Lecture Notes in Networks and Systems

Volume 482

The series "Lecture Notes in Networks and Systems" publishes the latest developments in Networks and Systems—quickly, informally and with high quality. Original research reported in proceedings and post-proceedings represents the core of LNNS.

Volumes published in LNNS embrace all aspects and subfields of, as well as new challenges in, Networks and Systems.

The series contains proceedings and edited volumes in systems and networks, spanning the areas of Cyber-Physical Systems, Autonomous Systems, Sensor Networks, Control Systems, Energy Systems, Automotive Systems, Biological Systems, Vehicular Networking and Connected Vehicles, Aerospace Systems, Automation, Manufacturing, Smart Grids, Nonlinear Systems, Power Systems, Robotics, Social Systems, Economic Systems and other. Of particular value to both the contributors and the readership are the short publication timeframe and the world-wide distribution and exposure which enable both a wide and rapid dissemination of research output.

The series covers the theory, applications, and perspectives on the state of the art and future developments relevant to systems and networks, decision making, control, complex processes and related areas, as embedded in the fields of interdisciplinary and applied sciences, engineering, computer science, physics, economics, social, and life sciences, as well as the paradigms and methodologies behind them.

Indexed by SCOPUS, INSPEC, WTI Frankfurt eG, zbMATH, SCImago.

All books published in the series are submitted for consideration in Web of Science.

For proposals from Asia please contact Aninda Bose (aninda.bose@springer.com).

More information about this series at https://link.springer.com/bookseries/15179

Francesco Calabrò · Lucia Della Spina ·
María José Piñeira Mantiñán

Editors

New Metropolitan Perspectives

Post COVID Dynamics: Green and Digital
Transition, between Metropolitan and Return
to Villages Perspectives

Set 2

 Springer

Editors
Francesco Calabrò
Dipartimento PAU
Mediterranea University of Reggio Calabria
Reggio Calabria, Reggio Calabria, Italy

Lucia Della Spina
Mediterranea University of Reggio Calabria
Reggio Calabria, Italy

María José Piñeira Mantiñán
University of Santiago de Compostela
Santiago de Compostela, Spain

ISSN 2367-3370   ISSN 2367-3389   (electronic)
Lecture Notes in Networks and Systems
ISBN 978-3-031-06824-9   ISBN 978-3-031-06825-6   (eBook)
https://doi.org/10.1007/978-3-031-06825-6

This Springer imprint is published by the registered company Springer Nature Switzerland AG
The registered company address is: Gewerbestrasse 11, 6330 Cham, Switzerland

Preface

This volume contains the proceedings for the fifth International "NEW METROPOLITAN PERSPECTIVES. Post COVID Dynamics: Green and Digital Transition, between Metropolitan and Return to Villages' Perspectives", scheduled from May 25–27, 2022, in Reggio Calabria, Italy.

The symposium was promoted by LaborEst (Evaluation and Economic Appraisal Lab) of the PAU Department, Mediterranea University of Reggio Calabria, Italy, in partnership with a qualified international network of academic institution and scientific societies.

The fifth edition of "NEW METROPOLITAN PERSPECTIVES", like the previous ones, aimed to deepen those factors which contribute to increase cities and territories attractiveness, both with theoretical studies and tangible applications.

This fifth edition coincides with what is most likely the end of the COVID pandemic that began in 2020. The global health emergency, despite having been a phenomenon limited in time, has acted as an accelerator of some changes in behavior and in the organization of activities associated with the ever-increasing spread of ICT.

The phenomena are too recent and still ongoing to fully understand the implications they will have on settlement systems, but the conclusion reached at the previous edition of New Metropolitan Perspectives seems to be confirmed: from many of the works presented at the Symposium, a reduction in the relevance of the localization factor emerges with ever greater clarity, at least in the ways known so far from the times of the Industrial Revolution, bringing to light more and more a paradigm shift in the center-periphery dualism.

In fact, the phenomenon that in the past led to the birth of the modern city, the need to concentrate people and activities in small areas, seems to be decreasing: the progressive spread of smart working and the digital modality for the provision of services (just think, e.g., of the digital services of the Public Administration or online commerce) significantly reduces the gaps in terms of accessibility to goods and services between metropolitan cities and marginalized areas, such as inland areas.

But this edition of the symposium also coincides with the start of a new phase for European policies, guided toward the green and digital transition, for the period 2021-27, by the European Green Deal, especially through the tool of the Next Generation EU.

The links between new technologies and sustainability tend to focus on the role played and that can play the city at EU level in fighting climate change.

Many of the contributions collected in this volume address the issue of the green transition through multidisciplinary points of view, dealing with very different issues such as, for example: infrastructures and mobility systems, green buildings and energy communities, ecosystem services and the consumption of soil, providing interesting information on the main trends in progress.

The changes in individual behavior and social organization, associated with the digital transition, are illustrated by the contributions that have addressed the issue of rules and of social innovation practices that are prefiguring new forms of governance for the regeneration of settlement systems. In this context, the issues of the new declinations of the concept of citizenship were also addressed, also with reference to the need to create favorable contexts for individual initiative and entrepreneurship, especially for young people, as a possible response to the challenge of employability for the new generations.

In this context, territorial information systems take on a leading role, together with apps capable of making territories increasingly smart.

The substantial investments planned by the EU to support the green and digital transition in the coming years require multidimensional evaluation systems, capable of supporting decision makers in selecting the interventions most capable of pursuing the objectives. The financial resources used for the implementation of the policies are borrowed from future generations, to whom we will have the obligation to be accountable for our work.

Unfortunately, at the time of writing we must also register serious concerns for the future of humanity, stemming from the risks of the spread of the conflict between Russia and Ukraine. In addition to the obvious concerns about the suffering that wars always cause to civilian populations, this situation makes future scenarios even more uncertain: It is clear that the circulation of goods, people and ideas will be increasingly conditioned by future geopolitical balances.

The ethics of research, in the disciplinary sectors that the Symposium crosses, invites us to feed, with scientific rigor, policies and practices that make the territory more resilient and able to react effectively to catastrophic events such as the pandemic or the war: We hope to know the outcomes of these courses in the next editions of the New Metropolitan Perspectives symposium.

For this edition, meanwhile, the more than 300 articles received allowed us to develop 6 macro-topics, about "Post COVID Dynamics: Green and Digital Transition, between Metropolitan and Return to Villages' Perspectives" as follows:

1. Inner and marginalized areas local development to re-balance territorial inequalities

2. Knowledge and innovation ecosystem for urban regeneration and resilience
3. Metropolitan cities and territorial dynamics. Rules, governance, economy, society
4. Green buildings, post-carbon city and ecosystem services
5. Infrastructures and spatial information systems
6. Cultural heritage: conservation, enhancement and management.

And a Special Section, Rhegion United Nations 2020-2030, chaired by our colleague Stefano Aragona.

We are pleased that the International Symposium NMP, thanks to its interdisciplinary character, stimulated growing interests and approvals from the scientific community, at the national and international levels.

We would like to take this opportunity to thank all who have contributed to the success of the fifth International Symposium "NEW METROPOLITAN PERSPECTIVES. Post COVID Dynamics: Green and Digital Transition, between Metropolitan and Return to Villages' Perspectives": authors, keynote speakers, session chairs, referees, the scientific committee and the scientific partners, participants, student volunteers and those ones that with different roles have contributed to the dissemination and the success of the Symposium; a special thank goes to the "Associazione ASTRI", particularly to Giuseppina Cassalia and Angela Viglianisi, together with Immacolata Lorè, for technical and organizational support activities: without them the Symposium couldn't have place; and, obviously, we would like to thank the academic representatives of the University of Reggio Calabria too: the Rector Prof. Marcello Zimbone, the responsible of internationalization Prof. Francesco Morabito, the chief of PAU Department Prof. Tommaso Manfredi.

Thank you very much for your support.

Last but not least, we would like to thank Springer for the support in the conference proceedings publication.

Francesco Calabrò
Lucia Della Spina
Maria José Pineira Mantinan

Organization

Programme Chairs

Francesco Calabrò	Mediterranea University of Reggio Calabria, Italy
Lucia Della Spina	Mediterranea University of Reggio Calabria, Italy
María José Piñeira Mantiñán	University of Santiago de Compostela, Spain

Scientific Committee

Ibtisam Al Khafaji	Al-Esraa University College of Baghdad, Iraq
Shaymaa Fadhil Jasim Al Kubasi	Koya University, Iraq
Pierre-Alexandre Balland	Universiteit Utrecht, Netherlands
Massimiliano Bencardino	Università di Salerno
Jozsef Benedek	RSABabes-Bolyai University, Romania
Christer Bengs	SLU/Uppsala Sweden and Aalto/Helsinki, Finland
Adriano Bisello	EURAC Research
Mario Bolognari	Università degli Studi di Messina
Nico Calavita	San Diego State University, USA
Roberto Camagni	Politecnico di Milano, Presidente Gremi
Sebastiano Carbonara	Università degli Studi "Gabriele d'Annunzio" Chieti-Pescara
Farida Cherbi	Institut d'Architecture de TiziOuzou, Algeria
Antonio Del Pozzo	Università degli Studi di MessinaUnime
Alan W. Dyer	Northeastern University of Boston, USA
Yakup Egercioglu	Izmir Katip Celebi University, Turkey
Khalid El Harrouni	Ecole Nationale d'Architecture, Rabat, Morocco
Gabriella Esposito De Vita	CNR/IRISS Ist. di Ric. su Innov. e Serv. per lo Sviluppo

Fabiana Forte   Università degli Studi della Campania "Luigi
                                     Vanvitelli"
Chro Ali HamaRadha   Department of Architectural Engineering/Faculty
                                     of Engineering/Koya University, Iraq
Christina Kakderi   Aristotelio Panepistimio Thessalonikis, Greece
Karima Kourtit   Open University, Heerlen, Netherlands
Olivia Kyriakidou   Athens University of Economics and Business,
                                     Greece
Ibrahim Maarouf   Alexandria University, Faculty of Engineering,
                                     Egypt
Lívia M. C. Madureira   Centro de Estudos Transdisciplinares para o
                                     DesenvolvimentoCETRAD, Portugal
Tomasz Malec   Istanbul Kemerburgaz University, Turkey
Benedetto Manganelli   Università degli Studi della Basilicata
Giuliano Marella   Università di Padova
Nabil Mohäreb   Beirut Arab University, Tripoli, Lebanon
Mariangela Monaca   Università di Messina
Bruno Monardo   Università degli Studi di Roma "La Sapienza"
Giulio Mondini   Politecnico di Torino
Pierluigi Morano   Politecnico di Bari
Grazia Napoli   Università degli Studi di Palermo
Fabio Naselli   Epoka University
Antonio Nesticò   Università degli Studi di Salerno
Peter Nijkamp   Vrije Universiteit Amsterdam
Davy Norris   Louisiana Tech University, USA
Alessandra Oppio   Politecnico di Milano
Leila Oubouzar   Institut d'Architecture de TiziOuzou, Algeria
Sokol Pacukaj   Aleksander Moisiu University, Albania
Aurelio Pérez Jiménez   University of Malaga, Spain
Keith Pezzoli   University of California, San Diego, USA
María José Piñera Mantiñán   University of Santiago de Compostela, Spain
Fabio Pollice   Università del Salento
Vincenzo Provenzano   Università di Palermo
Ahmed Y. Rashed   Farouk ElBaz Centre for Sustainability and
                                     Future Studies
Paolo Rosato   SIEV Società Italiana di Estimo e Valutazione
Michelangelo Russo   SIU Società Italiana degli Urbanisti
Helen Salavou   Athens University of Economics and Business,
                                     Greece
Stefano Stanghellini   INUIstituto Nazionale di Urbanistica
Luisa Sturiale   Università di Catania
Ferdinando Trapani   Università degli Studi di Palermo

Robert Triest   Northeastern University of Boston, USA
Claudia Trillo   University of Salford, UK
Gregory Wassall   Northeastern University of Boston, USA

Internal Scientific Board

Giuseppe Barbaro   Mediterranea University of Reggio Calabria
Concetta Fallanca   Mediterranea University of Reggio Calabria
Giuseppe Fera   Mediterranea University of Reggio Calabria
Massimiliano Ferrara   Mediterranea University of Reggio Calabria
Tommaso Isernia   Mediterranea University of Reggio Calabria
Giovanni Leonardi   Mediterranea University of Reggio Calabria
Tommaso Manfredi   Mediterranea University of Reggio Calabria
Domenico E. Massimo   Mediterranea University of Reggio Calabria
Marina Mistretta   Mediterranea University of Reggio Calabria
Carlo Morabito   Mediterranea University of Reggio Calabria
Domenico Nicolò   Mediterranea University of Reggio Calabria
Adolfo Santini   Mediterranea University of Reggio Calabria
Simonetta Valtieri   Mediterranea University of Reggio Calabria
Giuseppe Zimbalatti   Mediterranea University of Reggio Calabria
Santo Marcello Zimbone   Mediterranea University of Reggio Calabria

Scientific Partnership

SIEV - Società Italiana di Estimo e Valutazione, Rome, Italy
SIIV - Società Italiana Infrastrutture Viarie, Ancona, Italy
SIRD - Società Italiana di Ricerca Didattica, Salerno, Italy
SIU - Società Italiana degli Urbanisti, Milan, Italy
SGI, Società Geografica Italiana, Roma, Italy

Organizing Committee

ASTRI Associazione Scientifica Territorio e Ricerca Interdisciplinare
URBAN LAB S.r.l.
ICOMOS Italia, Rome, Italy

Contents

Contents

Metropolitan Cities and Territorial Dynamics. Rules, Governance, Economy, Society

Sustainable Policies in a Latin-American Context

Daniela Santana Tovar$^{(\boxtimes)}$, Sara Torabi Moghadam, and Patrizia Lombardi

Interuniversity Department of Regional and Urban Studies and Planning, Politecnico di Torino,
Torino, Italy
daniela.santana@polito.it

Abstract. The rapid process of urbanization in the world is generating several harmful impacts. This is especially true in a Latin American context, where cities have a high growth expansion rate in terms of population. Hence, it is necessary to take action to shift cities into resilient and sustainable environments. The present study aims to develop an existing sustainable assessment framework in the city of Cali, Colombia. The framework proposes new indicators to be integrated to the current assessment system for monitoring the progress of sustainability achieved in the city. The COVID-19 pandemic has negatively affected the development of the United Nations 2030 Agenda, and also generated the need to reassess the SDG Index indicators. COVID-19 has resulted in some cases to a lack of information and in others, to the formulation of policies to reverse regression in the fulfillment of the goals. The result is a framework consisting of 23 indicators to measure sustainability, which consider SDG11 indicators. The scope of this analysis is limited to assessment at urban scale, which facilitates the implementation of an evaluation framework and the modification of urbanistic regulations and policies.

Keywords: Sustainable Development Goals · Sustainable assessment tools ·
Case study approach · Indicators · Latin-American context

1 Introduction

Urbanization constitutes one of the most important processes of change in landscapes. It is acknowledged that 2008 was the turning point in which more than 50% of the world's population lived in urban areas rather than rural ones. This trend is increasing every year, and it has become even more drastic in some contexts, such as in Latin America, where it is estimated that 81% of the population lives in urban settlements [1]. Therefore, what sociologists call "urban society" [2] constitutes one of the major concerns for urban planners and institutions, because of the high concentration of people that these settlements imply and key role this plays when implementing policies that have impacts on the environment on a broader scale.

Several concepts that contribute to sustainability which are considered a measurable dimension of "sustainable development", such as wellbeing [3]. Development of rapid and effective evaluation frameworks and monitoring systems is crucial in order

© The Author(s), under exclusive license to Springer Nature Switzerland AG 2022
F. Calabrò et al. (Eds.): NMP 2022, LNNS 482, pp. 935–945, 2022.
https://doi.org/10.1007/978-3-031-06825-6_89

to assess implementation strategies and design new solutions towards the sustainable development.

The crucial role of cities in facing the main challenges of our time is also reflected in the Agenda 2030 adopted by the United Nations in 2015 [4]. In fact, among the 17 Sustainable Development Goals (SDGs) to guide future development, SDG11 is completely dedicated to cities with the aim of making them more inclusive, safe, resilient, and sustainable. Though the Sustainable Development Goals (SDGs) represent a global consensus on the characteristics of a desired sustainable future, Costanza et al. 2016 nevertheless suggest that defining a methodology capable of measuring progress to achieve sustainable development is necessary. Several researchers have proposed diverse methodologies to measure the wellbeing and sustainability of communities and ecosystems, but all of them start by defining indices [5].

Terraza et al. [6] identify some key issues that affect the urban scale such as the use of soil (e.g., disorganized urban sprawl, lack of definition between urban and rural soil, low density and presence of urban void, a high percentage of unused soil), the deficit of public transport and inequitable distribution of public space and green areas, the strong socio-spatial segregation, the proliferation of informal settlements in vulnerable unoccupied areas of the city, and the high risk of natural disasters caused by climate change. At the same time, a "sustainable city" can be described by the following features: compactness (i.e., clear limits between urban and rural occupation); social cohesion (i.e., spaces for social interaction, presenting a good index of public spaces and green areas); and resilience towards natural disasters (which encourages social activities and allows the community to develop). In 2016, the Inter-American Development Bank BID (IDB) analysed the main challenges that Latin-American cities are facing [7]. In this study, data from different cities have been compared side by side and examined considering similarities. The basic indicators which have been selected are focused on improvement of the quality of life of the inhabitants. Latin-American cities can also be considered as new cores of investment and opportunities. They present similar problems, and consequently share possible causes and solutions. However, considering specificity of the local context for sustainability assessment is fundamental in order to identify the needs and priorities of the place [8].

Since most Latin-American cities present similar characteristics and challenges, they can be analysed using same criteria and design strategies to guide future urban planning and action plans toward the implementation of sustainable development principles. For public urban planning departments, it has been proved that changes in governance models, in development strategies and the policies that regulate them are necessary, by considering sustainable development as the condition of achievement of equal conditions and opportunities for all. However, the rates of inequality in Latin-America are significantly high. In Colombia, initial steps toward the development of a more sustainable society were already made by integrating the UN's Sustainable Development Goals (SDGs) [4] in national and local development plans in the urban planning sector and national environmental policies. Problems persist in relation to the lack of implantation of the goals at different scales (from national to local). Every scale of implementation represents an important level of analysis in order to design proper strategies and monitor progress in the achievement of SDGs.

The ongoing pandemic of Covid-19 is a strong reminder that urbanization has changed the way that people and communities live, work and interact, and even more than in the past a multidisciplinary approach is necessary to develop systemic operational skills capable of dealing with complexity. Humanity has been pushed into an existential crisis, showing the vulnerability of our response systems to sudden shocks. According to the SDG 2020 report, the crisis has exacerbated existing inequalities and injustices, deeply affecting the most vulnerable. It is key to continue making efforts to improve data availability amongst all countries, and especially in Latin-American countries, in order to direct development in correspondence with the SDG's. The present study aims at further developing an existing sustainable assessment framework in the city of Cali, Colombia. The developed framework proposes new indicators to be added to the current assessment system for monitoring the progress of sustainability achieved in the city. The main objectives are: (i) Initiate the process of integration between development plans and Land-use Plans (POT) over time, establishing common indicators for the future assessment of progress and (ii) Propose an evaluation framework to complete and improve the current evaluation system of the POT with a focus on SDG11 targets and indicators.

This paper is organized as follows: Sect. 1 introduces the importance of considering the assessment and adjustment of globalization strategies towards sustainability. This section also gives an insight into current assessment methodologies to evaluate sustainability from different points of view. The methodological approach is reported in Sect. 2. In this section the process of generating a list of sustainability assessment indicators is illustrated. Section 3 illustrates the results of the application of the methodology on the case study. The paper finishes with conclusive remarks and some future developments (Sect. 4).

1.1 Case Study

The selection of the case study was based on the idea of developing an assessment framework that could help improve sustainable development in Latin-American cities, inspired by the application of the UN's SDGs. The selection and contextualization of an evaluation framework was made in Cali, Colombia. Cali is a Colombian middle-sized city, the third most populated of the country with 2.400.000 inhabitants. Located in the southeast of Colombia, officially named Special District and is the main city in the metropolitan area. In Colombia, the concept of "sustainability" is introduced in the constitution of 1991, which gave Colombian law a modern perspective that considered human rights and obligations regarding the environment. Later in 1993, the "Environment and Sustainability ministry" was established as a response to the UN Conference in Rio de Janeiro regarding environment and development. Sustainability is a complex concept which is open to various interpretations. Especially when combined with the concept of development, its focus changes progressively to economic development rather than to the overall integrated concept of sustainable development [9]. The concept of sustainable development in Colombia contrasts with the idea of economic growth since it was introduced globally in the '80s. In fact, it became evident that industrialization started damaging significantly natural resources and environments. Sustainable development was defined by the Law of 1993 art.3 as the integrated guide to economic growth, elevation of quality of life, social wellbeing, avoiding the exhaustion of renewable resources in which it is grounded and the deterioration of the environment [10]. Even if the concept of sustainability was introduced in planning instruments and normativity since 1993, it was not until the 2000's that specific environmental policies were officially introduced by CONPES (Concejo Nacional de Política Económica y Social, The superior organ in charge of the planning the development of the country). Issues in relation to the integration of local strategies and projects into national policies introduced by CONPES still persist. National regulations about sustainable development mainly refer to the protection of natural areas and the conservation of their environment and biodiversity. Still, guidelines for sustainability related planning policies at the city scale are still missing, even if it is recognized as the favoured scale for implementing sustainable development strategies. The Territorial Ordinance Plan (POT) is the technical instrument that each municipality uses to organize the territory. The "Expedient Municipal" (Municipal record) is an instrument developed to assess the implementation and progress of projects included in the POT. The Expedient Municipal is an instrument designated to assess the fulfilment and progress of the projects mentioned in the POT and includes indicators designated to monitor the progress and development of the city. These indicators aim at simplifying the complex dynamics in the field of land management and progress achieved by municipal development policies. The indicators are centralized in the site of the System of Social Indicators (SIS)[1].

2   Methodology

The methodology adopted to improve the assessment framework of the POT with a new measurement framework of sustainability consists of 3 steps of analysis as summarized in (Table 1).

Table 1. Methodological steps of the analysis

Methodological step	Operational activities
Step 1: Preliminary analysis	Comparing and selecting the most common indicators of two well-known assessment tools - BREEAM Communities and LEED for Neighbourhood Development
Step 2: Filtering process	Filtering the selected indicators, considering the Sustainable Development Goal (SDG) 11 indicators
Step 3: Development proposal	Developing and integrating the new indicators to the current framework supported by the Cali city- ESC- (Emerging and Sustainable Cities program)

Step 1 consists in the comparison and analysis of indicators from two sustainability assessment protocols used at the urban level: LEED for Neighbourhood Development

---

[1] The SIS Cali, has been conceived as a set of human and technological resources that integrates statistical information, storing, processing, analyzing and sharing data produced by the administration. Also allows an effective and automatized data exchange to support governance processes.

and BREEAM Communities. Relationships between them have been analysed and a single set of indicators created.

Step 2 consists in filter the selected indicators from step 1 according to the targets defined in the SDG11 and completing the set with missing indicators from SDG 11. In conclusion, step 3 compares the lists of indicators resulted from step 2 with the relevant topic and challenges in the Latin-American context, using as reference the ESC (Emerging and Sustainable Cities Program) methodology. This step results with an accurate framework for measuring the level of sustainability of Latin-American cities, which is simple to use. Accordingly, a review of existing approaches to sustainability measurement at multiple scales was carried out to understand the influence indicators applied for the context and the scale of analysis, and to orientate the selection of methodologies and approaches to be adopted.

Following the 3 steps, the resulting list of indicators is based on the LEED and BREEAM protocols, including principles of SDG11 and specific for the context of Latin-America cities. BREEAM, is an environmental assessment tool that was established in the UK. It is the first commercially available tool, and it focuses on the mitigation of impacts generated by design projects in the built environment. For the assessment of a larger scale, the company created the protocol called "BREEAM Communities". In this methodology, the categories of infrastructure and transport are the most important ones. LEED (Leadership in energy & environmental design) is a rating system developed by the United States Green building council. The assessment is based mainly on 2 features of a design project, the site selection and the construction elements. The "LEED for neighbourhood development" connects these elements with the infrastructure. The scoring system of LEED works with a set of mandatory criteria that must reach a specific score to obtain the certification. Unlike BREEAM that gives a weight to each category in connection with the other credits, LEED defines a series of criteria that are mandatory to classify the project as certified and indicate criteria that the project fulfil or not. The definition of the points in each criterion is made according to the relevance and positive effect that generates in the project and its environment [11].

The United Nation's SDGs are presented as a guide for countries to develop policies to collaborate with global commitments but is finally left to each government the development of plans, policies, and programmes. In this specific case, the methodology used to contextualize the framework is called ESC (Emerging and sustainable cities program). ESC is a program applied in multiple Latin-American cities to establish general aspects that must be prioritized to improve living conditions in the cities and develop strategies towards sustainable development. The Methodological guide ESC is a fast application methodology and diagnosis to help emergent cities in the realization and application of an action plan that structure interventions that guide sustainability goals in the short, medium and long-term period, developed by the Inter-American Development Bank (IBD).

The proposed methodology consists of three main analytical phases followed by a final validation phase in which the final list of indicators is discussed with experts. The result is a set of indicators classified in categories and criteria that can be integrated in the municipal expedient which is the evaluation instrument of the POT. Existing sustainability protocols and frameworks are used as a reference to create a framework that can be used in Latin-American cities. Most approaches used to orient urban projects in the Latin-American context have mainly an environmental focus, without an integrated approach between different dimensions of sustainable development. For this reason, the proposed indicators have been grouped in three categories corresponding to three pillars of sustainability: environmental, social and economic. In the next paragraphs, the developed activities are described more in detail.

2.1   Step 1: Preliminary Analysis

The first methodological step (preliminary analysis) selects the preliminary indicators according to the following operations: 1. Listing categories of each assessment tool (BREEAM and LEED), 2. Definition of categories that relate to both assessment tools, 3. Classification of indicators in the selected categories, 4. Find links and similarities between indicators, 5. Condensation of indicators by relations, 6. Selection of mandatory and high weighted indicators.

In the first step, all indicators from LEED and BREEAM were analyzed. Different features described in each assessment tool regarding the definition of indicators and the categories used to group them were considered. To simplify, the selection of the categories to organize the indicators was adopted from the ESC framework. This approach will be described more in detail in step 3; however, it was used since the beginning of the analysis to allow a more efficient organization of the indicators according to the Latin-American context. The dimensions of sustainability in which the ESC guide is framed are: (i) Environmental sustainability and climate change, (ii) Urban Sustainability and Fiscal sustainability and (iii) Governability.

51 criteria from LEED-ND and 53 from BREEAM were analyzed and then divided into five categories, responding to different topics. Successively, indicators from the two assessment tools that have the same purpose were linked and overlapped. One of the limitations of the approach is the neighborhood scale for which the indicators were designed, since the resulting framework will be applied at the city scale. However, some indicators that were created considering a smaller scale still have coherence and can be applied on a larger scale. Indicators were then compared to the categories of the ESC to identify connections, as shown in Fig. 1. The output of step 1 is a list of indicators classified grouped into categories. The selected categories of this step are water, energy, waste, risk vulnerability, land-use, mobility, and business and employment.

2.2   Step 2: Filtering Process

Step two, named filtering process, aims at defining targets according to the following activities: 1. List categories and indicators, 2. Define categories that relate to SDG 11, and 3. Find links and similarities between indicators and sub-indicators.

The objective of the analysis is to establish an effective measurement framework of sustainability on the city scale. For this reason, the UN SDGs are used as a reference and specifically SDG11 that is focused on sustainable cities and communities. SDG11 establishes the importance of accommodating the growing population in affordable and quality housing, with access to public spaces, green areas, and efficient public transport, encouraging sustainable urban planning and management. Indicators from step 1 are

Fig. 1. Summarizing scheme of the methodological development step 2.

filtered according to targets and indicators defined by SDG11. Categories representing SDG11 targets are compared to indicators categories defined in step 1 as illustrated in Fig. 1. Some indicators of SDG11 that have a more specific focus are used as a complement description and definition of the sub-indicators.

The strategies for the achievement of the targets are left unclear by the UN as is left to each government to implement legally binding standards to ensure the compliance of national policies that can favour sustainability. Also, to create an economic environment that promotes rewards for institutions that contribute to sustainable development. Hajer et al. [12] states that SDG cannot be achieved in charge of intergovernmental organizations, but be mobilized by new agents such as business, cities and civil society. Concluding that SGD is a tool to guide governments into a vision that promotes sustainable development and be applied to introduce effective policies that represent an advantage to work towards sustainable development. The result of this step is a new set of indicators that measure sustainability in a neighborhood or city scale and that are related to SDG11. Moreover, the relation with the context of study is still missing, even if it is recognized as an important feature to ensure the accuracy of the assessment.

2.3 Step 3: Development Proposal

Step three (development proposal) intends to contextualize the indicators in Latin-American cities, was developed according to the following activities: 1. Analyse ESC indicators, 2. Define fundamental issues for Latin-American cities with reference to ESC, 3. List indicators from ESC that are related to the categories selected in step 2, and 4. Complete the set of indicators.

The ESC indicators are aimed at evaluating and rate the progress of a city and prioritizing the projects that should be developed, focused on people's basic needs and the quality of life that the city and local government should provide. Thus, the indicators can define the focus of the measurements of the criteria from the steps before.

The result of the last step is the final list of indicators contextualized in Latin-American cities. To achieve a more feasible set of indicators, some extra information to help the measurement of each criterion are added. The inclusion of information about: Assessment method, parameter for assessment, and type of calculation (qualitative or quantitative) have the objective of clarify the availability of data and data collection.

3   Results and Discussion

The result is a Sustainability Measurement Framework, formed by 11 indicators and 23 sub-indicators grouped in 5 categories: Waste, Risk vulnerability, Land-use, Mobility, and Environment. Those categories are mainly referring to SDG11 targets. Table 2 reports the evaluation framework.

Table 2. Final list of indicators to support decision-making in Latin-America urban planning activities.

Sustainability dimension	Category	Indicator	Sub-indicator	Measurement unit
Environmental sustainability and climate change	Waste	Solid waste management	Meet waste disposal demand	%
			Solid waste treatment	%
			Waste production from construction	%
	Risk vulnerability	Management of risks	Risk management and assessment	
	Environment	Air quality	Concentration of air pollutant	PM10 in μg/m^3 in 24 h
Urban and fiscal sustainability	Land-use	Housing provision	Meet housing demand	%
			Housing affordability	%
		Demographic needs and priorities	Access to green suitable green spaces	Ha/100.000 inh
			Access to suitable public spaces	Ha/100.000 inh
			Jobs availability	Ratio
			Delivery of services and facilities	
			Ensure urban safety	%
			Inclusive design	
		Land-use strategy	Compact development	Ratio

(*continued*)

Table 2. (*continued*)

Sustainability dimension	Category	Indicator	Sub-indicator	Measurement unit
		Utilities	Meet public services demand	%
		Enhancement of ecological value	Protect existing natural habitats	
	Mobility	Transport assessment	Meet transport demand	%
			Access to public transport	%
			Public transport facilities	
			Cycling network	km
			Cycling facilities	
		Access to quality transit	Transit facilities	
		Walkable streets	Safe and appealing street	km

Even if the aim of the assessment is to create an interaction between the scales of application of strategies of sustainable development, the city scale is recognized as the focal point for the creation of policies, strategies and projects, and monitoring frameworks for their implementation.

In step 3, indicators were filtered according to ESC priorities and, simultaneously, the indicators of the municipal record were filtered and integrated in the framework according to the contents of SDG 11. Also, only seven indicators from the municipal record can be used to measure sustainable urban planning according to SDG 11. This filter at this stage shows the importance of contextualizing the framework to identify existing connections between local strategies of sustainable development and the Agenda 2030 framework.

The resulting framework must be included into existing local assessment and monitoring frameworks, since it is aligned with the Agenda 2030. After the analysis of the case study and the revision of the results of the methodology, there is a lack of coherence between the models of development that are proposed by the multiple planning and management instruments. As well as a clear manifestation of the importance of considering sustainable strategies to adjust current projects and to implement them for future developments.

4 Conclusions

The present study proposes a methodology, which consists of the development of three steps that result in a final framework of assessment of sustainability based on SDG 11. This includes a review of existing sustainability assessment tools like LEED-ND, BREEAM for Communities and the ESC methodology in order to obtain a list of indicators that are contextualized in concordance with the case study.

The case study is Cali, Colombia, which is a middle-sized Latin-American city with almost 3 million inhabitants; characterized for its favorable geographic and natural conditions which give it a major advantage by it economic and industrial development: this, still needs a boost equally favoring investors, owners, clients and inhabitants. Considering sustainable development proposals and policies the best alternative is to develop the city and enhance the natural values that it possesses. The two tools that can support and encourage the proposal and implementation of ideas for sustainable development are the POT, which is the local land-use plan, and the municipal record, where indicators are used to measure the effectiveness and accuracy of the projects presented in the POT. For further developments, the proposal is to encourage local authorities and entities to invest in proposals for urban planning focused on sustainable development, through the consolidation of one system of indicators that can measure progress simultaneously. The purposes of the proposal are the evaluation of progress of the 2030 Agenda, and the formulation of urban planning and emergency policies that were not considered before the crisis caused by COVID-19 and can trace a route to fight inequality and climate change.

The application of the framework in multiple case studies can also help determine the factors and indicators that can be standardised, and which need to be analysed within a context, such as the articulation of the indicators in a current policies regulation plan or the generation of alternatives to introduce the measurement of sustainability in the urban development of the cities. To conclude, the current pandemic is a perfect scenario for rethinking resilience in a more transformative way, this could include better data management and evaluating the impact of the shock for future city resilience development.

References

1. UN-Habitat. Planning Suatainable Cities: Policy Directions: Global Report on Human Settlements 2009 (Abridged Edition) (2009)
2. Lefebvre, H.: De la Ciudad a la Sociedad Urbana. Revista de Estudios Culturales Urbanos, Bifurcaciones (2014). ISSN-e 0718-1132
3. Bakar, A.A., Mohamed Osman, M., Bachok, S., Ibrahim, M., Zin Mohamed, M.: Modelling economic wellbeing and social wellbeing for sustainability: a theoretical concept. Proc. Environ. Sci. **28**, 286–296 (2015)
4. UN. Transforminng our World: The 2030 Agenda for Sustainable Development (2015)
5. Cutaia, F.: The Use of Landscape Indicators in Environmental Assessment BT-Strategic Environmental Assessment: Integrating Landscape and Urban Planning (2016)
6. Terraza, H., Rubio Blanco, D., Vera, F.: De ciudades emergentes a ciudades sostenibles. Comprendiendo y proyectando las metròpolis del siglo XXI (2016)
7. Inter American Development Bank. Guía Metodológica Iniciativa ciudades emergentes y sostenibles (2016)

8. Mohammed Ameena, R.F., Mourshed, M.: Urban sustainability assessment framework development: the ranking and weighting of sustainability indicators using analytic hierarchy process. Sustain. Cities Soc. **44**, 356–366 (2019)
9. Verma, P., Raghubanshi, A.: Urban sustainability indicators: challenges and opportunities. Ecol. Indicat. **93**, 282–291 (2018)
10. Sánchez Pérez, G.: Desarrollo y medio ambiente: una mirada a Colombia. Econ. Desarrol. **1**(1), 79–98 (2002)
11. Zeinal Hamedani, A., Huber, F.: A comparative study of DGNB, LEED and BREEAM certificate systems in urban sustainability. The Sustainable City VII: Urban Regeneration and Sustainability (2012)
12. Hajer, M.N., et al.: Beyond cockpit-ism: four insights to enhance the transformative potential of the sustainable development goals. Sustainability (Switzerland) **7**, 1651–1660 (2015)
13. Prescott-Allen, R.: The Well-Being of Nations (2001)
14. Costanza, R., et al.: Modelling and measuring sustainable wellbeing in connection with the UN sustainable development goals. Ecol. Econ. **130**, 350–355 (2016)
15. Spangenberg, J.H.: Hot air or comprehensive progress? A critical assessment of the SDGs. Sustain. Dev. **25**(4), 311–321 (2016)
16. Yan, Y., Wang, C., Quan, Y., Gang, W., Zhao, J.: Urban sustainable development efficiency towards the balance between nature and human well-being: connotation, measurement, and assessment. J. Clean. Prod. **178**, 67–75 (2018)

The Never-Ending Story of the Metropolitan Area of Vigo (Galicia, Spain)

Alejandro Otero-Varela$^{(\boxtimes)}$ (iD) and Valerià Paül (iD)

Universidade de Santiago de Compostela, Santiago de Compostela, Spain
alejandrootero.varela@usc.es, v.paul.carril@usc.gal

Abstract. Both the city of Vigo and its metropolitan area, understood respectively as a municipality and a functional region, have the largest populations in Galicia: almost 300,000 inhabitants in the first case and more than half a million in the second. However, the boundaries of the metropolitan area of Vigo are ambiguous, with a dispersion of opinions observable in the research carried out to date and a profusion of non-concurrent political approaches. Moreover, the institutionalisation of the metropolitan area has suffered two problems. Firstly, as a result of its own delimitation, which is continuously changing and under constant discussion. Secondly, because of the confrontation between the Local Government of Vigo and the Regional Government of Galicia. Through the analysis of the configuration and institutionalization process of this metropolitan area, it seeks to advance in issues of governance and urban development.

Keywords: Metropolitan area · Vigo · Political geography

1 Introduction

For a little over a decade now, more than half of the world's population has been considered to live in cities. Somewhat simplistically, these cities are assumed to be governed, managed and planned by their respective local governments. However, at least since the beginning of the 20th century, cities no longer tend to correspond as a rule to the strict scope of a single local government. Thus, metropolitan areas emerge, understood as heterogeneous combinations of built-up and non-built-up areas consisting of cores and suburban and outer fringes with high commuting dynamics, commonly embracing several municipalities [1]. Although a priori it can be argued that the creation of a metropolitan political-administrative scale makes sense in spatial development, planning and governance terms, there is a governmental fragmentation due to political reasons [2]. Thus, the debates around metropolitan institutionalisation revolve around two issues: on the one hand, the role that the pre-existing government levels must play; on the other hand, the "loss" of powers and competences of these [3].

The metropolitan scale is presented as one of the most appropriate levels for a better economic, social, and territorial cohesion and, therefore, ideal to achieve a more sustainable development. This is in line with SDG 11 (target 11.a) about "Support positive economic, social and environmental links between urban, peri-urban and rural areas

© The Author(s), under exclusive license to Springer Nature Switzerland AG 2022
F. Calabrò et al. (Eds.): NMP 2022, LNNS 482, pp. 946–956, 2022.
https://doi.org/10.1007/978-3-031-06825-6_90

by strengthening national and regional development planning". However, metropolitan spaces, considered those with the largest population and GDP, have little decision-making capacity. Spain is included in the group of countries where the EU cohesion policy is carried out at the regional level, but where synergies with other territorial levels are limited by the poor quality of governance [4]. In this sense, the ESPON (2021) project points out the fragmentation of local power as one of the main obstacles to achieving an integrated territorial development strategy [5].

The city of Vigo, with 296,692 inhabitants, is the most populous municipality in Galicia. Within Spain as a whole, it is in 14th place and is also the most populous city that is not a provincial capital [6]. This fact, together with its economic and industrial role, makes it a prominent city. In this paper, we propose to analyse the process of configuration and institutionalization of the metropolitan area of Vigo. We seek to reflect on the relationship between territorial dynamics and administrative structures, in terms of the possibility of offering a better and greater quality of development and governance. After this introduction and the theoretical framework, we present the results of the research work. In the results section, we make an exhaustive analysis of academic proposals about the metropolitan area of Vigo, and we review the political debate produced in the Galician Parliament. The last section is the discussion and conclusions.

2   Multiple Ways of Metropolitan Areas Construction

Most metropolitan areas lack their own political institution and legal framework to facilitate inter-municipal cooperation [3]. In fact, cyclical patterns of 'life and death' of metropolitan areas have been described, with periods when metropolitan areas were institutionalised and then abolished, as in the case of London in 1986 and Barcelona in 1987 [7, 8]. These reforms led to a refragmentation of metropolitan areas, with the assumption of powers by municipalities that were generally too small [9]. They have even reappeared in a third phase, as was the case in London in 2000 and in Barcelona in 2010, to continue with the same examples. In this context of problematic institutionalisation of metropolitan areas, there is a long-standing debate on the most appropriate institutional structure. According to Savitch and Vogel (2004) and Tomàs (2012) three options can be distinguished and are described below.

Firstly, the *Metropolitan Reform Tradition*. The existence of many local governments within a metropolitan area is seen as an obstacle to supra-municipal planning and efficient provision of services [10]. Thus, the creation of a new institution of inter-municipal cooperation is recommended as the best way to solve these problems. This institutionalisation can take place in two ways. On the one hand, through the amalgamation by the central municipality of the other metropolitan municipalities, which would result in a single level of government. On the other hand, through the creation of a metropolitan government that respects the autonomy of the municipalities, which would result in a two-tier government: metropolitan government and various local governments [11].

Secondly, the *Public Choice Perspective* rejects the idea of institutionalisation as a way to solve metropolitan problems. The fragmentation of metropolitan areas among local governments is considered to ensure greater efficiency and effectiveness [11]. This path is also advocated on the grounds of subsidiarity by assuming that smaller governments have the appropriate scale to better serve their citizens. Metropolitan-wide problems are assumed to be addressed through purpose-oriented coordination and cooperation networks involving municipalities.

However, both of these forms are criticised. On the one hand, several experiences of metropolitan institutionalisation are discouraging as they fail to solve metropolitan-wide problems [2]. On the other hand, it is criticised that inter-municipal cooperation is not always completely voluntary, which sometimes results in undermining local autonomy itself [12]. Furthermore, it is noted that, under inter-municipal cooperation procedures, too much competition can develop between local authorities [3].

A third possibility of metropolitan institutionalisation has been developed within *New Regionalism* [13]. It assumes that metropolitan areas are made up of interdependent municipalities that, together with other public and private actors, must cooperate flexibly to address common problems [2]. Governance offers the possibility of working in the interest of the metropolitan area without the need to implement its own government with static boundaries and a bureaucratic structure. Thus, metropolitan governance would be achieved through agreements involving very heterogeneous actors, even beyond the metropolitan boundaries [11, 14]. However, there are some voices that point out that the lack of transparency in *New Regionalism* may imply a threat to local democracy [11]. In another sense, recent research points to the metropolitan areas' search for greater degrees of devolution within their respective countries [16]. Nevertheless, in most cases all these are "light" metropolitan initiatives, i.e., they assume and/or promote strategic and non-binding planning, transport, and environmental agendas [3].

3   The Search for a Metropolitan Area for Vigo

Several proposals, both from the academic and institutional sources, have been presented regarding the metropolitan area of Vigo since the 1980s. In addition, these works have been accompanied, since the beginning of the current century, by an intense political debate on their institutionalization.

3.1   Academic Proposals for the Metropolitan Area

In the 1980s and 1990s, geographical research in Galicia embraced the sub-discipline of Urban Geography in general, and the metropolitan areas, in particular [17]. Galician Geography had mainly focused on the study of rural areas, with practically no work on urban areas. In this sense, in the 1980s the metropolitan dimension of Vigo was detected by Galician geographers with different delimitation proposals (Fig. 1).

One of the first geographers to propose a metropolitan area for Vigo was X. M. Souto. In 1984, he proposed a three-level division: a "neighbourhood area" made up of seven municipalities; a "proximity area" made up of 18 municipalities; and an "urban region", made up of 29 municipalities [18]. A year later, in 1985, he presented another proposal for an "urban region" in which he included many more municipalities to a total of 38. This increase was due to the fact that he considered municipalities on the outskirts of the city of Pontevedra to be part of the "urban region" of Vigo [19]. More recently, he has continued to make proposals for metropolitan delimitation. However, his delimitation projects have varied over time, including and excluding municipalities. Thus, in 2002, he referred again to an "urban region" made up of 19 municipalities [20], while in 2009 he raised it to 23 [14].

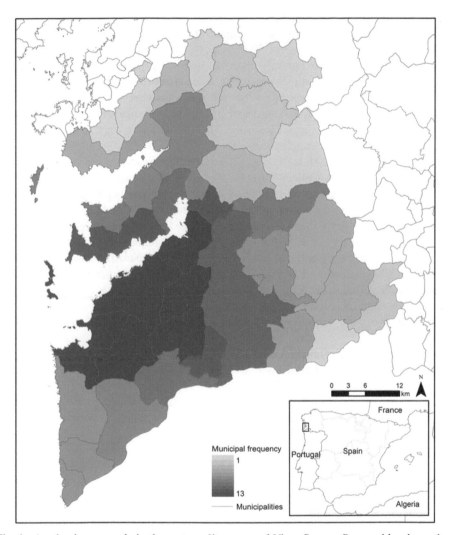

Fig. 1. Academic proposals in the metropolitan area of Vigo. Source: Prepared by the authors using a base map provided by IGN (http://www.ign.es/).

Another researcher, A. Precedo, considered in 1989 two levels: the "area of direct influence", with 29 municipalities; and the "metropolitan area", with 11 municipalities [21]. However, both within the one and the other, this geographer proposed the so-called "*comarcas*" (roughly speaking, *counties* in English) [22]. These counties, of which he said there could be between four and nine, would have their respective urban centres (county seats), with relative autonomy in relation to Vigo. However, when the Galician

Government passed the official county map of Galicia in 1997, proposed by A. Precedo himself, the delimitation of the so-called "county of Vigo" included 11 municipalities, which, importantly, were not the same as those in the "metropolitan area" he had proposed eight years before [23].

In 2004, another geographer, R. C. Lois, proposed a "metropolitan area" consisting of 14 municipalities [24]. There are three main differences with some of the previous proposals by X. M. Souto and A. Precedo. Firstly, R. C. Lois does not distinguish between different metropolitan scales but only proposes one level. Secondly, R. C. Lois clearly states that Pontevedra and its direct area of influence are not part of the metropolitan area of Vigo. Thirdly, R. C. Lois considers that the municipalities located on the Northern shoreline of the *ria* (inlet in Galician) are undoubtedly part of the metropolitan area of Vigo. Likewise, in 2011, R. C. Lois and M. J. Piñeira proposed an "area of direct influence" made up of 20 municipalities [25].

Between 2005 and 2009, there was a particular effervescence of academic debate on metropolitan areas in Galicia animated by political discussion [23]. As a result, several academic publications were published. In them, we will highlight two chapters authored by geographers proposing further delimitations of the metropolitan area of Vigo. The first, by R. Rodríguez-González, proposed a "metropolitan area" consisting of 23 municipalities [26], so-called "of the Rías Baixas", subtly amalgamating Pontevedra without naming the metropolitan area "of Vigo". "Rías Baixas" is a physical geography region basically consisting of a group of inlets located in south-western Galicia. The second, by J. A. Aldrey and J. Vicente, carried out research analysing part of the previous metropolitan maps approved or debated at the institutional level [27]. Nevertheless, they do not propose a different delimitation to those presented by the Galician government — see Sect. 3.2.

Finally, from outside Galicia, J. M. Feria led a research from 2004 to 2015 to propose a map of Spanish metropolitan areas using the US census method based on commuting between place of residence and place of work/study with data from the official 2001 and 2011 population censuses [28]. These studies indicate that the functionally defined metropolitan area of Vigo consists of 37 municipalities with a total population of over 700,000 inhabitants, which places it in 11th position in Spain.

The analysis of all these 13 academic studies reveals a huge dispersion, with delimitations ranging from 7 to 38 municipalities, and as many as 42 municipalities. Moreover, several of them use expressions such as "region", "county" or "area of influence", among others, which, strictly speaking, have different nuances to those of "metropolitan area" and therefore introduce a permanent lack of definition. Finally, some of these delimitations choose to include Pontevedra in the Vigo area, but others do not. Pontevedra is a municipality of 83,260 inhabitants [6], clearly urban and with an urban area of direct influence that globally exceeds 100,000 inhabitants. Its interrelation with Vigo is remarkable, but its autonomy is a debatable issue on which scholars have not agreed.

3.2   The First Steps for the Construction of the Metropolitan Area

In the absence of an institutionalised metropolitan area, supra-municipal cooperation has developed within a community of municipalities. In the metropolitan area of Vigo, the Community of the Intermunicipal Area of Vigo (in Galician, *Mancomunidade da*

Área Intermunicipal de Vigo) has been operating since 1992 for the joint provision of certain services [29]. Its boundaries have also changed over time. Initially, it was made up of 11 municipalities, those of the 1997 county, although later there were 12 (Fig. 2). At present, the Community is only made up of eight municipalities due to disagreements between the different local governments (Fig. 2). A significant aspect is that it was within the Community that the municipalities agreed to move towards the creation of an institutionalised metropolitan area for Vigo — i.e., political declarations held in Soutomaior in 1999 and Nigrán in 2000.

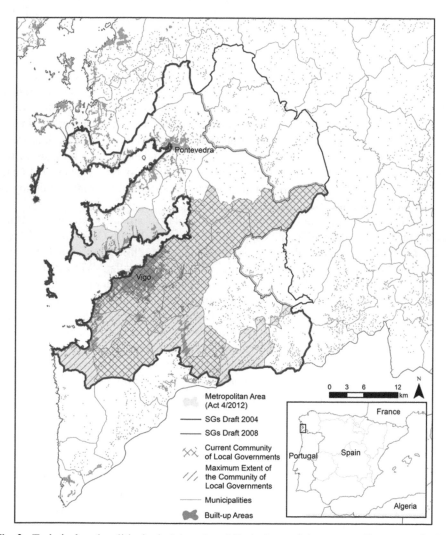

Fig. 2. Technical and political-administrative delimitations of the metropolitan area of Vigo. Source: Prepared by the authors using a base map provided by IGN (http://www.ign.es/).

In 2005, the Galician Government proposed the creation of a metropolitan area in Vigo made up of 14 municipalities by means of a bill (Fig. 2). This was the first official project of this kind in Galicia [23, 30]. The 12 municipalities of the Community were joined by Cangas and Moaña — located to the north of the inlet, but closely integrated with the south of the inlet, where Vigo is located, both by a motorway bridge and by sea transport. However, the project did not materialise because it was presented at the end of the legislature and the ruling party —the People's Party (*Partido Popular*, PP)— did not retain its parliamentary majority in the 2005 elections.

The new government, formed by a left-wing coalition between the Galician-Spanish Labour Party (*Partido dos Socialistas de Galicia*, PSdeG-PSOE) and the Galician Nationalist Bloc (*Bloque Nacionalista Galego*, BNG), agreed to establish metropolitan areas in Galicia. However, this policy was never developed. In March 2007, the PP, now in opposition, reintroduced the bill in Parliament to create the Metropolitan Area of Vigo, although the PSdeG-PSOE and the BNG opposed the bill, arguing that they envisaged "more ambitious" metropolitan areas, with their own competences and funding, and that they should be part of an in-depth reform of the political-administrative levels of government existing in Galicia [23]. Accordingly, both coalition parties considered that the provincial councils had to be reformed before implementing metropolitan areas, to avoid administrative overlapping. In short, provincial councils are a remnant of the old centralist Spanish State which subsist between the local governments and the Galician Government, without clear powers beyond the unspecific constitutional policy of assisting local governments.

The discussion on metropolitan areas also had its place in spatial planning [23]. Thus, in 2004, the first draft of the Spatial Guidelines (SGs; in Galician, *Directrices de Ordenación do Territorio*, the spatial master plan for Galicia as a whole) considered an extensive metropolitan area made up of 29 municipalities and so-called "Vigo-Pontevedra" (Fig. 2). In fact, the city of Pontevedra was included within the metropolitan area — unlike the 2005 project for the institutionalisation of the metropolitan area, which, as mentioned above, was limited to 14 municipalities, in a frequent gap between the spatial plans on paper and the political-administrative decisions. In the 2008 SGs draft, the "Vigo-Pontevedra" metropolitan area was reduced to 24 municipalities (Fig. 2). Finally, the SGs definitively passed in 2011 speak rhetorically of an "urban region of Vigo-Pontevedra", but avoid indicating the municipalities that compose it, possibly because the Galician Government avoids committing itself to its institutionalisation.

3.3 The Official Approval and Subsequent Discontinuation of the Institutionalised Metropolitan Area

At the beginning of 2012, the Galician Parliament passed Act 4/2012 institutionalising the Metropolitan Area of Vigo. This law is largely similar to the initial bill introduced in 2005. In fact, it includes the 14 municipalities from 2005 (Fig. 2), although its future expansion is not ruled out. The Act defines the following self-governing bodies: President, Metropolitan Government Board and Metropolitan Assembly, the first two assumed by the mayors of the metropolitan area, while the latter consists of some of the local councillors of the local governments in accordance with the population of each municipality. However, the metropolitan area is implemented without competences of its own and would only exercise competences delegated by other government levels, mainly the local governments and the Galician Government, amongst others potentially in the portfolio: economic promotion and employment; tourism and cultural promotion; mobility and passenger transport; environment, water and waste management; fire prevention and fire-fighting; civil protection; and cooperation in urban planning. The fact that its competences rest on the willingness of other governments to cede them led the PSdeG-PSOE and the BNG to position themselves against the law, which was passed with the votes only of the PP, once again with a majority in the Galician Parliament since 2009. The PSdeG-PSOE and the BNG considered that the Provincial Council of Pontevedra should be reformed to avoid overlapping, excluding those municipalities included in the metropolitan area from its sphere of action. Furthermore, PSdeG-PSOE and BNG did not agree with the boundaries of the metropolitan area and pointed out that there were other municipalities that should be included beyond the 14 municipalities proposed a decade earlier (the 2005 bill).

A major issue from the very moment the Act 4/2012 was passed was the refusal of the Local Government of Vigo to be part of it. Its mayor, from the PSdeG-PSOE, disagreed with the metropolitan governing bodies, considering that his municipality was underrepresented. This situation continued for several years until 2015, when the mayor of Vigo and the president of the Galician Government agreed to unblock the metropolitan area. Both leaders agreed to accommodate the demands of the Local Government of Vigo, which would increase its representation in the governing bodies. The metropolitan law was amended in 2016 with the unanimous support of all parties (Act 14/2016). Among the main changes was that the vote of the members of the Metropolitan Government Board would be based on the demographic weight of each municipality, which gave greater power to the Local Government of Vigo.

The consensus between different governments and political parties broke down shortly afterwards. The metropolitan area was constituted at the end of 2016 but the Galician Government argued that the Local Government of Vigo had not been incorporated into the Galician metropolitan transport plan and, therefore, the metropolitan area constitution was not valid. The mayors of the local governments in the hands of the PP did not participate in those first institutional steps. The mayor of Vigo was elected as Metropolitan President. In 2017, the Galician Government considered the implementation of the metropolitan area illegal and appealed to the courts to request its suspension. The High Court of Justice of Galicia decreed its precautionary suspension, while awaiting a final ruling to decide its future.

4   Discussion and Conclusions

This article has analysed, for the metropolitan area of Vigo, 13 academic delimitations formulated since 1984 and the evolution of political developments since 1999. None of them can be dissociated from the international debates on metropolitan areas. Thus, we note an overall wish towards the institutionalisation of the metropolitan area of Vigo, consistent with the *Metropolitan Reform Tradition* [10, 11]. This will is evident, above all, in Acts 4/2012 and 14/2016 of the Galician Parliament, which were to lead to the establishment of a metropolitan area consisting of 14 municipalities. However, the assumptions of the *Metropolitan Reform Tradition* are not fully met, because the competences of the metropolitan area of Vigo are not determined in these Acts of Parliament and, in fact, depend on other levels of government voluntarily devolving them. On the other hand, the local resistance in the Vigo area to abide by this legislation is evidence of the common perception by local governments that metropolitan institutionalisations undermine their autonomy [12]. Finally, the rigid parliamentary fixation of 14 municipalities, when in this article we prove that at some point a total of 42 have been considered (Fig. 1), is highly problematic and even arbitrary, and artificially shuts a much more complex discussion on the metropolitan boundaries down.

On the other hand, we have also noted the presence of the *Public Choice Perspective* in Vigo [10, 11], led by the local governments, namely Vigo, which are building a metropolitan scale based on voluntary collaboration: the Community of municipalities existing since 1992. However, this Community has lost municipalities (from the 12 it once had to the current eight), which leads us to state that the *Public Choice Perspective* is naïve, because local governments tend to compete among themselves and have difficulty in generating a common metropolitan-wide vision [3], not even voluntarily.

One possible way forward would be to follow the postulates of *New Regionalism* [11, 13]. In our case study, this would make it possible to combine the conflicts between local governments, the Galician Government, and the Provincial Government of Pontevedra [11]. Nevertheless, this is a path that opens many questions, given that it somewhat assumes that there will automatically be a confluence of interests between different political actors and governments. Furthermore, in the case of Spain, it should be noted that planning and governance have a lower level of quality than other EU countries such as France, Germany, or Sweden [4]. In our case study, it seems that Local Government of Pontevedra will never become part of a metropolitan area of Vigo, even though six of the 13 delimitations studied do include it (Fig. 1), as well as the three studied versions of the SGs also include it (Fig. 2) — in addition, the Spanish Ministry of Urban Agenda, in its statistical delimitation, considers Pontevedra part of metropolitan Vigo [31]. It is also unclear how key metropolitan policies such as transport or planning will be adequately developed on such a scale without metropolitan institutionalisation or, even worse, without a democratically elected and accountable metropolitan political authority [11, 15, 32].

References

1. Calzada, I.: Political regionalism: devolution, metropolitanization and the right to decide. In: Paasi, A., Harrison, J., Jones, M. (eds.) Geographies of Regions and Territories, pp. 231–242. Edward Elgar Publishing, Cheltenham/Northampton (2018)
2. Heinelt, H., Kübler, D. (eds.): Metropolitan Governance: Capacity, Democracy and the Dynamics of Place. Routledge, Abingdon (2005)
3. Tomàs, M., Martí, M.: La reconfiguración de la agenda urbana: el debate europeo. In: Iglesias, M., Martí, M., Subirats, J., Tomàs, M. (eds.) Políticas Urbanas en España: Grandes Ciudades, Actores y Gobiernos Locales, pp. 23–41. Icaria, Barcelona (2011)
4. Cotella, G., Janin-Rivolin, U., Pede, E., Pioletti, M.: Multi-level regional development governance: a european typology. Eur. Spatial Res. Policy **28**(1), 201–221 (2021)
5. ESPON. The Role and Future Perspectives of Cohesion Policy in the Planning of Metropolitan Areas and Cities. ESPON, Luxembourg (2021)

6. Instituto Nacional de Estadística (INE). https://www.ine.es. Accessed 1 Oct 2021
7. Nel·lo, O.: Las áreas metropolitanas. In: Gil-Olcina, A., Gómez-Mendoza, J. (eds.): Geografía de España, pp. 275–298. Ariel, Barcelona (2001)
8. Sorribes, J., Romero, J.: El fracaso de las experiencias de gobierno metropolitano en España. In: Farinós, J., Romero, J. (coords.): Gobernanza Territorial en España: Claroscuros de un Proceso a Partir del Estudio de Casos, pp. 397–414. UV, València (2006)
9. Kearns, A., Paddison, R.: New challenges for urban governance. Urban Stud. **37**(5–6), 845–850 (2000)
10. Savitch, H.V., Vogel, R.K.: Suburbs without a city: power and city-county consolidation. Urban Affair. Rev. **39**(6), 759–779 (2004)
11. Tomàs, M.: Exploring the metropolitan trap: the case of Montreal. Int. J. Urban Reg. Res. **36**(3), 554–567 (2012)
12. Eichenberger, R., Frey, B.S.: Functional, overlapping and competing jurisdictions: a complement and alternative to today's federalism. In: Ahmad, E., Brosio, G. (eds.): Handbook of Fiscal Federalism, pp. 154–181. Edward Elgar Publishing, Cheltenham/Northampton (2006)
13. Frisken, F., Norris, D.F.: Regionalism reconsidered. J. Urban Aff. **23**(5), 467–478 (2001)
14. Souto, X.M.: Un área metropolitana para as Rías Baixas: vellos localismos para novos problemas cidadáns. Glaucopis **14**, 337–364 (2009)
15. Kübler, D., Wälti, S.: Metropolitan governance and democracy: how to evaluate new tendencies. In: McLaverty, P. (ed.) Public Participation and Innovations in Community Governance, pp. 99–122. Routledge, Abingdon (2017)
16. Sellers, J.M., Kübler, D., Walter-Rogg, M., Walks, R.A. (eds.): The Political Ecology of the Metropolis. ECPR, Colchester (2013)
17. Lois, R.C.: La ciudad de Vigo en el centro de los análisis de Geografía Urbana en Galicia. Investig. Geográf. **10**, 135–142 (1992)
18. Souto, X.M.: Os problemas de rexionalización no sur de Galicia: o caso da rexión urbana de Vigo. In: III Coloquio Ibérico de Geografía, pp. 430–438. UB, Barcelona (1984)
19. Souto, X.M.: O problema da rexionalización en Galicia: encol da rexión urbana de Vigo. In: I Cuaderno de Xeografía, pp. 153–182. Ediciós do Castro, A Coruña (1985)
20. Souto, X.M.: Planeamento estratéxico e área metropolitana de Vigo. Glaucopis **8**, 203–219 (2002)
21. Precedo, A. (dir.): Vigo, Área Metropolitana. Fundación Caixa Galicia, A Coruña (1989)
22. Precedo, A.: O Plan de Desenvolvemento Comarcal de Galicia. Xunta de Galicia, Santiago de Compostela (1994)
23. Paül, V., Pazos, M.: Els darrers capítols del debat al voltant del mapa immutable de Galícia. Treball. Soc. Catal. Geograf. **67–68**, 199–229 (2009)
24. Lois, R.C.: Estructura Territorial de Galicia. In: Rodríguez-González, R. (dir.): Os Concellos Galegos Para o Século XXI, pp. 101–160. USC, Santiago de Compostela (2004)
25. Lois, R.C., Piñeira, M.J.: A rede urbana e a rápida urbanización do territorio. In: Piñeira, M.J., Santos, X.M. (coords.): Xeografía de Galicia, pp. 157–227. Xerais, Vigo (2011)
26. Rodríguez-González, R.: Hacer ciudad como acción pública. In: Rodríguez, R. (dir.): Ordenación y Gobernanza de las Áreas Urbanas Gallegas, pp. 153–195. Netbiblo, Oleiros (2009)
27. Aldrey, J.A., Vicente, J.: La Galicia de las mil ciudades: ordenación y funcionalidad de las áreas urbanas gallegas. In: Rodríguez-González, R. (dir.): Ordenación y Gobernanza de las Áreas Urbanas Gallegas, pp. 199–253. Netbiblo, Oleiros (2009)
28. Feria, J.M., Martínez-Bernabéu, L.: La definición y delimitación del sistema metropolitano español. Ciudad Territor. Estud. Territor. **187**, 9–24 (2016)
29. Rojo, A., López-Mira, Á., Varela, E.J.: Vigo. In: Iglesias, M., et al. (eds.): Políticas Urbanas en España, pp. 229–265. Icaria, Barcelona (2011)

30. López-Mira, Á.: Gobernanza local en Galicia e áreas metropolitanas: a superación da manco-munidade de Vigo. In: Rojo, A., Varela, E.J. (eds.): A Gobernanza Metropolitana, pp. 129–163. Xunta de Galicia, Santiago de Compostela (2007)
31. Ministerio de Agenda Urbana. http://atlasau.mitma.gob.es/. Accessed 21 Dec 2021
32. Paül, V.: Rural zones, parks, greenbelts, landscapes…? Assessing the shifting role and treatment of open spaces in metropolitan planning using the barcelona experience (1953–2019). J. Environ. Plan. Manag. **64**(2), 224–251 (2021)

Assessing the SDG11 on a Neighborhood Scale Through the Integrated Use of GIS Tools. An Italian Case Study

Francesca Abastante⬛ and Marika Gaballo$^{(\boxtimes)}$ ⬛

Politecnico di Torino, 10125 Turin, Italy
{francesca.abastante,marika.gaballo}@polito.it

Abstract. Sustainable Development Goal (SDG) 11 of the 2030 Agenda addresses issues of sustainable urban development and provides guidelines for monitoring progress, that should occur primarily at the neighborhood scale from which development begins. Thus, since the early 2000s many countries have developed Neighborhood Sustainability Assessment Tools (NSATs) as multi-criteria assessment tools for neighborhood sustainability using criteria and indicators, with the aim of encouraging sustainable urban planning.

Although NSATs can be useful tools for choosing sustainable measures, they are not yet able to return an integrated view of urban sustainability. In parallel, Geographic Information Systems (GIS) are tools used in planning to manage and spatialize urban data and are often combined with multi-criteria analysis tools to define the actual priorities to be pursued in terms of urban sustainability.

Contributing to the literature on urban planning, this article explores the combination of SNAT and GIS tools with respect to the assessment of the sustainability of cities, with the aim of investigating at the Italian national level if and in what terms SNAT could be an operational tool to support urban planning, in view of the achievement of SDG11.

Keywords: Neighborhood Sustainability Assessment Tools (NSATs) ·
Sustainable Development Goal 11 (SDG11) · Geographic Information System
(GIS) · Spatial analyses

1 Introduction

The United Nations (UN) emphasizes that achieving the global sustainability of the Sustainable Development Goals (SDGs) can only be possible starting from the local urban level [1], considering that human development is concentrated in cities.

During the 21st century, the concept of sustainable cities has gradually gained political consensus [2] with reference to the guidelines of the Sustainable Development Goal (SDG) 11 framework of the 2030 Agenda on how to make cities and human settlements sustainable and inclusive. However, knowledge on how to measure all dimensions of urban sustainability in an integrated way is still limited [3–5].

© The Author(s), under exclusive license to Springer Nature Switzerland AG 2022
F. Calabrò et al. (Eds.): NMP 2022, LNNS 482, pp. 957–967, 2022.
https://doi.org/10.1007/978-3-031-06825-6_91

Since the 1990s, the research in the field of sustainable construction has developed the Sustainability Assessment Tools (SATs) [6] aiming at assessing the performance of individual buildings [7]. During the first decade of the 2000s, the SATs started including a wider vision considering the performance of neighbourhoods with the Neighbourhood Sustainability Assessment Tools (NSATs) [8] so to properly integrate different urban elements [9, 10]. Despite the NSATs being the main tools for urban management to date [1] they: i) do not inform on how they contribute to the achievement of the SDG11 [7, 11]; ii) rely particularly on quantitative indicators, rarely considering qualitative values [12, 13] and failing to capture the interactions among different urban sustainability issues [1, 12].

The Geographic Information Systems (GIS) tools originated with the aim of associating data with their geographical position and processing them to extract different information that can be displayed in integrated maps [14]. Over time their role in spatial analysis to support urban decision-making problems has been increasingly recognized [15]. However, even though these tools are widely used to support urban planning, they fail to represent the priorities to be considered in different sustainability goals [16–18].

This contribution has a double aim: i) investigate how the SNATs could be operational tools to support local public decision-makers (DMs) in evaluating plans and projects with a view to achieving SDG11 [15, 19]; ii) how to integrate the achievement of SDG11 within the NSAT Green Building Council Neighbourhoods (GBC Neighbourhoods) [20] by using GIS starting from the analytical approach reported in [7].

Following this introduction, this paper is organized as follows: Sect. 2 clarifies what aspects of urban sustainability the SDG11 indicators focus on and how the GBC Neighbourhoods manages to capture them in its assessment framework; Sect. 3 delves into the research design used; Sect. 4 shows the results obtained on a case study of the city of Turin in Italy; Sect. 5 outlines the discussion and the conclusions of the methodology and the future developments of the research.

2   Assessing Urban Sustainability: Criteria and Indicators

The UN response to the new sustainability is expressed by the 2030 Agenda [21], which include 17 Sustainable Development Goals (SDGs), each of which is composed of specific targets that delve into the phenomena investigated and related statistical indicators, capable of expressing a specific phenomenon in the time frame examined [7, 11].

The 2030 Agenda recognizes the important role of cities in the achievement of sustainable development, since these are the main places where criticalities develop in both environmental, social and economic terms [7]. In fact, the SDG11 provides a reference model for the pursuit of urban sustainable development considering several interconnected issues [11, 22] articulated in specific targets and related indicators identified by the competent bodies, which change according to the global, European, and national scale.

Accordingly, for the SDG11 at the Italian level the ISTAT identifies 8 targets to which correspond 30 indicators, elaborated from the global model considering the specificities of the Italian territory [23]. Most of the indicators relate to "Air quality and temperature", followed by "Natural disaster", underlining an environmental focus within the evaluation framework, and almost completely omitting the social aspects that consider only 3.3% with respect to "Sexual harassment".

Since the 1990s, to assess the performance of urban products from a sustainability point of view many countries at the international level have started to develop the SATs at the single building scale [6]. Considering the continuous urban sprawl and the possible creation of closed communities that would fail in the pursuit of the SDG11 [24], achieving broader sustainability at the neighbourhood level has been progressively more fundamental than that of the single building [7]. In fact, a neighbourhood represents the basic element from which changes in terms of urban sustainability in a broader sense can take place, being able to observe its social, economic, institutional relationships, as well as monitoring its environmental conditions [11].

In this perspective, during the first decade of the 2000s, emerged the need for multi-criteria analyses that could assess the sustainability of urban products in a broader sense [25–27] leading to the development of the NSATs [8].

The NSATs assess the performance of neighbourhoods against a range of sustainability issues [28], encouraging planners and local decision makers to consider dimensions relevant to sustainable urban development [29].

Fig. 1. General structure of the NSATs

The general structure of the NSATs (Fig. 1) consists of a hierarchical system composed of sustainability categories specified in criteria or credits, assessed by indicators or by assigning weights, which represent the basic element for the assessment [6].

At the Italian national level, the most widely used NSATs are:

i. The GBC Neighbourhoods [20], composed of categories, credits and weights;
ii. The ITACA Urban Scale [30], composed of categories, criteria and indicators.

Both NSATs considered aim to provide guidelines on how to achieve sustainable urban development through neighbourhood performance assessment processes [31].

Although they do not make it explicit, these NSATs have a relationship with the SDG11, emphasizing a complementarity in terms of objectives, phenomena investigated and structure in terms of their criteria/credits and the ISTAT SDG11 indicators [7] (Fig. 2).

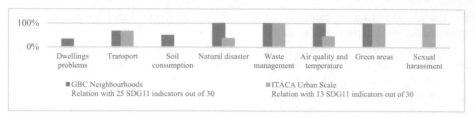

Fig. 2. NSATs GBC Neighborhoods and ITACA Urban Scale ability to capture the ISTAT SDG11 indicators

Figure 2 shows the ability of the GBC Neighbourhoods and the ITACA Urban Scale NSATs to capture the ISTAT SDG11 indicators, considering for each thematic area how many of the ISTAT SDG11 indicators are related to the criteria/credits of the NSATs. Figure 2 shows a prevalence of aspects related to environmental issues at the expense of social ones. Note that the GBC Neighbourhoods reports more relations than the ITACA Urban Scale, in fact covering almost all the thematic areas of the ISTAT SDG11 indicator's framework [7] and reporting overall with 25 out of 30 SDG11 indicators compared to 13 in the ITACA Urban Scale.

3 Research Design

This research considers the GBC Neighbourhoods SNAT since it shows most relationships with the SDG11 among the Italian SNATs (Fig. 2) [7].

The indicators were calculated and spatialized to test the combination of the GBC Neighbourhoods with the GIS tool and return an integrated assessment of the elements of the urban sustainability considered in the achievement of the SDG11.

The case study research method [32, 33] was used to collect and analyse the data of the ISTAT SDG11 indicators, to be able to represent the results obtained and outline new directions for future research developments [34]. As the case study had to allow for easy data retrieval [34], the city of Turin in Italy was considered fertile ground as it is the place where the research was developed. In particular, the city of Turin is the Regional County Seat of the Piedmont Region, which occupies 8th place in the ranking of the 20 Italian regions in the achievement of the SDGs and 4th place for SDG11 [35].

The assessment approach used included the unconventional use of the GBC Neighbourhoods to a specific neighbourhood but to the entire city of Turin [15], which was considered in terms of its 8 Districts, being the intermediate administrative units between the city and the statistical zones [36].

In this perspective, the phases of the methodology used with respect to the ISTAT SDG11 indicators having relations with the GBC Neighbourhoods included:

i. The choice and data collection phase, related to the indicators' choice and relative data considering both their cogency with the city of Turin in terms of the phenomenon measured and the availability of certain and updated data collected from open sources and taking into account the information provided by both the ISTAT metadata [37] and the GBC Neighbourhoods guidelines [20];

ii. The assessment and spatialization phase, which was developed directly in a GIS environment respectively to the phenomena measured by the chosen indicators, where they were assessed and then spatialized with respect to the administrative units related to the 8 Districts of the city of Turin.

4 Results

4.1 Choice and Data Collection Phase

For the purposes of the case study, the requirements of territorial cogency with the city of Turin and the availability of certain and updated data were applied for the 25 ISTAT SDG11 indicators' selection (Fig. 3).

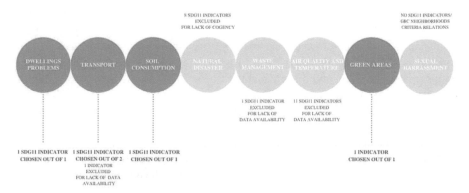

Fig. 3. ISTAT SDG11 indicators having relations with the GBC Neighborhoods

Figure 3 shows that among the 25 ISTAT SDG11 indicators reporting relations with GBC Neighbourhoods credits (Fig. 2), 21 were excluded considering the lack of cogency with respect to the case study. Therefore, the ISTAT SDG11 indicators chosen are 4 out of 25 (Table 1).

Table 1. Summary of the ISTAT SDG11 indicators chosen with the respective credits of the GBC Neighborhoods with which they relate.

	ISTAT SDG11 INDICATORS	**GBC NEIGHBORHOODS CREDITS**
DWELLINGS PROBLEMS	**11.1.1** *Percentage of people in homes with noise problems from neighbours or the street*	**Credit 16** *Acoustic Environment* (Category "Neighbourhood Organization and Planning")
TRANSPORT	**11.2.1** *Families by level of difficulty in connecting to public transport in the area where they live*	**Credit 3** *Accessibility to the public transport system* (Category "Site Location and Links")
SOIL CONSUMPTION	**11.3.1** *Waterproofing and per capita soil consumption*	**Credit 2** *Compact Development* (Category "Neighbourhood Organization and Planning")
GREEN AREAS	**11.7.1** *Incidence of urban green areas on the urbanized surface of cities*	**Credit 9** *Access to public spaces* (Category "Neighbourhood Organization and Planning")

For the spatialization of the 4 indicators on the case study, the retrieval of data has included open sources updated to 2020 and where possible to 2021 and has moreover considered both the information provided respectively by the metadata of the ISTAT [37] and by the guidelines of GBC Neighbourhoods [20]:

- the ISTAT indicator 11.1.1 "Dwelling problems" measures the percentage of people in dwellings exposed to noise problems [%] [37]. In parallel the GBC credit 16 "Acoustic Environment" aims to ensure suitable acoustic levels for people, measured by scores from 1 to 2 whether the acoustic exposure meets the maximum legal requirements in terms of dB(A) emission [20]. Accordingly, the shapefile updated to 2021 of the Acoustic Classification Plan of the province of Turin of ARPA Piemonte [38] and the data related to the resident population for each of the 8 Districts updated to 2020 [39] were considered;
- the ISTAT indicator 11.2.1 "Transport" measures the percentage of families with difficult access to public transport in residential areas [%] [37] and in parallel the GBC credit 3 "Accessibility to the public transport system" assesses in a catchment area less than 400 m from residential and non-residential buildings how many public transport stops are present and their frequency, assigning scores from 1 to 7 [20]. The open data of BDTRE updated to 2021 [40] and the shapefiles of the public transport stops updated to 2021 and downloadable in open mode from Open Street Map [41] were considered for this indicator;
- the ISTAT 11.3.1 indicator "Soil consumption" measures the per capita soil consumption [sqm/inhabitant] [37] while the GBC Credit 2 "Compact Development" aims to maintain a density that occupies less permeable area, assesses through scores from 1 to 6 the density of built-up area on the total reference area [20]. For this indicator the open data of the BDTRE and the resident population of the 8 Districts were taken into account;
- ISTAT indicator 11.7.1 "Green areas" measures the incidence of green areas on the urbanized area [%] [37] and in parallel the GBC credit 9 "Access to public spaces", aiming to increase urban public space with attention to green areas, evaluates their size in relation to the reference area and proximity to buildings by assigning 1 point [20]. For this indicator also the open data of the BDTRE was considered.

4.2 Assessment and Spatialization Phase

The collected data have been reported into the GIS environment, where the assessment functions of the tool have been used to assess and then spatialize the chosen indicators.

For the indicator 11.1.1, considering the assessment information of the GBC Neighbourhoods, only the Class II and III areas of the Acoustic Classification Plan have been identified (related to the predominantly residential and mixed type areas respectively) and in which an exceeding of the limits of dBA for a basic comfort occurs, fixed at 60 for the daytime period (06:00–22:00) and 50 for the night-time period (22:00–06:00). These data have been consequently crossed to the data relative to the resident population of the 8 Districts (Fig. 4).

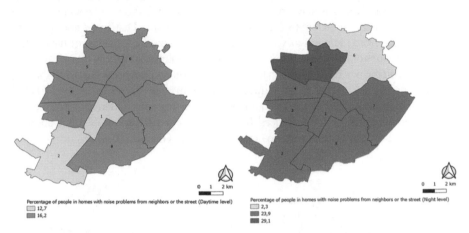

Fig. 4. ISTAT SDG11 indicator 11.1.1 (Daytime on the left and Night on the right)

Figure 4 shows that in general the noise problems occur during the night. In fact, during the night period only District 6 shows a percentage of people exposed to noise problems beyond the allowed dBA limits of less than 3% even though it has not the lowest resident population, while the other Districts exceed the 23%. During the day, Districts 1 and 2 report a population exposed over the limits of 12.7%, considering that they are respectively the ones with the lowest and highest resident population, while the other Districts report the 16.2%.

Regarding the indicator 11.2.1, the residential and mixed buildings (productive, commercial and offices) have been extracted from the BDTRE to which the shapefiles of the public transport stops have been crossed. Considering the guidelines provided by the GBC Neighbourhoods, to understand the density of the public transport stops the GIS calculation tool of Kernel Density Estimation [42] was applied, to report their diffusion within a 400 m radius (Fig. 5).

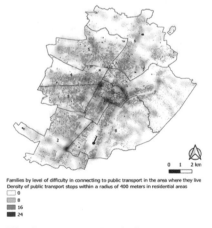

Fig. 5. ISTAT SDG11 indicators 11.2.1

Figure 5 shows how the density of stops is concentrated in District 1 (24 stops within 400 m) which corresponds to the city's centre. The District 2 shows a concentration in proximity of residences adjacent to the hospital area, while the Districts 7 and 8 are the most problematic from the point of view of connection to public transport of residential areas, probably due to the morphology of the hilly terrain that concerns them.

Moreover, on the one hand to obtain the per capita soil consumption [sqm/inhabitant] for the indicator 11.3.1, all the areas of the impermeable zones provided by the BDTRE were calculated and related to the resident population for each District; while on the other hand the indicator 11.7.1 considers the ratio between the surface of green areas of the BDTRE and the total urbanized area relative to each Districts (Fig. 6).

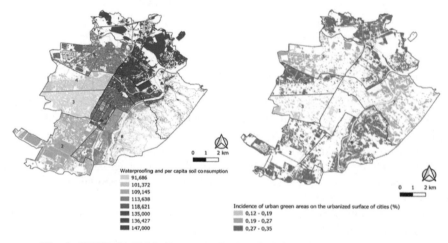

Fig. 6. ISTAT SDG11 indicators 11.3.1 (on the left) and 11.7.1 (on the right)

The integrated reading of the indicators 11.3.1 and 11.7.1 in Fig. 6 shows how, although District 6 has the highest per capita land use (147 sqm/inhabitant) it also reports the highest incidence of green permeable areas (0.27–0.35%). In parallel, District 3 despite having the lowest per capita land consumption (92 sqm/inhabitant) it also has the lowest percentage of green areas together with District 1, which however has also a high per capita land consumption of about 118 sqm/inhabitant.

5   Discussion and Conclusions

This paper has the general objective of investigating at the Italian level how the SNATs could be operational tools to support local public decision makers in evaluating plans and projects from a SDG11 attainment perspective, considering also the current international debate on the importance of localizing the SDGs to foster sub-national accountability in the pursuit of the 2030 Agenda [15, 19].

In this study, the GIS tool combined with GBC Neighbourhoods tested on the case study provides an operational basis for a possible integrated assessment of the effectiveness of plans and projects, in consideration of urban sustainability oriented to the achievement of the SDG11 within the planning practices. The methodology developed according to the two phases proposed succeeds in fact to return an integrated view of the different phenomena measured by the ISTAT SDG11 indicators with respect to each District. Indeed, the approach used passed in a complementary and simultaneously way from the visualization that provides easily readable outputs, to the results assessment that helps public decision makers to "get on the same page" [43] with an integrated view of the urban sustainability elements considered [42]. In fact, the GIS-based spatial analysis allows highlighting a comparison between the 8 Districts useful to outline intervention priorities for the development of sustainable urban policies [15].

However, this work underlined some challenges. First the lack of quality data in terms of updating and reliability required the combination of several datasets [15, 44], allowing at this stage of the research to analyse only four ISTAT SDG11 indicators and underlining a limitation of the work in integrating the different phenomena useful for achieving integrated sustainable urban development. Second, many of the GBC Neighbourhoods credits appear difficult to assess, since they refer only to guidelines making therefore particularly complicated the use of statistical data, that instead require more detailed information.

Future developments of the research will attempt to include the expansion of the datasets in consideration of additional ISTAT SDG11 indicators and the related comparison of results to define a global sustainability score, providing an evaluation model adaptable to different geographical contexts.

References

1. Boyle, L., Michell, K., Viruly, F.: A critique of the application of neighborhood sustainability assessment tools in urban regeneration. Sustainability **10**(4), 1005 (2018)
2. Dempsey, N., Bramley, G., Power, S., Brown, C.: The social dimension of sustainable development: defining urban social sustainability. Sustain. Dev. **19**, 289–300 (2011)
3. Garde, A.: Sustainable by design: Insights from U.S. LEED-ND pilot projects. J. Am. Plan. Assoc. **75**, 424–440 (2009)
4. Conte, E., Monno, V.: Beyond the buildingcentric approach: A vision for an integrated evaluation of sustainable buildings. Environ. Impact Assess. Rev. **34**, 31–40 (2012)
5. Komeily, A., Srinivasam, R.: A need for balanced approach to neighborhood sustainability assessment: a critical review and analysis. Sustain. Cities Soc. **12**, 32–43 (2015)
6. Castanheira, G., Bragança, L.: The evolution of the sustainability assessment tool SBToolPT: from buildings to the built environment. Sci. World J. **2014**, 1–10 (2014)
7. Abastante, F., Lami, I.M., Gaballo, M.: Pursuing the SDG11 targets: the role of the sustainability protocols. Sustainability **13**(7), 3858 (2021)
8. Cheshmehzangi, A., Dawodu, A., Song, W., Shi, Y., Wang, Y.: An introduction to neighborhood sustainability assessment tool (NSAT) study for China from comprehensive analysis of eight Asian tools. Sustainability **12**(6), 2462 (2020)
9. Benites, H.S., Osmond, P., Rossi, A.M.G.: Developing low-carbon communities with LEED-ND and climate tools and policies in Sao Paulo, Brazil. J. Urban Plan Dev. **146**(1), 04019025 (2020)
10. Sparshott, P.J., Darchen, S., Stjohn, D.: Do sustainability rating tools deliver the best outcomes in master planned urban infill projects? City to the Lake experience. Aust. Plann. **55**, 84–92 (2019)

11. Arslan, T.V., Durak, S., Aytac, D.O.: Attaining SDG11: can sustainability assessment tools be used for improved transformation of neighborhoods in historic city centers? In: Natural Resources Forum, vol. 40, no. 4, pp. 180–202. Blackwell Publishing Ltd., Oxford, UK (November 2016)
12. Sharifi, A., Dawodu, A., Cheshmehzangi, A.: Neighborhood sustainability assessment tools: a review of success factors. J. Clean. Prod. **293**, 125912 (2021)
13. Kaur, H., Garg, P.: Urban sustainability assessment tools: a review. J. Clean. Prod. **210**, 146–158 (2019)
14. Rogerson, P.A., Fotheringham, A.S.: Spatial Analysis and GIS. CRC Press (1994)
15. Pedro, J., Silva, C., Pinheiro, M.D.: Integrating GIS spatial dimension into BREEAM communities sustainability assessment to support urban planning policies. Lisbon case study. Land Use Policy **83**, 424–434 (2019)
16. Malczewski, J., Rinner, C.: Multicriteria Decision Analysis in Geographic Information Science, p. 331. Springer, New York (2015). https://doi.org/10.1007/978-3-540-74757-4
17. Ferretti, V., Montibeller, G.: Key challenges and meta-choices in designing and applying multi-criteria spatial decision support systems. Dec. Supp. Syst. **84**, 41–52 (2016)
18. Greene, R., Devillers, R., Luther, J.E., Eddy, B.G.: GIS-based multiple-criteria decision analysis. Geogr. Compass **5**(6), 412–432 (2011)
19. Barnett, C., Parnell, S.: Ideas, implementation and indicators: epistemologies of the post-2015 urban agenda. Environ. Urban. **28**(1), 87–98 (2016)
20. Green Building Council Italia: Manuale GBC Quartieri. Per progettare, realizzare e riqualificare aree e quartieri sostenibili (2015). https://www.gbcitalia.org/documents/20182/22088/Manuale+GBC+QUARTIERI+2015+def.pdf/b6cabb2a-200e-4404-b5d0-dffb9607b36c. Accessed 27 Oct 2021
21. UN-HABITAT: SDG goal 11 monitoring framework (2016). https://unhabitat.org/sdg-goal-11-monitoring-framework. Accessed 27 Oct 2021
22. Vaidya, H., Chatterji, T.: SDG 11 sustainable cities and communities. In: Franco, I.B., Chatterji, T., Derbyshire, E., Tracey, J. (eds.) Actioning the Global Goals for Local Impact. SSS, pp. 173–185. Springer, Singapore (2020). https://doi.org/10.1007/978-981-32-9927-6_12
23. Abastante, F., Lami, I., Mecca, B.: How Covid-19 influences the 2030 Agenda: do the practices of achieving the Sustainable Development Goal 11 need rethinking and adjustment? Valori e Valutazioni **26**, 11–23 (2020). https://doi.org/10.48264/VVSIEV-20202603
24. Candan, A.B., Kolluoğlu, B.: Emerging spaces of neoliberalism: a gated town and a public housing project in Istanbul. New Perspect. Turk. **39**, 5–46 (2008)
25. US Green Building Council (USGBC): LEED: Leadership in energy and environmental design (2016). https://www.usgbc.org. Accessed 27 Oct 2021
26. Haapio, A., Viitaniemi, P.: A critical review of building environmental assessment tools. Environ. Impact Assess. Rev. **28**, 469–482 (2008)
27. Berardi, U.: Sustainability assessment of urban communities through rating systems. Environ. Dev. Sustain. **15**(6), 1573–1591 (2013). https://doi.org/10.1007/s10668-013-9462-0
28. Sharifi, A., Murayama, A.: A critical review of seven selected neighborhood sustainability assessment tools. Environ. Impact Assess. Rev. **38**(1), 73–87 (2013)
29. Damen, R.G.: Evaluating urban quality and sustainability - presentation of a framework for the development of indicator assessment methods, by which the existing urban environment may be evaluated on quality and sustainability performance on a neighborhood scale. Master thesis, University of Twente (2014)
30. Istituto per l'innovazione e trasparenza degli appalti e la compatibilità ambientale and Ente Italiano di Normazione. Prassi di Riferimento (UNI/PdR 13.0:2019), Sostenibilità ambientale nelle costruzioni—Strumenti operativi per la valutazione della sostenibilità—Inquadramento generale e principi metodologici, Milano (2019). https://www.ediltecnico.it/wp-content/uploads/2019/07/UNI21000963_EIT.pdf. Accessed 27 Oct 2021

31. Aguiar Borges, L., Hammami, F., Wangel, J.: Reviewing neighborhood sustainability assessment tools through critical heritage studies. Sustainability **12**(4), 1605 (2020)
32. Gerring, J.: Case Study Research: Principles and Practices. Cambridge University Press, Cambridge, UK (2006)
33. Harrison, H., Birks, M., Franklin, R., Mills, J.: Case study research: foundations and methodological orientations. In: Forum Qualitative Sozialforschung/Forum: Qualitative Social Research, vol. 18. Freie University, Berlin, Germany (2017)
34. Abastante, F., Gaballo, M.: How to assess walkability as a measure of pedestrian use: first step of a multi-methodological approach. In: Bevilacqua, C., Calabrò, F., Spina, L.D. (eds.) NMP 2020. SIST, vol. 178, pp. 254–263. Springer, Cham (2021). https://doi.org/10.1007/978-3-030-48279-4_24
35. Regione Piemonte: Piemonte verso un presente sostenibile. Il posizionamento del Piemonte rispetto all'Agenda 2030 - Position Paper (Giugno 2021)
36. Ture, D.G.: Applicazione dei sistemi GIS nel verde urbano-L'esperienza della città di Torino. GEOmedia **21**(6), 1–4 (2017)
37. ISTAT: Italian Data for UN-SDGs-Sustainable Development Goals of the 2030 Agenda, 14 Maggio 2020. https://www.istat.it/storage/SDGs/SDG_11_Italy.pdf. Accessed 27 Oct 2021
38. ARPA Piemonte: Piani di Classificazione Acustica delle province di Torino, Novara ed Asti. Aggiornamento 2021. https://webgis.arpa.piemonte.it/geoportale/index.php/notizie-mob/22-rumore-e-vibrazioni/457-piani-di-classificazione-acustica-aggiornamento. Accessed 27 Oct 2021
39. Città di Torino: Popolazione registrata in anagrafe per genere e circoscrizione - Dati al 31/12/2020. https://dati.gov.it/view-dataset?groups=societa&organization=comune-di-torino&page=4. Accessed 27 Oct 2021
40. Regione Piemonte: GeoPiemonte. Territorial Reference Database of Bodies (BDTRE-2021). https://www.geoportale.piemonte.it/cms/bdtre/modalita-di-pubblicazione-e-fruizione. Accessed 27 Oct 2021
41. Open Street Map: Trasporti Pubblici di Torino (2021). https://www.openstreetmap.org/#map=12/45.0688/7.6642&layers=T. Accessed 27 Oct 2021
42. Abastante, F., Lami, I.M., La Riccia, L., Gaballo, M.: Supporting resilient urban planning through walkability assessment. Sustainability **12**(19), 8131 (2020)
43. Vennix, J.: Group Model Building: Facilitating Team Learning Using Systems Dynamics. Wiley, London, UK (1996)
44. Abin, S.M., Shaju, A.E., Jishnu Priya, K.C., Abhijith, V., Jose, T.: Role of GIS & analytics in transforming municipal administration. Int. J. Eng. Res. Technol. (IJERT) **09**(06), 85–91 (2021)

Methodological Proposal for the Sustainability of Sports Events

Tiziana Binda[1](✉), Sara Viazzo[2](✉) (iD), Marta Bottero[1](✉) (iD),
and Stefano Corgnati[2](✉) (iD)

[1] Interuniversity Department of Regional and Urban Studies and Planning,
Politecnico di Torino, Viale Mattioli 39, 10125 Torino, Italy
{tiziana.binda,marta.bottero}@polito.it
[2] TEBE-IEEM Research Group, Energy Department, Politecnico di Torino,
Corso Duca degli Abruzzi 24, 10129 Torino, Italy
{sara.viazzo,stefano.corgnati}@polito.it

Abstract. This paper provides a methodological proposal for the sustainability of sports events. Nowadays the topic of sustainability is of great interest due to the dangers of climate change the world is experiencing, mainly caused by human activities. The concept of sustainability has increasingly gained relevance in areas ranging from the efficiency of buildings, public transportation, and also in the field of major sporting events. However, at present there is no methodology that allows the level of sustainability of sports events to be estimated. This paper proposes a methodology that quantifies levels of sustainability by assigning a score to a number of carefully selected indicators, using a multicriteria approach, with the aim of producing a guideline for future sports events. The originality of this article lies in the creation of a methodology to quantify the level of sustainability of sporting events. Specifically, this novel methodology uses the existing methodologies which produce international building certifications, such as LEED or BREEM, or national certifications such as the Italian ITACA. A literature review of past sustainability reports was conducted in order to find quantifiable indicators for comparing the level of sustainability of different sporting events. The contribution of this article is to provide a methodology to support the figures involved in putting on such events, in order to create sports events that have a minimum environmental impact.

Keywords: Sports events · Sustainability · Certification

1 Introduction

Nowadays the topic of sustainability is very important, especially in light of the growing crisis of climate change. For this reason, it is necessary to act in different areas in order to meet the climate goals of the Paris Agreement and the Agenda 2030, reaching zero emissions of CO_2 by 2030 [1, 2]. One important action area is events organization. Numerous environmental certification systems have been developed in order to reduce the environmental impact of human activities, in particular in the field of construction

© The Author(s), under exclusive license to Springer Nature Switzerland AG 2022
F. Calabrò et al. (Eds.): NMP 2022, LNNS 482, pp. 968–978, 2022.
https://doi.org/10.1007/978-3-031-06825-6_92

and to some degree also in the organization of events. In fact, in the field of events, the United National Program has published a guideline that helps organizers of small and medium-sized events to make them more environmentally friendly [4].

At present, the international standards for environmental certification for the events sector are ISO 20121:2012 Event sustainability management systems—Requirements with guidance for use, and Green Meetings and Global Reporting Initiative-Events (GRI). ISO 20121 defines how to manage sustainable events assigning an environmental certification [4]. The first event which obtained this certification was the London Olympics games in 2012. On the other hand, GRI gives guidelines to develop standards to measure and report on the financial, environmental and social performance of different types and sizes of events [5].

These certifications provide useful guidelines for the sustainable management of events. However, they do not offer a standard methodology that allows the level of sustainability of different events to be compared [7, 8]. Instead, in the field of construction there are methodologies that measure the level of sustainability and assign a certificate. These systems aim to assess the level of sustainability of a building but also take into consideration the construction and maintenance phases. There are various sustainability certificates for buildings, such as the international sustainability model LEED (Leadership, Energy and Environmental Design) developed by the US Green Building Council, and also national sustainability models, e.g. ITACA for Italy [9, 10].

The innovation introduced by this paper is to propose a methodology for quantifying the level of sustainability of events. The Sustainable Sports Events (SSE) methodology proposed in this paper is based on the methodology used for building certification but is applied to sports events. The sustainability reports published by past events were used to develop the indicators for the methodology which would make it possible to measure the environmental, economic and social impact of sports events. A literature review of reports from the last 15 years was conducted in order to identify suitable indicators to measure the level of sustainability of sporting events.

The paper is divided into three parts. The first part describes how the SSE methodology was developed, while the second part shows how the SSE method could be applied. The last part is dedicated to the discussion of how to further apply this methodology in order to increase the sustainability of sports events.

2 Methodological Proposal

2.1 Literature Review

From the literature review of the sustainability reports of major sports events of the past it emerges that no standard report of sustainability to follow is present. In fact, Olympic sustainability reports are different from each other, using different categories, subcategories, and indicators. An example regards the "air quality" indicator, which is present in the sustainability report of Vancouver 2010, but is not present in the sustainability report of Turin 2006.

The aim of this paper is to propose a standard methodology which allows the level of sustainability of events to be estimated and compared. To achieve this goal, an analysis was conducted of the sustainability reports from past Olympic Games in order to establish subcategories and indicators. The analysis of these reports spans from the Turin Olympic Games 2006 to Pyongyang 2018. The last Olympic Games, Tokyo 2020, are not taken into account due to Covid19. Regarding the Covid19 pandemic, the World Health Organization has stipulated a document in order to organize sporting event safely for all the people involved in their management, but also participants and fans [11, 12]. The literature review was conducted in two phases. The first step aimed to identify the categories. These categories are subdivided according to the concept of sustainable pillars, which are: *environmental*, *economic*, and *social*. The second step was to identify the measurable subcategories, as reported in the following table (Table 1).

Table 1. Bibliographic analysis of sustainability reports to identify the subcategories [13–20].

	Environmental	Economic	Social
Torino 2006	Location	Tourism	Accessibility
	Waste Material		
	Mobility		
Beijing 2008	Location	Tourism	Accessibility
	Waste material		
	Mobility		
	Water efficiency		
	Sustainable material		
	Air quality		
	Energy performance		
Vancouver 2010	Location	Tourism	Accessibility
	Waste material		Gender equality
	Water Efficiency		
	Sustainable material		
	Air quality		
	Energy performance		
London 2012	Location	Tourism	Accessibility
	Waste material		Gender equality
	Water efficiency	Tourism	Accessibility
	Sustainable material		Gender equality
	Air quality		
	Energy performance		
Sochi 2014	Location	Tourism	Accessibility
	Water efficiency	Local business	

(*continued*)

Table 1. (*continued*)

	Environmental	Economic	Social
Rio 2016	Waste material	Tourism	Accessibility
	Water efficiency		Gender equality
	Air quality		
	Energy performance		
	Mobility	Tourism	Accessibility
Pyongyang 2018	Location		
	Mobility		
	Waste material		
	Water efficiency		
	Air quality		
	Energy performance		

2.2 Categories, Sub-categories and Criteria

In this section the subcategories and the criteria necessary in order to estimate the level of sustainability of the sports event are identified. As detailed in the previous paragraph, the model is divided in three categories based on the fundamental sustainable pillars, which are:

– *environmental*
– *economic*
– *social*

Each of the categories is composed of several subcategories. In detail, the *environmental* subcategory is composed of:

1. *Location:* the place is the first decision taken and plays a fundamental role in order to achieve sustainability [8]. The location choice is influenced by various issues, in particular transport connections and ease of reaching the area for the participants. The criterion used to measure the location is:

    1.1. Area: this index estimates the area occupied by the event. This criterion measures the reduction in percentage of the surface occupied by the event with respect to a past event.

2. *Mobility:* this index takes into account the internal travel arrangements of the people involved, with the aim of increasing sustainable transport, in detail:

    2.1. Percentage of travel on foot, on public transportation, by bicycle or carsharing.

3. *Water efficiency*: this subcategory aims to reduce the use of water estimated by different criteria; in detail:

    3.1. Water use reduction: the score is related to the percentage of water-saving, for example by installing a dual flush toilet.
    3.2. Water re-use: the percentage of reuse water, such as the sewage and the rainwater, is measured with respect to the model created.
    3.3. Reduction of plastic water bottles: the score is calculated by the percentage of plastic water bottle reduction.

4. *Energy performance*: the aspect of energy consumption is crucial during the events because they require a lot of energy for HVAC systems and for artificial lighting. The subcategories aim to reduce the environmental impact by:

    4.1. Reduction of energy consumption: the score is given by the percentage of energy consumption saved, for example using led lamps, or upgrading the HVAC system.
    4.2. Using renewable energy sources: the score is given by the percentage of energy created by renewable sources, such as photovoltaic panels or a heating pump.

5. *Air pollution:* this section aims to quantify the pollutants that are produced during the event, in particular to estimate the reduction of pollution, in detail:

    5.1. Reduction of CO_2 emissions.
    5.2. Reduction of PM emissions.
    5.3. Reduction of NOx emissions.
    5.4. Reduction of SOx emissions.

6. *Sustainable materials and recycling management.* This subcategory is composed of two criteria:

    6.1. Waste management: this has an important effect on the environment if it is not managed properly; the percentage is given by the material recycled compared to all the waste material produced during the event [19].
    6.1. Sustainable material: these criteria are estimated by the percentage of all the eco-friendly material with an EU environmental certification e.g. eco-label compared to all the materials.

The economic category is composed of two sub-categories, which measure the benefits deriving from the event, in detail:

7. *Enhancing and encouraging tourism*: the score is given by the percentage of tourism during the event compared with past tourism in the same period of the year [18].
8. *Local companies:* the local companies have to be encouraged in order to create a sustainable economy. In detail, two criteria to measure this subcategory are identified:

8.1. Zero km companies: the percentage relates to the local companies with respect to the total companies present during the event.

8.2. Catering: catering has an important impact on the sustainability. The score is given by the percentage of seasonal products or zero km products for food and beverage with respect to all the food and beverage used during the event [22].

The last category is the social one; normally in the literature this category is expressed by guidelines instead of indicators. For this category two subcategory are identified, expressed as follows:

9. *Gender equality*: this criterion measures the equity of gender (female and male) in the workplace during the event [23].

10. *Enhancement of accessibility:* this subcategory consists of three criteria:

10.1. Accessibility: this regards the architectural structure, in order that everyone can access it. Accessibility of the facilities is required by law [24], for this reason it is mandatory.

10.2. Seating places: this criterion refers to the percentage of seating places dedicated to people with disabilities [25].

10.3. Information accessibility: this indicator is based on the accessibility of information. Some elements are mandatory, such as signage and sign language. While braille language is not obligatory [26], its presence gives a higher score.

2.3 Methodology to Calculate the Score

The aim of the paper is to provide a methodology in order to evaluate the level of sustainability of the sports events (SSE). In detail, the SSE methodology is based on three categories and each of these is composed of different subcategories. Each subcategory can reach a maximum score equal to 10, and the total sum of categories give a total score equal to 100. The level of sustainability is different based on the total score achieved. In detail, there are 4 levels of sustainability, specifically:

1. Basic (40–49 points)
2. Sufficient (50–59 points)
3. Good (60–79 points)
4. Excellent (80–100 points)

The total score is based on the different range of scores which each criterion can reach, in detail, the range of scores are as follows: a) 0 to 10, b) 0 to 5, c) 0 to 4, d) 0 to 3 and e) 0 to 2. The scores are quantified in two different ways:

– Percentage (%): criteria can reach a maximum of points equal to 10, 5, 4, 3 and 2. A percentage of 100% corresponds to the maximum score x, for the assignment of the score ranging from 1 to x intervals of corresponding percentages defined by dividing 100% by x, as reported in Table 2. For the "gender equality" criterion the maximum percentage is equal to 50%, instead of 100%, therefore the percentage intervals are calculated by dividing 50% by x [21]. Thus, the "seating place" criterion has a starting percentage different to 1%, because 2.2% is the minimum percentage by law [23].

Table 2. Example of percentage classes for score definition (10 points).

Score	% Range
1	1%–10%
2	11%–20%
3	21%–30%
4	31%–40%
5	41%–50%
6	51%–60%
7	61%–70%
8	71%–80%
9	81%–90%
10	91%–100%

– Yes/no, in this case the percentage is given by the presence or the absence, specifically, yes corresponds to 100%, instead no correspond to 0%. These answers are translated as score, for example 3/0 or 0/3. Figure 1 represents an example of a weighting system for SSE referring to macro-areas, criteria (sub-categories) and indicators.

Figure 1 represents an example of a weighting system for SSE referring to macro-areas, criteria (sub-categories) and indicators.

Fig. 1. Examples of weighting system for SSE referring to macro-areas, criteria and indicators. Starting from the left: macro-areas, sub-categories and indicators regarding the subcategories "air quality & pollution".

3   Results

The methodology described in the previous sections, derived from the literature review, was used in order to identify the indexes which compose the SSE model. Table 3 shows the outcome of the SSE, which could be replicated for other events. Then, to understand better how the scoring works, a fictitious case study was created based on how sporting events are currently organized (see Fig. 2).

Fig. 2. Example of fictitious case study.

Table 3. Category, subcategory, criterion, and score of SSE methodology.

Category	Sub-category	Criterion	Unit	Score	
Environmental	1. Location	1.1. Area/biodiversity	%		10
	2. Mobility	2.1. On foot/bike/public transport/car sharing	%		10
	3. Water efficiency management	3.1. Water use reduction	%	3	10
		3.2. Water re-use		5	
		3.3. Plastic water bottle reduction		2	
	4. Energy performance	4.1. Reduction of energy consumption	%	5	10
		4.2. Using renewable		5	
	5. Air quality and Pollution	5.1. Reduction of CO_2 emissions	%	3	10
		5.2. Reduction of PM emissions		4	
		5.3. Reduction of NOx and SOx emissions		3	
	6. Material management	6.1. Waste management	%	5	10
		6.2. Sustainable material use		5	
Economic	7.Tourism	7.1. Enhancing and encouraging tourism	%		10
	8. Local company	8.1. Zero Km company	%	5	10
		8.2. Catering: food and beverages (zero km products/seasonal products)		5	
Social	8. Gender equality	8.1. Workplace equity	%		10
	9. Enhancement of accessibility	9.1. Accessibility to the event		Mandatory	
		9.2. Seating places	%	5	10
		9.3. Information Accessibility	Yes/No	5	

4   Conclusions and Future Developments

Today, the alarming situation of climate change has made it necessary to intervene to reduce CO_2 emissions caused by human activity before 2030. To reach this goal, numerous environmental certification systems have been developed to reduce environmental impact. Currently there are several international environmental certifications in the field of building construction. However, a similar certification has not yet been developed for the field of sports events.

The innovation of this paper was to develop a methodology in order to measure the level of sustainability of sports events (SSE), by assigning a score for various criteria. Part of the criteria were based on sustainability reports from past Olympic Games, which made it possible to evaluate the environmental, economic and social effects of sports events.

The SSE methodology is simple and direct, and it is easy to understand and to apply. The aim is to standardize the model in order to decrease the environmental impact created by sports events. This methodology could be an important contribution towards improving environmental certification. However, this is the first attempt at developing such a methodology. Therefore, the model still needs to be applied to future sports events in order to test it and to monitor the criteria used in the methodology. Trials on small and medium-sized sport events might be the most suitable context to test it.

From the future perspective, it will be important to add the criterion of Covid19, creating a specific index, such as the "Covid19 safety index", because it is mandatory to make sporting events safer for people in light of the pandemic. This index could be considered a negative value because it decreases the number of people who can participate, but thanks to technology, it could involve the fans virtually, developing "fan engagement". In this way, sports fans can be active during events and make a contribution to the development of sustainable sports events.

References

1. COP26 Homepages. https://ukcop26.org/it/la-conferenza/. Accessed 22 Dec 2021
2. United National Homepage. https://sdgs.un.org/goals/. Accessed 22 Dec 2021
3. Unite National Environmental Program (UNEP): Green Meeting Guide 2009. Roll out the Green Carpet for your Participants. United National Environmental Program, Nairobi, Kenya (2009)
4. International Organization for Standardization: Event sustainability management systems-requirements with guide for use. ISO 20121:2012. International Organization for Standardization, Geneva, Switzerland (2012)
5. GRI Homepage. https://www.globalreporting.org/standards/. Accessed 22 Dec 2021
6. Collins, A., Jones, C., Munday, M.: Assessing the environmental impacts of mega sporting event, two option? Tour. Manage. **30**, 828–837 (2009)
7. Muller, M., Wolfe, S.D., Gaffney, G., Gogishvili, D., Hug, M., Leick, A.: An evaluation of sustainability of the Olympic Game. Nat. Sustain. **4**, 340–348 (2021)
8. Boggia, A., Massei, G., Paolotti, L., Rocchi, L., Schiavi, F.: A model for measuring the environmental sustainability of events. J. Environ. Manage. **206**, 836–845 (2018)
9. USGBC U.S.: Green building council-green building rating system. USGBC U.S., Washington DC, US (2011)

10. ITACA Homepages: Managing Sport and Leisure (2020). http://www.itaca.org/certifica zione_ed_sost.asp. Accessed 22 Dec 2021
11. World Health Organization (WHO): Sporting events during the COVID-19 pandemic Considerations for public health authorities. World Health Organization, Geneva, UE (2021)
12. Ludvigsen, J.A.L., Hayton, J.W.: Toward COVID-19 secure events: considerations for organizing the safe resumption of major sporting events. Managing Sport Leisure **27**(1–2), 135–145 (2020)
13. Turin Organization Committee (TOROC): XX Olympic Winter Games. Torino 2006. LA84 Foundation, Los Angeles, US (2007)
14. United Nations Environment Program (UNEP): Independent environmental assessment, Beijing 2008 Olympic Games. United National Environmental Program, Nairobi, Kenya (2009)
15. Vancouver Organizing Committee (VANOC): Vancouver 2010 - Sustainability report 2009–2010. LA84 Foundation, Los Angeles, US (2010)
16. London Organizing Committee of the Olympic & Paralympic Games London (LOCOG): London 2012 Post- Games Sustainability Report – A legacy of change. London Organizing Committee of the Olympic & Paralympic Games London (LOCOG), London, U.K. (2012)
17. Sochi Olympic and Paralympic Organizing Committee: Official Report Sochi 2014 Olympic winter games, vol. 3. LA84 Foundation, Los Angeles, US (2014)
18. Rio Organizing Committee for the Olympic Games (ROCOG): Post -games sustainability – Report Rio 2016. Rio Organizing Committee for the Olympic Games (ROCOG), Rio, Brazil (2018)
19. International Olympic Committee (IOC): IOC Annual Report 2019. Credibility, Sustainability, Youth. International Olympic Committee (IOC), Lausanne, Switzerland (2018)
20. Raj, R., Musgrave, J.: The Economics of Sustainable Event. CABI, Oxfordshire, U.K. (2009)
21. Cavallo, R., Rosio, E., Bosio, L., Pavan, A., Ardito, L., Fenocchio, G.: La gestione sostenibile di grandi eventi sportivi. Ingegneria dell'ambiente **6**(1), 60–70 (2019)
22. Munesue, Y., Masui, T., Fushima, T.: The effects of reducing food losses and food waste on global food insecurity, natural resources, and greenhouse gas emissions. Soc. Environ. Econ. Policy Stud. -SEEPS **17**, 43–77 (2015)
23. European Institute for Gender Equality (EIGE): Giustificazione del quadro concettuale dell'indice sull'uguaglianza di genere per l'Europa. European Institute for Gender Equality, Vilnius, Lithuania (2012)
24. U.S. Access Board Homepage. https://www.access-board.gov/law/aba.html. Accessed 22 Dec 2021
25. International Paralympic committee (IPC): Accessibility guide – October 2020. International Paralympic committee (IPC), Bonn, Germany (2020)
26. European Network for Accessible Tourism (ENAT): Rights of Tourists with Disabilities in the European Union Framework. European Network for Accessible Tourism (ENAT), Brussels, Belgium (2007)

Sustainable Development Goals (SDGs) Evaluation for Neighbourhood Planning and Design

Valeria Saiu$^{(\boxtimes)}$ ⑩ and Ivan Blečić

Department of Civil and Environmental Engineering and Architecture, University of Cagliari,
09129 Cagliari, Italy
{v.saiu,ivanblecic}@unica.it

Abstract. The UN 2030 Agenda and the Sustainable Development Goals (SDGs) highlight the central role of cities and urban settlements for the implementation of sustainable development policies and actions at the local level. In this paper we investigate how neighbourhood sustainability contribute to the achievement of SDGs through the analysis of their correlations with project evaluation criteria used in existing assessment tools like LEED Cities and Communities (LEED v4.1). In particular, we illustrate a method for evaluating these correlations, both in qualitative and in quantitative terms. The results of this analysis highlight some open questions in the SDGs implementation at the neighbourhood level, and the potential role of current assessment practices in SDG localization and operationalization procedures. In conclusion, we argue that despite these challenges and open questions, the NSA frameworks have the potential to guide neighbourhood planning and design towards the sustainability transition but only if they are better aligned with the SDG framework.

Keywords: SDGs localisation · Urban sustainability · Neighbourhood sustainability · Assessment tools · Evaluation models

1 Introduction

The development and use of urban sustainability indicators and assessment tools have gained significant attention during the last decade, especially since the 2030 Agenda for Sustainable Development and their 17 Sustainable Development Goals (SDGs) entered into force. National governments are invited to establish SDGs monitoring systems and to report "Voluntary Local Reviews" (VLRs) in order to evaluate the overall implementation effectiveness and to give indications to review and follow up the progress towards these objectives. In this context, an increasing number of city administrations around the world, that have the primary responsibility for achieving many SDGs targets, are engaging with their VLR through the collection of different data sets concerning many urban issues that should be consistent with local needs and priorities [1].

© The Author(s), under exclusive license to Springer Nature Switzerland AG 2022
F. Calabrò et al. (Eds.): NMP 2022, LNNS 482, pp. 979–987, 2022.
https://doi.org/10.1007/978-3-031-06825-6_93

However, these reviews did not yet produce significant data and information for policy-making, planning and design at the neighbourhood level which is the most adequate spatial unit for the assessment of many sustainability aspects [2–4]. The main practical problems include the prevailing interest in urban questions that can be addressed through long-term policies and large investment projects, while the role of shorter term and smaller sized transformation processes are often neglected. This requires to define development objectives according to daily life problems that can bring direct tangible benefits to the quality of life of many people [5–7]. These transformations take place on the scale of the neighbourhood where, not surprisingly, efforts have been concentrated in recent decades to define common guidelines for the implementation of sustainability initiatives in communities [8, 9]. Certification rating systems and protocols, such as the Neighbourhood Sustainability Assessment (NSA) tools, have spread rapidly, as tools to measure performance of urban sustainability at suburban and neighbourhood level. In particular, NSA tools propose clear, relevant, representative and measurable indicators for an objective assessment, useful to simplify complex urban phenomena into easily understandable and valuable elements, by integrating different aspects in the design process, including environmental, economic, social and institutional issues [10–12].

This approach would seem to be consistent with the systemic framework of targets and indicators defined by the 17 SDGs. Despite one of the SDGs, the Goal 11 "Sustainable Cities and Communities", is specifically focused on urban perspective, "urban goals" go far beyond the sole Goal 11 because many sustainability dimensions covered by other goals, – such as SDG 1 (poverty reduction), SDG 3 (Health and wellbeing), SDG 6 (clean water and sanitation), SDG 7 (affordable and clean energy), SDG 8 (economic growth) and SDG 13 (climate action) – have implications for urban planning and policy. Thus, a more integrated approach is crucial to express the transformative potential of SDGs and for their effective implementation [13, 14].

Following these considerations, the aim of this paper is to evaluate the potential contribute of the NSA Tools to SDGs localization. For this, we report our analysis of one of the most prominent NSA tool, the last version of the Leadership in Energy and Environmental Design (LEED) rating system for "Cities and Communities" (i.e., LEED v4.1) developed by the US Green Building Council (USGBC), that allows to test the proposed methodology to a relevant case study. With this analysis, he main question this paper wishes to address is: which LEED planning and design criteria have impacts on the achievement of different SDGs, and what is the weight of those impacts?

The paper is structured as follows: next section describes the materials and methods used in this study; Sect. 3 presents the applied methodology and shows the results of data analysis. Then, Sect. 4 discusses these results in light of the objective of the paper and offers some concluding remarks.

2   LEED ND v4.1 and SDGs Localization

2.1   Materials

To contribute to the exploration of explore the potential employment of NSA tools for SDGs localization, we have evaluated the degree of correspondence between selected

SDGs targets and the selected NSA tool. We have analyzed the LEED scheme, a voluntary rating system among the most widely used tools to evaluate and certify the sustainability of the built environment, both building and neighborhood level. The first pilot version for neighbourhoods – LEED ND – was launched in 2007 whilst its present version, named "LEED for Cities and Communities" (LEED v.4.1), was released in April 2019 [15–18]. Compared to the previous editions, the current release focuses more on improved social equity, quality of life and standard of living. In general, the number of criteria has been reduced but the number of thematic areas has been expanded, offering a broad spectrum of sustainability issues, according to SDGs framework. The system was adapted for city and community (neighbourhood) level and, furthermore, for two different types of interventions, regeneration projects (existing) and new developments (plan and design). In this paper we consider the scheme adapted for existing neighbourhoods.

2.2 Methods

The assessment procedure was divided in three main steps (Fig. 1): (1) the analysis of the SDGs targets in order to select the most related with planning and design issues, defined as "urban targets"; (2) the analysis of the selected NSA tool – LEED v4.1 – credits and prerequisites; (3) the evaluation of the degree of correspondence between LEED v4.1 and selected SDGs "urban targets".

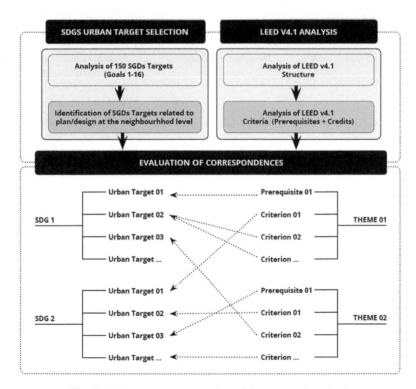

Fig. 1. Schematic representation of the proposed method.

3   Results

In the Step 1, we have analyzed the 150 SDGs targets, excluding SDG 17 which is related to the means of implementation of all the others targets. Then, we have selected the 52 most related with relevant project theme areas present in SDGs framework using the key words method (Table 1). We defined these "urban targets".

Table 1. SDGs "urban targets".

SDGs urban targets	SDGs project theme areas	Main planning and design issues
1.5, 11.5, 11.6, 11.B, 13.1, 13.2, 13.3	Resilience and risks	Climate change mitigation and adaptation
2.4, 6.6, 11.4, 14.4, 14.5, 15.1, 15.4, 15.5, 15.9	Ecosystem protection	Agriculture; water-related, mountain, coastal, terrestrial and inland freshwater ecosystems; reforestation; degraded land and soil, Biodiversity
6.4, 6.5, 7.1, 8.4, 9.1, 9.4, 11.C, 12.1, 12.2, 12.5	Resources efficiency	Water, Energy, built environment footprint, consumption and production patterns, waste
3.4, 3.6, 6.3, 11.2, 12.4, 14.1	Health and Safety	Mental health and wellbeing, Road traffic accidents, Pollution, chemical, waste, land base activities
4.4, 4.7, 4.a, 8.3, 8.9, 9.5, 12.8, 12.b	Education, Awareness, Job	Skills, Sustainable lifestyles, Education facilities, Job opportunities, Innovation, Sustainable tourism
1.4, 1.5, 2.3, 5.a, 6.b, 10.2, 10.3, 11.1, 11.3, 11.A, 16.4, 16.6	Equity and Justice	Basic services, Public Spaces, Land and other properties, Social inclusion, Equal opportunities, Participation

In the Step 2, we have analyzed the structure of LEED v4.1 (see Table 2). In total, there are 9 theme areas and 40 criteria, divided in 5 prerequisites (the mandatory requirements that must be fulfilled for a project before proceeding to the certification) and 35 credits that were designed to address impacts on specific project theme areas. Then, we analysed LEED v4.1 credits and prerequisites in order to evaluate their distribution across different theme areas and the related weight, both in terms of number and of maximum score assigned.

Table 2. LEED v4.1. The structure of the scheme adapted for Existing Communities.

LEED v4.1 theme areas		Prerequisites	Credits	Criteria		Score	
				Total number	Weight	Total points	Weight
IP	Integrative process	0	2	2	5%	5	4.5%
NS	Natural System and Ecology	1	4	5	12.5%	9	8.2%
TR	Transportation and Land Use	0	6	6	15%	15	13.6%
WE	Water Efficiency	1	4	5	12.5%	11	10%
EN	Energy and GHG Emissions	1	5	6	15%	30	27.2%
MR	Materials and Resources	1	5	6	15%	10	10%
QL	Quality of Life	1	7	8	20%	20	18.1%
IN	Innovation	0	1	1	2.5%	6	5.4%
RP	Regional Priority	0	1	1	2.5%	4	3.6%
		5	35	40	100%	110	100%

Finally, in the Step 3, we verified the direct and indirect correspondences between the 40 LEED v4.1 criteria with SDGs urban targets (see Table 3).

Table 3. LEED v4.1 and SDGs correspondences.

Theme areas	Project issues	SDGs directly related	SDGs indirectly related
IP	Integrative Planning, Green Building Policy and Incentives	11, 16	4–9, 10, 12,13
NS	Ecosystems, Green Spaces, Natural Resources, Light Pollution, Resilience Planning	11, 15, 13	1, 3, 6, 8, 9, 14
TR	Public Transport, Mixed Use Development, Pollution and Environmental Impact of Transport, Priority Locations	9, 11	1, 3, 6, 8, 10, 12, 13, 15
WE	Equitable Access to Water and Sanitation Services, Water Management, Stormwater Management,	1, 6, 9, 13	3, 8, 10–12, 15
EN	Equitable Access to Power, Zero Energy and Emissions, Energy Efficiency, Renewable Energy, Low Carbon Economy, Consumers Participation	1, 7, 13, 16	3, 8, 9, 11, 12
MR	Waste Management, Material Recycle, Circular Economy	12	3, 8, 9, 11, 13
QL	Living Standards, Equitable Economic Prosperity, Equitable Access to Community Services, Environmental Pollutants, Affordable Housing, Community Participation, Civil and Human Rights	1, 3, 10, 11, 16	2, 4–6, 8, 12,13, 15
IN	Exceptional and innovative performances	9	3, 4, 6–8, 11–13
RP	Local socio-economic and environmental priorities	11	1, 5, 10, 13–16

Figure 1 shows the distribution of credits into various SDGs displayed as a histogram. There are two bars for each SDG, one for the score weight (black) and one for credits weight (color). The first value is calculated as the sum of points assigned by LEED to various credits correlated with different Goals divided by the maximum score possible and multiplied by 100 to attain percentage scores; the second as the number of credits correlated with different Goals divided by the total number of credits defined by the tool (Fig. 2).

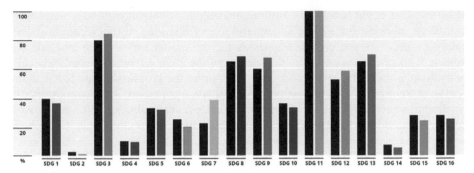

Fig. 2. Correspondences between SDGs and LEED v4.1 credits and score weights: the black bar shows the score weight and the colored bar the number of criteria weight.

4 Discussion and Concluding Remarks

The results of this study allow us to verify the possible impact of LEED v4.1 credits on different SDGs. There are many planning and design guidelines at the neighborhood scale, expressed by different credits and their related indicators, that can contribute to the achievement of Goals 3, 13, 8 and 9, while only few credits address issues related to Goals 2, 14 and 4. This means that in its current formulation LEED v4.1 provides fewer operational indications for the localization of these SDGs at the neighborhood scale, especially for SDG2 "zero hungry", SDG4 "quality education", SDG5 "gender equality", SDG14 "life below water". There is no credit that directly addresses these issues and few consider them indirectly. This limitation can be overcome by integrating specific criteria that outline the project performance in these areas.

Moreover, the study shows that SDGs have different weighting in the overall score. In general, a high number of criteria correspond to a higher score, but there are many cases in which this does not happen, such as for the Goal 7. In this case, the LEED theme area 5 "energy and greenhouse emissions" provides 6 criteria, the 15% of totals, and assigns 30 maximum points corresponding to the 27,2% of total score, as reported in the Table 2. Three of these criteria have the direct objective to increase energy efficiency, while the others focus on environmental and social goals that can be achieved through energy related measures, as well as the equitable access to power (EN 01), the reduction of the environmental and economic harms associated with fossil fuels and greenhouse gas emissions (EN 04), and the consumer participation in energy use optimization (EN 06).

Despite some specific considerations on the tool analyzed, this study reveals that the NSA tools can constitute an operational framework from which to start defining design guidelines for the SDGs achievement at the small-scale of the neighbourhood [19–21]. Further developments of this research must be focused on the comparison of different NSA tools, to define a comprehensive set of criteria useful for the better implementation of all SDGs. This could lead to the development of an assessment tool to be used in everyday practice at the local level. In particular, urban planning offices are expected to track and evaluate the impacts of their policies and interventions towards SDGs, in order to provide objective information on projects effectiveness and define improvement measures. This could support the institutionalization of neighbourhood sustainability assessment allowing for a more widespread use of this kind of tools, currently used mainly by agencies and private companies for the evaluation of a few and selected projects. This process is critical to the success of the 2030 Agenda that explicitly calls for evaluation methods for assessing and monitoring the progress on the SDGs at the local level. Such an institutionalization of the evaluation practices could also contribute to collaboration between public administrations, policy makers, planners, citizens' groups, businesses, and other local stakeholders in achieving sustainability goals. Changes in current urban planning and design practice, in fact, require a shared knowledge and a collective vision for the future of local places, and this implies to improve municipal statistical offices, awareness and education on urban sustainability, and to establish a permanent dialogue between different actors.

References

1. Fox, S., Macleod, A.: Localizing the SDGs in cities: reflections from an action research project in Bristol, UK. Urban Geogr. **0**, 1–21 (2021). https://doi.org/10.1080/02723638.2021. 1953286
2. UN-Habitat, UCGL: Guidelines for Local Voluntary Reviews. A comparative analysis of Existing VLRs (2020)
3. Sala Benites, H., Osmond, P., Rossi, A.M.G.: Developing low-carbon communities with LEED-ND and climate tools and policies in São Paulo, Brazil. J. Urban Plan. Dev. **146**, 04019025 (2020). https://doi.org/10.1061/(ASCE)UP.1943-5444.0000545
4. Sharifi, A., Dawodu, A., Cheshmehzangi, A.: Neighborhood sustainability assessment tools: a review of success factors. J. Clean. Prod. **293**, 125912 (2021). https://doi.org/10.1016/j.jcl epro.2021.125912
5. Vaidya, H., Chatterji, T.: SDG 11 sustainable cities and communities. In: Franco, I.B., Chatterji, T., Derbyshire, E., Tracey, J. (eds.) Actioning the Global Goals for Local Impact. SSS, pp. 173–185. Springer, Singapore (2020). https://doi.org/10.1007/978-981-32-9927-6_12
6. Saiu, V.: Evaluating outwards regeneration effects (OREs) in neighborhood-based projects: a reversal of perspective and the proposal for a new tool. Sustainability **12**, 10559 (2020). https://doi.org/10.3390/su122410559
7. Banchiero, F., Blečić, I., Saiu, V., Trunfio, G.A.: Neighbourhood park vitality potential: from Jane Jacobs's theory to evaluation model. Sustainability **12**, 5881 (2020). https://doi.org/10. 3390/su12155881
8. Dawodu, A., Akinwolemiwa, B., Cheshmehzangi, A.: A conceptual re-visualization of the adoption and utilization of the pillars of sustainability in the development of neighbourhood sustainability assessment tools. Sustain. Cities Soc. **28**, 398–410 (2017). https://doi.org/10. 1016/j.scs.2016.11.001
9. Saiu, V.: The three pitfalls of sustainable city: a conceptual framework for evaluating the theory-practice gap. Sustainability **9**, 2311 (2017). https://doi.org/10.3390/su9122311
10. Huovila, A., Bosch, P., Airaksinen, M.: Comparative analysis of standardized indicators for Smart sustainable cities: what indicators and standards to use and when? Cities **89**, 141–153 (2019). https://doi.org/10.1016/j.cities.2019.01.029
11. Sharifi, A.: A typology of smart city assessment tools and indicator sets. Sustain. Cities Soc. **53**, 101936 (2020). https://doi.org/10.1016/j.scs.2019.101936
12. Morano, P., Tajani, F., Guarini, M.R., Sica, F.: A systematic review of the existing literature for the evaluation of sustainable urban projects. Sustainability **13**, 4782 (2021). https://doi. org/10.3390/su13094782

13. Allen, C., Metternicht, G., Wiedmann, T.: Initial progress in implementing the sustainable development goals (SDGs): a review of evidence from countries. Sustain. Sci. **13**(5), 1453–1467 (2018). https://doi.org/10.1007/s11625-018-0572-3

14. Abastante, F., Lami, I.M., Gaballo, M.: Pursuing the SDG11 targets: the role of the sustainability protocols. Sustainability **13**, 3858 (2021). https://doi.org/10.3390/su13073858

15. Diaz-Sarachaga, J.M., Jato-Espino, D., Castro-Fresno, D.: Evaluation of LEED for neighbourhood development and envision rating frameworks for their implementation in poorer countries. Sustainability **10**, 492 (2018). https://doi.org/10.3390/su10020492

16. USGBC: LEED v4.1 Cities and Communities. https://www.usgbc.org/sites/default/files/2021-09/Guide%20to%20Certification%20in%20LEED%20v4.1%20for%20Existing%20Cities.pdf

17. Talen, E., et al.: LEED-ND as an urban metric. Landsc. Urban Plan. **119**, 20–34 (2013). https://doi.org/10.1016/j.landurbplan.2013.06.008

18. Dall'O', G., Zichi, A.: Green protocols for neighbourhoods and cities. In: Dall'O', G. (ed.) Green Planning for Cities and Communities. RD, pp. 301–328. Springer, Cham (2020). https://doi.org/10.1007/978-3-030-41072-8_13

19. Saiu, V., Blecic, I., Meloni, I.: Making sustainability development goals operational at neighbour-hood level: potentials and limitations of neighbourhood sustainability assessment tool. Environ. Impact Assess. Rev. (Forthcoming)

20. Saiu, V., Blecic, I., Cocco, G., Meloni, I.: Urban Sustainability and SDGs implementation between regional strategy and local practice: case of Sardinia. In: Filho, W.L. (ed.) Implementing the SDGs in the European Region. Springer (Forthcoming)

21. Saiu, V., Blečić, I., Meloni, I., Piras, F., Scappini, B.: Towards a SDGs based neighbourhood sustainability evaluation framework: a tool for assessing sustainability at the urban micro-scale. In: Abastante, F., et al. (eds.) Urban Regeneration Through Valuation Systems for Innovation. Springer (Forthcoming)

Around Cosenza

Francesca Moraci, Celestina Fazia, and Dora Bellamacina(✉)

Mediterranea University, Reggio Calabria, Italy
f.moraci@unirc.it

Abstract. The essay outlines the results of the research carried out in LabStUTeP (The experiences of the Laboratory of Urban and Territorial Strategies for Planning (LabStUTeP) belonging to the Department of Architecture and Territory of the Mediterranean University of Reggio Calabria, of which Prof. Francesca Moraci is coordinator, in the context of the research activities of the last years, have investigated some territorial systems of the province of Cosenza. In particular, the activities mainly involved the Cosenza-Rende conurbation system, the municipal area of Acri and finally Santa Sofia d'Epiro) regarding the role of territorial centers and internal areas in the wider regional context. The role attributed is that of "polycentrism towards resilience", understood as an aggregation of small urban centers with strong historical connotation and unexpressed potential (territorial dialogue, marginalized and undervalued linguistic minorities and identities, etc.). Propulsive lies in the new "aggregative" logics that look with interest to dynamic and propulsive network systems. Systems ready to welcome ideas and innovation, prepared to face the ecological transition and the resilient territory project. The pro-active dynamics tendentially and programmatically affect the pre-hilly and hilly contests of the Calabrian coast: they are areas with a strong historical and ethno-anthropological connotation that have preserved their distinctive treaties and traditions, which have protected the territory and the testimonial aspects of settled cultures and civilizations. The paper, in addition to summarizing the experiences conducted up to now on the hypothetical future scenarios (trend and programmatic) of Cosenza, proposes an examination of unresolved questions that slow down the reorganization and relaunching processes of the aforementioned emerging strategic areas. It identifies the new organizational cornerstones and a different vision of the territory, overcoming conceptual constructs and theoretical assumptions that have proved to be ineffective and at times obsolete.

Keywords: Polycentric aggregation · Internal areas · Small municipalities

1 Introducing the "Cosentino" System

1.1 Between Civitas et Urbs

The construction of the "new" territorial project takes into account the place, in terms of context and uniqueness, and the material and intangible relationships that come from it. The place is linked to the presence of men and their relationships in space. Civitas and urbs are two closely connected systems: one would be, without the other, just a

F. Calabrò et al. (Eds.): NMP 2022, LNNS 482, pp. 988–998, 2022.
https://doi.org/10.1007/978-3-031-06825-6_94

community of people who "suffer" from the lack of a defined and organized settlement and relational system. Territories are the organized set of places in which relationships emerge and are enhanced, through which the community dialogues, in an epistemic form, without giving in to the narcissistic vision of being "individual" and not collectivity.

The interpretation is guided by the identity and cultural relations of the centers in question and by the bet that new territorial aggregations, outlined by the implementation of innovative urban planning forms, can assume greater centrality and in the context of studies on sovereignty-urban planning and policies economic, and on widespread urban planning, such as the re-composition of the territorial plots of the landscape, within macro relationships of a large area.

Studies and projects aim to connect internal areas, as well as to enhance smaller centers and marginalized - or marginal - realities through broader strategies of recovery of large areas, urban regeneration and resilient practices.

1.2 Research Objectives and Methodological Path

The research proposes a methodological approach which starting from the "polycentrism of necessity" (orography, accessibility, urban fragmentation) comes to value a "polycentrism of opportunity and proximity" whose strengths are:

- local identities;
- sense of community;
- unexpressed potential.

It is a form of common strategy that must be the basis of any territorial transformation intervention.

1.3 Future Strategic Scenarios for "Cosentino" System

The city of Cosenza - center of gravity of the neighboring municipalities, has a multipolar centrality, characterized by the medium-sized metropolitan agglomeration that connects the northern part of Rende with the south-eastern part of Acri and Corigliano. The "QTRP", indeed, classifies the municipality of Cosenza as an urban center at the regional level, which represents a functional attraction pole for the entire regional territory.

Several municipalities in the foothills of Cosenza are part of the APTR "La Sila e la pre-Sila Cosentina", one of the great naturalistic complexes of the Calabria Region. The Cosenza-Silane system is a potentially functional reality to support the strategic role of Cosenza in its new dimension as a medium-sized metropolis.

The possible lines of development concern the development of innovative criteria and methods, which are capable of recognizing multipolar centrality systems and relating to areas of gravity with variable geometry. A priority is to define, especially in this historical moment of the "ecological transition", the strategic role of the micro and macro polarities and centrality of the urban and extra-urban or provincial level. Without wanting to create hierarchies free from "supply chain" contamination, it is believed that it is appropriate to consider objectives and actions and the strategic effects on the territory.

Therefore, from a strategic point of view, it is considered necessary:

– to overcome the exclusive use of socio-economic statistical indicators; the demographic dimension of the agglomeration under study, alone, is not significant for the potential of the area, any trend scenario does not confirm the demographic increase;
– the acquisition of "new forms of medium-sized metropolitan agglomeration", based on new centrality systems with variable geometry, organized by rhythms of use and themes of interest; it is the logic of the functioning and functionality of the hypothesized system that may possibly prevail over other discriminating factors;
– the recognition of sub-joints; as metropolitan endowments, metropolitan centralities, metropolitan cities, metropolitan areas, it tends to identify the roles of territorial realities with respect to alternating variables or visions that do not have to be rigid and static, but can complement and balance themselves according to the gravitations and driving forces in the system "multipolar" so defined.

Studies on territorial communities have defined a series of objectives useful for achieving the aforementioned strategies, such as: the protection of the cultural identities of the territory; the conservation and enhancement of the landscape and valley "ecomosaic"; the adoption of innovative and sustainable governance models of territorial transformations and the activation of self-sustainable local development dynamics; the orientation towards sustainability, the uses of territorial resources and rivers (in particular the Crati); directing downstream production towards high quality objectives (in particular the agri-food sector); limitation of land consumption; the prevention of hydrogeological instability; the strengthening of the processes of strategic aggregation between local authorities, through innovative tools and models, on a voluntary basis (cooperation agreements, understandings, associations, etc…); the reconfiguration of the relationships between city and urban area with the agricultural, rural and peri-urban contexts of the valley.

2 Case Studies in "Cosentino" System (2015–2021)

Cosenza is the northernmost capital of the Calabrian province, and it stands on a hill - Colle Pancrazio - at an altitude of 238 m. s.l.m.. The city is crossed by the river Crati and it extends into the valley. The point of confluence of the rivers delimits the historical area of the city from the new one.

The historic city is located in the highest part of the Pancrazio hill, while the new settlements have developed alongside the the stream bed of the Crati river. The historic core called "Cosenza Vecchia" is an ancient hilly settlement, whose urban fabric is formed by steep, narrow and winding alleys. The shape of the city is given by the noble palaces as well as by the small houses.

The entire settlement is protected to the west by the southern coastal chain that separates the city from the Tyrrhenian Sea, and to the east by the high wooded plain of the Sila.

The city of Cosenza extends over an area of 7.86 km^2, that climbs up a vertical drop of 402 m in the municipal area, with the highest point located to the south-east 589 m. s.l.m. and the lowest to the north 187 m. s.l.m.

The modern city is the *fulcrum* of the urban agglomeration, which gravitates above all to the conurbation formed with the municipality of Rende, where the university center is located.

The historic center of Cosenza embodies, in its architectural forms, the apex of the expansion and culture of the Bruzi, with monumental buildings, manors and prestigious palaces, but also the urban design characterized by a maze of alleys that wind around to the fabric. The whole urban space is forced between rivers (Fig. 1).

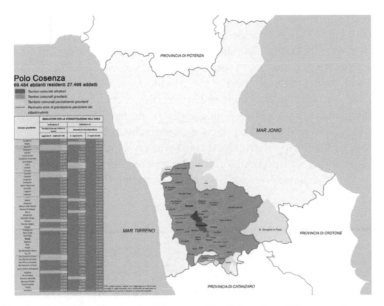

Fig. 1. The "Cosentino System" in a graphic elaboration of the research group.

In the early seventies, the new Vittorini Town Plan and the birth of the University of Calabria in the narrow municipality of Rende represented the centre of the modern urban expansion of the city. The city, with the aim of redeveloping the urban layout, improving the road layout, reducing the population density, has intensified the development towards the north, thus determining it on the urban axis that constitutes the agglomeration with Rende and Castrolibero, tourning into a polycentric city. The urban development of the capital has had a notable boost above all with the entry into force of the general variant to the master plan (1996), an urban tool that together with the policy of public works and integrated interventions made possible by the "Urban" and from the use of European Union funds, it has changed the face of the urban fabric.

The Cosenza-Rende road axis (2006) ensured the direct connection between the different centers of the Cosenza urban area and, at the same time, it modified the configuration of the urban fabric of the city, through the connection between the districts of the city. Indeed, the modern city, today presents itself as an orderly and regular network, with wide and straight streets in a north-south direction.

In the context of sustainable mobility, Cosenza has framed the urban cycle within a wider project of sustainable mobility of the city, it represents a system of metropolitan cycle paths about 30 km long, which allow the connection of the entire system of urban agglomeration.

Although the annual ranking of Il Sole 24 ORE sees Cosenza in 93rd place for the quality of life in the province, it is worth it to report the data from Legambiente which classifies it in 4th place for urban ecosystem, on a national basis.

2.1 Acri's Case, 2015–2018

Acri is a mountain municipality - average altitude: 720 m. s.l.m. - 193 m s.l.m. < 1 379 million. s.l.m. - with a considerable territorial extension - over 200 km^2 of territory and just over 20,000 inhabitants. Acri is a complex and widespread settlement system that promoted the drafting of the Municipal Structural Plan PSC, at the Planning Office and with the scientific support of the Mediterranean University of Reggio Calabria. This initiative was undertaken in order to relaunch the level of project integration with strategic planning tools and economic-social planning at municipal, provincial and regional level, to ensure the methodological coordination of the initiatives undertaken and to be activated in the legislative context in order to intercede an integrated planning of territorial resources with particular reference to intergenerational values referring to landscape-environmental and historical systems –wedding (Figs. 2 and 3).

Fig. 2. Territorial location of Acri.

In recent years we have witnessed profound changes in the economic, productive, social and environmental systems, which, for the Municipality of Acri, it was deemed necessary to combine land management models based on a plan and a articulated and sectorally specialized operational instrumentation, in particular, as regards the immense patrimony of natural and anthropic resources that constitute (Moraci 2018).

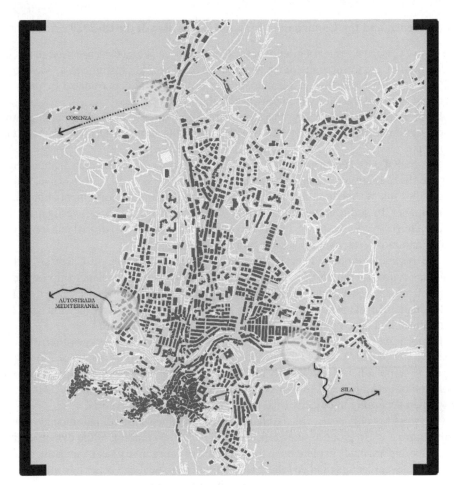

Fig. 3. Territorial location of Acri.

The effectiveness of sectoral policies, the interrelationships that intervene between territorial, economic, environmental and urban development issues, require the achievement of high levels of effectiveness in the integrated and coordinated approach with strategies and programming tools., aimed at supporting the vocations and specificities of the individual municipal areas in order to favor high levels of qualification and enhancement of existing resources and an overall and harmonious strengthening of territorial economic development.

The Municipality of Acri falls within the territory of the Regional Landscape Territorial Area (APTR) n. 12 - La Sila and Presila Cosentina. This territorial area is dominated by the Sila massif which, located in the northern part of the region, represents the largest mountain system, and borders the Presilana belt, the Crati valley and the Cosentino Ionian Sea.

2.2 Cosenza and the Conurbation System Coseza-Rende, 2018–2020

The city of Cosenza, center of gravity of the Cosenza-Rende and Cosenza-Acri districts, is identified by the Community of the National Urban Planning Institute as a "Vast Area and macro-regional dimension"; is classified in a recent study coordinated by prof. R. Mascarucci one of the Thirty medium-sized Italian metropolises.

It has a multipolar centrality, characterized by a medium-sized metropolitan conurbation that connects the northern part of Rende with the south-eastern part of Acri and Corigliano.

The goal of the research group was to test the efficiency of classification methodology proposed by the national group, in the context of an urban system, the Cosenza one, which rises to the rank of a medium-sized metropolis, demonstrating the fulfillment of physical requirements. In addition the research analyses the urban dynamics, investigating the Cosenza territory of a vast area, in order to identify appropriate territorial enhancement actions and more effective forms of territorial governance.

The research, supported locally by the Mediterranea di Reggio Calabria by the regional section of the INU, addressed aspects relating to the relationship between Cosenza, medium-sized metropolitan agglomerations, and the two Cosenza systems of Rende and Acri-Cosenza, strengthening the dialogue and defining the strategies for the relaunch and development of the Silan territory and the vast area.

It was a question of studying realities, such as that relating to the vast territory of Cosenza, not included in the 14 metropolitan cities referred to in law 56 of 2014 with respect to some indicators previously identified by the national group to hypothesize the definition of new intermediate forms of cities, and of spatial and spatial models that can be typified according to objective and numerically comparable criteria.

The work not only deals with the characterization of new types of intermediate cities, furthermore it try to propose some reflections. The critical reading of the dynamics that take place and are made explicit on it, the presence of "attractive places", and other qualitatively relevant factors (and no longer and not just statistical and economic indicators), determine an institutional redesign through the criteria that make also referring to the contents of the Sustainable Urban Agenda (Fig. 4).

In recent years in Calabria, as in the rest of Italy, we have witnessed profound transformations of the territorial and urban systems but, also, significantly of the economic-productive, social and environmental systems. The municipalities of Acri, Spezzano della Sila and San Giovanni in Fiore have found it necessary to combine land management models based on programming and an articulated and sectorally specialized operational instrumentation, in particular, with regard to the immense wealth of natural and anthropogenic resources, which constitute The Sila massif, in its various articulations, is perhaps the element that most structures the eco-morphological structure of the region, its breadth gives rise to very different contexts. The environment of the Sila Grande is mainly mountainous, while the Sila Piccola and the Greca have large stretches of landscape on the slopes and hills. Today tourism and activities linked to cultural and environmental resources are struggling to assume the role once occupied by mountain economic structuring, even if significant traits of productive agriculture remain. The system of the Silan plateau is characterized above all by the mountainous area around

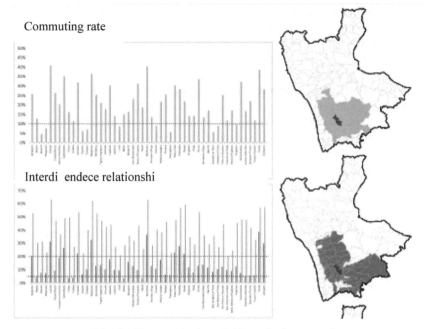

Fig. 4. The conurbation of "Cosentino" system. Data.

the lakes affected by a dense network of paths and urban areas consisting of numerous centers and villages for winter tourism with the relative accommodation facilities.

2.3 The Arberia Union and Santa Sofia D'Epiro, 2021-Incoming

The research - in the preliminary phase - aims to identify the territorial relationships between the smaller centers of the Calabrian internal areas in relation to the territorial vision and strategy capacity of the urban plans in associated form. There have been few successful cases of this practice even if LUR 19/2002[1] amplifies it and hopes for its implementation.

The interpretation is guided by the identity and cultural relations of the centers in question and by the bet that new territorial aggregations, outlined following the implementation of innovative urban planning forms, can assume greater centrality and in the context of studies on economic planning and policies, and on widespread urban planning, such as the recompositing of the territorial textures of the landscape, within the macro-relationships of a vast area.

The object of the proposal is to investigate and identify the internal attractive forces of the urban archipelago characterized by five centers, with a strong ethno-anthropological identity: the Italian-Albanian cities settled on the hilly side of the province of Cosenza. The municipalities of S. Sofia d'Epiro, S. Demetrio Corone, S. Cosmo A., S. Giorgio A., Vaccarizzo A.; a group of Calabrian foothills of origin arbëreshë.

---

[1] "Legge Urbanistica Regione Calabria" n.19 2002.

The theme is that of "Urban archipelagos. Connecting internal areas, enhancing, regenerating urban centers: between resilience and interrupted landscapes".

The objective aims to connect the internal areas as well as to enhance the smaller centers and the marginalized - or marginal - realities through broader strategies of recovery of large areas, urban regeneration, and resilient practices.

The construction of the territorial project considers the place, in terms of context and uniqueness, and the material and intangible relationships that derive from it. The place is linked to the presence of men and their relationships in space. Civitas and urbs are two closely connected systems: one without the other would be just a community of people who "suffer" from the lack of a defined and organized settlement and relational system. Territories are the organized set of places in which relationships emerge and are enhanced, through which the community dialogues, in an epistemic form, without giving in to the narcissistic vision of being "individual" and not collectivity.

The Albanian centers are born by replicating the modules of "proximity" and dialogue; in particular S. Sofia d'Epiro presents rules for the physical-formal structuring of the space starting from 3 blocks, Gjitonia, rules that connote urban identity, replicating and characterizing full-empty relationships up to defining their mutual contamination between the places of living, the sociability and the urban landscape that permeates the structure (Fig. 5).

Fig. 5. Typological scheme of Arbëreshë housing cells

Through reflection, the territorial and urban conditions underlying the planning tool are highlighted; centrality of interpretation to the reconstructed or still hidden plots of the small contiguous urban realities, administratively different, that PSA[2] organizes by design and politically.

The current PSA can be effective in enhancing the Albanian minorities present in Calabria by removing administrative discontinuities.

The design dimension of the transformations and of the historical conservation project, in sustainable terms with the local culture, requires at the same time an innovative, regenerative and attractive push. The needs of urban well-being and care of the existing and of the landscape are intertwined with mechanisms of redistribution of quality and involvement of individuals.

However, some questions arise that motivated the choice of the proposed theme regarding the fulfillment of strategic choices for the development and the territorial structure: the innovative result of the PSA, the choice to protect identities (otherwise dissolutive) natural aggregation.

---

[2] "Piano Strutturale Comunale in forma Associata".

It means overturning the crisis of the hinterland, to be understood no longer as a land of waste but as models of resilient development that sees the five smaller centers as protagonists. In particular, the rationalization of urban networks and archipelagos aims to reorganize these areas according to a logic of relations that is free from uncontrolled speculation, a protagonist of the last decades of the last century.

The strong characterization of latent union determined by planning tools that strengthen the cities in the logic of the natural territorial network is evident, the possible developments of which are expected precisely from a congruous and associated planning, which cannot transcend an overall vision framework.

The fil rouge on which the bet is based tends towards an innovative approach that thinks of an integrated model between urban planning and tourism development strategies, pursuing total sustainability dictated by urban, economic and social regeneration interventions, so that the little ones centers stop being penalized and are solid realities of development and quality of life respecting their ethno-anthropological connotation.

3 Final Observations

From what has been analyzed, it emerges that there is a profound need for renewal, change of course and modernization of the (static) policies implemented for years with the assumption that everything was manageable with a good imposition of rules, rules and roles to be respected. The challenges to be faced are instead the speed of transformations and the unexpected, time and nature, aspects of the latter that no one is able to foresee, stop, postpone or control. In the case of the provincial territory of Cosenza (and of the case studies analyzed) the (historically) driving role of the Cosenza-Rende conurbation and the destiny of marginalization - "announced" by old policies and aberrant logic - which results from the label "smaller towns" sewn on to all the small and pulverized towns in the area which had been given the role of "towed". This vision is obsolete and it is clear, that each reality is in itself "productive" for the territory, having only to implement strategic and functional choices for the enhancement of unexpressed talents and vocations. Acri, the neighboring municipalities, and still S. Sofia d'Epiro and the Alberesche communities, have a strong identity connotation and can actively participate, with different and always important contributions, in the construction of a territory of innovation in the era of ecological transition and green. Internal areas and smaller centers, with their historical, naturalistic, and wooded heritage, are areas with a high concentration of unique, "non-repeatable" territorial values, they can be actively involved in the network by implementing territorial mending policies using financing and measures of the strategies for internal areas, accompanying community planning and PNRR[3].

With the research, the working group was able to define some cornerstones. First, is the scarce infrastructural endowment and the "physical/functional" marginalization of the internal areas: the absence of an infrastructural network connecting the main arteries and connecting the centers often becomes an expedient to manage investments elsewhere and programming based on the legitimate but sometimes pretext that a work is carried

---

[3] "Piano Nazionale di Ripresa e Resilienza".

out if there is a strong demand and the ability to bear its economic and environmental burdens.

The second is innovation and new policies. In order to innovate and give impetus to the territorial realities, it is necessary to reverse the point of view: there are no longer any minor towed centers but poly-centralities capable of being an integral part of the territory.

The next point is the role of the *cosentino* aggregations and internal area strategies (with PNRR). We need to aim for aggregations of affinity and opportunity. Exploit the role that each center or community has or can have by offering a contribution to the ecological transition, also with respect to the major issues of urban and territorial resilience.

The third concerns the concept of polycentrism, the role of inland areas with respect to the vast area. Infrastructure, provision of services and connections are necessary prerequisites.

Today it is necessary to review the "geographies of opportunity" and redesign strategic assets by interpreting the demands that move from the territory; only by decoding an unheard, latent or unanswered question, can the right directions be found to redeem places and communities.

References

1. Dani, F.: Piano strutturale comunale: conformatività e salvaguardia. Istituzioni del federalismo. Rivista di studi giuridici e politici, Rubettino, Milano (2021)
2. Moraci, F., Passarelli, D., Errigo, M.F., Fazia, C., Bellamacina, D., Bartucciotto, A.: Cosenza. In: Mascarucci, R. (ed.) Città medie e metropoli regionali, INU Edizioni, Roma, vol. 16, pp. 221–231 (2020)
3. Lami, I.M.: Shapes, rules and value. In: Lami, I.M. (ed.) Abandoned Buildings in Contemporary Cities: Smart Conditions for Actions. SIST, vol. 168, pp. 149–162. Springer, Cham (2020). https://doi.org/10.1007/978-3-030-35550-0_9
4. Kourtit, K., Nijkamp, P.: In praise of megacities in a global world. Reg. Sci. Policy Pract. 5(2), 167–182 (2013)
5. Harvey, D.: La crisi della modernità, Il saggiatore, Milano (2010)
6. Moraci, F., Fazia, C.: Il Documento d'Indirizzi e l'Instant Report. In: Fazia, C., Il piano strutturale comunale nell'attuazione della legge urbanistica regionale della Calabria, Centro Stampa d'Ateneo, Reggio Calabria (2008)
7. Bauman, Z.: Modernità liquida. Laterza, Roma-Bari (2003)
8. Martinotti, G.: Matropoli. La nuova morfologia sociale della città. Il Mulino, Roma (1993)
9. Gottmann, J.: Megalopoli, funzioni e relazioni di una pluricittà, Einaudi, Torino (1970)
10. Martinotti, G.: Metropoli. La nuova morfologia sociale della città, Il Mulino, Bologna (1993)

Tackling the COVID-19 Pandemic at the Metropolitan Level. Evidence from Europe

Giancarlo Cotella[(⊠)] [iD] and Erblin Berisha[iD]

DIST | Politecnico di Torino, Viale Mattioli, 39, 10125 Turin, Italy
{giancarlo.cotella,erblin.berisha}@polito.it

Abstract. Metropolitan areas progressively joined cities as catalysts of European development, putting traditional governance models into crisis and triggering episodes of institutional experimentation. Also the European Union has progressively adapted its cohesion policy to cater to the needs of metropolitan areas, and arguments in this direction have multiplied as a consequence of the role that the latter could play in tackling the pandemic. Aiming at shedding some light on this issue, the contribution draws on the results of the ESPON METRO project, discussing how selected metropolitan areas in Europe have reacted to the pandemic, and to what extent they have been able to use the EU cohesion policy. Overall, whereas metropolitan areas appear well positioned to react to the pandemic and to contribute to plan its aftermath, the collected evidence shows that the scope and magnitude of their activity has been rather limited: a situation that may contribute to threaten European social, economic and territorial cohesion.

Keywords: Metropolitan areas · EU cohesion policy · COVID-19 pandemic

1 Introduction

Metropolitan areas have progressively joined cities as catalysts and drivers of global development. This puts traditional spatial governance models into crisis, and a growing number of institutional experimentations have emerged to address the metropolitan dimension [1, 2]. The importance of metropolisation has been also recognised by the European Union (EU), which progressively adapted the EU cohesion policy to cater to metropolitan needs [3]. Arguments in favour of a growing role for metropolitan areas in territorial development have increased incrementally in the last two years due to the role that they could in the aftermath of the COVID-19 pandemic [4]. Metropolitan territories have been those affected the most [5–7], and are called to mitigate the impact of the virus on their territories [8]. At the same time, metropolitan institutions seem be better positioned than traditional governance arrangements to govern the pandemic impacts and to plan for its aftermath [9, 10]. Drawing on the results of the ESPON

© The Author(s), under exclusive license to Springer Nature Switzerland AG 2022
F. Calabrò et al. (Eds.): NMP 2022, LNNS 482, pp. 999–1008, 2022.
https://doi.org/10.1007/978-3-031-06825-6_95

METRO project[1], this contribution discusses the extent to which this is actually happening in selected metropolitan areas in Europe – Metropolitan City of Turin, Barcelona Metropolitan Area, Lisbon Metropolitan Area, Brno Metropolitan Area, Metropolitan Area of Gdańsk-Gdynia-Sopot, Metropolitan City of Florence, Lyon Metropolitan Area and Riga Metropolitan Area [11, 12]. After this brief introduction, the following section sketches out the main coordinates characterising metropolitan governance in Europe, with particular reference to the metropolitan dimension of the EU cohesion policy and of the COVID-19 pandemic. Section 3 discusses how have the metropolitan areas under scrutiny reacted to the pandemic and, to what extent they have been able to take advantage of the EU cohesion policy in their action. The role played by metropolitan areas in the programming and implementation of the Recovery and Resilience Facility is given account of in Sect. 4. Finally, a concluding section rounds off the contribution, summing up its main argument and speculating on what the future may entail in relation to the contribution that metropolitan areas will be able to offer in the post-pandemic scenario.

2 Metropolitan Governance, the EU Cohesion Policy and the COVID-19 Pandemic

Metropolitan areas are responsible for the production of almost 70% of the EU GDP. However, metropolitan phenomena remain hard to address, also due to the complex relations among the centres, the suburban areas and the large peripheries that characterise metropolitan territories [13]. Since almost three decades, metropolitan areas in Europe have been both the scope of and the reason for institutional experimentation, with public authorities that have progressively engaged in the development of metropolitan strategic visions seeking to integrate spatial developments at different scales [1]. Whereas this often occurred via informal inter-municipal cooperation, a number of governance structures have been institutionalised from the bottom-up. At the same time, in some countries, formal administrative bodies have been established top-down and provided with the responsibility to manage the development of metropolitan territories. The exact nature of the cooperation is often unique, and different arrangements may depend on the different spatial governance and planning systems that characterise Europe [14]. Additional complexity emerges when comparing the institutional arrangement in place to the metropolitan functional dimension, as traditional governance and planning practices struggle to deal with phenomena that go beyond existing administrative jurisdictions [1]. In this concern, recent studies highlighted how hard it may be to adapt traditional planning practices to urbanisation trends that go beyond the jurisdictions of a single administrative authority, and the emergence of 'soft spaces with fuzzy boundaries' for planning and policy approaches that are more liquid and process-oriented [2, 15]. The key challenge seems to find the right problem 'owner(s)', that is/are able to address the metropolitan conundrum at the right scale and with the relevant tool(s). That is to say that the functional, political and representational relations within a given metropolitan area need to be understood in their institutional context before taking action [13].

---

[1] The ESPON METRO project (The role and future perspectives of Cohesion Policy in the planning of Metropolitan Areas and Cities. https://www.espon.eu/metro) analyses the role of the EU Cohesion Policy in the planning and implementation of metropolitan policies.

2.1   The Metropolitan Dimension of the EU Cohesion Policy

The importance of metropolitan areas is also recognised by the EU, as it is witnessed by their increasing relevance within supranational spatial development strategies and guidance document, as well as by the growing share of funds dedicated to urban development goals [16, 17]. The centrality of metropolitan areas has been recently reaffirmed in the renewed Leipzig Charter on Sustainable Cities [18] and in the EU Territorial Agenda 2030 [19]. More in detail, the latter advocates place-based territorial development and multilevel policy coordination as overarching principles for all places and policy sectors, while the former provides guidance for applying these principles in cities and their functional regions. These arguments have been further reinforced by the fact that metropolitan areas seem best positioned to manage the consequences of the COVID-19 pandemic and to plan for its aftermath [4]. In parallel to the consolidation of the metropolitan dimension within the European spatial planning discourse [20], the EU cohesion policy has been progressively adapted to cater to their needs. In the programming period 2014–20, at least 5% of the ERDF allocation was dedicated to sustainable urban development strategies, a share that raised to 8% in the present programming period. Also the ESF co-finances employment-related projects and investments targeting workers, young people and unemployed at a metropolitan scale. Since 2014, new instruments were introduced to ensure greater flexibility in tailoring funds' allocations to territorial needs. Integrated Territorial Investments (ITI) were used to favour the development and implementation of integrated metropolitan development strategies, addressing the challenges of given areas from priority axes of one or more operational programmes. Despite the described efforts, the adoption of suitable metropolitan governance and multi-scalar institutional arrangements remains a challenge. Metropolitan areas do not yet play a primary role neither in the design of the operational programmes, nor in the decision to use new instruments such as ITI, often due to the multilevel governance tensions resulting from the variable power relations that characterise each member states' administrative structures, that in turn influence that the design and management of the EU cohesion policy [21, 22].

2.2   The COVID-19 Pandemic in Metropolitan Areas

The COVID-19 pandemic triggered the third and greatest economic and social shock of the 21st century, after 9/11 and the 2008 global financial crisis. Over the past months, stringent measures have been applied to contain the virus and minimise pressure on healthcare infrastructure. Global economy has contracted unprecedentedly, with GDP that dropped on average by around 10 percent in the OECD countries in 2020, this representing the largest economic dip since the 1930s great depression [23]. While situations vary greatly across and within countries, most metropolitan areas have registered sharp downward trends in GDP, employment levels and fiscal revenues [5]. A survey conducted in May 2020 by the Council of European Municipalities and Regions on the impact of the crisis on local and regional finance confirms that, in addition to their significant losses in income, local governments are on the frontline of the crisis response and faced with large increases in expenditure[2]. Moreover, the OECD note on the territorial impact of the COVID-19 crisis across levels of government points out that budget constraints among subnational governments are expected to be long lasting[3]. When looking closer at some large European metropolitan areas, Paris saw its economic activity decrease by 37% since mid-March 2020, in contrast to the 34% at the national scale, whereas Barcelona estimates a drop of 14% in GDP, 4 times worse than during the 2008 financial crisis. The core cities of the United Kingdom estimated that the crisis had incurred costs amounting to GBP 1.6 billion among their cities alone by 22 May 2020. From the above, it emerges that the pandemic hit particularly strongly metropolitan areas, where density is associated with poverty, poor housing conditions, etc. In order to fight the direct health impacts and also the indirect economic and social consequences, preventive, protective or containment measures had to be introduced. However, the reaction to the pandemic has followed a certain degree of centralisation of policy making on national level. Cities and metropolitan areas have faced unprecedented levels and new forms of social and economic problems to which they had to react, in many occasions without the necessary powers and financial resources.

3   Metropolitan Areas Tackling the COVID-19 Pandemic

Metropolitan actions related to the pandemic are reported in most of the case studies analysed in the METRO project. These actions are conditioned by the legal status of each metropolitan authority and by the resources at their disposal. In some cases, important initiatives have been adopted that however did not take advantage from the EU cohesion policy. This is the case of the Barcelona Metropolitan Area, that put in place the most comprehensive set of measures among the analysed case studies, and also adapted two existing instruments to the scope: The Environmental Sustainability Plan and the Program of actions to improve the natural and urban landscape. When it comes to the measures directly supported through the EU cohesion policy, the cases of the Lyon Metropolitan Area and of the Lisbon Metropolitan Area report positive findings. In 2021, the Lyon Metropolitan Area has launched a call for projects focusing on mental health, financed through the leftovers of the ESF programme 2014-20 programming. Also in the case of the Lisbon Metropolitan Area, a number of measures have been implemented, this time in the field of transport. Moreover, the Lisbon Regional Operational Programme financed the adaptation of the metropolitan area's health care facilities so that they could better respond to the pandemic needs. The two Italian metropolitan cities investigated in the project show a rather different attitude. On the one hand, the Metropolitan City of Florence exploited the opportunities offered by the National Operational Programme (NOP) METRO to face the pandemic challenge by investing on sustainable mobility, housing, energy efficiency, urban forestry and digitalisation. On the contrary, in the case of Turin the metropolitan dimension of the intervention implemented through the NOP

---

[2] https://ccre.org/img/uploads/piecesjointe/filename/200629_Analysis_survey_COVID_local_finances_EN.pdf.

[3] http://www.oecd.org/coronavirus/policy-responses/the-territorial-impact-of-covid-19-man aging-the-crisis-across-levels-of-government-d3e314e1/.

METRO has been limited to the urban core. In other case studies, the role of metropolitan institutions has been more limited. In the case of Brno there has been no specific metropolitan instrument dedicated to react to the COVID-19 pandemic at the metropolitan scale. Also in the case of the Metropolitan Area of Gdansk-Gdynia-Sopot no specific metropolitan instrument is dedicated to react to the pandemic at the metropolitan level. However, two ITI projects have been enlarged in scope to accommodate reactions to COVID-19, by the decision of Region.

3.1 The Level of Metropolitan Engagement

When assessing the level of metropolitan engagement in the reaction to the pandemic, it differs across the metropolitan areas under investigation. More in detail, it is possible to distinguish between (Table 1): (i) no engagement, metropolitan authorities did not play any role in the implementation of COVID-19 mitigation measures; (ii) scarce engagement, metropolitan authorities played a limited role in addressing the emergency; (iii) sectoral engagement, metropolitan authorities have implemented some COVID-19 mitigation measures, mainly focusing on sectoral issues and, (iv) comprehensive engagement, metropolitan authorities have adopted a more integrated approach. According to the collected evidence, no case reports a full engagement of the metropolitan authority. In a number of cases (i.e. Gdańsk-Gdynia-Sopot, Lyon and Barcelona), metropolitan authorities have been involved in the development of sectoral interventions addressing the pandemic emergency and its impacts. The Gdansk-Gdynia-Sopot metropolitan area has undertaken several initiatives related to COVID-19, mainly of 'soft' nature, such as measures in support to the local tourism industry, cultural initiatives, social initiatives in support to local restaurants, and social inclusion measures. Lyon Metropolitan Area has directly acted on COVID-19 related issues through a set of dedicated measures: in April 2020 it launched a €100 million emergency fund in favour of local businesses, to then adopt a series of actions supporting health and social facilities, services dedicated to child protection and social inclusion. The Barcelona Metropolitan Area has lunched the most structured set of interventions, although also of a sectoral nature and detached from the EU cohesion policy framework. Two extraordinary investment programmes and a New Mobility Pact were approved, aimed at accelerating the sustainable mobility and energy transition. A number of initiatives, although of a less structured nature, have been put in place in the cases of the Metropolitan areas of Lisbon and Florence. In Lisbon the pandemic was tackled by the metropolitan authority and the regional government through the adjustment of certain priorities of the Regional Operational Programme, namely those focusing on social inclusion, education and training. Here the pandemic crisis put under stress the metropolitan administration's capacity to deal with emergencies, at the same time highlighting a gap between the institutional metropolitan competences and the expectations, hence causing frustration among the different stakeholders. As said, also Florence used the NOP METRO funds to strengthen its sustainable mobility system. However, due to their urban-centric nature, it is hard to read the metropolitan added value of the implemented interventions. Finally, in three out of eight cases (Brno, Riga, Turin), the metropolitan authorities did not play any substantial role in addressing the pandemic emergency.

Table 1. Level of Metropolitan engagement in dealing with the pandemic emergency

Metropolitan area	Metropolitan engagement			
	No	Scarce	Sectoral	Comprehensive
Metropolitan City of Turin	X	–	–	–
Metropolitan Area of Barcelona	–	–	X	–
Lisbon Metropolitan Area	–	X	–	–
Brno Metropolitan Area	X	–	–	–
Riga Metropolitan Area	X	–	–	–
Gdańsk-Gdynia-Sopot Metro Area	–	–	X	–
Metropolitan City of Florence	–	X	–	–
Lyon Metropolitan Area	–	–	X	–

3.2 The Nature of Policy Responses

Although in the majority of cases the metropolitan areas have had limited room for action, where this has been possible the measures that have been undertaken are rather heterogeneous (Table 2): (i) reactive, metropolitan authorities promote short-term measures aiming at giving an immediate response to the pandemic issue; (ii) containment driven, metropolitan authorities take strict decision in order to contain the pandemic by, for example, reallocating ordinary budget or EU funds and, (iii) proactive, metropolitan authorities support the implementation of long-term measures where decisions are taken based on a strategic view in order to overcome the pandemic impact. When examining the cases where the metropolitan authorities have somehow played a role in addressing the pandemic, only the Metropolitan Area of Barcelona seems to have adopted a proactive approach. Although of sectoral nature, the New Mobility Pact is aimed at influencing the mobility system of the metropolitan area in the post-pandemic scenario. On the other hand, in the cases of Lisbon, Lyon and Florence, a readjustment of funds has been made in the light of containing the pandemic emergency instead, as for instance strengthening soft mobility, or reinforcing health and social facilities. Finally, in the case of the Gdańsk-Gdynia-Sopot metropolitan area the majority of initiatives have been reactive to the pandemic, trying to provide support to local businesses and tourist activities in a way that they could survive the pandemic emergency, instead of being driven by more long-term oriented strategies and priorities. Overall, the heterogeneity of policy responses shows how articulated and challenging the pandemic has been for each territory. Perhaps because of their limited room for action due to the lack of competences, dedicated instruments and resources, the role that metropolitan areas could have played in addressing the pandemic has remained largely unexpressed. It will be interesting to see whether and how the situation will change with the implementation of the actions funded under the Recovery and Resilience Facility, and included in the National Recovery and Resilience Plans that some of the METRO stakeholders have in one way or another contributed to substantiate.

Table 2. Nature of Policy Responses of metropolitan areas dealing with COVID-19

Metropolitan area	Nature of policy responses		
	Reactive	Containment	Proactive
Metropolitan City of Turin	–	–	–
Metropolitan Area of Barcelona	–	–	X
Lisbon Metropolitan Area	–	X	X
Brno Metropolitan Area	–	–	–
Riga Metropolitan Area	–	–	–
Gdańsk-Gdynia-Sopot Metropolitan area	X	–	–
Metropolitan City of Florence	–	X	–
Lyon Metropolitan Area	–	X	–

3.3 The Type of Policy Responses

The actions put in place in each of the analysed contexts also differs in relation to their type (Table 3). It is possible to identify three categories of responses according to the diversity of instruments that they adopted to deal with the emergency: (i) incentives, metropolitan areas have activated *ad hoc* incentives to support specific sectors or social categories; (ii) strategies, metropolitan areas have introduced medium and long term sectoral strategies aiming at reducing the impact of the pandemic as well as increasing the quality of life of their territories and, (iii) projects, metropolitan areas have implemented specific *ad hoc* initiatives to limit or mitigate the impact of the pandemic.

Table 3. Type of Responses of metropolitan areas dealing with COVID-19

Metropolitan area	Type of responses		
	Incentives	Strategies	Projects
Metropolitan City of Turin	–	–	–
Metropolitan Area of Barcelona	X	X	X
Lisbon Metropolitan Area	X	–	X
Brno Metropolitan Area	–	–	–
Riga Metropolitan Area	–	–	–
Gdańsk-Gdynia-Sopot Metropolitan area	X	–	X
Metropolitan City of Florence	–	–	X
Lyon Metropolitan Area	X	–	X

The above categorisation highlights that no specific regulations have been adopted by metropolitan areas to deal with the pandemic. The majority of normative restrictions have been introduced by the national, regional and local authorities, following the distribution of responsibilities on health matters or the principle of subsidiarity. Among the five metropolitan areas that have been active in addressing the pandemic, all of them seem to have promoted incentive-based initiatives. Those mainly consist on allocating funds to support sectors and/or social groups particularly affected by the pandemic as for instance in Lyon where cohesion funds have been used for targeted interventions on some problematic areas such as those related to social cohesion and unemployment. These incentive-based initiatives are seen as the way to alleviate contingent emergency problems instead of supporting post-pandemic reconstruction. Examples of proactive and future oriented measures (i.e. long-term strategies) are very few across the METRO metropolitan areas. At this regard, is worth to mention the New Mobility Pact adopted by the Metropolitan Area of Barcelona, which took the momentum to go beyond the emergency towards the definition of a post-pandemic scenario where mobility will definitely play a crucial role. Finally, almost everywhere there was a proliferation of ad hoc projects that supported metropolitan areas to implement short-term initiatives aiming at softening the impact of the COVID-19.

4   Metropolitan Areas and the Recovery and Resilience Facility

After discussing the reaction of the analysed metropolitan areas to the pandemic, it is important to reflect on the role that they are playing in relation to the Recovery and Resilience Facility. The collected evidence shows that their involvement varies widely, due to the different approaches adopted in the programming of the National Recovery and Resilience Plans. The Italian metropolitan cities seem well positioned, as they were required to deliver to their representation in ANCI (the National Association of Italian Municipalities) a number of metropolitan flagship projects. The Mayor of the Metropolitan City of Florence has activated a working group collecting proposals from local actors. The Metropolitan City of Turin has proposed 20 projects related to green transition, digital transition for the public administration, cohesion, sustainability, inclusion and mobility. At the same time, Italian metropolitan cities will be responsible for the development and implementation of so-called Urban Integrated Plans, aimed at favouring the landing of the programmed interventions on the ground. In France, Lyon Metropolitan Area proposed 23 projects to the REACT-EU programme, concerning the thermal renovation of schools and social housing, the development of inclusive digital projects, and the purchase of personal protective equipment. In addition, five projects have been presented by the City of Lyon, concerning the development of a new vaccination centre and the thermal renovation of schools. Also the Barcelona Metropolitan Area has presented a number of preliminary projects to the regional and national governments. The Latvian National Recovery and Resilience Plan devotes a large amount of resources to climate objectives, and in particular to sustainable mobility. The greening of the Riga Metropolitan Area transportation system will be implemented through the plan, and involve Riga city and the neighbouring municipalities. The remaining cases do not seem to have been involved yet in the programming of the Recovery and Resilience Facility to any relevant extent. In the case of Brno, the main cities that compose the metropolitan areas and the holders of the ITI have gained since early 2021 some opportunities to comment upon the preparation of the National Recovery and Resilience Plan.

The Portuguese plan is seen as an opportunity to make heavier investments and change the paradigm in the transport and mobility domain in the metropolitan area of Lisbon, while it also includes priorities dedicated to the health sector, digital transition, and housing. A similar situation concerns the Metropolitan Area of Gdansk-Gdynia-Sopot where, despite the expectations the metropolitan influence in the programming phase has been so far limited.

5 Concluding Remarks and Future Research Perspectives

Whereas metropolitan areas are well positioned to react to COVID-19 pandemic and to contribute to planning its aftermath, due to the crucial role they could play in the promotion and coordination of intermunicipal strategies and actions, the collected evidence reports until now a limited engagement. This situation is further worsened by the fact that the Recovery and Resilience Facility is mostly managed at the central level in the member states, in partial contradiction to the partnership principle and, most importantly, to the fact that, across Europe, large urban and metropolitan areas have been the ones hit hardest by the pandemic. Drawing on the collected insights, some observations can be made in terms of metropolitan resilience and capacity to recover from shocks. Firstly, it appears clear that a wide range of place-based responses is ongoing, from measures to protect citizens to business support aiming at longer-term impact. In the first stage, most actions were short-term, but some metropolitan areas seem to have understood that the pandemic can be turned into an opportunity to make cities more resilient, circular, smarter, and better connected with rural areas, via the way goods are produced, the energy consumed and transport and other services organized [24–26]. To this end, they are undertaking policies aiming at providing efficient social and community services for disadvantaged groups such as health care and home care, through the design and implementation of ambitious social innovation strategies. The pandemic is slowly fading away and European metropolitan areas are faced with an unprecedented responsibility. At the same time, the chance has clearly opened for them to rethink from the ground up their development paradigms. If nothing else, the provided evidence shows that some of them are trying to seize the opportunity.

References

1. Albrechts, L., Balducci, A., Hillier, J.: Situated Practices of Strategic Planning. Routledge, London (2017). https://doi.org/10.4324/9781315679181
2. Zimmermann, K., Galland, D., Harrison, J. (eds.): Metropolitan Regions, Planning and Governance. Springer, Cham (2020). https://doi.org/10.1007/978-3-030-25632-6
3. Medeiros, E. (ed.): Territorial Cohesion: The Urban Dimension. Springer, Cham (2019). https://doi.org/10.1007/978-3-030-03386-6
4. Metropolis - Metropolitan governance and health: The Experience of COVID-19. Seminar organized in the framework of the World Metropolitan Day (2020). Metropolitan Governance and Health: The Experience of Covid-19 | Metropolis. Accessed 28 Dec 2021
5. Kapitsinis, N.: The underlying factors of the COVID-19 spatially uneven spread. Initial evidence from regions in nine EU countries. Reg. Sci. Policy Pract. **12**(6), 1027–1045 (2020). https://doi.org/10.1111/rsp3.12340

6. Cotella, G., Brovarone, E.V.: Questioning urbanisation models in the face of Covid-19. TeMA, 105–118 (2020). https://doi.org/10.6092/1970-9870/6913
7. Cotella, G., Vitale Brovarone, E.: Rethinking urbanisation after COVID-19. What role for the EU cohesion policy. Town Plan. Rev., 1–8 (2021). https://doi.org/10.3828/tpr.2020.54
8. Yahagi, H., Abe, D., Hattori, K., Cotella, G., Bolzoni, M.: Will Cities Change with COVID-19? Gakugei Publishing, Kyoto (2020)
9. Deslatte, A., Hatch, M.E., Stokan, E.: How can local governments address pandemic inequities? Public Adm. Rev. **80**(5), 827–831 (2020)
10. Moore-Cherry, N., Pike, A., Tomaney, J.: City-regional and metropolitan governance. In: Callanan, M., Loughlin, J. (eds.) A Research Agenda for Regional and Local Government. Edward Elgar Publishing, pp. 63–77 (2021). https://doi.org/10.4337/9781839106644.00010
11. ESPON METRO – The role and Future Prospective of Cohesion Policy in the Planning of Metropolitan Areas and Cities (2021a)
12. ESPON METRO – The role and Future Prospective of Cohesion Policy in the Planning of Metropolitan Areas and Cities. Annex II (2021b)
13. Salet, W., et al.: Planning for the new European metropolis: functions, politics, and symbols. Plan. Theory Pract. **16**(2), 251–275 (2015). https://doi.org/10.1080/14649357.2015.1021574
14. Berisha, E., Cotella, G., Janin Rivolin, U., Solly, A.: Spatial governance and planning systems and the public control of spatial development: a European typology. Eur. Plan. Stud. **29**(1), 181–200 (2021). https://doi.org/10.1080/09654313.2020.1726295
15. Allmendinger, P., Haughton, G., Knieling, J., Othengrafen, F.: Soft Spaces in Europe. Routledge, London (2015). https://doi.org/10.4324/9781315768403
16. Atkinson, R., Zimmermann, K.: Cohesion policy and cities: an ambivalent relationship. In: Piattoni, S., Polverari, L. (eds.) Handbook on Cohesion Policy in the EU. Edward Elgar (2016)
17. Cotella, G.: The urban dimension of EU cohesion policy. In: Medeiros, E. (ed.) Territorial Cohesion. TUBS, pp. 133–151. Springer, Cham (2019). https://doi.org/10.1007/978-3-030-03386-6_7
18. DE Presidency: The New Leipzig Charter. The transformative power of cities (2020a)
19. DE Presidency: Territorial Agenda 2030. A future for all places (2020b)
20. Adams, N., Cotella, G., Nunes, R.: Spatial planning in Europe: the interplay between knowledge and policy in an enlarged EU. In: Adams, N., Cotella, G., Nunes, R. (eds.) Territorial development, cohesion and spatial planning, pp. 1–25. Routledge, London (2011)
21. Cotella, G.: How Europe hits home? The impact of European Union policies on territorial governance and spatial planning. Géocarrefour **94**(3) (2020). https://doi.org/10.4000/geocarrefour.15648
22. Cotella, G., Dabrowski, M.: EU cohesion policy as a driver of Europeanisation: a comparative analysis. In: Rauhut, D., Sielker, F., Humer, A. (eds.) EU Cohesion Policy and Spatial Governance Territorial, Social and Economic Challenges, Edward Elgar, pp. 48–65 (2021)
23. OECD Economic Outlook, Volume 2020 Issue 1, No. 107, OECD Publishing, Paris (2020)
24. Bottero, M., Caprioli, C., Cotella, G., Santangelo, M.: Sustainable cities: a reflection of potentialities and limits based on existing eco-districts in Europe. Sustainability **11**(20), 5794 (2019). https://doi.org/10.3390/su11205794
25. Rotondo, F., Abastante, F., Cotella, G., Lami, I.: Questioning low-carbon transition governance: a comparative analysis of European case studies. Sustainability **12**(24), 10460 (2020). https://doi.org/10.3390/su122410460
26. Valkenburg, G., Cotella, G.: Governance of energy transitions: about inclusion and closure in complex sociotechnical problems. Energy Sustain. Soc. **6**(1), 1–11 (2016). https://doi.org/10.1186/s13705-016-0086-8

Regulating Urban Foodscapes During Covid-19 Pandemic. Privatization or Reorganization of Public Spaces?

Anita De Franco(✉)

Department of Architecture and Urban Studies (DASTU), Polytechnic University of Milan, Via Bonardi 3, 20133 Milan, MI, Italy
anita.defranco@polimi.it

Abstract. This paper describes how the municipality of Milan incentivized the occupation of public spaces by private activities. From a methodological viewpoint, this study adopts a neo-institutionalist view on "foodscapes". The approach is mainly conceptual but empirically illustrated, to discuss the spatial implications at stake. The first section (§ 1) provides an introduction on how Covid-19 pandemics made the relations that link objects and people in urban spaces more visible. To do this, the focus of our attention will be on restaurant businesses, which are among those that have had to review their way of working radically in the face of specific epidemiological issues but also the more general changes taking place in urban areas during the pandemic. The second section (Sect. 2) is devoted to the spatial analysis of emerging foodscapes in Milan, highlighting the distribution of restaurants outdoors (i.e. on urban streets). The third section (Sect. 3) discusses the issues affecting local businesses and restaurant entrepreneurs in Italy in the aftermath of the first year of the pandemic. Section (Sect. 4) concludes.

Keywords: Urban foodscapes · Regulation · Covid-19 · Restaurants

1 Introduction: Urban Foodscapes During Covid-19

During the first year of the Covid-19 pandemics, food consumption drastically changed the street outlook of many metropolitan cities [1, 2].

This paper[1] focuses on how restaurants adapted their business to arrange outdoor dining during Covid-19. With the term "urban foodscape", we mean those street installations identifiable as the outdoor activity of local restaurants. Therefore, we do not consider street food or street vendors even if some elements here discussed may be relevant also for them [3].

The "foodscape" can be simply defined as urban areas devoted to food and beverage consumption [4]. Considering everyday urban life, note that such environments are

---

[1] This research is part of the research project "Norms, Uncertainty and Space (NOUS): Cities in the age of Hyper-Complexity" funded by the Department of Architecture and Urban Studies (DASTU), Polytechnic University of Milan (Milan, Italy).

© The Author(s), under exclusive license to Springer Nature Switzerland AG 2022
F. Calabrò et al. (Eds.): NMP 2022, LNNS 482, pp. 1009–1017, 2022.
https://doi.org/10.1007/978-3-031-06825-6_96

possible everywhere, as food consumption also, and quite often, occurs in open and undesignated spaces. For instance, wherever there is a place to sit or stand [5]. This possibly naive and generic conception of foodscapes was at the centre of attention in the first year of Covid-19 pandemics in Italy and especially in metropolitan cities such as Milan. As the disease rapidly spreads indoors and at a short distance (through droplets and aerosol [6, 7]), restaurants were the most endangered and contested activities in urban areas.

While it is clear that aspects such as density, street connectivity and mixed uses are important for food consumption in urban contexts [8], these very aspects became particularly problematic during the pandemics. Cafés, bars and restaurants had to revise their way of working radically, also considering the "responsibility" of what was happening in the immediate surroundings. Mainstream media, in many countries, fuelled negative perceptions of urban social life, leading many to sustain a greater presence of the police forces and stricter regulations for citizens and restaurant owners [9].

Italy was among the nations that implemented the most restrictive measures earlier and longer than many Western states [10]. This had enormous impacts on the economy in general and for the restaurant sector in particular. To the numerous restrictions on mobility, gatherings and so on additional tasks were added for restaurateurs to remain open (when this was allowed). In sum, local restaurant entrepreneurs had to (a) reshape business activities as a *forced choice* in the light of the lockdowns; (b) manage their spaces as *prescribed* by the authorities; (c) monitor the spread of the virus as a *formal request* [6]. In Italy, all these additional measures were not subsidized by the governments, nor directly beneficial in the light of the higher running costs and decreased number of clients (especially where home confinement and mobility prescriptions were stricter, e.g. in Italy and France, and differently from Germany and other European countries [11]).

Over time, the general debates shifted from the *"density pathology"* narrative, typical of the first wave (based on the prohibition of any type of activity, at least in Italy), to an *"organising densities"* narrative, especially from the second wave [9]. Restaurant entrepreneurs had to undertake new business choices linked to both indoor activities (e.g. reducing seats, placing tables in new layouts, channel entry/exit flows, providing additional health procedures, using contactless technologies) and outdoor activities (i.e. creating outside seating areas).

Some cities welcomed this opportunity to experiment with new designs and business ideas making open public spaces directly available for retail re-openings and so fuelling a renewed interest in streetscapes design [1, 2]. The next section will explore the case of Milan.

2   The General Problem and the Case of Milan

2.1   Preliminary Aspects

From a methodological viewpoint, the case study is based on a qualitative and quantitative analysis, focused on: (i) a review of normative records and documents (e.g. local ordinances, codes, rules), sectorial reports from public and private agencies (i.e.

FIPE: *Federazione Italiana Pubblici Esercizi*), and municipal open data; (ii) on-the-spot observations. In terms of approach, this study adopts a neo-institutionalist view on "foodscapes" during Covid-19. The following sub-sections will provide an overview of the problem in Italy and preliminary spatial analysis (e.g. GIS mapping) and detailed information on the occupation of public land by local restaurants in the city of Milan.

2.2 General Overview

In Italy, as elsewhere, cafes, bars, pubs and restaurants have been considered as "non-essential" activities during the pandemics, even though it was evident that their closure had tremendous impacts on the local economy, social life and wellbeing [12–14]. They have been part of that relatively small part of the economy that was completely closed during the first-wave lockdowns and reopened intermittently in subsequent periods. The lack of people, of certain noises and smells coming out of restaurants and cafes, created mixed feelings of nostalgia and anxiety for city users [15]. The problem lies not so much in having disregarded certain cultural/identity values behind food, leisure, etc.; but, rather, in having completely ignored the entrepreneurial issues at stake.

According to the analyses of Italian national associations of retailers [16], due to the restrictive and often confusing rules for restaurateurs, the sector has lost more than 514,000 jobs and 31 billion euros only between 2020 and 2021. In the same period, 97.7% of restaurants in Italy recorded a sharp decline in profits (at least 60% suffered a decline of 50%) and, more generally, a decline in values for food consumption (household food spending was particularly low in budget and quality, covering less than 20% of the total loss in food and beverage sector). Further enquiries, underline that for Italian entrepreneurs, the decline in the sector is mainly due to the costs of people's mobility (88%), the reduction of the internal capacity of the restaurants (35%) and the reduction of tourist flows (31%). State subsidies were intended for those who could prove a decrease (of at least 30%) in turnover, but no incentives were given to those who managed to keep the business afloat. Taxes were not eliminated but extended at most (by 30 days, then by a few months) and without particular discounts. In the face of all this, 89.2% of entrepreneurs felt that the State aids were little (47.9%) or not at all (41.3%) effective. This, in Italy and other countries, tremendously affected family-owned businesses, small and medium enterprises as well as young adults either as customers, employees or future entrepreneurs [17, 18].

Lockdowns, either gradual or instantaneous, resulted in a decrease in output and inputs (e.g. investments) for restaurant retailers [19] and created a fear of food insecurity (e.g. unavailability and price fluctuations; [20]), unemployment, bankruptcies and recession [14]. All these problems are stimulating debates in other countries (like Germany); for instance, on the necessity to lift certain restrictions and to decentralize and make management of the far-reaching effects of the (post)pandemics [13, 20] more effective. Specifically for the food sector, many scholars appear to be quite supportive of the progressive approaches to urban food policies and the emerging new-municipalist era [21], although the discussions on how Covid-19 affect food production and distribution in cities require more critical attention to fully understand the types of innovations that the Covid-19 pandemic has made necessary and timely.

2.3 Milan's Response

As regards the city of Milan, it seems appropriate to recall the main events related to the pandemic: (i) February 2020: outbreak of the Covid-19 in Lombardy; (ii) March 2020: whole-city lockdown; (iii) April 2020: selective re-opening (e.g. bookshops, certain industrial activities); (iv) May 2020: mobility re-opening (e.g. inside and across cities); (v) June 2020: gradual re-opening of all urban activities (e.g. restaurants, bars, beauty centres, gyms included).

Restaurants have been among the latest businesses to return active through a wide renewal of normative frameworks (supralocal and local).[2] In Milan, the possibility to increase seat capacity by occupying outdoor public areas depended on three crucial steps taken by local governments.

Firstly, the adoption of Milan's adaptive strategy (i.e. *Strategia di Adattamento Milano 2020*, by May-April 2020 [22]). This included a collection of citizens' opinions and ideas for the re-openings (a total of 3,000). Outdoor dining collected 130 suggestions, showing, however, a certain degree of opposition especially on the concession of public spaces to private actors, fear of gatherings, noises, litter, etc.[3]

Secondly, the creation of a new ad-hoc tax regime (i.e. *COSAP: canone per l'occupazione di spazi ed aree pubbliche;* by December 2020 [23]). This consisted of the unification of municipal taxes concerning the occupation of public land, after the first experiments of the reopenings (from May 2020). This tax regime unified all types of fees concerning the installation of any elements on public land (from posters to kiosks). Local entrepreneurs colloquially call it the "tax on the shadow" (*tassa sull'ombra*). This reform consisted of lower rates for occupation and simplified procedures.

Thirdly, the publication of design guidelines for outdoor installations (aimed especially at bars, cafes and restaurants [24]). The key criteria were that such occupations should be easily removable, not completely closed and excluding any damage to public land (e.g. with fixing operations). The total cost of the occupation is made up of (i) *basic rates*, multiplied by (ii) a *coefficient* that evaluates the economic benefit of the area, (iii) the *road category* and (iv) the *square meters* and/or *days* of occupation (if these are temporary). The types of design affect the overall tariff that entrepreneurs have to pay (see Tab 1) whether they create occupations with closed structures (e.g. *dehors*) adhering

---

[2] Among which, at the European level: The State Aid Temporary Framework (19th March 2020). At the national level, a series of law decrees: DL n. 9/2020 (2nd March 2020); DL n.18/2020 "*Cura Italia*" (17th March 2020); DL n. 23/2020 "*Liquidità*" (8th April 2020); DL n. 34/2020 "*Rilancio*" (19th May 2020); "*Ulteriori misure urgenti per fronteggiare l'emergenza epidemiologica da COVID-19*" (L. 14th July 2020); DL n. 104/2020 "*Agosto*" (14th August 2020); DL n. 137/2020 "*Ristori*" (28th October 2020); DL n. 149/2020 "*Ristori-bis*" (9th November 2020); DL n.154/2020 "*Ristori-ter*" (23rd November 2020); DL n. 157/2020 "*Ristori-quater*" (30th November 2020). At the local level, between 11th March 2020 and 30th December 2021 43 ordinances were issued by the mayor of Milan (34 in 2020 and 9 in 2021; corresponding to one ordinance every 15 days). Available online: https://www.comune.milano.it/home/corona virus-informazioni-e-link (accessed 28 December 2021).

[3] Available online: https://www.comune.milano.it/documents/20126/95930101/Milano+2020+ risultati+finali.pdf/ed03823a-94d9-9ef8-f8ce-2ec84a486e59?t=1593436694366 (accessed 2 December 2021).

to or adjacent to buildings (rows 1 and 2), occupations delimited with tables and chairs (e.g. by screens: row 3) or not delimited (row 4), occupations on driveways (row 5).

Table 1. Overview of the coefficients affecting the costs of temporary and permanent occupations of public land in the city of Milan. From: [23]

Elements for the occupation	Coefficient for permanent occupations (2012 av.)	Coefficient for permanent occupations (2015)	Coefficient for temporary occupations
Closed structures adhering to buildings	2.25	2.00	2.00
Closed structures adjacent to buildings	3.85	4.00	4.00
Structures delimited with tables and chairs	1.60	1.50	1.00
Occupations with tables and chairs	1.10	1.00	1.00
Occupations on driveways	1.40	0.70	0.70

At the moment, the municipality of Milan is using the 2015 rates (almost the same for permanent and temporary occupation). The average rates have tended to decrease since 2015 but not homogeneously. In particular, since 2012 the tariffs for the occupation of driveway spaces decreased sharply (−50%), while those for the occupation of spaces adjacent to the building are the only ones that have increased (+4%). The concessions issued in 2020 (which took advantage of the transitional regulations) and those issued in 2021 now in place, are renewed until 31 March 2022.

2.4 Spatial Analysis

The spatial effects of the adaptive strategy, the new tax regime and local guidelines had interesting effects on the context (see Fig. 1). The local government's objectives were to prepare for a progressive resumption of activities to stabilize temporary light occupations, also taking into account residential parking and the interests of pedestrians [22–24]. From the first months of experimentation, thousands of businesses have joined the initiative. Looking at the municipal data, in 2020 (May-December) 777 licences were issued; in 2021 (June-September) there were more than 2700 licences (+250%).

There is a fair distribution of activities city-wide. Between 2020 and 2021, the areas with the largest number of licences were the semi-central ones (from 445 to 1533), followed by central areas (from 210 to 885) and peripheries (from 122 to 368). The vast majority of licences concerned small spaces for occupation (i.e. below 25 m^2). Breaking down the data, it is possible to understand that the new fiscal regime repurposed mainly pavements and parking spaces (respectively, 46% and 23% of the total licences) for

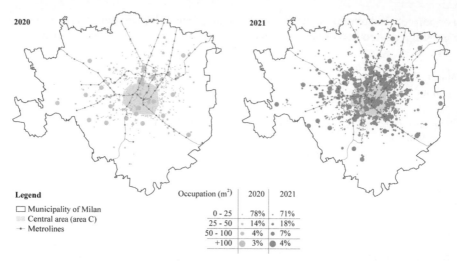

Fig. 1. Spatial distribution of licenses for the occupation of public spaces in Milan. Data for 2020 (May-December) and 2021 (June-September). From [25, 26]

outdoor dining; few licences concerned other, possibly already compatible, open spaces such as pedestrian and green areas (respectively, 14% and 17% of the total licences).

While similar studies emphasise the correlations between higher concentration of licences and higher income [2], for Milan's case it seems more appropriate to recognize that the higher concentration in the city centre is due to the already existing urban tissue that favoured the spreading of outdoor dining (e.g. areas with higher accessibility, more pedestrian spaces, presence of limited traffic zones).

The fair spread of occupations also in peripheral areas suggests that such activities, when made accessible, can change the appearance and functional endowment of urban fringes.

3 Discussion

In 2021, many restaurants spread and occupied public spaces in cities. Many (urban) scholars critically debated this trend pointing to an alleged end of public spaces, due to the ongoing privatization of areas and "commodification" of urban experience [27, 28]. Considering the specific case of the Covid-19 adaptive strategy of Milan, it is however necessary to clarify three main points.

First of all, what happened was not necessarily a deregulation of entrepreneurial activities, but a *recalibration* of human costs in general. The fear of contagion has raised the operative and bureaucratic running costs of many workplaces [1]. Thinking about restaurants, the interiors are already largely shaped by various standards, codes and laws (e.g. health, security, hygiene [27]) that add to the rules for the exteriors (e.g. land-uses, zoning, ordinances, decorum standards, business hours). Considering Covid-19 regulations, new strata of rules, and deontic meanings, are added to everyday objects, shaping also contemporary urban foodscapes. The critical point here is to understand which ones will be useful in the future; to keep them viable (and credible) these cannot be too costly or too numerous.

Second, the new activities were not necessarily exclusionary or exclusive to wealthier areas. To be clear, "privatization" implies the acquisition (i.e. ownership) of a piece of land bought from the public authority or the market. This should be no surprise (especially when legitimately done) and yet it does not coincide with the type of phenomena here described. In Milan, the focus was on licences for private entrepreneurs. Either in the case of "temporary" or "permanent" occupations, the collective benefits deriving from higher revenues and clustering effects should be regarded with particular interest, especially where there is a greater need to enlarge local entrepreneurial opportunities (e.g. in Milan's peripheries).

Third, what happened was not necessarily a commodification of urban foodscapes, but enrichment through the *diversification of experiences*. It is obvious that certain private activities, even the smallest ones, truly contribute to the overall enjoyment of local living. Other cities in Europe explicitly tried a policy of "non-closure" for restaurants (e.g. in Madrid) fully subsidizing the occupations and support of recreational activities for citizens [2]; Milan's approach was more moderate, less generous in this regard. Take for instance how diminishing parking spaces seems more appealing than establishing appropriate dining spaces (e.g. the rate for closed structures adjacent to buildings costs almost five times more). In conclusion, the narrative on deregulation, privatization and commodification in neo-liberal times does not always seem convincing. It is, in particular, questionable in the case considered.

4   Final Remarks

The peculiarity of the epidemics coupled with the need to establish a "new normality" has brought food and beverage consumption out in the open. The density of urban spaces and businesses was undoubtedly problematic for urban activities in general and dining in particular. To remain economically viable, on the one hand, many entrepreneurs heavily relied on public subsidies but, on the other hand, they had to prove their ability to adapt to emerging challenges [17].

Considering the case of Milan, the challenge has been interpreted mainly as "bring your activities in the open" by using fiscal incentives to accommodate certain practices more than others. Diminishing parking spaces is not necessarily a good business idea for local retailers and clients [29], but it is much in line with the mobility strategy that the Municipal Council is pushing forward (e.g. walkable 15 min city). This trend is common also in other metropolitan areas and highly consensual for the renewed interest in sustainable mobility [2]. On the other side of the coin, the sacrifice of parking and roadway spaces may be to the detriment of the *diversity* of resource flows (intra/extra-urban), types of activities (think for instance of drive-in) and customers (e.g. especially those with limited mobility), on which especially quality restaurateurs tend to focus.

This paper suggested critical reflections on (i) the overall *costs* of the pandemics and how to redirect some of them to foster innovation (i.e. recalibration of human energies, not necessarily deregulation); (ii) the possibility of enlarging the *opportunities for entrepreneurship* (especially in the periphery); (iii) the diversity of *experiences* inhibited and fostered in light of the pandemics.

References

1. Florida, R., Rodríguez-Pose, A., Storper, M.: Cities in a post-COVID world. Urban Stud., 1–23 (2021). https://doi.org/10.1177/00420980211018072
2. Pérez, V., Aybar, C., Pavía, J.M.: COVID-19 and changes in social habits: restaurant terraces, a booming space in cities. The case of Madrid. Mathematics 9(17), 1–18 (2021)
3. Zeb, S., Hussain, S.S., Javed, A.: COVID-19 and a way forward for restaurants and street food vendors. Cogent Bus. Manag. 8(1), 1–10 (2021)
4. Vonthron, S., Perrin, C., Soulard, C.T.: Foodscape: a scoping review and a research agenda for food security-related studies. PLoS ONE 15(5), 1–26 (2020)
5. Whyte, W.H.: The Social Life of Small Urban Spaces. Project for Public Spaces, New Work (1980)
6. Rozanova, L., Temerev, A., Flahault, A.: Comparing the scope and efficacy of COVID-19 response strategies in 16 countries: an overview. Int. J. Environ. Res. Public Health 17(24), 1–17 (2020)
7. Teller, J.: Urban density and Covid-19: towards an adaptive approach. Build. Cities 2(1), 150–165 (2021)
8. Black, J.L., Carpiano, R.M., Fleming, S., Lauster, N.: Exploring the distribution of food stores in British Columbia: associations with neighbourhood socio-demographic factors and urban form. Health Place 17(4), 961–970 (2011)
9. McFarlane, C.: Repopulating density: COVID-19 and the politics of urban value. Urban Stud. (2021). https://doi.org/10.1177/00420980211014810
10. COVID-19 Government Response Tracker. Blavatnik School of Government & University of Oxford. https:www.bsg.ox.ac.uk, Accessed 27 Dec 2021
11. Kuhlmann, S., Hellström, M., Ramberg, U., Reiter, R.: Tracing divergence in crisis governance: responses to the COVID-19 pandemic in France, Germany and Sweden compared. Int. Rev. Adm. Sci. 87(3), 556–575 (2021)
12. Kaufmann, K., Straganz, C., Bork-Hüffer, T.: City-life no more? young adults' disrupted urban experiences and their digital mediation under Covid-19. Urban Planning 5(4), 324–334 (2020)
13. Steinmetz, H., Batzdorfer, V., Bosnjak, M.: The ZPID lockdown measures dataset for Germany. ZPID Sci. Inf. Online 20(1) (2020). https://doi.org/10.23668/psycharchives.3019
14. Onyeaka, H., Anumudu, C.K., Al-Sharify, Z.T., Egele-Godswill, E., Mbaegbu, P.: COVID-19 pandemic: a review of the global lockdown and its far-reaching effects. Sci. Prog. 104(2), 1–18 (2021)
15. Young, A.: The limits of the city: atmospheres of lockdown. Br. J. Criminol. 61(4), 985–1004 (2021)
16. FIPE: Federazione Italiana Publici Esercizi, Ristorazione a pezzi dopo un anno di pandemia, 2021 in profondo rosso. Comunicato stampa. https://www.fipe.it, Accessed 23 Oct 2021
17. Gkoumas, A.: Developing an indicative model for preserving restaurant viability during the COVID-19 crisis. Tour. Hosp. Res. (2021). https://doi.org/10.1177/1467358421998057
18. Milne, G.J., Xie, S., Poklepovich, D., O'Halloran, D., Yap, M., Whyatt, D.: A modelling analysis of the effectiveness of second wave COVID-19 response strategies in Australia. Sci. Rep. 11(1), 1–10 (2021)
19. Kaszowska-Mojsa, J., Włodarczyk, P.: To freeze or not to freeze? epidemic prevention and control in the DSGE model using an agent-based epidemic component. Entropy 22(12), 1–33 (2020)
20. Mustafa, S., Jayadev, A., Madhavan, M.: COVID-19: need for equitable and inclusive pandemic response framework. Int. J. Health Serv. 51(1), 101–106 (2021)

21. Morley, A., Morgan, K.: Municipal foodscapes: urban food policy and the new municipalism. Food Policy **103**(August), 1–10 (2021)
22. Comune di Milano, Strategia di Adattamento Milano 2020 [Milan's Adaptative Strategy 2020]. https://www.comune.milano.it/documents, Accessed 01 Dec 2021
23. Comune di Milano, Disciplina del diritto ad occupare il suolo, lo spazio pubblico o aree private soggette a servitù di pubblico passo. Resolution no 132 of 4th December 2020. https://www.comune.milano.it/documents, last accessed 2021/10/26
24. Comune di Milano, Linee guida per la progettazione delle occupazioni di Suolo Pubblico Leggere e Temporanee. https://www.comune.milano.it/documents, Accessed 26 Oct 2021
25. Comune di Milano, Occupazioni suolo temporanee straordinarie (2020). https://dati.comune.milano.it, Accessed 23 Sept 2021
26. Comune di Milano, Occupazioni suolo temporanee straordinarie (2021). https://dati.comune.milano.it, Accessed 23 Sept 2021
27. Honey-Rosés, J.,et al.: The impact of COVID-19 on public space: an early review of the emerging questions–design, perceptions and inequities. Cities Health, 1–17 (2020)
28. Paköz, M.Z., Sözer, C., Doğan, A.: Changing perceptions and usage of public and pseudo-public spaces in the post-pandemic city: the case of Istanbul. Urban Design Int. **27**, 1–16 (2021) https://doi.org/10.1057/s41289-020-00147-1
29. Credit, K., Mack, E.: Place-making and performance: the impact of walkable built environments on business performance in Phoenix and Boston. Environ. Plan. B Urban Anal. City Sci. **46**(2), 264–285 (2019)

How Covid-19 Pandemic Has Affected the Market Value According to Multi-parametric Methods

Laura Gabrielli$^{(\boxtimes)}$ ⓘ, Aurora Greta Ruggeri ⓘ, and Massimiliano Scarpa ⓘ

University IUAV of Venice, Dorsoduro 2206, 30123 Venice, Italy
laura.gabrielli@iuav.it

Abstract. This research paper aims to discuss the changes that the Covid-19 pandemic has brought to the demand in the real estate market in Padova. Two databases are compared: database A dates back to a pre-Covid scenario, while database B represents the actual situation. A multi-parametric approach, based on the use of Artificial Neural Networks, is used to create a forecasting algorithm to predict the market value of the properties as a function of their characteristics. This multi-parametric perspective allows isolating each attribute's singular influence on the price. Comparing the two databases makes it possible to see how the demand preferences have changed during the pandemic. Some characteristics are now more appreciated than before, such as the external spaces, while others are less appreciated, such as the location. These changes in preferences can be attributed to the new lifestyle, habits and working schedule that the pandemic has led to.

Keywords: Market value · Artificial Neural Networks · Covid-19 pandemic

1 Introduction

As the Authors are writing this research paper, it is early 2022, and two years have passed after the first Covid-19 alerts, when the pandemic started in China in December 2019 [1]. This pandemic has shown evident consequences not only in the health sector but also in the economy, labour or trade, including, among the others, the real estate sector [2]. Real estate professionals are now questioning how the Covid-19 pandemic has affected (and will affect) the real estate market, and they will have to adapt their job to this new scenario. Researchers are also discussing this very same issue, analysing new tendencies and dynamics [3–7].

1.1 A Shift in Real Estate Market Demand

A significant transformation that the housing market has experienced during the pandemic is a change in demand preferences.

During the lockdown, many Italians became aware of the shortcomings of their homes. As a result, they shifted their interest to less central areas in exchange for more square metres, natural light, gardens, terraces and balconies, and top floors.

F. Calabrò et al. (Eds.): NMP 2022, LNNS 482, pp. 1018–1027, 2022.
https://doi.org/10.1007/978-3-031-06825-6_97

There is no doubt that the suburbs and hinterlands of large cities have become of considerable interest to buyers, mainly due to smart working and online education on digital platforms. The new demand is now looking for multifunctional, larger rooms with modular spaces for smart working and online education, providing a private area for each family member.

This means that everything sold until before the pandemic because it was small, easy to manage, and cheaper has become inadequate to meet the new housing needs. People are now more willing to look for larger housing solutions in the hinterland of smaller towns connected to their workplaces by adequate transport infrastructures. Buyers now seem to prefer peripheral areas as there is more space available, they are away from possible contagions, from city traffic, and there is the possibility of having more green space around them to walk in case of lockdowns while remaining close to home.

However, there is also a counter-trend: some families are looking for a house much closer to the city centre than to the suburbs because this brings them closer to services or supermarkets and pharmacies. Thus, they will have everything at hand in case of a new lockdown.

Another aspect concerns the quality of services and common spaces in multi-family houses. There is a need to have common spaces adequately designed and distributed within the building to meet the needs that demand cannot satisfy in the private area of the house. Finally, the 110% Superbonus [8] has led to a renewed appeal for property solutions to be renovated, for example, in period houses in more or less central locations in the city. In addition, sales are promoting properties with low energy performance and low maintenance levels, with the prospect of zero-cost renovations.

This could be a predominantly emotional response that presupposes an expectation of the structural nature of the changes that have taken place over the past two years, the resilience of which seems all but certain, as the bite of Covid-19 loosens its grip.

1.2 The Purpose of the Research

The Covid-19 pandemic certainly cannot be considered an insignificant phenomenon from a real estate market perspective. However, we cannot state whether it will be a **permanent force** or a **temporary** (but indeed huge) event. We can see by now that the pandemic has somehow affected the real estate sector, therefore changing its internal dynamics.

The open question is *how* the Covid-19 pandemic has altered the real estate market.

This paper aims to investigate this issue from the **market value** perspective. Our goal is to determine if, after almost two years from the first Covid-19 alerts, the factors that contribute to the price formation have changed. For example, does the demand now appreciate different buildings features than before? Nevertheless, on the other hand, are some of the most appreciated characteristics less influential in assessing the market value? Besides, considering a smart working perspective, the lockdown experience, and the increased amount of time spent at home, are other buildings' characteristics now more appreciated than before the Covid pandemics?

As further illustrated in next Sect. 2, the research scope is pursued by comparing two databases. The first one dates back to before the pandemic, while the second represents the actual situation. A multi-parametric forecasting algorithm is consequently developed based on each database to predict the market value as a function of its features. Finally, each feature that has a positive effect on the price is isolated so that it is possible to see how the demand side has changed after the Covid-19 pandemic.

2 A Methodological Approach

This discussion has employed two different methodologies to answer the research questions: **multi-parametric** market value assessment techniques [9, 10], and artificial intelligence techniques [11, 12]. Specifically, we have developed an **Artificial Neural Network** (ANN) to define a mathematical algorithm that can forecast the property market value as a function of its intrinsic and extrinsic characteristics.

Multi-parametric market value assessment techniques differ from mono-parametric market value assessment techniques because they associate the market value forecast to not only one building characteristic but multiple building characteristics [13]. In mono-parametric approaches, the market value is most commonly assessed as a function of the dwelling's area (sqm). Instead, multi-parametric procedures analyse multiple attributes of a building to create a forecasting function of the market value. Each independent variable inside the function contributes differently to the estimate of the price (the dependent variable).

Buildings descriptive attributes could be synthesised as follow:

I.   **Structural/physical attributes**: the quality of the building, such as its maintenance conditions, living area, its systems and facilities, the energy class, the presence of balconies, basements, garages or others.
II. **Neighbourhood quality**: the quality of the block or suburb, as outdoor environment, pollution, public parks, schools, services, shops, or others.
III. **Locational characteristics**: proximity to the employment centres, bus stops, stations, etc.

At the basis of multi-parametric procedures, there is the following assumption: the value of a property is based on (is a function of) a bundle of features, i.e. structural attributes, neighbourhood, and locational characteristics of a housing unit. Many factors and characteristics are crucial in influencing property values. The demand considers almost all those features before purchasing a dwelling, so reflecting them in the property prices. Different characteristics are valued differently by the demand side, depending on the type of the housing market, cycle, real estate, spending capacity, social status, etc. Features are also appreciated differently over time, so research always identifies their different weights in price formation. For example, the buyer's attention to the technological aspects of energy saving, the quality of neighbourhood and environment, and safety has increased over time.

Among multi-parametric approaches, there are the hedonic prices techniques, where the most common is the regression analysis [14]. A marginal (and implicit) price is estimated and associated with each building attribute in hedonic prices. The hedonic pricing method is founded on the principle that the property market value is built on a bundle of characteristics [15].

Among other multi-parametric procedures, we suggest employing in this paper an ANN to elaborate a forecasting tool to predict the market value of a property as a function of its descriptive characteristics [16]. ANNs are very flexible and precise statistical instruments whose results are highly reliable. ANNs, in fact, are able to capture patchy behaviours, non-linear interactions, and complex relationships among a considerable number of variables.

In this case, the input variables of the ANN will be the characteristics of the buildings, while the ANN output will be the market value. In particular, our goal is to compare an ANN developed on a **pre-Covid database** against another ANN established on the basis of a **post-pandemic database**. This way, it will be possible to compare a pre-pandemic scenario versus a post-pandemic one, understanding each characteristic's influence on the market value and highlighting any change in market preferences.

3 A Case-Study

This study is a part of a more comprehensive research line about real estate market investigations and analysis in Italy. For a few years now, the Authors of this paper have been collecting a considerable number of information about properties on sale in Padova in order to keep a historical track of that specific real estate market.

In particular, we developed a Python language crawling software to automatically download the asking prices of a set of real estate properties on sale (associated with the corresponding characteristics of the buildings) from specific selling websites [17].

The databases downloaded with the web crawler have been thoroughly checked and corrected by manually re-downloading a control group of the same data and information. This huge and time-consuming operation was extremely important since it calibrated the web crawler and tested the data obtained, estimating the error and excluding those data that contained excessively high misleading information.

Since we have followed this research line, we are now able to compare two different databases that refer to the **real estate asking prices market in Padova**. The first database we will analyse, called **database A**, dates back to before the emergency caused by the Covid-19 pandemic (II semester 2019). The second database, called **database B**, was downloaded in November 2021 (II semester 2021), maintaining the same web crawler structure, search domain and crawling features to allow for a significant comparison. As far as the definition of the web searching domain is concerned, we had limited the online search to residential properties on sale in Padua, including all the fourteen areas the Municipality is divided into. New constructions are considered in the domain, as well as existing buildings, comprising, among the others, apartments, terraced houses, penthouses, lofts, and farmhouses.

The procedure employed to extract the information from each sale advertisement using the web crawler has been developed in Python language with the support of the library *"Beautiful Soup"* (so as to parse and extract data from HTML documents), as well as the data analysis library *"Pandas"*, which is tailored for extracting and organizing data in the form of a.xls table. The buildings characteristics extracted with the web-crawler developed are listed in Table 1.

Table 1. Buildings attributes extracted with the web-crawler

Variable	Unit	Variable	Unit	Variable	Unit
Web URL	text	Private Garage	0/1	Optical Fiber	0/1
LOCATION AND NEIGHBOURHOOD		Private Garage Area	sqm	Fireplace	0/1
Zone	text	Common Parking Space	0/1	Lift	0/1
Address	text	Basement	0/1	Solar Panels	0/1
Latitude	coordinate	Basement area	sqm	Heat Pump	0/1
Longitude	coordinate	Terrace	0/1	Fireplace	0/1
MARKET VALUE		Terrace Area	sqm	**BUILDING TYPOLOGY**	
Price	€	Top Floor	0/1	Apartment	0/1
Floor area	sqm	**MAINTAINANCE CONDITIONS**		Apartment in a Villa	0/1
Price per sqm	€/sqm	Maintenance level	1/2/3/4	Penthouse	0/1
BUILDING CHARACTERISTICS		Energy Class	A4 --> G	Farmhouse	0/1
no. of bathrooms	number	Construction year	Year (number)	Loft	0/1
no. of rooms	number	**SYSTEMS AND TECHNOLOGIES**		Attic	0/1
Floor	number	Building Automation	0/1	Multi-storey single-family home	0/1
no. of internal floors	number	Central Heating	0/1	Single-family home	0/1
Common garden	0/1	Photovoltaic System	0/1	Terraced house	0/1
Private Garden	0/1	Mechanical Ventilation	0/1	Two-family villa	0/1
Private Garden Area	sqm	Air Conditioning	0/1	Multi-family villa	0/1

4 Application

If we compare database A against database B, at a first look, we can state that the **asking prices** have slightly increased after almost two years since the first Covid alerts. The mean market value produced by database A is 1792 €/sqm, while the mean value produced by database B is 2094 €/sqm. This result can be considered consistent with other sources of information, such as web historical archives [18], *"Nomisma spa – Servizi di analisi e valutazioni immobiliari"* [19] and also *"Agenzia delle Entrate - Osservatorio del mercato immobiliare"* [20].

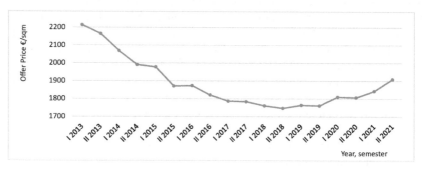

Fig. 1. Asking prices historical series in Padova (immobiliare.it)

As shown in Fig. 1, the recorded **asking prices** of properties on sale have decreased until 2019. From the II semester of 2019, in correspondence with the first pandemic alerts, the asking prices have begun to keep constant values and then slightly increase. This increment seems relatively stable during the two-year period 2019–2021.

Considering, instead, another source of information, Fig. 2 and Table 2 represent the historical series of market values (actual prices of real transactions) in Padova recorded by the Nomisma real estate observatory. From 2019 till the I semester of 2021 (since II semester data are still not available), the transaction prices kept relatively constant values. In fact, the response to the increase in the asking prices will be registered with a time-lag on the transaction prices.

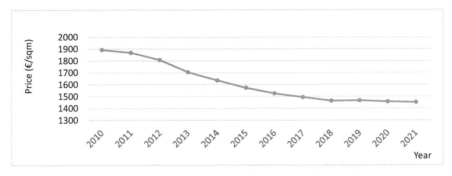

Fig. 2. Average transaction prices, historical series in Padova (Nomisma)

In order to deepen this comparative analysis, we develop two different ANN to predict the property market value as a function of its characteristics. ANN(A) is designed specifically on database A, reflecting a pre-pandemic market, while ANN(B) is developed on database B to depict the current situation.

In general, ANNs act as computational systems, very similar to human brains during biological learning procedures. ANNs are constituted of two elements: artificial neurons and connections. The artificial neurons are the computational units, while the connections between neurons are like artificial synapsis. The neurons are organised into multiple separated layers, where we distinguish the input layer, the output layer, and the inner hidden layers. The **input layer** is formed of the input neurons, which, in this study, are the descriptive characteristics of the properties. The **output layer** consists of one or more output neurons. In this research, the output neuron is the market value. Between the input layer and the output layer, there is/there are one/some **hidden layer**/s that define the specific structure of the ANN. The information flow from the input layer to the output layer through the connections. Specifically, all the connections entering a neuron bring information, which is combined inside the neuron through a weigh function and a bias function. An activation function converts the neuron value into a response value, and a numerical output is thus produced and exits the neuron through the next connection.

An ANN "learns" how the input neurons are related to the output neurons during a **training process**. ANNs, in fact, are able to analyse very complex input-output databases and define the most suitable forecasting algorithm. The input neurons and the output neuron used in this research are listed in Table 3. Unused variables have been tested through a feature-selection process and consequently excluded from the training because of too much missing information, outliers, or unlikely data that misled the ANN structure.

At this stage, we applied an **optimization selection procedure** developed in Python code to identify the optimal ANN(A) structure, i.e. the one ANN leading to the most

Table 2. Average transaction prices in Padova considering location and maintenance conditions (Nomisma)

Semester, Year	Maintenance conditions	Prestigious areas		Centre		Semi-centre		Suburbs	
		min	max	min	max	min	max	min	max
I 2021	New constructions	3482	4209	2793	3518	1846	2159	1192	1496
	Good maintenance conditions	2583	3038	2041	2496	1273	1685	925	1197
	Poor maintenance conditions	1573	2062	1442	1707	835	1140	608	824
I 2020	New constructions	3520	4323	2761	3440	1857	2168	1201	1507
	Good maintenance conditions	2612	3170	2075	2524	1285	1664	901	1183
	Poor maintenance conditions	1571	2155	1440	1718	847	1154	603	835
I 2019	New constructions	3451	4226	2745	3485	1848	2175	1212	1542
	Good maintenance conditions	2571	3108	2042	2583	1307	1672	919	1194
	Poor maintenance conditions	1577	2142	1483	1788	855	1147	618	841

Table 3. Input-output neurons

Variable	Use	Variable	Use	Variable	Use
Web URL	-	Private Garage	Input	Optical Fiber	Input
LOCATION AND NEIGHBOURHOOD		Private Garage Area	-	Fireplace	-
Zone	-	Common Parking Space	Input	Lift	Input
Address	-	Basement	Input	Solar Panels	-
Latitude	Input	Basement area	-	Heat Pump	-
Longitude	Input	Terrace	Input	Fireplace	Input
MARKET VALUE		Terrace Area	-	**BUILDING TYPOLOGY**	
Price	-	Top Floor	Input	Apartment	Input
Floor area	Input	**MAINTAINANCE CONDITIONS**		Apartment in a Villa	Input
Price per sqm	Output	Maintenance level	Input	Penthouse	Input
BUILDING CHARACTERISTICS		Energy Class	Input	Farmhouse	Input
no. of bathrooms	Input	Construction year	-	Loft	Input
no. of rooms	Input	**SYSTEMS AND TECHNOLOGIES**		Attic	Input
Floor	Input	Building Automation	Input	Multi-storey single-family home	Input
no. of internal floors	-	Central Heating	Input	Single-family home	Input
Common garden	-	Photovoltaic System	Input	Terraced house	Input
Private Garden	Input	Mechanical Ventilation	Input	Two-family villa	Input
Private Garden Area	-	Air Conditioning	Input	Multi-family villa	Input

reliable forecast of the market value. Resulting from the optimization process, the trained ANN(A) has a total of 6 layers, 1 input layer with 33 input neurons, 4 hidden layers with 32 hidden neurons each, and 1 output layer with 1 output neuron. Besides, the hyperbolic tangent is the activation function used, the training strategy is based on the mean squared error function, and a mean standard deviation scalarization is employed for the data scaling and unscaling process. As a result, the error produced by the optimal ANN(A) on the testing set is 0.04%.

The same optimization selection procedure is also developed on database B to define ANN(B). This second artificial neural network, very similar to the first one, has again a total of 6 layers, 1 input layer with 33 input neurons, 4 hidden layers with 34 hidden neurons each, and 1 output layer with 1 output neuron. The activation function used is the hyperbolic tangent. The training strategy is based on the mean squared error function, and the data scaling and the unscaling process is based on a mean standard deviation scalarization. The error produced by the optimal ANN(B) on the testing set is 0.06%.

5   Results and Discussion

After ANN(A) and ANN(B) have been developed on their corresponding databases, it is possible to compare the influence that any input produces on the output. To this purpose, the two correlation coefficients are analysed, as shown in Table 4.

Table 4. Correlation chart ANN(A)

Variable	II semester 2021	II semester 2019	Difference	Variable	II semester 2021	II semester 2019	Difference
LOCATION AND NEIGHBOURHOOD				Floor	0.0437	0.0459	0.2%
Latitude	- 0.1288	- 0.1323	0.3%	Top floor	0.2618	0.2373	2.4%
Longitude	- 0.0546	- 0.0785	2.4%	Private Garden	0.1746	0.1573	1.7%
BUILDING TYPOLOGY				Private Garage	- 0.0135	- 0.0507	3.7%
Apartment	- 0.3753	- 0.3310	4.4%	Common Parking Space	- 0.0148	- 0.0194	0.5%
Apartment in a villa	- 0.0019	- 0.0102	0.8%	Basement	0.1086	0.1151	0.6%
Penthouse	0.2618	0.2373	2.4%	Terrace	0.1438	0.0939	5.0%
Farmhouse	0.0123	- 0.0027	1.5%	**MAINTAINANCE CONDITIONS**			
Loft	0.0132	0.0013	1.2%	Energy Class	0.2091	0.1759	3.3%
Attic	- 0.0057	- 0.0107	0.5%	Maintenance level	0.2341	0.1827	5.1%
Multi-storey single-family home	0.2301	0.2257	0.4%	**SYSTEMS AND TECHNOLOGIES**			
Single-family home	0.0154	0.0027	1.3%	Building Automation	0.1021	0.0719	3.0%
Terraced house	0.0151	- 0.0013	1.6%	Central Heating	- 0.1934	- 0.1732	2.0%
Two-family villa	- 0.0686	0.0335	10.2%	Photovoltaic system	0.0746	0.0452	2.9%
Multi-family villa	0.0735	0.0729	0.1%	MCV	0.0536	0.0387	1.5%
BUILDING CHARACTERISTICS				Air Conditioning	0.1395	0.1380	0.2%
Floor Area	0.7279	0.6893	3.9%	Optical Fiber	0.0493	0.0201	2.9%
no. of bathrooms	0.6139	0.5957	1.8%	Fireplace	0.3153	0.3142	0.1%
no. of rooms	0.4712	0.4460	2.5%	Lift	0.0230	0.0106	1.2%

Let's compare the correlations in the database (A) versus (B). It is clear that several variations in the marginal appreciation attributed to the characteristics of the buildings have already occurred.

It seems that, after two years since the first Covid alerts, the market demand is prone to appreciate more characteristics of the buildings, such as a wider surface area, the number of rooms, the presence of external spaces like a private garden or a terrace.

Also, buildings technologies and installations seem to give a more significant contribution to the market value than before, such as building automation, the photovoltaic system, the mechanical ventilation, or the optical fibre.

As far as the building typology is concerned, we can state that the market preferences seem changed too. If the smaller typologies were the most appreciated before, now villas, detached and semi-detached houses and farmhouses seem more appealing for the demand.

These results can be considered consisted with other research carried out in different Italian cities, such as the analysis performed by Tajani et al. in Rome to assess the changes in the marginal prices of several building characteristics, comparing a pre and post-Covid situation [7].

6   Conclusions

In conclusion, this research aimed to analyse the first changes that the Covid-19 pandemic has produced on real estate market preferences.

Two databases that referred to the real estate market in Padova have been compared: database(A) depicted a pre-Covid market. In contrast, database(B) captured a two-year post-Covid alerts situation. Therefore, two forecasting multi-parametric models have been produced based on an Artificial Neural Network approach. The forecasting models relate the estimate of the price of a property (dependent variable) to 33 buildings features (independent variables). Since 33 characteristics have been analysed, it was possible to assess the influence of each building feature on the market value. Therefore, it was clear how market preferences have changed in two years, reflecting how the demand has reacted to the pandemic.

These variations may be attributed to the substantial changes in daily life due to the Covid-19. People have spent, and now still spend, more time at home than before. In addition, smart working has created new needs at home, such as quiet and isolated spaces to work, the creation of an efficient and comfortable workstation, requiring spacious and bright environments, as well as the presence of systems and technologies such as optical fibre and mechanical ventilation.

Even though this is an "in itinere" kind of analysis, since Covid-19 consequences will keep influencing the real estate market in the future, this comparison already allows us to understand the very first changes and outlines a development trend.

References

1. World Health Organization. https://www.who.int/, Accessed 15 Nov 2021
2. Nicola, M., et al.: The socio-economic implications of the coronavirus pandemic (COVID-19): a review. Int. J. Surg. **78**, 185–193 (2020). https://doi.org/10.1016/j.ijsu.2020.04.018
3. Quaglio, C., Todella, E., Lami, I.M.: Adequate housing and COVID-19: assessing the potential for value creation through the project. Sustainability **13**(19), 10563 (2021). https://doi.org/10.3390/su131910563
4. Tokazhanov, G., Tleuken, A., Guney, M., Turkyilmaz, A., Karaca, F.: How is COVID-19 experience transforming sustainability requirements of residential buildings? a review. Sustainability **12**(20), 8732 (2020). https://doi.org/10.3390/su12208732
5. Tucci, F.: Pandemia e Green City. Le Necessità Di Un Confronto per Una Riflessione Sul Futuro Del Nostro Abitare. In: Pandemia e sfide green del nostro tempo. Roma, Green City Network e Fondazione per lo Sviluppo Sostenibile (2020)
6. De Toro, P., Nocca, F., Buglione, F.: Real estate market responses to the COVID-19 Crisis: which prospects for the metropolitan area of Naples (Italy)? Urban Sci. **5**, 23 (2021). https://doi.org/10.3390/urbansci5010023
7. Tajani, F., Morano, P., Di Liddo, F., Guarini, M.R., Ranieri, R.: The effects of Covid-19 pandemic on the housing market: a case study in Rome (Italy). In: Gervasi, O., et al. (eds.) ICCSA 2021. LNCS, vol. 12954, pp. 50–62. Springer, Cham (2021). https://doi.org/10.1007/978-3-030-86979-3_4
8. Superbonus 110%. https://www.governo.it/superbonus, Accessed 17 Dec 2021
9. Wang, A., Xu, Y.: Multiple linear regression analysis of real estate price. In: Proceedings of the International Conference on Robots and Intelligent System, pp. 564–568. IEEE, Changsha (2018)
10. Feng, J., Zhu, J.: Nonlinear regression model and option analysis of real estate price. Dalian Ligong Daxue Xuebao/J. Dalian Univ. Technol. **57**, 545–550 (2017). https://doi.org/10.7511/dllgxb201705016
11. Yakub, A.A., Hishamuddin, M.A., Kamalahasan, A., Abdul Jalil, R.B., Salawu, A.O.: An integrated approach based on artificial intelligence using anfis and ann for multiple criteria real estate price prediction. Plan. Malaysia **19**, 270–282 (2021). https://doi.org/10.21837/PM.V19I17.1005

12. Kalliola, J., Kapočiūte-Dzikiene, J., Damaševičius, R.: Neural network hyperparameter optimization for prediction of real estate prices in Helsinki. PeerJ Computer Science **7**, 1–25 (2021). https://doi.org/10.7717/peerj-cs.444
13. Simonotti, M.: Metodi Di Stima Immobiliare. Flaccovio, Palermo (2006)
14. Amrutphale, J., Rathore, P., Malviya, V.: A novel approach for stock market price prediction based on polynomial linear regression. In: Shukla, R.K., Agrawal, J., Sharma, S., Chaudhari, N.S., Shukla, K.K. (eds.) Social Networking and Computational Intelligence. LNNS, vol. 100, pp. 161–171. Springer, Singapore (2020). https://doi.org/10.1007/978-981-15-2071-6_13
15. Rosen, S.: Hedonic prices and implicit markets: product differentiation in pure competition. J. Polit. Econ. **82**, 34–55 (1974)
16. Štubňová, M., Urbaníková, M., Hudáková, J., Papcunová, V.: Estimation of residential property market price: comparison of artificial neural networks and hedonic pricing model. Emerg. Sci. J. **4**, 530–538 (2020). https://doi.org/10.28991/esj-2020-01250
17. Gabrielli, L., Ruggeri, A.G., Scarpa, M.: Using artificial neural networks to uncover real estate market transparency: the market value. In: Gervasi, O., et al. (eds.) ICCSA 2021. LNCS, vol. 12954, pp. 183–192. Springer, Cham (2021). https://doi.org/10.1007/978-3-030-86979-3_14
18. Quotazioni Immobiliari Nel Comune Di Padova. https://www.immobiliare.it/mercato-imm obiliare/veneto/padova/, Accessed 17 Dec 2021
19. Nomisma Spa – Servizi Di Analisi e Valutazioni Immobiliari. https://www.nomisma.it/ser vizi/osservatori/osservatori-di-mercato/osservatorio-immobiliare/, Accessed 17 Dec 2021
20. Agenzia Delle Entrate - Osservatorio Del Mercato Immobiliare. https://www.pd.camcom. it/gestisci-impresa/studi-informazione-economica/quotazioni-immobili-1, Accessed 17 Dec 2021

Adapting Anti-adaptive Neighborhoods. What is the Role of Spatial Design?

Elena Porqueddu[✉]

via Giuseppe Bardelli 8, 20131 Milan, Italy
elena.porqueddu@archiworld.it
https://www.researchgate.net/profile/Elena-Porqueddu,
https://independent.academia.edu/elenaporqueddu

Abstract. The recent theory of planning and urban design highlights how healthy vibrant cities and neighborhoods behave as complex systems that are capable of constantly evolving and renewing over time, by endlessly adapting to new emergent needs and unexpected changes. Although adaptivity is increasingly considered an essential quality in these uncertain times, our urban areas are still dotted with post-war modernist settlements which have been widely described as Anti-Adaptive Neighborhoods (AANs) with a low degree of openness to processes of incremental adaptation and continuous adjustment. Since the anti-adaptive nature of these settlements often led to their obsolescence and consequent decline over time, an urgent question for planners and designers is: How can the adaptive capacity of AANs be increased, thereby turning them into structures that can evolve and self-regenerate over time? This question is becoming even more relevant in this historic period, in which the sudden and unexpected changes generated by the COVID 19 pandemic have often exacerbated pre-existing problems linked to anti-adaptivity, thus accelerating cycles of decline in AANs. By illustrating the renovation project for "Cité du Grand Parc" in Bordeaux by the architects Lacaton & Vassal, Druot and Hutin, the present paper highlights how designers can play a crucial role in turning rigid definitive structures into open adaptive systems and how their site-specific practice can offer valuable insights for planners and policy makers.

Keywords: Spatial design · Adaptability · Complexity · Modernist neighborhoods · Urban regeneration · Housing

1 Regenerating Anti-adaptive Neighborhoods: Challenges for Designers and Planners

The recent theory of planning and urban design highlights how healthy vibrant cities or neighborhoods behave as complex living systems that are capable of constantly adapting, evolving, and renewing over time [2, 4, 8, 17, 24, 25]. Such complex adaptive environments can adjust to unexpected conditions and welcome a diversity of uses and lifestyles.

E. Porqueddu—Independent researcher.

F. Calabrò et al. (Eds.): NMP 2022, LNNS 482, pp. 1028–1038, 2022.
https://doi.org/10.1007/978-3-031-06825-6_98

Their nature is emergent and evolving: after their initial construction, they provide sufficient room for progressive, unforeseen adaptations [6, 15, 21]. Their overall shape is, to a certain extent, unpredictable and in a constant state of becoming because it keeps transforming in relation to the multiple creative actions of their inhabitants rather than being entirely pre-defined and designed from the top down [10–13, 17, 21, 28].

This conception of cities and neighborhoods as dynamic entities in a constant state of becoming has led to a shift from planning and designing for presumed 'static' situations toward planning and designing in dynamic processes of change and has fostered the emergence of alternative forms of adaptive planning and design [8, 21]. In this respect, the unexpected transformations developed during the COVID 19 pandemic have highlighted the need to take the unknown seriously and to learn to deal with uncertainty, in order to increase our ability to progress together with unpredictable changes.

While adaptivity is increasingly considered a crucial quality in cities, our urban areas are still dotted with modernist neighborhoods, which have been widely described as anti-adaptive structures [13] that are incompatible with the emergent self-organizing nature of complex cities [1, 6]. At the time they were built, after the second world war, Anti-Adaptive Neighborhoods successfully responded to the housing pressures, and they generated opportunities for individuals and families to dwell at affordable prices. Nonetheless, they were designed and built all-at-once and from scratch and conceived as finished projects, where everything was over-defined from the micro to macro scale. Their initial structure was not designed to provide room for progressive adaptation to unpredictable uses, thus preventing continuous adjustments from their inhabitants both on the architectural and urban scale. In this sense, over time these settlements revealed a low propensity to adapt to emergent needs and lifestyles and to respond to social, cultural, technological, and environmental changes (sudden or gradual, expected, or unexpected), thus soon becoming obsolete and unattractive. This anti-adaptivity has been further highlighted by the COVID 19 pandemic, which profoundly and suddenly affected the relationship between people's behaviors and residential spaces. By exacerbating pre-existing problems and revealing hidden potential, the unexpected situations generated by the lockdown have raised further questions about how to assess the responsiveness of the existing residential stock [27].

In this respect, since AANs are widespread in cities all over the world and many of them have fallen into dramatic cycles of decline, planners and designers face the difficult challenge of inverting the process of decline and fostering their capacity to self-regenerate. In this respect, the present paper addresses the following question: How can planners and designers increase the adaptive capacity of AANs and turn them into structures that can evolve and self-regenerate over time? Although this question involves several aspects related either to planning or design theory and practice, the present paper mainly focuses on the role of spatial design (see Sect. 2). It is structured as follows: the first section explores the role of designers in turning rigid spatial configurations into adaptive structures and identifies a possible design approach. The second section presents a design strategy that is useful for exemplifying and further exploring this design approach. The third section traces some final remarks highlighting how site-specific design practice can offer valuable insights for planners and policy makers.

2   Transforming Rigid Spatial Configurations into Adaptive Structures: The Role of Spatial Design

Carter and Moroni [5] identify three groups of constraints that influence the potential for individual choice for the inhabitants, and therefore the degree of urban adaptability: planning constraints (mainly on 'normative freedom'), organizational constraints (mainly on 'normative powers') and design constraints (mainly on 'purely physical freedom'). Although these three types of constraint are interrelated, this paper will focus mainly on the third group: on the physical constraints which condition the possibility of each inhabitant to adapt an existing spatial structure over time. In general, the physical constraints that influence the adaptive capacity of a certain spatial configuration are determined directly in the design process, which can concern either the construction of a new neighborhood or the regeneration of an existing one.

As I illustrated elsewhere [21–23], in the case of the construction of a new building or neighborhood, the role of designers who aim to maximize adaptivity consists of shaping the initial spatial conditions which can structure a dynamic open evolution over time without overdetermining its formal outcome. In this case, the robust initial structure, which is designed from the top down, (1) provides room for unpredictable individual actions (which might emerge from the bottom up), (2) preserves the collective interest without overcontrolling and predefining every single intervention, and (3) provides the physical constraints that prevent the socio-spatial system from falling into chaos, without blocking individual creativity [23].

In the case of the regeneration of an existing building or neighborhood, the role of designers consists of altering the actual spatial structure, with the objective of increasing its adaptive capacity through minimal (and possibly progressive) spatial manipulations. For example, a design action might (1) strengthen or insert a robust spatial structure when the system lacks a cohesive layer and tends to fall into chaos, or rather (2) reduce some physical constraints in overdetermined rigid structures [23].

This second type of design action is particularly suitable for the regeneration of AANs because it makes it possible to increase their adaptability without resorting to a process of complete demolition and reconstruction that would probably create the conditions for the development of a new AAN. Indeed, a process of complete demolition would leave the residents without a dwelling and would generate the urgent need to design and reconstruct a new neighborhood that could provide them with a new house. This urgency would probably lead to the same all-at-once approach which was crucial in generating the Anti-Adaptive Neighborhoods. Furthermore, the entire demolition of an existing settlement recreates a condition of *tabula rasa* which is historically associated with the modern approach that tended to replace an existing urban fabric with a new one entirely designed by one or more designers and built from scratch [21].

Instead, an approach based on the modification of the existing urban fabric through minimal progressive adjustments focuses on the question of time. Such an approach entails treating AANs not as definitive inalterable physical structures, but rather as unfinished configurations which need to be modified, completed, and extended over time. By removing, or decreasing, the physical constraints which prevent AANs from adapting over time, designers can contribute to re-activating an evolving cycle which was interrupted after their initial construction. By creating new room for the spontaneous initiative of each resident, this approach increases the potential of these neighborhoods to self-regenerate over time. Moreover, by reducing economic costs, it helps to maintain or increase the available number of affordable dwellings, thus limiting the risk of gentrification (see Sect. 3).

In this respect, designers face the following challenges:

- What are the minimum spatial manipulations which can turn a rigid anti-adaptive structure into an open dynamic space that is capable of co-evolving with its inhabitants?
- What are the physical constraints that need to be removed (or decreased) in order to (1) provide space for unpredictable individual interventions, (2) limit the possibility for emergent chaos and (3) foster an overall upgrade of the neighborhood over time?

The next section presents a design strategy that is useful for addressing these questions, as well as for exemplifying and further exploring this approach.

3 Regeneration of the 'Cité du Grand Parc' in Bordeaux

The first phase of the project for the regeneration of the "Cité du Grand Parc" AAN in Bordeaux, which was developed by the architects Lacaton & Vassal, Druot and Hutin between 2011 and 2016, was based on a more general design strategy [9]. This strategy was elaborated by the architects in response to the National Program for Urban Renewal (PRNU-Programme Nationale de Rénovation Urbaine), which was launched in 2004 by the Agence Nationale de Rénovation Urbaine (ANRU - France's Urban Regeneration Agency). The aim of the PRNU was to restructure those neighborhoods that had been classified as sensitive urban areas. It included urban improvement operations by means of development plans and interventions involving the replacement or renovation of social housing, and the reorganization of spaces intended for economic and commercial activities [9].

While the government allocated billions of euros for the complete demolition and reconstruction of modernist housing estates, Lacaton & Vassal and Druot proposed a design strategy which consisted of adding, transforming, reusing, and expanding, rather than demolishing, the existing buildings. In their report Plus + [9], they established and outlined a precise inventory of site conditions and assets for transforming these modernist buildings successfully and cost effectively, by working with the existing structure. In this respect, they argued that many of these buildings were still functional from a structural point of view.

Furthermore, they highlighted the fact that the PRNU planned to rebuild fewer dwellings than the number it intended to demolish, thus causing a loss of social units in a context in which there is ever-growing demand for homes. Instead, a process of adaptation would consistently reduce economic costs, thereby making it possible to increase the available number of refurbished dwellings. Finally, the strategy they proposed would enable the dwellers to remain in these settlements, thus also reducing social costs whereas, with a process of full demolition and reconstruction of these ageing estates, tenants would have had to be moved out (an upheaval that would also have endangered the community).

After their general report Plus+ [9], the architects had the opportunity to experiment their ideas in several projects. This section will focus on the first phase of the regeneration of the Cité du Grand Parc, a post-war settlement, built in the early 1960s on reclaimed marshland. It covers 60 ha in the heart of city of Bordeaux and comprises 4,000 housing units (100% social rents) [14]. The settlement currently comprises four tower blocks of 22 storeys each and 18 other buildings of between 10 and 15 storeys. The main settlement landlord was Aquitanis, the Public Office for Housing of the Bordeaux Metropolitan Area. The project was also made possible thanks to Bernard Blanc (who became director of Aquitanis in 2008), who shared with Lacaton & Vassal, Druot and Hutin a strong belief in renewing housing stock rather than demolishing it. In this respect, Aquitanis (1) negotiated with the local Bordeaux authority for the UNESCO site boundaries to be extended in order to include the Grand Parc estate (thus protecting the neighborhood from any demolition plan) and (2) launched a competition for the regeneration of the neighborhood, which the trio won [26].

The first phase of the project concerned the transformation of three inhabited buildings that hosted 530 dwellings. Here the designers faced several challenges. The first consisted of turning the rigid overdetermined spatial layout of the obsolete dwellings into a more open architecture that can provide space for unexpected uses and the residents' appropriation. This proved to be particularly difficult in buildings that were characterized by a standardized fixed structure. Indeed, post-war construction techniques generated a final layout that was almost entirely pre-determined with little margin of flexibility [20].

Since modifying this original skeleton would have been both difficult and expensive, the designers chose to add a new flexible layer to the original robust structure. This layer materializes into a winter garden (up to 44 m^2 extra-surface per unit) and a balcony (up to 14 m^2 extra-surface), which are added as an extension of all existing apartments [14]. This is an open unprogrammed space (3,80 m deep), which alters the fixed overdetermined plan layout of the apartment and prompts different ways of using the space. Each existing room becomes accessible from the winter garden. This facilitates movement within the apartment, the inner circulation becoming freer, more flexible, redundant, and dynamic. By providing unprogrammed extra space, the architects increase the degree of behavioral freedom of the inhabitants, who could use the winter gardens in different unexpected ways, since the new structure provides space for free infill and new temporary activities [16] (Figs. 1, 3).

Furthermore, the partial demolition of the existing façade and the addition of the winter gardens and balconies turned the obsolete apartments with ordinary windows into luminous open and flexible spaces whose new glass façade provides a great view towards Bordeaux's city center (Fig. 2).

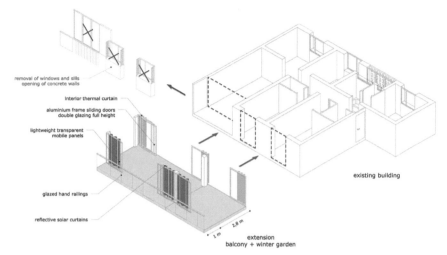

Fig. 1. Minimal demolitions of the existing façade and extensions (addition of winter gardens and balconies). © Lacaton & Vassal, Druot and Hutin. (Courtesy of Lacaton & Vassal).

Fig. 2. From tiny balconies and ordinary windows to a new glass façade with winter garden, balcony, and panoramic view. © Lacaton & Vassal, Druot and Hutin. (Courtesy of Lacaton & Vassal).

At the end of the project, it was evident that every family used the winter garden in a different way: some of the dwellers turned it into a studio, some of them into a workshop or a garden, others simply used it as an extension of the living room, etc. In this respect, the architects provide a free space that can adapt to different and evolving needs, activities and uses: the new additions unlock the form-function overdetermination

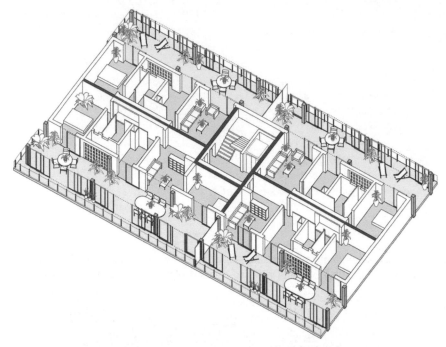

Fig. 3. The addition of an open undetermined extra-space to the original overdetermined structure. © Lacaton & Vassal, Druot and Hutin. (Courtesy of Lacaton & Vassal).

that is typical of these types of buildings and settlements, and increase their capacity to co-evolve with their dwellers and to adapt to unpredictable changes in their lifestyle.

The second challenge consisted of providing the maximum amount of extra-space by reducing construction costs. Here cost reduction is considered crucial to preventing an increase in rent, which would force many tenants to move out because they could no longer afford it. On the contrary, the idea is to allow the residents to remain in their apartment, not only after, but also during, the construction process. This would cancel the social and economic costs due to the displacement of entire families during and after the renovation process, thereby reducing the risk of gentrification. In this respect, the increase in rent was avoided by indexing it to building costs rather than to floor areas. The displacement of the inhabitants during the construction process was avoided by applying a specific conception of building site logistics and construction techniques and by involving the residents in the project [3].

The extension of the apartments was realized through prefabricated reinforced con-crete floors placed on a self-bearing structure (with its own foundations) that was attached to the façades. Reinforced concrete was preferred to steel because it is cheaper, fire-resistant and easier to implement. Part of the old façade was demolished, and the old windows were replaced by sliding full-height windows overlooking the new winter gardens, which were protected by polycarbonate sliding fixtures and thermal curtains (Fig. 1). This design strategy and construction process, which took three weeks per flat for a total amount of 24 months [20], entailed the collaboration of the inhabitants who were involved in the project from the beginning. The local authorities set up a show flat where residents could experience the spatial upgrade and learn about all the phases and implications of the project. Furthermore, they were supplied with a guide for living within the building during the renovation work and were helped in preparing their flat for refurbishment. Residents were also equipped with ear defenders that helped them to deal with noise problems. Lastly, three fully fitted flats were made available during the day to residents who were unable to cope with the disruption [26].

The rent increase was also avoided by enhancing the building's thermal performance. Indeed, better energetic performance led to a considerable reduction of the expenses, so that the total annual sum paid by residents could remain the same since energy costs were reduced thanks to the winter garden. The new unheated semi-external spaces, which were added on the south façade of building HI and on the east-west facades of building G, act as a thermal buffer, as 'inhabitable insulation' that stores solar heat during the winter and shades the interior spaces in the summer [20].

4   Final Remarks and Open Lines of Research

The design strategy presented here contributed to inverting a cycle of decline and obsolescence by adding an adaptive layer that was capable of turning existing un-adaptive buildings into flexible structures that can be constantly transformed by their dwellers according to their evolving needs and lifestyles. In this respect, the project developed by Lacaton & Vassal, Druot and Hutin demonstrates that even the most overdetermined rigid structure, which is shaped by a top-down formal approach to urban planning and design, can be turned into a more adaptive organism, and it shows that spatial designers can play a crucial role in this process.

In this case, the design intervention also has a strong symbolic power as it turns obsolete apartments consisting of tiny rooms, ordinary windows, by a rigid circulation and limited functional adaptivity, into luminous flexible open spaces with huge glass windows and wide luxury winter gardens and balconies offering a great view of the city (Fig. 2). The increase in the attractiveness of the single apartments and the façades of the buildings contributed to creating a positive perception of the neighborhood, a crucial factor in triggering further processes of self-regeneration [23] and increasing the economic value of the residential stock. In this case, the focus on cost reduction did not lead to a decrease in "ambition, pleasure, comfort, or freedom" [19]. On the contrary, the proposed design strategy made it possible to "do the maximum with less" [19].

In this initial phase of the "Cité du Grand Parc" regeneration, the increase in the adaptive capacity was limited to the potential for each dweller to endlessly rearrange the inner space of every single unit. Nonetheless, this is particularly relevant in this historic period in which the pandemic has increased the widespread need (and desire) for a house that acts as a hybrid space, and that is able to welcome a wide range of activities. In this respect, the addition of an unprogrammed extra-space contributes to broadening the *person's overall opportunity set* [5]: the extension of the apartments could also offer new room for micro-entrepreneurial activities (a seamstress workshop, a small lab for repairing micro-works, a home office, a micro gym, a set for web-yoga teaching, a meeting room, or other uses that we cannot even imagine).

In addition to increasing the level of spatial adaptability, the strategy developed in "Cité du Grand Parc" also provides direct access to balconies and terraces and increases the thermal comfort and the (indoor) air quality of each unit. In this respect, the illustrated intervention, even though it was developed long before COVID 19, fulfils three requirements which, since the lockdown, have been identified as being crucial to increasing the quality of life of the inhabitants and the economic value of the dwellings during the post-pandemic era [7, 27].

Furthermore, the strategies presented by the architects in the report Plus + [9] extend beyond the boundaries of each dwelling, including transformations concerning the open public space. In this respect, these spatial strategies highlight how designers who aim to increase adaptivity in anti-adaptive neighborhoods need to develop a multi-scale approach [23]. In their report, the architects show how the re-arrangement of the ground floors, and the densification of the non-occupied areas, could provide room for services, amenities, shops, and other emergent unpredictable activities, thus fostering proximity between housing and other types of emergent uses. The de-densification of some intermediary floors could provide new extra-spaces that are suitable for services and specific facilities for the residents, or for unpredictable activities deriving from the initiative of single residents or groups of them. In this respect, these new hybrid spaces could also help broaden the *person's overall opportunity set* [5], thus enhancing the adaptive capacity of the whole neighborhood.

Finally, the present paper shows how the discipline of spatial design can elaborate and experiment creative solutions that could possibly inform, to a certain extent, the decisions of planners and policy makers. While designers are supposed to act within the framework established by the current regulations and policies, in this case they contested the policy and proposed a different approach that was potentially useful for upgrading the policy itself. In this perspective, the project presented here also highlights the need for an increasing collaboration and exchange of ideas between policy makers, planners, and designers.

In this respect, it is important to underline that the physical configuration of the space is just one of the factors that influence the degree of adaptability of AANs: even when a spatial structure is provided with flexible unprogrammed extra-space suitable for a wide range of uses, the potential for unexpected activities could be limited by planning or organizational constraints: for example, by stringent planning rules that *a priori* establish and lock the functional program [5]. Another additional factor that might influence the degree of adaptability of AANs is property distribution [6]: the fact that usually in AANs the residents are not the owners of their unit also affects their potential to modify, update and upgrade their apartment over time.

In this regard, further research focusing on the complex relationships between design, planning and organizational constraints, could have a crucial role in unlocking new potential for adaptivity across our existing urban fabric, far beyond the specific topic of the regeneration of AANs.

References

1. Alfasi, N., Amitai, R.S., Davidson, M., Kahani, A.: Anti-adaptive urbanism: long-term implications of building inward-turned neighborhoods in Israel. J. Urban. **13**(4), 387–409 (2020)
2. Allen, P.M., Sanglier, M.: Urban Evolution, self-organization and decision making. Environ. Plan. A **13**(2), 169–183 (1981)
3. Ayers, A.: Retrospective: lacaton & vassal. Archit. Rev. **1463**, 77–86 (2019)
4. Batty, M.: Complexity in city systems: understanding, evolution and design. In: De Roo G., Silva, A. (eds.) A Planner's Encounter with Complexity. Routledge, London (2007)
5. Carter, I., Moroni, S.: Adaptive and anti-adaptive neighborhoods: investigating the relationship between individual choice and systemic adaptability. Urban Anal. City Sci. (2021). https://doi.org/10.1177/23998083211025542
6. Cozzolino, S.: The (anti) adaptive neighborhoods. Embracing complexity and distribution of design control in the ordinary built environment. Environ. Plan. B: Urban Anal. City Sci. **47**(2), 203–219 (2019)
7. D'Alessandro, D., Gola, M., Fara, M.F., Rebecchi, A., Settimo, G., Capolongo, S.: COVID 19 and living space challenge. Well-being and Public Health recommendations for healthy, safe, and sustainable housing. Acta Biomed. **91**, 61–75 (2020)
8. De Roo, G., Rauws, W.: Adaptive Planning: generating conditions for urban adaptability. Lessons from dutch organic development strategies. Environ. Plan. B: Plan. Des. **43**(6), 1052–1074 (2016)
9. Druot, F., Lacaton, A., Vassal, J.P.: Plus +. La Vivienda colectiva. Territorio de exception. Gustavo Gili, Barcclona (2007)
10. Habraken, N.J.: Design for flexibility. Build. Res. Inf. **36**(3), 290–296 (2008)
11. Habraken, N.J.: Cultivating complexity: the need for a shift in cognition. In: Portugali, J. Stolk, E. (eds.) Complexity, Cognition, Urban Planning and Design, pp. 55–74. Springer, Berlin (2016). https://doi.org/10.1007/978-3-319-32653-5_4
12. Ikeda S.: The city cannot be a work of art. Cosmos+Taxis **4**(2), 79–86 (2017)
13. Jacobs, J.: The Death and Life of Great American Cities. Vintage book, New York (1961)
14. Lacaton&Vassal Homepage. https://www.lacatonvassal.com/index.php?idp=80. Accessed 25 Feb 2022
15. Leupen, B., Heinjne, R., Van Zwol, J.: Time-based Architecture. 010 Publishers Rotterdam (2005)
16. Moratilla, J.M.: Open Conditions for Permanent Change. Interview with Anne Lacaton. Materia Arquitectura **18**, 22–29 (2018)
17. Moroni, S., Cozzolino, S.: Action and the City. Emerg. Complex. Plan. Cities **90**, 42–51 (2019)
18. Moroni, S., Cozzolino, S.: Multiple agents of self-organisation in complex cities: the crucial role of several properties. Land Use Policies **103**. https://doi.org/10.1016/j.landusepol.2021.105297 (2021)
19. Oswalt, P.: Designing the Brief. Jean-Philippe Vassal in conversation with Philipp Oswalt. Arch +. J. Archit. Urban. 64–73 (2019)
20. Pedrotti, L.: Gounod. Haendel and Ingres refurbishment. Arketipo **119**, 2–9 (2018)
21. Porqueddu, E.: Toward the Open City. Design and Research for Emergent Urban Systems. Urban Des. Int. **23**(3), 236–248 (2018), https://doi.org/10.1057/s41289-018-0065-0
22. Porqueddu, E.: Designing for the Open City. Directing rather than Mastering Emergent Transformations. Trialog. A J. Plan. Build. Global Context **136**(1), 20–23 (2020)

23. Porqueddu, E.: Triggering Spontaneous Self-regeneration in Cities. Towards a systemic approach to spatial design. In: Proceedings of Relating Systems Thinking and Design (RSD10) 2021 Symposium, Delft, The Netherlands, 2–6 November, Systemic Design Symposium (2021)
24. Portugali, J.: What makes cities complex? (2013). http://www.spatialcomplexity.info/files/2013/10/Portugali.pdf. Accessed 10 Sept 2021
25. Portugali, J.: Complexity theories of cities: Achievements, criticism and potentials. In: Portugali, J., Meyer, H., Stolk, E., et al. (eds.) Complexity Theories of Cities Have Come of Age, pp. 47–62. Springer, Berlin, Germany (2012)
26. Publica: Vital Neighborhoods - Lessons from international housing renewal, London (2017). https://www.lacatonvassal.com/publications.php. Accessed 25 Feb 2022
27. Quaglio, C., Todella, E., Lami, I.M.: Adequate housing and COVID-19: assessing the potential for value creation through the project. Sustainability **13**(19) (2021). https://doi.org/10.3390/su131910563
28. Sennett, R.: The Open City. Urban Age (2013). https://urbanage.lsecities.net/essays/the-open-city. Accessed 5 May 2021

Types of Uncertainty:
Cities from a Post-pandemic Perspective

Daniele Chiffi[(✉)] and Francesco Curci

Department of Architecture and Urban Studies, Politecnico di Milano, Milan, Italy
{daniele.chiffi,francesco.curci}@polimi.it

Abstract. Almost all decisions occur under conditions of uncertainty. Understanding uncertainty is thus an essential prerequisite for effective decision-making. In this chapter, we started by recalling the classic distinction between probabilistic risk and severe uncertainty. We ground our analysis in Hansson's recent classification of eight types of uncertainty: *factual uncertainty, possibilistic uncertainty, metadoxastic uncertainty, agential uncertainty, interactive uncertainty, value uncertainty, structural uncertainty*, and *linguistic uncertainty*. Based on this classification, we investigate and demarcate some of the determinants of each type of uncertainty by taking into account their different sources and scales. Finally, we apply our analysis of these determinants to urban decisions that might occur in a post-pandemic context and propose future lines of research.

Keywords: Uncertainty · COVID-19 · Cities · Urban planning · Philosophy of science

1 Introduction

The COVID-19 pandemic is a highly complex and uncertain phenomenon. Although many pandemics in the history of humankind have shown some *known* recurrent features—frequent attribution of the pandemic's origin to Asia; westward spread to Europe, directional spread within Europe, and acceleration of global spread; and presentations of different 'waves' (Morens and Taubenberger 2011)—the severe uncertainty associated with COVID-19 and our failure to understand the determinants of present and past pandemics have kept us from important information that could help to prevent and control modern pandemics in general. The spatial diffusion of COVID-19 is also sensitive to urbanization. This context is highly significant due to the disease's respiratory transmission mechanism, which is enhanced by the proximity of people to one another. Further, marginalized and vulnerable groups living in urban areas are disproportionately affected by COVID-19 and its related impacts (Sharifi and Khavarian-Garmsir 2020).

Following the classic distinction between forms of uncertainty represented by types of quantifiable (probabilistic) *risk* and the *severe uncertainty* that may resist (probabilistic) quantification (Knight 1921), we want to investigate the truly complex structure of cities and confirm the need to differentiate approaches to urban planning based on the

F. Calabrò et al. (Eds.): NMP 2022, LNNS 482, pp. 1039–1047, 2022.
https://doi.org/10.1007/978-3-031-06825-6_99

type of uncertainty present and analyzing its main determinants. Even if the distinction between risk and severe uncertainty cannot always be clearly delineated, we can compute the probability of future events and their consequences under conditions of risk. In contrast, under conditions of severe uncertainty, this calculation is not always possible, as we may face forms of uncertainty that are unquantifiable because we do not know—or even ignore—the statistical distribution of the events (Hansson 2022; Carrara et al. 2021; Chiffi and Chiodo 2020). Moreover, under conditions of severe uncertainty and even *unknown unknowns,* all the potential events we wish to evaluate may not be easily listable, in the sense that unexpected situations and truly novel scenarios may occur (Moroni and Chiffi 2022).

This interplay between conditions of risk and severe uncertainty has been recently acknowledged in connection with the outbreak of the COVID-19 pandemic (Hofmann 2020). Test accuracy—sensitivity, specificity, predictive values—in different contexts, the effects and side effects of new promising treatments, and the prevalence of the disease are all clearly connected with conditions of probabilistic risk. Further, viral spread, levels of immunity, diffusion of new variants of a virus, the effectiveness of intensive care treatments, and so on occur under conditions of severe uncertainty. As we will see, both the COVID-19 pandemic and contemporary cities are complex phenomena that require planning policies targeted toward conditions of risk and severe uncertainty. Unfortunately, the interaction between these two complex phenomena may pose socio-spatial and environmental challenges that are difficult to reconcile. This hardship is due to the spatial unlimitedness of the present interaction between highly uncertain systems (Hansson 2011), such as cities, and the novel problems the pandemic poses; this combination should invite us to be particularly alert to the *unknown.* While probabilistic trends, urban statistics, and so on are, of course, useful for understanding the impact of the pandemic on cities, urban planning strategists cannot rely solely on them. Instead, they must be open to critically managing several different types of severe uncertainty.

COVID-19 is not a *black swan* (Taleb 2012), i.e., a highly uncertain event; rather, it is a phenomenon with aspects ranging from probabilistic risk to severe uncertainty. We propose that the same can be said for cities. Indeed, the interaction between the pandemic and cities is extremely complex. In order to be prepared to deal with such complexity, we need to better understand different types of uncertainty surrounding the impact of COVID-19 on cities. Section 2 critically discusses different types of uncertainty, while Sect. 3 proposes a new approach aimed at understanding the main determinants of these types of uncertainty. Section 4 illustrates how each type of uncertainty may affect cities in a post-COVID-19 perspective underlying the relevance of local determinants of uncertainty. Finally, Sect. 5 concludes the chapter.

2   Types of Uncertainty

Uncertainty, together with the related concepts of risk and ambiguity, has long been a key concept in psychology, economics, decision-making, and planning processes (Lipshitz and Strauss 1997). Uncertainty comes in different forms, and several studies have proposed specific taxonomies based on the various factors that contribute to its formation, management, and treatment, especially within decision-making and planning processes.

We discuss here some interesting classifications and conceptualizations that may help us to understand the specific features of uncertainty present within cities during COVID-19.

An initial taxonomy is based on the analysis of hundreds of decision-making self-reports and differentiates three main causes of uncertainty: *inadequate understanding*, *incomplete information*, and *undifferentiated alternatives* (Lipshitz and Strauss 1997). Inadequate understanding may depend on equivocal information due to novelty, fast-changing or unstable situations. Incomplete information may depend on a partial or complete lack of information or on unreliable information. The third cause of uncertainty (undifferentiated alternatives) refers to the fact that, even when information is perfect, decision-making can be affected by the conflict among alternatives owing to equally attractive outcomes or to incompatible role demands.

A second taxonomy is founded on the *nature* and *object* of uncertainty (Bradley and Drechsler 2013). According to Bradley and Drechsler, the nature dimension relates to the kind of judgement being made. In this case, it is possible to distinguish three forms of uncertainty: *modal uncertainty* about what is possible or what could be the case; *empirical uncertainty* about what is the case (or has been or would be the case); and *normative uncertainty* about what is desirable or what should be the case. The object dimension relates to the features of reality that agents' judgements are directed toward. Here it is possible to distinguish two forms of uncertainty: *factual uncertainty* about the way things are and *counterfactual uncertainty* about the way things could or would be if things were other than the way they are.

Furthermore, in attending to dynamically adaptive systems, some authors have proposed taxonomies based on the sources of uncertainty across the three distinct levels of the management and decision-making process: the *requirements level*, the *design level*, and the *run-time level* (Ramirez et al. 2012). Uncertainty at the first level is owed to the idealization, misunderstanding, and incompleteness of functional and non-functional requirements (missing or ambiguous requirements, falsifiable assumptions). The second level is uncertain primarily due to unexplored alternatives and untraceable design. Uncertainty in the third occurs primarily because of environment unpredictability.

A comprehensive and transversal taxonomy that embraces previous proposals has been outlined recently by Hansson (2022). According to this taxonomy, we can list the following types of uncertainty:

(1) *Factual uncertainty*. This uncertainty surrounds the facts of the physical world and may usually be quantified and formalized;
(2) *Possibilistic uncertainty*. This form of uncertainty is about what can possibly be known. In this case, uncertainty depends on many factors, such as (i) the constraints on the information that an agent may obtain in a specific context at a given time and (ii) the very nature of the decision, which may deal with forms of logical, physical, biological, and social possibility;
(3) *Metadoxastic uncertainty* (or *uncertainty of reliance*). Our beliefs may be uncertain. This uncertainty is a second-level judgment about the accuracy of one's beliefs and is expressed, for instance, by second-order probabilities or confidence intervals;
(4) *Agential uncertainty*. This type of uncertainty considers individual future decisions and actions. It cannot be formalized or quantified, as there is no suitable decision

method. For instance, there is no proper methodology to formalize or compute the consequences of whether one will get married in two years. This uncertainty is thus related to the decisions and behaviors of individuals.

(5) *Interactive uncertainty.* Uncertainty may be the result of the interaction between individuals or between individuals and institutions or companies. This form of uncertainty can usually be formalized by means of (epistemic) game theory, even if it cannot be quantified.

(6) *Value uncertainty.* Philosophers and economists have recently started discussing the normative component of uncertainty, which goes beyond its factual forms. They have thus introduced the notions of *moral uncertainty* and *normative uncertainty.* (Lockhart 2000; Sepielli 2013 and 2014; MacAskill et al. 2020). The latter is a much broader concept than the former. Moral uncertainty is "uncertainty about what we all-things-considered morally ought to do" (MacAskill et al. 2020, p. 2). Normative uncertainty involves norms in the legal sense, but it "also applies to uncertainty about which theory of rational choice is correct and uncertainty about which theory of epistemology is correct" (MacAskill et al. 2020, pp. 2–3). It focuses on the value-based dimensions of conditions of inexactness and unpredictability (Taebi et al. 2020). Thus, normative uncertainty involves valuative considerations in those aspects of decision-making related to epistemology, ethics, law, and politics. By *value uncertainty* we mean an umbrella term for both moral uncertainty and normative uncertainty.

(7) *Structural uncertainty.* The true structure, limitations, and impacts of complex decisions are almost always unknown. This uncertainty may be caused by a number of factors: (i) the delimitation of the issue covered by the decision may not be fixed or known; (ii) the scope of the decision may be unclear; (iii) it may be unclear who is going to make the decision; (iv) the timing of the decision may be uncertain; and (v) the consequences of the decision may be difficult to conceive and evaluate.

(8) *Linguistic uncertainty (ambiguity).* This uncertainty is due to linguistic ambiguity and is mainly related to the semantics of the terms involved in decisions.

3   The Determinants of Uncertainty

The different forms of uncertainty classified by Hansson (2022) have different weights according to the context, the moment in which they emerge, and the events that produce them. This variability does not prevent us, theoretically, from analyzing the determinants of uncertainty. To understand what these determinants are regardless of the uncertainty's specific "objects", we opted to rely on two elements:

a.   the *scale* (local or global) at which uncertainty occurs or manifests itself.
b.   the *source* (impersonal, individual, or collective) of uncertainty.

In contrast to Ramirez et al. (2012), we do not intend "source" as an operational level within the decision-making process but rather as the "subject" contributing to the production of the various forms of uncertainty.

From this analysis, the following picture emerges, which is summarized in Figs. 1 and 2.

1) *Factual uncertainty* has an impersonal nature and can be generated both locally and globally. Indeed, it does not depend on preferences but on facts: that is, on events that are not linked directly to the choices of individuals even when they depend indirectly on human actions.

2) *Possibilistic uncertainty* is the broadest conception of uncertainty because it relates to everything that could be known or is technically feasible thanks to new events and forms of knowledge. For this reason, it can relate to all scales and sources.

3) *Metadoxastic uncertainty* is essentially an individual form of uncertainty, as it concerns "the veracity or accuracy of one's own beliefs" (Hansson 2022). Although it is mainly a form of uncertainty based on what individuals believe, it can nonetheless also manifest itself on a large scale when masses of individuals simultaneously rely in the same way on certain information.

4) *Agential uncertainty* depends on the degree to which individuals believe they can control themselves and their future actions with present decisions. Since it is a quite psychological form of uncertainty, it is purely personal and local.

5) *Interactive uncertainty* involves the ways agents expect other agents to react to their choices and decisions. Based as it is on human interactions, it is subject to social imitation or contagion. Thus, it is a purely collective form of uncertainty working at local or global scale.

6) *Value uncertainty* is based on comparisons made among different and often incommensurable values. It depends on the ranking people (individuals or collectives) wish to assign to alternatives whose relative values are uncertain. This type of uncertainty is ubiquitous and can manifest itself on both a local and global scale.

7) *Structural uncertainty* is strongly connected with problems that need to be clarified and systematized before considering a decision's outcome. Thus, it surrounds the structure of the decision, which includes the delimitation of the relevant issues, roles and responsibilities, timing, options, and criteria. Since it is a type of uncertainty linked to specific agents (individuals or collectives) and their ways of organizing themselves to deal with a problem, it stands predominantly at the local level.

8) *Linguistic uncertainty* is a purely collective uncertainty, given that languages are socio-cultural and context-dependent artefacts. Apart from English and a few other languages, linguistic uncertainty usually has a local nature instantiated in specific countries or regions.

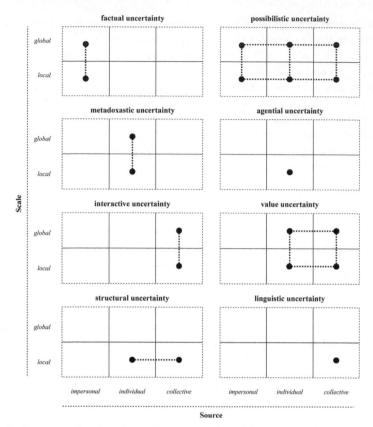

Fig. 1. Sources and scales of the different types of uncertainty (authors' elaboration)

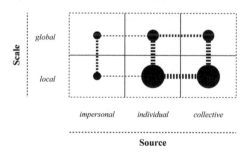

Fig. 2. Determinants of uncertainty (authors' elaboration).

4   Uncertainties and Cities in the Post-COVID-19 Perspective

In this section, we apply Hansson's taxonomy (Hansson 2022) to a more specific object of investigation: the relationship between uncertainty and the city from a post-pandemic perspective. We intend to identify and explore how different types of uncertainty are interrelated with each other when evaluating such a complex phenomenon. Our aim is to clarify the ways in which different types of uncertainty affect the decisions and planning processes related to the future of cities after the current pandemic.

1) *Factual uncertainty*, on the one hand, is linked to what has been happening during the (ongoing) pandemic phase, including all the aspects related to epidemiological data and statistics, with a special emphasis on differences between countries, regions, and cities. On the other hand, this type of uncertainty will involve: (i) the persistence of different forms of the virus, (ii) the way in which variants of the virus will develop, (iii) how contagious and lethal they will be, and (iv) to what extent they will spread in cities compared to less densely populated areas. Science will play a decisive role in this type of uncertainty and will be called upon to evaluate what happens and what to expect from COVID-19 or new viruses.

2) *Possibilistic uncertainty* is mainly connected to everything we might discover but that is presently deeply unknown (*unknown unknowns*). The so-called long-term effects of COVID-19 will feed uncertainty, along with the combinations of viruses and other unexpected phenomena that may occur both locally and globally. Problems relating to vaccines can also be included in this category, with reference not only to their effectiveness but also to their possible negative long-term effects (if any), or problems related to new forms of urban transportation based on disruptive technologies like autonomous vehicles.

3) *Metadoxastic uncertainty* involves, for instance, the cogency of various forecasting models that will be used on several fronts in the post-pandemic era. The effect of this form of uncertainty on previsions regarding the life of future cities is essential in order to understand the post-pandemic city, in which epidemiological indicators are going to remain crucial aspects of urban life.

4) *Agential uncertainty* is connected to how likely it is that people will consider cities less habitable or safe than other geographic and settlement contexts while also anticipating new types of housing, jobs, and transportation.

5) *Interactive uncertainty* will particularly influence the way in which precautionary limitations maintained even after the pandemic emergency will be accepted, transgressed, or possibly subjected to forms of social imitation. This type of uncertainty also encompasses the risks of organized forms of protest by groups of people against decision-makers, political leaders, and public institutions.

6) *Value uncertainty* is mainly linked to the principles placed at the basis of national and international political agendas and constitutional frameworks. By way of example, we may ask whether environmental sustainability will still guide public agendas and policies or whether the pandemic may contribute to shifting them towards new forms of social and territorial justice. For instance, widespread new hospitals, both large and small, with intensive care units could go against environmental and landscape protection.

7) *Structural uncertainty* is related to the following questions: what legal competences will be demanded of central or local governments making decisions on the measures to be applied in urban areas after COVID-19? Do we need new institutions to cope with the uncertainty of future cities? Will cities require new public structures specialized in preventive measures and epidemiological management? Will there be an abundance of technical committees supporting political choices?

8) *Linguistic uncertainties* concern the ways in which certain terms specifically linked to the containment and management of the pandemic might be perceived and interpreted. For example, we can consider the definitions of "green pass", which is the EU Digital COVID Certificate Regulation, "super-green pass", "lockdown", "quarantine", "curfew", "red zones", etc. It is clear that linguistic ambiguity connected to these terms may have a great impact on the life of a city.

From the examples listed above, it is clear that the interplay between these types of uncertainty will affect planning strategies and decisions for post-pandemic cities. Unfortunately, uncertainty is often treated as a uniform phenomenon; our analysis reveals its multifaceted nature, showing that based on the type of urban uncertainty that we are facing, we can isolate specific determinants and take decisions based on a better understanding of complex urban phenomena. It does not mean separating or fragmenting the decision-making and planning processes to deal with the different forms of uncertainty, but building more aware and complete interpretative frameworks, even though more complex. For urban planning, this means crossing the multidimensionality of uncertainty with the diversity and relationality of the urban spaces and phenomena.

5   Conclusion

In this chapter, we started by recalling the classic distinction between probabilistic risk and severe uncertainty. We grounded our analysis of the determinants of uncertainty in Hansson's recent classification of the eight types of uncertainty: *factual uncertainty, possibilistic uncertainty, metadoxastic uncertainty, agential uncertainty, interactive uncertainty, value uncertainty, structural uncertainty,* and *linguistic uncertainty.* While Hansson questioned the possibility of formalizing and quantifying all types of uncertainty, we investigated and delineated some of the determinants of each type of uncertainty by taking into account their different sources and scales. Then, we applied our analysis of these determinants to the urban decisions that might occur in a post-pandemic context.

Our analysis is intended to shed light on the relationship between uncertainty, COVID-19, and cities. We have pointed out that although the pandemic is a global phenomenon, its implications for uncertainty management also deeply affect the local level as well; likewise, it covers the individual and the collective dimensions with a strong relational dimension. In short, issues concerning the impact of the COVID-19 pandemic on cities seem to align with what emerges from our analytical scheme (see Fig. 2) of the determinants of uncertainty. This framework can be used as a suitable tool for communicating and managing urban uncertainty. Finally, since some types of uncertainty cannot be quantified or even formalized, future lines of research might investigate how to deal in an urban planning perspective with these types of uncertainty whose understanding and management is particularly challenging and involving the specific determinants that we have identified.

Acknowledgements. The research is supported by the RIBA project "Norms, Uncertainty and Space (Nous): Cities in The Age of Hyper-Complexity", Department of Architecture and Urban Studies, Excellence Department - "Fragilità Territoriali", Politecnico di Milano. The research of

Daniele Chiffi is also supported by the Italian Ministry of University and Research under the PRIN Scheme (Project no. 2020SSKZ7R).

References

Bradley, R., Drechsler, M.: Types of uncertainty. Erkenntnis **79**(6), 1225–1248 (2013)

Carrara, M., Chiffi, D., De Florio, C., Pietarinen, A.-V.: We don't know we don't know: asserting ignorance. Synthese **198**(4), 3565–3580 (2021). https://doi.org/10.1007/s11229-019-02300-y

Chiffi, D., Chiodo, S.: Risk and uncertainty: foundational issues. In: Balducci, A., Chiffi, D., Curci, F. (eds.) Risk and Resilience. SAST, pp. 1–13. Springer, Cham (2020). https://doi.org/10.1007/978-3-030-56067-6_1

Hansson, S.O.: Coping with the unpredictable effects of future technologies. Philosophy Technol. **24**(2), 137–149 (2011)

Hansson, S.O.: Can uncertainty be quantified? Perspect. Sci. **30**(2), 210–236 (2022). https://doi.org/10.1162/posc_a_00412

Hofmann, B.: The first casualty of an epidemic is evidence. J. Eval. Clin. Pract. **26**(5), 1344–1346 (2020)

Knight, F.H.: Risk, Uncertainty, and Profit. Hart, Schaffner & Marx; Houghton Mifflin Company, Boston, MA (1921)

Lipshitz, R., Strauss, O.: Coping with uncertainty: a naturalistic decision-making analysis. Organ. Behav. Hum. Decis. Process. **69**(2), 149–163 (1997)

Lockhart, T.: Moral Uncertainty and its Consequences. Oxford University Press, Oxford (2000)

MacAskill, W., Bykvist, K., Ord, T.: Moral Uncertainty. Oxford University Press, Oxford (2020)

Morens, D.M., Taubenberger, J.K.: Pandemic influenza: Certain uncertainties. Rev. Med. Virol. **21**(5), 262–284 (2011)

Moroni, S., Chiffi, D.: Complexity and uncertainty: Implications for urban planning, pp. 319–330. In: Portugali, J. (ed.) Handbook on Cities and Complexity. Edward Elgar Publishing, Cheltenham (2021)

Moroni, S., Chiffi, D.: Uncertainty and planning: cities, technologies, and public decision making. Perspect. Sci. **30**(2), 237–259 (2022). https://doi.org/10.1162/posc_a_00413

Ramirez, A.J., Jensen, A.C., Cheng, B.H.: A taxonomy of uncertainty for dynamically adaptive systems. In: 2012 7th International Symposium on Software Engineering for Adaptive and Self-Managing Systems (SEAMS), pp. 99–108. IEEE (2012)

Sepielli, A.: Moral uncertainty and the principle of equity among moral theories. Philos. Phenomenol. Res. **86**(3), 580–589 (2013)

Sepielli, A.: What to do when you don't know what to do when you don't know what to do... Noûs **48**(3), 521–544 (2014)

Sharifi, A., Khavarian-Garmsir, A.R.: The COVID-19 pandemic: Impacts on cities and major lessons for urban planning, design, and management. Sci. Total Environ. **142391**, 1–14 (2020)

Taebi, B., Kwakkel, J.H., Kermisch, C.: Governing climate risks in the face of normative uncertainties. Wiley Interdisciplinary Reviews: Climate Change **11**(5), e666 (2020)

Taleb, N.N.: The black swan: The impact of the highly improbable, vol. 2. Random house (2012)

Urban Change in Cities During the COVID-19 Pandemic: An Analysis of the Nexus of Factors from Around the World

Hussaen A. Kahachi[(✉)] [iD], Marwah Abdulqader Ali [iD],
and Wahda Shuker Al-Hinkawi [iD]

University of Technology, Baghdad, Iraq
kahhhtchi@gmail.com, Wahda.S.Mahmoud@uotechnology.edu.iq

Abstract. This research is concerned with understanding the context of influences the COVID-19 pandemic had/has on urban change in cities. It focuses on understanding the nexus of relations through which the pandemic impacted the urban space, urban place and public realm; specifically, the economic, environmental, social, human behavior, administrative and health factors. Using a mixed-methods approach and examples from around the world, an analysis was conducted to highlight the most important factors influencing urban change as a result of the pandemic. The research concluded that there are a multitude of direct and indirect connections between the factors that resulted in long-term and short-term impact on urban change and public realm. An understanding of this nexus of factors constitutes an important and essential step in creating more comprehensive and responsive urban planning/design policies. Thus, mitigating any undesirable urban change and better governing the public realm during similar situations and pandemics in the future.

Keywords: Urban change · Urban transformation · Public realm · COVID-19 · City planning · Economy · Social · Environment · Human behavior · Urban services · Health · Urban management

1 Introduction

Cities around the world faced many pandemics over the past centuries. Since the Athens pandemic in 430 BC, the plague that stretched all over Europe and then the world in the Middle Ages, the rapid spread of the Ebola pandemic in Africa, until recently the Corona pandemic in 2019. The COVID-19 pandemic rapidly spread throughout the world starting from China and significantly affected various sectors and aspects including education, health, economy, tourism, and services [1–4]. The COVID-19 pandemic, much like previous pandemics, had a great impact on cities from several aspects, resulting in a multitude of urban transformations/change. The effects ranged from urgent, direct and short-termed, to gradual, indirect and long-termed effects. The subject of urban change during and post pandemic has quickly gained great importance in academia, especially for highly populated and dense cities. The high density was considered by

© The Author(s), under exclusive license to Springer Nature Switzerland AG 2022
F. Calabrò et al. (Eds.): NMP 2022, LNNS 482, pp. 1048–1058, 2022.
https://doi.org/10.1007/978-3-031-06825-6_100

many as the main driver for the rapid spread of the pandemic especially when the necessary precautions are not taken [5, 6]. Also, because such pandemics could have dire consequences, effects and risks have and play an inevitable role in people's lives and well-being, hence, present and future planning of the cities and urban areas and public realm [7–9].

The effects of the pandemic on cities, urban space and public realm were on two levels, direct and indirect [10–12]. For example, due to the conditions and the worldwide shock associated with the pandemic and its rapid spread in 2019 when there was a lack of full-understanding of its nature and the way to deal with it, many cities and countries incorporated a first-stage set of precautionary measures, procedural and preventive emergency plans to limit/restrict its spread. This included lockdown, public curfew, health restrictions in public spaces, limiting movement and transportation, closing many non-essential public spaces and services including shopping centers, malls, cultural and religious centers, service and civic centers, setting strict controls to limit large and medium gatherings, and other measures to discourage the spread of the pandemic. A second stage of a more-relaxed precautionary measures followed in many countries and cities that regulated the public space and public realm in a way that enabled some activity while ensuring a reduced direct human contact. For example, by shifting towards working systems that do not require the presence of employees, customers, users and guests or reduce contact between them as far as possible - i.e. work-from-home and service-from-home systems-. In some cases, entire sectors that depend on direct human interaction or non-essential services, such as are tourism, cultural and recreational services, were completely paralyzed during this stage.

Some researchers argue that these conditions and measures directly affected the urban space and the public realm in many cities, constraining various activities and restricting movement of residents, transportation to and from cities and urban areas, and many other activities and services [13–16]. On the other hand, many researchers have discussed the existence of indirect effects of the pandemic on cities, urban space and public property. These effects extend to cities, urban space and public realm indirectly through other intermediately systems/aspects which generally takes longer to recover from the effects of the pandemic, and therefore medium/long-lasting effects on urban space, cities and public realm [17, 18]. Among the most important of these aspects that have been affected by the pandemic, and through which the pandemic has affected urban space and public realm, are economic, environmental, health, human behavior, services and administrative aspects [4, 12, 18–21]. Although many researchers and specialists have been interested in writing about the effects of the pandemic on urban space and public realm, there is gap in the academic literature related to the lack of a comprehensive understanding of nexus of COVID-19 based factors affecting urban change in cities within both the short and long timeframe.

For this reason, this research paper is concerned with urban change resulting from the COVID-19 pandemic and its effects on cities and activities in the urban space and public space both in short-term (usually less than two months), and long-term (usually more than one month). It will analyze and discuss the factors affecting urban change as a consequence to the pandemic from several aspects; economic, environmental, health, human, service, and administrative. The research will use this to draw a conceptual diagram of these factors' relationships network in the conclusion.

2 Aspects Through Which COVID-19 Influenced Urban Change

2.1 The Economic Aspect

Perhaps one of the most visible and important effects of the worldwide COVID-19 pandemic is its impact on the economy in many cities, countries, and even worldwide [22–24]. As the changes brought about by the pandemic and the subsequent major economic transformation in important sectors such as tourism, travel, recreational and cultural services, has led to transformations in the urban and inactive parts of cities alike [25]. Figure 1 shows how much economic activity has shifted in some countries as a result of the pandemic. It clearly shows the decline of the tourist movement and its associated activities such as transportation, hospitality and some administrative activities and services that surrounded a number of tourist attractions, especially closed ones in many cities in addition. This matter was directly affected by commercial activities in these areas or that are linked to them, such as transportation services, food and drink and housing. This change has led to the disappearance of some businesses and activities and the emergence of others. For example, people's visits to museums has decreased significantly (See Fig. 2), while the visit of local tourists in particular to open areas outside the city or in its suburbs, where population density is lower, has increased relatively. This resulting decrease in the profitability of some businesses led some of them to stop or change. Additionally, the decline in the purchasing and economic movement in general due to the loss of many self-employed and earners of their livelihoods, has also led to the cessation or erosion of a number of activities that were recognized in the cities, and this in turn led to the transformation of the economic movement towards the basic aspects of living Such as food, medicine and water during and after the pandemic.

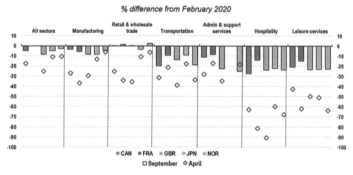

Note: Monthly GDP in Canada, Norway and the United Kingdom, monthly output in Japan and France. Data based on national industrial classifications. Data on all sector output are not available for France, and data on administrative and support services are not available for Japan. Transportation data for Norway exclude ocean transport.

Source: Office of National Statistics, United Kingdom; Ministry of Economy, Trade and Industry, Japan; Insee, France; Statistics Canada; Statistics Norway; and OECD calculations.

Fig. 1. The impact of the pandemic on economy in five countries by sector Source: [26]

Fig. 2. Change in activities' type and intensity during COVID-19 Source: [27]

2.2 The Environmental Aspect

Another important effect of the pandemic is the environmental aspect [28–30]. As a result of the cessation or decline of some commercial traffic and the closing of factories, for example, and the subsequent reduction in dependence on mechanized means of transportation, all of this has led to a reduction in environmental, audio and visual pollution in a number of places and public spaces and the city as a whole.

On the one hand, the low intensity of use of some green open spaces within the city has given nature an opportunity to recover in these spaces. This and the subsequent culture of social distancing have led to important transformations in urban space and public property. We note through Fig. 3, which shows the difference before and during the pandemic on the same urban space, that the environment was able to recover clearly due to the low intensity of use, which usually had a negative impact on the environment due to its height. Where the use of urban areas changed from general use by any person in the city, for example, to general use for residents of areas close to space and for necessary or essential cases at least. This effect was temporary in most cases, as the movement of industry and transport quickly regained its activity. However, these restrictions strongly emphasized the harmful effects of excessive/abnormal use of the environment, which rethought the systems, administrative and planning related to the urban environment.

Fig. 3. Comparison of urban space's environment before/during the pandemic Source: [31]

2.3 The Health and Hygiene Aspect

Another important aspect of linking urban transformation as a result of the pandemic is the health aspect [20, 32, 33]. The spread of the culture of public hygiene and personal protection has led to the emergence of new and different symptoms on the urban space as a whole. In certain regions and countries, there was great interest in public cleanliness and achieving the cleanliness and health of the public space, both through intensifying the cleaning movement, changing the nature of uses in the spaces to reduce or increase certain activities, and others.

On the other hand, in some regions and countries, the movement of interest in personal protection, unfortunately, had counterproductive results that led to an increase in waste due to poor handling of masks and sterilization for example (Fig. 4). Moreover, there are long-term effects on public health and social awareness. It became clear how the excessive population density in some urban areas influenced the spread of germs. Hence, highlighting the importance of incorporating the correct regulation to organize and control the use of urban space.

Fig. 4. Two models of the impact of COVID-19 on the health aspect within urban space Right: Tossed Masks Source: [34] - Left: public sterilization queue Source: [35]

2.4 The Human-Behavior Aspect

Human behavior is one of the aspects that the pandemic has affected, and in turn, it has affected urban space and public property. According to surveys conducted on a number of people, [28, 33, 36, 37], the majority of people are looking for an open urban public space that can be accessed by means that usually depend on muscular effort (cycling or bicycles). Not only that, but to protect themselves and their families, they began searching for new public spaces that they had not previously resorted to ensure social distancing, especially if the usual public spaces were overcrowded or where safety conditions could not be applied. The shift towards means of working and communicating remotely using communication services via the World Wide Web aided this as it saved time usually used for transportation, and allowed close interaction with families.

According to the Kubler-Ross diagram of change (Fig. 5), societies during the COVID-19 period went through stages of change of behavior. Societies first went through a stage of shock that led to exaggerated actions and movements by some, for example, buying needs in large quantities, storing them in homes and staying at home. Denial as they try to find different reasons and explanations for what is happening. Nervousness, followed by depression, which severely limit non-essential activities. Adaptation where societies begin to adapt, finding alternatives. These phases had an impact on the urban space, even if it was temporary and gradual, but the collective memory will keep these phases in the user's imagination in preparation for something similar in the future.

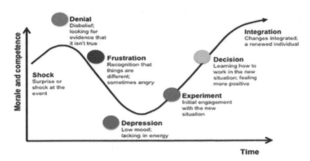

Fig. 5. The Kubler-Ross change curve source: [38]

2.5   The Urban Services Aspect

One of the most important aspects that have been affected in many cities is the service aspect. For example, there is less dependence on public transport in general, and even on the private car in many cases, especially with the increase in dependence on the means of work, shopping and providing services from a distance [32, 36]. On the other hand, the need for other services such as hygiene and sterilization and health services such as hospitals and examination and prevention centers has increased. The level of utilization of religious and cultural services such as mosques, public libraries, cinema, theater and others decreased. All this led to an important transition and a direct challenge to designers and users alike [28], as the use of surrounding spaces or their use associated with these activities decreased, such as metro and bus stations, and public spaces near theaters, shopping centers and religious buildings, and their occupancy and utilization decreased, which became limited to the use of a few. of people adopting healthy personal protective equipment. This change led to the emergence of a heterogeneous mixture between the interest in open green spaces and the movement paths for the commuters and bicycles at times to reduce movement outside the house and achieve home isolation at other times and according to conditions such as the availability of green spaces and appropriate movement paths and their design.

2.6   The Administrative Aspect

One of the important changes brought about by the pandemic is an important change in the mechanisms of managing cities and urban areas [1, 19]. For example, the increase in the need for open spaces, especially in overcrowded areas in city centers, the increase in the need to achieve a mixed use of services and activities at the urban level, and the lack of concentration of services in hotspots, reducing the need for mechanized transport in general and in general in particular, the development of mechanisms and alternatives to revitalize certain sectors that suffered during the pandemic, Increased interest in more flexible urban management mechanisms so that they can respond to changes and crises when they occur in the future [19, 32, 33]. This would bring about fundamental changes in cities, urban spaces and public properties in the future by adopting the means of technological control, artificial intelligence, real-time analysis and future simulation of urban areas, with a return to the theories and methodologies of future cities such as smart cities, responsive cities and others in order to give cities a higher ability to identify variables and adapt with her.

3   Results

The issue of urban transformations due to the pandemic and the accompanying transformations at various levels in general, is, as it appeared to us, one of the complex issues that have dimensions and links with several aspects. Also, this particular topic is still current, and many researchers, planners and designers are still conducting studies to determine the effects of this pandemic on the urban level and at various levels. Nevertheless, based on the foregoing, the urban transformations resulting from COVID-19 pandemic can be summarized as follows:

The economic aspect: It includes 1) the atrophy of a number of activities in city centers, especially with regard to the movement of tourism and tourist facilities. 2) The rise and emergence of a number of new activities in the city, especially health and treatment. 3) Moving away from centralization and moving to decentralization sometimes and sometimes relying on electronic or remote economic exchange means.

The environmental aspect: It includes 1) a gradual (often temporary) improvement in the environment and a decrease in pollution in general. 2) Reducing the intensity of the use of a number of open green spaces and giving nature a chance to recover. 3) The performance of some urban spaces has improved due to the accompanying environmental improvement.

The health aspect: includes 1) the spread of a culture of personal and public health. 2) Increasing attention to hygiene and sterilization of public facilities and regulating their use. 3) The increased use of sterilization and personal protection methods has sometimes generated problems in dealing with them.

The human behavior aspect: It includes 1) moving away from areas and spaces that are crowded with users in many city centers, especially closed ones. 2) Orientation towards open urban areas and spaces, which include new and diverse activities. 3) Increasing the free time available to the residents due to the absence of the need to move to and from the workplace.

The urban services aspect: It includes 1) a decrease in reliance on mechanized transportation, especially public ones, and an increase in reliance on means of movement with muscular effort, even if this is a relative matter according to the availability of the appropriate requirements. 2) A significant decrease in the demand for cultural, scientific, religious and recreational public services in general, and a shift in the type of services required and their standards. 3) Achieving home isolation at times (especially at the beginning of the pandemic) generated an increasing desire to achieve social communication in other ways, the most important of which was electronic platforms, which reduced dependence on public spaces.

The administrative aspect: It includes 1) important changes in the mechanisms and visions of managing cities and urban areas that took the emergence of such pandemics into account. 2) The emergence of the importance of achieving mixed use in urban areas and distributing activities in a homogeneous manner, while staying away from specialized urban centers. 3) Increasing interest in modern and future theories and approaches such as artificial intelligence and urban resilience that will have a significant impact on urban space in the future if applied.

4   Conclusions

In conclusion to what was discussed above, there is a direct and indirect impact of COVID-19 pandemic on Urban Change in many cities and regions of the world. The impact varied in type, size, and duration depending on a number of factors. The research discussed a multitude of COVID-19 factors impacting Urban Change from various aspects; economic, environmental, health, human, service, and administrative. Aside from the relations connecting the COVID-19 to Urban Change, there is a clear connection between these factors and aspects themselves.

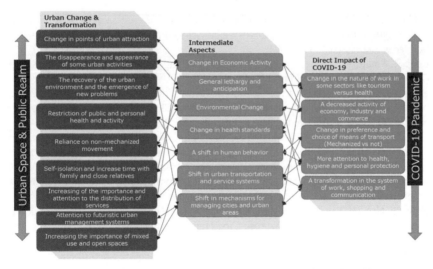

Fig. 6. Diagram of the different factors COVID-19 has on Urban Change

Figure 6 shows the effects of the COVID-19 pandemic on urban space and public property, passing through the economic, environmental, health, service, administrative, and human-behavior aspects. It appears from the figure that most of the transformations that resulted or result from the pandemic and the changes associated with it are temporary to medium transformations in terms of time, and this is a logical thing since urban transformations generally take a long-medium time, and therefore most of the effects of the temporary pandemic will only affect an emergency temporary And if things go back to normal for a long time, this includes, for example, changing urban attractions, recovering the urban environment, reducing dependence on mechanized movement, isolation and increasing time with family and others. Despite this, there is an important exceptional case, which is the administrative aspect, which, although associated with some emergency interim effects, may have long-term effects on city planning and management, especially during periods of similar crises and achieving urban resilience. Among these influences, for example, is the emergence of the importance of mixed use, distribution of services, and responsive urban management systems. As the effects of the pandemic contributed greatly to pushing for solutions, policies, and planning and urban systems capable of dealing with these emergency conditions, or at least limiting their impact through theories such as future cities. However, most of the available studies on the subject and on these systems are studies that depend on visions that may be logical, but they depend on specific situations in place and time that often lead to expectations, predictions and speculations that may occur and may not occur on the ground in the future, especially since cities and societies differ among themselves from Where the active activities in it and the dimensions affecting it socially, economically, administratively and even politically.

References

1. Uddin, M.: Covid19 Impact & Post Pandemic Public Spaces: A case of Indian Cities | Follow Your Rhythm. Placy (2020). https://www.placy.city/en/post-quarantine-urbanism/india-public-places/. Accessed 11 Oct 2021
2. Low, S., Smart, A.: Thoughts about public space during covid-19 pandemic. City Soc. **32**(1) (2020). https://doi.org/10.1111/CISO.12260
3. Zhou, G., Li, C., Zhang, J.: Identification of urban functions enhancement and weakening based on urban land use conversion: a case study of Changchun, China. PLoS One, **15**(6), e0234522 (2020). https://doi.org/10.1371/JOURNAL.PONE.0234522
4. Garrido, C., Tuduri, M., Gonzàlez, A., Giorgi, E.: Cities after Covid-19: How will public spaces and social life change? (2020). https://www.citiestobe.com/covid-19-how-are-public-space-and-social-life-going-to-change/. Accessed 12 Oct 2021
5. Kahachi, H.A.H.: Spatial planning contribution to addressing large scale multi-national urban problems and challenges in practice. J. Green Eng. **10**(10), 7693–7708 (2020)
6. Kahachi, H.A.H.: The impact of the 2008 world financial crisis on homelessness in developed countries : the case study of New York City. Hum. Soc. Sci. **44**(3), 269–280 (2017). https://doi.org/10.12816/0040582
7. Richards, B.: Poverty and housing in Chile: the development of a neo-liberal welfare state. Habitat Int. **19**(4), 515–527 (1995). https://doi.org/10.1016/0197-3975(95)00043-F
8. Kiss, G., Jansen, H., Castaldo, V.L., Orsi, L.: The 2050 city. Procedia Eng. **118**, 326–355 (2015). https://doi.org/10.1016/j.proeng.2015.08.434
9. Kahachi, H.A.H.: The practice of sustainable spatial planning between an essentialist and relational conceptions. J. Green Eng. **10**(9) (2020)
10. United Nations. COVID-19 | United Nations (2021). https://www.un.org/en/desa/covid-19. Accessed 12 Oct 2021
11. United Nations. Everyone Included: Social Impact of COVID-19 | DISD (2020). https://www.un.org/development/desa/dspd/everyone-included-covid-19.html. Accessed 12 Oct 2021

12. Honey-Rosés, J., et al.: The impact of COVID-19 on public space: an early review of the emerging questions – design, perceptions and inequities. Cities Health, 1–17 (2020)
13. Wray, A., Fleming, J., Gilliland, J.: The public realm during public health emergencies: exploring local level responses to the COVID-19 pandemic. Cities Health, pp. 1-4 (2020)
14. Barron, C., Emmet, M.-J.: Back gardens and friends: the impact of COVID-19 on children and adolescents use of, and access to, outdoor spaces. Ir. Geogr. **53**(2), 173–177 (2020)
15. Cheshmehzangi, A.: Social and public life during disruptive times: a public realm perspective. In: Urban Health, Sustainability, and Peace in the Day the World Stopped, pp. 123–128 (2021). Springer. https://doi.org/10.1007/978-981-16-4888-5_14
16. Burzynska, K., Contreras, G.: Gendered effects of school closures during the COVID-19 pandemic. Lancet **395**(10242), 1968 (2020)
17. Lusk, K., Einstein, K., Glick, D., Palmer, M., Park, S., Fox, S.: Urban parks and the public realm: equity & access in post-COVID cities (2021)
18. Stevens, N.J., Tavares, S.G., Salmon, P.M.: The adaptive capacity of public space under COVID-19: exploring urban design interventions through a sociotechnical systems approach. Hum. Fact. Ergon. Manufact. **31**(4), 333 (2021)
19. Gang, J., Hawthorne, C., Sennett, R.: Greg Lindsay. The Big Rethink: Cities After COVID-19 - Public Space. Worldwide: NewCities (2021)
20. Wortzel, J.D., et al.: Association between urban greenspace and mental wellbeing during the COVID-19 pandemic in a U.S. Cohort. Front. Sustain. Cities, **3** (2021)
21. Hollander, J.B.: Biometrics and the Public Realm: Urban Sustainability During the COVID Pandemic (2021)
22. Ijjasz, E., Mukim, M.: How competitive cities are coming through COVID-19 | World Economic Forum (2021). https://www.weforum.org/agenda/2021/03/covid19-cornavirus-cities-economic-recovery. Accessed 12 Oct 2021
23. United Nations. No winners but fewer losers in global economy from COVID than expected | | UN News (2021). https://news.un.org/en/story/2021/03/1087712. 12 Oct 2021
24. BBC. Coronavirus: How the pandemic has changed the world economy - BBC News (2021). https://www.bbc.com/news/business-51706225. Accessed 12 Oct 2021
25. Nigel Pain. The impact of the COVID-19 pandemic on sectoral output – ECOSCOPE. OECD Economics (12 December 2020). https://oecdecoscope.blog/2020/12/12/the-impact-of-the-covid-19-pandemic-on-sectoral-output/. 20 Dec 2021
26. OECD. OECD Economic Outlook, 2020(2). OECD (2020)
27. Gehl People. Public Space & Public Life during COVID-19 (2020). https://gehlpeople.com/announcement/public-space-public-life-during-covid-19/
28. Mateer, T.J., Rice, W.L., Taff, B.D., Lawhon, B., Reigner, N., Newman, P. Psychosocial factors influencing outdoor recreation during the COVID-19 pandemic. Front. Sustain. Cities **3** (2021). https://doi.org/10.3389/FRSC.2021.621029/FULL
29. Ali, H.M., Dom, M.M., Sahrum, M.S.: Self-Sufficient community through the concepts of collective living and universal housing. Procedia. Soc. Behav. Sci. **68**, 615–627 (2012). https://doi.org/10.1016/J.SBSPRO.2012.12.253
30. Sivam, A., Karuppannan, S., Mobbs, M.: How open are open spaces: evaluating transformation of open space at residential level in Adelaide - a case study. Local Environ. **17**(8), 815–836 (2012). https://doi.org/10.1080/13549839.2012.688734
31. Bashir, M.F., MA, B., Shahzad, L.: A brief review of socio-economic and environmental impact of Covid-19. Air Qual. Atmos. Health **13**(12), 1403–1409 (2020)
32. Shen, M., et al.: Assessing the effects of metropolitan-wide quarantine on the spread of COVID-19 in public space and households. Int. J. Infect. Dis. **96**, 503–505 (2020)
33. Song, X., Cao, M., Zhai, K., Gao, X., Wu, M., Yang, T.: The effects of spatial planning, well-being, and behavioural changes during and after the COVID-19 pandemic. Front. Sustain. Cities **3**, 47 (2021). https://doi.org/10.3389/frsc.2021.686706

34. Lakey, J.: Tossed masks, gloves reveal the disgusting side of human nature under COVID-19 (2020)
35. World Health Organization. COVID-19 and Urban Health (2019). https://www.who.int/teams/social-determinants-of-health/urban-health/covid-19
36. Jia, J.S., Lu, X., Yuan, Y., Xu, G., Jia, J., Christakis, N.A.: Population flow drives spatio-temporal distribution of COVID-19 in China. Nature **582**(7812), 389–394 (2020)
37. Johns Hopkins Medicine. Coronavirus, Social and Physical Distancing and Self-Quarantine. Johns Hopkins Medicine (2020)
38. Wemakeplaces. The Grief & Joy of Lockdown (2020). https://wemakeplaces.org/the-grief-joy-of-lockdown/

Local Authorities and Pandemic Responses in Perspective. Reflections from the Case of Milan

Carolina Pacchi[✉]

DAStU Politecnico di Milano, Milan, Italy
carolina.pacchi@polimi.it

Abstract. While in general the effects of the Covid-19 pandemic have been analysed in academic literature from many points of view, such as the medical, occupational, social, economic and psychological ones, much less attention has been paid
to the systems of norms which, in an emergency situation, have been elaborated
and applied, and to their relationship with space.

In order to explore this gap, the paper will propose to open a discussion on
the role and effects of systems of pandemic regulation at local level, proposing a
research framework and some very first applications to a specific case, the City of
Milan in Northern Italy, an area badly hit by the pandemic.

The paper thus moves from three research questions, related to questions
of framing, governance, and timescale of the effects, as the main interpretive
perspectives to understand and conceptualise the relationship between norms,
actors, and decision-making processes, with a specific focus on the spatial realm.
While the case in point chosen is the City of Milan, the research questions open to
a wider debate, and to possible cross-context comparisons with other urban areas.

Keywords: Covid-19 restrictions · Pandemic regulation · Local authorities
preparedness

1 Pandemic Restrictions and Local Scale

The pandemic has caused effects that can be read at different scales, but with a significant
concentration at the local scale, the scale of people's daily lives. While the pandemic
diffusion is indeed global and has reminded all of us of the inherent interconnectedness of
contemporary life, based on flows and mobility, the actual effects of the main restriction
strategies chosen by most governments across the world have been very much linked to
the actual spaces inhabited, used and negotiated in daily life.

While, in general, action to curb, or at least to slow down the pace of the pandemic has
been largely based on medical and pharmaceutical intervention, such as hospitalization,
hospitalization in ICUs for the most serious cases, and pharmaceutical treatment at
home (Giordano et al. 2020), we can highlight, in parallel, how non-pharmaceutical
intervention has been consistently gaining ground, even if in different ways and with
different pace in different countries regions of the world (Bo et al. 2020; Bendavid

F. Calabrò et al. (Eds.): NMP 2022, LNNS 482, pp. 1059–1064, 2022.
https://doi.org/10.1007/978-3-031-06825-6_101

et al. 2021). In this field, we can list and discuss restrictions to individual behaviour and restrictions to activities, both of which we will examine more in depth in the next sections.

The effects of the many restrictions imposed have been analysed from multiple points of view, at the personal, occupational, social, economic, psychological levels. On the contrary, the systems of norms which, in an emergency situation, have been elaborated and applied has been much less investigated (Di Mascio et al. 2020). There are indeed useful repositories of different types of norms (e.g. the *Covid-19 Government Response Tracker*, Blavatnik School of Government, University of Oxford[1]), but in general literature on the specific role of norms, and their relationships with the use of space (Pasqui 2022), is quite scant.

This paper thus aims to open a discussion on the role and effects of systems of pandemic regulation at local level, proposing a research framework and some very first applications to a specific case. Even norms which regulate behaviors visible at the local scale, on the other hand, are located at different scales, and are therefore linked to different institutional levels, to their interactions, to the models of government (and governance) experienced in a situation of strong emergency.

In order to investigate such regulatory aspects, the paper proposes a path towards a critical reconstruction of the debate, the decisions taken, and the system of rules applied by the Municipality of Milan, an area heavily affected by the pandemic, especially in the first phase during the spring of 2020. In order to read such responses in a dynamic and evolutionary way, the analysis will concern the period between February 2020 and September 2021. While the former date represents the very start of the pandemic diffusion in Italy, the latter one enables to reflect on the emerging elements with some time distancing.

With respect to this background, the paper moves from three research questions, related to as many interpretive perspectives on the relationship between norms, actors, and decision-making processes, with a specific focus on the spatial realm. While the case in point chosen is the City of Milan, in Northern Italy, the research questions open to a wider debate, and to possible cross-context comparisons with other urban areas.

2   Framing, Governance, Effects: Building a Conceptual Map

2.1   Local Government Responses to Covid-19: Research Questions

The first question is related to *framing* (Schön, Rein 1994) and it investigates how framing has been contributing to the definition of the problem to tackle, and therefore to shaping the regulatory and policy responses. The paper aims at better understanding how the Municipality of Milan, in its political and technical/bureaucratic components, did understand and conceptualise the pandemic; in order to do so, it is important to explore which interpretative, normative and operational frameworks were the basis of City of Milan's intervention in the different phases, and what was the role played by emergency in it, and by the ways to address it. Moreover, we might further investigate if emergency itself has been used more or less instrumentally.

---

[1] https://www.bsg.ox.ac.uk/research/research-projects/covid-19-government-response-tracker.

Along these same lines, it will be important to investigate the way in which radical uncertainty has been conceptualised in the discourses and interventions brought about by this Local Administration, both in general (which means, in relation to different sets of unexpected or disruptive occurrences), and in the specific case of the Covid-19 pandemic. What conceptual and normative frameworks have they drawn on to address it? Was the specific type of pandemic emergency present in policy documents, discourses, administrative culture? Was there some form of preparedness towards these emergencies, and how was it defined in organizational terms?

In relation to the conceptualization of uncertainty, it is useful to further examine the issue of emergency, the conceptualization of the pandemic as a state of emergency, and thus the justification/legitimation of norms proposed and designed as responses to unprecedented and exceptional situations.

The second question concerns the *models of government and governance* that have been tested in practice to manage the pandemic situation (Capano 2020). In what way was the production of ad hoc norms the combined effect of vertical (EU-Government-Region Lombardia-Municipality of Milan) and horizontal (with other local agencies, and other public and private stakeholders) forms of comparison and interaction? How much, therefore, are the ordinances and other regulations produced the result of exogenous processes of interaction, decision-making and evaluation and how much endogenous? What forms and modes of regulation were chosen and why?

The third question has to do with the *short- and medium-term effects of the norms* applied. We want to discuss whether, and in what way, the Milanese municipal adminis-tration has evaluated the specific effects of the regulations produced both on a national and local scale, in the short and in the medium term. The type of regulation issued, the different texts, ordinances, strategic documents, etc., have had very different effects on both citizen behaviour and spatial transformation, and the ongoing research aims at critically assessing them, in their dynamic effects (Fig. 1).

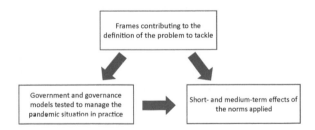

Fig. 1. Conceptiual map of research questions

2.2   Main Research Steps

To discuss these aspects, the research, still underway, is based on the analysis of a mix of secondary and primary sources.

Operationally, it is organized in three steps. The first one is a *punctual collection of data on the progress of the pandemic* in Italy and in Milan between February 2020 and

September 202, concerning the development and spread of the pandemic, as well as it spatial distribution, mortality, hospitalization and ICUs, etc.

The second one is the *collection and critical interpretation of the norms produced*. They include both the Italian national decrees and the local ordinances, sectoral policies and strategic documents issued by the Municipality of Milan, looking at how many and which ones have been issued; how often they have been issued; whether local ordinances simply implemented the national decrees or if they integrate, specify, and make the national norms more restrictive or locally targeted.

Finally, the third step will include *in-depth interviews* with politicians and technical bureaucrats of the Milan municipal administration, as well as, depending on the evidence that emerges, with other local players (experts in the health domain, representatives of other agencies, decentralized state administrations, private interests, etc.).

The first research steps concern the collection and analysis of the documents produced by the Municipality of Milan in response to the pandemic, since February 2020. As we have seen, the Municipality has responded to the pandemic emergency and its aftermath through different type of documents, such as Mayoral ordinances, strategic documents, and sectoral policies.

Mayoral ordinances, which are immediately applicable, concern a wide range of issues, from constraints on individual behaviour to constraints to activities. In the first phases, such ordinances concern also procedural streamlining and simplification, as well as the suspension of sets 'ordinary' rules and practices, such as vehicle traffic limitations. Finally, some of them contain first support measures to individuals and economic activities. While in 2020 the Mayor issued 34 ordinances, in 2021 he issued 8. As far as the local regulatory texts are concerned, the ordinances issued by the Municipality include measures introducing restrictions to personal behaviour and activities; bureaucratic simplification and reduction of red tape; temporary suspension of ordinary measures, rules and uses; support and welfare measures, aimed both at citizens in fragile situations and at economic activities in distress.

A second type of document issued has less coercive power compared to ordinances, because it mainly holds a strategic value, which means that it acts as a guideline for future policy design or rule-making. A Strategic Document on pandemic adaptation has first been issued by the Municipality of Milan in April 2020 (Comune di Milano 2020) and it has been opened up to extended public debate through a public participation process aimed at collecting different points of view and possible suggestions from citizens and their organisations.

Finally, the Municipal offices have issued a number of sectoral documents, with more or less cogent power, concerning different domains of their action, such as school, public space, mobility, etc. Among them we can mention the *Patto Milano per la Scuola*, jointly promoted by the Prefect of Milan (the local representative of the National Government) and the Mayor; (Prefettura di Milano 2020; 2021) the Guidelines for the Occupation of public land; different policy measures concerning mobility, and a number of policies in the area of social assistance.

These strategic or sectoral documents propose a wide range of more or less coercive regulatory tools, such as policy recommendations and different type of procedural reorganization, or short- and medium-term incentives. Moreover, looking towards the pandemic aftermath, they propose recovery and adaptation strategies, through the development of possible future scenarios, as well as piloting policy innovations. These last ones, in particular, suggest that the Municipality may have used the emergency situation to propose and test (more or less) disruptive policy measures, such as the ones linked to forms of slow mobility or the so called 'city of proximity'. These measures, related to the pandemic adaptation, but wider in scope, may have been implemented in any case, while the emergency situation can be seen as an occasion to modify the political opportunity structure (Tilly, Tarrow 2007), by exploiting the state of emergency and speeding up decisions that would have otherwise required extensive confrontation with the municipal structure, and outside.

3   Local Rules and Urban Effects. Emerging Research Directions

3.1   First Reflections on New Regulation Introduced During the Pandemic

In the wake of the framework outlined, it is possible to propose some first considerations, even if our research is still underway.

One first issue concerns the conceptualisation of emergency by the City of Milan: here, following considerations on the national response to the pandemic, it is possible to highlight that emergency situations tend to unveil the ordinary mismatches and disfunctions of institutional action (Capano 2020). More than the specific response or preparedness to the exceptional situation, therefore, the type of institutional response and reveals the ordinary capacity and robustness of institutions, local ones in this case.

Secondly, the analysis of local regulatory responses critically highlights the complexity of the relationship with national and regional systems of norms and their evolution, with the risk of loss of clarity, ambiguity and possible duplications. Moreover, it is possible to highlight how the measures included in local ordinances (restrictions, simplification, suspensions, support) have a totally different degree of cogency and imply different levels and intensity of control, even if they tend to be mixed in the same texts and therefore may generate confusion and ambiguity.

Lastly, as we have seen, emergency regulation can be seen as an opportunity to bring forth policy innovations which are already in the political agenda, but maybe difficult to promote in 'ordinary' times. In this sense, the changing structures of political opportunities pushed by the emergency situation enables the piloting and testing of forms of policy innovations. Some interventions on public space and on the road network, aimed at fostering and facilitating slow mobility (walking and cycling) have been publicly legitimized and justified as linked to the pandemic emergency, even if they would have probably been proposed regardless, maybe with a different timing.

3.2   Further Research Directions

Finally, we can highlight that many of these aspects need further exploration, since they are interesting suggestions, that have to be further probed through deeper conceptual and empirical investigation. Such investigation may lead to unpack the dimension of institutional learning. In order to understand whether the Municipality of Milan has learned from the pandemic experience, specifically in the area of emergency and recovery regulation, it will be necessary to examine through what processes this learning has taken place (interaction with external actors, internal review and evaluation processes, etc.), how has this learning been institutionalised and, finally, how it may become a guide for the production of regulations in other emergency situations.

Looking at different sectoral domains, it will be important to unpack the temporal dimensions, and distinguish short terms effects related to pandemic mitigation from longer terms institutional learning effects. This distinction will be relevant to interpret more precisely the developments in domains such as the reorganization of school spaces, and of their interface with their surroundings (Prefettura di Milano 2021), or the city of proximity and the new relevance of the neighbourhood level (Comune di Milano 2020).

References

Bendavid, E., Oh, C., Bhattacharya, J., Ioannidis, J.P.: Assessing mandatory stay-at-home and business closure effects on the spread of COVID-19. Eur. J. Clin. Invest. **51**(4), 1–9 (2021)

Blavatnik School of Government, University of Oxford Covid-19 Government Response Tracker. https://www.bsg.ox.ac.uk/research/research-projects/covid-19-government-response-tracker. Accessed January 2022

Bo, Y., et al.: Effectiveness of non-pharmaceutical interventions on COVID-19 transmission in 190 countries from 23 January to 13 April 2020. Int. J. Infect. Dis. **102,** 247–253 (2021)

Capano, G.: Policy design and state capacity in the COVID-19 emergency in Italy: if you are not prepared for the (un)expected, you can be only what you already are. Policy Soc. **39**(3), 326–344 (2020)

Comune di Milano: Milano 2020 Strategia di adattamento. Documento aperto al contributo della città (2020)

Di Mascio, F., Natalini, A., Cacciatore, F.: Public administration and creeping crises: insights from COVID-19 pandemic in Italy. Am. Rev. Pub. Adm. **50**(6/7), 621–627 (2020)

Giordano G., et al.: Modelling the COVID-19 epidemic and implementation of population-wide interventions in Italy. Nat. Med. Lett. **26**(6), 855-860 (2020)https://doi.org/10.1038/s41591-020-0883-7

Pasqui G.: Coping with the Pandemic in Fragile Cities, Spriger Briefs. Springer Nature, Cham (2022)

Prefettura di Milano: Piano Operativo per il coordinamento degli orari delle attività e del servizio di trasporto pubblico locale, December 2020

Prefettura di Milano: Patto Milano per la Scuola, September 2021

Schön, D., Rein, M. (eds.): Frame Reflection. Basic Books, New York (1994)

Tilly, C., Tarrow, S.: Contentious Politics. Paradigm Publishers, Boulder (2007)

Credentials as Regulatory Tools in the COVID Era

Giuseppe Lorini and Olimpia G. Loddo(✉)

University of Cagliari, Cagliari, Italy
lorini@unica.it, olimpia.loddo@gmail.com

Abstract. Credentials are the documents we are required to have in our possession to achieve certain ends—for instance, driving licenses, identity cards, passports, and credit cards. Credentials play an important role in COVID regulations. This paper aims to reconstruct and analyse their role during the pandemic. In particular, the paper will stress how credentials may act as an innovative non-prescriptive tool of social control.

Keywords: Credentials · COVID-19 · Document act · Constitutive rules · Pandemic

1 What Are Credentials?

1.1 A Definition of Credentials

The specific types of documents known as credentials can be defined as follows: 'Credentials are documents we are required to have in our possession to reach certain ends' [1]. Credentials are the documents that we must have in our possession when we perform certain actions—they are, for example, identity cards, driving licences, passports, or credit cards.

Examples of credentials include, in this sense, the boarding pass we need to board a plane, the badge we need to access the building in which we work, or the supermarket club card we use for shopping. Credentials are also required at airports and international borders to cross certain areas within state borders (Fig. 1).

F. Calabrò et al. (Eds.): NMP 2022, LNNS 482, pp. 1065–1071, 2022.
https://doi.org/10.1007/978-3-031-06825-6_102

Fig. 1. A passport is an example of credentials. Photo by Henry Thong on Unsplash.com

1.2   The Characteristics of Credentials

Any credential is a 'document or certificate proving a person's identity or qualifications'. Connection to a bearer, portability, and inspectability are the three fundamental characteristics of a credential. 'Connection' means that each credential is related to its specific bearer, 'portability' indicates that the bearer should be able to carry the credential easily on his or her person (thus, that it should be be small and light), while inspectability requires that the credential can be made publicly visible (to a human being, barcode scanner, or chip reader). Generally, credentials should also be immediately recognisable and identifiable as being a credential of the relevant type [1].

Credentials are status indicators—in this sense, credentials have an 'epistemic function'. In Searle's lexicon, credentials are '*status indicators*'. In this sense, they are 'official representations' required by institutional facts [2]. It is important to point out that since 'institutional realities exist in virtue of social recognition, then the tools/instruments that make it possible for us to recognize these realities also play a role in maintaining them in existence' [1].

A credential can be forged. A forged credential is either a new or a modified object designed to look like an authentic credential, in order to illegally carry out the same function. Authentic credentials can also be used in illegal ways, for example, when a twin uses the driving licence of his brother.

Finally, credentials are institutional objects: they presuppose institutions and are a product of social reality. Constitutive rules shape their characteristics in different ways.

2   The Constitutive Rules of COVID Credentials

The traditional idea of rules is that of prescriptive rules—that is, rules that impose prohibitions and obligations. For example, according to Norberto Bobbio [3], norms are prescriptive propositions; their function is to make people act in a certain way.

During the pandemic, lawmakers introduced numerous prescriptive rules, such as the ban on shaking hands and the obligation to maintain social distancing and wear a mask. However, prescriptive rules do not exhaust the typology of rules.

Meanwhile, prescriptive rules are not the only tool of regulation in the context of the pandemic; indeed, constitutive rules play a fundamental role.[1] Constitutive rules do not impose duties or obligations; rather, they shape the conditions and characteristics of social reality.

2.1   A Special Kind of Constitutive Rule: The Anankastic-Constitutive Rule

A particular type of constitutive rule, the anankastic-constitutive rule, has become particularly relevant for management of the pandemic crisis. The neologism 'anankastic-constitutive rule' originally appears in the work of Amedeo G. Conte [5]. The research of Amedeo G. Conte and Giampaolo Azzoni [6] on the concept of anankastic-constitutive rules shows two different characterisations of the concept: (i) ontological and (ii) semantic.

According to the ontological characterisation, anankastic-constitutive rules are 'rules which set a necessary condition, *a condicio sine qua non*', on the objects to which they relate.

According to the semantic characterisation, anankastic-constitutive rules are rules that determine the extension of the term to which they refer [7].

Examples of anankastic-constitutive rules:

(i)   The holographic will must be signed by the hand of the testator.
(ii)   The donation must be made by public deed.
(iii) The referees must be Italian.
(iv)   Those who received three doses are eligible for a Green Pass.
(v)   The Green Pass is valid for no more than six months from the last vaccination day.

2.2   Rules Can Modify the Extension of What People Can Do Without Being Prescriptive

Some anankastic-constitutive rules indirectly impose limits on activities that require credentials. Anankastic-constitutive rules establish credential-related conditions concerning the performance of certain acts, such as crossing a border (the traveller must carry a passport), travelling by car (the driver must carry a driving licence), and so on. As we have seen, possessing a credential gives us the deontic power to exercise one or more

---

[1] The concept of constitutives rule has been the subject of lively debate in social philosophy and in the philosophy of law. Some authors deny the existence of these rules and reduce everything to prescriptions; on this subject, see the very recent essay by Roversi [4].

associated rights. However, at the same time, the associated anankastic-constitutive rules impose limits on the exercising of those rights.

During the pandemic, an interesting aspect emerged in the relationship between anankastic-constitutive rules and credentials: anankastic-constitutive rules can affect the regulated behavior, without prescribing anything.

Anankastic-constitutive rules governing the Green Pass concern the document itself, alter its conditions of validity, and indirectly affect what can and cannot be done. However, they are not true prescriptions.

Their object is not human behaviour but a document that acquires new functions or loses validity under certain conditions set by the rules.

3   Two Kinds of Credentials that Played a Role During the Pandemic

In this paragraph, we distinguish between two different kinds of credentials that played important roles in the different phases of the pandemic: self-produced credentials and digital credentials.

3.1   Self-produced Credentials

In recent history, a particular example of such documents played a role in managing the COVID pandemic. It is well known that in early 2020, numerous governments instituted lockdown regimes to prevent the spread of infection. At that time, citizens had the right to move from their homes only in special cases (for example, health reasons or essential tasks linked to the workplace).

In some countries (such as France, Italy and Greece), to exercise their very limited freedom of movement, citizens had to carry a form of self-certification verifying the existence of the circumstances that legitimised their movement, based on the provisions of the legislation in force.

This document, the self-certification, did not determine an increase in rights; rather, it allowed its free exercise in case of police control [8]. Self-certification can be updated by the bearer (Fig. 2).

AUTODICHIARAZIONE AI SENSI DEGLI ARTT. 46 E 47 D.P.R. N. 445/2000

Il/La sottoscritto/a _____ , nato/a il ____ . ____ . _____

a _____ (____), residente in _____

(____), via _____ e domiciliato/a in _____

(____), via _____ , identificato/a a mezzo _____

nr. _____ , rilasciato da _____

in data ____ . ____ . _____ , utenza telefonica _____ , consapevole delle conseguenze penali

previste in caso di dichiarazioni mendaci a pubblico ufficiale **(art. 495 c.p.)**

DICHIARA SOTTO LA PROPRIA RESPONSABILITÀ

➢   **di essere a conoscenza delle misure normative di contenimento del contagio da COVID-19 vigenti alla data odierna, concernenti le limitazioni alla possibilità di spostamento delle persone fisiche all'interno del territorio nazionale;**

➢   **di essere a conoscenza delle altre misure e limitazioni previste da ordinanze o altri provvedimenti amministrativi adottati dal Presidente della Regione o dal Sindaco ai sensi delle vigenti normative;**

➢   **di essere a conoscenza delle sanzioni previste dall'art. 4 del decreto-legge 25 marzo 2020, n. 19, e dall'art. 2 del decreto-legge 16 maggio 2020, n. 33;**

➢   **che lo spostamento è determinato da:**

◯ **- comprovate esigenze lavorative;**

◯ **- motivi di salute;**

◯ **- altri motivi ammessi dalle vigenti normative ovvero dai predetti decreti, ordinanze e altri provvedimenti che definiscono le misure di prevenzione della diffusione del contagio;**
   (specificare il motivo che determina lo spostamento):

_____ ;

➢   **che lo spostamento è iniziato da** *(indicare l'indirizzo da cui è iniziato)*

_____ ;

➢   **con destinazione** *(indicare l'indirizzo di destinazione)*

_____ ;

➢   **in merito allo spostamento, dichiara inoltre che:**

_____ .

Data, ora e luogo del controllo
Firma del dichiarante   L'Operatore di Polizia

Fig. 2. Self-certification in use during the COVID-19 pandemic in Italy

3.2　Digital Credentials

Credentials are currently made of paper or plastic. However, in a process still in progress, many paper and plastic credentials are replaceable by electronic documents that we can store inside our phones. This does not mean the extinction of credentials— the credentials themselves survive, but now in electronic form. Furthermore, we will still need to carry them around with us, even after our wallets have been replaced by a wallet app [4]. Therefore, the term 'credential' refers to 'paper and plastic documents and their digital substitutes' [1] (Fig. 3).

Fig. 3. A digital credential in use during the pandemic. Photo by R. Grachev on unsplash.com

The Green Pass is an example of a digital credential. A credential's digital nature depends not on its display but on how it is created and updated. Italy's 'Green Pass' certificate comes in digital or paper versions showing that people have received their anti-COVID vaccination, tested negative, or recovered from COVID-19. Even though this document can be displayed on paper, it is created and updated through an entirely digital procedure. Often, updates of the Green Pass are even independent of the activity of the bearer. For instance, the lawmaker can change the anankastic-constitutive rule that determines the validity of the document.

4　Conclusion: The Function of Credentials in the COVID Era

Many rules governing credentials do not prescribe any behaviour but set conditions for exercising a preexisting right.

For instance, the right to travel is established by both national and international legislation. However, nowadays, only possessing a passport or other similar credentials allows the possibility of exercising these rights concretely [9].

Interesting uses of credentials as instruments of non-prescriptive regulation [10] emerged during the COVID-19 pandemic. In this sense, digital credentials such as the Green Pass introduce an important innovation in regulating social behaviour, independent of the direct use of coercion and the direct imposition of obligations [11].

References

1. Smith, B., Loddo, O.G., Lorini, G.: On credentials. J. Soc. Ontol. **6**(1), 47–67 (2020). https://doi.org/10.1515/jso-2019-0034
2. Searle, J.R.: The Construction of Social Reality. Allen Lane, London (1995)
3. Bobbio, N.: Teoria della norma giuridica. Giappichelli, Torino (1958)
4. Roversi, C.: In defence of constitutive rules. Synthese **199**(5–6), 14349–14370 (2021). https://doi.org/10.1007/s11229-021-03424-w
5. Conte, A.G.: Materiali per una tipologia delle regole. Materiali per una storia della cultura giuridica **15**, 345–368 (1985)
6. Azzoni, G.: Il concetto di condizione nella tipologia delle regole. CEDAM, Padova (1988)
7. Lorini, G.: Anankastico in deontica. LED, Milano (2017)
8. Smith, B., Lorini, G., Loddo, O.G.: Le credenziali: parole, disegni e poteri deontici. Teoria e critica della regolazione sociale 59–74 (2020)
9. Muchmore, A.I.: Passports and Nationality in International Law; SSRN Scholarly Paper ID 1649122; Social Science Research Network: Rochester, NY (2004)
10. Żełaniec, W.: Create to Rule: Studies on Constitutive Rules. LED, Milano (2013)
11. Lorini, G., Morioni, S.: Ruling without Rules. Regulation beyond Normativity. Glob. Jurist 1–11 (2020). https://doi.org/10.1515/gj-2019-0051

Healthy City for Organizing Effective and Multifaceted Actions at the Urban Level

Roberto De Lotto(✉) ⓘ, Caterina Pietra ⓘ, Elisabetta Maria Venco ⓘ,
and Nastaran Esmaeilpour Zanjani ⓘ

DICAr– University of Pavia, Via Ferrata 3, 27100 Pavia, Italy
uplab@unipv.com

Abstract. The presented paper consists of two main aims: the first one is to introduce the concept of Healthy City (HC), while the second one proposes the organization of a public action striving for the healthiness of cities. For this purpose, authors exploit research based on the experience approach concerning a specific topic related to HC (urban mobility) and the administrative procedure from developing a policy to its implementation phase. HC represents a highly complex system, and the practical actions emerge as a synthesis of different layers composing the public administration (at least considering the local scale). Generally speaking, managing the entire healthy process is fundamental to avoid losing the multifaceted attributes of objects, subjects, roles, and objectives. The grounded theory approach highlights the elements that compose the complex system substantially and clearly; moreover, it permits to keep control of the process while the singular steps are defined. Initially, the authors describe the HC idea and its main elements; then, they introduce a precise topic and organize a public action using the experiential approach.

Keywords: Healthy City · Grounded theory method · Public action

1 Introduction

The actual context that developed from the spreading of Covid-19 has certainly generated numerous impacts on more fields and at different scales. Notably, the city's complex dimension has been put in evidence, showing a real difficulty in maintaining control over each layer and confirming that urban health outcomes cannot be predicted with complete accuracy due to constant changes and individuals' actions [1–3].

Thus, the complex environment is requesting accurate politics to face the urban phenomenon and its consequences. A big problem lies in the deterministic character characterizing current procedures, which works efficiently when the interdependence level is low, and the local stability is high. In this regard, the Italian context appears to be strictly connected to the relational dynamics deriving from political-bureaucratic positions. The participatory processes suffer the consequences of these decisions, even though various experiences have shown that these processes produce better and, above all, shared solutions. Indeed, it is demonstrated that bottom-up interventions can significantly

F. Calabrò et al. (Eds.): NMP 2022, LNNS 482, pp. 1072–1081, 2022.
https://doi.org/10.1007/978-3-031-06825-6_103

increase the possibility of achieving what is planned, and this is made possible both for small-scale and large and complex issues such as the environmental one.

Bertuglia and Staricco effectively summarize the fundamental characteristics that determine what can be defined as the phenomenology of cities as complex systems: fragmentation of effective authority [4, 5]; structure organized on several levels [5]; inter-temporal multiplicity [5]; consequences devoid of valid reasons and surprising [4]. Knowing that planning activity is not a deterministic process [6], the definition and the management of policies that involve different administration layers need to be organized in e theoretical way but also to be tested practically.

Although the Healthy City system proves to be complex, the elements that compose it can be easily schematized and made formally intelligible. For example, an ontology, going beyond its philosophical meaning, represents an efficient tool that can effectively contribute to this operation by allowing for more accessible communication and integration of information among different sectors and actors. An opposite approach, entirely practical, is the experiential one. Regarding public administration, the more the goal is broad and multifaceted, the more diverse subjects with different roles and technical skills (that are not always complimentary) have to interact to translate policies into practical actions. The ontological approach can define the role of each element which is part of the comprehensive system abstractly, without however losing the general design (it constitutes a sort of deductive method); the grounded theory method, that is based on the awareness of the method's ontological, epistemological, and methodological perspectives [7–10] is here focused on a specific HC action. Moreover, it is used to mark the interferences among the subjects involved (it establishes a sort of indictive method). Also, the grounded theory method, which is based on the awareness of the method's ontological, epistemological, and methodological perspectives [7–10], is here focused on a specific HC action. The purpose of using "grounded theory" is to move beyond the description and production or finding of a theory, an abstract analytical scheme, and a process or action or interaction.

In the paper, the authors present the Healthy City concept, a related specific topic (urban mobility), and the organization of the steps from policy to actions concerning in particular the Municipality of Segrate (Milan province, Italy).

2 Healthy City: A Short Description

The principles defined by the Healthy City concept promote models and guidelines towards this direction by formally recognizing the local action's fundamental role to develop health issues, the great value of urban environments in terms of health and well-being, and the responsibility of local governments to foster beneficial environments for healthy living for all [11].

The modern meaning of "Healthy City" (HC) is attributed to Leonard Duhl and Trevor Hancock, who were both involved from the very beginning in the healthy cities project that emerged in 1984 during a conference entitled "Beyond Health Care", held in Toronto. They described an HC as "one that is continually creating and improving those physical and social environments and expanding those community resources which enable people to mutually support each other in performing all the functions of life and in developing to their maximum potential" [12]. Furthermore, a healthy city does not correspond to one that has obtained a particular state of health; rather, it constitutes a specific system aware of the health factor that strives to improve and promote it over time.

Again, Duhl and Hancock clarified the main feature of HCs and defined them as: "the example par excellence of complex systems: emergent, far from equilibrium, requiring enormous energies to maintain themselves, displaying patterns of inequality and saturated flow systems that use capacity in what appear to be barely sustainable but paradoxically resilient networks" [12].

Indeed, a complex systems framework to be applied to HCs will comprehend the physical, social, economic, and political environments. However, the set of relations that follow the principle of non-linearity and causation is multidirectional, thus even simple causal relations occurring between dependent and independent factors are difficult to isolate. Causes at the basis of a decision determine outcomes in the same way. Dynamicity extends the time sequence that exists from cause to effect; thus, looking for causal relations becomes a more difficult task using conventional analytic methods. Another complex layer is represented by intraurban diversity: cities are a mix of communities, including those that own a geographical localization and those that do not have one. Many stakeholders act on communities, and their different interests heavily affect diversity [14]. Moreover, the complexity of HCs is not a static feature and varies with scale. Interconnectedness among different levels is often not considered by existing models that end up studying systems in a too general way.

The corona times illustrate the new patterns of people's mobility in different countries, for example, changes in traffic peak time or the rising popularity of slow-motion forms of traffic like walking and biking; and an overall shift in route and mode of transportation, for example, the decrease of using public transportation. The spatial proximity among people is seen as a potential health threat, and it brings many consequences for spatial mobility and spatial interaction choices. This health threat shows itself in two ways:

1. A relative shift to individualized modes of transport, especially cars, and bicycles.
2. An absolute decline in general mobility for example distance working which is all thanks to technology [16].

3 Healthy City Topics and Urban Mobility

In city planning and management, the most relevant topics dealing with HCs concern: Air quality; Lifestyle; Urban green spaces.

Considering that the role of the public administration is, first of all, to guarantee public health and intervene in emergencies, the HC concept and its development are perfectly aligned with the duties that public bureaus should chase.

At the local scale, such as at the Municipality scale, many policies and actions can significantly improve all the main three cited topics. Moreover, there is a straight connection between the improvement of green spaces and the enhancement of air quality, between air quality and perspective to be active, between specific actions of a promising lifestyle, and the need for spaces where to improve them.

About air quality, an HC promotes the use of active and public transport: the aspect of walkability, meant as the possibility of reaching the work, commercial or social space on foot, by bicycle, or by public transport, is of particular interest. Active mobility has an important impact on health: besides improving physical health, it reduces depression, anxiety, and other mental health problems. Furthermore, cycling and walking help promote a sense of community inclusion.

In addition, this topic is at the center of numerous projects within the Italian Healthy Cities Network, a program established in 1995. Initially, it was born as a municipalities' initiative; later, in 2001, it became a non-profit association. To date, over 70 municipalities have taken part in it: the common goal is to increase health for all, equity, and sustainability. The Italian Network is accredited in the European Healthy Cities Network and has active conventions with several national bodies; among the others, it is worth mention mentioning the Italian Federation of the Environment and Bicycle (FIAB), with which the Network has recently stipulated a protocol to collaborate for the European project "Bike2Work – a smart choice for commuters". Notably, the project points to shifting the mobility quota from cars to bicycles, with a program aimed at modifying the collective vision of commuter workers (and companies) towards daily journeys by bicycle.

After all, the need to link EU mobility with more sustainable criteria has been affirmed in Europe for at least 30 years, moving forward the ecological transition. Starting from the 1990s, the key principles for urban development defined by the Aalborg Charter (1994) promoted the adoption of sustainable mobility forms to favor walking, cycling, and public transport, assigning priority to ecologically compatible means of transport. In 2004, during the fourth European Conference of Sustainable Cities, Aalborg Commitments were approved. The main strategies and goals to achieve are the reduction of private motorized transport and promotion of new efficient and accessible alternatives; the increasing of journeys made by public transport, on foot or by bicycle; the promotion of low emission vehicles; the development of integrated and sustainable urban mobility plans; the reduction of the impact of transport on the environment and public health.

Subsequently, as stated by the European Commission in the Green Paper "Towards a new culture for urban mobility" (2007), it is necessary to move from a culture of urban mobility dependent on the car to a new culture centered on the person-oriented towards sustainable transport, to an optimization in the use of all means of transport (collective such as train, tram, subway, and individual such as car, motorcycle, bicycle, on foot) through a better organization of the multi-modality between them.

In addition, during the last years, numerous international initiatives have arisen: the European Green Deal of 2019, the calls of Horizon 2020 – Work Program 2018–2020: Smart, green, and integrated transport (European Commission Decision C (2020) 6320 of 17 September 2020), and the Guidelines for Developing and Implementing a Sustainable Urban Mobility Plan of 2019.

Moreover, similar results are desirable for any city that wants to ensure healthy and sustainable environments, as clearly demonstrated by the 17 Sustainable Developments Goals (SDGs) developed by the United Nations in 2015. One year later, in 2016, the UN adopted the New Urban Agenda (NUA) and it was proposed as the primary vehicle for the direct strengthening of the SDGs, particularly in the context of urban settlements. For the first time, health was established as a desirable outcome deriving from sustainable urban development and explicitly recognized as a central component of urban governance and planning, beyond the exclusive provider of health care services [17].

3.1 Soft Mobility and COVID-19 Pandemic

The contingent situation created by the COVID-19 pandemic has produced a spread of individual soft mobility: at the European scale, some analyzes referring to 2020 report how the number of bicycles was higher than in the same period of the previous year. For example, in Italy, there was an increase of + 48.4%; moreover, the analysis of monthly data shows significant peaks in the use of bicycles in the periods of May (+81%) and September/October (+73%); finally in 2020, there was almost 200 km of new cycle paths (in most cases bike lanes).

Soft mobility can be defined as a special form of sustainable mobility capable of optimizing urban livability, improving individual and collective health, safeguarding the individual right to mobility. It includes any non-motorized means of transport that uses only human energy, with zero impact as true active mobility: pedestrians, bicycles, roller skates, scooters, skateboards, etc. Moreover, its definition can be extended to all environmentally friendly means of transport (including sharing mobility): electric cars and assisted pedal-assisted vehicles (e-bikes, e-scooters, electric unicycles, hoverboards, etc.). Following private cars and public transport, soft mobility is the third most used means of transport for daily commuting. Developed for short journeys in urban settlements, it owns multiple advantages: it saves time and money, allows training, and reduces the overall environmental impact. Many cities are encouraging this kind of mobility to reduce polluting emissions, thereby decreasing the environmental costs of motorized transport and reducing traffic congestion on main roads.

Generally, the Covid 19 crisis has brought about at least some temporary environmental benefits in reaching sustainable development goals (SDG). Emissions of CO_2 worldwide have dropped significantly due to reduced industrial activity, lower energy consumption, and reduced transportation of material and people [18].

4   Methodology

In the simplest possible form, Grounded theory represents the process of constructing a codified theory through organized data collection and inductive data analysis to answer new questions of those qualitative research that lack sufficient theoretical foundations in the field of study. Moreover, it is one of the methods that allow the regular recognition of views and meanings from the perspective of individuals in a particular situation [19]. Here this method is intrinsic in the role of a City Executive Council member when he/she is also an expert or a technician. The political vision and the technical competencies let actors have a comprehensive view of the phenomena occurring in the Municipality. Moreover, the qualitative nature of political decisions can be adequately weighted throughout the measure of the achievable results.

On the other side, the ontological tool can be built ad hoc to provide support while defining the policies' structure and relations among actions. Indeed an ontology constitutes a technological artifact since it enables a correct and formal semantic information exchange between humans and systems. A common vocabulary can be defined, and the set of concepts and inter-concept relations can be described following a formal logic representation to generate an abstract view of a specific domain[20]. The experimentation within the urban context is not new, and some cases were specifically aimed at facilitating the communication between planners, stakeholders, and information systems [21]. Overall experimentations have enabled a formal comprehension of the urban planning domain by establishing specific concepts and relationships. Different domains were covered regarding road systems, urban mobility, urban renewal [22], urban communication [23], land use [24], smart city [25] the city and its services [26], cultural heritage management, and transportation system [27]. Nevertheless, by keeping in mind the potentiality of the ontology tool and the complex urban layers, it is possible to affirm that ontology's development in the urban planning domain still has room for improvement and investigation.

As an example of the usefulness of the practical experience in managing complex systems and in promoting formal HC actions, we consider one author's practical skill acquired as a City Executive Council member for urban planning and mobility in the city of Segrate (Italy, Lombardy Region, Milan Province) from 2015 and 2020.

5   Application

Concerning the specific case of Segrate Municipality, the sectors involved in HC policies implementation are, at least: urban security, city planning, public works, environment management, financial management, culture, education, inter-sectoral topics such as mobility (that involve: urban planning, viability management, public works).

In 2020, Segrate was defining the new Urban Plan of Sustainable Mobility (Piano Urbano della Mobilità Sostenibile, PUMS); the Transport Plan (Piano Urbano del Traffico, PUT) was approved in 2010 (formally decayed). In 2020, the existing cyclable path system was coherently organized depending on the destination in a unique network called "micropolitan".

According to the cited regulatory framework about city planning and to the Italian legal framework about the organization of Public Administrations (D.Lgs. 18 August 2000 n. 267 and subsequent modifications and integrations), the hierarchy of panning instruments is:

A. Regional Plan (PTR)
B. Province Plan or Metropolitan Plan (PTCP or PTCM);
C. City plan (Piano di Governo del Territorio);
D. Actuation plans (i.e. PII).

The city plan contains the strategic goals and the normative indications for direct and established convention-building activities.

The actuation plan may have a specific regulative structure that defines public and private interests. The approval of the actuation plan denotes a separate procedure from the City Plan. The subjects involved are politicians, public and private technicians, and private stakeholders.

The Municipality executive council or Municipality Council approve the final act in which they are specified: Public commitments; Private concession; Design details (private building scheme, public works, and design scheme according to the forecasts of the public necessities that have been defined in the City Plan and the sectorial plan such as the Municipality Sustainable Mobility Plan); Economic evaluation; Economic guarantees for public works; Gantt of the public and private works.

Considering the specific actuation plan PII, the private stakeholder proposes the plan, and the Municipality (political and technical bureaus) approves the final plan. Once the actuation plan is approved, the private stakeholder has to start building the public works (or paying the equivalent amount of money) according to the Gantt scheme.

In the Lombardy Region context, the regulatory framework about city planning is based on the law 11 March 2005 n. 12 that introduced a new city plan model called Territorial Government Plan (Piano di Governo del Territorio, PGT). This plan is divided into three documents: the strategic component (Documento di Piano, DDP), the public city government (Piano Dei Servizi. PDS), the private and the normative system (Piano delle Regole, PDR).

Given that Segrate Municipality has a city plan approved in 2017 (PGT), the authors analyze one of the development areas, Milano4You actuation plan (Programma Integrato di Intervento - PII), and the related soft mobility actions such as bicycle path building which include the following steps:

1. Presentation of the detailed project by the stakeholder;
2. Check by the Municipality technicians;

   a) Coherence with the urban mobility strategies;
   b) Coherence with the Services Plan of the PGT;
   c) Coherence with the actuation plan;
   d) Coherence with the mobility plans;
   e) Economic balance and guarantees analysis;
   f) GANTT scheme analysis;

3. Approval by the Municipality executive council;
4. Request of building permit by the stakeholder;
5. Issue of the building permit by the Municipality technicians;
6. Beginning of work in the building site.

From the ontological point of view, the representation of the steps and subjects involved could be as follows:

Table 1. Phases and sectors involved in city planning and implementation steps.

Different sectors		Private sector		Public sector			% of involvement
Type of plan	Sectors / Actions	Private stakeholder	Private technician	City council	Executive council	Municipality technician	
C city plan				X	X	X	60%
D actuation plan		X	X	X	X	X	100%
D actuation plans	Presentation of the detailed project	X	X				40%
	checking		X			X	40%
	approval				X		20%
	Request of a building permit	X	X			X	60%
	Issue of the building permit					X	20%
	Beginning of work in the building site	X	X			X (supervision)	60%
	Number of steps	3	4	0	1	4	

6 Discussion and Conclusions

The authors highlighted the following aspects: the complex nature of HC; the opportunity to invest in soft mobility infrastructures and facilities to improve HC; the post-pandemic links with soft mobility. The proposed practical example becomes emblematic considering the following statements: to improve HC, it is necessary implementing general policies into actions involving and linking different subjects; the actions require a practical organization that is partially defined by the regulative framework and partially by the subjects involved. To manage the process, both technical knowledge and soft skills are required. The grounded theory approach can be synthesized in specific administrative figures (such as the technician-politician); the organizational scheme is not a permanent and deterministic process because it has ranges of approximation or modification (hand in hand with the complex nature of cities).

The authors aimed to underline the various interrelations among policies, topics, subjects, rules, and practical actions that emerge in the territorial management toward a healthier city. The small-scale sample (a public cyclable path in a private actuation plan) was also chosen to emphasize the need for public administrators to have perfect knowledge of the bureaucratic steps to organize the procedures and reach the goals in the best ways. In the comparison between City plans and Actuation plans, the research found out the involvement of all sectors (private and public) to implement an Actuation plan or city plan, and the result shows this involvement, 100% for an actuation plan. Private sectors include private stakeholders and private technicians and the public sector consists, of the city council, executive council, and municipality technicians.

Table 1 illustrates the same comparison of the involvement of different sectors(private and public) in the step of a soft mobility action, here is the cycling path. As it has shown, in the steps of " a request of building permit" and "beginning of work in the building site" with 60% involvement which has the most involved are the public and private sectors. The other result is about the role of municipality technicians and private technicians in the different steps of doing an actuation plan by obtaining numbers 4 out of 6 steps. They are similar in 3 steps, checking, request of building permit, and the beginning of work in the building site. In the next rank are private stakeholders that have an important role in the implementation of Actuation plans with one difference. It is important to mention that the city council does not play a role in the implementation of Actuation plans and they are just involved in the first approval of the plan.

The authors suggest similar research at different scales of urban, regional, and territorial development for future research.

References

1. Annells, M.: Grounded theory method: philosophical perspectives, the paradigm of inquiry, and postmodernism. Qual. Health Res. **6**(3), 379–393 (1996)
2. Barton, H., Grant, M.A.: Health map for the local human habitat. J. R. Soc. Promot. Health **126**, 252–253 (2006)
3. Batty, M.: Cities as complex systems: scaling; interaction, networks, dynamics and urban morphologies. In Encyclopedia of Complexity and Systems Science. Springer, New York (2011)
4. Barnes, D.M.: An analysis of the grounded theory method and the concept of culture. Qual. Health Res. **6**(3), 429–441 (1996)
5. Bertuglia, C.S., Vaio, F.: La città e le sue scienze. Milan-Rome: FrancoAngeli Editore (1997)
6. Bertuglia, C.S., Staricco, L.: Complessità, autoorganizzazione, città. Milan-Rome: FrancoAngeli Editore (2000)
7. Charmaz, K.: Constructionism and the grounded theory method. Handb. Constr. Res. **1**(1), 397–412 (2008)
8. Duhl, L., Hancock, T.: Promoting health in the urban context (WHO Healthy Cities Papers, n. 1), Copenhagen: PADL Publishers (1988)
9. Glouberman, S., Gemar, M., Campsie, P., et al.: A framework for improving health in cities: a discussion paper. J. Urban Health **83**, 325–338 (2006)
10. Grimm, N.B., Faeth, S.H., Golubiewski, N.E., et al.: Global change and the ecology of cities. Science **319**, 756–760 (2008)
11. Pandit, N.R.: The creation of theory: a recent application of the grounded theory method. Qual. Rep. **2**(4), 1–15 (1996)
12. Pietra, C., De Lotto, R., Bahshwan, R.: Approaching healthy city ontology: first-level classes definition using BFO. Sustainability **13**, 13844 (2021). https://doi.org/10.3390/su132413844

13. Portugali, J.: What makes cities complex? In: Portugali, J., Stolk, E. (eds.) Complexity, Cognition, Urban Planning and Design. SPC, pp. 3–19. Springer, Cham (2016). https://doi.org/10.1007/978-3-319-32653-5_1

14. Pumain, D.: Ricerca urbana e complessità. In: Bertuglia, C.S., Vaio, F. (eds.) La città e le sue scienze. Le scienze della città, vol. 2, pp. 1–45. FrancoAngeli, Milan-Rome (1997)

15. Ramirez-Rubio, O., Daher, C., Fanjul, G., et al.: Urban health: an example of a health in all policies approach in the context of SDGs implementation. Glob. Health **15**, 87 (2019). https://doi.org/10.1186/s12992-019-0529-z

16. https://doi.org/10.3390/su13074023 Marcu, S.: Towards sustainable mobility? the influence of the COVID-19 pandemic on romanian mobile citizens in Spain. Sustainability **13**(7), 4023 (2021)

17. Sterman, J.: Business Dynamics. System Thinking and Modeling for a Complex World. The McGraw-Hill Companies, USA (2000)

18. Corbin, J.M., Strauss, A.: Grounded theory research: procedures, canons, and evaluative criteria. Qual. Soc. **13**, 3–21 (1990). https://doi.org/10.1007/BF00988593

19. Sachs, J., Schmidt-Traub, G., Kroll, C., Lafortune, G., Fuller, G., Woelm, F.: The Sustainable Development Goals and COVID-19. Sustainable Development, Report 2020. Cambridge University Press, Cambridge (2020)

20. Gruninger, M.: Ontologies: Principles, methods, and applications. Knowl. Eng. Rev. **11**(02), 93 (2009). https://doi.org/10.1017/S0269888900007797

21. Teller, J., Billen, R., Cutting-Decelle, A.F.: Bringing urban ontologies into practice. J. Info. Tech. Constr. **15** (2010)

22. Berdier, C., Roussey, C.: Urban ontologies: the towntology prototype towards case studies. In: Teller, J., Lee, J.R., Roussey, C. (Eds.), Ontologies for Urban Development. Studies in Computational Intelligence, vol. 61, pp. 143–155. Springer, Berlin, Heidelberg (2007). https://doi.org/10.1007/978-3-540-71976-2_13

23. Métral, C., Falquet, G., Vonlanthen, M.: An ontology-based model for urban planning communication. In J. Teller, J.R. Lee, C. Roussey (Eds.), Ontologies for Urban Development. Studies in Computational Intelligence, vol. 61, pp. 61–72. Berlin, Heidelberg: Springer (2007). https://doi.org/10.1007/978-3-540-71976-2_6

24. Montenegro, N., Gomes, J.C., Urbano, P., Duarte, J.P.: A land use planning ontology: LBCS. Future Internet **4**, 65–82 (2012). https://doi.org/10.3390/fi4010065

25. Otero-Cerdeira, L., Rodríguez-Martínez, F.J., Gómez-Rodríguez, A.: Definition of an ontology matching algorithm for context integration in smart cities. Sensors **14**(12), 23581–23619 (2014). https://doi.org/10.3390/s141223581

26. Bellini, P., Benigni, M., Billero, R., Nesi, P., Rauch, N.: Km4City ontology building vs data harvesting and cleaning for smart city services. J. Vis. Lang. Comput. **25**, 827–839 (2014). https://doi.org/10.1016/j.jvlc.2014.10.023

27. Nandini, D., Shahi, G.K.: An ontology for transportation system. In: Proceedings of the 2nd International Joint Conference on Rules and Reasoning (RuleML + RR 2018), Luxembourg, 18–21, September 2018 (2018)

Real Estate Values and Urban Quality: Definition of an Indicator

Sebastiano Carbonara[1] (iD), Lucia Della Spina[2] (iD), and Davide Stefano[1(✉)] (iD)

[1] Department of Architecture, G. d'Annunzio University, 65127 Pescara, Italy
{s.carbonara,davide.stefano}@unich.it
[2] Department of Cultural Heritage, Architecture, Urban Planning,
Mediterranea University of Reggio Calabria, 89124 Reggio Calabria, Italy
lucia.dellaspina@unirc.it

Abstract. Urban quality and real estate values represent different but inextricably correlated and reciprocal aspects. The interventions and transformations induced by urban plans and policies have a significant impact on market dynamics and are reflected in the values which, consequently, can be interpreted as a measure of the quality of the urban spaces that house the buildings.

Starting from these premises, the paper proposes the construction of an urban quality index defined through the development of four macrosystems: environmental, infrastructural, settlement and services. Its statistical significance was tested through a multiple linear regression model. An indicator of this nature is certainly useful in mass estimates but can also be considered as a functional tool for city government decisions.

Keywords: Urban quality index · Real estate value

1 Introduction

An essential representation of the concept of city governance can be circumscribed to the development and implementation of tools relating to the transformation of the physical space and policies that affect the offer of services.

Both aspects end up decisively conditioning the real estate market, which is extremely reactive to the changes they cause.

Consequently, real estate values can represent a key to reading the variation in the quality level of urban areas and offer, in some way, a measurement [1, 2].

Starting from these assumptions, a multi-parameter model of multiple regression was developed which, through the reading of the real estate market, led to the definition of a representative indicator of the quality of the different urban contexts [3].

It is known, in fact, that real estate values derive both from the characteristics of each real estate unit (easier to identify), but also - and above all - from those characteristics that the estimation literature defines as extrinsic characteristics, namely those relating to each specific urban area [4–6].

F. Calabrò et al. (Eds.): NMP 2022, LNNS 482, pp. 1082–1090, 2022.
https://doi.org/10.1007/978-3-031-06825-6_104

In fact, while the intrinsic characteristics, linked to the material and architectural properties of the buildings [7], are easily identifiable (surface area, time of construction or renovation, level of the floor, etc.), the extrinsic ones are difficult to identify.

Therefore, the index developed will allow to recalculate the market values of the properties to be placed at the basis of taxation, thus trying to resolve the inequality still existing today in the tax field.

The first issue to be resolved is the definition of the boundaries of each urban area to which the quality indicator refers.

The starting reference was found in art. 2 of Presidential Decree 138/1998, i.e. the legislative provision that initiated the reform of the Land Registry in Italy, which is still largely unfinished, introducing the concept of micro-zone: "The micro-area represents a portion of the municipal area (in many cases, co-incident with the entire Municipality) that presents homogeneity in the characteristics of position, urban planning, historical-environmental, socio-economic characteristics, as well as in the provision of urban services and infrastructures. In each micro-zone, the real estate units are uniform in terms of typological characteristics, construction period and prevailing destination."

Most of the Italian municipalities have provided for the definition of these areas (often in collaboration with the cadastral offices), indicatively in the period 2000–2002.

Such a territorial demarcation is undoubtedly interesting for the purposes of estimation, although the possibility cannot be excluded that the error that is always made in transactions of this nature may be too high, and twenty years have passed since the first definition. On the other hand, the strong institutional value should not be overlooked considering that from the procedure provided for by Presidential Decree 138, once fully operational, the taxation of properties should follow.

In any case, once the polygons into which the city can be subdivided have been assumed, it is a question of defining the elements that allow the indicator to be processed.

2 The Extrinsic Variables: The Urban Quality Index

For the definition of the Urban Quality Index (U_{QI}), it was necessary to develop a Geographic Information System (GIS) [8] capable of containing all the information deemed useful for the objective representation of the urban contexts analysed [9]. Starting from the Regional Numeric Technical Map (CTRN), the system was implemented not only with the maps of the Abruzzo Region Web-GIS and the Cadastral Maps of the Revenue Agency, also with specific elements deriving from both a "virtual" survey (through internet services and geographic software such as Google Maps, Google Street View, Google Earth) which gives a direct survey.

For the definition of the Urban Quality Index (U_{QI}), four macro-systems have been identified in each reference area (micro-zone): Environmental, Infrastructure, Settlement, Services, each of which is broken down into subsystems [10].

The quality index was obtained as the sum of the contributions provided by each macrosystem:

$$U_{QI} = \sum_{i=1}^{n} I_i \tag{1}$$

To make it possible to compare the 10 different micro-zones, we proceeded through the normalization of the data, relating to the unit the maximum value found for each characteristic (benchmark).

All the measurements carried out (both in terms of surface and in linear meters) were related to the territorial surface of the i-th micro-zone. Subsequently, the normalization was carried out by identifying the benchmark (identifiable in the maximum value) among the values obtained within the same category. The sum of the normalized values expresses the share of the urban quality index attributable to the environmental system.

2.1 Environmental System

It represents the extent of the surfaces and the linear development of natural areas whose influence on the quality of public spaces and on the landscape is to be considered as improving [11]. The environmental system was then broken down and analysed into two subsystems: green areas and specific environmental characteristics of the city analysed.

Green areas include all public surfaces such as parks and gardens (Parks and Public spaces - PP), peripheral agricultural areas (cultivated, uncultivated or wooded) (Agricultural Lands - AL) and the linear development of tree-lined avenues (Tree- Lined streets - TL).

The specific environmental characteristics of Pescara are represented by the riverfront (River Front - RF) and the coastline (Sea shoreline - S) whose influence has been defined as being within the limit of 250 m.

2.2 Infrastructural System

The infrastructural system includes areas, routes, and the number of services subject to travel by public, private or cycle-pedestrian transport. The infrastructural system was then broken down and analysed into three subsystems: the road network (RN), local public transport (Local Public Transport - LPT) and interchange nodes (Junctions - J).

The road network (RN) includes vehicular roads (Main Roads - MR), secondary roads (Secondary Roads - SR), sustainable mobility networks (Cycling Lanes - CL and routes in parks - Routes in Parks - PR), limited traffic zones (LTZ) and parking areas (PK).

Local public transport (LPT) by road (in Pescara there is no rail transport system) was analysed through two parameters: the development in linear meters of the urban lines (Bus Lines $_{(1)}$ - BL $_{(1)}$) in the i-th micro-zone in relation to the territorial extension of the same micro-zone (in hectares) and the number of lines (Bus Lines$_{(2)}$ - BL$_{(2)}$) that run through the single i-th micro-zone. The normalization was carried out by identifying the benchmark where the greatest supply of transport services (max) among the values obtained within the same category was revealed.

Lastly, the nodes (Junctions - J), or the access points to the main networks: railway (Railway Junctions - RWJ) and road traffic (Road Junctions - RJ), were identified, considering their presence, in absolute number, for the i-th micro-zone. The normalization was carried out by identifying the benchmark where the greatest supply of transport services (max) among the values obtained within the same category was revealed.

2.3 Settlement System

For the construction of the settlement system index (SS) it was necessary to analysed in detail all 17,603 properties in the city. The overall surfaces, the construction/maintenance quality of the properties (Building Quality Score - BQS), the type (Building Typology Score - BTS) and the intended use (Building Designed Use - BDU) were calculated.

For the two sub-indices of construction/maintenance quality and type, reference was made to a previous study developed by the authors in 2012 [12] in which the criteria for the attribution of scores were set.

Direct observation of the facades of the buildings made it possible to detect the presence of precious materials or workmanship or, vice versa, the presence of forms of deterioration [13, 14].

As regards the building typologies, [15] 8 were identified, assigning defined scores on the basis of interviews subjected to a sample of the population, according to the scheme:

- industrial buildings (score = 1),
- tower buildings, multi-storey, non-residential block (score = 3),
- terraced, block or courtyard buildings (score = 5),
- villas and cottages (score = 7).

The normalization was carried out by identifying as a benchmark the highest score (max) among the values obtained within the same category.

The intended use sub-index was constructed by noting the intended use of each unit present within the individual buildings. Three building categories have been established:

- exclusively residential (Residential Building - RB),
- mixed use (Mixed use Building - MB),
- for the exclusive use of tertiary activities or public offices (Services Building - SB).

Based on the real estate values provided by the IMO (Real Estate Market Observatory), it was shown that the average real estate values (referred to residences) of Micro-zone no. 4 are the highest in the entire municipal area: this figure has led to considering as a reference benchmark, the combination of intended use present in Micro-zone no. 4 (30% exclusively residential, 59% mixed use and 11% tertiary, public and private offices) to carry out the normalization of the values found in the other micro-zones.

2.4 Service System

The greater presence of services within the analysed micro-zone was assessed as a favorable condition both in terms of public and private services (Public/Private services System - PPS). The survey was carried out considering the services present for the City of Pescara in order to the following categories:

- basic education (nursery schools, kindergartens, primary and secondary schools) and upper school (upper secondary, universities and research centers),
- hospitals, nursing homes, health facilities,
- libraries, museums, cinemas, theaters, auditoriums, exhibition halls,
- places for religious expression,
- facilities for sport and free time, commerce, and private services (large and small distribution, catering, bars, commercial areas, hotels, crafts, tertiary sectors, banks, and agencies), public squares and meeting places.

The values obtained are an expression of the ratio between the number of services present for the i-th micro-zone, without considering the range of influence that proximity to the service can have on the formation of real estate prices, which can reasonably be evaluated positively [16]. The normalization was carried out by identifying as a benchmark the maximum number of services (max) by comparing the values of all the micro-zones. By applying Formula (1) to the values obtained for each of the systems studied (Fig. 1), it was possible to determine the urban quality variable for each of the 10 micro-zones of the city of Pescara (Fig. 2).

micro-zone	Agricultural Lands - AL	Parks and Public spaces - PP	Tree-Lined streets - TL	River Front - RF	Sea shoreline - S	Building Designed Use - BDU	Building Quality Score - BQS	Building Typology Score - BTS	Main Roads - MR	Secondary Roads - SR	Cycling Lanes - CL	Routes in Parks - PR	Parking Areas - PK	Limited Traffic zones - LTZ	Local Public Transport - LPT	Junctions - J	Public/Private services System - PPS	Urban quality index
1	0,00	1,00	0,52	0,00	1,00	0,69	1,00	0,57	1,00	1,00	0,44	0,38	0,47	0,00	0,97	0,00	0,08	**9,11**
2	0,01	0,30	0,52	0,00	0,64	0,73	1,00	0,56	0,61	0,53	0,28	1,00	0,54	0,00	1,73	0,00	0,36	**8,80**
3	0,38	0,51	0,26	0,14	0,00	0,65	0,92	0,26	0,42	0,37	0,04	0,00	0,20	0,00	0,73	0,14	0,66	**5,66**
4	0,00	0,07	1,00	0,00	0,96	1,00	0,95	1,00	0,71	0,74	1,00	0,47	1,00	1,00	1,95	0,14	0,42	**12,41**
5	0,00	0,16	0,54	0,76	0,61	0,86	0,94	0,84	0,69	0,79	0,36	0,00	0,61	0,14	1,38	0,86	1,00	**10,51**
6	0,20	0,36	0,10	0,40	0,00	0,68	0,88	0,65	0,23	0,41	0,14	0,00	0,12	0,00	0,49	1,00	0,71	**6,36**
7	0,15	0,62	0,34	0,00	0,00	0,75	0,94	0,63	0,78	0,62	0,20	0,00	0,46	0,00	0,70	0,14	0,54	**6,86**
8	0,08	0,40	0,13	1,00	0,75	0,64	0,94	0,21	0,21	0,50	0,35	0,00	0,19	0,00	0,83	0,14	0,15	**6,52**
9	0,83	0,90	0,05	0,00	0,00	0,61	0,96	0,16	0,23	0,25	0,00	0,00	0,04	0,00	0,41	0,29	0,07	**4,79**
10	1,00	0,54	0,00	0,23	0,00	0,60	0,92	0,13	0,17	0,19	0,00	0,00	0,00	0,00	0,19	0,14	0,16	**4,27**

Fig. 1. Data normalization

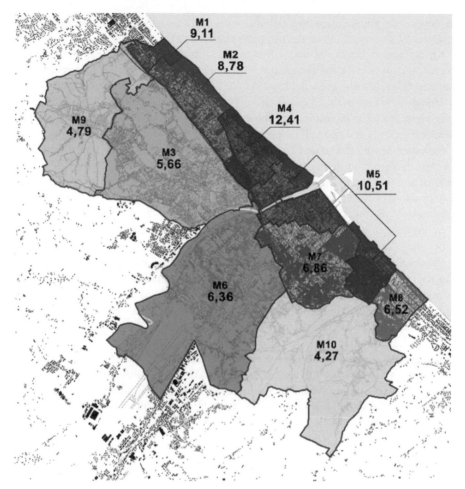

Fig. 2. Urban Quality Index (UQI) aggregated in the 10 micro-zones of the City of Pescara.

3 Conclusion

In conclusion, it is useful to recall some elements that animated this work. The research from which it arises pursued the objective of developing a mass appraisal model that would allow to estimate and update real estate values quickly, in order to proceed with the taxation of properties with reference to real values and not conventional ones. (Cadastral values) still used today in Italy.

In order to pursue this objective, it was necessary to have an indicator that would allow the characteristics of the different urban contexts to be represented.

Hence the need to define an indicator of urban quality.

There is no doubt that the concept of urban quality can be declined starting from different observation points and perspectives, which originate from the different disciplinary visions. The latter often end up privileging only some of the elements at stake and, beyond the appreciation on a cultural level [17] of some of them, they can prove to be the test of slippery facts on the implementation level.

An obvious point of weakness lies in the difficulty of attributing to the concept of urban quality a political dimension which, in the opinion of the writer, cannot be separated from its measurement [18–23].

This does not mean that the operation is in itself easy and decisive, since the choice of the elements to be measured and the construction of the indicator itself is part of a dialectical process that can still be extremely complex [24–28].

However, having elaborated and inserted the quality index within a multiple regression model, allowed to verify its significance, at least on a statistical level. If the constructed equation, including the variable relating to urban quality, allows for a good approximation of the value of the properties, with widely comforting statistical tests, its usefulness for the purpose can be deduced [29, 30].

In any case, a whole series of questions remain open that can in any case generate doubts and perplexities.

Meanwhile, the built indicator ultimately represents an indicator of density: where services, equipped spaces, residential construction, commercial activities, etc. are concentrated, the level increases. This is a position that some debate, since the higher density would correspond to all those inconveniences associated with dense agglomerations, not least a reduced quality of environmental parameters. Yet, in the cities of the world, the highest real estate values are almost always found in densely structured and inhabited areas, demonstrating that consumer preferences are oriented in this direction and their willingness to pay follows the trend. The Kibera slum in Nairobi is perhaps as densely populated as a division of Manhattan in New York, but all urban facilities are substantially absent in the first case, while they are found to be significantly concentrated in the second.

A further element of discussion is represented by the social characteristics attributable to urban areas [31, 32], including, for example, the frequency of crimes against persons and property and the sensation of risk perceived by the population in the various urban areas. In the built indicator, this aspect was deliberately omitted as it is non-structural, therefore variable over time, not perfectly territorially circumscribable and above all indirectly represented in the conformation of the different urban spaces.

In the end, what is important is to have a sort of census of the characteristics of the different urban areas and to have a measure that demonstrates the different levels in each of them. The consequent step on the political level is to proceed with a rebalancing that tends to offer more equitable and balanced services [20–23, 25–28, 32].

From this point of view, the environmental characteristics of urban areas represent a priority element of intervention. The pandemic events of this last two years will sooner or later tend to run out, but it will be difficult to get rid of the idea that Paul Virilio visually represented in the Innertie polaire, imagining us as inert poles of a global network and consequently more attentive and caring in the comparisons of the nearest space.

Author Contributions.   The paper is the result of the joint work of the three authors. Scientific responsibility is attributable equally to Sebastiano CARBONARA Lucia DELLA SPINA, and Davide STEFANO.

References

1. Carbonara, S., Stefano, D., Di Prinzio, A.: "Transforming surface rights into property rights: an analysis of current estimation procedures and a comparison with an alternative," Valori e Valutazioni. Teorie ed Esperienze **24**, 5–17 (2020)
2. Carbonara, S., Stefano, D.: The Transformation of Surface Rights into Property Rights. A Financial Resource for Rebalancing Municipal Budgets. The Case of Pescara. Appraisal and Valuation. Contemporary issues and new frontiers, 2021, pp. 91–101 (2021) https://doi.org/10.1007/978-3-030-49579-4_7
3. Carbonara, S., Faustoferri, M., Stefano, D.: Real estate values and urban quality: a multiple linear regression model for defining an urban quality index. Sustainability **13**(24), 13635 (2021). https://doi.org/10.3390/su132413635
4. Rosen, S.: Hedonic prices and implicit markets: product differentiation in pure competition. J. Polit. Econ. **82**(1), 34–55 (1974)
5. Mangialardo, A., Micelli, E.: Grass-roots participation to enhance public real estate properties. Just a fad? Land Use Policy **103**, 105290 (2021). https://doi.org/10.1016/j.landusepol.2021.105290
6. Faka, A., Kalogeropoulos, K., Maloutas, T., Chalkias, C.: Urban quality of life: spatial modeling and indexing in Athens Metropolitan area, Greece. ISPRS Int. J. Geo Inf. **10**(5), 347 (2021). https://doi.org/10.3390/ijgi10050347
7. Forte, C., De'Rossi, B.: Principi di economia ed estimo, Etas libri (1974)
8. Mollica, E., Massimo, D.E.: Valutazione degli strumenti del territorio e strumenti GIS. Aestimum **32**, 710–757 (2002). https://doi.org/10.13128/Aestimum-6782
9. Faisal, K., Shaker, A.: Improving the accuracy of urban environmental quality assessment using geographically-weighted regression techniques. Sensors **17**(3), 528 (2017). https://doi.org/10.3390/s17030528
10. Garau, C., Pavan, V.M.: Evaluating urban quality: indicators and assessment tools for smart sustainable cities. Sustainability **10**(3), 575 (2018)
11. Zhang, Y., Dong, R.: Impacts of street-visible greenery on housing prices: evidence from a hedonic price model and a massive street view image dataset in Beijing. ISPRS Int. J. Geo Inf. **7**(3), 104 (2018). https://doi.org/10.3390/ijgi7030104
12. Carbonara, S.: The effect of infrastructural works on urban property values: the asse attrezzato in Pescara, Italy. In: Murgante, B., et al. (eds.) ICCSA 2012. LNCS, vol. 7334, pp. 128–143. Springer, Heidelberg (2012). https://doi.org/10.1007/978-3-642-31075-1_10
13. Gavrilidis, A.A., Ciocănea, C.M., Niţă, M.R., Onose, D.A., Irina Iulia, N.: Urban landscape quality index – planning tool for evaluating urban landscapes and improving the quality of life. Procedia Environ. Sci. **32**, 155–167 (2016). https://doi.org/10.1016/j.proenv.2016.03.020
14. Wilson, J.Q., Kelling, G.L.: Broken windows: the police and neighborhood safety. Atl. Mon. **249**, 29–38 (1982)
15. Rong, H.H., Yang, J., Kang, M., Chegut, A.: The value of design in real estate asset pricing. Buildings **10**(10), 178 (2020). https://doi.org/10.3390/buildings10100178
16. Liebelt, V., Bartke, S., Schwarz, N.: Urban green spaces and housing prices: an alternative perspective. Sustainability **11**(13), 3707 (2019). https://doi.org/10.3390/su11133707
17. Carbonara, S., Stefano, D., Faustoferri, M.: Public Real Estate's in Italy: from Decommissioning to Valorization. Legislative Evolution Future Perspectives LaborEst **21**, 39–46 (2021)
18. Della Spina, L.: Multidimensional assessment for "Culture-Led" and "Community-Driven" urban regeneration as driver for trigger economic vitality in urban historic centers. Sustainability **11**(24), 7237 (2019). https://doi.org/10.3390/su11247237

19. Della Spina, L., Giorno, C., Galati Casmiro, R.: Bottom-up processes for culture-led urban regeneration scenarios. In: Misra, S., et al. (eds.) ICCSA 2019. LNCS, vol. 11622, pp. 93–107. Springer, Cham (2019). Doi: https://doi.org/10.1007/978-3-030-24305-0_8

20. Della Spina, L., Giorno, C., Galati Casmiro, R.: An Integrated Decision Support System to Define the Best Scenario for the Adaptive Sustainable Re-Use of Cultural Heritage in Southern Italy, pp. 251–267 (2020). https://doi.org/10.1007/978-3-030-52869-0_22

21. Della Spina, L.: Cultural heritage: a hybrid framework for ranking adaptive reuse strategies. Buildings 11(3), 132 (2021). https://doi.org/10.3390/buildings11030132

22. Della Spina, L.: A multi-level integrated approach to designing complex urban scenarios in support of strategic planning and urban regeneration. In: Calabrò, F., Della Spina, L., Bevilacqua, C. (eds.) ISHT 2018. SIST, vol. 100, pp. 226–237. Springer, Cham (2019). https://doi.org/10.1007/978-3-319-92099-3_27

23. Della Spina, L., Ventura, C., Viglianisi, A.: A multicriteria assessment model for selecting strategic projects in urban areas. In: Gervasi, O., et al. (eds.) ICCSA 2016. LNCS, vol. 9788. Springer, Cham (2016). https://doi.org/10.1007/978-3-319-42111-7_32

24. Della Spina, L., Rugolo, A.: A multicriteria decision aid process for urban regeneration process of abandoned industrial areas. In: Bevilacqua, C., Calabrò, F., Della Spina, L. (eds.) NMP 2020. SIST, vol. 178, pp. 1053–1066. Springer, Cham (2021). https://doi.org/10.1007/978-3-030-48279-4_99

25. Carbonara, S., Stefano, D.: How the Italian state finances post-seismic reconstruction: the 2009 Abruzzo Earthquake. In: Morano, P., Oppio, A., Rosato, P., Sdino, L., Tajani, F. (eds.) Appraisal and Valuation. GET, pp. 249–267. Springer, Cham (2021). https://doi.org/10.1007/978-3-030-49579-4_17

26. Carbonara, S., Stefano, D.: Building recovery, property values and demographic decline after the 2009 Abruzzo earthquake, New Metropolitan Perspectives. Knowledge Dynamics and Innovation-driven Policies Towards Urban and Regional Transition Volume 2, vol. 2, no. 178, pp. 779–790 (2021). https://doi.org/10.1007/978-3-030-48279-4_73

27. Carbonara, S., Faustoferri, M., Stefano, D.: Abruzzo post-seismic reconstruction: an exploratory study on the investments outcomes. LaborEst, 20, 23–29 (2020)

28. Carbonara, S.: La stima dei costi del patrimonio edilizio privato nella ricostruzione post-sismica abruzzese: un'analisi critica delle procedure utilizzate. Territorio 70, 119–125 (2014). https://doi.org/10.3280/TR2017-070019

29. Della Spina, L.: Integrated evaluation and multi-methodological approaches for the enhancement of the cultural landscape. In: Gervasi, O., et al. (eds.) ICCSA 2017. LNCS, vol. 10404. Springer, Cham (2017). https://doi.org/10.1007/978-3-319-62392-4_35

30. Della Spina, L., Viglianisi, A.: Hybrid evaluation approaches for cultural landscape: the case of "Riviera dei Gelsomini" Area in Italy. In: Bevilacqua, C., Calabrò, F., Della Spina, L. (eds.) NMP 2020. SIST, vol. 178, pp. 1369–1379. Springer, Cham (2021). https://doi.org/10.1007/978-3-030-48279-4_128

31. Della Spina, L.: Strategic planning and decision making: a case study for the integrated management of cultural heritage assets in southern Italy. In: Bevilacqua, C., Calabrò, F., Della Spina, L. (eds.) NMP 2020. SIST, vol. 178, pp. 1116–1130. Springer, Cham (2021). https://doi.org/10.1007/978-3-030-48279-4_104

32. Della Spina, L.: Revitalization of inner and marginal areas: a multi-criteria decision aid approach for shared development strategies. Valori e Valutazioni 2020(25), 37–44 (2020)

Co-evolutionary, Transformative, and Economic Resilience During the COVID-19 Pandemic Crisis. Evidence-Based Experiences of Urban Community Design in Turin (Italy)

Coscia Cristina[1]($\boxtimes$) (iD) and Voghera Angioletta[2] (iD)

[1] Dipartimento di Architettura e Design, Politecnico di Torino-DAD, c/o Castello del Valentino, viale Mattioli 39, 10125 Turin, Italy
cristina.coscia@polito.it
[2] Dipartimento Interateneo di Scienze, Progetto e Politiche del Territorio - DIST, c/o Castello del Valentino, viale Mattioli, 39, 10125 Turin, Italy
angioletta.voghera@polito.it

Abstract. This article seeks to interpret co-evolutionary and transformative resilience in a broad sense, with the aim of understanding how it may come into the practices of urban planning and project-making, innovating project procedures and generating economic effects. This article studies this through the case of *Bottom up!,* the Turin-based *Festival of Architecture*, and observes the procedures through which resilience takes action in different territories, interpreting territorial problems and crises such as the pandemic, viewing them as opportunities to innovate the system, suggesting integrated action on the natural, cultural, financial and social capital, experimenting with new practices, and holding institutions accountable.

Keywords: Co-evolutionary resilience · Urban community design · Economic resilience

1 Introduction

1.1 La Resilience Metaphor: An Overview

Recently, the scientific community has made multiple multidisciplinary research attempts to (more or less) circumscribe the ample parameters of resilience. On an international level, the concept of resilience is acknowledged and used endemically as an umbrella term or a semantic utopia [1]; it can be analysed according to several definitions and variations in relation to its ability to evoke themes and multi-disciplinary actions, in addition to the vast possibilities of cultural and social aggregation around its vague description [2–4]. Therefore, there is a relevant attractiveness towards resilience as an approach in terms of urban and territorial issues [5] and towards the features of adaptability and complexity of socio-economic systems [6, 7]. Resilience seen as

F. Calabrò et al. (Eds.): NMP 2022, LNNS 482, pp. 1091–1101, 2022.
https://doi.org/10.1007/978-3-031-06825-6_105

a metaphor of urban action still seems to be under investigation in terms of aspects connected to urban communities and the related economic consequences and systems in place to monitor impacts and "reactions". In this sense, and starting from the rich literature on the theme, the sections that follow highlight certain analyses supported by a Turin-based experience, which sheds light on some facts.

1.2 The Research Aims: Resilience in the Field of Urban Community Co-design

As mentioned above, this article aims to draw attention to issues which have not yet been thoroughly investigated. These issues are social and community resilience [8] and some of its specific interpretations, such as economic resilience (in particular, measuring and monitoring impact and reactions) in urban regeneration practices [9, 10]. Furthermore, this article investigates certain emergent practices in terms of evidence "of success", which may constitute factors of scalability and replicability of resilience in action. As a matter of fact, community resilience encompasses notions of well-being, adaptability, and resourcefulness in the face of adverse conditions through the activation of a network of actors to maintain the structure of the system, and its social capital for defining territorial and strategical governance [11, 12].

An analysis in the literature of how, in recent planning processes and urban projects, the connection between planning community design and the economy has provided theoretical and practical answers to contemporary challenges (including the pandemic) in an innovative and resilient key has been provided.

More specifically, in the broader landscape, we assume the definition of territorial resistance [13] as a reference to explore the practices of community-based innovation. Through different disciplinary contributions, territorial resilience is promoted as an operative theory, which interprets co-evolutionary dynamics as the property of a system, an emerging concept to overcome vulnerabilities affecting the system, considering the relationship of the communities, the heritage, the governance system and its learning capacity for the resilient co-evolution of the community and the territorial system [13].

Territorial resistance assumes the nonlinear adaptive ability of a system which is influenced by phenomena that is difficult to measure, but which also sees the co-evolutionary dimension as an essential component of resilience, paying attention to procedures of change rather than to the state of the system. The resilience of the territorial system [14] is defined through the assessment of vulnerability, and the multidisciplinary integration of different fields of investigation, with a focus on urban territorial architectural aspects and financial impacts and the use of a combination of quantitative and qualitative features to assess and interpret the system, with a view of guiding its transformation. From this perspective, resilience can be intended as a generative metaphor of new approaches, as is clear from the study by Pickett [15], useful for the innovation of territorial governance. Resilience as a metaphor or as an act of imagination can inspire specific technical meanings, which could be represented also in an informal and non-technical way, explaining the success of the concept on an international level.

From this angle, resilience can become the theoretical tool necessary to build operative models of representation of the system, capable of creating new metaphors, new cognitive approaches and open perspectives of action, which in turn can make experts interact with communities and strengthen social responsibility and self-organization. A model capable of representing the system and its components through specific temporal and spatial rules of functioning and interaction between the different components of the system (ecological, social, economic, and of governance), essential in order to define trajectories of urban development and transformation.

With regard to the financial definition of resilience, three macro-trajectories of research emerge: 1) social resilience connected to preserving the memory of the lands and communities [15, 16] and the specific urban and territorial features connected to processes of local development [17, 18]; 2) the measurable effects, even financial and organizational, of resilience in times of crisis, climate emergencies, and beyond [11, 19–22]; 3) the importance of the supply chain through the synergy and leverage effect of new types of public-private partnerships, which reveal themselves to be extremely flexible, paying attention to the financial components of impact and ethics, sometimes in support of the processes of urban metabolism [23–29].

The research background that has just been illustrated was instrumental in reading the case study of a "resilient community" (see Sects. 2 and 3), which saw the inception of 12 "resistant" projects within the framework of the *Bottom up! Festival of Architecture: When a city is transformed from the bottom up* of the Architecture Foundation of Turin and of the Order of Architects, Landscape Architects, Planners and Conservationists of Turin, which concluded in the month of May 2021. The practice of *Bottom Up!*'s resilient communities is paradigmatic in highlighting some indispensable theoretical assumptions for the reinterpretation of the concept of resilience in territorial, urban and architectural planning and project-making practices.

2   An Experience in Urban Community Design in Turin: A *Bottom Up!* Experimentation

2.1   *Bottom Up!* in Turin (2020–2021)

The case study of the *Bottom up! Festival* in Turin [30] proved to be a successful and paradigmatic practice to highlight response factors in terms of urban, economic and community resilience, that is to say flexibility and innovation, thanks to the "crowd" approach of the actions that accompanied the proposals.

It was a transformational "experiment" of the city at the service of the urban communities, starting from Turin and taking into account Turin's 2030 Action Plan. It was coordinated by the Architecture Foundation in Turin (a private not-for-profit body of the Order of Architects, Landscape Architects, Planners and Conservationists of Turin, a public law organisation) in its 2019–2021 mandate in collaboration with the Order of Turin, also taking into consideration the request of the MIBAC [30] regarding the importance of fostering good architectural practices in the process of regenerating spaces. As a matter of fact, the format places architecture at the centre as a shared valued practice of relations and construction, and – in an attempt to create a virtuous process of transformation and recovery – it gives way to regeneration or "generative" projects of certain underused urban areas, particularly those located in more peripheral areas, through experimentation and innovation, promoting higher quality of life and of the environment across the territory. The bottom-up model of accompaniment and social participation has revealed itself to be virtuous in terms of listening to and transforming the needs of the places and the people who inhabit them into processes in which architecture leads to change. The involvement of a public, public-private or private benefactor, and a network of communities with strong and heterogenous formal and informal groups of citizens has laid the basis for a long-term commitment, even on behalf of the Public Administration (the municipality of Turin) in supporting this bottom-up public participation project model. The role of the Public, as well as that of the sponsoring organisations, of the architects who turn dreams into reality, and of the citizens who made "donations", also proved to be strategic in raising awareness among sponsors and civil society at large of the crowdfunding campaign aimed at retrieving the resources required to carry out and carry forward the projects suggested. At the heart of all 11 projects, there were analyses and the qualitative monitoring of the effects of the organisation in urban contexts, even in relation to the climate crisis, the planning of living spaces and the life of the community, focusing on integration and inclusion.

The final results of the Festival have been published in the *Bottom up!* "User's Manual", of which a small portion is presented here (Fig. 1.a): 12 projects in Turin and 2 in Milan, whose initial goals of collecting funds to start the "core" projects were all reached for a total of 142,386.00 Euro (100,226.00 Euro of special donations) with 105 communities involved, 929 donors, and an average donation of 33 euros. In addition, it was a "widespread" Festival (Fig. 1.b) because the geographical distribution of the products highlighted the multi-centrality of the concept of urban transformation.

Fig. 1. (a) (to the left) and (b) (to the right). The Festival's numbers and the geographical distribution of the projects (Source: www.bottomup.it, see *Credits and knowledge*)

2.2 Innovative Elements Related to Urban Community Design and Economic Resilience

The practice proved to be representative and has highlighted recurrences in terms of aspects of innovation for territorial projects. These aspects are:

– creative diversity [31, 32] (of actions that are promoted by the community [33, 34] which produce innovations that affect multiple aspects of urban action (social, environmental, landscape, economic), acknowledging the interconnections and inter-dependencies between the many levels of the components of the socio-ecological system;

– attention to time variables associated to thresholds, or in other words, the strategies connected to recovery scenarios, adaptation, and evolution [35, 36].

– flexibility and innovation seen as the ability of local project-makers to learn, experiment and develop [37–39], to embrace changes, recognize or acknowledge the memory of the system, and strengthen it.

Focusing our attention on the aspect of urban resilience, community resilience and the economic and evaluative aspect of resilience, an extremely important nexus appears to be the brief and design stage of the accompaniment project, in which the structure and writing in the tenders in terms of the criteria for selection and assessment of the projects taking part in Bottom up! [40] become the guiding tools of the project guidelines and of the strategic factors of response to emergencies. As a matter of fact, in the tender during the pre-assessment phase, which allowed for a subsequent monitoring of the projects in terms of resilience, the following qualitative and quantitative requirements to present the proposed projects were highlighted:

A) the assessment of project proposals using **impact criteria** in descriptive-qualitative terms: 1) the quality of reading and interpreting the social fabric of the context; 2) the level of innovation of the proposal in relation to the ability to listen; 3) evidence of environmental sustainability, elements of scalability and replicability of the practice in analogous urban realities; 4) the inclusive ability and social quality of the community proposing the project;

B) the sustainability of economic resilience through **impact criteria** in qualitative-quantitative terms: 1) elements of circular economy; 2) the growth of ability for self-organization; 3) the empowerment of citizens; 4) elements of transparency and accountability accompanied by preliminary business plan models; 5) sharing goods and services.

Planning ability entails the ability to enact specific actions of resilience on an urban and territorial scale, acting on communities as a motor of social and architectural innovation, and as an actor in the processes, also in support of the financial feasibility of the project [40, 41].

3   Evidence from the *Bottom-Up!* Case: Resilience as a Metaphor of the Project for Bottom-Up Co-evolution

The festival is a community-based resilient transformation of the city, boosted by the active participation of the local community and the ability of the local community to self-organize. Resilience is a driver of local change, that is to say change in communities, institutions, organizations and social networks [41]. The practices strengthen the robustness of the system, strictly linked to social capital (trusts and social networks), to the capacity for institutional learning, to social memory and self-organization to build an adaptable governance that can guide change.

3.1 Bottom Up: The Adaptive and Co-evolutionary Model

Bottom up! (see Sect. 2.1) manages to promote an urban regeneration process on different scales and on different timelines to strengthen the ability for institutional learning, social memory, and self-organization viewed as essential factors for adapting and guiding change. The case study allows us to interpret resilience in relation to urban self-organization.

In fact, considering that cities are non-linear systems, and open adaptive complex systems [42, 43], a possible feature is the idea of self-organization. Cities are open because they exchange matter, energy, information and people, with their environment. Being open implies that the system continuously reacts to external changes, in an adaptive process. Secondly, cities are complex because «their parts are so numerous, and changing, that there is no way to describe them in terms of cause an effect (as did the urbanists of the 1950s and 1960s), nor in terms of probabilities (as did the urbanists since the end 1960s and the regional scientists of the 1970s and 1980s)» [43] (p. 46). Thus, an important property of cities as open and complex systems is related to self-organization process: cities could self-organize their internal structure independently of external causes.

Cities, considered as open adaptive complex systems, led to the development of a modified Panarchy Model [43], where cities are characterized by self-organization and therefore able to gradually adapt, learning from their past experiences and preserving their memory In other words, as *Bottom up!* in the COVID 19 period, the reaction of cities to external emergencies (system disturbances) depends on a certain degree of their self-organization. They can self-regulate and create innovative solutions to urban development. Resilience and self-organisation are therefore strictly related, in term of dynamic processes of renewal and urban and architectural design, through constant adaptation and innovation. Within this framework, the practices we discussed here centralize the need to make institutions and societies accountable. Through the Bottom up! Festival, they learn to self-organize, to develop from crises and view these as windows of opportunity of the system. They are practices which represent models that develop the ability to "stay in the game" [15] and operate in crises, seeing risks and threats as "windows of opportunity" [44]: COVID-19 was treated as such.

Considering economic resilience, the "new" feasibility of "accompanied" processes, particularly those combined with bottom-up processes and crowdfunding, highlight elements of innovation, which can be viewed as levers of resistance and "success" for the realization of said processes, thanks to the ability to reach the objectives of urban community.

The following are deemed strategic and are developed through some aspects of the Bottom up! Initiative and discussed in relation to other selected case studies:

– the resilience and authority of the network (formal/informal communities, project designers and beneficiaries) who make the crowdfunding action trustworthy and one that people get on board with also according to a principle of equity, which means higher resilience of the community (such as the specific contribution of architect's community similarly to other experiences) [43];
– the direct engagement of the beneficiaries who, in addition to authority, participated by making small contributions financing "their desires", or the "desires that convinced

them" (such as the role of local institutional actors as Turin Municipality or the same role assumed in the civic crowdfunding project of the city of Milan and today it is largely diffusing (https://www.produzionidalbasso.com/network/di/comune-di-mil ano#comunedimilano-initiative);

– co-generations of values, identity and memory as shared and "heritage" factors, which become permanent factors of the investment and of the co-participation in the risk of the investment operations (such as the values recognized by community in Bottom up! Designs, and in other case studies such as in the well-known process generated in Favara Farm Cultural Park (see https://www.farmculturalpark.com/) [25, 27, 46, 47];

– social finance [45, 46] and ethical impact in the processes of investment and ethical responsibility of the economic operators and beneficiaries: "Equity means higher resilience for the community" (such as the crowdfunding as a model of social co-responsabilisation in each Bottom up project. This is the principle underlying other processes and activities also implemented by subjects such as foundations, such as the Cottino Foundation (see https://www.cottinosocialimpactcampus.org/) [34, 46, 47]; assessment of the processes and projects for impacts accompanied by a process of continuous monitoring [9]: beyond the adoption of robust analysis models from a statistical and pre-visual perspective, or of qualitative analyses or of the identification of synthetic indicators, tools that monitor the impacts appear to be "resilient"; these, through preliminary assessment grids, measure impacts ex ante through qualitative-quantitative criteria or composite synthetic indicators (such as the widespread of evaluation in the application of Next Generation EU designs DNSH EU Model) [48, 49];

– the accompaniment of the project phases of the business plan and proportional budgets with incremental steps based on reaching milestones. Often, this accompaniment foresees a different organization of the system in terms of partnerships, which identify common timelines, shared objectives and innovative financial tools, according to an "inclusive" finance approach (this principle is being consolidated, despite some critical issues, also in the private sector, for example in the Platforms "Produzioni dal Basso"(see https://www.produzionidalbasso.com) e "Real Estate Equity Crowdfunding Italy (see https://www.concreteinvesting.com) [48, 49];

These factors were deemed essential also in the case of the Turin study, and they are behind its success, as illustrated in Sect. 2.1.

4 Conclusions

The festival is a *site-specific* innovation model connected to Turin, to the quality of the local fabric and associations available and the centrality of the local professional and cultural system. However, it can be used to activate new approaches of territorial governance based on the synergy between the institutions and the community and based on a vision which is proactive towards policies and governance, in which the communities play a vital role for active learning, robustness, and the ability to adapt and innovate in the face of change (Mehmood, 2015). In addition, communities have played a central role in establishing the risk linked to the new social and environmental vulnerabilities caused by COVID. They have managed to predict and manage social, institutional, and economic macro-changes to create projects accompanying the processes of co-evolution of the system and its innovation/transformations from the bottom up through the elaboration of adaptive strategies and dynamic processes of planning and design capable of interacting and responding to local needs. The actions encourage bottom-up processes able to promote procedures of innovation in planning, financing processes and to carry out actions, which are adaptable, multi-objective and trans-sectorial, and can produce positive outcomes or multiple components of the system through "multi-objective" strategies and "multi-scalar" strategies both in relation to the temporal dimension and vast-to-local spatial dimension, using processes and new decision-making and action-taking "construction tools". Project and procedural innovation which, as mentioned by Carta di Peccioli (Article 8 Biennale of Architecture 2020), celebrates the interface spaces project: private and public spaces, open or closed, with permanent co-generation practices, capable of building resilient communities and a new alliance between living species, space and society, individuals and the community.

Credits and Knowledge. "Bottom up! Festival of Architecture: when a city is transformed from the bottom up" is a project promoted by Fondazione per l'Architettura/Torino and Ordine Architetti Torino, edited by Maurizio Cilli e Stefano Mirti, coordinated by Raffella Bucci, Maurizio Cilli, Cristina Coscia, Eleonora Gerbotto, Raffaella Lecchi, Stefano Mirti, Serena Pastorino e Alessandra Siviero.

References

1. Gabellini, P.: Le mutazioni dell'urbanistica. Principi, tecniche, competenze, Carocci Editore, Roma (2018)
2. Star, S.L., Griesemer, J.R.: Institutional ecology, translations' and boundary objects: amateurs and professionals in Berkeley's Museum of Vertebrate Zoology, 1907–39. Soc. Stud. Sci. **19**(3), 387–420 (1989). https://doi.org/10.1177/030631289019003001
3. Brand, F.S., Jax, K.: Focusing the meaning (s) of resilience: resilience as a descriptive concept and a boundary object. Ecol. Soc. **12**(1) (2007)
4. Meerow, S., Newell, J.P., Stults, M.: Defining urban resilience: a review. Landsc. Urban Plan. **147**, 38–49 (2016). https://doi.org/10.1016/j.landurbplan.2015.11.011
5. White, I., O'Hare, P.: From rhetoric to reality: which resilience, why resilience, and whose resilience in spatial planning?. Environ. Plan. C Gov. Policy **32**(5), 934–950 (2014)
6. Batty, M.: The size, scale, and shape of cities. Science **319**(5864), 769–771 (2008)
7. Godshalk, D.R., Brody, S., Burby, R.: Public participation in natural hazard mitigation policy formation: challenges for comprehensive planning. J. Environ. Plann. Manag. **46**(5), 733–754 (2003). https://doi.org/10.1080/0964056032000138463
8. Andersson, E.: "Reconnecting cities to the biosphere: stewardship of green infrastructure and urban ecosystem services" - where did it come from and what happened next? Ambio **50**(9), 1636–1638 (2021). https://doi.org/10.1007/s13280-021-01515-z
9. Cheshire, L., Esparcia, J., Shucksmith, M.: Community resilience, social capital and territorial governance. Ager: revista de estudiossobre despoblacion y desarrollo rural (18), 7–38 (2015). https://doi.org/10.4422/ager.2015.08
10. Hallegatte, S., Economic resilience: definition and measurement. In: World Bank Policy Research Working Paper, vol. 6852 (2014)

11. Rose, A.: Measuring economic resilience to disasters: an overview. In: An Edited Collection of Authored Pieces Comparing, Contrasting, and Integrating Risk and Resilience with an Emphasis on Ways to Measure Resilience, vol. 197 (2016)
12. Prati, G., Pietrantoni, L.: Resilienza di comunità: definizioni, concezioni e applicazioni. Psychofenia, 20, vol. XII, pp. 1–26 (2009)
13. Coyle, S.J., Duany, A.: (a cura di): Sustainable and Resilient Communities. A Comprehensive Action Plan for Towns, Cities and Regions. Wiley, Hoboken (2011). https://doi.org/10.1108/ijshe.2011.24912daa.010
14. Brunetta, G., et al.: Territorial resilience: toward a proactive meaning for spatial planning. Sustainability 11(8), 2286, 17 (2019). https://doi.org/10.3390/su11082286, www.mdpi.com/journal/sustainability
15. Pickett, S.T.A., Cadenasso, M.L., Grove, J.M.: Resilient cities: meaning, models, and metaphor for integrating the ecological, socio-economic, and planning realms. Landsc. Urban Plan. 69, 369–384 (2004). https://doi.org/10.1016/j.landurbplan.2003.10.035
16. Voghera, A., Giudice, B.: Green infrastructure and landscape planning in a sustainable and resilient perspective. In: Arcidiacono, A., Ronchi, S. (eds.) Ecosystem Services and Green Infrastructure. Perspectives from Spatial Planning in Italy, pp. 213–224, Springer, Cham (2021). ISBN 978-3-030-54345-7. https://doi.org/10.1007/978-3-030-54345-7_16
17. Wilson, G.A.: Community resilience and social memory. Environ. Values 24(2), 227–257 (2015)
18. Battaglini, E., Masiero, N.: Sviluppo locale e resilienza territoriale: un'introduzione. Sviluppo locale e resilienza territoriale: un'introduzione, 5–22 (2015)
19. Saporiti, G., Echave, C., Scudo, G., Rueda, S.: Strumenti di valutazione della resilienza urbana. TeMA-J. Land Use Mob. Environ. 5(2), 117–130 (2012)
20. Fadda, N., Pischedda, G., Marinò, L.: Sustainable-oriented management come fattore di resilienza organizzativa. Un caso di studio. Manag. Control (2021)
21. Kim, H., Marcouiller, D.W.: Natural disaster response, community resilience, and economic capacity: a case study of coastal Florida. Soc. Nat. Resour. 29(8), 981–997 (2016). https://doi.org/10.1080/08941920.2015.1080336
22. Rose, A.Z.: Economic resilience to disasters (2009)
23. Rose, A.: Economic resilience in regional science: research needs and future applications. In: Jackson, R., Schaeffer, P. (eds.) Regional Research Frontiers - Vol. 1. ASS, pp. 245–264. Springer, Cham (2017). https://doi.org/10.1007/978-3-319-50547-3_15
24. Coscia, C.: The Ethical and Responsibility Components in Environmental Challenges: Elements of Connection Between Corporate Social Responsibility and Social Impact Assessment Chapter Intech (2020)
25. Coscia, C., Mukerjee, S., Palmieri, B.L., Quintanal Rivacoba, C.: Enhancing the sustainability of social housing policies through the social impact approach: innovative perspectives form a "Paris affordable housing challenge" project in France. Sustainability 12(23), 9903 (2020). https://doi.org/10.3390/su12239903
26. Coscia, C., Rubino, I.: Fostering new value chains and social impact-oriented strategies in urban regeneration processes: what challenges for the evaluation discipline? In: Bevilacqua, C., Calabrò, F., Della Spina, L. (eds.) NMP 2020. SIST, vol. 178, pp. 983–992. Springer, Cham (2021). https://doi.org/10.1007/978-3-030-48279-4_92
27. Coscia, C., Rubino, I.: Unlocking the social impact of built heritage projects: evaluation as catalyst of value? In: Bisello, A., Vettorato, D., Haarstad, H., Borsboom-van Beurden, J. (eds.) SSPCR 2019. GET, pp. 249–260. Springer, Cham (2021). https://doi.org/10.1007/978-3-030-57332-4_18
28. Russo, M.: Ripensare la resilienza, progettare la città attraverso il suo metabolismo. Techne 15, 39–44 (2018). https://doi.org/10.13128/Techne-23200

29. Stewart, G.T., Kolluru, R., Smith, M.: Leveraging public-private partnerships to improve community resilience in times of disaster. Int. J. Phys. Distrib. Logist. Manag. (2009). https://doi.org/10.1108/09600030910973724
30. Bottomuptorino. https://www.bottomuptorino.it/
31. MIBAC, Bando Festival dell'Architettura (2019). https://creativitacontemporanea.benicultu rali.it/
32. Moulaert, F., Sekia, F.: Territorial innovation models: a critical survey. Reg. Stud. **37**(3), 289–302 (2003). https://doi.org/10.1080/0034340032000065442
33. Moulaert, F., Mehmood, A.: Spaces of social innovation. In: Handbook of Local and Regional Development, pp. 234–247. Routledge, Abingdon (2010)
34. Cottino, P., Zeppetella, P.: Creatività, sfera pubblica e riuso sociale degli spazi. Forme di sussidiarietà orizzontale per la produzione di servizi non convenzionali. Cittalia - Fondazione Anci Ricerche, Roma (2009). http://www.anci.it/Contenuti/Allegati/Paper2.pdf
35. Melis, A.: Radical Creativity. Drawing the future. Italian Design Day 2020, Associazione Per Il Disegno Industiale, pp. 124–125 (2020)
36. Newman, P., Beatley, T., Boyer, H.: Resilient Cities. Responding to Peak Oil and Climate Change. Island Press, Washington (DC) (2009)
37. Otto-Zimmermann, K., (eds.): Resilient cities. Cities and Adaptation to Climate Change - Proceedings of the Global Forum 2010. Springer, London (2011). https://doi.org/10.1007/978-94-007-0785-6
38. Davoudi, S., et al.: Resilience: a bridging concept or a dead end? "Reframing" resilience: challenges for planning theory and practice interacting traps: resilience assessment of a pasture management system in Northern Afghanistan urban resilience: what does it mean in planning practice? Resilience as a useful concept for climate change adaptation? The politics of resilience for planning: a cautionary note. Plan. Theory Pract. **13**(2), 299–333 (2012). CTRL SE 2013
39. De Filippi, F., Coscia, C., Cocina, G.G.: Piattaforme collaborative per progetti di innovazione sociale. Il caso Miramap a Torino. Techne **14**, 218–225 (2017). https://doi.org/10.13128/Tec hne-20798
40. https://www.bottomuptorino.it/il-bando/
41. Jordan, E., Javernick-Will, A.: Measuring community resilience and recovery: a content analysis of indicators. In: Construction Research Congress 2012: Construction Challenges in a Flat World, pp. 2190–2199 (2012)
42. Berkes, F., Folke, C. (eds.): Linking Social and Ecological Systems: Management Practices and Social Mechanisms for Building Resilience. Cambridge University Press, Cambridge (1998)
43. Portugali, J.: Self-organization and the City. Springer, Heidelberg (2000). https://doi.org/10.1007/978-3-662-04099-7
44. Folke, C.: Resilience: the emergence of a perspective for social-ecological systems analyses. Glob. Environ. Chang. **16**(3), 253–267 (2006)
45. Coscia, C., Russo, V.: The valorization of economic assets and social capacities of the historic farmhouse system in peri-urban allocation: a sample of application of the corporate social responsible (CSR) approach. In: Bisello, A., Vettorato, D., Laconte, P., Costa, S. (eds.) SSPCR 2017. GET, pp. 615–634. Springer, Cham (2018). https://doi.org/10.1007/978-3-319-75774-2_42
46. Calderini, M., Chiodo, V.: La finanza sociale: l'impatto sulla dinamica domanda-offerta. Rivista Impresa Sociale-4/11 (2014)
47. Melis, A., Melis, B.: Community resilience through exaptation. notes for a transposition of the notions of exaptation into a design practice to promote diversity and resilience as an alternative to planning determinism during crisis. Interdiscipl. J. Arch. Built Environ. (2021)

48. Jordan, E., Javernick-Will, A: measuring community resilience and recovery: a content analysis of indicators. In: Construction Research Congress 2012: Construction Challenges in a Flat World, 2190–2199 (2012). https://doi.org/10.1061/9780784412329.220
49. Zampieri, M., Weissteiner, C.J., Grizzetti, B., Toreti, A., van den Berg, M., Dentener, F.: Estimating resilience of crop production systems: from theory to practice. Sci. Total Environ. **735**, 139378 (2020). https://doi.org/10.1016/j.scitotenv.2020.139378

Institutional Logics and Digital Innovations in Healthcare Organizations in Response to Crisis

Stefania De Simone(✉) and Massimo Franco

Università degli Studi di Napoli Federico II, Naples, Italy
{stefania.desimone,mfranco}@unina.it

Abstract. In all countries with advanced welfare systems, healthcare organizations operate in complex institutional systems, which define their space of autonomy in relation to health policy choices and affect their strategic choices, organizational design and management. Healthcare systems are a particularly fruitful context in which to examine how a range of contrasting norms and practices shape innovation. They are in fact confronted with multiple values and demands and the challenge is to simultaneously enhance the quality and reduce the costs of care. The paper examines how competing institutional logics shape innovation development through the use of digital technologies in healthcare organizations responding to emerging events, such as Covid-19 pandemic. Specifically, we adopt an institutional logics perspective to provide insight into the process of innovation, with a focus on the role of telemedicine. The need for social distancing and minimal physical contact challenged and interrupted hospital practices and, in response, digital technologies lead to new processes and services. Remote audio-visual functionality of digital technologies were appropriated in different ways, as stakeholders (state actors, managers, health professionals, and family members) sought to improvise and enhance the protection of persons concerned. Through remote monitoring of patients, telehealth works as a preventative measure to avoid admissions and is therefore a carrier of the managerialist logic of reducing costs by enabling, at the same time, a fast and accurate response to patients' needs.

Keywords: Institutional pressures · Healthcare organization · Digital innovation

1 Introduction

In all countries with advanced welfare systems, healthcare organizations operate in complex institutional systems, which define their space of autonomy in relation to health policy choices and affect their strategic choices, organizational design and management. Over 20 years, hospitals in Italy as well as in other European countries have evolved and changed in response to institutional pressures. The entrance of new logics and governance structures contributed to a transformation of the health system, with a distinction between productive and managerial dimension. The need for rationalization was stimulated by a number of factors, such as the rising health care costs, due to the use of increasingly sophisticated technologies, the progressive aging of the population, the general rise in the cultural level of the population, which led to growing needs to meet [1].

The changes in political and institutional mechanisms defining the resource allocation, and the increasing complexity of relationships among internal and external stakeholders create new challenges to hospital administrators. Governments introduce measures to save costs, patients express growing demands, and citizens demand greater transparency on the functioning and resource utilization. Healthcare organizations must be able to develop appropriate traits of flexibility and innovation to deal with these pressures [2].

The institutional approach focuses on social processes of construction of reality, and how the existence of social interactions tends to stabilize reality through processes of legitimation and to define constraints on the range of possible actions, reducing the variability and unpredictability of individual behavior. The context in which organizations operate is composed of shared and ingrained cultural elements that act as template to organize activities and the modification of this template determines potential organizational changes [3]. If organizations want to survive, their organizational choices must consider external pressures [4]. An organizational form can be adopted not because it is more efficient in terms of transaction costs [5] or adaptation through differentiation-integration (theory of structural contingencies), but because it is considered the appropriate way to organize activities as legitimized in the institutional context. Thus, an organization can adopt an innovation not just for the need to improve its performance, but to get legitimacy. Organizations in a population adapt to their environment, which is constituted by other organizations adapting to it [6]. This reflects the tendency to isomorphism, useful concept to understand innovative processes in health sector.

Thus, healthcare systems are a particularly fruitful context in which to examine how a range of contrasting norms and practices shape innovation. In particular the paper examines how a crisis event shapes innovation development through the use of digital technologies in a healthcare organisation. The crisis is an high-impact and unexpected situation that is perceived by stakeholders to threaten the survival of the organization [7, 8]. The crisis can be defined as "creeping" characterized by the uncertainty of its temporal boundaries, which hinders rapid detection, and by the unforeseen changes that create problems for managers and politicians [9]. Crisis is open to rapid innovation, with established ways of working replaced by alternative practices guided by institutional logics.

The research question of this study is how does the presence of multiple logics affect the adoption of an innovative practice in health care? In the following sections, we discuss institutional logics and how they can be linked to digital technology innovation in healthcare and conclude with implications for practice and directions for future research.

2 Institutional Logics and Digital Innovations

In time of crisis, organizations often introduce innovation practices in order to appear legitimate towards influential external actors. Institutional logics constitute the norms and beliefs that regulate the behaviour of individuals and the selection of technologies [10]. Due to their normative power, logics can constrain human action and be a source of resistance to change [2] and digital innovation [11]. New practices and technologies carry with them new institutional logics which, in turn, challenge the dominant logic of an organisational field. In the institutional approach, the analyses unit is the organizational field including different typologies of actors of a recognized area of institutional life: key suppliers, resource and product consumers, regulatory agencies, and other organizations that produce similar services or products [6]. In the healthcare field the main actors are: hospitals, health professionals, pharmaceutical companies (key suppliers); patients and potential patients (resource and product consumers); national and regional government, medical associations (regulatory agencies); medicine providers, social service provider (organization producing similar services or products) [12]. New logics challenging the dominant logic of an organisational field can become a source of new meanings and practices that actors may enact to bring about change [13]. An organizational field is commonly accepted as a unit of analysis characterized by the dynamics of its institutional logics comprehensive of common beliefs and values both formalized in the law and not [14].

Institutional logics, the beliefs and practices that guide and shape individual or organisational identities and actions [9], are important filters of attention activating goals and schema for individuals to act on, thus guiding how technology is taken up. Logics can thus be important in directing actions during the organisational recovery and adjustment of work practices and influence digital innovation.

During the Covid-19 pandemic, stakeholder groups are connected around digital technology use, including state level actors (the Ministries) corporate actors (technology firms, industry partners and hospital executives), professionals (medical doctors and nurses), and end users (patients and their family). Italian hospitals rapid responses to the COVID-19 crisis demonstrate how telemedicine technologies, serving as digital personal protective equipment, were reutilized for diverse purposes, such as surveillance, control of operations, inpatient safety and quality of care, and family support, as guided by the different institutional logics [15].

Stakeholders with a dominant *state logic* directing their recovery to the emerging pandemic focused their attention on affording national pandemic control, with a goal of protecting citizens within the national boundaries. Drawing on their interests of national governance within the healthcare arena, improvised use of telemedicine technologies traced disease progression and the movement of individuals to contain the disease. Stakeholders with a dominant *managerial logic* focused their attention on affording control of hospital operations. The sudden influx of patients, rapid redistribution of new spaces for hospital beds and new work routines directed their recovery to regain efficiency by integrating real time logistics information with telemedicine technologies. These adjustment practices enabled them to reclaim visibility over their dispersed operations without the need to physically visit contaminated zones. Stakeholders with a dominant *professional logic* addressed their attention towards collaboration and communication with colleagues, overcoming the interferences of protective barriers. Drawing on their need to coordinate expertise, improvised enactment of telemedicine in care delivery enabled them to make use of safe spaces to interact with colleagues and patients from a distance. Their adjustments to the new protective requirements imposed by the pandemic directed their recovery to innovate new ways of sharing expertise and monitoring patient care. The professional logic has been characterised as basing its norms in guilds and associations, following a strategy of increasing personal reputation, and deriving its authority through professional expertise [9]. Stakeholders directed by this logic would be expected to focus their attention on the possibilities for using technology to maintain their expert autonomy and to increase their knowledge, or projecting this knowledge-ability to others. Stakeholders with a dominant *family logic* focused their attention on affording encouragement and commitment to kin. Patients respond positively to family involvement and recovery of contact was improvised using telemedicine technology. The digital innovation of electronic visiting hours became an important crisis response to encourage grieving families [15].

3   Innovation in Healthcare: Telemedicine

Healthcare information technology is the application of information systems and technology to clinical and administrative work in healthcare facilities [16]. It refers to a wide range of clinical, operational, and strategic systems used in hospitals, such as electronic medical records, computerized physician order entry, and patient billing systems.

In Covid-19 time the disease presented the hospital with a marked influx of patients requiring respiratory related treatment, and needing ICU beds. All incoming patients had to be screened. Further, Covid-19 patients could infect others including staff, putting the workforce at risk. In response, directors at Italian hospital constructed rapid design solutions to increase the inpatient bed capacity, pointing to their logic of extending corporation size. Telemedicine is a modality of providing health care services through the use of information technology in situations where health professionals and patients are not in the same location, with transmission of secure information and medical data in the form of texts, sounds, images, or other forms necessary for the prevention, diagnosis, treatment, and patient follow-up. Telemedicine solution included: medical care for isolated coronavirus inpatients, home hospitalisation for coronavirus patients, and continuity of care for non-coronavirus ambulatory patients [17]. Operations were further adjusted to separate coronavirus patients from most of the medical staff. All the healthcare units were designed with separation between 'clean' and 'contaminated' zones, distinguished circulation routes, and a special control room to supervise remotely the operations of the unit. Digital technologies can afford the management of patients through audio-visual communication between a 'control room' in the 'clean' zone and the patients and necessary staff in the 'contaminated' zone. As stakeholders sensitised by corporate logic direct their attention to gaining control, the innovative use of technology provides real-time information supporting workforce and patient safety. Using the telemedicine, operational directors of the hospital guided by the logic of keeping managerial control and authority remotely adjusted their practices.

Telemedicine technologies were set up to care for patients in the hospital, particularly Covid-19 patients in critical care. These patients required intense monitoring, which was difficult for staff to sustain wearing the stifling personal protective equipment. The Covid-19 pandemic presented a crucial challenge to family as they were suddenly unable to visit their hospitalised relatives, and thus separated from their ill loved ones. In response, practices were adjusted through remote care technologies to enable patients to communicate with family. Sensitized by the family logic of making family interests part of the hospital's processes, visiting practices were adjusted using telemedicine technologies to afford encouragement between patients and family members in difficult times.

Italy as well as other several countries increased their intensive care units (ICU) capacity response by converting general ICU in dedicated Covid-19 facilities. In the ICU of University of Naples Federico II logistics and staff organizations were fundamental to avoid in hospital the spread of the virus while creating dedicated Covid-19 facilities [18]. Each ICU bed was equipped with a full monitoring of vital parameters and a mechanical ventilator. Each monitor is duplicated in the centralized control unit equipped with microphones and glasses to allow the communications between the staff. During the 12-h shift, the nursing and medical working was organized as follow: the most experienced ICU physician is the work shift coordinator and stays in the green area to control the compliance of the staff with the procedures and to check the patients from the centralized monitoring area. Medical staff review the medical records of each patient, and then a briefing with the whole staff is made to plan the actions of the shift. A simple logistic project and clear organizational plan may be the keys to the success of surging the ICU capacity with dedicated facilities during the COVID-19 outbreak [18].

4   Information Technology Innovation as Carrier of Multiple Logics

Information Technology (IT) innovations are complex initiatives involving various stakeholders and professions, with divergent expectations of what an IT innovation should do and how it should be deployed, often, retarding its adoption and implementation [10, 19]. Multiple interpretations of an IT innovation resonate different institutional logics [20]. namely, the cultural resources and norms that shape the way individuals perceive their social reality and, therefore, guide their behaviours and decisions [21].

IT innovation promote the logic of managerialism, a set of principles and practices that value cost-efficiencies, performance, and accountability, in contrast with the logic of medical professionalism, which safeguards the autonomy of clinical practice in the provision of patient care [22]. More recently, healthcare policies and IT innovations promoting health self-management and home-based monitoring have contributed to the diffusion of the logic of patient-centred care. This logic promotes a care model that empowers patients to make informed decisions giving them more control over their own health [23]. It thus challenges medical professionalism by diminishing the authority of medical practitioners over patients' decisions. Hence, medical professionals with managerial positions and in charge of IT innovation have to integrate new technologies into day-to-day work by safeguarding the integrity of medical practice.

IT innovations can generate tensions among competing logics. Clinical management information systems often respond to the managerialist logic of performance and efficiency in healthcare resource management and clinical practice [24, 25]. The introduction of these systems creates tension with medical professionalism by disrupting established patterns of work and challenging the professional autonomy of clinicians [26–28].

Telehealth can thus be the carrier of multiple logics. Through remote monitoring of patients, telehealth works as a preventative measure to avoid admissions and is therefore a carrier of the managerialist logic of reducing costs. At the same time, telehealth can enable a fast and accurate response to patients' needs, by improving the quality of care [24].

The success of IT innovation in healthcare depends on how stakeholders shape and are shaped by the tensions among competing logics and the misalignment of interests and values that such logics entail [19].

5 Conclusions

This paper has important implications for theory, particularly in the interaction between logics, digital technology and crisis. We add to understand the role of institutional logics in directing the attention of stakeholders to find diverse action possibilities through digital technology in times of crisis. Understanding the dominant logic held by a stakeholder group is important in shedding light on how digital innovation emerges in response to a crisis.

For management, awareness of the different institutional logics informing innovation processes can help decision-making to become more proactive, and reflect on priority issues, as well as the necessary trade-offs between stakeholders.

The health care sector is a striking example of a public organizational field where multiple values and demands are at play. Hospitals in many countries are confronted with the challenge to simultaneously enhance the quality and reduce the costs of care.

In times of transformative environmental changes, only those organizations, matching their capabilities to the changing environment, will survive and learn. Organizations must find the way to get external legitimacy in order to achieve knowledge, financial and intellectual resources [29].

Clinical managers play a crucial role in securing the implementation and sustainability of IT innovation in healthcare. Yet, not all clinical managers are willing and able to support IT innovation, particularly when the institutional logics of an IT innovation challenge their professional practice [24].

Institutional pressures stimulate the development of innovations and organizational learning, the process by which organization must adapt to environmental changes, modifying its behavior to meet both internal and external demands [1]. The interaction between environment's demands and organization's capabilities can create innovative processes not planned before (involuntary isomorphism). If the hospital organization is able to maintain its legitimacy, the acquisition of knowledge will lead to knowledge creation and ensure that the hospital organization will fit its changing institutional environment. Differently, organizations can decide to imitate other innovator actors to get external legitimacy (voluntary isomorphism).

Given the possible trajectories and uses for a digital technology, a practical implication is to encourage decision makers to develop greater awareness and openness to the multiple logics that are relevant for the success of their organisation.

A research question is to further examine the role of digital innovation in sustaining organisational responses to crises over the longer term and how these might be enabled through a culture of innovation both within the organisation and across a wider range of stakeholders embodying a complex array of institutional logics.

References

1. De Simone, S.: Isomorphic pressures and innovation trends in Italian health care organizations. Int. J. Bus. Manag. **12**(6), 26–32 (2017)
2. Franco, M., De Simone, S.: Organizzazioni sanitarie: dal design al management. McGraw-Hill, Milan (2011)
3. DiMaggio, P.J.: Constructing an organizational field as a professional project. In: Powell, W.W., DiMaggio, P.J. (eds.) The New Institutionalism in Organizational Analysis, pp. 183–203. University of Chicago Press, Chicago (1991)
4. Oliver, C.: Strategic responses to institutional processes. Acad. Manag. Rev. **16**, 145–179 (1991)
5. Williamson, O.E.: Comparative economic organization: the analysis of discrete structural alternatives. Adm. Sci. Q. **36**, 269–296 (1991)
6. DiMaggio, P.J., Powell, W.W.: The iron cage revisited: institutional isomorphism and collective rationality in organizational fields. Am. Sociol. Rev. **48**(2), 147–160 (1983)
7. Pearson, C.M., Clair, J.A.: Reframing crisis management. Acad. Manag. Rev. **23**(1), 59–76 (1998)
8. Williams, T.A., et al.: Organizational response to adversity: fusing crisis management and resilience research streams. Acad. Manag. Ann. **11**(2), 733–769 (2017)
9. Troisi, R., Alfano, G.: Is regional emergency management key to containing COVID-19? A comparison between the regional Italian models of Emilia-Romagna and Veneto. Int. J. Publ. Sect. Manag. (2021). https://doi.org/10.1108/IJPSM-06-2021-0138
10. Thornton, P.H., Ocasio, W., Lounsbury, M.: The Institutional Logics Perspective: A New Approach to Culture, Structure, and Process. Oxford University Press, Oxford (2012)
11. Sandeep, M.S., Ravishankar, M.N.: The continuity of underperforming ICT projects in the public sector. Inf. Manag. **51**(6), 700–711 (2014)
12. Reay, T., Hinings, C.R.B.: The recomposition of an organizational field: health care in Alberta. Organ. Stud. **26**(3), 351–384 (2015)
13. Lounsbury, M.: A tale of two cities: competing logics and practice variation in the professionalizing of mutual funds. Acad. Manag. J. **50**(2), 289–307 (2007)
14. Troisi, R., Alfano, G.: Towns as safety organizational fields: an institutional framework in times of emergency. Sustainability **11**, 7025 (2019). https://doi.org/10.3390/su11247025
15. Oborn, E., et al.: Institutional logics and innovation in times of crisis: telemedicine as digital 'PPE'. Inf. Organ. **31** (2021)
16. Greenhalgh, T., et al.: Tensions and paradoxes in electronic patient record research: a systematic literature review using the meta-narrative method. Milbank Q. **87**(4), 729–788 (2009)
17. Leshem, E., Klein, Y., Haviv, Y., Berkenstadt, H., Pessach, I.M.: Enhancing intensive care capacity: COVID-19 experience from a Tertiary Center in Israel. Intensive Care Med. **46**(8), 1640–1641 (2020). https://doi.org/10.1007/s00134-020-06097-0
18. Vargas, M., De Marco, G. , De Simone, S., Servillo, G.: Logistic and organizational aspects of a dedicated intensive care unit for COVID-19 patients. Crit. Care May 18, **24**(1), 237 (2020)
19. Bunduchi, R., et al.: When innovation fails: an institutional perspective of the (non) adoption of boundary spanning IT innovation. Inf. Manag. **52**(5), 563–576 (2015)

20. Boonstra, A., Eseryel, U.Y., van Offenbeek, M.A.G.: Stakeholders' enactment of competing logics in IT governance: polarization, compromise or synthesis? Eur. J. Inf. Syst. **27**(4), 415–433 (2017)
21. Friedland, R., Alford, R.R.: Bringing society back in: symbols, practices, and institutional contradictions. In: Powell, W.W., DiMaggio, P.J. (eds.) The New Institutionalism in Organizational Analysis, pp. 232–263. University of Chicago Press, Chicago (1991)
22. Reay, T., Hinings, C.R.: Managing the rivalry of competing institutional logics. Organ. Stud. **30**(6), 629–652 (2009)
23. Shaw, J.A., et al.: The institutional logic of integrated care: an ethnography of patient transitions. J. Health Organ. Manag. **13**(1), 82–95 (2017)
24. Bernardi, R., Exworthy, M.: Clinical managers' identity at the crossroad of multiple institutional logics in IT innovation: the case study of a healthcare organisation in England. Inf. Syst. J. **30**(3), 566–595 (2020)
25. Currie, W.L., Guah, M.W.: Conflicting institutional logics: a national programme for IT in the organisational field of healthcare. J. Inf. Technol. **22**(3), 235–247 (2007)
26. Abraham, C., Junglas, I.: From cacophony to harmony: a case study about the IS implementation process as an opportunity for organizational transformation at Sentara Healthcare. J. Strateg. Inf. Syst. **20**(2), 177–197 (2011)
27. Davidson, E.J., Chismar, W.G.: The interaction of institutionally triggered and technology-triggered social structure change: an investigation of computerised physician order entry. MIS Q. **31**(4), 739–758 (2007)
28. Exworthy, M.: The iron cage and the gaze: interpreting medical control in the English health system. Prof. Professionalism **5**(1) (2015)
29. Zimmerman, M., Zeitz, G.: Beyond survival: achieving new venture growth by building legitimacy. Acad. Manag. Rev. **27**, 414–431 (2002)

Sustainable Strategic Mobility Plans Towards the Resilient Metropolis

Bruno Monardo$^{(\boxtimes)}$ and Chiara Ravagnan

Department of Planning, Design, Technology of Architecture (PDTA),
Sapienza University of Rome, Rome, Italy
{bruno.monardo,chiara.ravagnan}@uniroma1.it

Abstract. Recent times see cities at the forefront in the fight against the pandemic in the framework of the harmful effects of climate change and urban inequalities issues. This territorial and urban condition is emphasizing the need for a 'holistic' approach to urban resilience and the importance to focus on sustainable mobility policies and planning towards green and inclusive metropolises. The main goal of this paper is to investigate and highlight innovative approaches in European Sustainable Mobility Plans, aimed at overcoming the sectoral critical aspects within a metropolitan resilience perspective. Bologna represents a paradigmatic case of a sustainable metropolitan area with new plans based on resilience, cohesion, connectivity. Starting from the concept of 'new urbanity' to be recognized as "the reciprocal adaptation of urban fabric morphology and conviviality form" in the contemporary metropolitan city, findings and lessons are expected to be useful to extract relevant suggestions for the specific interpretation styles of resilience in planned strategies and specific projects to be applied, notably, in the European context.

Keywords: 'Sustainable mobility plan' · Resilience · 'Metropolitan cities' · 'Urbanity'

1 Introduction: Conceptual Framework, Goals, Methodology

In recent times the policymaking sensitiveness and the scientific disciplinary debate around urban policies and settlement evolution have mostly focused on the present and future role of contemporary cities as ramparts against the harmful effects of climate change disasters, increasing urban inequalities, and - last but not least - threats to collective health. It has been argued about the increasing levels of global warming and heat islands within the urban environment, soil waterproofing and flood risks, water and air pollution generated by location criteria of production activities, increasing mobility flows, progressive lack of biodiversity in the green urban and metropolitan corridors. Furthermore, all these issues are provoking frequent extreme events within the general framework of the global and local impact of human activities on climate changes [1, 2]. Since the beginning of the XXI century, there has been a vibrant and caustic debate about the capacity of urban realities to face, tackle and embrace adaptation and regeneration strategies in front of recurrent systemic crises.

F. Calabrò et al. (Eds.): NMP 2022, LNNS 482, pp. 1110–1121, 2022.
https://doi.org/10.1007/978-3-031-06825-6_107

Looking at the condition of contemporary metropolitan areas, the debate has pointed out the necessity to manage urban sprawl and the related fragmentation of public space, urban fabrics, and local communities, exacerbated in the new millennium by the economic crises of 2001 and 2008, and the present pandemic phenomenon. In particular, the last years see big cities at the forefronts in the fight against the pandemic, facing the spatial, environmental, and social issues of poor accessibility to public facilities, lack of open green and blue spaces and infrastructures, unequal distribution of commons, discontinuity of pedestrian and cycle networks, all issues worsened by social distancing measures related to Covid-19 [3, 4].

These problems, worsened by social distancing measures, have increasingly emphasized the need for an authentically 'holistic' perspective to urban resilience [5], previously fostered by the UN Sustainable Development Goals as well as the European Union Policies.

In the last decades, the term 'resilience' has become an incredibly 'cool concept' due to the explosion of the contemporary global phenomena and emerging crucial issues mentioned before. However, these ancient and new plagues, despite changing dramatically everyone's lifestyles, paradoxically represent a powerful opportunity to boost new paradigms in urban and metropolitan planning strategies. Arguing about 'resilience', unfortunately, when a term is so successful to permeate the common international lexicon, it runs the risk of becoming trivial and suggests a clarification in order to define specific theoretical and operational references.

"Urban Resilience is the capacity of individuals, communities, institutions, businesses, and systems within a city to survive, adapt, and grow no matter what kinds of chronic stresses and acute shocks they experience" [6]. "Urban Resilience is the measurable ability of any urban system, with its inhabitants, to maintain continuity through all shocks and stresses, while positively adapting and transforming toward sustainability." [7].

In the increasing universe of the term resilience applied to urban and territorial settlements, these two definitions appear quite compelling as they are not just limited to highlighting the condition of survival and adaptation, but they underline the capability to grasp the adverse events and related crises as an evolutionary potential.

Starting from this context, the paper's purpose is to highlight how resilience can be considered the key to rethinking the multiple dimensions of regeneration within a holistic approach framework, combining spatial, environmental, social, and institutional issues. In consistency with the recent theoretical and operational references for urban resilience, this concept is reflected in the integrated approach of mobility strategies between public, green and movement spaces in the framework of climate changes and pandemic issues (Sect. 2). Furthermore, the concept is supported by a strategic set of planning tools, as in the case study of the Bologna Metropolitan City, selected by the authors in the Italian context as one of the most vibrant and thriving urban ecosystems as well as one of the most advanced local public administrations. Concerning the case study development, the research method follows the inductive approach (from the particular representative case to the general lessons) and the classic interpretation keys based on the 'descriptive' category [8] mostly developed with a qualitative approach and supported by public reports-analysis, on-site direct investigation sources and interviews (Sect. 3); hence the possibility to argue some main references for plans able to entangle the structuring choices for urban and environmental sustainability, tactical urbanism and participatory approach trying to reflect and highlight the 'urbanity' concept and its dimension, related with the emerging mobility paradigm, concluding the contribution through some critical remarks developed as an open issue (Sect. 4).

2   Exploring the Holistic Approach to Urban Resilience

From a theoretical perspective, the polysemous nature of the term "resilience" in urban policies is enabling innovative multi-disciplinary entanglement, implementing a virtuous dialogue between several knowledge domains (such as health, ecology, environment, socio-economy, law, planning).

Resilience is an answer to urban complexity and interactions, guiding all these sectors towards a sustainable urban metabolism, the use of smart technologies, the implementation of eco-friendly and adaptive urban spaces and networks, as well as the improvement of institutional cooperation.

Moreover, resilience, deepened in the framework of an ecosystemic perspective [9], is related to the concept of anti-fragility [10, 11] that fosters the capability of adaptation to external perturbations, facing vulnerability and preventing risks, offering multiple and coordinated actions and ways of interventions that enable improvements of systems within rapid stresses and long-lasting changes. This concept thus fosters a proactive character of dynamism and adaptation of transformation choices to environmental, economic, and socio-cultural changes and pays attention to the uncertainty of the scenarios and the scarcity of resources, and the need for data analysis, flexibility, and reversibility. At the same time, it affirms the importance of being rooted in the *milieu* and place-based approaches, focusing on the overall and multi-scale quality of the networks of physical, cultural, economic and social relationships.

With such an objective, it is evident that urban resilience requires a holistic approach to urban equity, efficiency, safety and security in cities strengthening the interactions between material networks (infrastructures and transports as well as green corridors) and immaterial networks (ICT, regulated social interactions, and institutional cooperation) [12] considered strategic vectors for the 'right to the city' [13] toward a 'smart city' [14].

Urban resilience requires, thus, new references for a shared cultural project, such as the new European Bauhaus [15] and for an integrated methodological framework. The present methodological frameworks point out the complexity of interactions between networks, agents and factors [16], in order to put in synergy new forms of smart governance of spaces, services and processes [17] combining infrastructure and ecosystems, leadership and strategy, health and wellbeing, economy and society [18].

From an operational perspective, the post-pandemic recovery policies, launched in 2020–21 through the allocation of huge public resources at the international level (i.e. the USA 'American Jobs Plan' or the 'Next Generation EU'), emphasized the resilience concept (Italian 'National Recovery and Resilience Plan', 2021) that fosters an integrated strategy on infrastructure and digital networks (from the ecological transition of rails to MaaS), in consistency with the cohesion principles of territories and civic communities. Furthermore, the documents point out the importance of the strategic dimension for planning to coordinate the different interventions and actions coming from the stakeholders and the city users and to mend the separation between top-down policies and bottom-up practices.

2.1 Mobility Strategies and Urban Regeneration

The post-COVID-19 phase brings with it the potential to build "a new normal" [19] in cities, placing issues related to the health and social distress of citizens but also to the vulnerability of economic systems at the centre of regeneration strategies. Urban planning, which was born in the industrial age as a discipline aimed at addressing sanitation problems and the organization of urban networks and services, took on new responsibilities in the twentieth century, starting from the awareness of the complexity of the contemporary city, and of the deep interrelations between anthropic and natural dynamics in the Anthropocene era [20].

In this context, mobility models and infrastructures assume a central role, starting from data that confirm not only the contribution of road transport to greenhouse gas emissions in European urban areas (about 25%) but also the related economic and health effects [21].

The global and European agendas offer a reference framework for national, regional and local governments to promote a new paradigm of sustainable development, giving priority to investments and resources that hold together objectives intervention (environment and landscape, mobility and infrastructures, public space and urban services), tools (policies, plans, programs, projects) and scales of intervention (European, national, regional, local) placing the theme at the centre of a multilevel governance framework of resilience.

Urban resilience finds particular concreteness in the choices aimed at strengthening sustainable mobility and ecosystem services in the construction of urban networks, reconfiguring the methodological references for the urban space planning, design and management. Integration of urban mobility strategies and tools are the basis for innovation and the keys of best practices. Looking at the 'space of movement', new planning tools can overcome ancient separations with the land use design and the open space system by concretely implementing integrated regeneration strategies [3, 4].

First of all, the EU Sustainable Urban Mobility Plan (SUMP) was officially introduced through the 'Mobility Urban Package' [22] and progressively implemented by the 27 member states. It represents the ambition of combining mobility and transport infrastructures with the urban space design in order to implement the ecological transition through the coordination of different infrastructure networks and urban spaces, overcoming a sectoral approach. The main good practices have been held in Bologna as well as in Bruxelles Capital Region, Grenoble, and Great Manchester.

Furthermore, the combination of cycle, green and public spaces networks is the specificity of Spanish practices experimented in the context of the arising role of the *Estrategia Nacional de Infraestructura Verde y de la Conectividad y Restauración Ecológicas* through new tools that integrate the construction of mobility, public spaces and green networks: emblematic examples are the tools put in place in Vitoria-Gasteiz such as the integrated *Mobility and public space Plan* and the *Plan de Acción Territorial de la Infraestructura Verde del Litoral de la Generalitat Valenciana* [23] where the cycle system is a backbone for the construction of a multiscalar green infrastructure.

Additionally, many Mobility Plans are developing 'local mobility grids' [24] in order to improve local accessibility of facilities and centralities. This goal is supported by the theoretical concept of the "ville du quart d'heure" [25] consolidated within years of studies and pointed out in the phase of Pandemic in the Paris case study. The 15 min-city promotes a reorganization of local accessibility with compact fabrics and services, in order to enable an increase in the quality of life in the ordinary phases and risk reduction during environmental and health crises. A proposal for a local grid is also developed in the Good Move Mobility Plan for the Capital-Region of Bruxelles 2020–2030 (awarded as best SUMP in 2020), where the design strategy of the 'Espace rue' proposes an hypothesis to organize relationships, interactions and conflicts between public spaces and mobility space at the local and urban scale, highlighting the importance of an integrated approach to streets, in consistency with the indicators of the "healthy street" defined by Transports for London [26].

Finally, the flexible and reversible expansion of the space dedicated to pedestrians and local greenery in the framework of "tactical urbanism", from the 'Superillas' in Barcelona [27] to 'Piazze aperte' in Milan is paying growing attention to cycle paths, as a method for the implementation of temporary bike lanes or as experimentation for future structural projects of cycle systems in SUMP, in order to test the interest of citizens and the possible synergies and conflicts with other forms of mobility and public spaces [19].

These strategies are the common keys of numerous experiments and studies at an international level that reveal an acceleration and timeliness of practices in some metropolises: Bologna, Milan, Paris, Brussels, Barcelona, Madrid, New York, Bogotà, Lima [4]. This timeliness is closely linked to the consolidation of structural choices on sustainable mobility considered a priority both by the administration and by the citizens, who have been involved for years in participatory processes and debates on urban regeneration. Among these, Bologna represents an emblematic case in Italy, also for being the first metropolitan city to approve the Sustainable Urban Mobility Plan following the EU format.

3 The Bologna Metropolitan City: Towards a Holistic Approach to Mobility

Bologna represents a paradigmatic case of a sustainable, thriving community, a surprising cradle of policies, plans and projects conceived, developed and implemented following the idea of an emerging identity of 'small metropolis' based on resilience, cohesion, attractiveness, and connectivity to be pursued through the construction of innovative tools in which mobility networks are not conceived as a sectoral dimension but as a crucial bridge connecting 'polis' to 'civitas' and 'urbs'.

Within a few years the local public institutions, the Metropolitan City and the Municipality of Bologna in particular, have been able to pursue a rich and effective path conceiving, discussing and approving numerous integrated and coherent tools in order to face the emerging challenges of contemporary urban communities. Bologna Municipality has recently approved an intriguing new General Urban Plan (July 2021) pursuing an advanced strategic profile according to the innovation principles introduced by the

Emilia Romagna Regional law (n. 24/2017). Looking at the inter-municipal level, recent planning tools as the Metropolitan Strategic Plan, the Metropolitan Territorial Plan and the Sustainable Urban Mobility Plan (SUMP) are proving to be original interpretations of integrated and inclusive planning processes, particularly consistent with the holistic approach promoted by European policies. In the following lines, it is particularly highlighted the crucial role played by mobility and its plans in pursuing and interpreting urban resilience.

3.1 Supporting the Emerging Metropolis Through the Planning Framework

In Italy, metropolitan areas are still a young, 'in progress' juridical reality. The first legislative measure that introduced them more than 30 years ago (l. 142/1990) didn't find mature conditions to be implemented. The relaunch occurred only a few years ago by the re-introduction of the 'Metropolitan City' institution (l. 56/2014), whose primary mission was the identification and coordination of development strategies for the whole metropolitan area through a Strategic Plan.

Within the framework of the UN 'Agenda 2030' objectives and the 'Bologna Charter for Environment', Bologna - the first new Metropolis constituted in Italy (2015) - approved its Metropolitan Strategic Plan (PSM 2.0) in 2018 pursuing three fundamental dimensions:

– sustainability in its environmental, economic and social dimensions nourished by the culture of legality and education in civic values;
– inclusiveness, interpreted as the ability to enhance differences and peculiarities, transforming them into common assets and wealth;
– attractiveness as openness to the original, unexpected, different issues, aware of how to increasingly strengthen its international and cosmopolitan identity.

The Metropolitan Territorial Plan (PTM) fosters a sustainable, resilient and attractive territory, in which the protection of the environment, the beauty of urban and natural places, together with work and innovation can find unitary and propulsive synthesis.

The plan, approved in 2020, within full Covid-19 emergency, is the main vector of new tasks: promoting urban regeneration, enhancing ecosystem services, managing the progression towards 'zero new land consumption', redistributing in an equalized way, on a metropolitan scale, the resources generated by the main urban transformations. The 'territorialization' of development strategies represents the specific object and added value of PTM with particular attention to territorial and urban resilience, risk prevention, service accessibility, welfare system, quality of production areas, and above all adequacy of mobility networks and infrastructural connections.

3.2 Sustainable Mobility Plan: Cultural Roots and Metropolitan Interpretation

At the end of 2019, Bologna was also the first metropolitan city to approve the 'Piano Strategico della Mobilità Sostenibile' (PUMS), the 'Italian interpretation' of the Sustainable Urban Mobility Plan (SUMP), EU strategic tool designed to meet the mobility demand of residents, economic activities, and city-users for the quality-of-life improvement. The SUMP promotes the innovation of traditional approaches between 'settlement space' and 'movement space' through the principles of resilience, integration, participation. Its formalization at the EU level represents the climax of an evolutionary path starting from the Action Plan on Urban Mobility [28] and the Transport White Paper (2011), as well as through a consultation conducted on behalf of the European Commission from 2010 to 2013 with the involvement of numerous experts and sector players; the work finally led to the 'Urban Mobility Package' [29] which recognized the SUMP as a new strategic tool for integrating mobility, accessibility and the city realm all over EU urban and metropolitan areas. In some countries as France or Italy, it has become compulsory for cities or polycentric areas with at least 100,000 inhabitants. The joint work has merged into the first and second edition of SUMP Guidelines [30, 31], official documents by the European Commission addressing public and private stakeholders towards the collective conception, implementation and management of the plan with the ambition of integrating mobility networks, transportation systems and urban planning strategies.

Unlike the more traditional approaches to mobility planning, the SUMP philosophy hinges on main axes as the participatory involvement of citizens and diffused stakeholders, the coordination of administrations at different levels, the harmonization of sectoral strategies enhancing the synergy between existing and in-progress tools. The SUMP aims at participatory democracy, processualism, prefiguration and evaluation of evolutionary scenarios, careful monitoring, and remodelling of the implementation phases.

In Italy, the mobility plan concept represents the mature evolution of the 'Mobility Urban Plan' (PUM), originally introduced in 2000 drawing inspiration from the French *'Plan de Déplacements Urbains'* (PDU), born in 1982 with the law LOTI (*Loi d'Orientation des Transports Intérieurs*). Originally the French model was based on the principle of 'right to transport' mostly meant in its technical-functional dimension; afterward, the plan profile was enriched with themes and contents related to the emerging ecologic-environmental dimension (law LAURE, *Loi sur l'Air et l'Utilisation Rationnelle de l'Énergie*, 1996) and 'urban welfare' policies (law SRU, *Solidarité et Renouvellement Urbain*, 2000). In the last decades, the PDU has reached an explicit organic integration with urban and inter-municipal planning tools, strengthening its 'strategic' role overcoming the sectoral dimension and expanding urban and metropolitan identity. Therefore, if the mobility plan allows the virtuous integration of social, environmental and symbolic dimensions with land use and infrastructural design, it could give substance to the original Lefebvrian idea of the 'right to the city' [32], recovered and interpreted with growing awareness in the literature that discussed the metamorphosis of urban lifestyles [33–35].

Indeed, the French model of 'movement space' and 'network urbanism' [36], which has inspired the EU idea of sustainable mobility plan, is not limited to pursuing the efficiency (and safety) of displacement vehicles, the rethinking of parking system or the rational circulation of people and goods, but proposes a more inclusive idea of urban welfare policies, prefiguring the passage from *'droit au transport'* (LOTI, 1982) to *'droit à la mobilité'* (LOM, *Loi d'Orientation des Mobilités*, 2019), tackling every form of inequality, marginalisation and isolation in the city.

In the first national Sustainable Mobility inter-municipal plan, the vision aims to "make Bologna metropolitan area more attractive through high levels of urban quality and liveability in order to enhance the cohesion and attractiveness of the territorial system as a whole and strengthen the role of its capital as international city".

The tool pursues the objectives of territorial development and regeneration by placing the crucial focus on values, rights and primary needs of the community, from health to safety, from accessibility to essential services and social inclusion, from education to work and leisure.

The holistic approach evoked in the disciplinary debate finds concreteness in the macro-objectives that outline the pillars of urban and territorial sustainability. The mobility and accessibility issues stand out in their kaleidoscopic interpretations: from the physical-spatial dimension of the reconnection between centrality and peripheries, to the 'environmental imperative' of tackling emissions and fostering resilience to climate change.

The accessibility ensured by collective transport networks and by encouraging micro-mobility is then seized as an opportunity to restore urbanity, social cohesion, proximity to local facilities and a 'sense of belonging' to the communities widespread across the territory [37]. Bologna, even in times of pandemic, confirms the rule that requires the administrations traditionally active in outlining integrated policies and open processes to be resilient and embrace adaptive flexible geometries facing striking times and undesired events.

In the Bologna sustainable mobility plan the resilience approach is highlighted by the "Biciplan", a sort of ecological cycle metro (inspired by the Réseau Vélo of the Paris region), conceived 'ex-ante' and integrated into the new plan, a precious resource for its capacity to create an organic framework, the 'Bicipolitana', bike structural and interconnected network that proved to be very effective both for the tactical interventions solicited by the health emergency and for the long term strategic relationship system.

4   Synergic Mobility Planning as 'New Urbanity'. Open Issues

In the pandemic season, cities and metropolitan areas have been identified as the privileged domain for rethinking plans, programs, and projects useful to manage the community health issues looking at specific problems such as the correlation between settlement densities, public transportation and virus spread or the pathological rise of structural inequalities at social and economic level [38].

According to emerging principles in the scientific-disciplinary debate, despite the persistent crisis, this condition represents the trigger to speed up processes of urban regeneration requiring an integrated approach to urban planning and mobility, in order to create the conditions for more inclusive, green and resilient cities [19], mostly aiming at proximity displacements [25], non-polluting active mobility, flexible and inclusive public spaces [7, 39].

Are local administrations proving to be equipped to manage conditions and opportunities suddenly opened up by the crisis and design virtuous scenarios of transformation? Ideas and initiatives paint a multicolour landscape and reveal the pendulum between many tactical, pop-up initiatives and rare strategic scenarios. Urban and metropolitan actors are called to face a terrific challenge: issues at stake are relevant and the cultural 'dna' of mobility styles, more or less sensitive to sustainable displacements, is making the difference even in emergency contingencies. The temptation to encourage micro-mobility with 'pret à porter' tactical urban planning has proved unavoidable, but international cases clearly demonstrate that the short-term approach cannot be enough.

The local administrations that have responded more promptly and effectively to the pandemic challenge seem to largely coincide with those traditionally active in designing integrated policies, and therefore equipped with new plans, programs and other tools consistent with a vision of overall strategic transformation. In fact, the implementation of coordinated actions between short and long-term choices requires a wide range of strategic and operational tools able to manage emergency and temporary interventions within balanced relationship frameworks, maximizing the virtuous effects in space and time [40]. The Bologna experience shows how the cities that have adopted timely and convincing measures are those with a consolidated cultural background, capable of promoting and implementing tools characterized by holistic and strategic approaches able to re-boost synergistically physical and intangible networks for urban and metropolitan resilience.

The integrated and inclusive planning process in the metropolitan area of Bologna appears particularly advanced due to its consistency with the holistic approach promoted in EU policies and implemented by significant metropolitan areas in western Europe. Its recent planning path shows the integration of mobility networks and public transportation systems with urban patterns, green-blue corridors and public spaces to be planned through participatory democracy's steps. Recent Bologna metropolitan tools represent the essential matrix useful to rethink and adapt spaces and forms of mobility, struck by unpredictable emergencies.

Original principles of new mobility plans escape from the sectoral dimension, assuming a strategic role through the ambition to integrate infrastructural space and land use design at a metropolitan scale. It's the attempt to prefigure a 'new urbanity' based on the synergy between the dimension of movement and the space of stasis that we all have experimented during lockdowns.

What is meant by "new urbanity"? It is one of the most complex and polymorphic concepts: prestigious schools of thought speak of urbanity as "the reciprocal adaptation of urban fabric morphology and conviviality form" [41, 42]. It represents the intertwining of the social '*mixité*' which promotes integration, and the public space in all its forms, which becomes the privileged place for its development. And the mobility spaces, with particular attention to the pedestrian and cycle paths, offer original and creative interpretations of urbanity [43, 44]. The lesson that emerges from the change in lifestyles as a result of the pandemic cannot, therefore, concern only the optimization of what already exists: it is necessary to rethink the infrastructural sites in the intertwining with urban functions, reflect on the consequences of the density remodulation, not only residential, but that of workplaces, university towns, urban services for education, free time, consumption and above all the flow densities in the space of movement, with a regulated downsizing of the capacity of collective transport carriers to be made more attractive and safe, preventing at the same time dangerous crowds during peak hours. All features that must find a new interpretative key in terms of public health protection.

Mobility is not just an opportunity of creating relationship spaces but embodies the 'place' essence wherever it is produced. Not just a technique to connect nodes and areas, but the vector of an everlasting new 'urbanogenesis'.

Research Context and Authors' Contribution. This paper illustrates the first investigation paths within the 'Sapienza' University of Rome international research project "Mobility Infrastructures. Towards new interpretation paradigms and operational tools for the resilience of European Metropolitan cities" (Coord. Monardo B. and Ravagnan C.). These reflections represent an authors' evolution of a preprinted paper in the 57th ISOCARP World Planning Congress Proceedings. Both authors conceived and developed organically the paper, however, Sects. 1 and 2 are by C. Ravagnan, Sects. 3 and 4 by B. Monardo.

References

1. IPCC: Climate Change 2021. The Physical Science Basis. Contribution of WG1 to the Sixth Assessment Report of IPCC. IPCC, Geneva (2021)
2. IPCC: Climate Change 2014. Synthesis Report. Fifth Assessment Report of the Intergovernmental Panel on Climate Change. IPCC, Geneva (2014)
3. Un-Habitat: Un-Habitat Guidance on Covid-19 and public space (2020). www.unhabitat.org/sites/default/files/2020/06/un-habitat_guidance_on_covid-19_and_public_space.pdf. Accessed 10 Feb 2022
4. OECD: Respacing our cities for resilience (2020). www.itf-oecd.org. Accessed 10 Feb 2022
5. UNDRR: Making cities resilient (2020). www.unisdr.org/campaign/resilientcities/. Accessed 10 Feb 2022
6. The Rockefeller Foundation. https://www.rockefellerfoundation.org/100-resilient-cities/. Accessed 10 Feb 2022
7. Un-Habitat Urban Resilience Hub. https://urbanresiliencehub.org/. Accessed 10 Feb 2022
8. Yin, R.K.: Case Study Research: Design and Methods. Sage Publications, Beverly Hills (1984)
9. Acierno, A.: La visione sistemica complessa e il milieu locale per affrontare le sfide. In: AAVV, Le sfide per la resilienza urbana, TRIA, vol. 15 (2015)
10. Taleb, N.N.: Antifragile: Things That Gain from Disorder. Random House Publishing, New York Group, New York (2012)
11. Blecic, I., Cecchini, A.: Verso una pianificazione antifragile. Come pensare al futuro senza prevederlo. Franco Angeli, Milano (2016)
12. Gargiulo, C., Maternini, G., Tiboni, M., Tira, M.: New scenarios for safe mobility in urban areas. TeMA (Special issue), 1 (2022)
13. Amato, C., Cerasoli M., de Ureña J.M., Ravagnan C.: Percorsi di resilienza in Italia e Spagna. Fenomeni insediativi contemporanei e nuovi modelli di mobilità. In: Talia M. (ed.), La città contemporanea: un gigante dai piedi d'argilla. Planum, Milano (2019)
14. Lauri, C.: Smart mobility. Le sfide regolatorie alla mobilità urbana. Rivista Trimestrale di Scienza dell'Amministrazione, 1 (2021)
15. EC. New European Bauhaus (2021). https://europa.eu/new-european-bauhaus/about/about-initiative_en. Accessed 10 Feb 2022
16. Troisi, R., Alfano, G.: Towns as safety organizational fields: an institutional framework in times of emergency. Sustainability (Switzerland) **11**(24), 7025 (2019)
17. Frantzeskaki, N. Urban resilience. A concept for co-creating cities at the future, Resilient Europe (2016). https://urbact.eu/sites/default/files/resilient_europe_baseline_study.pdf. Accessed 10 Feb 2022

18. Rockefeller Foundation 2013, City resilient framework. https://www.rockefellerfoundation.org/wp-content/uploads/City-Resilience-Framework-2015.pdf. Accessed 10 Feb 2022
19. OECD, City Policies Responses, in Tackling Coronavirus. Contributing to a Global effort. www.oecd.org/coronavirus/en/. Accessed 10 Feb 2022
20. Crutzen, P.J.: The "Anthropocene". In: Ehlers, E., Krafft, T. (eds.) Earth System Science in the Anthropocene. Emerging Issues and Problems. Springer, Cham (2000). https://doi.org/10.1007/3-540-26590-2_3
21. European Environment Agency EEA: Greenhouse gas emissions from transport in Europe. https://www.eea.europa.eu/data-and-maps/indicators/transport-emissions-of-greenhouse-gases/transport-emissions-of-greenhouse-gases-12. Accessed 10 Feb 2022
22. European Commission: Together towards competitive and resource-efficient urban mobility. COM 913, Brussels (2013)
23. Ravagnan, C.: Rigenerare la città e i territori contemporanei. Prospettive e nuovi riferimenti operativi per la rigenerazione. Aracne, Roma (2019)
24. Ravagnan, C., Cerasoli, M., Amato, C.: Post-Covid cities and mobility. TeMA – J. Land Use Mob. Environ. 87–100 (2022). https://doi.org/10.6093/1970-9870/8652. Accessed 10 Feb 2022
25. Moreno, C.: Droit de cité: de la "ville-monde" à la "ville du quart d'heure." Editions de l'Observatoire, Paris (2020)
26. Transport for London, Healthy streets. www.london.gov.uk/sites/default/files/healthy_streets_explained.pdf. Accessed 10 Feb 2022
27. Rueda, S.: La supermanzana, nueva célula urbana para la construcción de un nuevo modelo funcional y urbanístico de Barcelona (2016). www.bcnecologia.es/sites/default/files/proyectos/la_supermanzana_nueva_celula_poblenou_salvador_rueda.pdf. Accessed 10 Feb 2022
28. European Commission: Action Plan on Urban Mobility, COM 490, Brussels (2009)
29. European Commission: White Paper, Roadmap to a Single European Transport Area - towards a competitive and resource efficient transport system, Brussels (2011)
30. Rupprecht Consult (ed.): Planning for People. Guidelines. Developing and Implementing a Sustainable Urban Mobility Plan, European Commission, Directorate-General for Mobility and Transport, Brussels. www.eltis.org/sites/default/files/guidelines-developing-and-implementing-a-sump_final_web_jan2014b.pdf. Accessed 10 Feb 10
31. Rupprecht Consult (ed.) Guidelines for Developing and Implementing a Sustainable Urban Mobility Plan, Second Edition, European Commission, Directorate-General for Mobility and Transport, Brussels. www.eltis.org/sites/default/files/sump_guidelines_2019_interactive_document_1.pdf. Accessed 10 Feb 2022
32. Lefebvre, H.: Droit à la ville. Anthropos, Paris (1968)
33. Mitchell, D.: The Right to the City: Social Justice and the Fight for Public Space, 1st edn. The Guilford Press, New York (2003)
34. Harvey, D.: Rebel Cities: From the Right to the City to the Urban Revolution. Verso Books, London (2013)
35. Secchi, B.: La città dei ricchi e la città dei poveri. Laterza, Bari (2013)
36. Dupuy, G.: L'urbanisme des reseaux. Théories et methods. A. Colin, Paris (1991)
37. Monardo, B.: il ruolo delle reti della mobilità dolce per una nuova urbanità post-Covid 19. In: Moccia, D., Sepe, M. (eds.) XII giornata internazionale di studi INU, Benessere e/o salute? 90 anni di studi, politiche, piani. Urbanistica Informazioni, S.I. 289 (2020)
38. Nomisma, 3° Rapporto sul mercato immobiliare in Italia. www.nomisma.it/presentati-i-dati-del-3-rapporto-sul-mercato-immobiliare-2020. Accessed 10 Feb 2022
39. Honey-Rosés, J., et al.: The Impact of COVID-19 on Public Space: An Early Review of the Emerging Questions – Design, Perceptions and Inequities. Cities & Health. Taylor and Francis, Cambridge (2020)

40. Lydon, M., Garcia, A.: Tactical Urbanism. Short-term Action for Long-term Change. Island Press, Washington DC (2015)
41. Choay, F.: Le Règne de l'urbain et la mort de la ville. In: Dethier, J., Guiheux, A. (eds.) La Ville, art et architecture en Europe, 1870–1993, pp. 26–35. Editions du Centre Pompidou, Paris (1994)
42. Choay, F.: Urbanité. In: Merlin, P., Choay, F. (eds.) Dictionnaire de l'urbanisme et de l'aménagement. PUF, Paris (1996)
43. Levy, J.: La mesure de l'urbanité. Urbanisme **296**, 58–60 (1997)
44. Lévy, J.: Modèle de mobilité, modèle d'urbanité. In: Institut pour la ville en mouvement (ed) Les sens du mouvement. Modernité et mobilité dans les sociétés contemporaines. Belin, Paris (2004)

A GIS Application for the Hospitalization of COVID-19 Patients

Michele Mangiameli[✉] and Giuseppe Mussumeci

Department of Civil Engineering and Architecture, University of Catania, Catania, Italy
{michele.mangiameli,giuseppe.mussumeci}@unict.it

Abstract. During the COVID-19 pandemic period, it is often necessary to hospitalize a patient positive for the virus and in serious health conditions in suitable hospitals. The difficulty in managing these emergencies arises from the fact that it is often not possible to know which is the nearest hospital with beds available for hospitalization of the COVID-19 patients. In this work, we present a GIS application based on a relational database that allows to determine an optimal path for the patient transport from a starting point to the nearest hospital with free places for hospitalization. The developed application reduces the patient's transport time, decreasing the exposure time of the medical staff in the ambulance in contact with the positive patient. The application was developed in the urban area of Catania where hospitals gather a large pool of users and therefore it is essential to have a system that in real-time provides the available beds and thus optimize the distribution of COVID-19 positive patients who need an admission to a hospital.

Keywords: Covid-19 · Spatial database · Desktop GIS · WebGis

1 Introduction

Almost two years have passed since Chinese authorities identified a deadly new coronavirus strain, SARS-CoV-2 (January 7, 2020), causing more than 100,000 people infected and thousands of deaths. The whole world scientific community has realized that the pandemic problem would soon affect the whole world.

This important and dangerous virus must be fought by everyone and with all the weapons at the disposal of science with research activities involving all sectors from medicine, geography, engineering, etc. [1–9]. Obviously, many works are affecting the research activities of medicine that must deal with the virus biologically, even with its changes and variants.

New information technologies have provided valid tools for studying and fighting the virus. Among all, certainly the GIS technology has been widely used because it is a valid tool to decision support systems. Indeed, GIS technology is used in various fields providing valid applications, ranging from energy management, risk management, mobile robotics, socio-economic studies, land planning and more.

Several GIS applications have been developed to manage the COVID-19 pandemic phenomenon thanks also to the ability to manage big data. In this way, the GIS technology has been used in identifying the spatial transmission of the epidemic, in spatial prevention and control of the epidemic, in spatial allocation of resources, and in spatial detection of social sentiment, among other things [10]. South Korea has given the best example in controlling COVID-19 outbreak thorough GIS applications, in order to identify, track and monitor the infected people, and the places that each detected patient had visited before being certified as positive to the virus [11, 12]. Many studies have also involved the development of GIS applications for socio-economic analysis in order to understand how the socio-economic aspect can contribute to the spread of the virus. For example, from a socio-economic analysis carried out on the territory of Iran with GIS technology, it was found that the reduction of population concentration in some urban land uses is one way to prevent and reduce the spread of COVID-19 disease. The results obtained using the GIS application have shown that the central and the eastern regions of Tehran are more at risk and public transportation stations and pharmacies were the most correlated with the location of COVID-19 patients in Tehran [13]. Finally, several studies have been addressed to understand the spatiotemporal dynamics of COVID-19 to define decision-making processes, planning and community action to mitigate and clarify the extent and impact of the pandemic [14–16].

In this work, a GIS application was developed that allows reducing the patient's hospitalization time by providing the optimal path according to the COVID-19 beds available in hospitals. The application is characterized by an architecture where the data are collected and structured within a spatial RDBMS (Relational Database Management System) and linked to a GIS platform. We tested the GIS by using the city of Catania (Sicily, Italy) as test case.

2   The GIS Application

The transport of a positive COVID-19 patient to a hospital must be carried out as quickly as possible both to avoid worsening the patient's health conditions and to reduce the time exposure of the transporters. To reduce the patient's travel time, it is necessary to know the nearest hospital with free covid beds and the shortest route to reach them. To these purposes, we developed a GIS platform to manage the transport of a COVID-19 patient using a database containing a road graph structured in arcs and nodes and a vector layer containing the hospitals capable of managing COVID-19 positive patients. The GIS application, which can be implemented in the WEBGIS version, allows to choose a starting point and automatically provides the fastest route to the nearest hospital with free beds. The automatic calculation of the starting path in the GIS environment is performed using the network analysis tools, which is able to consider different variables, including the road interruptions or the traffic conditions, if available as vector or raster data.

The relational database has been implemented in a Relational Database Management Systems (RDBMS) extern to the GIS platform [2], in which the trigger functions have been implemented to automatically discard hospitals that have run out of beds and therefore will not be used as possible destinations for the patient.

Using this hardware architecture and the dedicated RDBMS software, it is possible to avoid problems of data redundancy and inconsistency, problems of competition for access to data by multiple simultaneous users, loss of data integrity, security problems, and problems of efficiency from the point of view of data search and updating.

The hardware architecture is characterized by a server where the RDBMS and the spatial database are located, while all GIS applications (Desktop and Web) are in the external device (tablet, PC, smartphone, etc.). With this architecture, all the characteristics of the DBMS and the spatial database are exploited, and the speed for updating the database modification and query is improved, thanks to a totally dedicated machine.

The deployment diagram of Fig. 1 shows how and where the architecture proposed is to be deployed, reporting the relationships between the hardware and software components used for the implementation of the developed GIS application.

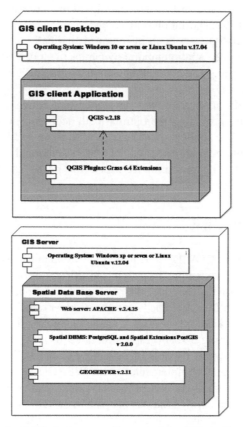

Fig. 1. UML deployment diagram of the GIS application

3 The Spatial Database

The spatial database has been developed using a relational data structure in an open-source external RDBMS. The use of this data structure allows managing different types of geometric information, spatial relationships, topological relations, directional relations, and proximity relations within the GIS application developed.

The design of the data structure to be implemented within the RDBMS has followed three main phases:

1. identification of hospitals and road infrastructures belonging to the area under study;
2. the relationships between the entities previously identified;
3. the conceptual design of the data structure using the entity relationship diagram;
4. the physical implementation of the spatial database within in the spatial RDBMS.

The structure of spatial relational database is characterized by three entities: nodes, arches, hospitals.

The arc-node relation represents the road graph allowing to reach hospitals. The Arc entity is characterized by the attributes Arc_ID and Name, which are respectively the primary key and an attribute. With respect to the architecture arc-node of the road graphs, the entity Arc is the entity related to Node.

The Nodes entity is characterized by the attributes Nodes_ID, Node_type and cost as respectively the primary key and simple attributes. In particular, the node type attribute contains the access or street_node value if it belongs to the road graph or access node to a hospital.

The hospitals entity is characterized by the attributes Hospital_ID, Name and Beds. This entity includes a trigger function that assigns a high cost to the access node corresponding to the hospital that has run out of beds. In this way the access node is discarded by the path search algorithm, providing the user with another destination with available beds.

Figure 2 shows the E-R diagram relating to the entities described above.

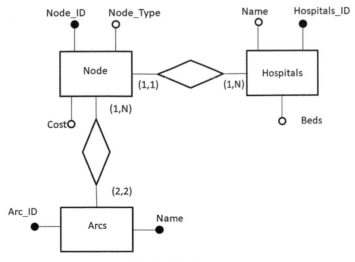

Fig. 2. E-R diagram

4 Test Case: The City of Catania

The developed GIS application was tested on the metropolitan area of Catania (Sicily, Italy), where six main public hospitals are located.

As raster cartographic support, we chose a geo-referenced orthophoto in EPSG 6708, managed in the GIS environment as a WMS service using the web link to "Sistema Informativo Territoriale Regionale - Regione Siciliana" (Fig. 3).

Using this cartographic basis, the arcs and nodes were digitized as vector thematisms with linear and puncture geometry, respectively. These two themes make up the roadgraph shown in Fig. 4.

Fig. 3. Raster cartographic support

Fig. 4. Road graph of the city

The node thematisms contain both the nodes belonging to the road graph and the access nodes to the hospitals and includes the starting and ending points for the optimal path search algorithm. Therefore, a graphic style was chosen to differentiate the two types of nodes (Fig. 5).

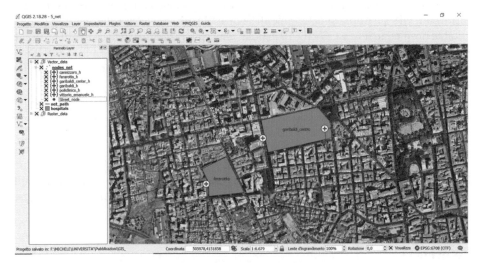

Fig. 5. Access nodes to hospitals

Finally, all the hospitals in the metropolitan city of Catania with COVID-19 beds have been digitized.

The application automatically selects all the access nodes to the hospitals with the lower cost (i.e. with more beds available) and subsequently the shortest path is calculated between the starting point and the node selected at the lowest cost. In case of same costs, the closest point to the starting node is chosen (Fig. 6).

Fig. 6. Best path calculated from a starting point and the nearest hospital with available COVID beds

5   Conclusions

In this work, we present a GIS application that is able to provide a valid tool for the fight against the COVID-19 pandemic.

The methodology we have proposed exploits a spatial database with relational support to the GIS platform to determine the optimal path between a starting point and a suitable hospital to host a COVID-19 patient. The path is optimized because all hospitals with exhausted beds are excluded. This result reduces the travel times of the COVID-19 patient by reducing the exposure times of the medical staff and the arrival times in the hospital.

The big advantage of the developed application is the zero cost, the speed calculation of the shortest path and the expandability of the platform on the web for accessing the application to multiple users at the same time.

The limit of the application is the direct access to the computer system of the hospitals, which should update the database with the actual number of beds available in real time. Thus, our approach can be implemented in governance procedures if the regional health companies can provide the information relating to the beds available through dedicated web portals.

References

1. Famoso, D., Mangiameli, M., Roccaro, P., Mussumeci, G., Vagliasindi, F.G.A.: Asbestiform fibers in the Biancavilla site of national interest (Sicily, Italy): review of environmental data via GIS platforms. Rev. Environ. Sci. Bio Technol. **11**, 417–427 (2012)
2. Mangiameli, M.; Mussumeci, G.: GIS approach for preventive evaluation of roads loss of efficiency in hydrogeological emergencies. Int. Arch. Photogrammetry Remote Sens. Spat. Inf. Sci. ISPRS Arch. (2013)
3. Mangiameli, M., Mussumeci, G., Roccaro, P., Vagliasindi, F.G.A.: Free and open-source GIS technologies for the management of woody biomass. Appl. Geomat. **11**, 309–315 (2019)
4. Troisi, R., Castaldo, P.: Technical and organizational challenges in the risk management of road infrastructures. J. Risk Res. (2022). https://doi.org/10.1080/13669877.2022.2028884
5. Gennaro, A., Mangiameli, M., Muscato, G., Mussumeci, G., Sgarlata, M.: Geomatic techniques for surveying and mapping an archaeological site. IEEE International Conference on Metrology for Archaeology and Cultural Heritage (Cassino 2018), MetroArchaeo 2018 - Proceedings, pp. 277–281, 9089803 (2018)
6. Mangiameli, M., Mussumeci, G., Cappello, A.: Forest fire spreading using free and open-source GIS technologies. Geomatics **1**(1), 50–64 (2021). https://doi.org/10.3390/geomatics1010005
7. Mangiameli, M., Mussumeci, G., Oliva, S.: Free and open source GIS technologies for the assessment of tsunami hazards in the ionic sea. In: Misra, S.., et al. (eds.) ICCSA 2019. LNCS, vol. 11622, pp. 216–224. Springer, Cham (2019). https://doi.org/10.1007/978-3-030-24305-0_17
8. Ganci, G., et al.: 3D lava flow mapping in volcanic areas using multispectral and stereo optical satellite data. In: AIP Conference Proceedings, vol. 2293, p. 300003 (2020)
9. Mangiameli, M., Candiano, A., Fargione, G., Gennaro, A., Mussumeci, G.: Multispectral satellite imagery processing to recognize the archaeological features: the NW part of Mount Etna (Sicily, Italy). Math. Methods Appl. Sci. **43**(13), 7640–7646 (2020)

10. Gatta, G., Bitelli, G.: A HGIS for the study of waterways: the case of Bologna as ancient city of waters. In: IOP Conference Series: Materials Science and Engineering, International Conference Florence Heri-tech: the Future of Heritage Science and Technologies, 14–16 October 2020, vol. 949, Online Edition (2020)
11. Korea's Fight against COVID-19 (2020). http://www.mofa.go.kr/eng/brd/m_5674/view.do?seq=320048
12. Rezaei, M., Nouri, A.A., Park, GS., Kim, DH.: Application of geographic information system in monitoring and detecting the COVID-19 outbreak. Iran. J Publ. Health **49**(Suppl.1), 114–116 (2020)
13. Razavi-Termeh, S.V., Sadeghi-Niaraki, A., Farhangi, F., Choi, S.: COVID-19 risk mapping with considering socio-economiccriteria using machine learning algorithms. Int. J. Environ. Res. Publ. Health **18**(18), 9657 (2021). https://doi.org/10.3390/ijerph18189657
14. Franch-Pardo, I., Napoletano, B.M., Rosete-Verges, F., Billa, L.: Spatial analysis and GIS in the study of COVID-19. A review. Sci. Total Environ. **739**(15 October 2020), 140033 (2020)
15. Troisi, R., Di Nauta, P., Piciocchi, P.: Private corruption: an integrated organizational model. Eur. Manag. Rev. (2021). https://doi.org/10.1111/emre.12489
16. Troisi, R., Alfano, G.: Is regional emergency management key to containing COVID-19? A comparison between the regional Italian models of Emilia-Romagna and Veneto. Int. J. Publ. Sect. Manag. (2021). https://doi.org/10.1108/IJPSM-06-2021-0138

Influence of Near Fault Records on the Optimal Performance of Isolated Continuous Bridges

Elena Miceli$^{(\boxtimes)}$ (iD)

Politecnico di Torino, Corso Duca degli Abruzzi, 10129 Turin, Italy
elena.miceli@polito.it

Abstract. This study aims at evaluating the optimal value of the friction coefficient in case of multi-span continuous deck bridges equipped with single concave friction pendulum devices. The bridge is modelled with a six-degree-of-freedom system considering the presence of the isolator on top of both the abutment and the pier. The friction pendulum device behaviour is modelled by including the dependency of the friction coefficient on the velocity. The equation of motions have been solved by adopting a nondimensionalization with respect to the peak ground acceleration-to-velocity ratio, which is a measure for the ground motion period. A parametric analysis has been performed by using different values for the friction coefficient, for the pier and deck periods and the masses of the deck and of the pier. The uncertainty in the seismic input is included by considering 40 near fault records. Finally, an optimal value for the friction coefficient able to minimize the substructure peak response is calculated as function of the peak ground acceleration-to-velocity ratio and the period of the deck.

Keywords: Continuous deck bridges · Friction coefficient · Optimal value

1 Introduction

The seismic isolation is one of the largely adopted techniques to improve the seismic safety of civil structures [1, 2]. In particular, the goal of the seismic isolation in case of bridges is to increase the period of the isolation system with the aim to reduce the inertia forces acting on the deck and subsequently transmitted to the substructure, i.e., the piers [3]. The advantage of using the friction pendulum systems (FPS) devices is to make the isolation period independent from the mass of the deck [4, 5]. Big efforts have been made within the researchers to study how to model the FPS devices behaviour [6–10]. The existence of an optimal value of the friction coefficient, able to minimize the seismic response of the substructure, was first introduced by Jangid in [11, 12]. To this aim, the optimal friction coefficient is evaluated in [13, 14] through a parametric analysis (i.e., by varying many properties of the structure and the seismic input). However, in these works the optimization is not independent from the seismic input.

In this study, the optimal friction coefficient of multi-span continuous deck reinforced concrete (RC) bridges, isolated with single concave FPS devices, is analysed. The bridge is modelled through a six-degree-of-freedom (dof) system, where 5 dofs are used to

© The Author(s), under exclusive license to Springer Nature Switzerland AG 2022
F. Calabrò et al. (Eds.): NMP 2022, LNNS 482, pp. 1130–1139, 2022.
https://doi.org/10.1007/978-3-031-06825-6_109

model the lumped masses of the pier and 1dof for the deck. The latter is considered infinitely rigid. Two FPS bearings are placed on top of the elastic RC pier and the rigid RC abutment, which is considered as a fixed support. The bridge is subjected to a set 40 of natural near fault records. Furthermore, many bridge models are considered by varying: the pier fundamental period, the ratio between the masses of the deck and the pier, the friction coefficient of the FPS devices and the ratio between the fundamental periods of the deck and of the ground motion. Nondimensional equation of motions have been solved such that the response becomes independent from the characteristic of the records, expressed in terms of peak ground acceleration-to-velocity ratio (PGA/PGV). The final result of this work is the computation of a linear regression in order to compute the nondimensional friction coefficient of the isolator as function of the deck period and the ground motion period, useful in the design phase of the FPS devices.

2 Bridge Model and Equations of Motion

The seismic response of the bridge is evaluated by modelling a six-degree-of-freedom (dof) system, accounting for 5 dofs for the lumped masses of the pier and 1 additional dof to model the rigid RC deck. Previous studies have demonstrated that the use of 5 lumped masses to model the pier allows to obtain enough accuracy of the results avoiding larger computational efforts [13, 14]. The friction pendulum system devices (FPS) are placed on top of the pier and the abutment, which is modelled as rigid and fixed. Figure 1 shows the 6 dofs model adopted and their horizontal relative displacements.

When the structure is subjected to a seismic input, the equation of motion of the 6 dofs, evaluated along the horizontal direction, are:

$$m_d\ddot{u}_d(t) + m_d\ddot{u}_{p5}(t) + m_d\ddot{u}_{p4}(t) + m_d\ddot{u}_{p3}(t) + m_d\ddot{u}_{p2}(t) + m_d\ddot{u}_{p1}(t) + c_d\dot{u}_d(t)$$
$$+ F_p(t) + F_a(t) = -m_d\ddot{u}_g(t)$$
$$m_{p5}\ddot{u}_{p5}(t) + m_{p5}\ddot{u}_{p4}(t) + m_{p5}\ddot{u}_{p3}(t) + m_{p5}\ddot{u}_{p2}(t) + m_{p5}\ddot{u}_{p1}(t) - c_d\dot{u}_d(t) + c_{p5}\dot{u}_{p5}(t)$$
$$+ k_{p5}u_{p5}(t) - F_p(t) = -m_{p5}\ddot{u}_g(t)$$
$$m_{p4}\ddot{u}_{p4}(t) + m_{p4}\ddot{u}_{p3}(t) + m_{p4}\ddot{u}_{p2}(t) + m_{p4}\ddot{u}_{p1}(t) - c_{p5}\dot{u}_{p5}(t) - k_{p5}u_{p5}(t) + c_{p4}\dot{u}_{p4}(t)$$
$$+ k_{p4}u_{p4}(t) = -m_{p4}\ddot{u}_g(t)$$
$$m_{p3}\ddot{u}_{p3}(t) + m_{p3}\ddot{u}_{p2}(t) + m_{p3}\ddot{u}_{p1}(t) - c_{p4}\dot{u}_{p4}(t) - k_{p4}u_{p4}(t) + c_{p3}\dot{u}_{p3}(t) + k_{p3}u_{p3}(t)$$
$$= -m_{p3}\ddot{u}_g(t)$$
$$m_{p2}\ddot{u}_{p2}(t) + m_{p2}\ddot{u}_{p1}(t) - c_{p3}\dot{u}_{p3}(t) - k_{p3}u_{p3}(t) + c_{p2}\dot{u}_{p2}(t) + k_{p2}u_{p2}(t) = -m_{p2}\ddot{u}_g(t)$$
$$m_{p1}\ddot{u}_{p1}(t) - c_{p2}\dot{u}_{p2}(t) - k_{p2}u_{p2}(t) + c_{p1}\dot{u}_{p1}(t) + k_{p1}u_{p1}(t) = -m_{p1}\ddot{u}_g(t)$$

$$(1a,b,c,d,e,f)$$

where u_d is the deck displacement relative to the pier top, u_{pi} is the relative displacement of the ith lumped mass of the pier with respect to the lower one, m_d is the mass of the deck, m_{pi} is the mass of the ith lumped mass of the pier, while k_{pi} is the corresponding stiffness. The viscous damping coefficient for the device and for the pier masses are, respectively, c_d and c_{pi}; t is the instant of time and the dots are used to indicate differentiation. In the end, the resisting forces of the FPS bearings located on top of the abutment and on the pier are, respectively, $F_a(t)$ and $F_p(t)$. These forces can be expressed as the sum of an elastic component, coming from the pendular behaviour of the FPS device, and a viscous component, as follows [4]:

$$F_a(t) = \frac{m_d g}{2}\left[\frac{1}{R_a}\left(u_d(t) + \sum_{i=1}^{5} u_{pi}\right) + \mu_a\left(\dot{u}_d + \sum_{i=1}^{5} \dot{u}_{pi}\right)\mathrm{sgn}\left(\dot{u}_d + \sum_{i=1}^{5} \dot{u}_{pi}\right)\right]$$

$$F_p(t) = \frac{m_d g}{2}\left[\frac{1}{R_p}u_d(t) + \mu_p(\dot{u}_d)\mathrm{sgn}(\dot{u}_d)\right] \tag{2a,b}$$

where R_a and R_p are the radii of curvature of the FPS devices placed on the abutment and on the pier, respectively and assumed equal, the stiffness of the deck is computed as $k_d = W/R = m_d g/R$, half for the bearing on the abutment and half for the pier; g is the gravity constant; μ is the sliding friction coefficient of the bearings. It is noteworthy that the fundamental period of the deck can be expressed as $T_d = 2\pi\sqrt{m_d/k_d} = 2\pi\sqrt{R/g}$, hence, it only depends on the geometrical property of the isolator (the radius of curvature), and not on the mass of the deck [4]. In literature [4, 6, 10, 15], many experimental evidences have emphasizes the dependence of the sliding friction coefficient on different parameters (e.g., sliding velocity, cumulative movements, temperature). In particular, the dependency of the sliding friction coefficient on the velocity is such that:

$$\mu(\dot{u}_d) = f_{\max} - (f_{\max} - f_{\min})\cdot\exp(-\alpha|\dot{u}_d|) \tag{3}$$

where $f_{\max}$ and $f_{\min}$ define, respectively, the sliding friction parameter at large and null velocity, α governs the transition from low to large velocities. Experimental results suggest to assume $f_{\max} = 3f_{\min}$ and $\alpha = 30$ [15].

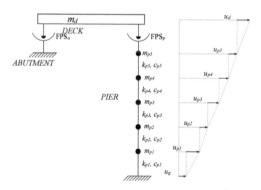

Fig. 1. Schematic representation of the six-degrees-of-freedom bridge model.

The equation of motions in (1) are then expressed in a nondimensional form, according to the Buckingham's Π-theorem, as illustrated in [16–18]. In particular, a time scale and a length scale are introduced and adopted for the nondimensionalization. In this study, the former is represented by $1/\omega_g$, whit $\omega_g = 2\pi/T_g = PGA/PGV$ indicating the circular frequency of the ground motion and herein evaluated as the peak ground acceleration-to-velocity ratio. The latter is assumed as a_0/ω_g^2, where a_0 is an intensity measure for the seismic input. The time scale is used to pass from the time t to $\tau = t\omega_g$ such that the ground motion input of Eq. (1) is expressed as:

$$\ddot{u}_g(t) = a_0 l(t) = a_0 \ell(\tau) \tag{4}$$

where $l(t)$ is a nondimensional function of the seismic input indicating its variation over the time t, while $\ell(\tau)$ contains the same information but in the time τ.

Finally, dividing the equations in (1) for the deck mass m_d and introducing the time and length scales, the equations in terms of nondimensional parameters are:

$$\ddot{\psi}_d(\tau) + \ddot{\psi}_{p5}(\tau) + \ddot{\psi}_{p4}(\tau) + \ddot{\psi}_{p3}(\tau) + \ddot{\psi}_{p2}(\tau) + \ddot{\psi}_{p1}(\tau) + 2\xi_d \frac{\omega_d}{\omega_g} \dot{\psi}_d(\tau) +$$

$$+ \left[\frac{1}{2} \frac{\omega_d^2}{\omega_g^2} \psi_d(\tau) + \frac{\mu_p(\ddot{u}_d)g}{2a_0} \mathrm{sgn}(\dot{\psi}_d) \right] +$$

$$+ \left[\frac{1}{2} \frac{\omega_d^2}{\omega_g^2} \left(\psi_d(\tau) + \sum_{i=1}^{5} \psi_{pi}(\tau) \right) + \frac{g}{2a_0} \mu_a (\ddot{u}_d + \sum_{i=1}^{5} \ddot{u}_{pi}) \left(\mathrm{sgn}\left(\dot{\psi}_d + \sum_{i=1}^{5} \dot{\psi}_{pi} \right) \right) \right] = -\ell(\tau)$$

$$\lambda_{p5} \left[\ddot{\psi}_{p5}(\tau) + \ddot{\psi}_{p4}(\tau) + \ddot{\psi}_{p3}(\tau) + \ddot{\psi}_{p2}(\tau) + \ddot{\psi}_{p1}(\tau) \right] - 2\xi_d \frac{\omega_d}{\omega_g} \dot{\psi}_d(\tau) + 2\xi_{p5} \frac{\omega_{p5}}{\omega_g} \lambda_{p5} \dot{\psi}_{p5}(\tau) +$$

$$+ \lambda_{p5} \frac{\omega_{p5}^2}{\omega_g^2} \psi_{p5}(\tau) - \left[\frac{1}{2} \frac{\omega_d^2}{\omega_g^2} \psi_d(\tau) + \frac{\mu_p(\ddot{u}_d)g}{2a_0} \mathrm{sgn}(\dot{\psi}_d) \right] = -\lambda_{p5}\ell(\tau)$$

$$\lambda_{p4} \left[\ddot{\psi}_{p4}(\tau) + \ddot{\psi}_{p3}(\tau) + \ddot{\psi}_{p2}(\tau) + \ddot{\psi}_{p1}(\tau) \right] - 2\xi_{p5} \frac{\omega_{p5}}{\omega_g} \lambda_{p5} \dot{\psi}_{p5}(\tau) + 2\xi_{p4} \frac{\omega_{p4}}{\omega_g} \lambda_{p4} \dot{\psi}_{p4}(\tau) +$$

$$- \lambda_{p5} \frac{\omega_{p5}^2}{\omega_g^2} \psi_{p5}(\tau) + \lambda_{p4} \frac{\omega_{p4}^2}{\omega_g^2} \psi_{p4}(\tau) = -\lambda_{p4}\ell(\tau)$$

$$\lambda_{p3} \left[\ddot{\psi}_{p3}(\tau) + \ddot{\psi}_{p2}(\tau) + \ddot{\psi}_{p1}(\tau) \right] - 2\xi_{p4} \frac{\omega_{p4}}{\omega_g} \lambda_{p4} \dot{\psi}_{p4}(\tau) + 2\xi_{p3} \frac{\omega_{p3}}{\omega_g} \lambda_{p3} \dot{\psi}_{p3}(\tau) - \lambda_{p4} \frac{\omega_{p4}^2}{\omega_g^2} \psi_{p4}(\tau) +$$

$$+ \lambda_{p3} \frac{\omega_{p3}^2}{\omega_g^2} \psi_{p3}(\tau) = -\lambda_{p3}\ell(\tau)$$

$$\lambda_{p2} \left[\ddot{\psi}_{p2}(\tau) + \ddot{\psi}_{p1}(\tau) \right] - 2\xi_{p3} \frac{\omega_{p3}}{\omega_g} \lambda_{p3} \dot{\psi}_{p3}(\tau) + 2\xi_{p2} \frac{\omega_{p2}}{\omega_g} \lambda_{p2} \dot{\psi}_{p2}(\tau) - \lambda_{p3} \frac{\omega_{p3}^2}{\omega_g^2} \psi_{p3}(\tau) +$$

$$+ \lambda_{p2} \frac{\omega_{p2}^2}{\omega_g^2} \psi_{p2}(\tau) = -\lambda_{p2}\ell(\tau)$$

$$\lambda_{p1} \ddot{\psi}_{p1}(\tau) - 2\xi_{p2} \frac{\omega_{p2}}{\omega_g} \lambda_{p2} \dot{\psi}_{p2}(\tau) + 2\xi_{p1} \frac{\omega_{p1}}{\omega_g} \lambda_{p1} \dot{\psi}_{p1}(\tau) - \lambda_{p2} \frac{\omega_{p2}^2}{\omega_g^2} \psi_{p2}(\tau) + \lambda_{p1} \frac{\omega_{p1}^2}{\omega_g^2} \psi_{p1}(\tau) = -\lambda_{p1}\ell(\tau)$$

$$\tag{5a,b,c,d,e,f}$$

where $\psi_d = u_d \omega_g^2 / a_0$ and $\psi_{pi} = u_{pi} \omega_g^2 / a_0$ are the nondimensional displacements, $\omega_d = \sqrt{k_d/m_d}$ and $\omega_{pi} = \sqrt{k_{pi}/m_{pi}}$ are the circular vibration frequencies, $\xi_d = c_d/2m_d\omega_d$ and $\xi_{pi} = c_{pi}/2m_{pi}\omega_{pi}$ are the damping factors (respectively for the deck and for the i-th lamped masses of the pier) and $\lambda_p = \lambda_{pi} = m_{pi}/m_d$ is the mass ratio of the i-th lumped mass (assumed equal for all the lumped masses). Furthermore, the nondimensional parameters Π of the problem are:

$$\Pi_{\omega_p} = \frac{\omega_p}{\omega_d}, \quad \Pi_{\omega_g} = \frac{\omega_d}{\omega_g}, \quad \Pi_{\lambda} = \lambda_p, \quad \Pi_{\xi_d} = \xi_d, \quad \Pi_{\xi_p} = \xi_{pi}, \quad \Pi_{\mu} = \frac{\mu(\dot{u}_d)g}{a_0}$$

$$(6a,b,c,d,e,f)$$

To discard the dependency of Π_{μ} on the velocity, we considered $\Pi_{\mu}^* = f_{max}g/a_0$.

3   Parametric Analysis

3.1   Ground Motion Records

According to the Performance Based Earthquake Engineering approach (PBEE) [19, 20], the seismic intensity a_0 is represented by the parameter PGA (i.e., peak ground acceleration). Furthermore, a set of 40 near fault (NF) ground motion records is considered, with peak ground acceleration-to-velocity ratios between 0.21 and 0.97 (i.e., low ranges of PGA/PGV), magnitude between 6.19 and 7.62 and soil types B and C.

3.2   Deterministic Parameters

The performance of the bridge isolated with FPS bearings is studied by including different values of the parameters involved in the problem. In particular, 2 pier periods are considered $T_p = 0.05s, 0.2s$, 15 period ratios in the range $T_d/T_g = 0.1 \div 1.6$, 3 mass ratios $\lambda_p = 0.1, 0.15, 0.3$ and 85 values for the nondimensional friction coefficient Π_{μ}^* between 0 and 1.5. The remaining parameters, i.e., Π_{ξ_d} and Π_{ξ_p} are set equal to 5% and 0% respectively [17, 21–24]. All the analyses have been solved in Matlab-Simulink [25]. To evaluate the performance of the bridge, the response parameter are chosen in terms of normalized relative peak displacement the pier (i.e., $\psi_{p,max} = u_{p,max}\omega_g^2/a_0$)

where the maximum pier response is intended as $u_{p,max} = \left| \left(\sum_{i=1}^{5} u_{pi} \right) \right|_{max}$. In the following, results are shown in terms of geometric mean (GM) and dispersion (β) of the response parameter D (i.e., $\psi_{p,max}$) considering the lognormality of the data [17, 26–38] as follows:

$$GM(D) = \sqrt[N]{d_1 \cdot \ldots \cdot d_N};$$

$$\beta(D) = \sigma_{\ln}(D) = \sqrt{\frac{(\ln d_1 - \ln(GM))^2 + \ldots + (\ln d_N - \ln(GM))^2}{N - 1}} \qquad (7a,b)$$

where D is the generic response parameter, d_j is the j-th realization of the response parameter and $j = 1,...,N$ with $N = 40$ the total number of NF inputs. Then, the k-th percentile of the response parameter is given by $d_k = GM(D)\exp[f(k)\beta(D)]$, where $f(k)$ [39] is equal to 0, 1 and −1 for the 50th, 16th and 84th percentile, respectively.

4 Optimal Sliding Friction Coefficient

The response parameter $\psi_{p,\max}$ corresponding to the normalized value of the maximum pier top displacement is illustrated in terms of geometric mean and dispersion as function of the normalized friction coefficient Π_μ^* and the ratio T_d/T_g for the two values of the pier fundamental periods T_p and the three quantity of λ_p. The geometric mean of the normalized response of the pier (Fig. 2a–b) increases for lower value of T_d/T_g since the lower the period of the deck, the larger are the forces acting on it and thus transmitted to the substructure. Furthermore, $GM(\psi_{p,\max})$ tends to decrease for larger values of the mass ratio. In addition, there is a slight decrease followed by an increase in the maximum normalized response of the pier top with the growth of the normalized friction coefficient, suggesting the evidence of an optimal value of Π_μ^* able to minimize the substructure response. Regarding the dispersion $\beta(\psi_{p,\max})$, the trend tends to slightly decrease in correspondence of the optimal value of the friction coefficient and turns out to be almost independent form the other parameters (Fig. 2c–d). The above mentioned considerations have resulted in the calculation of the optimal value for the normalized friction coefficient $\Pi_{\mu,opt}^*$ as function of the ratio T_d/T_g for the two values of T_p and the three quantity of λ_p, computed in terms of the 16th, 50th and 84th percentiles. In Fig. 3, the results corresponding to only the 50th percentile are shown. It can be noted that $\Pi_{\mu,opt}^*$ decreases with larger T_d/T_g ratios. Furthermore, Fig. 3 shows that there is not a large influence of both the pier fundamental period and the mass ratio in the computation of $\Pi_{\mu,opt}^*$. This

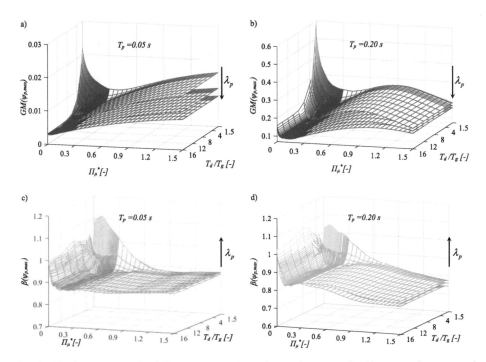

Fig. 2. Maximum normalized displacement of the pier top in terms of a–b) geometric mean and c–d) dispersion as function of Π_μ^* and T_d/T_g for $T_p = 0.05 - 0.2$ s and $\lambda_p = 0.1, 0.15, 0.2$.

suggests to perform a linear regression of the data to obtain the value of $\Pi^*_{\mu,opt}$ as only function of the parameter Π_{ω_g} as follows (Fig. 4):

$$\Pi^*_{\mu,opt} = c_1 + c_2 \Pi_{\omega_g} \geq 0 \tag{8}$$

where the coefficients c_1 and c_2 together with the *R-squared* values are in Table 1.

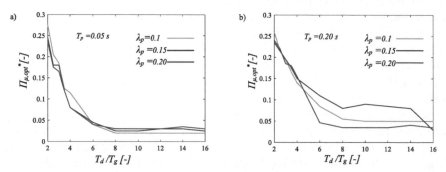

Fig. 3. Optimal value of the normalized friction coefficient as function of T_d/T_g for a) $T_p = 0.05$ s and b) $T_p = 0.2$ s and for the three values of $\lambda_p = 0.1, 0.15, 0.2$.

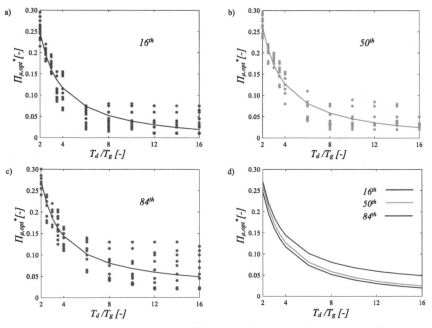

Fig. 4. Linear regression for the optimal value of the friction coefficient able at minimizing the a) 50th percentile b) 16th percentile and c) 84th percentile of the substructure response.

Table 1. Linear regression parameters for the optimal normalized friction coefficient

	50th	16th	84th
R-squared	0.9292	0.8872	0.8588
c_1	−0.0093	−0.0130	0.0172
c_2	0.5430	0.5190	0.5069

5 Conclusions

This study analyses the optimal value of the friction coefficient in case of seismic isolation of bridges with FPS devices. Many bridge models were considered within a parametric analysis. Then, each bridge model has been subjected to 40 near fault seismic records. The response of the substructure (i.e., the pier) has been evaluated in terms of nondimensional peak displacement normalized with respect to the ground motion characteristics through the acceleration-to-velocity (PGA/PGV) ratio. Results have shown the evidence of an optimal value of the friction coefficient able to minimize the substructure peak response. Hence, a linear regression of the data has led to an expression to compute the normalized optimal friction coefficient as function of the ratio between the fundamental period of the ground motion and of the deck.

References

1. Troisi, R., Alfano, G.: Towns as safety organizational fields: an institutional framework in times of emergency. Sustainability **11**, 7025 (2019). https://doi.org/10.3390/su11247025
2. Troisi, R., Castaldo, P.: Technical and organizational challenges in the risk management of road infrastructures. J. Risk Res. **2022** (2022). https://doi.org/10.1080/13669877.2022.202 8884
3. Ghobarah, A., Ali, H.M.: Seismic performance of highway bridges. Eng. Struct. **10**(3), 157–166 (1988)
4. Zayas, V., Low, S., Mahin, S.: A simple pendulum technique for achieving seismic isolation. Earthq. Spectra **6**(2), 317–333 (1990)
5. Su, L., Ahmadi, G., Tadjbakhsh, I.: Comparative study of base isolation systems. J. Eng. Mech. **115**(9), 1976–1992 (1989)
6. Mokha, A., Constantinou, M.C., Reinhorn, A.M.: teflon bearings in base isolation. I: testing. J. Struct. Eng. **116**(2), 438–454 (1990)
7. Constantinou, M.C., Mokha, A., Reinhorn, A.M.: Teflon bearings in base isolation. II: modeling. J. Struct. Eng. **116**(2), 455–474 (1990)
8. Almazàn, J.L., De la Llera, J.C.: Physical model for dynamic analysis of structures with FPS isolators. Earthq. Eng. Struct. Dyn. **32**(8), 1157–1184 (2003)
9. Mosqueda, G., Whittaker, A.S., Fenves, G.L.: Characterization and modeling of Friction Pendulum bearings subjected to multiple components of excitation. J. Struct. Eng. **130**(3), 433–442 (2004)
10. Jangid, R.S.: Computational numerical models for seismic response of structures isolated by sliding systems. Struct. Control. Health Monit. **12**, 17–137 (2005)
11. Jangid, R.S.: Optimum frictional elements in sliding isolation systems. Comput. Struct. **76**(5), 651–661 (2000)

12. Jangid, R.S.: Optimum friction pendulum system for near-fault motions. Eng. Struct. **27**(3), 349–359 (2005)
13. Castaldo, P., Amendola, G.: Optimal sliding friction coefficients for isolated viaducts and bridges: a comparison study. Struct. Control Health Monit. **28**(12) (2021)
14. Castaldo, P., Amendola, G.: Optimal DCFP bearing properties and seismic performance assessment in nondimensional form for isolated bridges. Earthq. Eng. Struct. Dyn. **50**(9), 2442–2461 (2021)
15. Constantinou, M.C., Whittaker, A.S., Kalpakidis, Y., Fenz, D.M., Warn, G.P.: Performance of seismic isolation hardware under service and seismic loading. Technical report MCEER-07-0012 (2007)
16. Makris, N., Black, C.J.: Dimensional analysis of inelastic structures subjected to near fault ground motions. Technical report: EERC 2003/05. Earthquake Engineering Research Center, University of California, Berkeley (2003)
17. Castaldo, P., Tubaldi, E.: Influence of FPS bearing properties on the seismic performance of base-isolated structures. Earthq. Eng. Struct. Dyn. **44**(15), 2817–2836 (2015)
18. Castaldo, P., Tubaldi, E.: Influence of ground motion characteristics on the optimal single concave sliding bearing properties for base-isolated structures. Soil Dyn. Earthq. Eng. **104**, 346–364 (2018)
19. Aslani, H., Miranda, E.: Probability-based seismic response analysis. Eng. Struct. **27**(8), 1151–1163 (2005)
20. Porter, K.A.: An overview of PEER's performance-based earthquake engineering methodology. In: Proceedings of the 9th International Conference on Application of Statistics and Probability in Civil Engineering, ICASP9, San Francisco, California (2003)
21. De Iuliis, M., Castaldo, P.: An energy-based approach to the seismic control of one-way asymmetrical structural systems using semi-active devices. Ingegneria Sismica-Int. J. Earthq. Eng. **29**(4), 31–42 (2012)
22. Castaldo, P., De Iuliis, M.: Optimal integrated seismic design of structural and viscoelastic bracing-damper systems. Earthq. Eng. Struct. Dyn. **43**(12), 1809–1827 (2014)
23. Castaldo, P., Palazzo, B., Ferrentino, T., Petrone, G.: Influence of the strength reduction factor on the seismic reliability of structures with FPS considering intermediate PGA/PGV ratios. Compos. B Eng. **115**, 308–315 (2017)
24. Castaldo, P., Palazzo, B., Ferrentino, T.: Seismic reliability-based ductility demand evaluation for inelastic base-isolated structures with friction pendulum devices. Earthq. Eng. Struct. Dyn. **46**(8), 1245–1266 (2017)
25. Math Works Inc.: MATLAB-High Performance Numeric Computation and Visualization Software. User's Guide. Natick (MA) USA (1997)
26. Troisi, R., Alfano, G.: Firms' crimes and land use in Italy. An exploratory data analysis. Smart Innov. Syst. Technol. SIST, **178**, 749–758 (2020)
27. Garzillo, C., Troisi, R.: Le decisioni dell'EMA nel campo delle medicine umane. In: EMA e le relazioni con le Big Pharma - I profili organizzativi della filiera del farmaco. G. Giappichelli, pp. 85–133 (2015)
28. Golzio, L.E., Troisi, R.: The value of interdisciplinary research: a model of interdisciplinarity between legal re-search and research in organizations. J. Dev. Leadersh. **2**, 23–38 (2013)
29. Nese, A., Troisi, R.: Corruption among mayors: evidence from Italian Court of Cassation judgments. Trends Organ. Crime **22**(3), 298–323 (2019)
30. Troisi, R., Golzio, L.E.: Legal studies and organization theory: a possible cooperation. In: 16th EURAM Conference Manageable Cooperation. European Academy of Management, Paris, 1–4 June (2016)
31. Troisi, R., Guida, V.: Is the appointee procedure a real selection or a mere political exchange? The case of the italian health-care chief executive officers. J. Entrepreneurial Organ. Divers. **7**(2), 19–38 (2018). https://doi.org/10.5947/jeod.2018.008

32. Troisi, R.: Le risorse umane nelle BCC: lavoro e motivazioni al lavoro. Progetto aree bianche. Il sistema del credito cooperativo in Campania 1, pp. 399–417 (2012)

33. Nesticò, A., Moffa, R.: Economic analysis and operational research tools for estimating productivity levels in off-site construction [Analisi economiche e strumenti di Ricerca Operativa per la stima dei livelli di produttività nell'edilizia off-site]. Valori e Valutazioni, n. 20, pp. 107–128 (2018). ISSN 2036-2404. DEI Tipografia del Genio Civile, Roma

34. Nesticò, A., Maselli, G.: Declining discount rate estimate in the long-term economic evaluation of environmental projects. J. Environ. Account. Manag. 8(1), 93–110 (2020). https://doi.org/10.5890/JEAM.2020.03.007

35. Castaldo, P., Gino, D., Marano, G., Mancini, G.: Aleatory uncertainties with global resistance safety factors for non-linear analyses of slender reinforced concrete columns. Eng. Struct. 255, 113920 (2022). S0141-0296(22)00078-5

36. Troisi, R., Alfano, G.: Is regional emergency management key to containing COVID-19? A comparison between the regional Italian models of Emilia-Romagna and Veneto. Int. J. Publ. Sect. Manag. (2021). https://doi.org/10.1108/IJPSM-06-2021-0138

37. Troisi, R., Di Nauta, P., Piciocchi, P.: Private corruption: an integrated organizational model. Eur. Manag. Rev. (2021)

38. Nese, A., Troisi, R.: Individual preferences and job characteristics: an analysis of cooperative credit banks. Labour 28(2), 233–249 (2014)

39. Ang, A.H.S., Tang, W.H.: Probability Concepts in Engineering-Emphasis on Applications to Civil and Environmental Engineering. Wiley, New York (2007)

Safety Management of Infrastructures Through an Organizational Approach: Preliminary Results

Roberta Troisi[1] (ID), Paolo Castaldo[2](✉) (ID), and Monica Anna Giovanniello[3]

[1] University of Salerno, Via Giovanni Paolo II 132, 84084 Fisciano, SA, Italy
rtroisi@unisa.it
[2] Politecnico di Torino, Corso Duca degli Abruzzi, 10129 Turin, Italy
paolo.castaldo@polito.it
[3] Department of Economia Aplicada, University of Balearic Islands, Palma, Illes Balears, Spain
ma.giovanniello@uib.cat

Abstract. The safety of the road infrastructures is a very relevant topic within the governance of a territory characterized by several structural systems built some decades ago. In fact, since these structural systems are affected by not negligible safety problems, the governance actors are called to define appropriate plans and interventions to reduce undesirable social impacts. It derives that the actors, i.e., public or private stakeholders, are responsible of both the administrative and economic procedures. The present study discusses this topic by means of some preliminary suggestions and results in relation to the safety management of the infrastructures from an organizational perspective through two models. In fact, the study highlights the stakeholders domain as the virtual place where the organizational principles should guide all the processes. In detail, as for a group of Italian infrastructures in Rome, the author suggest to adopt the monitoring by means of satellites for a territorial analysis. This technology-based data can lead to the elaboration of alert maps as well as risk analyses useful to upgrade the organizational features within the management and governance procedures.

Keywords: Organizational features · Infrastructure safety · DInSAR

1 Introduction

The safety of the infrastructures is a very relevant topic within the governance of a territory characterized by a large number of structural systems built some decades ago. In fact, since these structural systems are affected by not negligible safety problems, the governance actors are called to define appropriate plans and interventions to reduce undesirable social impacts, as commented in [1–5].

It derives that infrastructure safety is an important issue that involves numerous actors such as authorities, central and local actors (i.e., private or public stakeholders), who are responsible of both the administrative and economic procedures [6, 7], in addition to the researchers and technology experts.

© The Author(s), under exclusive license to Springer Nature Switzerland AG 2022
F. Calabrò et al. (Eds.): NMP 2022, LNNS 482, pp. 1140–1147, 2022.
https://doi.org/10.1007/978-3-031-06825-6_110

In fact, all plans and strategies provided by these actors obviously affect the safety of the infrastructures. Moreover, as marked in [8–10], the execution and design standards represent very important aspects when planning activities are delineated in a territory subjected to extreme natural events. These aspects, often missing [11], are essential in the planning with the purpose to reduce the risk improving the resilience.

In studies [12, 13], public policies have been proposed to ensure protection from natural events as well as risk-based plans combined with the use of the GIS (Geographical Information System) technique have been commented in [14, 15]. The studies [16, 17] describe the proposals of managerial approaches, inspired by organizational issues, for safety and security of road infrastructures.

Public and private actors have elaborated some strategies, defined as Bridge Management Systems, through the use of the Structural Health Monitoring (SHM) techniques [18, 19], such as, the one by means of satellites, i.e., the Differential Interferometry Synthetic Aperture Radar (DInSAR) [20, 21].

The present study aims at discussing this topic by means of some preliminary suggestions and results in relation to the safety management of the infrastructures from an organizational perspective. In fact, the study highlights the stakeholders domain as the virtual place where two organizational models should guide all the processes. In particular, two competing models can be useful in this case. Firstly, there is the centralized model based on a command-and-control approach. This is supported by a strict use of bureaucratic coordination rules and works well under two main conditions: 1) a limited number of agents to coordinate and 2) a low degree of alert.

Secondly, a flexible model [22] where centralized decisions go hands to hands with those decentralized based on a "loosely coupled" approach. Coordination rules turn into forms of partnership together with a significant leeway at the local level [23] particularly suitable under conditions of 1) high alert and 2) greater numbers of actors involved.

In detail, as for a group of Italian infrastructures in Rome, the authors suggest to adopt the monitoring through satellites for a territorial analysis to be supported through the organizational models.

This remote sensing activity is a very powerful monitoring technique to observe the infrastructure responses to some natural events (e.g., earthquakes, landslides, subsidence phenomena, temperature variations, degradation processes and seasonal phenomena) at territorial scale.

These technology-based data can lead to the elaboration of alert maps as well as risk analyses useful to upgrade the organizational issues in the management and governance procedures.

2   The DInSAR Technique Useful to an Organizational Approach Within the Safety Management of the Infrastructures

The DInSAR technique [24, 25] is based on the use of data from a satellite constellation (e.g., the Italian COSMOSkyMed) to achieve measurements in terms of both velocities and displacements on the topographic surface. These measurements obviously depend on time due to the occurrence of many natural or anthropic events.

The DInSAR technique exploits, in its basic form, the phase difference of (at least) two complex-valued SAR images, acquired by different sensors along so called Lower Earth Orbits (LEO), between 500–800 km, following polar (ascending and descending - with opposite ground looking directions) orbits to ensure a global coverage. As for the SAR sensors, they may be different mainly for maximum measurable displacement, band, acquisition period, revisiting time, line of sight and resolution [24]. All these factors can influence the quality of final data.

These data, referred to a territorial context where the infrastructures are built, can be properly elaborated to achieve alert thresholds as well as perform risk analyses. Successively, the results can be detailed in the GIS environment leading to the definition of territorial maps illustrative of the infrastructures safety.

These satellite-based maps can be adopted as common data to all actors and, therefore, are useful to define an organizational framework (Fig. 1) in the management and governance processes of infrastructures safety. In fact, the stakeholders are easily recognized and involved in the institutional processes.

The domain of the stakeholders should englobe all the analytical and management processes. This domain should be seen as the virtual place where the organizational principles should guide all the processes.

In this way, a synergic behavior between the different institutions, public actors and the various administration levels can be increased according to the guidelines described in the previous section. Specifically, the displacement maps will address for the suitability of the two models by considering the number of actors involved as well as the degree of alert.

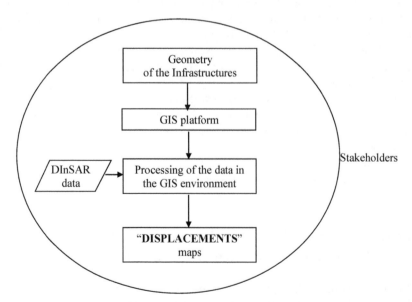

Fig. 1. Organizational framework.

3   Italian Infrastructures in Rome

Referring to some Italian road infrastructures in Rome, in compliance with the research project [26], this section illustrates some preliminary results achieved from the elaboration of the DInSAR data.

Table 1. Infrastructures in addition to the corresponding stakeholders.

Infrastructures	Stakeholders
"Autostrada del Sole - A1"	"ASPI - AutoStrade Per l'Italia"
"A91"	"ANAS S.p.A."
"Autostrada Azzurra - A12"	"ASPI - AutoStrade Per l'Italia"
"Circonvallazione"	Municipalility
"Grande Raccordo Anulare"	"ANAS S.p.A."
"A24"	"Strada dei Parchi S.p.A."
"Lungotevere"	Municipalility
"Maremmana - SP 216"	Province

Figure 1 depicts a thematic map in the GIS environment showing the physical boundaries of Rome Municipality in addition to the infrastructures extrapolated for the present analysis.

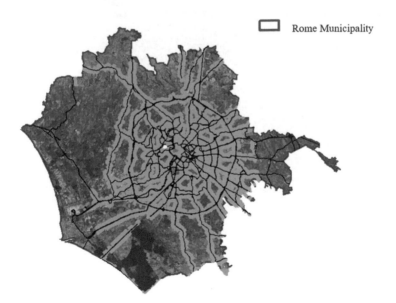

Rome Municipality

Fig. 2. The infrastructures with the points monitored over time.

Table 1 reports the corresponding stakeholders: "Strada dei Parchi S.p.A.", "ASPI - AutoStrade Per l'Italia", "ANAS S.p.A.", Municipality and Province.

After that, we have properly managed the DInSAR data, regarding the SAR sensor images (COSMO-SkyMED) along the ascending orbit within a timeframe of the last eight years. Selecting 0.6 as reference value for the coherence [27], we have monitored more than six millions of points, as illustrated always in Fig. 1.

With the purpose to carry out the computation in terms of the displacements, all the infrastructures have been divided by means of "cells" with dimensions equal to 50x50m [24, 28]. Next, the horizontal displacements have been calculated, as suggested by [24, 28], and are shown in Fig. 2.

These preliminary results indicate that some infrastructures are affected by potential damages due to high values of the horizontal displacements. Precisely, some infrastructures, managed, respectively, by "ANAS S.p.A.", "Autostrade per l'Italia - ASPI", Province and Municipality (Figs. 1 and 2) present severe alerts as regards the horizontal displacements. In this case, there is not a prevailing model that can be applied.

On one hand, any infrastructures can be referred to different authorities both when considered the hierarchical level and the nature (private vs public). On the other one, there is a similar risk that does not suggest a model inspired to flexibility as decisions should come with similar timeframes and thus are hard to be coordinated through common guidelines. A possible solution could refer to a hybrid model where both some features of the top-down model and the flexible one can be applied. We are thinking to a command

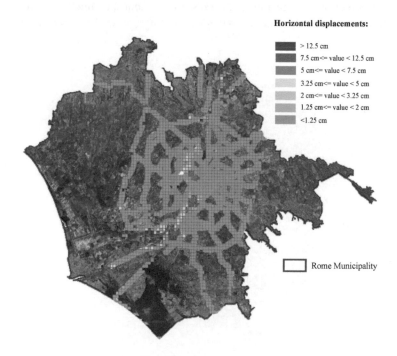

Fig. 3. Horizontal displacements.

chain among public authorities and the use of more flexible solutions when private authority is involved.

This propose should be better implemented considering a certain rigidity of the competences and of the rules, particularly for the public field. However, it could be useful for two reasons: 1) theoretical model needs reality to be adjusted, and a hybrid model better catches the complexity of the infrastructure system 2) one must carefully consider the trade-off between the rigidity of systems and the efficiency of the solution purposed.

It is important to specify that further and more detailed analyses need to be performed to assess the safety conditions of the infrastructures especially when severe damages are fearsome.

Although these results are preliminary, they represent a useful contribution for territorial analyses. In fact, they demonstrate the actual way to relate the infrastructures and the stakeholders for a more appropriate and sustainable planning in terms of structural safety at a territorial scale.

4 Conclusions

The study discusses some preliminary suggestions and results in relation to the safety management of the infrastructures from an organizational perspective.

A preliminary framework is illustrated in order to put in evidence the role of the stakeholders to be coordinated according to two main models. In fact, the domain of the actors represents the virtual place where the organizational rules should guide all the processes, where the institutional and administrative processes should be synergic and systemic in relation to the safety of the infrastructures.

Furthermore, some technical instruments (i.e., satellite data and GIS) are discussed for their features useful to the organizational aims at a territorial scale.

In detail, as for a group of Italian infrastructures in Rome, the results, derived from the use of the monitoring through satellites, demonstrate their effectiveness for a territorial analysis from an organizational perspective.

The preliminary results indicate that some infrastructures are affected by potential damages due to high values of the horizontal displacements. Precisely, some infrastructures, managed, respectively, by "ANAS S.p.A.", "Autostrade per l'Italia - ASPI", Municipality and Province present severe alerts as regards the horizontal displacements.

These technology-based data can lead to the elaboration of alert maps as well as risk analyses useful to upgrade the organizational features within the management and governance processes. These latter ones are not flexible enough to justify different models of coordination that are not respectful of hierarchy. This is particularly true in case of public stakeholders.

References

1. Blochl, B., Braun, B.: Economic assessment of landslide risks in the Schwabian Alb, Germany - research framework and first results of homeowners and experts surveys. Nat Hazards Earth Syst Sci **5**, 389–396 (2005)

2. Cotecchia, V.: The Second Hans Cloos Lecture. Experience drawn from the great Ancona landslide of 1982. Bull. Eng. Geology Environ. **65**, 1–41 (2006)
3. Iovine, G., Petrucci, O., Rizzo, V., Tansi, C.: The March 7th 2005 Cavallerizzo (Cerzeto) landslide in Calabria – SouthernItaly. In: Proceedings of the 10th IAEG Congress, 6–10 September 2006, 785, pp. 1–12. Nottingham, Great Britain (2006)
4. Mansour, M.F., Morgenstern, N.R., Martin, C.D.: Expected damage from displacement of slow-moving slides. Landslides **7**, 117–131 (2011)
5. Castaldo, P., Calvello, M., Palazzo, B.: Probabilistic analysis of excavation-induced damages to existing structures. Comput. Geotech. **53**, 17–30 (2013)
6. Nesticò, A., Moffa, R.: Economic analysis and Operational Research tools for estimating productivity levels in off-site construction [Analisi economiche e strumenti di Ricerca Operativa per la stima dei livelli di produttività nell'edilizia off-site]. Valori e Valutazioni **20**, 107–128 (2018). ISSN: 2036-2404. DEI Tipografia del Genio Civile, Roma
7. Nesticò, A., Maselli, G.: Declining discount rate estimate in the long-term economic evaluation of environmental projects. J. Environ. Acc. Manage. **8**(1), 93–110 (2020). https://doi.org/10.5890/JEAM.2020.03.007
8. Turner, R.K., Kelly, P.M., Kay, R.C.: Cities at Risk. BNA International, London (1990)
9. Smyth, C.G., Royle, S.A.: Urban landslide hazards: incidence and causative factors in Niterói, Rio de Janeiro State. Brazil. Appl. Geogr. **20**, 95–117 (2000)
10. Alcantara-Ayala, I., et a.: Disaster risks research and assessment to promote risk reduction and management. In: Ismail-Zadeh, A., Cutter, S. (eds.) ICSU-ISSC Ad Hoc Group on Disaster Risk Assessment (2015)
11. Briceño, S.: Looking back and beyond Sendai: 25 years of international policy experience on disaster risk reduction. Int. J. Disaster Risk Sci. **6**(1), 1–7 (2015)
12. McWilliam, W., Brown, R., Eagles, P., Seasons, M.: Evaluation of planning policy for protecting green infrastructure from loss and degradation due to residential encroachment. Land Use Policy **47**, 459–467 (2015)
13. Kubal, C., Haase, D., Meyer, V., Scheuer, S.: Integrated urban flood risk assessment – adapting a multi-criteria approach to a city. Nat. Hazards Earth Syst. Sci. **9**(6), 1881–1895 (2009)
14. De Mendonca, M.B., Gullo, F.T.: Landslide risk perception survey in Angra dos Reis (Rio de Janeiro, southeastern Brazil): a contribution to support planning of non-structural measures. Land Use Policy **91**, 104415 (2020)
15. Hossein, Y., Sachio, E.: Geothermal power plant site selection using GIS in Sabalan area, NW IRAN/Natural Resources and Environment (NRE). In: Proceedings of 6th Annual International Conference on Geographical Information Technology and Applications, pp. 1–18. GIS Development Press, Kuala Lumpur (2007)
16. Troisi, R., Alfano, G.: Towns as safety organizational fields: an institutional framework in times of emergency. Sustainability **11**(24), 7025 (2019)
17. Troisi, R., Castaldo, P.: Technical and organizational challenges in the risk management of road infrastructures. J. Risk Res. (2022). https://doi.org/10.1080/13669877.2022.2028884
18. Figueiredo, E., Moldovan, I., Barata Marques, M.: Condition Assessment of Bridges: Past, Present and Future. A Complementary Approach. Universidade Católica Editora, Unipessoal, Lda (2013). ISBN: 978-972-54-0402-7
19. Italian Infrastructure and Trasport Ministry (MIT): Guidelines for classification and management risk, security assessment and monitoring of existing bridges (2020)
20. Crosetto, M., Biescas, E., Duro, J.: Generation of advanced ERS and Envisat interferometric SAR products using the stable point network technique. Photogramm. Eng. Remote Sens. **4**, 443–450 (2008)
21. Crosetto, M., Monserrat, O., Cuevas-González, M., Devanthéry, N., Crippa, B.: Persistent scatterer interferometry: a review. ISPRS J. Photogrammetry Remote Sens. **115**, 78–89 (2016)

22. Janssen, M., Van Der Voort, H.: Agile and adaptive governance in crisis response: lessons from the COVID-19 pandemic. Int. J. Inf. Manage. **55**, 102180 (2020)
23. Guo, X., Kapucu, N.: Examining stakeholder participation in social stability risk assessment for mega projects using network analysis. Int. J. Disaster Risk Manage. **1**(1), 1–31 (2019)
24. Peduto, D., Cascini, L., Arena, L., Ferlisi, S., Fornaro, G., Reale, D.: A general framework and related procedures for multiscale analyses of DInSAR data in subsiding urban areas. ISPRS J. Photogramm. Remote. Sens. **105**, 186–210 (2015)
25. Berardino, P., Fornaro, G., Lanari, R., Sansosti, E.: A new algorithm for surface deformation monitoring based on small baseline differential SAR interferograms. IEEE Trans. Geosci. Remote Sens. **40**(11), 2375–2383 (2002)
26. ReLUIS: Research project between the Italian Civil Protection Department and the Italian Universities: WP 11 – Task 11.4: Monitoring and satellite data (2019–2021)
27. Yang, Y., Pepe, A., Manzo, M., Bonano, M., Liang, D.N., Lanari, R.: A simple solution to mitigate noise effects in time-redundant sequences of small baseline multi-look DInSAR interferograms. Remote Sensing Lett. **4**(6), 609–618 (2013)
28. Calvello, M., Cascini, L., Mastroianni, S.: Landslide zoning over large areas from a sample inventory by means of scale-dependent terrain units. Geomorphology **182**, 33–48 (2013)

Evaluation of Seismic Reliability for Isolated Multi-span Continuous Deck Bridges

Guglielmo Amendola(⊠) 📵

Politecnico di Torino, Corso Duca degli Abruzzi, 10129 Turin, Italy
guglielmo.amendola@polito.it

Abstract. The main goal of the study is evaluating the seismic reliability of isolated multi-span continuous deck bridges considering the influence of the friction pendulum isolators and of the structural properties. The adopted system is modelled by considering five degrees of freedom accounting for five vibrational modes of the elastic reinforced concrete pier whereas the response of the infinitely rigid deck equipped with the isolators is also considered with an additional degree of freedom. To take into account the interaction pier-abutment, the latter has been modelled as a fixed support. The inherent uncertainty related to the friction coefficient is assumed to follow a normal probability density function. The friction itself has been assumed velocity-dependent by means of a non-linear law. Record-to-record variability is taken into account through a set of different natural records. With the main purpose of evaluating the seismic reliability of the abovementioned system, fragility curves of both the pier and isolation system supporting the deck are evaluated firstly, and then, by means of the convolution among the hazard curve the corresponding seismic reliability curves are obtained.

Keywords: Reliability assessment · Continuous deck bridges · Friction coefficient

1 Introduction

Seismic isolation of bridges is a widely adopted strategy to uncouple the seismic response of the deck from the substructure, permitting a significant reduction in terms of both the deck acceleration and forces, transmitted to the pier, as demonstrated in [1, 2]. This leads to improvements in safety of infrastructures [3, 4] with important consequences within the territorial and urban planning. The main advantage of using the friction pendulum systems (FPS) devices consists in making the isolation period independent from the mass of the deck [5, 6]. As far as the variability along the friction coefficient is considered, in this study the model adopted is a velocity-dependent law [7–10].

Some studies have proposed the seismic reliability-base design (SRBD) for systems isolated with the FPS isolators, such as [11]. In [12], a probability-based reliability assessment method to consider the variations of the seismic isolation properties under different conditions is presented. Specifically, referring to a base-isolated two span reinforced concrete (RC) bridge, key structural properties such as temperature, seismic

© The Author(s), under exclusive license to Springer Nature Switzerland AG 2022
F. Calabrò et al. (Eds.): NMP 2022, LNNS 482, pp. 1148–1157, 2022.
https://doi.org/10.1007/978-3-031-06825-6_111

hazard, the dimensions and the material mechanical properties are modelled as random variables. Other studies have mainly focused the attention in finding an optimal value of the friction coefficient, able to minimize the seismic response [13, 14], throughout a parametric analysis (i.e., by varying many properties of the structure and the seismic input).

In this study, the seismic reliability of isolated multi-span continuous deck bridges, equipped with FPS isolators is evaluated for a wide range of bridge properties. The isolated bridges have been modelled through a five-degree-of-freedom (dof) model representative of the elastic RC pier behaviour and an additional dof is adopted to analyse the response of the infinitely rigid RC deck isolated by the FPS devices. The RC abutment is modelled as a rigid support. By means of that, the pier-abutment-deck interaction is also considered, even if the abutment-backfill soil interaction is neglected [15].

Within a probabilistic approach, the main random variable coincides with the friction coefficient, and the Latin hypercube sampling method (LHS) is adopted as a sampling technique [16]. Moreover, a set of 30 natural seismic records with different characteristics has been scaled to different increasing intensity levels in line with the seismic hazard of the reference site (i.e., L'Aquila (Italy)). Incremental dynamic analyses (IDAs) [17] have been carried out to define the response statistics in terms of peak deck displacement and peak pier displacement with the aim to derive the seismic fragility curves [18]. The above-mentioned fragility curves, in a cascaded flow, have been useful to assess the seismic reliability of bridges isolated with FPS bearings, according to the PEER-like modular approach [19], assuming the hazard curves of the site and a specific design life.

2 Equations of Motion

The seismic response of the bridge is herein modelled throughout a six-degree-of-freedom (dof) system in which 5 dofs account for the lumped masses of the pier and 1 additional dof relates to the rigid RC deck. Previous studies have demonstrated a great effectiveness in the use of the adopted 6dof model as demonstrated in [13, 14]. The friction pendulum system devices (FPS) are placed both on the top of the pier and the abutment, which is modelled as rigid and fixed. Figure 1 shows the 6 dofs model.

The corresponding equations of motion are:

$$m_d \ddot{u}_d(t) + m_d \ddot{u}_{p5}(t) + m_d \ddot{u}_{p4}(t) + m_d \ddot{u}_{p3}(t) + m_d \ddot{u}_{p2}(t) + m_d \ddot{u}_{p1}(t) + c_d \dot{u}_d(t) + f_b(t) + f_a(t) = -m_d \ddot{u}_g(t)$$

$$m_{p5} \ddot{u}_{p5}(t) + m_{p5} \ddot{u}_{p4}(t) + m_{p5} \ddot{u}_{p3}(t) + m_{p5} \ddot{u}_{p2}(t) + m_{p5} \ddot{u}_{p1}(t) - c_d \dot{u}_d(t) + c_{p5} \dot{u}_{p5}(t) + k_{p5} u_{p5}(t) - f_b(t) = -m_{p5} \ddot{u}_g(t)$$

$$m_{p4} \ddot{u}_{p4}(t) + m_{p4} \ddot{u}_{p3}(t) + m_{p4} \ddot{u}_{p2}(t) + m_{p4} \ddot{u}_{p1}(t) - c_{p5} \dot{u}_{p5}(t) - k_{p5} u_{p5}(t) + c_{p4} \dot{u}_{p4}(t) + k_{p4} u_{p4}(t) = -m_{p4} \ddot{u}_g(t)$$

$$m_{p3} \ddot{u}_{p3}(t) + m_{p3} \ddot{u}_{p2}(t) + m_{p3} \ddot{u}_{p1}(t) - c_{p4} \dot{u}_{p4}(t) - k_{p4} u_{p4}(t) + c_{p3} \dot{u}_{p3}(t) + k_{p3} u_{p3}(t) = -m_{p3} \ddot{u}_g(t)$$

$$m_{p2} \ddot{u}_{p2}(t) + m_{p2} \ddot{u}_{p1}(t) - c_{p3} \dot{u}_{p3}(t) - k_{p3} u_{p3}(t) + c_{p2} \dot{u}_{p2}(t) + k_{p2} u_{p2}(t) = -m_{p2} \ddot{u}_g(t)$$

$$m_{p1} \ddot{u}_{p1}(t) - c_{p2} \dot{u}_{p2}(t) - k_{p2} u_{p2}(t) + c_{p1} \dot{u}_{p1}(t) + k_{p1} u_{p1}(t) = -m_{p1} \ddot{u}_g(t) \qquad (1a,b,c,d,e,f)$$

where u_d is the deck displacement relative to the top of the pier, u_{pi}, with $i = 1 : 5$, is the relative displacement of the i-th lumped mass of the pier with respect to the lower one, m_d is the mass of the deck, m_{pi} is the mass of the i-th lumped mass of the pier, while k_{pi} is the corresponding stiffness. The viscous damping coefficient for the device and for the pier masses are, respectively, c_d and c_{pi}; t represents the instant of time and the dots are used to indicate differentiation. As for the resisting forces of the FPS bearings located on either the abutment or the pier, they are herein indicated respectively as $f_a(t)$ and $f_p(t)$. These corresponding reactions apply, as follows [5]:

$$f_a(t) = \frac{m_d g}{2}\left[\frac{1}{R_a}\left(u_d(t) + \sum_{i=1}^{5} u_{pi}\right) + \mu_a\left(\dot{u}_d + \sum_{i=1}^{5}\dot{u}_{pi}\right)\mathrm{sgn}\left(\dot{u}_d + \sum_{i=1}^{5}\dot{u}_{pi}\right)\right]$$

$$f_p(t) = \frac{m_d g}{2}\left[\frac{1}{R_p}u_d(t) + \mu_p(\dot{u}_d)\mathrm{sgn}(\dot{u}_d)\right] \tag{2a,b}$$

where R_a and R_p are the radii of curvature of the FPS devices placed on the abutment (pedix a) and on the pier (pedix b) and they are assumed to be equal; the stiffness of the deck, coming from the elastic component of the reaction force over the FPS, is evaluated as $k_d = W/R = m_d g/R$, where half of the total weight W is carried from the bearing on the abutment and half by the pier's bearing; g is the gravity constant; μ the sliding friction coefficient of the isolator on the abutment (subscript a) or of the isolator on the pier (subscript p), which depends on the bearing slip velocity $\dot{u}(t)$ (that is relative to the ground for the isolator on the abutment or to the pier top for the isolator on the pier), and $\mathrm{sgn}(\cdot)$ denotes the sign function.

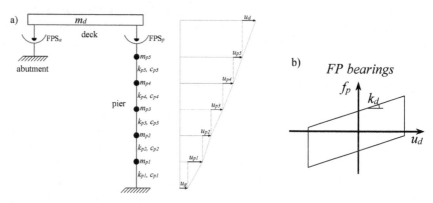

Fig. 1. 6dof model of a bridge isolated by FPS devices (a); response of the FPS device on the pier (b).

By implementing the FP devices, the fundamental period of the deck can be expressed as $T_d = 2\pi\sqrt{m_d/k_d} = 2\pi\sqrt{R/g}$ [5].

In this work, the dependency of the sliding friction coefficient on the velocity has been considered, as proposed by [7, 8, 20]:

$$\mu(\dot{u}_d) = f_{\max} - (f_{\max} - f_{\min}) \cdot \exp(-\alpha|\dot{u}_d|) \tag{3}$$

where $f_{\max}$ and $f_{\min}$ define, respectively, the sliding friction parameter at large and null velocity, α governs the transition from low to large velocities. Experimental results suggest to assume $f_{\max} = 3f_{\min}$ and $\alpha = 30$ [21, 22].

Dividing all equations of the system (1) by the deck mass m_d, dimensionless equations derive similarly to [13, 14].

3   Main Aleatory and Deterministic Variables

3.1   Seismic Records and Intensity Measure

Coherently with the Performance Based Earthquake Engineering approach (PBEE) [23, 24], this study adopts a scale factor, i.e. an intensity measure (*IM*) to scale a set of 30 natural ground motions, chosen from different databases (i.e., PEER, ITACA, ISESD-Internet Site for European Strong-Motion Data), see also [14]. The intensity measure is the spectral-displacement $S_D(T_d, \xi_d)$, in correspondence of the isolated period of the bridge system, $T_d = 2\pi/\omega_d$ with a damping ratio ξ_d equal to zero [14, 25], due to its effectiveness [26, 27]. Therefore, the corresponding *IM* is hereinafter denoted to as $S_D(T_d)$ and is assumed ranging from 0.10 m to 0.45 m (Table 2) to perform the IDAs in line with the seismic hazard of the reference site (L'Aquila (Italy)) [28].

Table 1. Selected values of the intensity measure $S_D(T_d)$.

IM	1	2	3	4	5	6	7	8
$S_D(T_d)$[m]	0.10	0.15	0.20	0.25	0.30	0.35	0.40	0.45

3.2   The Friction Coefficient and Structural Parameters

As for the uncertainty on the friction coefficient [29, 31], a truncated normal PDF, from 0.5% to 5.5% with an average value equal to 3%, has been used to model the sliding friction coefficient at large velocity f_{max}; starting from this PDF, the LHS method [21] has been used to generate 15 values.

Regarding the structural properties, a wide parametric analysis has been conducted encompassing different types of isolated bridges, specifically, the isolated superstructure period T_d varies in the range between 1 s and 4 s; the RC pier period T_p equal to 0.05 s; $\lambda = \sum_{i=1,5} m_{pi}/m_d$, that represents the overall mass ratio related to the sum of the *i*-th mass ratios (assumed equal), varies in the range between 0.1, 0.15 and 0.2.

4   Incremental Dynamic Analysis (IDA)

For each one of the combinations between the deterministic (T_d, λ and T_p) and aleatory parameters assumed, the differential equations of motion (Eq. (1)) have been repeatedly solved for the 30 ground motion records, scaled to the increasing 8 values of $S_D(T_d)$ (Table 1). In this way, IDAs [17] are performed in Matlab [32–35].

The IDAs have been interpreted assuming the following engineering demand parameters (EDPs): $u_{d,max}$ and $u_{p,max} = \left| \left(\sum_{i=1}^{5} u_{pi} \right) \right|_{max}$. As a consequence, a set of samples is obtained for each EDP at each *IM* and is assumed to follow a lognormal probability density function. The lognormal PDF can be estimated for each EDP by calculating the sample lognormal mean $\mu_{\ln}(EDP)$ and the sample lognormal standard deviation, or dispersion, $\sigma_{\ln}(EDP)$ through the maximum likelihood technique [18, 21, 22, 36–50]. So, it is possible to determine the 50th, 84th and 16th percentiles of each PDF [21].

5 Seismic Fragility Assessment

Assessing the seismic reliability of the system under investigation needs the evaluation of the seismic fragility, herein introduced as the probabilities P_f exceeding different limit states (LSs) conditioned to each level of the IM assumed in the previous analyses. For this reason, the LS thresholds need to be defined. Regarding the LS thresholds related to the isolation system, nine different values of the radius in plan of the single concave surface have been assumed [30, 31], as reported in Table 2. As for the pier, four discrete LSs (LS1, LS2, LS3 and LS4), corresponding respectively to "fully operational", "operational", "life safety" and "collapse prevention", are provided by [51] and expressed in terms of pier drift index (PDI) (Table 3) [11].

Table 2. Limit State thresholds for the isolation system

	1	2	3	4	5	6	7	8	9
r[m]	0.10	0.15	0.20	0.25	0.30	0.35	0.40	0.45	0.50

Table 3. Limit State thresholds for the pier.

	LS1 fully operational	LS2 operational	LS3 life safety	LS4 near collapse
PDI_{IB}	$PDI_{IB} = 0.23\%$	$PDI_{IB} = 0.5\%$	$PDI_{IB} = 0.83\%$	$PDI_{IB} = 1.67\%$
P_f	$5 \cdot 10^{-1}$	$1.6 \cdot 10^{-1}$	$2.2 \cdot 10^{-2}$	$1.5 \cdot 10^{-3}$

5.1 Seismic Fragility Curves

In this sub-section, the probabilities P_f exceeding the abovementioned LS thresholds with regard to the isolation level (consequently to the deck) and to the RC substructure/pier have been numerically calculated. Figure 2 and Fig. 3 show the fragility curves in relation to both the values of T_p and T_d, for each LS and each value of λ, referring to the isolation system. Generally, the seismic fragility decreases for increasing the LS thresholds. Moreover, the seismic fragility decreases as T_d increases. For low T_d values, an increase of T_p leads to slightly higher values of the seismic fragility.

As far as the seismic fragility referred to the RC pier is concerned, Fig. 4 and Fig. 5 show a more marked variability with the variation of λ: the pier is less fragile (vulnerable) when this parameter increases for fixed T_p. Furthermore, for low T_p values and medium-high T_d values, it is possible to observe how the failure probabilities assume negligible

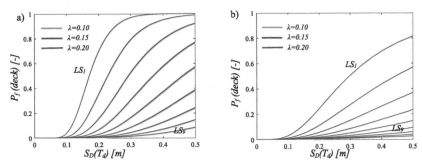

Fig. 2. Seismic fragility curves of the deck response for $T_p = 0.05$ s and a) $T_d = 1$ s, b) $T_d = 4$ s.

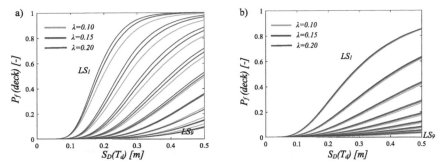

Fig. 3. Seismic fragility curves of the deck response for $T_p = 0.2$ s and a) $T_d = 1$ s, b) $T_d = 4$ s.

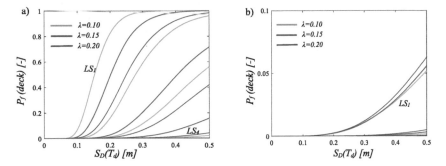

Fig. 4. Seismic fragility curves of the pier for $T_p = 0.05$ s and a) $T_d = 1$ s, b) $T_d = 4$ s.

values, due to the effectiveness of the seismic isolation technique combined with the high rigidity of the substructure (pier) and to the low seismic demand of the superstructure (deck). An increase of T_p causes an increase of the seismic fragility.

6 Seismic Reliability Curves

Integrating the seismic fragility curves with the seismic hazard curves, expressed in terms of the same *IM*, related to the specific site (L'Aquila (Italy)), permits to calculate

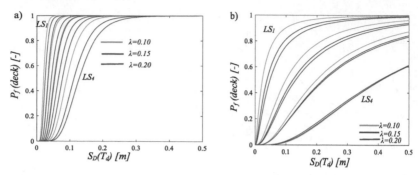

Fig. 5. Seismic fragility curves of the pier for $T_p = 0.2$ s and a) $T_d = 1$ s, b) $T_d = 4$ s.

the mean annual rates exceeding the *LSs*. Then, by using the Poisson distribution, the probabilities of exceedance in 50 years [17] have been calculated. The seismic reliability curves for the isolation system are obtained, as shown in Fig. 6. These curves can be used for the preliminary design of the dimensions in plan (i.e., radius in plan, *r*, of the concave surface) of the friction pendulum devices.

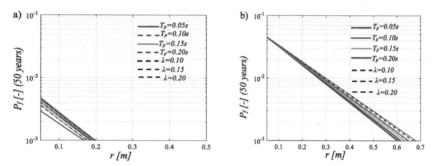

Fig. 6. Seismic reliability curves of the deck response for a) $T_d = 1$ s, b) $T_d = 4$ s.

7 Conclusions

The study aims at analyses the seismic reliability of multi-span continuous deck bridges, equipped with single concave friction pendulum (FPS) throughout an extensive, involving both the isolators properties and the bridge properties. The results are very important and useful for territorial and urban planning.

The inherent aleatory uncertainty involving the friction itself has been accounted through an appropriate PDF. As for the seismic uncertainty, a set of 30 records have been selected and then scaled to increasing seismic intensity, herein considered coincident with the spectral displacement at the isolated period of the system. The most representative response parameters of the system have been subsequently extrapolated. Successively, the fragility curves, referred to both the isolated deck and the RC pier, and then the reliability curves have been obtained. What comes out of more interesting from the analysis, is how the seismic reliability of the isolated deck decreases with the increase of the curvature radius of the isolator due to the high seismic hazard of the site. When the pier is concerned, it could be observed that the seismic reliability decreases for both higher pier periods and higher isolated periods due to the aleatory uncertainty of the sliding friction coefficient and to the high seismic hazard of the site.

References

1. Jangid, R.S.: Seismic response of isolated bridges. J. Bridg. Eng. **9**(2), 156–166 (2004)
2. Ghobarah, A., Ali, H.M.: Seismic performance of highway bridges. Eng. Struct. **10**(3), 157–166 (1988)
3. Troisi, R., Alfano, G.: Towns as safety organizational fields: an institutional framework in times of emergency. Sustainability **11**, 7025 (2019). https://doi.org/10.3390/su11247025
4. Troisi, R., Castaldo, P.: Technical and organizational challenges in the risk management of road infrastructures. J. Risk Res. (2022). DOI:https://doi.org/10.1080/13669877.2022.2028884
5. Zayas, V., Low, S., Mahin, S.: A simple pendulum technique for achieving seismic isolation. Earthq. Spectra **6**(2), 317–333 (1990)
6. Su, L., Ahmadi, G., Tadjbakhsh, I.: Comparative study of base isolation systems. J. Eng. Mech. **115**(9), 1976–1992 (1989)
7. Mokha, A., Constantinou, M.C., Reinhorn, A.M.: Teflon bearings in base isolation. I: Testing. J. Struct. Eng. **116**(2), 438–454 (1990)
8. Constantinou, M.C., Mokha, A., Reinhorn, A.M.: Teflon bearings in base isolation. II: modeling. J. Struct. Eng. **116**(2), 455–474 (1990)
9. Almazàn, J.L., De la Llera, J.C.: Physical model for dynamic analysis of structures with FPS isolators. Earthquake Eng. Struct. Dynam. **32**(8), 1157–1184 (2003)
10. Mosqueda, G., Whittaker, A.S., Fenves, G.L.: Characterization and modeling of Friction Pendulum bearings subjected to multiple components of excitation. J. Struct. Eng. **130**(3), 433–442 (2004)
11. Castaldo, P., Alfano, G.: Seismic reliability-based design of hardening and softening structures isolated by double concave sliding devices. Soil Dyn. Earthq. Eng. **129**, 105930 (2020)
12. Nassar, M., Guizani, L., Nollet, M.J., Tahan, A.: A probability-based reliability assessment approach of seismic base-isolated bridges in cold regions. Eng. Struct. **197**, 109353 (2019)
13. Castaldo, P., Amendola, G.: Optimal sliding friction coefficients for isolated viaducts and bridges: a comparison study. Struct. Control Health Monit. **28**(12) (2021)
14. Castaldo, P., Amendola, G.: Optimal DCFP bearing properties and seismic performance assessment in nondimensional form for isolated bridges. Earthquake Eng. Struct. Dynam. **50**(9), 2442–2461 (2021)
15. Mitoulis, S.A.: Seismic design of bridges with the participation of seat-type abutments. Eng. Struct. **44**, 222–233 (2012)
16. Celarec, D., Dolšek, M.: The impact of modelling uncertainties on the seismic performance assessment of reinforced concrete frame buildings. Eng. Struct. **52**, 340–354 (2013)
17. Vamvatsikos, D., Cornell, C.A.: Incremental dynamic analysis. Earthq. Eng. Struct. Dynam. **31**(3), 491–514 (2002)
18. Castaldo, P., Amendola, G., Ripani, M.: Seismic fragility of structures isolated by single concave sliding devices for different soil conditions. Earthq. Eng. Eng. Vib. **17**(4), 869–891 (2018). https://doi.org/10.1007/s11803-018-0481-6

19. Cornell, C.A., Krawinkler, H.: Progress and challenges in seismic performance assessment. PEER Center News **4**(1), 1–3 (2000)
20. Constantinou, M.C., Whittaker, A.S., Kalpakidis, Y., Fenz, D.M., Warn, G.P.: Performance of Seismic Isolation Hardware Under Service and Seismic Loading. Technical Report MCEER-07-0012 (2007)
21. Castaldo, P., Tubaldi, E.: Influence of FPS bearing properties on the seismic performance of base-isolated structures. Earthq. Eng. Struct. Dynam. **44**(15), 2817–2836 (2015)
22. Castaldo, P., Tubaldi, E.: Influence of ground motion characteristics on the optimal single concave sliding bearing properties for base-isolated structures. Soil Dyn. Earthq. Eng. **104**, 346–364 (2018)
23. Aslani, H., Miranda, E.: Probability-based seismic response analysis. Eng. Struct. **27**(8), 1151–1163 (2005)
24. Porter, K.A.: An overview of PEER's performance-based earthquake engineering methodology. In: Proceedings of the 9th International Conference on Application of Statistics and Probability in Civil Engineering, ICASP9, San Francisco, California (2003)
25. Ryan, K., Chopra, A.: Estimation of seismic demands on isolators based on nonlinear analysis. J. Struct. Eng. **130**(3), 392–402 (2004)
26. Iervolino, I., Cornell, C.A.: Record selection for nonlinear seismic analysis of structures. Earthq. Spectra **21**(3), 685–713 (2005)
27. Pinto, P.E., Giannini, R., Franchin, P.: Seismic Reliability Analysis of Structures. IUSS Press, Pavia (2003)
28. NTC18. Norme tecniche per le costruzioni. Gazzetta Ufficiale del 20.02.18, DM 17.01.18, Ministero delle Infrastrutture
29. Wei, B., Fu, Y., Li, S., Jiang, L., He, W.: Scaling errors of a seismic isolation system with a shear key. Soil Dyn. Earthq. Eng. **139**, 106382 (2020)
30. Castaldo, P., Palazzo, B., Ferrentino, T., Petrone, G.: Influence of the strength reduction factor on the seismic reliability of structures with FPS considering intermediate PGA/PGV ratios. Compos. B Eng. **115**, 308–315 (2017)
31. Castaldo, P., Palazzo, B., Ferrentino, T.: Seismic reliability-based ductility demand evaluation for inelastic base-isolated structures with friction pendulum devices. Earthq. Eng. Struct. Dynam. **46**(8), 1245–1266 (2017)
32. De Iuliis, M., Castaldo, P.: An energy-based approach to the seismic control of one-way asymmetrical structural systems using semi-active devices. Ingegneria Sismica-Int. J. Earthq. Eng. **29**(4), 31–42 (2012)
33. Amendola, C., De Silva, F., Vratsikidis, A., Pitilakis, D., Anastasiadis, A., Silvestri, F.: Foundation impedance functions from full-scale soil-structure interaction tests. Soil Dyn. Earthq. Eng. **141**, 106523 (2021)
34. Castaldo, P., De Iuliis, M.: Optimal integrated seismic design of structural and viscoelastic bracing-damper systems. Earthq. Eng. Struct. Dynam. **43**(12), 1809–1827 (2014)
35. Math Works Inc. MATLAB-High Performance Numeric Computation and Visualization Software. User's Guide. Natick (MA) USA (1997)
36. Troisi, R., Alfano, G.: Firms'crimes and land use in Italy. An exploratory data analysis. Smart Innovation, Systems and Technologies 178 SIST, pp. 749–758 (2020)
37. Garzillo, C., Troisi, R.: Le decisioni dell'EMA nel campo delle medicine umane. In EMA e le relazioni con le Big Pharma - I profili organizzativi della filiera del farmaco. G. Giappichelli, pp. 85–133 (2015)
38. Palazzo, B., Castaldo, P., Marino, I.: The dissipative column: a new hysteretic damper. Buildings **5**(1), 163–178 (2015)
39. Golzio, L.E., Troisi, R.: The value of interdisciplinary research: a model of interdisciplinarity between legal re-search and research in organizations. J. Dev. Leadersh. **2**, 23–38 (2013)

40. Nese, A., Troisi, R.: Corruption among mayors: evidence from Italian Court of Cassation judgments. Trends Organized Crime **22**(3), 298–323 (2019)

41. Troisi, R., Golzio, L.E.: Legal studies and organization theory: a possible cooperation. In: 16th EURAM Conference Manageable cooperation. European Academy of Management, Paris, 1–4 June (2016)

42. Troisi, R., Guida, V.: Is the appointee procedure a real selection or a mere political exchange? the case of the italian health-care chief executive officers. J. Entrepreneurial Organ. Diversity **7**(2), 19–38 (2018). https://doi.org/10.5947/jeod.2018.008

43. Troisi, R.: Le risorse umane nelle BCC: lavoro e motivazioni al lavoro. Progetto aree bianche. Il sistema del credito cooperativo in Campania **1**, 399–417 (2012)

44. Troisi, R., Alfano, G.: Is regional emergency management key to containing COVID-19? a comparison between the regional Italian models of Emilia-Romagna and Veneto. Int. J. Public Sect. Manag. (2021). https://doi.org/10.1108/IJPSM-06-2021-0138

45. Castaldo, P., Nastri, E., Piluso, V.: FEM simulations and rotation capacity evaluation for RHS temper T4 aluminium alloy beams. Compos. B Eng. **115**, 124–137 (2017)

46. Nesticò, A., Moffa, R.: Economic analysis and Operational Research tools for estimating productivity levels in off-site construction [Analisi economiche e strumenti di Ricerca Operativa per la stima dei livelli di produttività nell'edilizia off-site], Valori e Valutazioni n. 20, pp. 107–128 (2018). ISSN: 2036-2404. DEI Tipografia del Genio Civile, Roma

47. Nesticò, A., Maselli, G.: Declining discount rate estimate in the long-term economic evaluation of environmental projects. J. Environ. Acc. Manage. **8**(1), 93–110 (2020). https://doi.org/10.5890/JEAM.2020.03.007

48. Castaldo, P., Gino, D., Marano, G., Mancini, G.: Aleatory uncertainties with global resistance safety factors for non-linear analyses of slender reinforced concrete columns. Eng. Struct. **255**, 113920 (2022). S0141-0296(22)00078-5

49. Troisi, R., Di Nauta, P., Piciocchi, P.: Private corruption: an integrated organizational model. Eur. Manage. Rev. (2021)

50. Nese, A., Troisi, R.: Individual preferences and job characteristics: an analysis of cooperative credit banks. Labour **28**(2), 233–249 (2014)

51. SEAOC Vision 2000. Committee. Performance-based seismic design engineering. Report prepared by Structural Engineers Association of California, Sacramento, CA (1995)

The Effect of Local Emergency Policies on the Performance of the Italian Regional Health Care System

Gaetano Alfano[(⊠)] [iD]

Department of Political and Communication Sciences, Territorial Development Observatory (OST), University of Salerno, Fisciano, Italy
galfano@unisa.it

Abstract. This study aims to investigate the efficiency of the Italian regional health care system and how the emergency policy adopted by the local government can influence those performances. For this purpose, a two-step empirical analysis was performed. At first, a directional distance frontier model was used to assess the efficiency of the regional health care systems. Successively through a fixed-effect panel method, the effects of the policy on the performances were evaluated. The article presents evidence on how local policy influences the performance of the health care system. In particular, the two-step method proposed highlights how the reduction of the workload on the hospitals, obtained preferring home isolation of the infected to hospitalizations, and more stringent control on the public activities, by means of local ordinances, lead to higher performance of the health care system. Those results are considered useful to local government to identify more specific and efficient solutions for the management of creeping crisis as the Covid-19.

Keywords: Health care systems · Emergency policies · Efficiency

1 Introduction

The Covid-19 outbreak showed the worldwide lack of preparedness of the traditional emergency management systems to a "creeping crisis" [1]. The uncertainty linked to its spatial spread and the difficulty of defining its temporal boundaries makes this type of crisis even more difficult to deal with by governments [2]. As a result, the local government had to experiment with new strategies to ensure a quick and effective response to deal with the non-linear path of these events [3]. During the pandemic response, most countries have experienced new emergency management strategies involving jointly national and sub-national governments. In addition, their strategies may be characterized by the objectives considered as priorities when choosing the measures to be taken [4]. On this line, the decision was taken at a sub-national level (e.g. by the regional government) has a not negligible impact on the management of the Covid-19 endemic [5, 6]. Every local government takes their major decisions, including the measures to adopt for social distancing and isolation and treatment of the infected, to reduce the Covid-19

F. Calabrò et al. (Eds.): NMP 2022, LNNS 482, pp. 1158–1165, 2022.
https://doi.org/10.1007/978-3-031-06825-6_112

spread [7, 8]. Thus, in this situation, all the decision that influences the response of the local health system assumes great importance. Various studies have highlighted how European countries have adopted policy responses to contain the virus, mainly resulting in heavy limitations on public life. Only during the second phase of the endemic, the governments have switched to a more balanced system of measures with the objective to partially resume ordinary activities [9].

The main objective of this study is to evaluate the effect that some strategies adopted by the local governments have on the performance of the health care system.

Comparing the measures adopted by different regional emergency management allows assessing the effectiveness of different policies regarding the health care system [10]. The aim is to help policymakers to cope with local crisis management.

The focus is on the Italian scenario from March to October 2020. Italy was the country after China that has been most severely hit by the Covid-19 pandemic [11, 12]. It was also the first European country reached by the pandemic and with a significant increase in the number of cases in just a few weeks.

The exponential increase of infected cases in March and April 2020 showed how inadequate was the health care system to cope with a continuous flow of patients [13]. The ongoing crisis from March to November 2020 was characterized by two waves. The first wave, from March to May, forced the central government to adopt a general Lockdown. After the first wave, the central government adopted a different strategy providing more autonomy to the local government (i.e. regions) on the decisions to adopt than during the first wave.

The analysis proposed concerns the efficiency of regional health care systems and the effect of local policy adopted by the regions on them. The two-step method proposed highlights how the reduction of the workload on the hospitals, obtained preferring home isolation of the infected to hospitalizations, and more stringent control on the public activities lead to higher performance of the health care system.

2   Data and Method

A two-step method was used to assess the effects of the Regional Emergency Measures on the performance of the health care system. The first step is the evaluation of the regional health care system performance through the use of a Directional Distance Function (DDF) model. The second steps evaluate the effect over time of the emergency measures adopted by the regions on the efficiency score, obtained in the first step, by a fixed-effect panel model.

2.1   Data Sources

The data utilized in the DDF model are obtained mainly from the daily reports of the Minister of Interior and Minister of Health. In particular, the Minister of Interior provides information on the monthly number of confirmed cases, the monthly number of dismissed/healed cases and deaths. The Minister of Health gave information about the hospital staff.

The operational measures adopted by the Regional Emergency Management are provided by the regional ordinances and the Public Health and Health care plans enacted from March 2020 to October 2020. The study was stopped to October 2020 because until that time the effect of the national restriction can be hypnotized equal for all the regions since the national ordinances were the same for all regions. Starting November 2020 Italy adopted a new tier system according to the regional degree of risk, consisting of different restrictions for each region based on the regional risk indicator defined by the Minister of Interior.

In addition to the ordinances, the daily report of the Minister of the interior and Minister of Health was useful for obtaining information on the number of hospitalizations, number of intense care units and distribution of medical equipment.

2.2 Methods

The first step of the analysis is the evaluation of the efficiency of the health care system using a Directional Distance Function (DDF) model.

The model is a variation of the classical Data Envelope Analysis widely used in previous works on the assessment of the performance of hospitals [14]. Both models consent to evaluate the efficiency of each Decision-Making unit (DMU), in our case the regions.

The DEA model finds the efficiency of homogeneous units resolving the following linear model proposed initially by Charnes et al. [15]:

$$\min_{\theta, \lambda} \theta \ Z \tag{1}$$

$$s.t. \ \theta x_i \geq X \lambda$$

$$Y \lambda \geq y_i$$

$$\lambda \geq 0$$

where $X = \{x_1, x_2, \ldots, x_k\}$ and $Y = \{y_1, y_2, \ldots, y_k\}$ are, respectively, the inputs and outputs of the i-th DMU, λ is an optimal weight vector and $\theta \leq 0$ is the optimal score wanted. A DMU is defined as technically efficient when it lies on the frontier ($\theta = 1$).

The model has two advantageous proprieties that justify its diffusion in many works on public sectors and, in particular, health care systems [14].

As a first advantage, there is no need to specify the formal relationship between inputs and outputs, consenting to simplify the analysis, especially in the cases when the relationship is unknown. As a second advantage, it can be used also when the amount of DMU is low.

As proposed by other work [16–18], in this paper the efficiency of the health care regional systems are evaluated through the use of a DDF model. This variant of the DEA consents to evaluate the efficiency of the DMU considering any undesirable output inevitably related to the production (e.g. nuclear energy production is inevitably related to the production of nuclear waste).

As stated by Färe and Grosskopf [19], the use of traditional DEA when bad outputs are jointed to the production leads to efficiency evaluation biased.

The DDF model is analytically similar to the traditional DEA, but consent to define the efficiency of each DMU maximizing the desirable output and simultaneously minimizing the undesirable one. Formally: $P_{CRS} = \{(c, y^d, y^u)|x \geq \lambda X, y^d \leq \lambda Y, y^u = \lambda Y, \lambda \geq 0\}$, where y^d and y^u are respectively the desirable and undesirable outputs.

In particular, in this work, the number of medical staff and the number of infected are considered as the inputs [14]. The number of dismissed cases and cases of death are considered respectively as the desirable and undesirable output variables [16].

The second step of the procedure is the econometric analysis of panel data using R-software package glm. This analysis consents to analyze the relationship between the independent variable (i.e. the performance scores obtained in the previous step) with a set of explanatory variables over time. Generally, a panel model can be written as follows:

$$Y_{it} = \beta_{it} x_{it} + u \tag{2}$$

where y_{it} is the dependent variable considered, β_{it} is the vector of parameters to be estimated, x_{it} is the vector of explanatory variables considered and u is the error term.

The explanatory variables considered in this works are: 1) The number of regional ordinances on health care was considered as a measure of the level of restriction imposed by the regions. Local government during the crisis adopted the "redundancy" strategy consisting in adding to the national restrictions local restriction to further restrict public activities. This strategy even if it is not a cost-effective solution allows facing uncertainties quickly [20]. They adopt an increasing number of restrictions to reduce the uncertainties related to the diffusion of the Covid-19. 2) The percentage of hospitalized cases. The variable is a proxy of the workload that medical staff had to support during the crisis. Some studies in human resource management on the performance in the working place highlight how high workload in time leads to a lower level of performance [21]. During the crisis, some regions (e.g. Veneto) gave priority to quarantine at home cases and hospitalizing only the most severe cases, while other regions made a major use of hospitalization. 3) The number of intense care units over 1 million inhabitants in the region [14]. The variable measures the variation over time of the number of intense care units. 4) The amount of medical equipment delivered to the hospitals during the time considered, expressed in thousands of pieces. [22]. The variable proxies the amount of economic investment done by the regions during the crisis.

An Hausman Test (chisq = 9.9008, df = 4, p-value = 0.04213) suggest that a fixed-effect model is the more suitable for this work. Fixed-effects models have the advantage that omitted variables constant over time (e.g. the characteristic of the population, the socio-economic characteristic of the environment) do not influence the variation of the dependent variable over time, thus any variation of the dependent variable is related only to time-depending variables [23, 24].

3 Results

This section presents and discusses the results. Table 1 reports the full set of results for the DDF models, while Table 2 reports the results of the fixed-effect panel model.

The monthly scores have an average score of around 0.53 denoting a non-negligible inefficiency. The monthly efficiency scores for each Italian region highlight a remarkable difference among the regional health care systems. In particular, the analysis shows how the health care system of Lombardia and Trento have the highest level of performance among the Italian regions. In particular, Lombardia is the region that more often is on the frontier of efficiency than other regions.

Considering the geographical position of the regions, the scores in Table 1 point out how north regions are characterized by higher values of the performance score.

Table 1. DDF results.

Region	03/20	04/20	05/20	06/20	07/20	08/20	09/20	10/20
Abruzzo	0,32	0,12	0,18	0,07	0,48	0,48	0,44	0,36
Basilicata	0,10	0,25	0,89	0,46	0,35	0,45	0,38	0,23
Calabria	0,27	0,18	0,50	0,46	0,52	0,53	0,62	0,61
Campania	0,51	0,20	0,45	0,24	0,45	0,55	0,52	0,30
Emilia-Romagna	0,53	0,65	0,32	0,50	0,44	0,24	0,47	0,96
Friuli Venezia Giulia	1,00	0,75	0,47	0,35	0,55	0,53	0,58	0,59
Lazio	0,78	0,25	0,11	0,09	0,12	0,15	0,25	0,26
Liguria	0,67	0,66	0,51	0,90	0,87	0,85	0,63	0,90
Lombardia	1,00	1,00	0,88	1,00	1,00	1,00	1,00	1,00
Marche	0,02	0,61	0,35	0,80	0,15	0,18	0,22	1,00
Molise	0,98	0,41	1,00	0,80	0,84	0,66	0,54	0,58
P.A. Bolzano	0,72	0,64	0,86	1,00	0,90	0,90	0,12	0,82
P.A. Trento	0,63	0,88	0,78	1,00	1,00	1,00	0,98	1,00
Piemonte	0,17	0,41	0,21	0,70	0,78	0,77	0,76	0,86
Puglia	0,23	0,10	0,21	0,12	0,19	0,48	0,40	0,32
Sardegna	0,48	0,20	0,52	0,43	0,58	0,40	0,55	0,26
Sicilia	0,55	0,17	0,35	0,26	0,35	0,18	0,27	0,22
Toscana	1,00	1,00	1,00	1,00	0,87	0,51	1,00	0,09
Umbria	0,48	1,00	0,98	0,47	0,37	0,33	0,33	0,30
Valle d'Aosta	0,45	0,22	0,15	0,37	0,31	0,70	0,55	0,45
Veneto	0,88	0,68	0,29	0,30	0,08	0,07	0,19	0,72
Average score	**0,56**	**0,49**	**0,52**	**0,54**	**0,53**	**0,52**	**0,51**	**0,56**

Table 2 report the results of the fixed-effect panel model and its diagnostic tests.

Table 2. Fixed-effect regression results.

Variable	Coefficient (SE)	P-value
Local operational measures	2.20e−02 (2.20e−02)	1.194e-12 ***
Hospitalizations	−5.35e−03 (1.99e−03)	0.007942 **
Intense care units	−8.00e−04 (8.53e−04)	0.349459
Medical equipment	7.59e−06 (3.31e−06)	0.023185 *
R-square	0.28	
F-statistic	17.90 p-value: 2.36e−12	
Significance codes: *** p < 0.001, ** p < 0.01, * p < 0.05		

The local operational measures adopted by regional ordinances are significant and positively related to the performance scores of the health care system (2.20e−02, p < 0.001). This suggests that in regions where the local government added a higher number of further restrictions to the national restrictions, the health care systems performed better than regions with a fewer number of restrictions.

The results highlight a significant negative relationship between the percentage of hospitalizations and the performance score of the health care system over time (−5.35e−03, p < 0.01). Thus, in regions where the local government preferred to hospitalize only the more severe cases of Covid-19 a higher level of performance is achieved.

At last, the amount of medical equipment distributed to the hospitals is positively related and significant to the performance scores over time (7.59e−06, p < 0.05). This results in line with other studies [22], shows how a higher level of resources invested increase the overall performance level of the public health care systems.

The number of intense care units and its variation over time has been found not significant.

4 Conclusions

This study investigates the relationship between the policies adopted by local/regional government and the performance of the regional health care systems.

The initial directional distance frontier model highlights the difference in performance of the health care system of the Italian regions. In particular, Lombardia has been found to have the most efficient health care system during the entire crisis.

Successively, the effect of the local policies on the performance of the health care systems has been analyzed. The results show that regions where the national restrictions to public life have been further restricted by the local government through the enacting of regional ordinances generally perform better. The redundancy strategy applied seems to give a better response to the uncertainty related to the Covid-19, in line with the literature [20]. In addition, the model shows how policies oriented to quarantine the less serious cases of Covid-19 and hospitalize only the more severe cases, increase the overall performance level of the health care system. When this policy is applied, the health care system is generally subjected to a lower workload that ends in a better performance of the medical staff. At last, the study confirms that the amount of resources invested, in particular on medical equipment, highly influences the response of the health care system.

Finally, the study has some political implications. The knowledge of the effect of some policies on the efficiency of the health care system performances is useful to identify the specific and efficient solutions for the management of creeping crisis as the Covid-19.

References

1. García-Basteiro, A., et al.: The need for an independent evaluation of the COVID-19 response in Spain. Lancet **396**(10250), 529 530 (2020)
2. Bernanke, B.S., Geithner, T.F., Paulson, H.M.: Firefighting: the financial crisis and its lessons. Penguin Books (2019)
3. Handmer, J., Dovers, S.: Handbook of disaster policies and institutions: Improving emergency management and climate change adaptation. Routledge (2013)
4. Troisi, R., Alfano, G.: Towns as safety organizational fields: an institutional framework in times of emergency. Sustainability **11**(24), 7025 (2019)
5. Troisi, R., Alfano, G.: Is regional emergency management key to containing COVID-19? a comparison between the regional Italian models of Emilia-Romagna and Veneto. Int. J. Public Sector Manage. (2021)
6. Mangiameli, M., Mussumeci, G.: A gimbal platform stabilization for topographic applications. In: AIP Conference Proceedings, vol. 1648, p. 780011 (2015). https://doi.org/10.1063/1.4912991
7. Mohanta, K.K., Sharanappa, D.S., Aggarwal, A.: Efficiency analysis in the management of COVID-19 pandemic in India based on Data Envelopment Analysis. Current Res. Behav. Sci. **2**, 100063 (2021)
8. Tortia, E., Troisi, R.: The resilience and adaptative strategies of Italian cooperatives during the COVID-19 pandemic. Foresight STI Gov. **15**(4), 78–88 (2021)
9. Toshkov, D., Carroll, B., Yesilkagit, K.: Government capacity, societal trust or party preferences: what accounts for the variety of national policy responses to the COVID-19 pandemic in Europe? J. Eur. Public Policy, 1–20 (2021)
10. Wankhade, P.: Performance measurement and the UK emergency ambulance service: unintended consequences of the ambulance response time targets. Int. J. Public Sect. Manag. **24**(5), 384–402 (2011)
11. Mangiameli, M., Mussumeci, G.: Real time transferring of field data into a spatial DBMS for management of emergencies with a dedicated GIS platform. In: AIP Conference Proceedings 1648, p. 780012 (2015). https://doi.org/10.1063/1.4912992
12. WHO Director-General's opening remarks at the media briefing on COVID-19-11 March 2020. https://www.who.int/dg/speeches/detail/who-director-general-s-opening-remarks-at-the-media-briefing-on-covid-19---11-march-2020

13. Cesari, M., Proietti, M.: COVID-19 in Italy: ageism and decision making in a pandemic. J. Am. Med. Dir. Assoc. **21**(5), 576 (2020)
14. O'Neill, L., Rauner, M., Heidenberger, K., Kraus, M.: A cross-national comparison and taxonomy of DEA-based hospital efficiency studies. Socioecon. Plann. Sci. **42**(3), 158–189 (2008)
15. Charnes, A., Cooper, W.W., Rhodes, E.: Measuring the efficiency of decision-making units. Eur. J. Oper. Res. **2**, 429 (1978)
16. Du, J., Cui, S., Gao, H.: Assessing productivity development of public hospitals: a case study of Shanghai, China. Int. J. Environ. Res. Public Health **17**(18), 6763 (2020)
17. Gennaro, A., Candiano, A., Fargione, G., Mangiameli, M., Mussumeci, G.: Multispectral remote sensing for post-dictive analysis of archaeological remains. A case study from Bronte (Sicily). In: Archaeological Prospection, vol. 26(4), pp. 299–311 (2019)
18. Deng, Z., Jiang, N., Pang, R.: Factor-analysis-based directional distance function: the case of New Zealand hospitals. Omega **98**, 102111 (2021)
19. Färe, R., Grosskopf, S.: Modeling: undesirable factors in efficiency evaluation: comment. Eur. J. Oper. Res. **157**(1), 242–245 (2004)
20. Weick, K.E., Sutcliffe, K.M.: Managing the unexpected, vol. 9. Jossey-Bass, San Francisco (2001)
21. Ning, X., Wang, Q., Wu, B.: The Influence of Schedule Target on Project Performance. Shanghai University, China (2004)
22. Chowdhury, H., Zelenyuk, V.: Performance of hospital services in Ontario: DEA with truncated regression approach. Omega **63**, 111–122 (2016)
23. Stock, J.H., Watson, M.W.: Introduction to econometrics (3rd updated edition). Age (X3), 3(0.22) (2015)
24. Troisi, R., Di Nauta, P., Piciocchi, P.: Private corruption: an integrated organizational model. European Management Review (2021)

Fundamental Rights Implications of Covid-19. Religious Freedom and Resilience During the Pandemic

Angela Iacovino(✉) ⓘ and Milena Durante ⓘ

University of Salerno, Via Giovanni Paolo II 132, 84084 Fisciano, SA, Italy
{aiacovino,mdurante}@unisa.it

Abstract. The article aims to highlight some institutional and constitutional problems that have arisen during the COVID-19 health emergency. The issues concern the functioning of the State, the relations between the powers of the State, the compliance with the principle of legality, and the protection of fundamental rights. A relevant part of the doctrine has already addressed this particular strand of legal literature so, without any claim to exhaustiveness, we will try to make the point on some guidelines emerged, clarifying a series of doubts related to the constitutional compatibility, and compliance with the principles and rules of constitutionalism, regarding a set of measures adopted by the government. In particular, we focus on the restrictive measures and the pandemic effects on religious freedom in Italy, as a case study, and to related solutions that the Italian institutions have provided to legitimate the freedom restrictions. Finally, the study has the purpose to assess whether the adopted solutions come within the "reasonable accommodation" although we are aware that "post-Covid-19" phase, will be the real test bench of the resilient institutional capacity to relaunch and give new life to the protection of the fundamental rights currently in "standby".

Keywords: Sars-CoV 2 · Health emergency · Religious freedom

1 Introduction

The Covid-19 health pandemic is, to date, a global challenge requiring extraordinary measures to protect public health. The latter has become the fundamental social value and the political objective which complex government strategies have been focused on. The health emergency that we are experiencing represents an unprecedented scenario when compared with the crisis faced in the past. This extraordinary condition could justify the peculiar procedures followed to manage the consequent state of crisis. Thus, the Coronavirus health emergency has led to the adoption of diligent measures aimed at the containment of infections which in the emergency logic have been developed by the executive power [1]. It deals with various areas of legal standardization, in particular certain areas relating to fundamental freedoms, which are considered as the backbone of constitutional democratic structures. More specifically, a significant set of constitutionally guaranteed freedoms – freedom of movement, expatriation, assembly, but also

F. Calabrò et al. (Eds.): NMP 2022, LNNS 482, pp. 1166–1175, 2022.
https://doi.org/10.1007/978-3-031-06825-6_113

religious freedoms, the right to defense, the right to education, freedom of private economic initiative, the right to work– has suffered a drastic limitation in the name of the health protection. The latter is further supported by the chrism of universality [2], in term of fundamental right of the individual and the interest of the community, ex Art. 32 Cost. Emergency discipline is in our legal system fragmented, underpinned by a rationale that emphasizes urgency legislation instead of the emergency legislation [3]. All emergencies have a critical element that is counterbalanced with a necessary weakening of the rights of freedom, the form of government, and the whole system of sources. All of them involve a major risk for the maintenance of democracy [4]. The coronavirus has hit the structural institutional weaknesses by enhancing the crisis that " endemically afflicts the system of sources" [5]. In short, the current epidemiological emergency raises the following questions: to what extent inviolable rights ensuring the democratic stability (Article 2, Constitution) can be limited? In particular, we will focus on the restrictive measures going over the current crisis for their implications on the freedom of expression the religion. It is a fundamental right, which concerns the innermost private dimension together with its outward manifestation [6].

The territorial institutions involved are called to adopt solutions of "reasonable accommodation" between conflicting freedoms and the emergence of good practices in "this different seasons of history" [7], trying to understand "whether and in what way the religious factor can contribute to support the legal response to the needs raised during the emergency, without weakening but rather strengthening the resilience of values-purposes of the system" [8]. We will show how in times of pandemic the religion can regain its constitutional role, in compliance with the constitutional state of emergency [9]. Our study also identifies those actions in line with duties of solidarity that enforces collaborative strategies aimed at protecting public health [7].

2   Emergency and Constitution: The Balancing of Rights at the Time of Covid-19 and (Still) Unresolved Issues

The age of freedom seems to be replaced by the age of contagion [10], and the fear of the emergency [11] thus, the interpreter is engaged in investigating the kind of deviations occurred in the sphere of the effectiveness, to report any ongoing constitutional transformations and their surreptitious alteration of Constitutional values [12]. To what extent does the Constitution adapt to such an unprecedented emergency? There is no doubt that our common home is "suffering" since it is forced to step aside in some of its expressions reflecting the fundamental rights of the legal system. The emergency does not simply reverse the balance between freedoms and the protection of individual and collective health, rather it balances the Constitution with the same emergency "fatally destined to benefit the latter with heavy costs for the first" [13]. Since the emergency has jeopardized the balance between civil, economic, and ethical-social relations, constitutionally protected [14], how are freedoms, inviolable rights, and health protection reconciled? Protecting public health and human rights are not mutually exclusive choices. Therefore, how can we ensure the balance between these two protections? The limitation of our rights, which could be acceptable in a definite time horizon, has led to a sort of their suspension by means of acts that the Constitution does not recognize as suitable.

A suspension of rights is justified by the need to hinder the violation of the right to health *ex* art. 32 Cost. Through the emergency decree, freedom of movement, assembly, worship and business were sacrificed "according to the synthetic expression which lacks any definite terms – at least in technical-legal terms –lockdown" [15]. The prevalence of the right to health over other individual prerogatives has resulted in the compression of our freedoms. After the first Covid wave, it has slightly weakened, passaging from a radical suspension of fundamental freedoms to a restoration through behavioral control dynamics, however, capable to limit the right to privacy. Additionally, what about the impact of the pandemic on parliamentarism particularly for its difficult compatibility with states of exception? In times of crises, the constitutional categories usually adopted are equally in crisis, with implications on the form of government as well as on the statute of rights [5]. They have been "hit hard" as for centrality of parliamentary debate, the confrontation between political forces requiring- by definition-time, and the complexity of parliamentary decision-making processes. What is under pressure is the same communicative action underpinning the dynamics of representative democracy. If emergencies can be challenging for the political decision-maker – in particular for the "expert bureau" does a space for Parliament decision-making still survive? Otherwise, are we facing another constitutional era [16], where the parliament is deprived of authority as it is bypassed by the political decision-maker? How valid, from the legitimacy perspective, are the emergency measures adopted by the Government and local authorities (in the form of decree-laws, decrees of the Prime Minister, ministerial and civil protection ordinances, and circulars, regional ordinances). The measures imposed by the Government to face the emergency have defined a complex regulatory framework. Nevertheless, it is possible to catch relevant elements that deserve attention. First of all, administrative acts, namely emergency ordinances, (rectius DPCM) provide limitations to fundamental freedoms to ensure the protection of public health. Despite being covered by decree-laws, they are problematically evanescent in their implementation of the most restrictive provisions [17]. Criticisms are multiplied when one considers that they affect fundamental freedoms as in the case of movement and residence, (Article 16 Constitution) as well as the personal freedom, (Article 13 Constitution), according to the principle of reserve law [18]. Legal scholars are skeptics about how a decree-law can guarantee adequate legitimacy of a governmental administrative act, for the lack of explicit parameters and criteria. Conversely, DPCM provides "an extra-ordinem self-legitimization arising from the state of exception in itself rather than from the decree-law" [19]. Increasingly, atypical government sources, such as the DPCM, are questioning the principle of substantive legality and the principle of absolute reservation of law where fundamental rights are affected; then, the indispensable intervention of parliamentary law in the conversion of the decree-law. It is at stake" the observance of the procedure necessary for the correctness of the relationship between primary and secondary sources" [20] and which allows the comparison of rules of governmental origin with general and abstract law to ensure effective judicial protection. The sequence of acts by the Government, in time of "infinite emergency" [21], undermines the constitutional logic sublimating itself in a state of exception, Furthermore, emergency on the one hand, and request for safety and protection of public health on the other, do not remain stable but continuously change, favoring the dialectic between the unpredictable replication of the virus and the ordinal system of constitutional guarantees and sources. In this sense, uncertainty and the evanescence of the emergency regulations, as well as the sudden change in the regulation of public and private life, is disconcerting: moreover implementing the citizens distrust in institutions [22]. Italian Constitution, which establishes restrictions on freedoms "for reasons of health and safety (art. 16.1), appears incomplete since it does not contain provisions on states of emergency. Article 78 relating to the conferral on the Government of the necessary powers in the event of deliberation of the state of war by the Chambers has to be excluded. It governs a different situation not to be invoked in the current circumstances. Nevertheless, it contains the right and duty of the State to defend the institutions and protect the community. It deals with a defense that leads to a duty of integrity whose essential core is found in that principle of political solidarity (Article 2 Constitution). It entitles the legislator to establish both reasonable limits to individual rights, and the veto of conducts which infringe the rights of others [23]. Moreover, the emergency power, while explicitly provided in the Charter, can find its implicit foundation in the supreme guarantee of the preservation of the Republic [1]. In the specific pandemic, therefore, our Constitution could show resilience because it provides a discipline capable of ensuring a balance between governability and representativeness – between timely choices and parliamentary involvement. However, in considering the measures put in place their constitutional adherence can be discussed. The national emergency declaration, adopted by the Council of Ministers on a proposal from the Prime Minister – resolution of 31 January 2020 containing " Declaration of the state of emergency as a result of the health risk associated with the onset of diseases arising from transmissible viral agents" - while finding its justifications [24] in the document adopted on 30 January 2020 by the WHO defining the Covid-19 epidemic as "a public health emergency of international importance" and in the exponential increase in the infection curve, has expanded the legitimizing basis of the subsequent DPCM," substantially ratified from time to time by certain decree-laws" [11]. Indeed, their formal legitimacy has come under attack from who do not consider possible to take exceptional measures in derogation from primary sources and restricting rights through such invasive administrative acts (9). Nevertheless, a significant set of normative acts has been developed to face the emergency. In particular decrees-laws that have conferred an additional legitimation to the orders of the Prime Minister [25]. This turns into worrying restriction of parliamentary prerogatives and a risk of implementing the power of the Prime Minister. It should be noted that the legal nature of the DPCM is mainly administrative, "classifiable in the category of emergency orders that find their presupposition in the source of primary rank, the Single Text of Civil Protection" [1]. It is, therefore, an *extra-ordinem* administrative measure, whose legitimacy may be assessed according to the parameters laid down by law. The right of the emergency must then conform with the constitutional framework and must also be limited in time: "the emergency in its most specific sense an anomalous condition even if temporary. It legitimizes unusual measures but they lose legitimacy if unjustifiably prolonged in time" (Corte Cost., 15/1982). Even if the restrictions provided with the administrative measures are justified by the spread of the epidemic, the guarantee of health protection must always comply with constitutional principles. Any limitation of the fundamental freedom must necessarily be defined by the elective assemblies, which are required to supervise the executive power, to avoid flowing towards a "subtle" presidentialism. The wild recourse to the decree-law, in fact, not only jeopardizes the constitutional system, but also makes the urgent decree the rule and parliamentary law the exception [26]. In emergency times in which the health crisis addresses choices hindering rights and freedoms, Parliament – "the great absent in the time of the Coronavirus" [16] must, rather, return to play its role to ensure the democratic resilience of the country, showing "an additional solidity" without evading "the fulfillment of the functions pre-ordered to the guarantee of rights" [1]. In this sense, it would be appropriate to provide for a parliamentary right of emergency to better manage both crisis as well as exceptional periods. Many comparative experiences offers important lessons for the management of the pandemic, as well as a further rethinking of the role of Parliament, reduced to a forum [27] during the Covid crisis, could reaffirm its specific legitimacy as the main arena of popular sovereignty [28, 29]. Only Parliament has a degree of the discretion in balancing the principles, freedoms, and constitutional rights involved, whose plurality requires mutual interaction and limitation. As stated by the Constitutional Court, all fundamental rights protected by the Constitution are in a relation of mutual integration, then it is not possible to identify the absolute prevalence of one of them; their protection must always be systemic to avoid conflict (Corte cost. 264/ 2012) and also to prevent "the unlimited expansion of one of the rights, since any protected legal situations constitutes, as a whole, an expression of the dignity of the person" [30]. In this direction, the point of balance must be assessed by the legislator "in the decision of the rules and by the judge in the sentences - according to criteria of proportionality and reasonableness, such as not to allow a sacrifice of their essential core" (Corte cost. 85/2013). On the contrary we have witnessed to a set of administrative measures that in providing severe restrictions, limit fundamental freedoms, sometimes in areas covered by absolute legal reserves, arising compatibility issues with constitutionally guaranteed rights [31]. They turn into temporary suspensions of constitutional requirements, in the name of the protection of public health. The derogation from the ordinary constitutional rules, on the one hand, must be provisional and, on the other hand, the limiting measures must be only those that are unavoidable and proportionate to the risks, and must therefore be the result of a correct assessment of the balance between security and freedom [32]. The primacy of the right to health over other constitutional rights is controversial: competing positions are, on one hand about the idea that the protection of health and life represents a value, a necessary precondition of the other rights [33]; on the other it is considered the danger of an authoritarian drift [19, 34]. The primacy of the value of life and health has characterized the common thread of pandemic legislation: it seems to be a priority "to ensure the preservation of the effectiveness of the legal system for health reasons"- since the right to life threatens to collapse like a sandcastle, and forever, all rights, from individual freedoms to the right to work, to education to economic freedoms. "the right to health is transformed into a meta-positive "value" that orients all other rights and that cannot be balanced with them" [35]. However, the measures against the spread of contagion, which affected the maintenance of the constitutional provisions, also jeopardize the religious freedom (Article 19 of the Constitution) and the discipline of connection between the State and religious confessions (Articles. 7 and 8 Cost.). In the pandemic crisis some provisions "have shown unsuspected interactions with the protection of human health"

[36]. Let us, therefore, try to understand how the restrictive measures have affected the right to religious freedom and the pursuit of the objective of protecting human health.

3   Religious Freedom in Times of Covid-19: A Right that is Compressible but not Limited

In recent decades we are witnessing a generalized return of faith in the public space [37–39] a new phenomenon of protagonism of religions that in the sociological field has been defined as "deprivatization of religion" [40, 41]. In this connection, in such an emergency, where some freedoms are limited in favor of the wider protection of public health, it is appropriate to investigate the consequences of the fundamental right of health, whose main effects deal with the innermost sphere of the individual. We are talking about religious freedom, which are in the realm of the absolute rights. They can be defined freedoms that cannot be restricted as dealing with the most intimate dimension of the person, going beyond the jurisdiction of the State, belonging to so-called "forum internum". However, this right has equally an external dimension, the so-called "forum externum". The provision recognizes the right to concretely exercise one's religious faith through worship, teaching, practices, observance of rites, etc., that is to say, its practical expressions which, considered as relative rights, may legitimately be restricted [42]. Therefore, the religious freedom is not a limited right since it can be limited in some of its manifestations, in compliance with precise time constraints and based on measures that are proportionate and based on real needs and urgency for the protection of constitutional values [43]. To fully understanding the effects of the Covid-19 pandemic on freedom of religion in Italy, the starting point is in the constitutional provision, namely Article 19 which recognizes the right of all the people to profess their religious faith freely in an individual or associated form, to make propaganda and to practice its worship in private or in public, provided that it is not contrary to common decency. Art. 19 Cost., however, in the economy of such reflection, takes particular importance since, on the one hand, recognizes the protection of religious sentiment as an inalienable right of the person, on the other, identifies the only hypothesis of legitimate limitation. In fact, for all inalienable and fundamental human rights, the Constitution expressly provides for limits the possibility of freely enjoying them, recognizing a particular relevance especially to the limits relating to the protection of health, safety, and security. The art. 19 does not, however, expressly endorse such categories of limits to religious freedom, but merely the provision of the objective limit of morality. However, it is not conceivable to think of being faced with an absolute and unlimited right [44]; we must not be forgotten that 'fundamental freedoms' are not absolute, since they may be subject to restriction by the State authorities to achieve a fair balance between the protection of general interest and the request for the protection of fundamental rights of individuals. At the same time, it is clear that to legitimize such restrictions, it is necessary to assess certain parameters, through which to assess the reasonableness of any restrictive measures. In this regard, it should be recalled that the limit of health to freedom of religion is expressly provided for by the European Convention on Human Rights which, according to the provisions of Art. 117 Cost., is a binding supranational source for the Italian legislator [44]. In the ECHR, the protection of health is one of the reasons which, established by law, allow

Member States to restrict the freedom of expression, assembly, association, movement, and, precisely, the expression of a religion or belief (art. 9 of the ECHR) [36]. At the same time, however, the ECHR stipulates that, in limiting the scope of the freedom to express one's religion or belief, only those restrictions which are established by law may intervene, which constitute necessary measures in a democratic society and which are proportionate to the objective to be pursued and, to verify the presence of these criteria, the European Court conducts a check though the so-called proportionality test, which allows it to determine the compatibility of the limitation of the right with the conventional obligations, or prepare to assert a breach of the European Convention. What has just been said makes it possible to establish that even the limited discipline of freedom of worship, introduced by government decrees to contain the contagion from covid-19 on the Italian territory, does not in itself constitute a violation of religious freedom as it is functional to the protection of health as the supreme interest of the community. It is legitimate to think, therefore, that in conditions of a health emergency, in a democratic system that professes to be secular and pluralist, the external dimension of religious freedom, or the possibility of participating in religious functions or ceremonies, can be called to bear temporary forms of compression [7]. At this point, however, it is good to wonder how the Italian institutions have provided the legitimate possibility of limiting this freedom, and especially if they have adopted the solutions of "accommodation". The D.P.C.M. of 23 February 2020, concerning Urgent measures in the field of containment and management of the epidemiological emergency from Covid-19, published in the Official Gazette - General Series no. 45 of 23 February 2020 and converted, with amendments, with law no. 13 of 5 March 2020, has foreseen the "suspension of manifestations or initiatives of any nature, events and any form of reunion in a public or private place, also of a religious character" (…) even if carried out "in closed places open to the public" (art. 1, paragraph 2, lett. c). After a few days, a further measure reiterated the operation of the measure just mentioned and, in art. 2, paragraph 1, lett. d), provided that "the opening of places of worship is subject to the adoption of organizational measures such as to avoid gatherings of people, taking into account the size and characteristics of the places, and such as to ensure the further possibility of respecting the distance between them of at least one meter". The regulatory framework of the so-called Phase 1 was further defined, then, with the adoption of Decree-Law no. 19 of 2020 with which it was reiterated once again: the suspension of religious ceremonies including funeral ceremonies; the suspension of religious events in every place; conditional access to places of worship. To interpret and mitigate the effects of the provisions of the DPCM and the decrees-law mentioned then intervened the Central Directorate of the Affairs of Cults, which in a note specified that liturgical celebrations are not "in themselves prohibited, but can continue to take place without the participation of the faithful, to avoid groupings that could become potential opportunities for contagion" (Ministry of Interior – Department for Civil Liberties and Immigration – Central Directorate of Cult Affairs, 27 March 2020 [6, 43, 45]. In response to these measures, the CEI (Italian Episcopal Conference), reporting the positions adopted by the majority religious confession in Italy, through a statement of 5 March 2020, declared to "share (the) situation of hardship and suffering of the country" and to "take joint initiatives with which count the spread of the virus" [6]. It is evident, in short, that at the time of the so-called Phase 1, the Catholic Church and the other ecclesiastical authorities, have taken a collaborative attitude towards government measures, adapting to restrictive measures and modifying how certain religious activities are carried out. Only later voices were raised, particularly among the authorities of the Catholic Church, to claim a much higher level of organizational autonomy of their religious community, even in exceptional times [6], although the restrictive measures adopted by the State do not produce any violation of the sovereignty and independence of the Catholic Church, since the defense of the physical integrity of its citizens not only does not constitute an interference in the proper order of the Church (as of any other confessional order) but rather is an expression of the duty of the State to secure the supreme principles of the constitutional order [7, 44].

4 Conclusions

What emerges from this brief reflection is that the restrictive measures applied to religious freedom, as well as to other constitutional rights, were an immediate and necessary response to the need to protect the health of the community, to avoid the collapse of the national health system and at the same time to continue to ensure the stability of the entire democratic system, thus "avoiding democratically convoluted drifts" [6, 36].

The proof is that, by recognizing each fundamental right as primary, but not absolute, the balancing ensures systemic protection, sacrificing one or more rights in favor of the greater achievement of another constitutional interest, infact, "in balancing operations there cannot be a decrease in the protection of a fundamental right if it is not matched by a corresponding increase in protection of other interest of equal rank" (Corte cost. 143/2013). In the future months it will be possible to verify whether, in the conflict between all rights and interests at play, the right balance has been applied, and it will also be possible to ascertain the degree of resilience achieved by public institutions in managing to prevent or manage serious and exceptional national emergencies. Only in a future "post-Covid-19" phase, will we be able to measure the effective capacity of our democracy to be able to fall into the "ranks of normality", to relaunch and give new life to the suspended path of protection and development of those fundamental rights currently remained in "standby".

References

1. Nicotra, I.A.: Stato di necessità e diritti fondamentali. Emergenza e potere legislativo. AIC. **1**, 98–165 (2021)
2. Alicino, F.: Introduzione. Salute umana e tradizioni religiose di fronte alle emergenze sanitarie. Daimon. Diritto comparato delle religioni. Numero speciale, pp. 3–13 (2021)
3. Clementi, F.: Il lascito della gestione normativa dell'emergenza: tre riforme ormai ineludibili. Osservatorio Costituzionale **3**, 33–47 (2020)
4. Cabiddu, M.A.: Emergenze abissali. Cosa resterà di diritti e processo dopo la pandemia. AIC **1**, 303–326 (2021)
5. Ruggeri, A. Il coronavirus, la sofferta tenuta dell'assetto istituzionale e la crisi palese, ormai endemica, del sistema delle fonti. Consulta Online. I, pp. 209–216 (2020)
6. Michetti M.: La libertà religiosa e di culto nella spirale della emergenza sanitaria Covid-19, in Dirittifondamentali.it, n. 2/2020, pp. 526–546 (2020)

7. Macrì G.: La libertà religiosa alla prova del Covid-19. Asimmetrie giuridiche nello "stato di emergenza" e nuove opportunità pratiche di socialità, in Stato, Chiese e pluralismo confessionale, n.9/2020, pp. 46–48 (2020)
8. D'Angelo, G., Pasquali Cerioli, J.: L'emergenza e il diritto ecclesiastico: pregi (prospettici) e difetti (potenziali) della dimensione pubblica del fenomeno religioso. Stato, Chiese e pluralismo confessionale **19**, 25–78 (2021)
9. Raffiotta, E.: Sulla legittimità dei provvedimenti di governo a contrasto dell'emergenza virale da coronavirus. BioDiritto, On Line first (2020)
10. Giordano, P.: Nel contagio. Einaudi, Torino (2020)
11. Rescigno, F.: La gestione del coronavirus e l'impianto costituzionale. Il fine non giustifica ogni mezzo. Osservatorio Costituzionale **3**, 253–271 (2020)
12. Mangia, A., Bin, R.: (a cura di) Mutamenti costituzionali. Diritto costituzionale Rivista quadrimestrale **1**, 1–133 (2020)
13. Ruggeri, A.: Il Coronavirus contagia anche le categorie costituzionali e ne mette a dura prova la capacità di tenuta. Diritti Regionali **1**, 368–378 (2020)
14. Covolo, S.: Il difficile bilanciamento tra la salute come diritto fondamentale dell'individuo e interesse della collettività e gli altri diritti inviolabili, ai tempi dell'emergenza coronavirus. Soltanto il parlamento può essere garante contro l'arbitrio del potere esecutivo. Diritto.it (2020)
15. D'Antonio, V., Scocozza, S.L.: Emergenza sanitaria e sistemi di contact tracing: modelli normativi e tutela della privacy. In: Castagna, L., Conte, A., Parrella, R. (cds.) La storia senza aggettivi. Studi in onore di Luigi Rossi. Rubbettino, Soveria Mannelli, pp. 165–180 (2021)
16. Tripodina, C.: La Costituzione al tempo del coronavirus. Costituzionalismo.it **1**, 77–88 (2020)
17. Baldini, V.: Emergenza sanitaria e Stato di prevenzione. Diritti fondamentali.it. www.dirittifondamentali.it **1**, 561–565 (2020)
18. Piccirilli, G.L.: "riserva di legge" Evoluzioni costituzionali, influenza sovranazionali. Giappichelli, Torino (2019)
19. Venanzoni, A.: L'innominabile attuale. L'emergenza Covid-19 tra diritti fondamentali e stato di eccezione. Forum Quaderni Costituzionali (2020)
20. Brunelli, G.: Sistema delle fonti e ruolo del Parlamento dopo (i primi) dieci mesi di emergenza sanitaria. AIC **1**, 384–398 (2021)
21. Simoncini A.: Tendenze recenti della decretazione d'urgenza in Italia e linee per una nuova riflessione. A. Simoncini (a cura di), L'emergenza infinita. EUM, Macerata, pp. 19–52 (2006)
22. Cassese, S.: Troppe norme scritte senza buon senso. Così si alimenta la sfiducia dei cittadini. Il Messaggero 20 dicembre 2020
23. Giuffrè, F.: voce Solidarietà. Digesto delle discipline pubblicistiche. Aggiornamento. Utet, Torino (2021)
24. Luciani, M.: Il sistema delle fonti del diritto alla prova dell'emergenza. AIC. www.rivistaai c.it **2**, 109–141 (2020)
25. Mazzarolli, L.A.: "Riserva di legge" e "principio di legalità" in tempo di emergenza. Federalismi.it. 1 (2020)
26. Chinni, D.: Decretazione d'urgenza e poteri del Presidente della Repubblica. Editoriale Scientifica, Napoli (2014)
27. Paterniti, F.: La Corte "pedagogista" di un legislatore colpevolmente inerte. Riflessioni critiche su una svolta problematica della recente giurisprudenza costituzionale. Federalismi.it (2020)
28. Troisi, R., Di Nauta, P., Piciocchi, P.: Private corruption: an integrated organizational model. Eur. Manag. Rev. (2021). https://doi.org/10.1111/emre.12489
29. Troisi, R., Castaldo, P.: Technical and organizational challenges in the risk management of road infrastructures. J. Risk Res. (2022). https://doi.org/10.1080/13669877.2022.2028884

30. Spadacini, L.: I diversi bilanciamenti tra principio democratico, libertà di voto e diritto alla salute nei rinvii delle scadenze elettorali a causa della situazione pandemica. Consulta Online. II, 615–649 (2021)

31. De Marco, E.: Situazioni di emergenza sanitaria e sospensione di diritti costituzionali. Consulta online. II, 368–377 (2020)

32. Formisano, A.: Limiti e criticità dei sistemi costituzionali a fronte dell'emergenza Covid-19. NOMOS. 1 (2020)

33. Morana, D.: Sulla fondamentalità perduta (e forse ritrovata) del diritto e dell'interesse della collettività alla salute: metamorfosi di una garanzia costituzionale, dal caso ILVA ai tempi del coronavirus. Consulta online (2020)

34. Guzzetta, G.: E se il caos delle norme anti-contagio fosse un trucco per toglierci la voglia di libertà). Il Dubbio (2020)

35. Grandi, F.: L'art. 32 nella pandemia: sbilanciamento di un diritto o "recrudescenza" di un dovere? Costituzionalismo.it 1, 82–136 (2021)

36. Alicino F.: Costituzione e religione in Italia al tempo della pandemia, in Stato, Chiese e pluralismo confessionale, n.19/2020, pp. 2–18 (2020)

37. Casanova J.: Oltre la secolarizzazione. Le religioni alla riconquista della sfera pubblica, Il Mulino, Bologna (2000). 77 ss

38. Aldridge A., La religione nel mondo contemporaneo, Il Mulino, Bologna (2005),.15 ss

39. Dessì U.: Religioni e globalizzazione. Un'introduzione, Carocci Editore, Roma (2019)

40. McCrudden, C.: Quando i giudici parlano di Dio. Fede, pluralismo e diritti umani davanti alle Corti, Il Mulino, Bologna, pp. 31–32 (2019)

41. Roy O.: L'Europa è ancora cristiana? Cosa resta delle nostre radici religiose, Feltrinelli, Milano (2019)

42. Blando G.: Libertà religiosa e libertà di culto ai tempi del Covid-19: una questione di bilanciamento, in Federalismi.it, 5 maggio 2020

43. Fuccillo, A.: La religione "contagiata" dal virus? La Libertà religiosa nella collaborazione Stato Chiesa nell'emergenza Covid-19, in in Mazzoni G, Negri A., Libertà religiosa e Covid-19: tra diritto alla salute e salus animarum, in I focus del dossier Olir "Emergenza coronavirus", maggio 2021, 75

44. Colaianni, N.: La libertà di culto al tempo del coronavirus, in Stato, Chiese e pluralismo confessionale, n.7/2020, pp. 32–38

45. Consorti, P.: La libertà religiosa travolta dall'emergenza, in Forum di Quaderni Costituzionali, 2/2020

Multi-stakeholder Governance in Social Enterprises: Self-organization, Worker Involvement and Client Orientation

Ermanno C. Tortia[✉] [iD]

Trento University, 38122 Trento, Italy
ermanno.tortia@unitn.it

Abstract. The paradigm of self-organization is used to explain the evolutionary patterns observed in social enterprises (SEs), understood as multi-stakeholder, non-profit organizations, in Italy. Data from a survey on Italian social cooperatives are used to analyze the structure of governance and ability of SEs to pursue social, non self-seeking aims. Coherently with the paradigm of self-organization, the most salient results are as follows: (i) different stakeholder tend to follow persistent patterns in self-positioning themselves within the governance structure; (ii) paid workers appear to be the dominant insider stakeholder in social cooperatives even if this outcome is not prescribed by law, since they are present in almost all the organizations, and are the only stakeholder in a large share of them; (iii) the multi-stakeholder nature of SEs is confirmed by client orientation, since clients' wellbeing appears prominently among the objectives of the organization, also in its distributive patterns.

Keywords: Social enterprises · Multi-stakeholder governance · Client orientation · Self-organization

1 Introduction

This paper looks at the multi-stakeholder orientation of social enterprises (SEs) in terms of emergent pattern of client orientation and worker control, considering also the impor-

F. Calabrò et al. (Eds.): NMP 2022, LNNS 482, pp. 1176–1187, 2022.
https://doi.org/10.1007/978-3-031-06825-6_114

tant impact of volunteer work and charitable donations.[1] Background theory and empirical research show that the issue of the viability and potential of multi-stakeholder governance (MSG) and related control rights in different proprietary forms (investor owned, non-profit, cooperative and social enterprise) in the social economy is not settled yet (Cornforth 2012).

To achieve the objective, the potential of multi-stake governance is studied in two different directions: (i) theoretically, it applies the paradigm of self-organization: different stakeholder-patron self-position themselves into the governance structure of the non-profit enterprise to devise the organizational solutions most effective to reach the stated social missions; (ii) MSG can turn out to be the most effective and in some cases efficient organizational solution when different groups need to take an active role to limit contractual and governance failures (financiers, employees, clients, volunteers),. At the empirical level, on the other hand, national survey data are used to test descriptively what is the composition of the membership and board of directors, and the impact of multi-stakeholdership on the distribution of resources in favor of clients' wellbeing.

Self-organization requires that the institutional framework does not identify ex-ante the controlling stakeholder (e.g. investors in investor owned companies, workers in worker co-ops etc.....), but only defines the broad features of MSG without ordering the importance and relative position of different stakeholders. New avenues for the development of MSG in non-profit firms are envisaged and analyzed using Italian survey data on the composition of the social base and BoD in social enterprises (192 social cooperatives, SCs). Descriptives show that SCs are mainly multi-stakeholder. However, in 33% of cases they are indeed mono-stakeholder organizations controlled by paid workers. Hence, paid workers emerge spontaneously as the dominant stakeholder group in both the membership base and the board of directors, even if the majority of SCs have more than one stakeholder in the governance bodies, including clients, beneficiaries, volunteers, associations and local authorities.

---

[1] The Italian law on social cooperatives was passed by the Italian Parliament in 1991 (law 381/91), while the law on social enterprises was passed in 2006 (law 155/06) and reformed in 2016 within the overall reform of the third sector (law 106/16). In Italy, SEs can take different proprietary forms: investor owned companies, non-profit organization and cooperative enterprises (social cooperatives). In this work, data on labor relations in and governance of social cooperatives are used. All Italian SEs can be characterized as non-profit enterprises, since both IOFs and SCs undergo a partial non-profit distribution constraint that makes them similar to non-profit organizations. SCs are SEs ex-lege, while instead IOFs and non-profit organization need to apply to governmental agencies to become SEs. The total of the social economy in Italy included in 2015 about 380 thousand organizations, producing a total value added of almost 50 billion euro, and employing slightly less than 1.5 million people, while the number of volunteers is about 5.5 million. 14263 active SCs existed in 2015, producing a total value added higher than 8 billion euro and employing about 380000 people. Also, about 44 thousand volunteers were involved in the operation of SCs.

2 Theoretical Background

Institutional theory of the firm states that the low diffusion of organizations controlled by more than one group of stakeholder-patrons is doomed to failure due to high ownership costs related to heterogeneity of preferences both within each stakeholder group and between groups (Hansmann 1996, 2013). The latter category is especially characteristic of MSG and derives from heterogeneity of objectives and preferences engendering high costs of coordination between different patrons. Coordination costs are expected to grow exponentially as the internal complexity and heterogeneity of organizational governance and of different groups interacting within it grows. These elements would be particularly damaging on efficiency grounds.

Other authors (Sacchetti and Borzaga 2021) criticize this approach showing the existence and viability of different kinds of multi-stakeholder organization. For example, non-profit organizations of an associative kind ore often characterized by a composite and heterogeneous membership. This evidence is taken to show that coordination costs between different stakeholder groups do not need to be uncontrollably high while and can be kept under control by suitable governance solutions. (working rules instituting rewards and penalties, obligations, permissions, allowances etc....; cfr. Commons 1950; Ostrom 1990). At the same time, when several markets fail at one and the same time (e.g. both sale contracts with clients, voluntary donations and labor contracts with employees are characterized by pronounced asymmetries in information) contractual costs can become higher than the costs of governing plural relations between different stakeholders inside organizational boundaries (Sacchetti and Tortia 2020). In this kind of situation, the Coasean argument (Coase 1937) can become relevant again and multi-stakeholder organizations can emerge as more efficient in terms of minimization of transaction costs as opposed to mono-stakeholder ones.

2.1 The Self-organization Paradigm

In this paper, self-organization is seen as a concept related to autopoiesis and explicitly drawn from complexity theory and social systems theory (van Meerkerk et al. 2013; Euler and Heldt 2018). In the social sciences, the concept of self-organization, as imported from complexity science, is used to refer to the emergence of stable patterns through autonomous and self-reinforcing dynamics at the micro-level (Anzola et al. 2017). Ashby's (1947) first used explicitly the concept "self-organization" referring to the self-organizing dynamics in complex systems. Recent contributions such as Gilbert et al. (2015) suggest there are four factors that are common across definitions: pattern formation, autonomy, robustness and resilience, and dynamics. The process of self-organization contemplates several kinds of patterns and measurements such as cooperation, segregation, stratification, normalization. Autonomy deals with processes that do not have pre-determined coordination or central control. Robustness and resilience point to the existence of self-organizing dynamics that displays stability at any point in time (robustness) and over time (resilience). Robust social phenomena resist change, while resilient phenomena achieve endurance despite change. Finally, the dynamics of self-organizing systems refer to change, and tends to disregard individual states (Anzola et al. 2017).

In this paper, the concept of self-organization is applied to the self-positioning of different stakeholders within the boundaries of the social enterprise in terms of a cooperative process requiring that each stakeholder group seeks to reach both self-seeking and the social objectives at one and the same time. Effective self-positioning implies efficient production. All stakeholder will tend to recognize and approve those organizational solutions that are most likely to lead to efficient outcomes in terms of its own objectives. Relative self-positioning of different stakeholders implies a complex process of reciprocal recognition and identification of the most suitable and differential roles within organizational governance and at the community level (Edelenbos e al. 2018). When more than one stakeholder is found in the controlling position at one and the same time, second level structures are expected to emerge, such as arbitration and conflict resolution processes etc.... These second layer structures are geared to reduce the expected costs of conflict and ineffective decision-making processes (Rupasingha et al. 1999; Buijs 2010).

When self-organization and self-positioning is set in operation, the interplay between self-seeking and social objectives produces an open-ended process (as it does not have a clear end-point, like optimal solutions or equilibrium), in which individual and collective motivations are at play. The outcomes of the process cannot be predicted in advance, but instead it may be possible to observe and study relevant organizational patterns. Change over time is expected to improve adaptive fitness to internal and external conditions. The best interpretation of the process is not in terms of optimality, but instead in terms of effectiveness and adaptive fitness in reaching the both self-seeking and social objectives. Outcomes crucially depend on the dominant motivations of the relevant stakeholders. The function of the institutional framework is to let potential behaviors (motivations) become effective (Ruzavin 1994; Hannachi et al. 2020).

2.2 Multi-stakeholder Social Enterprises

The existence of multi-stakeholder organizations, even if they represent a minority phenomenon, points to the possibility that when several markets fail in a significant way at one and the same time, they can generate high contractual costs, as in Sacchetti and Borzaga (2021). The substitution of these markets with administration within organizational boundaries can be functional just to reduce these costs. Also, multi-stakeholder organizations, for example social cooperatives, are often observed to be non-profit enterprises that undergo an at least partial non-profit constraint and accumulate their assets in non-appropriable, indivisible reserves of capital (Tortia 2018). These institutional elements can be explained with the necessity to limit appropriation in the presence of several groups of stakeholder patrons, in a way similar to non- profit organizations in the Hansmann's approach, to protect client/users, who as a norm do not hold decision making power, against the risk of opportunistic behaviors by the organization, that is by those stakeholders (investors, paid workers, volunteers, other producers etc....) that take strategic and operational decisions. This configuration of MSG, which is close to the Italian model of the social cooperative, would derive from the presence of high transaction costs on several markets, which would lead to the necessity to internalize different contractual relation within organizational boundaries and to share control rights. On the other hand, when client involvement in decision making proves not feasible, the non-profit characterization would add guarantees against the possibility that the other stakeholders exploit opportunistically information and other advantages deriving from decision making power. Failures on multiple markets due to asymmetric information, incomplete contracts and the high relational content of the services provided are more widespread just in those sectors that are more intensively populated by non-profit organizations and multi-stakeholder social enterprises, such as social cooperatives (social services, health care, education and the performing arts). Indeed, these two organizational solutions (non-profit organization, NPOs and multi-stakeholder social enterprise, MSSE) can be considered two different ways to solve the same problem. While NPOs have much longer history, MSSEs represent newer instances of organizational design and innovation that explicitly involve stakeholder active participation, which implies better circulation of information and creation of new organizational knowledge. While NPOs simply base their operation on the non-profit constraint and governance steered by a board of trustees, MSSE make stakeholder interrelations explicit and formalize them to certain degrees in the attempt to achieve better adaptation, interaction and coordination. The process of creation of MSSE led to an organizational model in which the possibility of active participation is explicitly recognized by law, while, at the same time, the non-profit and socially oriented nature of the organization adds further guarantees in favor of those stakeholder-patrons that may not hold decision making power, especially donors, beneficiaries and client/users.

3   From Non-profit Organizations to Multi-stakeholder Social Enterprises

In the institutionalist literature (Hansmann 1996, 2013), the possibility of MSG is linked to the possibility that several markets fail at the same time, increasing this way the costs of contracting with different stakeholders (e.g. investors, employees, clients etc....) can open new room for the development of MSG, whose relevance is under-estimated and under-researched to date. When multiple markets fail at one and the same time, contractual costs are high in more than one market and mono-stakeholdership may not guarantee efficiency (Sacchetti and Borzaga 2021).

Hansmann (1996) defines non-profits as organizations without owners. Given contract failures in the sectors populated by non-profits (especially asymmetric information in favor of the organization) owners would have an incentive to exploit such imperfection to their advantage, to increase profits. In order to prevent this happening the non-profit distribution constraint (NPC) is imposed, an asset lock or trust fund are used to store the assets of the organization, and ownership rights in terms of appropriation rights are barred.

In this framework, the exclusion of property rights in terms of both residual control rights and appropriation rights allows this kind of organization not only to receive charitable donations, but also to reduce other potential failures in their relations with other stakeholder groups, especially clients. Contractual failures relates to asymmetric information in the production of services whose quality cannot be predicted in advance and is not easy to evaluate by clients (Blandi 2018). In contemporary non-profits, donations and volunteer work account for a small fraction of the economic value produced.

Factors of production (employees and capital) are paid their market price, while service fees represent the price for the services received by clients (clients are not beneficiaries). This implies that the NDC does not serve the simple function of preventing private appropriation of donations, but, more subtly, to repair contractual failures especially in terms of asymmetric information, when the quality of the product is difficult to evaluate. This way, the focus of analysis shifts towards the study of contractual costs, which are expected to be particularly high in the case of care, health and educational services because of asymmetric information and of the relational and non-standardized nature of such services. The NDC, by preventing private appropriation of surpluses, has the additional positive feature of favoring strengthened trust between the organization and its clients (Hansmann 1996; Blandi 2018).

The economic value of trust rests with the reduction of actual and potential contractual costs at the expenses, however, of creating an organization that is not explicitly governed by any stakeholder group, that is an organization, in Hansmann's interpretation, devoid of control rights.

A different stream evidences that non-profits are not only created by actors on the demand side (clients), but also on the supply side. Non-profit entrepreneurs can select those projects that are not guided by private or self-seeking gains, but by social motivations and objectives (Young 1983; Stryjan 2006). Supply side stakeholders, especially entrepreneurs and associations of workers, can have a fundamental role in instituting and achieving non-profit objectives, by adding knowledge and ability to carry out targeted programs. Easy examples come to mind, since in orthodox non-profit institutions such as universities and hospitals, substantial degrees of decision-making power is granted to staff (medical workers and academics).

The growing focus on the explicit positive role of different stakeholder groups in the making of the non-profit enterprise can be interpreted in new institutionalist terms as a process directed to reduce the negative imports of contractual failures in achieving socially beneficial aims. When the costs deriving from contractual failures increase, while the costs of participation of different groups of patrons and of their interaction in the governance of the organization (membership costs) are not too high, MSG understood as presence in the membership and active participation of different groups of patrons in representative bodies. can become viable. MSG is. The increase in membership costs can be limited, possibly, by the development of suitable working rules and other governance solutions (Commons 1950). The costs of contractual failures are reduced through a process of internalization of contractual relations and effects within governance boundaries (Borzaga and Tortia 2017).

Other authors (Sacchetti and Borzaga 2021; Sacchetti and Tortia 2020) point out that also social costs, understood as the costs of exclusion of weaker and marginalized stakeholder groups can be reduced thanks to their direct or indirect involvement as beneficiaries of the organization. Multi-stakeholder organizations enjoy a recognized ability to generate new social surplus not only by producing new services, but also by reducing negative external effects, such as the "costs of exclusion". This is attained by involving different groups of patrons or "publics" in their governance. Weak social groups, for example hard to employ and disabled workers, are marginalized by the mainstream socio-economic system. The involvement of weak social groups as beneficiaries or users of the organization can reduce social costs for example by reducing the amounts of sub-sidies directed to help these groups. In this way, MSG is able to reduce the total amount of costs (including social or exclusion costs) attached to the operation of the organization and can become more efficient than the conventional mono-stakeholder solutions.

Complementary, in balancing different objectives and disputes resolution, active par-ticipation does not only increase membership costs related to decision making processes, but can also increase economic and social value thanks to better involvement, circulation of information and knowledge creation. Again, well-known examples come to mind. The German system of codetermination in industrial relations is reported to increase coordination costs between the three dominant stakeholders (employers, workers' rep-resentatives and the national government) in terms of lengthy and complex contractual procedures (which increase governance costs), but, at one and the same time, it is also reduces costs related to contractual failures (due to industrial action, dispute and con-frontation), and increases economic surplus by creating stable patterns of growth for entire industries mainly through a more sustained accumulation of human capital in a long run perspective.

The stringent conditions are likely to be more pronounced in the sectors populated by non-profit firms and non-profit organizations, especially in social and welfare services, health care, education, culture and the arts. Indeed, MSG is more often observed in these sectors than in the rest of the economy. In the production of semi-public and meritorious goods which characterizes these sectors, contractual failures related to asymmetric infor-mation, contract incompleteness, production of positive externalities, or the relational and non-standardized nature of the services delivered are observed in a marked way. This evidence can led one to conclude that the net surplus (economic and social) produced by MSG, after adding increased membership costs, but subtracting lower costs related to contractual failures and exclusion, increases when compared to mono-stakeholder gov-ernance. In the most extreme cases, multi-stakeholder organizations in the third sector are observed to produce general interest services in situations in which neither private enterprises nor public authorities are able or willing to intervene. This is due, in the former case, to low profitability of these activities, and, in the latter case, to lack of public finance or inability to organize production in a timely and innovative way.

4   Empirical Evidence

The ICSI survey describes the most important dimensions of MSG in terms of presence of different stakeholder groups in the membership and board of directors (BoD). In addition, we address the issue of customer orientation, as customers are most often absent in membership and representative bodies, but still seem to be a crucial stakeholder in this type of organization. SCs appear to pay substantial attention to improving client/user welfare, and also to distribute resources either under market value or for free in favor of their clients and beneficiaries. These elements confirm their multi-stakeholder nature and the fact that stakeholder welfare is not necessarily promoted by voting rights and decision-making power in formal governing bodies (membership base and BoD). It can be fostered by the nonprofit and socially oriented nature of the organization, and by elements of corporate social responsibility, such as mission statements in statutes, codes of ethics, and social accounting (social reporting is mandatory for SCs in Italy). More generally, by social motivations, non-self-seeking motives and collective decision making in favor of stakeholder groups not sitting in organizational bodies.

4.1 Multi-stakeholder Governance

To test the actual features of governance in multi-stakeholder social enterprises (MSSE), data on Italian social cooperatives, a socially oriented typology of membership-based organization, as defined by law 381/1991, are used. The data are from the 2008 survey on Italian Social Cooperatives (ICSI) completed in 2007. A nationally representative sample of 310 SCs was extracted and questionnaires compiled by directors and managers. The descriptive statistics presented in the following paragraphs refer to SCs producing social and care services (Type A cooperatives).

Description of the membership base starts by exhaustively listing all 10 possible stakeholder groups: paid workers, clients/users; volunteer workers; generic supporters; financial members; private non-profit institutions; private for-profit institutions; public institutions; financial institutions. Paid workers are the most important stakeholder group in the membership, as they are present in 98% of the 192 Type A cooperatives for which data are available. Volunteers represent the second most relevant stakeholder after paid workers (present in 54% of organizations). They are predominantly active workers employed in other enterprises (Marino and Schenkel 2018). The third most relevant stakeholder group is financial members, as they are present in about 1 out of 4 organizations. As required by law, however, they never control the organization (they can't elect more than 1/3 of members of the BoD).

SCs are prevalently multi-stakeholder organizations, even if they are mono-stakeholder (they have one single stakeholder group represented in the BoD) in 33% of cases, while the remaining 66% have two or more group of patrons in the membership. Paid workers are the only stakeholder group in 32% of cases. Instead, 39.5% of organizations have 2 stakeholder groups represented in their membership (this is the modal outcome), 17.1% three groups, 7.3% four groups, and 1.6% five groups. No organization has more than 5 groups (Depedri 2007). As for the BoD, the most represented stakeholder group are again paid workers, who are present in 91% of cases, followed by volunteers who are recorded in 32% of cases. Users/clients are recorded in 4% of cases, and financial members in 9% of cases. The presence of directors external to the membership is recorded in 9% of cases, while institutional members are found in 8.5% of cases. Also, the BoD testifies to the multi-stakeholder nature of SCs: paid workers are the only represented stakeholder in 54.5% of organizations, 2 stakeholder groups are recorded in 35% of organizations, 3 groups in 8.5%, and 4 groups in 1.5%. No organization has more than four groups in its BoD.

4.2 Client Orientation

Even if clients are rarely present in the membership and BoD of Italian SCs (only in about 2% of cases), the fundamental importance of client orientation can be shown in several ways, but first of all by referring to law 381/1991, whose Article 1 defines SCs as businesses created with the aim of "pursuing the general interest of the community in human promotion and social integration of citizens". The general interest of the community, which must be reflected in their statutory bylaws and mission statement, can be understood to include the interests of the users of their services.

Several pieces of related evidence show the importance of client orientation. A first question shows that the organizational form (the non-profit social cooperative) has a fundamental role in satisfying users (64% of respondents say that the organizational form is important or very important). 2 Furthermore, when asked whether the inclusion of clients/users is a positive thing because it improves social inclusion, on a 1 to 7 Likert scale the average score is 5.2, while the modal (highest frequency) score is 7.3 When asked if the quality of services is one of the important elements in the social mission of the organization, 68.3% answered affirmatively. Especially, when asked if interaction with users/clients is important for the organization in terms of "trust", "quality of relations" and "mutual understanding" on a 1 to 7 Likert scales, scores were, respectively, 6.49, 6.61, and 6.43. In all three cases the modal and median answer is 7.4 On a 1 to 4 Likert scale, the quality of the services provided as element of strength for the cooperative receives a score of 3.65, and both the modal and median outcomes are 4.5 Finally, in terms of outcomes, when asked how directors evaluate the results reached by the cooperative concerning its relations with user/clients, on a 1 to 10 Likert scale the average score is 8.17, while both the modal and median scores are 8.6 In other contributions, users' wellbeing has been shown to be the main determinant of both paid workers' and volunteers' job satisfaction (Michelutti and Schenkel 2009).

As for participation in decision-making, results concerning client/user involvement are not univocal. On a 1 to 3 Likert scale, ICSI data show that clients' involvement in strategic and operational decision-making is limited (average score equal to 1.68), even if not completely absent. However, clients are regularly informed about the decisions taken by the cooperative (score 2.18) and, as a norm, the cooperative creates stable formal or informal channels for clients' involvement (score 1.79). This confirms that clients, as a rule, do not hold decision-making power, even if they can be directly involved in some cases. However, the same results show that clients are accounted among the main stakeholders of the organization.

These pieces of empirical evidence taken together show that, even if social coops are in most cases controlled by their workers, client orientation is recognized as a fundamental dimension of the activity of the organization. Besides pursuing workers' objectives (wages are low, but employment is stable and the organization's procedures are considered very fair by workers), worker control does not appear to be incompatible with other social goals and welfare increasing measures in favor of Unser/clients. Worker-members, who are the strongest patron both in the membership and in the BoD, appear to act as "collective entrepreneurs" (Stewart 1989; Spear 2006; Burress and Cook 2009; Ribeiro-Soriano and Urbano 2010).

4.3 Distributive Function

Once we have ascertained stakeholder participation and propensity to pursue clients' welfare, we now focus on the intensity of their "distributive function", which is defined as the amount of resources, in terms on overtime or volunteer labor, and in terms of services delivered below market value or for free to client/users and/or beneficiaries (Borzaga et al. 2011). These resources can be though to embody client orientation by increasing the benefits received by non-controlling stakeholders without or with partial monetary or in-kind compensation. A further mechanism allowing distribution of resources in favor of clients/users is price discrimination: the non-profit nature of SCs can induce clients to disclose more truthful information concerning their ability and willingness to pay for the service since they do not risk that the organization exploits this information opportunistically to increase its profits. In turn, following a pattern of positive reciprocity, the organization can use this information to set lower prices for individuals or groups characterized by lower ability to pay (cfr. Ben-Ner and van Hoomissen 1991; Grillo 1992).

Qualitative results (self-ratings on Likert scales) from the ICSI survey, show that SCs distribute some extra services free of charge to all their clients in more than 52% of cases, distribute some services free of charge to the poor individuals in society in 40% of cases, and sell their services at less than market price in one third of cases (Borzaga et al. 2011). Furthermore, a non-negligible portion of SCs distributes resources in favor of society in general (35.5%). Finally, a high portion of SCs states that the services supplied are explicitly developed to protect user/clients and satisfy their needs (50% occasionally, and 33% systematically, 83% in total).

When the origin of the additional services delivered free of charge is examined, the most relevant elements appear to be, in decreasing order of importance, resources accumulated to the asset lock (34% of cases), voluntary work (23%), other resources obtained thanks to cost savings (19%), overtime or underpaid work (partial work dona-tions, 12.5%). Finally, cooperatives with a stable and significant distributive function more frequently pursue social benefit aims (83 vs 70%) and are characterized by a democratic managerial style (in 53% vs 27% of cases). Hence, the broader the social missions, and the more democratic the style of management, the stronger the distributive function and the wider the effects on social well-being. The distribution of resources in excess of paid prices or fees emerges as one of the main channels through which SCs increase social welfare without an exchange equivalent. On the bases of the calculation of correlations coefficient, the same data allow to state that the extent of the distribution of resources is positively correlated with: (i) the intensity of actions taken in favor of client/user benefit (5% significance level); (ii) the number of stakeholders partaking the membership of the organization (1% sig. level). This last result can show that the multi-stakeholder nature of SCs favors increased user/client welfare through the distribution of resources on top of market exchanges.

To sum up the whole empirical section: (i) SCs are multi-stakeholder, even if they are not controlled by user/clients; (ii) SCs are worker run, multi-stakeholder organizations; (ii) SCs actively pursue users/clients welfare by: (iiia) factoring in clients' welfare and service quality into their objectives; (iiib) distributing resources (services and other resources) to clients and beneficiaries below market value or for free; (iii) volunteers' well-being and users' wellbeing are positively correlated.

References

Anzola, D., Barbrook-Johnson, P., Cano, J.I.: Self-organization and social science. Comput. Math. Organ. Theory **23**(2), 221–257 (2017). https://doi.org/10.1007/s10588-016-9224-2

Ashby, W.R.: Principles of the self-organizing dynamic system. J. Gen. Psychol. **37**(2), 125–128 (1947)

Ben-Ner, A., Van Hoomissen, T.: Nonprofit organizations in the mixed economy: a demand and supply analysis. Ann. Public Cooperative Econ. **62**(4), 519–550 (1991)

Blandi, V.: Customer uncertainty: A source of organizational inefficiency in the light of the modularity theory of the firm. Ph.D. dissertation. Trento Doctoral School of Social Sciences, Online (2018). http://eprints-phd.biblio.unitn.it/3056/. Accessed 26 Dec 2021

Borzaga, C., Depedri, S., Tortia, E.C.: Testing the distributive effects of social enterprises: the case of Italy. In: Sacconi, L., Degli Antoni, G. (eds.) Social Capital, Corporate Social Responsibility, Economic Behaviour and Performance, pp. 282–303. Palgrave, New York (2011)

Borzaga, C., Tortia, E.C.: Co-operation as coordination cechanism: a new approach to the economics of co-operative enterprises. In: Michie, J., Blassi, J., Borzaga, C. (eds.) The Oxford Handbook of Mutual, Co-operative and Co-owned Business, pp. 55–75. Oxford University Press, Oxford (2017)

Buijs, J.M.: Understanding connective capacity of program management from a self-organization perspective. Emergence Complex. Organ. **12**(1), 29–38 (2010)

Burress, M.J., Cook, M.L.: A primer on collective entrepreneurship: A preliminary taxonomy. University of Missouri. Department of Agricultural Economics, WP 2009-4 (2009). https://ideas.repec.org/p/ags/umcowp/92628.html. Accessed 16 Mar 2022

Coase, R.H.: The nature of the firm. Economica **4**(16), 386–405 (1937)

Commons, J.R.: The Economics of Collective Action. Macmillan, New York (1950)

Cornforth, C.: Nonprofit governance research: limitations of the focus on boards and suggestions for new directions. Nonprofit Volunt. Sect. Q. **41**, 1116–1135 (2012)

Depedri, S.: Le cooperative sociali tra single e multi-stakeholder. Impresa Sociale **3**, 69–82 (2007)

Edelenbos, J., van Meerkerk, I., Schenk, T.: The evolution of community self-organization in Interaction With Government Institutions. Am. Rev. Public Adm. **48**(1), 52–66 (2018)

Euler, J., Heldt, S.: From information to participation and self-organization: visions for European river basin management. Sci. Total Environ. **621**, 905–914 (2018)

Gilbert, N., Anzola, D., Johnson, P., Elsenbroich, C., Balke, T., Dilaver, O.: Self-organizing dynamical systems. In: Wright, J.D. (ed.) International encyclopedia of the social & behavioral sciences, pp. 529–534. Elsevier, London (2015)

Grillo, M.: Cooperative di consumatori e produzione di beni sociali. In: Granaglia, E., Sacconi, L. (eds.) Cooperazione, Benessere e Organizzazione Economica, Milao: Franco Angeli, pp. 95–138 (1992)

Hannachi, M., Fares, M., Coleno, F., Assens, C.: The "new agricultural collectivism." J. Co-operative Organ. Manage. **8**, 100111 (2020)

Hansmann, H.: The Ownership of the Enterprise. Harvard University Press, Harvard (1996)

Hansmann, H.: All firms are co-operative, and so are governments. J. Entrepreneurial Organ. Diversity **2**, 1–10 (2013)

Michelutti, M., Schenkel, M.: Working for nothing and being happy. In: Destefanis, S., Musella, M. (eds.) Paid and Unpaid Labor in the Social Economy, pp. 81–96. Springer, Heidelberg (2009)

Marino, D., Schenkel, M.: Recent Trends in Volunteerism. In: Gil-Lafuente, J., Marino, D., Morabito, F.C. (eds.) Economy, Business and Uncertainty, pp. 58–83. Springer, Heidelberg (2018)

van Meerkerk, I., Boonstra, B., Edelenbos, J.: Self-organization in urban regeneration: a two-case comparative research. Eur. Plan. Stud. **21**(10), 1630–1652 (2013)

Ostrom, E.: Governing the Commons. The Evolution of the Institutions for Collective Action. Cambridge University Press, New York (1990)

Ribeiro-Soriano, D., Urbano, D.: Employee-organization relationship in collective entrepreneurship: an overview. J. Organizational Change Manage. **23**(4), 349–359 (2010)

Rupasingha, A., Wojan, T.R., Freshwater, D.: Self-organization and community-based development initiatives. J. Community Dev. Soc. **30**(1), 66–82 (1999)

Ruzavin, G.: Self-organization and organization of the economy and the search for a new paradigm in economic science. Probl. Econ. Transit. **37**(6), 67–81 (1994)

Sacchetti, S., Borzaga, C.: The foundations of the "public" organization: strategic control and the problem of the costs of exclusion. J. Manage. Gov. **25**, 731–758 (2021)

Sacchetti, S., Tortia, E.: Social responsibility in non-investor-owned organizations. Corporate Gov. Int. J. Bus. Soc. **20**(2), 343–363 (2020)

Spear, R.: Social entrepreneurship: a different model? Int. J. Soc. Econ. **33**, 399–410 (2006)

Stewart, A.: Team Entrepreneurship. Sage, Newbury Park (1989)

Stryjan, Y.: The practice of social entrepreneurship: theory and the Swedish experience. J. Rural Cooperation **34**(2), 195–224 (2006)

Tortia, E.C.: The firm as a common. Non-divided ownership, patrimonial stability and longevity of cooperative enterprises. Sustainability **10**, 1023 (2018)

Young, D.R.: If not for profit, for what? Lexington Books, Lexington (1983)

Topsis Tecniques to Select Green Projects for Cities

Antonio Nesticò[1]([email]) [ID], Piera Somma[1] [ID], Massimiliano Bencardino[1] [ID], and Vincenzo Naddeo[2] [ID]

[1] Department of Civil Engineering, University of Salerno, Fisciano, SA, Italy
{anestico,mbencardino}@unisa.it
[2] Department of Political, Social and Communication Sciences, University of Salerno, Fisciano, SA, Italy
vnaddeo@unisa.it

Abstract. The issue of sustainability is crucial today. Resolving environmental problems also requires radical choices and actions for cities, which need more green areas capable of providing a variety of ecosystem services. But the scarcity of monetary resources does not allow all the projects proposed to be implemented. This raises the question: how can we select new green space initiatives that best meet the objectives of urban sustainability? The research proposes a Technique for Order Preference by Similarity to Ideal Solution (TOPSIS). It is a multi-criteria approach capable of considering the totality of the effects – environmental, social, cultural and financial – that the investment determines, through algorithms that are easy to read and easy to implement. The paper also provides indicators that the evaluator can use for economic analyses. Research perspectives concern the application of the model to real cases, as well as the construction of a wider dataset of evaluation indices, in order to provide a tool that can be widely used in spatial planning processes.

Keywords: Economic evaluation · Multicriteria techniques · TOPSIS · Green Areas

1 Introduction

Sustainable urban development, recognised as a subset of sustainable development, is complex [1]. The pursuit of sustainability cannot address issues from an exclusively technical point of view, but must be aware of the fundamental contribution made by environmental, social, intellectual and cultural aspects concerning the city itself and its inhabitants and users [2–4].

The city can be defined as an 'urban ecosystem', made up of built-up areas and open areas that require, in parallel and in a balanced way, adequate planning for a complete urban regeneration [5–9].

A. Nesticò et al.—Contributed equally to this work.

© The Author(s), under exclusive license to Springer Nature Switzerland AG 2022
F. Calabrò et al. (Eds.): NMP 2022, LNNS 482, pp. 1188–1196, 2022.
https://doi.org/10.1007/978-3-031-06825-6_115

In the perspective of urban sustainability, the theme of planning has acquired new perspectives: in addition to being enriched by aspects increasingly linked to the recovery of already built-up or urbanised areas, it has focused on the valorisation and safeguarding of areas not yet compromised [10].

In this context, public green areas represent one of the main urban planning standards for sustainable planning [11]. In particular, urban greenery is a tool for improving the quality of life and an opportunity to redevelop the natural environment. The transformations carried out over time have had a negative impact on the balance of the ecosystem and have contributed to making large portions of the territory unusable for the community [12, 13].

Ensuring the development of urban areas in these terms implies the implementation of adequate planning strategies [14]. Multi-criteria analyses can guide decision-makers in their choices, in order to solve complex problems for which the data to be analysed are heterogeneous and the criteria to be considered are conflicting [15, 16].

Several multicriteria evaluation techniques are proposed in the literature. For its peculiarities, the TOPSIS method is particularly effective for the optimal selection of "green projects" in cities [17–19]. Indeed, TOPSIS is: rational, i.e. it has a clear logical structure, which makes it easy to check the performance of each alternative; versatile, as it is applicable in the presence of a large number of both alternatives and evaluation criteria, and the number and programming effort of the steps remain the same regardless of the number of criteria and alternatives; simple to handle, in the presence of quantitative rather than qualitative criteria. Furthermore, TOPSIS is able to aggregate monotonically increasing or decreasing numerical information; it does not require independence between evaluation criteria; it needs only little input from the decision-maker and its output is a cardinal ranking of alternatives that is very easy to understand.

The aim of this paper is to propose the TOPSIS approach as a valid multi-criteria analysis tool for the identification of areas to be prioritised for urban greening.

The article is structured as follows. Section 2 outlines the algorithms on which the TOPSIS technique is based. Section 3 proposes a TOPSIS model for the selection of urban green projects. Section 4 sets out the conclusions of the study, highlighting the limitations and potential of the analysis model and outlining research prospects.

2 TOPSIS Method

TOPSIS is a Technique for Order Preference by Similarity to Ideal Solution [20, 21]. Of all the possible solutions, the TOPSIS algorithms select the one with the shortest distance (and therefore the closest) to the positive ideal solution (PIS) and the longest distance to the negative ideal solution (NIS). The ideal solution maximises the benefit criteria and minimises the cost criteria; the anti-ideal solution minimises the benefit criteria and maximises the cost criteria (Fig. 1).

The inputs required by the TOPSIS method are: a decision matrix (Table 1) of order $n \times m$, where one dimension represents the n alternatives, the other the m judgement criteria and the generic element a_{ij} (attribute) expresses the performance of the generic alternative A_i ($i = 1, 2,..., n$) with respect to the generic criterion C_j ($j = 1, 2,..., m$); a vector of relative weights w that provides information on the preferences of the decision makers [22, 23].

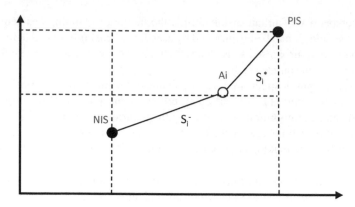

Fig. 1. TOPSIS logic schematization.

Table 1. Decision matrix.

	C_1	C_2	...	C_m
A_1	a_{11}	a_{12}	...	a_{1m}
A_2	a_{21}	a_{22}	...	a_{2m}
...	...	...	...	...
...	...	...	...	...
A_n	a_{n1}	a_{n2}	...	a_{nm}

To determine the weights of the criteria, the following are proposed: subjective approaches, which are based on the preference information given by the decision makers; objective approaches, which refer to data present in the decision matrix and therefore not influenced by the inconsistency of the decision makers' choices.

Among the objective approaches the most widely used in the literature is the entropy method [24]. Entropy is a measure of uncertainty in information and refers to probability theory, according to which a broad distribution expresses more uncertainty than one with a sharp peak. In order to determine the objective weights using the entropy measure, the decision matrix must be normalised:

$$r_{ij} = \frac{a_{ij}}{\sqrt{\sum_{i=1}^{n} a_{ij}^2}} \quad i = 1, 2, \ldots, n \; j = 1, 2, \ldots, m \tag{1}$$

The entropy E_j is given by:

$$E_j = -k \cdot \sum_{i=1}^{n} r_{ij} \cdot \ln(r_{ij}), \text{ con } k = \frac{1}{\ln(n)} \tag{2}$$

k ensures that $0 \le E_j \le 1$.
The weights to use are:

$$w_j = \frac{d_j}{\sum_{i=1}^{n} d_i} \text{con } d_j = 1 - E_j \tag{3}$$

To carry out an analysis according to TOPSIS [25–27], the steps are as follows:

1. Construction of the decision matrix $A = [a_{ij}]$;
2. Determination of the normalised decision matrix $R = [r_{ij}]$, whose elements are defined as in (2);
3. Definition of the weighted normalised decision matrix $V = (v_{ij})$ whose elements are

$$v_{ij} = r_{ij} \cdot w_j \quad i = 1, 2, \ldots, n \quad j = 1, 2, \ldots, m \tag{4}$$

4. Determination of the positive ideal solution A^* and the negative ideal solution A^-

$$A^* = \left\{ (\max_i v_{ij} \text{ con } j \in J_b), (\min_i v_{ij} \text{ con } j \in J_c), \text{ per } i = 1, \ldots, n \right\} = \{v_1^*, v_2^*, \ldots, v_n^*\}. \tag{5}$$

$$A^- = \left\{ (\min_i v_{ij} \text{ con } j \in J_b), (\max_i v_{ij} \text{ con } j \in J_c), \text{ per } i = 1, \ldots, n \right\} = \{v_1^-, v_2^-, \ldots, v_n^-\}. \tag{6}$$

5. Calculation of the Euclidean distance S_i^* and S_i^- of the alternative A_i ($i = 1, 2, \ldots, n$) from the ideal alternative A^* and the negative-ideal alternative A^- respectively:

$$S_i^* = \sqrt{\sum_{j=1}^m (v_{ij} - v_j^*)^2} \quad \text{per } i = 1, \ldots, n \tag{7}$$

$$S_i^- = \sqrt{\sum_{j=1}^m (v_{ij} - v_j^-)^2} \quad \text{per } i = 1, \ldots, n \tag{8}$$

6. Calculation of the relative closeness C_i^* of the alternative from the negative ideal solution by means of the ratio

$$RC_i = \frac{S_i^-}{S_i^- + S_i^*} \tag{9}$$

7. Classification of alternatives according to their proximity to the ideal alternative: the larger the value of RC_i, the better the alternative A_i.

3 A TOPSIS Approach for the Selection of Urban Green Projects

The objective of guaranteeing urban sustainability also includes the implementation of projects aimed at increasing the total area of green areas available to the city [29]. The localisation of green areas makes it possible to increase urban quality and to requalify the so-called 'urban voids', providing the territory with important ecosystem services. Well, how to select among the different projects that can be financed?

TOPSIS algorithms can be used to solve this problem. They make it possible to assess the suitability of an investment project in terms of its financial, environmental, social and cultural effects. According to a multi-criteria scheme, TOPSIS leads to a ranking of project alternatives, from best to worst.

To implement urban sustainability in a broad spatial policy framework, a system of indicators must be defined to assess the actual level of sustainability of a city and to identify policies and actions to improve this level of sustainability [28]. These indicators must be relevant to the objectives set, applicable in different spatial contexts and over time, and geo-referenced in order to observe how results are distributed over the territory [30–32].

Based on the analysis of the reference literature, the indicators in Table 2 apply to the issue under consideration and in relation to the chosen methodological approach.

Table 2. Evaluation criteria and indicators.

Criterion	Description	Indicator
Extencion	Ratio between intervention area and city area	%
Accessibility	Ratio between population living within 300 m of the intervention site and the population of the city	%
Usability	Ratio between surface area dedicated to services on the intervention site and its total surface area	%
New employees	Number of jobs generated as a result of new activities established on the intervention site or as a result of increased employment of activities in the immediate vicinity	Number
Quality of life	Assessment of the rehabilitation of degraded areas, sites of petty crime and health and hygiene risks	Values 1 to 7
Investmentcost	Total investment spending	€
Financial flows	Growth of economic activity as a result of increased number of users on the site	€
Increase in real estate value	Revaluation of properties in the vicinity (200 m) indicating the perception of living in a healthy, quality environment	m^3

The TOPSIS algorithm requires that the evaluation criteria are all monotonically increasing, i.e. advantageous. For this reason, the investment cost criterion, being the only inversely proportional one, should be considered according to the reciprocal ratio.

As an example, if there are four possible alternatives, according to the criteria mentioned, the decision matrix is of the type shown in Table 3.

Applying (2) we obtain the normalised decision matrix in Table 4.

We now determine the weights w_j from (3) in order to construct the weighted normalized decision matrix in Table 5.

Table 3. Decision matrix.

	C_1	C_2	C_3	C_4	C_5	C_6	C_7	C_8
A_1	a_1	b_1	c_1	d_1	e_1	f_1		g_1	h_1
A_2	a_2	b_2	c_2	d_2	e_2	f_2	g_2		h_2
A_3	a_3	b_3	c_3	d_3	e_3	f_3	g_3		h_3
A_4	a_4	b_4	c_4	d_4	e_4	f_4	g_4		h_4
	$\sqrt{(\sum a_i{}^2)}$	$\sqrt{(\sum b_i{}^2)}$	$\sqrt{(\sum c_i{}^2)}$	$\sqrt{(\sum d_i{}^2)}$	$\sqrt{(\sum e_i{}^2)}$	$\sqrt{(\sum f_i{}^2)}$	$\sqrt{(\sum g_i{}^2)}$	$\sqrt{(\sum h_i{}^2)}$

Table 4. Normalized decision matrix.

	C_1	C_2	C_3	C_4	C_5	C_6	C_7	C_8
A_1	a'_1	b'_1	c'_1	d'_1	e'_1	f'_1	g'_1	h'_1
A_2	a'_2	b'_2	c'_2	d'_2	e'_2	f'_2	g'_2	h'_2
A_3	a'_3	b'_3	c'_3	d'_3	e'_3	f'_3	g'_3	h'_3
A_4	a'_4	b'_4	c'_4	d'_4	e'_4	f'_4	g'_4	h'_4

Table 5. Weighted normalized decision matrix.

	C_1	C_2	C_3	C_4	C_5	C_6	C_7	C_8
A_1	$w_1 \cdot a'_1$	$w_2 \cdot b'_1$	$w_3 \cdot c'_1$	$w_4 \cdot d'_1$	$w_5 \cdot e'_1$	$w_6 \cdot f'_1$	$w_7 \cdot g'_1$	$w_8 \cdot h'_1$
A_2	$w_1 \cdot a_2$	$w_2 \cdot b'_2$	$w_3 \cdot c'_2$	$w_4 \cdot d'_2$	$w_5 \cdot e'_2$	$w_6 \cdot f'_2$	$w_7 \cdot g'_2$	$w_8 \cdot h'_2$
A_3	$w_1 \cdot a_3$	$w_2 \cdot b'_3$	$w_3 \cdot c'_3$	$w_4 \cdot d'_3$	$w_5 \cdot e'_3$	$w_6 \cdot f'_3$	$w_7 \cdot g'_3$	$w_8 \cdot h'_3$
A_4	$w_1 \cdot a_4$	$w_2 \cdot b'_4$	$w_3 \cdot c'_4$	$w_4 \cdot d'_4$	$w_5 \cdot e'_4$	$w_6 \cdot f'_4$	$w_7 \cdot g'_4$	$w_8 \cdot h'_4$

For each criterion, the ideal solution A^* and the ideal negative solution A^- are found using formulae (5) and (6). Thus, for criterion C_1 we have:

$$A_1^* = \max (w_1 \cdot a_1; \, w_1 \cdot a_2; w_1 \cdot a_3; \, w_1 \cdot a_4)$$

$$A_1^- = \min (w_1 \cdot a_1; \, w_1 \cdot a_2; w_1 \cdot a_3; \, w_1 \cdot a_4)$$

For each alternative under consideration, the Euclidean distances S_i^* and S_i^- are calculated as per (7) and (8). For alternative A_1:

$$S_1^* = \sqrt{\begin{array}{l} (w_1 \cdot a_1 - A_1^*)^2 + (w_2 \cdot b_1 - A_1^*)^2 + (w_3 \cdot c_1 - A_1^*)^2 + (w_4 \cdot d_1 - A_1^*)^2 \\ + (w_5 \cdot e_1 - A_1^*)^2 + (w_6 \cdot f_1 - A_1^*)^2 + (w_7 \cdot g_1 - A_1^*)^2 + (w_8 \cdot h_1 - A_1^*)^2 \end{array}}$$

$$S_1^- = \sqrt{\begin{aligned}&\left(w_1 \cdot a_1 - A_1^-\right)^2 + \left(w_2 \cdot b_1 - A_1^-\right)^2 + \left(w_3 \cdot c_1 - A_1^-\right)^2 + \left(w_4 \cdot d_1 - A_1^-\right)^2 \\ &+ (w_5 \cdot e_1 - A_1^-)^2 + (w_6 \cdot f_1 - A_1^-)^2 + (w_7 \cdot g_1 - A_1^-)^2 + (w_8 \cdot h_1 - A_1^-)^2\end{aligned}}.$$

From (9) the RC_i ratio is calculated for each alternative. The alternative with the highest value of this ratio is the optimal alternative.

4   Results and Conclusions

In the light of the structure of the analysis algorithms, the TOPSIS model proposed for the selection of green projects is: rational, i.e. clearly able to evaluate the performance of each project alternative; versatile, as it can be easily applied even in the presence of numerous alternatives and multiple evaluation criteria; simple to use, as it requires little input data and provides an immediately comprehensible output, i.e. a cardinal ranking of the alternatives.

The model is a useful tool in urban planning processes, as it guarantees the optimal allocation of the financial resources available to the public operator. In fact, when it is necessary to establish which areas are to be dedicated to public green spaces according to pre-established projects, the TOPSIS approach makes it possible to establish which investments are to be financed, according to the capacity of each initiative to determine higher ecosystem services. According to the evaluation scheme, these ecosystem services are estimated through appropriate indicators, capable of capturing the financial, but also the social, cultural and environmental effects generated on the territory.

The construction of a broader dataset of valuation indicators and the application of the model to different real cases determine interesting research prospects.

References

1. Dobrovolskiene, N., Pozniak, A., Tvaronaviciene, M.: Assessment of the sustainability of a real estate project using multi-criteria decision making. Sustainability **13**, 4352 (2021). https://doi.org/10.3390/su13084352
2. Morri, E., Santolini, R.: Valutare I servizi ecosistemici: un'opportunità per limitare i disturbi al paesaggio (2013)
3. Losasso, M., Lucarelli, M.R., Rigillo M., Valente R.: Adattarsi al clima che cambia. Innovare la conoscenza per il progetto ambientale. MAGGIOLI EDITORE, Santarcangelo di Romagna (2020)
4. Troisi, R., Castaldo, P: Technical and organizational challenges in the risk management of road infrastructures. J. Risk Res. 1-6 (2022). https://doi.org/10.1080/13669877.2022.2028884
5. Lak, A., Sharifi, A., Khazaei, M., Aghamolaei, R.: Towards a framework for driving sustainable urban regeneration with ecosystem services. Land Use Policy 105736 (2021). https://doi.org/10.1016/j.landusepol.2021.105736
6. Li, F., Guo, S., Li, D., Li, X., Li, J., Xie, S.: A multi-criteria spatial approach for mapping urban ecosystem services demand. Ecol. Ind. **112**, 106119 (2020). https://doi.org/10.1016/j.ecolind.2020.106119
7. Andreucci, M. B.: Progettare Green Infrastructure. Tecnologie, valori e strumenti per la resilienza urbana. Wolters Kluwer, Milano (2017)

8. Calabrò, F., Cassalia, G., Lorè, I.: The economic feasibility for valorization of cultural heritage. the restoration project of the reformed fathers' convent in Francavilla Angitola: the Zibìb territorial wine cellar. In: Bevilacqua, C., Calabrò, F., Della Spina, L. (eds.) NMP 2020. SIST, vol. 178, pp. 1105–1115. Springer, Cham (2021). https://doi.org/10.1007/978-3-030-48279-4_103

9. Della Spina, L., Giorno, C., Galati Casmiro, R.: An integrated decision support system to define the best scenario for the adaptive sustainable re-use of cultural heritage in Southern Italy. In: Bevilacqua, C., Calabrò, F., Della Spina, L. (eds.) NMP 2020. SIST, vol. 177, pp. 251–267. Springer, Cham (2020). https://doi.org/10.1007/978-3-030-52869-0_22

10. Martinelli, L., Battisti, A., Matzarakis, A.: Multicriteria analysis model for urban open space renovation: an application for Rome. Sustain. Cities Soc. 14, 10–20 (2015). https://doi.org/10.1016/j.scs.2014.07.002

11. Malscevichi, S., Bisogni, G. L.: Green Infrastructures and ecological reconstruction in urban and peri-urban areas. J. Tech. Arch. Environ. 11, 33–39 (2016). https://doi.org/10.13128/Techne-18398

12. Liu, H., Hu, Y., Li, F., Yuan, L.: Associations of multiple ecosystem services and disservices of urban park ecological infrastructure and the linkages with socioeconomic factors. J. Clean. Prod. 174, 868–879 (2018). https://doi.org/10.1016/j.jclepro.2017.10.139

13. Santolini, R.: Servizi ecosistemici e sostenibilità. Ecoscienza 3, 20–23 (2011)

14. Bencardino, M., Nesticò, A.: Demographic changes and real estate values. a quantitative model for analyzing the urban-rural linkages. Sustainability 9, 536 (2017). https://doi.org/10.3390/su9040536

15. Maselli, G., Nesticò, A.: L'Analisi Costi-Benefici per progetti in campo ambientale. La scelta del Saggio Sociale di Sconto. LaborEst 20/2020 (2020). https://doi.org/10.19254/LaborEst.20

16. Nesticò, A., Maselli, G.: Declining discount rate estimate in the long-term economic evaluation of environmental projects. J. Environ. Accout. Manag. 8, 93–110 (2020). https://doi.org/10.5890/JEAM.2020.03.007

17. Zapolskyte, S., Vabuolyte, V., Burinskiene, M., Antucheviciene, J.: Assessment of sustainable mobility by MCDM methods in the science and technology parks of vilnius. Lithuania Sustain. 12, 9947 (2020). https://doi.org/10.3390/su12239947

18. Vakilipour, S., Sadeghi-Niaraki, A., Ghodousi, M., Choi, S.: Comparison between multi-criteria decision-making methods and evaluation the quality of life at different spatial levels. Sustainability 13, 4067 (2021). https://doi.org/10.3390/su13074067

19. Zhenga, W., Shenb, G.Q., Wangc, H., Hongd, J., Li, Z.: Decision support for susyeinable urban renewal: a multi-scale model. Land Use Policy 69, 361–371 (2017). https://doi.org/10.1016/j.landusepol.2017.09.019

20. Hwang, C.L., Yoon, K.: Multiple Attributes Decision Making Methods and Applications. Springer, Berlin (1981)

21. Tzeng, G.H., Hwang, J.J.: Multiple Attributes Decision Making Methods and Applications. CRC Press Taylor & Francis Group, USA (2011)

22. Triantaphyllou, E., Mann, S.H.: An examination of the effectiveness of multi-dimensional decision-making methods: a decision making paradox. Int. J. Decis. Supp. Syst. 5, 303–312 (1989). https://doi.org/10.1016/0167-9236(89)90037-7

23. Triantaphyllou, E.: Multi-Criteria Decision Making Methods: A Comparative Study. Kluwer Academic Publisher, Holland (2002)

24. Shanian, A., Savadogo, O.: A methodological concept for material selection of highly sensitive components based on multiple criteria decision analysis. Expert Syst. Appl. 36, 1362–1370 (2009). https://doi.org/10.1016/j.eswa.2007.11.052

25. Dammak, F., Baccour, L., Alimi, A.M.: Crisp multi-criteria decision making methods: state of the art. Int. J. Comput. Sci. Inf. Secur. 14(8), 252 (2016)

26. Deng, H., Yeh, C., Willis, R.J.: Inter-company comparison using modified topsis with objective weights. Comput. Oper. Res. **27**, 963–973 (2000). https://doi.org/10.1016/S0305-054 8(99)00069-6

27. Ren, L., Zhang, Y., Wang, Y., Sun, Z.: Comparative Analysis of a Novel MTOPSIS Method and TOPSIS. Oxford University Press, Oxford (2007)

28. Dolores, L., Macchiaroli, M., De Mare, G.: A dynamic model for the financial sustainability of the restoration sponsorship. Sustainability **12**, 1694 (2020). https://doi.org/10.3390/su1204 1694

29. Nesticò, A., Elia, C., Naddeo, V.: Sustainability of urban regeneration projects: novel selection model based on analytic network process and zero-one goal programming. Land Use Policy **99**, 104831 (2020). https://doi.org/10.1016/j.landusepol.2020.104831

30. Meadows, D.H.: Indicators and Information System for Sustainable Development. The Sustainability Institute, Lynedoch (1998)

31. Toli, A.M., Murtagh, N.: Environmental sustainability indicators in decision-making analysis on urban regeneration projects: the use of sustainability assessment tools. ARCOM CONFERENCE (2017)

32. Pileri, P.: Interpretare l'ambiente – Gli indicatori di sostenibilità. Alinea Editrice, Firenze (2002)

Human Smart Landscape: An Integrated Multi-phase Evaluation Framework to Assess the Values of a Resilient Landscape

Lucia Della Spina$^{(\boxtimes)}$ (iD) and Claudia Giorno

Mediterranea University of Reggio Calabria, 89125 Reggio Calabria, RC, Italy
lucia.dellaspina@unirc.it

Abstract. Starting from the most recent international debate on the theme of the Human Smart Landscape, the study explores the theme of the landscape as a "common good", as a field of investigation and experimentation of an innovative long-term sustainable cultural-tourism development model. In this context, the research aims to define a multi-phase integrated assessment framework aimed at helping local actors and users to understand the multifunctional resources and values present in the study landscape and neighboring territories and to stimulate their collaboration in management and design of new tourist enhancement strategies. The research activities are still in the initial stages. The research illustrates the first results relating to the first cognitive phase relating to the case study conducted in the Sila National Park (Southern Italy), a UNESCO site of excellence.

Keywords: Human Smart Landscape · Resilient landscapes · Space Decision Support System · Tourist-cultural routes

1 Introduction

In over time, the concept of landscape has explored the man-nature, built-environment bond as both dimensions necessary for the protection and life of a portion of the territory. This perspective has led to the overcoming of an aesthetic, photographic vision of the landscape towards a dynamic and multilevel vision [1], in which the multidisciplinary analyzes that investigate the interaction between environmental and cultural aspects take on great interest, economic, social, for the evolution and conservation of a vision of the territorial context [2].

In recent years, landscape culture has grown, and various disciplines have made significant contributions, especially separately [3]. However, the Historical Urban Landscape (HUL) approach [4], aspires to a more general and operational concept, which aims to preserve the qualities of the landscape, improving the productive and sustainable use of resources thanks to a territory and the development of tools capable of improving the active role of a community in the management and planning of landscape resources. From this point of view, landscape and social innovation are an inseparable pair. The most recent orientations look on the one hand to local development as a key theme of innovation [5] on the other to social innovation as "revolutionary innovation" able to start a new cycle for the management of common goods, in which for-profit and non-profit entities take on a leading role in the management of resources with respect to the public administration [6].

Therefore, the HUL approach aspires to provide dynamic tools, able to adapt to different contexts, to define actions based on shared value judgments that are able to initiate processes to enhance the identity values of the material and immaterial culture of a place [7]. The research therefore analyzes and experiments an approach that allows users/inhabitants of the landscape to participate in the development and testing of innovative solutions for a specific territory.

Research in this field underlines the need for evaluative approaches based on complex adaptive systems, that is, focused on the response (adaptation) to different types of feedback. Exploring a broad decision-making context, it can be seen that the values of the resources and those of the subjects involved (direct, indirect, potential, future generations) are interdependent and interconnected, in the perspective of Complex Social Value (CSV) [2, 8].

Starting from the most recent international debate on the theme of cultural landscape, the study explores the theme of landscape as a "common good", as a field of investigation and experimentation of an innovative model of long-term sustainable tourism development. In this context, the document illustrates a multi-stakeholder territorial decision-making process, based on an evaluation approach useful to support decision-makers in defining improvement strategies for resilient landscapes. The research activities are in the initial stages, this contribution illustrates the first results of the cognitive phase relating to the case study conducted in the Sila National Park (Southern Italy), a UNESCO site of excellence.

The aim of the experimentation is the identification of new paths for the touristic-cultural use of the park, capable of restoring a 'new affective geography', in which the Park landscape is reconstructed starting from the experiences and users.

2 Case Study

The structuring of an interpretative and evaluative model has been implemented in the cultural landscape of the Sila National Park (SNP), in Southern Italy.

The SNP is a site included in the World Network of UNESCO sites of excellence which houses one of the richest systems of biodiversity, with a rich heritage of unique geological qualities and varieties. The park extends over a large and heterogeneous territory, consisting of mainly mountainous areas that attract many tourists especially in the summer; it also includes inner areas that are sparsely populated, and which seek to leverage their cultural and environmental heritage to promote their land and thus face the phenomenon of de-population.

The International Coordination Council of the MAB Program (Man and the Biosphere Program), during its 26th session in Jönköping in Sweden, approved the registration of Sila as the 10th Italian Biosphere Reserve in the UNESCO World Network of Sites of Excellence.

Currently, the protected area of the park is a candidate to be included in the world list of UNESCO Global Parks.

It must be said that the Sila National Park not only includes a vast territory rich in naturalistic and landscape beauties, but also contains within it a concentrate of typical flavors, aromas and high-quality knowledge that absolutely deserve to be experienced and enhanced through experiential tourism routes.

3   Materials and Methods

The application of integrated methodologies to the case study is based on a Space Decision Support System (SDSS) for a multifunctional landscape aimed at helping local actors and users to understand the multifunctional resources and values present in the park and in the neighboring territories and to stimulate their collaboration in the management and design of new tourism-cultural enhancement strategies.

The methodological approach is structured in four phases: (i) Intelligence, (ii) Design, (iii) Choice and (iv) Outcome [9], in order to take into account the needs of the community and implement widely shared enhancement strategies. For the elaboration of each phase, will be chosen some evaluation methods suitable for representing and managing the complexity of the interests and objectives at stake, stimulating territorial co-creativity towards regenerative models [10].

In the first phase (Intelligence phase), through laboratories and surveys with the park community, the activities in the field are aimed at identifying the most significant places in the park and understanding the characteristics that determine an 'affective' meaning for the inhabitants and for users who frequent the park landscape, even temporarily.

Through the use of geo-statistics techniques and with the aid of GIS instruments, the 'places of value' indicated by the Park community will be traced and will be defined new tourist routes (Design phase).

In the Choice phase, the alternative paths identified will be evaluated according to a multi-criteria approach, in order to define shared strategies for the enhancement of the park.

Finally, in the Outcome phase the composite maps relating to the multifunctionality of the landscape will be represented. The information collected, processed and geo-referenced in the GIS environment will return summary maps capable of providing a synthetic cognitive framework useful for classifying the various Municipalities of the Park according to the different level of density of the services offered.

Research activities are still in progress. Therefore this contribution illustrates the first results of the Intelligence phase.

4   Intelligence Phase

In accordance with the Cultural Values Model (CVM) [11], the Intelligence phase was marked by focus groups and in-depth interviews, in order to identify a sample of individuals interested in working together to achieve shared objectives.

According to the CVM, the cultural values of the landscape are the foundation of the identity of a local community, as the result of its relationship with the environment they have inhabited and changed over time. To determine which values are shared by the users of the landscape and to decipher their meanings in relation to the places perceived as most significant, were conducted studies and surveys aimed at the perception of the landscape by specific groups of people and their expectations.

This allows us to recognize the general predisposition of users towards the choice of a certain type of landscape, the reasons for the choice, the importance of the experience, the main positive or negative factors related to the personal perception of the landscape [12–14].

In particular, the studies on the perception of the landscape of the SNP examined the personal experience of some groups of people and their expectations on the landscape of the park, recognizing the general predisposition of users towards the study landscape: why it was important to live it, what led them to do it, the key factors and the positive or negative characteristics related to personal perception [12–14] of the park landscape.

To define these meanings, the CVM [11, 13] chooses an open and in-depth survey, which is aimed at a sample of dominant figures recognized [15] as the community of the analyzed landscape and to people from different places different.

The main result of the survey is aimed at identifying the "places of value" of the park and at understanding the characteristics that determine their values for the inhabitants and for those who frequent the park's landscape, even temporarily.

Therefore, starting from the paths suggested by people, it is possible to analyze human behavior in exploring territory and attribute values to a stratified reality such as the landscape.

3.1.1 Investigation Tools. Interviews and Questionnaires

In order to identify the places of value of the Park, in the research is experimented an approach to landscape according to the indications of the European Convention, where people have been stimulated to express the vision of the landscape in which they live.

The identification of values involves the exploration and research of ways to detect subjective fields, the value of a place may not necessarily be attributed for objective requirements (site, heritage, humanity) but can depend on subjective parameters (ability to tribute meaning, to the need for community).

Following the approach of the CVM [11], there are two main categories of actors linked to the landscape: the insiders who live, work, or have their roots in a given context and therefore express an internal point of view of collective knowledge; outsiders who are the external visitors, usually tourists or experts.

The sample of actors who participated in the survey are the users of the landscape (insiders and outsiders) potentially interested in the alternative promotion activities of the park landscape under study.

The sample was identified in the field, during the activities and events organized on the territory of the Park. In particular, for the insiders, the following were selected: i) the inhabitants present in the places where the activities took place, who do not play an active role; ii) the inhabitants recognized by the community as experts. For outsiders, the following were selected: i) the experts and scholars who have participated in scientific promotion events in the area or who have a consolidated knowledge of years of research on fieldwork; ii) the regular tourists of the area, interested in expanding their knowledge of the park.

The specific objective of the survey is to improve the involvement of insiders in the activities promoted by local groups, experimenting with different methods of collecting information, functional to the interaction with different users and the return of data that can be analyzed. Inhabitants, tourists, experts and park scholars were then interviewed in three distinct ways: semi-structured interview, direct questionnaire and online questionnaire, from which were obtained different results. The choice between the semi-structured interview or the face-to-face questionnaire was made based on the interest of the interviewee and his or her knowledge of the area. While the online questionnaire was disseminated through the Facebook social channels.

The semi-structured interviews and questionnaires, with simple and general open questions, aim to identify the places in the park and to understand the characteristics that determine their values for the inhabitants and for those who frequent the territory even temporarily. An approach to the landscape is therefore experimented close to the indications of the European Convention, stimulating people to express the vision of the environment in which they live and structuring ways and tools to detect this type of information. Indeed, reasoning by values implies the exploration and research for ways to detect subjective fields. The value of a place is not necessarily attributed for objective requirements (site, heritage, humanity) but it can depend on subjective parameters (ability to contribute to the meaning, to the need for community), it can reside in the cognitive, confidential relationship between place and person, demonstrated, for example by the punctuality and topographical precision with which the places are marked: a house, a school, a church, a tree, a precise path, a clearly recognizable open space, a natural setting. The relationship, and therefore the value, is more often constructed by points, by microcosms [16].

The administration of the semi-structured interviews is aimed at returning "thematic paths of value of the park" starting from the paths indicated by the interviewees and the detailed and precise subjective explanation they give of the material and intangible resources they visited, and from which it is possible to derive useful information for the promotion and enhancement of the park's landscape. In fact, the interview identifies the reason for the lived experience, the potentially interested public, the preferable period for fruition, and the means of transport with which to travel the route. Finally, the interviewee is also asked to indicate a slogan that re-assumes a lifestyle for the promotion of the Park.

For the survey, two versions were developed: one direct frontal and one online. The direct frontal survey was used to indicate "places of value" [16]. The goal is to collect reports also from interlocutors who do not have in-depth/itinerant knowledge of the area, writed down the places that are significant for their life or which they would recommend to a friend for a visit to the park. Each place is associated with: reason, potential interested audience, preferable season, means of transport and slogan.

The online survey was developed starting from the indications obtained from the first results of the frontal one, which facilitated the selection and organization of multiple and single choice questions and provided useful information for the alternatives of closed answers. The online survey system used is LimeSurvay integrated with the Sila Labscape platform. The system features have allowed the use of maps and interactive systems to indicate the order of preference for multiple responses.

Both the semi-structured interview and the frontal questionnaire proved capable of establishing conversational and simple contact with the interlocutors. Different categories of users have in fact answered the questions with ease and availability, interacting with the research and acting as an intermediary with other interlocutors whose point of view and knowledge of the territory were considered relevant. The simplicity of interacting starting from places testifies that talking about places is a need that must only be stimulated and "cultivated" [16]. It should be noted that in some cases a place has been marked for the possibility of knowing a person or the story of an experimental project, helping to build the network of contacts and information.

3.1.2 Intelligence Phase: Intermediate Results

The processing of the soft data, obtained from the survey tools, made it possible to extract relevant indicators on the subjective perception of users regarding the "places of value". This has allowed the identification of tourist clusters/domains of shared landscape, linked to experiential tourism, representative of collective identity values, recognized by insiders and outsiders.

Therefore, starting from the paths and 'places of value' indicated by people, it was possible to analyze human behavior in exploring territory and attribute values to a stratified reality such as the landscape under study [17]. In fact, when human beings move in space, they keep the memory of their perceptions. The marked path, therefore, represent the connection of people to places of value and allow to read the dynamics of the different ways of experiencing the landscape.

The semantic processing of the information extracted from the interviews allows to highlight some recurring semantic fields and to geographically and spatially locate the places that insiders and outsiders have indicated in their paths. Each interview was then returned through a specific route (a succession of points in the geographical space), to which has been attributed a theme, according to a Semantic Analysis [18], in relation to some fields of the interview relating to: trip' reason, slogans and potentially interested audiences. A first semantic processing of the interview data has highlighted some recurring semantic fields. Furthermore, through subsequent frequency analysis, were then selected some relevant keywords associated with each main semantic field. The comparison between all the keywords that emerged in all the interviews made it possible to identify some subjective semantic fields, common to the two groups of users (insiders and outsiders) relating to: Identity, Community, Discovery, Well-being, Quality of Life. The subjective fields identified represent the identity values of the landscape shared by the interviewees.

Quite frequently emerges that the motivation of the route (the theme) it is very often the combination of several fields.

The survey through questionnaires, still ongoing online, returns, albeit partially, the postcards of 'places of value' for insiders and outsiders. What is required is to try to tell the landscape through the images, words, and advice by who have lived a travel experience, aiming to collect suggestions, perceptions, comments, and preferences.

To the online survey has been added a field "when" for a possible indication regarding the order of preferability of the seasons. This response method was also chosen to identify the categories of users potentially interested to visit the indicated place (lover of nature; lover of culture; lover of fun; inhabitant; inhabitant of the city; expert; sportsman).

Through appropriate corrections, combining open and closed, single and multiple responses (in order of preference), the survey was modified for the offline (front) and online channels, adding in the latter the possibility of attaching photos and comments. It is significant to highlight how the usefulness of the "photo name/comment" field has made it possible to trace a specific place, often not expressly indicated in the "place of value" field and in some cases providing information on the more emotional and perceptive aspects.

From the elaborations, therefore emerges an order of preference of the places of value reported by insiders and outsiders. The preference is the result of the frequency and order in which the places were named by users. Each interviewee was asked to indicate a maximum of three places, indicating additional information for each (trip' reason, type transportation, period, trip's recipients, slogans, photos, comments). On the multiple-choice fields, with alternatives to be sorted, the preference (frequency and sorting) has been elaborated; on the single answer fields, in which to select only one of the alternatives listed, the frequency has been processed. The semantic analysis was carried out on the open-ended fields, tracing the keywords.

Ultimately, through the examination of the values perceived by the interviewees, [19, 20] it was possible to recognize, through frequency analysis, some recurrent keywords, those more frequent, which express the emotionally subjective component of the experience travel and which belong to the same area of meaning/semantic field [18, 21]. This activity therefore allowed the selection of tourist clusters/domains of shared landscape [22–26], recognized by insiders and outsiders. It turns out that there are two main tourist clusters, identified with related associated keywords, which represent a wide range of aspects of the perceived landscape, appreciated and shared by the interviewees:

- Cultural Tourism (A1) which aims to promote and enhance cultural tourism and local identity through the enhancement of places of cultural interest, the tasting of typical products, as well as the use of naturalistic places in the Park. The set of keywords that represent the Cultural Tourism cluster are: i) Identity: Community relations, Landscape, Hospitality, Food; ii) Discovery: Known, Hidden, In extinction, Not well-known, Submerged, Heritage, Culture, Historic town, Towns in state of abandon, Water, Local food, Local crafts and knowledge Encounter, Landscape, Traditional holiday.
- Natural Tourism (A2) which aims to mainly promote naturalistic tourism, personal well-being, implementing the quality of life through slow mobility and the improvement of services, as well as promoting experiences related to "food routes". The set of keywords that represent the Natural Tourism cluster are: Well-being: Quality of life, Tranquility, Healthy life, Peaceful life, Well-being Involvement, Find oneself, Recover the meaning of life, Recover your dreams, Peace within, Freedom, Traditions, Gastronomy Relaxation, Sea Hospitality, Landscape, Culture, Uniqueness, Natural oasis.

5   Conclusions

In the context of the debate on landscape as a "common good", the research experiments the construction of a process of analysis of multi-stakeholder spatial decisions.

The methodology is still in a phase of development and experimentation within the 'Sila Labscape' project conducted in the Sila National Park, a UNESCO site of excellence.

Through the integration of different methodologies interpretative and evaluative tools [10, 22, 23, 26–30], the research aimed to develop a process for the enhancement of the park landscape and its internal resources, useful for reconstructing an 'affective geography' of the territory, from different points of view (insider and outsider).

In the subsequent phases of the research, the emotional ties with the landscape will be further explored through the spatial rendering of soft data and with the support of the analysis of the proximity relationships between specific geographical points.

The identification of perceived values, linked to specific places and the personal relationships of users with the territory, will allow the definition of new tourist-cultural routes for the use of the park: 'places of value' linked to the quality of life of individual and that reflect the concept of well-being perceived by a community [6]. This will allow, in the further development of research, the generation of new landscape maps capable of describing the emotional paths that represent the places where 'values and personal relationships' are concentrated, thus giving back the possibility of living a unique experience in the park [22, 24–28].

The methodological framework, aimed at searching for new ways of enhancing the Park and its resources, constitutes a research field that is still open, from which some relevant issues emerged. The most relevant results of the study concern the evident connection between the assigned meanings and the perceived values of the landscape, according to the components of the Cultural Value Model [11], the recognition of the subjectivity of the objective environment, its representation in space and further interpretations.

The process of structuring and subsequent evaluation of the perceived places of value, will make it possible to return maps of new tourist-cultural routes, the expression of shared intersubjective perception, useful for supporting the development of collaborative local strategies for the enhancement of local resources, in particular in fragile contexts such as the territory in which the Park is located.

The landscape values traced by users will provide to Park promoters useful information that can be used to develop innovative and sustainable strategies and develop a deeper knowledge and awareness of local qualities and potential.

The landscape becomes a field of research/action for bottom-up innovative processes, in which to experiment new ways of managing landscape resources. Sustainable processes in which the environmental, economic and social dimensions are an integral part of a path that recognizes values as an expression of cultural and identity ties between communities and places, which pays close attention to social cohesion, creativity and quality of life, and is able to create synergies between nature, rural and urban areas.

The vision of the landscape, in which the management of resources takes place thanks to collaboration with citizens, is close to the logic of the Human Smart City [31, 32], which aims to build a new sense of belonging and identity starting from improving and well-being of its inhabitants. People are thus at the center of a process of social innovation that aims to stimulate local development: new services arise from the real needs of people and are co-designed with the citizens themselves through an interactive, dialogic and collaborative process [31–37].

The Human Smart Landscape [7, 10] is a research field that is still open in which new approaches and tools for the representation, evaluation, monitoring, and management of the landscape and tools. It is therefore a complex space attentive to social cohesion, creativity and quality of life, in which citizens play a central role.

There is a need for actions capable of integrating the design of spaces and infrastructures, reducing consumption and waste of resources, to recover the cultural and environmental heritage in sustainable terms [10, 33–40], to create synergies with natural, rural and urban areas, to make a qualified use of ecological networks [28, 41, 42], to enhance in a conscious way the different forms of culture, the attractiveness of the landscape [22, 24, 25], in order to improve the quality of spaces, uses and relationships [43–47].

Author Contributions. The paper is the result of the joint work of the two authors. Scientific responsibility is attributable equally to Lucia Della Spina and Claudia Giorno.

References

1. Convenzione europea sul paesaggio (2000). http://www.convenzioneeuropeapaesaggio.ben iculturali.it. Accessed 13 Dec 2021
2. Fusco Girard, L.: Risorse architettoniche e ambientali. Valutazioni e strategie di conservazione. Angeli, Milano (1987)
3. Voghera, A.: Dopo la Convenzione Europea del Paesaggio. Politiche, Piani e Valutazione, Alinea, Firenze (2011)
4. UNESCO: Recommendation on the Historic Urban Landscape. UNESCO World Heritage Centre, Resolution 36C/23, Annex, Paris, France (2011)
5. Seravalli, G.: Innovazione e sviluppo locale. Concetti, esperienze, politiche. WP 7/2007 Serie: Economia e Politica Economica, Settembre 2007 (2007)
6. Zamagni, S.: Dal welfare della delega al welfare della partecipazione. Il Distretto di Cittadinanza come esempio evoluto di sussidiarietà circolare (2011)
7. Fusco Girard, L.: Toward a smart sustainable development of port cities/areas: the role of the historic urban landscape approach. Sustainability 5(10), 4329–4348 (2013)
8. Fusco Girard, L., Nijkamp, P.: Le valutazioni per lo sviluppo sostenibile della città e del territorio. Angeli, Milano (1997)
9. Simon, H.A.: The New Science of Management Decision. Prentice Hall PTR, Upper Saddle River (1977)
10. Della Spina, L.: Strategic planning and decision making: a case study for the integrated management of cultural heritage assets in southern Italy. In: Bevilacqua, C., Calabrò, F., Della Spina, L. (eds.) NMP 2020. SIST, vol. 178, pp. 1116–1130. Springer, Cham (2021). https://doi.org/10.1007/978-3-030-48279-4_104
11. Stephenson, J.: The Cultural Values Model: An integrated approach to values in landscapes. Landsc. Urban Plan. 84, 127–139 (2008). http://www.sciencedirect.com/science/article/pii/ S0169204607001661. Accessed 30 Jan 2022
12. Ståhlbröst, A.; Holst, M.: The Living Lab Methodology Handbook. Social Informatics at Luleå University of Technology and CDT, Centre for Distance-spanning Technology: Luleå, Sweden (2012)
13. MacFarlane, R., Haggett, C., Fuller, D., Dunsford, H., Carlisle, B.: Tranquillity Mapping: Developing A Robust Methodology for Planning Support. Report to the Campaign to Protect Rural England, Countryside Agency, North East Assembly, Northumberland Strategic

Partnership, Northumberland National Park Authority and Durham County Council, Centre for Environmental & Spatial Analysis; Northumbria University: Newcastle Upon Tyne, UK (2004)

14. Djamel, F.: Valuation Methods of Landscape. Int. J. Res. Methodol. Soc. Sci. **2**, 36–44 (2016)

15. De Marchi, B., Funtowicz, S.O., Lo Cascio, S., Munda, G.: Combining participative and institutional approaches with multi-criteria evaluation. An Empirical Study for Water Issue in Troina, Sicily", Ecological Econ, **34**(2), 267–282 (2000)

16. Zanon, S., Luciani, D.: Il concorso Luoghi di valore. Contributo alla conferenza internazionale "Paysages de la vie quotidienne, regards croisés entre la recherche et l'action Landscapes of everyday life. Intersecting perspectives on research and action. Perpignan (Francia) e Gerona (Spagna), 16–18.III.2011 - Fondazione Benetton Studi Ricerche, Treviso (2012). www.fbsr.it

17. De Luca, V., Bertolo, M.: Urban games to design the augmented city. Eludamos. J. Comput. Game Culture **6**(1), 71–83 (2012)

18. Goddard, C.: Semantic Analysis. A Practical Introduction. 2nd edn. Oxford University Press, Oxford (2011)

19. Kenter, J.O.R., et al.: Shared, plural and cultural values: a handbook for decision-makers. In: UK National Ecosystem Assessment Follow-On Phase; UNEP-WCMC: Cambridge, UK (2014)

20. Irvine, K., et al.: Ecosystem services and the idea of shared values. Ecosyst. Serv. **21**, 184–193 (2016)

21. Van den Brink, A., Bruns, D., Tobi, H., Bell, S.: Research in Landscape Architecture: Methods and Methodology. Routledge, New York (2017)

22. Della Spina, L. Giorno, C.: Cultural landscapes: a multi-stakeholder methodological approach to support widespread and shared tourism development strategies. Sustainability **13**, 7175 (2021). https://doi.org/10.3390/su13137175

23. Della Spina, L.: Cultural heritage: a hybrid framework for ranking adaptive reuse strategies. Buildings **11**, 132 (2021). https://doi.org/10.3390/buildings11030132

24. Campolo, D.: The sustainable development of inland areas through cultural landscape and cultural. [L'uso sostenibile delle aree interne attraverso il paesaggio culturale e le cultural routes.] LaborEst **12**, 80–84 (2016). https://doi.org/10.19254/LaborEst.12.13

25. Campolo, D., Schiariti, C., Tramontana, C.: To the origin of humanism: a cultural route for a competitive network of mediterranean cities. [Alle origini dell'umanesimo: un itinerario culturale per una rete competitiva di città del mediterraneo]. LaborEst **9**, 9–13 (2014). https://doi.org/10.19254/LaborEst.09.02

26. Della Spina, L.: Revitalization of inner and marginal areas: a multi-criteria decision aid approach for shared development strategies. Valori e Valutazioni **2020**(25), 37–44 (2020)

27. Della Spina, L.: Integrated evaluation and multi-methodological approaches for the enhancement of the cultural landscape. In: Gervasi O. et al. (eds) ICCSA 2017. LNCS, vol 10404. Springer, Cham. (2017). https://doi.org/10.1007/978-3-319-62392-4_35

28. Della Spina, L.: A multi-level integrated approach to designing complex urban scenarios in support of strategic planning and urban regeneration. In: Calabrò, F., Della Spina, L., Bevilacqua, C. (eds.) ISHT 2018. SIST, vol. 100, pp. 226–237. Springer, Cham (2019). https://doi.org/10.1007/978-3-319-92099-3_27

29. Giuffrida, S., Ventura, V., Nocera, F., Trovato, M.R., Gagliano, F.: Technological, axiological and praxeological coordination in the energy-environmental equalization of the strategic old town renovation programs. In: Mondini, G., Oppio, A., Stanghellini, S., Bottero, M., Abastante, F. (eds.) Values and Functions for Future Cities. GET, pp. 425–446. Springer, Cham (2020). https://doi.org/10.1007/978-3-030-23786-8_24

30. Assumma, V., Bottero, M., Monaco, R. Mondini, G.: An Integrated Evaluation Model to Assess the Values and the Pressures of the Vineyard Landscape of Piedmont. [Un approccio

integrato per la misurazione dei valori e delle pressioni dei paesaggi vitivinicoli del piemonte.] LaborEst **16**, 75–80 (2018). https://doi.org/10.19254/LaborEst.16.12

31. Oliveira, À.D.: The human smart cities manifesto: a global perspective. In: Concilio, G., Rizzo, F. (eds.) Human Smart Cities. ULP, pp. 197–202. Springer, Cham (2016). https://doi.org/10.1007/978-3-319-33024-2_11

32. Bonomi, A., Masiero, R.: Dalla smart city alla smart land. Marsilio Editori spa (2014)

33. Della Spina, L., Giorno, C., Galati Casmiro, R.: An integrated decision support system to define the best scenario for the adaptive sustainable re-use of cultural heritage in southern Italy. In: Bevilacqua, C., Calabrò, F., Della Spina, L. (eds.) NMP 2020. SIST, vol. 177, pp. 251–267. Springer, Cham (2020). https://doi.org/10.1007/978-3-030-52869-0_22

34. Della Spina, L., Giorno, C., Galati Casmiro, R.: Bottom-up processes for culture-led urban regeneration scenarios. In: Misra, S., et al. (eds.) ICCSA 2019. LNCS, vol. 11622, pp. 93–107. Springer, Cham (2019). https://doi.org/10.1007/978-3-030-24305-0_8

35. Della Spina, L.: Multidimensional assessment for culture-led and community-driven urban regeneration as driver for trigger economic vitality in urban historic centers. Sustainability **11**, 7237 (2019). https://doi.org/10.3390/su11247237

36. Della Spina, L.: The integrated evaluation as a driving tool for cultural-heritage enhancement strategies. In: Bisello, A., Vettorato, D., Laconte, P., Costa, S. (eds.) SSPCR 2017. GET, pp. 589–600. Springer, Cham (2018). https://doi.org/10.1007/978-3-319-75774-2_40

37. Bonomi, A., Roberto M.: Dalla smart city alla smart land. Marsilio Editori spa (2014)

38. Tajani, F., Morano, P., Locurcio, M., D'Addabbo, N.: Property valuations in times of crisis: artificial neural networks and evolutionary algorithms in comparison. In: Gervasi, O., et al. (eds.) ICCSA 2015. LNCS, vol. 9157, pp. 194–209. Springer, Cham (2015). https://doi.org/10.1007/978-3-319-21470-2_14

39. Manganelli, B., Tajani, F.: Optimised management for the development of extraordinary public properties. J. Property Investment Financ. **32**(2), 187–201 (2014)

40. Giuffrida, S., Trovato, M.R.: A semiotic approach to the landscape accounting and assessment. an application to the urban-coastal areas. Paper presented at the CEUR Workshop Proceedings, 2030, pp. 696–708 (2017)

41. Trovato, M.R., Giuffrida, S.: The protection of territory from the perspective of the intergenerational equity. In: Mondini, G., Fattinnanzi, E., Oppio, A., Bottero, M., Stanghellini, S. (eds.) SIEV 2016. GET, pp. 469–485. Springer, Cham (2018). https://doi.org/10.1007/978-3-319-78271-3_37

42. Mangialardo, A., Micelli, E.: New bottom-up approaches to enhance public real/estate property. In: Stanghellini, S., et al. (eds.) Appraisal: From Theory to Practice, Green Energy and Technology (2017). https://doi.org/10.1007/978-3-319-49676-4_5

43. Della Spina, L., Ventura, C., Viglianisi, A.: A multicriteria assessment model for selecting strategic projects in urban areas. In: Gervasi, O., et al. (eds.) ICCSA 2016. LNCS, vol. 9788, pp. 414–427. Springer, Cham (2016). https://doi.org/10.1007/978-3-319-42111-7_32

44. Della Spina, L., Viglianisi, A.: Hybrid evaluation approaches for cultural landscape: the case of "Riviera dei gelsomini" area in Italy. In: Bevilacqua, C., Calabrò, F., Della Spina, L. (eds.) NMP 2020. SIST, vol. 178, pp. 1369–1379. Springer, Cham (2021). https://doi.org/10.1007/978-3-030-48279-4_128

45. Della Spina, L.: Scenarios for a sustainable valorisation of cultural landscape as driver of local development. In: Calabrò, F., Della Spina, L., Bevilacqua, C. (eds.) ISHT 2018. SIST, vol. 100, pp. 113–122. Springer, Cham (2019). https://doi.org/10.1007/978-3-319-92099-3_14

46. Della Spina, L.: Historical cultural heritage: Decision making process and reuse scenarios for the enhancement of historic buildings. In: Calabrò, F., Della Spina, L., Bevilacqua, C., (eds.) New Metropolitan Perspectives; ISHT 2018; Smart Innovation, Systems and Technologies, vol. 101. Springer, Cham (2019)
47. Della Spina, L., Rugolo, A.: A multicriteria decision aid process for urban regeneration process of abandoned industrial areas. In: Bevilacqua, C., Calabrò, F., Della Spina, L., (eds.) New Metropolitan Perspectives; NMP 2020; Smart Innovation, Systems and Technologies, vol. 178. Springer, Cham (2021)

Hydrogeological Damage: An Overview on Appraisal Issues

Antonio Nesticò$^{(\boxtimes)}$ ⓘ, Gabriella Maselli ⓘ, and Federica Russo ⓘ

Department of Civil Engineering, University of Salerno, Fisciano, SA, Italy
anestico@unisa.it

Abstract. The growing frequency of hydrogeological instability reflects the fragility of many territories, due to their geological, geomorphological, and hydrographic conformation. In addition, human actions, such as continuous deforestation, intensive urbanisation, unauthorised building, and lack of maintenance of slopes and watercourses, contribute to and encourage these events. The resulting damage involves the population, infrastructure, real estate, and the environment. This has serious economic, social, and environmental repercussions. The analysis of a phenomenon of hydrogeological instability certainly can't elude an economic-estimative investigation. This study intends to address the issue of estimating the costs necessary to restore the *status quo ante*, with specific reference to a hydrogeological instability that involves an urban area. To this end, first, the cost approach is introduced. Then, the analytical estimation procedure and the synthetic-comparative estimation procedure are exposed, both useful to evaluate the reconstruction cost, highlighting their limits and advantages. Finally, a database of parametric reconstruction costs in the Campania Region (Italy) for different types of civil works is built. This represents a tool available to operators for applications aimed at expeditious estimation of damage from hydrogeological instability.

Keywords: Hydrogeological instability · Appraisal issues · Cost-approach

1 Introduction

Hydrogeological instability is an issue of considerable and growing importance for the international scientific community. Italy, in particular, due to its geological, geomorphological and hydrographic conformation, is naturally predisposed to this phenomenon now widespread throughout the territory. The broad term "hydrogeological instability" refers to «all those processes that range from moderate and slow erosion to more consistent forms of surface and subsurface degradation of slopes, up to the imposing and severe forms of landslides» [1]. It is therefore a question of phenomena connected with the ruinous flow of open water, on the surface and in the soil, which have a significant impact on the population, the economic fabric, and the environment.

The contribution to this paper is the result of the joint work of the three authors, to which the paper has to be attributed in equal parts.

© The Author(s), under exclusive license to Springer Nature Switzerland AG 2022
F. Calabrò et al. (Eds.): NMP 2022, LNNS 482, pp. 1209–1217, 2022.
https://doi.org/10.1007/978-3-031-06825-6_117

The orography of the Italian territory is certainly penalising in this respect, given the prevalence of mountainous and hilly areas [2]. The causes of these instability phenomena are to be found in the fragility of the soils, in the radical modification of the hydrogeological balances along watercourses and in the lack of maintenance by man. The latter is often both the cause and contributor to such catastrophic events.

The progressive phenomenon of intense urbanization, the abusive building, the abandonment of the mountain lands, the continuous deforestation, the lack of maintenance of slopes and watercourses, have increased the incidence of these phenomena defined as "man-induced natural disasters" [3–5].

An updated framework on hydrogeological hazard in Italy is in the National Report (2018) by the Institute for Environmental Protection and Research (ISPRA) [6], which shows that:

- 16.6% of the national territory falls into the highest hazard classes (about 50,000 km^2);
- 7,275 municipalities (91% of the total) are at risk from landslides and/or floods, an increase of 3% compared to 2015 data;
- almost 4% of Italian buildings, over 550,000, fall in areas of high and very high landslide hazard and more than 9%, over one million, in areas of medium flood risk;
- about 12,000 buildings of historical and cultural interest fall in landslide risk areas and 30,000 in flood risk areas;
- over 80,000 industries and services are affected by landslide risk and almost 600,000 by flood risk;
- 1.28 million inhabitants are exposed to landslide risk and over 6 million to flood risk.

The figures unequivocally show how the consequences of such geo-hydrological events can be translated into tragic disruptions in the social and environmental sphere with extremely serious socio-economic impacts, such as high levels of material damage, loss of life and livelihood for affected communities [7, 8]. In addition, if we look at the data concerning cultural heritage, we understand the emerging need to protect the historical-artistic heritage which, especially for some smaller centres, characterises their local identity and constitutes a a strength for their economy [9–11].

2 Damage from Hydrogeological Instability

The damage D, to be intended as the consequence of the occurrence of the feared hydrogeological instability events, can be expressed in terms of vulnerability V and exposed value E:

$$D = V \cdot E \tag{1}$$

The vulnerability V expresses the attitude of people and property, natural and anthropic, exposed to hazard, to resist the effects of the intensity of a given event.

The exposed value E, instead, represents the element that must bear the event, expressed in terms of number of people or value of natural and economic resources exposed to a certain hazard.

To be able to talk about risk (R) it is necessary to consider vulnerability (V), expose value (E) and hazard (H). The latter which expresses the probability of a certain phenomenon occurring in an area:

$$R = H \cdot V \cdot E \tag{2}$$

Risk thus expresses the expected number of casualties, injuries, damage to property, destruction of economic activities or natural resources, due to a particular damaging event.

With specific reference to damage, a first classification, commonly adopted, is that which distinguishes direct and indirect damage. In particular, direct damage is linked to a physical interaction between the calamitous event and the population, buildings, environment. Indirect damages are induced by the former and generally occur after the event, both in space and time: they include, for instance, an interruption of traffic, public utilities, and local businesses. Losses due to consequent business interruptions or non-use of infrastructure providing public services may approach, or even exceed, those due to damage to property and assets [12, 13].

An additional distinction, for both categories, is that between tangible and intangible damages: the first ones are easily expressed in monetary terms; the second ones refer to things and services that cannot be assigned a monetary value. These are damages to people's health, to the environment, to the cultural heritage; damages that are difficult or considered almost impossible to estimate and for this reason are often underestimated or ignored [14, 15]. Ideally, all these elements should be included in an assessment of total costs but, in practice, most often only direct and tangible damages are considered to estimate economic losses. Yet, it has been demonstrated that the total impact of hydro-geological phenomena is significantly higher than just material damage [16].

A further and similar classification defines three types of damage [17]:

– direct damage resulting from the total or partial loss of the elements of the impacted area, such as buildings and their contents, vehicles, infrastructures, environmental assets, people;
– indirect damage, which often occurs after the event and derives, for example, from the interruption of activities, electricity, water and transport, assistance;
– intangible damage, which is difficult to quantify or convert into a monetary equivalent. These are, for instance, detours or queues on the home-work route, psychological impairments, abandonment of the areas most exposed to risk.

The most recent Ispra data show that, in Italy, repairing damage related to hydrogeological hazards costs on average four times more than interventions to prevent it. From 1998 to 2018, Italy has spent around 20 billion euros, about one billion a year, to remedy the effects of instability, compared to 5.6 billion euros that have been invested in the planning and implementation of prevention works [18].

In addition, recent studies show that in the last few decades disasters have produced more victims and have been more costly for developing countries than for developed ones [19, 20]. Increased wealth causes relatively higher losses in high-income countries [21]. However, higher incomes increase private demand for security: more income allows people and countries to employ additional and costly precautionary measures [22].

3 Estimation of the Costs to Restore the *status quo ante*

Economically evaluating the damage caused by a phenomenon of hydrogeological instability means considering direct costs and indirect costs:

- direct costs represent the "most visible" economic consequence of such phenomena. With regard to buildings, infrastructure and urbanisation works, the estimate should be made in terms of the cost of restoring the original condition (for functional and/or aesthetic damage) or in terms of the cost of partial or total reconstruction (for structural damage);
- indirect costs, on the other hand, are generally linked to loss of income, contraction of real estate values, increased unemployment and other economic aspects related to the interruption or reduction of production, distribution, and consumption of goods. The estimation of these costs is complex and, according to some studies, would be of a similar order to direct damages [23].

When hydrogeological instability affects an urbanised territory, the first problem is to estimate the costs necessary to restore the *status quo ante*; that is, the costs to be sustained for the restoration of infrastructures, housing and production units involved in the tragic phenomenon. The cost approach is used for estimates, according to the depreciated reconstruction cost criterion. This approach is recognised both in estimation literature and by international valuation standards. The logic refers to the purposes of: reproduction, where attention is paid to the use of the original materials in reconstructing a replica of the property being estimated; replacement, if the use of new and current construction methods and materials is used to obtain a structure structure with the same initial function. In all cases, it is necessary to estimate the costs of new reconstruction of works having the same utility and equivalent function to be able to perfectly replace existing ones and damaged by the instability.

According to art. 32 of D.P.R. 207/2010, in Italy the Construction Cost CC is the sum of the Technical Construction Cost TCC, the General Costs GC and the Constructor's Technical Profit CTP:

$$CC = TCC + GC + CTP \tag{3}$$

The Technical Construction Cost TCC is given by the direct variable costs, which are directly proportional to the works carried out. These are the costs: MT for materials; L for labour; FT for freight and transport. The technical cost TCC has a significant impact on the total construction cost.

General Expenses GE include indirect costs, both variable and fixed: variable indirect costs are related to the opening, setting up and safety of the construction site; fixed indirect costs, on the other hand, are general site costs, independent of the site activity, such as the cost for the personnel employed, financial charges, etc.

The Constructor's Technical Profit CTP represents the remuneration for the organization of building production and work phases, as well as the premium for any risks related to an increase in the costs of production factors, penalties for delays or problems in the works. The following relationships apply:

$$TCC = MT + L + FT \tag{4}$$

$$GE = 13 - 15\% \cdot TCC \tag{5}$$

$$TCP = 10\% \cdot (TCC + GE) \tag{6}$$

For estimating the construction cost, based on the availability of data, the processing time and the degree of reliability required by the forecast, two procedures can be used:

– indirect (or analytical) procedure, focused on the analysis of the production process of the work and, therefore, on the identification and quantification of all the production factors involved in the process;
– synthetic-comparative procedure, based on a forecast of the construction cost through the comparison with known costs regarding similar works already realised.

The analytical estimation procedure involves the elaboration of an Estimation Metric Calculation (EMC) through which the production process is broken down into individual processes. By identifying and quantifying all the production factors, the construction cost is estimated as the total amount reported at the end of the calculation. The analytical procedure leads to a detailed estimate of the resources necessary for the realization of the intervention but, at the same time, requires the elaboration of a definitive or executive project to be referred to. In addition, if we want to apply this procedure to an urbanized area, it is immediate to understand the difficulties in drawing up a metric calculation for each civil work involved in the instability.

The synthetic comparative procedure is more rapidly used and therefore frequently used in expeditious estimates. This procedure uses: the identification of a sample for each of the types of building involved; the analytical estimate of the construction cost of each work/building sample, to establish a parametric unit cost; the extension of the parametric values to the entire area involved in hydrogeological instability.

In the application of the synthetic procedure a first phase is planned to find the unit costs for the construction of building or urbanization works similar to those to be estimated. Data sources can be construction companies, contracting authorities, chambers of commerce, national statistical institutes, public bodies, professional orders, builders' associations. The next step concerns the choice of the technical parameters able to measure the consistencies: for residential buildings, the parameter can be the volume, the surface area (gross or useful), the number of rooms. Finally, each consistency, expressed in the appropriate technical parameter, is to be multiplied by the unit cost of building production, so as to obtain the total cost of reconstruction.

To provide concrete indications of the costs to be incurred for the total reconstruction of a territorial area subject to hydrogeological instability, Table 1 shows a list of parametric costs valid for the Campania Region (Italy). These are CC construction unit costs derived from official sources and are specifically useful in expeditious estimates of damage due to hydrogeological instability [24, 25]. It is assumed that the instability has severely involved the area, to the point that all civil constructions have to be demolished and entirely rebuilt.

Table 1. Parametric costs of demolition, transport to dump and reconstruction of irreparably damaged civil works in the area exposed to hydrogeological instability (Campania Region, Italy).

Type of civil work	Unit cost of demolition and transport to dump	Unit cost of reconstruction
Residential tower building Medium-rise residential building. Reinforced concrete structure developed on 8 levels and provided with lift system	17 €/m^3	318 €/m^3
Multi-level residential building Social housing. Reinforced concrete structure developed on 6 levels and provided with lift system	17 €/m^3	245 €/m^3
Office building Reinforced concrete structure on 4 levels with underground level and lift system	17 €/m^3	352 €/m^3
Multi-level industrial building On-site reinforced concrete structure on 3 levels	17 €/m^3	783 €/m^2
School building Reinforced concrete structure on 3 levels with an underground level. Expected capacity of 750 students	17 €/m^3	225 €/m^3
Level car park Surface area with public lighting for a total of 367 parking spaces and 43 motorbikes	376 €/m^3	50 €/m^2
Public garden Area of 10,500 m^2 with public lighting and walking ways	34 €/m^2	35 €/m^2
Public road Road with a calibre of 16.50 m obtained with a 12 m wide carriageway and two pavements each 2.25 m wide	376 €/m^3	96 €/m^2
Motorway Road, in trench or raised, with two lanes in each direction and two carriageways each 22 m wide	726 €/m^3	$2,148 \text{ €/m}$

(*continued*)

Table 1. (*continued*)

Type of civil work	Unit cost of demolition and transport to dump	Unit cost of reconstruction
Viaduct Road with two lanes and a 10.50 m wide carriageway and platform	1,171 €/m^3	10,800 €/m

It should be noted that the approach described above for estimating the cost of restoring the *status quo ante* leads to costs of reconstruction of civil works, without considering the actual conditions of conservation, maintenance, age, and obsolescence of the structures subject to demolition due to hydrogeological instability.

4 Conclusions

Examining the economic-estimative aspects linked to the more frequent phenomena of hydrogeological instability, means better understanding the risk levels of a territory. Damage from hydrogeological instability involves multiple sectors, with serious consequences not only in terms of loss of life, but also on the productive tissue, infrastructures, and real estate [26].

Quantifying the impact of a hydrogeological phenomenon in monetary terms leads not only to a greater awareness of the consequences of our actions, but also to underline the necessity of urgent and effective actions for risk mitigation [27].

This research focuses on an important issue: the estimation of the cost of the structures required to restore the *status quo ante*. First, the possible procedures – analytical and synthetic-comparative – for the estimation of the reconstruction cost are described, highlighting the advantages and limitations of each. Then a database of parametric costs to be incurred in the Campania Region (Italy) in case of total reconstruction of an urbanized territorial area subjected to hydrogeological instability is built.

The information provided in terms of costs necessary to restore the conditions preceding such phenomena constitute terms of reference for case studies and real applications. In addition, this information shows that the costs of reconstruction are significant and certainly higher than the costs of planned risk prevention and/or mitigation measures. In this regard, Triglia *et al.* [6] point out that repairing the immense damage caused by hydrogeological risk costs on average four times more than prevention measures. It results that the attention of decision-makers with respect to the issues in question problems in question must be maximum, directing the allocation of public resources to protect the territory and also directing changes to the regulatory reference framework.

In this direction, moreover, public participation as well as information and training activities on these issues are certainly useful.

In view of the above, it's easy to understand the importance of implementing policies to protect the territory. These policies must involve all stakeholders, moving from the logic of repair to that of prevention and territorial requalification. This will also provide an opportunity for economic relaunch.

References

1. Commissione Interministeriale per lo Studio della Sistemazione Idraulica e la Difesa del Suolo (Commissione De Marchi): Atti della Commissione, vol. I, II, III, IV. Roma (1970)
2. Di Raimondo, S.: Hydrogeological instability: a perennial emergency for roads infrastructures. Strade Autostrade **141**, 142–147 (2020)
3. Gisotti, G., Benedini, M.: Il dissesto idrogeologico. Previsione, prevenzione e mitigazione del rischio. Carocci, Roma (2000)
4. Maselli, G., Nesticò, A.: L'Analisi Costi-Benefici per progetti in campo ambientale. La scelta del Saggio Sociale di Sconto. LaborEst (20), 99–104 (2020). https://doi.org/10.19254/LaborEst.20
5. Troisi, R., Castaldo, P.: Technical and organizational challenges in the risk management of road infrastructures. J. Risk Res. (2022). https://doi.org/10.1080/13669877.2022.202884
6. Trigila, A., Iadanza, C., Bussettini, M., Lastoria, B.: Dissesto Idrogeologico in Italia: pericolosità e indicatori di rischio. In: Edizione 2018. ISPRA (2018)
7. Guzzetti, F., Polemio, M.: Il rischio idrogeologico in Italia e il ruolo della ricerca scientifica. Convegno Nazionale sul Dissesto Idrogeologico. Geologia dell'ambiente 2 (2012)
8. Salvati, P., Bianchi, C., Rossi, M., Guzzetti, F.: Societal landslide and flood risk in Italy. Nat. Hazard. **10**, 465–483 (2010)
9. Calabrò, F., Cassalia, G., Lorè, I.: The economic feasibility for valorization of cultural heritage. the restoration project of the reformed fathers' convent in francavilla Angitola: the Zibìb territorial wine cellar. In: Bevilacqua, C., Calabrò, F., Della Spina, L. (eds.) NMP 2020. SIST, vol. 178, pp. 1105–1115. Springer, Cham (2021). https://doi.org/10.1007/978-3-030-48279-4_103
10. Della Spina, L., Giorno, C., Galati Casmiro, R.: An integrated decision support system to define the best scenario for the adaptive sustainable re-use of cultural heritage in southern Italy. In: Bevilacqua, C., Calabrò, F., Della Spina, L. (eds.) NMP 2020. SIST, vol. 177, pp. 251–267. Springer, Cham (2020). https://doi.org/10.1007/978-3-030-52869-0_22
11. Dolores, L., Macchiaroli, M., De Mare, G.: Sponsorship's financial sustainability for cultural conservation and enhancement strategies: an innovative model for sponsees and sponsors. Sustainability **13**(6), 9070 (2021). https://doi.org/10.3390/su13169070
12. Marsh & McLennan Companies: Sunk costs: The socioeconomic impacts of flooding. Rethinking Flood Series **1**, 14–15 (2021)
13. Troisi, R., Alfano, G.: Is regional emergency management key to containing COVID-19? A comparison between the regional Italian models of Emilia-Romagna and Veneto. Int. J. Public Sector Manage. (2021)
14. Schuster, R., Fleming, R.: Economic losses and fatalities due to landslides. Bull. Assoc. Geologists **XXII I**(1), 11–28 (1986)
15. Petrucci, O., Gullà, G.: A simplified method for assessing landslide damage indices. Nat. Hazards **52**(3), 539–560 (2009)
16. Balbi, S., Giupponi, C., Olschewski, R., Mojtahed, V.: The economics of hydrometeorological disasters: approaching the estimation of the total costs. BC3 Working Paper Series, pp. 2–8 (2013)
17. Gaschen, S., Hausmann, P., Menzinger, I., Schaad, W.: Floods: An Insurable Risk? A Market Survey. Swiss Reinsurance Company, Zurich (1998)
18. Gisotti, G., Masciocco, L.: Il ruolo economico del riassetto idrogeologico in Italia. Valutazione ambientale **23**, 74–78 (2013)
19. Dore, M., Etkin, D.: The importance of measuring the social costs of natural disasters at a time of climate change. Aust. J. Emergency Manage. **15**(3), 46–51 (2000)

20. Toya, H., Skidmore, M.: Economic development and the impacts of natural. Econ. Lett. **94**(1), 20–25 (2007)
21. Raschky, P.A.: Institutions and the losses from natural disasters. Nat. Hazard. **8**(4), 627–634 (2008)
22. Petrucci, O.: The impact of natural disasters: simplified procedures and open problems. In: Tiefenbacher, J. (eds.) Approaches to managing disasters- assessing hazards, emergencies and disaster impacts, pp. 110–133. IntechOpen, United Kingdom (2012). https://doi.org/10.5772/29147
23. Sterlacchini, S., Frigerio, S.: Landslide risk analysis: a multi-disciplinary methodological approach. Natural Hazards Earth Syst. Sci. **7**(6), 657–675 (2007)
24. Collegio degli Ingegneri e Architetti di Milano: Prezzi Tipologie Edilizie. QuineDEI Tipografia del Genio Civile, Milan (2019)
25. Regione Campania: Prezzario Regionale delle Opere Pubbliche (2021). https://box.regione.campania.it/data/public/prezzario2021. Accessed 10 Oct 2021
26. Bencardino, M., Nesticò, A.: Demographic changes and real estate values. A quantitative model for analyzing the urban-rural linkages. Sustainability **9**(4), 536 (2017). https://doi.org/10.3390/su9040536
27. Nesticò, A., Maselli, G.: Declining discount rate estimate in the long-term economic evaluation of environmental projects. J. Environ. Acc. Manage. **8**(1), 93–110 (2020). https://doi.org/10.5890/JEAM.2020.03.007

Urban Sustainability: Reporting Systems and Dataset in the European Union

Giorgia Iovino[(✉)]

University of Salerno, Via Giovanni Paolo II, 132 , 84084 Fisciano, SA, Italy
giovino@unisa.it

Abstract. The issue of urban sustainability has been at the center of the European political agenda for over a decade. Consequently, there is an increasing demand for geographic data and analytical tools that can usefully direct policy choices at different territorial scales, in order to select transformative options more careful in terms of environmental quality and social equity.

In this perspective, the paper proposes a brief review of the reporting systems, and data set developed at European level to monitor the urban environment and measure its sustainability, highlighting its functions and utilities for urban policies.

Keywords: Urban sustainability · Reporting systems · European union

1   Introduction

At the basis of the need to decline sustainable development on an urban scale lies the recognition of the preponderant role of the city in contemporary society. Currently over half of the planet's population lives in urban areas. In the European Union cities host two thirds of the population, use about 80% of energy resources and generate up to 85% of European GDP (EC, DG Regional Policy 2011).

Engines of the economy and catalysts of creativity and innovation, cities are also the places where environmental and socio-economic problems reach the most alarming levels. As recognized by the European Commission (EC, UN-Habitat 2016), urban areas today face major challenges ranging from climate change to the growth of poverty and social inequalities, from the consumption of natural resources (soil, energy, landscape) to the economic and financial crisis.

It is no wonder, therefore, that the urban question and more specifically the question of the sustainability of cities and urban settlements has been placed at the center of European and international policies, as evidenced by the signing in 2016 of the Amsterdam Pact for the European Urban Agenda (CE 2014) and the New urban Agenda adopted by the UN at the Habitat III Conference, held in Quito in the same year. As is well known, sustainable cities and communities are also one of the 9 priority objectives identified in the context of the 7th Action Program of the European Union (EC 2014) and one of the 17 Sustainable Development Goals (SDGs) that define the framework reference for the United Nations 2030 Agenda for sustainable development.

F. Calabrò et al. (Eds.): NMP 2022, LNNS 482, pp. 1218–1228, 2022.
https://doi.org/10.1007/978-3-031-06825-6_118

This renewed attention to the sustainability of the urban environment has revived interest in the analytical tools that can be used for its measurement and has increased the demand for geographic data, indicators and reporting systems that can usefully direct policy choices (at different territorial scales), directing them towards transformative options economically efficient and less burdensome in terms of environmental quality and social equity.

In order to respond to this need, the paper offers an overview of the reporting systems and cartographic tools developed at European level to monitor the urban environment and measure its sustainability. Specifically, after a section (par. 2) dedicated to outlining the main partitioning and classification methods of urban space, the main datasets or frameworks developed on a European scale in recent years are examined (par. 3): from the European Common Indicators to the Urban Metabolism Framework, from the European Green City Index to the Urban Sustainability Indicators. A brief description of each database is provided to highlight its functions and usefulness for urban policies.

2   The European Urban Territory: *Découpage* and Classifications

The identification of a system of indicators to measure urban sustainability calls into question the issue of *découpage*, i.e. the territorial units of analysis. There are today different ways of defining and delimiting the city, which reflect the complexification and reconfiguration of urban space[1] that has taken place in recent decades, both from a physical-morphological point of view (see the urban sprawl processes) and from a functional point of view (see the growth of interconnections and expansion of gravitation areas of cities).

At European level, the European Environment Agency (EEA 2009) identified three classification systems for urban areas in 2009. The first one is based on a purely administrative approach, which uses the political-administrative borders in each member state The second is based on a morphological approach (ESPON 2013, 2015), where the urban area (*UMZ Urban Morphological Zone*) is defined in purely physical terms (population density, extension of buildings, networks and infrastructures, presence of industrial areas ampersands, etc.). The third classification system uses a functional approach according to which the extent and shape of the city are given "by the attraction that the city exercises on the surrounding area in socio-economic, productive, service delivery terms" (EEA 2009, p. 11). Using the latter approach, Eurostat in collaboration with the OECD (OECD 2012; Eurostat 2016) in 2011 identified 911 *Functional Urban Areas* (Fig. 1) on a European scale (EU and EFTA countries). Each FUA is composed of a central or core area (with a population density between 1.500 and 50.000 inhabitants per sq. km) and a commuting area, identified through the commuting flows for work reasons (>15% of outgoing flows of each municipality to the core).

---

[1] These reconfigurations are the result of multiple interconnected phenomena such as the increase in individual mobility, the development of infrastructural networks, the fragmentation of the social fabric and businesses, the increase in urban income in central urban areas and the decentralization of residential functions and productive in peri urban areas. On the topic of urban sprawl and new settlement forms see EEA, FOEN 2016.

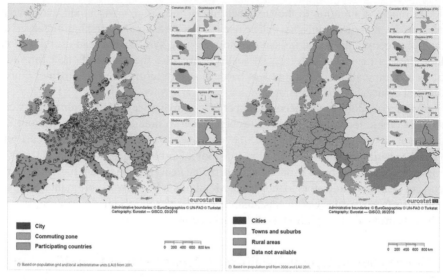

Fig. 1. Urban Functional Areas (a) and degree of urbanization (b) Source: Eurostat (2016).

This procedure follows the one already experimented in the Urban Audit area for the identification of the *Larger urban Zones* (LUZ)[2]. In Italy, Istat has long used a similar methodology for defining Local Labor Systems, based on commuting flows for work or study purposes[3].

Approximations of the FUAs are the *Metropolitan regions*, regions of over 250 thousand inhabitants made up of one or more Nuts 3 level regions that take their name from the main internal FUA within its borders[4]. As regards Metropolitan Regions, a distinction is made between capital City Regions, second-tier Metro regions and smaller Metro Regions. The first include capital cities, the second comprise the largest non-capital cities in the country, the third include smaller Metro Regions identified through a natural break in the data series.

The classification system of the Degurba database recently developed by Eurostat in collaboration with the OECD is, instead, based on a grid cell approach (EC, DG Regio

---

[2] The spatial levels of subdivision of the urban space proposed by Urban Audit are three: the core city, generally included in the administrative boundaries (LAU2) or in larger cities given by aggregations of LAU2; the sub-city districts, internal subdivisions of the city on a demographic basis; the Larger Urban Zones, functional areas based on daily movements for work reasons. See ESPON 2011.

[3] ISTAT (2107) has divided the 661 Italian SLLs identified in 2011 into three hierarchically articulated typologies: at the first level the local systems of the main urban realities, i.e. the 14 metropolitan cities and those SL with a population greater than 500 thousand inhabitants or with the population of the provincial capital of the SL over 200 thousand inhabitants (Busto Arsizio, Como, Bergamo, Verona, Padova, Taranto), at the second level the local systems belonging to medium-sized cities (86) and at the third level the remaining 504 local systems.

[4] If more than 50% of the population of an adjacent NUTS3 region lives within the FUA it is included in the metropolitan region.

2014, Eurostat 2018, 2019). The European territory is divided into cells of 1 sq. Km. By calculating the density of each cell, three types of areas are identified according to the degree of urbanization (Fig. 1): 1) *cities*, more densely populated areas (at least 50.000 inhabitants), where the majority of the population lives in the cells defined as urban centers with a density of at least 1.500 inhabitants/sq km; 2) *towns and suburbs*, intermediate density areas obtained by surrounding perimeter cells with a density of at least 300 inhabitants/sq km and a total population of at least 5.000 inhabitants (cells defined as urban clusters or urban clusters); 3) *rural areas*, located outside urban settlements. Using a grid-based approach, this classification system has the advantage of taking into account local administrative units (LAU2) and at the same time eliminating distortions related to the different size of the administrative units, improving the comparability between national contexts and different regional. In fact, only municipalities (LAU2) with at least half of the population living in the urban center are candidates to be classified as cities.

Where the urban area develops beyond the city border, an additional spatial level called greater city has been created. In Italy, Naples and Milan are an example of this.

For capital cities and cities that exceed 250 thousand inhabitants, a subdivision into sub districts is provided, with two levels: the first relating to the partitions already in force within the cities, the second that sets demographic threshold values (a population included between 5.000 and 40.000 inhabitants) for greater comparability.

In summary, the spatial levels currently used to classify the urban territory in Europe are the following: 1. Metropolitan region; 2. FUA-Functional Urban Area, (Larger Urban Zone -LUZ, in Urban Audit); 3. Greater City (Kernel in Urban Audit); 4. City (core city in Urban Audit); 5. Sub City Districts (level 1 and 2). Figure 3 provides further details about the building process of three of the five levels indicated above (Cities, FUA and Metro region).

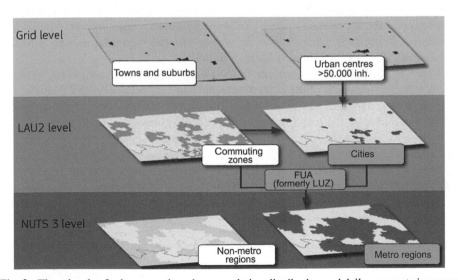

Fig. 2. Three levels of urban areas based on population distribution and daily commute journeys. *Source:* Eurostat (2018)(modified)

3 Reporting Systems and Urban Sustainability Indicators

In order to monitor urban sustainability, various scalable and ready-to-use reporting standards and dataset systems (Table 1) have been developed on a European scale (EC, DG Environment 2015; PBL 2016).

Table 1. The main reporting systems on urban sustainability

Dataset/indicator	Promoter
European Common Indicators	EC
Urban Ecosystem Europe	ICLEI; Ambiente Italia
Urban Sustainability Indicators	Eurofound
Urban Metabolism Framework	EEA
European Green Capital Award	EC
European Green City Index	EIU

Source: Author's elaboration

The European Common Indicators (ECI) were developed in the 1990s on the initiative of the European Commission by the Group of Experts for the Urban Environment (which in 1999 formed the Working Group on Sustainability Indicators) with the aim of supporting the implementation of the Local Agenda 21 and reducing the ecological footprint in urban areas (CE and Ambiente Italia 2003). Based on a bottom-up approach (local authorities are the main actors), the ECI can be considered as complex indicators of the quality of life in urban areas. Compared to the indicators used in environmental analysis that generally provide objective quantitative information (for example, levels of emissions into the atmosphere, quantity of waste produced, concentration of pollutants present in water bodies), the ECI also take into consideration qualitative aspects linked to the behaviors and perceptions by local communities.

They are useful as they contribute to the development of benchmarking systems between the performances of the cities at national and European level and facilitate the creation of environmental accounts. Of the 10 indicators identified and reported in the table below (Table 2), the first 5 represent the key indicators that should be used by all local authorities, while the remaining are specific optional indicators, which deepen and better describe some aspects.

The *Urban Ecosystem Europe* (UEE) is an evaluation of 32 European cities carried out by the Research Institute Ambiente Italia, which is part of the International Council for Environmental Initiatives (ICLEI) to monitor the progress made by cities towards the commitments signed in the Charter of Aalborg[5]. The survey (Dexia and Ambiente Italia

---

[5] The Aalborg Charter signed in 1994 by 80 European Local Administrations and 253 representatives of international organizations, national governments, scientific institutes, etc. it essentially develops into three parts that define, respectively, the declaration of principle for a sustainable urban model for European cities, the campaign of sustainable European cities, the commitment to implement local Agenda 21.

Table 2. The common European indicators

n.	Dimension and main indicator
1	Citizen satisfaction with reference to the local context *Satisfaction (general and average) with relation to the local context*
2	Local contribution to global climate change (and / or ecological footprint) *Per capita CO2 emissions*
3	Local mobility and passenger transport *Percentage of movements occurring with private motorized vehicles*
4	Accessibility of public green areas and local services *Percentage of citizens who live within 300 m of public green areas >5.000 mq*
5	Local air quality *PM10 net exceedances*
6	Home-school travel for children *Percentage of children going to school by car*
7	Sustainable management of the local authority and local businesses *Percentage of environmental certifications on to the total number of companies*
8	Noise pollution *Percentage of population exposed at night to noise levels >55 dB (A)*
9	Sustainable products *Percentage of people buying sustainable products*

Source: Commissione Europea e Ambiente Italia (2003)

2007) takes up the analysis prepared annually by Legambiente (Urban Ecosystem), comparing the European cities, which provided the data. The system consists of 25 indicators, grouped into 6 thematic areas to allow a more integrated reading (Table 3).

Table 3. The urban ecosystem europe dataset.

Thematic area
Local action for health and protection of our common goods
Responsible consumption and lifestyle
Planning, urban form and sustainable mobility
From local to global: energy and climate protection policies
Innovative, sustainable local economy and equity, justice and social cohesion
Local management towards sustainability, good governance and participation

Source: Dexia and Ambiente Italia (2007)

Specific attention is paid to local governance and the quality of life, but the indicators manage to touch further aspects of urban sustainability. The cities evaluated range from

150.000 inhabitants to more than 2 million, indeed scalability of the indicator guarantees a fairly good comparability across cities of very different size.

According to the commitments established in the Aalborg Charter, Eurofound, the European Foundation for the Improvement of Living and Working Conditions, has developed the framework of the *Urban Sustainability Indicators* (Mega and Pedersen 1998). The indicators were assigned to each political theme identified in the Charter, paying specific attention to environmental health measures. The dataset has been tested on a few European cities and includes 16 indicators. The geographical scale of each indicator depends on its nature: the indicators of global climate, acidification, toxicity of ecosystems and sustainable economy are relevant at city level, while the rest are more relevant at neighborhood level if they have to reflect the different rhythms of development within the same city. The originality of this set of indicators is represented especially by the last category, that of unique sustainability, an indicator defined by each city according to its uniqueness, such as particular local climatic conditions or specific practices or events. According to Mega and Pedersen (1998, p. 27) "this indicator should represent the degree to which these unique factors/events lead to urban sustainability".

The *Urban Metabolism Framework* is a system of indicators developed by Minx et al. (2011) for the European Environment Agency, in order to evaluate the sustainability of a city based on its metabolic flows[6] rather than on its performance or current status. This system was composed by collecting a wide range of indicators from different frameworks, such as Urban Ecosystem Europe, all based on publicly available municipal data. Starting from this, the authors generated a dataset of 15 quantitative key indicators (Table 4), which provides continuous and low-cost monitoring of urban metabolism. The strength of this framework of indicators is in its simplicity, in the easy accessibility data sources and in its ability to be applied to cities of various sizes. However, it does not provide the most complete measure of how sustainable a city is and is more informative at European level rather than at individual city level.

The *European Green Capital Award* is an annual award active since 2010 that recognizes the commitment of European cities to environmental issues, such as, local transport, and soft mobility, green areas and sustainable land use, nature and biodiversity, local contribution to climate change, air quality, noise pollution, waste production and management, water consumption, waste water treatment, eco-innovation and sustainable employment, environmental management of the local authority, energy performance. The evaluation of these issues is conducted by a panel of experts, who express an opinion based on a set of information about the performance by cities (European Green Leaf Award Secretariat, Phrenos, Arcadis, Expert Panel 2021). The twelve themes listed above are described through a set of about fifty indicators, both quantitative (CO_2 emissions per capita, inhabitants connected to the district heating network, public green areas per capita, etc.) and qualitative such as the policies put in place regarding the reduction of greenhouse emissions, consumption, waste, traffic, soil sealing or the measures taken to mitigate noise or improve air quality, water, accessibility to green areas, etc. The cities

---

[6] Just as living organisms transform food, water and oxygen into energy and waste, cities likewise need energy, materials and resources. Urban metabolism regulates the transformation and replacement of these materials to allow the production, self-production and reproduction of the city: this is facilitated by the governance policies, infrastructures and inhabitants of the city.

Table 4. The urban metabolism framework.

	Indicator	Dimension
H1	Per capita CO2 emissions from energy consumption	Urban Flows
H2	Energy efficiency of transport	Urban Flows
H3	Efficiency of the use of residential energy	Urban Flows
H4	Efficiency in urban use of water	Urban Flows
H5	Intensity of waste	Urban Flows
H6	Recycling	Urban Flows
H7	Urban land occupation	Urban Flows
H8	Access to green spaces	Urban Quality
H9	NO2 concentrations	Urban Quality
H10	PM10 concentrations	Urban Quality
H11	Unemployment rate	Urban Quality
H12	Land use efficiency	Urban Patterns
H13	Length of the public transport network	Urban Patterns
H14	Registered cars	Urban Drivers
H15	GDP per capite	Urban Drivers

Source: Minx et al. (2011)

that apply for the award must, in fact, compile a report that analyzes the current situation, past trends and future commitments for each of the indicated issues.

Another method for assessing the sustainability of cities is the *Green City Index*, developed by the Economist Intelligence Unit (2009). This index analyzes how 30 European capitals - ranging in demographic size from less than one million to more than three million inhabitants- face the challenges related to sustainable development and provides a ranking in terms of performance. As part of the evaluation, an expert panel developed a framework of 30 indicators to compare these cities. The dataset covers all the major areas of urban environmental sustainability and scores by evaluating 8 themes: greenhouse gas emissions, energy, construction, transport, water, waste and land use, air quality, environmental governance. The evaluation, which takes place through a weighted average of the 30 indicators, provides a final score, on a scale of 100 points, both overall and by single category. Of the 30 indicators identified 17 are quantitative and 13 qualitative, although the focus is more on measuring energy consumption or CO2 emissions, rather than on measures of health, happiness and quality of life.

In addition to the reporting systems analyzed so far, it is worth mentioning (Table 5) the battery of indicators developed to monitor the objective 11 dedicated to sustainable cities and communities within the SDGs (Eurostat 2021). Only two indicators have quantitative target (road deaths and recycling rate of municipal waste).

In recent years, various interactive applications, equipped with mapping tools, have also been developed in institutional European websites. Examples are the Urban Data

Table 5. Progress towards goal 11 of the SDGs.

Topic	Indicator
Quality of life in cities and communities	Crowding rate
	Population suffering from noise pollution
	Exposure to pollution
	Population living in degraded and unsafe homes
	Population that reports the occurrence of crimes, violence or vandalism in the area of residence
Sustainable mobility	People killed in traffic accidents
	Share of buses and trams on total passenger transport
Adverse environmental impacts	Settlement area per capita
	Municipal waste recycling rate
	Population linked to wastewater treatment

Source: Eurostat 2021

Platform, the Statistics Illustrated in the Eurostat website or the CityBench app. The first is an initiative of the Joint Research Center and the Directorate-General for Regional and Urban Policy of the European Commission. It collects data on over 800 cities derived from multiple sources, disaggregated into three spatial levels (807 Cities, 672 Functional Urban Areas, 271 Metropolitan Regions). Its goal is to provide direct access to information regarding the state and trends of European cities and surrounding regions, to support territorial policies and the implementation of the European urban Agenda. The Eurostat database contains various types of data concerning sustainability: environmental data in the "Environment and energy" folder and the system of monitoring indicators relating to the Europe 2020 strategy and the 17 Sustainable Development Goals. More specifically on the topic of sustainable urban development, the Urban Audit dataset is available, and it can be accessed through the Statistics Illustrated section. The CityBench is a webtool developed by Espon to support political decision-making processes or private sector investment choices. It allows the user to compare the performance of European cities on different themes, such as demography, the economy, the quality of life, the environment, etc., displaying the indicators in graphical and cartographic form, in order to bring out similarities and complementarities and evaluate weaknesses and potentials of the individual cities in a benchmarking perspective.

4   Concluding Remarks

In the context of the EU, the supply of reporting systems and datasets relating to urban sustainability has multiplied in recent years, to respond to the growing attention to the ecological transition issue by scholars and policy makers.

The reporting system and dataset analyzed in the paper can be a first essential step to increase knowledge on urban sustainability and to face the challenges posed by the

European and international programs and agreement on this issue. However, some limitations need to be highlighted. First, most systems follow a top-down expert led approach, neglecting a more participatory and place-based approach that involves citizens and local knowledge. Secondly, a fixed set of indicators is useful for comparative analyzes at the European level, but not very suitable for developing a more specific knowledge of the context that can be used on a local scale, for planning purposes. A third limit is linked to the imbalance between quantitative and qualitative indicators to the advantage of the former. In short, there is a weak territorialization of the processed datasets. A possible solution would be to integrate the fixed datasets with a more flexible set of context specific indicators, as cities, neighborhoods and urban communities have different perceptions, behaviors, needs and priorities.

References

CE-Commissione Europea: Verso un'Agenda urbana integrata per l'UE. Ufficio delle pubblicazioni dell'Unione europea, Lussemburgo (2014)

CE-Commissione Europea, Ambiente Italia: Indicatori Comuni Europei. Verso un Profilo di Sostenibilità Locale. Ambiente Italia, Milano (2003)

Dexia, Ambiente Italia: Urban Ecosystem Europe Report. Milano (2007)

EC: General Union Environment Action Programme to 2020. Living well, within the limits of our planet. Publications Office of the European Union, Luxembourg (2014)

EC - DG Environment: In-depth Report: Indicators for Sustainable Cities. Science for Environment Policy. Issue 12, Bristol (2015, revised 2018)

EC, DG REGIO: A harmonized definition of cities and rural areas: the new degree of urbanization. Working Paper 01 (Dijkstra L., Poelman H. eds) (2014)

EC, DG Regional Policy: Cities of tomorrow - Challenges, visions, ways forward. Publications Office of the European Union, Luxembourg (2011)

EC, UN-Habitat: The state of European cities 2016 Cities leading the way to a better future. Publications Office of the European Union, Luxembourg (2016)

Economist Intelligence Unit: European green city index. Assessing the environmental impact of Europe's major cities. Siemens AG, Munich (2009)

EEA: Ensuring quality of life in Europe's cities and towns. EEA Report 5/2009. Copenhagen (2009)

EEA, FOEN: Urban sprawl in Europe. EEA Report 11. Copenhagen (2016)

ESPON: LUZ specifications (Urban Audit 2004). Technical Report Luxembourg (2011)

ESPON: Naming UMZ: making them more operational for urban studies. Technical Report, Luxembourg (2013)

ESPON: ESPON Database 2013 Phase II (2011–2014) Multi Dimensional Database Design and Development (M4D). Final Report. Luxembourg (2015)

European Green Leaf Award Secretariat, Phrenos, Arcadis, Expert Panel: Technical Assessment Synopsis Report - European Green Leaf Award 2022, Phrenos (2021). http://www.ec.europa.eu/europeangreencapital. Accessed 10 Dec 2021

Eurostat: Urban Europe - Statistics on cities, towns and suburb. Publications Office of the European Union, Luxembourg (2016)

Eurostat: Methodological manual on city statistics 2017 edition. Publications Office of the European Union, Luxembourg (2018)

Eurostat: Methodological manual on territorial typologies 2018 edition, Publications Office of the European Union, Luxembourg (2019)

Eurostat: Sustainable development in the European Union. Monitoring report on progress towards the SDGs in an EU context 2021 edition. Publications Office of the European Union, Luxembourg (2021)

Istat: Forme, livelli e dinamiche dell'urbanizzazione in Italia. Istat, Roma (2107)

Mega, V., Pedersen, J.: Urban Sustainability Indicators. The European Foundation for the Improvement of Living and Working Conditions (Eurofound), Dublin (1998)

Minx, J., Creutzig, F., Ziegler, T., Owen, A.: Developing a pragmatic approach to assess urban metabolism in Europe? Final Report to the European Environment Agency. Project Reference EEA/NSV/09/001 (2011)

OECD: Redefining "urban": A new way to measure metropolitan areas. Paris (2012)

PBL-Netherlands Environmental Assessment Agency: Cities in Europe. Facts and figures on cities and urban areas. PBL Publishers, The Hague (2016)

Territorial Dynamics Emerged During the Festival "Terra2050 Credenziali per Il Nostro Futuro"

Maria Laura Pappalardo[✉]

Università degli Studi di Verona, Via San Francesco, 22, 37129 Verona, Italy
`maria.pappalardo@univr.it`

Abstract. On the occasion of the Festival "Terra2050 Credenziali per il Nostro Futuro", held in Verona and Mantova from October 13 to 24, 2021, academics from different disciplines have discussed the meaning of sustainability and, above all, how it is possible to combine this objective with both environmental and social but also economic development.

The debate saw on stage speakers from the world of business, medical personnel and "ordinary people", all united by the desire to succeed in drawing up "Credentials for Our Future" that aim to promote a new ethic.

Keywords: Sustainability · Ethics · Future

1 The Reason for the Festival "Terra2050 Credenziali per Il Nostro Futuro"

In order to introduce our topic of reflection, it is useful to reread the metaphor extracted from an old story by R. D. Bradbury that narrates the adventures of a hunter who is faced with an unusual safari offered by a travel agency of the future. Bradbury begins his story by describing the agency and its advertising techniques. The sign at the entrance reads, "… Hunt a dinosaur just before it dies a natural death. Our time machine will take you to the past".

What hunter-tourists need to do is very simple: listen to the guide, don't collect objects from the past era, notify the guide of any news, and don't leave a metal path that avoids visitor contact with the prehistoric ecosystem.

Unfortunately, the character in the story is rather clumsy, especially when he sees the Jurassic giant: in fact, he doesn't kill the dinosaur, he flees in fear, stumbles and exits the metal path. Nevertheless, he decides not to tell anyone, not even the guide, about what happened. And, in doing so, he violates one of the deliveries of that very particular trip. When the protagonist of the story returns to his own time, he not only notices a slight change in the writing announcing the trip, but also that the agency is no longer run by a young lady but by an almost human reptile, and he hears that the television announces the rise to world government of a tyrant who would dominate the planet with a very fierce policy. When he gets home, he realizes that the house is no longer the same but looks like a prison. Obviously, he doesn't understand anything anymore. When he wakes up and goes to sleep, he discovers that he has a Jurassic butterfly attached to the sole of his shoe. It is then clear to him that during his crazy escape he had brought it back to his era, and that the butterfly, with its absence, has changed the evolution of the world.

Beyond the metaphor, one of the purposes of the Festival "Terra2050 Credenziali per il Nostro Futuro" was to analyze from different points of view what is the "weight" exerted by each of us, and especially by the actions of each of us, on the Earth. In order to understand our "weight", we had to proceed with our reflections, remembering Renzo Piano's statement: "An ugly book you can't read; an ugly music you can't listen to; but the ugly building we have in front of us, we have to see it", because this is the only way to start some reflections on modern society. This, with its fashions and models, has taught us that we must abandon the idea that there is only one definition of *beauty* and that we can know, through fixed canons, what *beauty* is; even the *ugly* has its own role, that of making people feel pain, of being a denunciation of the world, a representation of what is different.

Brodel remembers: "The ugly has always been considered as the shadow of the beautiful, as its evil twin brother; therefore substantially, at the beginning of our civilization, the ugly has the analogous characteristic to that of the false or to that of the moral evil, that is to say, one wants to deny its positive existence. In Greek philosophy, especially from Plato to Plotinus, who most theorized these concepts, the ugly is presented precisely in the form of non-being. This, however, is the simplest form of expressing things. When, in the tradition that begins with Plotinus and arrives at the Italian Renaissance, ugliness is considered, it is always linked to the idea of a threat, of something that beauty manages to tame, but not completely. The ugly is the emergence, so to speak, of chaos in the order".

However, in order to clarify, it is necessary to remember how the concept of *beauty* depends on the various cultures and how in each of them, but especially in ours, *beauty* is the result of a series of stratifications only by virtue of which it is possible to define it, using and connecting the main variants, the answers that have been provided over time. The term *beauty* derives from Latin *bellus*, diminutive from the root *duenulus bonulus*, something *good in small*. The concept of *beauty* is united, therefore, in our as in many other cultures, to that of *good*. Also in Greece, for example, in ancient times the term *kalós - beauty* was often found connected to the term *good*. In modern Greek, *kalós* no longer means *beauty*, but *good*. In Western knowledge since the Pythagorean School, in the Magna Graecia of the sixth-fourth century BC, the concept of *beauty* was specified and entered in relation to the concept of *true* as *well* as *good*, coming to constitute a triad in which values are accumulated (*the beauty, the good, the true*) that have as characteristic the *measure*. With the passing of the centuries, however, we realize that beauty often does not coincide with order and regularity, that there is *something extra* that is neither calculable nor measurable, something that contributes to establishing what *beauty* is, something that is added to normality: there is no longer a shared criterion to which to appeal. One must have taste to be able to appreciate *beauty*.

As far as the *ugly* is concerned, it has always been considered, as we mentioned earlier, the shadow of the *beauty*, the false, the moral evil, which nevertheless exists in nature. In nature, in fact, we have always been aware that there is also the formless, the deformed, the *ugly but necessary*. From these premises comes the conviction, as scholars, that every landscape includes something sensitive (from the colors of cultivated fields, to the shapes and lines of urban architecture): consequently, the knowledge of the landscape itself becomes, for the simple spectator as well as for the careful observer, a kind of detonator of higher emotional charges, which can lead to the sensitive beauty and, provoking an aesthetic sensation, allow to appropriate the intelligible beauty. In the landscapes, however, in different and complex forms and ways, the *ugly* and the *beauty*, the *bad* and the *good*, the *useful* and the *useless* coexist, giving life to a composite and complex reality.

2 The Landscape and Its Issues

At the center of the research activities of urban planners, architects, historians, economists, sociologists, anthropologists - and always of geographers - the landscape is now recognized as a primary collective good and the foundation of virtuous processes of construction of the common good. During the various opportunities for debate offered by the Festival, it was stressed that the relationship between the landscape and the population is not completed in the influence played by the dynamics belonging to the society in the construction of landscapes: while the population deeply marks the landscape, at the same time the landscape suggests the population by inspiring emotions and feelings, spurring the composition of meanings and values, thus becoming an important element of the quality of life of the populations themselves. It can therefore be said that this is a relationship of reciprocity, indeed, of circularity. Man and society behave towards the territory in which they live in two ways: as actors who transform, in an ecological sense, the environment in which they live, giving it the mark of their action, and as spectators who know how to look at and understand the meaning of their work on the territory. In this sense, the landscape becomes *the interface between doing and seeing what is done*, the roles of actor (the one who builds) and spectator (the one who observes) cannot be separated since one observes what one builds, but, equally, one builds on the basis of what one observes and how one observes it.

Given these premises, during the days of the Festival it was more urgent than ever to impose new interpretative keys and methods for the identification of more innovative forms of *governance*, oriented to a multi-sectoral approach not only integrated but also aware of the importance of involving multiple territorial actors in the construction of their own landscape project. Among them, those paths related to sustainable development, which have highlighted complex interweaving of interactions between economic and social problems of local systems and the needs of ecological and environmental protection, acquire an added value.

Today we speak of megacities, metropolises and mega-cities: a series of terms with different, nuanced and often controversial definitions. In any case, these are "poles", with varying densities of construction, which expand their influence on the surrounding space, a space that is sometimes very large.

Faced with such a defined reality, both scientific-cultural and political-normative attitudes are increasingly directed towards the recognition of new territories, in an attempt to find different forms of management focused on the enhancement of ordinary landscapes.

From the European Landscape Convention onwards, the idea of the necessary attention and care of everyday places has spread, as well as of extraordinary landscapes, without natural or cultural exceptionalities, but significant for the communities that live in them. The new landscape project, therefore, must start from here: "From the awareness that in these landscape realities material and immaterial values are sedimented, meanings that contribute to the vitality of the places that are declined in them. Projects that specifically address the redevelopment of, for example, peri-urban landscapes: marginal open spaces, the result of city-country hybridism, or abandoned areas, uncultivated land that stands between land used for agriculture.

These spaces are only apparently *empty* and meaningless, in which in truth it is worthwhile to intervene since they show an already existing possibility to experience *another temporality* linked to the rhythms of nature and the body, which in the world of technology remains confined almost only in the space of nature often close to our homes. Only by virtue of the formation of a more innovative culture of the landscape, they will be able to transform themselves into territories of social association in which to find the condition of urbanity and rurality today totally confused, and, in so doing, proceed to a complex and complete valorization of the landscape.

3   Some Key Words

Key words that emerged in this Festival were: sustainability, preservation, protection, enhancement and management of the resources of our planet.

In particular, active protection, with more subjects involved, not plastered landscapes and museums, but landscapes seen in evolution and transformation, not only constraints, but participatory and shared projects.

During the Festival, the city (which grows and transforms) was the focus of specific analyses, and to most people it seems to stagnate outside of creativity, increasingly opposed, by ecological insurgencies, to a *beautiful nature* (to be protected) and to a *historical center* (to be preserved). Often, in the approach to the reading and interpretation of the urban center, there is a tendency to constitute categories in which value attributions such as *ancient, historical, to be preserved, emerging*, referring both to the built heritage and to the urban fabric and form, stand out. In this process of separation, the characters of degradation, the phenomena that we consider negative, are *left out* (and many times outside of research and survey), the peripheral parts of the city, as if there was a possibility of extracting (or perhaps *abstracting*) a protected area or forgetting the contemporary urban landscape that incessantly reproduces and regenerates itself with unimaginable quantities and variations, often constituting the background and the budding horizon in which Architecture stands out, instead, today presents new ways to re-propose an urban space for living.

Too many times on our planet we operate only with the desire to typify or classify the landscape to find areas to preserve and protect, and we do not stop to reflect on the fact that now are dissolving all the margins and boundaries, both historical and natural, which allowed to perceive the landscape in the contrast of its diversity.

Before there was a landscape outside the walls and a landscape inside the walls, there was the city and the countryside, the hill and the forest.

Now we live in a planet where the city is continuous, incessant, where it is increasingly rare to find places in which it is possible to interpolate building facts and infrastructural interventions. Unfortunately, however, it is often an incomplete, indeterminate, *hybrid* urban landscape, whose identity must be rediscovered above all through the study of all those *negative phenomena* that make it appear in a state of progressive degradation and, therefore, make us fear the irreversibility of the phenomenon.

Landscape is culture but also historical memory: an inseparable alchemy that offers us wonderful beauties on the planet but also, unfortunately, monstrously ugly works.

During the debates that took place from 13 to 24 October 2021 in Verona and Mantua, we also discussed another term, neither *beauty* nor *ugly*, but which certainly makes *useful* even what is too often considered useless: *sustainable*, which can be sustained. This is a concept that has unfortunately become alien to man, the only creature on the planet that does not act according to nature, that does not follow those invisible laws that govern the world so that a balance is maintained between living beings and the surrounding environment.

We are all aware of the continuous destruction carried out in the name of the *beauty*, the *just*, the *useful* but not the *necessary*; every day we assist to acts of fury against the Earth, apparently unaware of the fact that resources are not inexhaustible.

While, in fact, men are growing, both in number and in their needs, and the traditional primary energy sources are always the same and are dramatically thinning, the landscape bears the marks of this irresponsible comportment, whether they are represented by landslides, pollution of different kinds of air, water and soil and, more generally, by the different and complex stresses that our societies produce on Earth.

From these statements, during the three days of the Scientific Conference: "In search of balance: towards a Sustainable Development" and the two days of the International Scientific Conference: "I don't even know where Canada is: Cartography yesterday, today and tomorrow", came the confirmation of the usefulness of the useless, not only of all those knowledges whose essential value is completely free from any utilitarian purpose, but also and above all of geography and cartography that, for their nature free and disinterested, have a fundamental role in the safeguard of our planet.

It has been written, "Is life widespread? Or is the Earth special, not only for us who live there, but for the cosmos in the broadest sense? As long as we know only one biosphere, our own, we cannot rule out that it is unique: complex life could be the result of a chain of events so unlikely to have taken place only once in the entire observable universe, on the planet that became ours. On the other hand, life could be widespread and be developed on every Earth-like planet (and perhaps in many other cosmic environments). We still know too little about how life began and evolves to decide between these two extreme possibilities…".

So it is not just a question of answering the question of how much oil is left to exploit or whether we will be able to stop global warming, whether *artistic beauty* will win or whether *ugliness* will have its revenge! When considering such a vast problem it is easy to feel confused, unable to make any changes. However, we must avoid reacting in this way: all crises, including the one our planet is experiencing and to which the signs on the landscape bear witness, can only be resolved if individuals take responsibility, at least in part.

Only by educating ourselves and others, by doing our part to reduce degradation and pollution, by valuing the useful, geographically speaking, can we make a difference.

Geography, with its method of analyzing reality, and cartography as a tool of fundamental representation of reality itself, can help us heal from that partial blindness in the way we consider the effect of our decisions on the natural world that is a major obstacle to efforts being made to formulate sensible responses to the threats currently facing the environment.

Studying the landscape and *reading* the objects present in it not in terms of mere *beauty* and *usefulness* economically understood, but as the theater of human action, are a vital condition to find the right balance in the relationship between man and his living environment, to reverse the now widespread trend that sees us decided only to be indecisive, resolute only to be irresolute, motionless in movements, firm in instability, omnipotent in the determination to be powerless.

Geography and cartography have always had a close link: Dematteis writes, for example, that maps are the tools most commonly associated with geography, simple but powerful objects that allow us to represent and visualize the different parts of the world.

4   Who Reads the Law?

During the Festival, the awareness emerged from many speakers and the audience that the landscape policies implemented according to the directives of the Code of Cultural Heritage and Landscape and the European Landscape Convention, as well as regional urban planning laws, to be effective require the construction of widespread knowledge. In order to guarantee the drafting of plans and programs for the transformation and protection of the territory that are also shared by local communities, it is necessary to develop an adequate knowledge of the landscape as a material and cultural expression of the urban and rural territory, a landscape heritage with strong identity connotations. Interesting were, in particular, the references to environmental law, recently conceived if compared to other areas and subjects. So recent as not to have found an express reference in our Constitution that directly protects the landscape and only mediately, through the protection of health, the right to live in a healthy environment. Moreover, it has been underlined how the environmental law revolves around fundamental principles whose elaboration, progressive in time, demonstrates the evolution of knowledge, of awareness and consequently of the protection ensured by new laws. These are, in particular, the principles: "polluter pays", prevention and precaution. Since a couple of years, however, it has been observed, as the new challenges such as the implementation of the Circular Economy, a new development model that helps to combat and resist the climate crisis, as well as the incentive towards renewable energy sources, require a reflection to try to identify, at this precise moment in history, the priorities.

Other reflections came from the study of the Intergovernmental Panel on Climate Change in the 5th Report published in October 2013, prodromal to what was then debated in the Paris Conference of 2015. "Global warming is unequivocal and, since the ´50s, many of the phenomena observed in recent decades have not occurred for hundreds, sometimes thousands, of years. The atmosphere and oceans have warmed, the supply of snow and glaciers has decreased, sea levels have risen, and the concentration of greenhouse gases has increased. The human influence on climate is evident. This is evidenced by the increase in the concentration of greenhouse gases and radiation in the atmosphere, the increase in warming and the more intense climate variability. (…). It is highly likely that human influence has been the dominant cause of the global warming that has been taking place since the middle of the last century. (…). Continued emissions of greenhouse gases will cause further increases in temperatures and changes in all average atmospheric conditions. Limiting climate change will require a substantial and sustained reduction in greenhouse gas emissions". These findings have led to an increasing awareness, even if this has not happened for many countries, of the need to put in place effective structural measures to regulate polluting emissions that cause the increase in temperature and, consequently, climate change. Endless are the signals that, at least in the last decades, our planet is trying to send us through the occurrence of countless tragedies announced that are affecting every corner of it with the tropicalization of temperate areas, the melting of ice with its rise in sea levels and oceans, long periods of drought with famine and migration, extreme events that cause death and destruction, in short, our Earth system can no longer withstand the constant increase in global temperature.

As geographers we have also underlined, in the various occasions of discussion during the Festival, how the equal relationship existing between man and environment has been destroyed, with man, society, and economic-political choices that have decided to be able to manipulate what surrounds us at will, so much so as to destroy, hopefully not definitively, that natural greenhouse effect that guaranteed the perfect thermal balance on Earth. It is rightly said that planet Earth has been given to us on loan by our children and grandchildren and we should give it back to them at least in the conditions in which we received it. Instead, we are destroying it with pollution, the main cause of what everyone now calls climate change.

Several contributions have shown that this issue has been debated for a long time, even if it is not the case to go back to the Swedish scientist Arrhenius supporter, already in 1896, that the thermal increase was due to the higher concentration of carbon dioxide due to the use of fossil fuels, even if, generally, it was believed that human activities had a minimal impact compared to natural ones.

It is now well known that mankind has an increasing influence on the climate and on the temperature of the earth due to the massive use of fossil fuels responsible for additional quantities of greenhouse gases that are added to those already present in the atmosphere, implementing the consequences. Just think of carbon dioxide, a greenhouse gas produced mainly by human activity and responsible for 63% of global warming. Its concentration in the atmosphere currently exceeds by more than 50% the level recorded at the beginning of the industrial era. Other greenhouse gases must be added to it, which, although released in smaller quantities, also have the power to generate heat, so that, for example, methane is responsible for over 19% of anthropogenic global warming and nitrogen oxide for 6/7%.

Often, when reasoning with traditional models, it is thought that the consideration of the environmental variable presupposes non-development. And yet, if environmental sustainability criteria were to be introduced into all planning, policies and projects, it would be possible to achieve lasting benefits over time; unfortunately, however, the projections of human actions into the future are often not taken into account.

We all remember how development models that have to do with environmental sustainability presuppose the possibility of satisfying the needs of present generations while taking future generations into account. This paradigm today, although respected and considered constitutional in many countries of the world, is difficult to apply.

Certainly, changes in ecosystems are inevitable and in accordance with the progress of societies and may or may not produce harmful situations. The problem arises from the awareness that both human groups, in order to survive, have always adapted nature to their needs, and that development cannot and must not be stopped.

Therefore, it is necessary as soon as possible to achieve a true intersectoral integration between nature conservation policies, development plans that only concern the economic aspect and social projects. Certainly, protecting the environment or using non-energy-consuming technologies may imply limiting some economic areas in the short term. But it is now necessary to become aware of the fact that these restrictions will bring real, long-term benefits in the future. Some examples presented in the discussions that took place during the Festival can clarify what has now been stated. If, at the time of planning new urbanization, we had the foresight to consider the environmental variable, we could avoid the classic problems of flooding produced by poor land use and land modification. The same happens with pollution caused by waste that does not have adequate space where to be placed. Undoubtedly, however, the most complex aspect lies in finding solutions to contamination and threats of long standing. Is it possible to reintroduce fauna that, due to habitat modifications, has moved away? How much does it cost to recover a river that, due to human action, has its bed covered with heavy metals? And how does this contaminate river populations?

5 The Challenges Ahead

The people of the twenty-first century should dare, should accept the challenge to "think" about a sustainable use of ecosystems.

At this point, the question has also arisen as to what are the fundamental management tools.

We must, in this sense, recall environmental planning and regulation, the establishment of a specific regulatory structure, environmental education, the action of the mass media as an engine for social change and Environmental Balance Studies.

In order to achieve real and lasting development, a series of principles must be accepted that, at the very moment they are incorporated into sectoral policies, favor economic growth in harmony with the territory. We cite first of all the "precautionary principle" (ie: the absence of scientific certainty of absolute character on some phenomena is not an excuse to postpone measures that are considered good in themselves) and the "principle of prevention" (according to which it is necessary to try to prevent environmental risks before the damage itself is produced), although we cannot keep silent the "principle of the polluter-payer" (largely, However, we cannot omit the "principle of the polluter-payer" (now largely insufficient), that of "intergenerational equity" and that of "internalizing the environmental cost" (when developing products or setting projects in motion, the use of nature must be included as a cost factor and eco-nomically advantageous prices must be calculated). These criteria, elaborated for most of the 20th century and included in International Law, have given life to Environmental Law, promoting a management of resources that, while aiming at development, does not forget conservation.

From what has been said, therefore, it is indispensable both to spread environmental quality certifications, given by the competent authorities to companies or activities that comply with environmental regulations (these certifications, in fact, are interesting management tools that require a study of activities from the environmental point of view), and to implement the protection of natural and cultural heritage. Protection of fauna and flora, protection, preservation and enhancement of cultural heritage, but also, for example, creation of protected areas and implementation of policies oriented to the sustainable use of environmental and cultural heritage.

The planet is a territorial reality occupied by human beings (but not exclusively by human beings!) and these, in order to contribute to their own survival and develop a local, regional, national or international economic system, must foresee the consequences of their actions on the natural balance, but they cannot be limited in their development programs by prejudices of any kind.

The concept of ecological impact is therefore certainly very important, but it cannot - if not in theory - be considered "the measure of things". It is enough to reread the past, more or less recent, to be convinced, in fact, that there have been developments of different types and forms in different eras.

It is therefore necessary to work for the well-being of the Earth and not on the Earth for our own well-being, as often happens when we talk about sustainability, especially by some who claim, in the last instance, that the Environmental Impact Assessment is the *conditio sine qua non* to review any economic activity. The Environmental Impact study is immanent and not transcendent, and as such must be carried out, in a prudent way or, better, according to us geographers, it must be replaced by the Environmental Balance.

Another interesting aspect was pointed out by the speakers: considering a community as the creator of the construction of its territory leads to reconstruct the objects and the plots that have supported them with the support of historical memory. This reconstruction of the object teaches us to remake the object itself and it is therefore possible to set the cognitive and interpretative premises to intervene and safeguard, enhance and requalify the set of historical stratifications of the territory.

We must become aware that every humanized landscape that has been created over the centuries on Earth, seen through the mechanism of perception and collective memory, becomes a spatial code, a way of organizing space and experiencing it on a daily basis. The landscape of the barracks (in some of them some events of the Festival took place), of the military use of the city space (even if for the past) of Verona, for example, is culture, both in its diachronic dimensions and in the synchronic interpretation given by the community living these places. And since the ability of modern society to ensure a future for the past represents the possibility of ensuring the future of mankind through and by means of a proper relationship with nature… from this correlation arises the need to ensure, through proper management of the signs of the past, a link of continuity to the relationships that can be established between human groups, productive activities and resources, achieving an adequate relationship with the environment because, by intervening on the past of reality, positive effects can be exerted on the process of economic growth and production on the same cultural promotion.

As far as the meaning attributable to these humanized landscapes is concerned, for "their tomorrow", it is necessary that a policy of management of the present resources be concretized in increasingly defined forms, moving towards a new territorial fabric in which the functions and possibilities of use and modification are organically coordinated. It will then be truly possible to implement the real preservation and enhancement of these cultural assets, restoring, where it has gradually deteriorated over time, their value of testimony and documentation and their potential to communicate the semiotic value of the ancient, which in this way will continue to convey valid messages. What is hoped for is, in other words, the implementation of an "active" preservation of these landscapes, compatible, therefore, with the socio-economic reality, with the production needs and with the problems of use of the local environment.

Hence the conviction that the preservation of the signs that so characterize these realities, through a functional reuse of environmental components, can concretely take place by assigning to these signs a collective value, of fruition of goods and services economically convenient.

And this is because the today of a territory is not only the result of its intrinsic prerogatives, which alone could place it, compared to the more general system in which it is inserted, in a dominant or marginal category, but also, and in many cases unfortunately, of the situations produced by neighboring realities.

Already in 1992, the European Union approved the Fifth Environmental Action Plan in which it hoped for a change in the behavior patterns of society by promoting the participation of all sectors and strengthening the spirit of co-responsibility that extended to the public administration, businesses and the community. There was a definite awareness of how environmental protection should be integrated into the definition and implementation of the various Community policies, not only for the good of the environment but also for the good and progress of all sectors. But in order to achieve this, it is necessary to overcome the individualistic vision of the solution of problems in order to move towards the logic proposed by a vision of these entities as united in a systemic region.

Indeed, at this point we must make a general consideration. The different humanized landscapes that characterize the Earth contribute to compose a complex reality that, although articulated in multiple and different parts, must be felt, understood and managed, especially when talking about such complex issues as sustainability, as a single entity (a geographic region) in which the coherence of interventions make it a "functional unit".

Interpreting these landscapes as a unitary systemic reality, even if articulated in territorial ambits diversified by the qualitative-quantitative variety of the components, means planning a geographical reality that depends on the intrinsic characteristics of the places, rather than on the typological characteristics and expansive force of the "external" territories, which have often annulled the birth of a true interactive and synergic relationship between the various parts of the Earth as a whole. Only by cancelling the particularisms will it be possible to concert a true sustainable development of our Planet.

6 In Search of Sustainable Development

In the spirit of the search for a sustainable development, it is necessary to start the realization of a system in which the different components, whether they are the protection of the environment and local communities, the organization of the existing and potential eco-nomic realities, the careful and accurate information to all actors, the solidarity between the different realities present in different territories, the local culture, are all linked together in a logic of globalization of acting in and for the protection of the environmental, human and economic heritage present and future.

Convinced that a true sustainable development can be considered as such when the rate of regeneration of resources exploited by economic activity is higher than the rate of exploitation of the resources of the same activity, it is necessary that not only legislators but also each of us in our choices invest in projects that enhance, at an increasing rate, the regeneration of resources that are used: only in this way our future can be defined as sustainable and useful both for those who inhabit the Earth today, but also, and above all, for future generations.

A proposal to solve the problems outlined in the previous pages is based on the concepts of integration and tradition. By integration we mean the phenomenon that aims to insert the individual "strong realities" of a landscape in integrated and coherent territorial entities, not identified only on the homogeneity of some elements taken as a parameter but, rather, by that complementarity that generates gravitation and weaves strong relational links. These entities need their own global planning, able to coordinate and aggregate even those composite and fragmented realities that combine different characteristics. The concept of tradition is linked, instead, to the transformations carried out or that can be carried out: in accordance with tradition is what is part of the typicality of the landscape without generating imbalances and contradictions.

These realities, weaker because of the location, infrastructure or production situations, despite their consolidated population are now subject to an exodus of the population that has "desertified", so as to make problematic that quali-quantitative demographic recovery on which every harmonious evolution depends. In this way, choices have sometimes been made that have trivialized the typology of the places, making entire realities become problem-areas, not having used the potential (sometimes hidden, but nevertheless existing) of the single places and not having therefore given life to those innovations of orientation from which only the possibility of development springs. However, it is necessary to remember that no place is useless and absolutely devoid of potential; if nothing else, it has the function of intermediary and complement. The depressed areas of our planet are in fact such either because they have been inert or because they have been burdened by activities in contrast with each other and/or with the potential of the place. These are spaces that need to see their main spaces remodeled in a logic of the whole, including through the promotion of greater sensitivity.

And so it is that the implementation of a practice that has been indicated many times, but as many times ignored, comes back with force. It is necessary to carry out a complete analysis of the single components of the existing situation; an evaluation of the real potentialities and predispositions of the different territorial realities; a programmatic choice with a consequent verification of the positivities and negativities that it generates in the territories and in their components. In order to do this, interventions that innovate the pre-existing realities of the physical and human environment should not be feared; therefore, artificial choices should not be feared, that is, those that can change the previous evolutionary model. Artificial choices, i.e., those that can change the previous evolutionary model, should not be feared. However, artificial interventions should be avoided, i.e., those that are in contrast with the equilibrium of and between environments that have been so painstakingly achieved.

Much more could be said about the debates on the future of our Planet, but before concluding, we would like to underline how the problem of social sustainability has also considerably occupied the time and reflections of the Festival. In particular, the great wound of the lack of inclusion of women in many countries with the disfigurement of their fundamental rights.

Before introducing ourselves into the theme, let's consider the meaning of the term "inclusion", a word that underlies our entire reflection, offering us a vision of life, of man, of politics. A word, a process that makes us human, witnesses to our fragility and our infinite potential. "Inclusion" requires the denunciation of violence, hypocrisy, manipulation, with respect for the honest analysis of the facts. To do this requires commitment, study, confrontation, courage, training, a culture that is not selfish, not good. The process of "inclusion" requires hope, it requires building "communication bridges" between different cultures, it requires courage, to face situations... together, like the person who, for a stretch of road, takes off the shoes of his own culture and learns to walk barefoot, to put himself in the shoes of others! It is not possible to talk about Earth, Environment, Climate, without taking into account the serious problem of violence against women, children, immigrants, the elderly, often alone and abandoned!

The Earth, "a flowerbed that makes us so fierce", remembering what the Supreme Poet wrote in Canto XXII at verse 151 of Paradise, too watered with blood: the Mediterranean Sea, a cemetery of faces and names that cry out their pain and call us to our responsibility.

With the thought of women and children and all those who, in various ways, suffer violence, not excluding the many, too many, absurd victims at work - during the days of the Festival we made our own the words of the Collaborators of Gino Strada: "... We will continue to be extremely realistic but, at the same time, to cultivate utopia".

During the seminars it was recalled that democracy is not won once and for all, it requires a shared culture and ethics, a condition that the current moral degradation is putting at serious risk. Enclosing oneself in a fortress to defend oneself from whom? From the last ones, from the poor people who escape from wars, persecution, hunger, despair? We must have the strength, all indiscriminately, in every facet of our lives, in all these segments of our existence we must always keep in mind, daily, our values and our ethics acting consequentially to it. And moreover, we must have the courage to use words in their correct sense - avoiding, for example, euphemisms, calling those who defend the small interests of their own garden to the detriment of the entire world population selfish and not frugal - because it is also from the correct use of language that change must start.

The new culture, or rather, the new ethic that is hoped for the years to come cannot be reduced to a series of urgent and partial responses to the problems that gradually arise regarding environmental degradation, the depletion of natural reserves, pollution, social problems, immigration, the difficult economic relations between the various parts of the planet,....

What is needed is a different outlook, a way of thinking, a policy, an educational program, and a lifestyle that gives shape to a resistance to the advance of the technocratic paradigm, opting, for example, for less polluting production systems, supporting non-consumer models of life, happiness and conviviality, when technology is primarily oriented towards solving the concrete problems of others, with the commitment to help them live with more dignity and less suffering, when the search for beauty and its contemplation manage to overcome the power of objectification.

Humanity has changed profoundly and the accumulation of continuous novelties consecrates a fleetingness that drags us to the surface in a single direction. It becomes difficult to stop and recover the depth of life. If architecture reflects the spirit of an epoch, megastructures and mass-produced houses express the spirit of globalized technology, in which the permanent novelty of products is combined with a heavy boredom, conceptual art can be a means through which to work for a sustainable development by recovering the values and the great aims destroyed by a megalomaniac unrest that no longer recognizes nature either as a valid norm or as a living refuge.

Authentic sustainable development has a moral character and presupposes full respect for the human person, but must pay attention to the natural world, taking into account the nature of each being and its mutual connection in an ordered system. Environmental problems are also ethically and spiritually rooted, and it is therefore unthinkable to look for solutions not only in technology, but also in a change in human beings, because otherwise only the symptoms will be addressed.

If therefore, as Valéry wrote back in 1931: "The future is no longer what it once was", we need a cultural revolution that makes us all aware that for a sustainable development of society and the territory, we need to embrace the concept of "common good", in other words, to practice a far-sighted vision, to invest in the future, to care for the community, to pay attention to every being of creation, subordinating to it every single interest that is in contrast with it.

Paraphrasing Popper, the future is very open, and depends on us, on all of us, on what you and I and all other men do and will do, today, tomorrow and the day after tomorrow. And what we do and will do depends in turn on our thinking and our desires, our hopes and our fears. It depends on how we see the world and how we evaluate the possibilities of the future that are open to us. A future that, as Einstein said, "comes so soon" and that: "belongs to those who believe in the beauty of their dreams".

Unfortunately, we are indebted to the future to pay the debts of the past, but the past cannot be recreated. We can pretend, we can delude ourselves, but what is over does not come back, does not come back anymore. Therefore, we must get used to thinking about the future according to an approach that takes into account the complexity of the landscape and the need for interdisciplinary interventions, as our colleagues present at the Festival have well pointed out.

The task of the geographer is to work attentively, lucidly and punctually on difference: it is necessary to work on the recognition of the individual parts and in this sense to work rigorously to bring aesthetics back to being the daughter of ethics. In this "*media potpourri*" of images, the temptation of those who work in and on the landscape, whether

they be architects, landscape architects, or jurists, often slips into self-celebratory nota-
tion, they make themselves strong with great special effects, espousing the demand
for marketing of sad "designer flavor", which in its self-referential language becomes
franchising: we therefore lose the narrative of a specificity, of a story attentive to the
difference of a territory, in its individual stories described.

It is important to learn to recognize, protect and enhance the richness of the differences present in each landscape, which must become common cultural capital, in the
construction of a new reality, even if we now live in a liquid society in which everything
changes quickly, consumption becomes the engine, and the idea of time and space play
a role of continuous mobility.

Much more could be written to recount the important results achieved during the
days of the Terra2050 festival, but we prefer to end our reflections at this point, making
an appointment for next fall when the second edition will be held, and quoting this
reflection by Calvino: "Cities are like dreams: everything imaginable can be dreamed,
but even the most unexpected dream is a rebus that hides a desire or its reverse, a fear.
Cities like dreams are built of *desires* and *fears*".

References

1. Cera, M.: Le mappe raccontano il mondo. Storia della cartografia dalle prime esplorazioni al
   mondo d'oggi, Cairo Ed., Milano (2021)
2. Davico, L., Sviluppo sostenibile. Le dimensioni sociali, Carocci, Roma (2004)
3. De Masi, D.: L'età dell'erranza, Il turismo del prossimo decennio, Marsilio Ed., Venezia
   (2018)
4. De Vecchis, G.: Da problema a risorsa: sostenibilità della Montagna Italiana, Kappa, Roma
   (1996)
5. Documenti CNEL: Sviluppo Ambiente Occupazione, CNEL, Roma (1994)
6. Festival Terra2050 Credenziali per il Nostro Futuro. Libro dell'evento: Verona e Mantova
   13–24 ottobre 2021, Testi di Pappalardo M.L., Global Map, Verona, 2022
7. Marshal, T.: Il potere delle mappe. Le dieci aree cruciali per il futuro del nostro pianta,
   Garzanti, Milano (2021)
8. Pappalardo, M.L.: Itinerari di odissee veronesi, Global Map, Verona (2021)
9. Pappalardo, M.L.: L'oasi dei bisbigli, il deserto delle grida, Global Map, Verona (2021)
10. Pappalardo, M.L.: Mappe e carte antiche, Global Map, Verona (2021)
11. Settis, S.: Il paesaggio come bene comune, La scuola di Pitagora Ed., Napoli (2013)
12. Sviluppo locale: territorio, attori, progetti. Confronti internazionali, a cura di Faggi P.,
    Dipartimento di Geografia, Università di Padova, Padova (1986)
13. Tétart, F.: Il mondo nel 2021 in 200 mappe: atlante di geopolitica, LEG, Gorizia (2020)

The Role of the Institutional Dimension in Defining Sustainable Development Policies in Italy

Massimiliano Bencardino[1](✉) [iD], Antonio Nesticò[2] [iD], Vincenzo Esposito[1],
and Luigi Valanzano[3] [iD]

[1] Territorial Development Observatory (OST), Department of Political and Communication
Sciences, University of Salerno, 84084 Fisciano, SA, Italy
mbencardino@unisa.it

[2] Territorial Development Observatory (OST), Department of Civil Engineering, University of
Salerno, 84084 Fisciano, SA, Italy

[3] Department of Cultural Heritage Sciences, University of Salerno, 84084 Fisciano, SA, Italy

Abstract. By integrating the theoretical perspectives on local development and
those on sustainability, the study aims to overcome the traditional conception of
development as economic growth by expanding it to include other dimensions. In
particular, the study focuses its analysis on the institutional component of Sustainable Development. To this end, the component was explored through a set
of variables extracted from the open data offered by the Open Civitas platform
promoted by the Ministry of Economy and Finance and SOSE. The goodness of
correspondence of the variables is measured in the first case by the correlation
coefficient R^2, while in the second by the presence of high-high and low-low
clusters on the maps generated by the algorithm of Moran's I bivariate.

Keywords: Historical expenditure · Taxable income · Public policies · Cluster
analysis · Spatial correlation

1 Introduction

Recent studies highlight the peculiarity of the local dimension within the dynamics of
sustainable development [1, 2]. At the same time, the concept of sustainable development
itself has been enriched with new components of analysis, although that of the three
pillars (social, economic and environmental) remains dominant [3–6]. In fact, three main
approaches to the concept of sustainable development can be distinguished: a one-pillar
model; a three-pillar model [7]; a multi-pillar and inter-pillar model [8].

The social and economic aspects of development (three-pillar model) configure a
model that attempts to achieve social, economic and ecological goals in equal measure
[9, 10]. However, the social dimension remains the most conceptually elusive and vague

---

The contribution to this paper is the result of the joint work of the Authors, to which the paper has
to be attributed in equal parts.

© The Author(s), under exclusive license to Springer Nature Switzerland AG 2022
F. Calabrò et al. (Eds.): NMP 2022, LNNS 482, pp. 1243–1251, 2022.
https://doi.org/10.1007/978-3-031-06825-6_120

in the sustainable development discourse [11–13]. Difficulties in identifying exclusively social issues increase as there are significant over-positions and trade-offs between the pillars of sustainable development. These are particularly pronounced between economic and social pillars as many issues result relevant to both dimensions (e.g., employment and unemployment, equitable distribution of resources, etc.) [11]. With the aim of defining the social dimension of sustainable development, the concepts of social sustainability [12, 14–17] and sustainable social development [18] have been introduced. Several social dimensions emerge from this literature. Littig and Grießler Griessler identify quality of life, social justice, and social coherence as social dimensions of sustainability [7]. Chan and Lee identify social infrastructure, employment opportunities, urban landscape design, preservation of local characteristics, and satisfaction of psychological needs as factors that promote sustainability [16]. Cuthill classifies social capital, social infrastructure, social justice, and widespread governance (participatory democracy) as factors promoting social sustainability [17]. Dempsey et al. identify two categories of social sustainability: social equity and community sustainability [12].

Many subsequent contributions claim the need to consider other dimensions in addition to the three foundational dimensions. In particular, the importance of the political-institutional [19–21] and cultural [22–24] dimensions is recognized.

The consideration of the multidimensionality of sustainability has its roots in the capability approach as a reaction to the utilitarian model of development based on the availability of material resources [25, 26]. Material well-being is replaced by an idea of well-being, of human well-being, of the ability to transform available resources into achievements, goals, and results. Thus, in an evaluation exercise the focus should be on the processes - structural and personal conditions - that influence the individual's ability to choose. These conditions are conversion factors and include personal characteristics but also social and institutional characteristics [27, 28].

2   Methodology

This paper emphasizes the attention on the "institutional" dimension of sustainable development in a five-pillar model (Fig. 1), declining it through the function of public policies because they are able to influence the well-being of local communities. To this end, the component was explored through a set of variables extracted from the open data offered by the OpenCivitas platform [29]. OpenCivitas is a portal promoted by the Ministry of Economy and Finance and SOSE to ensure access to information on municipalities, provinces and metropolitan cities of the Italian Regions with ordinary statutes.

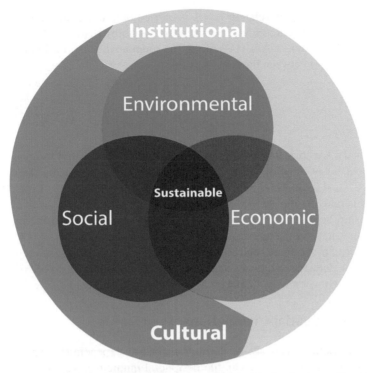

Fig. 1. Concept map of the analytical dimensions of sustainable development (Our elaboration).

In order to identify the "institutional" component in development policies on a local scale we identify a set of variables: the historical per capita spending (2017); the standard per capita spending (2017); the level of public services provided (as of 2017); and the taxable income for the purposes of additional IRPEF (three-year average 2015–2017).

The application of standard requirements to the amount of historic expenditure gives rise to standard expenditure. Historical expenditure is defined by the amount actually spent by the municipality in a year for the provision of services to citizens. Standard expenditure measures the financial requirements of an entity on the basis of territorial characteristics, the socio-demographic aspects of the resident population and the services offered. While the level of services provided measures the percentage difference in services offered by the municipality compared to the average of municipalities in the same population range [20]. The functions of the municipalities relate to general administration, roads and territory, waste management, social services and kindergartens, local police and, finally, public education. Income taxable for IRPEF surcharges is plausibly considered a measure of economic well-being, that explains several territorial dynamics [30–33]. In effect, all other things being equal, the lower a overall income produced would correspond to a lower tax base and therefore a lower tax revenue capable of financing local services. Hence, it follows that taxable incomes are capable of affecting the quantity and, indirectly, the quality of services offered by individual local governments.

The measurement of correlation levels is carried out, at the regional scale, on the basis of the linear regression model between the variables Historical Expenditure and Average Taxable Income - ATI (1) and, at the municipal scale, on Local Moran's I bivariate analysis (2) between the same variables:

$$ATI = a + b * HE \tag{1}$$

$$I_i = c * HE_i * \sum_j w_{ij} * ATI_j \tag{2}$$

where a, b and c are constants and w_{ij} are the elements of the spatial weights matrix.

The goodness of fit of the variables is measured in the former by the correlation coefficient R^2, and in the latter by the presence of High-High and Low-Low clusters on the maps generated by the algorithm. In essence, Moran's I bivariate analysis captures the relationship between the value of a variable at location x_j and the mean of the neighbouring values for another variable, i.e., its spatial lag $\sum w_{ij}*y_j$.

The choice of using this second estimator for analyses on a local scale is corroborated by the desire to identify clusters of spatial correlation, since a direct correlation between spending and income on a municipal scale appears less significant.

3   Linear and Spatial Correlations

The spatial distribution (Fig. 2) of the dimensions of expenditure (expressed in per capita terms) shows higher values in the peripheral municipalities, in correspondence of the Alpine crown and the Apennine arc and the smaller municipalities of Central Italy, in particular Umbria and Marche. In addition, both expenditure dimensions seem to decrease in intensity as one moves from the northern regions to the southern regions.

The difference between historical expenditure and standard expenditure can give different results, but it is not sufficient to evaluate the efficiency of an agency. A standard requirement difference from historical expenditure, therefore, is the result of both the efficiency of local services provision and the quality and quantity of services offered. Therefore, by analysing OpenCivitas data, if in large cities such as Milan, Turin, Rome, Florence, historical spending is greater than standard spending, the same is not true in the cities of Naples, Bari, Genoa and Reggio Calabria. The negative differential between historical expenditure and standard expenditure could depend on the lack of financial resources necessary for the provision of local services.

Similarly, Fig. 3 also shows an unequal distribution of the quantity of services provided by local government, with lower levels in the southern regions. The geography of taxable incomes (Fig. 4) marks a substantial unequal distribution among the various Italian regions. In fact, the mapping of taxable income shows higher values in the central-northern urban areas with respect to the size of spending. This is also linked to the capacity of the Public Administration to provide services. The per capita revenues of the Public Administrations in the Italian regions reflect the incomes and taxable bases that in the South are notoriously lower.

In order to verify the existence of significant relations between expenditure and levels of well-being, correlation algorithms are examined at the regional, provincial and municipal levels.

At the regional level, the correlation between each expenditure dimension (historical and standard expenditure in per capita terms) and average taxable income shows significant values. The Lombardy region was excluded from the calculation of the variables. This choice is a function of two observations relative to both the greater population, which in relative terms is greater than the other regions, and to the lower influence that public spending would cause on incomes, which are already conditioned by the existence of a stronger and more varied economic context. Therefore, net of Lombardy, the correlations show R2 values of 0.683 for standard expenditure-taxable income (Fig. 5) and 0.601 for historical expenditure-taxable income, respectively.

The Local Moran's bivariate analysis (Fig. 6) shows a good correlation with the high-high clusters in the central municipalities of the north and with the low-low clusters in large portions of the southern territory. Other areas show inverse correlation, highlighting how the interdependence of the variables is not always verified and that spending is not the only variable to be taken into consideration in this type of analysis. The same can also be seen when correlating standard per capita spending to average taxable income.

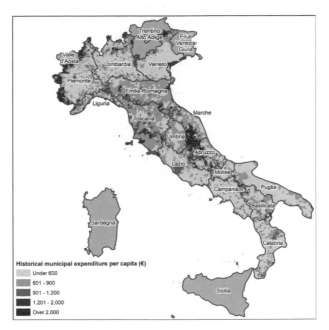

Fig. 2. Historical expenditure per capita (Our elaboration on OpenCivitas-MEF data).

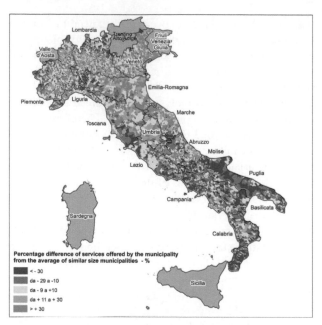

Fig. 3. Percentage difference of services offered by the municipality from the average of similar size municipalities (Our elaboration on OpenCivitas-MEF data).

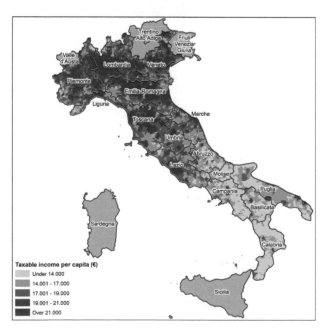

Fig. 4. Taxable income per capita (Our elaboration on OpenCivitas-MEF data).

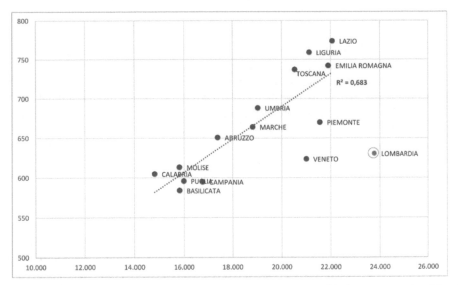

Fig. 5. Correlation between historical expenditure and average taxable income at regional level (Our elaboration on OpenCivitas-MEF data).

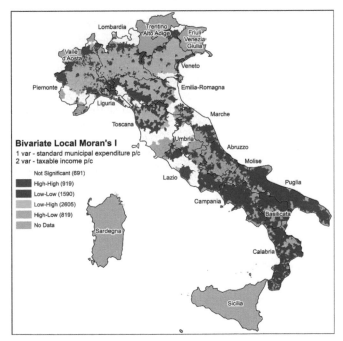

Fig. 6. Spatial correlation between the historical expenditure and average taxable income with local Moran's algorithm at municipal level (Our elaboration on OpenCivitas-MEF data).

4 Conclusions

The correlations show the existence of positive relationships between spending and income. These correlations are not, however, satisfactory for understanding the impact of municipal public spending on well-being. In fact, in areas such as the Lombardy Region, this correlation proves weaker. One could deduce here a more marginal role of public spending on local economic well-being, probably due to the existence of a more competitive regional productive context. Vice versa, the same correlation is strong in the peripheral areas of Northern Italy and in the southern regions. If, in the first case, the correlation is positive, on the other hand, it is negative in the southern areas where a lower value of expenditure corresponds to a lower value of taxable income. In the southern regions, the negative correlation seems to have an impact on the lower supply of services provided by the individual administrations, thereby inferring negative effects on the overall wellbeing of local populations.

Overall, the analytical municipal dimension does not appear to be the most suitable for representing correlations between spending and income. More relevant correlations can be demonstrated on a supra-communal scale. Likewise, spending does not seem to be the only variable to be taken into consideration in this type of analysis. This will be a direction for future research.

References

1. Moallemi, E.A., et al.: Achieving the sustainable development goals requires transdisciplinary innovation at the local scale. One Earth **3**(3), 300–313 (2020)
2. Szetey, K., Moallemi, E.A., Ashton, E., Butcher, M., Sprunt, B., Bryan, B.A.: Co-creating local socioeconomic pathways for achieving the sustainable development goals. Sustain. Sci. **16**(4), 1251–1268 (2021). https://doi.org/10.1007/s11625-021-00921-2
3. Purvis, B., Mao, Y., Robinson, D.: Three pillars of sustainability: in search of conceptual origins. Sustain. Sci. **14**(3), 681–695 (2019). https://doi.org/10.1007/s11625-018-0627-5
4. Alaimo, L.S., Maggino, F.: Sustainable development goals indicators at territorial level: conceptual and methodological issues—the Italian perspective. Soc. Indic. Res. **147**(2), 383–419 (2020)
5. Delli Paoli, A., Addeo, F., Mangone, E.: Sustainability and sustainable development goals (SDGs): from moral imperatives to indicators and indexes. A methodology for validating and assessing SDGs. In: Nocenzi, Mariella, Sannella, Alessandra (eds.) Perspectives for a New Social Theory of Sustainability, pp. 47–68. Springer, Cham (2020). https://doi.org/10.1007/978-3-030-33173-3_5
6. Delli, P.A., Addeo, F.: Assessing SDGs: a methodology to measure sustainability. Athens J. Soc. Sci. **6**(3), 229–250 (2019)
7. Littig, B., Grießler, G.E.: Social sustainability: a catchword between political pragmatism and social theory. Int. J. Sustain. Dev. **8**, 65–79 (2005)
8. Murphy, K.: The social pillar of sustainable development: a literature review and framework for policy analysis. Sustain. Sci. Pract. Pol. **8**(1), 15–29 (2012)
9. Wichaisri, S., Sopadang, A.: Trends and future directions in sustainable development. Sustain. Dev. **26**(1), 1–17 (2018)
10. Diaz-Sarachaga, J.M., Jato-Espino, D., Castro-Fresno, D.: Is the Sustainable Development Goals (SDG) index an adequate framework to measure the progress of the 2030 Agenda? Sustain. Dev. **26**(6), 663–671 (2018)

11. Thin, N., Lockhart, C., Yaron, G.: Conceptualising Socially Sustainable Development. Department for International Development and World Bank, London (2002)
12. Dempsey, N., Bramley, G., Power, S., Brown, C.: The social dimension of sustainable development: defining urban social sustainability. Sustain. Dev. **19**(5), 289–300 (2011)
13. Vifell, Å.C., Soneryd, L.: Organizing matters: how 'the social dimension' gets lost in sustainability projects. Sustain. Dev. **20**(1), 18–27 (2012)
14. Goodland, R.: Sustainability: human, social, economic and environmental. in: Munn T (ed) Encyclopaedia of Global Environmental Change, Wiley, Hoboken, NJ, pp. 488–489 (2002)
15. Turkington, R., Sangster, K.: From housing to social mix: housing's contribution to social sustainability. Town Country Plan. **75**(6), 184–185 (2006)
16. Chan, E., Lee, K.: Critical factors for improving social sustainability of urban renewal projects. Soc. Indic. Res. **85**(2), 243–256 (2008)
17. Cuthill, M.: Strengthening the 'social' in sustainable development: developing a conceptual framework for social sustainability in a rapid urban growth region in Australia. Sustain. Dev. **18**(6), 362–373 (2010)
18. Vavik, T., Keitsch, M.: Exploring relationships between universal design and social sustainable development: some methodological aspects to the debate on the sciences of sustainability. Sustain. Dev. **18**(5), 295–305 (2010)
19. Spangenberg, J.H.: Institutional sustainability indicators: an analysis of the institutions in Agenda 21 and a draft set of indicators for monitoring their effectivity. Sustain. Dev. **10**(2), 103–115 (2002)
20. Spangenberg, J.H., Pfahl, S., Deller, K.: Towards indicators for institutional sustainability: lessons from an analysis of Agenda 21. Ecol. Ind. **2**(1–2), 61–77 (2002)
21. Pfahl, S.: Institutional sustainability. Int. J. Sustain. Dev. **8**(1–2), 80–96 (2005)
22. Hawkes, J.: The fourth pillar of sustainability: culture's essential role in public planning; Common Ground Publishing Pty Ltd in association with the Cultural Development Network (Vic), Victoria (2001)
23. Nurse, K.: Culture as the fourth pillar of sustainable development. Small Stat. Econ. Rev. basic Stat. **11**, 28–40 (2006)
24. Barkin, D., Lemus, B.: Understanding progress: a heterodox approach. Sustainability **5**(2), 417–431 (2013)
25. Sen, A.: Inequality Re-examined. Clarendon Press, Oxford (1992)
26. Sen, A.: Development as freedom. Oxford University Press, Oxford (1999)
27. Robeyns, I.: The capability approach in practice. J. Pol. Philos. **14**(3), 351–376 (2006)
28. Frediani, A.A.: Sen's capability approach as a framework to the practice of development. Dev. Pract. **20**(2), 173–187 (2010)
29. OpenCivitas database. https://www.opencivitas.it/
30. Bencardino, M.: Squilibri territoriali nella distribuzione del reddito pro capite in regione campania: una sperimentazione alla scala delle frazioni censuarie. Bollettino dell'Associazione Italiana di Cartografia **165**, 59–73 (2019). https://doi.org/10.13137/2282-572X/29825
31. Bencardino, M., Nesticò, A.: Demographic changes and real estate values. A quantitative model for analyzing the urban-rural linkages. Sustainability, **9**(4), 536 (2017). MDPI AG, Basel, Switzerland https://doi.org/10.3390/su9040536
32. Bencardino, M., Nesticò, A.: Spatial correlation analysis among land values, income levels and population density. In: Calabrò, F., Della Spina, L., Bevilacqua, C. (eds.) ISHT 2018. SIST, vol. 100, pp. 572–581. Springer, Cham (2019). https://doi.org/10.1007/978-3-319-92099-3_64
33. Bencardino, M., Nesticò, A.: Urban sprawl, labor incomes and real estate values. In: Gervasi, O., et al. (eds.) ICCSA 2017. LNCS, vol. 10405, pp. 17–30. Springer, Cham (2017). https://doi.org/10.1007/978-3-319-62395-5_2

The Role of the Coordination Models in Urban Resilience Against Covid-19

Roberta Troisi[1]([✉]) [iD], Gaetano Alfano[1] [iD], and Rocío Blanco-Gregory[2] [iD]

[1] Department of Political and Communication Sciences, Territorial Development Observatory (OST), University of Salerno, Fisciano, Italy
{rtroisi,galfano}@unisa.it
[2] Departamento de Dirección de Empresas y Sociología, Facultad de Empresa, Finanzas y Turismo, Universidad de Extremadura, Badajoz, Spain
trblanco@unex.es

Abstract. This study analyzed the effects of the urban institutional actions and voluntary actions on COVID-19. We use two fundamental organizational model that underpin crisis management on local level: the top-down and the loosely coupled model to understand what choice works better.

Based on data in 64 Italian regions, we employed a random-effect panel model.

It was found that: 1) despite the non-significant effect of local police controls, urban planning has an important role in constraining the spread of the pandemic since context characterized by an update local plan are related to smaller values of COVID-19 incidence. 2) Vaccination on voluntary action also helps to reduce the COVID-19 incidence. 3) NPOs initiative are positively related to the incidence over time. In light of results, a mixed coordination model with authority in a central role and voluntary action inside a formal net of emergency seems to be the optimal urban resilient response to the pandemic.

Keywords: Coordination models · Urban resilience · Covid-19

1 Introduction

Since the early days of the COVID-19 crisis, social sciences scholars' interest focused both on its transmission mechanisms and the ways of avoiding the spread by leveraging on a series of people habits. Since it is well-known we are facing a transmissible disease, the great attention paid to citizens' behaviors has dealt with different scales. Particularly preventing and controlling the spread of COVID-19 has been largely attributed to the need to change a set of social rules. This study offers an understanding of how these changes have affected the control of the diffusion of COVID-19 on a local basis. Four main ideas are at play: 1) The attention at the local level, namely the urban level, is explained since cities by their nature are places with a high concentration of economic activities and population and, therefore, are more vulnerable to various stressors such as natural and man-made disasters. For the same reasons, they have been considered hot spots of COVID-19 infections [1]. Understanding local dynamics can help gain knowledge of an

F. Calabrò et al. (Eds.): NMP 2022, LNNS 482, pp. 1252–1261, 2022.
https://doi.org/10.1007/978-3-031-06825-6_121

important piece of the problem. 2) This analysis is based on two main driving forces at the basis of the change in citizens' behavior at the urban level. Substantially, it is up to citizens' or private organizations voluntary choices as well as it can be provided with regulation, particularly that on the local level. On one hand, it has been proven that for an unpredictable event in absolute terms, the spatial and temporal scale is conditioned by the level of urban preparedness mainly depending on public measures related to transport, public spaces, facilities as well as a set of controls to regulate citizens' behaviors [2]. On the other, minor interest has been addressed to the voluntary actions at the local level. They stem from the civic sense spamming from the voluntary support to the people in need, to the vaccination, at the time we are writing, still provided voluntarily. Civic sense and civic capital have been mainly explored at the macro level both on national and European level as driving forces to address the major incidence of voluntary actions [3, 4]. The overlooking of the local perspective is probably due to a general dearth of data at the urban level 3) Our focus is on the impact of two different kinds of actions (from now institutional and voluntary actions) on the control of the spread of the COVID-19 virus. They can be traced back to a fundamental organizational framework in the emergency management field that defines when and in what circumstances decision-making in times of crisis should be exclusively in the institutional actors' realm or it could be enlarged to a voluntary basis. In its basic sense, it refers to a model of hierarchical coordination and a loosely coupled one [5]. 4) In our idea, the control of the COVID-19 virus through local institutional and voluntary actions expresses urban resilience for two reasons. First, it is in line with a shared idea of resilience according to the recent literature on hazard and disaster. It frequently refers to the resilience concept as a guiding principle behind any effective responses to coping with the impact of the event. It stands in the intrinsic ability of a system, community or society, hit and shocked by a disaster, to adapt and endure through changing its non-essential features and reconfiguring itself [6]. Furthermore, it catches the more specific features of the urban resilience referring to the ability of an urban system "to face of a disturbance, to adapt to change and to quickly transform systems that limit current or future adaptive capacity" [7] by involving different agents within the urban area.

Therefore, an in-depth analysis of these two factors could provide a vision for urban resilience improvement concerning pandemics, particularly because this crisis is unprecedented. Although a vast body of research has been published on the impacts of a wide range of disasters on cities and the related urban resilience [8] all the lessons learned in the past could not be suitable for this case. Thus, any explanation aimed at understanding the control of the impact of the pandemic on cities can represent a new important lesson to share in these difficult times.

The main innovations and contributions of this paper are as follows: first, to more effectively reflect the actual situation of the pandemic at the local level we collect a set of data from different sources gathering a unique dataset on real-time pandemic data and urban characteristics. Our data are collected on an urban basis whereas the bulk of the analysis relies on national or at the most regional data. In this way, an important unit of the spread of COVID-19 is missed. The urban level represents a significant local unit since the city context is, in its nature, marked by systematic relationships between the local government, the citizens, and the local systems of entrepreneurs, civic associations, and group interests [9].

Second, to analyze the effects of institutional and voluntary actions on urban resilience we applied a random-effect panel model. This has the benefit to analyze over time the effects of time-variant and time-invariant explanatory variables on the dependent variable of interest. Third, as said, we consider our results as a lesson whose implications are two-fold. They can give some suggestions to policymakers as for the capacity of the single actions in addressing the prevention and control of COVID-19 and, at the same time, they can represent long term means for a future crisis, thus aimed to enhance urban resilience.

The rest of this paper is structured as follows: the second section briefly reviews the literature on emergency management and urban governance. The third section describes our research data and the econometric model. The fourth section is about the conclusions.

2   Facing the Crisis at the Local Level: The Coordination Models from the Emergency Management and Urban Governance Literature

The bulk of the studies both in the field of general emergency management and at the local level faces the problem of decision-making and actions in times of crisis through two basic coordination mechanisms.

The emergency management literature emphasizes the type of coordination together with the nature of involved agents, through three different kinds of emergency models. The first turns into a rigid paradigm where the agents are mainly public since governments are supposed to possess exclusive capabilities and resources to manage a range of disaster-related responses. In its original version, this paradigm is based on command and execution, coordinated through bureaucratic rules. Compliance must be enforced without any kind of leeway at the bottom level such as the citizens' level. Its main advantage stands in the capacity to avoid decision paralysis by implementing pre-defined schema and pre-determined action plans [10].

The second model focuses on a more adaptive response where the public agents are supplemented by further agents in a wider network. Reality has shown an increasing resort to private actors, based on the idea that complex problems can be better solved through multi-actor collaborations and joint solutions [11]. However, the public agents keep on being the key subjects, permanently belonging to the core of management, while the other agents are found in a peripheral role, that can change according to the nature of the event [12]. This paradigm recalls the loosely-coupled model [13], particularly by moving away from the core, coordination mechanisms loosen up, orders are replaced by guidelines within which a degree of leeway is viable, typically turning into decentralized decisions, as long as programmed. At the same time, private organizations are involved as affiliated, considered formal agents in formal plans.

The advantage of such a model stands in the ability to rapidly assess and adapt to a crisis which requires flexibility both in decision-making schemes and coordination between the different actors involved.

The third kind of emergency management does not provide a model rather it identifies a group of agents and their responses to a crisis. The agents act spontaneously and are not associated with any formal disaster response plan. Some studies underline how informal voluntary action, in the form of individuals and emergent groups, is an important resource and capacity for emergency response [14]. Since their contribution is characterized by high degrees of spontaneity, improvisation and ad hoc behavior it is marginally considered in the general emergency management frameworks.

Similar concepts can be found in local-level studies that can be brought to the general themes of urban governance and urban governance in times of crisis.

As for the first strand of research, the urban governance notion has been consolidated into two models which in many ways recall the coordination mechanisms mentioned above. The bulk of the contributions considers two main models characterized by hierarchy or loosely coupled, but are not focused on the voluntary actions outside the institutional net of emergency.

On the one hand, the urban governance characterized by a clear division of roles along the command chain is described: this kind of study typically consider the mayor as the local authority in the most strategic position [15]. On the other hand, a more integrated governance model is analyzed characterized by a sustainable perspective in coordinating many activities and multi-stakeholders in the long-term [16].

However, when referring the urban governance in times of crisis there is no unanimous consensus on the model to prefer to face emergencies. Some study shows how in some cities facing COVID-19 emergency through multi-governance tools has successfully prevented the spread of the pandemic since it allowed rapid and coordinated solutions among agents [17]. Others show that the centralization of decision-making powers, particularly referred to the mayors, has equally led to positive results in terms of prevention [18].

Differently, Chu et al. [19], show a series of cases in which the local experience seems to have produced excessive fragmentation, excessive autonomy of choice, with no appreciable result when considered in terms of whole national performance.

3 Data and Method

In this section, the data used in the analysis and the model are presented. The analysis evaluates the effect of institutional and voluntary actions on the incidence of COVID-19 through the use of a random-effect panel model.

3.1 Data Sources

The data used in the analysis were obtained from different sources. The number of daily new cases for each Province was obtained from the daily reports of the Ministry of Interior. In particular, those data are useful to define COVID-19 incidence for each Province analyzed. The variable defined as the number of daily new cases over 10000 habitants is a proxy of the risk of contracting the disease [20]. We consider the daily incidence as the dependent in the random-effect panel model employed.

Regarding the institutional actions, information on urban planning are given by the National Institute of Urban Planning reports [21], while data on the local police controls are gathered from the Minister of Interior and Prefectures' websites.

Regarding the voluntary actions, the activities of no-profit organizations (NPO) are reported on the NPOs' websites while info on the daily vaccination are gathered from the Department of Civil Protection's daily reports.

At last, ISTAT provides further information about provinces as population and population density.

The work is still in progress. In this paper, we report the initial results obtained considering 64 of the 107 Italian provinces randomly selected.

3.2   Method

The relationship between the institutional and voluntary actions is evaluated employing a random-effect panel model. A Hausman Test (p-value $= 0.12$) suggests that the random-effect models have to be preferred to a fixed- effect model. Analytically the model can be written as follows [22]:

$$Y_{it} = \alpha + \beta_{it}x_{it} + z_i\gamma + u_{it} \tag{1}$$

where Y_{it} is the dependent variable of interest, in this work the incidence over time. x_{it} is a set of time-varying explanatory variables and γ are the time-invariant explanatory variables. α, β_{it} and z_i are respectively the intercept and the parameters to be estimated while u is the idiosyncratic error term.

Variables. We consider four explanatory variables, two of which refer to the institutional actions and two to the voluntary actions.

1) Urban planning. Two main reasons lead us to consider urban planning as the main institutional action that orient citizens' behavior in mitigating the effect of COVID-19 virus. First, the presence of large parks in cities, the balanced distribution of housing avoiding, or, at least controlling areas of a high density of population, the presence of the necessary facilities and the regulation of transport for each urban area, are features associated with citizens' greater well- being during the period of lockdown [23]. At the same time, this kind of urban planning has the side effect of facilitating social distancing, avoiding that a series of imbalances among areas can lead to non-compliance behaviors particularly in the most disadvantaged areas. In essence, the plans developed considering the whole urban quality of life central, are in themselves more suitable plans to deal with any type of emergency [24]. In Italy, in the last two decades, urban legislation has imposed a series of constraints both in terms of construction and maintenance respecting a general principle of quality of life and sustainability (e.g. DL 380/2001). In this direction, up-to-date plans present more constraints in comparison with urban plans preceding national laws. Therefore, the variable used distinguishes between plans entered into force after 2010 and plans previously entered into force. The variable is operationalized as the percentage of municipalities with an urban plan adopted after 2016.

2) Local controls. In many countries, local policies have been considered the main enforcer of the social restrictions that citizens have experienced during COVID-19. Sargeant et al. [25] show that there is a positive relationship between police effectiveness, collective efficacy, and public willingness to intervene when others violate lockdown restrictions. Differently in countries with endemic problems of overpopulation, the police work was shown as little appreciated and as an additional burden to a situation already very complicated [26]. The variable in this study has been operationalized as the number of daily violations reported by the Prefectures.

3) Vaccination. Vaccination promotion is a crucial strategy to achieve herd immunity against COVID-19. Since in many countries it is voluntary, one of the risks underlined is that some citizens can have an incentive to free-ride as they could benefit from the reduction in COVID-19 spread thanks to the majority of citizens' choice to be vaccinated [27]. Races, gender, age, socio-economic conditions have been tested to understand their impact on the assessment of the severity of the COVID-19 virus and the related importance of being vaccinated. Nevertheless, whereas these results are contrasting, there is more evidence in terms of the right motivation in being vaccinated that can be summarized in the need to protect individual, family or community health [28] this variable has been operationalized as the number of daily vaccinations.

4) NPOs initiatives. Participating in the community response to tackle emergencies is difficult to catch in such a complex scenario. However, some study underlines how this kind of response has been mainly addressed by non-profit-organizations [29]. Their contribution matters in creating system resilience both in coping with a health crisis and its social effects. COVID-19 seems to be an important test bench for these organizations for their role as engines of social innovation. Furthermore, a high-degrees of spontaneity, ingenuity, improvisation and ad hoc behavior seems to represent their added value in supporting people in need more promptly than the public interventions. It is due to their close and enduring relationships with their embedding communities despite the scenario characterized by strict personal restrictions [30]. The variable has been operationalized as the weekly number of initiatives undertaken during the COVID-19 crisis by NPOs.

The population density is considered as a control variable. At last, a dummy variable considering the Italian regions was also added. This variable, as suggested in [31], is useful to deal with the effect that the outlying incidence of some regions may have on the estimation of the parameters. The variable also consents to take count of the fixed-effect that the regional emergency policies have on the incidence of COVID-19 [10]. Finally, as said, we consider COVID-19 daily incidence in terms of new cases over 10000 inhabitants as the dependent variable. The reduction of the trend analytically expresses the containment of the virus, and theoretically, the capacity of urban resilience.

4 Results

Table 1 reports the results of the random-effect panel model and the diagnostic tests.

Table 1. Fixed-effect regression results.

Variable	Coefficient (SE)	P-value
Intercept	−3.28e+01 (7.45e+00)	1.07e−05 ***
Urban planning	−9.32e+01 (7.31e+00)	<2.2e−16 ***
Local police control	1.36e−05 (1.82e−04)	0.94
Vaccinations	−1.58e−03 (3.80e−04)	3.17e−05 ***
NPOs initiatives	1.55e−01 (1.33e−02)	<2.2e−16 ***
Population density	2.47e−01 (3.54e−03)	<2.2e−16 ***
Basilicata	4.14e+00 (7.84e+00)	0.59
Calabria	7.89e+00 (6.91e+00)	0.25
Campania	−1.86e+01 (7.51e+00)	0.01 *
Emilia-Romagna	7.41e+01 (6.46e+00)	<2.2e−16 ***
Friuli Venezia Giulia	8.61e+00 (7.87e+00)	0.27
Lazio	6.48e+01 (6.23e+00)	<2.2e−16 ***
Liguria	−1.47e+01 (1.02e+01)	0.14
Lombardia	5.56e+01 (6.22e+00)	<2.2e−16 ***
Marche	−2.91e+01 (6.98e+00)	3.10e−05 ***
Molise	4.14e+00 (7.85e+00)	0.59
Piemonte	1.58e+02 (7.88e+00)	<2.2e−16 ***
Puglia	3.92e+01 (6.48e+00)	1.43e−09 ***
Sardegna	1.07e+00 (6.41e+00)	0.86
Sicilia	−1.30e+00 (6.24e+00)	0.83
Toscana	1.24e+01 (5.87e+00)	0.03 *
Umbria	6.44e+01 (1.01e+01)	1.89e−10 ***
Valle d'Aosta	1.71e+01 (1.01e+01)	0.09
Veneto	9.23e+01 (6.29e+00)	<2.2e−16 ***
R-square	0.25	
F-statistic	9234.4 on 23 DF, p-value: <2.22e−16	
Significance codes: *** p < 0.001, ** p < 0.01, * p < 0.05		

The value of the R-squared could seem small (0.25): for this reason, a statistical power analysis of the result was performed [32, 33]. The analysis consents to evaluate the statistical power of the study considering the relationship among the four variables involved in the statistical inference: the sample size, the significance criterion, the population effect size, and the statistical power. Following the procedure proposed by Cohen [34], we evaluated that the value of 0.25 is very close to the value proposed for large R-squared (0.2595, for regression analysis with a significance criterion of 0.01, a medium population effect size and our sample size). Regarding the institutional variables, urban planning is significant and negatively related to the regional incidence ($-9.32e+01$, $p < 0.001$). The results confirm that an updated urban plan is useful to mitigate the incidence of COVID-19. In contrast, local police controls are not significant.

Regarding the voluntary action variables, vaccinations are significant and negatively related to COVID-19's incidence ($-1.58e-03$, $p < 0.001$). Thus, the communities with a higher number of vaccinations, therefore closer to herd immunity, are characterized by a smaller value of the incidence over time. In contrast, the number of NPOs initiatives are positively related to the incidence ($1.55e-01$, $p < 0.001$). This means that in local contexts where there are a higher number of NPOs' initiatives the incidence is higher than in places whit a smaller number of initiatives.

The population density is positively related to the incidence ($2.47e-01$, $p < 0.001$). Therefore, more densely populated contexts are characterized by a higher value of the incidence over time. At last, the regional dummy confirms the outlying effect that some regions have on COVID-19 incidence (e.g. Lombardia and Veneto are positively related to the incidence while Marche and Campania are negatively related).

5   Conclusion

What is the most suitable decision-making and action coordination model that is resilient to this pandemic?

The combination of voluntary actions with institutional actions to mitigate the effects of COVID-19 comes down to the mixed model, according to the loosely coupled criterion. Urban planning contributes to resilience, by encouraging virtuous behavior instead of imposing rules of conduct: it moves away from the idea of a rigid chain of command where authority orders and the citizens execute. In addition, local controls, which are closely linked to a strict hierarchy model since they verify the citizens' compliance, show that they are not able to mitigate COVID-19. As for the two voluntary actions, the choice to vaccinate shows to mitigate the effects COVID-19 although within the public vaccination plans. This confirms a model that works well when provides institutional guidelines and relies on citizens for their spontaneous implementation. Contrarily, the non-profit organizations voluntary actions do not produce mitigation effects. The result is not counter-intuitive: despite their social value, improvisation and spontaneity probably solve problems of need, but at the same time increase contacts between people.

Finally, there is a lesson to learn. Urban resilience cannot be intended as a specific response for an unprecedented event. The current response to pandemics put together institutional actions with voluntary although inside the net of the emergency management. Such an integrated approach could facilitate developing appropriate long-term visions considering a "planned integration" a good solution to face future crisis. In this sense, resilience is not a contingent response as well as the lesson that founds long-term strategies to minimize the impacts of future disruptive events.

References

1. Borjas, G.J.: Demographic determinants of testing incidence and COVID-19 infections in New York City neighbourhoods (No. w26952). National Bureau of Economic Research (2020)

2. Mouratidis, K.: How COVID-19-19 reshaped quality of life in cities: a synthe-sis and implications for urban planning. Land Use Policy **111**, 105772 (2021)
3. Barrios, J.M., Benmelech, E., Hochberg, Y.V., Sapienza, P., Zingales, L.: Civic capital and social distancing during the COVID-19-19 pandemic☆. J. Public Econ. **193**, 104310 (2021)
4. Di Marco, G., Hichy, Z., Sciacca, F.: Attitudes towards lockdown, trust in institutions, and civic engagement: a study on Sicilians during the coronavirus lockdown. J. Public Affairs e2739 (2021)
5. Troisi, R., Castaldo, P., Arena, L.: Maintenance management of infrastructure systems: orga-nizational factors in territorial planning. In: IOP Conference Series: Materials Science and Engineering, vol. 1203, no. 3, p. 032098. IOP Publishing (2021)
6. Tiernan, A., Drennan, L., Nalau, J., Onyango, E., Morrissey, L., Mackey, B.: A review of themes in disaster resilience literature and international practice since 2012. Policy Des. Pract. **2**(1), 53–74 (2019)
7. Meerow Meerow, S., Newell, J.P., Stults, M.: Defining urban resilience: a review. Landsc. Urban Plan. **147**, 38–49 (2016)
8. Sharifi, A., Yamagata, Y.: Urban resilience assessment: multiple dimensions, criteria, and indicators. In: Yamagata, Y., Maruyama, H. (eds.) Urban resilience. ASTSA, pp. 259–276. Springer, Cham (2016). https://doi.org/10.1007/978-3-319-39812-9_13
9. Troisi, R., Alfano, G.: Towns as safety organizational fields: an institutional framework in times of emergency. Sustainability **11**(24), 7025 (2019)
10. Troisi, R., Alfano, G.: Is regional emergency management key to containing COVID-19? A comparison between the regional Italian models of Emilia-Romagna and Veneto. Int. J. Publ. Sector Manag. (2021)
11. Jung, K., Song, M., Park, H.J.: The dynamics of an inter-organizational emergency man-agement network: interdependent and independent risk hypotheses. Public Adm. Rev. **79**(2), 225–235 (2019)
12. Robinson, S.E., Eller, W.S., Gall, M., Gerber, B.J.: The core and periphery of emergency management networks. Public Manag. Rev. **15**(3), 344–362 (2013)
13. Perrow, C.: Systems and Accidents. Accident At Three Mile Island: The Human Dimensions (2019)
14. Whittaker, J., McLennan, B., Handmer, J.: A review of informal volunteerism in emergencies and disasters: definition, opportunities and challenges. Int. J. Disaster Risk Reduct. **13**, 358–368 (2015)
15. Gissendanner, S.: Mayors, governance coalitions, and strategic capacity: drawing lessons from Germany for theories of urban governance. Urban Affairs Rev. **40**(1), 44–77 (2004)
16. Kokx, A., Van Kempen, R.: Dutch urban governance: multi-level or multi-scalar? Eur. Urban Regional Stud. **17**(4), 355–369 (2010)
17. Sharifi, A.: The COVID-19 Pandemic: Lessons for Urban Resilience. In: Linkov, I., Keenan, J.M., Trump, B.D. (eds.) COVID-19: Systemic Risk and Resilience. RSD, pp. 285–297. Springer, Cham (2021). https://doi.org/10.1007/978-3-030-71587-8_16
18. Earl, R.: Vietnam Living with authoritarianism: Ho Chi Minh city during COVID-19 lockdown city Soc., 32 (2020)
19. Chu, Z., Cheng, M., Song, M.: What determines urban resilience against COVID-19: City size or governance capacity? Sustain. Cities Soc. **75**, 10330 (2021)
20. Shields, L., Twycross, A.: The difference between incidence and prevalence: this paper is one of a series of short papers on aspects of research by Linda Shields and Alison Twycross. Paediatr. Nurs. **15**(7), 50–55 (2003)
21. Istituto nazionale di urbanistica: Rapporto dal territorio 2016, INUed, Roma (2016)
22. Stock, J.H., Watson, M.W.: Introduction to Econometrics, vol. 3. Pearson, New York (2012)
23. Mouratidis, K., Yiannakou, A.: COVID-19 and urban planning: Built environment, health, and well-being in Greek cities before and during the pandemic. Cities **121**, 103491 (2021)

24. Mouratidis, K.: How COVID-19 reshaped quality of life in cities: a synthesis and implications for urban planning. Land Use Policy **111**, 105772 (2021)
25. Sargeant, E., Murphy, K., McCarthy, M., Williamson, H.: The formal-informal control nexus during COVID-19: what drives informal social control of social distancing restrictions during lockdown?. Crime & Delinquency, 0011128721991824 (2021)
26. Wasdani, K.P., Prasad, A.: The impossibility of social distancing among the urban poor: the case of an Indian slum in the times of COVID-19-19. Local Environ. **25**(5), 414–418 (2020)
27. Sasaki, S., Saito, T., Ohtake, F.: Nudges for COVID-19 voluntary vaccination: how to explain peer information? Soc. Sci. Med. **292**, 114561 (2022)
28. Giubilini, A.: Vaccination ethics. Br. Med. Bull. **137**(1), 4–12 (2021)
29. Tortia, E., Troisi, R.: The resilience and adaptative strategies of Italian cooperatives during the COVID-19 pandemic. Foresight STI Govern. **15**(4), 78–88 (2021)
30. Diab, A.: The accountability process during the time of COVID-19 pandemic and the emerging role of non-profit associations. Acad. Strategic Manag. J. **20**(1) (2021)
31. Bell, A., Fairbrother, M., Jones, K.: Fixed and random effects models: making an informed choice. Qual. Quant. **53**(2), 1051–1074 (2018). https://doi.org/10.1007/s11135-018-0802-x
32. Kraemer, H.C., Blasey, C.: How many subjects?: Statistical power analysis in research. Sage Publications (2015)
33. Faul, F., Erdfelder, E., Lang, A.G., Buchner, A.: G* Power 3: a flexible statistical power analysis program for the social, behavioral, and biomedical sciences. Behav. Res. Methods **39**(2), 175–191 (2007)
34. Cohen, J.: Statistical power analysis. Curr. Dir. Psychol. Sci. **1**(3), 98–101 (1992)

The Financial Sustainability a Cultural Heritage Adaptive Reuse Project in Public-Private Partnership

Lucia Della Spina[1]([⊠]) [iD], Sebastiano Carbonara[2] [iD], and Davide Stefano[2] [iD]

[1] Mediterranea University of Reggio Calabria, 89125 Reggio Calabria, RC, Italy
lucia.dellaspina@unirc.it
[2] G. d'Annunzio University, 65127 Pescara, PE, Italy

Abstract. In recent decades, there has been a greater attention and awareness of the key role of the functional reuse of cultural heritage for the creation/sustainable development of cities, in a perspective of circular economic growth. However, interventions related to the enhancement of heritage involve complex decisions for the interests at stake and a persistent scarcity of public resources. The need to preserve the values of assets and to meet the rules relating to their protection, entailing high costs that discourage public and private investments for reuse interventions. It therefore becomes essential to support the decisions of reuse through adequate evaluation methodologies for the evaluation of the multidimensional impacts of the projects of functional reuse of cultural heritage in the circular economic perspective. This would allow public and private entities to recognize the value of heritage and their possible integration into the local economic system. In this context, the contribution proposes a methodological framework based on the integration of different decision support and evaluation methodologies, but which does not neglect the values historical and cultural, as well as economic and financial feasibility ones, to verify the involvement of private capital in the implementation and management of the intervention. The application of the methodology to a case study allows evaluating in the final phase the financial sustainability for the private sector, or if the investment for the redevelopment and reuse of the asset will be repaid through the management of the new configuration, selected through a path shared with the local community in the preliminary evaluation phase.

Keywords: Cultural heritage · Functional reuse · Financial sustainability · Public-private partnership · Discounted Cash Flow Analysis (DCFA)

1   Introduction

Cultural heritage is recognized as a key driver in local sustainable development processes, as it contributes to the identity of the territories and the cultural diversity of local communities.

As expressed by the Faro Convention [1], the "heritage communities" represent the fundamental actors able to guide and implement the exercise of civic responsibility and (inter) cultural policies, to build shared and sustainable development scenarios [2–4].

© The Author(s), under exclusive license to Springer Nature Switzerland AG 2022
F. Calabrò et al. (Eds.): NMP 2022, LNNS 482, pp. 1262–1272, 2022.
https://doi.org/10.1007/978-3-031-06825-6_122

In this perspective, many real estate assets which are recognized as having a public utility function should be chosen as reference elements to initiate or consolidate urban, territorial and environmental redevelopment processes. We should start from those which, by definition, represent the places of people, of daily life, of culture, of communication and of exchange. These spaces, appropriately enhanced, could become the driving force for a new renaissance of the city and the territory, in line with the pressing need to recover a new public ethical tension and promote livability on a human scale [5].

An enhancement action should have the purpose of conferring new values on public goods, through the recovery of lost merits and the recognition of the peculiarities considered [6, 7]. In this way, would be increased and improved the public visibility of assets and their ability to provide new benefits, both tangible and intangible, would be increased and improved [8]; the possibility of their use and fruition, in the forms deemed most advantageous by the community; the activity for their maintenance and conservation. It would mean acting on a heterogeneous set of potentials of the good capable of increasing its social, cultural, environmental and economic performance by procuring direct and indirect, immediate and deferred, both on a local scale and on a wider territory [9–12].

All this would presuppose an integrated enhancement intervention, based on the many and different characteristics and potentials of the asset (physical, functional, historical-cultural, aesthetic) and on a heterogeneous flow of reciprocal inferences (positional, environmental, socio-economic) between the asset and the context that welcomes it.

It is believed that the integrated enhancement of public goods, if creatively and rationally related to the system of needs, vocations and objectives expressed by the relative territorial and socio-economic context, is able to effectively contribute to the stable regeneration of the areas concerned and to improve the relational relationships between the same context and the subjects who act in it [13–15]. This is due to the benefits induced on large swaths of territory, over extended periods of time, on the quantity and quality of the interventions that can be carried out, on the physical and socio-economic components of the places.

Putting at the center of attention those assets which, for various reasons, are considered suitable for intercepting the interests of a community and, sometimes, for identifying particular values, means affecting the founding elements of sociality, all essential factors for guiding development processes towards shared objectives and to prepare public policies suitable for achieving them.

The considerations made so far show that the scenario in which to operate for the enhancement of public goods is very problematic and complex.

Today more than in the past, the scarce availability of public money pushes local technicians and administrators to evaluate possible investments in advance; this, from a purely theoretical point of view, would avoid the frequent creation of public works that are not strictly necessary, which offer services on an occasional and non-continuous basis and which burden public budgets in terms of management costs [16].

Evaluating public investments at this time means verifying on the one hand the convenience for the community, if the functional reuse program really responds to a specific public need. On the other hand, it is also necessary to verify the convenience for the private subject to participate in the investment project, in order to identify for which works and to what extent the involvement of private capital in the implementation and management of the intervention can be envisaged.

The increasingly topical issue of the Public-Private Partnership (PPP) [17, 18] therefore becomes closely linked / connectable to the economic-social and economic-financial convenience, because only in conditions of "equilibrium" can be established a collaboration between the two subjects who have different purposes (social utility and financial profit) in implementing operations for the reuse of underused or discarded assets through appropriate functional reconversions, aimed at satisfying the needs of new spaces of collective interest [19, 20].

The contribution proposes a methodological framework based on the integration of different decision support and evaluation methodologies. In particular, the application of the methodology to the case study makes it possible to evaluate in the final phase the financial sustainability of a PPP initiative for the private sector: that is, whether the investment for the requalification and reuse of the asset will be repaid through management of the new configuration, chosen as the highest and best destination for reuse, selected through a path shared with the local community in the preliminary phase through a Multi-Criteria Analysis.

2 Methodological Framework

The present research fits into the context outlined above. According to a multiphase decision-making process, a distinctive feature of the methodology followed in this study is the combined use of different evaluation tools, capable of supporting the choice relating to the reuse of a public cultural asset located in the city of Catanzaro (Italy), through a "shared strategy", built on the needs of the local community, but which does not neglect historical and cultural values, as well as economic and financial feasibility [20].

In particular, the first step of the proposed decision-making process consists in the development of a Stakeholder Analysis (SA) [21] with the aim of identifying the actors involved in the problem, as well as their values and objectives. The system of objectives identified in the preliminary phase is used for the development of a Multi-Criteria Decision Analysis (MCDA) [13–15, 18, 22] which aims to choose the best alternative project for the reuse of the asset in question. Finally, the final step of the process develops a Discounted Cash Flow Analysis (DCFA) [17–19, 23, 24] for the financial sustainability of a PPP initiative for the private sector, selected in the second phase of the procedure through a MCDA methodology [13–15, 18, 22].

The study will illustrate the final phase of the evaluation process, in which the DCFA is applied [17–19, 23, 24], for the verification of the sustainability and financial feasibility of a PPP initiative, relating to the reuse of a public cultural asset through the "concession of enhancement" procedure [5, 19]. For an in-depth study and application of the MCDA, in relation to the selection of the best alternative project for the reuse of the asset in question (Phase 1), please refer to other studies and research by the authors applied to historical heritage [13, 14, 25–29].

3 Case Study

The case study concerns the reuse of a former monastery complex linked to the church of S. Nicola Coracitano, known as Palazzo Stella. It is located in the historic center of Catanzaro (Southern Italy), the main regional capital with regional public activities and services.

As part of the strategic plan, the Municipality of Catanzaro has planned a series of integrated interventions, of which development strategy provides for the enhancement of the historic center, together with the creation of new public services.

The quadrangular complex of the former monastery has undergone numerous interventions, transformations, additions, and alterations over time, induced by the stratification of uses, which have partially distorted its original characteristics. The PA acquired the monastery in 2010, although it has remained unused since 2003.

As mentioned, the property is located in the historic center along a secondary road, away from the main communication routes. The complex consists of three main parts: the former monastery, the church, the theater, and the cloister.

The monastery is over three levels, around a central courtyard. On the ground floor (about 1000 sqm), there is the entrance, the reception, the porch, the common rooms, the kitchen, and the canteen; on the first floor (about 850 sqm), there are the bedrooms; and on the second floor (about 800 sqm), there are service areas, common areas, and other bedrooms.

It is the objective of the Municipality of Catanzaro, owner of the property, to proceed with its enhancement through a PPP procedure, according to the hypothesis of Enhancement Concession (EC), in order to verify its financial sustainability. In the current economic situation, characterized by a persistent scarcity of public resources, the Public–Private Partnership (PPP) constitutes a fundamental tool, capable of bridging the gap between the limited public financial resources available and the need to enhance assets in public real estate [23, 24].

In particular, for the enhancement of Palazzo Stella, the PA assumes, according to the hypothesis of Enhancement Concession, that the recovery and restructuring interventions for new functions of the historic asset are carried out by a private investor in exchange for the concession for a limited period of time. For the entire period of the concession, the private investor is responsible for the costs of modernization and all new works deriving from the consequent management. Upon expiry of the concession, the PA returns to the full availability of the good with the acquisition of the good, complete with any transformation and improvement carried out.

4 Assessment of Financial Sustainability

The objective of this phase of the evaluation process is to evaluate the financial sustainability of the private sector, through a procedure of "Enhancement Concession" [5, 19, 23, 24, 29], or rather whether the investment for the requalification and reuse of the asset will be repaid through the management of the asset with the destination: "Social and health residences for non-self-sufficient elderly", new configuration chosen as the highest and best reuse destination selected through an MCDA procedure.

The verification of the financial sustainability of an initiative for the enhancement of a public property carried out through a PPP procedure (in individual or corporate form), is carried out through the development of a Discounted Cash Flow Analysis (DCFA).

The feasibility conditions of the project are verified in terms of the Net Present Value (NPV), the Weighted Average Cost Of Capital (WACC), Debt Service Cover Ratio (DSCR), and Loan Life Cover Ratio (LLCR) [16, 17]. The willingness of the private subject accepting the cooperation mechanism with the Public Administration is based on the private convenience of the intervention and, therefore, on the ability of the initiative to repay the initial monetary outlay, to remunerate the management, and generate a financial surplus.

4.1 Appraisal of Investment Costs

The costs of the initiative concern: i. the investment for the renovation and arrangement of the property and the outdoor area, and ii. the direct and indirect management for the intended uses leased to third parties.

The parametric estimation construction costs were taken from the lists of public and private works currently in force in the Calabria Region and validated through formal investigations conducted at construction companies operating in the Province of Reggio Calabria.

The investment paid by the private operator is therefore equal to 1.8 million euros. The plan also assumes (on a ten-year basis) future interventions, such as extraordinary maintenance and replacement of parts of the furnishings.

For extraordinary maintenance, the estimated amount (equal to 3% of the overall value of the structure, approximately 3.3 million euros) is approximately 100,000 euros; regarding the furnishings, however, it is assumed that the dealer needs to change only one-third of the furniture every ten years, equal to about 110,000 euros. In both cases, the values refer to the moment of the estimate and are included in the plan increased for each year based on the inflation forecasts.

4.2 Appraisal of Management Costs

The overall management costs framework, calculated based on a period of activity of the established functions equal to 365 days/year, includes the costs for healthcare and nursing staff; expenses for other staff; hotel costs for users; general operating expenses; expenses for staff training courses; ordinary maintenance costs; and the annual rent for the property for the lease of the respective surfaces.

The estimate of the healthcare and nursing staff refers to the regional regulatory requirements and starts from the detail of the staff managed by the concessionaire, divided by qualification and specific skills, with an indication of each minimum service to be provided according to the number of users hosted in the facility.

The nursing and personal protection fees have been sized according to the minimum benefits to be provided to users of the facility.

The costs for healthcare and nursing staff were therefore determined based on these organizational hypotheses and based on the average gross wages derived for each profes-sional figure from the respective national employment contracts. The costs for personal protection assistance were instead estimated at 75% of the costs of the nursing staff. The total annual cost is 470,000 euros. Regarding the other structure personnel, it was decided to consider the coverage of three functions as the administration, the secretariat, and the concierge. Still, regarding regional regulations, the concessionaire is also required to provide users with a basic hairdressing service, as regards the minimum activities related to personal hygiene. The total annual cost is 75,000 euros.

In estimating the costs for RSA users, daily meals were identified (calculated in days of food, taking into consideration that the dealer decides not to produce meals directly in the structure due to the lack of adequate spaces), the laundry (which includes cleaning and flat linen, both strictly personal), and the cleaning of the structure. The total cost is 430,000 euros.

The operating expenses also estimated the annual expenses for heating, the expenses for drinking water, the supply and consumption of electricity, telephone expenses, expenses for insurance, miscellaneous expenses, costs for two mini- professional refresher courses of 20 h / year for the duration of the concession. The total annual cost is 100,000 euros. The ordinary maintenance cost was estimated at 0.50% of the structure value at the end of the works; this expenditure, as well as the other costs considered so far, also increases every year with inflation. The total annual cost is 16,500 euros.

4.3 Appraisal of Operating Revenues

To estimate revenues, it is necessary to establish the type of assistance provided; with a view to greater flexibility, it was decided to consider an equitable division of patients into the three social healthcare categories (high, medium, and low).

As a precaution, the plan provides coverage of 90% of the available places (54 users against the 60 available beds) equally divided, as mentioned, among the three levels of social and health services. Finally, it was assumed that the places occupied were 100% under an agreement with the Calabria Region and with the Local Health Authority of reference. Therefore, the indications referred to in the regional regulations were used to define the unit tariffs. Total annual revenues are 1,650,000 euros. The plan then prefigures an update of tariffs every six years, equal to 100% of the increase suffered by costs due to inflation (estimated at 2% per year).

4.4 The Construction of the Financial Plan

The development of the DCFA allows to verify the feasibility of the investment from the point of view of the private investor. With reference to the case study analyzed, in terms of duration, a 30-year concession was assumed, in addition to the year necessary for the implementation of the initial recovery and requalification.

Regarding the financial structure of the investment, it was estimated that the mix of financial resources most suitable for financing the entire structure was made up of 30% risk capital (financing supported by the concessionaire) and 70% debt capital (financing sustained by taking out a loan from a credit institution). The duration of the loan for the initial capital loan has been assumed to be 30 years, which is the entire duration of the management concession. To deal with subsequent investments (extraordinary maintenance and replacement of furnishings) the dealer contracts new mortgages of twenty and ten year durations. The applicator interest rate is equal to 4.75%, inclusive of spread.

The amortization of the initial building investment was distributed over 30 years, with an annual rate of 3.33%; as said, 70% of the furnishings purchased with the initial investment are assumed to have low technical and functional obsolescence, that is, the dealer is not forced to bear additional costs for their replacement during the duration of the concession. These purchases are in any case amortized over 10 years, taking into account the maximum percentage shares prescribed by the reference ministerial tables. The remaining 30% of the new acquisition, on the other hand, behaves according to an average level of obsolescence (the useful life of 10 years). From the tenth year onwards, the additional rate calculated among depreciation relates to investments for extraordinary maintenance.

The direct taxes considered are the corporate income tax and production tax. For indirect taxation, the recovery of Value Added Tax (VAT) was assumed referring to the initial investment and costs incurred during the duration of the concession.

4.5   Financial Sustainability and Evaluation of Results

In terms of duration, a 30-year concession was assumed, in addition to the year necessary for the implementation of the initial intervention.

Against an initial investment by the private investor of 1.8 million euros (of which only 30% is of equity capital and the remaining 70% of loan capital), the cash flows developed in the following 30 years (management phase) lead to a positive NPV. The economic convenience of the project or the ability of the intervention to create value and generate a level of profitability for the invested capital, adequate to the expectations of the private investor, is confirmed by the result of the indicators considered: the project, based on the assumptions made, can cover the investment and management, with an NPV higher than 742,000 euros (covers the invested equity capital) and a WACC equal to 8.04% (weighted average cost of capital).

Financial sustainability, expressed in terms of bankability, or the ability of the project to generate sufficient cash flows to guarantee the repayment of the loans activated, was analyzed through two main coverage coefficients: the DSCR and LLCR. In this specific case, both indices have values far greater than unity.

5   Conclusion

The applied research for the case study confirms the advantages of a multi-methodological evaluation process capable of supporting the entire decision-making process, characterized by numerous elements of complexity (from the phase of defining the problem, to the choice of the highest and best alternative of reuse, up to the evaluation of the feasibility and financial sustainability of the investment) [30–36].

In particular, the development of the DCFA allows to verify if the investment for the private concessionaire/manager of the intervention is feasible and sustainable, and therefore if the investment by the concessionaire/manager of the asset for the its reuse (including the costs for ordinary and extraordinary maintenance) will be repaid through its management [5, 17–19, 23, 24, 29, 36].

Ultimately, the analysis carried out highlights the significant support that can derive from a preventive evaluation in the initial planning phase of defining investments in PPP for the enhancement of the public cultural heritage. The identification of the various subjects involved in the reuse project makes it possible, systematize and clarify the roles of each private investor, the related costs to be incurred and the revenues deriving from the management of the asset [7].

The implementation of a DCFA and the evaluation of the performance indicators made it possible to verify the ability of the investment to adequately compensate the risks of the initiative for the concessionaire/manager, as well as generating a positive financial surplus [5, 17–19, 23–29], and this without any other charge to be borne by the Public Administration (PA), owner of the asset. The choice of the functions envisaged by the reuse project and the management method also allowed, on the one hand, to satisfy the need for convenience for the concession investor, and on the other, to transfer most of the investment risks real estate to private investors most competent for each market sector.

Ultimately, the study is a valid support for decisions, through an integrated approach between the SA, MCDA and DCFA, aimed at the evaluation and selection of alternatives for the reuse of historical heritage [5, 7, 25].

Today, the new challenge for the PA is to regenerate the building heritage by involving the community and local stakeholders to create new governance models [31, 37–41], capable of guaranteeing the enhancement of the historical heritage and cultural values, which, as in the case study, represents a symbolic asset for the community and strategic investment for sustainable urban development for the area [30–34, 41–48].

In this perspective, future research in this field is oriented, according to the concept of "heritage as a common good", to prefigure sustainable experiments for the management of common goods to support the decisions of the PA [1, 5, 6, 15, 18, 28, 30–34, 42–49].

Author Contributions.   The paper is the result of the joint work of the three authors. Scientific responsibility is attributable equally to Lucia Della Spina, Sebastiano Carbonara and Davide Stefano.

References

1. Convention, F.: The Framework Convention on the Value of Cultural Heritage for Society. Faro, Portugal (2005)
2. United Nations General Assembly: Transforming Our World: The 2030 Agenda for Sustainable Development; United Nations, Department of Economic and Social Affairs: New York, NY, USA (2015)
3. Bebbington, J., Russell, S., Thomson, I.: Accounting and sustainable development: reflections and propositions. Crit. Perspect. Account **48**, 21–34 (2017)
4. Brandon, P.S., Lombardi, P., Shen, G.: Future Challenges for Sustainable Development within the Built Environment. Wiley, Chichester (2017)
5. Della Spina, L.: Cultural heritage: a hybrid framework for ranking adaptive reuse strategies. Buildings **11**, 132 (2021). https://doi.org/10.3390/buildings11030132

6. Della Spina, L., Calabrò, F. (eds.): Enhancement of Public Real-estate Assets and Cultural Heritage Management Plans and Models, Innovative Practices and Tools in Supporting the Local Sustainable Development; MDPI: Basel, Switzerland (2020). https://www.mdpi.com/books/pdfview/book/2731. Accessed Dec 2021

7. Trovato, M.R., Giuffrida, S.: The monetary measurement of flood damage and the valuation of the proactive policies in sicily. Geosciences (Switzerland) **8**(4) (2018). https://doi.org/10.3390/geosciences8040141

8. Fusco Girard, L.: Capitale culturale intangibile e sviluppo locale "circolare". In: Colletta, T. (ed.) Festività Carnevalizie, Valori Culturali Immateriali e Città Storiche. Una Risorsa per lo Sviluppo Turistico di Qualità del Mezzogiorno. Franco Angeli: Milano, Italy (2018)

9. Fusco Girard, L. Energia, Bellezza, Partecipazione: La Sfida della Sostenibilità. Valutazioni Integrate tra Conservazione e Sviluppo; Franco Angeli: Milano, Italy (2005)

10. Latham, D.: Creative of Buildings. Donhead Publishing, Shaftesbury (2000)

11. MacArthur, E.: Towards the Circular Economy, Economic and Business Rationale for an Accelerated Transition; Ellen MacArthur Found: Cowes, UK (2013)

12. Ellen MacArthur Foundation and McKinsey Center for Business and Environment: Growth within: A circular Economy Vision for a Competitive Europe (2015). https://www.ellenmacarthurfoundation.org/assets/downloads/circular economy/Growth-Within-Report.pdf . last accessed 2021/12/18

13. Della Spina, L.: Multidimensional assessment for culture-led and community-driven urban regeneration as driver for trigger economic vitality in urban historic centers. Sustainability **11**, 7237 (2019). https://doi.org/10.3390/su11247237

14. Della Spina, L., Giorno, C., GalatiCasmiro, R.: Bottom-up processes for culture-led urban regeneration scenarios. In: Misra, S., et al. (eds.) ICCSA 2019. LNCS, vol. 11622, pp. 93–107. Springer, Cham (2019). https://doi.org/10.1007/978-3-030-24305-0_8

15. Della Spina, L., Giorno, C., Galati Casmiro, R.: An integrated decision support system to define the best scenario for the adaptive sustainable re-use of cultural heritage in Southern Italy. In: Bevilacqua, C., Calabrò, F., Della Spina, L. (eds.) NMP 2020. SIST, vol. 177, pp. 251–267. Springer, Cham (2020). https://doi.org/10.1007/978-3-030-52869-0_22

16. Roscelli, R. (ed.): Manuale di Estimo: Valutazioni Economiche ed Esercizio della Professione; UTET Università: Milano, Italy (2014)

17. Prizzon, F.: Gli Investimenti Immobiliari. Analisi di Mercato e Valutazione Economico-Finanziaria degli Interventi; Celid: Torino, Italy (1995)

18. Della Spina, L.: The integrated evaluation as a driving tool for cultural-heritage enhancement strategies. In: Bisello, A., Vettorato, D., Laconte, P., Costa, S. (eds.) SSPCR 2017. GET, pp. 589–600. Springer, Cham (2018). https://doi.org/10.1007/978-3-319-75774-2_40

19. Tajani, F., Morano, P., Liddo, F.: À roles and risks of the subjects involved in public-private partnerships: the feasibility analysis of an enhancement investment in the city of Rome (Italy). [Complementarieta'dei ruoli dei soggetti coinvolti in procedure di partenariato pubblico privato per l'efficacia degli interventi e la diversificazione dei rischi di mercato: Analisi di fattibilita'di un progetto di valorizzazione nella citta'di Roma]. LaborEst 18/2019, 27–33 (2019). https://doi.org/10.19254/LaborEst.18.04

20. Nuti, F.: La valutazione economica delle decisioni pubbliche. Dall'analisi costi-benefici alle valutazioni contingenti. Giappichelli, Torino (2001)

21. Dente, B.: Understanding Policy Decisions. Springer, Cham (2014). https://doi.org/10.1007/978-3-319-02520-9

22. Figueira, J.R., Greco, S., Ehrgott, M. (eds.): Multiple Criteria Decision Analysis: State of the Art Surveys. Springer, Heidelberg (2005). https://doi.org/10.1007/b100605

23. Tajani, F., Morano, P., Di Liddo, F., Locurcio, M.: An innovative interpretation of the DCFA evaluation criteria in the Public-Private Partnership for the enhancement of the public property assets. [Un'interpretazione innovativa dei criteri di valutazione della DCFA nel partenariato pubblico-privato per la valorizzazione del patrimonio immobiliare pubblico] Labor-Est 16/2018, pp. 53–57 (2018). https://doi.org/10.19254/LaborEst.16.09

24. Calabrò, F., Della Spina, L.: La fattibilità economica dei progetti nella pianificazione strategica, nella progettazione integrata, nel cultural planning, nei piani di gestione. Un modello sperimentale per la valorizzazione di immobili pubblici in Partenariato Pubblico Privato. LaborEst 16/2018, pp. 1–49 (2018). https://doi.org/10.19254/LaborEst.16.IS

25. Della Spina, L.: Historical cultural heritage: Decision making process and reuse scenarios for the enhancement of historic buildings. In: Calabrò, F., Della Spina, L., Bevilacqua, C. (eds.) ISHT 2018. SIST, vol. 101, pp. 442–453. Springer, Cham (2019). https://doi.org/10.1007/978-3-319-92102-0_47

26. Sdino, L., Rosasco, P., Magonia, S.: The Financial Feasibility of a Real Estate Project: the Case of the ex Tessito. [La fattibilità finanziaria di un progetto immobiliare: il caso dell'ex Tessitoria Schiatti] LaborEst 13/2016, pp. 34–37 (2016). https://doi.org/10.19254/Lab

27. Della Spina, L., Rugolo, A.: A multicriteria decision aid process for urban regeneration process of abandoned industrial areas. In: Bevilacqua, C., Calabrò, F., DellaSpina, L. (eds.) NMP 2020. SIST, vol. 178, pp. 1053–1066. Springer, Cham (2021). https://doi.org/10.1007/978-3-030-48279-4_99

28. Della Spina, L.: Strategic planning and decision making: A case study for the integrated management of cultural heritage assets in Southern Italy. In New Metropolitan Perspectives; NMP 2020; Smart Innovation, Systems and Technologies; Bevilacqua, C., Calabrò, F., Della Spina, L., Eds.; Springer: Cham, Switzerland, 2020, Volume 178 (2021)

29. Calabrò, F., Della Spina, L.: The Public–Private Partnership for the Enhancement of Unused Public Buildings: An Experimental Model of Economic Feasibility Project. Sustainability **11**, 5662 (2019)

30. Carbonara, S., Stefano, D., Di Prinzio, A.: Transforming surface rights into property rights: an analysis of current estimation procedures and a comparison with an alternative. Valori e Valutazioni. Teorie ed Esperienze **2020**(24), 5–17 (2020)

31. Carbonara, S., Stefano, D.: The valorisation of public real estate assets in italy: a critical reconstruction of the legislative framework. In: Bevilacqua, C., Calabrò, F., Della Spina, L. (eds.) NMP 2020. SIST, vol. 178, pp. 475–485. Springer, Cham (2021). https://doi.org/10.1007/978-3-030-48279-4_45

32. Della Spina, L.: Cultural heritage: a hybrid framework for ranking adaptive reuse strategies. Buildings **11**, 132 (2021). https://doi.org/10.3390/buildings11030132

33. Della Spina, L. Giorno, C.: Cultural landscapes: a multi-stakeholder methodological approach to support widespread and shared tourism development strategies. Sustainability **13**, 7175 (2021). https://doi.org/10.3390/su13137175

34. Della Spina, L.: Revitalization of inner and marginal areas: a multi-criteria decision aid approach for shared development strategies. Valori e Valutazioni **2020**(25), 37–44 (2020)

35. Nesticò, A., Moffa, R.: Economic analysis and Operational Research tools for estimating productivity levels in off-site construction [Analisi economiche e strumenti di Ricerca Operativa per la stima dei livelli di produttività nell'edilizia off-site], Valori e Valutazioni n. 20, pp. 107–128, ISSN: 2036–2404. DEI Tipografia del Genio Civile, Roma (2018)

36. Nesticò, A., Maselli, G.: Declining discount rate estimate in the long-term economic evaluation of environmental projects. J. Environ. Acc. Manag. **8**(1), 93–110 (2020). https://doi.org/10.5890/JEAM.2020.03.007

37. Manganelli, B., Tajani, F.: Optimised management for the development of extraordinary public properties. J. Prop. Invest. Fin. **32**(2), 187–201 (2014)

38. Tajani, F., Morano, P., Locurcio, M., D'Addabbo, N.: Property valuations in times of crisis: artificial neural networks and evolutionary algorithms in comparison. In: Gervasi, O., et al. (eds.) ICCSA 2015. LNCS, vol. 9157, pp. 194–209. Springer, Cham (2015). https://doi.org/10.1007/978-3-319-21470-2_14

39. Manganelli, B., Morano, P., Tajani, F.: The risk assessment in Ellwood's financial analysis for the indirect estimate of urban properties. AESTIMUM **55**, 19–41 (2009)

40. Dolores, L., Macchiaroli, M., De Mare, G.: Sponsorship's financial sustainability for cultural conservation and enhancement strategies: an innovative model for sponsees and sponsors. Sustainability **13**(6), 9070 (2021). https://doi.org/10.3390/su13169070

41. Dolores, L., Macchiaroli, M., De Mare, G.: A Dynamic model for the financial sustainability of the restoration sponsorship. Sustainability (Switzerland) **12**(4), 1694 (2020). https://doi.org/10.3390/su12041694

42. Dolores, L., Macchiaroli, M., De Mare, G.: A model for defining sponsorship fees in public-private bargaining for the rehabilitation of historical-architectural heritage. In: Calabrò, F., Della Spina, L., Bevilacqua, C. (eds.) ISHT 2018. SIST, vol. 101, pp. 484–492. Springer, Cham (2019). https://doi.org/10.1007/978-3-319-92102-0_51

43. Benintendi, R., De Mare, G.: Upgrade the ALARP model as a holistic approach to project risk and decision management. Hydrocarb. Process. **9**, 75–82 (2017)

44. Della Spina, L.: A multi-level integrated approach to designing complex urban scenarios in support of strategic planning and urban regeneration. In: Calabrò, F., Della Spina, L., Bevilacqua, C. (eds.) ISHT 2018. SIST, vol. 100, pp. 226–237. Springer, Cham (2019). https://doi.org/10.1007/978-3-319-92099-3_27

45. Della Spina, L., Ventura, C., Viglianisi, A.: A multicriteria assessment model for selecting strategic projects in urban areas. In: Gervasi, O., et al. (eds.) ICCSA 2016. LNCS, vol. 9788, pp. 414–427. Springer, Cham (2016). https://doi.org/10.1007/978-3-319-42111-7_32

46. Della Spina, L., Viglianisi, A.: Hybrid evaluation approaches for cultural landscape: the case of "Riviera dei Gelsomini" area in Italy. In: Bevilacqua, C., Calabrò, F., Della Spina, L. (eds.) NMP 2020. SIST, vol. 178, pp. 1369–1379. Springer, Cham (2021). https://doi.org/10.1007/978-3-030-48279-4_128

47. Della Spina, L.: Scenarios for a sustainable valorisation of cultural landscape as driver of local development. In: Calabrò, F., Della Spina, L., Bevilacqua, C. (eds.) ISHT 2018. SIST, vol. 100, pp. 113–122. Springer, Cham (2019). https://doi.org/10.1007/978-3-319-92099-3_14

48. Della Spina, L.: Strategic planning and decision making: a case study for the integrated management of cultural heritage assets in Southern Italy. In: Bevilacqua, C., Calabrò, F., Della Spina, L. (eds.) NMP 2020. SIST, vol. 178, pp. 1116–1130. Springer, Cham (2021). https://doi.org/10.1007/978-3-030-48279-4_104

49. Della Spina, L.: Integrated evaluation and multi-methodological approaches for the enhancement of the cultural landscape. In: Gervasi, O., et al. (eds.) ICCSA 2017. LNCS, vol 10404, pp 478–493. Springer, Cham (2017). https://doi.org/10.1007/978-3-319-62392-4_35

Migrants, Retail Properties and Historic Centre. Urban and Economic Resilience in Palermo (Italy)

Grazia Napoli$^{(\boxtimes)}$ (iD) and Simona Barbaro (iD)

University of Palermo, 90133 Palermo, Italy
grazia.napoli@unipa.it

Abstract. Migratory flows arriving in the countries of the European Union over the recent years have generated significant urban, social and economic transformations, especially in metropolitan areas. Although with different intensity, the increase of new cultures and ethnic groups living in Italy has led to changes in both social value system and urban economic dynamism, generating consequences also on the real estate market. In order to analyze urban dynamics linked to migration flows, this study focuses on the retail properties used for inner-city shops owned/managed by immigrants. In particular, the city of Palermo (Italy) is chosen as a case study due to its multicultural stratification that is particularly complex in the historic centre. This study highlights also the resilience of foreign business activities after the Covid-19 pandemic emergency and their contribution to the production of social wealth.

Keywords: Multicultural city · Retail real estate · Migration flows

1 Introduction

In the European Union, migration flows of foreign populations have settled especially in the main metropolitan areas where the coexistence of different cultural identities, lifestyles and aspirations has led to transformations in the urban fabric. These transformations are particularly manifest in the neighbourhoods with a high concentration of people of foreign origin, in public spaces such as parks, squares, streets inhabited by multicultural communities, and also in the proliferation of retail shops owned/run by immigrants. As a result, new challenges arise to deeply understand new trends in the development of urban space that must be "considered not only as a container for migrants' activities, but as a mobile and fluid entity that transforms itself based on individual and collective practices in which migrants are protagonists, and can constitute a real relational and strategic resource for them" [1] (p. 4).

Moreover, this phenomenon produces significant changes in social value systems and urban economy, as well as urban spatial hierarchy, and also affects real estate markets [2, 3]. In recent years, several studies have analysed the economic, political, social and demographic aspects of immigration [4], concerning specific territorial contexts, such as inland areas [5] or district of provincial capitals [6], and particular communities of foreigners [7, 8], highlighting the impact of migration flows [9].

It is worth reporting that the European Commission recognised the contribution of immigrant entrepreneurs to the European economy in the "Entrepreneurship 2020 Action Plan", attributing them a fundamental role in the economic revival of the EU [10]. In Italy, for example, the value of wealth produced by foreigners was €134 billion in 2020, with an incidence on the national Gross Domestic Product (GDP) of 9.0%, although below its 9.5% in 2019 [11, 12].

Starting from these considerations, this research examines the case of Palermo (Italy) as a representative example of a Mediterranean and multicultural city and focuses on the dynamics of the historic centre that are related to the retail properties used for shops owned/managed by foreign citizens. This research also shows how resilient foreign business activities have been in recent years, characterized by the crisis in the economic system, due to the lockdown periods and the Covid-19 pandemic.

2   Methodology

The proposed methodology consists of the following steps:

- Demographic analysis of the distribution of migrants among the city's neighbour-hoods;
- Identification of urban areas with a large number of migrants and selection of the main high streets;
- Direct survey of data describing how retail properties are used and what type of retail shops are owned/run by immigrants;
- Analysis of the real estate market.

3   The Case Study of the City of Palermo (Italy)

The city of Palermo, located in southern Italy, is chosen as a case study [13]. There were 25,445 foreign residents in Palermo as of January 1, 2020, out of a population of 640,720 [14]. In addition, 4,327 foreigners had acquired Italian citizenship during 2019

Table 1. Data of foreigners in Palermo in 2017–2020.

Year	Total Population (No.)	Foreigners (No.)	Foreigners/ Population (%)	Annual variation in foreigners (No.)	Annual variation in foreigners (%)	Acquired Italian citizenship (No.)
2017	668,405	25,607	3.83	−1,119	−4.37	4,323
2018	652,720	25,753	3.95	+146	+0.57	4,320
2019	660,048	25,522	3.87	−231	−0.91	4,002
2020	640,720	25,445	3.97	−77	−0.30	4,327

(Table 1). The share of foreigners in the total population in 2020 was 4% as in 2015, although in the latter case the absolute values were different: 26,647 foreigners out of a population of 674,435.

Foreigners live mainly in the historic centre and they are 2,158 and 2,869 respectively in the "Tribunali-Castellammare" and "Palazzo Reale-Monte di Pietà" neighbourhoods, corresponding to 17.60% and 20.33% of the inhabitants. There is also a high concentration of foreigners in other nearby neighbourhoods such as "Oreto-Stazione", "Zisa", "Noce" and "Politeama". In 2019, 69.26% of all foreigners lived in these six neighbourhoods, while the remaining 30.74% was distributed in other areas of the city (Table 2 and Fig. 1).

Immigrants living in Palermo are from more than 120 different countries, but the most numerous communities rooted in the urban fabric are those from Bangladesh (5,341), Sri Lanka (3,270), Romania (3,205), Ghana (2,611), the Philippines (1,756), Tunisia (1,042), China (984), Morocco (979), Mauritius (832) and Nigeria (630).

Table 2. Foreigners living in the neighbourhoods of Palermo in 2019 (source: Comune di Palermo).

Neighbourhood No.	Neighbourhood name	Total population (No.)	Foreigners (No.)	Foreigners/ Population (%)
1	Tribunali - Castellammare	12,261	2,158	17.60
2	Palazzo Reale - Monte di Pietà	14,113	2,869	20.33
3	Oreto - Stazione	39,916	4,209	10.54
6	Zisa	34,244	2,538	7.41
7	Noce	27,856	2,271	8.15
10	Politeama	32,029	3,631	11.34
	Others	499,629	7,846	1.57
	Total	**660,048**	**25,522**	**3.87**

The location of migrants in the central districts and, especially, in the historic centre has led to the use of a great share of the residential housing stock which, often, is of low construction quality because it is affordable for low-income households such as those of many immigrants [15, 16]. The use of public spaces also reflects interactions within and outside foreign communities in terms of habits, timeslot, and type of users, forming new hierarchies of inclusive or exclusive spaces [17]. Immigrants also influence the use of retail properties, where many shops offer products and services to their own community, as well as to all residents and even tourists, strengthening the rootedness of migrants in neighbourhoods.

Because there has been an oversupply in the retail real estate market under current conditions, surveys directly conducted in recent years reveal that migrants have actively contributed to the resilience of the local real estate market and have provided a continuous income flow to property owners. In order to verify the resilience and role of migrants'

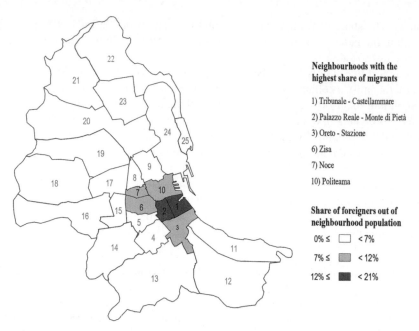

Fig. 1. Shares of foreigners living in the neighbourhoods of Palermo in 2019.

shops in this market segment, the retail shops located in the high streets of the historic centre were analyzed from 2017 to 2020. The streets considered were as follows:

- Via Vittorio Emanuele;
- Via Maqueda, which crosses Via Vittorio Emanuele and divides the historic centre into four "Mandamenti" (areas);
- Via Roma, built in the 19th century as a partition of the historic fabric.

3.1 Retail Properties in the Main Streets of the Historic Centre

A survey of the retail properties and shops located in the three high streets previously mentioned was directly conducted between 2017 and 2020 during workshops with students of the degree of Town Planning and Urban Studies. The collected data were subsequently checked and processed to obtain a consistent database, despite some missing data. The analysis consists of two different levels:

- *Level 1 – Use of retail properties*;
- *Level 2 – Retail shops owned/run by foreign residents.*

In *Level 1 – Use of retail properties*, the retail properties were analyzed to record presence/absence and type of retail shop, and were categorized as follows:

- *"food"*, retail shops which sell food-related products (e.g. cafes, restaurants, supermarkets, grocery, etc.);

- *"non-food"*, all other types of retail shops (e.g. clothing shop, souvenir shop, etc.);
- *"empty"*, unused shop property;
- *"for lease"*, retail property that is for rent;
- *"for sale"*, retail property that is for sale.

The analysis of the collected data –almost 2,000 (Table 3)– shows that the use of retail properties for "food" shops increased in number over time at the expense of "non-food" shops. This phenomenon happened especially in Via Maqueda, where the share of "food" shops had increased from 25.9% in 2017 to 38.6% in 2020, probably as a result of the municipal decision to pedestrianize the street to support its tourist vocation. The share of empty properties had increased significantly over the years, up to 25.72% in 2020, in all the high streets, but the highest number of unused retail properties was found in Via Roma (56.0% in 2020).

Table 3. Type of use of the retail shops.

Type of use	2017		2018		2019		2020	
	No.	%	No.	%	No.	%	No.	%
Food	50	15.53	134	24.45	105	25.30	173	24.86
Non-food	211	65.53	335	61.13	268	64.58	342	49.14
Vacant	45	13.98	66	12.04	42	10.12	179	25.72
For rent	12	3.73	13	2.37	0	0.00	2	0.29
For sale	4	1.24	0	0.00	0	0.00	0	0.00
Total	**322**	**100.00**	**548**	**100.00**	**415**	**100.00**	**696**	**100.00**

Table 4. Data on retail shops owned/run by foreign residents (2017–2020).

Location	Shops owned/run by foreigners	2017		2018		2019		2020	
		No.	%	No.	%	No.	%	No.	%
Total	Total shops	46	17.6	71	15.1	68	18.2	97	18.8
	Food	14	30.4	23	32.4	24	35.3	37	38.1
	Non-food	32	69.6	48	67.6	44	64.7	60	61.9
	Asia	40	87.0	64	90.1	62	91.2	87	89.7
	Africa	6	13.0	7	9.9	6	8.8	10	10.3
Via Vittorio Emanuele	Total shops			14	6.5	18	10.8	20	9.1
	Food			5	35.7	7	38.9	7	35.0
	Non-food			9	64.3	11	61.1	13	65.0

(*continued*)

Table 4. (*continued*)

Location	Shops owned/run by foreigners	2017		2018		2019		2020	
		No.	%	No.	%	No.	%	No.	%
	Asia			14	100	17	94.4	19	95.0
	Africa			0	0	1	5.6	1	5.0
Via Maqueda	Total shops	38	32.8	53	30.3	43	37.4	66	33.5
	Food	13	34.2	16	30.2	15	34.9	23	34.8
	Non-food	25	65.8	37	69.8	28	65.1	43	65.2
	Asia	35	92.1	49	92.5	39	90.7	61	92.4
	Africa	3	7.9	4	7.5	4	9.3	5	7.6
Via Roma	Total shops	8	5.5	4	5.0	7	7.6	11	11.2
	Food	1	12.5	2	50.0	2	28.6	7	63.6
	Non-food	7	87.5	2	50.0	5	71.4	4	36.4
	Asia	5	62.5	1	25.0	6	85.7	7	63.6
	Africa	3	37.5	3	75.0	1	14.3	4	36.4

Level 2 – Retail shops owned/run by foreign residents. Data on shops owned/run by foreigners were collected noting the continent and, if available, the retailers' country of origin. The *Level 2* database was obtained from the *Level 1* database by extracting all data related to properties used by foreign retailers at least once in the 4 years considered. This also made it possible to detect the frequency and location of all the changes in ownership/management, closures or openings of retail activities. Almost 300 data out of 2,000 of *Level 1* were related to foreign retailers (Table 4).

The highest share of shops owned/managed by foreign residents was in Via Maqueda where, in fact, in 2020 it was about one-third of retail properties (33.5%). Via Roma and Via Vittorio Emanuele, on the contrary, had the lowest percentage, 11.2% and 9.1% respectively in 2020. The 18.8% average presence in the three high streets of foreign retailers demonstrates the active contribution of immigrants in supporting housing market demand, even in the retail segment.

In terms of nationality, almost 90% of foreign retailers is of Asian origin, mainly from Bangladesh, and only a few are of African origin, most of them from Tunisia. No retailers from America, Oceania or other non-EU countries were detected. However, it was not always possible to know the countries of origin of all retailers.

In general, the presence of foreign shops along Via Vittorio Emanuele and Via Maqueda has been rooted for a long time and contributes to the commercial real estate dynamism of the high streets of the historic centre and also of its cross streets, as analysed by other studies [17]. For example, in Via Maqueda six Italian shops were replaced by retail businesses run by immigrants. At the same time, however, two previously Asian shops were replaced by Italian retail activities (see Table 5).

However, the presence of foreign shops along Via Maqueda, especially in the first section from Piazza Giulio Cesare to Piazza Vigliena, seems to be more constant and rooted than in the other axes of the historic centre.

Table 5. Turnover in retail shops owned/run by foreign residents (2017–2020).

Retail shops from ▶ to			Via Vittorio Emanuele	Via Maqueda	Via Roma	Total
Foreign	▶	Italian	2	4	3	9
Italian	▶	Foreign	2	6	7	15
Foreign	▶	Foreign	0	3	1	4
Foreign	▶	Empty	0	5	1	6
Empty	▶	Foreign	1	2	2	5
Permanence of shops			14	41	2	57

During the period under consideration, the number of foreign businesses (15) that took over Italian ones was greater than the number of Italian businesses (9) that took over foreign ones, while 5 new shops are offset by 6 closed retail businesses leaving retail properties vacant. In via Roma, the retail properties used by immigrants underwent frequent turnover and the number of foreign retailers taking the place of Italian ones was prevalent (see Fig. 2), they are of African and, in particular, Tunisian origin.

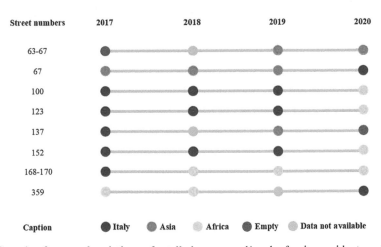

Fig. 2. Sample of punctual variations of retail shops owned/run by foreign residents and located in Via Roma.

The comparison of the data related to the high streets, Via Vittorio Emanuele, Via Maqueda and Via Roma, allows making some considerations on the outcomes of the pedestrianization and the establishment of Limited Traffic Zones (ZTL) both decided by the Municipality of Palermo. Although all the three streets fall within the perimeter of the ZTL, Via Vittorio Emanuele and Via Maqueda were largely pedestrianised, while Via Roma has always remained a vehicular street. These decisions have created the conditions for an increase in demand for the rental of retail properties and the opening of shops in Via Vittorio Emanuele and Via Maqueda. On the contrary, Via Roma has been penalized, accentuating the trend of decline already underway as a commercial axis that has caused an increase in unused properties and vacancy periods exceeding one year.

Indeed, although the percentage of empty commercial properties in Via Roma was very high (56% in 2020), there were still a few properties rented to foreigners. In Via Maqueda, on the other hand, the share of vacant properties is low (10.5% in 2020) and the presence of foreigners was considerable as previous mentioned (33.5% in 2020). Moreover, data show that there was a prevalence of Asians (particularly Bengalis), who intercepted the consumption needs of young people, tourists and residents by selling ethnic fast-food, clothing and other products with opening hours extended into the evening. However, it is worth noticing that these shops were mainly located in the first section of Via Maqueda where the rents were lower than those in the second section –from piazza Vigliena to piazza Giuseppe Verdi– and, according to a direct survey in 2018, the most frequent rents were between 100 and 200 €/sq.m per year [17].

3.2 Commercial Real Estate Market

After carrying out the analysis of data on retail properties owned/run by immigrants, the OMI (Osservatorio del Mercato Immobiliare) quotations [18] of this type of real estate were collected according to the OMI's zones B14, B16, B17, B18 and B19, in which the previously selected streets fall (Fig. 3).

In 2020 there was a significant decrease in the sale values compared to 2017, ranging from −9% to −16%, in all the zones. The only exception was zone B16 –where the second section of Via Maqueda falls– which bucked the trend with slightly higher sales values (see Table 6).

Based on these values and assuming an average size of 30 sq.m per property, it can be supposed that in 2020 the total amount of annual revenue, corresponding to the rental of retail properties to immigrants along the three considered high streets, ranged from a minimum of €230,000 to a maximum of €420,000.

Fig. 3. The OMI's zones and the main streets in the historic centre of Palermo.

Table 6. Values of retail properties according to the OMI's zones in Palermo.

OMI's zones	Street name	Year	Semester	Min sales value (€/mq)	Max sales value (€/mq)	Min rental value (€/mq year)	Max. rental value (€/mq year)
B14	Vittorio Emanuele	2017	2	1,550	2,150	102.0	168.0
		2020	2	1,150	1,950	102.0	180.0
B16	Maqueda (2nd section)	2017	2	1,050	1,800	79.2	120.0
		2020	2	1,000	1,850	72.0	138.0
B17	Maqueda (1st section)	2017	2	930	1,300	72.0	92.4
		2020	2	790	1,100	72.0	96.0

(*continued*)

Table 6. (*continued*)

OMI's zones	Street name	Year	Semester	Min sales value (€/mq)	Max sales value (€/mq)	Min rental value (€/mq year)	Max. rental value (€/mq year)
B18	Roma (2nd section)	2017	2	930	1,500	73.2	110.4
		2020	2	880	1,250	72.0	116.4
B19	Roma (1st section)	2017	2	1,000	1,500	73.2	110.4
		2020	2	860	1,300	73.2	120.0

4 Conclusions

This study analyzed some implications associated with the presence of migrant-owned/run retail shops along the main axes of the historic centre of Palermo. Knowledge of these implications is crucial to understand the economic, spatial, social, but also real estate dynamics that characterize many neighborhoods and to support processes of urban regeneration [19, 20], that promote the inclusion of migrant communities within the urban fabric. In addition, knowing the potential durability and resilience of certain processes is useful for studying the characteristics of specific segments of the real estate market, such as retail, and estimating their future performance.

The pandemic has dramatically highlighted the importance of flexibility and adaptability of public and private spaces in our cities, including workplaces, businesses, and retail activities. The permanence of retail shops owned/run by immigrants in the historic centre of Palermo, even after the first waves of the pandemic, is an important indicator of the inclusion of migrant populations in the economic, social and urban fabric of the city and their contribution to the production of social wealth.

References

1. Schmoll, C.: Spazi insediativi e pratiche socio-spaziali dei migranti in città. Il caso di Napoli. Studi Emigrazione, 699–719, halshs-00212027 (2006)
2. Napoli, G.: Values spaces migration. Appraisal scenarios for an intercultural society. Valori e Valutazioni **28**, 49–58 (2021). https://doi.org/10.48264/VVSIEV-20212805
3. Cochrane, W., Poot, J.: Effects of immigration on local housing markets. In: Kourtit, K., Newbold, B., Nijkamp, P., Partridge, M. (eds.) The Economic Geography of Cross-Border Migration. FRS, pp. 269–292. Springer, Cham (2021). https://doi.org/10.1007/978-3-030-48291-6_12
4. Micelli, E.: The intercultural city: real estate markets, migratory trends and social dynamics. Valori e Valutazioni **28**, 59–66 (2021)
5. Oppio, A.: Migrants and Italian inner areas for an anti-fragility strategy. Valori e Valutazioni **28**, 93–100 (2021)
6. Kalantaryan, S., Alessandrini, A.: Housing values and the residential settlement of migrants: zooming in on neighbourhoods in Italian provincial capitals. Spatial Demography **8**(3), 293–350 (2020). https://doi.org/10.1007/s40980-020-00068-1

7. Rosasco, P., Sdino, L., Sdino, B.: Immigration in Genoa: real estate demand survey in the historic centre. Valori e Valutazioni **28**, 67–80 (2021)
8. Forte, F., De Biase, C., De Paola, P.: The multicultural territory of domitian coast: housing condition and real estate market. Valori e Valutazioni **28**, 81–92 (2021)
9. De Biase, C., Manna, M.: The city as a place of ethnic and cultural fusion - how the growth of foreign population influenced the transformations of the territory. In: Gambardella, C. (ed.) World heritage and Legacy. Culture, Creativity, Contamination, Architecture, pp. 211–215. Gangemi Editore International, Roma (2019)
10. European Commission: Entrepreneurship 2020 Action Plan. Reigniting the entrepreneurial spirit in Europe (2013). https://eur-lex.europa.eu/LexUriServ/LexUriServ.do?uri=COM: 2012:0795:FIN:en:PDF. Accessed 20 Dec 2021
11. Moressa, F.L.: Rapporto annuale sull'economia dell'immigrazione. Edizione 2021. 1st edn. Il Mulino, Bologna (2021)
12. Ministero del Lavoro e delle Politiche Sociali: XI Rapporto annuale. Gli stranieri nel mercato del lavoro in Italia (2021). https://www.lavoro.gov.it/documenti-e-norme/studi-e-statistiche/ Pagine/default.aspx. Accessed 20 Dec 2021
13. Bonafede, G., Napoli, G.: Palermo multiculturale tra rischi di gentrification e crisi del mercato immobiliare nel centro storico. Archivio di Studi Urbani e Regionali **113**, 123–150 (2015)
14. Comune di Palermo, Archivio Open data (2019). https://opendata.comune.palermo.it/ope ndata-dataset.php?. Accessed 20 Dec 2021
15. Napoli, G., Giuffrida, S., Trovato, M.: Fair planning and affordability housing in urban policy. The case of syracuse (Italy). In: Gervasi, O., Murgante, B., Misra, S., Rocha, A.M.A.C., Torre, C.M.M., Taniar, D., Apduhan, B.O.O., Stankova, E., Wang, S. (eds.) ICCSA 2016. LNCS, vol. 9789, pp. 46–62. Springer, Cham (2016). https://doi.org/10.1007/978-3-319-42089-9_4
16. Napoli, G.: Housing affordability in metropolitan areas. The Application of a Combination of the Ratio Income and Residual Approaches to Two Case Studies in Sicily, Italy. Buildings **7**(4), 95, 1–19 (2017). https://doi.org/10.3390/buildings7040095
17. Napoli, G., Bonafede, G.: The urban rent in the multicultural city: retail shops, migrants and urban decline in the historic centre of Palermo. Valori e Valutazioni **27**, 67–76 (2020). https:// doi.org/10.48264/VVSIEV-20202707
18. Osservatorio del Mercato Immobiliare – OMI. https://www.agenziaentrate.gov.it/portale/ web/guest/schede/fabbricatiterreni/omi. Accessed 20 Dec 2021
19. DellaSpina, L., Giorno, C., GalatiCasmiro, R.: Bottom-up processes for culture-led urban regeneration scenarios. In: Misra, S., et al. (eds.) ICCSA 2019. LNCS, vol. 11622, pp. 93–107. Springer, Cham (2019). https://doi.org/10.1007/978-3-030-24305-0_8
20. DellaSpina, L., Giorno, C., GalatiCasmiro, R.: An Integrated decision support system to define the best scenario for the adaptive sustainable re-use of cultural heritage in Southern Italy. In: Bevilacqua, C., Calabrò, F., Della Spina, L. (eds.) NMP 2020. SIST, vol. 177, pp. 251–267. Springer, Cham (2020). https://doi.org/10.1007/978-3-030-52869-0_22

Dam Break-Induced Urban Flood Propagation Modelling with DualSPHysics: A Validation Case Study

Salvatore Capasso[1](✉) , Bonaventura Tagliafierro[1] , and Giacomo Viccione[2](✉)

[1] Department of Civil Engineering, University of Salerno, 84084 Fisciano, Italy
[2] Environmental and Maritime Hydraulics Laboratory (LIDAM), University of Salerno, 84084 Fisciano, Italy
gviccion@unisa.it

Abstract. Sustainable, safe and healthy growth of urban areas is currently achieved by a series of measures and initiatives, including the planning, management and mitigation of natural and anthropic possible hazards. Among potential menaces, extreme precipitation events of short duration, inducing severe urban flooding, may pose a significant threat for residents, especially if no counter-measures are planned to tackle with. Urban development, if not integrated with flooding mitigation strategies, generally goes with an increase of the impermeable surface, yielding a decreasing of infiltration and water evaporation as a result, and ultimately an increase of runoff peaks and a decrease of concentration times. These kinds of phenomena tend to worsen due to the ongoing climate change. In this framework, the sudden release of water through a simplified scaled urban configuration, following an abrupt dam breaching, is here numerically investigated. The open-source Lagrangian-based DualSPHysics solver was used for this aim. The aligned square city layout of 5 x 5 buildings case of Soares-Frazão and Zech (2008) was adopted as a validation case study. DualSPHysics post-processed free surface vertical profiles were compared with data recorded at 17 water-level gauges and numerical results of a finite-volume shallow water scheme, showing a reasonable agreement.

Keywords: Dam-break · Urban flooding · SPH · DualSPHysics

1 Introduction

Urban floodings, that is the submerging of populated areas underwater, are among the most significant threats with devastating consequences in terms of human casualties, economic losses, damages to infrastructures and buildings [1, 2]. The World Bank's 2020 flagship report Poverty and Shared Prosperity [3] revealed that floods keep on being the most economically devastating disaster in the world. Contaminants carried by floodwaters, e.g. heavy metals and pesticides, are responsible for soil contamination, an aspect rarely considered for urban areas [4, 5]. Progressive surface impermeabilization and climate change yield an alteration of the urban basin hydrological response, i.e.

© The Author(s), under exclusive license to Springer Nature Switzerland AG 2022
F. Calabrò et al. (Eds.): NMP 2022, LNNS 482, pp. 1284–1292, 2022.
https://doi.org/10.1007/978-3-031-06825-6_124

the way the urban fabric responds to precipitations, reducing soil infiltration and water evaporation [6–8]. Consequently, flow rate peaks keep increasing while the concentration times going reducing [9, 10], a worrying combination that urban areas are facing more and more [11–13]. Impermeable components, such as streets, roofs, parking lots etc., enhance flow propagation, yielding sudden level rises and dangerous flow velocities. Nature-based solution strategies, consisting of blue-green infrastructure (BGI), can provide environmental, social and economic benefits while helping urban resilience [14–18]. Permeable pavements, green roofing, and rain gardens are among the possible pursuing strategies to mitigate urban flooding [19–21].

Numerical simulation of dam breaking wave propagation and impact on structures can provide valuable insights on the kinematics and dynamics response, once proper initial and boundary conditions are given [22–29]. Furthermore, the non-stopping growth of computing power, the availability of more and more advanced and accurate solution schemes makes the use of numerical solvers as a viable alternative to costly experimental setups.

In this work, the fast-transient peculiarity associated with dam-breaking flow waves in urban areas is numerically analyzed with DualSPHysics [30], a Smoothed Particle Hydrodynamics (SPH) [31, 32] based solver for free surface flow problems, already tested for a wide range of applications: wave energy harvesting [33–36], wave propagation and overtopping [25, 37], fluid-structure interaction [39–40]. The experimental set-up presented by Soares-Frazão and Zech [41] and related numerical results are adopted for numerical validation.

The paper is structured as follows. In Sect. 2, materials and methods are given. In Sect. 3, obtained results and related comparisons are provided, finally, the conclusions are drawn in Sect. 4.

2 Materials and Methods

2.1 The Experimental Set-Up

The experiments here considered refer to the dam-break measurements recorded at the wave flume (Fig. 1) located at Hydraulic Laboratory of the Civil and Environmental Engineering Department of the Université Catholique de Louvain, Belgium published in [41].

The initial water level at the upstream tank was 0.40 m. A centrally placed sluice gate, 1-m wide, was instantly removed to simulate the sudden release of water downstream. Channel bottom was made initially wet, providing a 0.011 m water height. The idealized city layout here considered, corresponds to a square of 5 ×5 buildings of parallelepiped shape, each of horizontal and vertical edge length of 0.30 m, and 0.40 m high, aligned with the incoming flow direction (Fig. 2).

Seventeen resistive level gauging devices, used to record local water heights in time, were considered, the latter compared with the results of DualSPHysics, as next discussed. The comparison, later presented in the paper, concerns the water elevation over the streets within the proposed city, at different instants of time.

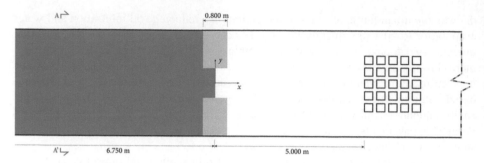

Fig. 1. Plan view of the experimental set-up and channel dimensions in meter.

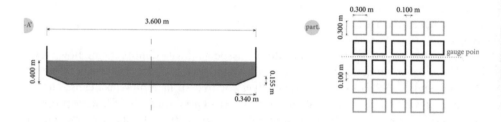

Fig. 2. Left. Lateral cross section of the channel. Right. Zoomed view of the city layout (dimension in meter).

2.2 Numerical Models

SWE Model. The results of the unsteady, two-dimensional (vertical effects neglected) shallow-water numerical model proposed in [41], were taken as well per comparison. Spatial discretization was performed with adaptive unstructured triangular meshes. The ruling Saint-Venant equations, written in the hypothesis of hydrostatic pressure distribution over the generic vertical axis so that the streamlines are straight in the downstream direction, are next provided:

$$\frac{\partial}{\partial t}\begin{bmatrix} h \\ uh \\ vh \end{bmatrix} + \frac{\partial}{\partial x}\begin{bmatrix} uh \\ u^2h + gh^2/2 \\ uvh \end{bmatrix} + \frac{\partial}{\partial y}\begin{bmatrix} vh \\ uvh \\ v^2h + gh^2/2 \end{bmatrix} = \begin{bmatrix} 0 \\ gh(S_{bx} - S_{fx}) \\ gh(S_{by} - S_{fy}) \end{bmatrix} \quad (1)$$

where h is the local flow depth, uh and vh the unit flow rates in the x- and y-directions, respectively, S_{bx} and S_{by} the bed slopes, respectively, and S_{fx} and S_{fy} the components of the friction slope, modelled according to the Manning's resistance formula:

$$S_{fx} = \frac{n^2 u \sqrt{u^2 + v^2}}{n^{\frac{4}{3}}}; S_{fy} = \frac{n^2 v \sqrt{u^2 + v^2}}{n^{\frac{4}{3}}} \quad (2)$$

being $n = 0.010 \text{ sm}^{-1/3}$ the experimentally Manning friction coefficient derived in [41].

DualSPHysics Model. In the SPH method, the domain is discretized into a set of particles (i.e. the computing nodal points), where the values of the physical properties (position, velocity, density, pressure, etc.) are obtained by local interpolation of the properties of the surrounding particles. The Navier-Stokes equations are solved during the particle interactions. The contribution of the neighbouring particles is given using a kernel (or weighting) function, for which a smoothing length or radius of interaction is defined.

The SPH method is mathematically built up on a convolution integral approximation: any function F can be defined by:

$$F(r) = \int F(r')W(r - r')dr' \tag{3}$$

where W is the kernel function, r is the position of the point where the function F is computed, r' is the position defined within the compact support. The F value is approximated by taking into account the contributions of a finite set of surrounding particle. A summation is performed all over the particles within the compact support of the kerne, that isl:

$$F(r_a) = \sum_n F(r_a)W(r_a - r_b, h)\frac{m_b}{\rho_b} \tag{4}$$

where r_a is the position of the interpolated position, r_b is the position of the j-th neighboring particles, $j = 1,...,n$, m and ρ being the mass and the density, respectively, m_b/ρ_b the volume associated with the neighbouring particle b, and h is the smoothing length. The kernel functions W fulfils several properties, such as positivity on the compact support, normalization, and monotonically decreasing with distance. The weighing function used in this work is the piecewise polynomial Quintic Wendland (QW) kernel:

$$W(q) = \alpha_D\left(1 - \frac{q}{2}\right)^4(2q + 1)\, 0 \leq q \leq 2 \tag{5}$$

where α_D is a real number that ensures the kernel normalization property, $q = r/h$ is the non-dimensional distance between pointed and surrounding particle. In this work, the QW kernel is used to compute interactions between particles at a distance r up to the value of $2h$.

The differential system of equations (NS) can be thus written in the discrete SPH formalism [42] as:

$$\frac{dv_a}{dt} = -\sum_b m_b\left(\frac{P_b + P_a}{\rho_b \cdot \rho_a} + \Pi_{ab}\right)\nabla_a W_{ab} + g \tag{6}$$

$$\frac{d\rho_a}{dt} = \rho_a\sum_b m_b v_{ab}\nabla_a W_{ab} + 2\delta hc\sum_b(\rho_b - \rho_a)\frac{v_{ab}\nabla_a W_{ab}}{r_{ab}^2}\frac{m_b}{\rho_b} \tag{7}$$

where t is the time, v is the velocity vector, P pressure, g is the gravitational acceleration, ∇_a is the gradient operator, W_{ab} the kernel function, whose actual value depends on the distance between particles a and b, $r_{ab} = r_a - r_b$ with r_k being the position of the particle k, and c is the speed of sound, δ is a parameter that governs the diffusive term.

The artificial viscosity term, Π_{ab} is added in the momentum equation based on the Neumann–Richtmeyer artificial viscosity, aiming to reduce oscillations and stabilize the SPH scheme.

Boundary conditions are defined following the reference paper by English and co-workers [43], in which the modified Dynamic Boundary Conditions (mDBC) are presented. The basic formulation (DBC) was proposed in [44], in which an enhanced scheme was developed to remove the unphysical gap when fluid particles approach an initial-dry surface.

3 Model configuration and results

In the following, the model setup is presented along with some details on the use of the boundary conditions applied to various city group elements. All the surfaces that are in touch with the fluid phase are treated by using the mDBC technique, which entails the definition of a normal for each particle. This is done by using automatic routines that come bundled to the code.

3.1 Setup

The geometry shown in Figs. 1 and 2, is used to create the initial setup to be used for the SPH simulation, representing it at a 1-to-1 scale. The reservoir in Fig. 1 comprises 9.80 cubic meters of fluid, discretized using two initial inter-particle distances, i.e., $dp_1 = 0.020$ m and $dp_2 = 0.015$m, giving 10 M and 22 M computing particles overall, respectively. The fluid column was kept in place by a mobile gate into the experimental setup which, however, is not modelled for the purpose of these simulations, thus assuming that the water column is instantly released when the simulation starts. Another essential detail regards the wet surface over which the flow propagates. As above specified, there could be a thin layer of water to avoid uncertainties in the interpretation of the results. Nevertheless, due to the particle sizes used in this work, the initial layer of water would have a distorted contribution to the overall fluid dynamics, thus it was not included. Downstream of the building block (city), an outlet is set up to allow the fluid flow out of the simulation domain, thus avoiding extra computation for the part of the domain that is not of interest. This last assumption is fully compliant with the physical test, for which there is no downstream wall.

3.2 Numerical Validation

The model accuracy is tested against the surface profiles available in the experimental reference paper [41], see Fig. 2. Figure 3 compares the numerical and experimental water-surface profiles for two different resolutions. The four panels a), b), c) and d) compare the response of the numerical model at different instants of time of the simulation. Note that the shaded areas in each plot represent the buildings locations. Panel a) reports an initial underestimation of the surface elevation that could be explained by the lack of the initial layer of water. The remaining part of this case proves to be in good agreement with experimental points. The pattern described for panel a) can be observed in Panels b), c), d) as well, demonstrating the consistency of the model throughout the whole benchmark.

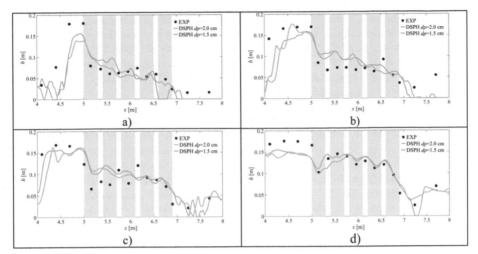

Fig. 3. Experimental and numerical free-surface profiles in central longitudinal street for a) t = 4.00 s, b) t = 5.00 s, c) t = 6.00 s, and d) t = 10.00 s.

4 Conclusions

The proposed SPH-based numerical model has proved to be able to accurately reproduce the fluid-solid interaction that occurs when violent flows are considered within a complex-shaped geometry. The number of particles, needed to achieve a satisfactory agreement between experimental and numerical results is relatively low, if one considers the ease with which a similar setup can be defined. Furthermore, the extension of the validated setup to more interesting applications, such as, for example, the dissipative capabilities of an ensemble of trees upstream of human settlement, could be investigated with relative ease, including the debris transport phenomenon as well. The authors believe that the SPH technique has an edge over other modelling approaches when free-surface flows are interacting with objects.

References

1. Cornwall, W.: Europe's deadly floods leave scientists stunned despite improvements, flood forecasts sometimes failed to flag risks along smaller streams. Science **373**(6553), 372–373 (2021). https://doi.org/10.1126/science.373.6553.372
2. Blöschl, G., et al.: Changing climate both increases and decreases European river floods. Nature **573**(7772), 108–111 (2019). https://doi.org/10.1038/s41586-019-1495-6
3. World Bank: Poverty and Shared Prosperity 2020: Reversals of Fortune. Washington, DC (2020). https://doi.org/10.1596/978-1-4648-1602-4
4. Rözer, V., et al.: Impact-based forecasting for pluvial floods. Earth's Future **9**(2), 2020EF001851 (2021). Doi: https://doi.org/10.1029/2020EF001851
5. Stoppiello, M.G., Lofrano, G., Carotenuto, M., Viccione, G., Guarnaccia, C., Cascini, L.: A comparative assessment of analytical fate and transport models of organic contaminants in unsaturated soils. Sustainability **12**, 1–24, 2949 (2020). Doi: https://doi.org/10.3390/su1207 2949

6. Bian, G., Wang, G., Chen, J., Zhang, J., Song, M.: Spatial and seasonal variations of hydrological responses to climate and land-use changes in a highly urbanized basin of Southeastern China. Hydrol. Res. **52**(2), 506–522 (2021). https://doi.org/10.2166/nh.2021.087

7. Wu, J., Wu, Z.-Y., Lin, H.-J., Ji, H.-P., Liu, M.: Hydrological response to climate change and human activities: a case study of Taihu Basin China. Water Sci. Eng. **13**(2), 83–94 (2020). https://doi.org/10.1016/j.wse.2020.06.006

8. Gao, Y., Chen, J., Luo, H., Wang, H.: Prediction of hydrological responses to land use change. Sci. Total Environ. **708**, 134998 (2020). https://doi.org/10.1016/j.scitotenv.2019.134998

9. Sarchani, S., Awol, F.S., Tsanis, I.: Hydrological analysis of extreme rain events in a medium-sized basin. Appl. Sci. (Switzerland) **11**(11), 4901 (2021). https://doi.org/10.3390/app111 14901

10. Wang, Y., et al.: Quantifying the response of potential flooding risk to urban growth in Beijing. Sci. Total Environ. **705**, 135868 (2020). https://doi.org/10.1016/j.scitotenv.2019.135868

11. Fekete, A., Sandholz, S.: Here comes the flood, but not failure? Lessons to learn after the heavy rain and pluvial floods in Germany 2021. Water (Switzerland) **13**(21), 3016 (2021). https://doi.org/10.3390/w13213016

12. Francipane, A., Pumo, D., Sinagra, M., La Loggia, G., Noto, L.V.: A paradigm of extreme rainfall pluvial floods in complex urban areas: The flood event of 15 July 2020 in Palermo (Italy). Nat. Hazard. **21**(8), 2563–2580 (2021). https://doi.org/10.5194/nhess-21-2563-2021

13. Arrighi, C., Castelli, F.: The 2017 flash flood of Livorno (Italy): lessons learnt from an exceptional hydrologic event. In: Fernandes, F., Malheiro, A., Chaminé, H.I. (eds.) Advances in Natural Hazards and Hydrological Risks: Meeting the Challenge. ASTI, pp. 117–120. Springer, Cham (2020). https://doi.org/10.1007/978-3-030-34397-2_23

14. Magliocchetti, M., Adinolfi, V., Viccione, G., Grimaldi, M., Fasolino, I.: Small rivers and landscape NBS to mitigate flood risk. Sustainable Mediterranean Construction (Special Issue 5), 32–36 (2021). ISSN 2385-1546

15. Hobbie, S.E., Grimm, N.B.: Nature-based approaches to managing climate change impacts in cities. Philos. Trans. Roy. Soc. B: Biol. Sci. **375**(1794), 20190124 (2020). https://doi.org/ 10.1098/rstb.2019.0124

16. Ourloglou, O., Stefanidis, K., Dimitriou, E.: Assessing nature-based and classical engineering solutions for flood-risk reduction in urban streams. J. Ecol. Eng. **21**(2), 46–56 (2021). https:// doi.org/10.12911/22998993/116349

17. Schiavo, V., Fasolino, I., Viccione, G. An ecosystem services-based model for the definition of decision support scenarios for urban flood hazard mitigation: a case study. Sustainability (Switzerland). Under review

18. Viccione, G., Nesticò, A., Vernieri, F., Cimmino, M.: A pilot plant for energy harvesting from falling water in drainpipes. technical and economic analysis. In: Misra, S., et al. (eds.) ICCSA 2019. LNCS, vol. 11624, pp. 233–242. Springer, Cham (2019). https://doi.org/10.1007/978-3-030-24311-1_16

19. Longobardi, A., D'Ambrosio, R., Mobilia, M.: Predicting stormwater retention capacity of green roofs: an experimental study of the roles of climate, substrate soil moisture, and drainage layer properties. Sustainability (Switzerland) **11**(24), 6956 (2019). https://doi.org/10.3390/ su11246956

20. Osheen, M., Singh, K.K.: Rain garden—a solution to urban flooding: a review. In: Agnihotri, A., Reddy, K., Bansal, A. (eds.) Sustainable Engineering. Lecture Notes in Civil Engineering, vol. 30, pp. 27–35. Springer, Singapore (2019). https://doi.org/10.1007/978-981-13-6717-5_4

21. Hernández-Crespo, C., Fernández-Gonzalvo, M., Martín, M., Andrés-Doménech, I.: Influence of rainfall intensity and pollution build-up levels on water quality and quantity response of permeable pavements. Sci. Total Environ. **684**, 303–313 (2019). https://doi.org/10.1016/j. scitotenv.2019.05.271

22. Pugliese Carratelli, E., Viccione, G., Bovolin, V.: Free surface flow impact on a vertical wall: a numerical assessment. Theoret. Comput. Fluid Dyn. **30**(5), 403–414 (2016). https://doi.org/10.1007/s00162-016-0386-9

23. Viccione, G., Bovolin, V., Carratelli, E.P.: A numerical investigation of liquid impact on planar surfaces. In: ECCOMAS Congress 2016 - Proceedings of the 7th European Congress on Computational Methods in Applied Sciences and Engineering, vol. 1, pp. 627–637 (2016). https://doi.org/10.7712/100016.1842.10842

24. Viccione, G., Bovolin, V., Carratelli, E.P.: Simulating fluid-structure interaction with SPH. AIP Conf. Proc **1479**(1), 209–212 (2012). https://doi.org/10.1063/1.4756099

25. Viccione, G., Bovolin, V.: Simulating triggering and evolution of debris-flows with SPH. In: International Conference on Debris-Flow Hazards Mitigation: Mechanics, Prediction, and Assessment, Proceedings, pp. 523–532 (2011). https://doi.org/10.4408/IJEGE.2011-03.B-058

26. Kocaman, S., Evangelista, S., Guzel, H., Dal, K., Yilmaz, A., Viccione, G.: Experimental and numerical investigation of 3d dam-break wave propagation in an enclosed domain with dry and wet bottom. Appl. Sci. (Switzerland) **11**(12), 5638 (2021). https://doi.org/10.3390/app11125638

27. Kocaman, S., Güzel, H., Evangelista, S., Ozmen-Cagatay, H., Viccione, G.: Experimental and numerical analysis of a dam-break flow through different contraction geometries of the channel. Water (Switzerland) **12**(4), 1124 (2020). https://doi.org/10.3390/W12041124

28. Amicarelli, A., et al.: SPHERA v.9.0.0: A Computational Fluid Dynamics research code, based on the Smoothed Particle Hydrodynamics mesh-less method. Comput. Phys. Commun. **250**, 107157 (2020). https://doi.org/10.1016/j.cpc.2020.107157

29. Viccione, G., Izzo, C.: Three-dimensional CFD modelling of urban flood forces on buildings: a case study. J. Phys. Conf. Ser. **2162**, 012020 (2022). https://doi.org/10.1088/1742-6596/2162/1/012020

30. Domínguez, J.M., et al.: DualSPHysics: from fluid dynamics to multiphysics problems Comput. Particle Mech. (2021). https://doi.org/10.1007/s40571-021-00404-2

31. Vacondio, R., et al.: Grand challenges for Smoothed Particle Hydrodynamics numerical schemes. Comput. Particle Mech. **8**(3), 575–588 (2020). https://doi.org/10.1007/s40571-020-00354-1

32. De Padova, D., Ben Meftah, M., De Serio, F., Mossa, M., Sibilla, S.: Characteristics of breaking vorticity in spilling and plunging waves investigated numerically by SPH. Environ. Fluid Mech. **20**(2), 233–260 (2019). https://doi.org/10.1007/s10652-019-09699-5

33. Tagliafierro, B., et al.: A new open source solver for modelling fluid-structure interaction: case study of a point-absorber wave energy converter with a power take-off unit. In: Proceedings of the 11th International Conference on Structural Dynamics, Athens, Greece, 23–26 November 2020, pp. 656-668 (2020), ID 165382. ISSN: 23119020

34. Crespo, A.J., et al.: SPH modelling of extreme loads exerted onto a point absorber WEC. In: Developments in Renewable Energies Offshore - Proceedings the 4th International Conference on Renewable Energies Offshore, RENEW 2020, 206-213, 172350 (2021). https://doi.org/10.1201/9781003134572-25

35. Ropero-Giralda, P., et al.: Efficiency and survivability analysis of a point-absorber wave energy converter using DualSPHysics. Renewable Energy **162**, 1763–1776 (2020). https://doi.org/10.1016/j.renene.2020.10.012

36. Ropero-Giralda, P., et al.: Modelling a heaving point-absorber with a closed-loop control system using the DualSPHysics Code. Energies **14**, 760 (2021). https://doi.org/10.3390/en14030760

37. Altomare, C., Tagliafierro, B., Dominguez, J.M., Suzuki, T., Viccione, G.: Improved relaxation zone method in SPH-based model for coastal engineering applications. Appl. Ocean Res. **81**, 15–33 (2018). https://doi.org/10.1016/j.apor.2018.09.013

38. O'Connor, J., Rogers, B.D.: A fluid–structure interaction model for free-surface flows and flexible structures using smoothed particle hydrodynamics on a GPU. J. Fluids Struct. **104**, 103312 (2021). https://doi.org/10.1016/j.jfluidstructs.2021.103312

39. Capasso, S., et al.: On the development of a novel approach for simulating elastic beams in dualsphysics with the use of the project Chrono library. In: COMPDYN Proceedings, 174550 (2021). ISSN: 26233347

40. Capasso, S., Tagliafierro, B., Martínez-Estévez, I., Domínguez J.M., Crespo A.J.C., Viccione, G.: A DEM approach for simulating flexible beam elements with the Project Chrono core module in DualSPHysics. Comput. Particle Mech. 1–17 (2022). https://doi.org/10.1007/s40 571-021-00451-9

41. Soares-Frazão, S., Zech, Y.: Dam-break flow through an idealised city. J. Hydraul. Res. **46**(5), 648–658 (2008). https://doi.org/10.3826/jhr.2008.3164

42. Capasso, S., et al.: A numerical validation of 3D experimental dam-break wave interaction with a sharp obstacle using DualSPHysics. Water (Switzerland) **13**(15), 2133 (2021). https://doi.org/10.3390/w13152133

43. English, A., et al.: Modified dynamic boundary conditions (mDBC) for general-purpose smoothed particle hydrodynamics (SPH): application to tank sloshing, dam break and fish pass problems. Comput. Particle Mech. (2021). https://doi.org/10.1007/s40571-021-00403-3

44. Crespo, A.J.C., Gómez-Gesteira, M., Dalrymple, R.A.: Boundary conditions generated by dynamic particles in SPH methods. Comput. Mater. Continua **5**(3), 173–184 (2007). ISSN: 15462218

Transport Infrastructures and Economic Development of the Territory

Antonio Nesticò[✉] [iD] and Federica Russo[iD]

Department of Civil Engineering, University of Salerno, Fisciano, SA, Italy
{anestico,frusso}@unisa.it

Abstract. Transport infrastructures offer a fundamental contribution to the economic and social development of a country. These, branching out over the territory, allow the movement and transport of people and goods with a potentially significant reduction in mobility time. Saving time has a positive impact on productivity, on the labour market and on the quality of life of citizens. The objective of this paper is to highlight the importance of the economic impacts generated by transportation infrastructures, particularly in the urban real estate market. Among the benefits induced by investments in transport infrastructures, the increase in territorial accessibility is certainly significant: increasing the attractiveness of properties in relation to their location, improved accessibility also leads to an increase in the relative real estate values. Hedonic pricing method has long been the most widely used reference for studying and quantifying these increases in value. The study also reports on this, along with research perspectives that must concern: the selection of useful variables to express the levels of urban accessibility; the characterization of a regression function for the quantitative estimation of the effects that accessibility determines on the formation of real estate prices

Keywords: Transport infrastructures · Real estate market · Territorial economy

1 Introduction

The economic and social role of the infrastructures is an issue of increasing interest. Indeed, the Economy has long demonstrated the positive correlation between infrastructure and income capacity of an area: infrastructure is a precondition and, at the same time, an essential ingredient for economic development, influencing the territorial, social, economic and political cohesion of a community [1, 2].

The objective of this work is to highlight the importance of the economic impacts generated by transport infrastructures and, particularly, the significant incidence that these infrastructures have on the real estate market and on the formation of urban real estate prices: by improving accessibility, in fact, transport infrastructures contribute to an increase in the attractiveness of real estate in relation to its location.

A. Nesticò and F. Russo contributed equally to this work.

F. Calabrò et al. (Eds.): NMP 2022, LNNS 482, pp. 1293–1302, 2022.
https://doi.org/10.1007/978-3-031-06825-6_125

Research prospects concern the definition of a value function that explicitly includes the powerful interaction between transport infrastructure and real estate values through a variable capable of expressing the level of accessibility of the area.

This article is structured in five sections.

This section introduces the concept of infrastructure and the respective classifications adopted over the years, with insights into the effects that economic and social infrastructures have on the territory. Section 2 reports the interactions between the territorial system and the system of transport infrastructures, focusing on the concept of accessibility as a potential opportunity for interaction between economic operators. Section 3 highlights the economic impacts generated by investments in the transport sector. Specifically, the results of various studies confirm the positive connection between economic development and transport infrastructures using an aggregate production function. Section 4 underlines the close connection between infrastructure and the real estate market. In particular, the results obtained from various studies, with reference to different countries, show how the presence of an efficient transport infrastructure increases the values of urban real estate it serves. In conclusion, in Sect. 5, there is a reflection on the importance of the issues discussed with the future objective of characterizing a regression model useful for a quantitative estimate of the effective influence of transportation infrastructure on the formation of urban real estate prices.

1.1 Definition and Classification of Infrastructures

In modern economic literature, the main reference to the concept of infrastructure has been provided by the German economist Hirschman, who describes infrastructures as «those basic services without which primary, secondary and tertiary productive activates cannot function».

Moreover, A. Mari defines them as «works necessary and instrumental to the performance of economic activities or indispensable for urban developments and, ultimately, aimed at enabling the life of people and communities and improving their conditions»[3]; a definition also adopted by Nubi [4], who describes infrastructures as the set of all structures that allow a city to function efficiently.

The wide variety of works identified by the term "infrastructure" explains the necessity, over the years, to use a classification to distinguish them.

Biehl [5] makes a distinction between network infrastructures and nucleus infrastructures: network infrastructures spread widely over the territory and are characterized by a series of interconnected points, including road, rail and communication networks and energy and water supply systems; nucleus infrastructures present the characteristic of being useful as single units [6] and among these are included schools, hospitals, museums.

Di Palma et al. [7] divide infrastructures into tangible and intangible: tangible infrastructures are characterized by their "physicality" such as transport, water and energy networks; intangible infrastructures are aimed at development, innovation and training, such as research centers or business services.

Furthermore, Aschauer [8] classifies infrastructures as core and non-core infrastructures, including, among the first, roads and motorways, airports, public transport, electricity and gas networks, water distribution networks and sewage networks; types of infrastructures also defined as closely linked to the production process. The non-core infrastructures represent, instead, a residual component such as hospitals and public buildings.

Similar distinction is proposed by Hansen, who classified infrastructure into Social Overhead Capital *SOC* and Economic Overhead Capital *EOC* [6, 9]: «Those items classified as *EOC* are primarily oriented toward the support of directly productive activities or toward the movement of economic goods. *SOC* items [...] may also increase productivity, the way in which they do so is much less direct than in the case for *EOC* items» [10].

Particularly, economic infrastructures contribute to lowering production costs, increasing the exchange of information and reducing freight transport times, which in turn enable the market to become more functional, expand its borders and allow for greater specialization [11]. This category includes road, rail, electricity and sewage networks, ports, aqueducts and airports.

Social infrastructures, on the other hand, have an indirect effect on the economy of the territory, being primarily aimed at increasing social and collective wealth. In this case, Hansen refers to works such as green areas, remediation and urban regeneration interventions, schools, training centers, sports facilities, retirement homes, etc.

As he points out, support for this type of infrastructures is strategic, especially in "lagging regions", areas characterized by a low standard of life due to small-scale agriculture or small industries. Hansen, in this case, considers it appropriate to give priority to investments in social infrastructures since it would affect the education and culture of society, which, in turn, would support the growth of an enterprising spirit necessary for the processes of industrialization. In such areas, economic infrastructures would risk being left unused, even if in abundance, due to the absence of developmental prerequisites.

Investment in economic infrastructures, instead, is advantageous for "intermediate regions", much more so than social regions. These areas, in fact, are characterized by an abundance of well-trained labour, low-cost energy, raw materials and an environment conducive to the presence of this type of infrastructures.

Indeed, it's common thought that the efficient functioning of infrastructures is a necessary condition for a country's competition, but the impact that these generate is strongly characterized by territorial specificities [12].

In this sense, according to Rostow [13], the main objective of the public operator is to implement economic infrastructures in a coordinated way on the national territory. The type of these will depend on the level of development achieved: only when the country's economy has reached a "mature" state, then the attention will shift to social infrastructures.

In contrast, Glaeser, Lolko and Saiz [14] show that for sustainable regional growth and development, it's important for a location to be attractive to both residents and businesses; therefore, they give social infrastructures equal importance to economic ones.

2   Interactions Between Transport Infrastructures and Territory

Infrastructures, especially transport infrastructures, are the subject of various studies that over the years have investigated urban dynamics and the economic impacts generated by them.

Transport infrastructures, branching out over the territory, allow, in different ways, the movement and transport of people and products.

In line with what the economists Wegener and Fürst [15] had already said in 1999, many studies have demonstrated a mutual interconnection between the territorial system and the transport system: the territorial system represents the relationship between the physical, the built and the anthropic environment that generates the identity of a place; the transport system is composed of two elements that influence each other, namely offer and demand.

The demand for transport derives from the necessity of interaction between different activities located in different points of the territory and depends on the number of people and products that move for different reasons, at different times of the year and with different modes of transport.

Instead, the transport offer is constituted by the set of infrastructures and organizational elements, such as tariffs, timetables, and road rules, which make travel possible. This offer has a limited capacity, allowing only a limited number of users to move within a given time interval. An increase in demand, beyond the capacity of the offer, can generate the phenomenon of congestion, with the formation of queues, long travel times and low travel speeds.

In response to changes in the territorial and transport systems, decisions on the location and mobility of private actors, such as families and industries, vary.

In this regard, Wegener and Fürst (1999) highlight the existence of a cyclical relationship between the transportation system and the spatial system, summarized as follows:

– distribution of "land uses", such as residential, industrial, or commercial activities, determines the location of human activities, such as work, education, shopping, and leisure;
– distribution of human activities in space generates the necessary displacements to overcome the distances existing between the places where the different activities are located;
– distribution of transportation infrastructures, necessary to support travel, creates opportunities for spatial interactions that can be measured in terms of accessibility;
– distribution of accessibility in space influences, in turn, the location choices involving changes in the territorial system.

One of the main results induced by investments in transport infrastructure is the increase in territorial accessibility, which can be defined as a potential opportunity for interaction between economic agents located or not in the same region [16]. Specifically, the relationship between accessibility and transport infrastructures is identifiable, more than in terms of physical or temporal distance, as a level of connectivity between regions:

that is, as the ability to establish economic and social ties that are more attractive to economic and commercial activities.

The latter, taking advantage of the accessibility provided by the infrastructure system, will enjoy greater returns than other businesses located elsewhere.

In general, it has been shown that areas with good accessibility to shops, schools, jobs and places of entertainment have a greater influence on the location of residences and develop faster with higher land prices. At the same time, areas with better accessibility to motorways, places for stopover of products, generate a catalytic effect towards economic activities such as businesses, industrial establishments.

It's clear, then, that the mobility system should be thought of as an incorporated element of the spatial distribution of land use and accessibility opportunities provided by transportation infrastructures [15, 17].

Despite the obvious interrelationship between the two systems, however, land use and transportation planning are often separate operations in practice.

3   Transport Infrastructures and Territorial Economy

Transport infrastructures are recognized as strategic and essential factors in the economic and social development of a country and its competitiveness and productivity. These infrastructures, in fact, allow above all the reduction, even potentially relevant, of the times of mobility of men and goods. This saving of time, in turn, positively influences aspects such as the increase in productivity, due to a simplification in the distribution of products, and the expansion of the labour market. The latter can, in turn, affect the improvement of production and encourage an increase in leisure time, translating into an improvement in the quality of life [18, 19].

What still remains a topic of debate is the causality and estimation of the economic impact generated by transportation infrastructure [6].

Does investment in transport promote economic growth or does growth encourage increased demand for transport, and therefore further investment? The conventional view is that the relationship is bidirectional, with transportation acting both as an important facilitator of economic development and as an important outlet for capital investment resulting from economic growth.

Adam Smith, in his 1776 paper "The Wealth of Nations", is one of the first to note the evident positive connection between economic development and transportation infrastructures. He declares for certain the existence of a cause-and-effect relationship between infrastructures, provided they are efficient and effectively allocated, and economic development.

Subsequently, Fogel and Fishlow [20], economists of the last century, highlight the impact of the American railway network on income increases. Still Aschauer [8], in a study on the US economy in the 1950s and 1960s, confirms the importance and strong impact of infrastructure investments.

The same conclusions are also widely shared by contemporary economists, such as Krugman [21], who states that the development of the railway had brought about a transformation in the American economy, initially based on agriculture and livestock, and later industrialized and facilitated by the reduction of transport costs.

The most common approach used in analysing the relationship between investment in transport infrastructure and economic development is to develop a production function model. In the latter, transport infrastructure is treated as public capital that has an impact on output, as do other inputs such as primarily private capital and labour.

A relevant contribution comes from the pioneering work of the previously mentioned economist Aschauer who, based on data relating to the U.S. economy (1949–1985), uses the production function approach estimating a 4% increase in productive output – measured as an increase in GDP – because of a 10% increase in public capital endowment.

In line with this estimate, many others highlight, albeit with different values, the positive influence of public capital on output: Sturm and de Haan [22] for the Netherlands and Stephan [23] for Germany estimate an output elasticity that even exceeds 6%; for Spain, Bajo Rubio and Sosvilla-Rivero [24] obtain an elasticity of 2%; for France, the elasticity measured by Cadot et al. [25] is about 1%, while for Germany, Kemmerling and Stephan [26] estimate an effect of 2%.

In Italy, although with different results, the positive impact generated by transport infrastructure is confirmed, especially for the South where, due to lower initial endowment of infrastructures, they produce greater productivity gains: Paci and Saddi [27] estimate an elasticity of about 2% for the entire Italian territory, in line with the results of Bronzini and Piselli [28], while Bonaglia et al. [29] do not find significant effects except for the South with an elasticity of about 4%.

In general, research shows higher productivity effects for road networks more than for other transport infrastructures. This relates more to the long than to the short and medium term.

4 Transport Infrastructures and Real Estate Market

In addition to directly influencing production and consumption, transport infrastructures improve accessibility and, consequently, have an impact on location choices and the local real estate market. Increasing accessibility, in fact, involves a reduction in travel time and a consequent positive influence on the attractiveness of real estate in relation to its location.

The real estate sector plays a fundamental role in the economy of a country as it brings together different economic agents: construction companies, property owners, tenants, banks. A large percentage of household's assets is invested in real estate and a third of bank lending is represented by loans to households for real estate mortgages and financing to businesses in that sector.

A change in real estate prices, therefore, has important consequences for the entire economic system.

The study of the relationship between transport infrastructures and property values, especially residential, has been the focus of much research over the years. As early as 1964, Alonso's urban location model of residential activities provided a theoretical basis for such studies [30]. For the city of Madrid was estimated the effect on housing prices generated by the proximity to the metro stations Metrosur: a house 1,000 m from the nearest metro station costs between 2,18% and 3,18% less than an equal house, however, located next to the station [31].

Furthermore, for the city of Sydney, it is shown that the impact of subway transit on real estate prices is negative during the announcement phase and positive during the construction of the infrastructure: the price increase is on average 0.037% for each 1% reduction in the distance to the nearest subway station [32].

For Slovenia, the construction of a new highway infrastructure shows a significant and positive impact on the real estate market, represented by an increase in value of about 11%, consistent with the results of research by Efthymiou and Antoniou [33] with reference to the city of Athens and Seo et al. [34] for the city of Phoenix.

The most used estimation model for this type of research, as well as for the studies mentioned, is that of hedonic prices.

Originally used by Court [35] for the analysis of price changes in the automobile market, this method was later taken up by Lancaster [36] who developed a first theoretical basis. Rosen, however, is responsible for the theory of hedonic prices, according to which a good can be valued as the sum of the utilities generated by its characteristics that contribute to its price [37].

A classification of the characteristics that affect the market value of a residential property is proposed by Carlo Forte [38], who distinguishes:

– extrinsic positional characteristics, which relate to the urban context (habitat quality, noise, accessibility to transport services, provision of public services);
– intrinsic positional characteristics, which describe the relationship that the property has with the context in which it is located (luminosity, accessibility, floor level);
– typological and technological characteristics, which include the architectural and material properties of the building (age of construction, state of preservation, quality of finishes);
– productive-economic characteristics, which concern the aspects of the property that affect profitability and determine its usability in an economic sense (possibility of changing the intended use or size, presence of constraints).

In its simplest form, an hedonic equation is a multiple regression, linear or non-linear, able to quantify, for each of the real estate characteristics, the relative hedonic price. Specifically, the independent variables in the equation represent individual housing characteristics and the regression coefficients measure their relative implied prices.

Variation in the measure of one of these variables is reflected in the value of the property. Through the application of this method, it is possible, therefore, to study the connection between transport infrastructures and property values.

Only the introduction in a regression model of appropriate variables able to consider the levels of accessibility to the real estate unit can allow the estimation of the incidence that the infrastructural endowment has on the formation of prices.

5 Conclusions

The aim of this work is to highlight the importance of the economic impacts generated by transport infrastructures, with particular attention to the impact on the formation of urban property prices.

The above-mentioned studies confirm the fundamental and strategic role that infrastructures, especially transport infrastructures, play in the economic and social development of a country. An improvement in infrastructure endowment has a strong influence on the productivity of the entire economic system as well as on the location choices of companies in the area and on the lives of citizens. The quality of infrastructure reflects, at the same time, that of institutional and market mechanisms.

The planning and implementation of transport infrastructures have strong economic and social repercussions in the long term, which is the reason why estimating the economic impacts of such investments appears increasingly important to guide decision-making processes. The approach commonly used for such estimates is to develop an aggregate production function model in which infrastructure is understood as public capital and included as an input into the function equal to other factors such as private capital and labour.

It is also clear that one of the major problems connected with infrastructure investments has always been that of their financing [39]. Therefore, an issue of great importance is the finding of new and greater resources for infrastructure, both from the public sector and by experimenting with new forms of private involvement [40, 41].

The results of numerous studies show that investments in transport infrastructure improve accessibility by generating increases in the value of land and buildings located near them or at their access points. In this regard, many economists suggest the possibility of financing these infrastructures through the "capture" of increases in real estate value, allowing property owners who benefit from them to contribute to their construction costs.

Hedonic pricing method is certainly the most widely used reference for estimating the benefits generated by transportation in terms of accessibility. These benefits can be capitalized by property owners, giving rise to a policy of "capturing property values" to finance public transport and improve economic sustainability.

Highlighted the strong interaction between infrastructures and real estate values, research perspectives concern first the careful selection of variables useful to express levels of urban accessibility, then the characterization and therefore the concrete implementation of a regression function for the quantitative estimation of the effects that accessibility determines on the formation of real estate prices.

References

1. Rubino, P.: I settori infrastrutturali di servizio pubblico: caratteristiche economiche e loro regolazione. In: Biancardi, A. (ed.) L'eccezione e la regola. Tariffe, contratti e infrastrutture, pp. 43–46, Bologna (2009)
2. Troisi, R., Castaldo, P.: Technical and organizational challenges in the risk management of road infrastructures. J. Risk Res. (2022)
3. Mari, A.: Infrastrutture. In: Cassese, S. (ed.) Dizionario di diritto pubblico, ad vocem, vol. IV, p. 3134. Giuffrè, Milano (2006)
4. Nubi, T.O.: Procuring, managing and financing urban infrastructure: towards an integrated approach. In: Omirin, et al. (eds.) Land Management and Property Tax Reform in Nigeria Department of Estate Management. University of Lagos, Akoka (2003)
5. Biehl, D.: The role of infrastructure in regional development. In: Vickerman, R.W. (ed.) Infrastructure and Regional Development. Pion, London (1991)

6. Torrisi, G.: Public infrastructure: definition, classification and measurement issues. Econ. Manag. Financ. Markets **4**(3), 100–124 (2009)
7. Di Palma, M.D., Mazziotta, C., Rosa, G.: Infrastrutture e sviluppo. Primi risultati: indicatori quantitativi a confront (1987–95). Quaderni sul Mezzogiorno e le politiche territoriali (4). Confindustria, Roma (1998)
8. Aschauer, D.A.: Is public expenditure productive? J. Monet. Econ. **23**, 177–200 (1989)
9. Paradisi, F., Brunini, C.: Le infrastrutture in Italia. Un'analisi provinciale della dotazione e della funzionalità (7), ISTAT (2006)
10. Hansen, N.M.: The structure and determinants of local public investment expenditures. Rev. Econ. Stat. **2**, 150–162 (1965)
11. Castaldo, P., Amendola, G.: Optimal sliding friction coefficients for isolated viaducts and bridges: a comparison study. Struct. Control. Health Monit. **28**(12), e2838 (2021)
12. Bencardino, M., Nesticò, A.: Demographic changes and real estate values. A quantitative model for analyzing the urban-rural linkages. Sustainability **9**(4), 536 (2017)
13. Rostow, W.W.: The take-off into self-sustained growth. Econ. J. **66**(261), 25–48 (1956)
14. Glaeser, E.L., Kolko, J., Saiz, A.: Consumer city. J. Econ. Geography **1**, 27–50 (2001)
15. Wegener, M., Fürst, F.: Land-Use Transport Interaction: State of the Art. Berichte aus dem Institut für Raumplanung 46. Institute of Spatial Planning, University of Dortmund, Dortmund (1999)
16. Rietveld, P., Bruinsma, F.: Is transport infrastructure effective? Transport Infrastructure and Accessibility: Impacts on the Space Economy. Springer, Berlino (2012)
17. Papa, E.: Urban transformations and rail stations system: the study case of Naples. 45th Congress of the European Regional Science Association, Amsterdam (2005)
18. Maselli, G., Nesticò, A.: L'Analisi Costi-Benefici per progetti in campo ambientale. La scelta del Saggio Sociale di Sconto. LaborEst (20), 99–104 (2020). https://doi.org/10.19254/LaborEst.20
19. Nesticò, A., Maselli, G.: Declining discount rate estimate in the long-term economic evaluation of environmental projects. J. Environ. Acc. Manag. **8**(1), 93–110 (2020). https://doi.org/10.5890/JEAM.2020.03.007
20. Fogel, R., Fishlow, A.: Railroads and American Economic Growth: Essays in Econometric History. Johns Hopkins Press, Baltimora (1964)
21. Krugman, P.: Increasing return and economic geography. J. Polit. Econ. **99**(3), 483–499 (1991)
22. Sturm, J., De Haan, J.: Is public expenditure really productive? New evidence for the US and the Netherlands. Econ. Model. **12**(1), 60–72 (1995)
23. Stephan, A.: Regional infrastructure policy and its impact on productivity: a comparison of Germany and France. Appl. Econ. Q. **46**(4), 256–327 (2000)
24. Bajo Rubio, O., Sosvilla Rivero, S.: Does public capital affect private sector performance?: An analysis of the Spanish case, 1964–1988. Econ. Model. **10**(3), 179–185 (1993)
25. Cadot, O., et al.: Contribution to productivity or pork barrel? The two faces of infrastructure investment. J. Public Econ. **90**(6–7), 1133–1153 (2006)
26. Kemmerling, A., Stephan, A.: the contribution of local public infrastructure to private productivity and its political economy: evidence form a panel of large German Cities. Public Choice **113**(3–4), 403–424 (2002)
27. Paci, R., Saddi, S.: Capitale pubblico e produttività nelle regioni italiane. SR Scienze Regionali **1**, 5–26 (2002)
28. Bronzini, R., Piselli, P.: Determinants of long-run regional productivity with geographical spillovers: the role of R&D, human capital and public infrastructure. Reg. Sci. Urban Econ. **39**(2), 187–199 (2009)
29. Bonaglia, F., et al.: Public capital and economic performance: evidence from Italy. Giornale degli Economisti e Annali di Economia **2**, 221–244 (2000)

30. Alonso, W.: Location and Land Use. Harvard University Press, Cambridge (1964)
31. Dorantes, L., Paez, A., Vassallo, J.M.: Analysis of house prices to assess economic impacts of new public transport infrastructure. J. Transp. Res. Board **2245**(1), 131–139 (2011)
32. Chen, Y., Maziar, Y., Mohammad, M., Sidney, N.: The impact on neighbourhood residential property valuations of a newly proposed public transport project: the Sydney Northwest Metro case study. Transp. Res. Interdiscip. Perspect. **3**, 100070 (2019)
33. Efthymiou, D., Antoniou, C.: How do transport infrastructure and policies affect house prices and rents? Evidence from Athens, Greece. Transp. Res. Part A Pol. Pract. **52**, 1–22 (2013)
34. Seo, K., Aaron, G., Michael, K.: Combined impacts of highways and light rail transit on residential property values: a spatial hedonic price model for Phoenix, Arizona. J. Transp. Geogr. **41**, 53–62 (2014)
35. Court, A.T.: Hedonic price indexes with automotive examples. The Dynamics of Automobile Demand. General Motor Corporation, 99–117 (1939)
36. Lancaster, K.: A new approach to consumer theory. J. Polit. Econ. **82**, 34–55 (1964)
37. Rosen, S.: Hedonic price and implicit markets: product differentiation in pure competition. J. Polit. Econ. **82**(1), 34–55 (1974)
38. Forte, C.: Elementi di Estimo urbano. Etas Kompass, Milano (1968)
39. Della Spina, L., Giorno, C., Galati Casmiro, R.: Bottom-up processes for culture-led urban regeneration scenarios. In: Misra, S., et al. (eds.) ICCSA 2019. LNCS, vol. 11622, pp. 93–107. Springer, Cham (2019). https://doi.org/10.1007/978-3-030-24305-0_8
40. Dolores, L., Macchiaroli, M., De Mare, G.: A dynamic model for the financial sustainability of the restoration sponsorship. Sustainability **12**(4), 1694 (2020). https://doi.org/10.3390/su12041694
41. Calabrò, F., Cassalia, G., Lorè, I.: The economic feasibility for valorization of cultural heritage. The restoration project of the reformed fathers' convent in Francavilla Angitola: the Zibìb territorial wine cellar. In: Bevilacqua, C., Calabrò, F., Della Spina, L. (eds.) NMP 2020. SIST, vol. 178, pp. 1105–1115. Springer, Cham (2021). https://doi.org/10.1007/978-3-030-48279-4_103

Spatial Patterns of Blue Economy Firms in the South of Italy

Massimiliano Bencardino[1]([✉]) [iD], Vincenzo Esposito[1], and Luigi Valanzano[2]

[1] Territorial Development Observatory (OST), Department of Political and Communication Sciences, University of Salerno, 84084 Fisciano, SA, Italy
mbencardino@unisa.it
[2] Department of Cultural Heritage Sciences, University of Salerno, 84084 Fisciano, SA, Italy

Abstract. This study aims to measure the Blue Economy (BE) through the methodology of cluster analysis with a multiscalar approach. For this purpose, the Orbis database is used, building a sample of firms subjected to aggregation operations. The sample investigated is enriched by additional firm indicators offered by the platform (such as Value Added), where these are available.

The Authors have identified the supply chains of the sea, consisting only of consolidated sectors, through the lines dictated by the European Commission. Moreover, in order to bring out the territorial dynamics of the maritime clusters, the Authors have selected individual business activities with details of firm information at the NUTS 2 and NUTS 3 scale, ascribing the analysis exclusively to the Mezzogiorno area.

The contribution to this paper is the result of the joint work of the Authors, to which the paper has to be attributed in equal parts.

Keywords: Blue economy · Fishing · Shipbuilding · Value chain · Cluster analysis

1 Introduction and Purpose of the Work

Recent trends place the Blue Economy (BE) at the center of new policies for pia-nification of land-sea space [1]. The European Commission (EC) has long been engaged in measuring the trends and performance of maritime supply chains within member countries. In fact, the sum of the maritime supply chains generates 1.5% of the total value added (VA) (176 billion euros) and 2.3% of the total employment (4.5 million) of the EU-27 [2: page 6].

Several authors [3–6] point out that there is still no universal definition of the BE concept. However, it certainly represents an alternative economic paradigm to the model of linear growth by imprinting the lines of regional development both on the promotion of a strategy of conservation of marine ecosystems and on the enhancement of local resources that contribute to the economy of the sea [7].

The maritime cluster survey is a useful exercise to understand the spatial dynamics of the different dimensions that contribute to the Blue Economy [8, 9]. In this study, BE

F. Calabrò et al. (Eds.): NMP 2022, LNNS 482, pp. 1303–1312, 2022.
https://doi.org/10.1007/978-3-031-06825-6_126

is assessed through the methodology of cluster analysis with a multiscalar approach. The study aims to bring out the relationships between companies that, in a logic of supply chain, operate in the sectors of BE in the eight regions of southern Italy, ascribing the research only to the sectors defined by the European Commission as "consolidated": Marine living resources; Marine non-living resources; Marine renewable energy; Ports activities; Shipbuilding and repair; Maritime transport; Coastal tourism [2].

In the literature [10, 11] the perimeter of BE sectors follows different classifications and is quite heterogeneous. The maritime sector encompasses numerous activities and includes both maritime and land-based production areas. For this reason, their participation in BE supply chains prescinds from the direct use of the sea resource. Many international organizations [12, 13] have tried to formalize a schematic framework of the sectors that represent the main backbone of the Maritime Economy. The Ecorys study [14] breaks down the BE sectors according to their level of development: mature stage (Shortsea shipping, Offshore oil and gas, Coastline tourism and yachting, Coastal protection); growth stage (Marine aquatic products, Offshore wind, Cruise shipping including port cities, Maritime monitoring and surveillance); pre-development stage (Blue biotechnology, Ocean renewable energy, Marine minerals mining).

On this basis, the Blue Growth Strategy promoted by the European Commission in 2012 [15] clearly highlights the difference between established sectors and emerging sectors. Thus, emerging and innovative BE sectors are defined in the latest European Commission Report [2] in: Ocean Energy; Bioeconomy and Blue Biotechnology; Desalination; Marine Minerals; Defense, Security and Maritime Surveillance; Research and Education; Marine Infrastructure and Works (submarine cables, robotics, etc.).

2 Sample Construction and Methodology

In the survey, a sample of companies was constructed based on information provided by the Orbis database of the Bureau van Dijk company - a Moody's Analytics company - whose level of coverage and representativeness of economic, financial and corporate micro-data of companies is rated among the most complete [16, 17]. The database includes information on over 200 million companies, not including the agricultural sector and public services. Orbis is, in fact, the main source for the construction of the sample of companies that usefully contribute in the survey to the definition of the various segments of the Maritime Economy. However, a number of premises underlying the effectiveness of the database is necessary.

The level of data availability and the degree of depth of the information contained in the database can vary by country and by reference year, being based on different national sources. Furthermore, for the same number of variables, the information contained in Orbis at the level of individual business is frequently incomplete. There are, in fact, numerous cases in which data on value added (VA), the number of employees or the value of production are missing. In fact, the lack of this information would require an indirect construction of the incomplete values; this detailed operation has not been addressed in this study, which is intended to be preliminary [17]. Finally, the surveys in Orbis do not include individual firms.

That said, the potential offered by the Orbis database is entirely consistent with the objective of this investigation. In fact, rather than evaluating the economic incidence of the BE sectors in the individual regional contexts, the study intends to bring out the level of intensity of the relationships between companies that interact in the land-sea space. It is possible to achieve this objective only through clustering operations of maritime supply chains. For this reason, Orbis is a valid tool, allowing the extrapolation of useful business information to meet the requirements of geo-localization.

In particular, for the construction of maritime supply chains, the sample of companies was defined following the methodological approach adopted by "The EU Blue Economy Report 2021", drawn up annually by the European Commission to measure the performance of the BE at the scale of a single Member State [2].

Along the lines of the Report, the Orbis survey allowed for the extrapolation of all active companies surveyed on the basis of the statistical classification NACE rev.2 (Eurostat, 2008), with details corresponding to the fourth level of articulation of the European classification. In particular, the Authors have identified the economy of the sea in 46 classes of activity, corresponding only to the consolidated sectors of the Maritime Economy. Successively, the companies were aggregated into 16 sub-sectors, which in turn were grouped into 7 maritime sectors or chains.

In order to bring out the territorial dynamics of maritime clusters, the Authors have selected individual business activities with details of business information at the NUTS 2 and NUTS 3 scale. Specifically, this analysis is limited to the regions of Southern Italy, including islands: Abruzzo, Molise, Campania, Puglia, Basilicata, Calabria, Sicily and Sardinia. In addition, the depth of the firm information provided by the database, in particular the geographical information has enabled a more effective sectoral clustering operation. The multi-scalarity of the methodological approach adopted by the analysis is a fundamental element of differentiation with respect to the blue economy report, whose evaluation focuses exclusively on the performance of national macroeconomic aggregates.

Additional firm indicators offered by the database, where available, enrich the sample. These concern the number of employees in the last year, the value of production and added value, with both figures expressed in euros and taken from the last year available, and the firm size class. The size of the individual firm is recorded in accordance with the Orbis scheme, which compares firm data with that relating to the respective class: Very large companies (VL); Large companies (L); Medium sized companies (M); Small companies (S).

According to these findings, the sample is composed of 229,171 active firms. These contribute to the formation of the supply chains of the Marine Economy within the 8 regions of Southern Italy (in brackets the percentage of active companies): Campania (32%); Sicily (23%); Apulia (17%); Calabria (10%); Sardinia (8%); Abruzzo (7%); Basilicata (2%); Molise (1%).

Respectively, Very large companies (VL) represent the 0.04% of the sample, Large companies (L) the 0.3%, Medium sized companies (M) the 5.1% and Small companies (S) the 94.6% of the sample analyzed. The sum of the added values generated by the individual active companies is equal to 9 billion euros, while the sum of the production values is equal to 75 billion euros. Finally, they generate employment for over 600,723 people, 58.1% of whom are employed in Small companies (S); 26.2% in Medium sized companies (M); 9.9% in Large companies (L); 5.8% in Very large companies (VL).

It is necessary to make some clarifications. As already mentioned, the southern maritime sector is the result of the aggregation of the 229,171 active companies recorded in the fourth digit in the respective maritime sub-sectors and sectors or supply chains. Unlike the blue economy report, the construction of the maritime supply chains addressed was not subject to any weighting exercise between the classes of activity in this survey and the actual degree of intensity of the maritime resource within the production processes is not identifiable. This would have required a more detailed taxonomy of business activities beyond the fourth digit, and also the use of additional, alternative and complementary statistical sources. Orbis, in fact, does not allow a survey beyond the fourth level of the NACE rev.2 classification. Precisely in an attempt to enhance the exercise of clusterization of the maritime supply chains in the regions and provinces of Southern Italy, the values (such as VA, the value of production and the number of employees) refer exclusively to those associated by the Orbis database to the individual firm surveyed. It is not possible, therefore, to conduct an exercise of their indirect reconstruction.

All this implies that the large amount of unavailable firm information makes the sample result less consistent. Moreover, the analysis has shown how the coherence of the sample of firms associated with some sub-sectors of the maritime economy is sometimes mixed or even residual, falling within them companies that have no direct relationship with the maritime economy. The consequence of these clarifications is that the real size of the southern maritime economy is here "raw" and overestimated in its values. The overestimation finds confirmation in the economic dimension recorded by the Blue Economy Report, which associates the national BE with a VA of 23.7 billion euros in 2018. An incidence of 1.5% of the national VA (the report does not report on the regional dimension). Differently, the dimension of the economy of the sea "enlarged" in the present survey attributes to the southern economy alone a VA of 9 billion (Table 1).

All of the above, with the aim of verifying in which territories the incidence of the specific sector of the blue economy is higher and where the concentration of VA is greater, the procedure developed by Getis and Ord [18] and known as High/low Clustering is used.

3   Cluster Analysis

The supply chains that contribute to the Southern Maritime Economy concentrate more companies within Coastal Tourism. This sector alone represents 87% of the sample. The "Marine living resources" sector represents 6% of the companies making up the entire sample. The sector contributes to 9% of the VA and 6% of the employment of the Maritime Economy in the Southern regions.

The production areas that contribute to the formation of the chain are three: "Primary sector", "Processing of fish products" and "Distribution of fish products". In particular, the "Distribution of fish products" component reveals the presence of companies operating in very heterogeneous activities. On the other hand, the "Primary sector" component shows a production process totally characterized by the use of the sea resource. Here fishing and aquaculture account for 15% of the total VA of the sector.

The regional distribution of the supply chain reveals how in Campania, Puglia and Sicily concentrate the 72% of the supply chain companies, equal to 75% of the total

Table 1. Summary values of the maritime supply chains.

Maritime supply chains	Firms number	Employees	VA (€ x 1000)	Sample coherency[1]
Accommodation	18.732	53.132	1.138.764	mixed
Other expenditures	175.417	391.999	3.028.778	mixed
Transport	5.074	35.352	1.389.704	mixed
Coastal tourism	*199.223*	*480.483*	*5.557.246*	
Cargo and warehousing	1.560	21.444	515.384	mixed
Port and water projects	1.301	5.890	310.651	whole
Port activities	*2.861*	*27.334*	*826.035*	
Freight transport	53	1.608	159.631	whole
Passenger transport	579	5.464	281.188	whole
Services for transport	2.532	23.469	680.066	mixed
Maritime Transport	*3.164*	*30.541*	*1.120.885*	
Equipment and machinery	4.771	11.624	290.540	residual
Shipbuilding	1.545	8.134	219.967	whole
Shipbuilding and repair	*6.316*	*19.758*	*510.507*	
Offshore wind energy	3.629	1.981	280.371	residual
Marine renewable Energy	*3.629*	*1.981*	*280.371*	
Oil and gas	31	962	78.693	mixed
Other minerals	387	1.803	84.049	mixed
Marine non-living resources	*418*	*2.765*	*162.742*	
Distribution of fish products	7.370	15.312	371.961	mixed
Primary sector	3.501	12.443	124.167	whole
Processing of fish products	2.688	10.106	337.260	mixed
Marine living resources	*13.559*	*37.861*	*833.388*	
Total	*229.170*	*600.723*	*9.291.174*	

[1] Sample coherency. This item indicates the level of intensity of the maritime component within the production process. It is expressed through three levels: whole, mixed, residual.

sectoral VA. In particular, the cluster analysis identifies a greater intensity of firms (with complete data in terms of VA) in the metropolitan area of Naples, in the predominantly northern Puglia, in Trapani and in Ragusa (Fig. 1). Moreover, in Sardinia, Olbia emerges as the headquarters of Generale Conserve spa. The company, with its Asdomar brand, is the second Italian firm among canned fish producers and the first for tuna production generates the highest VA value, with 32 million euros, within the supply chain and has 762 employees.

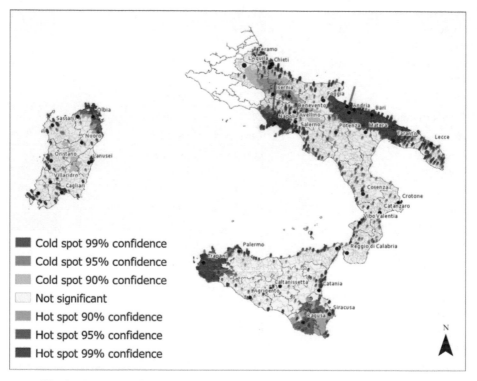

Fig. 1. Clustering of Marine leaving resource VA (Our elaboration on Orbis DB).

In Campania and Puglia, the distribution category has the greatest impact on their respective supply chains with 68% and 55%. However, Puglia shows an enlarged data given by the significant presence of land-based activities operating in olive cultivation. As mentioned, the segment includes a large number of production areas: in Campania, for example, next to Shedir Pharma srl operating in the nutraceutical market, there are companies such as Eurofish Napoli s.r.l. and Al Delfino s.r.l.

In Sicily, the "Processing of fish products" segment emerges as significant, showing an incidence of around 50% in terms of VA records on the respective regional supply chain. For example, in Trapani the Nino Castiglione s.r.l., with the Auriga brand, generates a VA of 16.7 million euros and employs 236 people. While in Palermo the canning firm, Flott s.p.a., shows a VA of 9.3 million euros and 737 employees. The map shows how in many cases the territorial dynamics of Marine living highlight the formation of smaller clusters This is the case of Calabria, where a cluster of lesser intensity exists in Vibo Valentia. This area is the setting of major firms such as Callipo Conserve Alimentari s.r.l., operating in the production of tuna and canned fish.

The transformation processes induced by a new land-sea space planning are strategic, integrated and systemic, and certainly place port areas as a tool for rethinking and overturning current growth models [1]. The ports of Southern Italy plays a central role in European, North African and Asian traffic [7, 19]. It affects the total national freight traffic and container handling by 47% and 45% respectively and generates 33% of the added value of the maritime sector [20].

In the study, the "Port activities", "Maritime Transport" and "Shipbuilding and repair" supply chains represent 5% of firms, 26% of VA and 13% of employment in the blue economy in Southern Italy. Of the total number of companies that make up the supply chain, 87% are small and 12% are medium-sized companies.

The geo-referenced analysis shows a high concentration of supply chain activities along the Naples-Salerno route (Fig. 2). The sectors instrumental to the development of the maritime economy represent an indispensable vehicle for the growth of the regional industrial apparatus (construction and repair shipbuilding, storage depots for energy products in port, development of project cargo services for assembly in logistics areas). In fact, the two provinces are respectively the first and the second in terms of number of companies and the first and the fifth in terms of VA. In these terms, the two provinces represent the 27% of the total of the three sectors in the entire southern territory. A greater intensity of the supply chains is also observed in the areas of Bari, Messina and Palermo, while, exclusively in terms of number of businesses, Sassari and Catania also stand out.

In the Blue Economy Report 2020 [21], the data on shipbuilding, construction and repair, report an expanding, highly innovative and dynamic sector despite having suffered the impacts of the great recession of 2009. According to the report, the European shipbuilding industry - with more than 300 shipyards, a production value of 42 billion per year and a total generation of direct employment of over 300,000 units - occupies a global market share of 15% in terms of compensated gross tonnage and 34% of the global share in terms of value. The "Shipbuilding and repair" supply chain incorporates the "Shipbuilding" and "Equipment and machinery" components. While in the first the marine resource is clearly identifiable, in the second it fades away in favor of a heterogeneous complex of activities, as occurs in the Teramo and Chieti areas. Although the Tecnimpianti s.p.a. in Palermo has the highest added value, the leadership of the Neapolitan area in the shipbuilding industry is highlighted by a significant number of firms in the area. In the sub-sector, among the first 50 firms sampled in terms of VA, 25 are located in the Neapolitan area. Thus, the Naples port cluster emerges with Cantieri del Mediterraneo s.p.a. (43 employees and a VA of 5 million euros) and Nuova Meccanica Navale s.r.l. (81 employees and a VA of 4.9 million), but above all it confirms its leadership of the historical cluster of the Stabia area with the CA.TU.NA. s.r.l. in Scafati (168 employees and a VA of 8 million), C&C Costruzioni s.r.l. in Pompei (168 employees and a VA of 4.6 million), Tecnaval s.r.l. in Santa Maria La Carità (134 employees and a VA of 4.6 million), Arcadia Yachts s.r.l. in Torre Annunziata and other medium-sized companies. Outside the Naples area, other significant clusters are found in Ragusa,

Taranto, Lecce and Catanzaro, where two important firms, Aschenez s.r.l. and Motonautica Fratelli Ranieri s.r.l., are located with 77 and 47 employees respectively and a VA of over 3 million euros for both.

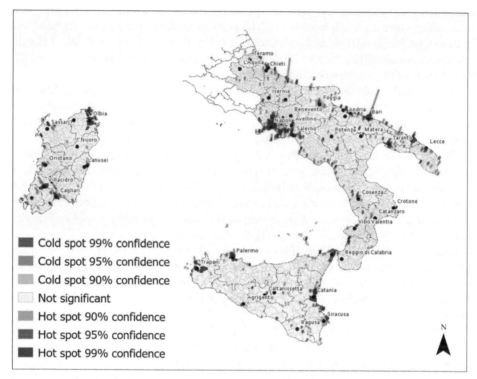

Fig. 2. Clustering of Shipbuilding VA (Our elaboration on Orbis DB).

4   Conclusions and Research Perspectives

This study shows how the methodology of cluster analysis with a multi-scalar approach proves useful in bringing out the intensity of the sea-related supply chains within territorial dynamics. Despite the shortcomings presented by the Orbis database, which did not allow for a more detailed measurement of the incidence of corporate values on the overall southern economy, the tool nevertheless proved useful for the geo-referencing of the extrapolated companies. This operation has made it possible to show the existence of territorial clusters with different degrees of intensity, bringing to light specific productive features worthy of future study.

In particular, the analytical detail of the "Marine living resources" sector reveals a particular intensity of companies in the regions of Campania, Puglia and Sicily, with levels of differentiation in the territorial distribution of sub-sectors. Similarly, the "Port activities", "Maritime Transport" and "Shipbuilding and repair" sectors show a greater concentration of activities on the Naples-Salerno route. Depending on whether one represents, the concentration of companies or the production of BE expressed in terms of VA, there is a change in the ranking at the NUTS3 scale. However, the first in terms of number of companies and VA account for 27% of the total of the three sectors in the Mezzogiorno area.

The presence within the clusters of consolidated territorial excellence is particularly evident. The competition of these excellences to the creation of value of the BE in the South seems to confirm the existence of strongly representative segments under the profiles of the district and the territoriality in the relationships between enterprises. The peculiar connotation of spatial relationships responds to a "short chain" model.

Future works will be oriented to the monitoring and diachronic analysis of emerging sectors and individual territorial clusters and the measurement of their contribution to the formation of the wider added value of regional economies.

References

1. Prezioso, M.: New harbouring and Blue Growth. Italian challenges. Bollettino dell'Associazione Italiana di Cartografia, **163**, pp. 11–26 (2018)
2. European Commission: The EU Blue Economy Report. 2021. Publications Office of the European Union, Luxembourg (2021)
3. Smith-Godfrey, S.: Defining the blue economy. Maritime Affairs: J. Natl. Maritime Found. India **12**(1), 58–64 (2016)
4. Voyer, M., Quirk, G., McIlgorm, A., Azmi, K.: Shades of blue: what do competing interpretations of the Blue Economy mean for oceans governance? J. Environ. Pol. Plan. **20**(5), 595–616 (2018)
5. Keen, M.R., Schwarz, A.M., Wini-Simeon, L.: Towards defining the Blue Economy: practical lessons from pacific ocean governance. Marine Pol. **88**, 333–341 (2018)
6. Garland, M., Axon, S., Graziano, M., Morrissey, J., Heidkamp, P.: The blue economy: Identifying geographic concepts and sensitivities. Geography Compass **13**(7), e12445 (2019)
7. Bencardino, M.: L'altra sponda. Prospettive di Crescita Blu nel Mediterraneo centro–occidentale, Roma, Aracne, Geografia economico-politica, pp.168 (2021)
8. Doloreux, D.: What is a maritime cluster? Marine Policy **83**, 215–220 (2017)
9. Pinto, H., Cruz, A.R., Combe, C.: Cooperation and the emergence of maritime clusters in the Atlantic: analysis and implications of innovation and human capital for blue growth. Marine Policy **57**, 167–177 (2015)
10. Surís-Regueiro, J.C., Garza-Gil, M.D., Varela-Lafuente, M.M.: Marine economy: a proposal for its definition in the European Union. Mar. Policy **42**, 111–124 (2013)
11. Katila, J., Ala-Rämi, K., Repka, S., Rendon, E., Törrönen, J.: Defining and quantifying the sea-based economy to support regional blue growth strategies–Case Gulf of Bothnia. Mar. Policy **100**, 215–225 (2019)
12. OECD: The Ocean Economy in 2030. OECD Publishing, Paris (2016)
13. World Bank: The Potential of the Blue Economy. United Nations, Washington (2017)
14. ECORYS: Blue Growth Scenarios and drivers for Sustainable Growth from the Oceans, Seas and Coasts, Third Interim Report, Rotterdam/Brussels, 13/3/2012
15. European Commission, Crescita blu. Opportunità per una crescita sostenibile dei settori marino e marittimo, COM (2012) 494 final, Bruxelles, 13/9/2012 (2012)
16. ORBIS, database. https://orbis.bvdinfo.com/

17. Bajgar, M., et al.: Coverage and representativeness of Orbis data. OECD Science, Technology and Industry Working Papers, **No. 2020/06**, OECD Publishing, Paris (2020)
18. Getis, A., Ord, J.K.: The analysis of spatial association by use of distance statistics. Geogr. Anal. **24**(3), 189–206 (1992)
19. Bencardino, M., Esposito, V.: Tanger MED SEZs: a logistic and industrial hub in the Western Mediterranean. In: Gervasi, O., et al. (eds.) ICCSA 2020. LNCS, vol. 12255, pp. 40–50. Springer, Cham (2020). https://doi.org/10.1007/978-3-030-58820-5_4
20. SRM: Italian Maritime Economy. Porti, rotte, noli e shipping: specchio di un cambiamento globale. Sostenibilità e logistica sfide per essere competitivi nel Mediterraneo. 8° Rapporto annuale, Studi e Ricerche per il Mezzogiorno (SRM), Napoli, Gianni Editore (2021)
21. European Commission: The EU Blue Economy Report. 2020, Office of the European Union, Luxembourg (2020)

Electrification of Commercial Fleets: Implementation Practices

Franco Corti[(⊠)] (iD)

Department of Management and Engineering, University of Padova, 36100 Vicenza, Italy
franco.corti@unipd.it

Abstract. The electrification of commercial fleets is one of the most promising processes for transition toward a more sustainable mobility. The role of companies that provide services in the urban area is relevant, since they manage the usage and circulation of means of transport that can have a huge impact on cities. We put in place preliminary surveys with firms potentially interested in these investments, to study the opportunities and challenges related to the adoption of electric mobility solutions in the provision of services. We have highlighted some considerations emerged from their experience, on the one hand the benefits on the environmental, maintenance and usage side of the pilot vehicles that are being tested, on the other hand the barriers to implementation related to costs, safety, and infrastructure constraints that the subjects involved find in approaching these changes. The vehicles involved in the last-mile logistics represent good candidates for these kinds of investments.

Keywords: Urban sustainable transport · Electric-vehicles · Sustainable logistics · PNRR

1 Introduction

1.1 Context

To promote the decarbonization process, the Transport Sector is crucial, given its major impact on Greenhouse Gases (GHG) emissions. Over a total of 49bln CO_2 emissions worldwide in 2018, the sector emitted 8.2bln, 75% of which related to road transports. In Italy, the latter has become the most polluting in terms of CO_2 emissions, with 104mln of CO_2 emitted in the same year. Urban areas are mainly responsible for these emissions, contributing to 75% of the country's CO_2 emissions, and consume 75% of the produced energy in the country [1]. This decarbonization challenge is addressed by Regulation 2021/1119 of the European Parliament approved in June 2021, which targets the goal of climate neutrality by 2050 and the reduction in emissions by 55% by 2030, upgrading the previous target of 40%. In July 2021, the program "Fit for 55" presented by the EU Commission was approved to align all European policies to the abovementioned target. Policymakers at the European levels are interested in promoting the post-pandemic economic growth with three targets: provide a more digital, ecological and resilient

© The Author(s), under exclusive license to Springer Nature Switzerland AG 2022
F. Calabrò et al. (Eds.): NMP 2022, LNNS 482, pp. 1313–1318, 2022.
https://doi.org/10.1007/978-3-031-06825-6_127

Europe [2]. In the national context, the National Recovery and Resilience Plan (PNRR) addresses the disruptions brought by the COVID-19 pandemic with an unprecedented amount of 230 bln€ investments, of which 38 bln€ specifically related to Sustainable Mobility.

To support an economy that is strongly linked to trading routes, Italy must invest in a more sustainable logistics infrastructure. Within the national innovation strategy (Strategia Nazionale di Specializzazione Intelligente - SNSI), the targets that are mainly addressed are: i) Systems of intelligent urban mobility for logistics and people; ii) Systems for security in urban environment, environmental monitoring; and iii) Prevention of critical events. Moreover, the focus of the strategy is on materials, components and vehicles, sensors, information, and communication technologies (ICT), applications for Intelligent Transport Systems (ITS). These are the enabling elements to reach the innovation objectives, through systems for the management of demand and mobility services of people and goods. For these reasons, many European countries have prioritized the electrification of transportation, considering the importance of adopting Electric Vehicles (EVs) with collaborative sharing solutions to promote a greener transportation and reduce environmental downsides.

1.2 Electrification of Fleets

The electrification of mobility services has been addressed by many scholars as one of the most promising challenges of the next years. The concept of electric Mobility as a Service (eMaaS) has been developed in recent years regarding optimization, interconnectivity and integration of smart services [3]. Moreover, the growing interest on last-mile delivery in scientific literature has led to the specific definition of Last-mile-as-a-service (LMaaS). These concepts integrate different modes of transportation and storage to provide an answer to challenges linked to on-time deliveries as well as to critical events and crises that have particularly affected long supply chains, such as the COVID-19 pandemic [4].

To leverage the full potential of eMaaS, some researchers [5] have proposed data management models with an interoperability approach to improve operations in smart cities through big-data collected from a diversity of stakeholders. Some scholars have raised questions on the expected impacts of eMaaS business models on city sustainability [6]. There are three types of barriers to the implementation highlighted in literature: economic difficulties, safety-related barriers, and operational obstacles. Within the first group, it is interesting to mention the cost of purchasing, producing electric power as well as of batteries usage. Concern for safety issues, the high risk of accidents due to silent engines and the problems related to battery maintenance are the most evidenced. Studies have also found some operational barriers related to charging time and stations and limited luggage space of EVs [7].

2   Preliminary Surveys

2.1   Data

The aim of our work is to conduct an exploratory case study using a qualitative approach, with a review and feedback mechanism on the different agents involved in the provision of services. Therefore, we evaluate of great interest to develop the study on some grounded realities, following the methodology developed by Yin [8]. We chose two companies operating with Light Commercial Vehicles (LCV) for the provision of their services of last-mile delivery and urban waste management and that could be interested by the possibility to convert their fleet partially or entirely in the next years. Data have been collected through questionnaires and interviews. The following results come from these preliminary surveys about the propension to converting commercial fleets and preliminary feedbacks on the pilot tests that they have conducted in the last two years. Then, the main findings are divided in two groups, one related to the vehicles and issues emerged during the testing phase, and the other one presents the aspects and considerations about the electrification of fleets in the view of the subjects involved.

Data Sources. A summary of people interviewed is presented in Table 1.

Table 1. Interviews

Current Role – Firm	Education	Experience
New Plant and Program Manager F2	Ph.D	>15 years
Controller F2	MSc	5 < 15 years
Service Planning Manager F2	MSc	5 < 15 years
Logistic Director F1	MSc	>15 years
Operations Manager F1	High School	5 < 15 years

2.2 Main Findings

Pilot Tests

The two firms have recently tested some LCV, considering that the numbers of these means of transport and the possibility to customize the vehicles according to the specific needs are considerably growing. The pilot tests were made in urban context with favourable conditions of flat level and not too narrow streets. The test was made with the standard of two shifts a day, morning, and afternoon, but also with three daily shifts, a great challenge for operation (battery life and severity index).

The priority for interviewed subjects is safety, ensuring security on mobility within and outside the firm activity, but mainly for the operator who must work in a safe and comfortable way in both warm and cold seasons. The level of autonomy of trucks was sufficient for a normal usage and the experience was positive for Firm 1, which provide a more regular kind of service. While Firm 2, which provides delivery and a more variable kind of service, has experienced more technical issues and difficulties in receiving assistance and put the vehicle quickly back on track. For these reasons it is still more sceptical about the technological readiness of certain vehicles.

Table 2. Firms feedback to electric LCV tests

Firm	Positive feedback	Negative feedback
Firm 1	Ecological	Difficulty on traffic bumps
	Good autonomy (2 shifts)	No underground parking
	Comfortable	Weak lights (safety issues)
	Urban Agility	Cost of purchasing
	Increasing variety of commercial vehicles	Fear of market uncertainty on materials provision
	Maintenance costs	
	Noiseless	
Firm 2	Ecological	Cold chain delivery requires more batteries
	Noiseless	Cost of purchasing
	Good autonomy (2 shifts)	Technical assistance

The following Table 2 provides a summary of the positive and negative feedbacks that have emerged.

Approaches to Electrification of Fleets

First, the service must be guaranteed at the present level and the principal need of firms is to understand how these vehicles will impact operative business. The three priorities asserted for both services are security, optimization of the services, adoption of low environmental impact vehicles.

The attention of both subjects to the environmental consequences was high before the recent policy suggestions and requirements. Moreover, they are leading firms with above market standards in their respective sectors, for both environmental and safety strategies.

The electrification of fleets (or less polluting fleets e.g., Euro6 vs Euro3) is incentivized, because it gives higher score in tenders. But right now, the interviewed firms have no electric vehicles in their operative fleets, and they are in a testing phase.

Hybrid technology is the one that is more interesting for the firms interviewed, but there are not hybrid LCVs available to maintain the level of service. Fast charge technology in these cases was not so relevant, given that these LVC most of the times could charge over the night and have second small charge in between of the shifts. For similar reasons, the charging infrastructure was not important at this level of testing, because the idea is to provide the large amount of charge inside the firm warehouse rather than open public/private parking slots.

There are some other aspects related to the impact of battery usage that arise some scepticism: decay/degradation, battery rental preferred to purchase option, possibility of fire and safety risks, charging anxiety on private or medium-length journeys.

Since the targets of these services are traditional in the urban context, right now the high-performance logistics are granted more by human operators rather than informatization. Usually, a city evolves very slowly, and some services are repetitive. After few months they are almost the same, so the difference is made when the operators know how to manage at best. In the experience of the firms, digitalization is crucial for data reporting and for analysis, but tracking information are more useful for its legal and insurance reasons, rather than live information of the service provisions.

3 Conclusion

Last mile logistics and service provision seems to be ideal for electrification. Most of vehicles make two shifts a day, with half charge in the middle of the day and full charge at night. Additionally, they don't need specific park areas in the city center during the day.

There must be a dedicated charging infrastructure in the warehouse, but the perception of the interviewed subjects is that the existing technology is not ready to maintain the level of services and the economic viability of business models.

Some aspects regarding the evaluation of electrification of the fleets have emerged during the interviews, suggesting some potential leads to future research on which are the most relevant barriers to these processes. In this case we faced a situation which is preliminary to potential investments in fleets renewal. From the feedbacks we have received we see that in the testing phase some safety-related and operational barriers have been highlighted, while the economic difficulties have not yet been addressed.

As for this, in the experience of the firms involved, the interest in the funds announced by the PNRR for the investments in a greener mobility seems to be distant from their actual situation, due to the preliminary testing phase of technical viability of services.

Acknowledgments. Financial support was provided by a grant from the Ministry of Education, University and Research, PON Scholarships R&I for 37° Cycle PhD program.

References

1. Chiaroni, D., Frattini, F.: Smart Mobility Report 2021. La sostenibilità nei trasporti: le sfide per una mobilità sostenibile nello scenario post-Covid. Milano (2021). https://www.energystrategy.it/
2. European Commission: Recovery plan for Europe (2021). https://ec.europa.eu/info/strategy/recovery-plan-europe_en
3. Dijk, M., Orsato, R.J., Kemp, R.: The emergence of an electric mobility trajectory. Energy Policy **52**, 135–145 (2013). https://doi.org/10.1016/j.enpol.2012.04.024
4. Correia, D., Teixeira, L., Marques, J.L.: Last-mile-as-a-service (LMaaS): an innovative concept for the disruption of the supply chain. Sustain Cities Soc. (2021). https://doi.org/10.1016/j.scs.2021.103310
5. Anthony, J.B., Abbas, P.S., Ahlers, D., Krogstie, J.: Big data driven multi-tier architecture for electric mobility as a service in smart cities: a design science approach. Int. J. Energy Sect. Manag. **14**, 1023–1047 (2020). https://doi.org/10.1108/IJESM-08-2019-0001

6. Le Pira, M., Tavasszy, L.A., Correia, G.H. de A., Ignaccolo, M., Inturri, G.: Opportunities for integration between Mobility as a Service (MaaS) and freight transport: a conceptual model. Sustain. Cities Soc. (2021). https://doi.org/10.1016/j.scs.2021.103212
7. Paddeu, D., Parkhurst, G., Fancello, G., Fadda, P., Ricci, M.: Multi-stakeholder collaboration in urban freight consolidation schemes: Drivers and barriers to implementation. Transport **33**, 913–929 (2018). https://doi.org/10.3846/transport.2018.6593
8. Yin, R.K.: Case Study Research and Applications: Design and Methods, 6th edn. SAGE Publications Ltd., Thousand Oaks (2017)

Medium-Long Term Economic Sustainability for Public Utility Works

Luigi Dolores[1] , Orlando Giannattasio[2], Maria Macchiaroli[1](✉) ,
Gianluigi De Mare[1] , and Rosa Maria Caprino[1]

[1] University of Salerno, Via Giovanni Paolo II, 132, Fisciano, SA, Italy
{ldolores,mmacchiaroli,gdemare,rcaprino}@unisa.it
[2] Studio Tecnico S.In.Tec., Via C. Sorgente N.18, Salerno, SA, Italy
o.giannattasio@studiosintec.it

Abstract. The conditions dictated by the pandemic and by the support measures adopted to fight it and relaunch the production systems, creating the conditions for the relaunch of the economy at a global level and, in particular, of the Italian one with average growth expected in the two-year period 2021/22 by 5%. However, the very actions dictated by the European Community, with capital flows concentrated in favor of member Countries in little more than five years, risk tarnishing investors' lucidity, affecting their forecasting capacity. In detail, economic operators historically accustomed to counting on their own resources have to deal with competitors who could have more convenient operating conditions, damaging the principle of free and effective competition. The most evident case concerns entrepreneurs who operate project financing who build the solidity of the investment hypothesis on the self-sustainability of the project idea, technically translated through the business plan. This instrument is obviously based on the reliability of the market and scenario forecasts which, in fact, today become rather uncertain given the persistence of the health crisis and the whirlwind of news on the financing conditions for the sectors concerned. This study deals with the case of the construction of a multi-story car park in project financing with respect to the critical issues deriving from a business plan that is found to be unreliable due to the change in the macroeconomic horizon and due to conditions depending on complex or not observed contracted rules.

Keywords: Project financing · Business plan · Public works ·
Economic-financial sustainability · PEF

1 Introduction

The Italian National Plan for Recovery and Resilience (PNRR) [1] reserves € 25.40 billion for public infrastructural works, i.e. 13.26% of the total. The main interventions concerning high speed on the Salerno-Reggio Calabria, Naples-Bari, Palermo-Catania lines and the extension of the Turin-Venice high speed line. Confirmed objectives are the

---

L. Dolores—Contributed equally to this work.

© The Author(s), under exclusive license to Springer Nature Switzerland AG 2022
F. Calabrò et al. (Eds.): NMP 2022, LNNS 482, pp. 1319–1327, 2022.
https://doi.org/10.1007/978-3-031-06825-6_128

reduction of travel times, the greater capillarity of the main services, the strengthening of freight transport.

Investments for the dynamic monitoring of the networks and for their connection to the terminal hubs (ports and airports) will be supported. With cascading effects on regional lines and modernization of the sections.

By combining these projects with funds destined for urban regeneration, a framework of territorial and city mobility is determined that tends decisively to contain the use of private vehicles, consequently favoring the strengthening of exchangers in which to switch from personal to public transportation.

Private individuals will be able to participate by financing a maximum of 25% of the total cost of the project.

In this condition there is therefore a clear reference to the construction models of public works introduced in the national legislation since law 415/98 and then rationalized with the system outlined by art. 7 of Merloni quater (n. 166/02) and, finally, reformed with Article 183 of the Public Contracts Code (Legislative Decree 50/16).

This type of intervention characterized by the direct investment of the private entrepreneur [2–6] which returns from the capital invested and earns a certain return, is based on multi-year concessions - often over thirty years - with unquestionable needs for a correct forecast of the microeconomic scenario and of the macroeconomic one.

These forecasts converge in the business plan, annexed to the concession agreement and regulatory instrument par excellence for the economic sustainability of the intervention [7].

The weight of the business plan is poured out in full in the credit-guy loan investigation that the entrepreneur must bear for the capital advances necessary until the service that can be provided by the work being carried out is fully operational. These advances are concessions to the vehicle company based on the cash flows envisaged in the business plan.

In this historical moment, even in the presence of interest rates that are still very low (at least in Europe, given that the first maneuvers to tighten the credit system are registered in the United States), the problem arises of correctly assessing the socio-economic scenarios in the medium term and to consider the advantages and costs of close collaboration with the public, based on the available funds but also the inherent criticalities in the management and reporting of the same.

The managerial, administrative, and fiscal [8–12] skills as well as technical-planning ones required of companies lead to their merger in more articulated and complex legal subjects than the current most common configuration of the well-known Italian small-average companies.

2   Business Plan

The business plan is a document aimed at portraying by a prospective way of the entrepreneurial project with the intent of evaluating its feasibility - in relation both to the company structure in which the project is, and to the context in which the proposing company operates - and to analyze the possibilities that have an impact on the main company decisions and on the economic and financial expected results. The business plan represents the descriptive synthesis and translation in economic-financial terms of the entrepreneur's business proposal, including operational and financial details, marketing opportunities and management strategies, skills, and competences [13].

The functions of a business plan are many:

- trace a strategic and operational path, supporting the assessment of business opportunities, also simulating possible alternatives both in terms of opportunities and threats with the aid of strategic planning models such as the SWOT matrix [14];
- define specific objectives for company management, following the indications dictated by the Project Management discipline and, therefore, in line with the SMART criteria (specific, measurable, achievable, related, time-based) [15];
- sharing in a clear and detailed way with all stakeholders the general strategy, the priorities of the company and the specific actions to be implemented; note how the inadequate communication of the business plan and the future objectives to be achieved can become one of the main causes of the failure of corporate strategies [16];
- to support company growth and to be the necessary tool for finding financial resources;
- allow the financial management of cash flows;
- support potential exit strategies through the analysis of exit barriers [17].

A business plan always combines a descriptive and an economic-financial part [18]. The description one contains qualitative information about the business, the strategic business area of interest [19], the analysis of the reference sector and the business strategy to be adopted. The economic-financial area, on the other hand, contains information of a quantitative nature, illustrated and described in the forecast economic-financial plan of the business activity. The same economic-financial plan is then accompanied by assessments regarding the economic feasibility of the plan and its financial sustainability. The preparation of the investment plan is, in this phase, preparatory to the subsequent elaboration of the economic-equity structure of the entrepreneurial project. In particular, it is important to detail:

- investments in fixed assets and in circulating assets. Distinguishing investment decisions in multi-year use activities with respect to ordinary management costs to be remunerated during the year is important both in order to establish the most appropriate source of financing and for the correct estimate of the parameters based on to whom the performance of an investment is generally evaluated, and which then affect the overall evaluation of the entire business project [20–22];
- the multi-year economic plan;
- the financial plan, with the estimate of the needs and the relative coverage procedures;
- the formalization of the financial plan and the cash budget.

Finally, is also necessary the assessment of the plan's earnings feasibility and its financial sustainability through tools [23, 24] such as the payback time or recovery time of the investment, the estimate of the Net Present Value (NPV) and of the Internal Rate of Return (IRR), the evaluation of the Return on Equity (ROE) and the Return on Investment (ROI) or, for the financial sustainability of the plan, the Debt service cover ratio (DSCR) and the Long Life coverage ratio (LLCR). Assessment tools also provided for by the Public Contracts Code (Legislative Decree 50/2016) and detailed in the n.9 National Authority for anticorruption (ANAC) guidelines.

3 Study Case

An example of how the complexity of the contractual elements and the changes in the macro and microeconomic scenario can affect the effectiveness of investments in project financing derives from a city in the Southern of Italy where in the early years of the century a private company built a multi-story car park in central area with a view to strengthening the traffic plan and improving the living conditions of the urban areas concerned.

The overall investment and, therefore, the financial exposure to which the vehicle company has subjected itself over a period of fifteen years is equal to approximately € 11 million.

The most critical elements for the concessionaire are of double nature: the administration's failure to comply with clauses in the agreement and relating to the areas of influence (as regards their correct management in terms of regulation of parking procedures and monitoring the application of these methods); the unpredictable evolution of living conditions dictated by the pandemic events of early 2020.

Evidently, the verification of the existence of the conditions of economic-financial equilibrium is not limited to the proposal or award phase, since the administration is required to verify their permanence during the concession period. This is governed by specific contractual clauses which guarantee the parties with respect to exogenous risk but also against endogenous behaviors aimed at activities that tend to have conflicting objectives [25–27]. In particular, on the basis of the law (articles 165 and 182 of Legislative Decree 50/16) and the ANAC resolution 318/2018, the agreement must provide for the causes and methods for revising the economic-financial plan. Therefore, for the case study, the condition is configured according to which the failure to respect the transfer of the areas of influence in the management of the vehicle company entails a strong impact on the economic and financial plan (PEF) due both to the effects induced on the non-regulated parking on the entrances of the managed neighboring car parks and to the lost revenues connected with the potential revenues deriving from the parking rates in the areas of influence and from the penalty costs for cars left out of place in the same.

As can be seen directly, the pandemic and the related lock-down periods have determined absolutely reduced vehicle flows for over a year, leading to a further compression of the revenues for the financial statements.

Another complication generated by the pandemic was the difficulty of data collection, necessary to allow effective analysis; for this purpose, a method was studied that should respect the following characteristics:

- continuous access to data;
- possibility of retrieving historical data;
- good quality of data collection and their verification;
- compliance with the health conditions in force;
- operator safety.

Referring to the techniques used in the transport systems engineering sector, it is possible to use the mobile observer technique [28] alongside new systems for data collection, in order to obtain data that respect the characteristics listed above.

The cash-flow analysis (in Table 1) model for the work is based on the ex-ante simulation that can ordinarily be developed with the use of sensitivity checks, outlining the foreseeable risk and the impact of contractual defaults as well as pandemic events on economic and financial plan with the possible consequences for the Company and the Administration.

Table 1. Cash flows from parking ten years after construction.

| Year | Invest.* | Revenue | | | Outflows | Cash balance | Present value |
		Sales	Management	Condominium management	Total	Management		
2001	774,69							
2002	774,69							
2003	774,69							
2004	774,69							
2005	774,69	947,80	2,09		949,89		949,89	625,72
2006		1.130,87	45,78		1.176,65		1.176,65	598,28
2007		90,09	87,35		177,44		177,44	94,87
2008		494,02	111,24		605,26		605,26	291,53
2009		64,60	108,37		172,97		172,97	75,06
2010		28,00	115,20		143,20		143,20	55,98
2011			98,81		98,81		98,81	34,80
2012		18,10	90,25	7,84	116,19		116,19	36,87
2013			75,93	22,15	98,08		98,08	28,04
2014		19,00	83,29		102,29	56,65	45,64	11,75
2015			67,78		67,78	20,30	47,48	11,02
TOT	3.873,45	2.792,48	886,09	29,99	3.708,56	76,95	3.631,61	1.863,92

The values are in thousands of €.

Time for construction (Year) 5

*Starting investment/years for construction

As evident from the Fig. 1 representing the flows, starting from 2007 the financial trend is outlined without the first years in which the sales of the boxes determine a distorting effect of the typical management. With continuously decreasing cash balances, starting from 2007 (excluding the year 2008 which once again thrives on extraordinary items) it becomes unthinkable to achieve the performance objectives assumed in the thirty years of concession, with evident need to realign the PEF through initiatives shared with the Administration. The data, of course, can only worsen starting from 2020 following the pandemic events that hit the entire country.

The method of surveying traffic data and parking conditions typical of the study areas assumes particular importance in the analysis conducted. Using historical information available on google maps, it was possible to derive the trend lines of the use of parking

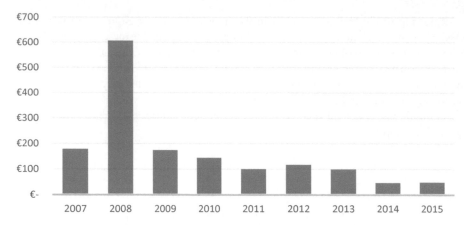

Fig. 1. Cash flows from parking. Values are in thousands of €.

spaces by motorists, verifying their behavior and consequences on the use of spaces adjacent to urban roads. In overall terms, the public-private binomial is unbalanced due to the breach of the contract but also to the occurrence of external events with respect to which the principle of risk for the private entrepreneur must certainly be reviewed also with the use of new insurance aids. nature and methods.

Overall, diagram in Fig. 2 shows the shift in the pay-back period beyond the thirty-year concession period.

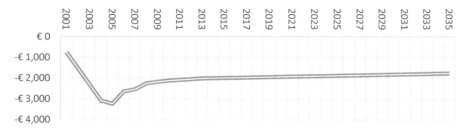

Fig. 2. Pay-back period beyond the thirty-year concession period. Values are in thousands of €.

4   Conclusions

The massive change of the economic-financial bargaining rules in public investments with private partnerships and the macroeconomic scenario completely in fibrillation due to the pandemic conditions as well as the strong reaction of national and continental governmental bodies, induce the need to reflect on entrepreneurial skills to make reliable predictions over medium-long periods. Such capabilities are particularly necessary for investments, such as those in project financing [29–32], which generate convenience over more than ten-year time intervals lasting.

Therefore, the definition of verification and control protocols of the permanence of the conditions of economic-financial equilibrium envisaged in the PEF during the negotiation phase, becomes particularly useful in support of the technical-economic feasibility of public utility works.

The case study addressed highlights how unobserved contractual clauses and natural phenomena of global significance can lead to significant losses for the entrepreneur and substantial repercussions on the quality of the public services provided.

The need to operate with forms of rebalancing of the conditions of sustainability and to introduce insurance tools (inherent to business interruption [33]) that look at scenarios unimaginable only a few years ago is significant.

References

1. https://www.governo.it/sites/governo.it/files/PNRR.pdf (2021)
2. Calabrò, F., Della Spina, L.: The public-private partnerships in buildings regeneration: a model appraisal of the benefits and for land value capture. In: 5th International Engineering Conference 2014 (KKU-IENC 2014). Advanced Materials Research, vol. 931–932, pp. 555–559. Trans. Tech. Publications, Switzerland (2014). https://doi.org/10.4028/www.scientific.net/AMR.931-932.555
3. De Mare, G., Di Piazza, F.: The role of public-private partnerships in school building projects. In: Gervasi, O., et al. (eds.) ICCSA 2015. LNCS, vol. 9156, pp. 624–634. Springer, Cham (2015). https://doi.org/10.1007/978-3-319-21407-8_44
4. Grimaldi, M., Sebillo, M., Vitiello, G., Pellecchia, V.: An ontology based approach for data model construction supporting the management and planning of the integrated water service. In: Misra, S., et al. (eds.) ICCSA 2019. LNCS, vol. 11624, pp. 243–252. Springer, Cham (2019). https://doi.org/10.1007/978-3-030-24311-1_17
5. De Mare, G., Granata, M.F., Forte, F.: Investing in sports facilities: the Italian situation toward an Olympic perspective. In: Gervasi, O., et al. (eds.) ICCSA 2015. LNCS, vol. 9157, pp. 77–87. Springer, Cham (2015). https://doi.org/10.1007/978-3-319-21470-2_6
6. Campanile, L., Cantiello, P., Iacono, M., Marulli, F., Mastroianni, M.: Risk analysis of a GDPR-compliant deletion technique for consortium blockchains based on pseudonymization. In: Gervasi, O., et al. (eds.) ICCSA 2021. LNCS, vol. 12956, pp. 3–14. Springer, Cham (2021). https://doi.org/10.1007/978-3-030-87010-2_1
7. Madera, R., Pisano, A.: Risparmio, Investimenti a lungo termine e crescita sostenibile in Long-term investing in Europe: re-launching fixed, network and social infrastructure, Rome (2014)
8. De Mare, G., Nesticò, A., Macchiaroli, M., Dolores, L.: Market prices and institutional values. In: Gervasi, O., et al. (eds.) ICCSA 2017. LNCS, vol. 10409, pp. 430–440. Springer, Cham (2017). https://doi.org/10.1007/978-3-319-62407-5_30
9. González-Méndez, M., Olaya, C., Fasolino, I., Grimaldi, M., Obregón, N.: Agent-based modeling for urban development planning based on human needs. Conceptual basis and model formulation. Land Use Policy **101** (2021). https://doi.org/10.1016/j.landusepol.2020.105110
10. Dolores, L., Macchiaroli, M., De Mare, G.: A model for defining sponsorship fees in public-private bargaining for the rehabilitation of historical-architectural heritage. In: Calabrò, F., Della Spina, L., Bevilacqua, C. (eds.) ISHT 2018. SIST, vol. 101, pp. 484–492. Springer, Cham (2019). https://doi.org/10.1007/978-3-319-92102-0_51
11. De Mare, G., Nesticò, A., Macchiaroli, M.: Significant appraisal issues in value estimate of quarries for the pubblic espropriation. Valori e Valutazioni (14), 47–62 (2017). ISSN: 2036-2404. DEI Tipografia del Genio Civile, Roma

12. Mallamace, S., Calabrò, F., Meduri, T., Tramontana, C.: Unused real estate and enhancement of historic centers: legislative instruments and procedural ideas. In: Calabrò, F., Della Spina, L., Bevilacqua, C. (eds.) ISHT 2018. SIST, vol. 101, pp. 464–474. Springer, Cham (2019). https://doi.org/10.1007/978-3-319-92102-0_49

13. Zimmer, T.W., Scarborough, N.M.: Essential of Enterpreneurship and Small Business Management, 4th edn. Perason Prentice Hall, Upper Saddle River (2002)

14. Giacobbe, G., La, S.W.O.T.: Analysis quale strumento di pianificazione strategica, StreetLib, EAN: 9788892567351 (2016)

15. Norma UNI ISO 21500:2013 Linee Guida per la gestione dei progetti

16. Mankins, M.C., Steele, R.: Turning great strategy for great performance. Harv. Bus. Rev. **83**(7–8), 64–72 (2005)

17. Porter, M.E.: Competitive Strategy: Techniques for Analyzing Industries and Competitors. The Free Press, New York (1980). ISBN 0-684-84148-7

18. Finch, B.: How to Write a Business Plan, 2nd edn. Kogan Page, London (2006)

19. Abell, D.F.: Defining the Business: The Starting Point of Strategic Planning. Prentice Hall, New York (1980)

20. Cuomo, M.T., et al.: Dalla strategia al piano. Elementi informativi e di supporto. G. Giappichelli Editore, Torino (2016). ISBN/EAN: 978-88-921-0365-8

21. Benintendi, R., De Mare, G.: Upgrade the ALARP model as a holistic approach to project risk and decision management. Hydrocarb. Process. **9**, 75–82 (2017)

22. Nesticò, A., Maselli, G.: Declining discount rate estimate in the long-term economic evaluation of environmental projects. J. Environ. Account. Manag. **8**(1), 93–110 (2020). https://doi.org/10.5890/JEAM.2020.03.007

23. Metallo, G.: Finanza sistemica per l'impresa, Terza edizione. G. Giappichelli, Torino (2013). ISBN/EAN: 978-88-348-9951-9

24. Nesticò, A., Maselli, G.: A protocol for the estimate of the social rate of time preference: the case studies of Italy and the USA. J. Econ. Stud. **47**(3), 527–545 (2020). https://doi.org/10.1108/JES-02-2019-0081

25. OECD: Society at a Glance. In: OECD Social Indicators. OECD Publishing (2014)

26. D'Alpaos, C.: The privatization of water services in Italy: make or buy, capability and efficiency issues. In: Mondini, G., Fattinnanzi, E., Oppio, A., Bottero, M., Stanghellini, S. (eds.) Integrated Evaluation for the Management of Contemporary Cities. SIEV 2016. Green Energy and Technology, pp. 223–231. Springer, Cham (2018). https://doi.org/10.1007/978-3-319-78271-3_18

27. Gerundo, R., Fasolino, I., Grimaldi, M.: ISUT model. A composite index to measure the sustainability of the urban transformation. In: Papa, R., Fistola, R. (eds.) Smart Energy in the Smart City. GET, pp. 117–130. Springer, Cham (2016). https://doi.org/10.1007/978-3-319-31157-9_7

28. Wardop, J.: Some theoretical aspects of road traffic research. Proc. Inst. Civ. Eng **2**, 325–378 (1952)

29. Dolores, L., Macchiaroli, M., De Mare, G.: Sponsorship's financial sustainability for cultural conservation and enhancement strategies: an innovative model for sponsees and sponsors. Sustainability **13**(6), 9070 (2021). https://doi.org/10.3390/su13169070

30. Hodge, G.A., Greve, C.: Public-private partneships: an international performance review. Public Adm. Rev. **67**(3), 545–558 (2007)

31. Dolores, L., Macchiaroli, M., De Mare, G.: A dynamic model for the financial sustainability of the restoration sponsorship. Sustainability **12**(4), 1694 (2020). https://doi.org/10.3390/su12041694

32. Hodge, G.A., Greve, C., Boardman, A.E.: International Handbook on Public-Private Partnership. Edward Elgar Publishing, Cheltenham (2010)
33. https://www.eiopa.europa.eu/document-library/other-documents/eiopa-staff-paper-measures-improve-insurability-of-business_en

A GIS-BIM Approach for the Evaluation of Retrofit Actions in Urban Planning. A Methodological Proposal

Gabriella Graziuso⬭, Michele Grimaldi(✉)⬭, and Carla Giordano⬭

University of Salerno, Via Giovanni Paolo II, 84084 Fisciano, Italy
{ggraziuso,migrimaldi,cagiordano}@unisa.it

Abstract. In the last years, the increase in urbanization has caused the deterioration of urban environments. Consequently, a series of evaluation frameworks has been developed to monitor the sustainability of urban transformations. Considering the complexity of the urban and territorial scale, professionals need a tool capable of satisfying the planning, design and management needs of urban space. The large amount of data and the possibilities of managing the multiple information contained in a BIM model can, indeed, be integrated usefully and declined at higher scales than the single building one, since they can be extended from the sphere of pure architectural design to the spatial field. In such a wide context, the GIS-BIM approach can represent a real shift of paradigm aimed at managing the complexity of urban processes more effectively. In this paper a methodology focused on the integration of GIS and BIM to design sustainable spatial scenarios of urban transformations will be presented and applied to a case study in south of Italy.

Keywords: Geographic Information System (GIS) · Building Information Modelling (BIM) · Urban planning

1 Introduction

In 2008, more than half of the world population already lived in the urban areas and it is expected to increase to 68% in 2050 [1, 2]. This expansion of urbanization in cities will be characterized by the reduction of the size of housing plots and green spaces, and the consequent increment of densities and needs for new types of houses, which reflect also the new composition of the population [3]. Although new infrastructures are designed according to sustainable principles, the use of innovative and heterogeneous materials, colors and techniques are elements evaluated, above all, for the indoor comfort. The implications for the outside environment cannot be managed simply with cooling and heating systems and each urban transformation can be crucial and modify the temperature making it uncomfortable for pedestrians and residents. The increase of urban air temperatures and the urban heat island (UHI) effects, indeed, can produce generally an objective deterioration of cities environment [4, 5].

F. Calabrò et al. (Eds.): NMP 2022, LNNS 482, pp. 1328–1336, 2022.
https://doi.org/10.1007/978-3-031-06825-6_129

In order to monitor the sustainability of constructions and urban transformations, it is necessary a more integrated approach, able to deepen the relationship between urban, building, building systems, and material. It is a one-to-one relationship, because the microclimates, which are governed by the urban systems, on the one hand, could have major impact on the energy, thermal, and lighting performance of buildings, and, on the other, more attention should be payed throughout the building delivery process from inception and design, to construction, operation, and maintenance of the built environment [6–9].

Currently, the representative models of digital cities are based on geospatial data deriving from a Geographic Information System (GIS), which can be considered an indispensable tool for the spatial development. At the same time, Building Information Modelling (BIM) can support the detailed semantics of parts of buildings or other functional parts of the city [10]. However, BIM tools can be also useful for urban planning because they allow the analysis of the physical environment, related changes, and the interaction with cities environment and the population.

Consequently, both these tools are involved in the knowledge of the territory and are characterized by their own specificities, focusing on the data, the objects, the building and spatial features and the overall modelling differently. Specifically, a model, that is based on a complete information repository, can be used both for the characterization of the spatial evolution and the building lifecycle and the data of the model can be updated and shared readily [11].

The relation between urban and building systems, i.e. the connection between the urban and the building scale could be facilitated by a better integration between GIS and BIM. This could be made through the interoperability of the two tools, namely their ability to exchange data and information each other [12]. The data deriving from a GIS system and those deriving from a BIM model, indeed, can be integrated with the consequent advantage of managing project information in a coordinated manner [13, 14].

Thanks to the interoperability feature, it is possible to move from the simple concepts of GIS and BIM to the new one of City Information Modeling (CIM), which is characterized by a multidisciplinary union of all the data of the model [15]. Furthermore, the CIM can involve a multiplicity of factors to collaborate in the development of sustainable, participatory and competitive cities [16]. It can be structured into sub-modules, each representing essential part of the city.

The combination of GIS and BIM systems allows the creation of a software tool useful for the analysis of retrofit actions and the support of decision-making process related to the urban planning. Taking into account the geometry of the urban area and its interaction with the built environment, it is possible to simulate the consequences the new intervention will have on the urban fabric [17]. Starting from these considerations, this paper analyses a possible working methodology focused on the integration of GIS and BIM to design sustainable spatial scenarios of urban transformations.

2   Methodology

In order to create a parametric model useful for urban planning, a methodology that allows the association of the classic urban indicators with BIM system was defined.

This parametric model, integrated with the data coming from a GIS system, permits the creation of a database able to facilitate the stakeholders in the management of information and in the evaluation of the urban transformation processes [18].

The methodology can be divided into three phases (see Fig. 1)., i.e.:

1. GIS environment modelling;
2. BIM environment modelling;
3. Scenarios construction and assessment of urban transformations.

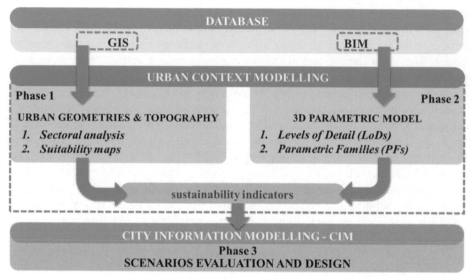

Fig. 1. Methodology for the CIM definition and the construction of dynamic scenarios.

Each phase is organized in several steps. While the first two phases are related to the urban and the buildings contexts respectively, the third phase contribute to define the model for the generation of different scenarios. Since a CIM can handle heterogeneous data, it is possible to construct a unique database containing detailed information on buildings as well as other relevant entities for urban planning [19, 20]. The data are geo-referenced and each item is placed in a proper system of co-ordinates, being associated with a geometric entity. This data is related to different scales, i.e. building, district and city. Specifically, while the large scale information can be easily implemented in a GIS system, the building characteristics are already stored in the BIM environment and only few data can be collected at the urban level. At the district scale, moreover, a balanced mix between GIS and BIM data is required in order to improve the decision processes and the information system [21].

The first phase, which is implemented in GIS environment, can be divided into two steps. The first step concerns the acquisition and systematization of all the information deriving from the planning system of the study area, from the cognitive layers about the uses, the morphology and geology of the soil. The second step involves the synthesis with overlay map operations of the information layers and the construction of suitability maps of the transformation in terms of distribution of the lots and their appropriate urban functions.

The second phase can be, also, divided into two steps and includes the definition of levels of detail (LoD), for step 1, and definition of parametric families (PFs), with regards to step 2. In the BIM environment, the geometric representation of the construction elements can vary in granularity, depending on the design state, with the definition of several LoDs, such as the symbolic (LoD A), generic (LoD B), defined (LoD C), detailed (LoD D), specific (LoD E), executed (LoD F) and updated (LoD G) objects (see Fig. 2). To support the dynamic nature of the design process, step 2 of the second phase is connected with the generative modelling approach of BIM, which allows the changes of object models quickly and efficiently by creating parametric families. For example, the thickness of a wall component can be simply changed by adjusting the only width parameter, while the change of geometry is implicit [17]. The parametric model, that is the output of the second phase, allows the modelling of objects intended as datasets, interchangeable through the Industry Foundation Classes (IFC) data model, i.e. the international BIM standard for interoperability with more chance of success.

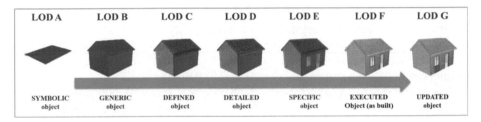

Fig. 2. Definition of Levels of Detail (LoDs).

In order to evaluate the sustainable characteristics of an urban system, starting from the model generated in the previous phases, it is possible to define a set of sustainability indicators (see Table 1): thermal dispersion index (IDt), inverse average reflection coefficient ($ICrm$) and aspect ratio (Ar). They can be selected according to the available known information of a territory, which generally coincides with the one acquired for the creation of urban planning tools [22] on a municipal basis and permits to measure the morphometric characteristics of the urban system that affect the anthropogenic heat flow stored in the urban structure.

Finally, the third phase is characterized by two levels of implementation, with reference to the design and the approval phases. In the design phase, indeed, it provides for the simulation of different configurations in compliance with the set threshold limits while, during the approval phase by the public administration, the model allows the verification of compliance with the technical implementation standards.

Table 1. Sustainability indicators.

Indicator name	Formula	Description
Thermal dispersion index	$IDt = \frac{S}{V}$	S = dispersive surface V = heated volume
Inverse average reflection coefficient	$ICrm = \frac{1}{\sum Ci} \cdot \frac{Ai}{\sum Ai}$	Ci = coefficient of refraction of a given material Ai = surface area of a specific material
Aspect ratio	$Ar = \frac{\sum Hi \cdot Di}{2 \sum Wi \cdot Di}$	Hi = height of the i-th front of the building facing the street Di = linear extension of the i-th front of the building facing the street Wi = distance between the i-th front of the building and the street

3   Case Study

The methodology has been applied in the municipality of Palma Campania, in the province of Naples, in South Italy (Fig. 3). Palma Campania has been the object of study and research as part of an agreement stipulated by the University of Salerno and the municipality for the preparation of the municipal urban plan.

The dataset is based on a vector cartographic support, scaled 1:2000, modelled on an aerial photogrammetric survey, done in 2010, integrated with an alphanumeric database containing information on volumes, floor to area ratio, and land use.

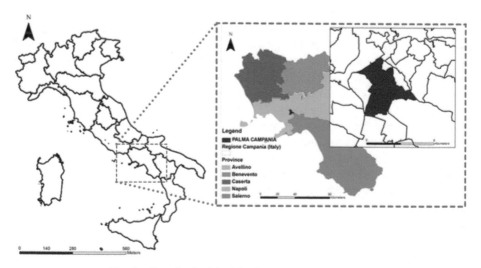

Fig. 3. Case Study: Municipality of Palma Campania.

With regard to the measurement of indicators, representing the morphometric characteristics of the urban structure, a first division of the components of the urban fabric in horizontal and vertical surfaces was carried out, transforming the polygon feature class in the multi-patch feature class, resulting in the fields alignment. Horizontal surfaces included the entire road surface (asphalt or similar) and shells of buildings that were all considered flat. All façades of buildings, with exception of the contact surfaces between them, fell into the category "vertical surfaces". In this way it was possible to calculate for each surfaces the *IDt* indicator (see Fig. 4a). As regard to the calculation of the *ICrm*, it was considered eligible to adopt, for each building, a prevailing reflection coefficient according to the surfaces extension (see Fig. 4b). For the calculation of the *Ar*, both the average distance between the façades facing the street and the average height for each road arch was measured (see Fig. 4c).

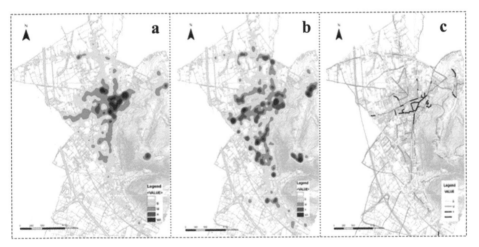

Fig. 4. Calculation of the sustainability indices: a. *IDt*; b. *ICrm*; c. *Ar*.

The obtained maps allow the evaluations of different design alternatives relating to specific transformation areas identified in the Municipal Urban Plan. Once completed the factor map for each parameter, it was possible to obtain the summary map that permits to define intensity thresholds of the studied phenomenon in the urban fabric (see Fig. 5a). As respect to this map, different design solutions can be evaluated and modeled in a BIM environment, relating to urban districts, verifying the change that those solutions can make to the current state in compliance with the aforementioned intensity values.

As regard to the case study, two alternatives relating to an urban regeneration area were compared (see Fig. 5b). The values of each map were aggregated through zonal statistics operation relating to the perimeter of the study area, obtaining three sustainable indices: *Isa, Isn, Isu*. Moreover, it was possible to define also a synthesis index (Is) related to the summary map. For each design alternative, indeed, the values of each map were recalculated and aggregated. The percentage variation with respect to the starting situation was then calculated for each of them. The comparison between the values of the two different alternatives (see Fig. 5c) shows how the solution with the greatest compactness and with different distribution of green spaces cause an increase of the *Is* index, permitting to attribute a judgment on the goodness of the two solutions.

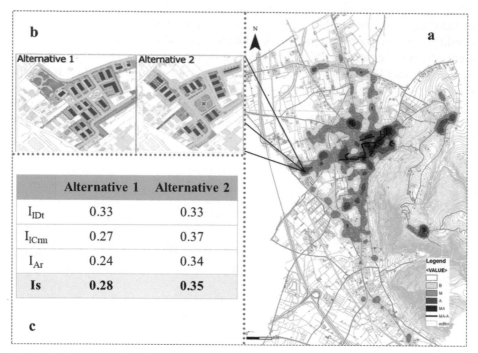

Fig. 5. Project proposal: a. Synthesis index representation in the summary map; b. Selection of the study area and design of two alternatives; c. Sustainable indices calculated for the two designed alternatives.

4 Conclusions

In this paper, a methodology for the creation of city information modeling has been defined. It is useful for the design and evaluation of sustainable urban scenarios through the combination of two tools, that connects different aspects at different scales [23]. The integration of GIS and BIM environments, indeed, provide a graphical database able to improve the visual monitoring and collection of data about landscape, grid, buildings and city. The methodology has been tested on some project applications to evaluate their efficacy, as instrument that supports the urban project of settlement and allows the choice between different project solutions for the same area.

In conclusion, the creation of a CIM model defined with a spatial decision support system could be useful for the strategic environmental assessment procedures of the areas in relation to their detailed forecasts defined by the municipal urban plan. Furthermore, the integration of GIS and BIM systems could provide quantitative information on the effects induced by urban choices on the environment, stimulate participatory contributions, which can become an indispensable element to support decision-making processes in defining priorities and selecting possible alternatives for the urban transformations.

References

1. United Nations, Department of Economic and Social Affairs, Population Division. World Urbanization Prospects: The 2018 Revision (2018). https://population.un.org/wup/Publicati ons/. Accessed 21 Dec 2021
2. Santamouris, M., Synnefa, A., Karlessi, T.: Using advanced cool materials in the urban built environment to mitigate heat islands and improve thermal comfort conditions. Sol. Energy **85**, 3085–3102 (2011)
3. Oke, T.R.: The energetic basis of the urban heat island. Q. J. R. Meteorol. Soc. **108**, 1–24 (1982)
4. Yang, W., Wong, N.H., Jusuf, S.K.: Thermal comfort in outdoor urban spaces in Singapore. Build. Environ. **59**, 426–435 (2013)
5. Dolores, L., Macchiaroli, M., De Mare, G.: Sponsorship's financial sustainability for cultural conservation and enhancement strategies: an innovative model for sponsees and sponsors. Sustainability **13**(6), 9070 (2021)
6. Dolores, L., Macchiaroli, M., De Mare, G.: A model for defining sponsorship fees in public-private bargaining for the rehabilitation of historical-architectural heritage. In: Calabrò, F., Della Spina, L., Bevilacqua, C. (eds.) ISHT 2018. SIST, vol. 101, pp. 484–492. Springer, Cham (2019). https://doi.org/10.1007/978-3-319-92102-0_51
7. Coppola, F., Fasolino, I.: The axial analysis for defining neighborhoods' crime vulnerability. A methodological proposal. In: Gervasi, O., et al. (eds.) ICCSA 2021. LNCS, vol. 12958, pp. 457–473. Springer, Cham (2021). https://doi.org/10.1007/978-3-030-87016-4_34
8. Coppola, F., Fasolino, I., Grimaldi, M., Sebillo, M.: A model to construct crime risk scenarios supporting urban planning choices. In: La Rosa, D., Privitera, R. (eds.) INPUT 2021. LNCE, vol. 242, pp. 123–130. Springer, Cham (2022). https://doi.org/10.1007/978-3-030-96985-1_14
9. Azhar, S.: Building information modeling (BIM): trends, benefits, risks, and challenges for the AEC industry. Leadersh. Manag. Eng. **11**, 241–252 (2011)
10. Dantas, H.S., Sousa, J.M.M.S., Melo, H.C.: The importance of City Information Modeling (CIM) for cities' sustainability. IOP Conf. Ser. Earth Environ. Sci. **225**, 12074 (2019)
11. Del Giudice, M., Osello, A., Patti, E.: BIM and GIS for district modeling. In: Christodoulou, S., Scherer, R. (eds.) Ework and Ebusiness in Architecture, Engineering and Construction 2014, pp. 851–854. CRC Press—Taylor & Francis Group, Boca Raton (2015)
12. Osello, A., et al.: BIM GIS AR per il Facility Management, Dario Flaccovio Editore, Palermo (2015)
13. Irizarry, J., Karan, E.P., Jalaei, F.: Intergrating BIM and GIS to improve the visual monitoring of construction supply chain management. Autom. Constr. **31**, 241–254 (2013)
14. Xu, X., Ding, L., Lou, H., Ma, L.: From building information modeling to city information modeling. J. Inf. Technol. Constr. **19**, 292–307 (2014)
15. Almeida, F., Andrade, N.: A integração entre BIM e GIS como ferramenta de gestão urbana. In: Proceedings of the VII Encontro de Tecnologia de Informação e Comunicação na Construção, pp. 371–383 (2015)
16. Sacks, R., Eastman, C., Lee, G., Teicholz, P.: BIM Handbook: A Guide to Building Information Modeling for Owners, Managers, Designers, Engineers and Contractors, 3rd edn. Wiley, Hoboken (2018)

17. Kolbe, T.H., Donaubauer, A.: Semantic 3D city modeling and BIM. In: Shi, W., Goodchild, M.F., Batty, M., Kwan, M.-P., Zhang, A. (eds.) Urban Informatics. TUBS, pp. 609–636. Springer, Singapore (2021). https://doi.org/10.1007/978-981-15-8983-6_34
18. Grimaldi, M., Giordano, C., Graziuso, G., Barba, S., Fasolino, I.: A GIS-BIM approach for the evaluation of urban transformations. A methodological proposal. WSEAS Trans. Environ. Dev. **18**, 247–254 (2022)
19. Liu, X., Wang, X., Wright, G., Cheng, J.C.P., Li, X., Liu, R.: A state-of-the-art review on the integration of Building Information Modeling (BIM) and Geographic Information System (GIS). ISPRS Int. J. Geo Inf. **6**(2), 53 (2017)
20. Zhu, J., Wright, G., Wang, J., Wang, X.: A critical review of the integration of geographic information system and building information modelling at the data level. ISPRS Int. J. Geo Inf. **7**(2), 66 (2018)
21. Torabi Moghadam, S., Ugliotti, F.M., Lombardi, P., Mutani, G., Osello, A.: BIM-GIS modelling for sustainable urban development. In: NEWDIST, pp. 339–350 (2016)
22. Gerundo, R., Grimaldi, M.: The measure of land consumption caused by urban planning. Procedia Eng. **21**, 1152–1160 (2011)
23. Giannattasio, C., Papa, L.M., D'Agostino, P.: Bim-oriented algorithmic reconstruction of building components for existing Heritage. Int. Arch. Photogramm. Remote Sens. Spatial Inf. Sci. **XLII-2/W15**, 513–518 (2019)

Economic-Financial Sustainability and Risk Assessment in the Water Sector in Italy

Maria Macchiaroli⬤ and Luigi Dolores(✉)⬤

University of Salerno, Via Giovanni Paolo II, 132, Fisciano, SA, Italy
{mmacchiaroli,ldolores}@unisa.it

Abstract. The National Recovery and Resilience Plan (PNRR) is the document prepared by Italy to relaunch the economy after the COVID-19 pandemic and enable the country's green and digital development. One of the many objectives of the PNRR is to ensure the sustainability of water resources and the improvement of the environmental quality of the water.

With this work, we intend to define a procedural process for the assessment of the risk of lack of economic and financial sustainability in projects relating to Mission 2 Category 4 (Protection of the territory and water resources) of the PNRR. Specifically, the probabilistic tools for economic-financial risk assessment are integrated with the As Low As Reasonably Practicable (ALARP) principle, traditionally used in high-risk sectors to estimate acceptable and tolerable levels of health and safety. The work is part of the recent stream of literature that uses the ALARP principle to evaluate the economic and financial convenience of projects in the fields of construction and civil engineering. The model is applied to a potential project, financed through funds allocated to the PNRR, aimed at the construction of a new purification plant and the expansion of the existing sewer network in a medium-sized city. The application allows you to select the best design alternative in terms of both expected return and the probability of failure.

Keywords: Integrated Water Service · Risk assessment · ALARP principle · PEF · Economic-financial sustainability · PNRR

1 Introduction

By aligning itself with the other member states of the European Union, in recent years Italy has been gradually introducing the principles of the green economy in the context of the Integrated Water Service. This is because it is necessary to define tools and actions aimed at safeguarding and sustainable management of the water resource as soon as possible. To start the water sector on the path of the green economy, adequate investments and coherent institutional governance are required. Unfortunately, many of the actions necessary for sustainable management of water resources are still very late, especially in the South. For example, the following issues have not yet been adequately resolved: the containment of water losses, the availability of the resource in drought periods, the

---

M. Macchiaroli and L. Dolores—Contributed equally to this work.

© The Author(s), under exclusive license to Springer Nature Switzerland AG 2022
F. Calabrò et al. (Eds.): NMP 2022, LNNS 482, pp. 1337–1346, 2022.
https://doi.org/10.1007/978-3-031-06825-6_130

reduction of consumption, the adaptation of infrastructures, the abatement of evasion of the tariff, the streamlining of risk assessment and management procedures in the water supply chain [1]. About the last point, there are many risk factors, uncertainties, and conflicts in the water sector. Four main categories of risks can be identified:

- **Risks associated with the initial assignment phase**: risks concerning the selection of the manager when the service is awarded. The main difficulty concerns the information asymmetries that characterize the collaboration relationship established between the grantor and the potential concessionaire of the service. Most of the Area Plans (PdA) and Economic-Financial Plans (PEF) attached to the agreements do not reach the level of processing required by private lenders.
- **Risks associated with investments and service standards**: in Italy, the allocation of investment risks is still unclear. The Area Management Body (EGA) is responsible for defining the investments, while the individual water manager has the task of making them by responding to specific service standards. Given the correlation between investments and standards, the manager inevitably takes on a risk that he or she does not have full control over.
- **Risks associated with the early termination of the concession**: an indemnity is paid by the concessionaire in the event of early termination of the concession [2].
- **Risks associated with economic and financial sustainability**: risks that can depend on various factors. The main ones are the following: errors in the recognition of the works, errors in the evaluation of the necessary interventions, errors in the estimation of the costs and timing of the investments, exogenous increases in construction costs, delays due to lack of the necessary authorizations, technological change, inefficient management of the construction process, incorrect estimate of the operator's demand and revenues, errors in the tariff structure, failure to achieve technical, qualitative and managerial standards. Finally, the risks associated with inflation and structural changes in the financial markets should not be overlooked [3–5].

The last category of risk is the subject of study in this work. The mitigation of the risks associated with the economic and financial sustainability of the infra-structural networks of the Integrated Water Service is a highly topical topic [6]. In fact, the issue is addressed in the National Recovery and Resilience Plan (PNRR), a document that the Italian government has prepared to illustrate to the European Commission how the country intends to invest the funds allocated under the Next Generation Eu program. Issues related to efficiency in the management of water services are included in Component 4 (C4 - Protection of the territory and water resource) of Mission 2 (M2 - Green revolution and ecological transition). Specifically, the investments contained in M2C4 aim to ensure the safety, supply, and sustainable management of water resources throughout the entire cycle. Extraordinary maintenance work is planned on the reservoirs and completion of the water networks. These interventions are to be carried out in such a way as to improve the state of chemical and ecological quality of the water as well as ensure the efficient allocation of the water resource between the various sectors (urban, agriculture, hydroelectric, industrial). The planned interventions concern the following four investment categories:

- **Investment 4.1**: investments in primary water infrastructures for the security of water supply. Due to climate change, water crises are increasingly frequent, so it is necessary to make aqueduct infrastructures efficient and resilient, guaranteeing water supply to all sectors of interest.
- **Investment 4.2**: reduction of water losses through the digitalization and monitoring of networks. The goal is to promote optimal resource management by reducing the waste of drinking water and limiting inefficiencies.
- **Investment 4.3**: investments in the resilience of the irrigation agricultural system for better management of water resources. The main objective is to improve irrigation efficiency and effectively monitor the use of the resource.
- **Investment 4.4**: investments in sewerage and purification. Often the Italian sewer and purification network does not comply with the European Directives. This lack is widespread above all in the South, where networks are often obsolete and sometimes even absent. Because of this, the EU has initiated four infringement procedures against Italy in the last few years. Effective and technologically innovative water purification is therefore necessary. In this sense, it is planned to carry out projects that allow the recovery of energy and sludge and the reuse of purified water for agricultural and industrial purposes.

The objective of the paper is to define a protocol for assessing the risk of lack of economic and financial sustainability of the projects relating to M2C4 of the PNRR. In particular, the risk management process is applied to a potential project of the Investment 4.4 (Investments in sewerage and purification). The novelty of the work consists in integrating the As Low As Reasonably Practicable (ALARP) principle, generally used in high-risk sectors to estimate acceptable and tolerable levels of health and safety, in the traditional assessment of the economic and financial risks of Integrated Water Service projects. The work is part of the recent stream of literature that uses the ALARP principle to evaluate the economic and financial convenience of projects in the fields of construction and civil engineering [7–9]. In this work, the ALARP principle is applied for the first time to a potential investment of the PNRR to contribute to the sustainable development of the territory and to the effective management of resources.

2   Method

This section briefly describes the main approaches used for the economic and financial risk assessment of an investment project. A more in-depth study is reserved for the Monte Carlo simulation (MCS), a technique that was applied to the case study. Below is a description of the ALARP principle which, as anticipated, was integrated into the risk assessment phase of the risk management process.

2.1   Traditional Techniques for Risk Assessment in Investment Projects

The classic approaches for risk assessment in investment projects are as follows:

- **Statistical approach** (or mean-variance): with this approach, a probability distribution is associated with cash flows. Once the expected return and the standard deviation (volatility) have been estimated, it is consequently possible to express an opinion of economic convenience.

- **Methods for adjusting the net present value**: in this case, there are two different roads. In the first, it is possible to transform risky cash flows into certain equivalent flows through an appropriate risk rate. The second way is to increase the discount rate by considering a risk premium (Capital Asset Pricing Model - CAPM).
- **Probabilistic tools**: consists in identifying the sensitive variables of the project and in defining an appropriate probability distribution for each of them. The final step is to estimate the cumulative distribution function of the performance indicator (NPV or IRR). Monte Carlo simulation (MCS) is usually used to estimate the cumulative probability distribution of the chosen indicator [10].

In this paper, probabilistic tools were used, integrating the MCS with the ALARP principle in the risk assessment phase.

MCS, one of the most used techniques for the assessment of investment risk, was devised in Los Alamos in the 1940s by Ulam, von Neumann, and Fermi to solve complex deterministic equations. Since then, the technique has undergone continuous evolution and has been applied in many research fields [11]. MCS is a statistical analysis technique that consists first in defining a probability distribution for each sensitive variable. Having done this, through many iterations, making the sensitive variables vary randomly within their definition range, it is possible to estimate the cumulative probability distribution of the chosen performance indicator [12].

2.2   The ALARP Principle

Acronym of the expression As Low As Reasonably Practicable, the ALARP is that portion of the frequency of occurrence-number of fatalities diagram, between the level of acceptability and the level of tolerability of the risk, within which the cost-benefit analysis is applied as a guiding criterion in making risk management decisions in the presence of uncertainty for a given structure. The extreme acceptability and tolerability thresholds delimit the risk tolerance region.

The ALARP approach was introduced by the Health and Safety Executive in the United Kingdom to estimate the probability of loss of human life in industrial sectors at risk of a major accident. In summary, the following three risky regions are considered (see Fig. 1):

- **Widely acceptable area**: the risk is negligible and widely accepted by most people, which is why it is not necessary to reduce it.
- **Tolerable area or ALARP**: the risk, although not desired, is tolerated in view of the benefits that can be obtained by accepting it. In this region, the risk is tolerable either because it is impossible to reduce it or because the mitigation costs are disproportionate to the future benefits.
- **Unacceptable area**: The risk is so high that no one is willing to take it. The losses far outweigh the benefits of accepting this risk. It is necessary to reduce the risk by making it at least tolerable [13].

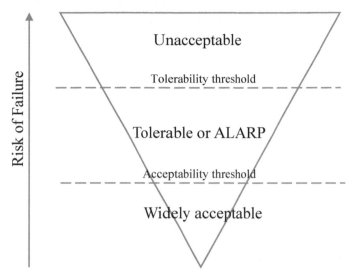

Fig. 1. Risk areas in the ALARP principle (own elaboration).

In this work, the ALARP approach was used by defining the tolerability and accept-ability thresholds of the economic and financial risk of an investment project aimed at the construction of a new purifier (project compliant with Investment 4.4 of the PNRR). ALARP is therefore integrated with the traditional MCS technique.

3 Application and Results

To apply the model defined in the previous section, it is assumed that a new purifier will be built to serve a medium-sized city (380,000 inhabitants). The goal is to align the wastew-ater treatment system with EU standards to ensure greater protection of the environment, in accordance with the planned investments of M2C4 of the PNRR. Interventions are also planned on the sewage system to reduce infiltrations and to allow wastewater to reach the new purification plant. In addition, it is planned to connect 10,000 new users to the sewer system. Two project scenarios are proposed for the purification plant:

- **Scenario 1**: Drying and incineration of sludge and deposit of ashes in landfills.
- **Scenario 2**: Reuse in agriculture and/or for energy crops (after dehydration).

In addition to the two project scenarios, scenario 0 should also be considered, i.e., the current scenario characterized by the absence of the new purifier. It is assumed that 97% of the population's wastewater is currently collected in the existing sewer system and discharged into the river that crosses the city without having been previously treated. The assumptions made are representative of the actual situation of some Italian cities, especially in the South.

Table 1 shows the basic parameters necessary to draw up the Economic-Financial Plans (PEF) for each of the three scenarios. These parameters were defined following the indications suggested by the European Commission's cost-benefit analysis guide [14].

Table 1. Starting parameters.

Reference year of the evaluation	2021
Analysis period	30 years
Duration of the construction phase	3 years
Price system	All amounts are defined at constant prices in euros
Discount rate	4%
Terminal value	14,8 million euros

Table 2 shows the estimated values of the main economic and financial performance indicators (NPV and IRR) for each of the three hypothesized scenarios.

Table 2. Estimated performance indicators for the three reference scenarios.

	NPV [MEUR]	IRR [%]
Scenario 0	2.51	6.25
Scenario 1	0.82	4.20
Scenario 2	8.45	6.14

The following sensitive project variables have been identified: investment costs, tariff revenues, the volume of water supplied to users, increase in post-project operating costs. Sensitive (or critical) are defined as those variables which, making them vary by 1%, lead to a consequent variation in the performance indicator of more than 1% [14]. Following an in-depth analysis of the literature and after hearing the opinion of sector experts, it was possible to attribute a triangular probability distribution to each variable of scenarios 1 and 2, defining the minimum, maximum, and most probable values (see Table 3 and Table 4).

Once the ranges of the variability of the critical parameters were defined, it was possible to implement MCS. The IRR was chosen as an economic-financial performance indicator. The tolerability and acceptability thresholds characteristic of the ALARP approach have been defined as follows:

- **Tolerability threshold**: the probability of having an IRR greater than or equal to the project discount rate (equal to 4.00%) must not be less than 50%.
- **Acceptability threshold:** the probability of having an IRR greater than or equal to the IRR of scenario 0 (equal to 6.25%) must not be less than 50%.

For both design scenarios, the tolerability requirement is met. In fact, for scenario 1 the probability of having an IRR greater than or equal to the discount rate is 53%, therefore greater than 50%. For scenario 2, this probability is even close to 100%. Hence, both investments do not fall within the unacceptable region.

Table 3. I hypothesis on the probability distribution of critical variables (scenario 1).

Critical variable	Most probable	Min	Max
First year investment costs [MEUR]	−18.5	−20.4	−16.5
Second year investment costs [MEUR]	−22.5	−24.8	−20.0
Third year investment costs [MEUR]	−23.5	−25.9	−20.9
Tariffs (and therefore revenues) [$€/m^3$]	1.5	1.4	1.7
Annual volumes of water (demand) [hm^3]	3.8	3.2	4.3
Annual incremental operating costs following the project [MEUR]	−3.5	−3.9	−3.1

Table 4. I hypothesis on the probability distribution of critical variables (scenario 2).

Critical variable	Most probable	Min	Max
First year investment costs [MEUR]	−17.5	−17.6	−17.4
Second year investment costs [MEUR]	−22.0	−22.1	−21.9
Third year investment costs [MEUR]	−23.0	−23.1	−22.9
Water tariff + Soil improver tariff [$€/m^3$]	1.6	1.5	1.7
Annual volumes of water (demand) [hm^3]	3.8	3.7	4.0
Annual incremental operating costs following the project [MEUR]	−3.4	−3.5	−3.3

Figure 2 shows that for scenario 1 the acceptability requirement is not met (the probability of having an IRR greater than or equal to 6.25% is 5.6%), so the investment is in the ALARP area. Differently, from Fig. 3 it is possible to verify how the investment of scenario 2 is in the widely acceptable area, being the probability of having an IRR greater than or equal to 6.25% higher than 50% (precisely, equal to 51%).

The analysis carried out shows that the most convenient investment for the integrated water service manager is that relating to scenario 2. This is not only because this solution is able to guarantee higher NPV and IRR than those of scenario 1, but above all, because the probability of failure is low enough and considered acceptable.

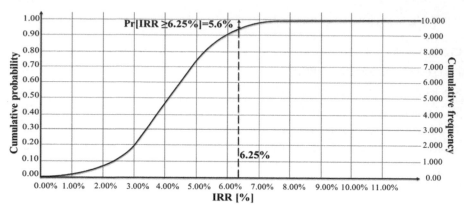

Fig. 2. Risk assessment (scenario 1): the project is in the ALARP area (own elaboration).

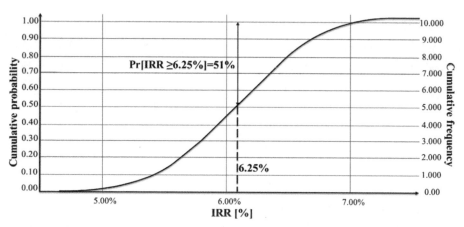

Fig. 3. Risk assessment (scenario 2): the project is in a widely acceptable area (own elaboration).

4   Discussion and Conclusions

The sustainable development of the territory is threatened by the rapid expansion of metropolitan areas and the consequent increase in the demand for water and energy resources. These interdependent resources are fundamental for the socio-economic growth of a country [15]. On purely energy issues, the National Integrated Plan for Energy and Climate (PNIEC) was recently introduced in Italy. The aim of the plan is to achieve and exceed the EU objectives in terms of energy efficiency and security, the use of renewable sources, and the development of the internal energy market [16]. The objectives of the PNIEC were considered in the preparation of the investments and reforms in the field of Green Transition contained in the National Recovery and Resilience Plan (PNRR). Unfortunately, of the 59.47 billion euros allocated by the PNRR to the M2 mission (Green Revolution and ecological transition), only 4.38 billion are assigned to the water sector (funding that will then be further divided between integrated water service and irrigation). Specifically, only 600 million are intended for the completion of the sewage and purification infrastructure, and therefore for the elimination of situations of non-compliance with European legislation [7]. Therefore, there is insufficient attention to the delicate issue of wastewater treatment, which represents the major critical aspect of the national water service. For this reason, in this work it was decided to implement the risk assessment protocol for a project aimed at the construction of a new treatment plant and the expansion of the existing sewerage network, taking as a reference a medium-sized Italian city.

The approach adopted integrates the traditional risk assessment expressed in probabilistic terms with the ALARP logic, identifying thresholds of tolerability and risk acceptability. From the application, it emerges that both design alternatives respect the tolerability condition. However, only for the second alternative is the condition of full acceptability satisfied. Although the deterministic IRR of scenario 2 is lower than that of scenario 0 (current situation), the probability of obtaining for scenario 2 an IRR greater than that of scenario 0 is greater than 50%. Furthermore, a not secondary aspect, the second solution is also the most sustainable in terms of impact on the natural environment.

The approach adopted could also be applied to the other missions, categories, and investments of the PNRR, such as those relating to the areas of urban regeneration and sustainable transport, as well as to all the projects necessary to achieve the objectives of the PNIEC [17–20]. The proposed model is a useful tool for policy makers and private managers to monitor the evolution of projects as the main sensitive parameters change. It is, therefore, an useful decision support tool and could find a wide use in real applications [21–23].

References

1. Stati Generali della Green Economy. https://www.statigenerali.org/2016/07/il-pianeta-acqua-alla-sfida-della-green-economy/
2. Anwandter, L., Rubino, P.: Rischi, incertezze, e conflitti d'interesse nel settore idrico italiano: analisi e proposte di riforma. Unità di Valutazione degli Investimenti Pubblici, Roma (2006)
3. Mazzei, A., Peruzzi, P.: La regolazione dei servizi idrici e l'allocazione dei rischi. Associazione Nazionale Autorità ed Enti d'Ambito, 2008/4 (2008)
4. Maselli, G., Macchiaroli, M.: Tolerability and acceptability of the risk for projects in the civil sector. In: Bevilacqua, C., Calabrò, F., Della Spina, L. (eds.) NMP 2020. SIST, vol. 178, pp. 686–695. Springer, Cham (2021). https://doi.org/10.1007/978-3-030-48279-4_64
5. Macchiaroli, M., Dolores, L., Pellecchia, V., De Mare, G., Nesticò, A., Maselli, G.: Application to a player operating in Italy of an AHP model for the identification of the most advantageous technical alternatives in the management of the integrated water service. In: Gervasi, O., et al. (eds.) ICCSA 2020. LNCS, vol. 12255, pp. 146–161. Springer, Cham (2020). https://doi.org/10.1007/978-3-030-58820-5_11
6. Macchiaroli, M., Pellecchia, V., D'Alpaos, C.: Urban water management in Italy: an innovative model for the selection of water service infrastructures. WSEAS Trans. Environ. Dev. **15**, 463–477 (2019)
7. Council of Ministers of the Italian Republic: Piano Nazionale di Ripresa e Resilienza (2021)
8. Nesticò, A., Macchiaroli, M., Maselli, G.: An innovative risk assessment approach in projects for the enhancement of small towns. Valori e Valutazioni **2020**(25), 91–98 (2020)
9. Benintendi, R., De Mare, G.: Upgrade the ALARP model as a holistic approach to project risk and decision management. Hydrocarb. Process. **9**, 75–82 (2017)

10. Burja, C.A., Burja, V.: The risk analysis for investments projects decision. Ann. Univ. Apulensis, Ser. Econ. **11**(1), 98–105 (2009). http://www.oeconomica.uab.ro/upload/lucrari/112 0091/09.pdf

11. Tazid, A., Boruah, H., Dutta, P.: Modeling uncertainty in risk assessment using double Monte Carlo method. Int. J. Eng. Innov. Technol. (IJEIT) **1**(4), 115–118 (2012)

12. Berk, J., De Marzo, P., Venanzi, D.: Capital Budgeting, 1st ed., Pearson Paravia Bruno Mondadori, Milan, Italy (2009). ISBN 9788871925875

13. Melchers, R.E.: On the ALARP approach to risk management. Reliab. Eng. Syst. Saf. **71**(2), 201–208 (2001). https://doi.org/10.1016/S0951-8320(00)00096-X

14. Sartori, D., et al.: Guide to Cost-Benefit Analysis of Investment Projects–Economic Appraisal Tool for Cohesion Policy 2014–2020. European Commission, Brussels, Belgium (2014)

15. Fang, D., Chen, B.: Linkage analysis for the water–energy nexus of city. Appl. Energy **189**, 770–779 (2017). https://doi.org/10.1016/j.apenergy.2016.04.020

16. Italian Ministry of Economic Development, Italian Ministry of the Environment, Italian Ministry of Infrastructure and Transport: Piano Nazionale Integrato per l'Energia e il Clima (2019)

17. Nesticò, A., De Mare, G., Frusciante, B., Dolores, L.: Construction costs estimate for civil works. A model for the analysis during the preliminary stage of the project. In: Gervasi, O., et al. (eds.) ICCSA 2017. LNCS, vol. 10408, pp. 89–105. Springer, Cham (2017). https://doi.org/10.1007/978-3-319-62404-4_7

18. Dolores, L., Macchiaroli, M., De Mare, G.: Sponsorship's financial sustainability for cultural conservation and enhancement strategies: an innovative model for sponsees and sponsors. Sustainability (Switzerland) **13**(16), 9070 (2021). https://doi.org/10.3390/su13169070

19. De Mare, G., Granata, M.F., Forte, F.: Investing in sports facilities: the Italian situation toward an Olympic perspective. In: Gervasi, O., et al. (eds.) ICCSA 2015. LNCS, vol. 9157, pp. 77–87. Springer, Cham (2015). https://doi.org/10.1007/978-3-319-21470-2_6

20. De Mare, G., Di Piazza, F.: The role of public-private partnerships in school building projects. In: Gervasi, O., et al. (eds.) ICCSA 2015. LNCS, vol. 9156, pp. 624–634. Springer, Cham (2015). https://doi.org/10.1007/978-3-319-21407-8_44

21. Macchiaroli, M., Dolores, L., Nicodemo, L., De Mare, G.: Energy efficiency in the management of the integrated water service. A case study on the white certificates incentive system. In: Gervasi, O., et al. (eds.) ICCSA 2021. LNCS, vol. 12956, pp. 202–217. Springer, Cham (2021). https://doi.org/10.1007/978-3-030-87010-2_14

22. Dolores, L., Macchiaroli, M., De Mare, G.: A model for defining sponsorship fees in public-private bargaining for the rehabilitation of historical-architectural heritage. In: Calabrò, F., Della Spina, L., Bevilacqua, C. (eds.) ISHT 2018. SIST, vol. 101, pp. 484–492. Springer, Cham (2019). https://doi.org/10.1007/978-3-319-92102-0_51

23. Dolores, L., Macchiaroli, M., De Mare, G.: A dynamic model for the financial sustainability of the restoration sponsorship. Sustainability **12**(4), 1694 (2020). https://doi.org/10.3390/su1 2041694

Awareness Campaigns and Sustainable Marketing for an Efficient Use of Territorial Resources

Luigi Dolores$^{(\boxtimes)}$ ⓘ, Maria Macchiaroli ⓘ, and Gianluigi De Mare ⓘ

University of Salerno, Via Giovanni Paolo II, 132, Fisciano, SA, Italy
{ldolores,mmacchiaroli,gdemare}@unisa.it

Abstract. Economic development and population growth are clear signs of the progressive depletion of natural resources. It is, therefore, necessary to encourage more efficient use of territorial resources. Citizens need to be better informed on green issues to encourage a positive attitude about the conscious use of resources. Many companies are involved in the process of informing and raising citizens' awareness of environmental issues. The use of green marketing strategies is becoming more and more widespread as consumers are intent on making purchasing choices such as to reward companies committed to the ecological front.

This paper intends to propose a model through which to maximize the degree of utility that users of essential public services attribute to the green marketing strategies undertaken by private managers. To this end, the Cobb-Douglas function is used to estimate the optimal budget that the companies in the analysis should allocate to awareness campaigns on green issues to maximize consumer satisfaction and achieve an optimal level of corporate reputation. The model is applied to a private Integrated Water Service Manager in southern Italy. The application shows that more than 550 users are predisposed to perceive in a favourable manner the green marketing activity promoted by the Manager. Investments in awareness campaigns, green communication, and sustainable marketing can undoubtedly bring significant benefits for managers of essential services both in terms of reputation and financial sustainability.

Keywords: Green marketing · Sustainable marketing · Resource management · Integrated Water Service · Cobb-Douglas function

1   Introduction

One of the primary objectives of modern society is to promote the growth of the efficiency with which natural resources are used in the socio-economic process. In recent decades, the idea has been strengthened according to which, for the human system to move towards ecological sustainability, it is necessary to reduce the resources taken from the natural environment. Classical economic theory holds that the purpose of any economic activity is to satisfy the needs of individuals. Therefore, the creation of economic value

L. Dolores, M. Macchiaroli and G. De Mare—Contributed equally to this work.

presupposes the use of the resources necessary for the satisfaction of needs. Only on the fringes of classical economic theory are the so-called negative externalities analysed, i.e., those unwanted effects deriving from improper use of resources [1]. This gap, now largely overcome, has also had repercussions at the institutional level in many countries for many years. For example, in Italy, the governance of the commons has only recently acquired a central role in guaranteeing the protection and safeguarding of natural resources. In the Italian context, starting from 2011 the term «common» became popular thanks to the press and digital media that disseminated the slogan «water as a common good» during the referendum on the privatization of the water service. From that moment on, in many of the campaigns carried out to raise public awareness on specific issues, the «common» label has been associated with each of the areas of interest, such as the environment, school, culture, art, work, public transport, health, state property, etc. [2]. Thanks to the mobilization of civil society, citizens are acquiring greater awareness of the failure of the development models followed so far, which are now increasingly less sustainable. For this reason, companies, whose main task is to offer goods and services to citizens, are forced to review their business models. This is to identify new medium-long term development strategies, verify their correspondence with environmental issues and make the new strategic conducts operational. Marketing practices have often contributed to the creation of negative externalities such as environmental pollution, the waste of non-reproducible resources, the uncontrolled production of waste, the unbridled consumption of goods and services [3]. Fortunately, today marketing often plays an antithetical role, being used by many companies from a green perspective. According to Polonsky, "sustainable marketing" or "green marketing" consists of a series of activities carried out to generate and facilitate any exchanges necessary to satisfy human needs or desires in ways that have the least negative impact on the natural environment [4]. Sustainable marketing involves the entire business organization and aims to identify more sustainable business models. Its main purpose is to stimulate sustainable consumption styles, as well as to increase the corporate reputation. Through green marketing, we intend to ensure both the economic sustainability of the company (with the generation of profit) and the environmental and social one (through the creation of a long-term advantage for the community).

Remaining in the Italian context, sustainable marketing activities should also be increasingly carried out by private companies that offer public goods and services, such as companies that deal with the production and distribution of electricity and natural gas and those engaged in the management of the Integrated Water Service (SII). Their main objective should be to reduce the gap between individual and collective needs, between rich and poor areas, and between the exploitation of resources and their availability [3]. This objective is even more priority if contextualized with respect to the recent National Integrated Plan for Energy and Climate (PNIEC), a document that establishes Italy's 2030 objectives on energy efficiency, renewable sources, and the reduction of CO_2 emissions [5]. A further push towards the ecological and environmental transition in Italy comes from the National Recovery and Resilience Plan (PNRR). This is a document that each Member State of the European Union must prepare to access the funds of the Next Generation EU (NGEU), to guarantee the recovery after the Covid-19 pandemic and to relaunch the economy of the Member States by making them greener and digital [6]. By virtue of the institutional and social change taking place, companies that offer goods and services of the public utility can no longer ignore green communication methods. This is because consumers now tend to reward companies committed to the ecological front and to prefer products and services that have the characteristic of environmental, ethical, and social sustainability.

In this work, attention is focused on the Integrated Water Service (SII). The water sector is often considered to be sustainable, since the mission of every good manager is to guarantee access to the resource for present and future generations, ensuring the protection of quality and quantity. Each managing body is faced with a double challenge: to effectively manage the infrastructures given to them in concession and to guarantee the uninterrupted supply of the service in compliance with quantitative and qualitative standards. Regarding the last point, however, we must deal with the frequent water crises and with the waste and unconscious use of the resource. Awareness-raising campaigns aimed at users are therefore necessary to inform them on issues such as climate change and anthropogenic actions capable of reducing the availability of water. In addition to information campaigns, the water managers should also take an active role, providing material and financial support to all those initiatives promoted by bodies, public administrations, non-profit organizations, and private citizens aimed at protecting and safeguarding the water resource. By making use of green marketing and sustainable communication strategies, water managers can improve their reputation, obtaining multiple benefits. Among these, there could be a reduction in the level of user insolvency or a reduction in tax evasion and avoidance.

The aim of the work is to demonstrate how, by resorting to green marketing, sustainable communication strategies, and awareness campaigns on the rational use of resources, companies that offer public goods and services or of public utility (commons) can build a positive reputation with stakeholders and customers. To this end, a model for maximizing the utility generated by the initiatives based on the Cobb-Douglas function is proposed [7]. Specifically, the virtual profits attributed to the green marketing activity are considered as a proxy variable of the utility that stakeholders attribute to this strategy. The model in question, which takes its cue from previous studies on sponsorship in the cultural field [8–12], is applied to a water manager of Southern Italy. It allows estimating the optimal budget that the manager should allocate to sustainable marketing initiatives to maximize the level of satisfaction of the service users.

2   Method

The Cobb-Douglas production function is a mathematical function formulated by C.W. Cobb and P.H. Douglas in 1928, widely used in economic analysis. It describes how the output varies in relation to the variation of production factors [7]. This function is the subject of numerous applications because of its mathematical properties that allow subsequent processing, such as the calculation of the marginal productivity of factors, but also because its econometric estimate offers a rather convincing representation of reality. In maximization problems, the Cobb-Douglas function is used to calculate the quantity of one factor of production capable of maximizing profits. It is clear how, with the same unit cost of the production factor considered, as its quantity increases, both revenues, and overall production costs can increase. Therefore, the maximum profit is obtainable only when the condition of economic optimum is reached, that is when the quantity of a good is produced in correspondence with which the marginal revenues equal the marginal costs [13].

For the case study, the production function is represented by the following equation:

$$R_v = A^\alpha \cdot W^{1-\alpha} \cdot G^{1-\alpha}. \tag{1}$$

This function expresses the virtual revenues R_v that the Integrated Water manager obtains in the current year. Specifically, A is the set of capital available to the manager (assets), W is the number of workers available to the company (blue-collar and white-collar workers) and G is the number of green marketing initiatives undertaken in the analysis exercise (including awareness campaigns and investments in sustainable communication). The constant $\alpha < 1$ allows to measure the marginal return of A, while the constant $1 - \alpha$ < 1 offers a measure of the marginal returns of both W and G. Since both constants are less than one, the marginal returns of the three factors of production they are all decreasing. The same exponential constant is attributed to the factors W and G since the Cobb-Douglas function considered is like that proposed by Romer and used to explain the endogenous growth generated in an economic system starting from the human capital factor [14]. The latter factor is here replaced with G. Revenues R_v are to be considered virtual as a portion of them are presumed to derive from G. But in the case of the Integrated Water Services, with the same water tariff, there is almost never a substantial change in revenues, since the number of users (and therefore the demand) is roughly constant in the reference area. Only the manager's reputation level and the perception of the quality of the service offered to the user can vary. Therefore, the higher revenues generated by the production factor G are to be considered as a proxy variable of the number of users and stakeholders who have positively received the green marketing activity undertaken by the manager. At this point, you can define the cost function as follows:

$$C = aA + wW + gG, \tag{2}$$

where a is the average return on assets, w is the unit cost of the labour factor and g is the unit cost, assumed constant, of the green marketing activity. Virtual profits can be represented as follows:

$$P_v = A^\alpha \cdot W^{1-\alpha} \cdot G^{1-\alpha} - aA - wW - gG. \tag{3}$$

The parameter α can be estimated starting from the log-linearity property of the Cobb-Douglas function as follows:

$$\alpha = (\ln R_v - \ln W - \ln G)/(\ln A - \ln W - \ln G) \tag{4}$$

Marginal revenues equal marginal costs when the first derivative of P_v with respect to G is zero. By imposing this condition, it is possible to obtain the number of optimal green marketing initiatives G^* capable of maximizing virtual profits, i.e., the degree of user satisfaction:

$$G^* = \left(\frac{g}{(1-\alpha)A^\alpha W^{1-\alpha}} \right)^{-\frac{1}{\alpha}}. \tag{5}$$

The optimal budget to invest in sustainable marketing is $g \cdot G^*$. The maximum virtual profits $P_{v,max}$ is instead equal to:

$$P_{v,max} = A^\alpha \cdot W^{1-\alpha} \cdot G^{*(1-\alpha)} - aA - wW - gG^*. \tag{6}$$

Finally, the number of users u who positively perceive the manager's green marketing and awareness campaigns is equal to the ratio between the maximum virtual profits P_v, max and the average annual cost per household of the water resource c_w:

$$u = \frac{P_{v,max}}{c_w}. \tag{7}$$

The model is applied to the case study in the following section.

3   Results and Discussion

As anticipated, the model is applied to a water manager in Southern Italy. The Manager is a joint-stock company that deals with the collection, supply, and distribution of drinking water and other services of public interest originally managed by the municipal administration (now majority shareholder). The number of end-users served is approximately 5,000. The amount of water supplied to users each year is approximately 650,000 m³/year. Table 1 shows the data necessary for the application of the model deriving from the latest available financial statements (the year 2020) [15].

Table 1. Financial statements data of the water manager (Source: AIDA database, 2020)

Data	Symbols	Values
Virtual revenue [EUR]	R_v	1,559,225
Production costs [EUR]	C	1,426,015
Virtual Profits [EUR]	P_v	133,210
Cost of assets [EUR]	aA	915,390
Labour cost [EUR]	wW	460,625
Cost of green marketing activities [EUR]	gG	50,000
Assets [EUR]	A	4,152,342
Number of employees	W	11
Number of green marketing activities	G	1
Average return on assets [%]	a	0.22
Unit labour cost [EUR]	w	41,875
Unit cost of green marketing activities [EUR]	g	50,000

Suppose that for the year of analysis (the year 2021) the revenues and costs remain constant (ceteris paribus). Furthermore, we assume a unitary investment in green marketing equal to € 50,000, which flows into the total cost of production. The average annual cost per household of the water resource c_w is set at € 307 (estimated consumption for the year 2020 of an average household of three people [16]). By applying (4), (5), (6), and (7), the results of Table 2 are obtained.

Table 2. Results of the maximization model (Source: own elaboration, 2021).

Results	Symbols	Values
Marginal rate of return [%]	α	0.92
Optimal number of investments in green marketing	G*	2.56
Optimal budget to invest in green marketing [EUR]	gG*	€ 127,753
Maximum virtual profit [EUR]	$P_{v,max}$	€ 171,121
Number of users who favourably perceive the green Marketing activity	u	557

The model shows that the Manager, by investing an amount of € 130,190 in green marketing, sustainable communication, and awareness campaigns, can obtain the highest level of user satisfaction. In particular, the investments made are favourably perceived by as many as 557 households (it is assumed that each household is on average composed of three members [16]). Thanks to the investments made, the Manager can strengthen its reputation with stakeholders and with customers. Finally, not necessarily the profits deriving from the marketing activity are all to be considered virtual. In fact, the Operator's commitment in terms of Corporate Social Responsibility could lead to a better perception of the water service, which in turn could result in a decrease in the level of user insolvency, and therefore in the abatement of tax evasion and avoidance of the tariff. Furthermore, a reward mechanism could also be hypothesized for those managements aimed at promoting environmental and ecological issues. In fact, through the tariff multiplier mechanism, a portion of the investment cost can be transferred to the user of the water service. This in turn could trigger further virtuous behaviour, encouraging users to reduce water consumption to bring tariffs back to a lower level.

4   Conclusions

Economic development and demographic growth are clear signs of the progressive exhaustion of natural resources. Common goods such as water, clean air, ecosystem services, and soil are essential for both the quality of life and human health. It should not be overlooked that natural goods are available in limited quantities. The unbridled competition in grabbing these resources is not only a cause of scarcity but could lead to an increase in prices that could generate a serious economic crisis worldwide. It is, therefore, necessary to encourage more efficient use of resources by both businesses and private citizens. As far as businesses are concerned, it is necessary that they voluntarily contribute to the progress of civil society and the protection of the environment, integrating social and ecological assessments in entrepreneurial transformation and in governance relations with stakeholders [17]. On the other hand, private citizens need to be better informed on green issues to encourage a positive attitude about the conscious use of resources. The Green Revolution is therefore also influencing the world of communication. In recent years, companies are seeing full compliance with sustainability policies. Many of the companies involved in the process of informing and raising awareness of their customers on environmental issues acquire greater competence and awareness than the recipients of their communication. New and more efficient green marketing strategies are spreading as consumers are increasingly intent on making purchasing choices aimed at rewarding companies committed to the ecological front [18]. Given the growing privatization of essential services such as water, energy, waste, and local transport, it becomes necessary to extend the green marketing strategies to private management.

This paper intends to propose a model through which to maximize the level of utility that users of essential public services attribute to the green marketing strategies undertaken by the companies that provide these services. To this end, the Cobb-Douglas function is used to estimate the optimal budget that these companies should allocate to sustainable marketing and awareness campaigns on green issues to maximize consumer satisfaction and achieve an optimal level of corporate reputation. From the application of the model to a private Integrated Water Service manager operating in a Southern Italian city of about 12,000 inhabitants, it can be deduced that more than 550 users are predisposed to perceive in a favourable manner the green marketing activity promoted by the manager. Therefore, investing in awareness campaigns, green communication, and sustainable marketing can certainly bring significant benefits for managers of essential services in terms of both reputation and money (for example, triggering a mechanism to reduce the level of insolvency). In this way, management is not only greener but also more sustainable from a financial point of view [19–24].

References

1. Femia, A.: Cambiare le priorità: dalla produttività del lavoro alla efficienza nell'utilizzo delle risorse naturali (2010)
2. Gattullo, M.: Implicazioni geografiche sulla natura dei beni comuni. Bollettino della Società Geografica Italiana 12(2), 179–199 (2015)
3. Scott, W.G.: Sostenibilità del marketing e marketing sostenibile. Micro Macro Market. 12(2), 171–182 (2003)
4. Polonsky, M.J.: An introduction to green marketing. Electron. Green J. 1(2), 1–10 (1994)
5. Italian Ministry of Economic Development, Italian Ministry of the Environment, Italian Ministry of Infrastructure and Transport. Piano Nazionale Integrato per l'Energia e il Clima (2019)
6. Council of Ministers of Italy: Piano Nazionale di Ripresa e Resilienza (2021)
7. Cobb, C.W., Douglas, P.H.: A theory of production. Am. Econ. Rev. 18(1), 139–165 (1928)
8. Dolores, L., Macchiaroli, M., De Mare, G.: Sponsorship's financial sustainability for cultural conservation and enhancement strategies: an innovative model for sponsees and sponsors. Sustainability 13(16), 9070 (2021). https://doi.org/10.3390/su13169070
9. Dolores, L., Macchiaroli, M., De Mare, G.: A dynamic model for the financial sustainability of the restoration sponsorship. Sustainability 12(4), 1694 (2020). https://doi.org/10.3390/su12041694

10. Dolores, L., Macchiaroli, M., De Mare, G., Nesticò, A., Maselli, G., Gómez, E.M.: The estimation of the optimal level of productivity for sponsors in the recovery and enhancement of the historical-architectural heritage. In: Gervasi, O., et al. (eds.) ICCSA 2020. LNCS, vol. 12253, pp. 285–299. Springer, Cham (2020). https://doi.org/10.1007/978-3-030-58814-4_20

11. Nesticò, A., De Mare, G., Frusciante, B., Dolores, L.: Construction costs estimate for civil works. a model for the analysis during the preliminary stage of the project. In: Gervasi, O., et al. (eds.) ICCSA 2017. LNCS, vol. 10408, pp. 89–105. Springer, Cham (2017). https://doi.org/10.1007/978-3-319-62404-4_7

12. Dolores, L., Macchiaroli, M., De Mare, G.: A model for defining sponsorship fees in public-private bargaining for the rehabilitation of historical-architectural heritage. In: Calabrò, F., Della Spina, L., Bevilacqua, C. (eds.) ISHT 2018. SIST, vol. 101, pp. 484–492. Springer, Cham (2019). https://doi.org/10.1007/978-3-319-92102-0_51

13. Varian, H.R.: Microeconomia. Libreria Editrice Cafoscarina, Venezia, Italy (2012). ISBN 9788875433079

14. Romer, P.M.: Increasing returns and long-run growth. J. Polit. Econ. **94**, 1002–1037 (1986)

15. Analisi Informatizzata Delle Aziende Italiane (AIDA). https://aida.bvdinfo.com/Report.serv?_CID=61&context=2S5WC2KM1F60WAU

16. ADNKRONOS.   https://www.adnkronos.com/acqua-317-euroanno-spesa-media-per-fam iglia-tipo-di-3-persone_veD5kBptmTman3UVO2gqd

17. Gazzola, P.: CSR e reputazione nella creazione di valore sostenibile. Economia Aziendale Online- **2**, 27–45 (2012)

18. Marzulli, L., Pazienza, R., Zammartini, S., Macaddino, V.: Green marketing. XVII Master in Marketing Management, Fondazioneistud, pp. 1–27 (2014). http://service.istud.it/up_media/pwmaster13/tesina_green_marketing.pdf

19. Nesticò, A., Endreny, T., Guarini, M.R., Sica, F., Anelli, D.: Real estate values, tree cover, and per-capita income: an evaluation of the interdependencies in Buffalo City (NY). In: Gervasi, O., et al. (eds.) ICCSA 2020. LNCS, vol. 12251, pp. 913–926. Springer, Cham (2020). https://doi.org/10.1007/978-3-030-58808-3_65

20. Morano, P., Tajani, F., Guarini, M.R., Sica, F.: A systematic review of the existing literature for the evaluation of sustainable urban projects. Sustainability **13**(9), 4782 (2021). https://doi.org/10.3390/su13094782

21. Macchiaroli, M., Dolores, L., Nicodemo, L., De Mare, G.: Energy efficiency in the management of the integrated water service. a case study on the white certificates incentive system. In: Gervasi, O., et al. (eds.) ICCSA 2021. LNCS, vol. 12956, pp. 202–217. Springer, Cham (2021). https://doi.org/10.1007/978-3-030-87010-2_14

22. Benintendi, R., De Mare, G.: Upgrade the ALARP model as a holistic approach to project risk and decision management. Hydrocarb. Process. **9**, 75–82 (2017)

23. De Mare, G., Granata, M.F., Forte, F.: Investing in sports facilities: the Italian situation toward an olympic perspective. In: Gervasi, O., et al. (eds.) ICCSA 2015. LNCS, vol. 9157, pp. 77–87. Springer, Cham (2015). https://doi.org/10.1007/978-3-319-21470-2_6

24. De Mare, G., Di Piazza, F.: The role of public-private partnerships in school building projects. In: Gervasi, O., et al. (eds.) ICCSA 2015. LNCS, vol. 9156, pp. 624–634. Springer, Cham (2015). https://doi.org/10.1007/978-3-319-21407-8_44

Infrastructure Accessibility Measures and Property Values

Gabriella Maselli⬛, Stefano de Luca⬛, and Antonio Nesticò$^{(\boxtimes)}$⬛

Department of Civil Engineering, University of Salerno, Fisciano, SA, Italy
{gmaselli,sdeluca,anestico}@unisa.it

Abstract. The aim of this paper is to investigate the relationship between accessibility to services and transport infrastructure and property prices. Empirical evidence shows that an increase in accessibility levels tends to result in a positive impact on property values. However, assessing this potential benefit is not straightforward.

In this research, we first clarify which are the methods mainly employed to measure accessibility to services and infrastructures; then, we build a dataset of indicators useful to define the price function. The output is the characterization of a Hedonic Pricing Model (HPM) able to evaluate the effect of accessibility on residential properties, seldom considered in estimates. Two main findings emerge from the study. The first is that an HPM should be a function not only of the traditional intrinsic and extrinsic characteristics generally used to explain property values, but also of specific accessibility indicators, distinguishing between local and system accessibility. The second is that HPM, generally based on the use of multiple regression models, fails to consider the spatial correlation that is often particularly significant for the accessibility variable. Therefore, in the case of high levels of spatial heterogeneity, regression models must be supported by spatial econometric models.

The study conducted represents a starting point for applications to real case studies that will allow to test the defined model.

Keywords: Accessibility · House pricing · Hedonic Pricing Models · Spatial Econometric Models

1 Introduction

The impact of accessibility to the urban system services and activities on land use and real estate prices is extensively studied in the relevant literature and by transport planners [1]. As early as 1959, the American planner Hansen systematically analyzed the relationship between accessibility and land use [2]. Classical theories of urban economics have also investigated the trade-off between accessibility and space [3]. In 1969, Muth had found that dwellings with better accessibility to urban system services and activities generally have higher property values per unit area [4]. These theories show that investments

G. Maselli, S. de Luca and A. Nesticò—Contributed equally to this work.

© The Author(s), under exclusive license to Springer Nature Switzerland AG 2022
F. Calabrò et al. (Eds.): NMP 2022, LNNS 482, pp. 1355–1365, 2022.
https://doi.org/10.1007/978-3-031-06825-6_132

in transportation systems result in an increase in the level of accessibility, so that the resulting incremental benefits can be capitalized by property owners.

To understand how to measure the impact of accessibility on urban house prices, it is necessary to first specify what is meant by the term "accessibility". According to the classical urban model, the city structure can be interpreted as a trade-off function between access to jobs, generally located in the town centre, and housing prices. This results in a more densely populated form of city with higher land values in the centre, and steadily decreasing densities and prices as you approach the peripheral areas [5]. Nowadays, the shape of metropolitan areas has changed dramatically. First, cities are no longer monocentric. Empirical studies document the decentralization of both jobs and population, as well as the existence of multiple concentrations [6]. In addition, the literature shows that accessibility is also related to other factors, such as the opportunity to access scarce public resources such as high-quality schools, shops and services, bus stops, metro and railway stations, urban parks, and green streets. As a result, the level of infrastructure significantly affects accessibility and, consequently, property values. Generally, dwelling prices analysis employs Hedonic Price Models (HPMs) according to which the dependent variable (house price) is a function of a bundle of characteristics that households place values on, including transport accessibility [7].

Although the relationship between accessibility and property values has been widely studied, there are still several challenges related to the topic that need further investigation. First, it is necessary to investigate how to consider different levels of accessibility when characterising HPM. Often, the shortest distance or travel time from a home to a facility of interest is chosen as a proxy variable but neglecting the differences between facilities and their scarcity. Second, the housing market is heterogeneous, so the housing price may be related to unobserved characteristics. Therefore, it is necessary to carefully select the extrinsic features - in addition to the intrinsic ones - that characterize the model, and sometimes also to evaluate the combined effects of multiple factors. Third, accessibility is a typical spatial attribute, so HPMs often fail to adequately account for it. In fact, in a typical HPM, all assumptions of the linear multiple regression are assumed to be satisfied. However, in the analysis of neighbourhood attributes, spatial correlation between observations can be found. Therefore, it is essential to check for the presence of spatial autocorrelation in unadjusted HPM and to evaluate its effects if necessary [8].

This paper first provides a literature review of the possible measures of accessibility that can be included in the characterization of an HPM. Then, we focus on the models generally employed to assess the effect of accessibility on property values, highlighting their limitations and advantages. Finally, we characterize a price function, in which we define the independent variables, the relative units of measurement and the indicator or *proxy* variable useful for estimating them. The output is a new methodological approach based on HPM to assess the impact on housing prices of both intrinsic and extrinsic characteristics and accessibility.

2 Literature Review and Conceptual Background

2.1 Intrinsic Characteristics, Extrinsic Attributes, and Housing Prices

The literature shows a wide use of HPM to determine the price function of real estate. This function is defined by a set of implicit or hedonic prices, in relation not only to the characteristics proper to the dwellings (called intrinsic), but also according to the location and the peculiarities of the place (extrinsic or zonal characteristics) [9]. According to Wittowsky et al. [10], dwellings attributes can be divided into: (i) dwelling-specific characteristics; (ii) neighbourhood characteristics; and (iii) accessibility characteristics. The last two can also be understood as local characteristics. Intrinsic characteristics include, for instance: dwelling type and floor area; number of rooms, bathrooms, and balconies; exposure and brightness; presence of elevator; floor level [11]. Extrinsic (or zonal) characteristics contain social, economic, and environmental factors of the neighbourhood. Social factors include, but are not limited to: crime and proximity to noxious facilities; quality of nearby schools; racial or ethnic composition of the population. Economic variables consist of average per capita income and municipal tax level. Finally, environmental factors involve proximity to urban parks and green spaces, which determine a positive effect on housing prices, but also the level of environmental and noise pollution and proximity to ecosystem services, which instead lead to a decrease in housing market value [12–14]. Regarding accessibility characteristics, the literature shows that residential property values are a function of access to services, urban facilities, jobs, but also of distance to the Central Business District (CBD), major infrastructure and public transport in general [6]. The next section focuses on key urban accessibility measures.

2.2 Measuring Accessibility

The term "accessibility", widely used in urban and transport planning, refers to the ability to reach a potential destination i from a given location j by a particular transport system [15]. Thus, the accessibility of a place results mainly from two factors: the transportation system and the land use pattern [16]. Geurs and van Wee [17], instead, identified four components that should be included in the accessibility measure: (i) land use, which refers to the spatial distribution and quality of opportunities; (ii) transportation, which considers the disutility related to moving from a given place to a relevant opportunity; (iii) the temporal component due to the availability of opportunities at different times of the day; and (iv) the individual component that refers to the needs and preferences of different individuals.

There are many empirical studies showing that accessibility is a key 'external factor' that significantly influences house prices [18]. Generally, it has been shown that increased accessibility related to improved transport results in higher property values. However, some studies have found neutral or negative effects, particularly in railway station accessibility due to the negative externalities associated with these facilities [19]. Regarding strictly the measurement of accessibility, different approaches exist in the literature in relation to the purpose of the study. Handy and Niemeier [16] classified the indicators useful to measure accessibility into three main groups: (a) gravity type or Hansen-gravity; (b) accumulated opportunities; and (c) based on random utility theory.

Gravity-type or Hansen-gravity indicators, which measure accessibility to a given service, have been the most widely used in practice and are considered the most robust. Their general formulation is:

$$A_i = \sum_j f(E_j, C_{ij}) \tag{1}$$

In (1): A_i represents the accessibility of opportunities in each zone i; E_j measures the attractiveness of zone j (in terms of business opportunities, jobs, etc.); C_{ij} is a measure of the travel cost between zones i and j.

The Hansen method is suitable for measuring accessibility to employment, as generally employment opportunities are proportional to the size of an area or the number of potential people. In contrast, accessibility to services such as education can be measured by using the travel time to the nearest school destination as a proxy variable for accessibility, as each child will tend to attend the nearest school.

The main difference between gravity-type indicators (a) and accumulated opportunities indicators (b) is that the former is able to differentially weight opportunities according to the cost of travel; in the latter case, instead, costs can only be weighted in a binary way, i.e., they can take a value of 1 or 0 depending on whether the opportunities are inside or outside a given range [20].

Finally, compared to utility-based indicators (c), gravity-type indicators (a) have a zonal nature that makes them more appropriate for research focused on an intermediate scale [18].

Several authors use gravity-type (a) indicators [7, 11, 21, 22]. Among them, Cascetta [23], Coppola and Nuzzolo [24], Cordera et al. [18] apply the Hansen-gravity model distinguishing between active and passive accessibility. Active accessibility is understood as the ability of an area to reach opportunities present in other districts, while passive accessibility represents the ability of an area to be reached by populations in other neighbourhoods. In formula:

$$ACC\_ACT_i = \sum_j [\exp(\alpha_2 \cdot C_{ij}) \cdot E_j^{\alpha_1}] \tag{2}$$

$$ACC\_PAS_j = \sum_j [\exp(\alpha_4 \cdot C_{ij}) \cdot P_i^{\alpha_3}] \tag{3}$$

In which: P_i is the population, or the number of households present in zone i. E_j and C_{ij} measure the attractiveness of zone j and the cost of travel between zones i and j, respectively. α_1, α_2, α_3, and α_4 are estimated by Ordinary Least Sqaures (OLS), expressing both terms of expressions (2) and (3) in logarithmic terms.

Du and Mulley [7] used travel time as a measure of accessibility. To assess accessibility to education by public transport they used the closest method, i.e., they calculated the travel time to the nearest school. To evaluate the accessibility of employment they used the weighted Hansen accessibility measure, calculated with a gravity-based formula. Martínez and Viegas [21] distinguish system-level accessibility attributes and local accessibility attributes. System-level accessibility indicators were evaluated by means of a gravity model. Local accessibility is measured by means of the all-or-nothing and continuous approach, considering the influence (total or null) resulting from the proximity to public transport entry points or nerve centres. In this case, accessibility is estimated through a set of dummy variables, which are assigned the value 1 if the property is close to the public transport line or road, 0 if it is far away or not affected by the accessibility benefit. The all-or-nothing approach fails to capture system accessibility but, being of simpler practical implementation, is adopted by several authors [21, 25]. Finally, other authors evaluate local accessibility not only through dummy variables (1 or 0) but also through quantitative variables: distance in meters from metro stations, bus lines, railway stations; distance or time in minutes from Centre District Business (CDB) or main urban services; access to shopping, gardens, bus lines etc. (walking time in minutes) [22, 26, 27].

2.3 Hedonic Price Models and Spatial Econometric Models

Hedonic price modelling is mainly based on Lancaster's theory of consumer behaviour, according to which it is not the good itself that creates utility but its specific characteristics [28]. That is, the value of the outputs – in this case the value of residential properties – is given by the sum of the values that the consumer gives to each of the different attributes that constitute it. In this way, it is possible to estimate the prices of those characteristics that are not explicitly exchanged in observable market transactions [29].

Generally, HPM uses OLS, or Multiple Linear Regression (MLR) models. The general model is typically expressed through the following formulation:

$$Y = X \cdot \beta + \varepsilon \tag{4}$$

In (4): Y is a vector (n × 1) of individual house prices, generally specified in logarithmic terms; X is a matrix (n × k) of independent variables, generally represented by property characteristics (intrinsic and extrinsic); β is the vector of k coefficients to be estimated; ε is a vector (n × 1) of independent and identically distributed errors [16].

In some studies, the non-stationarity between different areas in the relationship between accessibility and housing values has been demonstrated [7, 23]. This means that depending on the socio-economic characteristics of the investigated area, accessibility may result in a positive effect on housing prices in some neighbourhoods and neutral or negative effects in others. This spatial non-stationarity may result, on the one hand, from an incorrect characterization of the model, especially when some data are not available or when some variables are neglected in the analysis. On the other hand, variables such as those defining accessibility are spatially heterogeneous. Precisely to account for spatial dependence and heterogeneity, linear regression models can be supported by spatial econometric models [7]. Such spatial dependence can be accounted for by including "lagged" spatial variables in the model [29].

The most widely used Spatial Econometric Model (SEM) to account for spatial dependence in linear regression observations is the Simultaneous Autoregressive Model (SAR). In this case, the autoregressive process is applied to the variable Y which is modelled as a "lagged" variable:

$$Y = \rho \cdot W \cdot Y + X \cdot \beta + \varepsilon \tag{5}$$

In (5) ρ is the spatial autocorrelation parameter, W is the spatial weights matrix of dimension n × n, with n number of observations.

In the Spatial Error Model (SEM), instead, the spatial dependence is applied to the error term:

$$Y = X \cdot \beta + \lambda \cdot W \cdot u + \varepsilon \qquad (6)$$

where: λ is the spatial autoregression coefficient, u is the vector of the spatial error term, W is the matrix of weights, ε is the spatially independent error term.

In the Spatial Durbin Model (SDM), spatial autoregression is applied to all variables:

$$Y = \rho \cdot W \cdot Y + X \cdot \beta + W \cdot x \cdot \gamma + \varepsilon \qquad (7)$$

X is the matrix of independent variables (n $\times$ k), γ is the autoregression coefficient applied to the X matrix.

The Spatial Autocorrelation Model (SAC) uses two matrices of weights, one for the dependent variable and one for the spatial error:

$$Y = \rho \cdot W_1 \cdot Y + X \cdot \beta + u, \quad \text{where } u = \cdot W_2 \cdot u + \varepsilon \qquad (8)$$

Finally, Geographically Weighted Regression (GWR), which can be regarded as the 'local' version of the linear regression model using OLS, has also been extensively tested in practice [23]. Unlike SLMs, SEMs and SDMs, which provide a single marginal price for each independent variable, with GWR it is possible to assess the local variation of the implicit marginal prices for each characteristic introduced by the model. It emerges that crucial aspects concern: (a) the identification of the best regression model; (b) the understanding of the coefficients of SEM. Regarding point (a), Anselin [30] characterises a methodology to select the best performing regression model. With reference to point (b), Golgher et al. [31] explain how to interpret coefficients of SEM.

3   Research Setting and Methodology

Below we define the logical-operational steps to be followed to assess the effect of accessibility on property values.

Step 1: Characterization of the Hedonic Price Model (HPM). It is necessary to define the variables on which the property price function Y depends. In general terms, Y is a function of intrinsic characteristics C_i of the property, extrinsic factors C_e, indicators of local accessibility A_l, indicators to assess system accessibility A_s.

$$Y = f(C_i, C_e, A_l, A_s) \qquad (9)$$

The most widely used model for assessing the effect of accessibility is the semilogarithmic model, according to which:

$$\ln Y = \beta_0 + \beta_1 \cdot X_1 + \cdots + \beta_i \cdot X_i + \cdots + \beta_k \cdot X_k + \varepsilon \qquad (10)$$

In such a model, a change in an explanatory variable (X_i) results in a percentage change $(100 \times \beta i)$ in the dependent variable Y and represents the elasticity of that variable. For the indicator variables - or dummy variables - the elasticities can be calculated as follows:

$$E_i = [\exp(\beta_i) - 1] \cdot 100 \qquad (11)$$

Table 1 below reports a panel of variables useful to specify the model described by formula (10). The analyst will be able to select the useful indicators to calibrate the model based on the specific social, economic, and environmental characteristics of the study area and based on the availability of data. These variables are deduced from the bibliographic references of the sector.

Step 2: Verifying the Goodness of the Model. Once the model has been calibrated and implemented, it is necessary to verify that the assumptions on which the multiple regression is based are verified. First, we need to verify the acceptability of the results through the indices of determination R^2, R^2_{corr}, Akaike's Information Criterion (AIC). Then it is verified: (a) the normality of the conditional distributions and the linearity of the relationships between the variables by means of the Q-Q plot method; (b) the hypothesis of homoscedasticity by means of residual analysis; (c) the existence of a significant relationship between the dependent variable P and the set of explanatory variables, by means of the global significance F-test; (d) the significance of each of the explanatory variables of the model, by implementing the *t-test* on the individual regression coefficients [12, 17].

Step 3: Autocorrelation Analysis and Implementation of Spatial Econometric Models. MLR assumes that the regression coefficients are spatially homogeneous. However, the use of spatial data - such as those related to accessibility analysis - does not always satisfy the basic assumptions of OLS regression. It is necessary to analyse the spatial autocorrelation by means of indices such as: the Moran I index, which makes it possible to check for the presence of residual autocorrelation; the Lagrange multiplier (LM) test, which is useful for detecting specification errors not considered by MLR models [25]. If the spatial autocorrelation present in the residuals of the MLR models turns out to be significant, it is necessary to support the analysis with spatial econometric models, introduced in Sect. 2.3, to provide more accurate estimates.

Table 1. Factors influencing the price of real estate properties.

Independent variable	Description
Internal characteristics	
Flat	Dummy (1 = yes; 0 = no)
Terraced property	Dummy (1 = yes; 0 = no)
Detached property	Dummy (1 = yes; 0 = no)
Apartments	Proportion of apartments (%)
Bedroom	Total number in the house

(*continued*)

Table 1. (*continued*)

Independent variable	Description
Bathroom	Total number of bathrooms
Construction years	Dummy (0,1) for time intervals or for single years
Age	Average age of the buildings (dummy)
Surface Area	In m^2 or as a dummy variable considering various surf. ranges
Garage, parking, elevator, terrace, garden, safety door, auto heat	Presence or absence (1 = yes; 0 = no)
Improvement	Major improvement needed (1 = yes; 0 = no)
Floor	Floor level Dummy (1st floor, 1;0; 2nd floor, 1;0...)
Orientation	sunny, corner and front (1 = yes; 0 = no)
External characteristics	
House	Dwellings/Km^2 - Built square meters/square meters
Density	Measure of the zone's population density
Restaurant, ATM, medical services, neighbourhood shopping	Number of services within a certain radius (500 – 1.000 m) or dummy (presence or absence within a certain radius)
Pupil	Pupil-to-teacher ratio
Student Income	€-$
Elementary school	Dummy (presence or absence within a certain radius)
Quality school	Average score of the nearest secondary school
Educational Index	N. of undergraduate persons over 20 years old (500 m radius)
Ethnic minority	% of ethnic minority
Higher professional occupations	% of higher professional occupations
Unemployment	% of unemployment
Green Area (GA)	GA % in the neighbourhood/Distance from the nearest GA
Jobs	(a) Number of employments present in the area where the property is located; (b) Jobs/Km^2
Agriculture surface in zone	Ratio of agriculture surface in zone
Socio-economic attributes	Dummy (property located in mediocre, good or excellent area)
Foreign population	Percentage of population born outside the state

(*continued*)

Table 1. (*continued*)

Independent variable	Description
Individual or household income	Median individual income (or household) per week $
Automobiles	Number of automobiles per dwelling
Ecosystem Disservices (EDs)	Distance from the nearest ED (Km)
Environmental Polluters (EP)	Value of measured Eps (e.g. CO, NOX, PM10 in $\mu g/m^3$)
Conservation State (CS)	N. of buildings with a "very good", "good", "poor" CS
Local accessibility	
Distance to coast, from centre, airport, rail station, metro station, main road, bus lines	Km or through a set of dummy variables (for instance, property within a 10 min-walk to station/bus line etc.)
CDB distance	Time in minutes which it takes at morning rush hour to reach the city's CBD from the property using the road network
Bus lines/railways/metro interaction	Number of internal lines bus (metro o train) serving the zone /Cumulated frequency of all bus lines (metro o trains)
Being in the centre/near the beach/in a commercial zone/in a zone of prestige	Dummy (1 = yes; 0 = no)
Access to services or transport	Walking time in minutes to these activities or services
Road	Road density (km/km^2)
Footpath	Footpath density (km/km^2)
Streetlight	Streetlight density (100 streetlight/km^2)
Systemwide accessibility	
Public Travel Time (PTT)	PTT (minutes) to reach secondary school
ACC_attive	Hansen type measure of employment active accessibility
ACC_passive	Hansen type measure of employment passive accessibility
Job accessability	Car travel time or public transport travel time (min) to reach employment (Hansen-gravity model)

4 Conclusions and Research Perspectives

This paper analyses the effect of transport accessibility on real estate prices. We first focus on the methods to measure accessibility and the most suitable HPMs to evaluate it. Then, we define the logical-operational steps to follow to assess the impact of accessibility on housing prices. It also provides a dataset of factors/indicators useful to define and calibrate the HPM.

From the analyses, the following main results emerge. First, the choice of independent variables is a crucial step in the characterization of HPM. Neglecting some variables in the analysis, would mean invalidating the result of the elaborations. Thus, to define a reliable model, four types of attributes must first be included in the HPM: (i) intrinsic characteristics of the property; (ii) extrinsic characteristics; (iii) local accessibility; (iv) system accessibility. Local accessibility indicators refer primarily to the distance of the property from public transportation. Systemwide accessibility indicators are generally gravity-type or Hansen-gravity and measure accessibility to a particular service or job. Evidently, testing the goodness of fit of the results returned by HPM is necessary to identify variables to be excluded from the analyses and to characterize progressively more refined models. Second, since spatial data are employed, it turns out that the assumptions on which multiple regression is based are not always satisfied. It follows that to handle spatial autocorrelation, MLR must be supported by spatial econometric models (SEM, SAR, GWR). This is precisely to obtain more accurate results in case the autocorrelation levels in space are significant.

Research prospects concern the application of the defined model to real case studies.

References

1. Grace, R., Saberi, M.: The value of accessibility in residential property. In: Australasian Transport Research Forum 2018 Proceedings, Darwin, Australia, 30 October–1 November (2018). http://www.atrf.info. Accessed 22 Nov 2021
2. Hansen, W.G.: How accessibility shapes land use. J. Am. Inst. Plan. **25**(2), 73–76 (1959)
3. Alonso, W.: Location and Land Use: Toward a General Theory of Land Rent. Harvard University Press, Cambridge (1964)
4. Muth, R.F.: Cities and Housing: The Spatial Pattern of Urban Residential Land Use. University of Chicago Press, Chicago (1969)
5. Mills, E.S.: Studies in the Structure of the Urban Economy. Johns Hopkins University Press, Baltimore (1972)
6. Giuliano, G., Gordon, P., Pan, Q., Park, J.: Accessibility and residential land values: some tests with new measures. Urban Stud. **47**, 3103–3130 (2010)
7. Du, H., Mulley, C.: Relationship between transport accessibility and land value: local model approach with geographically weighted regression. Transp. Res. Rec. **1977**(1), 197–205 (2006). https://doi.org/10.1177/0361198106197700123
8. Yuan, F., Wei, Y.D., Wu, J.: Amenity effects of urban facilities on housing prices in China: accessibility, scarcity, and urban spaces. Cities **96**, 102433 (2020)
9. Wen, H., Zhang, Y., Zhang, L.: Assessing amenity effects of urban landscapes on housing price in Hangzhou, China. Urban For. Urban Green. **14**, 1017–1026 (2015)
10. Wittowsky, D., Hoekveld, J., Welsch, J., Steier, M.: Residential housing prices: impact of housing characteristics, accessibility and neighbouring apartments – a case study of Dortmund, Germany. Urban. Plan. Transp. Res. **8**(1), 44–70 (2020)

11. Chiarazzo, V., dell'Olio, L., Ibeas, A., Ottomanelli, M.: Modeling the effects of environmental impacts and accessibility on real estate prices in industrial cities. Procedia Soc. Behav. Sci. **111**, 460–469 (2014). https://doi.org/10.1016/j.sbspro.2014.01.079

12. Nesticò, A., La Marca, M.: Urban real estate values and ecosystem disservices: an estimate model based on regression analysis. Sustainability **12**(16), 6304 (2020). https://doi.org/10.3390/su12166304

13. Nesticò, A., Maselli, G.: Declining discount rate estimate in the long-term economic evaluation of environmental projects. J. Environ. Account. Manag. **8**(1), 93–110 (2020). https://doi.org/10.5890/JEAM.2020.03.007

14. Dolores, L., Macchiaroli, M., De Mare, G.: A dynamic model for the financial sustainability of the restoration sponsorship. Sustainability **12**(4), 1694 (2020)

15. Troisi, R, Castaldo, P.: Technical and organizational challenges in the risk management of road infrastructures. J. Risk Res., 1–16 (2022)

16. Handy, S.L., Niemeier, D.A.: Measuring accessibility: an exploration of issues and alternatives. Environ. Plan. A **29**, 1175–1194 (1997)

17. Geurs, K., van Wee, B.: Accessibility evaluation of land-use and transport strategies: review and research directions. J. Transp. Geogr. **12**, 127–140 (2004)

18. Cordera, R., Coppola, P., dell'Olio, L., Ibeas, Á.: The impact of accessibility by public transport on real estate values: a comparison between the cities of Rome and Santander. Transp. Res. Part A Policy Pract. **125**, 308–319 (2019)

19. Bowes, D.R., Ihlanfeldt, K.R.: Identifying the impacts of rail transit stations on residential property values. J. Urban Econ. **50**, 1–25 (2001)

20. Koenig, J.G.: Indicators of urban accessibility: theory and application. Transportation **9**, 145–172 (1980)

21. Martínez, L.M., Viegas, J.M.: Effects of transportation accessibility on residential property values: hedonic price model in the Lisbon, Portugal, Metropolitan Area. Transp. Res. Rec. **2115**(1), 127–137 (2009)

22. Ibeas, Á., Cordera, R., dell'Olio, L., Coppola, P., Dominguez, A.: Modelling transport and real-estate values interactions in urban systems. J. Transp. Geogr. **24**, 370–382 (2012). https://doi.org/10.1016/j.jtrangeo.2012.04.012

23. Cascetta, E.: Transportation Systems Analysis: Models and Applications, 2nd edn. Springer, New York (2009). https://doi.org/10.1007/978-0-387-75857-2

24. Coppola, P., Nuzzolo, A.: Changing accessibility, dwelling price and the spatial distribution of socio-economic activities. Res. Transp. Econ. **31**, 63–71 (2011)

25. Munoz-Raskin, R.: Walking accessibility to bus rapid transit: does it affect property values? The case of Bogotá, Colombia. Transp. Policy **17**(2), 72–84 (2010)

26. Pan, H., Zhang, M.: Rail transit impacts on land use: evidence from Shanghai, China. Transp. Res. Rec. **2048**, 16–25 (2008)

27. Yang, L., Chau, K.W., Szeto, W.Y., Cui, X., Wang, X.: Accessibility to transit, by transit, and property prices: spatially varying relationships. Transp. Res. Part D Transp. Environ. **85**, 102387 (2020). https://doi.org/10.1016/j.trd.2020.102387

28. Lancaster, K.J.: A new approach to consumer theory. J. Polit. Econ. **74**(2), 132–157 (1966)

29. Efthymiou, D., Antoniou, C.: Measuring the effects of transportation infrastructure on real estate prices and rents. Investigating the potential current impact of a planned metro line. EURO J. Transp. Logist. **3**, 179–204 (2013)

30. Anselin, L.: Exploring spatial data with GeoDaTM: A workbook. Center for Spatially Integrated Social Science (2005)

31. Golgher, A.B., Voss, P.R.: How to interpret the coefficients of spatial models: spillovers, direct and indirect effects. Spat. Demogr. **4**, 175–205 (2016)

The Smart City NEOM: A Hub for a Sustainable Raise of Economy and Innovation

Elena Merino Gómez[1] (iD), Renato Benintendi[2], and Gianluigi De Mare[3](✉) (iD)

[1] Universidad Nebrija, Calle Pirineos, 55, Madrid, Spain
emerino@nebrija.es
[2] Megaris Ltd., 19 Cirrus Drive, Reading, Berkshire RG2 9FL, UK
renato.benintendi@megaris.co.uk
[3] University of Salerno, Via Giovanni Paolo II, 132, Fisciano, SA, Italy
gdemare@unisa.it

Abstract. Saudi Arabic project NEOM has been announced in 2017 as a living laboratory and hub for innovation. Unlike same typology projects, the aim is essentially to produce a model of comprehensive sustainability for life quality and innovative work. The peculiarity of this massive initiative, consisting of 500 billion US dollars, is not just the implementation of new technologies within a new economic and architectural frame, but rather to set a up a new living model, underpinning an innovative modality to conceive the multifaceted aspects of the economy, the environment and the urban asset. The project is to be related to the particular nature of Saudi economy and society and their trend, which is a fundamental driver to identify the development factors of the initiative and to enable one to define a more general picture of the sustainability concept of the future urban development. This study focuses on identifying and analysing the features of the NEOM projects, with the aims to highlight those founding factors which might be developed further to define new sustainability models for urban economy and architecture.

Keywords: NEOM · Smart city · Urban models · Sustainability

1 Introduction

Megacity NEOM project was announced in Riyadh on 24th October 2017 by Crown Prince Mohammed bin Salman within the conference Future Investment Initiative [1]. The purpose of the project in the founders' idea is included in its name, with is an heterogenous acronym of ancient and modern, notably ancient Greek νέοand modern Arabic مستقبل, *mustaqbal*, which stands for *future*.

The announcement was accompanied by the presentation of a multifaceted picture of opportunities and purposes, strictly related to the plan to implement a strong change of the economic and social Saudi frame, involving life quality, urban architecture, resources, energy and waste management, air quality, within a so effective sustainability plan to

---

E. M. Gómez—Contributed equally to this work.

F. Calabrò et al. (Eds.): NMP 2022, LNNS 482, pp. 1366–1372, 2022.
https://doi.org/10.1007/978-3-031-06825-6_133

result in a tremendous modification of the nation. The NEOM brochure [2] defines the Arabic smart city as a living laboratory and hub for innovation, a sustainable ecosystem for living and working and a model for the new future. It is also impressive the humanistic, ethic and social meanings the founders have given to the initiative, especially for a so technocratic and traditional society and economic scenario like Saudi Arabia.

The concept of an Ideal Edenic City is quite old and widely treated in philosophy, art and urban architecture. Since 19th century B.C., Egyptians produced the concept of the symbolic towns, such as *El Lahun*, with the aim to reflect social and hieratic values. Babel town and Platonic ideal society structure have remained in the western and eastern culture as underpinning bricks of human consciousness development, meaning the aim for a societal model where all contributing factors were properly harmonised to allow people to reach the Edenic condition including wellbeing, ethics, development, environmental sustainability, science and philosophic knowledge.

Based on the principles of Neoplatonism [3, 4], the Renaissance saw the creation of the most significant example of western ideal cities, which reflected classical and renewed principles, such as the theoretical framework of Leon Battista Alberti, based, in turn on Vitruvius'. Figure 1 depicts the outline of the geometrical perspective of an Ideal City, probably conceived at the Italian court of Federico da Montefeltro.

Fig. 1. *The Ideal City*, 1480 c. Unknown author. Marche National Gallery, Italy.

The emphasis given by Arabic government and the conceptual frame underpinning the announcement of the project are consistent with the historical *esprit* of traditional architecture. This is particularly important due to the course of Arabic economy undertaken in the lates decades and provides a strong basis of thinking and planning for the development of urban architecture.

2 Background

The Arabic economy has undergone a fast diversified behaviour in the latest decades. The oil-based development has tremendously changed a traditionally herding and agricultural multi-layered society, producing an impressive growth compressed in a relatively short time (Table 1). This has regarded the whole gulf region, even if Saudi Arabia has been by far the major player. Table 1 shows oil and gas production data for Gulf countries and world producers [5–7]. It demonstrates how heavy and intense was the role played by Saudi Arabia. For the purpose of the present study, it is worthy considering that the injection of oil technology into the traditional Arabic economy created an apparent growing gap in the structure of the polarized Arabic society. At one end withstood the millenary tribal structure and at the other end grew a new generation of economists and technologists, based on the British and American knowledge transfer.

Table 1. Oil and gas reserves in the Arab Economies at End 2000 (UNDP, 2012).

	Oil			Natural gas		
	Proved reserves (bbl)	Share of world reserves	R/P ratio	Proved reserves (Tcm)	Share of world reserves	R/P ratio
The GCC states	495.0	29.9%	69.5	42.4	20.3%	121.0
Bahrain	0.1	<0.05%	7.0	0.3	0.2%	26.8
Kuwait	101.5	6.1%	97.0	1.8	0.9%	>100
Oman	5.5	0.3%	16.9	0.9	0.5%	35.8
Qatar	24.7	1.5%	39.3	25.0	12.0%	>100
Saudi Arabia	265.4	16.1%	65.2	8.2	3.9%	82.1
UAE	97.8	5.9%	80.7	6.1	2.9%	>100
Other major oil producers	202.4	12.2%	110.5	9.6	4.6%	114.2
Algeria	12.2	0.7%	19.3	4.5	2.2%	57.7
Iraq	143.1	8.7%	>100	3.6	1.7%	>100
Libya	47.1	2.9%	>100	1.5	0.7%	>100
Other oil producers	16.2	1.0%	26.8	12.6	1.4%	159.8
Egypt	4.3	0.3%	16.0	2.2	1.1%	35.7
Sudan and S. Sudan	6.7	0.4%	40.5	0.1	<0.05%	-
Syria	2.5	0.2%	20.6	0.3	0.1%	34.3
Yemen	2.7	0.2%	32.0	0.5	0.2%	50.7
Total Arab World	713.6	43.2%	74.4	55.0	26.3%	107.2
Total World	1652.6	100%	54.2	208.4	100%	63.6

An additional key driver for the understanding of the evolution of the Arabic economy until the conception of the NEOM project is the particular weather conditions and the territorial orography. The western-like futuristic development of the Emirates, which produced strong modifications, have been a first approach towards the direction to NEOM project. However, to consider the former as an embryonic attempt of the latter would be a big mistake.

Regardless of this, the further jump of the Gulf economy has been the switching to the petrochemistry-based economy, which was inducted by the planet oil scenario, where many players entered the match and, last but not least, the *shale oil* presented a completed new scenario, fragmenting a monopolistic platform in the world energy resources balance [8].

This advancement must not be considered just as a technological change. In reality, the Gulf countries and, specifically, Saudi Arabia realized that a much more intense scientific growth impacting over all of the aspects of social life was in place. The latest petrochemical projects have been partnered with world centre of excellence groups, and the production scenario moved from the basic refineries cuts to complex engineered chemicals employed to improve life quality, energy related equipment, leisure items and new materials. The authors of these publications have been involved in complex Arabic projects as costly as 30 billion US dollars and have witnessed the significant change occurred in the Arabic economy and society.

This essential background frames the NEOM challenging project. Arabic government realized that a strong change was due, and NEOM has to be considered a technological, educational, architectural, ethic, environmental, in one-word sustainable laboratory hub.

3 Analysis

NEOM site is shown in Fig. 2. The contextualization is very important to capture the nature of the initiative. The localisation on the east part of the kingdom, West Coast of the Red Sea is a clear indication of the new vocational approach of the Saudi government. The proposed area is meant to cover more than 25.000 km^2, a size comparable to that of some small countries [9].

The NEOM futuristic kaleidoscopic vision (Fig. 3) is made of the shown main-coloured glasses and structural parts.

This image is very effective in demonstrating the inspiring though of the project devisers.

- FUTURE OF ENERGY
    NEOM's climate allows unrivalled complementary solar and wind potential through competitively priced renewable energy. New industries will drive the next wave of the energy transition by producing green hydrogen and carbon free resources.
- FUTURE OF WATER RESOURCES
    Located on the Red Sea with more than 450 km of coastline, NEOM has a large supply of seawater, with a huge potential of desalinated water.
- FUTURE OF MOBILITY
    Urban and regional connections will be via sea, air and land. The port will be the most advanced, sustainable, efficient and logistic zone of the world. NEOM will have direct connections with the rest of the world.

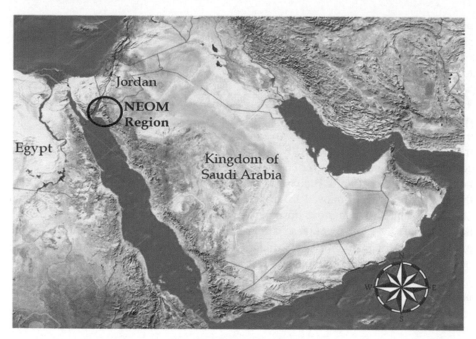

Fig. 2. NEOM localization.

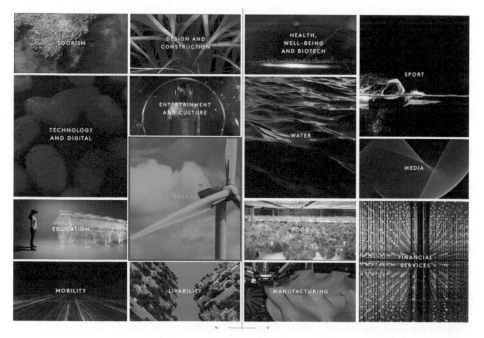

Fig. 3. NEOM kaleidoscope tiles (Government of Saudi Arabia).

It will operate on zero carbon emissions by offset. The architecturally iconic airport will feature biometric processing to eliminate bottlenecks through a seamless curb-to-gate experience. NEOM will stand as a world reference for safety, security and environmental compliance.

- FUTURE OF DIGITAL AND MEDIA

  Humans and machines will live in harmony, according to the principles of the digital design. Key words are artificial intelligent software, transfer between physical and virtual world, full digital information share. Industry education is expected to be given a never seen impulse. Universities, cultural media, books and data base will be accessible for the international community.

- FUTURE OF FINANCIAL SERVICES

  Universal access to financial services will be provided by means of new technologies which will support the economic growth.

- FUTURE OF TURISM

  This aspect should be prioritised in this item list. The kingdom aim is to convert the region into a world top attraction, which is relatively new for Saudi Arabia. The westernisation of the site choice, as well as the sea location work in this direction.

- FUTURE OF WELL-BEING, SPORT AND BIOTECH

  No comparable urban initiative will be found providing the integration of this indicators, according to the announced program [10]. Healthy lifestyle, sport structures and world-class prevention disease programs will be implemented.

- FUTURE CONSTRUCTION AND MANUFACTURING

  This part of the program has been conceived by foreseeing the adoption of new material, smart construction and manufacturing technologies, software operations and data management.

  This can be an embryonic source of the industrialized world, both Asian and Euromerican, not neglecting Africa.

This necessarily short itemization of the main NEOM tiles is effective in identifying the sustainability aspects of the program [11, 12].

It is evident the big change undertaken by the Saudi government which has in their mind a huge program of modernization, that is not only technological and environmental, but that will necessarily impact on the Arabic society as a whole.

4 Conclusions

The founding factors that define new sustainability models for urban economy and architecture are closely linked to correct economic and urban patterns strongly rooted in historical models and recent experiences, the lessons of which must be taken into account [13–16].

The NEOM project still has details to be defined, even on a large and medium scale. It must find in case studies, precedents and contemporaries, examples of success and failure [17] that must be thoroughly analysed to ensure success in a project whose ambition might represent its own doom.

References

1. Farag, A.A.: The story of NEOM city: opportunities and challenges. In: Attia, S., Shafik, Z., Ibrahim, A. (eds.) New Cities and Community Extensions in Egypt and the Middle East. Springer, Cham (2019). https://doi.org/10.1007/978-3-319-77875-4_3
2. NEOM Brochure. An Accelerator for Human Progress (2018). https://static1.squarespace.com/static/5bb3e839b9144925687dfc3f/t/60c32bc9e7f3934297b1c192/1623403481915/NEOM+Brochure-compressed.pdf
3. Campanella, T., La Citta'del Sole, Aeodnia Edizioni (2020)
4. More, T.: Libellus vere aureus, nec minus salutaris quam festivus de optimo rei publicae statu, deque nova insula, Utopia (1516)
5. Fattouh, B., El-Katiri, L.: Arab Human Development Report, Research Paper Series, United Nations Development Programme, Regional Bureau for Arab States (2012)
6. González-Méndez, M., Olaya, C., Fasolino, I., Grimaldi, M., Obregón, N.: Agent-based modeling for urban development planning based on human needs. Conceptual basis and model formulation. Land Use Policy **101** (2021). https://doi.org/10.1016/j.landusepol.2020.105110
7. Graziuso, G., Mancini, S., Francavilla, A.B., Grimaldi, M., Guarnaccia, C.: Geo-crowdsourced sound level data in support of the community facilities planning. A methodological proposal. Sustainability. **13**(10), 5486 (2021). https://doi.org/10.3390/su13105486
8. MEES, VOL. LIV, No 9, 28 February 2011
9. Aly, H.: Royal Dream: City Branding and Saudi Arabia's NEOM. Middle East - Topics & Arguments, Bd. 12, no. 1, pp. 99–109, June 2019. https://doi.org/10.17192/meta.2019.12.7937
10. De Mare, G., Granata, M.F., Forte, F.: Investing in sports facilities: the Italian situation toward an Olympic perspective. In: Gervasi, O., et al. (eds.) ICCSA 2015. LNCS, vol. 9157, pp. 77–87. Springer, Cham (2015). https://doi.org/10.1007/978-3-319-21470-2_6
11. Campanile, L., Cantiello, P., Iacono, M., Marulli, F., Mastroianni, M.: Risk analysis of a GDPR-compliant deletion technique for consortium blockchains based on pseudonymization. In: Gervasi, O., et al. (eds.) ICCSA 2021. LNCS, vol. 12956, pp. 3–14. Springer, Cham (2021). https://doi.org/10.1007/978-3-030-87010-2_1
12. Salameh, T., et al.: Optimal selection and management of hybrid renewable energy System: Neom city as a case study. Energy Convers. Manag. **244**, 114434 (2021). https://doi.org/10.1016/j.enconman.2021.114434
13. Dolores, L., Macchiaroli, M., De Mare, G.: Sponsorship's financial sustainability for cultural conservation and enhancement strategies: an innovative model for sponsees and sponsors. Sustainability **13**(6), 9070 (2021). https://doi.org/10.3390/su13169070
14. Dolores, L., Macchiaroli, M., De Mare, G.: A dynamic model for the financial sustainability of the restoration sponsorship. Sustainability **12**(4), 1694 (2020). https://doi.org/10.3390/su12041694
15. Dolores, L., Macchiaroli, M., De Mare, G.: A model for defining sponsorship fees in public-private bargaining for the rehabilitation of historical-architectural heritage. In: Calabrò, F., Della Spina, L., Bevilacqua, C. (eds.) ISHT 2018. SIST, vol. 101, pp. 484–492. Springer, Cham (2019). https://doi.org/10.1007/978-3-319-92102-0_51
16. De Mare, G., Di Piazza, F.: The role of public-private partnerships in school building projects. In: Gervasi, O., et al. (eds.) ICCSA 2015. LNCS, vol. 9156, pp. 624–634. Springer, Cham (2015). https://doi.org/10.1007/978-3-319-21407-8_44
17. Benintendi, R., De Mare, G.: Upgrade the ALARP model as a holistic approach to project risk and decision management. Hydrocarb. Process. **9**, 75–82 (2017)

Where is the City? Where is the Countryside? The Methods Developed by Italian Scholars to Delimit Urban, Rural, and Intermediate Territories

Valentina Cattivelli[1,2]([✉])

[1] Comune di Cremona, Cremona, Italy
valentina.cattivelli13@gmail.com
[2] Uninettuno University, Rome, Italy
valentina.cattivelli@uninettunouniversity.net

Abstract. The present text illustrates the methods developed by Italian scholars to delimit urban, rural, and intermediate territories in Italy over the last 15 years. While for official statistical investigations and governmental decisions the harmonized TERCET methods remain the favorite, within territorial studies more sophisticated and articulated methods are preferred. Based on a desk research, 4 different methods are considered and characterized in relation to the adopted spatial unit of reference and variables, their purpose, beyond the used statistical method and the territorial typologies. Findings reveal that the municipalities and provinces are preferred as spatial units, the economic indicators are chosen to delineate the basic characterization of the method. The definition of several territorial typologies is a sign of attention for the intermediate territories and their differentiation.

Keywords: Rural areas delimitation · Urban areas delimitation · Intermediate areas · Italy · Italian scholars · TERCET · EUROSTAT

1 Background

The pressure to adopt the method that best describes the territories accurately has increased in recent years (Cattivelli 2021b). It originates from the need to distribute efficiently European and national funds and to identify accurately the programme areas for specific planning decisions. It also stems from the need to map new emerging territories (the intermediate ones) as well as territories performances in comparative terms.

Within the vast scholarly terrain there are two opposing strands of thought to manage this pressure: one questions the effectiveness of articulated methods for international comparisons; the other extolls the need to represent thoroughly the local diversity as prerequisite for more adequate and territorial-based policies (ibid.).

The first strand is supported by EUROSTAT-European Commission. Within the TERCET initiative, they propose harmonized urban-rural typologies for international official statistics and standards at European level. These typologies are essentially based on

© The Author(s), under exclusive license to Springer Nature Switzerland AG 2022
F. Calabrò et al. (Eds.): NMP 2022, LNNS 482, pp. 1373–1383, 2022.
https://doi.org/10.1007/978-3-031-06825-6_134

demographic indicators calculate at grid level and the setting of only three thresholds for delimiting each territory: urban, rural, and intermediate (EUROSTAT 2020, 2021). Beyond European Commission, this strand is also supported by FAO, ILO, OECD, UN-Habitat and World Bank which have developed a new harmonised definition that can be applied to every country in the world, called the Degree of Urbanisation (2020). Instead of relying on only two classes, the new method uses three classes to capture the urban-rural continuum: 1) Cities, 2) Towns and semi-dense areas and 3) Rural areas.

However, both methods do not describe accurately territorial diversity at the local level.

In line with the second strand, national statistical offices and regional governments elaborate their own territorial definitions to measure progress towards local policies. These methods integrate an appreciable number of dimensions of analysis and indicators and delineate several territories as intermediate in addition to urban and rural ones (e.g., Bencardino and Nesticó 2017; Champion and Hugo 2017; Ortiz-Baez et al. 2021). However, they represent differently urban, rural, and intermediate delimitations, thereby preventing possible comparisons among homogeneous territorial categories (EUROSTAT 2021b). For Italy, a review of these methods is included in Cattivelli (2021a).

To settle this dispute, several scholars propose some original methods. These adopt a multi-scalar approach and focus on the patterns of territory diversity, thus challenging the framework based on the urban/rural dichotomy across all European regions. Being often a byproduct, they depend strictly on the research aims and are rarely replicated in other projects. Based on Cattivelli (2021b), 80 of these methods have been produced across Europe in the last years.

2   The Method of Investigation

The starting point of the present investigation is the identification of the methods developed by Italian scholars. While Cattivelli (2021a) describes those elaborated by Italian government and statistical institutions, this paper completes the investigation considering also those formulated by scholars. This identification takes place considering the 80 methods identified by the same scholar (ibid.) and those among them that are of interest. This has identified 4 methods to analyze (Table 1), i.e.:

Table 1. The considered method in the present investigation, based on Cattivelli (2021a, b, c).

Method	Bibliographical reference
Anania & Tenuta	Anania and Tenuta (2006). Ruralità, urbanità e ricchezza nelle Italiee contemporanee. *Agriregionieuropa*, 2(7)
Barbieri & Cruciani	Barbieri and Cruciani (2007). Caratteristiche dei sistemi locali urbani. In Esposito, G. Contabilità nazionale, finanza pubblica e attività di controllo. Scritti per il Cinquantenario ISCONA (p. 259–280). Roma: ISCONA
Boscacci	Boscacci (2010). Urban-rural relations. A methodology to classify rural areas. RUFUS/TRUST workshop "Diversities of rural areas in Europe and beyond". Hannover

Each method is then analyzed by considering some key characteristics, such as the statistical method, the purpose, the spatial unit, the variables, and the territorial typologies. The statistical method used can be a basic statistical procedure like the use of simple indicators, or more articulated like the application of statistical or econometric methods. The purpose is the aim inspiring the formulation of the method, for policy support or scientific detail. The spatial unit is the unit to which the variables refer. It can be administrative if it coincides with an administratively defined area (i.e., NUTS 3, LAU 2) or statistical one (grid) or political one (e.g., macro-region or business district). Variables can be economic, social, demographic or include land-use and distance indicators. They can be used individually or together. The territorial typologies are the number of categories into which the territory is divided. If two, scholars support the traditional urban/rural dichotomy. If more than two, they promote a further detail of the territory, including the identification of intermediate areas and their differentiation.

3   The Description of the Considered Methods

The first considered method is that formulated by Anania and Tenuta (2006). These scholars assume that the characterization of a municipality is determined not only by demographic factors, but rather also depends on urbanization degree and economic specialization. As such, in formulating their rurality/urbanity indicator (RUI) within a specific research project, they use some data concerning demographic density, population dispersion, population employed in agriculture and public services, degree of urbanization and availability of living spaces. These indicators are elaborated through applying a principal component analysis. Based on the values assumed by RUI, the authors define each municipality as extremely rural, rural, weakly rural, weakly urban, urban, and extremely urban. RUI application gives rise to the representation of a more heterogeneous Italy in the north, while in the south it identifies larger urban areas than other classifications (especially in Puglia, Basilicata, and Sicily). Figure 1 provides more details of what emerges from the application of this method[1].

---

[1] We have decided to replicate the Anania and Tenuta method exactly. However, this method dates back to 2006, and in the intervening years many municipalities have changed name, ISTAT code or have merged or been suppressed. Most municipalities in white have precisely recorded these variations.

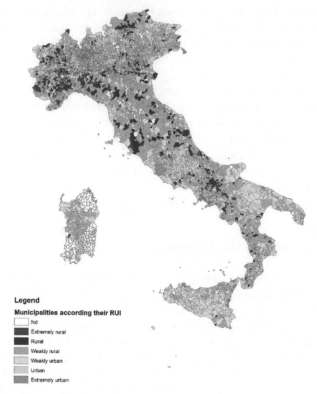

Fig. 1. A map of Italy obtained by applying Anania & Tenuta method. Source: Anania and Tenuta 2006.

Barbieri and Cruciani introduce their own method the following year, which is based on a different territorial unit: the Sistemi Locali del Lavoro (SLL) – Local Labor System (LLS). The Local Labor System (LLS) represents a territorial grid whose boundaries are defined using daily home/work travel flows (commuting). According to the 2011 census survey, there are 686 Local Labor Systems in Italy. The two scholars apply multivariate analysis techniques to these systems, to identify those with typically urban functions. In this way, Barbieri and Cruciani identify 4 different types of urban areas (highly specialized urban areas, low-skilled urban areas, unspecialized urban areas, urban areas and shipyards), but ignore the delimitation of any rural area. Their mapping exercise suggests that the vast majority of LLSs are not urban. Urban LLSs are scattered throughout the territory and differ in terms of their economic specialization (Fig. 2).

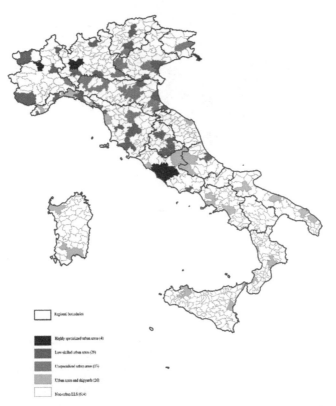

Fig. 2. A map of Italy according to the Barbieri & Cruciani method. Source: Barbieri and Cruciani 2007.

The two scholars later apply the same method to the municipalities but introduce additional variables concerning the degree of urbanization. Specifically, they measure the degree of urbanization by considering the incidence of urban agglomerations. In greater detail, they consider the proportion of the morphological urban agglomeration surface in the total municipal area, as well as the share of urban agglomeration population over municipal population. After performing the method using these two additional variables, the authors identify the urbanized municipalities only in terms of population density, the urbanized municipalities in terms of surface, the municipalities that are not urbanized and the highly urbanized municipalities. As demonstrated by Fig. 3, urban areas appear to be more extensive. While in the north, municipalities are mostly urbanized by surface area intensity, in the south they are mostly urbanized by population intensity.

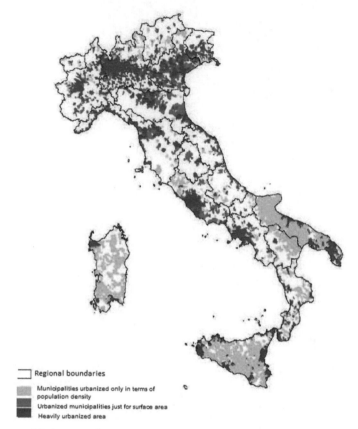

Fig. 3. A map of Italy according to the Barbieri & Cruciani method (2). Source: Barbieri and Cruciani, 2007.

Boscacci (2010) adopts a different territorial unit of reference, replacing munici-palities with provinces. Unlike other methods, he excludes any demographic variables when mapping these territories, as he assumes that the territories differ only in terms of economic vocation. As such, he only considers strictly economic indicators such as the productivity of the agricultural sector, the relevance of the agricultural sector to the provincial economy proxied by the comparison of used agricultural areas and the total provincial area, economic diversification measured by the number of employees in small firms, and the total agricultural labor force as well as urban sprawl. By combin-ing these indicators and applying them to the provinces, Boscacci identifies 5 territorial areas: strong province, province under pressure, province under pressure/weak and weak province. The areas defined as urban through application of almost all other methods considered (for example the Po Valley around Milan and Via Emilia) are also defined as urban here. The large size of the provinces as statistical unit precludes the possibility to analyze the great territorial diversity within them. This is illustrated in Fig. 4.

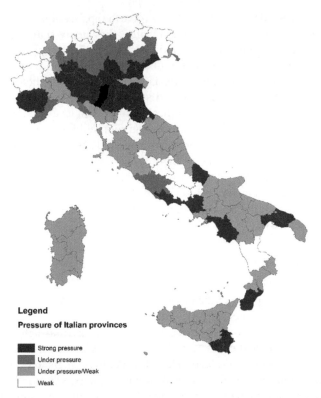

Fig. 4. A map of Italy resulting applying the Boscacci's method. Source: Boscacci 2010.

4 Discussion

The present paper describes the methods formulated by Italian scholars in the period 2005–2020. Research was fruitful in the earlier years of that period, while work in later years was limited to adapting or revising the methods tested in previous years. The obligation to adopt the TERCET methods to produce official statistics has certainly discouraged the development of new methods. Although the number of the methods considered appears limited, this analysis represents an occasion to better understand the methodological efforts to identify the territorial peculiarities. These methods in fact provide a framework for national and regional policy-makers to target economic and social policies at a territorial level and question the effectiveness of articulated methods for international comparisons.

The methods adopted by Italian scholars involve the application of different statistical units of reference. In addition to the municipality, they apply statistical techniques to the LLS Local Labor System (Barbieri and Cruciani) and to the provinces (Boscacci). This approach is driven by the conviction that territorial dynamics are based on social and economic relations, which extend over larger territories than the municipality. It is also facilitated by the availability of data that are only collected at a municipal level (and that can therefore be regrouped at a local labor system level) and/or at a provincial level. This is specifically true for economic, and morphological data, which, together with demographic data, are the most commonly used. These methods identify different territorial categories; however, none of them specifically frames peri-urban territories. They do not detail rural areas meticulously and this precludes their replication in rural planning. Finally, these methods work according to both sophisticated statistical analysis (multivariate analysis, Barbieri and Cruciani) than simpler with single or more indicators (the remaining methods).

5   Conclusion

This article offers an overview of the urban-rural methods formulated and implemented by Italian scholars over the period 2005–2020.

Some years ago, these actors began questioning the urban/rural dichotomy, proposing a multi-scalar approach, and challenging the framework based on traditional administrative boundaries. However, results of TERCET methods are the most widely used for official statistics, and this has discouraged Italian scholars, who in recent years have stopped developing new methods. To relaunch the debate in light of the incessant peri-urbanization of the Italian territory (Cattivelli 2021c), it is worth bearing in mind that each method offers a different territorial representation that depends, in turn, on the chosen statistical unit and method, as well as on the defined territorial categories and variables. The choice of municipality as statistical unit highlights a preference for simple, but accurate, representations or the availability of data. Opting for others such as LLSs reflects the aspiration to map territories, but also the economic relations that insist in them. This option can be also reinforced by the decision to adopt economic variables. This is a solution adopted several times, also in addition to demographic ones. This underlines that territorial delimitation depends on the economic structure or flows and demographic trend, rather than on distance or land-use measures. However, the choice of the variables results influenced by their availability. Sometimes, the unavailability of data hinders the development of more refined classification methods capable of better describing economic-social-territorial dynamics. The definition of territorial categories is also crucial. The adoption of a large number of categories is a sign of the capacity of the method to capture territorial diversity, while a small number indicates a desire to take a dichotomous approach. The choice of simple or complex statistical methods depends on the availability of data for the variables and the degree of understanding of those who apply it.

Differences among methods depend on the combination of choices related to these elements and it is not possible to establish a priori which method is best or most representative of the territorial characteristics because it depends on the general aims of the investigation (more complex methods in case of developing territorial targeted policies, simpler when producing statistics or performing comparisons among territories).

Future studies should investigate the differences between the different representations and calculate how much of the Italian territory is mapped differently as well as considering the implications of such differences (Table 2).

Table 2. A summary of the considered methods. Source: own elaboration based on literature review 2021

Method	Spatial unit	Variables	Statistical method	Territorial typologies
Anania and Tenuta	Municipalities	Demographic density, population dispersion, population employed in agriculture and public services, urbanization, availability of living spaces	Simple indicator, calculated after applying principal component analysis	*Rural *Weakly rural *Weakly urban *Urban *Extremely urban
Barbieri & Cruciani	LLS (Local labor system)	Economic specialization	Multivariate analysis techniques	*Highly specialized urban areas *Low-skilled urban areas *Unspecialized urban areas *Urban areas and shipyards
Barbieri & Cruciani	Municipalities	Economic specialization and degree of urbanization	Multivariate analysis techniques	*Urbanized municipalities only in terms of population density *Urbanized municipalities in terms of surface *Municipalities that are not urbanized *Highly urbanized municipalities

(continued)

Table 2. (*continued*)

Method	Spatial unit	Variables	Statistical method	Territorial typologies
Boscacci	Provinces	Productivity of agriculture, weight of agricultural area, economic diversification, urban sprawl	Simple indicators	*High productivity and High weight agriculture: *Strong province* *High productivity, Low weight agriculture and low urban sprawl: *Strong province* *High productivity, Low weight agriculture, high urban sprawl: *Province under pressure* *Low productivity, High weight agriculture low diversification: *Province under pressure/weak* *Low productivity, High weight agriculture, high diversification and high urban sprawl: *Province under pressure/weak* *Low productivity, High weight agriculture, high diversification and low urban sprawl: *Weak province* *Low productivity, Low weight agriculture: *Weak province*

References

Anania, G., Tenuta, A.: Ruralità, urbanità e ricchezza nelle Italiee contemporanee. Agriregionieuropa **2**(7) (2006)

Barbieri, G., Cruciani, S.: Caratteristiche dei sistemi locali urbani. In: Esposito, G. (ed.) Contabilità nazionale, finanza pubblica e attività di controllo. Scritti per il Cinquantenario ISCONA, pp. 259–280. ROMA, Iscona (2007)

Bencardino, M., Nesticó, A.: Demographic changes and real estate value. A quantitative model for analyzing the Urban-rural linkages. Sustainability **9**(4), 563 (2017)

Boscacci, F.: Urban-rural relations. A methodology to classify rural areas. In: RUFUS/TRUST Workshop "Diversities of Rural Areas in Europe and Beyond", Hannover (2010)

Cattivelli, V.: Institutional methods for the identification of urban and rural areas—a review for Italy. In: Bisello, A., Vettorato, D., Ludlow, D., Baranzelli, C. (eds.) SSPCR 2019. Green Energy and Technology, pp. 187–207. Springer, Cham (2021a). https://doi.org/10.1007/978-3-030-57764-3_13

Cattivelli, V.: Methods for the identification of urban, rural and peri-urban areas in Europe: an overview. J. Urban Regeneration Renewal **14**(3), 240–246 (2021b)

Cattivelli, V.: Planning peri-urban areas at regional level: the experience of Lombardy and Emilia-Romagna (Italy). Land Use Policy **103**, 105282 (2021c)

Champion, T., Hugo, G.: Introduction: moving beyond the urban-rural dichotomy. In: New Forms of Urbanization, pp. 3–24. Routledge, New York (2017)

Dijkstra, L., Papadimitriou, E.: Using a New Global Urban-Rural Definition, Called the Degree of Urbanisation, to Assess Happiness, UE and JRC report (2020)

EUROSTAT. Degree of urbanization. EUROSTAT, Bruxelles (2020)

EUROSTAT. Applying the Degree of Urbanisation, 2021 edition. EUROSTAT, Bruxelles (2021)

EUROSTAT. Demographic data - Rural and urban areas: differences. EUROSTAT, Bruxelles (2021b)

Ortiz-Báez, P., Cabrera-Barona, P., Bogaert, J.: Characterizing landscape patterns in urban-rural interface. J. Urban Manag. **10**(1), 46–56 (2021)

A Multicriteria Evaluation of Blockchain-Based Agrifood Chain in the New Scenario Post-Covid 19

Alessandro Scuderi[1]([✉]), Roberta Selvaggi[1], Luisa Sturiale[2], Giovanni La Via[1], and Giuseppe Timpanaro[1]

[1] D3A, University of Catania, 95127 Catania, Italy
{alessandro.scuderi,giovanni.lavia,giuseppe.timpanaro}@unict.it
[2] DICAR, University of Catania, 95127 Catania, Italy
luisa.sturiale@unict.it

Abstract. The agrifood supply chain is one of the largest in the world and includes all the businesses aimed at the production, processing, distribution and sale of products. The purpose of this article is to clearly explain what the blockchain technology really is, how it performs and whether it can really help our agriculture. It seems particularly useful to bring a knowledge support to institutional operators, agricultural enterprises, rural communities, activists, in the belief that it is essential to introduce in the current and future debate a critical vision that remains open to comparison and observation of the evolution of the 4.0 revolution that has just begun. In this context, blockchain technology can play an important role in the exchange of goods and services, which allows to democratize the procurement process, thus making the relationship between small farmers and large buyers more equitable. The objective of the research is to analyze which scenarios will benefit most from the implementation of blockchain in the citrus chain. The multi-criteria indices are representative of the variability in the citrus supply chain based on confirmed trends is safety, security, and traceability in the overall process for orange juice, from fruit to bottle.

Keywords: Resilence · Innovation · Value

1 Introduction

The agrifood industry is undergoing profound changes in relation to population growth, food demand, climate change, and increasing consumer demands for information. Through digital innovation and Agriculture 4.0, the agrifood sector can be the way to ensure greater competitiveness to the entire supply chain: from production in the field to distribution, decreasing the risk of fraud and adulteration. For the protection and enhancement of Italian agrifood and for the enhancement of Made in Italy can be useful to trace the products along the whole the production chain through a special technology: the blockchain.

F. Calabrò et al. (Eds.): NMP 2022, LNNS 482, pp. 1384–1399, 2022.
https://doi.org/10.1007/978-3-031-06825-6_135

New technologies will have to provide new agriculture production paradigms: the Internet of things, big data and artificial intelligence could be capable of renewing not only the products but also the production processes, the business organization and the company's approach to the market and consumers [1–5].

The reason for the application of blockchain-based technology in the food market relates to the challenges that this market is currently facing: food security, the creation of value through quality, the fight against fraud and counterfeiting, and, most importantly, the value of the information and experience embedded in the product [6, 7]. Consumers have become more aware of the origin and quality of the products they buy and food-safety regulations have also become stricter [8, 9]. The procedures initially established by Reg CE n.178/2002 define the fundamental importance of an integrated approach of the supply chain and regulations regarding food traceability, indicating the obligation to show the path of food destined for food production through all its phases of production, transformation, and distribution [10, 11]. These rules have been further updated with the EU Reg. 1169/2011, which regulates the information that must be made known to consumers, and with EU Reg. 625/2017, which concerns official controls in the supply chain [12].

The need to have information about the supply chain of quality food products can also be traced back to the country of origin [13, 14]. The real social, economic, and environmental values incorporated in the supply chain are increasingly complex to ascertain due to the intensification of the typical dynamics of an increasingly globalized economy [15–18].

The main motivations for this research into the European Blockchain Partnership (EBP) are to evaluate the information asymmetry, to try to decrease the distance between producers and consumers, and to establish the "value" [8] of the supply chain of citrus products through distributed ledger technologies. However, there are some phases of the supply-chain process that cannot officially codified, which represents a gap that needs to be filled to protect consumers [19].

In this context, the present research addresses the economic and social importance of citrus growing and, in particular, of the production and consumption of oranges.

The production and consumption of citrus fruits globally has been of considerable dynamism, deriving both from the growth of world production from globalization, with the relative "death of distance" [1] and from the evolution of consumption models. The present study analyzes the blockchain-based citrus supply chain to identify new approaches and opportunities in the agri-food sector that could be used to develop guidelines, both politically and as local planning tools, to enhance production, consumer protection, and the analysis of the value chain in the citrus supply chain.

2 Materials and Methods

2.1 2.1. Citrus Fruits Production and orange Supply Chain

The area dedicated globally to citrus fruits has increased in the last decade from 7.3 million hectares to 7.9 million hectares (and increase of c. 8%); this increase can be attributed to new geographical areas being used to cultivate citrus fruits, especially in developing countries [20]. Among the various citrus species, the cultivation of oranges is becoming increasingly important, covering an area of 3.8 million hectares in 2019, followed by small citrus fruits (tangerines, clementines, and mandarins) with 2.5 million hectares, and lemons with 1.1 million hectares, as well as other minor species.

The global production of citrus fruit in 2017 stood at 132.7 million tonnes, an increase of 19.6% from 2007 [21], showing a greater dynamism in production compared product and process innovations occurring in the same period. From the analysis of the geographical distribution relative to the world production of citrus fruits (Fig. 1), the Asian continent is the main global producer (42%), followed by the Americas (36%), and Africa (13%). Europe only produces 8% of the total production.

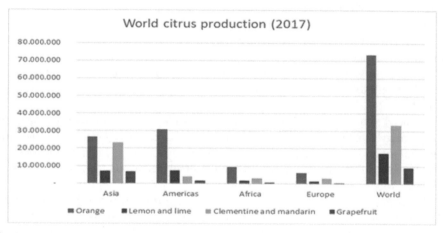

Fig. 1. Citrus production for main species in 2017 (tonnes). Source: Adapted from FAOSTAT (2019).

Given that oranges account for almost half of the world's citrus fruit production, the present study focuses on oranges. According to previous research [22, 23], the European Union has suffered a decrease in production to the point of becoming a net importer of oranges. Based on United States Department of Agriculture (USDA) data, in 2017, the European Union imported more than one million tons of oranges, 3% more than in the previous season, with a significant cost to the European market (USD 780 million); the largest exporters were South Africa and Egypt, followed by Morocco and Argentina [24].

The amount of oranges consumed worldwide in 2017 was 73.3 million tonnes, similar to the previous year [20]. In general, however, orange consumption shows a modestly increasing trend. The largest increase was in 2008, when consumption increased by 4.6% from the previous year. Over the period under review (2007–2017), global orange consumption achieved its peak volume of 73.6 million tonnes in 2016, decreasing slightly in the following year.

The global orange market revenue amounted to USD 44 billion in 2017, representing an increase of 3.6% from the previous year.

Regarding the consumption of oranges globally in recent years, the country with the highest consumption is Brazil, followed by China and India (Table 1) [24]. In total, orange consumption was 73.3 million tons in 2017, representing a decrease of 1.2% over the last decade. Average per capita consumption was 9.7 kg per person in 2017, representing a decrease of 0.2% from 2015.

Table 1. Per capita consumption of oranges in the main countries of the world, 2015–2017 (millions of tonnes, kg/year).

Country	Consumption, million tonnes			Population, million persons			Per Capita Consumption, kg per person			CAGR of Consumption, %	CAGR Per-Capita Consumption, %
	2015	2016	2017	2015	2016	2017	2015	2016	2017	2007-2017	2007-2017
Brazil	16.9	17.3	17.4	206.0	207.7	209.3	82.2	83.1	83.4	-0.7%	-1.6%
China	8.1	8.3	8.6	1428.4	1435.0	1441.1	5.7	5.8	6.0	9.6%	9.0%
India	7.7	7.6	7.7	1309.0	1324.2	1339.2	5.9	5.7	5.7	6.1%	4.8%
Mexico	4.5	4.6	4.6	125.9	127.5	129.2	35.7	35.9	35.4	0.8%	-0.7%
USA	5.4	5.0	4.3	319.9	322.2	324.5	16.9	15.5	13.2	-4.4%	-5.1%
Egypt	2.7	2.2	2.4	93.8	95.7	97.6	28.7	22.9	24.1	2.8%	0.7%
Indonesia	1.9	2.2	2.3	258.2	261.1	264.0	7.2	8.2	8.8	-1.4%	-2.6%
Spain	1.2	2.3	1.9	46.4	46.3	46.4	25.4	49.2	41.6	2.8%	2.6%
Italy	2.0	1.5	1.6	59.5	59.4	59.4	33.7	24.9	27.3	-4.3%	-4.3%
Turkey	1.5	1.5	1.6	78.3	79.5	80.7	19.5	18.7	19.9	2.2%	0.7%
Iran	1.5	1.8	1.6	79.4	80.3	81.2	19.4	22.9	19.8	-2.6%	-3.8%
Pakistan	1.7	1.6	1.6	189.4	193.2	197.0	8.9	8.5	8.0	-0.2%	-2.2%
Others	18.0	17.7	17.7	3220.1	3266.1	3312.2	9.9	9.8	9.7	1.2%	-0.2%
Total	73.1	73.6	73.3	7414.1	7498.2	7581.6	9.9	9.8	9.7	1.0%	-0.2%

These data confirm the importance of the citrus-fruit-production chain in terms of: value and quantity in the field of agri-food production; the considerable level of international trade; the growing interest in the traceability and characteristics of products by all actors in the chain; and, most importantly, consumers who are looking for healthy citrus fruits, deliberately seeking to establish the quality of the product at its origin.

2.2 Blockchains and Agri-Food System

The food market's significant interest in technology sector [25, 26], in particular blockchain-based projects [27], relates to the challenges this market faces [28] in the fight against fraud and counterfeiting [29–31]. Food fraud and the lack of transparency and traceability throughout the supply chain, as well as generating significant economic losses, cause potentially serious damage to the health of consumers [16].

Attention to food practices and the quality and safety of goods sold is also becoming a major problem [32] in relation to the growth of the world population, which is estimated to reach nine billion people in 2050. This will lead to increasing pressure on available resources [29], will be accompanied by the increasing presence of highly illegal practices throughout the supply chain. [33, 34]. The key focus of EU interventions has been to reduce information asymmetry, aiming to reduce, at least theoretically, the distance between producers and consumers [10].

Blockchain technology and their applications could provide a range of powerful tools that can be used both in the fight against fraud and counterfeiting and, above all, can allow greater transparency [35] related to business operations and the traceability of consumed products [36]. The emergence of large retailers and large-scale retailers has radically changed the dynamics of power within the value chain, further reducing the competitiveness of small producers [25]. Within this context, blockchain technology takes on strategic significance through its distributed ledger, which allows the creation of decentralized platforms for the exchange of goods and services that allow the democratization of the supply process, making the relationship between small farmers and large buyers fairer [16, 37–39].

However, an anti-innovation mentality and a poor technological culture can be significant obstacles to the expansion of technology [40, 41]. Therefore, it will be the task of early adopters to disseminate knowledge and effectively communicate innovation in order to ensure its correct and complete implementation, potentially revolutionizing the internal agri-food sector. [6, 42]. The blockchain builds consensus within the supply chain by exploiting a mechanism based on distributed and shared registers. This innovation entails a profound change for modern societies, which have been built and developed around centralized systems, while, through the blockchain, the maximum decentralization of the process is achieved.

Blockchain technology makes it possible to create an infra-structure that can solve many of the challenges that characterize today's supply chains (Fig. 2) [5].

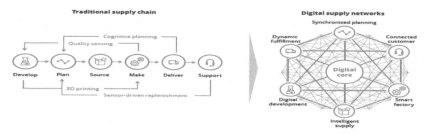

Fig. 2. From supply chain to Digital Supply Chain. (Source: Mussomeli et al. (2016).

The strategic aspect lies in the ability of the blockchain to decentralize authority and distribute power in a more democratic way, reducing, at least at a theoretical level, the concentration of information [38], of those who possess more resources [8].

There are two main categories of blockchain technology: private and public. The distinction is based on access: who can read and send transactions to a blockchain and participate in the validation process. In a public blockchain, anyone can access and take part in transactions while, in a private blockchain, only selected parties can access and make changes.

In the public transactions are transmitted to each individual participant and each node thus maintains a complete record of the entire transaction history. In addition to the public blockchain, many efforts have been made to create private ones, often used by industry consortia, which, due to privacy, regulatory problems, and system performance, restrict access to the blockchain only to those organizations that have been admitted to the network. Both public and private blockchains must have their transactions verified, and this is achieved by consensus. According to some authors [5, 38, 39, 43], the blockchain can be segmented by distinguishing between different permission models [44] (Table 2).

Table 2. Functional matrix of public and private blockchain technology.

	Property	Who can access and view transactions ?	Who can generate transactions and send them to the network ?	Who can update the network ?
Blockchain open	Publish without permission	Open to all	Anyone	Anyone
	Publish with permission	Open to all	Authorized participants	Group of authorized participants
Blockchain closed	Consortium	Restricted to authorized participants	Authorized participants	Group of authorized participants
	Enterprice	Completely private	Network node manager	Network node manager

Source: Our elaborations

3 Research Design and Methodology

The present study analyzes the blockchain-based citrus supply chain to identify new approaches and opportunities in the agri-food sector that can be used to develop guidelines, both politically and as local planning tools, to enhance production, consumer protection, and the analysis of the value chain in the citrus supply chain.

The proposed approach is based on integrating participatory planning (based on the establishment of focus groups of the various stakeholders) and the novel approach to imprecise assessment and decision environments (NAIADE) method for multi-criteria social assessment (SMCE) [45, 46] as a possible methodological structure to acquire and evaluate the "complex" information collected (quantitative and qualitative data) on possible alternative scenarios in relation to blockchain application.

The methodology can be considered a social experiment that is able to produce collective opinions, to detect communication barriers, to study conflictual behavior, to acquire local information, and to create acceptable options [45, 47].

The aim is to develop a methodological structure using suitable tools to acquire first, and process second, qualitative and quantitative information concerning the possible alternative scenarios of the problem under study. The opinions were collected through specific focus groups with local stakeholders, operators, consumers, and producers interested in the issue in question (including security, certification, innovation, interoperability, fraud repression, and the value chain), but from different points of view, in the presence of two researchers, one with the role of moderator and the other with the role of recording the responses of the individual subjects.

The adoption of this approach was limited to problems of territorial planning referable to SMCE, while there are more numerous articles that have employed SMCE for the resolution of problems related to the enhancement of production, consumer protection, and the control of value distribution in the supply chain.

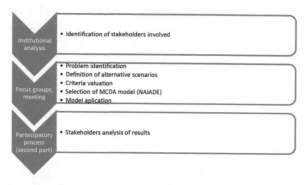

Fig. 3. Structure of the SMCE model (Source: adapted from Munda, (2006)).

The proposed model is based on:

- the individualization of the stakeholders involved (80 questionnaires);
- the definition of the alternative scenarios (definition of the three hypotheses of scenario: producers; consumers; and markets);
- the definition of the context of evaluation, namely the decisional criteria (citrus chain in Italy for shared projects);
- the evaluation of the impact of alternative scenarios relative to the criteria;
- the final creation of the impact matrix.

The model used focus groups as a social research methodology, aiming to acquire information on the opinions of stakeholders regarding a variety of scenarios for future development within the chain [23]. The choice of focus groups and, therefore, of the interaction among the actors involved, aims to support the phase of the choice and evaluation of the different aspects that will be included in the equity matrix. The matrices of impact and equity constitute the basis for the use of the discrete multicriteria evaluation NAIADE model, which is able to manage qualitative and quantitative data in order to evaluate the measures of intervention. This instrument supports the classification of the alternative scenarios proposed on the basis of determined decisional criteria and considerations of possible "alliances" and "conflicts" between the groups of stakeholders for the proposed scenarios, thus measuring their acceptability. The NAIADE model is an appropriate tool for the planning of problems characterized by great "uncertainty" and "complexity" regarding existing territorial, social, and economic structures and their interrelations. The basic input in the NAIADE method consists of: alternative scenarios to be analyzed; different decisional criteria for their evaluation; and different stakeholders who express opinions about the scenarios in question. One of the strengths of this tool in its application to the planning of interventions on green spaces is based in its ability to collect the conflicting perspectives of the stakeholders and to address the compromises among the environmental, social, and economic dimensions.

The objective of this study is to analyze the principal priorities, using as its methodology the model of blockchain application in the citrus supply chain. The evaluation through the focus groups was divided into three phases, referring in this specific case to the potential repercussions for the Italian citrus supply chain being evaluated.

Phase 1 involved "planning" the meetings. During this phase, the following were established:

- the number of sessions and the time dedicated to each of them, as an expression of the individual categories considered (producers, producers' trade union associations, dealers, retailers, consumers, consumer associations, institutions, and scientific groups);
- the creation of an interview guide to help conduct the discussion (scientific and dissemination materials, research papers, photos, and maps relative the different scenario);
- the selection of participants (stratified selection for homogeneous groups (age, gender, income).

The questionnaire used for the interviews was designed to explore the perception of traceability issues in the citrus-supply-chain context and to evaluate the real needs of actors in the supply chain. It comprised 21 questions aiming to collect information and opinions useful for the research related to three hypotheses proposed (for producers, consumers and markets):

P (producers): application of blockchain is hypothesized for the valorization of the agricultural productions on the basis of the quality of the product.

C (consumers): application of the blockchain is aimed at protecting the health of the consumer.

M (markets): application of blockchain in order to gain control over prices along the chain and its value chain.

Phase 2 comprised "carrying out" the entire activity, based on the guide to the preestablished interview. It began with the presentation of the topic relating specifically to the blockchain action strategy in the citrus supply chain, using the support material (articles, results, and photographs) prepared specifically to introduce the issue under consideration and stimulate discussion and interaction among participants. During this phase, various ideas and opinions were acquired that represent the reactions of the participants involved to the issues raised.

Phase 3 involved elaborating the "qualitative results" and producing of the final report.

The key advantage of focus groups dedicated to defining intervention strategies for enhancing the citrus supply chain, compared to other participatory techniques, lies in the deep interaction between the participants, becoming a "social network." The participants become fundamental tools to support a "mutual learning process" for the question examined. This participatory comparison technique makes it possible to reveal new dimensions of the issue under discussion, thus underlining the possibility for focus groups to highlight different opinions in this regard rather than produce generalized results. The analysis phase of the results of the focus groups was followed by the multi-criteria analysis, in which the basic input of the NAIADE method comprised: alternative scenarios to be analyzed; different decision criteria for the relative evaluation; and different stakeholders that expressed opinions on the scenarios in question. Based on this method, two types of analysis can be performed:

- a multicriteria analysis that, on the basis of the impact matrix, leads to the definition of the priorities of alternative scenarios with regard to certain decision-making criteria; and
- an analysis of equity that, on the basis of the equity matrix, analyzes possible "alliances" or "conflicts" among different interests in relation to the scenarios in question.

In this regard, the multicriteria analysis, according to the NAIADE methodology, aims to classify alternative scenarios based on the preferences of individual groups based on certain decision criteria.

The input of the NAIADE method is constituted by the impact matrix (criteria/alternative matrix), including scores that can take the following forms: crisp numbers; stochastic elements; fuzzy elements; and linguistic elements (such as very poor, poor, good, very good and excellent). To compare alternative scenarios, the concept of distance is introduced. In the presence of crisp numbers, the distance between two alternative scenarios with respect to a given evaluation criterion is calculated by subtracting the respective crisp numbers.

The classification of alternative scenarios is based on data from the impact matrix, used for:

comparison of each single pair of alternatives for all the evaluation criteria considered;

calculation of a credibility index for each of the aforementioned comparisons that measures the credibility of one preference relation, e.g. alternative scenario (a) is better/worse, etc. than alternative scenario (b) (preference relationships were used);

aggregation of the credibility indices produced during the previous stage leading to a preference intensity index $[\mu * (a, b)]$ of an alternative (a) with respect to another (b) for all the evaluation criteria, associated the concept of entropy $[H * (a, b)]$ as an indication of the variation in the credibility indices; and classification of alternative scenarios on the basis of previous information.

The final classification of the alternatives is the result (intersection) of two different classifications: the classification $\Phi + (a)$ (based on the "best" and "decidedly better" preference relationships); and the classification $\Phi - (b)$ (based on the "worst" and "decidedly worse" preference relationships).

In relation to the objective of the present study, the analysis will be applied to the main priorities, for the assessment of the scenario that benefits most from blockchain implementation in the citrus supply chain.

4 Results

The results of the present study provide a further multidisciplinary contribution to research on the management of blockchain technologies. Specifically, the analysis was conducted to address the research question: What are the opportunities that blockchain technologies can bring to the citrus-fruit supply chain? In order to evaluate the three hypotheses mentioned above, evaluation criteria were defined that "… a measurable aspect of judgment that can characterize a dimension of the various choices that are taken into consideration." The objectives of the evaluation activity were: security; certification; innovation; interoperability; fraud repression; and the value chain. Specifically, for each objective, the related evaluation criteria were considered (see Table 3).

Table 3. Objectives and evaluation criteria for blockchain technology applied to the citrus supply chain.

Objectives	Evaluation criteria
Security	Method of production, workers, productive exploitation, energy consumption
Certification	Origin, Product characteristics, Chemical residues, Sustainability
Innovation	Decentralization of the register, automation, ICT, Big data
Interoperability	Consent decentralization, supply chain verification, encryption
Fraud Repression	Immutability of data, traceability, production verification
Value-chain	Price origin, logistics, value chain

Source: our elaborations

Twenty one evaluation criteria were used (Table 4). These criteria were defined on the basis of the purpose and objectives of the evaluation of the analyzed case, which can be considered representative of the citrus supply chain.

P (producers) was revealed to be the best option for sharing, closely followed by C (consumers) and M (markets), but all three hypotheses had positive evaluations.

Then the stock matrix was then developed. This provided the views of interested parties on the three suggested hypotheses. The results show that a large number of stakeholders and groups of selected operators agreed with the assessment of the three hypotheses (Table 5).

The results of the multi-criteria analysis revealed that P (producers) hypothesis was the predominant hypothesis, followed closely C (consumers), while M (markets) acquired only a marginal rating (Table 6).

Table 4. Evaluation of the impact matrix of the different alternatives.

Evaluation criteria	Scenario Productions "P"	Scenario Consumers "C"	Scenario Markets "M"
Security			
Method of production	Very good	Exellent	Good
Workers	Good	Very good	Good
Productive exploitation	Poor	Very good	Good
Energy consumption	Poor	Good	Very poor
Certification			
Origin	Good	Exellent	Very poor
Product characteristics	Exellent	Very good	Good
Chemical residues	Very good	Exellent	Poor
Sustainability	Very good	Good	Poor
Innovation			
Decentralization of the register	Very good	Exellent	Exellent
Automation	Good	Very good	Exellent
ICT	Good	Very good	Very good
Big data	Very good	Good	Very good
Interoperability			
Consent decentralization	Good	Exellent	Exellent
Supply chain control	Very good	Very good	Good
Encryption	Exellent	Poor	Good
Fraud Repression			
Immutability of data	Good	Very good	Exellent
Traceability	Very good	Exellent	Exellent
Production verification	Exellent	Good	Good
Value-chain			
Price origin	Good	Very poor	Exellent
Logistics	Poor	Good	Very good
Value chain	Good	Good	Exellent

Source: our elaborations

Table 5. Equity matrix: opinions of the operators and stakeholders on the proposed scenarios.

Groups and stakeholders	Scenarios	Scenario Producers "P"	Scenario Consumers "C"	Scenario Markets "M"
Producers	A1	Very good	Good	Exellent
Producers' trade union associations	A2	Exellent	Very good	Exellent
Dealers	A3	Good	Very good	Good
Retailers	A4	Very good	Good	Good
Consumers	A5	Very good	Exellent	Very good
Consumer associations	A6	Good	Exellent	Good
Institutions	A7	Very good	Very good	Poor
Scientific association	A8	Exellent	Exellent	Good

Source: our elaborations

The results obtained through the analysis of the single answers were used to examine possible alliances or conflicts between the opinions of the interested parties regarding the decision on which hypothesis to adopt. The results in Table 7 show that a large number of interested parties, in addition to agreeing on the classification of the different hypotheses to be applied, agreed with P (producers), while noting that there were also significant consequences for the consumer scenario.

Table 6. Classification of the scenarios at the highest consensus level.

Groups and stakeholders		Scenario **Producers** "P"	Scenario **Consumers** "C"	Scenario **Markets** "M"
A1	Producers	0,7938	0,5118	0,8128
A2	Producers' trade union associations	0,8129	0,5339	0,7318
A3	Dealers	0,6218	0,4108	0,3739
A4	Retailers	0,5369	0,4217	0,4557
A5	Consumers	0,5924	0,7879	0,6137
A6	Consumer associations	0,6374	0,8631	0,7928
A7	Institutions	0,6139	0,7218	0,5436
A8	Scientific association	0,8218	0,8605	0,7161

Source: our elaborations

Table 7. Consensus and related prioritization of scenarios.

		Consent levels					
		0,6351	0,5429	0,6942	0,5236	0,6238	0,5537
Scenario classification		P	C	P	P	M	C
		C	M	C	M	C	M
		M	P	M	C	P	P
Groups of "alliances" at each level of consensus		All groups	All groups except A1 e A2	All groups except A4	All groups except A5 e A6	All groups except A7 e A8	All groups except A1 e A4

Source: our elaborations

The efficiency of this type of approach is based on the possibility of establishing a "learning platform" that facilitates participation, exchange of information, and mutual understanding among participants, which mutually stimulate each other to share the technology. The results allowed us to include different perspectives of the problem being studied, as demonstrated by the different groups involved, highlighting the perception of the actors in the supply chain concerning the acceptability of the proposed alternatives, which can lead to improving strategic decisions and, therefore, creating innovative ideas and new solutions to enhance and protect, based on the possibilities offered by these participatory processes (Fig. 4).

Overall, the results obtained from this model, developed through the integration of a participatory tool and a multi-criteria analysis, become strategic for the choices of investments in the agrifood system, in particular in relation to the current situation in which the producer, consumer, and retailer are trying to define a fundamental role in the supply chain.

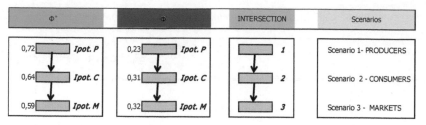

Fig. 4. Classification of alternative hypotheses and multicriteria assessment.

5   Conclusions

Blockchain-based technologies offer countless opportunities for the citrus chain that can directly impact aspects of the supply chain [35, 48, 49]. The ability to digitally decode a new concept of trust makes this technology potentially suitable to take on economic, political, and social value [43]. The blockchain, more than a technology, can be considered a new paradigm, that is, a new way of interpreting the large theme of decentralization and participation [5, 50]. Blockchain has started to change some of the key activities within supply chain management [17], facilitating greater transparency, product traceability, protection against fraud and counterfeiting, financing supply chains, democratizing supply relationships (making them more equitable), improving partner relationships, managing collaborative activities related to analytics, and automating decision making [8]. Producers represent the scenario with the greatest consensus in the citrus industry. This is perhaps due to the fact that the citrus supply chain is characterized by a lack of clarity and linearity, so much so that it has been defined as "tortuous" (no linear supply chain) with an increasing demand for transparency and role definition from production [37, 38, 51]. Regarding the regulatory aspects, of the blockchain application, the regulatory difficulties mainly concern data security and privacy [52, 53]. Finally, it is necessary that all actors in the supply chain spread knowledge about technology in order to stimulate investment, but it is essential that everyone understands the benefits. Only by creating a culture that businesses and, more broadly, the communities in which they live, can they hope to have a transparent and shared process [43].

These are key challenges that need to be addressed if we are to hope that blockchain will be applied in the agrifood system of the future [54–57]. For distributed ledgers to be adopted, however, these objectives must be achieved as soon as possible. If not, against all expectations, we will see the decline of this new technology that still has the power to transform substantially the entire system in which we live.

In conclusion, this technology is certainly revolutionary, but it cannot be considered the solution to all problems. Although it is not possible to predict how blockchain technology will be implemented over time, it is certain that the dynamics of the market will change, and perhaps it can be assumed that the blockchain [58–60] is not in all cases the best solution for the food system and that, perhaps, the traditional databases, if well-managed, will remain the dominant technology if we do not have the active involvement of all actors and the related information of the individual supply chains.

References

1. Armitage, C., Baillie, B., Polson, R., Wang, R., Roberts, J., Green, H.: Food fraud vulnerability assessment and mitigation: Are you doing enough to prevent food fraud?, PwC Report, November 2016
2. Scuderi, A., Pecorino, B.: Protected designation of origin (PDO) and protected geographical indication (PGI) Italian citrus productions. ActaHorticulturae **1065**, 1911–1917 (2015)
3. Ge, L., Brewster, C.A.: Informational institutions in the agrifood sector: meta-information and meta-governance of environmental sustainability. Curr. Opin. Environ. Sustain. **18**, 73–81 (2016)
4. Gupta, V.: Building the hyperconnected future on blockchains. World Government Summit report (2017). https://www.worldgovernmentsummit.org/kiosk/reports/building-the-hyperconnected-future-on-blockchains
5. Abeyratne, S.A., Monfared, R.P.: Blockchain ready manufacturing supply chain using distributed ledger. Int. J. Res. Eng. Technol. **5**(9), 1–10 (2016)
6. Lee, H., Mendelson, H., Rammohan, S., Srivastava, A.: Technology in agribusiness: Opportunities to drive value. White paper, Stanford Graduate School of Business (2017)
7. Pappalardo, G., Chinnici, G., Selvaggi, R., Pecorino, B.: Assessing the effects of the environment on consumers' evaluations for wine. Wine Econ. Policy **9**(1), 31–42 (2020)
8. William, N., Lijo, J., Harish, K.: How blockchain can shape sustainable global value chains: an evidence, verifiability, and enforceability (EVE) framework. Sustainability **10**(11), 3926 (2018)
9. Arena, E., et al.: Exploring consumer's propensity to consume insect-based foods. Empirical evidence from a study in Southern Italy. Appl. Syst. Innov. **3**(38) (2020). https://doi.org/10.3390/asi3030038
10. Chen, L., Lee, H.L.: Sourcing under supplier responsibility risk: the effects of certification, audit, and contingency payment. Manag. Sci. **63**, 2795–2812 (2016)
11. Vindigni, G., La Terra, G., Bellia, C.: Rethinking IPRs on agro-biotechnological innovations in the context of food security. Rivista di StudisullaSostenibilità **2**, 69–77 (2016)
12. Scuderi, A., Sturiale, L., Timpanaro, G.: The importance of "origin" for on line agrifood products. Quality - Access Success **16**, 260–266 (2015)
13. Benedetto, G., Sini, M.P.: Considering the social sustainability of rural landscape in Sardinia, Italy: the role of local identities. In: Island Landscapes: An Expression of European Culture, pp. 184–189 (2016)
14. Timpanaro, G., Foti, V.T., Scuderi, A., Schippa, G., Branca, F.: New food supply chain systems based on a proximity model: the case of an alternative food network in the Catania urban area. ActaHorticulturae **1215**, 213–217 (2018)
15. De Molli, V., et al.: Sfide e priorità per il settore alimentare oggi. Strumenti e approcci per la competitività. The European House – Ambosetti (2017). https://www.ambrosetti.eu/ricerche-e-presentazioni/settore-alimentare-oggi/
16. Lin, Y.P., et al.: Blockchain: the evolutionary next step for ICT. E-Agric. Environ. **4**(3), 50 (2017)
17. Tapscott, D.: Blockchain Revolution. Random House, New York (2016)
18. Pappalardo, G., Selvaggi, R., Pecorino, B., Lee, Y.Li., Nayga, R.M.: Assessing experiential augmentation of the environment in the valuation of wine: evidence from an economic experiment in Mt. Etna, Italy. Psychol. Mark. 1–13 (2019)
19. Kempe, M., Sachs, C., Skoog, H.: Blockchain use cases for food traceability and control: A study to identify the potential benefits from using blockchain technology for food traceability and control. Axfoundation, SKL Kommentus, Swedish Country Councils and Regions, Martin & Servera, and Kairos Future (2017)

20. FAOSTAT, FAO Statistical Databases (2019)
21. Scuderi, A., Zarbà, A.S.: Economic analysis citrus fruits destined to markets. Italian J. Food Sci. **23**, 34–37 (2011)
22. Carrà, G., Peri, I., Scuderi, A.: Euro-Mediterranean agricultural trade agreement: wichreal impact on the EU citrus protection? Quality - Access Success **15** (2014)
23. Scuderi, A., Sturiale, L.: Multicriteria evaluation model to face phytosanitary emergencies: the case of citrus fruits farming in Italy. Agric. Econ. **62**, 205–214 (2016)
24. USDA - United States Department of Agriculture. Citrus: World market and trade (2019). https://www.fas.usda.gov/data/citrus-world-markets-and-trade
25. Love, J., Somerville, H.: Retailer Carrefour using blockchain to improve checks on food products. Reuters (2018). https://www.reuters.com/article/carrefour-blockchain/retailer-carrefour-using-blockchain-to-improve-checks-on-food-products
26. MIPAAF - Ministero delle Politiche Agricole Alimentari e Forestali. Report di Attività 2017. Report Department of the Central Inspectorate and Quality Control and Fraud Repression of Agro-Food Products Department (2018)
27. Sachs, C., Exman, R., Nohrstedt, Göthberg, P., Kempe, M.: Blockchain use cases for food-traceability and control: A study to identify the potential benefits from using blockchain technologyfor food traceability and control. Axfoundation, SKL Kommentus, Swedish Country Councils and Regions, Martin & Servera, and Kairos Future, February 2017
28. Cagnina, M.R., Cosmina, M., Gallenti, G., Marangon, F., Nassivera, F., Troiano, S.: The role of information in consumers' behavior: a survey on the counterfeit food products. Economia Agro-Alimentare **20**(2), 221–231 (2018)
29. Banerjee, A.: Blockchain technology: supply chain insights from ERP. In: Pethuru, R. (ed.) Advances in Computers, vol. 11, pp. 69–98. Elsevier (2018)
30. Ge, L., Brewster, C., Spek, J., Smeenk, A., Top, J.: Blockchain for agriculture and food: Findings from the pilot study. Wageningen Economic Research, Report, 112 (2017)
31. Janvier-James, A.M.: A new introduction to supply chains and supply chain management: definitions and theories perspective. Int. Bus. Res. **5**(1), 194–198 (2012)
32. Kouhizadeh, M., Sarkis, J.: Blockchain practices, potentials, and perspectives in greening supply chains. Sustainability **10**(10), 1–16 (2018)
33. European Commission: Green Book: Promoting a European framework for Corporate Social Responsibility. European Commission, Brussels (2001)
34. Argandoña, A., von Weltzien Hoivik, H.: Corporate social responsibility: one size does not fit all. Collecting evidence from Europe. J. Bus. Ethics **89**, 221–234 (2009). https://doi.org/10.1007/s10551-010-0394-4
35. Mussomeli, A., Gish, D., Laaper, S.: The Rise of the Digital Supply Network. Deloitte University Press, New York (2016)
36. Pergola, M., et al.: Sustainability evaluation of Sicily's lemon and orange production: an energy, economic and environmental analysis. J. Environ. Manag. **128**, 674–682 (2013)
37. Maslova, A.: Growing the garden: How to use blockchain in agriculture.Cointelegraph (2017). https://cointelegraph.com/news/growing-the-garden-how-to-use-blockchain-in-agriculture
38. Subramanian, H.: Decentralized blockchain-based electronic marketplaces. Commun. ACM **61**(1), 78–84 (2017)
39. Zheng, Z., Xie, S., Dai, H., Chen, X., Wang, H.: Blockchain challenges and opportunities: a survey. Int. J. Web Grid Serv. **14**(4), 352–375 (2016)
40. Raimondo, M., Caracciolo, F., Cembalo, L., Chinnici, G., Pecorino, B., D'Amico, M.: Making virtue out of necessity: managing the citrus waste supply chain for bioeconomy applications. Sustainability **10**(12), 4821 (2018)
41. Valenti, F., et al.: Use of citrus pulp for biogas production: a GIS analysis of citrus-growing areas and processing industries in South Italy. Land Use Policy **66**, 151–161 (2017)

42. Tse, D., Zhang, B., Yang, Y., Cheng, C., Mu, H.: Blockchain application in food supply information security. In: International Conference on Industrial Engineering and Engineering Management, pp. 1357–1361 (2017)
43. Wang, J., Li, L., He, Q., Yu, X., Liu, Z.: Research on the application of block chain in supply chain finance. In: 3rd International Conference on Electronic Information Technology and Intellectualization (ICEITI) (2017). http://dpi-proceedings.com/index.php/dtcse/article/view/18857/18353
44. Peck, M.E.: Blockchain world - do you need a blockchain? This chart will tell you if the technology can solve your problem. IEEE Spectrum **54**(10), 38–60 (2017)
45. Munda, G.: A NAIADE based approach for sustainability benchmarking. Int. J. Environ. Technol. Manage. **6**, 65–78 (2006)
46. Munda, G.: Social Multicriteria Evaluation for a Sustainable Economy. Springer, Heidelberg (2008)
47. Munaretto, S., Siciliano, G., Turvani, M.: Integrating adaptive governance and participatory multicriteria methods: a framework for climate adaptation governance. Ecol. Soc. **19**(2), 74–87 (2014)
48. Scuderi, A., Foti, V.T., Timpanaro, G.: The supply chain value of POD and PGI food products through the application of blockchain. Quality - Access Success **20**, 580–587 (2019)
49. Chinnici, G., Pecorino, B., Scuderi, A.: La percezione della qualità dei prodotti tipici da parte del consumatore in Sicilia. Economia Agro-Alimentare **14**(1), 143–172 (2012)
50. Ganeriwalla, A., Casey, M., Shrikrishna, P., Bender, J.P., Gstettner, S.: Does your supply chain need a blockchain? Boston Consulting Group/MIT Media Lab report (2018)
51. Chinnici, G., Pecorino, B., Scuderi, A.: Environmental and economic performance of organic citrus growing. Qual. - Access Success **14**(S1), 106–112 (2013)
52. Krishnan, H., Winter, R.A.: The economic foundations of supply chain contracting. Found. Trends Technol. Inf. Oper. Manag. **5**(3–4), 147–309 (2012)
53. Kshetri, N.: Blockchain's roles in meeting key supply chain management objectives. Int. J. Inf. Manag. **39**, 80–89 (2018)
54. Alizadeh, M., Andersson, K., Schelen, O.: A survey of secure Internet of Things in relation to blockchain. J. Internet Serv. Inf. Secur. **10**, 47–75 (2020)
55. Leung, K.H., Lau, H.C.W., Nakandala, D., Kong, X.T.R., Ho, G.T.S.: Standardising fresh produce selection and grading process for improving quality assurance in perishable food supply chains: an integrated Fuzzy AHP-TOPSIS framework. Enterp. Inf. Syst. **15**, 651–675 (2020)
56. Ghadge, A., Er-Kara, M., Mogale, D.G., Choudhary, S., Dani, S.: Sustainability implementation challenges in food supply chains: a case of UK artisan cheese producers. Prod. Plan. Control (2020)
57. Yadav, V.S., Singh, A.R., Raut, R.D., Govindarajan, U.H.: Blockchain technology adoption barriers in the Indian agricultural supply chain: an integrated approach. Resour. Conserv. Recycl. **161**, 104877 (2020)
58. Ben-Daya, M., Hassini, E., Bahroun, Z., Banimfreg, B.H.: The role of internet of things in food supply chain quality management: a review. Qual. Manag. J. **28**, 17–40 (2020)
59. Mogale, D. G., Kumar, S., Tiwari, M.: Green food supply chain design considering risk and post-harvest losses: a case study. Ann. Oper. Res. **295**(1), 257–284 (2020). https://doi.org/10.1007/s10479-020-03664-y
60. Blockchain for organic food traceability: case studies on drivers and challenges. Front. Blockchain **3**, 43 (2020)

Experiences of Online Purchase of Food Products in Italy During COVID-19 Pandemic Lockdown

Luisa Sturiale[1]([✉]), Alessandro Scuderi[2], Biagio Pecorino[2], and Giuseppe Timpanaro[2]

[1] DICAR, University of Catania, 95127 Catania, Italy
luisa.sturiale@unict.it
[2] D3A, University of Catania, 95127 Catania, Italy
{alessandro.scuderi,pecorino,giuseppe.timpanaro}@unict.it

Abstract. The lockdown that affected our planet as result of the COVID-19 health emergency caused an important change in consumers' relationship with digital technologies and e-commerce. The paper introduces a brief overview of ecommerce in recent years and the main changes due to the pandemic. The aim is to analyze the purchasing experiences during the lockdown, in general and in Italy, with specific reference to food products, and the impact of the digital transformation due by the pandemic on consumer habits. The analysis was carried out by administering a specific questionnaire to a sample of 400 Italian consumers during the lockdown (February–May 2020). The results showed that this sad emergency was taken as an opportunity for change in lifestyles and in, the reconfiguration of corporate management.

Keywords: E-commerce · COVID-19 · Purchasing behavior · Digital transformation · Virtual market · Omnichannel

1 Introduction

The health emergency due to COVID-19, which began in Wuhan Province (China), caused, as we all know, a very heavy impact on national health systems (Allcott et al. 2020; WHO 2020). The governments of the countries affected by the pandemic have implemented measures of isolation and social distancing (sometimes of varying intensity). Workers were asked to work from home, schools were closed and lessons were held remotely online, shops were closed (except for essential services), and cities were emptied. But it has not been the only effect because it has involved the economic and environmental systems worldwide(Hudson 2020; Phillipson et al. 2020; Zambrano-Monserrate et al. 2020), although at different times and in different ways, the economic activities related to the process of the purchase of goods or services (Hall et al. 2020; Nicola et al. 2020).

The magnitude of the downturn and the level of recovery will be more difficult in the retail market than during the Great Recession. According to the International Monetary

Fund (IMF 2021), global GDP (Gross Domestic Product) will contract by 3.3% in 2020 compared to a 0.1% decline in 2009.

The lockdown measures resulting from the COVID-19 pandemic have led to change in people's lifestyles (Deb et al. 2020; Chen et al. 2020).The digital technologies and the integrated digital media experiences have made possible to develop new relational activities, smart working, virtual support, play activities, virtual events, online art therapy, telemedicine, digital fitness, etc. (Hoffmann et al. 2020; Miller and McDonald 2020), in general, and new consumer purchasing behavior, in particular, especially for the confinement of almost the entire world's population in their homes (Baker et al. 2020; Chronopoulos et al. 2020; Dirgantari et al. 2020; Eriksson and Stenius 2020; Hobbs 2020).

In particular, two phenomena that have had repercussions on the use of digital technologies, linked to work activities and purchasing behavior, are highlighted (Di Renzo et al. 2020; Haleem et al. 2020; Nicola et al. 2020; Signorelli et al. 2020).

The smart working (both in businesses, schools and universities, communities) has led to a surge in video conferencing apps. It surprised the immediate growth of Zoom compared to other platforms, followed by Skype, WhatsApp, Google and Microsoft Teams.

The e-commerce activity, particularly related to health and grocery, is booming in all countries that have adopted the lockdown measure. The pandemic has increased retail sensitivity to e-commerce, even though the phenomenon had gradually increased in recent years (Pejic-Bach 2020; Santhiya and Elavarasan 2020). In fact, online sales have increased at an unexpected rate in various production sectors, including agri-food products, which have had a considerable weight.

During the quarantine, food in general showed an upward trend in sales. But interesting, there have been some changes in purchasing behavior (Grashuis et al. 2020; Roggeveen and Sethuraman 2020).

Italy was the first country in Europe (and in the world, after China) to face the terrible epidemic, which led the Italian government to activate the drastic action of lockdown and physical (and social) distancing for several months (from 9th March to 3rd May, although already at the end of February in the "red zones" was active).

According to recent studies, many Italian consumers, even the elderly, have overcome their suspicions about technology to shop online for the first time during the country's lockdown (RetailX 2020).

The paper introduces a brief overview of the evolution of the "web ecosystem" and of the e-commerce in recent years and the main changes that have occurred as a result of the pandemic. The aim is to analyze the purchasing experiences during the period of lockdown, in general, and in Italy, in particular, with particular reference to food, and the impact that the digital transformation induced by the coronavirus event has had on business strategies and consumer habits. The analysis was carried out through the administration of a special questionnaire to a sample of 400 Italian consumers during the blocking period (February–May 2020). The results showed that this sad health emergency was seized as an opportunity to change lifestyles, to modify online shopping behavior and to reconfigure business management towards a digital rethinking of business functions, especially those related to customer-care, delivery and social media management.

2   The "Web Ecosystem": Main Characteristics in the Global Economic Context Pre-pandemic and Dynamics of Ecommerce in Emergency Healthcare Covid-19

Digital technology and the spread of computer networks have transformed the processes of production, access, transfer and use of information. We can say that Information and Communication Technologies (ICTs) are having an impact as important as electricity and transportation a century ago, and have spawned innovative services such as eHealth, smart city smart-farming, Agriculture 4.0, Internet of Tinghs, Big Data, Cloud, Augmented and Virtual Reality (Jayaram et al. 2015; Scuderi et al. 2019a, b; Scuderi et al. 2020).

Over the last decades the web has become a "web ecosystem" in which users create and share experiences (through e-mail, blogs, forums, communities, chats, virtual platforms, marketspace, social media, etc.). The Internet has become the global platform for the exchange of information and trade, where an economic force of global dimensions operates, which has radically changed the social and economic behavior of the whole world (Huang and Benyouce 2013; Noguti and Waller 2020; Scuderi et al. 2018; Scuderi and Sturiale 2019; Verdouw et al. 2013).

It is not easy to give a consistency of the phenomenon because it is characterized by a globalism and complexity difficult to control, however, there are reliable statistical sources that can give us a significant picture in statistical terms.

In 2020, the number of internet users in the world (Table 1) is more than 4.8 billion at a global level, Asia has the highest number of users, followed by Europe and Africa (Internet World States 2020).

Table 1. Some statistics on internet use in the world (2020)

World areas	Internet users	Penetration rate (%)	Incidence on world (%)	Variation 2000-2020 (%)
Asia	2,525,033,874	58.8	52.2	2,109
Europe	727,848,547	87.2	15.1	592
Africa	566,138,772	42.2	11.7	12,441
Latin America	467,817,332	71.5	9.7	2,489
North America	332,908,868	90.3	6.9	208
Oceania/Australia	28,917,600	67.7	0.6	279
WORLD TOTAL	4,833521,806	62.0	100.0	1,239

Source: our elaborations on data available online: https://www.internetworldstats.com/

For online sales, China is still the country that holds the leadership (it is estimated that 62.6% of e-commerce is concentrated in the Asia-Pacific area), even if in the first months of 2020 it has reduced the growth trend, while North America and Western Europe follow, but at a distance, with 19.1% and 12.7% (Cramer-Flood 2020).

Focus on the European Union (E.U.), the percentage of the population with internet access (internet users) has increased in recent years, going from 78% in 2015 to 87% in 2019, according to data provided by Eurostat (Eurostat 2020; Ecommerce Europe 2020).

According to first estimates, the value of e-commerce in Europe is expected to be 717 billion euros in 2020. This would mean an increase of 13% compared to the situation in 2019 (Ecommerce Europe 2020). The most interesting aspect is that the pandemic has certainly led to an acceleration in online shopping and consumers' behavior.

The share of consumers using the internet who shopped online in the last year varies per country. The greater part of B2C e-commerce is registered in Western Europe (83%), followed at a short distance from Northern Europe (79%), while taillight is Eastern Europe (36%).

It is possible to provide an overview of online purchasing preferences of european consumers before the pandemic, both in terms of merchandise and age groups. Considering only the percentage of the total value, the record was held by Clothes and sport goods with more than 60%, followed by Travel, holiday accommodation with more than 50%, then by Household goods, Tickets for event, while Food recorded only 27%, Telecommunication services and Computer hardware around 20% and Medicines less than 20% (Eurostat 2020).

Certainly the health emergency due to COVID-19 has transformed our home into the "new ecosystem of life". In fact, the home has become a work space, with the intensification of smart working; a shopping space, with the development of ecommerce (Veeragandham et al. 2020; Thakur et al. 2020).

The impossibility of being able to leave home for lockdown measures and make purchases in physical stores has led to a major change in the lifestyles of consumers, who have transformed their home into the shopping space, through e-commerce (but also social commerce and mobile commerce).

During the pandemic, the agri-food products has registered the highest growth rate (even if different among the various countries) after the medicines and health products (Hall et al. 2020; Santhiya and Elavarasan 2020).

Quarantine has resulted in a significant boost to e-commerce of fresh food products, both in traditional shops and supermarkets, which have ensured fast and reliable deliveries of fruit and vegetables (Galanakis 2020; Grashui et al. 2020; Sturiale and Scuderi 2016; 2017). Individual farmers have also activated e-commerce through social media, maintaining continuous interaction with their customers and ensuring zero kilometer home deliveries of fresh fruit and vegetables. This phenomenon has been observed, albeit with different territorial specificities, in all EU countries. (Germany, Italy, Spain and others). The interesting aspect of the phenomenon is that the e-commerce for fresh fruit and vegetables has not had the same growth trend as the other production sectors until 2019. As research has shown, e-commerce in food products is growing more slowly than in other sectors due to certain limitations, including the preference of consumers to choose fresh products themselves. Cultural aspects, product safety and health are the main obstacles to the diffusion of ICTs in the agro-food sector as well as an adequate supply chain organization (Bond et al. 2009; Lehmann et al. 2012; Weiss and Wittkopp 2005; Scuderi and Sturiale 2014; Scuderi et al. 2015; Sturiale and Scuderi 2011).

Since March there has been a surge in online orders for food products at supermarkets, a little in all EU countries. Many retailers have found themselves unprepared for the unexpected increase in orders (in some cases 25 times higher than in 2019) and delivery times have obviously increased, with dissatisfaction of customers (although often sympathetic to the current health emergency) (REWE in Germany, Migros and Leshop in Switzerland, to name a few). Moreover, the experience of the pandemic has oriented some online stores to strengthen their logistics (this is the case of the Dutch online supermarket Nemlig.com, which recorded a 50% increase in sales), while others, despite the increase in online sales, prefer omnichannel, both physical and virtual (the Coop, which has also increased its online orders but does not plan to become an exclusively online store (Dannenberg et al. 2020; von Abrams 2020).

In the scenario of European e-commerce, Italy is the third largest economy in the European Union (E.U.), although there are still limits to the development potential of the phenomenon. In fact, on the one hand, Italian consumers prefer to shop in person and, on the other, only 79% of consumers have access to the internet. Despite the various limitations, the growth of e-commerce in Italy in recent years shows a market that offers significant opportunities to retailers interested in the digital sales channel, so much so that in 2019 e-commerce grew by 15% (Osservatori Digital Innovation 2019).

The emergency situation has made it possible to accelerate business investment in the ICTs, in general, and in digital commerce channels, in particular, moving towards a digital transformation of the management of companies, of any size, which have oriented towards omnichannel and customer interaction increasingly focused and careful. This phenomenon seems to have been seized by companies as an opportunity for change in their business strategies, to be integrated with the tools offered by digital technologies.

In Italy (Osservatori Digital Innovation 2019; 2020) the value of e-commerce has increased by 26% in 2020 compared to 2019 (although it has already been growing since 2016 both in percentage terms and in absolute value). All production sectors have grown, but the one that has marked a significant increase is Food & Grocery, which has grown by 56% in recent months, an emerging sector that now accounts for 20% of total e-commerce retail.

The pandemic and subsequent lockdown have created one of the most profound and radical changes in retail in the last decade. Government regulations, the transformation of certain consumer habits and the enhancement of e-commerce projects by numerous players in the sector have led to profound changes in the Italian online market:

The largest volume of purchases in Italy was concentrated in the lockdown months (March-May), with a drop in the period from July to September, only to rise again during the second wave of the emergency, particularly between November and December 2020.

3   Methodology

The research was carried out to analyze the purchasing experiences during the lockdown period in Italy and, in particular, the purchasing behaviors of consumers of agri-food products through ICTs platform and to analyze the impact that the digital transformation induced by the coronavirus event has had, in general, on the lifestyles of consumers.

The research was conducted by collecting data through a structured questionnaire and through the use of the Google Docs platform. A total of 400 consumer questionnaires were processed, out of a total sample of 543 citizens, located in Italy, with the only prerequisite that they had bought food products online (Baumgartner 2010; Scuderi et al. 2019a, b). The questionnaires not completed in all its parts were discarded so as to reach the sample of 400 consumers.

The information and data collected with the questionnaire have been processed according to a methodological scheme already used for the study of the consumption characteristics of some food products (Chinnici et al. 2002; Sturiale et al. 2017).

In the first phase, we proceeded to describe the sample as a whole, using the tools of univariate statistical analysis, in order to outline consumer behavior.

The next phase involved the application of multivariate statistical analysis techniques in order to highlight the main variables affecting purchases during the COVID-19 and identify homogeneous consumer groups.

The questionnaire was structured in three sections, ICTs knowledge, online shopping and consumer behavior during the COVID-19. In total there are 10 questions, which have been validated in a first phase through the face to face methodology, in order to verify with interviews in the presence of possible critical issues of the questionnaire, taking about 5 min to complete.

The survey is available online for a period of about 60 days, starting on 25 February 2020 and ending on 04 May 2020. This time frame has been considered as corresponding to the maximum lockdown period.

These data, even though they are not an expression of a statistically representative sample of consumers, provide useful food for thought to understand the dynamics, behaviors and interests that affect the development prospects of the online agri-food products market even in periods other than COVID-19.

The objective of the survey was to detect the confidence with ICTs, the type of purchases on e-commerce sites during the COVID-19, with particular reference to the purchase of agri-food products, the degree of penetration as frequency and the use of social networks, with particular interest in the agri-food sector and finally general data of the respondent.

The questionnaire allowed to collect the answers of various target consumers, having gender, cultural level, employment, ICTs propensity, etc., in order to understand, identify and evaluate the different areas of the agri-food sector through the use of online shopping during the COVID-19 period (Ahmad and Murad 2020).

The most important advantage of cluster sampling exists when each bunch is not homogeneous, i.e. the units that make up the bunch differ from each other by a mode detected on the same characteristic. Only then does such sampling provide more efficient estimates than simple random sampling. However, this type of sampling was chosen in this research for its ease of organization and low cost, both in economic terms and in terms of the rapidity of data collection, processing and dissemination of the results, considering four clusters.

4   Results and Discussion

The analysis of the characteristics of the sample considers demographic, social and economic factors: gender, age group, status, educational qualifications, purchasing manager, occupation, family size and family income (Table 2).

Table 2. Characteristics of the sample (%)

Sex		Occupation	
Males	64	Employee	21
Females	46	Freelancer	9
		Worker	14
Age classes		Unemployed	11
20-40 years	41	Housewife	15
40-60 years	52	Retired	13
over 60 years	7	Student	17
Civil status		*Family size*	
Single	21	1 component	21
Married	79	2 components	25
		3 components	29
Qualification		4 components	18
Secondary school license	7	5 or more components	7
Diploma	43		
Degree	39	*Annual family income*	
Post-graduate degree	18	up to 10 thousand euro	14
		10-20 thousand euro	37
Purchasing manager		20-40 thousand euro	31
Yes	87	over 40 thousand euro	18
No	13		

Source: elaboration on directly measured data.

For the identification of the main variables, the analysis of the main components was used, as described above, which allowed to identify, from the 4 groups of selected variables. The group of variables considered (Table 3) are the socio-economic variables and among those surveyed only 8 were able to explain 70.3% of the total variance, through the extraction of 4 components (A1; A4).

The first component extracted (A1) alone explains 22.9% of the variance and describes the characteristics of a part of the sample interviewed, represented by a group of individuals of the female sex, young, on average married, who carry out an employment of the clerical type; in fact, these variables are positively correlated (values > 0) between them and with a diploma level qualification and an average annual family income.

Table 3. Rotated matrix of components relative to the characteristics of the sample

Variables	Main components			
	A1	A2	A3	A4
Sex	0,015	0,687	0,424	0,387
Age class	0,085	0,087	0,217	0,142
Marital status	0,148	0,187	0,187	0,307
Qualification	0,583	0,254	0,072	0,218
Profession	0,638	0,127	0,475	0,237
Family-run numbering	0,384	0,201	0,381	0,175
Purchasing Manager	0,179	0,184	0,298	0,385
Family income	0,852	0,078	0,197	0,687
Explained variance (%)	22,9	23,1	13,6	14,5

Source: elaboration on directly measured data.

The second component (A2), with 23.1% of the variance explained, is characterized by the population of medium-high age, married and caring for the family, being responsible for food purchases. These variables are strongly correlated with each other in a positive way (values > 0).

The third component extracted (A3), with 13.6% of the total variance explained, shows the middle class of the sample, averagely numerous, characterized above all by the male component that takes care of the family as it is not responsible for purchases.

The fourth component (A4) explains 14.5% of the variance and highlights and links through positive correlations the professional condition of employees with a high level of education.

The application of cluster analysis allowed to identify 4 homogeneous groups of consumers, the groups obtained by the application of the K-medium method.

Group 1 - Potentially interested consumers: (29.3% of the sample) is characterised by women, with a medium level of education and medium-high income. This group prefers processed food products (known through offline and online channels), for which they are willing to pay a premium price for their consumption as an expression of higher quality.

Group 2 - Curious consumers: (26.6% of the sample) is characterised mainly by the female component as the person responsible for purchases, with a household consisting of a few units and a medium-high level. They tend to consume products online "by trend".

Group 3 - Aware consumers: (22.7% of the sample) is represented by young, unmarried, high level of education. They habitually consume food products purchased online.

Group 4 - Consumers who are not interested: (21.4% of the sample) is characterised by a high number of family members, low to medium income and medium level of education). Conditioned by advertising, they are not inclined to buy bio-certified products because they are not willing to pay more than conventional products.

The analysis of the responses shows that almost all respondents (95.6%) surf the Internet. Confirming this result emerges a substantial "heavy internet user" profile (Baumgartner 2010) and medium-high technological competence in terms of computer skills and the use of tools such as PCs, smartphones, and the network in general.

The context of use is mainly dedicated to study and -e-learning (87.7%); smartworking (62.6%), followed by reading newspapers, studying one's hobbies, banking operations and e-commerce (42.6%). Although the latter is lower than the other uses, it is significant both because its use in previous years was much lower (Scuderi and Sturiale 2016; Scuderi and Sturiale 2014) and because it was almost obligatory. The general trend is that of the use of ICT applications by 100% of the sample, albeit combined with traditional channels.

Few individuals do not make purchases online, only 12.3% of them claim not to do so mainly for two reasons: due to inability to control products (40.7%) and lack of opportunities (55.6%).

The remaining part of consumers (87.7%), say they have made online purchases. Among all the product categories analyzed, it emerges that personal products (70.3%) are the most purchased, followed by food products (67.4%), sports, electronics, clothing, and then values that tend to decrease for other types of products including travel.

Certainly, the lockdown has led to strong growth in online purchases of agri-food products (67.4% of respondents), which is important because it expresses the high future potential. This figure bodes well for the potential of agri-food ecommerce, if it is based on targeted strategies, omnichannel and quality products and services offered to customers.

The most part of the representative sample (85.3%) declared that they are oriented to buy a product online compared to an offline product, because of the need dictated by COVID-19, while 60.3% declared that it derives from the convenience of the service, until now not considered.

The research results confirm that purchases are mainly dictated by the COVID-19 requirement (87.7%) but show that this channel is interesting for the possibility of purchasing local products (69.8%) and convenience (62.4%), while quality of products, curiosity and economic savings are marginal. These responses highlight the main factors of choice in online purchases of agri-food products in the future (Fig. 1).

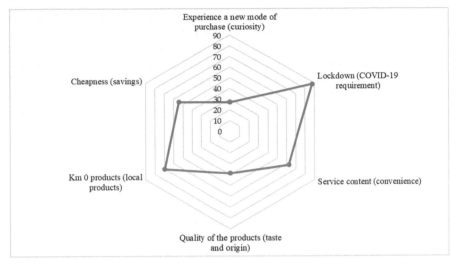

Fig. 1. Most important reasons for buying agri-food products online in COVID-19 lockdown Source: our elaboration on directly collected data.

The subjects who have never purchased food products online are 29.5% and states that mainly this happens for economic aspect, payment method and delivery time (Fig. 2).

Fig. 2. The mains limits on the purchase of agri-food products (%). Source: our elaboration on directly collected data.

Analysing the online purchases of agri-food products, it can be seen that, during the lockdown, 47.4% of the purchases were made by traditional local supermarkets with their online sites, 37.4% by the marketplaces of agri-food products, while only 5.8% were directly acquired by the producer. This data confirms that online purchases are distorted by the presence of supermarkets, confirming that offline and online in many cases represent the optimal synergy for the consumer at present.

Concerning the type of products purchased, the majority of the sample (72.9%) has purchased processed products with a long shelf life, while only 7.4% have sometimes purchased fresh fruit; 8.2% cheese and other types of dairy products; however low are the values for fresh meat and fish with a value of less than 5%. The majority of the sample bought: pasta (42.1)%, extra virgin olive oil 41.4)%, jams (37.2%), coffee (36.4%), liqueurs (29.7%) (wine, beer and spirits).

The determining factors influencing the purchase intention for agri-food products are the following: importance of the place of origin of the product (81% of the interviewees); preference for zero-km products (72%), preference for certified products (PDO, PGI and mainly organic products (57%).

They were asked to express an opinion on the factors that could influence in the future, after the lockdown period of COVID-19, positively on the purchase of agri-food products online, with an evaluation scale from 1 as minimum satisfaction value to 4 as the highest. The results confirm that if on the one hand the price, assumes a certain weight (72.4%), also the certification of products (60.4%), the positive reviews on the site (51.8%) and the origin are decisive.

The analysis confirmed that the consumer during the COVID-19 period approached the online market with less hostility, being in many cases forced, but at the same time he appreciated the possibility in terms of time-saving, the reasoned choice and the possibility to make differentiated purchases. In this sense, it was seen that for fresh fruit and vegetables, platforms on the main social networks have been introduced, which made it possible to order and receive products within 24 h directly from local producers.

Regarding the online user behavior during the COVID-19 lockdown, in the context of the relationship with brands in the agri-food sector, 76% of respondents stated that they had interacted at least once with agri-food companies through a social network.

5   Conclusion

The serious health emergency caused by COVID-19 has certainly changed our lives, in every aspect, social, economic and psychological.

Although unimaginable, what has been assisted has accelerated towards ICTs in the social context, redefining the organization of work, study and life in general and orienting choices towards what was forcedly applied during the lockdown period.

The research provided an analysis of the consumer's approach to the online channel in the COVID-19 pandemic and the potential for agri-food products. For food products there was a general interest in buying online both through the online platforms of traditional points of sale, and real online platforms dedicated to individual producers who offered their products both through online pages, through blogs and through social channels provided by Facebook, Instagram, WhatsApp, Telegram through which they offer the product portfolio and collect orders from individual consumers.

For agri-food products both a new "virtual" dimension has been considered until now at the margins of the traditional channel, today instead it has become strategic and fundamental to be able to reach a high number of consumers.

In Italy, the health crisis has generated an unprecedented digital evolution in the lifestyles of Italians, profoundly changing their purchasing behaviour. Today's customer journey is strongly influenced by the new "normality" in which we live: the frequency of e-commerce purchases is increasing (+79%), home delivery remains the preferred method of receiving products by over 93% of users and digital payments are increasingly preferred to cash payments (Osservatori Digital Innovation 2020).

Italian consumer demand has driven retailers and brands to adapt their services in physical shops, taking advantage of digital opportunities to offer increasingly satisfying and secure experiences. Unfortunately, the digitisation of Italian businesses is still far from the European average, but it is likely that, thanks to the new investments in digital innovation planned in the Recovery Found, this digital divide can be reduced. This orientation gives hope for a change of course towards a greater openness of Italian companies towards the digitization of the various business functions and, above all, of the commercial one. In particular, it is hoped that the opportunity offered by this sad period will be the opportunity to develop the omnichannel of sales, integrating offline and online, gathering the positive experience and adapting sales channels to the new needs of consumers, who have become increasingly demanding in their purchase choices.

This phenomenon can be seen as a great opportunity for companies that will be called to invest in omnichannel (i.e. strategic integration between offline and online), reducing the limits that were highlighted in the supply chain during the pandemic, especially in terms of logistics, transport and delivery modes and payment security.

References

Ahmad, A.R., Murad, H.R.: The impact of social media on panic during the COVID-19 pandemic in Iraqi Kurdistan: online questionnaire study. J. Med. Internet Res. **22**(5), e19556 (2020). https://doi.org/10.2196/19556

Allcott, H., Boxell, L., Conway, J., Ferguson, B., Gentzkow, M., Goldman, B.: Economic and Health Impacts of Social Distancing Policies during the Coronavirus Pandemic. https://ssrn.com/abstract=3610422. Accessed 8 July 2020

Baker, S.R., Farrokhnia, R.A., Meyer, S., Pagel, M., Yannelis, C.: How Does Household Spending Respond to an Epidemic? Consumption during the 2020 COVID-19 Pandemic. NBER Working Paper 26949, National Bureau of Economic Research, Cambridge, MA (2020)

Baumgartner, H.: Bibliometric reflections on the history of consumer research. J. Consum. Psychol. **20**(3), 233–238 (2010)

Bond, J.K., Thilmany, D., Bond, C.: What influences consumer choice of fresh produce purchase location? J. Agric. Appl. Econ. **41**, 1 (2009)

Chen, S., Igan, D., Pierri, N., Presbitero, A.: Tracking the Economic Impact of COVID-19 and Mitigation Policies in Europe and the United States. IMF Working Paper 20/125, International Monetary Fund, Washington, DC (2020)

Chinnici, G., D'Amico, M., Pecorino, B.: A multivariate statistical analysis on the consumers of organic products. Br. Food J. **104**(3/4/5), 187–199 (2002)

Chronopoulos, D.K., Lukas, M., Wilson, J.O.S.: Consumer Spending Responses to the COVID-19 Pandemic: An Assessment of Great Britain. https://ssrn.com/abstract=3586723. Accessed 30 Oct 2020

CNBC: Coronavirus is making grocery delivery services like Instacart really popular and they might be here to stay. https://www.cnbc.com/2020/05/13/coronavirus-making-grocerydelivery-services-like-instacart-popular.html. Accessed 24 June 2020

Cramer-Flood, E.: Global Ecommerce 2020. Ecommerce Decelerates amid Global Retail Contraction but Remains a Bright Spot. Executive Summary. https://www.emarketer.com/content/global-ecommerce-2020. Accessed 23 June 2020

Dannenberg, P., Fuchs, M., Riedler, T.: Digital transition by COVID-19 pandemic? The German food online retail. J. Econ. Soc. Geogr. **111**, 543–560 (2020)

Deb, P., Furceri, D., Ostry, J.D., Tawk, N.: The Economic Effects of COVID-19 Containment Measures. CEPR COVID Economics Vetted and Real-Time Papers 24, Centre for Economic Policy Research, London (2020)

Di Renzo, L., Gualtieri, P., Pivari, F., et al.: Eating habits and lifestyle changes during COVID-19 lockdown: an Italian survey. J. Transl. Med. **18**, 229 (2020)

Dirgantari, P.D., Hidayat, Y.N., Mahphoth, M.H., Nugraheni, R.: Level of use and satisfaction of e-commerce customers in Covid-19 pandemic period: an Information System Success Model (ISSM) approach. Indonesian J. Sci. Technol. **5**(2), 261–270 (2020)

Ecommerce Europe: Euro Commerce. Europe 2020 Ecommerce Region Report. Lauch Webinar, 08 July 2020

Ecommercenews: Ecommerce in Europe. https://ecommercenews.eu/ecommerce-in-europe/. Accessed 12 Jan 2021

Eriksson, N., Stenius, M.: Changing behavioral patterns in grocery shopping in the initial phase of the Covid-19 crisis—a qualitative study of news articles. Open J. Bus. Manag. **8**, 1946–1961 (2020)

Eurostat: Digital economy and society statistics. https://ec.europa.eu/eurostat/web/digital-economy-and-society/data/main-tables. Accessed 27 May 2020

Galanakis, C.M.: The food systems in the era of the coronavirus (COVID-19) pandemic crisis. Foods **9**, 523 (2020)

GlobalWebIndex: Coronavirus Research. Multi-market research wave 5, July 2020. www.globalwebindex.com. Accessed 15 July 2020

Grashuis, J., Skevas, T., Segovia, M.S.: Grocery shopping preferences during the COVID-19 pandemic. Sustainability **12**, 5369 (2020)

Haleem, A., Javaid, M., Vaishya, R.: Effects of COVID 19 pandemic in daily life. Curr. Med. Res. Pract. (2020). https://doi.org/10.1016/j.cmrp.2020.03.011

Hall, M.C., Prayag, G., Fieger, P., Dyason, D.: Beyond panic buying: consumption displacement and COVID-19. J. Serv. Manag. (2020). https://doi.org/10.1108/JOSM-05-2020-0151

Hobbs, J.E.: Food supply chains during the COVID-19 pandemic. Can. J. Agric. Econ. (2020). https://doi.org/10.1111/cjag.12237

Hoffman, G.J., Webster, N.J., Bynum, J.P.W.: A framework for aging-friendly services and supports in the age of COVID-19. J. Aging Soc. Policy **32**(4–5), 450–459 (2020). https://doi.org/10.1080/08959420.2020.1771239

Huang, Z., Benyoucef, M.: From e-commerce to social commerce: a close look at design features. Electron. Commer. Res. Appl. **12**, 246–259 (2013)

Hudson, D.S.: Chapter 5 The economic, social and environmental impacts of COVID 19. In: Hudson, D.S. (ed.) Goodfellow Publishers, Oxford (2020). https://doi.org/10.23912/978191 1635703-4429

IMF: World Economic Outlook Update. World Economic Outlook, April 2021. Managing Divergent Recoveries (imf.org). Accessed 2 May 2021

Internet World Stats: www.internetworldstats.com. Accessed 13 July 2020

Jayaram, D., Manrai, A.K., Manrai, L.A.: Effective use of marketing technology in Eastern Europe: web analytics, social media, customer analytics, digital campaigns and mobile applications. J. Econ. Financ. Adm. Sci. **20**, 118–132 (2015)

Lehmann, R.J., Reiche, R., Schiefer, G.: Future internet and the agri-food sector: state of the art in literature research. Comput. Electron. Agric. **89**, 158–174 (2012)

Miller, G., McDonald, A.: Online art therapy during the COVID-19 pandemic. Int. J. Art Ther. **25**(4), 159–160 (2020). https://doi.org/10.1080/17454832.2020.1846383

Nicola, M., et al.: The socio-economic implications of the coronavirus pandemic (COVID-19): a review. Int. J. Surg. **78**, 185–193 (2020)

Noguti, V., Waller, D.S.: Motivations to use social media: effects on the perceived informativeness, entertainment, and intrusiveness of paid mobile advertising. J. Mark. Manag. **36**(15–16), 1527–1555 (2020). https://doi.org/10.1080/0267257X.2020.1799062

Osservatori Digital Innovation.: L'ecommerce B2C. Motore di crescita e di innovazione del retail! OsservatorioeCommerce B2C, October 2019. www.osservatori.net. Accessed 13 May 2020

Osservatori Digital Innovation: Le infografiche 2020. I numeri chiave dell'innovazione digitale, January 2021. www.osservatori.net. Accessed 12 Jan 2021

Pejic-Bach, M.: Editorial: electronic commerce in the time of Covid-19 - perspectives and challenges. J. Theor. Appl. Electron. Commer. Res. **16**, 1 (2020)

Phillipson, J., et al.: The COVID-19 pandemic and its implications for rural economies. Sustainability **12**, 3973 (2020)

RetailX: Italy 2020: Ecommerce Country Report. https://retailx.net/countries/. Accessed 25 July 2020

Roggeveen, A.L., Sethuraman, R.: How the COVID-19 pandemic may change the world of retailing. J. Retail. **96**, 169–171 (2020)

Santhiya, R., Elavarasan, R.: A study on perceptive the Covid-19 effect on online shopping behavior. Acad. Leadersh. - Online J. **21**(7), 66–71 (2020)

Scuderi, A., Sturiale, L.: Evaluation of social media strategy for green urban planning in metropolitan cities. In: Calabrò, F., Della Spina, L., Bevilacqua, C. (eds.) ISHT 2018. SIST, vol. 100, pp. 76–84. Springer, Cham (2019). https://doi.org/10.1007/978-3-319-92099-3_10

Scuderi, A., Sturiale, L.: Analysis of social network applications for organic agri-food products. Int. J. Agric. Resour. Gov. Ecol. **10**, 176–189 (2014)

Scuderi, A., Bellia, C., Foti, V.T., Sturiale, L., Timpanaro, G.: Evaluation of consumers' purchaising process for organic food products. AIMS Agric. Food **4**(2), 251–265 (2019a)

Scuderi, A., Foti, V.T., Timpanaro, G.: The supply chain value of POD and PGI food products through the application of blockchain. Qual. - Access Success **20**, 580–587 (2019b)

Scuderi, A., La Via, G., Timpanaro, G., Sturiale, L.: Current and future opportunities of digital transformation in the agrifood sector. CEUR Workshop Proc. **2761**, 317–326 (2020). Proceedings of HAICTA 2020,9th International Conference on Information and Communication Technologies in Agriculture, Food & Environment, September 24–27, 2020, Thessaloniki, Greece

Scuderi, A., Sturiale, L., Timpanaro, G.: The importance of "origin" for on line agri-food products.Qual. – Access Success **16**, 260–266 (2015)

Scuderi, A., Sturiale, L., Timpanaro, G.: Economic evaluation of innovative investments in agri-food chain. Qual.-Access Success **19**, 482–488 (2018)

Signorelli, C., Capolong, S., D'Alessandro, D., Fara, G.M.: The homes in the COVID-19. How their use and values are changing. Acta Bio Med [Internet], 20 July 2020

Sturiale, L., Scuderi, A.: Information and Communication Technology (ICT) and adjustment of the marketing strategy in the agrifood system in Italy. CEUR Workshop Proc. **1152**(2011), 77–87 (2011)

Sturiale, L., Scuderi, A.: The digital economy: new e-business strategies for food Italian system. Int. J. Electron. Market. Retail. **7**, 287–310 (2016)

Sturiale, L., Scuderi, A.: The marketplaces and the integration between physic and virtual in the business models of fruit and vegetables e-commerce. In: CEUR Workshop Proceedings, HAICTA 2017 (2017)

Sturiale, L., Timpanaro, G., La Via, G.: The on-line sales models of fresh fruit and vegetables: opportunities and limits for typical Italian products. Qual.-Access Success **18**(S2), 444–451 (2017)

Thakur, A.C., Diwekar, B., Reddy, J., Gajjala, N.: A study of the online impulse buying behaviour during COVID-19 pandemic. Int. J. Res. Eng. Sci. Manag. 3(9), 86–90 (2020). https://doi.org/10.47607/ijresm.2020.294

Veeragandham, N., Patnaik, R., Tiruvaipati, Guruprasad, M.: Consumer buying behaviour towards e- during COVID-19. Int. J. Res. Eng. Sci. Manag. 3(9), 78–82 (2020). https://doi.org/10.47607/ijresm.2020.292

Verdouw, C.N., Vucic, N., Sundmaeker, H., Beulens, A.J.M.: Future internet as a driver for virtualization, connectivity and intelligence of agri-food supply chain networks. J. Food Syst. Dyn. 4, 261–272 (2013)

von Abrams, K.: What Does Online Grocery Shopping Look Like in Western Europe During COVID-19?. www.eMarketers.com. Accessed 14 Apr 2020

Weiss, C.R., Wittkopp, A.: Retailer concentration and product innovation in food manufacturing. Eur. Rev. Agric. Econ. 32(2), 219–244 (2005)

World Health Organization (WHO): Regional Office for Europe (2020). Coronavirus disease (COVID-19) pandemic. https://www.euro.who.int/en/health-topics/health-emergencies/coronavirus-covid-19/novel-coronavirus-2019-ncov. Accessed 7 July 2020

Zambrano-Monserrate, M.A., Ruano, M.A., Sanchez-Alcaldec, L.: Indirect effects of COVID-19 on the environment. Sci. Total Environ. 728, 138813 (2020)

Metropolitan Food Systems at the Test of Covid-19: Changes, Reactions, Opportunities Between Food Insecurity and New Needs

Valentina Cattivelli[1,2]([✉])

[1] Comune di Cremona, Cremona, Italy
valentina.cattivelli@uninettunouniversity.net,
valentina.cattivelli13@gmail.com
[2] Uninettuno University, Rome, Italy

Abstract. This paper briefly presents the effects of the COVID-19 pandemic and its related fallout on food metropolitan systems in the context of three main forces: (i) the interconnection among operators along the food chain; (ii) the increasing inequalities among population; (iii) the relations between the metropolitan areas and the near rural ones. The pandemic has made interconnections among the operators along the food chain as well as among territories along the urban-rural linkages vulnerable and has imposed the search of new solutions for building more resilient food systems. It also has increased inequalities in food access that could be alleviated through local government interventions and volunteering food provisioning initiatives at community level.

Keywords: Metropolitan food systems · Covid-19 · Food security · Resilience · New trends · Social innovation

1 Introduction

The Covid-19 pandemic has aggravated the risk of severe and/or extreme food insecurity. Its diffusion has affected the metropolitan food system severely and has induced potentially long-lasting transformations in terms of (i) interconnections among operators along the food chain; (ii) increase in inequalities among population; (iii) effects on linkages among the metropolitan areas and the near rural ones.

The present paper represents an attempt to contextualise the limited evidence on short-term trends related to these transformations in a longer-term dynamic perspective. To do this, the impacts on food metropolitan systems are analysed through the lens of the recent literature focusing specifically on these areas (2020–2022) and are generalized to those in developed countries. Although there is relevant uncertainty as regards this kind of analysis, the knowledge of these effects may educate policymakers to manage measures to prevent temporary difficulties in food access from becoming systematic and long-terms problems and prepare to manage long-lasting transformations (e.g., those induced by climate change).

© The Author(s), under exclusive license to Springer Nature Switzerland AG 2022
F. Calabrò et al. (Eds.): NMP 2022, LNNS 482, pp. 1415–1424, 2022.
https://doi.org/10.1007/978-3-031-06825-6_137

The paper is structured as follows. The second paragraph lists the documented immediate effects of the pandemic on metropolitan food systems, whereas the following ones
provide insights into the interconnection among operators along the metropolitan food
chain, increasing inequalities and the effects on urban-rural linkages. The last paragraph
discusses some possible implications and future developments.

2   Exploring the Effects of the Pandemic on Metropolitan Food Systems

In normal times, food systems contemporarily face three challenges: providing food security and nutrition for a growing and more demanding population, ensuring livelihoods
for people working along the food chain, and achieving environmental sustainability
moving from a linear to a circular economy. These challenges should not only be seen
through the lens of rural development. Conversely, there is an increasing recognition
that should be faced in terms of urban development. Urbanization is rapidly occurring
throughout the world, with burgeoning city populations and expanding peri-urban areas.
Globally, as of 2015, about 80% of rural residents live within three hours from an urban
center – a 57-percent increase since 2000 (IFRI 2019). Urban population currently consumes 70% of the world's food (Cabannes and Marocchino 2018) and live, in large part,
on calories produced in other areas and countries (The Economist 2020). By 2050, it is
estimated that 80% of all food produced around the world will have been consumed.

Metropolitan food system results are influenced by several factors, such as demographic change, urbanization, climate change and resource scarcity (Béné 2020). These
are managed by local institutions, which are increasingly required to find solutions
occurring within and beyond their boundaries, in the global and national context, across
an urban-rural continuum. Their functioning depends on the existence of functioning
institutions, private actors, instruments, resources, data, stakeholder engagement, and
multilevel coordination (ibid.).

The COVID-19 pandemic has had a negative impact on urban and metropolitan areas:
more than 90% of the contagions has been found in these territories (WHO 2020, 2021).
Its diffusion has particularly exposed urban and metropolitan food systems to unexpected
and increased vulnerability posing several challenges for cities and local governments in
terms of food availability, accessibility, and affordability for urban population (Garnett
et al. 2020; FAO 2020). The pandemic has brought about a food crisis, not for the sheer
of food, but due to disruption in the supply chain and farm trade activities (Torero 2020)
as well as evident food injustices across these territories (Bellamy et al. 2021). Mobility
and food importation restrictions have had important effects on food production and
supply and implied the mobilization of enough resources to respond to the difficulties in
managing food flows at a global and local level, e.g., dominance of small set of retailers,
reliance on just-in-time supply chains and the dependency on imported food (Zurek
et al. 2022). Short-term food shortages were fueled by panic buying and restricted supply
owing to stringent lockdown measures (UN 2020), resulting in food wastage, and volatile
prices from demand-supply feedbacks (Harris 2020); job losses, reduced incomes and
rising food prices worse food insecurity, exacerbating pre-existing inequalities among
people (with serious repercussions for the most vulnerable groups, Naidu-Ghelani 2020)

and territories (exacerbated by an uneven access and distribution of food, Wilkinson et al. 2020).

Local institutions are required to find solutions within and beyond metropolitan boundaries. In addition, they should recognize that dealing with food issues helps them address other urban issues, providing different perspectives, entry points and policy options for action.

2.1 The Interconnection Among Operators Along the Food Chain

The spread of the pandemic has imposed to all operators along the food chain an answer to unexpected disruptive changes and threats from the external environment. The traditional just-in-time characteristics of the food supply chain (Lang, 2020) aiming at offering fresh products and reducing food waste in short time require effectiveness and strict connections of each actor along the chain. When several actors are impacted, then the long-lasting adverse effects affect the entire chain (Ali et al. 2022).

Adverse effects for producers consist in the interruption of the input supply chain and the reduction in labour availability. Processors (both small/ medium-size and large companies) agonise from the demand drop of certain products, and shift food suppliers to ensure food flows from manufacturers. Like producers, they are engaged in the preservation of the workforce's health and face the negative effects of profitability drop or temporarily interruption.

Along the chain, retailers manage the instability of food traded for an unknown number of coming days, and the possible increase in the price of products. These conditions reduce business transactions and opportunities for consumers to food access. Answers include diversification, connectivity, and business substitution (Sing et al. 2021). Food retailers changed their just-in-time inventory management strategies to be able to deal with uneven shortages throughout their largely national distribution networks. Finally, they also adopt protective strategies to maintain safe distances between customers (Shah 2020) and to answer the requests for aid from food banks (e.g., Opinko 2020).

Local open-air and informal markets are forced to close. Among those operating here, one can presume that higher level of self-efficacy, better access to information and stronger cooperation are essential for their survival (Elbeyoğlu and Sirkeci 2021).

Consumers were forced to access food locally due to mobility restrictions. A growing number has experimented food insecurity due to income drops, the increase in costs related to food purchase (costs of transport, etc.) or to the limited local living offer in a food desert municipality. As consequence, its purchasing power decreased a lot (O'Meara et al. 2022). Consumers have also changed their food behaviours (Hansen 2022). They have drawn more attention to food safety and therefore on the infrastructure and workforce responsible for supplying reliable and safe food locally and internationally (Civero et al. 2021). For fear or to reduce supermarkets visits, they have also bought more food than they actually need and have increased their purchases at online groceries (Nicola et al. 2020). They have increased volume and types of food purchased without the stores having to adjust their supply chain (Goddard 2020). Their consumer food routine is also interrupted due to the infrequency of bars, restaurants for lunch break, as well as for boredom in the house which led to high fat, proteins and carbohydrate consumption. Stress derived from quarantine has pushed them towards junk foods or sugars for feeling happier, and this has increased the risk of obesity or other health-related diseases (Muscogiuri et al. 2020). Contemporarily, it has pushed new consumers to alternative local food systems in search of fresher, healthy and safe food (Nemes et al. 2021).

In the light of such pressures, there is enough room for investigating how operators have responded positively. Rowan and Galanakis (2020) consider COVID-19 as a possibility for the agri-food sector to transform itself towards a greener and more sustainable direction, enhancing food quality, and shortening food supply. Giudice et al. (2020) show its resilience, especially at local level, through reducing food waste. FAO (2020) reports the importance of coordinating cross-sectoral national and local level plans to allow the appropriate resource allocation at a local level, where needs and connection with key stakeholders may be rapidly identified. This offers the opportunity to establish multi-stakeholder and multi-scalar (from local to national) food governance mechanisms involving various local actors (e.g., community associations, slum associations, the informal food sector) that could become essential driving forces in emergency situations. Another key element to face these pressures is the promotion of local production through urban and peri-urban agriculture, thus preserving agricultural land in urban and peri-urban areas. To present, diverse types of urban and peri-urban agriculture within 20 kms of cities account for 60% of all irrigated cropland in the world, supplying up to 90% of the vegetables consumed in many cities (FAO 2020). Lastly, to face these pressures, the improvement of traditional markets, informal food networks of retailers around alternative food networks and urban gardening would be positive (e.g., Cattivelli 2020a, 2020b, Lal 2020). The informal economy generated by these food retailers and urban gardens cultivation should be better analysed and integrated in the metropolitan food planning as it contributes to diversify food provisions and distribute power along food chain more equally.

2.2 Increasing Inequalities

Access to food is an evident sign of how well society distributes it wealth, and the governments and citizens' level of commitment to assure the right to food (Dreze and Sen 1990).

Due to the Covid-19 virus proliferation, food insecurity has raised as a consequence of billions of people living under temporary lockdown or in analogous situations (Galanakis 2020). As such, the identification of ways to secure this access has come into sharp focus. Based on Klassen and Murphy (2020), "the pandemic is another reminder that having enough food in a food system does not protect people from hunger". One non-health impact of the pandemic has been the access to food, tied to slashed incomes, tightened borders, rapid shifts in demand, and business and market closures. This has taken place without any significant change to the amount of food produced yet, although future supplies are likely to be affected. Others refer to the lack of food in shops, the increased extension of food desert and isolation (Loopstra 2020). Outdoor markets were closed during initial lockdowns, resulting in food-access-related constraints for mid- to low-income families who, countering the typical image of alternative food systems as elitist niches, often supply fresh food in these markets. One health impact reveals the interconnections between social inequalities, food insecurity and hill-health. Diet-related diseases are a risk factor for some symptoms of the virus. In England, 62% of hospital deaths were obese individuals (Williamson et al. 2020; Public Health England 2020) and were prevalent in the most socially and economically disadvantaged communities, including populated metropolitan areas.

Resolving food access related problems is imperative in metropolitan areas, essentially for two main reasons.

The first one is related to health. A lack of food is the origin of deficiencies in critical nutrients and calories necessary to fight the onset of disease. Conversely, the surplus of nutrient-deficient food or poor quality leading to an increased consumption of junk food also to fight stress and anxiety of the period can ignite health problems, such as diabetes, obesity and hypertension that may compromise immune systems.

The second one is related to compromised social justice. To contrast food access difficulties, local governments have allocated assistance programs to support individuals or associations for food purchases and income support rather than to enhance productivity for farms. Actions have also included the distribution of food excess to centres feeding vulnerable people and food banks (Harvey 2020). In case of difficulty, several volunteering food provisioning initiatives at community level have been taken (e.g., Cattivelli and Rusciano 2020; Tarra et al. 2021). Through the connection of people in need, volunteers, voluntary associations, and small producers locally, these initiatives have re-organized the food chain and the food aid systems. They supply fresh food from local farmers and producers and then they mobilize canteens, supermarkets, restaurants proved to be mutually beneficial, increase food quality and quantity and alleviate the economic situation of medium-sized farms experiencing difficulties. By being socially innovative, they propose organizational changes in food provision, thus reconfiguring existing social practices. By adopting a more systemic perspective centred around communities' power, they encourage a rethinking of how food systems could be restructured across urban-rural linkages (ibid.). Its adoption also contributes to the enhancement of outcome and social well-being as well as improve the cooperation between the population, local associations, and food operators (Turetta et al. 2021).

2.3   The Effects on Urban-Rural Linkages

The COVID-19 crisis outlines how much food systems are globalized. With limits to international trades, food provisioning is becoming increasingly difficult. The national borders closure is one of the most relevant causes of its vulnerability. As a consequence, decentralization towards more localized systems at metropolitan level is desired as it redesigns the food chain, reduces its environmental impact and enhances local and regional production by building stronger links between consumers, retailers, and food producers.

Urban and peri-urban agriculture could represent an important food supply channel during the lockdown, for two main reasons: a greater flexibility and agility in moving and in handling goods and the possibility of remunerating local farms, thus contributing to the resilience of local agri-food systems. Proximity to production areas and shorter supply chains make towns less vulnerable to disruptions on food distribution networks as they consequently become more resilient to shocks (FAO 2020). UN Habitat (2020) reveals that food systems in small villages were more resilient to such shocks compared to larger urban areas due to proximity to production areas and shorter supply chains. This suggests that programs put in place to food recover should ensure that the food-related urban-rural linkages remain uninterrupted in time of crises and promote shorter supply chains. Some peri-urban and urban farmers have been thriving thanks to the direct selling, with own deliveries and online orders, of vegetable and fruit boxes. While costs are similar with those provided by supermarket, they enable direct access without queuing (e.g., Pitt 2020). However, being practiced in small areas, their production is very limited and contributes little to food self-sufficiency.

The COVID-19 crisis points to the relatively weak role of the farmers, despite their importance in the provision of raw agricultural products. Shorter supply chain means supporting farmers, improving their finance, which in turn enhance access to food at fairer conditions. Furthermore, as farmers are stewards and caretakers of the countryside, they promote environment-friendly agroecological practices and protect biodiversity (Marchetti et al. 2020).

The pandemic has also contributed to the replication and upscaling of alternative food systems by prompting agri-food systems actors and stimulating the participation of citizens and farms (Nemes et al. 2021). Their contribution was possible to address major social issues, such as food security and solidarity. Thanks to access, these systems are more likely to be perceived as safe, high-quality and healthy by consumers, to be seen to increase their power along the food chain, but their environmental impact is rarely put into question (Tefft et al. 2021).

Shorter supply implies a reduction in packaging and food miles.

3 Some Possible Implications and Future Developments

The COVID-19 outbreak has highlighted the fragility of metropolitan food systems. Of course, their impacts affect each metropolitan area differently. However, they are not always dependent from the virus itself, but from system failures, such as the high dependency of food imports, the instabilities of the relations among operators along the chain and across urban and rural areas. Strategies to mitigate these effects enhance the restructuring of actual food models at a metropolitan level to improve their resilience and safety as well as shorten distances among operators and with rural areas. They suggest reconsidering these systems on the regional and national agenda, minimizing their dependencies and vulnerability, applying social-ecological resilience with a focus on urban food systems (as already suggested in the past by Hodbod and Eakin 2015). This also implies an improvement in the relations with near rural areas and food operators. Reducing distances increases rural farmers' and consumers' power towards a stronger food democracy (Carlson and Chappell 2015; Petetin 2020). Consumers feel they are in control of food provenance without international food flows, whereas rural farmers increase their business feeding directly urban population at lower costs. The increasing interest for alternative food systems positively offer different perspectives on how food could be produced and consumed, and stimulate the creation of food hubs and e-commerce platform specifically for their informal networks.

Local administrations are required to promote the reorganization of food distribution at an urban level, also thanks to the support of communities and non-governmental actors, as well as to take action to improve the food environment (FAO 2020).

In other terms, evidence suggests a model for multilevel food governance at metropolitan level, which should be characterized by what consumers want i.e., "better food, more information and choices, and preference for local action and personal involvement" (Hamilton 2011). This new way of governing food systems supports the challenge of meeting population behaviors and growth as being strictly related challenges, e.g., food injustice, long-term malnutrition, climate change and soil consumption. Its guidelines include the promotion of people-centered consumption, the enhancement local democracy through a redistribution power for resilient supply chain, and the right to grow sustainable healthy food.

While the food system is the entry point for analysis and strategies, solutions cannot solely rely on food systems reinforcing linkages with multiple aims, scales (productive and territorial) and actors. To better prepare for future crises, governments should take this opportunity to invest in structural changes to reduce persistent inequities in food access due to poverty, health outcomes, decent work, and overall wellbeing, especially for people at risk of social exclusion and living in food deserts. Strategies should strengthen their capacity to promote efficient food system and strict links with other local development plans, also thanks to information, catalogue success, good practices collection and sharing across sectoral offices. This should be the starting point for an integration of food issues into urban and metropolitan development, land-use or sector-specific plans, as well as for an aim alignment with broader territorial goals and private and civil society involvement.

Multistakeholder platforms therefore become an essential instrument to assure effective collaboration among several actors within food issues and place where formal, informal organizations and networks could dialogue.

Recovery from the pandemic cannot only have to do with restoring the status quo but implies promoting the revision of all aspects of holistic food systems; stakeholders' engagement as key actors across metropolitan and near rural areas, and consequently also agro-innovations in a co-design process should be promoted. It would be well worth providing a more detailed and precise cost analysis which includes all external effects over a longer period, as well as those related to sudden events. This would represent a good exercise to refine the inclusion into food prices of certain long-term costs which are not usually taken into consideration in the calculation of real food price (e.g., the effects of climate change).

Recovery represents an opportunity to experiment initiatives that can shorten and diversify food chain and increase their resilience. Public investments must focus on re-building systems to uphold human rights and better protect people living in precarity.

References

Ali, I., Arslan, A., Chowdhury, M., Khan, Z., Tarba, S.: Reimagining global food value chains through effective resilience to COVID-19 shocks and similar future events: a dynamic capability perspective. J. Bus. Res. **141**, 1–12 (2022)

Sanderson Bellamy, A., Furness, E., Nicol, P., Pitt, H., Taherzadeh, A.: Shaping more resilient and just food systems: lessons from the COVID-19 Pandemic. Ambio **50**(4), 782–793 (2021). https://doi.org/10.1007/s13280-021-01532-y

Béné, C.: Resilience of local food systems and links to food security–a review of some important concepts in the context of COVID-19 and other shocks. Food Secur. **12**, 1–18 (2020)

Cabannes, Y., Marocchino, C.: Integrating Food into Urban Planning. UCL Press, London (2018)

Carlson, J., Chappell, M.: Restrictions on Genetically Modified Organisms. Global Legal Research Center (2015)

Cattivelli, V.: The motivations of urban gardens in rural mountain areas: the case of South Tyrol. Sustainability **12**, 4304 (2020a)

Cattivelli, V.: The urban gardens in South Tyrol (IT): spatial distribution and some considerations about their role on mitigating the effects of ageing and urbanization. Reg. Stud. Reg. Sci. **7**(1), 206–209 (2020b). https://doi.org/10.1080/21681376.2020.1770121

Cattivelli, V., Rusciano, V.: Social innovation and food provisioning during COVID-19: the case of urban–rural initiatives in the province of Naples. Sustainability **12**(11), 4444 (2020)

Civero, G., Rusciano, V., Scarpato, D., Simeone, M.: Food: not only safety, but also sustainability: the emerging trend of new social consumers. Sustainability **13**(23), 12967 (2021)

Dreze, J., Sen, A.: Hunger and Public Action. Clarendon Press, London (1990)

Elbeyoğlu, K., Sirkeci, O.: The effects of COVID-19 pandemic on street economy: solutions and proposals. In: A New Social Street Economy: An Effect of The COVID-19 Pandemic. Emerald Publishing Limited (2021)

FAO: Urban Food Systems and COVID-19: The Role of Cities and Local Governments in Responding to the Emergency. FAO, Rome (2020)

Galanakis, C.: The food systems in the era of the coronavirus (COVID-19) pandemic crisis. Foods **9**(4), 523 (2020)

Garnett, P., Doherty, B., Heron, T.: Vulnerability of the United Kingdom's food supply chains exposed by Covid-19. Nat. Food **1**, 315–318 (2020)

Giudice, F., Caferra, R., Morone, P.: COVID-19, the food system and the circular economy: challenges and opportunities. Sustainability **12**(19), 7939 (2020)

Goddard, E.: The impact of COVID-19 on food retail and food service in Canada: preliminary assessment. Can. J. Agr. Econ. **68**, 1–5 (2020)

Hamilton, N.: Moving toward food democracy: better food, new farmers, and the myth of feeding the world. Drake J. Agric. Law **117**, 118 (2011)

Hansen, T.: Consumer food sustainability before and during the Covid-19 crisis: a quantitative content analysis and food policy implications. Food Policy **107**, 102207 (2022)

Harris, J.: Diets in a time of coronavirus: Don't let vegetables fall off the plate. FPRI Covid-19 blogs (2020)

Harvey, A.: Canadian food banks struggle to stay open, just as demand for their services skyrockets (2020). https://www.theglobeandmail.com/canada/toronto/article-canadian-food-banks-struggle-to-stay-open-just-as-dem

Hodbod, J., Eakin, H.: Adapting a social-ecological resilience framework for food systems. J. Environ. Stud. Sci. **5**(3), 474–484 (2015). https://doi.org/10.1007/s13412-015-0280-6

IFRI: 2019 Global Food Policy Report. IFRI, Washington DC (2019)

Klassen, S., Murphy, S.: Equity as both a means and an end: lessons for resilient food systems from COVID-19. World Dev. **136**, 105104 (2020). https://doi.org/10.1016/j.worlddev.2020.105104

Lal, R.: Home gardening and urban agriculture for advancing food and nutritional security in response to the COVID-19 pandemic. Food Secur. **12**(4), 871–876 (2020). https://doi.org/10.1007/s12571-020-01058-3

Lang, T.: Feeding Britain: Our Food Problems and How to Fix Them. Penguin, London (2020)

Loopstra, R.: Vulnerability to food insecurity since the COVID-19 lockdown. The Food Foundation (2020)

Marchetti, L., Cattivelli, V., Cocozza, C., Salbitano, F., Marchetti, M.: Beyond sustainability in food systems: perspectives from agroecology and social innovation. Sustainability **12**(18), 7524 (2020)

Muscogiuri, G., Barrea, L., Savastano, S., Colao, A.: Nutritional recommendations for CoVID-19 quarantine. Eur. J. Clin. Nutr. **74**(6), 850–851 (2020)

Naidu-Ghelani, R.: Cost of living amid COVID-19 (2020). Ipsos, https://www.ipsos.com/en/cost-living-majority-say-cost-food-goods-and-services-have-increased-covid-19-began

Nemes, G., et al.: The impact of COVID-19 on alternative and local food systems and the potential for the sustainability transition: insights from 13 countries. Sustain. Prod. Cons. **28**, 591–599 (2021). https://doi.org/10.1016/j.spc.2021.06.022

Nicola, M., et al.: The socio-economic implications of the coronavirus pandemic (COVID-19): a review. Int. J. Surg. **78**, 185–193 (2020)

O'Meara, L., Turner, C., Coitinho, D., Oenema, S.: Consumer experiences of food environments during the Covid-19 pandemic: Global insights from a rapid online survey of individuals from 119 countries. Glob. Food Sec. **32**, 100594 (2022)

Opinko, D.: Save-On-Foods launches $1-million campaign to feed kids during COVID-19 pandemic (2020). https://lethbridgenewsnow.com/2020/04/02/save-on-foods-launches-1-million-campaign-to-feed-kids-during-covid-19-

Petetin, L.: The COVID-19 crisis: an opportunity to integrate food democracy into post-pandemic food systems. Eur. J. Risk Regul. **11**(2), 326–336 (2020)

Pitt, H.: C19 Horticulture Summit: Results from Edible Producers in Wales April 2020. Food Sense Wales (2020)

Public Health England: Disparities in the risk and outcomes of COVID-19. Public Health England, London (2020)

Rowan, N., Galanakis, C.: Unlocking challenges and opportunities presented by COVID-19 pandemic for cross-cutting disruption in agri-food and green deal innovations: Quo Vadis? Sci. Total Environ. **748**, 141362 (2020). https://doi.org/10.1016/j.scitotenv.2020.141362

Shah, M.: Grocery chains install checkout shields, raise wages in response to coronavirus pandemic (2020). Global News Retrieved from https://globalnews.ca/news/6713693/grocery-stores-coronavirus-canada/

Singh, S., Kumar, R., Panchal, R., Tiwari, M.: Impact of COVID-19 on logistics systems and disruptions in food supply chain. Int. J. Prod. Res. **59**(7), 1993–2008 (2021)

Tarra, S., Mazzocchi, G., Marino, D.: Food system resilience during COVID-19 pandemic: the case of roman solidarity purchasing groups. Agriculture **11**(2), 156 (2021). https://doi.org/10.3390/agriculture11020156

Tefft, J., Jonasova, M., Zhang, F., Zhang, Y.: Urban Food Systems Governance: Current Context and Future Opportunities. FAO and the World Bank (2021)

The Economist. The global food supply chain is passing a severe test (2020). https://www.economist.com: The Economist. How to Feed the Planet—The Global Food Supply Chain Is Passing a Severe Test. 9 May 2020. Available online: https://www.economist.com/leaders/2020/05/09/the-global-food-supply-chain-is-passing-a-severe-test, Accessed 2 July 2020

Torero, M.: How to stop a looming food crisis. Foreign Policy News (2020). https://foreignpolicy.com/2020/04/14/how-to-stop-food-crisis-coronavirus-economy-trade

Turetta, A., Bonatti, M., Sieber, S.: Resilience of community food systems (CFS): co-design as a long-term viable pathway to face crises in neglected territories? Foods **2021**(10), 521 (2021)

UN. COVID-19: The global food supply chain is holding up, for now. UN, New York (2020)

UN Habitat. Issue Brief: COVID-19 through the Lens of Urban Rural Linkages-Guiding Principles and Framework for Action (URL-GP). UN Habitat, New York (2020)

WHO. WHO Coronavirus (COVID-19) Dashboard. WHO (2020–2021)

Wilkinson, A., Tulloch, O., Ripoll, S.: Key considerations: COVID-19 in informal urban settlements. Social Science in Humanitarian Action (2020)

Williamson, E., et al.: OpenSAFELY: factors associated with COVID-19-related hospital death in the linked electronic health records of 17 million adult NHS patients. MedRxiv (2020)

Zurek, M., Ingram, J., Alexander, P., Barnes, A.P.: Food system resilience: concepts, issues, and challenges. Ann. Rev. Environ. Res. (2022)

Designing Food Landscape in the 15-Min Post-covid City. Imagining a New Scenario for Low-Density Spaces in Metropolitan Areas

Catherine Dezio[1]([✉]) and Mario Paris[2] [iD]

[1] DASTU, Politecnico di Milano, Milan, Italy
catherine.dezio@polimi.it
[2] DISA, Università degli Studi di Bergamo, Bergamo, Italy
mario.paris@unibg.it

Abstract. After the covid crisis, the territorial governance will be a testing ground for innovative approaches that involve spatial configurations and the sustainability of current living practices. In this contribution, we reflect on the potential role of food landscapes as catalysts for the regeneration of marginal urban systems. Therefore, we would explore if – and how -projects in rural contexts can become a tool for applying the 15-min city model in low-density spaces. To address these objectives, we considered three case studies of the Milanese periurban context: a project coordinated by the Municipality of Milan and conducted by a large partnership financed by European funds (OpenAgri) and two experimental simulations produced together with young colleagues for the Area of *Parco Agricolo Sud*. The method adopted for this investigation is the perspective of *"research-by-design,"* a type of academic investigation of the architecture field, through which design is explored as a method of inquiry. In all these experiences, the production, the transformation, and the consumption of local food became the engine of a larger spatial transformation, which involves social, economic, and cultural aspects. Starting from the lessons learned from these projects, the paper concludes by outlining scenarios for the role of the territorial project in a post-covid future.

Keywords: Urban food policies · Urban planning · Research-by-design · Post-covid model

1 Introduction

In a recent study, E. Casti, F. Adobati and I. Negri [1] pointed out that the current health crisis has led to an acknowledgment of social and territorial vulnerabilities. The answer to the pandemic involves a deep reconsideration of current models of living and urge to rethink them considering the emerged weaknesses.

Therefore, in their analysis, the researchers involved pandemic, environmental, and social issues as three inseparable clusters that we need to address to rework territorial governance.

F. Calabrò et al. (Eds.): NMP 2022, LNNS 482, pp. 1425–1436, 2022.
https://doi.org/10.1007/978-3-031-06825-6_138

Indeed, efforts produced by scholars, activists and the policy communities should support a progressive transition toward renewed ways of living territory, which may be seen as instruments for fulfilling the needs of humans, ensuring their quality of life, and achieving well-being and happiness.

The idea that it is possible to overcome a crisis starting from a physical territorial project is increasingly gaining ground [2], and the post-covid governance is a testing ground for innovative approaches to the territorial governance and for the re-imagination of current living practices. This opportunity is a push to review risks and occasions of metropolitan urban areas, even adapting the discussed 15-min city model to their scattered patterns [3]. This effort is original, and we identified a lack in the academic studies in the application of this model in the low-density spaces. There are theoretical and design references that propose interpretative keys for the landscape on board, legitimizing a vision of the city based on ecosystems and attributing value to the diversity of material and immaterial aspects [4–14]. Over the last two decades, and in Italy, thanks to pilot initiatives, we experimented with a cultural shift in urban planning that rediscovered the rural space as an opportunity [15]. Incrementally, policy communities recognized the role of agriculture and the rural landscape, not only as a sectoral domain, but also as an activating resource of possible virtuous triggers.

However, what escapes the policies is the role of the design project of rural territories, which, today, are the backcloth for a range of crucial planning issues [16]. In this contribution, we reflect on the role of food landscapes as catalysts for the regeneration of marginal urban systems, starting from the opportunities that the design project can become an application and integrated tool for various distant policies. The multiple objectives of this paper are: 1. to explore the tool of agriculture as a connection platform for material and immaterial aspects; 2. identify new lines of research for the regeneration of periurban territories and for the relationship between the city and its marginal context; 3. investigate the possibilities of the landscape project to become an instrument of dialogue between different skills; 4. outline restart scenarios for the post-covid era. To address the following objectives, we considered three case studies of the Milanese periurban context: one of these is a project coordinated by the Municipality of Milan and conducted by a large partnership financed by European funds (OpenAgri) [17]; while the other two are team-built projects as an experimental simulation. The method adopted for this investigation is the perspective of *"research-by-design"* [18], a type of academic investigation of the architecture field (in particular, landscape architecture and urban design), that uses design as a method of inquiry [19, 20]. This method is based on a succession of deconstructive operations of composition and decomposition. Through this process, it is possible to explore the relationship between the parts and the whole [21]. With this type of methodology, we will explore the projects chosen as case studies by reading them in their procedural dimension. After a brief reflection on state of the art in this field (1.), we analyze three experimental proposals (2.) and discuss their applications, expecting to recognize different convergences and criticalities in the development of these projects related to periurban regeneration (3.). In the conclusions (4.) we will indicate a set of potential developments of this research, useful for launching new insights on the landscape design project in the agro-food systems. We will focus on the interaction between the scenarios proposed within the projects, analyzing their spatial features and constraints and the current expectations of local governments. Starting from the analyses developed in the three case studies, we will highlight the difficulties encountered by planners designing projects on rural landscapes and food production as resources for local, sustainable regeneration. While recognizing the limitations of research based on academic simulations and not on real projects, we consider these experiments as pilot experiences and testing grounds. We evaluated the current boundaries and limits of the implementations of the actions. With them, we tried to offer solutions on an objective basis, addressing all the impediments and contradictions that mark the translation of the principles into actual actions.

2   Food Landscape Design in the Post-covid Periurban Areas

A recent Italian research claims that in the last nine years, 35,000 neighborhood shops have closed and, only in 2019 with a closing rate of 14 units per day [20]. Other data, presented by Istat [22] pointed out that almost 450.000 companies show fragilities and among them, 292.000 show a profile defined *"static and in crisis"* because they lack any strategy for answering the emergent issues due to the pandemic [21]. Half of them work on the services sector, and especially those are hotels and restaurants/cafe, tourist facilities, retail, personals services, and entertainment and they involve almost 1 mln of workers. A progressive loss of presence of local business could ends with a parallel loss of complexity and vibrancy of current urban settlements, especially in those areas where these companies function as a catalyst of social encounters and provide a set of basic services for the local communities. At the same time, this impoverishment could also affect the agricultural production chain, the quality of landscapes, the productiveness of local spinnerets and the food accessibility for inhabitants of certain urban spaces, especially when those are peripheral and not easily accessible.

2.1   The 15-Min City

The dramatic impact of the pandemic affects the social and economic reality of Italy, and in this paper, we will focus on the potential reaction to this condition for the most fragile spaces of the country: the low-density periurban ones. We choose those areas because after the lockdown due to the covid19, cities have experimented with a set of proposals that deal with two intrinsically connected components of a territorial project: public space and essential services [2]. These two elements support the *15-min city model* that gained ground in recent years. But often, the narrative – or the rhetoric – about these experimentations involves pilot experiences in urban, over-populated neighborhoods, marked by a high density of services and a high intensity of social relationships. Indeed, the opportunities and the risks connected to the application of this model in periurban environments must be overhauled and there is a lack of studies on this field.

The *'15-min city',* developed by C. Moreno in 2016 to support the new urban plan of Paris, is an iteration of the idea of *'neighborhood units'* developed by American planner C. Perry during the 1920s. The proposal improves the quality of life of people living in dense metropolitan areas, with a green, inclusive, resilient, and proximity urban organization based on the neighborhood scale and walkability [23]. In its TED Talking dedicated to the issue, the architect explained that in the 15-min city, dwellers can access their essential living needs that includes the workplaces and all the other facilities that they will need in their everyday life. The transition to this model is a challenge that concerns all urban stakeholders and requires everyone to reconsider their own role within urban life to open to alternative horizons, conveying high quality societal life where the inhabitants have an easy access to the essential services (grocery shops, healthcare and personal services, social and cultural equipment, public spaces). During and after the pandemic, policymakers assume the 15-min city as the model to develop interventions for supporting a shift in settlement paradigms and included this imaginary in the medium-term vision that orients the territorial governance. In this light, it is interesting that the new territorial approach to think, manage and provide services takes in account the design of public spaces and facilities as a driver for planning actions, not only for reacting the pandemic situation, but to rethink the urban/territorial pattern in a more inclusive and sustainable way. Therefore, over the last few years, it took place a lively plurality of experiences based on'*the back to the place*' through the reconstruction of a '*collective awareness*' of the territorial heritage value [24] in central or semi-central urban areas.

2.2 The Paradigm of Periurban Spaces Evolves Thanks to the 15-Min City Concept

Often, cities use periurban spaces as edges, backyards, or buffer areas between urban cores and the countryside. These spaces are fragmented, chaotic, and often they belong to different municipalities, which provide different visions and rules for their territorial governance.

As recognized by Vindigni et. Al. [25], periurban landscapes can be seen as a spatial and figurative broken network, characterized by fragmentation, lack of urban and ecologic continuity, hybrid (not rural, not urban) environments, thus lacking identity.

At the same time, they are the spaces that could "*contribute to urban sustainability transitions, offering an opportunity to invest in environmental safety, the improvement of ecological performance and the urban environment, also in terms of the quality of public space and in the economic dimension*" [25, p. 2].

The quality of these spaces is the critical aspect that affect the quality of life of their inhabitants. This impact appeared with a mayor impact during the pandemic, due to the lack of existing services, and the difficulties to the access to facilities – that normally are in the urban cores. The movement restrictions and the contagion risk connected to the public transport affected local dwellers. At the same time, rethinking the paradigms in a post-covid era, these are the spaces in which settlement, production and movement dynamics could evolve faster and that could respond better to these proposals, due to their dynamism.

In previous studies we conducted [26], we noticed that a reduced number of experiences deal with the implementation of the 15-min city in low-density spaces. We think that this concept could be relevant for the implementation of the quality and the sustainability of low-density urban fringe, even if we must re-decline and adapt the idea to the contexts.

There is room for innovate in the empowerment of local actors in the management of environmental and agricultural resources, envisioning a new scenario in which rural spaces will gain a new role due to the proximity to potential consumers and food production areas and – partially – contrast the progressive loss of economic activities that affects these spaces. At the same time, periurban agricultural landscape can become a driver for new ecosystem opportunities in those spaces, and the (re)activation of potential processes in those contexts. Innovations can emerge thanks to the role of nature and wilderness, the quality of the environment and the presence of economic opportunities connected to the tourism, the leisure and the demand for close enjoyable spaces, the qualitative productions and the new interest raised up during the pandemic for low-density environments as an alternative to the dense, polluted, overpopulated urban settlements.

These potential spaces ask for initiatives that enhance local productions [27] to strengthen local agri-food chains and, at the same time, reconquer rural areas as a sharable space able to provide alternative to the overcrowded and privatized urban spaces.

The 15'-min city model could be implemented in the low-density space. C. Moreno suggested to decline a 30'-min city in peripheral areas where the slow movements by bike and pedestrian could be integrated by the private traffic. There is a need to reflect about the provision of services and facilities in those environments and their accessibility more than the speed of the travel to reach them.

This declination imposes to the activists (designers, urban planners, agronomists, among others) and policy makers involve in the process a set of skills and abilities that nowadays are not included in their consolidate profile.

Nowadays, current pressures impose a need to generate profoundly innovative solutions, thanks to the presence of a plurality of actors capable of promoting ideas and practices.

There is a need for innovative forms of 'local project' [24], capable of combining tangible and intangible aspects in a single view: open spaces' project; support of small farms; healthy, accessible, and affordable food; new job places, also inclusive of vulnerable people [28]; citizen services planning; protection of cultural heritage and local biodiversity; and much more.

The need for public spaces that are well-designed, safe, and easily accessible is increasing. Often this is linked to choices concerning the economic dimension (tourism, food, lifestyle). All this meets a model of multifunctional agriculture which, by reviewing the role of the rural landscape, can meet multiple advantages [29]: (i) it can support the income of quality farms; (ii) can answer the demand for public open spaces; (iii) can provide essential new services, such as the provision of healthy and affordable food, to communities in marginal territories; (iv) it can shift the gravity's center of the people's flow to marginal and underused areas; (v) can generate off-season tourism; (vi) it can generate opportunities for protection and care of areas that are generally not very populated, and much more.

Often, current practices lack of creativity and knowledge to propose innovative solutions. We had the opportunity to evaluate our alternative approach in three different projects based on this vision and we will show in the next section we discuss the potential impacts of these proposals.

3   Three Milan Experiences: Re-imagining Planning Implication for Periurban Metropolitan Contexts.

The research aims to evaluate how concepts and tools connected to urban and territorial regeneration can produce opportunities in low-density metropolitan areas, exceeding the over-simplistic rhetoric often created by the policy communities in this domain. Therefore, in different cases, we reflected on what approaches we should improve in the post-covid metropolitan areas.

3.1   Data Collection/Why These Cases?

We proposed three case studies: one based on a European project coordinated by Municipality of Milan; two based on design simulations produced by young colleagues (Miriam Colombo for the case of Abbiategrasso [30] and Gloria Signorini for Rozzano [31]) in the context of Master's Architecture of AUIC School of Politecnico di Milano (in particular, the Master's in Architecture and Urban Design and the Master's in Landscape Architecture).

According to Table 1, the elaboration process of the three cases took a variable time between and collected a set of research questions linked to the different Milan contexts where they take place that came across the issues we proposed in the present article.

Table 1. Title, context, and elaboration process of the proposals.

Title of the project	Context	Elaboration process
What future for Abbiategrasso? Rural networks, tourism, and landscapes for a territorial regeneration project	Abbiategrasso (Milan, Italy)	May, 2020–Dec, 2020
The role of landscape design in promoting agroecological systems in periurban farms. A vision for Cascina Sant'Alberto	Cascina Sant'Alberto, Rozzano (Milan, Italy)	Jun 2021–Dec 2021
OpenAgri, within the "Urban Innovative Action" European Program	Cascina Nosedo, Milan (Milan, Italy)	Apr 2017–Apr 2020

Elaboration by the authors (2021)

They focus on low-density, periurban spaces marked by a different degree of fragility. The proposals activate regeneration processes working on the local scale and considering agricultural production as an asset, expecting metropolitan impacts. Therefore, all of them frame the analysis and the design schemes within original and non-institutional spatial dimensions. These dimensions exceed the municipal and administrative scales, and they are helpful to understand and describe the ongoing phenomena in a more sensitive and detailed way. In this light, the exploration moves from the identification of three conditions.

- Spatial constraints, linked to the geographical context and the morphological conditions of the area (topography, water, vegetation, land-uses).
- Socio-economic constraints, in which candidates recognize the role of existing spinnerets, sectoral logic that influenced the evolution of the space and its current territorialities, where the food production has a crucial role in the local economy and welfare.
- Relational constraints. They explored existing or potential relationships (productive systems, settlement systems), interactions, and networks (public and open spaces, slow mobility infrastructures, ecological corridors) among the elements recognized on the space at different scales.

All these analyses imposed new attention to the open spaces as a value. In our approach, the open and green spaces network is the underused or underestimated factor involved in the proposal as a catalyst for reactivation.

3.2 Approach to the Case Studies

All the proposals pointed out the opportunity to support the multifunctional transition [32, 33]. The rural landscapes shift from an exclusive agricultural production to a complex and overlapped mix of uses. As pointed out by Frank and Hibbard [34]: *"Multifunctionality does not entail the abandonment of agriculture and natural resource extraction, but rather a condition in which rural landscapes and their communities concurrently serve production, consumption, and protection functions"*. We agree to this approach, and we tried to transmit them into the four presented focuses. In addition, the authors proposed a set of tasks in both hardware and software domains to make this transition operational.

We used the same approach to propose a transversal reading of the three projects (Table 2). This exploration is the basis for design the integration of food production with other functions in a process where the values of rural-urban interfaces are the catalyst for a more considerable number of activities. It can generate work opportunities for the local population and attractiveness for Milanese citizens, looking for ecosystem services such as recreational open spaces and food at km0.

The three cases can generate many suggestions in terms of good practices for the implementation further projects at local scale. There are five lessons learned that may have an interesting replicability character for different contexts: i) links with existing corridors and rural axes; ii) recovery of buildings; iii) slow routes connected with the city center; iv) design of spaces open to the public; v) sale of locally produced food. In

Table 2. A comparison between the case studies

	Cascina Nosedo	Cascina Sant'Alberto	Abbiategrasso
Location	South-east of Milan (Parco Agricolo Sud)	South of Milan (Parco Agricolo Sud)	West Milan (Parco Agricolo Sud)
Area	34 ha	130 ha	4.778 ha

(continued)

Table 2. (*continued*)

	Cascina Nosedo	Cascina Sant'Alberto	Abbiategrasso
Ownership	Public (Comune di Milano)	Private: Brioschi Group	Public (Comune di Abbiategrasso (MI)
Proposal	European Project (UIA – OpenAgri)	Landscape architecture design simulation	Urban design simulation
Status	Consolidate settlement (Chiaravalle abbey and the network of ancient farmhouses) Historical rural settlement abandoned and degraded	Rural area partly reclaimed in 1999 marked by the presence of pathways, even their quality and the equipment are poor	Fragmented agricultural pattern The urban edge fades into the rural space without a designed transition
Condition to deal with	Reclamation of rural spaces and a farmhouse (*cascina*)	Regeneration/reactivation of an abandoned area including a disused historic farmhouse	Reuse of a brownfield (ex Iar-Siltal)
Design actions	Implementation of an Agricultural Park, representative of a multifunctional agro-ecosystem Activation of the existing farmhouse as an innovative agricultural hub, and the landscape design project includes slow routes, green and blue infrastructures, and productive fields The fields cultivate innovative agriculture products, proposed by young non-farmers	Reinforcement of the renaturation and ecological reconnection fostering existing corridors Implementation of two areas for experimenting with agroecology: the Food Research Hub and Agri-market and the Cascina Sant'Alberto, an Educational Hub with adjoining farmhouse. Other micro-functions (wetland park, birdwatching tower educational paths) integrate the functional program	Recovering the continuity of ecologic connections between the urban center and the outskirts Reconnection of the area to existing canals and consolidated trails. Definition of a green link that embraces the existing urban core and the rural land. The transformed area became an interface between the rural and urban spaces. Food production/transformation services, education buildings (high schools, services.) reinforce the rural vocation of the space

this sense, the reference to the 15-min city model defined by Moreno is significant, even designer should adapt to the conditions and constraints of rural areas. In fact, the 15-min city model is made up of relationships, based on services, spaces and connections. In this sense, the agri-environmental system can respond to all three components: supply, production spaces that become collective, and connections that become paths. Therefore, thanks to the attention paid to the slow-mobility and the accessibility to the facilities for agrarian functions, can enrich the rural space as economic and social ecosystem.

4 Conclusions

The three experiences have been configured as experimental laboratories and re-search opportunities on issues little addressed in Italy, both from the point of view of disciplinary approach to the periurban agri-environmental landscape, and as regards the implementation of 15'-min cities in the low-density, marginal areas.

In the territories of the fragmentation and of the (il)logic accumulation [35], the agri-environmental system is an opportunity to exceed the endless city, by virtue of its multiple contents and its natural predisposition as uninterrupted modular and connective surface. This approach, based on physical elements of the local scale, attempts a crucial step towards a multiscale dialectic between problems and potentials (as a representation of the conditions of ambiguity of the margin).

By adopting a different point of view, which can be a driving force for the induction of new life cycles for marginal territories, we support analytical and operational strategies. This investigating spirit, which uses mechanisms of disassembly and reassembly, seeks an internal logic to the disorder that can be justified and recognized only by referring to ecological processes. These are hypotheses and operational reflections capable of driving an approach to the territorial system that uses the field of Landscape Ecology as an analytical and propositional framework.

Approaching the marginal agri-environmental territories, therefore, will require an ecological thought that does not depend on quantitative or qualitative tools or on the scale of the project, but pertains to the level of disciplinary sharing of an idea.

This paper made it possible to elaborate multiple considerations about the potential role of an innovative approach in the urban planning field. From these thoughts, we frame a sort of domain for our action as planners, activists, and educators. Finally, we collected a set of three "directions" listed above, that we will explore in the future, which are also open to a set of insights.

The first direction concerns the relationship between agroecosystems and an antifragility and resilient development model. As also well stated by the 17 Goals of the United Nations, thinking about sustainable cities also means thinking about their communities and the socio-economic and cultural dynamics that concern them. It means *"interpreting the present and thinking about the future, reasoning in systems"* [36]. The agroecosystem is a complex dimension made up of interdependent elements that affect the economy and the territories' wellbeing and communities. The potential for sustainable development models capable of responding to unforeseen and devastating dynamics, such as the covid pandemic, is indisputable. Still, we need to work on it, producing a solid disciplinary approach that in the planning domain is just in an initial stage.

The second direction concerns the possible and perhaps necessary interactions between policies and projects relating to 15'-min cities and agroecosystems. In this light, activated/reactivated rural spaces became key and accessible facilities for low-density spaces. Thanks to their reticular distribution over the agricultural land and the different networks that they generate, it could be possible to re-think the pattern of equipment and public spaces and for the "edge city" [37] and its food accessibility. This aspect will provide a range of new opportunities also for those rural spaces and their inhabitants that are not involved in the agricultural activities. This point is fundamental, because among the rural spaces, the agricultural spaces show a certain tension and dynamism meanwhile, an apparently hopeless inertia affected the other spaces. Within the case studies, we proposed agricultural productions and local productive spaces as a real alternative to the consolidate production chains, and they also represent a booster for the local economy and labor market, to be implemented and that could support a different relationship between customers and producers in metropolitan spaces. At the same time, they could stimulate new forms of living and opportunities for the local scale where inhabitants could experiment new forms of rooting to those spaces.

The third direction concerns the potential of "research-by-design" method, which sees in the analysis of the physical space an opportunity to understand the coevolutionary processes behind socio-economic and political-cultural phenomena. We will keep experimenting with our approach in this light, applying the lessons learned on other cases at different scales and discussing their adaptation to different conditions. In this light, we will produce a continuous learning process that will enhance our ability to interact with rural environment, its actors, and its approaches.

The exploration of these three directions will support the production of a needed, alternative approach for spatial governance of marginal contexts. This spatial governance will be effective if oriented to real opportunities for these spaces, connected to food supply chains, to slow tourism and environmental quality. Taking these factors as reliable opportunities, planners can exceed the limitations of current practices. In the paper, we tried to show how rural areas would feed the post covid 15'-min city model, in an integrated and multidimensional way. They deal with local constraints and features connected to the geo-morphology, the history, the economy and productive processes, and the associated settlement development. In parallel, we need to establish a situated and integrated governance [38] for these spaces. On it, planners and designers play with sectoral logic and approaches, proposing convergent visions that take advantage of specialized expertise.

References

1. Casti, E., Adobati, F., Negri, I.: Mapping the epidemic: a Systemic Geography of Covid-19 in Italy. Elsevier, Amsterdam (2021)
2. Balducci, A.: I territori fragili di fronte al covid. Scienze del Territorio, Special Issue, Abitare il territorio al tempo del Covid, 169–176 (2020)
3. Glaeser, E.: The 15-minute city is a dead end—cities must be places of opportunity for everyone (2021). https://blogs.lse.ac.uk/covid19/2021/05/28/the-15-minute-city-is-a-dead-end-cities-must-be-places-of-opportunity-for-everyone/, Accessed 04 Jan 2022
4. Baumann, Z.: Nascono sul confine le nuove identità. Corriere della Sera, May 24 (2009)
5. Berger, A.: America's urbanized environments have over the past century evolved from dense, vertical, and architecturally dominated places to the horizontal opposite. Drosscape: wasting land in urban America. Princeton Architectural Press, New York (2006)
6. Clément, G.: Manifesto del terzo paesaggio. Quodlibet, Macerata (2005)
7. Corboz, A.: Le territoire comme palimpseste. Diogene, 121, gen-mar (1983)
8. De Matteis, G.: La città dappertutto. AA.VV. Geometria e natura, atti di convegno nazionale ANCSA, Bergamo (2007)

9. Donadieu, P.: Campagne urbane: una nuova proposta di paesaggio della città. Donzelli, Roma (1998)
10. Indovina, F.: Dalla città diffusa all'arcipelago metropolitano. FrancoAngeli, Milano (2009)
11. Ingersoll, R.: Sprawltown: cercando la città in periferia. Meltemi, Roma (2004)
12. Secchi, B.: Prima lezione di urbanistica. Laterza, Roma-Bari (2000)
13. Turri, R.: Megalopoli padana. Einaudi, Torino (2000)
14. Viganò, P.: La città elementare. Skira, Milano (1999)
15. Scott, A.: Rediscovering the rural–urban fringe: a hybrid opportunity space for rural planning. In: Scott, M., Gallent, N., Gkartizios, M. (eds.) The Routledge Companion to Rural Planning, pp. 469–484. Routledge, Abingdon (2019)
16. Frank, K., Hibbard, M.: Rural planning in the 21st century: context-appropriate practices in a connected world. J. Plan. Educ. Res. **37**(3), 299–308 (2017)
17. OpenAgri. https://open-agri.it/openagri/, Accessed 04 Jan 2022
18. Corner, J.: Terra Fluxus. In: Waldheim, C (ed.) The Landscape Urbanism Reader. Princeton Architectural Press, New York (2006)
19. Roggema, R.: Research-by-design: proposition for a methodological approach. Urban Sci. **1**, 2 (2017)
20. Frankel, L.; Racine, M.: The complex field of research: for design, through design, and about design. In: Proceedings of the design and complexity – DRS International Conference, Montreal, QC, Canada, 7–9 July 2010 (2010). https://dl.designresearchsociety.org/drs-confer ence-papers/drs2010/researchpapers/43, Accessed 15 Oct 2021
21. Confesercenti: Commercio: Confesercenti – SWG, continua la crisi dei negozi: un'attività su quattro si avvia a chiudere l'anno in perdita (2019). https://www.confesercenti.it/blog/ commercio-confesercenti-swg-continua-la-crisi-dei-negozi-unattiva-su-quattro-si-avvia-a-chiudere-lanno-in-perdita/, Accessed 04 Jan 2022
22. Istituto Nazionale di Statistica (Istat): I profili strategici e operativi delle imprese italiane nella crisi generata dal Covid-19 (2021). https://www.istat.it, last accessed 2022/01/04
23. C40: 15-minute cities: How to develop people-centred streets and mobility (2020). https:// www.c40knowledgehub.org/s/article/15-minute-cities-How-to-develop-people-centred-str eets-and-mobility?language=en_US, Accessed 06 Jan 2022
24. Magnaghi, A.: Il Progetto locale. Verso la coscienza di luogo. Bollati Berlinghieri, Torino (2010)
25. Vindigni, G., et al.: shedding light on peri-urban ecosystem services using automated content analysis. Sustainability **13**, 1–17 (2021)
26. Paris, M., Fang, Y.: Towards a new normality in via paolo sarpi (Milan, Italy)? social behaviors and spatial transitions during and after the lockdown. J. Human Behav. Social Environ. **31**(1–4), 305–324 (2021)
27. Morazzoni, M., Zavattieri, G.G.: Tutela attiva e sistemi agroalimentari nelle Aree Interne Italiane. Geography Notebooks, vol. 1 (2018)
28. INEA: Il capitale umano in agricoltura (2013)
29. Dezio, C.: Ripartire dalle risorse. Patrimonio rurale come Capitale Territoriale Antifragilità. Valori e Valutazioni, vol. 24 (2020)
30. Colombo M.: Quale future per Abbiategrasso? Reti rurali, turismo e paesaggi per un progetto di rigenerazione territoriale. Master's thesis in Architecture and Urban Design, Polytechnic of Milan (2020)
31. Signorini G.: The role of landscape design in promoting agroecological systems in periurban farms. A vision for Cascina Sant'Alberto. Master's thesis in Architecture and Urban Design, Polytechnic of Milan (2021)
32. Holmes, J.: Impulses towards a multifunctional transition in rural Australia: gaps in the research agenda. J. Rural. Stud. **22**(2), 142–160 (2006)

33. McCarthy, J.: Scale, sovereignty, and strategy in environmental governance. Antipode **37**(4), 731–753 (2005)
34. Frank, K., Hibbard, M.: Production, consumption, and protection. the multifunctional transition in rural planning. In: Scott, M., Gallent, N., Gkartizios, M. (eds.) The Routledge Companion to Rural Planning, pp. 469–484. Routledge, Abingdon (2019)
35. Walker, R.: Edgy Cities, Technoblurbs and Simulcrumbs: Depthless Utopias and Dystopias on the Suburban Fringe. University of California, Los Angeles (1994)
36. Meadows, D.: Thinking in Systems: A Primer by Donella H. Meadows. Chelsea Green Publishing, Chelsea (2008)
37. Garreau, J.: Edge City: Life on the New Frontier. Anchor Books, New York (1991)
38. Albrechts, L., Balducci, A., Hillier, J. (eds.): Situated Practices of Strategic Planning: An International Perspective. Routledge, Abingdon (2017)

Turning 'Food to Be Wasted' into Food Security and Multi-ethnic Integration:

The Example of *Refoodgees* in the New Esquiline Market

Alessandra Narciso(✉) ⓘ

Roma Tre University, Rome, Italy
alessandra.narciso@uniroma3.it

Abstract. Feeding the city is one of the most important challenges for modern production systems. Urban food networks are associated with food safety, the quality and healthiness of products, efficient use of resources of consumption, and the sustainability of the urban model in general.

According to the latest policy reports, the COVID-19 pandemic has negatively impacted the purchasing power of already marginalized consumers (e.g., immigrants, students, and the elderly), and their nutritional security has further deteriorated as a direct result. This paper presents the initial findings of a study on the role of voluntary associations in providing alternative food networks. The partnership of one of them, "Refoodgees," with one of the oldest food markets in Rome, "New Esquiline Market," is described as an example. This sort of partnership can offer resilient, alternative food networks better able to meet the needs of the urban population, by redesigning the foodscapes of an urban district in a time of emergency—through a combination of environmental approaches and a new way of "revalorizing" products—while filling the gaps of urban food policies.

Keywords: Alternative food networks · COVID-19 food security · Markets · Rome

1 Introduction

This article discusses an alternative food network in Rome at the New Esquiline Market, created by the voluntary association *Roma Salva Cibo – Refoodgees* (hereinafter *Refoodgees*), which operates in successful partnership with the cooperative that manages the market. *Refoodgees*' initiative is an example of a circular economy developed for social purposes. Modeled on the example of *Eco dalla Città* in Turin (EcodalleCittà n.d.), *Refoodgees* was born in 2017, and within two years had become a key stakeholder in food distribution to marginalized and vulnerable populations (Refoodgees n.d.). Originally founded to recycle perishable food, *Refoodgees* has understood the potential of its endeavor for contributing to the food security of many vulnerable people, even during the time of COVID-19. While fighting against food loss, they have also been supporting food security in the most ethnically diverse neighborhood of central Rome.

© The Author(s), under exclusive license to Springer Nature Switzerland AG 2022
F. Calabrò et al. (Eds.): NMP 2022, LNNS 482, pp. 1437–1447, 2022.
https://doi.org/10.1007/978-3-031-06825-6_139

This article demonstrates that community networks in Rome during the pandemic went beyond the scope of recycling perishable food that most likely would have been wasted otherwise: they satisfied the need of the most vulnerable categories among the Roman people and, through a participatory approach, created a valuable sense of community during the pandemic. The article also highlights the role and importance of markets as symbols of Roman social and cultural cohesion. The paper is built on the thesis that voluntary low-scale activities can better react to various aspects of urban life—including food security during emergencies—even though they were originally born to find solutions to the problem of wasted food. Furthermore, markets can be conceived of as model spaces in which urban populations meet, not only to find economic solutions for purchasing their food, but also to meet their needs for personal interaction and social cohesion.

The article is divided into five sections: an introduction, with a methodological subsection; Sect. 2 "Food security in Rome during Covid-19 pandemic"; Sect. 3 "The role of markets in food security during emergencies: the case of New Esquiline Market Markets and Refoodgees"; Sect. 4 "Voluntary organizations have more flexibility to respond to food security in times of emergencies"; a conclusion.

1.1 Methodology

The methodology used is an interdisciplinary analysis of policy papers, scientific articles, and ten one-on-one interviews with coordinators of the organization *Refoodgees*, volunteers, and simple customers, the director of the Esquiline's market cooperative. The interviews' main questions were about: a) group motivations, b) origins of the collaborations, c) obstacles and challenges, d) suggestions on how to better improve the proposed cooperation.

The theory of "care economy" (Tronto 2013) is applied to analyze the *Refoodgees* model of intervention. Although applied to diverse contexts, it explains why caring actions undertaken to maintain a peaceful "world" can help find positive solutions to the most critical problems. Further, the circular economy—with the adjustment of social focus (Cattivelli 2015; Moriggi et al. 2020)—applied to the market structure can provide various opportunities for social change and societal innovation.

2 Food Security in Rome During COVID-19 Pandemic

In recent years, the importance of food systems in an urban environment has become crucial, particularly to the improvement of the 'quality of life'. In fact, in many countries around the world, the topic has attracted the attention of researchers as it can affect the local economy, the environment, public health, and the quality of neighborhoods (Pothukuchi *et al.* 1999). The issue of food security is particularly felt in the city of Rome, where there is a need to satisfy a population of 2,763,804 (Istat 2021), among which 8,000 are homeless people according to the last census (Istat 2015). The problem of food vulnerability has increased in Rome among the range of marginalized groups during the COVID-19 pandemic (Caritas 2020).

As a consequence of the recent pandemic, the categorization of those marginalized expanded to also include more low-income people, as well as those who had lost their jobs during the pandemic and thus becoming "new paupers" (Community of Sant'Egidio). Caritas reports (Caritas 2020) that 19.4% of the local population could not afford the overall high cost of housing (including electricity/heating, condominium costs on top of other taxes). The overall costs of housing contribute to the level of inequality in Italy, which is higher than in most advanced economies (OECD 2021a).

Many soup kitchens and food banks organized by religious groups, including Caritas, have always been very active in the municipality of Rome. These organizations usually provide food to the poorest and most marginalized populations, most of whom live on the streets of the capital city. However, the greatest challenges during the COVID-19 pandemic have been: i) the new categories of people in need ii) reaching out to people, particularly during freedom of movement restrictions that have periodically been adopted to combat the transmission of the virus.

Volunteering initiatives that emerged to guarantee food security for the most vulnerable population have gone beyond their initial scope: they set up, consciously or unconsciously, networks of associations and people, where "the societal challenges and social innovations somehow came together" (Cattivelli 2020). Nevertheless, many organizations, including *Refoodgees*, found obstacles to reaching out to people during the emergency, where restrictions of movement were imposed, and the market was required to obey strict time constraints. When interviewed, the organizations' leaders and the market's director made clear the necessity of creatively and strategically rethinking access to the market, and consequently the distribution of the leftover food, during the COVID-19 pandemic. Two main approaches were adopted: limiting the number of volunteers and introducing a stand inside the market, rather than outside the market where it is usually positioned, to better control and limit the flux of people at the entrance of the market.

Secondly, *Refoodgees* increased partnerships with other organizations active in door-to-door food distribution. Exploring alternative food systems is crucial to reaching out to more needy people. Yet, according to the interviews with the association's leaders, the volunteer food distributors have experienced difficulty in approaching people who were already living in fragile economic settings. COVID-19 increased these people's sense of not belonging and their struggle to cope with an emergency that further deteriorated their living standards. Those who have not been able to access food for economic reasons lie in stark contrast to the "smart" society of consumers able to place orders for groceries through the internet. A general increase in isolation accompanied the pandemic, in Rome as in other major Italian cities. However, it was extremely evident in Rome, where a feeling of lost dignity was felt even more by low-income groups, as the cost of living is higher than in many other Italian cities.

The number of people who have been affected in their food choices (limited or altered eating habits) is still not well estimated. Not all organizations reported on how much food they distributed in the street at soup kitchens or door-to-door. However, retired people in Rome with low or minimum pensions have been severely affected by food insecurity during COVID-19 (Roma Capitale 2021). This is not surprising, since Italy has one of the oldest populations in Europe (OECD 2017b; OECD 2021c), a vulnerability that increases even more in times of disaster (WHO 2015). As emerged from the interviews, the elderly hid even more from society. This is one of the main obstacles when trying to analyze the impact of the pandemic on the aging population of Rome and their loss of purchasing power (Report Roma Capitale 2021). Overall, the negative effects of the COVID-19 pandemic have also impacted students, immigrants, and other vulnerable categories, exposing them to food insecurity. Restriction of movement has caused not only job loss, but also difficulties in food distribution.

The networks of restaurants, shops, supermarkets, and hotels in Rome that provide food to soup kitchens have greatly diminished. During the COVID-19 pandemic, charity associations and municipal authorities have had to find new ways to reach the populations to which they typically provide assistance, as well as to the "new invisible" to address the expansion of the food security emergency.

3   The Role of Markets in Food Security During Emergencies: The Case of New Esquiline Market and *Refoodgees*

Markets are a symbol of Rome and are present in each Roman district. They are usually cheaper than supermarkets, as they offer customers the possibility of choosing from among many items to purchase and this competition leads to lower prices. Roman markets can help us to better understand the development of society, its cultural changes, and social challenges. They are also a traditional expression of Mediterranean culture and a staple on the Italian foodscape:

> "Food, because of its cultural and historical place in the Mediterranean tradition, has a significant role in configuring the areas where exchanges take place, which are, therefore, specific places for meeting and forming relationships within the public spaces of a city. Markets are such places, together with the function they bring to city squares and streets. Food consumption, in the shape, for example, of street food, backed by better weather patterns, is closely linked to Mediterranean urban behaviour. Together with the network of personal space, these aspects help to define complex city geographies of food and food flows, functional in terms of relationships and space, that can re-write not only the form of space but also the types of behaviour that such space generates" (Cavallo et al. 2014, p. 217).

Markets are also a growing model of tourist attraction, a symbol of city life that has transformed over the years and has incorporated food into urban planning (Cabannes 2018). Markets have always contributed to the transformation of the urban landscape, to the introduction of new products, and to orient consumers' tastes. Not only have the social aspects changed but also the squares, the city streets, to create these spaces of not only food but also a meeting of cultures.

Markets in larger cities also represent the connection between city and countryside, between center and periphery, as well as between city dweller and farmer (Civero 2021). One key to their success, particularly in times of crises, is that they facilitate the meeting of consumers and (local) producers – and more urban policies should be promoted to develop this connection, including between institutional and local stakeholders (Pothukuchi *et al.* 1999). The recently approved Italian Circular Economy Strategy refers to a new approach to a circular economy, where the downstream and upstream movements should also be integrated (MITE 2021). Retail markets are the perfect synthesis of this approach that tries to integrate social and infrastructural aspects of life in an urban area.

Markets are the lens through which it is possible to analyze changes in many aspects of urban life: consumers' behavior, the demographic composition of customers, ethnic background, food security/safety, and food waste implemented policies. Markets can demonstrate a transition process in society that can be measured in a context often made up of products from different cultures, or simple mixtures between local farmers and foreign sellers, or between foreign producers and Italian sellers.

The consumer who chooses the market chooses in some way to adhere to this cultural change that combines tradition with innovation in perfect harmony. In the case of Rome, markets have historically symbolized the city's evolution and changes in social planning and structure. Rome's authorities view the city's markets as an attractive socio-economic and cultural element that can adapt to consumers' demands, which is why the mayor has launched a program to renovate the city's markets (Roma Capitale n.d.).

Dating to the Umbertino period, the New Esquiline Market was established in 1913 by then-Mayor Ernesto Nathan in Piazza Vittorio, where almost anything could be found at a reasonable price, and its screaming vendors have long been a fixture in the folklore of Rome, inspiring numerous artistic and literary works (Gadda 1957; Jaran 2014). It is no longer an open-air market, as it was moved from Piazza Vittorio Emanuele II in 2001 (Mercatidiroma 2021) to its present location in the former Sani police station—still in the Esquilino district in the center of Rome. Its present location has several entrances on roads not far from *Termini*, the city's main train station. Administered by a cooperative, it is open every day from 7 A.M. to 2 P.M. Divided into two sections: one fruit/vegetable, spices, meat/fish; and one with clothes and shoes, it covers 10,000 square meters which make it one of the biggest markets in Italy. The faces of the merchants and consumers have changed; they are no longer just Romans or Italians behind the counters, but often the first generation of immigrants who come from different countries. This is evident in kosher and halal food, and also oriental vegetables, and local regional specialties. Despite its position and historical background, the market is not yet organized to host street food as are other markets in Europe and it is also not yet a well-known tourist attraction.

Markets also serve as a tool for addressing new dynamics in a growing urban society with increased immigrants and elders (Lands Onlus 2019). The New Esquiline Market, better known as the ex-Piazza Vittorio market, distinguishes itself as an extraordinary melting pot of diversity that has become known beyond Italy's borders (Donadio 2013). In recent years, the current market administration is trying to implement alternative and circular food networks including producers, traders, customers, municipal and city authorities, and associations like *Refoodgees* in what could be described as a small-scale circular economy. Other partnerships like the one with the environmental association *RespiroVerde*, are helping to educate the public on important values like the respect for urban green practices and spaces.

Refoodgees is an example of a volunteer-based initiative that started with the aim of recycling food waste for food security and also fosters a microcosm of social inclusion for immigrants, refugees, women, unaccompanied minors and the elderly. The organization was established in 2017 by a group of volunteers, who also had experience working with immigrants' associations. Currently, *Refoodgees* has 20 volunteers who dedicate themselves, every Saturday after the closing market time, to the collection and distribution of unsold fruit and vegetables of the New Esquiline Market in Rome. After having established a relationship with the director of the cooperative managing the market, the association was authorized to have a stand in front of the market. Since then, all customers – all people but mostly the most vulnerable – can receive fresh food that otherwise would become rubbish every Saturday at the market's closing time. In the last few years, *Refoodgees* has collected thousands of tons of unsold food for Roman citizens, particularly for the elderly, women, immigrants, and refugees.

Charity is not, however, *Refoodgees'* primary focus. Rather than merely collecting food products for distribution to those in need, eliminating food waste and advancing food security—while embracing a vision for a more inclusive society—is the goal. They even organize culinary, community-based, and cultural events with the intent of spreading cultural diversity and creating human connections.

Increasingly during the COVID-19 pandemic, *Refoodgees* has built relationships with other public and private stakeholders who are centered around the New Esquiline Market. Despite the limited response from the city major authorities, as reported by *Refoodgees'* founders and the market's cooperative when interviewed, *Refoodgees* has established a good working relationship with Municipality I, as well as with Roman social organizations such as the *House of Social Rights*, *Binario 95*, *Piano Terra*, *A Buon Diritto*, to meet the needs of volunteers and beneficiaries who had been asking *Refoodgees* where they could find social and health services.

In fact, by starting with addressing food waste, *Refoodgees* tackled food security needs and expanded to cover many relevant necessities of the local and vulnerable population. Born with a focus on food recycling, *Refoodgees* is now a reference point for the neighborhood and works closely with medical and social services.

The partnership between markets and local/neighbor organizations can improve urban food policies and represent a valid tool to change the approach to food security/food waste in urban environments. Further, by 'revalorizing food products,' it contributes to socio-cultural cohesion (Parsons et al. 2021).

In Rome, there are also many local markets integrated into the urban, social and temporal space and increasingly inclusive of the immigrant experience, some with the addition of gourmet cooking. Nevertheless, the experience of the New Esquiline Market is unique for the collaboration between the market authorities and the voluntary networks of people that believe in contributing to a better world.

4   Voluntary Organizations Have More Flexibility to Respond to Food Security in Times of Emergencies

The current food regimes are regulated by the "empires of food" (Van Der Ploeg 2009), where the culture of food is mostly represented by supermarket chains. Although food markets do represent higher risks for food safety (as recently shown by the COVID-19 pandemic, which probably originated in the Wuhan market), they are a great link with an "unplasticized" face-to-face approach to food, and strengthen the consumer-seller relationship. While cities like Turin and Milan were already more advanced on food security planning (Calori et al. 2017), such as developed in the *Milan Urban Food Policy Pact* (Mufpp) (Dansero et al. 2016), Rome had and still has a long way to go to adequately address these policies. These new rules of food governance have introduced "new values" that bypass the time of emergency and that are different from those "capitalistic approaches" that currently prevail (Orlando 2018).

COVID-19 has presented a great challenge in many aspects of urban life in Rome and other Italian cities, but it also has brought new opportunities. During the pandemic, the EU *Green Deal* has been approved, and people have started to understand even more how the world is interconnected and that no economic development is possible without caring for the environment and its people.

An illustrative example is the practice of turning some "food to be wasted" into recyclable food to satisfy the food security needs of the most vulnerable when is still edible. The Esquiline Market is a circular economy coupled with the model of the economy of care – the link between the seller and consumer has no filters nor mediation. *Refoodgees* distributes still edible food at the closing time of the market. The alliance with the cooperative in charge of the market agreed to promote this initiative and offer its assistance. *Refoodgees* can recycle from 600 to 1,200 kg. of food every Saturday in only two hours. The food consists only of fruit and vegetables, is not sold outside the market area and conforms to regional and city regulations regarding hygienic and sanitary standards. The food is still edible and, not being packaged food, shelf life is not an issue. It is up to the volunteers to assess whether the fruit or vegetable can still be sold or not. They decide quickly and dismiss what they believe cannot be sold any longer. The food that has already passed quality control is sold, and the volunteers handle the food with care, wearing gloves and following the principle that they only distribute food that they would purchase for themselves.

The consumers are separated from the area where fruit and vegetables are kept (Art. 39, T.U.C. 2019). Most of the people queuing at *Refoodgees* are those who regularly buy products at the market "discounted" from their trusted sellers and then also wait for additional, no-cost supply after the market closes on Saturday, when *Refoodgees* starts to collect the leftover food to be distributed. In recent years, academics and experts have been talking about the relevant role played by cities in global food security, expressing a new vision of foodscapes. Previously, cities were not important players for food security because food production and food-waste processing took place outside the cities, while citizens responsible for food distribution and consumption were not relevant in the food supply chain. According to Fattibene (2018), the link between food security, political stability, and environmental crisis is especially relevant in the Mediterranean region. The increasing speed of the urbanization process in Mediterranean countries requires urban public stakeholders who understand and meet consumers' food needs. According to scholars Morgan and Sonnino, this "new food equation" is crucial in answering the global challenge of sustainable food security (Sonnino et al. 2014).

Particularly in Italy between 2000 and 2010, Rome's public policies focused heavily on food security and food-waste processing policies. In that decade, Rome successfully led food security and sustainable food system policies, especially in launching a school meals system (Sonnino *et al.* 2009) incorporating environmental and economic sustainability with social benefits. By bringing together customers, producers, municipal institutions, and NGOs, Rome organized an innovative procurement market for catering firms based on seasonality, variety, locality, and nutritiousness (Morgan 2008). Regrettably, this inclusive system has been dismantled in recent years, and the "Roman model" has been negatively impacted by bribery, scandals, and corrupt management of public resources (Fattibene 2018).

The example of *Refoodgees* is only one among many associations active in Rome engaged in fighting the food insecurity emergency during COVID-19. The Agri-food Center of Rome (CAR - Centro Agroalimentare Roma) has recovered tons of unsold food products from firms and redistributed them through Roman social organizations, including the *Food Bank*, the *Community of Sant'Egidio*, the diocesan *Caritas* of Rome and Tivoli, and the *Christian Associations of Italian Workers*; or the emergence of digital platforms, such as *Bitgood* or *Too Good to Go*, which bring together companies, unsold food, and beneficiaries.

5   Conclusion

As has been seen, *Refoodgees* is an example of a spontaneous and bottom-up voluntary association which has partnered with the New Esquiline Market cooperative during the pandemic to develop a flexible and innovative way of delivering food to Rome's most vulnerable. Furthermore, it has become a reference point of social services and medical care for those who request them. This work highlights the relevance of partnership between local institutions (in this case, the market cooperative) and voluntary associations, not only to educate the community on the importance of recycling food that is still edible, but also in reconnecting people with food practices that necessarily involve social interaction. This article presents an initial stage of research on the impact of voluntary food networks born or developed in the city of Rome during the COVID-19 pandemic. The collection of data and experiences of alternative food networks will help us to better understand the resilience of alternative food networks in times of emergency.

In the future, further research could include the analysis of other associations and local authorities to map socially-conscious alternative food networks, in the city of Rome and elsewhere. The level of the research could be improved through specific questionnaires distributed among the beneficiaries of food networks and through more detailed interviews with the leaders and volunteers to better understand the mission/vision, challenges, and way forward. The research could also take as reference different periods during the pandemic for comparing the data within this given timeframe.

Multiple effects and lessons learned can derive from the application of this economic model in combination with the "economy of care": turning food waste into food security for the most marginalized categories of society (e.g., immigrants, young people, and an aging urban population); decrease in the transport and energy costs of unused food; benefits for diverse groups of people for increased quality of life.

At a policy level, meeting people's needs, particularly the most vulnerable, is crucial, and the lack of a comprehensive policy in the city of Rome could be improved by favoring more creative food-network initiatives. The flexible structure created by *Refoodgees* and the New Esquiline Market, with a minimal level of bureaucracy, still conforms to sanitary provisions and can be a key player in times of emergency in interacting with different urban actors, including civil society organizations and institutional-local actors. Particularly, in this case, the collaboration creates a platform that also stimulates dialogue and fosters multi-ethnic integration.

Indeed, the role of city markets should be reevaluated and more efficient tools of collaborations even among markets, city authorities, and associations should be sought to better coordinate the food response in times of emergencies. As discussed, the socio-cultural relevance of food markets, which are traditional means of food distribution in the cities of Mediterranean counties, are places where circular low-scale economies can be easily implemented. In the urban setting of Rome, policies, governance and regulations are not adequately facilitating these needed changes, and bureaucracy and lack of coordination are still obstacles to be overcome.

Acknowledgments. The author would like to thank the anonymous reviewers for their useful feedback. The author is also grateful to Ms. Viola De Andrade, Mr. Salvatore Perrotta for their kind collaboration, and Ms. Alessia Ferrucci for her assistance in collecting some data.

References

Cabannes, Y., Marocchino, C.: Integrating food into urban planning. UCL Press, London (2018)

Calori, A., Dansero, E., Pettenati, G., Toldo, A.: Urban food planning in Italian cities: a comparative analysis of the cases of Milan and Turin. Agroecol. Sustain. Food Syst. **41**(8), 1026–1046 (2017). https://doi.org/10.1080/21683565.2017.1340918

Caritas: Rapporto 2020. La povertà a Roma: un punto di vista Nessuno si salva da solo. http://www.caritasroma.it/wp-content/uploads, Accessed 22 Oct 2021

Cattivelli, V. Not just food: the, new, importance of urban gardens in the modern socio-economic system: a brief analysis. In World Food Trends and the Future of Food; LeEdizioni: Milano, Italy, pp. 53–74 (2015)

Cattivelli, V., Rusciano, V.: Social innovation and food provisioning during COVID-19: the case of urban–rural initiatives in theprovince of Naples. Sustainability **12**(11), 4444 (2020)

Cavallo, A., Di Donato, B., Guadagno, R., Marino, D.: The agriculture in Mediterranean urban phenomenon: Rome foodscapes as an infrastructure. In: Conference: 6th AESOP Sustainable Food Planning conference Leeuwarden, The Netherlands 5–7 November 2014, vol. Finding Spaces for Productive Cities (2014)

Civero, G., Rusciano, V., Scarpato, D., Simeone, M.: Food: not onlysafety, but also sustainability. The emerging trend of new socialconsumers. Sustainability **13**(23), 12967 (2021)

Community of Sant'Egidio: https://www.santegidio.org/pageID/30468/langID/it/itemID/43253/Crescono-i-poveri-per-la-pandemia-il-pranzo-di-Natale-con-Sant-Egidio.html, Accessed 26 Dec 2021

Dansero, E., Di Bella, E., Peano, C., Toldo, A.: Nutrire Torino Metropolitana: verso una politica alimentare locale. Agriregionieuropa **12**, 44 (2016)

Donadio, R.: Vibrant market is heart of multiethnic capital, Section A, p. 4, The New York Times (2013)

Eco dalla Città n.d.: https://www.ecodallecitta.it, Accessed 23 Nov 2021

Fattibene, D.: From farm to landfill: how rome tackles its food waste. In: Woertz, E (ed.) Wise Cities in the Mediterranean? Challenges of Urban Sustainability, CIDOB, Barcelona, pp. 187–197 (2018)

Gadda, C.E.: Quer pasticciaccio brutto de via Merulana, Garzanti (1957)

Istat: Bilancio demografico anno 2021 (dati provvisori). https://demo.istat.it/, Accessed 28 Feb 2021

Istat: Le persone senza dimora - Homeless People in Italy (2015). http://www.fiopsd.org, Accessed 2 Mar 2022

Jaran, M.: Due gialli a Roma tra via Merulana e Piazza Vittorio: Gadda e Lakhous a confronto Quaderni d'italianistica, vol. XXXV(2), 193–213 (2014). https://mercatidiroma.com/nuovo-mercato-esquilino-ex-piazza-vittorio/esquilino, Accessed 11 Oct 2021

MITE Strategia Nazionale per l'Economia Circolare (2021). https://www.mite.gov.it/sites/default/files/archivio/allegati/economia_circolare/SEC_30092021_1.pdf, Accessed 21 Feb 2022

Morgan K., Sonnino R.: The School Food Revolution: Public Food and the Challenge of Sustainable Development, pp. 67–80. Routledge, Abingdon (2008). https://doi.org/10.5860/choice.46-6442

Moriggi, A., Soini, K., Bock, B.B., Roep, D.: Caring in, for, and with nature: an integrative framework to understand green care practices. Sustainability 12(8), 3361 (2020). https://doi.org/10.3390/su12083361

OECD : Economic Policy Reforms 2021: Going for Growth OECD Publishing, Paris, (2021a). https://www.oecd.org/economy/growth/Italy-country-note-going-for-growth-2021.pdf, Accessed 21 Nov 2021

OECD : Preventing Ageing Unequally, OECD Publishing, Paris (2017b). https://doi.org/10.1787/9789264279087-en

OECD : OECD Economic Surveys: Italy 2021, OECD Publishing, Paris (2021c)

Orlando G.: Le reti alternative del cibo dopo la crisi. Teoria, ipotesi di lavoro e un caso studio esplorativo, Archivio Antropologico Mediterraneo, ANNO XXI 20(1) (2018)

Parsons, E., et al.: Who really cares? introducing an 'Ethics of Care' to debates on transformative value co-creation. J. Bus. Res. 122, 794–804 (2021). https://doi.org/10.1016/j.jbusres.2020.06.058

Pothukuchi, K., Kaufman, J.L.: Placing the food system on the urban agenda: the role of municipal institutions in food systems planning. Agric. Human Values 16, 214 (1999). https://doi.org/10.1023/A:1007558805953

Refoodgees n.d.: https://www.facebook.com/RomaSalvacibo/, Accessed 18 Oct 2021

Roma Capitale (a cura di) Luca Ferrigni. Le nuove Povertà nel territorio di Roma Capitale. Aracne Edizioni, Roma (2021). https://www.aracneeditrice.eu/free-download/9791259941626.pdf, Accessed 20 Nov 2021

Roma Capitale, Mercati Rionali. https://www.comune.roma.it/web/it/scheda-servizi.page?contentId=INF690431&stem=commercio_su_aree_pubbliche, Accessed 26 Feb 2022

Sonnino, R.: Quality food, public procurement, and sustainable development: the school meal revolution in Rome. Environ. Plann. A Econ. Space 41(2), 425–440 (2009). https://doi.org/10.1068/a40112

Sonnino, R., Faus, M.A., Maggio, A.: Sustainable food security: an emerging research and policy agenda. Int. J. Sociol. Agric. Food, 173–188 (2014). https://doi.org/10.48416/ijsaf.v21i1.161

Terra! Onlus, Lands Onlus. Una Food Policy per Roma. Perché alla Capitale d'Italia serve una Politica del Cibo (2019). https://www.associazioneterra.it/wp-content/uploads/2017/03/Una-Food-Policy-per-Roma.pdf, Accessed 10 Dec 2021

Testo Unico del Commercio (T.U.C.), L. 22, 6 November (2019). BUR: 90 https://www.comune.roma.it/web-resources/cms/documents/Legge_Regione_Lazio_num_22_del_6_novembre_2019.pdf, Accessed 3 Feb 2022

Tronto, J.C.: Caring Democracy: Markets, equality, and justice. New York University Press, New York (2013)

Van Der Ploeg, J.D.: The imperial conquest and reordering of the production, processing, distribution, and consumption of food: a theoretical contribution, Sociologia Urbana e Rurale June (2009). https://doi.org/10.3280/SUR2008-087003

WHO: World Report on Ageing and Health, ISBN 978 92 4 069479 8 (ePub) Geneva (2015)

Food (in)security in a Nordic Welfare State: The Impact of COVID-19 on the Activities of Oslo's Food Bank

Julia Szulecka[1]([✉]) [iD], Nhat Strøm-Andersen[1] [iD], and Paula Capodistrias[2] [iD]

[1] TIK Centre for Technology, Innovation and Culture, University of Oslo, Blindern, P.O. Box 1108, NO-0317 Oslo, Norway
julia.szulecka@tik.uio.no

[2] Matsentralen Norge (Food Banks Norway), Ole Deviks vei 20, 0666 Oslo, Norway

Abstract. This chapter explores the impact of COVID-19 on metropolitan food redistribution chains in a Nordic welfare state by analyzing the case of Oslo's Food Bank. It assesses how alternative redistribution schemes can support the welfare state or reach out to vulnerable groups without sufficient assistance. The study finds that Oslo's Food Bank - *Matsentralen Oslo* - experienced an increase in food demand and new types of recipients. However, despite the COVID-related social challenges, it was able to redistribute more food in 2020 and 2021 compared to 2019 thanks to the combination of new strategies, new administrative and control systems, new internal structures, as well as the establishment of new partnerships. The increased food demand and the availability of surplus food combined with the food bank's capacity to adapt to sudden changes suggest that the provision of food security requires the presence of dedicated institutions with the flexibility to make quick adjustments and connect supply and demand. The study shows that food banks complement the welfare state in bringing social services and additionally reduce food waste. They thus combine socio-economic and environmental services, which are beyond the scope of public financial transfers.

Keywords: Norway · Oslo · Food banks · Food redistribution · Food security · Innovation

1 Introduction

The COVID-19 pandemic has brought unprecedented challenges to the food system, revealing the vulnerability of global food security [1, 2]. Lessons from this period should thus be taken seriously, considering possible new waves of the pandemic and preparing for future shocks of a different nature, such as making agri-food systems more resilient in the face of climate change and other unexpected adversities [3]. In this chapter, we study the effects of the COVID-19 outbreak on the activities of food banks and food redistribution at the metropolitan level.

What are the impacts of a sudden shock of COVID's magnitude on alternative metropolitan food chains in the context of robust welfare protection networks? Is the welfare state alone able to mitigate adverse effects, or are specific redistribution instruments needed?

F. Calabrò et al. (Eds.): NMP 2022, LNNS 482, pp. 1448–1461, 2022.
https://doi.org/10.1007/978-3-031-06825-6_140

To shed light on these questions, we look at Norway, which is widely seen as an example of the most robust social democratic welfare state, characteristic of the Nordic region, and specifically at the country's capital city—Oslo. To grasp the scale and nature of challenges to food security, we offer a case study of a critical institution operating on the frontier of food provision, reducing poverty and food waste—food banks. Operating on the interface of civil society, i.e., charity organizations and NGOs, food industry actors, and public institutions, food banks contribute to maintaining food security and make significant efforts to reduce food waste.

This chapter analyzes the activities of Matsentralen Oslo—the city's only and Norway's largest food bank—during the pandemic from several angles. First, we look at the volume of food redistributed by Matsentralen Oslo in the city's metropolitan area in 2019, 2020, and 2021 to assess the impact of the pandemic. We further take the perspective of frontline organizations providing food aid and study the change in the recipient group composition. Furthermore, we explore the demand for food at frontline organizations that provide hunger relief with foodstuffs received from the food bank. Lastly, we study innovations that have been developed by the Matsentralen Oslo under the crisis to face the challenges brought upon by the pandemic.

Our former study found that locally nested innovation in food bank practices mattered greatly in meeting the new challenges posed by the COVID-19 pandemic and maintaining efficiency under stress [4]. However, local dynamics and the role of the city level in food redistribution have not been widely diagnosed and studied. We also know little about the interaction of food bank activity with mature welfare state institutions. It can be hypothesized that welfare states provide the necessary socio-economic resilience making food redistribution via food banks a low priority concern. This case study interrogates this claim using city-level data from a capital city of a Nordic welfare state.

2   The Impact of COVID-19 on European Food Banks

Recent statistics show that in 2020, an estimated 21.9% of the EU population or some 96.5 million people were at risk of poverty or social exclusion [5] – a number which likely increased during the economic crisis fueled by the pandemic with a long unseen hike in energy prices and growing inflation. This dire socio-economic situation makes mechanisms of hunger relief through food redistribution immediately relevant. The European Federation of Food Banks (FEBA) is an umbrella organization that represents 335 European food banks located in 24 full member countries and 5 associate countries. In 2020, FEBA full members recovered, collected, sorted, stored, and redistributed 860,000 tonnes of food (a 12% increase compared to pre-COVID levels), helping 12.8 million people in need, an increase of 34.7% [6].

FEBA further identified a 90% increase in food demand, with 70% of new beneficiaries being people who have lost their jobs due to COVID-19. Finally, there was an increase in food demand from particular groups, such as 60% from families with children, 40% increase from elderly people and 10% increase from students, who reached out to food banks in Europe [6].

To meet immediate needs and provide a concrete response to the emergency, FEBA launched an open call for solidarity and established a COVID-19 Social Emergency

Fund to secure the activity of its members and ensure no good food was going to waste, while helping charities and frontline organizations that provide food aid to people in need [7–9].

The main challenges for European food banks thus include the growing demand for food, the increase in the number of people in need, the unstable flow of food donations, the disruption in the logistics and transport processes, the lack of volunteers, and unexpected costs and drop in financial resources [10].

One might expect that the social restrictions during the pandemic would be a considerable obstacle for European food banks and adversely affect their operations, especially those with older volunteers and complex logistics. However, a recent study indicated that, although the pandemic brought unprecedented challenges to food banks' operations, the European food banks associated with FEBA were resilient in food redistribution during the first year of the COVID crisis [4]. Their analysis of ten European countries revealed that, on average, European food banks redistributed 68% more food in 2020 than in 2019.

3   Methods

To study the way food redistribution and security practices are affected by sudden society-wide shocks in the context of a mature and robust welfare state, we look at Norway. We are particularly interested in how large cities are affected – as large metropolitan areas see the spatial condensation of various socio-economic challenges and often display higher levels of inequality, social vulnerability, and diversity. For that reason, we investigate the Oslo area: not limited to the city itself, with a population of just under 700,000 at the end of 2021 but encompassing the entire metropolitan area around the Oslo Fjord (estimated at over 1 million inhabitants). This constitutes one-fifth of the entire population of Norway, making the socio-economic stress on this area a primary challenge for the Norwegian welfare state system.

Our analysis focuses on Matsentralen Oslo, including frontline organizations from the Oslo area that collaborate with it. We analyze the Bank's internal statistics and documents, combining this with a desktop analysis of national and municipal statistics, surveys of food charities operating in the Oslo metropolitan area, and an expert interview with the managing director of the Matsentralen Oslo conducted on 18 November 2020. We relied on NVivo—a qualitative data analysis tool—to analyze the interview and the documents.

Data on food redistribution (Sect. 5.1) come from the internal food banks' statistics. Data on new beneficiaries (Sect. 5.2) results from two surveys assessing the pandemic's impact on charities collecting food from Matsentralen Oslo. The first survey was run by Matsentralen Oslo in June 2020. Among the 126 frontline organizations that collected food from the food bank in this period, 87 took part. The questionnaire included questions on new groups of beneficiaries and food demands. The second survey was run in September 2020. Among 134 frontline organizations that collected food from the food bank in this period, 114 took part. The details of the first survey have been discussed earlier in a report published by Matsentralen Norge [11]. Section 5.3 on innovation in food bank activities builds on an expert interview and internal food banks' materials. Innovation in food redistribution has also been studied in a comparative European framework [4].

4 Background: Poverty and Food Redistribution in Norway Before COVID-19

4.1 Poverty in Norway

Even though Norway, with a population of 5,425,270 people [12], is undoubtedly a wealthy nation, the poverty rate has increased since 2010, reaching a top of 12.7% in 2018 [13]. In 2016, 4 out of 10 children born to immigrant parents lived in poverty in Oslo [14], with the figure increasing to 6 out of 10 in some neighborhoods of the Norway's capital [15]. Over 100 000 children lived in households with persistent low income in 2016. Of these, 54.5% came from immigrant families. The proportion of immigrants or Norwegian-born with immigrant parents with constant low income was 28.3% in 2016; this is three times higher than the population's average. Many immigrant groups have lower occupational participation than the rest of the population and for that reason growing immigration accounts for much of the rise in poverty statistics, but there has also been an increase in the number of children in low-income families without an immigrant background [16].

Norway has a well-developed welfare system that ensures that people with lower financial resources have access to health care, schools, and other public services. Nevertheless, a living conditions survey done by the Norwegian Labour and Welfare Administration (NAV) shows that low household income contributes to material and social deficiencies. People with low income more often have poor health, live more often alone, and have less social contact than those above the low income threshold.

4.2 Norwegian Food Banks

Although some of the biggest charity organizations, such as the Salvation Army and the Church City Mission, have been offering food aid to people in need in Norway for over a hundred years, it was not until 2013 that they came together to fund the first food bank in the city of Oslo.

After a pilot project carried out by the grocery retail company NorgesGruppen, in cooperation with the Ministry of Agriculture and Food and the Church City Mission, a research project called ForMat was requested to take responsibility for further progress by establishing a food bank in Oslo as a means of reducing food waste in the food industry, while contributing to the food aid efforts of charity organizations working to reduce poverty.[1] The project received financial support from the grocery trade's environmental forum (DMF), the organization of Grocery Suppliers of Norway (DLF) and the Food and Beverages and Food and Agriculture sections of the Confederation of Norwegian Enterprise (NHO). In addition, the Kavli Foundation, NorgesGruppen, COOP, ICA, Tine, Nortura, BAMA and the Ministry of Agriculture and Food also contributed to the

---

[1] More process insights of working towards reducing food waste in Norway can be found [20].

project economically. Together with The Salvation Army, Blue Cross and the Church City Mission, a non-profit organization was established to run the food bank, which was officially opened on 2 September 2013 [17]. A year later, another food bank was opened in the northern city of Tromsø. Many other Norwegian cities followed and today, the network of Norwegian food banks includes eight banks, redistributing more than 4,000 tonnes of surplus food to charities providing food aid throughout the country. The ForMat network paved the way for industry self-regulation, with an Agreement of Intent to Reduce Food Waste signed on 7 May 2015 [18] and the Industry Agreement on Reduction of Food Waste signed on 23 June 2017 [19]. Norwegian food industry committed to specific reduction targets has become increasingly active in reducing food waste, including food donations to the growing food bank network.

The Norwegian food banks function as intermediaries that collect surplus food from the food industry and make it available to frontline organizations (charities) that offer food assistance to people in need. The network is organized under the umbrella organization Food Banks Norway (*Matsentralen Norge*) and collaborates with more than 450 non-profit organizations and more than 250 suppliers of surplus food. Frontline organizations that collect foodstuffs from the food banks provide food assistance by redistributing bags of food or preparing and serving meals. The suppliers of surplus food include actors from the whole value chain, including producers, distributors, and retailers.

Food redistributed through food banks can be seen as an alternative food chain. It needs to be noted that although partly supported by the Norwegian state (both from the state budget and through the municipalities) and food industry (donations of surplus food and financial donations), food banks remain an independent third-sector organization dependent on fundraising, volunteer work, and donations from individuals and charity foundations. Moreover, state support through the Ministry of Agriculture and Food (*Landbruks- og matdepartementet* - LMD) cannot be taken for granted and is subjected to annual state budget negotiations. The year 2017, when the government tried to cut financing for food banks, showed clearly that the relationship with the state and public welfare institutions is not entirely stable and state support is something the food banks need to constantly keep striving for.

4.3   The Impact of COVID-19 on Norway

The first case of SARS-CoV-2 coronavirus was registered in Norway on 26 February 2020, and the first wave of the pandemic hit in March. It led to a lockdown of the country, which included the closure of schools, kindergartens, and many businesses and non-essential services, leaving grocery stores and pharmacies as the only services remaining open for a period of several weeks. From May onwards, some restrictions were lifted, but the impact on Norway's economy was already large.

The pandemic brought upon an increase of 255% of people who received unemployment benefits in 2020 (371,100 people), compared with the 104,500 in 2019 [21]. By the end of August 2020, 212,700 were still unemployed. 7.2% of people employed at the end of 2019 were no longer employed in the fourth quarter of 2020. Furthermore, preliminary figures show that 11.2% of young people under the age of 30 were out of work, education, or work-oriented measures in 2020.

The proportion of non-employed is somewhat higher among those with an immigrant background than in the general population. Both among immigrants and Norwegian-born with immigrant parents, this proportion is around 11%, compared with 6.3% among those without an immigrant background [22]. A study by Nordregio shows that the pandemic increased social and economic inequalities in the Nordics, already facing challenges with labor markets integration of immigrants, especially those with low education [23]. The COVID-19 pandemic enhanced these challenges even further [24].

5   Analysis: Oslo's Food Bank Under Stress

The Norwegian food banks found themselves in a key position in between two significant challenges of the pandemic: food waste risk, resulting from disruptions in usual value chains and supply-demand patterns, and an increased need for food from groups and individuals who suddenly found themselves in a crisis situation. Never before in its short history was the work of the network of food banks as relevant as during this time.

The restrictions in the activity of schools, offices, events, and restaurants led to a situation where these actors found themselves with food that they could no longer sell. This led to an increase in food donations to the Norwegian food banks. The COVID-19 pandemic also affected frontline organizations providing food aid dramatically. With many volunteers and/or beneficiaries at risk groups (elderly, addicts, sick people, etc.) and restrictions in terms of serving food, 20% of the charities that collect food from the Food Banks Norway, had to suspend their activities[2].

5.1   Amount of Food Redistributed

As illustrated in Fig. 1, compared to 2019, the food redistributed by Matsentralen Oslo in 2020 and 2021 had seen a significant increase. Food redistribution was particularly high during the Spring of 2020 when the Norwegian government introduced key social restrictions (the total lockdown at the beginning of the pandemic in March). This result confirms the findings obtained by FEBA that the pandemic has impacted noticeably on the levels of food redistribution.

The second year of the pandemic (2021) shows more variation in food redistributed each month, with a deep decrease in the summer season (July is a traditional holiday month in Norway) but also sharp increase before Christmas, when a new wave of infections was noted leading to the reintroduction of some restrictions, e.g., obligatory home office and the closure of most restaurants and many hospitality sector enterprises.

In total, Matsentralen Oslo redistributed 1,149,944 kg of food in 2019, 1,547,700 kg in 2020 and 1,654,998 kg in 2021. Compared to the baseline year 2019, 2020 increased by 35%, while 2021 saw an increase by 44% (also compared to the pre-pandemic 2019).

---

[2] Expert interview with Matsentralen Oslo's managing director conducted on 18.11.2020.

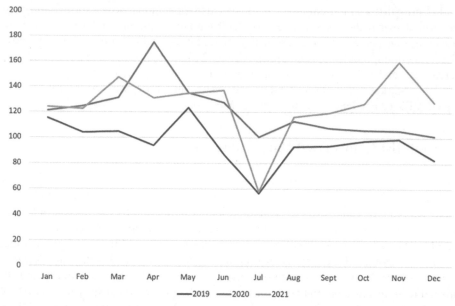

Fig. 1. Changes in the food redistribution by Matsentralen Oslo by year (2019–2020-2021) in tonnes (own elaboration).

5.2 New Beneficiaries

During the COVID-19 pandemic, changes in the number and composition of food recipients have been observed by the frontline organizations. While traditionally frontline organizations helped people struggling with addiction and mental health problems, refugees, immigrants, and former criminals, a survey conducted among frontline organizations collaborating with and acquiring food from Matsentralen Oslo showed a surge of new groups of people reaching out to food aid. Half of the charities reported new types of recipients, and similarly, nearly a half (49%) of all charities said that they needed more food than they received at that moment (Fig. 2).

Some of these new types of beneficiaries include people on *furlough* (people temporarily out of work, who remain on contracts but whose employment has been put 'on hold' with reduced pay, suspended for less than 3 months), low-income families and unemployed. As many as 27 frontline organizations reported an increase in new people on furlough reaching out for food aid. Meanwhile, 26 charities noticed an increase in new low-income families in need, whereas 26 reported new people who are unemployed.

Many of these people reached out to a charity organization for the first time, and because of decreased income, the loss of part time jobs, or unemployment. Some of them sought food aid as a result of being excluded in the welfare state's security scheme or economic relief packages launched during the pandemic.

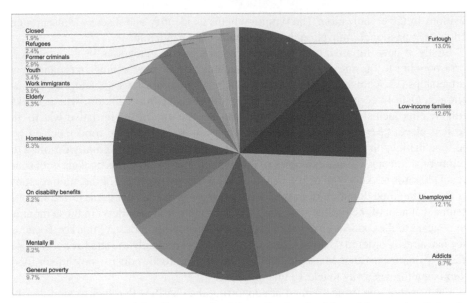

Fig. 2. The percentage of frontline organizations collecting food from Matsentralen Oslo who experience new types of recipients (own elaboration).

One frontline organization reported, "we see that there is a growing need for help among low-income families"; another respondent in the survey provided a local scale of this phenomenon "we experience many families in need. We had approx. 70–90 but now we have around 160–190 families who get food". Another comment read, "[we observe] lots of influx of immigrants and refugees during the corona period, but also many ethnic Norwegians who have been laid off or lost their jobs". A comprehensive discussion of the problem can be found in [11].

5.3   Innovation

Innovation, especially in adapting to new conditions and restrictions imposed by the COVID-19 pandemic, helped Matsentralen Oslo cope with and expand food redistribution work under the new constraints.

Tavassoli and Karlsson [25] distinguish six dimensions of organizational innovation that can serve as an informative framework to analyze Matsentralen Oslo's innovation practices developed in the pandemic. Those categories include (i) introduction and implementation of new strategies, (ii) introduction of knowledge management systems that improve the skills in searching, adopting, sharing, coding, storing, and diffusing knowledge among employees, (iii) introduction of new administrative and control systems and processes, (iv) introduction of new internal structures with their associated incentive structures including decentralized decision-making and team work, (v) introduction of new types of external network relations with other firms and/or public organizations including, vertical cooperation with suppliers and/or customers, alliances, partnerships, sub-contracting, out-sourcing and off-shoring, and (vi) hiring of new personnel for key positions in the organization. The typology helps us identify and discuss Matsentralen Oslo's organizational innovation developed under the pandemic.

Firstly, a new initiative can be seen as related to two categories in the typology, corresponding to the introduction and implementation of new strategies (i) and new partnerships (v). Known as "Matsentralen Kitchen," the project was developed by Matsentralen Oslo in cooperation with the non-profit organization Unikum, which works with creating inclusive employment opportunities. The aim of the initiative was to, in the first place, take advantage of the increased volumes of surplus food received by the food bank upon the COVID-19 restrictions, particularly food originally destined to commercial kitchens. Some examples of this type of food are 10 kilos buckets with sour cream, 30 kilos buckets with taco sauce or 10 kilos bags of rice. One of the main reasons for increasing food donation from the food industry was that, as our interviewee said, "I assume that a lot of the products we received from them [the suppliers] in those months were meant to the food service, to the horeca [the hospitality industry] market, because they had already ordered these products …., [which are] probably too late to cancel those orders." This means that Matsentralen Oslo helped the food industry save much food from being thrown away thanks to this initiative.

Furthermore, the project aimed at supporting the charities that had to temporarily stop serving food to people in need, by providing them with ready meals that they could distribute among their beneficiaries, especially those without access to a kitchen. Last but not least, the project provided the staff at Unikum with work at a time when all of their catering assignments were temporarily suspended.

Between April and December 2020, the project prepared over 20,000 ready packed meals. The Church City Mission's leader in Oslo commented on the importance of this initiative "the ready meals from Matsentralen Kitchen was the alpha and omega for our users, we do not know what we would have done without this offer" [26].

Although by the end of 2020, most charities were already preparing and serving food again, both Matsentralen Oslo and Unikum decided to continue with the project to keep supporting the charities, providing food relief to homeless or people without access to a kitchen. In 2021, in addition to using surplus food from the food bank, the project started to also collect surplus fruits and vegetables from selected supermarkets.

A second project also touches upon three categories in the organizational innovation typology, namely (i) introduction of new strategies, (iv) introduction of new internal structures and (v) introduction of new types of external network relations. The goal of the project "Internal Transport" was to increase the capacity of food banks to receive more surplus food from suppliers. While in the past, food banks were limited to receive volumes of food according to the local demand for food, establishing a system of internal transport allowed each food bank to rescue more food than locally needed and send the rest of the food to other food banks in the network. In addition to increasing the capacity of each food bank, this project also allowed a better distribution of food types among the food banks. To transport donated food, the food banks reached out to transport companies who offered the food banks a discounted rate. This was made possible because the transport companies could take advantage of, for example, returning empty trucks. The project's financing was a combination of funds donated by private actors and foundations.

This system was particularly attractive to the food suppliers with large volumes of surplus food available, who sometimes were not able to donate to the food bank because the food bank could only receive a fraction of the whole volume, which led to complex logistics for the suppliers. Thanks to this project, in 2020, Matsentralen Oslo was able to receive an additional 100 tonnes of food from its suppliers, which were sent to other food banks in the network, while receiving an additional 41 tonnes of food from different food banks (Fig. 3).

Fig. 3. Effects of the new internal transportation system between the Norwegian food banks on food received and food redistributed by Matsentralen Oslo (own elaboration).

With the success of this project, Matsentralen Oslo continued this practice in 2021 and is now one of the most essential tools to increase the volume of food redistributed.

The last initiative, we would like to highlight, corresponds to two categories in the typology, that is (i) implementation of new strategies and (iii) introduction of new administrative and control systems and processes. In addition to supermarkets and pharmacies, food banks were also allowed to continue operating despite the restrictions of the pandemic because they work under the category of food provision.

Nevertheless, the food banks had to comply with restrictions and limit the number of frontline organizations' visitors. While in the past, frontline organizations could spontaneously come to the food bank to collect food, a new system was put in place in the pandemic, through which charities had to book themselves into a 20-min spot to collect food from the food bank (the largest charities could book themselves into two time slots). This meant that only one organization could collect food at the food bank at a time. In Table 1 below we can see the schedule for Week 13 (March 23–29) of 2020.

Table 1. New scheduling system (March 2020) in Matsentralen Oslo (internal document, Matsentralen Oslo).

Fra	Til	Mandag 23/3	Tirsdag 24/3	Onsdag 25/3	Torsdag 26/3	Fredag 27/3
07:00	07:20		EOK Ev.senteret Oslo		EOK Ev.senteret Oslo	
07:20	07:40		EOK Ev.senteret Oslo		EOK Ev.senteret Oslo	
07:40	08:00					LIV Livsgiverne
08:00	08:20	KKA Kampen Omsorg++		BOK Blå Kors Kontakts.		LIV Livsgiverne
08:20	08:40		FLI FA Lillestrøm Korps			SLU FA Slumstasjon
08:40	09:00	Unikum (Kirkens Bym.)		Unikum (Kirkens Bymisjon)	Unikum (Kirkens Bymisjon)	SLU FA Slumstasjon
09:00	09:20	Unikum (Kirkens Bym.)	KOB Kongsberg Bibels.	Unikum (Kirkens Bymisjon)	Unikum (Kirkens Bymisjon)	MAS Maritastiftelsen
09:20	09:40	SLU FA Slumstasjon	KOB Kongsberg Bibels.	FAT Fattighuset	BVB Byggveien Bolig	FAT Fattighuset
09:40	10:00	SLU FA Slumstasjon	EMO Ev.senteret Moss	SLU FA Slumstasjon		
10:00	10:20	FID Evas matkasse		SLU FA Slumstasjon	VBK Vålerenga Baptistk.	FHA FA Halden Korps
10:20	10:40	FID Evas matkasse	MUE Matutdeling Eidsvoll	KBH Bybo Hollendergt.		
10:40	11:00	FDR FA Drammen Korps	MUE Matutdeling Eidsvoll		MEU Misjon Europa	FMO (FA Moss Korps
11:00	11:20			MER Mercyhouse		
11:20	11:40	FSB FA Sarpsborg korps	GUD Guts Drammen		MUE Matutdeling Eidsvoll	FAT Fattighuset
11:40	12:00			FAT Fattighuset		
12:00	12:20	FKO FA Kongsvinger korps	EVA Ev.senteret Varna		MAW Marita Women	FID Evas matkasse
12:20	12:40			SFM Senter flerkult. Moss	ELB Ellingsrudåsen Bolig	FID Evas matkasse
12:40	13:00	FEV FA Eidsvoll Korps	FOSA Fotballst. (Asker)			KLI Bymisjonen Lillestr.
13:00	13:20			FAM FA Askim Korps	THS Tyrkisk for. Oslo	FET FA Elevator
13:20	13:40	EFJ Ev.senteret Fjordtun	FFN Fengsels Felles.			
13:40	14:00	NRK Nesodden Ressursk.		FDØ FA Drøbak	MOS Misj.kirken Oslo Syd	GRI Grindestua Asker
14:00	14:20					
14:20	14:40	EOK Ev.senteret Oslo				
14:40	15:00					

A few weeks after the scheduling system was put in place, a new system of zones was established in Matsentralen Oslo's storage. By dividing the food bank into six zones, it was possible to receive up to 3–4 charities simultaneously while minimizing contact among visitors (Fig. 4).

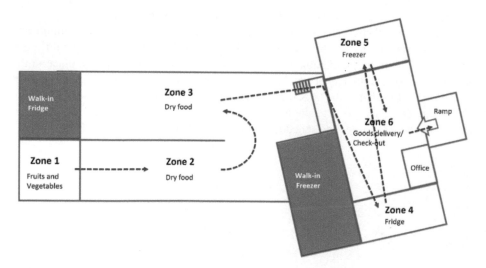

Fig. 4. New zones system in Matsentralen Oslo (internal document, Matsentralen Oslo).

The scheduling system was so successful that it remained in place even after the social distancing restrictions were lifted. Charities also expressed interest in keeping permanent time slots, what makes the food donation more effective. Current scheduling systems established for weeks 1–26, i.e., the first half of 2022, keep the new scheduling routines (Table 2).

Table 2. Refined scheduling system (January 2022) kept in Matsentralen Oslo (internal document, Matsentralen Oslo).

Time	Mandag 3. januar			Tirsdag 4. januar			Onsdag 5. januar			Torsdag 6. januar			Fredag 7. januar			Time
07:00				EOK	XL	Ev.s. Oslo				EOK	XL	Ev.s. Oslo				07:00
07:15																07:15
07:30	LIV	XL	LivsGiverne													07:30
07:45				Unikum												07:45
08:00	FEV	S	FA Eidsvoll korps	SLU	XL	FA Slumstasjonen Oslo	SPH	S	Seksjon Psykisk Helse				FOSA	S	Fotballst. Asker	08:00
08:15				Unikum			BOK	XL	Blå Kors Oslo Kontaktsenter	EMO	M	Ev.s. Moss	SLU	XL	FA Slumstasjonen Oslo	08:15
08:30	SLU	XL	FA Slumstasjonen Oslo	FOSK	XS	Fotballst. KFUM	SLU	XL	FA Slumstasjonen Oslo	Unikum			KBH	S	Bybo Hollendergata	08:30
08:45	VAB	S	Valle Bosenter	KLA	XS	Lauras hus				Unikum			Unikum			08:45
09:00	FON	S	Fontenehuset Asker	FOSV	S	Fotballst. Vålerenga	KAB	XS	Aktivitetshuset Bjerke	Unikum			FAT	XL	Fattighuset	09:00
09:15	KMM	M	Kafé med mening	EVA	L	Ev.s. Varna				FAT	XL	Fattighuset	Unikum			09:15
09:30	FRF	XL	FA Felleskjøkkenet	EMO	M	Ev.s. Moss	FAT	XL	Fattighuset	FSA	S	FA Sandvika korps	KKS	S	Kafé Søndre Nordstran	09:30
09:45	FLI	S	FA Lillestrøm korps	VBK	XS	Vålerenga Baptistkirke	MAW	XS	Marita Women	FFJ	XS	FA Fjellhamar	FDW	M	Firdaws	09:45
10:00	FID	L	Evas Matkasse	RKN	S	Nettverk e. soning Oslo	VTB	S	Veien Tilbake	RFS	S	Ringerike Frivilligsentra	FHA	L	FA Halden	10:00
10:15	MER	M	Mercy House	MUE	L	Matutdeling Eidsvoll	MER	M	Mercy House	IOO	XS	IOGT Oslo HK	FET	S	FA Elevator	10:15
10:30	FDR	L	FA Drammen	FAT	XL	Fattighuset	FAM	S	FA Askim korps (O)	FMO	S	FA Moss korps (O)				10:30
10:45	FOSH	O	Fotballst. HamKam	GAT	S	Gatebarnas Far	NRK	S	Nesodden ressurskafe	MEU	L	Misjon Europa				10:45
11:00	HJR	M	Hjerterommet	KOB	L	Kongsberg Bibelsenter				TFD	M	Tyrkisk for. Drammen	DYR	S	Dyrehjelperne	11:00
11:15	FSB	L	FA Sarpsborg	GUD	M	Guts Drammen (O)	MAS	S	Maritastiftelsen	MUE	L	Matutdeling Eidsvoll				11:15
11:30	FTE	XS	FA Templet korps	FFN	M	Fengsels Fellesskap Nor	FIV	S	Frivillige i Viken	ESV	M	Ev.s. Svene				11:30
11:45	BOK	XL	Blå Kors Oslo Kontaktsent	FST	XS	FA Stovner korps	EØS	L	Ev.s. Østerbo (O)	ÅEV	M	Ås Evangeliekirke				11:45
12:00	FOSÆ	XS	Fotballst. Stabæk	PSA	M	Salen - Det Grønne Hus	UFS	M	Ullensaker Fr.v.sentral	HJH	M	Hjelpende Hender				12:00
12:15	EOK	XL	Ev.s. Oslo	FHJ	S	FerieHjelpen	SFM	S	Senter Flerkult. Ungd. Moss	FID	L	Evas Matkasse				12:15
12:30	FVS	L	FA Varmestua Sandvika	BRT	S	Breakthrough (O)	KLI	M	KB Lillestrøm	THS	M	Tyrkisk forening Oslo				12:30
12:45	FAK	S	FA Asker korps	FMA	S	FA Majorstua	M2H	S	M&E 2nd Hand	STH	S	Streethope				12:45
13:00				FRO	S	FA Rusomsorg Stedet	FDØ	M	FA Drøbak korps (P)							13:00
13:15	LØG	S	Lørenskog Gatelag	FFR	M	FA Fredrikstad	SHF	S	St. Hansh. Friv.s. KB	SLØ	S	Sletteløkka				13:15
13:30	FOSS	XS	Fotballst. Strømsgodset	NKH	S	NaKuHel	RKØ	S	Nettverk e. soning Østf. (P)	ELB	S	Ellingsrudåsen boliger				13:30
13:45	MOS	M	Misjonskirken Oslo Syd	SHU	M	Selvhjelpens Hus	RKB	M	Bærum Røde Kors	H1N	S	Humanity 1st Norge (P)				13:45
14:00	FFY	L	FA Fyrlyset Oslo	FEL	XS	FA Ensjøtunet omsorg+	KKF	S	Kulturell Kvinneforening							14:00
14:15							WBO	S	Wayback Oslo							14:15
14:30	FRO	S	FA Rusomsorg Stedet													14:30
14:45																14:45

6 Discussion and Conclusions

Our analysis shows that Matsentralen Oslo, in collaboration with the frontline organizations, experienced an increase in food demand and new types of recipients in the two years of the pandemic—2020 and 2021. Low-income and vulnerable groups did not only rely on welfare state mechanisms (unemployment benefits and temporary layoffs) but also on food services from the charities during the pandemic. Despite the social restrictions and other related challenges, Matsentralen Oslo continued its operations without interruptions and increased its food redistribution in these two years. This was possible thanks to the various innovations being introduced by the food bank.

Our case study shows that, in such an emergency, the implementation of new strategies, the introduction of new administrative and control systems and processes, the creation of new internal structures, as well as the establishment of new types of external network relations were truly important factors, which helped the Bank not only to maintain its activities but also to reach out to more people in need. These initiatives, however, did not work alone but together, proving that Matsentralen Oslo was efficient, effective, and resilient in a crisis. What is even more interesting is that two of the initiatives, i.e., "Matsentralen Kitchen" and "Internal Transport", had been incubated before the pandemic, but were only activated during the crisis. Thus, the arrival of the pandemic was an opportunity for the food bank to take these ideas into practice.

Despite the efforts of the welfare state, poverty continued to increase in the last few years, and the shock of the arrival of COVID-19 had considerable repercussions. What we observe in this study is that even in a welfare state like Norway, the ability of dedicated institutions such as food banks to react with specific in-kind assistance to the people in need is necessary, and the Oslo case shows how quickly that reaction can occur within alternative metropolitan food chains. In addition, the results emphasize the vital role of the food banks in reducing food waste, especially in such a quick and effective manner. In this sense, food banks complement the welfare state in bringing social services and reducing food waste, thus combining socio-economic and environmental services, which are beyond the scope of financial transfers from the welfare state. This suggests that the provision of food security also depends on dedicated institutions with the flexibility to make quick adjustments and connect supply and demand efficiently.

References

1. Laborde, D., Martin, W., Swinnen, J., Vos, R.: COVID-19 risks to global food security. Science **369**(6503), 500–502 (2020)
2. Béné, C., Bakker, D., Chavarro, M.J., Even, B., Melo, J., Sonneveld, A.: Global assessment of the impacts of COVID-19 on food security. Global Food Secur. **31**, 100575 (2021). https://doi.org/10.1016/j.gfs.2021.100575. (in press)
3. Rowan, N.J., Galanakis, C.M.: Unlocking challenges and opportunities presented by COVID-19 pandemic for cross-cutting disruption in agri-food and green deal innovations: Quo Vadis? Sci. Total Environ. **748**, 141362 (2020)
4. Capodistrias, P., Szulecka, J., Corciolani, M., Strøm-Andersen, N.: European food banks and COVID-19: Resilience and innovation in times of crisis. Soc. -Econom. Plann. Sci. https://doi.org/10.1016/j.seps.2021.101187. (in press)
5. Eurostat, Living conditions in Europe. https://ec.europa.eu/eurostat/statistics-explained/index.php?title=Living_conditions_in_Europe. Accessed 20 Dec 2021
6. FEBA. https://asvis.it/public/asvis2/files/Eventi_Flash_news/FEBA_Report_Survey_COVID_Sept2021.pdf (2021). Accessed 18 Jan 2022
7. FEBA: https://lp.eurofoodbank.org/wp-content/uploads/2020/08/FEBA_Report_Survey_COVID_April2020.pdf (2020a). Accessed 18 Jan 2022
8. FEBA: https://lp.eurofoodbank.org/wp-content/uploads/2020/07/FEBA_Report_Survey_COVID_July2020.pdf (2020b). Accessed 18 Jan 2022
9. FEBA: https://lp.eurofoodbank.org/wp-content/uploads/2020/10/FEBA_Report_Survey_COVID_Sept2020.pdf (2020c). Accessed 18 Jan 2022
10. FEBA. https://lp.eurofoodbank.org/wp-content/uploads/2021/02/FEBA_Social_Forum_Report_2020_INTERACTIVE_FINAL.pdf, (2020d). Accessed 18 Jan 2022
11. Capodistrias, P.: Impact of the COVID-19 pandemic on the network of Norwegian Food Banks and partner organizations. Food Banks Norway, Oslo (2020)
12. SSB (2022). https://www.ssb.no/befolkning/folketall/statistikk/befolkning. Accessed 2 Mar 2022

13. Macrotrends: https://www.macrotrends.net/countries/NOR/norway/poverty-rate, (2022). Accessed 18 Jan 2022

14. SSB: Barn og unge voksne med innvandrerbakgrunn Demografi, utdanning og inntekt, https://www.ssb.no/befolkning/artikler-og-publikasjoner/_attachment/273616?_ts= 1562bfcd488 (2016). Accessed 18 Jan 2022

15. Brattbakk, I., et al.: På sporet av det nye Grønland. Sosiokulturell stedsanalyse av Grønland i Bydel Gamle Oslo. https://oda.oslomet.no/oda-xmlui/bitstream/handle/20.500.12199/6503/ r2017-04_Stedsanalyse%20Gr%c3%b8nland_RS.pdf?sequence=1&isAllowed=y   (2017). Accessed 18 Jan 2022

16. Furuberg, J., Grav, T., Lima, I.A.Å., Munch-Ellingsen, E.: Lavinntekt og levekår i Norge. NAV-rapport 2018, p. 3 (2018)

17. Hanssen, O.J., Møller, H.: Food Wastage in Norway 2013. Status and Trends 2009-13. Østfoldforskning, OR.32.13 (2013)

18. Regjeringen: Intensjonsavtale om reduksjon i matsvinn. https://www.regjeringen.no/conten tassets/e54f030bda3f488d8a295cd0078c4fcb/matsvinn.pdf (2015). Accessed 18 Jan 2022

19. Regjeringen: Industry Agreement on Reduction of Food Waste. https://www.regjeringen.no/ contentassets/1c911e254aa0470692bc311789a8f1cd/industry-agreement-on-reduction-of- food-waste_norway.pdf (2017). Accessed 18 Jan 2022

20. Szulecka, J., Strøm-Andersen, N.: Norway's food waste reduction governance: from industry self-regulation to governmental regulation? Scandinavian Political Studies. https://doi.org/ 10.1111/1467-9477.12219. (in press)

21. Nerdrum, A.H., Berge, A.: Antall dagpengemottakere økte med 255 prosent i 2020. SSB. https://www.ssb.no/inntekt-og-forbruk/artikler-og-publikasjoner/antall-dagpengemott akere-okte-med-255-prosent-i-2020 (2021). Accessed 19 Jan 2022

22. Olsen, B.: Innvandrere mest rammet av koronatiltakene. SSB, https://www.ssb.no/arb eid-og-lonn/artikler-og-publikasjoner/innvandrere-mest-rammet-av-koronatiltakene (2021). Accessed 19 Jan 2022

23. Gassen, N.S., Penje, O.: (eds.) Integrating immigrants into the Nordic labour markets. The impact of the COVID-19 pandemic. Nordregio. https://pub.norden.org/nord2021-050/#79390 (2021). Accessed 19 Jan 2022

24. Pettersen, M.: Trenden er brutt – flere unge utenfor i 2020. SSB, https://www.ssb.no/arb eid-og-lonn/sysselsetting/statistikk/tilknytning-til-arbeid-utdanning-og-velferdsordninger/ artikler/trenden-er-brutt--flere-unge-utenfor-i-2020 (2021). Accessed 19 Jan 2022

25. Tavassoli, S., Karlsson, C.: Persistence of various types of innovation analyzed and explained. Res. Policy 44(10), 1887–1901 (2015)

26. Capodistrias, P., Gerritsen, E.: Matsentralen Kjøkken, fra overskuddsmat til ferdigmat. Matsentralen Norge. https://uploads-ssl.webflow.com/5d5e9d19d7cea0cd72b43e4e/618 93c697f81d5d1334f4e48_Matsentralen%20Kj%C3%B8kken%20Rapport_Oppslag%20F ERDIG.pdf (2021). Accessed 19 Jan 2022

Green Buildings, Post Carbon City and Ecosystem Services

Valuation and Design for Economic and Social Value Creation

Isabella M. Lami, Beatrice Mecca[✉], and Elena Todella

Interuniversity Department of Regional and Urban Studies and Planning (DIST), Politecnico di Torino, Viale Mattioli 39, 10125 Turin, Italy
beatrice.mecca@polito.it

Abstract. The quality assessment of projects is becoming increasingly relevant in the estimative discipline, aiming at a greater control of the design effects of real estate interventions. If from an estimation point of view real estate development processes are traditionally linked to a concept of economic value creation, the current context requires reflections that also consider the assessment and creation of other values, arising from environmental, social, and cultural needs. In this context, the aim of the paper is to reflect on how the application of the estimative discipline to the field of design can support the development of sustainable architectural projects, reflecting on the centrality of values in the development of the project. In this sense, the paper intends to schematize which values should be considered, estimated, and evaluated in the ex-novo and transformation processes, giving exemplifications of the potential action of the architectural project on the space for the creation of these values. The reflection is carried out on economic and social values as a first step of a research that can be expanded to other specific values of the architectural context.

Keywords: Value creation · Project appraisal · Architectural sustainability

1 Introduction

The real estate development process is traditionally linked to the creation of economic value, as the predominant parameter to be considered. However, sustainability considerations on several dimensions, in addition to the economic one, are now framed at a global level by the objectives of the 2030 Agenda, at a European level by the Green New Deal (GND), and at a national Italian level by the National Recovery and Resilience Plan (PNRR), defined as a response to the Covid-19 pandemic, in which sustainability-related requests have been made operational. While the creation of economic value in real estate development processes remains undisputed, environmental, social, and cultural demands and needs are emerging to be reflected upon not only in the public but also in the private context. This new sensitivity directly influences and is registered by the market, thus raising measurement requests on which the valuation discipline must focus. In this sense, different values should be considered with effects on the future overall value of the real estate, and therefore this could lead to modifications on the aspects considered

F. Calabrò et al. (Eds.): NMP 2022, LNNS 482, pp. 1465–1475, 2022.
https://doi.org/10.1007/978-3-031-06825-6_141

in the valuation process. Sustainability is now one of the most debated topics, especially in the field of architectural and urban transformations [1–3].

The national context is currently affected by the operational fallout of the PNRR [4] prepared to boost the economy after the COVID-19 pandemic, to enable the green and digital development of the country. In accordance with European and global objectives, the plan defines a series of reforms and investments, among which the fundamental principles of reuse, urban regeneration and land consumption limitation are affirmed (p. 81), as well as historical and cultural enhancement interventions, enhancing the attractiveness, safety, and accessibility of places (p. 85). Furthermore, the PNRR emphasises the "social" dimension of policies, with particular attention paid to urban regeneration as a means of improving inclusion and urban well-being (p. 214).

In this context, the aim of the paper is a reflection on the potential of value creation with respect to the project. In this sense, the paper reflects on the centrality of values, and envisioning how they could be explicitly incorporated by the project and, consequently, pursued and measured (Sect. 2). A first output of the paper is therefore to outline which values need to be considered, estimated, and assessed in the processes of new development and transformation, depending on the public or private context of reference. Moreover, looking at the process of real estate estimation, we highlight some new forms of value that may give rise to new forms of measurement. We focus specifically on social value, to understand how this can be operationally useful in creating value through the project (Sect. 3). Finally, we provide some concluding reflections and possible directions for research (Sect. 4).

2 Evaluation "of" and "in" the Architectural Project

The quality assessment of projects has become increasingly relevant [5, 6]. Consequently, more importance is attributed to the analysis of the design process and the incidence of evaluation tools, in terms of their creative potential in the process [3, 7]; especially considering the inherently unpredictable, progressive, and iterative construction character of the project, strongly dependent on the individual choices implemented during the design process. The need to rethink some of the prerogatives of the estimation and evaluation disciplines in the training and practices of designers and urban planners is not a new theme [8–11], indeed it is part of a reflection started at national level on the roles that the disciplinary area plays with respect to the design process. Moreover, the effects of projects strongly depend on the interaction and concatenation between decisions and exchanges, which ultimately produce physical transformations through a set of actions [12]. So, the character of the evaluative discipline is understood here as potentially pervasive of the whole project process.

Alongside the traditional strands of research related to property appraisal and plan and programme evaluation, more and more importance is now being attached not only to quantitative requirements of a performance, functional and practical nature, but to the connotations of physical space in more qualitative terms, to other cultural and symbolic values. In this sense, if on the one hand the economic values of properties are traditionally assessed according to widespread practices, on the other hand their symbolic, perceptual meaning, as other aspects related to quality have generally remained unexplored – as in a "black box" [13]. Therefore, this research intends to take the opportunity to start from opening this "black box" and investigate the peculiarities through which the design process is articulated, to consider some precise parameters able to have effects on quality, as new and different values to pursue, particularly with respect to projects of architectural value. This does not mean rejecting the contribution of traditional evaluations but recognising the need to broaden reflection to include other aspects.

2.1 Between Valuation and Assessment

Project appraisal, in the context of estimation, is aimed at an evaluation of economic viability, however, it is also based on types of criteria/value bases/observed values different from the economic estimative ones. This aspect is crucial in the evaluation of projects, specifically architectural projects, since in the creation of value in a current context of necessary sustainability, it is necessary to consider different aspects that are not limited to economic aspects. In the realisation of an ex-novo project and in a reuse or transformation project, the estimation of the market value remains essential, concerning the evaluation of the market attractiveness of the architectural object, and of the cost value, concerning the economic sustainability of the project.

According to the Red Book of the Royal Institution of Chartered Surveyors (RICS) [14], market value is defined as "the estimated amount for which an asset or liability should be sold and acquired, at the valuation date, by an unrelated buyer and seller, both of whom are interested in the sale or purchase, on competitive terms, after proper marketing in which the parties have both acted in an informed, knowledgeable and uncoerced manner". The cost value is given as a function of the fact that every property needs to be created and therefore the construction process involves costs. Simonotti [15] highlights the differentiation of cost into two main distinctions according to two actors in the construction process: i) the production cost, which is the sum of the expenses that an entrepreneur needs to incur to realise the entire asset; ii) the construction or intervention cost, which is the sum of the expenses that the construction company needs to incur in order to realise the asset in material terms.

Alongside these two aspects of value, it is essential to consider other values, material and immaterial, that permeate objects, spaces, and architectural environments for the development of sustainable and valuable architecture. This demand arises from the request of Agenda 2030 and PNRR, and more in general from the goal to pursue sustainability.

This concept is not new, indeed, since the 1960s a field of study related to Multicriteria Analysis was introduced in the estimative discipline, due to the awareness that for the resolution of real problems, complex and characterized by the multidimensionality of objectives, it was necessary to consider multidisciplinary approaches to evaluate different effects (economic, social, environmental, political etc.) in an integrated way [16] The emergence of new issues, summarised in the concept of sustainability, led to the need for estimation to include complex project values that cannot be expressed in monetary terms [17]. Indeed, such analyses make it possible to assess quality through multiple criteria, which can also consider intangible effects that are difficult to express in monetary terms. The consideration of different criteria makes it possible to provide a rational basis for choice in complex and multidimensional social contexts. The criteria considered in real estate contexts are, indeed, of different nature: economic, social, physical-spatial, environmental, and cultural [18]. Furthermore, when dealing with and assessing cultural heritage, it is necessary to consider intrinsic values, constructed by the society that lives in the spaces. Fusco Girard and Vecco [19] report a classification of values in the public sphere into: cultural values, economic values, communication values and ecosystem values. Cultural values include historical, social, artistic, aesthetic, moral, scientific, cultural, spiritual/religious, and educational values. These values are reported in reference to the heritage, therefore to what is already existing, but they are particularly suitable for the public architectural context and therefore not only for the transformations of reuse of existing buildings, complexes, or areas but potentially to be considered valid also in a context of new construction. Indeed, in both contexts these values must be respected, enhanced, and created for the development of valuable projects.

The paper intends to focus not only on economic values but also on social values, as a first step of a wider research that can be expanded to the detailed analysis of the different values mentioned above. Indeed, in the last decades social sustainability has gained attention as a fundamental component of sustainable development and it is from this that we intend to start to test the potential of an enlargement of the dimensions of possible value creation through the project. For this reason, we observe and consider five key factors used in the literature to assess social sustainability (for more information refer to Lami and Mecca [3]), namely: social inclusion and participation, architectural identity, flexibility and adaptability of space and sense of security.

3 Architectural Design and Estimation Values

In this section we highlight the importance of evaluation in project design to support and have control over the creation of sustainable economic and social values and thus over the effects of the project. Indeed, in a process of change, the evaluation and assessment of the values created by the project is central to effective and fruitful strategies.

Starting from a private context we observe that the convenience of the investor is in the first instance economic-financial, as it is oriented to the maximisation of the profit. This objective can be achieved in the creation of an artefact that demonstrates the highest possible market value and the lowest possible cost value.

In this context, market value and cost value are considered "original" values, namely those from which the others can be "derived" [18]: in this sense we can affirm that we can limit the analysis to these two fundamental ones. Therefore, to create these values, the role of the valuation consists in supporting the definition of the most effective design solution for the investor through tools for assessing the factors that can positively or negatively influence the market and cost values, while the role of the project consists in the action on the spatial and material configurations necessary to provide the building with the qualities and performances that make it valuable. In this sense, from the point of view of value creation, the project cannot move and define itself outside the consideration and interaction with the estimation of these values, upon which the quality and sustainability of the building will be determined.

A design conceived to guarantee the fundamental characteristics of wellbeing, such as thermo-hygrometric, biophysical, acoustic, and optical aspects, through natural spatial and technological elements and strategies that are autonomous from energy mechanisms affecting the environment, can have a reflection on market value. Indeed, by acting on solar and climatic space planning in an efficient and healthy way, it allows to optimise the performance of the building favouring on the one hand the reduction of operating costs and consumption in the short term and on the other hand the reduction of the energy load of the building on the environment. This implies that higher performance corresponds to a higher market value. In addition, the design can act on a second factor, namely the flexibility and adaptability of the space. Indeed, a building designed with a view to be flexible to potential future transformations, brings with it a principle of circularity and reuse promoted as a practice of new constructions [20, 21]. This aspect of increased efficiency of the building in terms of transformability and adaptability to new uses can lead to savings on possible future transformative intervention costs and in this sense, this added value can be expressed in an increase in market value.

Observing the type of use and therefore of function located in the space, it is possible to consider the need to pay greater or smaller attention to the development of social values, understood as aspects of inclusion and social participation, architectural identity, flexibility and adaptability of the space and sense of security.

The development of an intervention for private use, therefore targeting few exclusive users for residential use or destined to a restricted public use as in the case of small or medium sized commercial premises, does not necessarily imply the necessity to consider the development of the aspects of social interaction and participation and architectural identity. However, the sense of security is an aspect to be considered and developed, and the design can intervene in this regard at the level of planning for controlled access and strategies of closure or semi-closure of boundaries.

Considering instead a private intervention with spaces targeting a wide range of users, be they tertiary workplaces involving many workers or large commercial premises, it is possible to observe the need to create further social values than sense of security, such as social interaction. For the latter, the project can act by paying attention to the design of the space to include specific places of interaction and aggregation, functionally connected to other spaces. In the light of the impacts due to the Covid-19 pandemic, the need for development and creation of social value understood as social interaction is even more inevitable [22, 23]. Indeed, isolation has made evident the need for such spaces, appropriately designed so that they respect the safety distances for the health of everyone, promoting individual psychological well-being.

The project action can concretise the creation of a lower cost value through the choice of materials that consider three requirements: i) suitable performance qualities; ii) cheapness; iii) and local derivation, to reduce transport costs and favour the development of the local economy. Moreover, considering what is promoted in the EC Action Plan, namely the promotion in the construction sector of careful and safe choices of materials with a view to reducing and recycling construction waste while ensuring efficient and safe performance [24, 25], it is possible to add two aspects to the three requirements to be considered. In terms of choice of materials that of iv) being recycled and/or recyclable materials, in favour of a circular and sustainable development. In terms of cost reduction, we can also consider (v) the possibility of adaptively reusing the spaces through a reconfigurable project.

In general, these concepts are valid in a private context of new construction but can also be valid for reuse interventions. In this case the observed value will be the transformation value: by acting with careful design mechanisms to increase the market value and decrease the costs it is possible to influence the creation of a higher transformation value.

In a public context, the convenience of investors is of an economic-social type, aimed at maximising the benefits for the community. Considering social utility not only as being dependent on the actual use of the good, but with a value independent of use, an overall value known as total economic value (TEV) has been defined [26, 27]. Therefore, the values to be considered are the direct and indirect use value, the option value, and the non-use value, understood as legacy and existence value. The economic value of public goods estimated and observed in terms of use leads us to consider that what can allow the creation of a use value is the usefulness and appeal of the architectural good itself. In this sense, the project can act in the development of these values in a way that is common to all five values estimated within the TEV. What we mean is that the potential for interesting use of a space, current or future, is determined by the attractiveness of the building, resulting from its forms, space design and function. The project indeed outlines the forms, connections, distributions and finishes that can make a building comfortable and functional. This connects with a concept of "beauty" [28]: the latter has the potential to attract the public and therefore this involves the creation of an economic value, namely a value of use, option, or non-use. Therefore, the project can act operatively, for instance by outlining iconic architectural objects in their lines and forms, thus making them unique and distinctive. In the case of a reuse intervention, the design action for the creation of value is instead outlined in the respect and maximisation of the main characteristics of the building, that is, what made it specific and recognisable. Therefore, the insertion of the new functions should not denaturalise the existing building with strong and impactful interventions.

Cino Zucchi, during the conference "The Value of Beauty" in October 2021, affirms that architecture has four states: first the ideation, followed by the conception and the realisation, and then it becomes the background of daily life and the stage of social exchange. This statement makes clear the importance of social value, especially in the context of public goods whose use involves the community, and how architecture plays an important role in its creation. Therefore, we highlight the importance of assessing and considering social aspects, such as social interaction and participation, architectural identity, flexibility and adaptability, and a sense of security, in the act of design. The latter can act, for instance, by the differentiation and distribution of different blocks (understood as a box-by-box design strategy) for public or public-private functions with a shared distribution and aggregation space, or by the act of connecting different portions of the building used for different types of functions to spaces designed to encourage social interaction. In this sense, there are spaces dedicated to public activities, such as cultural and recreational activities, which can increase social participation [3], effectively connected to each other and converging towards spaces designed for social interaction (e.g., covered squares, open spaces). Flexibility and adaptability of spaces are an important element due to the constant evolution of society and its needs, which imply the remodelling of buildings for new social uses. In this perspective, the design can be based on modular and reversible solutions in favour of a flexible design that allows changes over time. Architectural identity is an element that can run parallel to the beauty mentioned above, indeed it can be developed in new contexts defining iconic and representative places in their forms, finishes and materials, and in reuse contexts maintaining and highlighting the pre-existing features. Finally, the sense of security will have to be addressed to a greater or lesser extent depending on the level of security of the area in which the building in question is located. The design can act considering infrastructures and design devices that affect the perception of security: insertion of controlled and differentiated accesses for users and as already mentioned in the private case, the insertion of different strategies of closure or semi-closure of the boundaries.

Table 1 summarises schematically the different estimative values that are considered, estimated, and valued in the drafting of a private or public, new or reuse project, so that the design of the spaces can reason on these parameters in spatial terms to maximise them and increase their valence and quantity. It should be noted that as far as economic values are concerned, reference will only be made to the cost and market value as the primary values from which the other four derived, and to the TEV in general, not divided into direct and indirect use values and non-use values of legacy and existence.

Table 1. Estimation values enhanced and created through the project.

Estimation values	Project contribution	Example of output design strategies
Market value	Energy performance	Efficient and healthy solar, climatic, and acoustic planning, namely: ventilated facade, rainwater collection, photovoltaic and solar panel; solar shading (e.g. trees, brise soleil, curtains) to adequately manage solar inputs; natural acoustic shading (e.g. trees, bushes)
	Flexibility/Adaptability	Open free space configurations that allow for a variety of possible uses while maintaining both shape and size over time; Vertical closures made with "dry" construction technologies to transform open spaces into closed spaces and vice versa; Movable vertical partitions with sliding devices on rails or wheels to transform several units into a single spatial unit

(continued)

Table 1. (*continued*)

Estimation values	Project contribution	Example of output design strategies
	Sense of security	Differentiation of pathways and accesses according to private and public functions; For private functions avoid permeability of the site, i.e. privatise internal spaces with clear divisions between public and private (gates, green partitions, visual barrier systems limiting access); Technological video surveillance systems at boundaries; Raised floors which, due to their higher level, better protect privacy
	Social interaction	Specific places of interaction and aggregation, functionally connected to other spaces with different functions; Open shared spaces that with controlled and restricted access respect the safety distances imposed by Covid-19; Distributed spaces for a functional mix
Cost value	Material features	Features of building materials: made from recycled raw materials or abundantly available in the area; eco-sustainable (e.g. wood, glass, cork, stone, earthenware), which are recyclable or disposable, limiting their impact on the environment and ensuring maximum internal comfort, locally produced to cut transport costs; Adaptive reconfiguration of existing spaces limits costs of new builds
Total Economic Value	Beauty and/or architectural identity	Iconic architecture (e.g. pure and compact forms; forms inspired by nature, distinctive materials, colors and patterns); Emphasis on the new constructions, as new façades to increase visibility; Maximisation of the main characteristics of the pre-existing building

(*continued*)

Table 1. (*continued*)

Estimation values	Project contribution	Example of output design strategies
	Social interaction	Openness of the building to the public, favouring the permeability of the site through a non-sharp division between the public space pertinent to the building and the sidewalk and street; Outdoor and/or indoor squares open to the city and therefore to the public; Pedestrian pathways that invite and introduce the public to the places of interaction and aggregation
	Flexibility/Adaptability	The load-bearing structure, dimensioned with useful redundancies to allow potential extensions or elevations (e.g. frame structures and structural system on a modular grid, allowing different possible layouts in plan and elevation; wide spans and heights); Technological flexibility, in terms of demountability and constructive reversibility of partitions, assembled in a drywall manner; Wide open areas, articulated in multifunctional spaces through the box-in-box strategy
	Sense of security	Different strategies to access the different spaces according to the target users; Technological video surveillance systems at boundaries; Private indoor open spaces, accessible only by authorised users

4   Conclusions

The paper reflects on the centrality of different economic-estimative and social values in the design of sustainable architectural spaces. The combination of the reflection on values, understood as selection criteria or value bases on which evaluations are based, and the design process demonstrates a potential for value creation for the stakeholders, the community, and the environment. The document analyses the economic and social dimensions, observing how the convenience of both a private investor and a public investor can be fulfilled through the project act that considers certain parameters and aspects of the project with a view to economic and social sustainability. We observe that the values to be considered, estimated and valued are different mainly on the basis of two aspects: the private or public nature of the context of reference from which a different economic-estimative value arises, depending on the convenience of the investment; the typology of the functions that will be hosted by the spaces, as restricted and exclusive or vast and differentiated users can lead to consider different social needs and therefore different aspects of the social value. In the private case the project will have to reflect on the definition of the spaces to create an economic value on the basis of considerations in favour of a circular and sustainable concept leading to a positive reflection on the market value on the cost value; create a social value in order to fulfil the needs of security of the users and of adaptability of the spaces to the evolution of the society. From the public point of view the reflection at design level should take place: in terms of utility and appeal of the collective good, so that this can lead to the creation of an economic value intended as a higher potential of interest for current and/or future use; in terms of interaction, social participation, architectural identity, flexibility of the space and sense of security to meet the needs of the community, thus creating a positive social value.

We recognise that the considerations and conclusions are drawn at a theoretical level and require future practical and concrete feedback. However, this schematisation leaves open the potential for future development both in terms of the effective intersection of the disciplines of estimation and design through the conjugation of evaluation and design tools and in terms of expansion of research to include different values to be considered and created in an architectural context.

References

1. Capolongo, S., Buffoli, M., Oppio, A., Petronio, M.G.: Sustainability and hygiene of building: future perspectives. Epidemiol. Prev. **38**(6), 46–50 (2014)
2. Bassi, A., Ottone, C., Dell'Ovo, M.: I Criteri Ambientali Minimi nel progetto di architettura. Trade-off tra sostenibilità ambientale, economica e sociale. Valori e Valutazioni 22 (2019)
3. Lami, I.M., Mecca, B.: Assessing social sustainability for achieving sustainable architecture. Sustainability **13**, 1–21 (2021)
4. PNRR, Piano Nazionale di Ripresa e Resilienza (2021). https://www.governo.it/sites/gov erno.it/files/PNRR.pdf. Accessed 16 Dec 2021
5. Fattinnanzi, E., Micelli, E.: Valutare il progetto di Architettura. Valori e Valutazioni **23**, 3–14 (2019)
6. Berni, M., Rossi, R.: Pensare la qualità del progetto in relazione alla città come bene comune. Valori e Valutazioni **23**, 57–64 (2019)
7. Fattinnanzi, E., Acampa, G., Forte, F., Rocca, F.: La Valutazione complessiva della qualità nel Progetto di Architettura. Valori e Valutazioni **21**, 3–14 (2018)
8. Mondini, G.: La valutazione come processo di produzione di conoscenza per il progetto. Valori e Valutazioni 3 (2009)
9. Forte, F.: Il valore architettonico di un immobile: criterio o obiettivo? Valori e Valutazioni **8**, 105–117 (2012)
10. Bentivegna, V.: Editoriale. Valori e Valutazioni **17**, 1–3 (2016)
11. Lami, I.M., Mecca, B.: Architectural project appraisal: an active learning process. Valori e Valutazioni **28**, 3–20 (2021)

12. Armando, A., Durbiano, G.: Teoria del Progetto Architettonico. Dai disegni agli affetti. Carocci, Roma (2017)
13. Bentivegna, V.: Qui si parla di progetti, valutazioni e valutatori. Valori e Valutazioni 6 (2015)
14. RICS, Royal Institution of Chartered Surveyors (RICS), Valutazione RICS – Standard globali (2020). https://www.rics.org/globalassets/rics-website/media/upholding-professional-standa rds/sector-standards/valuation/rb-italian-2020.pdf. Accessed 1 Dec 2021
15. Simonotti, M.: La stima immobiliare. Con principi di economia e applicazioni estimative. UTET, Torino (1997)
16. Greco, S., Ehrgott, M. and Figueira, J.R. Multiple Criteria Decision Analysis: State of the Art Surveys. Springer, Berlin (2016). https://doi.org/10.1007/978-1-4939-3094-4
17. Del Giudice, V.: Estimo e valutazione economica dei progetti. Profili metodologici e applicazioni al settore immobiliare. Paolo Loffredo Iniziative Editoriali Srl, Napoli (2015)
18. Roscelli, R.: Manuale di estimo, valutazioni economiche ed esercizio della professione, De Agostini-UTET Università, Novara (2014)
19. Fusco Girard, L., Vecco, M.: Genius loci: the evaluation of places between instrumental and intrinsic values. BDC, Università degli Studi di Napoli 2, 473–495 (2019)
20. Ellen MacArthur Foundation and ARUP, Designing buildings for adaptable use, durability and positive impact (2019). https://www.ellenmacarthurfoundation.org/assets/downloads/2_Buil dings_Designing_Mar19.pdf. Accessed 24 Nov 2021
21. Kyro, R.K.: Share, preserve, adapt, rethink. A focused framework for circular economy. In: IOP Conference Series: Earth and Environmental Science 588, 1.11–1.14 (2020)
22. Quaglio, C., Todella, E., Lami, I.M.: Adequate housing and COVID-19: assessing the potential for value creation through the project. Sustainability 13, 10563 (2021)
23. Abastante, F., Lami, I.M., Mecca, B.: How Covid-19 influences the 2030 Agenda: do the practices of achieving the sustainable development goal 11 need rethinking and adjustment? Valori e Valutazioni 26, 11–23 (2020)
24. European Commission, Circular Economy Action Plan. For a cleaner and more competitive Europe (2020). https://op.europa.eu/en/publication-detail/-/publication/45cc30f6-cd57-11ea-adf7-01aa75ed71a1/language-en/format-PDF/source-170854112. Accessed 24 Nov 2021
25. Gravagnuolo, A., De Angelis, R., Iodice, S.: Circular economy strategies in the historic built environment: cultural heritage adaptive reuse. In: Proceedings of the STS Conference Graz (2019)
26. Girard, L.F.: I beni ambientali: valutazioni e strategie di conservazione, tra conflitto e cooperazione, Genio rurale – Estimo e Territorio 5, 38 (1994)
27. Stellin, G., Rosato, P.: La valutazione economica dei beni ambientali. CittàStudi Edizioni, Torino (1998)
28. SIDIEF: Società Italiana di Iniziative Edilizie e Fondiarie. Il valore del bello (2021). file:///C:/Users/tice_/Downloads/ricerca-il-valore-del-bello.pdf. Accessed 24 Nov 2021

Projecting the Underused. Increasing the Transformation Value of Residential Spaces Through Their Adaptive Reuse

Elena Todella[1]([⊠]), Caterina Quaglio[2], and Isabella M. Lami[1]

[1] Interuniversity Department of Regional and Urban Studies and Planning (DIST), Politecnico di Torino, Viale Mattioli 39, 20125 Turin, Italy
elena.todella@polito.it

[2] Department of Architecture and Design (DAD), Politecnico di Torino, Viale Mattioli 39, 20125 Turin, Italy

Abstract. What are the effects of the pandemic on the real estate market? How do we interpret the changes, in terms of spaces and services, to respond to new demands? The paper seeks to answer these questions in the Italian context: (i) by reviewing, through indirect sources on real estate market, the trends of the current demand in the residential sector in the light of Covid-19; (ii) by analyzing case studies, to show the added value of architectural solutions related to accessory spaces. A reflection is directed to the assessment of the potential for value creation through the project of those spaces, as the relationship between the demand analysis, the possibilities in terms of adaptive reuse of residential spaces and the transformation value.

Keywords: Transformation value · Underused spaces · Adaptive reuse

1 Introduction

The impacts of Covid-19 pandemic on the housing market in the short and long term and the potential of the existing urban fabric to respond to emerging needs [1–3] are current topics of research. The present paper tackles the issue of value creation through the project, integrating evaluation and design evidence to capture the specificities of the current contingency and highlight possible lines of action. To this end, the text draws on three perspectives: i) the key characteristics of emerging housing market demand; ii) the actual possibilities for adaptive reuse of residential spaces; iii) the transformation value of underused and undervalued housing spaces. These three perspectives have been related to a broader context to question. From the conjunction of these three dimensions, in a combined evaluation and project perspective, it was therefore possible to draw some operational considerations. Overall, this analytical process brings to the forefront several underused and undervalued housing spaces characterized by a potential value – and a transformation value – that could far exceed the one attributed to them by traditional estimation methods. Those spaces, normally considered as accessory to the residence – such as terraces, common courtyards, or cellars, are indeed currently estimated through

F. Calabrò et al. (Eds.): NMP 2022, LNNS 482, pp. 1476–1485, 2022.
https://doi.org/10.1007/978-3-031-06825-6_142

reductive coefficients or are completely excluded from the calculation of property value in real estate valuation practices in the Italian context, but present a high degree of adaptability to respond to emerging demands. How can this value be accounted and included in real estate evaluation? The paper addresses this question by suggesting a possible trajectory aimed, on the one hand, at making valuation tools a guide to design action and, on the other, at refining valuation methodologies by including a design perspective capable of considering a transformative dimension.

The structure of the paper reflects the logical pattern just described. After outlining the background, objectives, and methodology of the research (Sect. 2), the third section presents the results of the analysis of the new trends in residential demand (Sect. 3). The last section (Sect. 4) discusses the findings by introducing the concept of adaptive reuse in terms of most promising spaces to meet new needs from a transformation value perspective. Finally, some questions are addressed on the relationship between demand analysis and transformation value.

2   Adaptive Reuse and Real Estate Evaluation

Assessing the potential for value creation through designing the adaptive reuse of underused spaces aims at detecting spatial resources to be maximized through the project, in response to emerging needs, reflecting on the emerging potential for use in such spaces, on the one hand, and the current practices of evaluating them, on the other.

The pandemic has exacerbated some existing issues within residential spaces [1–3], underlining the inadequacy of some basic spaces and services of our homes. At the same time, some spaces pertaining to homes or buildings have emerged as unexpected sources of social sustainability for contingent needs, whether these be for socialization or privacy [4–6]. These needs produced several secondary impacts of Covid-19 [3, 7, 8] on the use of our homes, buildings, and neighborhoods. In this paper, the focus is on residential spaces pertaining to housing and buildings, reflecting on the potential for transformation related to the emergence of new uses both inside – and outside, as in balconies and terraces – of housing, and inside buildings in semi-private filter-spaces between the building itself and the external shared space – such as courtyards, condominium gardens, equipped condominium terraces. Facing the evidence that traditional houses are not sufficient to meet the new needs that see them as places of work, leisure, and relaxation, has emerged the need to think about the spaces in terms of adaptability and flexibility [1–3, 5, 7–10].

In this sense, the potential of some underused and "latent" residential spaces for creating added value arise from the adaptability and flexibility – and consequent intensified use – of such spaces, which are instead actually overlooked in the assessment of housing market and of transformation value. Currently, the market value of a property is calculated based on its commercial surface area [11], which may relate to an entire property, or part of it, and is understood to be based on the sum of the various weighted surfaces that make up the property. We talk about weighted surfaces because, to calculate the total commercial surface, it is not enough to simply add up the areas of the rooms and appliances, but appropriate reductive coefficients, expressed in percentage shares, must be used to homogenize the accessory surfaces to the main one (Table 1). Among the main rooms of a residential property are the rooms of daily use – e.g., kitchen, living room, bedroom – and among the accessory rooms those of smaller surfaces serving the real estate unit. These can be direct – e.g., bathrooms, closets, entrances, hallways, corridors – and indirect – e.g., cellars, garages, attics, warehouses – whether directly connected to the main or not. The reductive coefficients are relevant to the calculation of market value of a property because they imply that not all parts of a property have the same value. Therefore, these are considered in proportion to the importance of those spaces in the overall size of the house. The principle adopted by the Observatory of the Real Estate Market (the Italian abbreviation is OMI) for the calculation of the surface area of the main property uses, therefore including residential, is the assumption of the square meter of cadastral surface area as the unit of measurement of the consistency of property [12], also used in the professional practice of real estate agents and consultants.

Table 1. Examples of reductive coefficients adopted by the OMI [11].

Type of space	Reductive coefficient
Cellars and attics (communicating or not with the main rooms)	25–50%
Balconies and terraces (open or covered), patios and porches	25–35%
Own garden	10–15%
Verandas	60%
Habitable attics	80%
Habitable basement rooms	60%

Moreover, when calculating the commercial surface area of an apartment within a condominium, and its market and transformation value, the shares relating to the areas of common use are not included in the calculation relative to the single real estate unit – e.g., condominium meeting rooms, covered and uncovered areas, common stairs, landings, and balconies – neither garden, green areas nor walkways pertaining to the building [11]. In this sense, the potential added value of these spaces is not specifically highlighted in terms of both social and economic value.

What we are arguing is the need to rethink some modes of evaluation, in a reflection on the role of evaluation practice more generally in the design process of space transformation [13–15], in relation to the potential of creating value precisely through the project. Indeed, on the one hand, a demand for specific functions and spaces features, such as adaptability, emerged during Covid-19 both at the apartment and at the building levels; on the other hand, there are spaces that are worth more than what they are valued and that can be transformed in this direction. Moreover, since these accessory rooms are underused, they can be conceived as resources to be activated in response to emerging needs and demands, as non-static entities, with an adaptive reuse approach [3, 16–18].

In line with the aim, our methodology is structured as follows. First, assuming a field of analysis limited to a single European country (Italy), we propose an updated picture of the implications of the pandemic, in terms of demand for adaptable spaces, by reviewing, through indirect sources on the real estate market, the trends in the residential sector to highlight the current demand. Second, to identify some trends in the intensified use of spaces and to highlight what spatial, economic and ownership characteristics made them a particularly favorable adaptable space, we analyze 10 national case studies, to show the added value of specific architectural solutions related to accessory spaces, developed before and/or after the pandemic. In this sense, a reflection on the relation among evaluation practices and design strategies is directed to the assessment of the potential for value creation through the project of those spaces, as the relationship between the demand analysis, the possibilities in terms of adaptive reuse of residential spaces and the transformation value.

3 Reading the Real Estate Market Trends

3.1 The Emerging Demand

The global economy and the activities of real estate operators have slowed down overall at the beginning of 2020 [19–21], arising the question whether the impacts of Covid-19 on the Italian real estate market could permanently affect demand [22]. Investors have reacted in different ways, but in general there has been a reversal in decision-making processes: where they have not withdrawn, they have put negotiations on hold or asked for discounts on those advanced [20], thus affecting all economic activities related to the real estate market, despite the latter being the sector that has lost the least dynamism. In Italy, overall, the real estate landscape remained in fact evolving and 2020, even closed with 30% less than the previous year, reached almost 9 billion euros invested, still proving to be among the best results in the last twenty years. The living sector had an acceleration, whereby an increasing number of investors produced a volume that replicated four times that obtained in 2019 [20]. Relative to purchases and sales, looking at 2020, the health emergency and subsequent lockdown periods seem to have led to the need and desire of families to live in homes with a larger size [23]. The Economic Survey of the Italian Housing Market confirms this hypothesis, reporting an increase in purchases and sales for homes with a medium size [24]. Indeed, the 56.4% of the properties bought and sold after the outbreak of the epidemic had a size between 80 and 140 square meters, and properties with a size of less than 80 square meters accounted for 37.8% of brokerage [23].

During 2021, the trend remains constant [25] and, in the perception of real estate operators, the characteristics of the dwellings sought by potential buyers are increasingly oriented towards independent solutions and with outdoor spaces, complicit with the possibility of remote work, especially in metropolitan areas. Focusing on the type of demand for those planning to buy a home, the pandemic has changed habits and led to a reflection on what may be new concepts of living, because if the habits and needs of people change, inevitably change the spaces. Housing today must be designed to accommodate the different habits of each. Among the new requirements there is a need for greater flexibility, for houses that are increasingly adapted to heterogeneous needs, to new lifestyles and consumption, as "fluid" and multifunctional houses [26]. A more enriched demand for living emerges, linked to the search for a better quality of context and services [27], with different spaces, possibly larger, comfortable, and equipped with balconies, gardens, or terraces. Spaces, these, which have become very important evaluation criteria in defining the priorities of living, even at the level of the building, co-working condominiums, rooftops to be rediscovered as living and meeting spaces, landings with a role not of mere passage, but of filter [26]. In this sense, the offer of housing seems increasingly to have to concern a more complex and social living, of which there is still little trace both in the policies [27] and in the evaluation methodologies.

3.2 An Already Existing Offer

The second part of the analysis focuses on the qualitative case study of a series of recent housing projects (Table 2). The choice of the case studies was based on five key criteria. First, all the selected projects are in Italy to ensure a greater coherence with the bureaucratic system and consistently with what was previously discussed in relation to evaluation tools and methods. Second, since the objective of this part of the study is to investigate a time frame sufficiently recent but not restricted to the contingency of Covid-19 pandemic, they are all projects whose design phase has been completed within the last 15 years and whose construction phase has also been terminated – except for the most recent one. Third, from a typological point of view, they are exemplary of different forms of collective housing, while from the point of view of economic-management objectives, they correspond to affordable and/or social housing initiatives. Fourth, because of their social status, they necessarily respect a logic of constructive and managerial economy which favor simple and agile solutions and they have passed a particularly binding bureaucratic process implying public recognition. Finally, they are the result of design competitions for new construction, a criterion aimed at focusing the research on projects paying specific attention to urban and architectural design operating in favorable conditions from the point of view of contextual constraints.

Table 2. The selected case studies.

ID	Project name	Year	Location
1	Abitare nel parco	2008	Bolzano
2	Abitare a Milano - Via Gallarate	2009	Milano
3	Parma Social House	2010	Parma
4	Social Housing Casanova	2011	Bolzano
5	Cenni di Cambiamento	2013	Milano
6	Figino. Il borgo sostenibile	2014	Figino
7	Progetto Zoia	2014	Milano
8	Social Housing Romea	2015	Ravenna
9	Social Village Cascina Merlata	2016	Milano
10	House in Milanosesto	2019	Milano

Overall, the design solutions analyzed are easily translatable to a slice of the existing residential stock that is both particularly vulnerable but also the bearers of exceptional opportunities for transformation in recent times. From an operational point of view, the study was mainly based on design drawings, photographs, and written descriptions freely accessible online or in trade publications. In this sense, most of the features we observe and declare as present in the projects can therefore be verified through a documentary consultation on the projects. As a result of the in-depth analysis and comparison of the projects, we highlight some recurrences in the design priorities, which have been identified and classified with respect to three general spatial domains to identify some common trends (Table 3).

As regards the private domain of the single residential unit, it should be noted first that in all cases it is recognized the need to provide the apartments with accessory spaces – cellars; warehouses; car boxes, etc. – and private open spaces such as balconies, loggias, terraces, or, when possible, private gardens, obtained as a result of a general tendency to privatize space even in housing typologies traditionally organized only with a common court. The variety and adaptability of housing units represents another central theme, which assumes different declinations. While almost all cases pay particular attention to the heterogeneity of the residential offer – both in terms of type and size of the apartments and of housing tenures – and to the modularity of construction as a guarantee of possible and scalable subsequent changes, there are still relatively few projects that provide flexible solutions for the active modification of space by users. Energy efficiency and environmental sustainability are evidently issues of increasing priority that influence the design from the scale of the individual unit to that of the building and the neighborhood. As far as the communal domain is concerned, the most evident trend is the great attention paid to the increase of shared spaces and functions both as a strategy of efficiency and optimization and as a stimulus to sociability and community life. This tendency becomes visible in the care taken in the design of semi-public protected courtyards (often open to public access) or in solutions such as the oversizing of common distribution spaces and the provision of functional areas to accommodate community activities. Finally, the residential projects, object of the analysis, often extend into the public realm, with the aim to create actual pieces of the city in all their complexity. The residential project then becomes an opportunity to introduce new functions and public spaces, to provide new green spaces and urbanize the area of intervention with the design of urban furniture and public uses. While this implies an improvement in the quality of urban space, it is also emblematic of the recognition of the value of public spaces to the quality of residential supply.

Table 3. The emerging key factors in relation to the projects.

Domain	Emerging key factor	Case study (ID)
Private domain	Flexibility/dynamicity (e.g., partitionability)	1; 2; 4; 8
	High variety of the apartments offer	1; 2; 3; 5; 6; 7; 8; 9; 10
	Modular space	1; 2; 3; 4; 5; 6; 7; 8; 9; 10
	Balconies, loggias, terraces	1; 2; 3; 4; 5; 6; 7; 8; 9; 10
	Private garden (on the ground floor)	1; 3; 4; 5; 8

(*continued*)

Table 3. (*continued*)

Domain	Emerging key factor	Case study (ID)
	Accessory spaces (e.g., cellars, garages, storage rooms)	1; 2; 3; 4; 5; 6; 7; 8; 9; 10
	Energy efficiency systems (apartment scale)	1; 2; 5; 6; 8; 9 (after renovation); 10
Shared domain	Oversized distribution spaces (as a space for socialization)	1; 3; 5; 7; 8; 10
	Semi-public courtyards, courts, gardens	1; 3; 4; 5; 6; 7; 10
	Indoor communal premises (on the ground floor)	1; 3; 5; 6; 7; 8; 9; 10
	Communal services (e.g., bike parking/sharing; kitchen; laundries; concierge)	2; 3; 4; 5; 6; 7; 9; 10
	Communal terraces/green roofs	5; 8; 9; 10
	Energy efficiency systems (building scale)	1; 2; 4; 5; 6; 7; 8; 9; 10
Urban domain	New public services/Functional mix	2; 5; 6; 7; 9; 10
	Public greenery	2; 4; 5; 6; 7; 8; 9; 10
	Urban furniture/design	2; 4; 5; 6; 7; 8; 9; 10
	Energy efficiency systems (district scale)	2; 6; 7; 10

From a design standpoint, it is therefore possible to identify some tendencies in the considered case studies, which highlight partly innovative approaches and priorities and partly a revival and reinterpretation of solutions rooted in the Italian residential tradition. Many of these tendencies, in each case, anticipate needs that the pandemic has made even more evident and urgent. In synthesis, it means, on the one hand, equipping housing with dynamic spaces, from furniture to the possibility of splitting the spaces themselves for the coexistence of several functions at the same time; on the other hand, designing spaces capable of adapting to future changes, in a modular way, therefore reconfigurable [9]. In addition, the presence of balconies, loggias, terraces acquire an important added value [3, 7–9] as a possibility of external access and filter that preserves privacy and promotes coexistence with a general increase in shared spaces and activities. In fact, considering the communal domain, the greater attention paid to the provision and design of semi-private spaces at the scale of the building or estate enhance the transformability of these spaces, with the flexible adaptation of both indoor and outdoor spaces pertaining to the building [1, 3, 6–8], which become a source to accommodate functions and uses usually performed at the city level. With the possibility of rethinking such common spaces, it is possible to rediscover a surplus value related to flexible common services offered in larger environments than a housing arranged in strategic and barycentric points of the building – with the consequent opportunity of optimizing the use of private spaces for other functions [9]. Conversely, the greater care given to the design of public space, an aspect often neglected in the history of large-scale social housing construction, significantly improves not only the overall quality of the residential environment but also the quality of life it offers - with the introduction, for example, of proximity services.

4   Rethinking the Transformation Value of "Latent" Spaces

According to the results of the analyses of recent trends,, we draw on the concept of transformation value [28–30] as an appraisal criterion that evaluates an asset susceptible to transformation, and is based on certain requirements, which are: the property under appraisal can be transformed and/or changed in use; the transformation is deemed more profitable than the de facto conditions at the time of appraisal; the transformation entails a non-zero cost related to the transformation process. The transformation of an asset, obviously subject to change, is therefore given by the difference between the expected market value of the transformed asset and the cost of the transformation itself. In brief: $Vt = Vbt - Ct$, where: Vt = value of transformation; Vbt = market value of the transformed asset; Ct = cost of the transformation.

Consequently, we propose a reflection on the potential of enhancing the transformation value of residential building by: (i) increasing the potential market value of the transformed asset by re-discussing the reductive coefficient to some accessory or shared spaces, that can acquire value based on their flexibility and adaptability to different uses; (ii) reducing the cost of the transformation by proposing projects that conceive an adaptive reuse of such spaces to better respond to emerging demands in line with trends accentuated in the pandemic phase.

The first result of the research is therefore to have highlighted the transformation value of a certain category of spaces, which acquire particular interest in a perspective of adaptive reuse at low cost. It is therefore implicitly suggested the possible revision (i.e., by increasing the reductive coefficient) of residential valuation models for spaces such as: private open spaces; adaptable accessory spaces; modular and flexible areas; communal indoor and outdoor spaces; and even, in an extreme scenario, neighborhood services and public spaces. As a result, the research also highlighted how a project, implemented from the perspective of leveraging the potential for low-cost adaptability of these spaces, creates value, in itself. A series of design solutions developed from the scale of the apartment to that of the neighborhood can in fact potentially significantly increase the transformation value of a property in the sense described above. Houses where underused spaces can be allocated to new functions and transformed over time; where quality private open and closed spaces coexist with the enhancement of common areas and services in a logic of both optimization in the use of space and social sustainability; where the quality of the design of public, private and semi-private spaces contribute to the overall value of residential life.

Finally, these observations bring to light how especially low-value residential buildings and neighborhoods can acquire a much higher value from a transformation value perspective. This is the case of public and social, peripheral and/or degraded residential neighborhoods, which often present favorable characteristics for adaptive reuse by having (i) wide spaces outside the apartment; (ii) modular and standardized building systems; (iii) ample opportunity to invest in community and shared spaces and activities. Moreover, today these same neighborhoods in Italy benefit from special tax advantages for reuse.

5 Conclusions

The paper has highlighted a trajectory intended to emphasize the transformation value as a possible key to rethinking (i) the traditional models used to calculate the value of residential units; (ii) the priorities and strategies to be adopted in the design of such spaces; (iii) the potential value of a number of undervalued spaces and, in particular, of a specific segment of the existing residential stock.

If the need to discuss an alternative assessment model is outlined, the question of what the specific characteristics of a model of value creation through the project are remains open. Is it possible to imagine a model that allows quantifying the value related to this relationship between demand analysis, adaptive reuse of underused spaces, and transformation value? And how would it affect investment policies, for example as an incentive for institutional investors to invest in entire asset classes based on these new logics of real estate valorization through adaptive reuse? The research presented introduces these questions, leaving ample room for further investigation from the perspective of both design strategies and evaluation techniques.

References

1. Capolongo, S., et al.: COVID-19 and Cities: from urban health strategies to the pandemic challenge. A Decalogue of Public Health opportunities. Acta Biomed. **91**, 13–22 (2020)
2. Kang, M., et al.: COVID-19 impact on city and region: what's next after lockdown? Intern. J. Urban Sci. **24**, 297–315 (2020)
3. Quaglio, C., Todella, E., Lami, I.M.: Adequate housing and COVID-19: assessing the potential for value creation through the project. Sustainability **13**, 10563 (2021)
4. Bianchetti, C., Boano, C., Di Campli, A.: Thinking with quarantine urbanism? Space C **23**, 301–306 (2020)
5. Mehta, V.: The new proxemics: COVID-19, social distancing, and sociable space. J. Urban Des. **25**, 669–674 (2020)
6. Tokazhanov, G., Tleuken, A., Guney, M., Turkyilmaz, A., Karaca, F.: How is COVID-19 experience transforming sustainability requirements of residential buildings? a review. Sustainability **12**, 8732 (2020)
7. D'Alessandro, D., et al.: COVID-19 and living space challenge: well-being and Public Health recommendations for a healthy, safe, and sustainable housing. Acta Biomed. **91**, 61–75 (2020)
8. Megahed, N.A., Ghoneim, E.M.: Antivirus-built environment: lessons learned from COVID-19 pandemic. Sustain. Cities Soc. **61**, 102350 (2020)
9. Tucci, F.: Pandemia e Green City. Le necessità di un confronto per una riflessione sul futuro del nostro Abitare. In: Ronchi E., Tucci F.: Pandemia e sfide green del nostro tempo. Roma, Green City Network e Fondazione per lo Sviluppo Sostenibile (2020)
10. Jefferies, T., Cheng, J., Coucill, L.: Lockdown urbanism: COVID-19 lifestyles and liveable futures opportunities in Wuhan and Manchester. Cities Health, 1–4 (2020)

11. Osservatorio del Mercato Immobiliare: Manuale della Banca Dati delle Quotazioni dell'Osservatorio del Mercato Immobiliare Istruzioni tecniche per la formazione della Banca Dati Quotazioni OMI. Allegati. Agenzia delle Entrate, versione 2.0 aggiornata al 19 gennaio 2017

12. Decreto del Presidente della Repubblica del 23 marzo 1998 n. 138: Regolamento recante norme per la revisione generale delle zone censuarie, delle tariffe d'estimo delle unità immobiliari urbane e dei relativi criteri nonché delle commissioni censuarie in esecuzione dell'articolo 3, commi 154 e 155, della legge 23 dicembre 1996, n. 662

13. Mondini, G.: La valutazione come processo di produzione di conoscenza per il progetto. Valori e Valutazioni 3 (2009)

14. Mecca, S.: La valutazione nello scenario di cambiamento del progetto di architettura. Valori e Valutazioni 23, 15–17 (2019)

15. Fattinnanzi, E., Micelli, E.: Valutare il progetto di Architettura. Valori e Valutazioni 23, 3–14 (2019)

16. Douglas, J.: Building Adaptation, 2nd edn. Elsevier Ltd., Amsterdam (2006)

17. Lami, I.M.: Shapes, rules and value. In: Lami, I.M. (ed.) Abandoned Buildings in Contemporary Cities: Smart Conditions for Actions. SIST, vol. 168, pp. 149–162. Springer, Cham (2020). https://doi.org/10.1007/978-3-030-35550-0_9

18. Robiglio, M.: The adaptive reuse toolkit. how cities can turn their industrial legacy into infrastructure for innovation and growth (2016)

19. McKinsey & Company: Commercial real estate must do more than merely adapt to coronavirus. https://www.mckinsey.com/industries/private-equity-and-principal-investors/our-insights/commercial-real-estate-must-do-more-than-merely-adapt-to-coronavirus, Accessed 20 Dec 2021

20. Cushman & Wakefield. Italy Real-Estate Overview (2021). https://www.cushmanwakefield.com/it-it/italy/insights/italian-real-estate-overview, Accessed 20 Dec 2021

21. ISTAT: I e II trimestre. Mercato immobiliare: compravendite e mutui di fonte notarile (2020). https://www.istat.it/it/archivio/254283, Accessed 20 Dec 2021

22. Lami, I.M.: Editoriale. Valori e Valutazioni 28, 1–2 (2021)

23. PICTET: Mercato immobiliare: andamento e prospettive in era COVID-19. https://www.am.pictet/it/blog/articoli/mercati-e-investimenti/mercato-immobiliare-andamento-e-prospettive-in-era-covid-19, Accessed 20 Dec 2021

24. Banca d'Italia: Sondaggio congiunturale sul mercato delle abitazioni in Italia. IV trimestre (2020). https://www.bancaditalia.it/pubblicazioni/sondaggio-abitazioni/2020-sondaggio-abitazioni/04/index.html, Accessed 20 Dec 2021

25. Banca d'Italia: Sondaggio congiunturale sul mercato delle abitazioni in Italia. II trimestre (2021). https://www.bancaditalia.it/pubblicazioni/sondaggio-abitazioni/2021-sondaggio-abitazioni/02/index.html, Accessed 20 Dec 2021

26. Urban@it: Come abiteremo nel post Covid? https://www.urbanit.it/come-abiteremo-nel-post-covid/, Accessed 20 Dec 2021

27. Nomisma: La Casa e gli italiani. 14° Rapporto sulla Finanza Immobiliare (2021)

28. Simonotti, M.: La stima immobiliare. Con principi di economia e applicazioni estimative. UTET, Torino, (1997)

29. Gabrielli, L., Lami, I.M., Lombardi, P.: Il Valore di Mercato: note di lavoro per la stima di un immobile residenziale, Celid, Torino (2010)

30. Roscelli, R.: Manuale di estimo, valutazioni economiche ed esercizio della professione. De Agostini-UTET Università, Novara (2014)

Back to School. Addressing the Regeneration of the Italian School Building Stock in the Latent Pandemic Contingency

Caterina Barioglio$^{(\boxtimes)}$ (ID) and Daniele Campobenedetto (ID)

Department of Architecture and Design, Politecnico di Torino, Viale Mattioli, 39, 10125 Turin, Italy
{caterina.barioglio,daniele.campobene-detto}@polito.it
https://www.polito.it, https://full.polito.it

Abstract. In 2020, the health emergency put school buildings to the test, highlighting the limits of a dated heritage in need of urgent interventions. The approximately 40,000 buildings that make up the Italian school infrastructure are a layered and widespread legacy throughout the country, which now requires rethinking in light of the social, demographic, and pedagogical changes that have arisen in recent decades. The opportunity to invest in the existing school building stock, fuelled by national and European funds, requires an extended effort of exploration, analysis, and measurement of the school building heritage. In this framework, the *Re-school* research project was born to provide tools for systematising knowledge on the Italian school infrastructure. The main objective is to support the public authorities in strategic planning to activate a process of regeneration of the school building stock on a territorial scale. This article describes a proposal for a working method to explore the transformative potential of school buildings, thus allowing the adaptation of the existing infrastructure to face situations marked by sudden change (e.g., a pandemic or a drop in demographics).

Keywords: Educational building · Strategic planning · Urban regeneration · Heritage

1 Introduction

The COVID-19 pandemic put the global education system into jeopardy, resulting in an unprecedented disruption to education worldwide that has affected more than 1.6 billion students and amplified the pre-existing learning crisis [1, 2].

At least a third of the world's schoolchildren – 463 million children in 188 countries – were unable to access remote learning when COVID-19 shuttered their schools. Globally, from late February 2020 until early August 2021, education systems were on average fully closed for 121 instructional days and partially closed for 103 days [3]. Looking specifically at the Italian situation, according to the Italian National Institute of Statistics (ISTAT) evaluation, about three million children aged 6–17 had difficulties attending school during the lockdown [4].

F. Calabrò et al. (Eds.): NMP 2022, LNNS 482, pp. 1486–1495, 2022.
https://doi.org/10.1007/978-3-031-06825-6_143

In this situation, the pandemic put a significant strain on the spaces of existing school buildings. The management of activities in classrooms, the flow of people, the organisation of daily breaks, and the movement of students inside and outside the buildings are just some of the many challenges that the pandemic emergency brought to primary and secondary schools. School principals, teachers, and local authorities have also faced the urgency of rapidly adapting educational spaces to facilitate the return to in-person school activities after the end of multiple lockdowns.

Operating school facilities during a health emergency accentuated the pre-existing limits of a degraded and dated building heritage, which now requires urgent interventions. The approximately 40,000 buildings that constitute the Italian school infrastructure comprise a layered and widespread legacy throughout the country, which has required some rethinking based on the social, demographic, and pedagogical changes in recent decades.

Fuelled by national and European funds and programmes – the first of which was the "Piano Nazionale di Ripresa e Resilienza" – the country now has the opportunity to invest in the existing school building stock through strategic actions (on both the national and local scales) that require an extended effort of exploration, analysis, and measurement with respect to the existing school infrastructure.

In this framework, the *Re-school*[1] research project was founded. The project was developed to provide tools to systematise knowledge of the existing school infrastructure. The main objective is to support public authorities in strategic planning to activate a process of regeneration of the school building stock on a territorial scale [5]. This article, which summarises the early outcomes of the *Re-school* project, outlines a proposal for a working method to explore the transformative potential of existing school buildings in Italy.

2 A Transformative Model: From Typology Recognition to Design Action

During the health emergency, the challenges that local authorities face when dealing with school buildings are related to the management of the ordinary [6]. To develop territorial plans, it is necessary to overcome the logics that have oriented the main models of Italian heritage regeneration to date. The first model is based on *experimental projects* on individual buildings, which encourage qualitative transformations (and serve as a model for other projects) that are enacted on a "case-by-case" basis. Another well-established model is based on *extensive up-to-standard actions*, which have as a main aim compliance with minimum requirements, often without consideration of the intervention's overall quality [7].

The starting point of the *Re-school* project is the proposal of an alternative model of intervention that attempts to close the gap between the previous approaches. This

---

[1] *Re-school* is a research project carried out by the Future *Urban Legacy* Lab (FULL), based in Politecnico di Torino (http://www.full.polito.it). The *Re-school* research team members (2021) affiliated with FULL include the following: Matteo Robiglio (scientific director), Caterina Barioglio (coordinator), Daniele Campobenedetto, Marco Cappellazzo, Elena Guidetti, Caterina Quaglio, Giulia Sammartano, Nannina Spanò, Ilaria Tonti, and Emere Arco.

reset is based on the consideration of school buildings as territorial infrastructures. Such an approach entails moving from a "case-by-case" model to the adoption of a comprehensive perspective on the whole stock of buildings [8]. The model developed by the *Re-school* research team aims at considering both the identification of homogeneous groups of school buildings according to their physical consistency and the specificities and potentials of different territories.

2.1 Measure

Italian local authorities have access to an extraordinary amount of data on school building heritage, which has been collected in the national database *Sistema Nazionale dell'Anagrafe dell'Edilizia Scolastica* (SNAES) and in its regional "nodes" of the *Anagrafe Regionale Edilizia Scolastica* (ARES) [9]. The research team performs quantitative and qualitative analyses based on these datasets, identifying and quantifying recurring criticalities (e.g., with respect to the issue of sustainability and energy efficiency) and spatial resources (e.g., surplus space in schools located in non-urban areas). The main objectives of these actions are to spatialise and systematise knowledge concerning the school infrastructure and to provide an overview of the existing assets in terms of their transformative potential.

2.2 Assess

The second step involves classifying school buildings (defined by their layout, structure, massing, etc.) and their urban contexts (e.g., access points, relationship with the street network and the built environment in general, etc.) according to some recurring typologies. These typologies are the result of periods characterised by certain historical, legislative, and technical conditions [10, 11] that are reflected in the school building stock across Italy. The typological classification is also intertwined with the distribution of buildings on a territorial scale. The assessment of these two aspects combined aims at identifying statistically significant categories that can provide a concise but comprehensive description of the national school infrastructure. This approach allows the research team to explore the transformative potential of a building on an architectural level, with consideration of all the factors related to the territorial scale.

2.3 Select

By introducing assessment parameters (e.g., resources available for transformation, cost-benefit analysis, etc.), the results of the measurement can be evaluated and prioritised against the objectives of the local authorities. In this way, the analyses can provide guidelines to identify priorities for intervention on a territorial scale (Fig. 1).

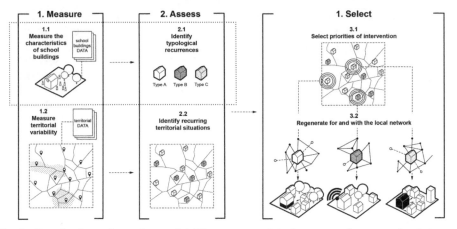

Fig. 1. *Re-school* transformative model. The present article focuses on the connection between 1.1 (Measure the characteristics of school buildings) and 2.1 (Identify typological recurrences), framed by the dotted line.

3 The "Ingredients" of Building Typologies: A Preliminary Exploration Through the Case Study of Turin

Does the Italian school building infrastructure have spatial resources that can be exploited to improve educational activities? Is it possible to identify the transformative potential of the buildings that make up the Italian school building stock? Could their lives be extended [12]?

To address these issues, it is useful to recognise some recurring types or groups of buildings with similar characteristics among the more than 40,000 Italian school buildings that are representative of this heritage. Such recognition can provide a description of the school building heritage through a limited number of architectural types and therefore constitutes a preliminary step for planning interventions on individual buildings. Each type of school building has a different transformative potential and can be described by four main elements: the layout, the structure, the position in the lot, and the relationship with the built environment.

The following taxonomy constitutes the first hypothesis that has been tested on the school building stock of the city of Turin, which, in terms of quantity, years of construction, and building techniques, is a consistent sample of the national stock [13] (Figs. 2 and 3).

3.1 Layout

The layout describes the organisation of spaces in the building. It is an important element in the recognition of transformative potential because it describes the relationship between functional spaces (e.g., classrooms, laboratories, canteens, gyms, etc.) and connective or multi-purpose spaces (e.g., corridors, atriums, and multifunctional spaces), as well as the massing of the building. Six main layout categories could be identified

Fig. 2. Abacus of the school buildings of the city of Turin. Source: "EDISCO Piemonte" database. Cartographic bases: "Carta Tecnica di Torino" (2019) and "BDTRE - base dati territoriale di riferimento degli enti" (2019) [5].

Fig. 3. Map representing the distribution of the school buildings in the city of Turin. Source: "EDISCO Piemonte" database. Cartographic bases: "Carta Tecnica di Torino" (2019) and "BDTRE - base dati territoriale di riferimento degli enti" (2019) [5].

according to a phenomenological analysis of the school building stock of the city of Turin, as well as some of the main studies on Italian school buildings [14]:

1. *Wing*: Single building developed in length, with corridor circulation;
2. *Court*: Wings organised around open space;
3. *Comb-shaped*: main wing on which smaller wings are grafted;
4. *Block:* One or more buildings developed in height, with a central common space;
5. *Pavilions*: Several autonomous buildings, connected by connective spaces, including external ones;
6. *Platform:* Building developed horizontally.

3.2 Structure

The structure describes the construction technology that characterises the load-bearing elements of the building. It is essential to the recognition of potential because it constitutes one of the main constraints for any transformation. With the same method applied before, two main categories can be identified that describe the construction systems of Italian schools:

1. *Continuous structures*: Masonry or concrete walls;
2. *Frame structures*: Pillars and beams with non-load-bearing walls.

3.3 Position in the Lot

The position in the lot characterises the way in which the buildings occupy the lot in which they are located [15]. This feature is particularly important for identifying the transformative potential of the school's open spaces. At least three main positions can be identified:

1. *Alignment to one or two sides;*
2. *Alignment to three or more sides;*
3. *Centred.*

3.4 Relationship with the Built Environment

This element refers to the built environment surrounding the school building. Three main categories can be identified:

1. *Dense fabric*: The building is surrounded by a dense and compact built fabric;
2. *Edge*: The building lot borders a dense built fabric on one or two sides, and the other sides extend towards an open space;
3. *Suburb*: The building lot is surrounded by open spaces or scattered buildings.

4 Recurring School Building Types and Their Transformative Potential[2]

The categories of *layout, structure, position on the plot,* and *context* are useful for identifying the main types of buildings that characterise the Italian school building

---

[2] All figures of this chapter are published in [5].

heritage. To describe this heritage in an almost exhaustive way, far fewer typologies are needed than those that would result from the combination of all the categories proposed. The specific characteristics of each type of building express a potential for transformation that is useful for identifying intervention priorities for parts of the overall heritage and/or actions to be taken on individual buildings.

4.1   Classroom-Based School

Description. These are wing-shaped, comb-shaped, or courtyard-shaped buildings with two or more floors. They are mainly characterised by corridors that connect a series of classrooms and are aligned on the edges of the lot. The load-bearing structure is continuous and is frequently made with masonry. In many cases, this typology describes mid-nineteenth-century buildings that are widespread in large cities and within the consolidated urban fabric of smaller centres.

Potential. Because of the masonry structure and the traditional technologies used in these buildings, rethinking the layout is particularly difficult. Concerning the interior, the corridors, as well as the large entrance halls, could be suitable for educational activities for individual students or small groups. The classrooms are characterised by above-average surfaces and heights.

4.2   Community-Based School

Description. This type includes courtyard- or comb-shaped buildings of two or more floors aligned with the edges of the lot. Inside the building, the classrooms and corridors are characterised by large windows to allow for the flow of natural lighting. Although very often built in dense urban environments, these buildings could have a courtyard; sports equipment (i.e., gym) is usually present within the buildings. In most cases, the frame structure is made of reinforced concrete. These schools were built mainly between the 1920s and 1940s; however, examples can also be found from the second half of the century.

Potential. The introduction of reinforced concrete allows for the reorganisation of interiors, enabling the unification of contiguous spaces or the reorganisation of connective spaces. The long corridors could be used as extensions of classrooms. The presence of more than one staircase, due to the articulation of the building wings, enables the diversification of paths within the building. Smooth surfaces and simple volumes, typical of *Rationalism*, ease the intervention process on the building envelope. The courtyards can be considered large open-air "rooms" that can potentially be adapted to host a wide range of educational activities.

4.3 Cluster-Based School

Description. This is one of the most recurring typologies in the Italian school building stock. Its definition is based on the organisational principle of the *cluster*, a group of classes belonging to the same section or year that share a common space. This typology includes block- or wing-shaped buildings with two or more floors, often with separate buildings hosting sports facilities. The structure is often made of reinforced concrete, and cases of prefabricated structures are not rare. Early constructions date back to the 1960s, but many others can be dated to the following decades, up to the 1990s.

Potential. This type is particularly interesting in terms of the transformative potential of the interiors. Wide connective spaces and the possibility of eliminating internal partition walls are conditions that allow the building to host public activities. The position of these buildings within the lot favours the identification of diversified entrances, both to the school lot and to the building.

4.4 Small-Group School

Description. This typology includes buildings divided into one or more groups of one- or two-story pavilions positioned in the centre of the lot. The structure is, in many cases, realised with prefabricated elements. These are mainly buildings dedicated to kindergarten or primary schools built during the 1960s or 1970s, which can be found in any type of built environment, but especially in large cities.

Potential. The small size and widespread distribution in the urban area favour the reuse of these buildings as territorial structures for various types of services. The pavilions, often directly accessible from the outside, favour diversified access for activities open to an enlarged community. Experimentation with building techniques and materials can pose maintenance and safety problems.

4.5 Platform School

Description. These are buildings developed horizontally, on one or more floors, and often positioned in the centre of the lot. The interiors are characterised by a complex network of circulation spaces, common areas, and services. The load-bearing structure is made of reinforced concrete or prefabricated systems. Many of the schools belonging to this typology were built in the 1970s as community centres in the expanding suburbs of large cities or educational centres in smaller cities (i.e., that host all the school grades in one building).

Potential. The dimensions and complex articulation of the spaces make these schools particularly interesting for the reorganisation of interiors. The oversized connective spaces allow for different types of educational activities to be hosted, while the horizontal extension makes it easy to identify alternative access points. The provision of large open spaces, which is very common in these cases, constitutes important potential for outdoor activities.

5 Conclusions

This article summarises a methodological approach to dealing with the regeneration of building stocks [16] and the evaluation of strategic actions to be taken on individual buildings. In dealing with built objects spread across wide regions and often designed and built by technical departments of public authorities, this approach is intended to facilitate understanding of how the transformation of the ordinary city could take place [17]. The focus of this first exploration is the school building stock; nonetheless, the method is conceived for different stocks that are widespread in a defined territory, with a limited number of managers/owners, and characterised by comprehensive dataset availability. Examples of such territorial infrastructure include social housing stock owned by public or private actors, one-company retail spaces, public or private health system facilities, and open public spaces (e.g., the network of green spaces within a dense urban environment). In all these cases, and within the field described above, it is possible to apply the method we described, especially the part related to the typological identification of buildings, to assess transformative potential.

The double-scale approach (from the territorial to the architectural scale) of this regeneration method is intended to address certain contingencies that affect the built environment (e.g., the pandemic) [18, 19]. In doing so, it can contribute to the development of evaluation tools that support decision-making processes, thus closing the gap between planning orientations and technical or architectural applications.

Acknowledgements. The authors contributed equally to the research and writing of this article.

References

1. UNESCO, UNICEF, World Bank: The State of the Global education crisis: A path to recovery (2021)
2. UNESCO, UNICEF, World Bank: What Have We Learnt?: Overview of Findings from a Survey of Ministries of Education on National Responses to COVID-19. Washington, DC (2020)
3. UNICEF: COVID-19: Are Children to Continue Learning During School Closures? A Global of the Potential Reach of Remote Learning Policies Data from 100 Countries. United Nations Children's Fund, New York (2020)
4. ISTAT: Rapporto annuale 2020. La situazione del Paese. Roma (2020)
5. Barioglio, C., Campobenedetto, D. (eds.): Re-school. Ripensare la scuola a partire dagli spazi. Politecnico di Torino, Torino (2021)
6. Fare spazio. Idee progettuali per riaprire le scuole in sicurezza. https://www.fondazioneag nelli.it/wp-content/uploads/2020/08/Fondazione-Agnelli-e-FULL-PoliTo-FARE-SPAZIO-030820.pdf. Accessed 29 Dec 2021
7. Barioglio, C., Campobenedetto, D.: The school as a model. two experimental urban school building in Turin. 1968–75. FAM Mag. **56**, 79–90 (2021)
8. Bertaud, A.: Order Without Design: How Markets Shape Cities. MIT Press, Cambridge MA (2018)
9. SNAES Homepage. https://dati.istruzione.it/opendata/opendata/catalogo/elements1/?area= Edilizia%20Scolastica&&pk_vid=1cbf0c7f2bcc104a1640794742a0d5c2. Accessed 29 Dec 2021
10. Hertzberger, H.: Space and Learning. 010 Publishers, Rotterdam (2008)
11. Hille, R.T.: Modern Schools: a Century of Design for Education. John Wiley and Sons, Hoboken (2011)
12. Wong, L.: Adaptive Reuse: Extending the Lives of Buildings. Birkhäuser, Basel (2016)
13. Fondazione Agnelli (eds): Rapporto sull'edilizia scolastica. Laterza, Roma-Bari (2019)
14. Deambrosis, F., De Magistris, A.: Architetture di formazione: note sull'edilizia scolastica italiana del Novecento. Territorio **85**, 103–113 (2018)
15. Haupt, P., Berghauser Pont, M.: Spacematrix: Space, Density and Urban Form. Nai010 Publishers, Rotterdam (2021)
16. Abramson, D.M.: Obsolescence: An Architectural History. University of Chicago Press, Chicago (2017)
17. Amin, A., Graham, S.: The ordinary city. Trans. Inst. Br. Geogr. **22**(4), 411–442 (1997)
18. Cuff, D. (eds).: Contingency. Design and the Challenge of Change. Ardeth, 6 I - Spring 2020, Rosenberg & Sellier, Turin (2020)
19. Quaglio, C., Todella, E., Lami, I.M.: Adequate housing and COVID-19: assessing the potential for value creation through the project. Sustainability **13**(19), 10563 (2021)

Post-covid City: Proximity Spaces, Sharing Economy and Phygital

Federica Marchetti[✉]

Politecnico di Milano, Milan, Italy
federica.marchetti@polimi.it

Abstract. The pandemic has accelerated a series of processes taking place within our cities, forcing people to live, work and socialize exclusively in their own home for several months and, even today, for large portions of time. On the one hand, the use of digital devices and web space for work, study and leisure has allowed to carry out most of the daily functions and work activities, on the other hand many others have had to open to new uses producing unprecedented spatial possibilities. All these processes are interesting because they define more hybrid, flexible, lean and phygital characteristics of the space. The aim of this research is to demonstrate how the proximity and sharing spaces of the city can be characterized by these new aspects. The methodology that was used is the selection of contemporary examples thanks to literature and web sources. In the end, the case studies outline possible scenarios describing opportunities for the post-covid age in the near future.

Keywords: Proximity space · In-between spaces · Sharing economy · Phygital

1 Introduction

1.1 Proximity and Sharing Spaces for the City

After the experience of the lockdowns due to the pandemic, but also in the current situation, it seems that the city must not be divided into specialized parts but should be more and more polycentric and diversified. The adaptive reuse of interstitial spaces, in the existing housing stock [1] or in the public realm, but also the importance of the communities in the use and appropriation of these spaces are important elements to understand these processes and to define a new quality of the contemporary city environment. In fact, the design of urban and architectural spaces are crucial aspects for the improvement of the social sustainability, in the sense of a human centered value, linked to the needs of people from a cultural and psychological point of view, but also referred to adaptability and growth [2]. In particular, the concept of sustainability is not considered as a defined and end state, but as a complex and contextualized factor [3]. The aim of this work is to demonstrate how, in the frame of a sustainable environment in the Post-Covid city, proximity and sharing spaces can be defined by new characteristics, more hybrid, flexible, lean and between online and offline reality. For these reasons the research main questions are: how these spaces become key elements after the pandemic? And, in which way online and offline tools can interact with these aspects?

© The Author(s), under exclusive license to Springer Nature Switzerland AG 2022
F. Calabrò et al. (Eds.): NMP 2022, LNNS 482, pp. 1496–1504, 2022.
https://doi.org/10.1007/978-3-031-06825-6_144

The Post-Covid cities needs new requirements that involves many characteristics for the space design including general health risks, new forms of consumptions of the resources [4] and the needs to define new indicators for the urban realm looking at the future years. Finding new ones, for example in the frame of the Sustainable Development Goal 11 – one of the 17 goals, titled "sustainable cities and communities", established by the United Nations in 2015, [5] – or in the proposals of the National Institute of Urban Planning for overcoming the emergency and relaunching the country (INU) in Italy [6], are important goals for the quality of the environment.

An environment where the mediation of technology is important more than ever [7] and each inhabitant can find what they need in the *home proximity* [8] considering that the proxemic in the indoor and outdoor places is changed because of the pandemic [9] and the dispositions related to the social distance by the World Health Organization (WHO).

For example, among the current urban visions, one of the most relevant is *the 15-min City* by Carlos Moreno, scientific director at the Sorbone, and it is promoted in Paris [10] by the mayor Hanne Idalgo. This is a city where people, by bike or on foot, can reach all the necessities, workplaces (also in co-working) and spaces for free time.

After all the new needs that emerged with the spread of Covid, Paris is not the only city to reflect on strategies for a more human-sized city. Indeed, even before these events, cities such as Portland and Melbourne spoke of *20 min neighborhood*, while in Sweden today the *one-minute city* is being experienced.

Returning to the pre-pandemic situation, the use of digital technology allowed the search for domestic aspects in public places, open spaces and workplaces. Indeed, co-working and co-living spaces were topics of interest as well as the co-design of public space, where people from simple users become protagonists in its definition together with other actors involved: municipalities, investors, brands, etc..

The pandemic has radicalized this situation by having to seek, in the open proximity spaces or in large, enclosed spaces for shared activities, those services offered in more distant spaces, having dimensions and spatial possibilities that do not make them accessible due to containment measures for Covid.

This trend is combined with that of seeking open space and working space at home, taking advantage of the in-between spaces in one's own apartments or inside condominiums. Therefore, proximity spaces are enriched in the post-covid era, the close ones, often created in threshold areas, interstitial places between the inside and the outside, those that bind memory, reality and fantasy together [11] (Fig. 1).

It seems that the circle started with the contemporary social transformations has closed and that today the characteristics of these spaces are **hybrid, flexible, lean, online/offline**, but also **creative** because they increasingly exploit the collective intelligence of people, codifying changes to depending on the uses of the inhabitants.

Fig. 1. Paris: the 15-min city. Source: Paris en commun.

1.2 Methodology and Structure of the Work

The methodology adopted to investigate the topic is based on a qualitative selection of examples from literature and web sources. This work traces some of the most interesting lines that have been undertaken from 2020 to date thanks to the examination of several case studies. In fact, starting from best practices is possible to define new strategies and instruments to design the post-covid city.

In particular, these are the elements that guided the collection of case studies, considering a type of place that allows: 1. to re-inhabit unused spaces, especially outdoors; 2. to add activities and uses of space, making the environments more and more multifunctional and to allow flexibility of use that guarantees social distancing and containment of the number of people 3. to expand domestic and non-domestic equipment thanks to greenery and digital solutions.

In the text these three factors are reflected into the three groups of examples of the next paragraphs: Reuse and Co-design of the Space for Work, Educational and Social Uses, Hybrid Spaces for Flexible Uses and Social Distancing, Greener and More Phygital.

It is important to note that in all these cases the concept of community, physical but at the same time virtual, becomes important for the use and management of space, linking itself to the concept of Sharing Economy or even more of *Community Economy* [12] where the groups of people to collaborate in defining responses to contemporary needs.

2   Case Studies

2.1   Reuse and Co-design of the Space for Work, Educational and Social Uses

The growing importance of communities in the co-definition of spaces, in the management and creation of new relationships and processes, between the physical and digital world, influences the vision of the city and its possibilities.

In Sweden, taking up the concept of the 15-min City, they started practicing the one-minute city. Through the *Street Moves project*, conceived by the Vinnova agency in partnership with ArkDes and Lundberg Design, they are testing the possibility of providing citizens with the use of wooden modules to modify the nearby public space according to their needs. A project that draws inspiration from the virtual world - the Minecraft video game – but also from the physical one – in the ease of assembly for users, like an IKEA furniture or a LEGO structure. New seats, parking spaces for bicycles and electric scooters, small playgrounds etc. extend the space between the building and the street, defining places that increase urban facilities.

A possibility that has a strong link with local communities, in the idea that more inclusive environments are produced from sharing and exchange.

A concept reminiscent of *parklets*, urban devices to be inserted in the parking stalls along the streets, born with bottom-up events and spread first in the United States – especially in San Francisco and Los Angeles – and then spread through the NACTO protocols also in Europe, Asia and South America [13] (Fig. 2).

Fig. 2. Street moves in Stockholm by Vinnova, ArkDes and Lundberg design. Source: globetrender.com.

These elements could also be interesting in the reuse of underused environments within buildings or in collective spaces, such as courtyards, roofs, distribution areas, to modify and enrich the spaces.

In fact, these kinds of uses can involve not only the space between the buildings, but also the architecture itself, in particular the distribution spaces and the roofs, making them pandemic-proof as in the project for the Markham College Lower School in Lima, Peru, by Rosan Bosch Studio and IDOM (2021).

In this case the concept of interior and exterior becomes more ambiguous because the lessons and activities can take place both indoors and outdoors – on the roofs or in passageways thanks to the distribution of spaces and the use of movable partitions whose conformation can be decided according to the needs of the users. The relationship between the building, green spaces and trees also becomes tighter and more integrated, making the environments more dilated and open than the outdoor space.

The school, designed to reflects the needs of the post-pandemic age, can represent an interesting scenario for other educational buildings, but also for public or office ones (Fig. 3).

Fig. 3. Markham college lower school in Lima di Rosan Bosch Studio e IDOM. Sources: rosanbosch.com.

2.2 Hybrid Spaces for Flexible Uses and Social Distancing

The domestic space was at the center of the debate during the lockdown periods and a source of analysis and design experiments. For example, during the Salone del Mobile in Milan 2021 some design installations were presented to reflect on the post-Covid home. The *fluid house* by Elisa Ossino Studio and Marco Bay with the partnership of Elle Decor and 3M presents sinuous and multifunctional spaces. These elements generate a sensorially welcoming place that allows different activities and not only single functions in the environments: the kitchen with a domestic garden, or the bed hidden by a curtain to define a relax, fitness and study area etc.. But the workplace also acquires new characteristics. From the ideas of installing microarchitectures – pods – in the garden or in interstitial spaces such as the project *My Room in the Garden* by the London studio Boano and Jonas Prišmontas, to a new way of seeing co-working spaces in large cities (Fig. 4).

Fig. 4. My room in the garden by Boano and Jonas Prišmontas. Source: boanoprismontas.com.

In fact, from the research conducted by Università Cattolica del Sacro Cuore, TRAILab - The Transformation of Milan Coworking in the Pandemic Emergency [14] - and from the study on the Geography of Coworking Spaces in Milan by DAStU [15] - Politecnico di Milano, emerges as after the closure of many offices in the city center, both companies and individual citizens have increased the demand for work spaces near their homes, looking for solutions in the co-working areas closest to their apartments. For this reason, it was born the term *nearworking*.

But post-pandemic reflections also appear in the ex-novo design. In the offices for the Yıldız Technical University in Istanbul (2020), the architect Salon Alper Derinbogaz designs the spaces according to the new post-pandemic goals. The rooms are connected by open-air paths which, in good weather, can be used as office spaces. The use of air conditioning is reduced by introducing a passive geothermal heating and cooling system, which uses pipes buried in the ground to sustainably regulate the temperature. In fact, air recirculation is just a risk factor during the pandemic. The distinction between exterior

Fig. 5. Offices for the Yıldız technical university in Istanbul by Salon Alper Derinbogaz. Source: salonarchitects.com

and interior becomes hardly noticeable also thanks to the use of plants both inside and outside the rooms (Fig. 5).

2.3 Greener and More Phygital

In addition to the concepts of proximity and multifunctionality of space, two other elements emerge: 1. the search for the outdoors by people in contrast to a purely closed domestic or work space; 2. the presence of digital devices to integrate the user experience between the physical and virtual world.

In the first case, it can be seen from the world of Real Estate that the price of houses with a terrace or garden is growing. Taking into consideration a study on the Italian case by Immobiliare.it, it represents 8% and, at the same time, the prices of those without open areas have collapsed [16].

So, the multifunctionality of the living space can be envisaged as the starting point of the design, allowing an increase in domestic gardens and green spaces through modular models inside the houses, in the interstitial spaces of housing complexes and on the roofs. In this way, even the current housing offer could expand its facilities both as a private level and as single units or condominiums. For example, the Canadian city of Edmonton has promoted the creation of new Community Gardens where the inhabitants can grow plants and vegetables independently. In addition to the possibility, guidelines have also been guaranteed to allow a self-design suitable for the Covid moment.

Another theme is the concept of **phygital** [17], born in marketing to indicate the hybrid user path between the online e-commerce world and the physical retail space, which, in this case, can be extended to actions involving sociality, work and leisure, not just the shopping experiences. In fact, the concepts of proximity and multifunctionality can now be supported and amplified thanks to the use of digital supports. Apps and web platforms, but also touchless devices (qr code, voice commands…) can predict a city beyond the smart city, leaner in its processes and safer; a city where technology communicates with the safety and the social distance provisions due to the Covid problem. These dynamics between online and offline also include ideas such as the "network of hybrid spaces" designed for Milan with the participation of the organization Collaboriamo.

The mapping of all those activities that have begun to perform a social function as well as a productive one: bookstores-coworking, neighborhood bar-concierge, shops that connect people to regenerate urban spaces. This digital mapping process is a way to research online and reach offline places that can serve communities.

3   Discussion and Conclusion

In conclusion, what emerges are scenarios that bring social value through the combined use of actions by the communities and the intermediate space inside or next to the homes. A type of city that reaches its own complexity within 15 to 20 min; not zoned and specialized but characterized by multiple aspects that collaborate between real and online dimensions, between closed and open spaces.

What appear to be individual issues – proximity to services, work, leisure – in such a system bring benefits to the entire urban center. In fact, it would lead to a decrease in congestion in the center and a decrease in dormitory suburbs. The difficult pandemic moment could be a factor of impetus for all communities of people towards a more sustainable, effective and shared life thanks to their active role in daily processes.

On the other hand, it is the governance policies that need to be structured allowing the expression of these new values and supporting citizens in this process. For this reason, the knowledge, preparation and implementation of digital management processes and systems becomes important together with a strategy and physical design of the spaces using co-design methodologies and practices.

Considering the case studies, it is possible to define some key findings that can contribute to the development of strategies and tools for future projects.

First of all, the importance of proximity spaces that originally are considered less relevant in the general hierarchy of a project: for example, thresholds between exterior and interior, roofs and distribution spaces. Then, the definition of places that allows many activities and not only single functions, looking at the domestic space and at the work space as more related than in the past. In the end, a new attention for the green areas, seen as proximity spaces of the residential units that can be developed by communities. All these elements can be aided by the use of digital technologies that encourage collaboration between people thanks to their management and mapping abilities. The individual aspects, but most of all their combinations and implementation can represent interesting instruments for architects and designers to answers the new challenges of the contemporary society.

References

1. Quaglio, C., Todella, E., Lami, I.M.: Adequate housing and COVID-19: assessing the potential for value creation through the project. Sustainability **13**, 10563 (2021)
2. Lami, I.M., Mecca, B.: Assessing social sustainability for achieving sustainable architecture. Sustainability **13**, 142 (2021)
3. Schroeder, T.: Giving meaning to the concept of sustainability in architectural design practices: setting out the analytical framework of translation. Sustainability **10**, 1710 (2018)
4. Tokazhanov, G., Tleuken, A., Guney, M., Turkyilmaz, A., Karaca, F.: How is COVID-19 experience transforming sustainability requirements of residential buildings? Rev. Sustain. **12**, 8732 (2020)
5. Abastante, F., Lami, I.M., Mecca, B.: How covid-19 influences the 2030 agenda: do the practices of achieving the sustainable development goal 11 need rethinking and adjustment. Valori Valutazioni **26**, 11–23 (2020)
6. Talia, M.: Le proposte dell'Istituto Nazionale di Urbanistica per il superamento dell'emergenza e il rilancio del Paese (2020)
7. Low, S., Smart, A.: Thoughts about public space during Covid-19 pandemic. City Soc. **32**(1) (2020)
8. Manzini, E.: Abitare la Prossimità. Idee per la Città dei 15 minuti. 1st edn. Egea, Milan (2021)
9. Mehta, V.: The new proxemics: COVID-19, social distancing, and sociable space. J. Urban Des. **25**, 669–674 (2020)
10. City of Paris web page, https://www.paris.fr/dossiers/paris-ville-du-quart-d-heure-ou-le-pari-de-la-proximite-37 , last accessed 2021/12/29
11. Nuvolati, G.: Interstizi della Città. Rifugi del Vivere Quotidiano. 1st edn. Moretti e Vitali, Bergamo (2019)

12. Mainieri, M.: Community Economy. Persone che Rivoluzionano Organizzazioni e Mercati. 1st edn. Egea, Milan (2020)
13. Comune di Milano news. https://comune-milano.img.musvc2.net/static/105044/assets/2/Slide_coworking%20Indagione%20condotta%20da%20Universit%C3%A0%20Cattolica.pdf. 29 Dec Accessed 2021
14. Collaboriamo.org media page. https://collaboriamo.org/media/2021/07/Coworking_a_Milano_politecnico.pdf. Accessed 29 Dec 2021
15. NACTO. https://nacto.org/publication/global-street-design-guide/. Accessed 29 Dec 2021
16. Guglielminetti, E., Loberto, M., Zevi G., Zizza R.: Living on my own: the Impact of the Covid-19 pandemic on housing demand. Banca d'Italia. Questioni di Economia e Finanza. Occas. Pap. **627**, 5–31 (2021)
17. Andreula, N.: Phygital. Il nuovo Marketing, tra Fisico e Digitale. 1st edn. Hoepli, Milan (2020)

Digital Platforms, Imaginaries and Values Creation: Opportunities for New Urban Dynamics

Maria Cerreta$^{(\boxtimes)}$, Fernanda Della Mura, and Eugenio Muccio

Department of Architecture, University of Naples Federico II, 80134 Naples, Italy
{maria.cerreta,fernanda.dellamura,eugenio.muccio}@unina.it

Abstract. Starting from 2015 in Naples, Italy, the Short-Term Rental season (STR) has been inaugurated with Airbnb. In 2019 the apartments of the STR panorama constituted 12.35% of the entire residential real estate stock in the historic centre of the city, starting a gradual process of airification. The observed condition has produced a progressive expulsion of low-income residents and students, accompanied by the increase in evictions for the transformation of properties that are not very attractive for the ordinary market into b&bs to be valued through the sharing platform. During the first months of the Covid-19 pandemic, the condition faced led to a downturn in listings on Airbnb due to the real estate reorganisation mechanisms, with the conversion of part of the b&bs into new uses. Furthermore, with the reopening of national borders and the possibility of travelling, tourism activities have restarted the mechanisms of territorial promotion, conveyed above all through the images published on the web, which influence the choices of users and the dynamics of the urban values.

The study contemplates the effects of digital platforms on the production of the Neapolitan imaginary, which, in turn, influences the dynamics of the real estate market, where tourism has made "waste properties" attractive. New processes encourage investments in properties that are scarcely desirable for the ordinary market, defining a new type of specificities and generating urban regeneration processes relevant for the production of the interstitial rent.

Keywords: Overtourism · Real estate market · Interstitial rent

1 Introduction

Tourism in the digital era uses new communication channels that ideally bring the tourist closer to the destination. Since 2007, Airbnb has opened the doors to the home-sharing season on a global scale, leveraging the push given to the sharing economy [1] in the dark years of the great economic crisis, thanks to the explosion of urban tourism [2]. In this context, due to the 2007–08 economic crisis, resulting from the collapse of subprime mortgages and the subsequent stagnation of the real estate market, the traditional postulates of economic and social growth and the entire capitalist system have been questioned. Instead, monetary values are derived from the possessions/capacities,

F. Calabrò et al. (Eds.): NMP 2022, LNNS 482, pp. 1505–1515, 2022.
https://doi.org/10.1007/978-3-031-06825-6_145

acquiring the awareness that each property can turn into a potential profit and each person into a potential entrepreneur [3, 4]. The short-term rental of a property is not a new phenomenon, but digital platforms, as "disruptive innovation" [5–10] fueled its transformation, diversification and expansion in an unprecedented way [11]. In this scenario, the platform has triggered a dizzying growth in the accommodation capacity of cities [6, 12], influencing the availability and accessibility of housing in the long-term rental market [13–15]. Indeed, the economic advantages envisaged by sharing practices, due to profit-oriented investors [16], have exacerbated the pre-existing conditions of urban hardship experienced by residents, linked both to the problems of social segregation [17, 18], and to the increase in property prices [15, 19], giving rise to a city that lends itself to investment and that for companies like Airbnb becomes the archetype of a city of tourism, of a short-term city.

In this context, overtourism has intensified the old social problems linked to contradictions of capitalism [20].

Cities find themselves living a theatrical dimension. Life itself and its relationships are the objects of tourist consumption [21]. A post-recession urban landscape has thus been configured, which must be investigated to understand how platforms produce urban space according to the logic of "platform urbanism" [22]. User interactions with the platform, processed through algorithms, contribute to the production of space [23], shaping the boundaries of the tourist city and influencing its perception, feeding the dynamics of "platform capitalism" [24–28]. In this way, the complex flow of information shapes a digital dimension of reality parallel to it. It acquires a real strength thanks to its power to globally convey an image that immediately becomes a substitute for actual reality. The impact is clear: the urban fabric, previously defined by the logic of planning and by the real needs of the population, is now slowly being modified by an erosive process that outlines the contours of new urban geography linked to the need to use the territories. If these needs are to meet the desires of tourists, we have the necessary ingredients to start that tourism process that changes the existing urban balances.

The research aims to investigate the production of the imaginary tourist city conveyed through the digital network, especially through Instagram, and its ability to create narratives capable of expressing the pregnant cultural values of the metropolitan area and influencing the dynamics of development and urban development transformation.

Especially during the Covid-19 pandemic, the social media environment has contributed to determining favourable narratives, often in support of local businesses, through digital work [29] aimed at countering the process of closures dictated by the health emergency, in the hope of restoring the vitality of the city itself. It, therefore, have redefined both the perceived value of domestic environments and the ones assigned to public space and its multiple uses [30].

The work presented in this paper is part of the PRIN 2017 "Short-term City" research activities, "Digital platforms and spatial (in)justice" [31], of which the authors are members of the research team. The paper's structure proceeds as follows: Sect. 2 explains the methodological approach by describing the context of the research and its development phases; Sect. 3 identifies the case study relating to the tourist city of Naples and implements the methodological process; Sect. 4 presents the discussion of the processed data and the conclusions with open issues about the research themes.

2 Materials and Methods

The research focused on studying the city of Naples (Italy), which assumes the connotations of an example to evaluate the effects of the production of the imagination in a city affected by overtourism.

The methodological process considers the following four main steps (see Fig. 1): 1. Main frame; 2. Main drivers; 3. Methods & Tools; 4. Output. The field of investigation is the platform economy and how this can influence urban values, generate new urban income, and leverage benign imaginaries linked to the tangible and intangible cultural values expressed by the city [32]. Alongside the imaginaries, other drivers of the transformations are constituted by the existing spatial and geographical hierarchies and by the modifications of the inhabited space, which often respond more to the needs of tourist use of the city than to the protection of residential uses.

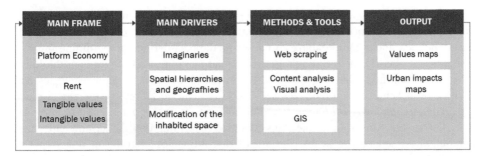

Fig. 1. The methodological framework

The extraction of data from the Instagram platform, in the form of photo and video content, made it possible to proceed with the content analysis and visual analysis and to spatialise the data processed through a reading framework in the GIS environment.

Values maps were obtained, expression of intangible values mediated by the community of Instagram users. Then, it was possible to get the maps of tangible values, the maps of short-term accommodation listings on Airbnb, and real estate dynamics integrating different data and related indicators. Finally, the set of maps made it possible to determine urban impact synthesis that constitutes a picture of the influence of imaginaries on urban values.

3 The Case Study

The subject of the investigation is that of the tourist city, identifying the city of Naples as a case study. Naples, a city in southern Italy (see Fig. 2), historically is a crossroads of peoples, for its strategic position and cultural resources, as well as the proximity to the ruins of the ancient civilisation of Pompeii, an exceptional stage since the time of the Grand Tour of the XVII AD. It is precisely from the Grand Tour that the distinctive look of the romantic tourist was born that led to the production of urban and artistic *topoi* of

Fig. 2. The city of Naples: the selected case study

the Bel Paese [33], which the research intends to analyse to identify the relationships between the production of imaginary and urban transformations.

The methodological process focuses on: investigating the symbolic production of the city; decoding the tourism city spaces; understanding the new geography of the tourist city; understanding if and how imaginaries and cultural services influence market dynamics.

The pandemic moment in progress has suggested the adoption of a digital exploratory methodology, selecting the images disseminated via Instagram through appropriate hashtags to define a meaningful database to describe the peculiarities of the different urban contexts and classify them through an image reading framework. Significant hashtags have been identified, able to express different points of view through which to look at the city to achieve the objectives set. These indicate a defined geographical place (*#quartierispagnoli #vicolidinapoli*) and/or abstract (*#paesagginapoletani #napolituristica*). The selection of these hashtags was also determined by the significantly high number of related photos in anticipation of discarding less significant images when structuring the database. Through platform scraping, a sufficient number of photos were collected for processing between April and May 2021. Of the collected images, 200 were analysed through a qualitative reading framework, described below.

An interpretative analysis was built (see Table 1) that investigates the Instagram community, outlining the profile of the person sharing content and the reason behind the sharing.

Table 1. Framework for reading the imaginary: Instagram community

Class	Data	Data description
Account type	N. photo	
	Link photo	
	Id account	
	Account name	
	Account type	Personal or Business
	Geo tag	Local or not-local
Content analysis	Materiality	Building; street; bodies; etc.
	Type of materiality	To specify: ex. hotel
	Setting	Urban, Nature, Landscape, Indoor

(continued)

Table 1. (*continued*)

Class	Data	Data description
	Type of setting	To specify: ex. square
	Social space practice	Special event; daily life, self-display, etc.
	Instagram digital objects	Emoticon; text; etc.
	Photo/Video	
	Hashtag	
	Co-hashtag	
	N. Like	

The interpretation of the selected images follows this analysis, carried out through the identification of Cultural Services (C.E.S.), one of the Ecosystem Services (E.S.) categories [34], expressed by them (see Table 2). The construction of a specific framework for reading the tangible and intangible benefits that can be enjoyed in the portrayed places starts from the reinterpretation of the most recent Service classification frameworks, the CICES (Common International Classification of Ecosystem Services) [35, 36], which considers the distinction of C.E.S. into two "Divisions", representing the tangible and intangible characteristics of these services.

Table 2. Framework for reading the imaginary: intangible values

Division	Group	Class	Indicator
Intangible	Spiritual	Sacred and religious	Place or type of representation
		Symbology and myth	Place or type of representation
	Emblematic	Inspiration	Place
		Sense of place	Place
		Aesthetic values	Place
		Public/Private space relations	Extension of domestic space in public space
			Extension of shop in public space
			View from the private space to public space
			View of urban landscape from a public facilities
			View of the interior private space

(*continued*)

Table 2. (*continued*)

Division	Group	Class	Indicator
		Informal practices	Itinerant commercial practices
			Street artist
			Street art
			Street game
			Self-built urban furniture
			Relaxing in the urban space
			Urban explorations
			Contemplation
		Identity	Cultural diversity
			Specificities
		Built environment	Picturesque
			Monumental
			Historical
			Public housing
			Density
			Porosity
			Interstitial

Operating on "Groups" and "Classes", the three domains and relative dimensions borrowed from the "Cultural and Creative Cities Monitor", a tool developed by the European Commission's JRC [37–39], were integrated as tangible services (Table 3).

Table 3. Framework for reading the imaginary: tangible values

Division	Group	Class	Indicator
Tangible	Cultural vibrancy	Cultural venues & facilities	Museums and galleries
			Monuments or monumental complexes
			Archaeological sites
			Libraries
			Theatres
			Parishes

(continued)

Table 3. (*continued*)

Division	Group	Class	Indicator
		Cultural participation & attractiveness	Museum visitors
			Cinema visitors
			Cultural event
	Creative economy	Creative jobs & activities	Cultural and creative operators
			Cultural and creative jobs
	Enabling environment	Human capital	Graduates in arts and humanities
			AFAM graduates
		Local connections	Train stop
			Bus stop
			Ferry stop
		Openness, tolerance & trust	Foreign graduates
			Foreign-born population
			Nonprofit organisations volunteers

4   Results and Conclusions

The representation of a comprehensive set of indicators and the urban narratives in GIS environment, combined with the relationships between them and the tourist offer, managed through the Airbnb platform, define how the imaginary [40, 41] determine material transformations, influencing the tourist hospitality market, strongly in friction with residential uses. The mapping of intangible values began with the qualitative reading of each image according to the selected indicators, detecting the presence or absence of the intangible services expressed by each of them. Therefore, the values map was obtained through the overlay of indicators' ones, which brings out the areas with the highest density of intangible C.E.S. connected to the selected hashtags. The level of attractiveness of these hashtags for representing the tourist city imaginary recognises the historic centre of Naples as a focus point. The map of tangible C.E.S. was generated with a similar overlap. This mapping is not affected by the selection of images and hashtags; it, therefore, photographs a situation that concerns the entire urban system. The common areas of concentration between tangible and intangible values represent how these hashtags can benefit from the greatest number of cultural services (see Fig. 3). It is, therefore, interesting to verify the points of contact with the mappings of market values and listings (see Fig. 4). It should be observed that while there was a phase of constant growth of listings in the pre-pandemic period (reaching peaks of 8600 listings), following the Italian lockdown (from March 9th to May 18th, 2020) there was a decline of listings, below 8000, due to the reorganisation of real estate assets. On this occasion, a part of the properties was intended for people in isolation, smart working and long-term residence. However, it emerged how the mechanisms of value creation of the built environment in pandemic contingencies are conditioned by those of value perception. Compared to the data of the listings on Airbnb, observed in April 2021 [42], of the 7961 listings present, most are concentrated within the Unesco perimeter, locating themselves near the places that the Neapolitan imaginary describes as functioning ecosystems. Hard data also specify this condition.

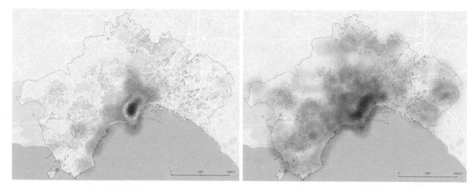

Fig. 3. Values map: intangible values and tangible values

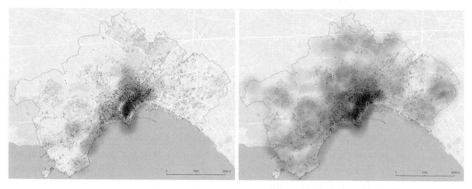

Fig. 4. Overlap between Airbnb's listings and intangible and tangible values

Comparing the data of the real estate market, deriving from the OMI prices of the first half of 2021, we observe that the area in which the greatest number of listings on the Airbnb platform persists coincides with that where the average market values range between 1275.00 €/sqm and 1775.00 €/sqm (see Fig. 5). A similar observation affects the long-term rental market compared with listings, with a preponderance of listings in areas where average rental values are between 3.80 €/sqm and 7.20 €/sqm. This area falls within the Unesco area and coincides with the historic centre of the city. Here there is an often degraded condition of the properties, which often lack the basic requirements of adequate housing [30] and therefore are poorly attractive for the ordinary market, but with a high positional value, which reflects the interest in placing rental properties on short-term rental platforms. Indeed, the imaginary, fueled by tourism development, helps to influence the choices and investments in properties that can be defined as "waste" for the ordinary market, but which represent a new type of specificity, capable of generating regeneration processes relevant for the production of interstitial rent, a surplus of rent produced starting from the difference between the rent extracted through short-term rental platforms and the rent captured by the property taxation regime [43].

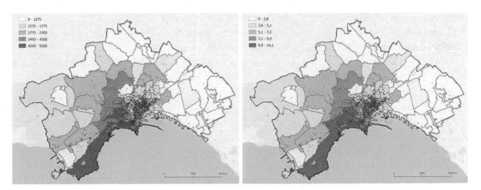

Fig. 5. Overlap between Airbnb's listings and market values and rental values

In conclusion, the structured methodology for this research makes it possible to consider complex socio-cultural variables to be quantified in the analysis of market values, thus relating imaginaries to short and long-term market dynamics.

The future objectives of the research aim to develop a scalable and adaptable tool to the specificities of other urban contexts, which allows assessing the increase in economic value due to imaginaries, which is the plus value of imaginaries.

References

1. Rossi, U.: Cities in global capitalism. Int. J. Urban Reg. Res. **43**(2), 394–396 (2017). Wiley Blackwell, Polity Press, Cambridge
2. D'Eramo, M.: Il selfie del mondo: Indagine sull'età del turismo. Feltrinelli, Milano (2017)
3. Langley, P.: The making of investor subjects in Anglo-American pensions. Environ. Plann. D Soc. Space **24**(6), 919–934 (2006)
4. Bernardi, M.: Un'introduzione alla Sharing Economy, Milano: Fondazione GianGiacomo Feltrinelli, Laboratorio Expo KEYWORDS (2015)
5. Hannam, K., Butler, G., Paris, C.M.: Developments and key issues in tourism mobilities. Ann. Tour. Res. **44**, 171–185 (2014)
6. Guttentag, D.: Airbnb: disruptive innovation and the rise of an informal tourism accommodation sector. Curr. Issue Tour. **18**, 1192–1217 (2015)
7. Dredge, D., Gyimóthy, S.: Tourism and Collaborative Economy. Springer, Cham (2017). https://doi.org/10.1007/978-3-319-51799-5
8. Oskam, J., Bowijk, A.: Airbnb: the future of networked hospitality businesses. J. Tour. Futures **2**(1), 22–42 (2016)

9. Prayag, G., Ozanne, L.K.: A systematic review of peer-to-peer (P2P) accommodation sharing research from 2010 to 2016: progress and prospects from the multi-level perspective. J. Hosp. Mark. Manag. **27**(6), 649–678 (2018)
10. Oskam, J.: The Future of Airbnb and the 'Sharing Economy.' Channel View Publications, Bristol (2019)
11. Colomb, C., Moreira de Souza, T.: Regulating short-term rentals. Platform-based property rentals in European cities: the policy debates. Property Research Trust, London (2021)
12. Zervas, G., Proserpio, D., Byers, J.W.: The rise of the sharing economy: estimating the impact of Airbnb on the hotel industry. J. Mark. Res. **54**(5), 687–705 (2017)
13. Lee, D.: How Airbnb short-term rentals exacerbate Los Angeles's affordable housing crisis. Harvard Law Policy Rev. **10**, 229–254 (2016)
14. Barron, K., Kung, E., Proserpio, D.: The sharing economy and housing affordability: evidence from Airbnb. In: Proceedings of the 2018 ACM Conference on Economics and Computation (2018)
15. Gainsforth, S.: Piattaforme digitali e spazio urbano. Il caso Airbnb. Crítica Urbana. In: Revista de Estudios Urbanos y Territoriales, vol.3, núm.10. Qué turismo. A Coruña: Crítica Urbana, Enero (2020)
16. McLaren, D., Agyeman, S.: Sharing Cities: A Case for Truly Smart and Sustainable Cities. MIT Press, Boston (2015)
17. Rypkema, D.: Heritage conservation and property values. the economics of uniqueness: investing in historic city cores and cultural heritage assets for sustainable development. In: Licciardi, G., Amirtahmasebi, R., World Bank (eds.) The Urban Development Series, pp. 107–142. The World Bank, Washington, DC (2012)
18. Koster, H.R.A., Van Ommeren, J., Rietveld, P.: Historic Amenities, Income and Sorting of Households; Vrije Universiteit Amsterdam—Spatial Econometrics Research Centre, Amsterdam (2013)
19. Nijkamp, P.: Economic valuation of cultural heritage. In: Licciardi, G., Amirtahmasebi, R., World Bank (eds.) The Urban Development Series, pp. 75–106. The World Bank, Washington, DC (2012)
20. Del Romero Renau, L.: Touristification, sharing economies and the new geography of urban conflicts. Urban Sci. **2**(4), 104 (2018)
21. Spillare, S.: L'esperienza turistica tra pratica di consumo e fattore di sviluppo locale sostenibile. Ph.D. dissertation, Alma Mater Studiorum Università di Bologna (2013)
22. Barns, S.: Platform Urbanism: Negotiating Platform Ecosystems in Connected Cities. Geographies of Media. Springer, Singapore (2020). https://doi.org/10.1007/978-981-32-9725-8
23. Farmaki, A., Christou, P., Saveriades, A.: A Lefebvrian analysis of Airbnb space. Ann. Tour. Res. **80**, 102806 (2020)
24. Olma, S.: Never Mind the Sharing Economy: Here's Platform Capitalism. Institute of Network Culture (2014)
25. Kenney, M., Zysman, J.: The rise of the platform economy. Issues Sci. Technol. **32**(3), 61 (2016)
26. Langley, P., Leyshon, A.: Platform capitalism: the intermediation and capitalisation of digital economic circulation. Finance Soc. **3**(1), 11–31 (2017)
27. Srnicek, N.: Platform Capitalism. Polity Press, Cambridge (2017)
28. Celata, F.: Territorio **86**, 48–56 (2018)
29. Bereitschaft, B., Scheller, D.: How might the COVID-19 pandemic affect 21st century urban design, planning, and development? Urban Sci. **4**, 56 (2020)
30. Quaglio, C., Todella, E., Lami, I.M.: Adequate housing and COVID-19: assessing the potential for value creation through the project. Sustainability **13**, 10563 (2021)

31. PRIN 2017 "Short-term City. Digital platforms and spatial (in)justice". https://www.stcity.it/. Accessed 14 Dec 2021
32. Carone, P., Panaro, S.: New images of city through the social network. Int. J. Web Inf. Syst. **10**(2), 209–223 (2014)
33. Boyer, M.: Il turismo, dal Grand Tour ai viaggi organizzati. Electa-Gallimard, Torino (1997)
34. MEA, Millennium Ecosystem Assessment: Ecosystems and Human Well-Being: Synthesis. Island Press, Washington, DC (2005)
35. Costanza, R., et al: Twenty years of ecosystem services: how far have we come and how far do we still need to go?.Ecosyst. Serv. **28**, 1–16 (2017)
36. Haines-Young, R., Potschin, M.: Common international classification of ecosystem services (C.I.C.E.S.) V5.1. Guidance on the Application of the Revised Structure. Fabis Consulting Ltd. The Paddocks, Chestnut Lane, Barton in Fabis, Nottingham, UK (2017)
37. Montalto, V., Tacao Moura, C., Panella, F., Alberti, V., Becker, W., Saisana, M.: The Cultural and Creative Cities Monitor: 2019 Edition, EUR 29797 EN, Publications Office of the European Union, Luxembourg (2019)
38. Cerreta, M., Muccio, E., Poli, G.: ValoreNAPOLI: la Valutazione dei Servizi Ecosistemici Culturali per un Modello di Città Circolare. BDC. Bollettino Del Centro Calza Bini **20**(2), 277–296 (2020)
39. Cerreta, M., Daldanise, G., La Rocca, L., Panaro, S.: Triggering active communities for cultural creative cities: the "hack the city" Play Rech mission in the Salerno historic centre (Italy). Sustainability, Switzerland **13**(21), 11877 (2021)
40. Kotler, P., Haider, D.H., Rein, I.: Marketing Places. The Free Press, New York (1993)
41. Urry, J.: The Tourist Gaze: Leisure and Travel in Contemporary Societies. Sage, London (1990)
42. Insideairbnb. http://insideairbnb.com/naples/. Accessed 02 Dec 2021
43. Cerreta, M., Mura, F.D., Poli, G.: Assessing the interstitial rent: the effects of touristification on the historic center of Naples (Italy). In: Gervasi, O., et al. (eds.) ICCSA 2020. LNCS, vol. 12251, pp. 952–967. Springer, Cham (2020). https://doi.org/10.1007/978-3-030-58808-3_68

Grid Governance Between Spatial Efficiency and Social Segregation: Chinese Gated Communities Socio-spatial Responses Amidst COVID-19 Outbreak

Edoardo Bruno[(✉)] and Francesco Carota

Department of Architecture and Design, Politecnico di Torino, China Room, Turin, Italy
{edoardo.bruno,carota.francesco}@polito.it

Abstract. The paper analyzes how the urban model of Chinese gated communities (GCs) is changing living spaces and administrative organization, starting from the implications caused by the COVID-19 pandemic. Gated Communities represent a consolidated sociospatial model within the urban landscape of contemporary China. Its wide diffusion, which has its roots both in the historic city and in the models of collectivization of the space of the socialist era, has allowed real estate investment companies to experiment with forms and solutions aimed at supporting a rapidly growing market. Likewise, their segregation has prompted government authorities to promote reforms to reduce their diffusion and encourage new forms of integration. The pandemic has projected the morphological pattern of GCs into a new urban role, proposing itself as a model of effective fight against the spread of the disease thanks to the activation of spatial structures capable of separating and hierarchizing homogeneous social groups, as well as aligning state administrative power at the local level and the spatial layout of the urban grid. Residents are thus transferring new hierarchies of values to GCs, influencing future market prices, implementation in the regulatory field and design practices.

Keywords: Chinese urbanization · Gated community · COVID-19

1 Introduction

The recent pandemic of COVID-19 led us to rethink and reframe many aspects of our life. Among the many speculations, the recent outbreak has surely opened a debate, both in the academia and outside of it, on the role that planning, urban design and housing forms have on people health and security, both during "normal" times and sanitarian crisis [14]. Particularly, in the field of urban design and planning, a fierce debate arose within the study, conception and assessment of dense cities, their functional organizations, planning systems and housing models [15]. In this frame, China represents an interesting field of investigation, not because the first outbreak spreading happened there, but because of the continuous overlapping between the urban housing spatial organization and the boundaries of its public governance.

© The Author(s), under exclusive license to Springer Nature Switzerland AG 2022
F. Calabrò et al. (Eds.): NMP 2022, LNNS 482, pp. 1516–1525, 2022.
https://doi.org/10.1007/978-3-031-06825-6_146

This paper, thus, investigates the effects of the COVID-19 pandemic on the perception and assessment of housing spaces in Chinese cities, with a particular emphasis on their physical attributes and urban form. Many scholars have already acknowledged the intertwined relations between housing spaces and social practices [21], as well as spatial structure and health [5]. These studies highlight that within a certain definition of the built environment, the potential exists for architectural and spatial forms to facilitate or constraint certain types of activity and thus to transform the essence of social relations, reflecting the way in which places are perceived and manifested. This paper aims to contribute to these debates, bringing some new reflections on the assessment of physical attributes of Chinese gated communities (hereafter GCs) before, during and after the first wave of COVID-19 outbreak. Several authors have already delt with the effects of the pandemic on this urban typology, raising questions whether and how GCs' spatial organizations are able to challenge external emergencies [20, 38], and to improve the sense of safety among what is perceived as "external" [39]. This literature assumes, therefore, a particular relevance if related to a large debate which discussed about the benefits and constraints of opening GCs in major Chinese cities [37,28], after they have been largely assumed as a socio-spatial dispositive able of reinforcing the social role of an emerging urban middle class [27].

After this type of development has been criticized both in academia and in institutional bodies, this paper asks whether and how, in a very contingent situation such as the COVID-19 outbreak, China's gated communities have been reconsidered; how their space have been adapted to new uses and possibilities; and how their design features are changing in the near coming future. To answer these questions the contribution, compare forms of use and perceived values of Chinese GCs before, during and after the first wave of the pandemic. Analysis have been carried out on secondary sources (particularly Chinese newspapers, online magazines and scientific articles), real estate datasets [17] (National Bureau of Statistic of China for the years 2016–2020) and long period of on-field research in China's GCs. The conclusions of the paper are based on the cultural shift that are detectable on the sources of this research, which count also on the direct observation of newest GCs architectural competitions occurring in China asking participants to imagine for the future housing solutions able to fight health emergencies.

2 Gated Community in China. Between Market Value and Urban Form

Although GCs have spread globally, their prevalence in China is worth of interest for many reasons. Indeed, CGs represent the basic model of urban development in the country, not only in terms of social organization and service management but also in terms of urban structure and design. According to Wu [35] Chinese GCs accounted in 2018 for the 80,61% of the housing type of urban population and, reaching a diffusion of 300,000 units across the country, becoming its market dominance [36]. After 1979, China has undergone a period of fast-paced urbanization, together with a profound reformation of its economical and planning system. This brought to an important process of land and housing commodification, in which the development of large-scale land plots represented the DNA of urban expansion. Chinese large and densely populated cities based their urban planning approach on the idea of the "super-block", a spatial construct that started to become popular in the western modernist urban planning discourse during the early stage of the 20th century [26].

Even though, the idea of the "super-block" is not a recent concept in the country, during the period of real estate market formation and transformation, the super-block remained the basic unit of urban planning and real-estate transaction, and thus it strongly defined the "new Chinese city" [16]. Similarly, to what happened in the American context during the post war period, the introduction of a grid plan model, developed under the western influences, brought the concept of block to partly loose its physical and social acceptation and to assume the meaning of a big parcel of land dived by giant roads, also before being transformed into a housing estate or a commercial cluster [13, 29]. Such type of real estate developments dilatated the urban fabric until turning urban monads into microcosms. Indeed, although often externally gated, inside they contain not only multiple places and functions to satisfy the needs of urban communities, but also all major public facilities [2] (Fig. 1).

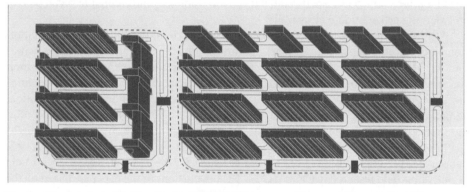

Fig. 1. General plan of Beijing Vanke COFCO holiday town (Source: Elaboration by the authors).

However, within the Chinese culture, gated living is not only a recent trend. According to Wu [33], it is possible to retrace different functions of gating in the Chinese history. Under socialism, gating reinforced political control and collective consumption organized by the state, while in the post-reform era, the gate demarcates emerging consumer clubs in response to the retreat of the state from the provision of public goods. In this regard, Huang [9] suggests how residents began to form collectives to protect their lifestyle and interests. Gating started to also have some influence on the privatization of service provision and governance, among the Chinese city and culture. New Chinese residents started to form homeowners' associations [4], in addition to other governance bodies such as residents' committees and community service centers [25], which work strictly with developers and community managers companies to provide and ensure basic private services to residents. Indeed, we need also to consider the influence of the rising awareness for privacy in the Chinese culture associated with a new form of private property ownership [18].

In addition to this, it is possible to see Chinese gated residential estates not just in terms of governance issues, but also in terms of symbolic value. After the system of public housing provision has been replaced by commodity housing programs, gated and architectural design features started to enter the Chinese culture and create a vehicle of desirability of newly built residential areas. Instead of simply providing security, the gate of newly built suburban estates can be considered a way to create a sense of magnificence and status among the residents. Large real estate developers, such as China Vanke, Poly and Longfor, adopted exclusive and innovative urban services, including both physical structures such as a laundry, a nursery, a swimming pool and a park with a games area and sports equipment, as well as social amenities such as educational structures, cultural activities and health care support, to market their estates [32] especially in the suburbs; this strategy became crucial for gaining a competitive advantage on the real estate industry (Fig. 2).

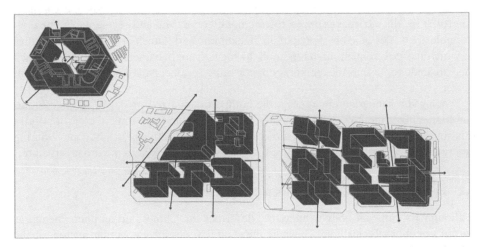

Fig. 2. Masterplan (Phase1 + Phase2) of Guangzhou Vanke cloud city (Source: Elaboration by the authors).

This fact makes the Chinese context different from that of western gated communities because the debate is not only focused on the theme of private governance [31], but rather about the specific building forms and their packaging practices [34]. Moreover, differently from the West, gating in China did not raise too much concern for such long time. It was only in 2016 that, because of a new national strategic direction within the "New Type of Urbanization Plan", China announced policy recommendations intending to end GCs. This policy, however, faced societal resistance and it was very little successful. Very few interventions were effectively developed under this direction (Fig. 3). As Chiu et al. [3] have largely acknowledge, indeed, privately produced communities continue to promote and justify exclusion, reinforcing socioeconomic and spatial inequalities. If this kind of values were the "normal" before the pandemic, the external health issue questioned spatial organization and local governance as discussed in the following paragraph.

3 Evaluating Chinese Gated Communities Socio-spatial Responses During COVID-19 Outbreak

3.1 The Reactivation of the Gates

The lockdown in Wuhan, where, according to statistics, 60% of all cases happened in China [22], reflects how Chinese urban areas have dealt with the pandemic, using their common planning grid as a spatial mean to curb population movements. The gated structure of GCs permitted to reactivate pre-existing surveilled entry points to contrast the diffusion of the virus. Only those who do not show symptoms of contagion were allowed to enter inside; entrance was permitted presenting digital certificates tracing previous activities, with relevant repercussions on the issue of social surveillance and how it worked during the pandemic. Outside the GCs, mobility infrastructure was again monitored by local authorities. This double mechanism, which contingently separated homogeneous social groups in space, allowed to activate a meticulous and capillary control. It recalls the frozen spaces, hierarchically organized and segmentable into controllable areas, described by Foucault in "Discipline and Punish" [6]: a form of control in which the government aligns its tools to local spatial characteristics, and in which the pandemic represents the temporary opportunity to conceive a disciplined exercise of power.

Facing the sanitarian emergency, the compartmentalized model of GCs separates and hierarchizes the flows of the population, allowing stricter forms of control [19]. In this sense, the GCs result as latent structures capable of activating tight control during emergencies, becoming administrative tools inextricably linked to the urban structure.

3.2 From Efficient GCs Responses to New Market Values

Although a significant body of literature describes the Chinese urban phenomenon as part of a neoliberal transition [8], where the capital drives local governance choices, the transformation of GCs as a control tool for facing external emergencies, suggests the re-emerging of the state in the management of local spaces and communities, partially overcoming the private management of the real estate sector. In this sense, Li [10] speaks of a pre-liberal disciplinary society, in which the policy of the "grid governance system", inaugurated in 2013, is implemented for building even more rigid control perimeters, activated by the government thanks to the distribution of dedicated personnel [24].

The efficient use of gates as a separation of the urban grid during the pandemic has challenged the 2016 tentative reform, which never took off due to residents' oppositions [7]. According to the values assigned by the study of Blakely and Snyder [1] to describe the preferences for living in GCs - lifestyle, prestige and safety - safety was surely the less significant on the desirability of Chinese gated enclaves [11], the implementation of control by property management companies and government bodies was not fully enforced, making them de facto open communities rather than fortified estates. Furthermore, the low level of criminality in Chinese cities defined the need for a fenced boundary only as a psychological device rather than an actual deterrent. On the contrary, a recent study by Li, Wan, and He [11] display that, in the case of some GCs in Beijing, the sanitary control activated at the entrances of CGs during the pandemic has strongly influenced the social perception of safety. This increased the social value of gating, because associated to a limitation of the effects of the pandemic, having thus a real impact on market prices of CGs.

Fig. 3. Entrance of a GC during COVID-19 outbreak surveilled by dedicated personnel (source: www.sixthtone.com)

The increasing positive perception of gating by the residents may solicit the introduction of restrictive measures of spatial governance even beyond the current contingency. In addition, real estate developers may take advantage from a real demand for increasing speculative profits, thus reducing the provision of new affordable housing spaces. Average selling price data (National Bureau of Statistics of China) from 2016 to 2020 [17] display how in the 5 first-tier cities of China the pandemic did not arrest the fast pace rising average selling price for residential building and confirmed local decreasing tendencies in already developed metropolitan areas [Table 1].

3.3 New Local Alliances

During the pandemic, GCs held also new forms of governance based on a stronger alliance between public institutions and homeowners' associations. The use of the "grid governance" was implemented thanks to the insertion of the state power at the local level, which reconfigured the presence of the public authority, increased the number of specially trained officials and cadres, and brought to experimenting new ways of cooperation between public bureaucracy and local private companies [40]. Indeed, the SARS crisis of 2003 had already demonstrated that the public administration could not entirely rely on private management companies to counter the emergency, because such firms were divided into a myriad of small realities and moved by profit-oriented mechanisms. During the new pandemic of COVID-19, a series of street offices, managed directly by the local government, had the opportunity to coordinate an increased number of volunteers, regaining in this manner the confidence of the emerging middle class living in CGs [22]. At the same time, the socio-spatial structure of GCs, formerly under a revision by some urban policies soliciting a better integration between the residential estate and the external urban realm [6], is now becoming more introverted.

Table 1. Average selling price of commercialized buildings [residential buildings] (yuan/sqm). Data available from the National Bureau of Statistics of China 2017, 2018, 2019, 2020, 2021. [http://www.stats.gov.cn/]

City	2016	2017	2018	Var%	2019	Var%	2020	Var%
Beijing	28489	34117	37420	+9,68	38433	+2,70	42684	+11,06
Tianjin	12879	15139	15924	+5,18	15423	−3,14	16391	+6,27
Shanghai	25910	24866	21582	+16,54	32926	+13,61	36741	+11,58
Guangzhou	16346	17685	21582	+22,03	24015	+11,27	27112	+12,89
Shenzhen	45498	48622	55441	+14,02	55769	+0,59	56844	+1,92

The result is a more significant growth of social inequalities within the spatial grid and a further limitation in accessing public services and goods [23]. In these regards, planners and architects are invited to envision how dedicated regulation may guarantee the collective enjoyment of public assets even in case of public health crisis.

3.4 Emerging Spaces in GCs Design Practices

Observing how GCs responded to COVID-19, it is possible to envision them as resilient socio-spatial devices within the urban realm, which bring to reflect more upon the future lines of development of the Chinese process of urbanization. Some physical elements of GCs are changing how humans interact with spaces responding to the contingent state of emergency.

As described in the previous paragraphs, the most evident is the entrance gate, which selects and distributes the external complexity by channeling it within a homogeneous and monitored context. The entrance device feeds social segregation, but when it becomes a market value expressed by a segment of consumers, it legitimizes planning action reinforced by the state of emergency. It is the same for the distribution of internal services, which guarantee the self-sufficiency of the community. If on the one hand, residential facilities urge to design GCs as multifunctional zones, on the other hand, they trigger the possibility that urban enclaves also become privileged places of consumption separated from the urban context. Even elevators are also changing the way in which they are used by local inhabitants. Allowing the vertical movement of masses inside residential towers, their use is becoming a second control barrier. Restrictions in the number of users and continuous sanitization, open the doors to both new technological means and a reconsideration of their dimensioning – larger and therefore more expensive – in future residential investments. Finally, apartment entrances represent the last sanitized threshold that must be as much as possible preserved. Thus, for instance, new spaces in other parts of the gated community are defined to accommodate delivery storages for online purchases, increasing more and more the distance between residents and external users of the space.

In the absence of a clear legislation that regulates the design of new GCs within the current pandemic contingency, spontaneous experimentations arise. Real estate developers, planners and local agencies are considering the mitigation of risks connected to pandemic and public health issues among the main criteria for assessing new design proposals. This was for instance the case of a design competition for the development of a multifunctional GC in Shenzhen, to which the authors have recently participated: innovative solutions, spanning from open space to interior design, were welcomed by the jury to evaluate the proposal under a holistic point of view. It is thus possible to record how the private sector is advancing in the standardization of new spatial features that might be delivered to an upcoming emergent market, in which the safe management of residents' flows might play a predominant role.

4   Concluding Remarks

This study displays how the emergence of COVID-19 had some significant implications on the urban model of Chinese GCs. Their spatial characteristics, even considering problems of segregation and separation between different social classes raised over time, gave an efficient response to the pandemic, easily re-activating latent structures, such as gated entrances and internal facilities for securing and monitoring residents' activities [30].

The rigid division between the inside and the outside of the estate, has, even more than before, become a market value for residents and prospective homebuyers. Therefore, firms operating in the real estate sector, in the absence of a specific legislation, could operate to intensify the socio-economic separation between the inhabitants of the GCs and urban citizens that from them are excluded. This drift would bring Chinese GCs closer to transnational models of enclave urbanism, which separate homogeneous social groups from the collectivity, far from representing the historical process of production of collective housing within Chinese society [12].

GCs thus find themselves representing a bias in the face of future public health challenges. They position themselves in between the efficiency shown in controlling health emergencies, thus legitimizing themselves from the point of view of public safety, and the acceleration of a process of social segregation. This study encourages further post-pandemic investigation to carry on in-depth analyses and fieldwork data collection on the physical and social impact of GCs on urban life, determining how COVID-19 may have defined new interpretative categories and values for framing, assessing and design contemporary Chinese urban development.

References

1. Blakely, E.J., Snyder, M.G.: Fortress America: Gated Communities in the United States. Brookings Institution, Washington (1999)
2. Carota, F.: City's fragments. In: Bonino, M., Carota, F., Governa, F., Pellecchia, S. (Eds.) China Goes Urban. The City to Come. Skira Editore: Milan (2020)
3. Chiu-Shee, C., Brent, D.R., Vale, L.J.: Ending gated communities: the rationales for resistance in China. Hous. Stud. 1–30 (2021)
4. Deng, F.: Is HOA capitalized in housing price? evidence from Chongqing, China. Int. J. Hous. Mark. Anal. 15(1), 94–107 (2022)
5. Forsyth, A., Salomon, E., Smead, L.: Creating Healthy Neighborhoods: Evidence-Based Planning and Design Strategies. Routledge, London (2017)

6. Foucault, M.: Discipline and Punish: The Birth of the Prison. Vintage Books, New York (1975)
7. Hamama, B., Liu, J.: What is beyond the edges? gated communities and their role in China's desire for harmonious cities. City, Territory Archit. **7**(1), 1–12 (2020). https://doi.org/10.1186/s40410-020-00122-x
8. Harvey, D.: A Brief History of Neoliberalism. Oxford University Press, Oxford (2005)
9. Huang, Y.: Collectivism, political control and gating in Chinese cities. Urban Geogr. **27**(6), 507–525 (2006)
10. Li, P.: Conceptualizing China's spatial lockdown during the COVID-19 pandemic: a neo-liberal society or a pre-liberal one? Soc. Trans. Chin. Soc. **17**(2), 101–108 (2021)
11. Li, L., Wan, W.X., He, S.: The heightened security zone function of gated communities during the COVID-19 pandemic and the changing housing market dynamic: evidence from Beijing. China. Land **10**(9), 983 (2021)
12. Li, M., Xie, J.: Social and spatial governance: the history of enclosed neighborhoods in urban China. J. Urban Hist. 00961442211040460 (2021)
13. Liu, H., Zhou, G., Wennersten, R., Frostell, B.: Analysis of sustainable urban development approaches in China. Habitat Int. **41**, 24–32 (2014)
14. Madden, D.: The urban process under covid capitalism. City **24**(5–6), 677–680 (2020)
15. Megahed, N.A., Ghoneim, E.M.: Antivirus-built environment: lessons learned from COVID-19 pandemic. Sustain. Cities Soc. **61**, 102350 (2021)
16. Monson, K.: String block vs superblock patterns of dispersal in China. Archit. Des. **78**, 46–53 (2008)
17. National Bureau of Statistics of China, http://www.stats.gov.cn. Accessed 28 Feb 2022
18. Pow, C.-P.: Gated Communities in China: Class, Privilege and the Moral Politics of the Good Life. Routledge, London (2009)
19. Qian, Y., Hanser, A.: How did Wuhan residents cope with a 76-day lockdown? Chin. Soc. Rev. **53**(1), 55–86 (2021)
20. Shi, C., Liao, L., Li, H., Su, Z.: Which urban communities are susceptible to COVID-19? an empirical study through the lens of community resilience. BMC Publ. Health **22**(1), 70 (2022)
21. Shove, E., Pantzar, M., Watson, M.: The Dynamics of Social Practice: Everyday Life and How It Changes. SAGE: Los Angeles (2012)
22. Song, Y., Liu, T., Wang, X., Guan, T.: Fragmented restrictions, fractured resonances: grassroots responses to Covid-19 in China. Crit. Asian Stud. **52**(4), 494–511 (2020)
23. Sun, C., Xiong, Y., Wu, Z., Li, J.: Enclave-reinforced inequality during the COVID-19 pandemic: evidence from university campus lockdowns in Wuhan. China Sustain. **13**(23), 13100 (2021)
24. Tang, B.: Grid governance in China's urban middle-class neighbourhoods. China Q. **241**, 43–61 (2020)
25. Tomba, L.: Middle classes in China: force for political change or guarantee of stability? PORTAL J. Multidiscipl. Int. Stud. **6**(2), (2009)
26. Vale, J.L.: Standardizing public housing. In: Ben-Joseph, E., Szold, T.S. (Eds.) Regulating Place: Standards and Shaping of Urban America. Routledge, New York (2013)
27. Wang, L., Gilroy, R.: The role of housing in facilitating middle-class family practices in China: a case study of Tianjin. Sustainability **13**(23), 13031 (2021)
28. Wang, H., Pojani, D.: The challenge of opening up gated communities in Shanghai. J. Urban Des. **25**(4), 505–522 (2020)
29. Wang, Z., Li, L., Li, Y.: From super block to small block: urban form transformation and its road network impacts in Chenggong. China. Mitig. Adapt. Strateg. Glob. Change **20**, 683–699 (2014)

30. Wang, Z., Liu, L., Haberman, C., Lan, M., Yang, B., Zhou, H.: Burglaries and entry controls in gated communities. Urban Stud. **58**(14), 2920–2932 (2021)
31. Webster, C., Glasze, G., Frantz, K.: The global spread of gated communities. Environ. Plann. Plann. Des. **29**(3), 315–320 (2002)
32. Wu, F.: Transplanting cityscapes: the use of imagined globalization in housing commodication in Beijing. Area **36**, 227–234 (2004)
33. Wu, F.: Rediscovering the gate under market transition: from work-unit compounds to commodity housing enclaves. Hous. Stud. **20**(2), 235–254 (2005)
34. Wu, F.: Gated and packaged suburbia: Packaging and branding Chinese suburban residential development. Cities **27**, 385–396 (2010)
35. Wu, X.L.: Utopian ideals and privatopia dilemmas in urban gated communities. [in Chinese.]. J. China Natl. Sch. Adm. **2**, 122–127 (2018)
36. Wu, X., Li, H.: Gated communities and market-dominated governance in urban China. J. Urban Plann. Dev. **146**(3), 04020025 (2020)
37. Wu, Z., Yang, L., Xu, K., Zhang, J., Antwi-Afari, M.F.: Key factors of opening gated community in urban area: a case study of China. Int. J. Environ. Res. Public Health **18**(7), 3401 (2021)
38. Zeng, P., Sun, Z., Chen, Y., Qiao, Z., Cai, L.: COVID-19: a comparative study of population aggregation patterns in the central urban area of Tianjin, China. Int. J. Environ. Res. Public Health **18**(4), 2135 (2021)
39. Zhong, T., Crush, J., Si, Z., Scott, S.: Emergency food supplies and food security in Wuhan and Nanjing, China, during the COVID-19 pandemic: evidence from a field survey. Dev. Policy Rev. dpr. 12575 (2021)
40. Zhu, T., Zhu, X., Jin, J.: Grid governance in china under the COVID-19 outbreak: changing neighborhood governance. Sustainability **13**(13), 7089 (2021)

Proposal for Mapping Social Housing Needs. The Apulia Region Case Study

Giulia Spadafina[✉] and Giovanna Mangialardi

Politecnico di Bari, Via Amendola 186, Bari, Italy
{giulia.spadafina,giovanna.mangialardi}@poliba.it

Abstract. The pandemic has affected society, transformed housing needs, and its issues, although the economic and structural changes had already been ongoing for a decade. General impoverishment has increased the segment of those living in deprivation conditions, especially those living in rented housing. These changes have not only widened the social range of those who find themselves in a condition of housing need, but they have also modified and broadened the targets. In Italy, there are no lists or mapping of such housing needs, except for the often-unmet rankings for Public Housing. The contribution, therefore, aims to investigate tools and normative references useful for mapping the social needs and the new targets, overcoming a gap in the literature and practice. Through the Apulia Region case study, a proposal for mapping social housing needs based on the 431/1998 National Law is presented, to improve the management of the relationship between supply and demand, including all parts of the social distress.

Keywords: Housing deprivation · Housing needs · Housing data · Support for rents

1 Introduction

The housing issue is universal. It is understood as the absence of quality of living generated by the lack of one or more elements including the adequacy of housing, its economic sustainability, and the guarantees given by the use rights (Di Biagi 2014). Universal in the sense that it cannot elect spaces, populations, or periods of life more relevant than others to investigate it. Starting from this statement, however, it can try to understand how certain factors affect to a greater or lesser extent the housing conditions of people living in certain places.

The recent Covid-19 pandemic that has affected the entire planet also heavily influenced the housing issue (Molinari 2020). This is shown by the data in the recent "Resolution of 21 January 2021 on access to decent and affordable housing for all"[1] of the European Parliament. The Resolution states that the "Covid-19 crisis has aggravated housing insecurity, over-indebtedness, and the risk of eviction and homelessness, and shown the precarious situations of many people, especially the elderly, but also migrant

---

[1] www.europarl.europa.eu/doceo/document/TA-9-2021-0020_EN.html.

F. Calabrò et al. (Eds.): NMP 2022, LNNS 482, pp. 1526–1535, 2022.
https://doi.org/10.1007/978-3-031-06825-6_147

workers and seasonal workers, who do not have access to housing that meets health and social-distancing requirements". Furthermore, the Housing Europe's new 2021 report "The State of Housing in Europe"[2] demonstrate that the global health crisis has made the "essential role of decent and affordable housing even more explicit, reinforcing the strong link between adequate homes and health".

In this context, it is crucial to question the current situation of the housing issue, by defining new targets and needs and assessing how the pandemic has affected and changed the society, although structural changes in society and thus in housing demands began even before the pandemic.

Indeed, the socioeconomic changes of the last decade have profoundly transformed housing and its issues (Lodi Rizzini 2013). General impoverishment has increased the segment of those living in a condition where housing-related expenses exceed the conventional share identified by the European Commission of 30% of income (Fregolent e Torri 2018). These changes have not only widened the social range of those who find themselves in a condition of housing need, but they have also modified and broadened the targets that find themselves in this condition. In Italy, there are no lists or mapping of housing needs, except for the often-unmet rankings for Public Housing. However, the public housing rankings are not sufficient to represent a useful sample for analysing these new needs (Storto 2018).

Furthermore, if on the one hand, property no longer guarantees a status of wealth, since it also impoverishes through the mortgage, on the other hand, the changes and the precariousness of work have put the spotlight back on the opportunities of renting.

The contribution, starting from these considerations and lacks, tries to investigate tools and normative references useful to reconstruct the social demand with particular attention to the Italian case. With a focus on the case of the Apulia Region, the research investigates how it is possible to reconstruct the mapping of social unease to better manage the relationship between supply and demand. To do this, the authors chose to use the data relative to the call for tenders' results "Support for rents" provided for by L. 431/1998 because they can represent, even if not in an exhaustive manner, a good starting point for understanding the characteristics of the housing needs of families living in rented accommodation, in conditions of housing need. The contribution starts with the description of the housing needs and deprivation. After the research method presentation, the Apulia Region case study is presented, using the description of the local call for tenders "support for rents" provided for by L. 431/1998 supported by the PUSH Platform. The last sections contain discussions and conclusions, work limitations, and suggestions for future research.

2   The Housing Needs: Definitions and Criticalities

Housing is a complex, multifactorial issue, shaped by the convergence of at least three dimensions: the dwelling, its tenants, and the context in which the dwelling is located (Olagnero 2008). The characteristics of these factors influence the living, generating when lacking in quality, conditions of housing deprivation. Depending on the severity of

---

[2] https://www.stateofhousing.eu/#p=3.

the problem, different housing deprivation situations can be distinguished: inadequacy, insecurity, or high cost (Tosi 2017).

Given the multifactorial nature of the housing issue, there are consequently many facets to be analysed to measure the quality of housing in a specific place.

At the European level, the Social Protection Committee (SPC) has developed a basic index of monitoring of the dimensions that characterize, on a territory, the presence of housing problems. The indicators used are housing cost overburden rate, overcrowding rate, housing deprivation, AROP (the at-risk-of-poverty) by accommodation tenure type, and share of housing costs in total disposable household income. These dimensions, although theoretically defined, are often difficult to apply and measure in practice (The Social Protection Committee 2014). This monitoring index from SPC is effective in identifying potential areas at risk of housing distress but does not characterize types of need.

On a national scale, taking Italy as a reference case, several factors make the measurement of the housing quality and consequently the characteristics of new needs extremely difficult. Specifically, the main difficulties are found first in a regulatory reason given the legislative and managerial autonomy of the regional government in the field of housing policies, it is often complex to produce homogeneous data. In addition, there are difficulties linked to administrative and bureaucratic issues, the reluctance of administrations in the digital transition, the lack of knowledge of the new housing issue, and therefore the inability to correctly analyse needs. More specifically, to date, except for the lists of beneficiaries of public housing, which often contain many families awaiting allocation, there is no tool for monitoring and mapping the quality of housing on a territorial scale.

In recent years, some Italian regions that are particularly virtuous in the field of housing policies, such as Emilia-Romagna and Tuscany, have produced reports monitoring housing conditions. Within these reports, data useful for monitoring housing needs are analysed.

The Emilia-Romagna Region, through ORSA, assesses the needs by detecting them at a provincial level to identify municipalities where housing policy interventions should be prioritised[3]. To estimate the extent of the needs, the Region conducts analyses on different fields such as families and resident population, living conditions, possible sources of economic hardship, and evictions.

Two years ago, on the other hand, the Region of Tuscany started to develop a real synthetic index of housing conditions[4] to provide a useful tool for authorities to interpret the phenomena related to housing (Cigolotti, et al. 2020). The index is developed based on the housing domains theorised by Palvarini (2006): the physical, legal, social, economic, and territorial domains. Twenty-two total indicators are identified based on these five domains.

Both indexes developed by Emilia-Romagna and Tuscany are complex and complete in their objective of identifying municipalities and areas with a potential situation of housing hardship to better address housing policies. Indeed, they come from different

---

[3] https://territorio.regione.emilia-romagna.it/osservatorio-delle-politiche-abitative/fabbisogno-abitativo.

[4] https://www.regione.toscana.it/documents/10180/13844663/Cond+abit+2021-web.pdf/fd3 a12f7-8dfb-8284-d147-8eb1350858ec?t=1638774810706.

databases[5] which are often not constantly updated and do not provide data at municipal scale (such as data on evictions or data on the use of buildings), which makes updating the index particularly expensive in the authors' opinion.

From literature, Baldini e Poggio (2013) analysed micro-data related to the call for tender on rental support in some large Italian cities, but without defining precise measurement indicators.

3   The Apulia Region Case Study

The contribution intends to overcome the absence of lists and rankings to define social deprivation, by means of the data coming from Law no. 431/1998, for planning housing policies more suitable to the real existing needs. In detail, starting from the question of the lack of up-to-date, uniform, and standardised data, the authors tried to analyse what could effectively monitor the general characteristics of the new housing needs on the one hand and identify a source of updated and updatable data on the other. To try to overcome these problems and to better understand the housing needs at the municipal scale, through data always available and updated yearly, the contribution reflects on the data coming from one of the most important housing support tools in the Italian context: "Support for rents" provided by art. 11 of Law n. 431 of 1998. To analyse the law application and its capabilities in mapping housing needs, the authors have chosen the Apulia Region case study. Apulia is a Region located in the south of the Italian peninsula. Regarding ISTAT data, it is included in the context of the South, which has a percentage of households in absolute poverty of 8.6% compared to 5.2% in the North and Centre of Italy (Nomisma 2020). Moreover, housing poverty is more prevalent in families living in rented accommodation than in those who own their homes. In the Apulian regional context, as in the whole of southern Italy, there is a deep negative imbalance in experimenting with new housing solutions such as social housing or cohousing (Mangialardi, Martinelli e Spadafina 2020). In detail, the case study presented concerns the analysis of data related to the call for tenders "Support for rents 2020" issued by the Housing Policies Section of the Apulia Region, managed through the PUSH platform. The data of Bari are analysed to reflect on their use to map and know the housing needs.

3.1   Law no. 431/1998

Law n. 431/1998 "Regulations governing leases and the release of properties for residential use"[6], in addition to regulating and legislating on leases and eviction measures, in article 11 "National Fund" establishes the National Fund for the support of access to rental housing. Through the fund, "supplementary contributions for the payment of rents are granted annually to tenants who meet the minimum requirements defined annually". The rent contribution established by art. 11 of Law 431/98, in a context of limited

---

[5] ISTAT, Region, INPS, Bank of Italy, Revenue Agency, various Ministries, Housing Agencies, Municipalities, etc.

[6] Legge 9 dicembre 1998, n. 431 "Disciplina delle locazioni e del rilascio degli immobili adibiti ad uso abitativo".

available housing, is a fundamental contribution for those who live in rented accommodation and who are unable, for economic reasons or due to requirements, to access public housing.

Subsequently, article no. 2 of Decree-Law no. 47/2014 has recognized the regions the possibility of contributing, in agreement with the municipalities, to the definition of the purpose of the fund, also coordinating with the Fondo defaulting tenants. At present, the resources must be distributed to the municipalities based on parameters that reward both the combination of accommodations at an agreed fee and households coming from subsidized public housing or subject to executive eviction procedures, and the number of lease contracts at an agreed fee overall brokered in the previous two years.

The fund distributes 90% of the annual resources among the Regions and the remaining 10% to the Regions that have made their resources available. In addition to the regional resources, since 2006, municipal resources have also contributed to the definition of the total amount allocated to support for rents. According to the monitoring carried out by the *Corte dei Conti* in 2020, the national funds have gradually decreased, delegating to the regions the task of financing these calls for proposals together with the municipalities, from around 310 million in 1999 to just under 10 million in 2011. As regards the requirements to be eligible for support, there are two income limits defined each year by the Regions, the first, the lower limit defines Band A beneficiaries, the second limit defines Band B beneficiaries[7]. In addition, it is necessary to be resident in rented accommodation, not to live in a 'valuable area' defined by each municipality and not to have used the rent deduction in the tax return. Regions may also identify special priority categories for the allocation of support.

3.2   The Regional Management of the "Support for Rent" Call for Tenders

To verify the usefulness of the data extrapolated from the findings of the call "support for rents" for mapping housing needs, this paragraph analyses in detail the features related to the Apulian call for tenders "Support for Rent" and the data collected by the PUSH Platform. Specifically, the call for proposals "Rent support 2020, income 2019" is analysed.

In 2017, the Apulia Region financed the implementation of the PUSH[8] (PUglia Social Housing) IT platform. The PUSH platform, by the provisions of the Digital Administration Code, implements the dematerialisation of administrative processes related to the housing policies section of the Apulia Region and addressed to ARCA[9] and Municipalities. It is also responsible for acquiring and managing databases on the public housing stock and on the results of the ranking lists of municipal calls for tenders relating to rent support. In this way, through the functions of the management dashboard, it is possible to monitor housing deprivation and better plan housing policies.

In 2020, the experimentation of the first call for the support for rents according to Article 11 of Law no. 431/98, approved by DGR n. 1999 of 4/11/2019, was launched

---

[7] Income limit for Band A is €13,338.26 and the limit for Band B is €15,250.00.

[8] https://push.regione.puglia.it/.

[9] Regional Agency for Housing and Living.

on the PUSH platform. The DGR introduced new procedures, constraints, and controls, which required major adjustments to the PUSH platform, which had already been designed and tested to implement the logic of execution of the notice of Law no. 431/98.

In line with national provisions, the regional announcement defines income limits for Band A and Band B, similar to those provided for ERP. After issuing the call for applications, the municipalities manage the reception of applications and, once the lists of beneficiaries and their specifications have been drawn up, upload them onto the Platform, where they are entered into the database and can then be analysed within the Management Dashboard of the Platform, through which it is possible to process and calculate the data in tabular form and export the information to spreadsheets.

Specifically, the Management Dashboard relating to the monitoring of beneficiaries of the Law no. 431/98 call allows the analysis of data using "dimensions" and "measures". Measures and dimensions can be analysed to aggregate and filter data from rankings to perform analysis and reflection on data that arrive in disaggregated, point-in-time form.

The "dimensions" are the filters that allow to group the collected data and are:

- Notice - Unique identifier (filtered by the call and year of the lease)
- Geographical dislocation (filtered by province, municipality, and Istat code)
- High housing tension[10]
- Beneficiary Type (filtered by Band A, Band B or excluded from rankings)
- Type of income (filtered by Self-employed/ Employee/ Mixed income/ Zero income/ Miscellaneous/ Not specified)

The "measures" are the numerical values expressing the target variables of the analysis. In the case of beneficiary monitoring, data are as follows, measurable in total value or average value:

- Call (no. of applications)
- Income information and contribution (Annual fee, Contribution, Conventional income, Taxable income, Employee income, Autonomous income, Othe income)
- Rent information (month lease, rent incidence, agreed rent, free rent)
- Dwelling information (surface of dwelling, rooms, lift presence, heating presence, prestige area)
- Beneficiary information (handicap, over65, immigrant, no. others component, no. dependent children)

The possibility of analysing and aggregating these data on a municipal, provincial, or annual scale is useful in order to identify, in the case of a territorial approach, potentially critical areas of housing deprivation, or, in the case of a temporal approach, the possibility of highlighting mutations and changes in housing needs.

3.3 Application to the Municipality of Bari

The data on the results of the call, collected by the computer system and available through the PUSH Management Dashboard, can be processed to provide more knowledge on

---

[10] Municipalities defined as High Tension Housing (ATA) according to CIPE Resolution 2003.

housing needs. In this paragraph, the data concerning the Municipality of Bari will be analysed and shown, as an example useful to support the thesis of the contribution. The sample of the Municipality of Bari, the regional capital with a population of 316 245 people, was chosen for the application of the data processing, as it is the most significant in the Apulia Region with a total of 3061 (1% of the total population) beneficiaries, of which 1901 in Band A and 1160 in Band B. Moreover, before analysing the data, it is interesting to underline the scope of the call which, with the contribution granted, manages to satisfy 78% of the needs for Band A (with a requirement of €4, 981, 031.66) and 34% for Band B (with a requirement of €2,220,207.80) for a total of 64% of the requirements satisfied.

The first analysis allows measuring the medium income, calculated with a weighted average, of the beneficiaries of the two bands. For Band A the average taxable income is 7610,83 euros and for Band B is 18117,71 euros[11]. It is interesting to underline the incomes distribution, which shows in Fig. 1 that most applicants have an income of between 5000 and 15000 euros per year, i.e., like public housing, showing a target group that is low, but not in absolute poverty.

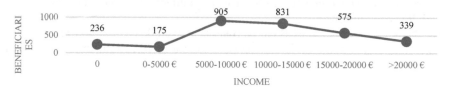

Fig. 1. Income distribution

Interesting is the incidence of housing costs on income, the so-called affordability. Data processing shows that most applicants are in a state of poverty compared to the conventional 30% limit set to assess housing affordability. However, this is to be expected considering that the sample analysed refers to beneficiaries of housing support. More interestingly, the distribution is overlapping for Band A and Band B, showing in Fig. 2 that even households with higher incomes, potentially excluded from housing policies, are in considerable housing poverty.

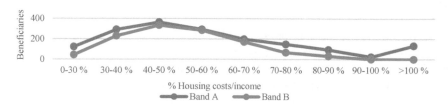

Fig. 2. Distribution of housing costs as a percentage of income

---

[11] It should be noted that, with respect to band B, the calculation used by the results is that of conventional income, calculated with a deduction formula based on dependent children. In this case, taxable income has been used to standardise the information.

The average age of applicants is broadly comparable between Band A and B with a higher incidence of applicants between 35 and 50 years old in Band A. Household composition generally shows a presence of small two/three members' households. Figure 3 shows that the poorest are single member households, which are most Band A applications, mostly single elderly people.

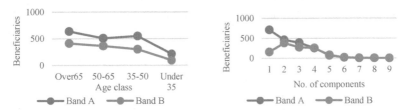

Fig. 3. Average age of applicants and distribution of households

In terms of households' origin, the presence of immigrants is around 15% of the total. It is interesting to note, as shown in Fig. 4, that the family composition of Italian households is mainly composed of households without a child or with only one dependent child, while immigrant households are more numerous and have more children.

Fig. 4. No. of dependent children in Italian and immigrants' families

Analysing the data relative to the availability of square meters per inhabitant, a widespread problem of overcrowding does not emerge. Emerge in Fig. 5, a condition of under-utilisation of dwellings emerges, due to the high number of elderly and single-person households, and a greater presence of serious conditions of overcrowding in foreign families, with more than one family out of ten not reaching the minimum threshold of 14 square meters/inhabitant.

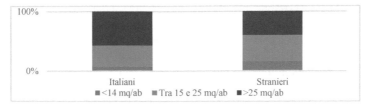

Fig. 5. Square meters for household

3.4 Proposal for Mapping Social Housing Needs

The mapping of housing needs through the data coming from the PUSH platform and analysed to know the characteristics of the housing demand responds to two needs. The first one is to widen the knowledge of the targets outside the public housing lists and the second one is to support in a concrete way choice concerning housing policies, directing them according to the real housing need in a specific territory. The application of the analysis to the data relating to the Municipality of Bari has made it possible also, to demonstrate the possibility of knowing the specific characteristics of this need and therefore being able to outline a housing supply starting from a known demand. Specifically, it is possible to describe the types of households and their composition, average age, housing conditions, and the incidence of housing costs on income. For this purpose, the data identified as useful in knowing the housing needs are income data, the incidence of housing expenses on income, family types and their composition, the average age of the beneficiaries, and the average square metres per person.

Concerning the index developed by the European SPC, indicators on the incidence of housing costs on income and overcrowding conditions are useful and usable. Regarding the number of people at risk of poverty for the title of tenure, it is not possible to apply it because the data refer to people all related to the type of rent; regarding housing deprivation, the findings do not collect the required data but only those related to the presence of heating and lift. The use of these data, in addition to those relating to public housing rankings, which have not yet been centralised and systematised on a regional scale, would allow a mapping of the housing needs of the poorest segments of the population, i.e., those most affected by housing problems.

4   Conclusions and Future Research

In Italy, a review of the rental system is needed, introducing legislation to improve guarantees for tenants and proprietors to increase the number of available homes for rent (Filandri, Olagnero e Semi 2020). In addition, as mentioned above, there is no systematic survey of the needs of the social groups with housing deprivation problems.

Therefore, with the intent of resolving gaps, the contribution highlighted the potentiality of the data collected by the regional application of Law no. 431/1998 for the social needs knowledge. In detail, the paper allows for the development of a repeatable processing methodology over time to update needs, going beyond the simple identification of areas of need but also characterizing it. The complete social housing needs mapping can be very useful in challenging current housing policies, which are unable to meet the large and diverse demand for housing. It can direct urban planning strategies, meeting different actors (public, private and third sector) and at different scales, offering more affordable and quality rental solutions (including public and social houses). This is even more important because of the huge financing given by the Italian PNRR (Piano Nazionale di Ripresa e Resilienza). This amount of money will be a great opportunity to innovate and implement the housing system and relative management methods. The research limitations are related to the development of the PUSH Platform, which is still ongoing, so some information and functionality are still being implemented. Also, the data needs further processing than how it comes out of the dashboard.

Future research should attest to the scope of even broader and more shared mapping at the regional and national scales. In addition, mapping of needs located within cities may help to better understand areas of greatest housing tension within the municipality.

References

Baldini, M.: e T Poggio. Il fondo sociale per l'affitto nell'esperienza di alcune grandi città italiane: un'analisi su microdati amministrativi. Quaderni della ricerca sociale 22 (Ministero del Lavoro e delle Politiche Sociali), 255–283 (2013)

Cigolotti, B., et al.: Un abitare complesso. Una proposta partecipativa per un indice della povertà abitativa. XIII COnferenza ESPAnet Italia (2020)

Di Biagi, P.: Dal Pubblico al Sociale. Nuove questioni abitative, città pubblica e spazi abitati. Edilizia Sociale e Urbanistica. La difficile transizione dalla casa all'abitare (2014)

Filandri, M.M.: Olagnero, e G Semi. Casa dolce casa? Italia un paese di proprietari. Bologna: Il Mulino (2020)

Fregolent, L.: e R Torri. L'Italia senza casa. Bisogni emergenti e politiche per l'abitare. I. Milano: FrancoAngeli (2018)

Lodi Rizzini, C.: Il Social Housing e i nuovi bisogni abitativi. In: Primo rapporto sul secondo welfare in Italia 2013, di Maurizio Ferrera Franca Maino. Torino: Einaudi (2013)

Mangialardi, G., Martinelli, N., Spadafina, G.: Abitare la Puglia, Criticità e sfide per nuovi modelli abitativi nel Mezzogiorno. QuAD (2020)

Molinari, L.: Le case che saremo. Abitare dopo il lockdown. Milano: Nottetempo (2020)

Nomisma. DIMENSIONE DEL DISAGIO ABITATIVO PRE E POST EMERGENZA Covid-19 (2020)

Olagnero, M.: La questione abitativa e i suoi dilemmi. Abitare (2008)

Palvarini, P.: Il concetto di povertà abitativa: rassegna in tre definizioni (2006)

Storto, G.: La casa abbandonata. Il racconto delle politiche abitative dal piano decennale ai programmi per le periferie. Roma: Officina Edizioni (2018)

The Social Protection Committee. Indicators and data to monitor developments in access to housing and housing exclusion (2014)

Tosi, A.: Le Case dei Poveri. Milano: Mimesis edizioni (2017)

Barcelona's Challenge to Supply Affordable Housing. Innovative Tenure Alternatives to Improve Accessibility

Maria José Piñeira-Mantiñán[✉], Ramón López-Rodríguez, and Francisco R. Durán Villa

University Santiago de Compostela, Pza. Universidade 1, 15782 Santiago Compostela, Spain
mariajose.pineira@usc.es

Abstract. Local governments in Spanish cities such as Barcelona are boosting alternative forms of tenure to facilitate housing accessibility. Specifically, Barcelona city council has developed co-living, cohousing, housing cooperatives or provisional proximity housing initiatives as an alternative to home ownership. In this chapter we critically analyze the housing tenure alternatives carried out by Barcelona city council, focusing on its spatial spread in the city.

Keywords: Housing policies · Housing accessibility · Rental prices · Barcelona · Spain

1 Introduction

Housing in Spain is a problem of multiple dimensions that is perpetuating itself over time. The actions derived from Spanish housing policy over the years have been unable to respond to the problems of access to housing for large sectors of society.

Since the creation of the Ministry of Housing in the 1950s to promote construction and respond to the housing shortage, the housing policy model has focused on promoting access to home ownership (regardless of household income) and understanding the acquisition of housing as an investment [1]. This model is widespread in southern European countries, where there is also a high stock of secondary and empty housing and a shortage of affordable social housing for rent [2]. In this regard, it should be noted that Spain is at the bottom of the OECD countries in terms of social rented housing, with barely 1.1% of the total, compared to countries such as the Netherlands, where this figure rises to 34.1%, the United Kingdom with 16.7% or France with 14% [3].

The bursting of the housing bubble in 2008 exposed the weakness of this model [1, 4], with an unprecedented increase in the volume of foreclosures and evictions. About 163 daily evictions are still recorded. Likewise, mortgage foreclosures and evictions derived from the Urban Leasing Law, which represent 56.2%, become more relevant.

Research in the framework of the projects (PID2019-108120RB-C31) & (Ref: CSO2016-75236-C2-1-R).

Likewise, the European Anti-Poverty Network has shown that more than 11 million people in Spain are in a situation of residential and housing emergency [5]. A problem that is not being addressed, as the volume of social housing built is the lowest since the 1950s (only 6,615 were built in 2019). As a result, the social housing stock in Spain only represents 0.96% of the main housing stock when in the European Union it is 9.3% [6]. A circumstance that causes a mismatch between supply and demand, as according to the registers of Applicants for Subsidized Housing of Spanish regions there are more than 400,000 people applying for subsidized housing.

In this context, there is an urgent need to promote a residential model that responds to the enforceability of housing as a right. Interviews with political leaders during the 2015–2019 legislature in several Spanish cities (Barcelona, Madrid, Cádiz, Valencia, Pamplona, Seville, A Corunna, Santiago de Compostela) confirm their limited capacity to promote subsidized housing (VPO), either due to budgetary problems or to the slowness of the bureaucratic process involved in its promotion and execution. On the other hand, some researchers [7] question to what extent VPO real estate can continue to be the main figure of housing policy in Spain, when during the entire crisis the number of subsidized housing transactions plummeted.

Local administrations are therefore obliged to promote innovative initiatives that facilitate, among other aspects, the development of public rental housing (new and refurbished) for direct promotion; encourage investment in the consolidated city as opposed to the production of new housing and promote refurbishment actions; incentivize the occupation of empty housing through public-private agreements; etc.

This paper first addresses the housing policies developed in Spain since the mid-twentieth century and the effects of a neoliberal residential model that came to light in the wake of the 2008 crisis, and then focuses on the case study of Barcelona, to study the initiatives promoted by the local administration to facilitate access to housing for the needy, in which social renting and public-private partnerships are the most noteworthy aspects.

2   The Housing Problem in Spain

2.1   An Approach to Housing Policies in Spain

"All Spaniards have the right to enjoy decent and adequate housing. The public authorities shall promote the necessary conditions and establish the relevant rules to make this right effective, regulating land use in accordance with the general interest to prevent speculation (…). Article 47 Spanish Constitution 1978

Housing in Spain is a problem of multiple dimensions that is perpetuating itself over time. It is paradoxical that in a country with an oversized housing stock it is becoming increasingly difficult to gain access to housing. This circumstance leads us to interpret article 47 of the 1978 Constitution to analyze whether the policies developed to date are in line with it. Let us start with the key idea, the "right to enjoy housing". Enjoying a home does not necessarily imply owning it, as has been interpreted until now by society, encouraged by policies that saw home ownership as the solution to boost the construction sector and thus recover the economy. These policies date back to the 1950s–1960s, when housing minister Arrese stated that "we do not want a Spain of proletarians, but a Spain of homeowners" [8]. This would be the turning point that would mark a new structure in the housing tenure regime. This is demonstrated by data from the Housing Observatory, which recorded that while in 1950 51% of the population lived in rented accommodation - reaching 90% in cities such as Madrid and Barcelona - by the time the housing bubble burst in 2007, 90% of the Spanish population owned a home [9]. For more than fifty years, the stock of owner-occupied housing has only grown, and this has been accompanied by a rise in the prices of homes (sale and rental) and land. On the contrary, the production and development of housing by the public sector has been reduced at the peak of the economic-real estate cycle and encouraged when it is contracting - albeit insufficiently -, generating a public intervention inversely proportional to the situation of the real estate market [10]. This brings us back to article 47 when it says "The public authorities shall promote the necessary conditions and establish the relevant rules to make this right effective, regulating the use of land in accordance with the general interest in order to prevent speculation". However, the measures adopted in Spain, mainly since 1985, promoted the opposite: (i) the declaration of the Boyer Decree underpinned the stimulation of the private purchase of housing thanks to the reduction of bank interest rates, easy access to mortgages, the unfreezing of rents and tax relief for the purchase of a home (whether it was to be occupied or not) [1]; ii) the approval of Law 6/1998 on the land regime and valuations, baptized as the "all developable land" law, determined that developable land reached its maximum market value (as if it were built)[4]. iii) The government's actions took the form of taxation that encouraged indebtedness through public aid aimed at subsidizing loans -both to individuals and to developers and builders- for the liberalization of land and the purchase of free housing, which would eventually end up in the hands of financial institutions. iv) Housing policies responded to a desire for development and speculation [11]. Instead of promoting affordable housing, they were subordinated to a financial and real estate lobby that attracted foreign currency, generated financing, increased the land market and promoted large real estate projects. The administration ceded the power of urbanization and control over the free housing stock.

This panorama led to the "prodigious" decade of the brick [4], the high point was 2006, when housing production in Spain was double that of countries such as France and Holland and quadruple that of Germany [12]. Housing had ceased to be a right to become a commodity and an object of speculation, and the nightmare of thousands of families who, because of the 2007 crisis, could no longer afford it, were evicted and were affected by processes of economic, social and residential vulnerability.

2.2 The 2007 Crisis: A Key Factor in the Change in the Housing Paradigm

The bursting of the housing bubble in 2007 exposed the weaknesses of the neoliberal growth model that had been in place until then. First, citizens became aware of the high volume of housing that was not intended for primary residence. According to the housing statistics of the Ministry of Transport, Mobility and Urban Agenda [13], in 2007 the housing stock was 24,034,966 units (14% more than in 2001), of which 30% were non-primary. This stock continued to increase to 25,882,055 in 2020, of which 25% were non-primary. They became aware of the high stock of unsold new housing - which reached its peak in 2009 with 649,780 units [14] and began to suffer the consequences of having prioritized the purchase of new free or subsidized housing over renting [7], unlike other European countries. To this must be added the sale by some administrations of public housing stock to new developers such as investment funds or vulture funds. The result is that Spain has one of the smallest stocks of social housing in Europe. While in 1985 the promotion of subsidized housing (51%) exceeded free housing, from 1988 onwards it began to fall to 7.2% in 2002 and currently stands at just over 1%, compared to almost 30% in the Netherlands, more than 20% in Austria and more than 15% in the UK and France [15]. This was a serious problem if we consider that in the period 2008–2019, two million people were affected by housing emergencies because of evictions [4]. Since the crisis, 684,385 evictions have been registered in Spain, reaching 186 per day in 2014 [16]. However, a mutation in their profile was observed, with an increase in those caused by non-payment of rent and a decrease in foreclosures since 2016.

According to the Association of Registrars in 2019, the average mortgage payment was 591€ and the average rental income was 966€; and the percentage of family income spent on buying and renting was 41% and 67%, respectively [16]. In the case of rents, the price increase was notable, rising from an average cost of $9.1€/m^2$ in September 2006 to $11.5€/m^2$ in September 2011 according to data from the real estate portal Idealista [17]. The causes were associated with an increase in demand (more and more people see renting as the only alternative for accessing housing), the shortage of rental housing at affordable prices, and an exponential increase in tourist rentals. Territorially, as of November 2021, the most expensive rents were in Madrid (13.6€/m2), Catalonia (13. €/m^2), the Basque Country ($12.4€/m^2$), the Balearic Islands ($12€/m^2$) and the Canary Islands ($10.3€/m^2$).

In this context, the Mortgage Victims' Platform (PAH) and the Tenants' Union began to express the urgency of promoting a public housing stock that would guarantee the right to housing for citizens. The response from the administration was the approval of the Sustainable Economy Law 2/2011 [18], which focuses on the construction of subsidized housing and rehabilitation; and the approval by the Council of Ministers on 26 October 2021 of the new Housing Law, which will come into force in 2022 and which will apply measures to regulate rental prices: promoting subsidized housing for rent at a limited price, promoting public rentals in new developments, preventing changes in the classification of public housing, offering tax advantages to small owners to lower rents and delimiting "stressed sectors" in the housing market. These stressed areas will be declared by the Autonomous Communities - as they are responsible for housing - and will have a multi-scale character (neighborhood, municipality or the whole region). They must meet at least the following requirements: average rental prices must have risen five points higher than the CPI over the last five years; and the average amount must represent more than 30% of the average income per household in the area to be declared [18].

However, much work remains to be done, especially bearing in mind that: in the field of housing there is a multiplicity of realities depending on the size of the city and the intrinsic characteristics of its neighborhoods (morphological, social and economic characterization); information on an intra-urban scale is scarce, obsolete and in some cases in the hands of private real estate portals. Only cities such as Madrid and Barcelona have a wider range of statistics available, promoted by the local administration, with a view to designing new housing access policies.

In this article we will focus our attention on the case of Barcelona, as it is the city with the highest population growth in recent years; it has been, together with Madrid, the Spanish city with the highest number of evictions and the highest rental prices; and where the municipal government is implementing innovative initiatives to promote access to rental housing at affordable prices.

3   Methodology

The literature on the housing situation in Spain, the real estate market, the evolution of policies since the mid-20th century and current legislation has been reviewed. The literature review has been accompanied by a mixed quantitative and qualitative methodology. Official statistical sources from the National Institute of Statistics, the Ministry of Public Works and Barcelona City Council were consulted. Likewise, in the period 2015–2021, semi-structured interviews were conducted with political leaders, platforms and citizen groups in ten Spanish cities: Madrid, Barcelona, Valencia, Pamplona, Seville, Cadiz, Ferrol, A Coruña, Santiago de Compostela, Las Palmas de Gran Canaria. Fieldwork that is part of two research projects funded by the Ministry of Science and Innovation R&D&I Projects: *New models of governance of Spanish cities, and intervention in urban spaces in the post-crisis period* (Ref: CSO2016–75236-C2–1-R) developed between 2015–2019; and *The housing problem in the fragmented metropolises of Spain. Perpetuation over time, new housing markets and solutions from the public adminis-tration* (PID2019-108120RB-C31) started in 2020. In both projects, a different interview model was designed for politicians and citizen platforms. For the Barcelona case study, the people interviewed in November 2021 were officials from the city council, the Tenants' Union and the Housing Observatory. All of them were asked about the factors limiting access to housing, what initiatives are being implemented to reverse this situation and how the tension arising from the increase in tourist rentals is being managed. This information was complemented by the seminar *Measurement and analysis of metropolitan inequalities. Themes, methods, hypotheses and results* in which experts from the Autonomous University of Barcelona and the Polytechnic School of Madrid took part to address the problem of housing as a factor of segregation, the problem of rising prices and tensions in the housing market.

4   The Challenge of Accessing Housing in the City of Barcelona

Accessing housing in Barcelona is becoming more complicated every day. In 2018, the Municipal Services Survey [19] showed that access to housing was one of the most serious problems in the city and this opinion is maintained in 2021. Among the factors causing the problem is the lack of housing available at affordable prices, which responds to the existing demand that is growing (1,496,268 in 2000 and 1,664,182 in 2020). Secondly, we must refer to the commodification of housing, which has led to an exorbitant rise in prices. With the recovery of the crisis, the city recorded a new rise in the price of flats for sale, which rose from 249,000€ in 2013 to 352,000€ in 2021, and the arrival of investors interested in buying properties, with 84% of the supply in sectors such as Barceloneta [20]. Urban sectors such as Ciutat Vella and Les Corts have suffered a brutal price increase, with the average price per m^2 built reaching 8,877€ in 2007 [21]. In other districts such as Sants-Montjuïc the cost per m^2 was reduced, but prices still high remain in central sectors such as Eixample and Les Corts and in districts further away from the center such as Gràcia and Sarrià-Sant Gervasi. These coincide with the districts where the highest rents in the city are concentrated.

In this context, citizens are finding it increasingly difficult to buy a home, and the situation is getting worse if we consider the growing gap between people's incomes and house prices. In Barcelona, incomes rose by 2.7% during 2018–2019, average house prices increased by 5.9% in new, 2.8% in second-hand and 5.3% in rented housing. The result has been a sharp fall in sales (from 15,676 in 2017 to 6,587 in 2021) and an increase in the volume of rentals, mainly among young people aged 16–34 (71.5% live in rented accommodation) and young adults aged 35–44 (42.3%). So, it is observed a significant change in the structure of the tenancy regime. Currently, 30% of Barcelonans have opted to rent, a percentage that is still far from the 60% recorded in Berlin, Amsterdam or Paris, where it is also protected by law. Since 2000 the average monthly rent has increased by 121.7% from 408.3€ to 905€. With these prices, according to data from the Housing Observatory, families must invest 40% or more of their income in housing, and young people see their emancipation limited, they have left their parents' homes 27% less than ten years ago. The most expensive sectors coincide with those where tourist activity is putting the greatest pressure on the housing stock, diverting flats intended for residents to visitors because it is more profitable for landlords. According to a report on the impact of tourist rental on residential rental in Barcelona [22], there are 15,881 tourist flats in the city, of which only 9,606 are licensed. This means that 39.5% of the supply is illegal. On the other hand, it indicates that tourist rentals represent 7.7% of the rental stock and that the income for the owner is between 2.3 and four times higher than that of conventional rentals, with profits between 7% and 13% compared to 3%–4%. This data explains why the stock of housing on the rental market is shrinking is alarmingly reduced. Thus, the high cost of housing prices is becoming: (i) the main factor in the expulsion of the population towards municipalities in the metropolitan area; (ii) a factor in residential exclusion and segregation, as the economic effort that families must make to access housing in the city is increasing, especially with the increase in the precariousness of the working conditions of the population.

Data from the 2017 survey on precariousness [23] showed that 47% of those surveyed were working under a temporary contract, 29.4% had a part-time contract; 50.1% had experienced at least one period of unemployment in the last ten years; 46% earned less than €1,000 per month. In the same year, the Foessa Foundation [24] highlighted that access to employment was insufficient to protect people from residential vulnerability. Consequently, we can affirm that in Barcelona, the income of the population is not sufficient to meet the costs of housing, leading to evictions (2,125 in 2019, 80.6% due to non-payment of rent). The population cannot live where they want, but where they can [25]. According to the Metropolitan Housing Observatory of Barcelona [26], in 2019 a family with an average gross household income of 48,859€ could afford a medium-priced home in Barcelona without exceeding the threshold of 30% of their income (27.6%

for new housing, 23.7% for second-hand and 21.2% for rental). However, a household with an annual income of 35,000€ would be excluded from the market in 7 of the 10 districts of the city and in 5 of the 35 surrounding municipalities; and if their income was 25,000€, their chances would be greatly reduced, only in 2 municipalities of the metropolitan area.

5   Innovation: The Alternative to Improve Access to Housing in Barcelona

The arrival of Mayor Ada Colau to the municipal government brought about a radical change in housing policies and the implementation of innovative initiatives to promote access to affordable housing. It was considered essential to draw up a Housing Plan 2016–2025 that focuses its interests on: housing emergency, social use of housing, affordable public and private housing stock and rehabilitation. The Plan assumes a leap in the public rental stock from just over 7,000 units (at the end of 2015) to 13,489 in 2025, and implies that Barcelona City Council has to meet the costs from its own funds and the nearly 180 million euros it has obtained from the European Investment Bank and the Council of Europe Bank.

To defend the right to housing, the city council has carried out three pioneering measures: i) it has bought entire buildings to avoid evictions, prevent speculation and stop tourist housing. ii) it has approved the reservation of 30% of subsidized housing in both new developments and major renovations - a interesting fact for the central neighborhoods that suffer speculation and gentrification-; iii) has declared the whole municipality as a housing market stressed area for 5 years and has passed the Law 11/2020 which establishes the Reference Price Index. This measure is accompanied by the application of penalties for abusive rentals (9,000–90,000€ if the rent exceeds the index by 20%; and 3,000–9,000€ if it exceeds it by less than 20%). These initiatives have been complemented by a census of empty homes, with the aim of mobilizing private housing to the affordable rental market quickly and dispersed by neighborhood. In 2019, a total of 10,052 dwellings that had been empty for 2 years in the city (1.22%) were registered, In the case of private owners, contact was made with them to offer the option of incorporating their homes into the city's rental market by providing guarantees on rental payments, facilities and benefits for the refurbishment of the homes. In the case of financial entities, management companies or companies, sanctioning procedures were initiated, since, in accordance with the Law on the Right to Housing, two years is the maximum time a property can be kept empty. Ciutat Vella is the district with the highest proportion of flats in which no one lives in relation to the total housing stock, although in absolute numbers, Eixmple (1,883), Sant Martí (1,313), Horta-Guinardó (1,199), Sants Montjuïc (1,128), Nou Barris (1,065), Gràcia (859), Sant Andreu (694), Sarrià-Sant Gervasi (655) and Les Corts (275) stand out.

In any case, despite all the initiatives promoted, the municipal government warned that the lack of affordable housing was a problem that could not be solved with public investment alone, and that it was necessary to tend towards a diversification of managers and promoters. It was time to promote public-private collaboration for housing construction under the umbrella of a non-speculative model of investment in the sector.

Under this premise, various initiatives were promoted. With the Habitat3 Foundation -social housing manager driven by the third sector- a process was started to capture and manage 250 empty homes on the private market to allocate them to families/persons in a situation of social exclusion. If the dwellings require refurbishment work, this is carried out by the program and by social organizations in the city. For its part, the owner receives the agreed rent for the duration of the rental contract, from which the financing of the refurbishment work (20% is non-refundable) will be deducted on a monthly basis. Among the actions is La Casa Bloc located in Sant Andreu neighborhood.

With the regional government it set up a rental housing exchange to offer mediation services between owners of empty homes and potential tenants. Also, it created a mixed company to set up the Metropolitan Housing Operator, in which the City Council and the Metropolitan Area of Barcelona owned 50% of the shares and the other half was in the hands of the private company that won the public tender. The objective was to build 4,500 affordable public rental flats in the Barcelona metropolitan area over a period of 8 to 10 years, at below-market prices of between 400 and 600 euros. The participants would have a profit limited to approximately 4% of the investment, and any profits in excess would be reinvested in Metrópoli Vivienda.

Collaboration with private developers also took the form of the construction of Provisional Proximity Housing (APROP) and cohousing. The former was a tool through which to offer fast, sustainable and quality housing in order to expand the public housing stock, prevent the expulsion of residents from the neighborhood and fight gentrification. The idea was to build through this process 4,000 homes in different neighborhoods and with different profiles of residents. As a pilot project, three highly gentrified urban sectors were chosen: Ciutat Vella, Sants-Montjuïc and Sant Martí.

These were residential buildings made from containers, but with the finishes of a conventional, high-quality building, and with facades covered to improve their thermal insulation. Overall, the building exceeded the standards of the highest level of sustainable building rating, AA. Among its advantages were the speed with which it could be built (nine months compared to 6–7 years for a normal house); and the fact that it was modular, transportable and demountable, which would allow empty spaces in the city to be mobilized quickly for periods of 5 or more years.

Finally, cohousing consisted of a formula for granting ownership of the property to a cooperative to be built without the land ceasing to be public. Barcelona City Council ceded 15 plots of land -distributed along the coastal districts of Sant Martí, Sants-Montjuic, Sarria-Sant Gervasi and the northernmost district of Gràcia- and 3 buildings to be built and rehabilitated by cooperatives. The surface right is ceded to the cooperative for a period of 75 years, extendable to 90. At the end of this period, ownership of the built property passes to the municipality. People enter the housing by paying a refundable down payment, so they are never owners but cooperative members. They can live there under a regime like indefinite renting, but at a much more affordable price and with collaborative management of the common spaces. In addition, the right to use the home can be indefinite, transferred and inherited. Among the advantages offered by this modality are: cut down bureaucracy, costs and time related to public procurement processes and improve access to financing [27].

6   Conclusions

The evolution of housing prices in Spain has caused the marked trend towards home ownership to shrink. Such a trend has been reinforced by the aforementioned initiatives carried out by Barcelona city council, where in principle social renting and co-management are the key aspects.

In this study we have focused on alternatives that are novel, such as cohousing, housing cooperatives or Provisional Proximity Housing. We have found that all these alternative forms of tenure facilitate urban renewal processes and the rehabilitation of historic and urban centers -not entailing price increases and gentrification- while allowing access to housing at affordable prices below the market average. Having analyzed the case of Barcelona, it has been observed that these alternative forms of tenure can be more viable if: i) the local governments declare their commitment to public housing and have publicly owned plots of land to build it on; ii) there are areas under stress because of rising housing -mainly historic centers-; iii) there is organized civil society interested in tenure alternatives.

However, these new forms of tenure can be useful for those local administrations with difficulties in promoting alternative forms of tenure due to the lack of complementary budgets from the national or regional government. For the future, it will be interesting to analyze the extent to which these initiatives are being developed in other Spanish cities, with less pressure on housing prices, and therefore what are the factors that favor their implementation in those cities.

References

1. Lois, R., Piñeira, M.J., Vives, S.: The urban bubble process in Spain: An interpretation from the theory of circuits of capital. J. Urban Reg. Anal. VIII(1), 5 – 20 (2016)
2. Maldonado, J.L., Martínez, A.: Tendencias recientes de la política de vivienda en España. Cuadernos de Relaciones Laborales. Monográfico: La cuestión de la vivienda **35**, 1 (2007) https://revistas.ucm.es/index.php/CRLA/article/view/54982
3. OCDE. Affordable Housing. https://www.compareyourcountry.org/housing/en/3/all/default
4. Burriel de Orueta E.L.: La década "prodigiosa" del urbanismo español (1997–2006). Scripta Nova **12** (270), 64 (2008)
5. EAPN.   https://www.eapn.es/noticias/1427/mas-de-11-millones-de-personas-sufren-exclus ion-residencial-en-espana
6. Cohispania. Estado de la vivienda protegida en España. https://www.cohispania.com/docume nts/Estudios_Mercado/informe_estado_vivienda_protegida_en_espana-CoHispania.pdf
7. Pareja, M., El Sanchez, M.T.: sistema de vivienda en España y el papel de las políticas: ¿qué falta por resolver? Cuadernos Económicos de ICE **90**, 4 (2015)
8. ABC,1959 No queremos una España de proletarios sino de propietarios. https://www. march.es/es/coleccion/archivo-linz-transicion-espanola/ficha/--linz:R-73814. Accessed 25 Nov 2021
9. Rodríguez Alonso, R., Espinoza Pino, M.: De la especulación al derecho a la vivienda. Más allá de las contradicciones del modelo inmobiliario español. 1st edn. Traficantes de Sueños, Madrid (2017)
10. García Pérez, E., Janoschka, M.: Derecho a la vivienda y crisis económica: la vivienda como problema en la actual crisis económica. Ciudad y Territorio. Estudios Territoriales XLVIII **188**, 213–228 (2016)

11. Trilla, C.: La política de vivienda en una perspectiva europea comparada. Colección Estudios Sociales 9. Fundación La Caixa. Barcelona (2001)
12. Gaja, F.: El tsunami urbanizador en el litoral mediterráneo. El ciclo de hiperproducción inmobiliaria 1996–2006. Scripta Nova. Revista Electrónica de Geografía y Ciencias Sociales **270**(66), Universidad de Barcelona (2008). http://www.ub.es/geocrit/sn/sn-270/sn-270-66. htm
13. MITMA Estimación parque de viviendas. https://www.mitma.gob.es/informacion-para-el-ciudadano/informacion-estadistica/vivienda-y-actuaciones-urbanas/estadisticas/vivienda-y-suelo. Accessed 28 Nov 2021
14. Ministerio de Fomento 2012 Informe sobre el stock de vivienda nueva 2012 (Serie 2008 - 2011 revisada). www.mitma.gob.es
15. The state of housing in Europe 2021. Housing Europe. https://www.stateofhousing.eu/The_ State_of_Housing_in_the_EU_2021.pdf. Accessed 20 Nov 2021
16. DESC & PAH. Informe Emergencia Habitacional en el Estado Español con datos inéditos sobre las ejecuciones hipotecarias. https://afectadosporlahipoteca.com/2013/12/17/informe-emergencia-habitacional/
17. Informe precios alquiler Idealista. https://www.idealista.com/sala-de-prensa/informes-pre cio-vivienda/alquiler/. Accessed 05 Dec 2021
18. Anteproyecto de Ley por el Derecho a la Vivienda. https://www.mitma.gob.es/el-ministerio/ sala-de-prensa/noticias/mar-26102021-1559. Accessed 18 Dec 2021
19. Ayuntamiento de Barcelona. Encuesta de Servicios Municipales de la ciudad de Barcelona. https://opendata-ajuntament.barcelona.cat/data/es/dataset/esm-bcn-evo
20. García Montalvo, J.: (Cood). XXXIII Informe sobre el mercado de la Vivienda. Tecnocasa y Universitat Pompeu Fabra. Barcelona
21. Ayuntamiento de Barcelona. Vivienda y mercado inmobiliario. https://ajuntament.barcelona. cat/estadistica/castella/Estadistiques_per_temes/Habitatge_i_mercat_immobiliari/index.htm
22. Duatis, J., Buhigas, M., Cruz, H.: Impacte del lloguer vacacional en el mercat de lloguer residencial de Barcelona. Ayuntamiento de Barcelona, Barcelona (2016)
23. Bolibar, M., Galí, I., Jódar, P., Vidal, S.: Precariedad laboral en Barcelona: un relato sobre la inseguridad. Universitat Pompeu Fabra, Barcelona (2017)
24. Caro, F.: Vulnerabilidad y empleo. Fundación Foessa, Madrid (2017)
25. Nello, O.: Transformar la ciudad con la ciudadanía. Criterios y reflexiones para el Plan de Barrios de Barcelona. Barcelona: Ayuntamiento de Barcelona (2017)
26. Observatorio Metropolitano de la Vivienda de Barcelona. L'habitatge a la metrópoli de Barcelona 2019. Barcelona (2019)
27. Cabré, E.: Barcelona Cohousing Programme. Barcelona, Ayuntamiento de Barcelona. https:// nws.eurocities.eu/MediaShell/media/05_Barcelona.pdf

Urban Planning and Urban Morphology Variables in Defining Real-Estate Sub-markets

Konstantinos Lykostratis[✉] and Maria Giannopoulou

School of Civil Engineering, Democritus University of Thrace, Campus Xanthi-Kimmeria,
67100 Xanthi, Greece
klykostr@civil.duth.gr

Abstract. One of the most important research subjects in urban economics concerns whether the real estate market is unified within the urban space, or consists of sub-markets. A review of the existing literature shows that there is no clear agreement on the definition of real estate sub-markets, with research focusing on defining their boundaries. Our effort aims to check and confirm the existence of real estate sub-markets with emphasis on the housing market applying both spatial and non-spatial approaches. For this purpose, urban planning parameters with geographic boundaries arising from the zoning system (building density and land uses), along with urban morphology parameters based on Space Syntax Theory (integration and choice) are used. Xanthi a medium size city of Northern Greece, is selected as a case study in order to check the research hypothesis. Based on hedonic model theory and spatial clustering techniques, our research confirms the hypothesis that the real estate market is not fixed across the urban space, highlighting the role of the selected parameters in shaping the boundaries, as well as in defining them. Analysis revealed that the segmentation method of the housing market based on geometric accessibility variables has a clear superiority over the urban zoning parameters.

Keywords: Housing market · Space syntax · Sub-markets

1 Introduction and Background

One of the basic assumptions of hedonic models is that the hedonic price equation should be applied in a market that has common housing and buyer characteristics [1, 2] defines the concept of real estate sub-markets as: "The urban housing market is a set of compartmentalized and unique submarkets with demand and supply influences likely to result in a different structure of prices in each", highlighting the existence of sub-areas within a city, each one having its own special characteristics, thus differentiating real estate prices.

A review of the existing literature shows that there is no clear agreement on the definition of real estate sub-markets [3], with research focusing on defining their boundaries. Administrative boundaries between regions, geographical boundaries or other criteria are traditionally used for the division in sub-markets [4] considering that the marginal

F. Calabrò et al. (Eds.): NMP 2022, LNNS 482, pp. 1546–1556, 2022.
https://doi.org/10.1007/978-3-031-06825-6_149

willingness to pay for additional units of a housing feature and the housing price per unit feature are constant [3, 5] summarizing previous research notes that the definition of sub-market boundaries can be based either on the structural characteristics of the building stock such as the housing typology (detached house, apartment, etc.) [6], on spatial parameters such as census tracts [7] and geographical boundaries resulting from real estate surveys [8], on buyers' characteristics such as income criteria [9], or even on the simultaneous influence of structural and spatial parameters within.

Both spatial and non-spatial approaches have been criticized for the accuracy of their results. For the former, the main argument is that the a-priori determination of the number and boundaries of sub-markets based on administrative and other criteria is insufficient to describe a complex urban system such as an ever-changing city [10], at least at a social level [11], especially when they are not combined with other features (e.g. structural) [12]. On the other hand, the ineffectiveness of a-spatial approaches is based on the fact that they do not take space into account [8].

Building regulations as imposed by the state are an important category of variables affecting the housing value [13]. In this context, variables used to express land uses [14–17] or the maximum allowable height of buildings and building density (as an expression of the restrictive or non-restrictive role of zones) appear statistically significant in real estate research [18, 19].

As generally accepted, the price of housing is shaped primarily by location parameters [20]. In real estate research, accessibility is usually expressed as distance from work centers, services, shops and infrastructure [21]. According to natural movement theory [22] open spaces' layout is responsible for people moving or understanding space [23]. The open spaces layout forming the urban grid, is represented through an axial map and then turned into an axial graph, the latter interpreted through graph theory [24]. The centrality measures generated by the graph quantifying accessibility for every segment of the network or the relationship between a network segment to all other segments of the network, are integration and choice. Integration measures mean depth (depth here should be sensed as the number of lines travelled to move from the origin line to the destination line) of a segment to every other segment of the network [25]. Hence, an axial line is highly integrated when it can be easily reached form other lines of the network, making integration an accessibility measure of the line [26]. Choice measures the participation degree of a line in all shortest paths between all network lines [26]. Spatial analysis based on Space Syntax theory can be conducted for the grid as a whole examining the relationship between every segment of the network to all other segments, but can also be conducted for grid parts defined by depth (in Space Syntax terms). Geometric accessibility, as it is alternatively called, emerges as a statistically significant parameter of land and housing value (e.g. [27–32]).

The influence of both urban planning parameters and urban morphology parameters in shaping housing values seems to spatially differentiate in urban space [33–36], being among the most influential variables of housing values at the local level [37]. This spatial differentiation of these parameters influence on housing value, make the presence of real estate sub-markets in urban space more than evident.

This research paper aims to check and confirm the existence of real estate sub-markets emphasizing on the housing market. For this purpose, urban planning parameters are used, highlighting their role in shaping the boundaries, as well as in defining them.

Fig. 1. Research area and housing sample.   **Fig. 2.** Allowable building area coefficient zones.

2   Data and Methodology

Xanthi a medium-sized city of Northern Greece, whose special characteristics are reflected in the diversity of its urban fabric (traditional and newer core) as well as in its building stock (modern, traditional and preserved constructions), is selected as a case study in order to check the research hypothesis. Web advertisements of housing sales are used as the sample of the study (313 observations, see Fig. 1), with the asking sale price per m2 being the dependent variable of the hedonic model. The dependent variable selection is in line with relevant research (e.g., [38, 39]). The sample concerns a specific period of time (February 2019 - September 2019) before the COVID-19 pandemic crisis, accounting for the necessary balance.

Size, age, floor, construction materials (dummy variable scored 1 if the bearing body is made of reinforced concrete, metal or wood and 0 otherwise), view (dummy variable scored 1 if the residence faces a street and 0 otherwise), building typology (scored 1 if the residence is a detached house and 0 if it is an apartment), minimum distances from school, bus-stop and park/square, population density (of the block in which the dwelling is located) and crime density (per square kilometer in a radius of 400 m) are used as independent variables for the model. DepthMap software is used to calculate the geometric accessibility variables (Space Syntax) while the spatial analysis is based on segment angular analysis. Variables regarding size, age and distances were log-transformed for linearity reasons.

The applied methodology is based on the framework introduced by [40]. The process begins by applying the hedonic model to the whole market (entire study sample). Then, statistically significant differences between the implicit prices of the examined sub-markets are checked by using the F-test [9, 41].

Fig. 3. Zones of allowed land uses.

Fig. 4. Choice (radius 400 m) measures.

Once the hedonic model is constructed and calibrated with OLS (Ordinary Least Squares), the hypothesis testing for market segmentation follows. The first 2 categories of sub-markets tested rise from the geographic boundaries of the zoning system applied in the city. The first concerns the "Allowable Building Area Coefficient (ABAC)" (see Fig. 2) and the second the "Number of Allowed Land Uses" (see Fig. 3). The land use zoning system applied in Greece is close to the hierarchical zoning, while ABAC is an indicator of the building density of the area.

The third category of sub-markets tested is related to accessibility in terms of Space Syntax theory. As there is no geographical definition of the sub-market boundaries in this category, the process of defining them is based on the choice and integration measures of the radius that best interpret the housing values of the sample, using a spatial cluster analysis algorithm integrated into GIS platforms. The Calinski-Harabasz Pseudo F-statistic is used to find the optimal number of subgroups, which is a measure of the relative ratio between the groups created and the relative similarity between the characteristics of each group. The creation/development of groups is based on the k-means algorithm, which attempts to minimize the differences between the characteristics of the groups [42].

The final stage of the framework is based on the comparison of the results of the 3 different sub-market categories, for which the standard error check, as proposed by [40], is used, checking whether the standard errors for each sub-market's model are reduced in relation to the whole model.

3 Results and Discussion

After completing the corresponding statistical tests, it seems that the hedonic model does not suffer from multicollinearity issues (Table 1). The OLS results indicate that the variance of the dependent variable y explained by the independent is 32.2%. The results of the analysis indicate that the variables relating to size, age, the floor where the residence is located, its type, as well as the distance from green spaces and bus stops and population density are statistically significant (significance level of a = 0.05). Having removed the insignificant variables, the final standard error of the model is 19.834, there are 7 significantly variables and the R2 is 0.314.

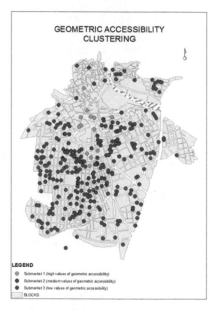

Fig. 5. Integration (radius 400 m) measures.   **Fig. 6.** Geometric accessibility clustering.

Following the methodology described above, the model is tested for submarkets. The first submarket category is based on the number of uses allowed separating the market into 3 groups (low, medium and high number of uses) with the latter coinciding with the city center of the study area. Moreover, the market is divided based on the Allowable Build Area Coefficient segregating the market into 3 groups indicating low, medium and high building density respectively.

Table 1. OLS Analysis results (* = statistical significance at p = 0.001 level).

Variables	Coef.	Std Coef.	Robust t	Robust Sig.	Adj R2	AIC	VIF
Constant	7,779		31,121	0,00*	0,322	48,245	
Size	−0,111	−0,205	−3,057	0,00*			1,426
Age	−0,245	−0,521	−9,691	0,00*			1,367
Floor	0,022	0,122	2,643	0,00*			1,246
Material	0,069	0,075	−0,594	0,55			1,165
View	−0,127	−0,057	1,489	0,14			1,367
Type	0,520	0,437	5,264	0,00*			2,012
Schools	−0,011	−0,022	−0,486	0,63			1,185
Bus_Stops	0,047	0,115	2,492	0,01*			1,120
Green	−0,046	−0,110	−2,532	0,01*			1,126
Pop_Density	−2,307	−0,116	−2,279	0,02*			1,310
Criminality	0,000	0,022	0,427	0,67			1,143

From the analysis presented in Table 2 it is evident that in each case, the part concerning the old city (sub-market 1) is characterized by the influence of the typology of the building. For the rest of the city, the influence of building age is important in any case. Special characteristics of influence emerge, depending on the prevailing situation in each sub-market. Thus, the values of the houses located on the city-borders (sub-market 2) are strongly influenced by the existence of green spaces and the low population density, highlighting these parameters as housing selection factors in these areas. In the areas identified spatially with the urban center (sub-market 3), the classification of the influence of the variables does not change between the 2 categories tested, with the population density not affecting statistically significantly.

The segmentation of the housing market based on urban morphology (geometric accessibility radius that strongly affects housing prices), is then tested. Previous research has proven that the radius that best interprets housing values is that of 400 m [37]. This radius corresponds (for this city size) to the walking scale [35, 43], indicating that buyers prefer homes that have increased accessibility at the neighborhood level. Space Syntax analysis results for the 400m radius are presented in Fig. 4 and Fig. 5. Following the spatial clustering (of the geometric accessibility) results, 3 sub-markets are created. It is worth noting that the form of the urban tissue appears to have a significant impact on the clustering (see Fig. 6). Thus, among the last 2 sub-markets created, the first settles on areas of organic tissue and the second on areas that grow in a rectangular grid.

Table 2. Submarket analysis results based on land uses and building density (* = statistical significance at p = 0.001 level).

	Submarket 1 (land uses)		Submarket 2 (land uses)		Submarket 3 (land uses)	
	R2	0,872	R2	0,298	R2	0,402
	RSS	0,250	RSS	11,053	RSS	6,568
Variables	Std Coef.	Sig.	Std Coef.	Sig.	Std Coef.	Sig.
Constant		0,00*	7,549	0,00*		0,00*
Size	−0,908	0,28	−0,036	0,42	−0,282	0,00*
Age	−0,572	0,39	−0,241	0,00*	−0,516	0,00*
Floor	−0,401	0,09	0,026	0,00*	0,135	0,07
Type	1,367	0,00*	0,435	0,00*	0,356	0,00*
Bus_Stops	−0,466	0,30	0,031	0,26	0,161	0,01*
Green	−0,142	0,45	−0,051	0,04*	−0,016	0,76
Pop_Density	0,585	0,00*	−2,821	0,01*	−0,058	0,43
	Submarket 1 (building density)		Submarket 2 (building density)		Submarket 3 (building density)	
	R2	0,365	R2	0,100	R2	0,358
	RSS	1,973	RSS	1,774	RSS	14,845
Variables	Std Coef.	Sig.	Std Coef.	Sig.	Std Coef.	Sig.
Constant		0,00*		0,00*		0,00*
Size	−0,397	0,48	−0,145	0,39	−0,203	0,00*
Age	−0,462	0,23	−0,572	0,00*	−0,493	0,00*
Floor	0,223	0,33	−0,056	0,62	0,139	0,01*
Type	0,968	0,02*	0,251	0,39	0,359	0,00*
Bus_Stops	0,543	0,28	0,021	0,86	0,122	0,01*
Green	−0,114	0,74	−0,241	0,03*	−0,136	0,01*
Pop_Density	−0,029	0,94	−0,399	0,00*	−0,075	0,14

As spotted by the results of the hedonic model applied in each sub-market (Table 3), the determining factor of housing price in the first sub-market is (as expected) the building typology. In the second sub-market, concerning the wider city center, the size of the house has a significant influence. The fact that this area is chosen mainly by students looking for small homes for their stay, seems to raise their prices. The third sub-market is mainly characterized by building age and typology parameters.

Results of the chow tests for all model sub-markets are presented in Table 4. The hypothesis is that the implicit prices of between every 2 examined sub-markets are equal. The statistically significant differences between certain pairs of sub-markets confirm the segmentation hypothesis. The case is more than evident for the sub-markets defined by geometric accessibility.

Table 3. Submarket analysis results based on geometric accessibility (* = statistical significance at p = 0.001 level).

	Submarket 1 (Accessibility 400 m)		Submarket 2 (Accessibility 400 m)		Submarket 3 (Accessibility 400 m)	
	R2	0,72	R2	0,372	R2	0,272
	RSS	1,436	RSS	5,971	RSS	10,056
Variables	Std Coef.	Sig.	Std Coef.	Sig.	Std Coef.	Sig.
Constant		0,00*		0,00*		0,00*
Size	−0,746	0,26	−0,236	0,00*	−0,100	0,20
Age	−0,971	0,07	−0,471	0,00*	−0,550	0,00*
Floor	0,273	0,57	0,130	0,02*	0,098	0,10
Type	1,566	0,03*	0,456	0,00*	0,287	0,00*
Bus_Stops	0,201	0,86	0,013	0,13	0,062	0,36
Green	0,243	0,39	−0,114	0,09	−0,127	0,05
Pop_Density	−0,050	0,28	−0,156	0,03*	−0,088	0,16

Finally, the results of the 3 different sub-market categories are compared. All models have a significant standard error reduction, greater than the critical value of 10% [40]

Table 4. Chow F-test results for all model submarkets.

Submarkets defined by land uses	F	Sig
Submarket 1 with 2	2,141	0,034
Submarket 2 with 3	2,274	0,023
Submarket 1 with 3	1,893	0,068
Submarkets defined by building density	F	Sig
Submarket 1 with 2	2,397	0,031
Submarket 2 with 3	1,506	0,159
Submarket 1 with 3	2,474	0,013
Submarkets defined by geometric accessibility	F	Sig
Submarket 1 with 2	2,455	0,018
Submarket 2 with 3	2,104	0,036
Submarket 1 with 3	3,214	0,002

Table 5. Estimation results of weighted standard error.

Model	Standard error	% reduction
Market-wide model	19,834	
Submarkets defined by land uses	8,712	56,08%
Submarkets defined by building density	12,003	39,48%
Submarkets defined by geometric accessibility	7,9333	60,00%

(Table 5). The segmentation method of the housing market based on the urban morphology parameters has a clear superiority. Applying the hedonic model to the sub-markets based on geometric accessibility significantly improves its estimation by 60%.

4   Conclusions and Perspectives

This research confirms the hypothesis that the real estate market is not unified across the urban space, but consists of sub-markets, highlighting the role of urban planning and morphology parameters in shaping their boundaries and defining them. The above analysis reveals that the segmentation of the housing market based on geometric accessibility parameters could be an important alternative non-spatial approach in identifying submarket boundaries. An interesting idea for future research could focus on a mixed approach, i.e. combining urban configuration variables and other urban parameters (building density, land uses) along with construction characteristics, in order to achieve a higher degree of reliability.

Acknowledgements. This research is co-financed by Greece and the European Union (European Social Fund- ESF) through the Operational Programme «Human Resources Development, Education and Lifelong Learning» in the context of the project "Reinforcement of Postdoctoral Researchers - 2nd Cycle" (MIS-5033021), implemented by the State Scholarships Foundation (IKY).

Ευρωπαϊκή Ένωση
European Social Fund

**Operational Programme
Human Resources Development,
Education and Lifelong Learning**

Co-financed by Greece and the European Union

References

1. Sirmans, G.S., Macpherson, D.A., Zietz, E.N.: The composition of hedonic pricing models. J. Real Estate Lit. **13**(1), 3–43 (2005)
2. Straszheim, M.: Hedonic estimation of housing market prices: a further comment. Rev. Econ. Stat. **56**(3), 404–406 (1974)
3. Watkins, C.A.: The definition and identification of housing submarkets. Environ. Plan. Econ. Space **33**(12), 2235–2253 (2001)
4. Feng, Y., Jones, K.: Comparing two neighbourhood classifications: a multilevel analysis of London property price 2011–2014. In: 22nd Pacific-Rim Real Estate Conference, Queensland, Australia (2016)

 5. Basu, S., Thibodeau, T.: Analysis of spatial autocorrelation in house prices. J. Real Estate Financ. Econ. **17**(1), 61–85 (1998)
 6. Gabrielli L., Giuffrida S., Trovato M.R.: Gaps and overlaps of urban housing sub-market: hard clustering and fuzzy clustering approaches. In: Stanghellini S., Morano P., Bottero M., Oppio A. (eds.) Appraisal: From Theory to Practice. Green Energy and Technology. Springer, Cham (2017). https://doi.org/10.1007/978-3-319-49676-4_15
 7. Christafore, D., Leguizamon, S.: Willingness to pay for hospital access in areas with high concentrations of blacks. Rev. Region. Stud. **45**(1), 87–104 (2015)
 8. Michaels, R.G., Smith, V.K.: Market segmentation and valuing amenities with hedonic models: the case of hazardous waste sites. J. Urban Econ. **28**(2), 223–242 (1990)
 9. Goodman, A.C., Thibodeau, T.C.: Housing market segmentation and hedonic prediction accuracy. J. Hous. Econ. **12**(3), 181–201 (2003)
10. Wu, C., Ye, X., Ren, F., Du, Q.: Modified data-driven framework for housing market segmentation modified data-driven framework for housing market segmentation. J. Urban Plan. Dev. **144**(4), 04018036 (2018)
11. Salvati, L., Ciommi, T.M., Serra, P., Chelli, F.M.: Land use policy exploring the spatial structure of housing prices under economic expansion and stagnation: the role of sociodemographic factors in metropolitan. Land Use Policy **81**, 143–152 (2019)
12. Gavu, E.K., Owusu-Ansah, A.: Empirical analysis of residential submarket conceptualization in Ghana. Int. J. Housing Mark. Anal. **12**(4), 763–787 (2019)
13. Glaeser, E.L., Gyourko, J.E.: The impact of building restrictions on housing affordability. Econ. Policy Rev. **9**(2), 21–39 (2003)
14. Asabere, P.K., Harvey, B.: Factors influencing the value of urban land: evidence from halifax-dartmouth. Canada Real Estate Econ. **13**(4), 361–377 (1985)
15. Kok, N., Monkkonen, P., Quigley, J.M.: Land use regulations and the value of land and housing: an intra-metropolitan analysis. J. Urban Econ. **81**, 136–148 (2014)
16. Schläpfer, F., Waltert, F., Segura, L., Kienast, F.: Valuation of landscape amenities: a hedonic pricing analysis of housing rents in urban, suburban and periurban Switzerland. Landsc. Urban Plan. **141**, 24–40 (2015)
17. Levkovich, O., Rouwendal, J., Brugman, L.: Spatial planning and segmentation of the land market: the case of the Netherlands. Land Econ. **94**(1), 137–154 (2018)
18. Duncan, M.: The synergistic influence of light rail stations and zoning on home prices. Environ. Plan **43**(9), 2125–2142 (2011)
19. Shimizu, C.: Estimation of hedonic single-family house price function considering neighborhood effect variables. Sustainability **6**(5), 2946–2960 (2014)
20. Kiel, K.A., Zabel, J.E.: Location, location, location: the 3L approach to house price determination. J. Hous. Econ. **17**(2), 175–190 (2008)
21. Orford, S.: Valuing locational externalities: a GIS and multilevel modelling approach. Environ. Plann. Plann. Des. **29**(1), 105–127 (2002)
22. Hillier, B., Penn, A., Hanson, J., Grajewski, T., Xu, J.: Natural movement: or, configuration and attraction in urban pedestrian movement. Environ. Plann. Plann. Des. **20**(1), 29–66 (1993)
23. Karimi, K.: A configurational approach to analytical urban design: space syntax methodology. Urban Des. Int. **17**(4), 297–318 (2012)
24. Penn, A., Turner, A.: Movement-generated land-use agglomeration: simulation experiments on the drivers of fine-scale land-use patterning. Urban Desing Int. **9**(2), 81–96 (2004)
25. Hillier, B.: Space is the Machine (e-edition). Space Syntax, London, UK (2007)
26. Hillier, B., Vaughan, L.: The city as one thing. Prog. Plan. **67**(3), 205–230 (2007)
27. Lykostratis, K., Giannopoulou, M.: Land value hot-spots defined by urban configuration. In: Calabrò, F., Della Spina, L., Bevilacqua, C. (eds.) ISHT 2018. SIST, vol. 100, pp. 590–598. Springer, Cham (2019). https://doi.org/10.1007/978-3-319-92099-3_66

28. Chiaradia, A., Hillier, B., Barnes, Y., Schwander, C.: Residential property value patterns in London: space syntax spatial analysis. In: Koch, D., Marcus, L., Steen, J. (eds.) Proceedings of the 7th International Space Syntax Symposium, pp. 015.1–015.12. Royal Institute of Technology (KTH), Stockholm (2009)
29. Topçu, M.: Accessibility effect on urban land values. Acad. J. 4(11), 1286–1291 (2009)
30. Narvaez, L., Penn, A., Griffiths, S.: Space syntax economics: decoding accessibility using property value and housing price in Cardiff, Wales. In: Greene, M., Reyes, J., Castro, A. (eds.) Proceedings of the Eighth International Space Syntax Symposium, pp. 1–19. Santiago de Chile (2012)
31. Law, S., Karimi, K., Penn, A., Chiaradia, A.: Measuring the influence of spatial configuration on the housing market in metropolitan london. In: Kim, Y.O., Park, H.T., Seo, K.W. (eds.) Proceedings of the Ninth International Space Syntax Symposium, pp. 121.1–121.20. Sejong University Press, Seoul (2013)
32. Law, S., Stonor, T., Lingawi, S.: Urban value: measuring the impact of spatial layout design using space syntax. In: Kim, Y.O., Park, H.T., Seo, K.W. (eds.) Proceedings of the Ninth International Space Syntax Symposium, pp. 061.1–061.20. Sejong University Press, Seoul (2013)
33. Lykostratis, K., Giannopoulou, M., Roukouni, A.: Measuring urban configuration: a GWR approach. In: Calabrò, F., Della Spina, L., Bevilacqua, C. (eds.) ISHT 2018. SIST, vol. 100, pp. 479–488. Springer, Cham (2019). https://doi.org/10.1007/978-3-319-92099-3_54
34. Matthews, J.W., Turnbull, G.K.: Neighborhood street layout and property value: the interaction of accessibility and land use mix. J. Real Estate Finance Econ. 35(2), 111–141 (2007)
35. Xiao, Y., Webster, C., Orford, S.: Identifying house price effects of changes in urban street configuration: an empirical study in Nanjing. China Urban Stud. 53(1), 112–131 (2016)
36. Shen, Y., Karimi, K.: The economic value of streets: mix-scale spatio-functional interaction and housing price patterns. Appl. Geogr. 79, 187–202 (2017)
37. Lykostratis, K.: Investigation and correlation of urban parameters and housing values by using spatial analysis methods in Geographic Infromation Systems environment. PhD Thesis (2018)
38. Hajnal, I.: An investigation of property value impairment caused by noise, in the case of the budapest ferenc liszt international airport, using a hedonic model. Period. Polytech. Soc. Manag. Sci. 25(1), 49–55 (2017)
39. Efthymiou, D., Antoniou, C.: How do transport infrastructure and policies affect house prices and rents? evidence from Athens, Greece. Trans. Res. Part Policy Pract. 52, 1–22 (2013)
40. Schnare, A.B., Struyk, R.J.: Segmentation in urban housing markets. J. Urban Econ. 3(2), 146–166 (1976)
41. Chow, G.: Tests of equality between sets of coefficients in two linear regressions. Econometrica 28(3), 591–605 (1960)
42. Jain, A.K.: Data clustering: 50 years beyond K-means. Pattern Recogn. Lett. 31(8), 651–666 (2010)
43. Song, Y., Knaap, G.J.: New urbanism and housing values: a disaggregate assessment. J. Urban Econ. 54(2), 218–238 (2003)

Housing Values Defined by Urban Morphology Characteristics

Konstantinos Lykostratis$^{(\boxtimes)}$ and Maria Giannopoulou

School of Civil Engineering, Democritus University of Thrace, Campus Xanthi-Kimmeria, 67100 Xanthi, Greece
klykostr@civil.duth.gr

Abstract. Different analysis methods are used in real estate studies, with hedonic models being the most frequently used. Although the vast majority of relevant studies conclude that factors related to the location of a property along with its neighborhood characteristics have a crucial role in housing value, urban parameters defining urban space have not yet been adequately considered in relevant literature. This research, based on hedonic model theory, aims at pointing out the importance of urban characteristics such as urban morphology characteristics (geometric accessibility) as parameters defining housing values. The proposed methodological framework with the selected variables entered in the hedonic model describing housing values is applied to Xanthi, a city with distinctive morphological features that are reflected both in its urban fabric and in its building stock, making it ideal for the implementation of the model. The results of the analysis confirm the primary hypothesis of the research that geometric accessibility significantly affects the housing value and thus should be taken into account in housing price models.

Keywords: Accessibility · Space syntax · Housing value

1 Introduction and Background

Hedonic regression models are econometric tools allowing the price analysis of a heterogenous commodity such as the housing market [1]. After reviewing urban economics literature, it is noticed that a wealth of variables are used as regressors in housing market research. It is also noted that neither the theoretical framework delineates which variables to be entered in the regression, nor there is an agreement on their choice [2]. It is generally suggested that the variable choice is based on structural characteristics of the property, its location, and its neighborhood characteristics [3].

[4] notes that "The three most important determinants of house price are location, location, location" making clear that variables related to the location of the property have a decisive impact on housing value. Accessibility variables have been consistently based on the city's development model. From the monocentric city model [5–7] and the distance from the Central Business District (CBD), to the polycentric model [8] and the distances to commerce, recreation and entertainment sub-centers, several variables have

© The Author(s), under exclusive license to Springer Nature Switzerland AG 2022
F. Calabrò et al. (Eds.): NMP 2022, LNNS 482, pp. 1557–1566, 2022.
https://doi.org/10.1007/978-3-031-06825-6_150

been tested for their impact on the housing value. Thus, besides accessibility (proximity) to road network and public transportation [9], researchers have tested accessibility to services and public facilities such as schools [10] sport facilities [11], and natural amenities [12].

Accessibility and proximity are based on the notion of centrality, meaning the fact that "some places are more important because they are more central" [13]. Accessibility estimated through Space Syntax theory is also known as spatial accessibility [14] or a special case of geometric accessibility, differentiated from the classic estimation methods of accessibility (geographic accessibility) [15]. Studying centrality as a spatial procedure [16], Space Syntax theory [17] elaborates accessibility measures of the urban core for the morphologic analysis of space in order to describe the relationship between human and built environment and its social aspect [18]. In other words the concept of spatial systems' accessibility along with its effects on human behavior and interaction is examined [19]. According to Natural Movement theory [20] the layout of open spaces is responsible for the way people move or conceive space [21], which is the reason why topological dependence of open spaces is used in the analysis. Geometric accessibility is pointed out as statistically significant parameter in recent land and housing value research (e.g. [22–26]).

Although it is commonly accepted that the location characteristics of a property are the most important factors shaping its value, geometric accessibility holds a very limited part in housing market literature [27] despite the fact that morphology plays an important role in the organization of property markets [28]. This research, based on hedonic model theory, aims at pointing out the importance of urban characteristics such as urban morphology characteristics (geometric accessibility) as parameters defining housing values.

2   Data and Methodology

The model developed herein is applied to Xanthi (see Fig. 1), a medium-sized town of 55,000 inhabitants, in northeastern Greece. The urban tissue consists of separate parts with an interesting variety in their form and density, traditional sections with coherent organic tissue, new extensions with rectangular grid or normal geometry and a large range in the sizes of building blocks [29]. Web advertisements of housing sales are used as the sample of the study (313 observations, see Fig. 2), with the asking sale price per m2 being the dependent variable of the hedonic model. The dependent variable selection is in line with relevant research (e.g. [30, 31]). The sample concerns a specific period of time (February 2019 - September 2019) before COVID-19 pandemic crisis, accounting for the necessary balance.

In order to test whether urban morphology characteristics define housing values, two models are constructed. Classic variables regarding construction, location and neighborhood characteristics are entered in the first model, while in the second geometric accessibility variables based on Space Syntax Theory are added in order for the models to be compared based on R-squared and AICc criterion.

Fig. 1. Research area.

Fig. 2. Housing sample.

Classic independent variables refer to the size of the house, the floor where it is located, the age and building materials of the building (characteristics derived from house ads), accessibility to services such as schools and public transport, the distance from green space, and the view it offers (parameters of the location and accessibility of the building and environmental characteristics related to the minimum distance from a place of interest based on the pedestrian network), population density and crime density (neighborhood quality characteristics resulting from police data and Hellenic Statistical Authority censuses).

The additional independent variables of the second model refer to the urban morphology (geometric) accessibility of the area in which the residence/observation is located. The variables used are "integration" and "choice" which describe the relationship that develops between an axial line (road segment) and all the other axial lines of the system [16] in terms of topology and geometry [32], based on the concept of distance and direction [33].

The calculation of these variables can refer to the network as a whole, but also to parts of it by looking at the relationship developed between a line and all the other lines of the subsystem (defined by a radius around each line of the system). Integration refers to how far a line is in relation to the others, with the higher integration axes being those that can be reached more easily than the other lines of the system. Choice refers to the extent to which a part of the network participates in all the shortest paths between all its segments [33] with the axes displaying a higher choice value being the ones to be selected more often for each possible move (for the particular analysis radius).

The model introduces local integration and local choice variables for 400 and 1200 m. Following [28] and [34] the radius selection is based on the fact that the 400–1200 m radii refer to the walking scale in a city. DepthMap software [35, 36] was used for analysis, while spatial analysis was based on segment angular analysis [33].

The results from the network analysis of the 400 m radius, reflecting short distance travel routes, are shown in Fig. 3, with the highest choice values appearing in the traditional part of the city (Old Town) and in the city center, which comprise the main core of the travel network. Figure 4 shows the analysis for the 1200 m radius which expresses larger but daily routine journeys. Integration maps (see Figs. 5 and 6) highlight the city center axes having the largest integration values. The radius of 400 m represents short radius of walkability highlighting small sub-centers consisted of individual "more integrated" axes (compared with those who surround them) in each of the city districts. While the study radius increases (1200 m) approaching the radius for the whole city the "more integrated" axes of the entire network are highlighted, which coincide with high commercial activity axes, the main traffic network segments of the city, confirming [20] who note that integration shows high correlation with pedestrian flows.

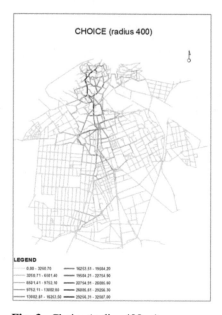

Fig. 3. Choice (radius 400 m) measures.

Fig. 4. Choice (radius 1200 m) measures.

Table 1 shows the independent variables entered to the hedonic model.
According to the above, the two models that are calibrated are:

$$y = b_0 + a_1x_1 + a_2x_2 + a_3x_3 + a_4x_4 + a_5x_5 + a_6x_6 + a_7x_7 + a_8x_8 + a_9x_9 + a_{10}x_{10} + a_{11}x_{11} \quad (1)$$

$$y = b_0 + a_1x_1 + a_2x_2 + a_3x_3 + a_4x_4 + a_5x_5 + a_6x_6 + a_7x_7 + a_8x_8 + a_9x_9 + a_{10}x_{10}$$
$$+ a_{11}x_{11} + a_{12}x_{12} + a_{13}x_{13} + a_{14}x_{14} + a_{15}x_{15} \quad (2)$$

where a1–a15 are the coefficients of the independent variables x1–x15 and b0 the constant term. For the linear models applied herein variables x1, x2, x7, x8, x9, x12, and x13 were log-transformed for linearity reasons. The models are calibrated using the Ordinary Least Squares (OLS) method and tested for multicollinearity, and heteroscedasticity.

Fig. 5. Integration (radius 400 m) measures.   **Fig. 6.** Integration (radius 1200 m) measures.

3 Results and Discussion

After completing the corresponding statistical tests, it seems that both models do not suffer from multicollinearity issues (Variance Inflation Factors) (Table 2), but from heteroscedasticity. The latter can be solved using the robust errors [37]. For the first model, the variance of the dependent variable y explained by the independent is 32.2%. Analysis results indicate that the variables related to age (x2), typology (x6), size (x1), the floor where the residence is located (x3) and the distance from green spaces (x9) and bus stops (x8) are statistically significant variables all having the expected signs.

For the second model (Table 2), R-squared rises to 35.4%. The comparative advantage of this model is its evidence when AICc values are compared. Regarding the coefficients, influence and signs of the previous model's parameters do not change, with accessibility measures for the 400 m (x12, x14) radius also being statistically significant. The statistical significance of the Space Syntax measures reveal that they are better accessibility estimators compared to classical accessibility variables, but also that prospective buyers take into account the smaller service radii when choosing their residence.

Table 1. Variable description.

Variable	Code	Type	Description
Price/m2	y	Continuous	The per square meter commercial price of the residence
Size	x1	Continuous	The area of the residence in m^2
Age	x2	Continuous	The age of the building hosting the residence, in years
Floor	x3	Continuous	The floor where the housing is located
Material	x4	Dummy	1 if the bearing body is made of reinforced concrete, metal or wood and 0 otherwise
View	x5	Dummy	1 if the residence faces a street and 0 otherwise
Type	x6	Dummy	1 if the residence is a detached house and 0 if it is an apartment
Schools	x7	Continuous	The minimum distance from primary school in m
Bus_Stops	x8	Continuous	The minimum distance from a bus stop to a public transport bus in m
Green	x9	Continuous	The minimal distance from the entrance of a park or square (green spaces) in m
Pop_Density	x10	Continuous	The building density of the block in which the dwelling is located
Criminality	x11	Continuous	The crime density per square kilometer in a radius of 400 m
CH_400	x12	Continuous	The choice value of the axis to which the house has direct access (for a 400 m radius)
CH_1200	x13	Continuous	The choice value of the axis to which the house has direct access (for a 1200 m radius)
INT_400	x14	Continuous	The integration value of the axis to which the house has direct access (for a 400 m radius)
INT_1200	x15	Continuous	The integration value of the axis to which the house has direct access (for a 1200 m radius)

Geometric accessibility variables for the different radii present opposite behavior. The choice coefficient for the 400 m radius (a12) is consistent with relevant literature figuring the fall of housing prices as the choice value increases, mainly associated with increased traffic and pedestrian flow volumes [38], and therefore with high noise levels [23]. The sign of the integration coefficient for the 400 m radius (a14) is also consistent with relevant literature [24, 28] revealing the buyers' willingness to acquire central houses in a neighborhood. Further analysis should be applied regarding the sign of the coefficient for the 1200 m choice (a13) and integration (a15) variable (although not statistically significant) because this could be an indication for local variation of geometric accessibility parameters.

Table 2. OLS analysis results (* = statistical significance at p = 0.001 level).

Variables	Model 1				Model 2			
	Std Coef.	Robust t	Robust Sig.	VIF	Std Coef.	Robust t	Robust Sig.	VIF
Constant		31,121	0,00*			28,884	0,00*	
Size	−,205	−3,057	0,00*	1,43	−,194	−2,851	0,00*	1,44
Age	−,521	−9,691	0,00*	1,37	−,541	−10,012	0,00*	1,43
Floor	,122	2,643	0,01*	1,25	,123	2,644	0,01*	1,28
Material	−,057	−0,594	0,55	1,37	−,062	−0,701	0,48	1,4
View	,075	1,489	0,14	1,17	,068	1,316	0,19	1,18
Type	,437	5,264	0,00*	2,01	,434	5,557	0,00*	2,09
Schools	−,022	−0,486	0,63	1,19	−,022	−0,497	0,62	1,21
Bus_Stops	,115	2,492	0,01*	1,12	,123	2,614	0,01*	1,19
Green	−,110	−2,532	0,01*	1,13	−,119	−2,659	0,01*	1,22
Pop_Density	−,116	−2,279	0,02*	1,31	−,095	−1,703	0,10	1,5
Criminality	,022	0,427	0,67	1,14	−,071	−1,150	0,25	1,67
CH_400					−,292	−2,222	0,03*	6,13
CH_1200					,110	1,029	0,30	4,27
INT_400					,300	2,287	0,02*	7,02
INT_1200					−,019	−0,178	0,86	4,58
R2	0,322				0,354			
AIC	48,245				42,052			

4   Conclusions and Perspectives

Decoding housing value by knowing the impact of various parameters on the housing value, which have not been taken into account, can be used by designers/urban planners in urban planning, but also help potential investors in decision making. The results of the analysis confirm the primary hypothesis of the research that urban parameters significantly affect the housing value and this is why they should be taken into account in housing price models. Small radii seem to better explain housing values, thus, when included, they empower hedonic models. Moreover, it is evident that they are better accessibility estimators compared to classical accessibility variables such as the least distance from schools, bus stops and green spaces. An interesting fact is that geometric accessibility suggests that buyers prefer homes that have accessibility in their neighborhood, accounting at the same time for their privacy. Future research should focus on a local scale analysis in order to both assess and confirm geometric accessibility local influence but also account for spatial dependence.

Acknowledgements. This research is co-financed by Greece and the European Union (European Social Fund-ESF) through the Operational Programme «Human Resources Development, Education and Lifelong Learning» in the context of the project "Reinforcement of Postdoctoral Researchers - 2nd Cycle" (MIS-5033021), implemented by the State Scholarships Foundation (ΙΚΥ).

Ευρωπαϊκή Ένωση
European Social Fund

Operational Programme
Human Resources Development,
Education and Lifelong Learning

Co-financed by Greece and the European Union

References

1. Rosen, S.: Hedonic prices and implicit markets: product differentiation in pure competition. J. Polit. Econ. **82**(1), 34–55 (1974)
2. Ceccato, V., Wilhelmsson, M.: The impact of crime on apartment prices: evidence from Stockholm, Sweden. Geografiska Annaler Ser. B-Hum. Geogr. **93B**(1), 81–103 (2011)
3. Taylor, L.O.: The hedonic method. In: Champ, P.A., Boyle, K.J., Brown, T.C. (eds.) A Primer on Nonmarket Valuation, vol. 3, pp. 331–393. Springer, Dordrecht Netherlands (2003). https://doi.org/10.1007/978-94-007-0826-6_10
4. Dubin, R.A.: Spatial autocorrelation: a primer. J. Hous. Econ. **7**(4), 304–327 (1998)
5. Alonso, W.: Location and Land Use: Toward a general Theory of Land Rent. Harvard University Press, Cambridge (1964)
6. Muth, R.: Cities and Housing. University of Chicago Press, Chicago (1969)
7. Mills, E.S.: Studies in the Structure of the Urban Economy. Johns Hopkins Press, Baltimore (1972)
8. Shin, K., Washington, S., Choi, K.: Effects of transportation accessibility on residential property values. Transp. Res. Rec. J. Transp. Res. Board **1994**(1), 66–73 (2007)
9. Sah, V., Conroy, S.J., Narwold, A.: Estimating school proximity effects on housing prices: the importance of robust spatial controls in hedonic estimations. J. Real Estate Finance Econ. **53**(1), 50–76 (2015). https://doi.org/10.1007/s11146-015-9520-5
10. Wen, H., Zhang, Y., Zhang, L.: Do educational facilities affect housing price? An empirical study in Hangzhou. China. Habitat Int. **42**, 155–163 (2014)
11. Ahlfeldt, G.M., Maennig, W.: Impact of sports arenas on land values: evidence from Berlin. Ann. Reg. Sci. **44**(2), 205–227 (2010)
12. Brueckner, J.K., Thisse, J.-F., Zenou, Y.: Why is central Paris rich and downtown Detroit poor? An amenity-based theory. Eur. Econ. Rev. **43**(1), 91–107 (1999)
13. Porta, S., Crucitti, P., Latora, V.: The network analysis of urban streets: a primal approach. Environ. Plann. B Urban Anal. City Sci. **33**(5), 705–725 (2006)
14. Charalambous, N., Mavridou, M.: Space syntax: spatial integration accessibility and angular segment analysis by metric distance (ASAMeD). In: Hull, A., Silva, C., Bertolini, L. (eds.) Accessibility Instruments for Planning Practice. COST Office, pp. 57–62 (2012)
15. Jiang, B., Claramunt, C., Batty, M.: Geometric accessibility and geographic information: extending desktop GIS to space syntax. Comput. Environ. Urban Syst. **23**(2), 127–146 (1999)
16. Cutini, V.: Centrality and land use: three case studies on the configurational hypothesis. Cybergeo: Eur. J. Geogr. Systèmes, Modélisation, Géostatistiques, Doc. **188** (2001)

17. Hillier, B., Hanson, J.: The Social Logic of Space. Cambridge University Press, Cambridge [Cambridgeshire] (1984)
18. Monokrousou, K., Giannopoulou, M.: Exploring methods and strategies for sustainable urban development. Int. J. Soc. Behav. Educ. Econ. Bus. Ind. Eng. 9(8), 2907–2910 (2015)
19. Monokrousou, K., Giannopoulou, M.: Interpreting and predicting pedestrian movement in public space through space syntax analysis. Procedia Soc. Behav. Sci. 223, 509–514 (2016)
20. Hillier, B., Penn, A., Hanson, J., Grajewski, T., Xu, J.: Natural movement: or, configuration and attraction in urban pedestrian movement. Environ. Plann. B. Plann. Des. 20(1), 29–66 (1993)
21. Karimi, K.: A configurational approach to analytical urban design: space syntax methodology. Urban Des. Int. 17(4), 297–318 (2012)
22. Lykostratis, K., Giannopoulou, M.: Land value hot-spots defined by urban configuration. In: Calabrò, F., Della Spina, L., Bevilacqua, C. (eds.) ISHT 2018. SIST, vol. 100, pp. 590–598. Springer, Cham (2019). https://doi.org/10.1007/978-3-319-92099-3_66
23. Law, S., Karimi, K., Penn, A., Chiaradia, A.: Measuring the influence of spatial configuration on the housing market in metropolitan London. In: Kim, Y.O., Park, H.T., Seo, K.W. (eds.) Proceedings of the Ninth International Space Syntax Symposium, pp. 121.1–121.20. Sejong University Press, Seoul (2013)
24. Law, S., Stonor, T., Lingawi, S.: Urban value: measuring the impact of spatial layout design using space syntax. In: Kim, Y.O., Park, H.T., Seo, K.W. (eds.) Proceedings of the Ninth International Space Syntax Symposium, pp. 061.1–061.20. Sejong University Press, Seoul (2013)
25. Narvaez, L., Penn, A., Griffiths, S.: Space syntax economics: decoding accessibility using property value and housing price in Cardiff, Wales. In: Greene, M., Reyes, J., Castro, A. (eds.) Proceedings of the Eighth International Space Syntax Symposium, pp. 1–19. Santiago de Chile (2012)
26. Chiaradia, A., Hillier, B., Barnes, Y., Schwander, C.: Residential property value patterns in London: space syntax spatial analysis. In: Koch, D., Marcus, L., Steen, J. (eds.) Proceedings of the 7th International Space Syntax Symposium, pp. 015.1–015.12. Royal Institute of Technology (KTH), Stockholm (2009)
27. Xiao, Y., Webster, C., Orford, S.: Identifying house price effects of changes in urban street configuration: an empirical study in Nanjing. China. Urban Stud. 53(1), 112–131 (2016)
28. Desyllas, J.: Berlin in transition: using space syntax to analyse the relationship between land use, land value and urban morphology. In: Major, M.D., Amorim, L., Dufaux, D. (eds.) Proceedings of the First International Space Syntax Symposium, pp. 04.1–04.15. University College London, London (1997)
29. Giannopoulou, M., Vavatsikos, A.P., Lykostratis, K.: A process for defining relations between urban integration and residential market prices. In: Calabrò, F., Della Spina, L. (eds.) Procedia - Social and Behavioral Sciences, vol. 223, pp. 153–159. Elsevier (2016)
30. Hajnal, I.: An investigation of property value impairment caused by noise, in the case of the Budapest Ferenc Liszt International airport, using a hedonic model. Period. Polytech. Soc. Manag. Sci. 25(1), 49–55 (2017)
31. Efthymiou, D., Antoniou, C.: How do transport infrastructure and policies affect house prices and rents? Evidence from Athens, Greece. Transportation Research Part A: Policy and Practice 52, 1–22 (2013)
32. Hillier, B., Iida, S.: Network and psychological effects in urban movement. In: Cohn, A.G., Mark, D.M. (eds.) COSIT 2005. LNCS, vol. 3693, pp. 475–490. Springer, Heidelberg (2005). https://doi.org/10.1007/11556114_30
33. Hillier, B., Vaughan, L.: The city as one thing. Prog. Plann. 67(3), 205–230 (2007)
34. Song, Y., Knaap, G.J.: New urbanism and housing values: a disaggregate assessment. J. Urban Econ. 54(2), 218–238 (2003)

35. Varoudis, T.: depthmapX – Multi-platform spatial network analyses software (2012). https://github.com/varoudis/depthmapX
36. Turner, A.: Depthmap 4, A Researcher's Handbook (2004). http://www.vr.ucl.ac.uk/depthmap/handbook/depthmap4.pdf
37. Verbeek, M.: A Guide to Modern Econometrics, 2nd edn. Wiley, Chichester (2004)
38. Turner, A.: From axial to road-centre lines: a new representation for space syntax and a new model of route choice for transport network analysis. Environ. Plann. B Urban Analy. City Sci. **34**(3), 539–555 (2007)

Integrated Evaluation Methodology for Urban Sustainable Projects

Pierluigi Morano[1], Francesco Sica[2(✉)], Maria Rosaria Guarini[2], Francesco Tajani[2], and Rossana Ranieri[2]

[1] Department of Science of Civil Engineering and Architecture, Polytechnic University of Bari, 126 Amendola Street, 70126 Bari, Italy
[2] Department of Architecture and Design, Sapienza University of Rome, 359 Flaminia Street, 00196 Rome, Italy
francesco.sica@uniroma1.it

Abstract. The projects for cities' development must comply with multiple instances defined according to the sustainability dimensions, i.e. economic, social and environmental ones. Even in the light of the most demanding international governmental and other current provisions for climate-change adaptation, initiatives of public and/or private interest must be evaluated on the basis of their performance level in the perspective of economic growth, environmental protection and social well-being.

The present paper focuses on the following issues: *i)* the economic-sustainable evaluation of urban design-solutions; *ii)* how to carry out integrated evaluations that systematically include multiple aspects while respecting the interests involved.

The use of multi-criteria analysis methods if, on the one hand, contributes to the achievement of commingling among dissimilar semantic-assessment fields, on the other hand, it is characterized by the difficulty of taking into account the interests of all the stakeholders able to affect the intervention priority among investment alternatives in urban context.

The work aims at describing an integrated evaluation methodology based, on the one hand, on the syntactic-operational formalism of the Analytic Hierarchy Process (AHP), and, on the other one, on the implementation of an analytical technique of a descriptive-participatory type (SWOT analysis + focus group) for the definition of evaluation criteria and relative weights for the sustainable evaluation of projects.

Keywords: Urban sustainable development · Integrated evaluation framework · AHP · SWOT analysis · Focus group

1 Introduction

Facing the climate changes is the most difficult challenge nowadays [1]. Globally, climatic conditions are rapidly evolving, worsening the effects on cities in terms of quality of life, mean urban temperature, capability to respond to natural disaster, etc. [2]. The attention given to the issue of implementing the sustainability in every sector and as global issue is risen in the past fifteen years. Indeed, the United Nations (UN) develop a specific Agenda for Sustainable Development, in which the issues addressed are multiple and interrelated, in order "[…] to achieve a better and more sustainable future for all". Specific objectives are recognised in seventeen goals (Sustainable Development Goals – SDGs) with a focus on cities in Goal 11 "Make cities and human settlements inclusive, safe, durable and sustainable" [3]. Furthermore, in a perspective of monitoring the vulnerability of the cities in 2015, the Third UN World Conference adopted the "Sendai framework", with the support of the United Nations Office for Disaster Risk Reduction (UNISDR) [4]. In the Sendai report the shift from "disaster management" to "disaster risk management" is highlighted through the statement of prevention activities consistent with the seventeen SDGs. Furthermore, the building's design process is characterized by new approaches with an emphasis on the integration and the protection of natural components in the project, such as biophilic design - as a strategy for promoting positive connections between people and nature -, or the *archinature*, i.e. the complex and articulated interweaving in which nature and artifice should collaborate in the construction of buildings and landscape [5–7]. This typology of design aims to sustain the resilience of natural systems over the time, modifying many components of the environmental conditions of a building or landscape in the short term, in order to support, in the long term, an ecologically robust and sustainable natural community [8, 9]. Furthermore, in the European Union (EU) there is a particular attention for the integration of environmental criteria into all phases of the purchasing process, encouraging the diffusion of technologies and environmentally sound products. The European Commission has developed the Green Public Procurement (GPP) [10], encouraging member States to adopt action plans for the integration of environmental requirements into public procurement and, subsequently, issued specific Guidelines for the drafting of a National Action Plans on GPP. In Italy, for example, the Procurement Code [11] has introduced the obligation to follow the minimum environmental criteria (CAM) – defined consistently with the GPP - in the planning of new interventions on the territory. Through this approach, the projects must consider the life cycle of products, services and concessions and has to be measured by objective criteria capable of weighing the environmental, energetically, social and economic performance on a medium and long analysis term.

Considering all the issues described above, it is possible to state that understanding the impact that climate change will have on cities is complex, because it involves many different factors hard to quantify [12]. In order to support and guide the planning and design of sustainable and resilient cities, it is necessary to develop methods, methodological approaches and tools able to assess urban sustainability criteria and appropriate performance indicators.

The research aims at developing an integrated evaluation methodology based on: *i*) the use of one of the main tools found in the literature for solving ranking problems – Analytic Hierarchy process (AHP) [13] – *ii*) the synergic implementation of a participatory process aimed at empirically quantifying the weights of the criteria involved in the evaluation flow of the examined projects.

The work is organized as follows: Sect. 2 contains a brief literature review in order to contextualize the research carried out; in Sect. 3 the evaluation methodology adopted is illustrated; in Sect. 4 the methodology is applied to a case study, in order to highlight its potentialities; in Sect. 5 the conclusions are discussed.

2   Sustainable Criteria and Supporting Assessment Methods

The selection of a set of proper criteria, indicators and/or index for the assessment of the sustainability of urban transformation projects is quite complex. Analyzing the current literature, a lack of consensus is noticeable, not only on the whole conceptual framework of reference, but on the selection and optimal number of indicators. Also, there is a lack of uniformity in identifying the relative weights to be attributed to sustainability criteria and indicators [14]. Early in 2002 Warhurst [15] - referring to the mining, metals and energy sectors - highlighted the need to structure links among sustainable development and corporate social, economic and environmental performance. Nowadays, a recent study carried out by Morano et al. [16] point out the heterogeneity both in the adoption of criteria and indicators, in the methodology approach and in the tools that could be used. Furthermore, Tanguay et al. [17] highlight that, according to the reference territory, there are different classifications and categorizations of Sustainable Development Indicators (SDI); moreover, the ambiguous definition of sustainable development makes hard the selection of methods and difficult the accessibility of qualitative and quantitative data.

In order to respond to this state of the art, many studies and research have been developed on the subject. With reference to the construction sector, Ayman and Wafaa [18] carried out a review in order to establishing an International Sustainability Index. In United Kingdom, Boyko et al. [19] defined a toolkit that establishes the relative sensitivity of sustainability indicators to facilitate the use of scenarios in any urban context and at any scale of reference. In the Indian context, Anand et al. [20] developed a study to assess various sustainability criteria in smart cities using a fuzzy-AHP method. With specific reference to the selection of sustainable materials in the construction sector, Figuereido et al. [21] also adopted a framework that involved the integration of Life Cycle Sustainability Assessment (LCSA), fuzzy-AHP, and Building Information Modeling (BIM) to choose suitable materials, by creating a system of 52 indicators of urban sustainable development that could address economic growth and efficiency, ecological and infrastructural construction, environmental protection, social and welfare progress. Other research tried to systematize the use of many indicators throughout the development of integrated tools like the Dashboard of Sustainability (DS) [22, 23], that is a mathematical and graphical tool that is able to create concise evaluations to support the decision-making process; or the methodology proposed by Fernández-Sánchez and Rodríguez-López [24], aimed at identifying a specific set of sustainability indicators in construction project management.

3   Integrated evaluation methodology for sustainable projects

In the light of literature review evidence in Sect. 2, it seems appropriate to be able to apply "mixed" evaluation methodologies in order *i*) to address, manage and translate multiple design-assessment aspects of projects, taking into account the logical-functional relationships in sustainable perspective, *ii*) to create a system of multi-scalar analysis that highlights the socio-economic and environmental characteristics of projects, as well as their sustainable effects for territorial growth. The proposed evaluation methodology allows for multi-dimensional assessments of alternative designs solutions based on a participatory and explicit information structure [25]. The novelty of the proposed methodological approach is identifiable in the mixture between the implementation of a participatory process according to a logic of direct functionality with respect to the multi-criteria evaluation method designed to solve the ranking problems between alternatives.

By considering a panel (I) of design solutions (i = 1, ..., n) to be evaluated from a sustainable perspective, the proposed methodology is constituted by the following sequential steps:

Step 1_SWOT analysis for sustainable criteria recognition: definition of the evaluation criteria (j = 1, ..., m) to be used to quantitatively and/or qualitatively measure the effects of the i-*th* intervention to be carried out in the urban area of reference from the point of view of sustainable development. The systematic process of the main performance criteria to be used in the evaluation phase of the i-*th* project alternative is carried out according to the logic of SWOT analysis. This highlights strengths, weaknesses, opportunities and risks that characterize the set of projects (I). The comparison among the different SWOT analysis carried out for each i-*th* alternative project is used to identify appropriate evaluation criteria in order to choose of a specific project consistent to the programmatic goals of sustainability at international level. The criteria (j-*th*) are directly derived by examining the SWOT analysis developed for each project, and identifying the keywords that are most frequently recurrent in the i-*th* alternative;

Step 2_Focus group for weighting-criteria: weighting in correspondence with the j-*th* evaluation criteria by Step 1 for the i-*th* project solution.

The weighting of the criteria is developed through the engagement of focus group practice, namely a group interview involving a small number of similar people or participants who have common experiences in urban sustainable studies;

Step 3_AHP analysis: measurement of the relative weights (w_{ij}) among design solutions *i* with respect to the j-*th* evaluation criterion by Step 1.

This with the Analytic Hierarchy Process that provides: *i*)the construction of the pairwise comparisons matrix between alternative projects; *ii*) the determination of relative local weights. The local weights of each project are multiplied by those of the corresponding via focus group activity. The products thus obtained are summed in global weights (or priorities value) of the analysis elements. To verify whether the weights obtained in the Step 2 are comparable to the judgments of the AHP, the Consistency Ration (CR) for each comparison matrix, as ratio between Consistence Index and Random Index [26] must be defined;

Step 4_Evaluation problem solving: the outcomes obtained by previous Steps could support the solving of multiple operative decision-making systems, e.g. the creation of the ranking lists, total and relative to each evaluation criterion, between project alternatives, or the selection of best project to finance for the urban sustainable development.

The proposed evaluation algorithm is tested on case study aimed to structure a ranking among alternatives related to urban regeneration projects located in different countries of the world. The application of the methodological framework is aimed at testing the flexibility of the proposed analysis scheme, even in the situation of having to evaluate examples of projects implemented in settlement contexts of different economic, social and environmental feature. In Sect. 4 the description of the case study under examination is presented.

4   A Ranking Case Between World-Wide Urban Renewal Projects

The case study takes into account a set of 13 projects, of diverse affiliation country, aimed to the urban renewal and recovery of property of historical and architectural significance. The projects are spread worldwide: five in Europe (two in Italy, one in France, one in Belgium, one in Slovakia), two in North Africa (Egypt and Morocco), two in middle east/Mediterranean (Lebanon, Turkey), four in western Asia (Iran). The identification of 13 case-studies was performed according to relative construction year, investment costs and capacity to pursue sustainability goals in the urban context where they are located.

The objective is to individuate which of this projects are able to determine the best financial, social, cultural and environmental benefits in sustainable perspective on their territory. Below descriptions of the steps of the proposed evaluation methodology with regard to the ranking case-study are reported.

Step 1: SWOT analysis for sustainable criteria recognition
The specificities of the several cities in the world where the i-th project is located, and the availability of information on the interventions examined, led to the description of each project according to the logic of SWOT analysis. The comparison among i-th SWOT analysis allows to identify key evaluation aspects – *sustainable evaluation criteria* – on the basis of which is possible to frame the ranking case-study. By the process of comparing alternative solutions, the evaluation of the projects set could be made according to the following sustainable evaluation criteria: C1) economic impact; C2) historical value; C3) environmental aspect; C4) location and relationship with the urban core of the city; C5) creation of open space; C6) private/public collaboration; C7) social sustainability; C8) cultural boost; C9) technological features; C10) correspondence to national and international strategies of sustainable development.

Step 2: Focus group for weighting-criteria
To establish weights-criteria for the i-th project examined, a focus group activity is carried out. A small-group discussion of 4 members that have knowledge on the project set and sustainability objectives, guided by a trained leader, is created. Through online means (video-chat, web-forum) the participatory evaluation among group-components is performed. Each component is called to express own judgment on importance degree for j-th criteria in correspondence to i-th project alternative. The evaluation process of importance degree attribution is made on the basis of the evaluation scale [0, 3, 5, 7] which stands as follows: not at all meaningful (0); low meaningful (3); quite meaningful (5); high meaningful (7). In Table 1 the outcomes of focus group activity for weighting-criteria are reported.

Table 1. Criteria-weights values

Project alternatives	Sustainable evaluation criteria									
	C1	C2	C3	C4	C5	C6	C7	C8	C9	C10
P1. Navile Market	3	5	7	7	7	5	3	3	0	5
P2. Regeneration of Maspero Triangle	5	5	5	7	5	5	3	0	3	3
P3. Progetto Manifattura	5	5	7	5	7	5	5	7	5	5
P4. Iran Mall	7	0	3	5	3	3	5	5	7	3
P5. Waterfront City	7	0	3	5	7	7	5	0	5	7
P6. Twin City	7	0	5	7	7	3	7	3	7	7
P7. Rabat City Train Station	7	7	0	7	5	7	3	0	5	3
P8. Galataport	7	0	0	7	5	5	7	5	7	3
P9. Chitgar Lake	3	3	3	7	5	3	5	5	0	0
P10. Naghsh E Jahan	5	7	0	5	3	3	5	3	3	0
P11. Charbagh Street	5	5	7	7	0	3	3	0	3	3
P12. Tivoli Greencity	3	0	7	3	3	3	7	0	0	5
P13. Philharmonie	0	0	3	5	0	3	3	7	7	3

Step 3: AHP analysis

The measurement of the relative weights among design solutions i with respect to the j-*th* evaluation criterion by Step 1 is made via AHP analysis. Ten comparison matrices among alternative projects for each evaluation criterion are performed. As evaluation scale of reference the follows is adopted: 3 (low meaningful versus other alternative); 5(quite meaningful versus other alternative); 7 (high meaningful versus other alternative). In spite of the well-established *Saaty* scale (1, 3, 5, 7, 9), this smaller range value scale has been adopted in order to preserve operational uniformity with the previous Step and corresponding evaluation process.

Table 2 reports an example of comparison matrix among project alternatives for the C1 (economic impact criterion).

To verify the judgments consistency of Step 3 and those underlying the AHP, the Consistence Ratio (CR) is calculated for j-*th* comparison matrix as in Table 3.

As noted via literature studies on AHP, a CR index value < 0.1 is considered permissible; when the CR value exceeds a threshold equal to 10%, the deviation from the perfect consistency condition is considered unacceptable [26]. For the case-study, the follows range values for CR index are considered (Table 4). The minimum and maximum CR values ($CR_{min} = 0.130$; $CR_{max} = 0.360$), as well as the intermediate benchmark ($\overline{CR} = 0.245$) were obtained by assuming the j-*th* comparison matrix consists entirely of the minimum, middle and maximum values (3, 5, 7) of the rating AHP scale. In accordance with the range values as in Table 4 the CR outcomes by AHP analysis of reference case-study are quite acceptable.

Table 2. Comparison matrix between project alternatives regarding economic impact criterion

C1.Economic impact	P1	P2	P3	P4	P5	P6	P7	P8	P9	P10	P11	P12	P13
P1	1,00	7	7	7	7	7	5	5	5	7	7	5	3
P2	0,14	1,00	5	7	7	7	7	7	3	5	5	3	3
P3	0,14	0,20	1,00	7	7	7	7	7	3	5	5	3	3
P4	0,14	0,14	0,14	1,00	5	5	5	5	3	3	3	3	3
P5	0,14	0,14	0,14	0,20	1,00	5	5	5	3	3	3	3	3
P6	0,14	0,14	0,14	0,20	0,20	1,00	5	5	3	3	3	3	3
P7	0,20	0,14	0,14	0,20	0,20	0,20	1,00	5	3	3	3	3	3
P8	0,20	0,14	0,14	0,20	0,20	0,20	0,20	1,00	3	3	3	3	3
P9	0,20	0,33	0,33	0,33	0,33	0,33	0,33	0,33	1,00	7	7	5	3
P10	0,14	0,20	0,20	0,33	0,33	0,33	0,33	0,33	0,14	1,00	5	3	3
P11	0,14	0,20	0,20	0,33	0,33	0,33	0,33	0,33	0,14	0,20	1,00	3	3
P12	0,20	0,33	0,33	0,33	0,33	0,33	0,33	0,33	0,20	0,33	0,33	1,00	3
P13	0,33	0,33	0,33	0,33	0,33	0,33	0,33	0,33	0,33	0,33	0,33	0,33	1,00

Table 3. Consistence Ratio (CR) values

	Sustainable evaluation criteria	CR
C1	Economic impact	0.349
C2	Historical value	0.322
C3	Environmental aspect	0.277
C4	Location	0.339
C5	Open spaces	0.290
C6	Public/Private collaboration	0.326
C7	Social sustainability	0.367
C8	Cultural boost	0.308
C9	Technological features	0.364
C10	Sustainable development	0.326

Table 4. CR range values

CR range value	Meanings
$0.130 \leq CR \leq 0.245$	High acceptable
$0.246 \leq CR \leq 0.360$	Quite acceptable
$CR \geq 0.361$	Not acceptable

Step 4: Ranking problem outcomes.
By previous steps total priority list and relative ones for j-*th* criterion are obtained, respectively in the Tables 5 and 6 following.

Table 5. Total priority list

	P1	P5	P8	P3	P2	P4	P6	P9	P10	P7	P11	P13	P12
i-th Global weight	0.184	0.139	0.130	0.101	0.096	0.084	0.072	0.059	0.046	0.040	0.034	0.028	0.022

Table 6. Relative priority lists

Relative priority lists									
C1	C2	C3	C4	C5	C6	C7	C8	C9	C10
P4	P1	P1	P1	P1	P1	P4	P3	P4	P1
P2	P2	P3	P2	P3	P5	P6	P4	P6	P5
P1	P7	P2	P4	P2	P2	P1	P1	P8	P6
P5	P3	P6	P7	P5	P7	P3	P9	P3	P3
P7	P10	P12	P9	P6	P3	P2	P8	P2	P2
P6	P11	P11	P3	P7	P4	P8	P13	P5	P4
P3	P9	P4	P6	P9	P8	P5	P6	P7	P12
P8	P4	P5	P5	P4	P6	P12	P10	P13	P7
P10	P5	P9	P8	P8	P9	P9	P2	P10	P8
P11	P6	P13	P11	P10	P10	P10	P5	P11	P11
P9	P8	P7	P10	P12	P11	P7	P12	P1	P13
P12	P12	P8	P13	P11	P12	P11	P7	P9	P10
P13	P13	P10	P12	P13	P13	P13	P11	P12	P9

5   Conclusions

Assessing the sustainability of projects requires the implementation of methodologies to express the projects performance according to the three dimensions of sustainability, namely economic, social and environmental ones. This should also be carried out by taking into account the decision-making context, in particular the interests related to the stakeholders involved in the life cycle of the projects to be assessed [27, 28]. The integrated evaluation methodology for urban sustainable projects [29, 30] described in this contribution attempts to comply, on the one hand, with the need to carry out assessments that respect the points of view of multiple stakeholders, and, on the other one, to formulate judgments according to a logical-operational process that is as rational and sequential as possible. The development of SWOT analysis, as first step, and then the implementation of focus group provides qualitative-quantitative input information data for the applicability of evaluative procedures (e.g., AHP), in order to support ranking cases among alternatives for the sustainable city development. The testing and checking phase of the applicability of the proposed evaluation framework on the set of 13 urban renewal projects in world-wide context allow to make some reflections. Namely, the proposed framework on which to set up the integrated evaluation of projects in sustainable key can change according to: *i*) the type of evaluation problem to be solved, *ii*) the category and number of stakeholders (public and/or private) involved in the project initiatives. The combination of these factors (type of evaluation problem, territorial analysis and stakeholders involved) defines the degree of detail and accuracy of the evaluation framework, as well as the level of accuracy of the evaluation system proposed, supporting the resolution of multiple evaluation cases of interest. Limitations of the research lie in the measurement of the weights of each criterion through a participatory process, and thus in the randomness of computing this data in the AHP evaluation process. Outlines of future researches will regard the integration of proposed methodological framework whit different multi-criteria decision tools, and its applicability to diverse type of investments in circumscribed contexts of analysis.

References

1. Bulkeley, H.: Cities and Climate Change, 1st edn. Routledge, Abingdon (2012)
2. National Geographic. https://www.nationalgeographic.com/environment/article/this-year-extreme-weather-brought-home-reality-of-climate-change. Accessed 21 Dec 2021
3. United Nations. https://www.un.org/sustainabledevelopment/sustainable-development-goals/. Accessed 21 Dec 2021
4. United Nations Office for Disaster Risk Reduction. https://www.undrr.org/publication/sendai-framework-disaster-risk-reduction-2015-2030. Accessed 21 Dec 2021
5. Kellert, S.R., Wilson, E.O.B.: Hypothesis (2008)
6. Xue, F., Gou, Z., Siu-Yu Lau, S., Lau, S., Chung, K.H., Zhang, J.: From biophilic design to biophilic urbanism: stakeholders' perspectives. J. Clean. Prod. **211**, 1444–1452 (2019)
7. Ippolito, A.M.: L'archinatura. Le diverse modalità di dialogo dell'architettura con la natura, Franco Angeli (2010)
8. Newman, P., Beatley, T., Boyer, H.: Resilient Cities: Overcoming Fossil Fuel Dependence, pp. 127–153. Island Press/Center for Resource Economics, Washington, DC (2017)
9. Torre, C.M., Morano, P., Tajani, F.: Saving soil for sustainable land use. Sustainability **9**, 350 (2017)
10. European Commission. https://ec.europa.eu/environment/gpp/index_en.htm. Accessed 21 Dec 2021
11. Legislative Decree No. 50 of April 18, 2016 Implementation of Directives 2014/23/EU, 2014/24/EU and 2014/25/EU on the award of concession contracts, public contracts and the procurement procedures of entities operating in the water, energy, transport and postal services sectors, as well as for the reorganization of the current regulations on public contracts for works, services and supplies
12. CCPI.v. https://ccpi.org/download/climate-change-performance-index-2022-background-and-methodology/. Accessed 21 Dec 2021

13. Darko, A., Chuen Chan, A.P., Ameyaw, E.A., Kingsford Owusu, E., Pärn, E., Edwards, D.: Review of application of analytic hierarchy process (AHP) in construction. Int. J. Construct. Manag. **19**(5) (2019)
14. Bell, S., Morse, S.: Sustainability Indicators: Measuring the Immeasurable?, 2nd edn. Routledge, Abingdon (2008)
15. Warhurst, A.: Sustainability Indicators and Sustainability Performance Management. International Institute for Environment and Development (IIED), United Kingdom (2002)
16. Morano, P., Tajani, F., Guarini, M.R., Sica, F.: A systematic review of the existing literature for the evaluation of sustainable urban projects. Sustainability **13**, 4782 (2021). https://doi.org/10.3390/su13094782
17. Tanguay, A., Rajaonson, J., Lefebvre, J.F., Lanoie, P.: Measuring the sustainability of cities: an analysis of the use of local indicators. Ecol. Ind. **10**(2), 407–418 (2010)
18. Ayman, O., Wafaa, N.: Towards establishing an international sustainability index for the construction industry: a literature review. In: First International Conference on Sustainability and the Future, Cairo, Egypt, vol. 1 (2010)
19. Nesticò, A., Endreny, T., Guarini, M.R., Sica, F., Anelli, D.: Real estate values, tree cover, and per-capita income: an evaluation of the interdependencies in Buffalo City (NY). In: Gervasi, O., et al. (eds.) ICCSA 2020. LNCS, vol. 12251, pp. 913–926. Springer, Cham (2020). https://doi.org/10.1007/978-3-030-58808-3_65
20. Anand, A., Rufuss, D.S., Rajkumar, V., Suganthi, L.: Evaluation of sustainability indicators in smart cities for India using MCDM approach. Energy Procedia **141**, 211–215 (2017)
21. Figueiredo, K., Pierott, R., Hammad, A., Haddad, A.: Sustainable material choice for construction projects: a life cycle sustainability assessment framework based on BIM and Fuzzy-AHP. Build. Environ. **196**, 107805 (2021)
22. Dolores, L., Macchiaroli, M., De Mare, G.: Sponsorship's financial sustainability for cultural conservation and enhancement strategies: an innovative model for sponsees and sponsors. Sustainability **13**(6), 9070 (2021). https://doi.org/10.3390/su13169070
23. Scipioni, A., Mazzi, A., Mason, M., Manzardo, A.: The dashboard of sustainability to measure the local urban sustainable development: the case study of Padua Municipality. Ecol. Ind. **9**(2), 364–380 (2009)
24. Fernández-Sánchez, G., Rodríguez-López, F.: A methodology to identify sustainability indicators in construction project management—application to infrastructure projects in Spain. Ecol. Ind. **10**(6), 1193–1201 (2010)
25. Haase, D., et al.: A quantitative review of urban ecosystem service assessments: concepts, models, and implementation. Ambio **43**(4), 413–433 (2014)
26. Saaty, T.L.: A scaling method for priorities in hierarchical structures. J. Math. Psychol. **15**(3), 234–281 (1977)
27. Calabrò, F., Mallamace, S., Meduri, T., Tramontana, C.: Unused real estate and enhancement of historic centers: legislative instruments and procedural ideas. In: Calabrò F., Della Spina, L., Bevilacqua, C. (eds.) ISHT 2018. SIST, vol. 101, pp. 464–474. Springer, Cham (2019). ISBN 978-3-319-92098-6, ISSN 2190-3018. https://doi.org/10.1007/978-3-319-92102-049
28. Della Spina, L., Giorno, C., Galati Casmiro, R.: Bottom-up processes for culture-led urban regeneration scenarios. In: Misra, S., et al. (eds.). ICCSA 2019. LNCS, vol. 11622, pp. 93–107. Springer, Cham (2019). https://doi.org/10.1007/978-3-030-24305-08

29. Della Spina, L., Giorno, C., Galati Casmiro, R.: An integrated decision support system to define the best scenario for the adaptive sustainable re-use of cultural heritage in Southern Italy. In: Bevilacqua, C., Calabrò, F., Della Spina, L. (eds.) NMP 2020. SIST, vol. 177, pp. 251–267. Springer, Cham (2020). https://doi.org/10.1007/978-3-030-52869-022

30. De Paola, P.; Del Giudice, V.; Massimo, D.E., Forte, F., Musolino, M., Malerba, A.: Isovalore maps for the spatial analysis of real estate market: a case study for a central urban area of Reggio Calabria, Italy. In: Calabrò, F., Della Spina, L., Bevilacqua, C. (eds.) ISHT 2018. SIST, vol. 100, pp. 402–410 Springer: Cham, (2019). https://doi.org/10.1007/978-3-319-920 99-346

Proposal of an Environmental-Economic Accounting System for Urban Renewal Projects

Maria Rosaria Guarini[1], Pierluigi Morano[2], Francesco Tajani[1], and Francesco Sica[1(✉)]

[1] Department of Architecture and Design (DIAP), University "Sapienza" of Rome, Rome, Italy
francesco.sica@uniroma1.it

[2] Department of Science of Civil Engineering and Architecture, Polytechnic University of Bari, Bari, Italy

Abstract. The multiple features of the urban systems require actions designed according to integrated logics. In line to current European dispositions on ecological transition and digital innovation, worldwide attention is focused on the programming-planning interventions in view of intergenerational equity. The valorization of existing built-natural environment by eco-systemic services is a strategic asset by European cities (and not ones) to become greening. Nowadays, the System of Environmental-Economic Accounting (SEEA) constitutes the reference framework at international level that integrates economic and environmental data to provide a more comprehensive and multipurpose view of the interrelationships between the economy and the environment in urban contexts. Few applications and logical-operational transpositions of the framework, like the SEEA one, in decision-making systems are found in the literature, especially to support the design of nature-solutions initiatives in urban contexts from the environmental, economic and social points of view.

In this view, the aim of this work is to favor urban interventions analyzed with an integrated logic namely that of ecosystem services. The objective consists in the descriptions of logic-operative framework of multi-criteria matrix for supporting the feasibility of design proposals evaluated in terms of trade-off concerning the environmental-social and economic frame of reference urban context. Expectations on possible operative transcription of the proposed ecosystem workflow assessment for urban projects will be outlined.

Keywords: Ecosystem services · Urban renewal projects · Environmental-economic accounting

1 Policy Framework

The decisions and policy-making related to the economic growth of the cities and the protection of its biodiversity degree are characterized by multi-level practices that arise from the interaction among multiple actors and processes. This is depicted in the form of the challenge of considering a wide range of interests and value systems in different political scales to legitimate processes that balance and negotiate such interests, ultimately seeking positive gains in economic valuation. This challenge is particularly important in contexts of urban transformation process [1].

The European Environmental Agency (EEA) published in 2019 *The EU State of the Environment Report 2020* where the biodiversity loss is one of the persistent problems Europe is facing [2]. As indicated by several international reports (e.g., from the Organization for Economic Co-operation and Development (OECD), Intergovernmental Science-Policy Platform on Biodiversity and Ecosystem Services (IPBES)), land use and cover change are the main threats to achieve positive biodiversity-related alterations [3, 4]. For this, the EU Territorial Agenda 2030 recognizes the need to create healthy environments in territories, and pursues to encourage integrated development and planning to safeguard a sustainable use of the soil and its ecological features [5]. Current realities are asking for a transition in the settlement transformation processes to decrease biodiversity loss, enhance biodiversity valuation while setting trajectories of social, economic and environmental integration towards meeting Sustainable Development Goals (SDG) [6, 7].

IPBES (2019) recognizes the relevancy of encouraging comprehensive environmental assessments, such as Strategic Environmental Assessment (SEA) and Environmental Impact Assessment (EIA), as support instruments to mitigate the impacts of development activities on biodiversity and Ecosystem Services Supply (ESS), and to promote intersectoral approaches in sustainable perspective [4]. Also, the EU Biodiversity Strategy 2030 has a similar recognition, when indicating that to enable transformative change on European biodiversity, it is imperative to commit, implement and enforce (and where necessary review and revise) EU environmental legislation, where both the EU SEA and EIA Directives are included [8]. The Energy & Financial Inclusion (E&FI) can then promote a more equitable redistribution of the benefits and costs related with land-use policies, by financing and policy action enabling territorial equitable biodiversity value sharing [9].

The EU strategy (2021) for financing the transition to a sustainable economy pursues to re-orient investments in four areas: transition finance, inclusiveness, resilience and contribution of the financial system and global ambition. This strategy adds new instruments to the policy options of addressing direct and indirect drivers of biodiversity loss in relation to settlements transformation processes. One of the core in current work will therefore be to evaluate how this strategy can safeguard Biodiversity and contribute to sustainable transformation of the cities [10].

2   Literature Review of Methods and Tools. Work Aims

As mentioned above, there are not specific EU laws that regulate the processes of spatial and urban planning, but the fact is that EU influences local development through sectoral policies [11]. Potential negative and positive impacts of these policies on the loss of biodiversity degree because the implementation of settlements transformation processes should ultimately be reduced by integrated evaluation instruments applied in combined manner according to the analysis type (social, economic and environmental) and assessment aim to achieve. In practice, it should be noticed the use of different methods and tools for supporting the integration of biodiversity value in the decision making systems for spatial and urban planning of the city. Many of them are based on multi-criteria logics for which indicator sets of multiple nature are considered in order to express, in quantitative and qualitative manner, the different impacts on biodiversity within the settlement transformation processes [12–14]. Among the most common known indicators, these could be considered during the evaluation phase of settlement transformation projects in biodiversity perspective: CITY keys indicators for smart city projects and smart cities, Urban Sustainability Index, The Reference Framework for Sustainable Cities (RFSC), The Sustainability Tools for Assessing and Rating Communities (STAR), and Urban Sustainability Indicators.

Evolving towards an integrated practice that protects and enhances biodiversity calls for a change among decision-makers involved in settlements process in urban contexts. The needs to reduce negative impacts of urban policies and processes, and enhance positive impacts, must be recognized.

Starting from these premises, the work aims to provide a methodology for evaluating in eco-systemic terms the effects that the settlements transformation processes can produce in urban contexts and aspired city. This methodology gives the possibility to support some evaluation problems linked to initiatives developed according to eco-systemic logics on the basis of specific performance indicators chosen according to the objective to be pursued, urban ecosystems and social-economic conditions of reference context.

The following Sect. 3 explains: an introduction to the issues to be addressed in Sect. 3.1 related the explanation of the verified framework of the System Environmental-Economic accounting (Sect. 3.2); the evaluation methods and tools for interventions in urban areas that include the conservation of biodiversity according to integrated eco-systemic logics (Sect. 3.3). In Sect. 4 the proposed system of environmental-economic accounting within the economic evaluation of urban projects is illustrated. At last, in Sect. 5 conclusions are reached and the potential for applying the proposed workflow assessment is discussed.

3 Material and Method

3.1 Premise

In order to be able to highlight multi-dimensional characters of initiatives by including in an integrated manner the many aspects of settlement transformation project, it is necessary to take into account, in the planning and design phase of the interventions, multiple aspects linked both to the prescriptions contained in the reference regulations in function of the specificity of the intervention to be carried out, and to the characteristics of the state of affairs of the object under examination, and also to the socio-economic conditions of the reference urban context [15].

Thus, preferring a logic of integration between the design aspects related to the same intervention, the following describes a multi-criteria evaluation approach in which to simultaneously include each of them in order to establish the priorities for action in compliance with the current urban system features. This methodology can also provide useful support in carrying out the procedures for accessing the funds allocated to the redevelopment and renovation of the cities. By favoring an integrated evaluation logic, it is possible to establish the type of intervention to be carried out to requalify according to an integrated problem based-solving approach inspired to SEEA.

3.2 The System of Environmental and Economic Accounting

The System of National Accounts is a statistical system used at international level. The United Nations has led development of international standards for environmental accounting, the first of which, the System of Environmental Economic Accounting Central Framework (SEEA CF), was officially published in 2012 [16]. The SEEA CF provides accounting standards for natural assets and inputs to economic production. It includes amounts of the stocks of ecosystem assets and the flows of services they bring to the economy and society.

On its basis, the DPSIR (drivers, pressures, state, impacts and response model of intervention) framework was adopted as driver for classifying the type of relations between environmental asset and human settlement. Focusing also on in which manner their links influence the variable production of ecosystem services in terms of Drivers, Pressures, State, impact on the Biodiversity degree of territory (DPSB).

Figure 1 shows linear relations between environmental asset, economic units and societal well-being by the SEEA. The SEEA foot-print is integrated whit DPSB workflow scale for highlighting the correspondence whit the relations between multiple systems (environmental, economic and social well-being) and various impacts degree likely in DPSB.

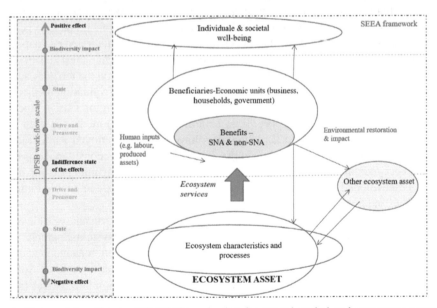

Fig. 1. Proper elaboration by EC (2018), valuation for natural capital and ecosystem accounting: synthesis paper. *Source*: https://ec.europa.eu/environment/nature/capital_accounting/pdf/Valuation_for_natural_capital_and_ecosystem_acounting.pdf (last accessed 29/11/2021).

3.3 Multi-criteria Models and Tools for Supporting the SEEA of Urban Renewal Projects

The different methodologies for the assessment of urban projects in ecosystem key can support the implementation of the mitigation hierarchy process based on the "no-net loss" concept. This is particularly relevant in the assessment of spatial planning projects, policies, plans whose impacts on biodiversity are balanced.

Biophysical methods describe how ecosystems contribute to the supply of ES to society through ecological processes. Examples of biophysical methods include direct measurements (field observations and surveys), indirect measurements (earth observation derivatives, statistical and spatial proxy data), and modelling tools, such as statistical and ecological models [17, 18]. Concerning social impacts, a set of indicators is aimed to be established, which take into consideration, adapt and upgrade the current most known urban social sustainability indicators. Concerning economic impacts, an assessment framework can be set-up that allows for assessing costs and benefits.

To assess the costs and benefits of ecosystems direct and indirect costs and benefits have been identified. Direct costs are for example those related to restoration of ecosystems, e.g., supporting ES such as provisioning of habitat for biodiversity. The inclusion of indirect effects is particularly important for benefits, as these are often secondary. For example, the implementation of ecosystems restoration may lead to a reduction in urban heat island effects, for which an economic effect can be calculated. Another example, restoration of woodland and agricultural areas can lead to biodiversity increases and, at the same time, to higher water quality and reduced carbon emissions. The use of multi-criteria evaluation methodologies makes it possible to take into account environmental, social and cultural aspects either separately or jointly with those of a financial type. This through the use of appropriate evaluation techniques that allow integrated evaluations to be carried out, referring contextually both to the morphological characteristics in which the area subject to intervention falls, and to the effects generated on the territory according to the evaluation question to be achieved [19–21]. A range of valuation methods are compatible with the System of Environmental-Economic Accounting (SEEA) methodological framework for this purpose. E.g., benefit transfer (BT) method is seen as an eligible method in SEEA, even if pilot studies have already employed value transfers (either in physical or monetary terms) in practice. BT should be reconsidered especially at this experimental stage of ecosystem accounting and explicit guidance for its use should be developed. It is a fast and cost-effective method that can enable empirical applications in situations of limited resources [22].

By the literature some of the most popular tools of multi-criteria analysis have been applied [23] to solve evaluation problems related to urban ecosystems, namely: Analytic Hierarchy Processes (AHP) for the selection between different management design options [24–27]; Techniques for Order of Preference by Similarity to Ideal Solution (TOPSIS) [28, 29] and Goal Programming [30] for ranking design alternatives in consideration of ideal solution to pursued; Operations Research optimization algorithms to answer, for example, financial questions for a distribution of available monetary resources among alternative investment projects.

Among the multi-criteria tools, those proposed by Operations Research are particularly useful for the using of logical-mathematical paradigms able to provide an optimal solution to the question posed. In particular, it is possible to solve many evaluation problems by structuring mathematical models of multi-objective optimization based on principles of linear programming, both continuous (PLC) and discrete (PLD).

4   Proposal of a System of Environmental-Economic Accounting Within the Economic Evaluation of Urban Projects

On the basis of the logical-functional relationships that trace the workflow of the SEEA, as likely in Fig. 2, the structuring of the proposed methodological approach can be synthetically articulated in an interactive integrated process made by the following steps:

- **Step 1:** definition of the specific objectives to be pursued and identification of possible strategies to achieve eco-systemic targets related to the bioclimatic, environmental, settlement, infrastructural, socio/economic conditions of the reference urban context;
- **Step 2:** analysis of the urban context condition, namely the bioclimatic, infrastructural, urban, economic and social features of the area subject to settlement transformation. This in order to collect data for describing the current state of the area and identify strengths, weaknesses, opportunities and risks related to the type of project to upload;
- **Step 3:** quantification, measurement, evaluation of the economic costs and benefits of the intervention; assessment of the impacts by single project produced in terms of ecosystem services supply; using of economic model for supporting decision-making systems by the view of public and private subjects in cases of judgements convenience process.

Specifically, on the basis of the objective of sustainability to be achieved and according to the interests of the stakeholders involved in the single initiative of transformation of the urban area of interest (Step 1), the proposed evaluation framework can guide public-private decision-makers to identify the best project alternative, taking into account the effects generated in terms of eco-system services on the territory. With this framework, it is possible to select and identify the most sustainable project option using a panel of performance indicators that cover multiple aspects of the single initiative, evaluated from an economic, financial, social and environmental point of view (Step 2). The indicators that can be used concern both the conditions representing the urban ecosystem in its naturalistic and biodiversity features, and the performance in terms of impact that the project solution is able to express with respect to the economic, social and environmental characteristics of the intervention area. Moreover, always with the proposed evaluation framework, the selection phase can be supported by the implementation of appropriate methods of analysis and evaluation of project proposals of multi-criteria matrix (Step 3). These methods make it possible to take into account multiple aspects of the project throughout the evaluation process, and to compare several alternatives with respect to one or more sustainability objectives to be achieved jointly.

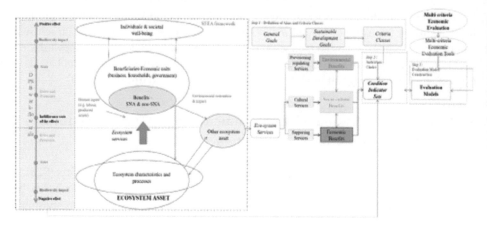

Fig. 2. Evaluation workflow proposed integrating the SEEA and DPSB framework

5 Conclusion

A systemic vision between the natural and built environment suggests the use of alternative strategies for a sustainable development of the city. Interventions based on integrated eco-systemic logic lend themselves to promote processes of transformation with a view to urban sustainability, taking into account the multiple effects they generate in the form of Ecosystem Services (ES) that can bring benefits related to the protection of existing biodiversity, economic growth and well-being of citizens. ES is essential to understand the ecosystems bio-value as component of decision-making systems related to the land-use policies. It is a crucial step towards sustainable policy and decision-making, important for both raising stakeholders' awareness and shaping decisions for both ecological processes and human activities.

In view of the interaction between environmental asset and value system of urban context, as possible to see in SEEA, it is important to use assessment models suitable to take into account the plurality of effects produced by this type of intervention, even in respect of the morphological characteristics of the reference context. The work proposes an evaluation protocol inspired by the SEEA with the aim of developing settlement transformation processes in the perspective of sustainable urban development. The use of evaluation protocols based on multi-criteria analysis offers the opportunity to build different evaluation models depending on the objective to achieve. This is also through the identification of appropriate condition indicators that differ according to the target to be achieved and the type of service. In this sense, the role of this instrument of investigation for the purposes of economic policy is clear, with a view to urban regeneration and environmental enhancement in accordance with integrated eco-systemic logic. Future perspectives will concern the construction of evaluation operative models bringing considerations related to the ecosystem services accounting and their influences on the decision-making processes in sustainable perspective. In this sense it will be deepened the implementation of the Cost-Benefits analysis concerning urban renewal projects where ecosystem services accounting will be computed.

References

1. Agger, A.: Towards tailor-made participation: how to involve different types of citizens in participatory governance. Town Plann. Rev. **83**(1), 29–45 (2012). www.jstor.org/stable/413 49079
2. European Environment Agency (EEA): The European Environment—State and Outlook 2020: Knowledge for Transition to a Sustainable Europe (2019). https://www.eea.europa. eu/publications/soer-2020. Accessed 26 Nov 2021
3. OECD: Policy Strategy: Towards Sustainable Land Use: Aligning Bio-diversity, Climate and Food Policies (2020)
4. IPBES: Summary for Policymakers: The global assessment report on Biodiversity and Ecosystem Services (2019)
5. European Commission (EC): Territorial Agenda 2030. A future for all places (2020). https://ec.europa.eu/regional_policy/sources/docgener/brochure/terririal_agenda_2030_en. pdf. Accessed 20 Nov 2021
6. Grima, N., Corcoran, W., Hill-James, C., Langton, B., Sommer, H., Fisher, B.: The importance of urban natural areas and urban ecosystem services during the COVID-19 pandemic. Plos ONE (2020). https://doi.org/10.1371/journal.pone.0243344
7. Venter, Z.S., Barton, D.N., Gundersen, V., Figari, H., Nowell, M.: Urban nature in a time of crisis: recreational use of green space increases during the COVID-19 outbreak in Oslo, Norway. Environ. Res. Lett. **15**(10) (2020). Published 6 October 2020. https://doi.org/10. 1088/1748-9326
8. EC: EU Biodiversity Strategy for 2030 Bringing nature back into our lives (2020). https://eur-lex.europa.eu/legal-content/EN/TXT/?qid=1590574123338&uri=CELEX:52020DC0380. Accessed 29 Nov 2021
9. Klapper, L., El-Zoghbi, M., Hess, J.: Achieving the sustainable development goals. The role of financial inclusion (2016). http://www.ccgap.org. Accessed 28 Nov 2021
10. Pettifor, A.: The Case for the Green New Deal. Verso Books, London (2020)
11. Dolores, L., Macchiaroli, M., De Mare, G.: A dynamic model for the financial sustainability of the restoration sponsorship. Sustainability **12**(4), 1694 (2020). https://doi.org/10.3390/su1 2041694
12. Spampinato, G., Malerba, A., Calabrò, F., Bernardo, C., Musarella, C.: Cork oak forest spatial valuation toward post carbon city by CO2 sequestration. In: Bevilacqua, C., Calabrò, F., Della Spina, L. (eds.) NMP 2020. SIST, vol. 178, pp. 1321–1331. Springer, Cham (2021). https:// doi.org/10.1007/978-3-030-48279-4_123
13. Della Spina, L.: Cultural heritage: a hybrid framework for ranking adaptive reuse strategies. Buildings **11**, 132 (2021). https://doi.org/10.3390/buildings11030132
14. Del Giudice, V., De Paola, P., Manganelli, B., Forte, F.: The monetary valuation of environmental externalities through the analysis of real estate prices. Sustain. Build. Environ. **9**, 229 (2017). https://doi.org/10.3390/su9020229
15. Morano, P., Tajani, F., Di Liddo, F., Amoruso, P.: The public role for the effectiveness of the territorial enhancement initiatives: a case study on the redevelopment of a building in disuse in an Italian small town. Buildings **11**(3), 87 (2021)
16. United Nations (UN): System of environmental economic accounting 2012—central framework. https://seea.un.org/sites/seea.un.org/files/seea_cf_final_en.pdf. Accessed 26 Nov 2021
17. Mondini, G.: Valutazioni di sostenibilità: dal rapporto Brundtland ai Sustainable Development Goal. Valori e Valutazioni (23) (2019)
18. Santos-Martin, F., et al.: Creating an operational database for ecosystems services mapping and assessment methods. One Ecosyst. **3**, e26719 (2018)

19. Sheppard, S.R., Meitner, M.: Using multi-criteria analysis and visualisation for sustainable forest management planning with stakeholder groups. For. Ecol. Manag. **207**(1–2), 171–187 (2005)
20. Diaz-Balteiro, L., Romero, C.: Making forestry decisions with multiple criteria: a review and an assessment. For. Ecol. Manag. **255**(8–9), 3222–3241 (2008)
21. Sica, F., Nesticò, A.: The benefit transfer method for the economic evaluation of urban forests. In: Gervasi, O., et al. (eds.) ICCSA 2021. LNCS, vol. 12954, pp. 39–49. Springer, Cham (2021). https://doi.org/10.1007/978-3-030-86979-3_3
22. Guarini, M.R., Morano, P., Sica, F.: Eco-system services and integrated urban planning. a multi-criteria assessment framework for ecosystem urban forestry projects. In: Mondini, G., Oppio, A., Stanghellini, S., Bottero, M., Abastante, F. (eds.) Values and Functions for Future Cities. GET, pp. 201–216. Springer, Cham (2020). https://doi.org/10.1007/978-3-030-23786-8_11
23. Morano, P., Tajani, F., Anelli, D.: Urban planning decisions: an evaluation support model for natural soil surface saving policies and the enhancement of properties in disuse. Prop. Manag. **38**(5), 699–723 (2020)
24. Nesticò, A., Endreny, T., Guarini, M.R., Sica, F., Anelli, D.: Real estate values, tree cover, and per-capita income: an evaluation of the interdependencies in Buffalo City (NY). In: Gervasi, O., et al. (eds.) ICCSA 2020. LNCS, vol. 12251, pp. 913–926. Springer, Cham (2020). https://doi.org/10.1007/978-3-030-58808-3_65
25. Calabrò, F., Cassalia, G., Lorè, I.: The economic feasibility for valorization of cultural heritage. The restoration project of the reformed fathers' convent in Francavilla Angitola: the Zibìb territorial wine cellar. In: Bevilacqua, C., Calabrò, F., Della Spina, L. (eds.) NMP 2020. SIST, vol. 178, pp. 1105–1115. Springer, Cham (2021). https://doi.org/10.1007/978-3-030-48279-4_103
26. Del Giudice, V., Massimo, D.E., De Paola, P., Forte, F., Musolino, M., Malerba, A.: Post carbon city and real estate market: testing the dataset of Reggio Calabria market using spline smoothing semiparametric method. In: Calabrò, F., Della Spina, L., Bevilacqua, C. (eds.) ISHT 2018. SIST, vol. 100, pp. 206–214. Springer, Cham (2019). https://doi.org/10.1007/978-3-319-92099-3_25
27. Vercellis, C.: Ottimizzazione. Teoria, metodi, applicazioni, pp. i-470. McGraw-Hill, New York (2008)
28. Manganelli, B., Tajani, F.: Optimised management for the develop-ment of extraordinary public properties. J. Property Investment Finance **32**(2), 187–201 (2014)
29. Tajani, F., Morano, P., Di Liddo, F.: The optimal combinations of the eligible functions in multiple property assets enhancement. Land Use Policy **99**, 105050 (2020)
30. Morano, P., Tajani, F., Guarini, M.R., Sica, F.: A systematic review of the existing literature for the evaluation of sustainable urban projects. Sustainability **13**(9), 4782 (2021). https://doi.org/10.3390/su13094782

The *Extended House* as Response to the Post-pandemic Housing Needs: Hints from the Real Estate Market

Francesca Torrieri[1]([✉]), Davide Di Ceglie[2], and Marco Rossitti[2]

[1] Università degli Studi di Napoli Federico II, Piazzale Vincenzo Tecchio 80,
80125 Naples, Italy
frtorrie@unina.it

[2] Politecnico di Milano, Via Bonardi 3, 20133 Milan, Italy

Abstract. The paper presents a study of new ways of living in response to the crisis from Covid 19-pandemic. In 2020, with the arrival of the pandemic and the repeated lockdowns, the limitations of apartments, conceived in a minimalist key that characterized the 20th century, were evident, especially due to the absence of common spaces; in fact, the concept of home has changed, becoming places used for the most varied activities previously carried out outside of them (e.g., work, study, culture, leisure, etc.). In the light of this, there is an increasing need for homes that are "More than living", i.e., equipped with free spaces (study rooms, terraces, gardens, gyms, etc.) that can allow these activities to take place and satisfy the changing needs of users. However, these homes are not always affordable for everyone, which is why many families are opting for more peripheral locations where, instead, it is possible to access larger homes with outdoor spaces. In this context, the contribution analyzes, based on a questionnaire submit to a sample of consumers, the willingness of users to live in houses where access spaces are shared. This could allow for more work and leisure space without resulting in an excessive increase in the rent or market price of the property. The first results obtained confirm the thesis that argues that the type of co-housing, or rather of the extended house, can be a valid response to new housing needs by making the use of accessory spaces accessible and affordable.

Keywords: Post-pandemic · Housing needs · Real estate market

1 From Common House to Co-housing: Shared Living as Crises Response in History

The twentieth century was characterized by continuous experimentations in architecture, especially related to the theme of residence. Population growth, urbanization, major economic crises, war destruction and subsequent reconstructions have forced designers to deal with the necessary transformations affecting the built environment and, among them, with the need to provide people with suitable residences. Different solutions were proposed, stemming from the leading paradigm of rationalist architecture, as the *Existenzminimum,* according to which spaces were designed to ensure the minimum level

F. Calabrò et al. (Eds.): NMP 2022, LNNS 482, pp. 1587–1595, 2022.
https://doi.org/10.1007/978-3-031-06825-6_153

of existence. This mainstream approach to 20th century's architecture was flanked by a further design response to the residence theme, less widespread than the previous one, but no less efficient: the collective residence.

The first declination of this theme can be identified in the Common house. The most relevant related experiments took place in the Soviet Union. In the 1920s avant-gardist architects proposed to build new social condensers. Their being Common derived from including in their design the logic of the Workers' Clubs [1].

Such an approach to the residence theme was shelved for several decades, until the 1970s, when the first examples of *Bofælleskab* were built in Denmark. The economic and social crises, affecting that historical period, together with the triumph of individualism, led to the fading of practices based on civic and community engagement, as collective residences are. In this context, the idea of *Living Communities* moved from recognizing the importance of recreating the environment and services provision characterizing the old villages' territorial dimension. With this aim, *Living Communities* represented the architectural attempt to reintroduce social relationships, typical of pre-industrial societies, into the 20th century's post-industrial reality, anonymous and impersonal: they can be considered as "an old idea, but a contemporary approach" [2].

At the end of the twentieth century, the mono-nuclear family stood out as the leading model in civic organization. However, starting from the mid-sixties, several change have affected the type-family's structure, both in terms of composition and habits [3]. This change in society doesn't find full correspondence in the design of residential buildings, that are often still conceived for a type-family, that no longer represent the major model. Moving from these premises, in the last decades Cohousing has started spreading out all over the world as an answer to the new society's needs.

In 2020 because of Covid 19-pandemic, the need for shared accessory spaces to homes for work and leisure has become a growing consumer need. In fact, the repeated lockdowns, have highlighted the limits of the apartments, conceived in a minimalist key, which characterized the 20th century mainly due to the absence of common spaces; in fact, the concept of home has changed, since homes have also become places used for the most varied activities previously carried out outside them (e.g., work, study, culture, leisure, etc.). By virtue of this, there is an increasing need for homes that are "More than living", i.e. equipped with free spaces (study rooms, terraces, gardens, gyms, etc..) able to allow the performance of such activities and meet the changing needs of users.

The residential market is therefore changing, especially for a new way of considering housing by the younger generations, and attention will be paid, in the future, more and more to those formats that, to date, in our Italian territory are present in an embryonic form, such as residential rental (multi-family) and other forms of residential as student housing, micro-living, senior living and co-housing. These are residential models that aspire to provide users with "additional" spaces that can be used in a variety of ways to combine sociality and utility. At the base is the idea of sharing: the choice of sacrificing a part of the private space in the apartments in favor of common areas that promote socialization.

However, these homes are not always affordable for everyone, which is why many families are opting for more peripheral locations where, instead, it is possible to access larger homes with outdoor spaces.

The real estate market, in fact, registers an increase in prices of properties equipped with open spaces, such as gardens or terraces, especially in central areas of large cities.

Social housing offers affordable housing and sustainable, social and shared living, thanks to quality, environmentally friendly interventions and many common and shared spaces [4]. While co-houses can use large, shared spaces (terrace, cinema room, etc.) and take advantage of services such as car and bike sharing. Not everyone can afford spacious accommodation, perhaps with a terrace, garden and a games room where they can spend their free time.

Starting from sector studies carried out by the main Italian real estate agencies, the present paper proposes to investigate if the typology of the social house, which, as previously described, has always represented an answer in times of crisis, can also today be a valid alternative to the changed needs of habitability, allowing to have more space for work and leisure without this leading to an excessive increase in rent or the market price of the property.

Based on a questionnaire submit to a heterogeneous sample of users, we analyzed the willingness to share accessory spaces and therefore the predisposition of users to live in this kid of "Extended houses".

The paper will therefore be structured as follows. First, the new housing needs and the effects on the real estate market will be analyzed based on studies conducted at the Italian leading real estate agencies. Then, the first results of the ongoing survey will be presented in order to highlight the willingness of users to share accessory spaces in their homes.

2   Covid-19 Pandemic and New Housing Needs: The Effects on the Real Estate Market

As we know, the real estate market is the place where supply and demand meet. De-mand changes faster than supply and for this it is necessary to analyze user demands to produce adequate supply. Since before the Covid 19-pandemic, the extension of economic and social progress has always accentuated the demand of quality and functional require-ments for housing, along with the demand for space and housing equipment. Events can happen that somehow accelerate progress, one of them is the health emergency that we are experiencing.

In Italy the balconies have been protagonists of the quarantine and thanks to the benefits they have allowed during the lockdown, now they may have acquired a different value.

As it is known, in fact, the market value of a property is a function of its extrinsic and intrinsic characteristics, the first referring to the environmental and infrastructural con-text, the second referred to the dimensional, compositive, technological and productive specificities of the property [5].

In the specific historical moment of the Covid 19-pandemic two studies conducted by *Immobiliare.it* and *Il Sole 24 Ore* [6, 7] have analyzed how the presence of open spaces both at the neighborhood level and at the single building level has had an influence in determining the market price of real estate in reference to different territorial contexts.

The study of *Immobiliare.it* (portal of real estate ads leader in Italy) in collaboration with Realitycs (proptech company specialized in market intelligence and automatic evaluations), has defined that the presence of a balcony or terrace in Italy makes increase the value of the property by 8% in the case of a high floor (from the fifth onwards). Carlo Giordano, CEO of *Immobiliare.it*, points out that this phenomenon varies from city to city. In Milan, due to the high price per square meter, heavy traffic and a harsher climate, terraces and gardens become a must-renounceable asset, while Rome and Naples are cities where you are more predisposed to spend in order to own a private space in the open air (Fig. 1).

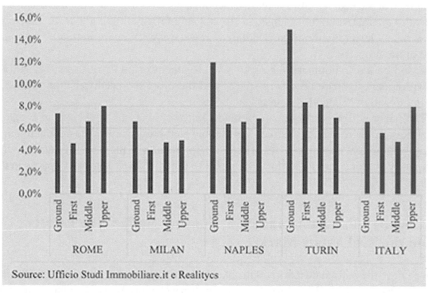

Source: Ufficio Studi Immobiliare.it e Realitycs

Fig. 1. Increase in value in the presence of an outdoor space compared to a property of the same characteristics.

Even *il Sole 24 Ore* has produced an elaboration to translate this trend. The search for real estate types with these "more than living" characteristics has led to a greater flow of buyers to the hinterland and to the provinces adjacent to the large Italian cities. Turin is the most expensive Italian city about solutions with outdoor spaces, so much so that the additional value reaches 9.5% on the first floor. "The increases depend a lot on the quality of the building fabric in the reference area" -explains Francesca Zirnstein, general manager of *Scenari Immobiliari*- for example, Turin is so expensive from this point of view because the most of buildings do not have balconies.

The additional value of the terrace naturally depends on the panoramic view and the floor on which it is located, as well as a whole other series of factors. In Italy on average the terrace on the first floor adds 5.8% value to the apartment, to the intermediate floors 6.4% and to the upper floors 6.7%. As we can see in the graph (Fig. 2), the percentage values of the two studies are approaching, the deviation is minimal. Therefore, we can say with certainty that, on average, in Italy, following the pandemic, an apartment with a terrace on a high floor will be worth about 7% more than before [8].

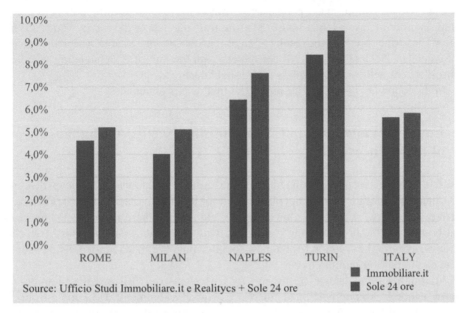

Fig. 2. Comparison between the increase in value found by immobiliare.it and *Sole 24 Ore* for the first floor only

The results of studies conducted by both *Immobiliare.it* and *Il Sole 24 Ore* show that the changing needs and experience of the lockdown have, at least in the short term, modified the habits and preferences of consumers and this phenomenon is clearly reflected in the trend of the real estate market. The increase in prices of houses with balconies and open spaces, especially in urban centers, has led on the one hand to a greater flow of buyers towards the hinterland and the provinces adjacent to large Italian cities, and on the other hand to the search for housing solutions that are more accessible to all.

The residential market is therefore changing, especially for a new way of considering housing by the younger generations, and attention will be paid, in the future, more and more to those formats that, to date, in our Italian territory are present in an embryonic form, such as residential rental (multifamily) and other forms of residential as student housing, micro-living, senior living and co-housing.

At the base is the idea of sharing: the choice to sacrifice part of the private space in the apartments in favor of common areas that promote socialization. Social housing offers affordable housing and sustainable, social and shared living, thanks to quality, environmentally friendly interventions with many common and shared spaces. One of the most interesting phenomena recorded between May and September 2021 (but the signs were already present during the lockdown, through online searches), was "the Covid 19-pandemic effect on demand with the optional green" [9]: the increase in requests for houses outside the city, but with outdoor spaces, more spacious and functional. We see it in the latest survey carried out by *Scenari Immobiliari* on families: "*La casa che vorrei*" [6], a survey conducted in September 2020, which involved about 22 thousand people. If the need is to have more space, people are willing to move to areas further away from the center in order to have a larger house. Starting from the data of the survey "*La casa che vorrei*" this paper presents the results of an online questionnaire submitted in the month of September 2021 to a sample of consumers with the aim of understanding whether, consumers are willing to live in less traditional housing.

In the following paragraph we report the first results obtained.

3   Covid-19 Pandemic and New Housing Needs: A Change of Preferences

In 2020, with the arrival of the pandemic and the repeated lockdowns, the concept of home has changed: homes have become places used for the most varied activities previously carried outside (work, study, culture, leisure, etc.). Now, there is an increasing need that homes are "more than living", equipped with free spaces (study rooms, terraces, gardens, gyms). The solution for the new social habits, post pandemic, must be sought in an evolution of minimum living.

Starting from the data of the survey "*La casa che vorrei*" [6] we created an online survey entitled "Abitare post-pandemico", responded a sample of 315 people from all Italy, including young people and the elderly. The aim of the survey is to understand whether, consumers, while not knowing cohousing and the different levels of sharing and privacy it offers, are willing to live in housing less traditional.

The survey showed that 61.6% of respondents are willing to share some accessory spaces while maintaining the autonomy and privacy of their apartment, in the face of a reduction in price/fee rent.

The balcony is the most popular accessory space (65.1%), but not everyone has it, 34.9% have lived the lockdown without an open space. Indeed, 17.1% do not have any kind of accessory space (Fig. 3).

According to 50.8% of respondents, the balcony is the space that proved most useful during the lockdown, followed by study room (28.3%), garden (22.2%) and passion room (17.8%). Again, 25.4% do not have any accessory space that proved useful during the lockdown.

The effects of repeated lockdowns on consumer tastes are evident in the choice of accessory spaces that they would like to have at home. In the survey "La casa che vorrei" in first place is the garden (58%) followed by the terrace (43%). In the survey "Abitare post-pandemico" immediately after the garden (47%) it is the room of passions to take preferences (41.9%), space that instead in the survey "La casa che vorrei" was at 19%. The two surveys were carried out about a year apart.

Making a cross-analysis between preferences expressed by the sample and personal data we can trace to more interesting information. For example, men are willing to share 7.3% more than women. The maximum openness to sharing is found in the under 25, which rises to 72.0%, 24.6% more than the over 25 (Fig. 4).

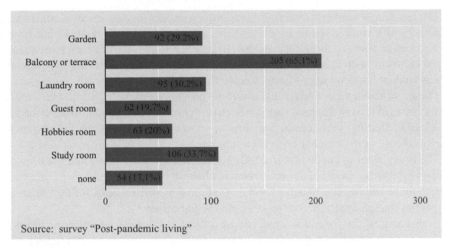

Source: survey "Post-pandemic living"

Fig. 3. Answer of the sample of 315 people to the question: "What accessory spaces are currently present in your home?"

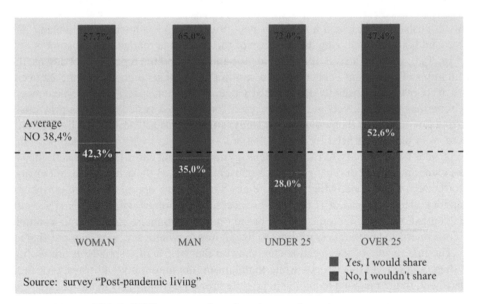

Source: survey "Post-pandemic living"

Fig. 4. Willingness to share based on gender and age group.

On the basis of the studies carried out and the results of the questionnaires it is possible to state that there has been an important passage on the side of the housing supply and demand, we have passed from the conception of the house seen only as a simple domestic space, to a house that becomes the protagonist in the other spheres of life daily, thanks to the inclusion of technological and environmental comforts as well as services in line with the new lifestyle.

The residential market is therefore changing, especially for a new way of considering housing by the younger generations, and the attention will be turned, in the future, more and more towards those formats that, to date, on Italian territory are present in an embryonic form, such as residential for rent (multi-family) and other forms of residential such as student housing, micro-living, senior living and co-housing [10].

These are residential models that aspire to provide users with "additional" spaces that can be exploited in the most varied ways to combine sociality and utility. At the base is the idea of sharing: the choice to sacrifice a part of the private space in the apartments in favor of common areas that favor socialization.

The point is not to oppose the shared home to the individual home and not even one family model to another, but to notice what is changing around us and try to seize it, to offer, to those who want, a housing solution that today is still a utopia, for the length and complexity that characterizes the process of new initiatives.

In this context the Extended house can be a valid response to new housing needs by combining the new space requirements with budgetary constraints of the families.

4   Conclusion

This contribution presented a study on new housing models such as co-housing and Extended housing, analyzing the response of the real estate market and consumers.

In 2020, with the arrival of the Covid 19-pandemic and the repeated lockdowns, the limitations of apartments, conceived in a minimalist key that characterized the 20th century, were evident, especially due to the absence of common spaces; in fact, the concept of home has changed, becoming places used for the most varied activities previously carried out outside of them (e.g., work, study, culture, leisure, etc.). In the light of this, there is an increasing need for homes that are "More than living", i.e., equipped with free spaces (study rooms, terraces, gardens, gyms, etc.) that can allow these activities to take place and satisfy the changing needs of users. However, these homes are not always affordable for everyone, which is why many families are opting for more peripheral locations where, instead, it is possible to access larger homes with outdoor spaces. In this context, the contribution analyzes, based on a questionnaire submitted to a sample of consumers, the willingness of users to live in houses where access spaces are shared.

The first results of the questionnaire showed that 61.6% of respondents are willing to share some accessory spaces while maintaining the autonomy and privacy of their apartment, in the face of a reduction in price/fee rent. Developments in the research can be directed towards a more in-depth analysis of both supply and demand. With regard to supply, a questionnaire has been prepared to be submit to real estate agencies in order to highlight the real market response.

References

1. Kopp, A.: Città e rivoluzione. Feltrinelli, Milano (1987)
2. McCamant, K.: Cohousing: A Contemporary Approach to Housing Ourselves. Ten Speed Press, Berkley (1989)
3. ISTAT – Annuariostatistico italiano, cap.3 "Popolazione e famiglie"

4. PNRR misura "M5C2 Rigenerazione Urbana e Housing Sociale"
5. Del Giudice, V., Passeri, A., De Paola, PF., Torrieri, F.: Estimation of risk-return for real estate investments by applying Ellwood's model and real options analysis: an application to the residential real estate market of Naples. Appl. Mech. Mater. **651–653**, 1570–1575 (2014). https://doi.org/10.4028/www.scientific.net/AMM.651-653.1570
6. Scenari Immobiliari, Rapporto Casa Italia (2021)
7. Analisi di Immobiliare.it e Realitycs sull'offerta di immobili residenziali, Comunicato stampa, Milano, luglio (2020)
8. Rogel, L., Corubolo, M., Gambarana, C., Omegna, E.: Cohousing. "L'arte di vivere insieme. Principi, esperienze e numeri dell'abitare collaborativo". Altreconomia, Milano (2018)
9. Manzini, E., Jegou, F.: Quotidiano sostenibile. Scenari di vita urbana. Edizioni Ambiente, Milano (1993)
10. UN-Habitat 2015 Housing at the center of new Agenda, pp. 3–4 (2015)

An Integrated Model for the Estimation of the Emissions of Pollutants from Traffic in the Urban Area

Domenico Gattuso, Gian Carla Cassone, and Domenica Savia Pellicanò[✉]

Mediterranea University, 89124 Reggio Calabria, Italy
domenica.pellicano@unirc.it

Abstract. The vehicular traffic still remains one of the main causes of air pollution, despite the progressive improvement, particularly in the urban areas. For this reason, several municipal administrations are forced to impose limitations and restrictions on the circulation of vehicles, especially the most polluting ones.

The paper proposes some specific emission models for CO, NOx and PM pollutants, per vehicle unit, derived from the composition of some literature models and capable of responding to the planning requirements for different road and traffic contexts. These tools appear interesting in relation to the need to verify the non-exceeding of the environmental capacity limits on the individual streets and on the global road network of a city.

Finally, a representative application of the simulation approach and the resulting model assessment are proposed on an Italian urban network case study.

Keywords: Models · Mobility · Urban traffic · Planning · Sustainability · Environmental Impacts

1 Introduction and Models Framework

Significant impact indicators are generally assumed for urban traffic simulations such as the average speed of flow, the road saturation degree, or other congestion measures such as average delays. It is also common practice to estimate impacts such as polluting emissions and energy consumption. In the case of polluting exhaust gas emissions, with reference to macroscopic runoff simulations, one approach is to make estimates using statistical-mathematical models. A large framework of gas emission models was drawn up by the authors [1].

A first classification of the emission models [2, 3], in relation to the level of complexity and the types of application, distinguishes:

- *average speed* models (COPERT, MOBILE, etc.), in which the emission factors are a function of the average travel speed;
- *traffic situation* models (HBEFA, ARTEMIS, etc.), where the emission factors correspond to particular traffic situations (i.e. stop-and-go-driving, freeflow);

- *modal* models (PHEM, CMEM, MOVES, etc.), with emission factors corresponding to specific operating conditions of the engine or vehicle.

In this paper, attention is focused on the Average Speed Models (ASM), at macroscopic level, which relate the Emission Factor (*EF*) with the travel time and indirectly with the vehicle speed.

The ASM are considered easy-to-apply models as tools to support the planning of transportation systems, which allow estimates of emissions at link, route and network level. A typical functional form of the ASM model assumed for estimating the EF, in *g/km*, is the following:

$$EF = a v^2 + b v + c$$

where v is the average speed in km/h; a, b and c are parameters calibrated by different researchers (Table 1).

Table 1. Parameters values in the ASM

Authors, year	Vehicle type	Pollutant	A	b	c
Wang et al., 2018	Small cars	CO	0.0032	−1.0866	92.46
		NOx	−0.0004	0.1093	−2.5028
	Medium cars	CO	0.0126	−1.811	111.19
		NOx	−0.0012	0.3534	−3.8
	Large cars	CO	0.0014	−0.2157	115.578
		NOx	0.0052	−0.453	29.78
Sharm and Mettew, 2016	Car	HC	0.0002	−0.211	0.6974
		NOx	0.000249	−0.00933	0.1405
		CO	0.0029	−0.288	10
		CO$_2$	0.5957	−50.71	1221
	SUV	HC	0.0006109	−0.05325	1.384
		NOx	0.001264	−0.118	3.448
		CO	0.001679	−0.1611	3.957
		CO$_2$	0.4979	−39.77	1074
	Truck	HC	0.0001957	−0.02933	1.139
		NOx	0.006815	−0.8451	27.6
		CO	0.0002483	−0.04091	1.698

Wang et al. [4] provided the parameter values for three different vehicle categories and for the CO and NOx pollutants. The small cars category includes mini-buses and vans; coaches and freight vehicles constitute medium cars; buses, large trucks and oversized freight vehicles are intended as large cars. *Sharma and Mettew* [5] proposed the calibration of the parameters for three types of vehicles (car, SUV and truck) and for the pollutants HC, NOx, CO and CO_2.

Other authors [6] proposed the following functional form:

$$EF = \frac{a}{v} + b + cv + dv^2$$

EF is the Emission Factor in g/km; v is the average speed in km/h; a, b, c, d are parameters to be calibrated. Table 2 shows the calibration proposed by *Song et al.* [6] for light duty and heavy duty vehicles.

Table 2. Parameters values in Song et al. models [6]

Vehicle type	Pollutant	a	b	c	d
Light duty	HC	10.8	−0.00711	0.000376	0.0000363
	NOx	2.0	−0.0449	−0.000336	0.0000349
	CO	80.8	1.16	0.00503	0.000535
	CO_2	4,780	111	−1.24	0.0237
Heavy duty	HC	15.5	0.392	−0.0072	0.0000531
	NOx	89.1	9.35	−0.136	0.000891
	CO	41.4	1.99	−0.011	0.0000299
	CO_2	3,670	534	−7.9	−0.0543

Some studies give the Emission Factor in relation to the average speed using the following expression:

$$EF = av^e + bv^f + cv + d$$

Corvàlan et al. [7] determined the values of the hot emission factor parameters for light gasoline vehicles (catalytic and non-catalytic cars), running around 2,000 tests on a sample of 166 vehicles. Calibration was also carried out for catalytic and non-catalytic commercial vehicles (Table 3).

Table 3. Parameters values in Corvàlan et al. models [7]

Vehicle type	Pollutant	a	b	c	d	e	f
Catalyst car	HC	0.3681	0	0	0	−0.4085	0
	NOx	$0.3 \cdot 10^{-6}$	$-3 \cdot 10^{-4}$	0.0068	0.4941	3	2
	CO	20.844	0	0	0	−0.7656	0

(continued)

Table 3. (*continued*)

Vehicle type	Pollutant	a	b	c	d	e	f
No catalyst car	HC	8.8083	0	0	0	−0.4792	0
	NOx	$9.5 \cdot 10^{-6}$	−0.0016	0.0738	1.2586	3	2
	CO	0	0.0243	2.5613	76.977	0	2
Catalyst commercial veh.	HC	0	0.0003	0.0277	0.7856	0	2
	NOx	$8 \cdot 10^{-6}$	0.001	0.03	0.4419	3	2
	CO	0	$-8 \cdot 10^{-5}$	0.0011	0.5633	0	2
No catalyst commercial veh.	HC	0	0.0006	0.0891	4.5941	0	2
	NOx	$3 \cdot 10^{-5}$	−0.0042	0.1669	1.738	3	2
	CO	0	0.0228	-2.4598	79.998	0	2

Joumard et al. [8] have proposed the following *EF* formulation for light vehicles:

$$EF = \frac{a + cv + ev^2 + f/v}{1 + bv + dv^2}$$

And for heavy vehicles, the expression:

$$EF = av^2 + bv + c$$

The parameters were calibrated by the authors in relation to different types of vehicle and pollutants; Table 4 shows the values relating to the most advanced EURO standards.

Table 4. Parameters values in Joumard et al. models [8]

Vehicle type		Poll	a	b	c	d	e	f
Light	EURO 4 Gasoline	HC	0.0118		$-3.47 \cdot 10^{-5}$		$8.84 \cdot 10^{-7}$	
		NOx	0.106		$-1.58 \cdot 10^{-3}$		$7.10 \cdot 10^{-6}$	
		CO	0.136	−0.0141	$-8.91 \cdot 10^{-4}$	$4.99 \cdot 10^{-5}$		
	EURO 3 Diesel	HC	0.0965^*	0.103^*	$-2.38 \cdot 10^{-4*}$	$-7.24 \cdot 10^{-5*}$	$1.93 \cdot 10^{-6*}$	
		HC	$0.0912°$		$-1.68 \cdot 10^{-3°}$		$8.94 \cdot 10^{-6°}$	
		NOx	2.82	0.198	0.069	$-1.43 \cdot 10^{-3}$	$-4.63 \cdot 10^{-4}$	

(continued)

Table 4. (*continued*)

Vehicle type		Poll	a	b	c	d	e	f
		CO	0.169		$-2.92 \cdot 10^{-3}$		$1.25 \cdot 10^{-5}$	1.10
		PM	0.0515		$-8.80 \cdot 10^{-5}$		$8.12 \cdot 10^{-6}$	
	Hybrid	HC	$5.50 \, 10^{-4}$		$-8.54 \cdot 10^{-6}$		$4.94 \cdot 10^{-8}$	
		NOx	0.0148		$4.20 \cdot 10^{-3}$		$4.29 \cdot 10^{-6}$	
		CO	$1.95 \cdot 10^{-4}$		$3.80 \cdot 10^{-5}$		$-2.64 \cdot 10^{-7}$	
Heavy	Gasoline N1-III	HC	$1.01 \cdot 10^{-5}$	$-5.56 \cdot 10^{-3}$	0.484			
		NOx	$1.20 \cdot 10^{-5}$	$-5.43 \cdot 10^{-3}$	0.535			
		CO	$7.40 \cdot 10^{-4}$	-0.122	6.16			
	Diesel N1-II	HC	$-5.49 \cdot 10^{-7}$	$-4.38 \, 10^{-4}$	0.0729			
		NOx	$1.85 \cdot 10^{-4}$	-0.0216	1.42			
		CO	$1.65 \cdot 10^{-5}$	$-5.19 \cdot 10^{-3}$	0.412			
		PM	$5.03 \cdot 10^{-6}$	$-7.00 \cdot 10^{-4}$	0.0490			

* vehicles with engine capacity < 2.0 l; ° vehicles with engine capacity > 2.0 l

A similar expression has been adopted by *EMEP/EEA* [9]:

$$EF = \frac{a v^2 + b v + c + d/v}{e v^2 + f v + g}$$

which provides the hot emission values of the COPERT model.

The parameters values are available by vehicle categories (Passenger Cars – PC; Light Commercial Vehicles – LCV; Buses – BS; Heavy Duty Vehicles - HDV), by fuel types (Gasoline - G, Diesel - D, Hybrid - H, etc.), by EURO standard, by different technologies and pollutants. Table 5 shows the values of the parameters by type of vehicle, for petrol and diesel engines, considering the EURO 5 and EURO 6 standards, for the pollutants CO, NOx and PM. For HDVs, a road grade of 0%, fully loaded with Selective Catalytic Technology (SCR), rigid segment of less than 7.5 t was considered.

Table 5. Parameters values in EMEP/EEA model [9]

Vehicle type	Poll	a	b	c	d	e	f	g
PC EURO 5 Gasoline	CO	4.45E−04	−0.1021	6.8769	10.3838	0.0016	−0.4376	30.3373
	NOx	−3.15E−04	0.1031	0.2391	−0.3393	0.0345	1.9860	1.2638
	PM	5.73E−07	−1.26E−04	0.00691	0.0642	3.30E−04	−0.0878	6.01
PC EURO 5 Diesel	CO	−0.00167	0.2126	−0.7450	0.7171	0.0274	1.5132	7.41E−09
	NOx	6.67E−05	−0.0114	0.946	1.92	−5.15E−05	4.26E−03	1.00
	PM	−2.69E−07	7.57E−05	0.0239	0.0827	−6.08E−04	0.211	4.46
PC EURO 6 Gasoline	CO	8.51E−04	−0.180	11.3	16.9	0.00264	−0.719	50.8
	NOx	−3.15E−04	0.103	0.239	−0.339	0.0345	1.99	1.26
	PM	7.36E−07	−1.58E−04	0.00928	0.0411	2.69E−04	−0.0779	6.02
PC EURO 6 Diesel	CO	−0.00167	0.213	−0.745	0.717	0.0274	1.51	7.41E−09
	NOx	6.67E−05	−0.0114	0.946	1.92	−5.15E−05	0.00426	1.00
	PM	4.67E−04	0.0664	−0.325	0.961	1.14	0.546	0.290
PC MIDI	PM	4.80E−04	0.683	−0.339	0.913	1.17	0.541	0.291
LCV EURO 5 Gasoline	CO	0.00151	−0.248	12.8	115	0.00174	−0.5	37.2
	NOx	1.59E−04	0.0546	−0.128	−0.122	0.0445	1.32	−4.09
	PM	8.73E−05	−2.38E−04	0.0416	−0.0863	−0.0245	4.87	0.0365
LCV EURO 5 Diesel	CO	−1.53E−05	0.00240	−1.13E−04	0.00135	0.0573	2.81	0.496
	NOx	1.54E−05	−0.00307	0.183	1.23	5.72E−05	−0.0145	1.00
	PM	9.09E−04	0.0384	−0.0148	0.0537	1.54	0.552	0.339
LCV EURO 6 Gasoline	CO	0.00151	−0.248	12.8	115	0.00174	−0.5	37.2
	NOx	1.59E−04	0.0546	−0.128	−0.122	0.0445	1.32	−4.09
	PM	8.73E−05	−2.38E−04	0.0416	−0.0863	−0.0245	4.87	0.0365
LCV EURO 6 Diesel	CO	−1.53E−05	0.0024	−1.13E−04	0.00135	0.0573	2.81	0.496
	NOx	1.54E−05	−0.00307	0.183	1.23	5.72E−05	−0.0145	1.00
	PM	9.09E-04	0.0384	−0.0148	0.0537	1.54	0.552	0.339
HDV EURO 5 Diesel	CO	0.00762	−0.0615	3.42	2.96	0.0128	−0.0143	1.09
	NOx	1.25E−04	−0.0197	0.842	2.01	−3.93E−05	0.00386	0.111
	PM	0.0157	0.394	−0.421	−0.0160	1.41	0.361	0.448
BUS MIDI EURO 5 Diesel	CO	0.00396	0.521	−2.16	4.47	0.0129	−0.0635	0.161
	NOx	2.56E−04	−0.0369	1.75	4.56	−4.42E−05	0.00588	0.0984
	PM	−5.38E−05	−0.00239	0.794	−0.542	−0.00781	0.717	1.72
BUS ST. EURO 5 Diesel	CO	0.00224	0.334	−1.27	3.01	0.00624	−0.0301	0.0891
	NOx	9.53E−04	−0.149	6.97	12	−2.19E−04	0.0232	0.232
	PM	0.00659	0.446	−1.46	1.71	0.406	−0.860	1.47

2   Proposal of an Integrated Model for the Estimation of Pollutant Emissions

Given the diversity of the literature models, it seemed appropriate to identify a set of easy-to-use models for traffic simulations, assuming a mathematical expression capable of summarizing the different expressions in relation to the type of pollutant and differentiating the emissions produced by light vehicles and heavy vehicles. A sort of envelope of the basic models was generated, through a statistical interpolation operation. The starting models were the model of *Joumard et al.* [8] and the models developed by *EMEP/EEA* [9]. The other models have not been considered due to a significant deviation of the relative trends compared to the trends revealed by the former and less reliable in relation to experience (some models did not appear credible due to evident over-estimation or in contrast in functional form). Specifically, the *EMEP/EEA* [9] models considered are the gasoline Passenger Cars (PC, B), EURO 5 and EURO 6 engines, and the EURO 6 gasoline Light Commercial Vehicles (LCV, B) for Light vehicles; Heavy Duty Vehicles (HDV) models with a segment below 7.5 t, MIDI (BS md) and standard (BS - sd) buses were used for heavy vehicles. The cases of EURO 5 diesel vehicles, fully loaded, with Selective Catalytic Technology (SCR), on a road with a slope of 0% have been analysed.

2.1   Standard Emission Models for CO

The model for estimating CO emissions, for Light Vehicles, has the following form:

$$EF = a v^2 + b v + c$$

where v is the average speed in *km/h*; a, b e c are model parameters.

It has been derived from the interpolation of four literature models (*Joumard et al. 2013*; PC, EURO 5, *EMEP/EEA* [9]; PC, EURO 6, *EMEP/EEA* [9]; LCV, EURO 6, *EMEP/EEA* [9]) and the calibrated parameters respectively have the following values: $a = 0.000179$ gh^2/km^3; $b = -0.0167$ gh/km^2; $c = 0.571$ g/km.

For Heavy Vehicles, the emission factor formula is:

$$EF = a v^b$$

where v is the average speed in *km/h*; a and b are model parameters.

The interpolation has made possible to calibrate a specific model and the last parameters values are *10.3* gh/sq.km and $- 0.701$ respectively.

This new models are proposed in red lines in Fig. 1; the same figure shows the original models. The trend of CO emissions for light vehicles (left) is parabolic with a minimum of *40* km/h. The parable is not symmetrical; for low speed values it has a modest slope instead, for speeds above *40* km/h, the slope is higher. Concerning heavy vehicles (right) the model assumes the trend of an equilateral hyperbola; for speeds of *10* km/h the emissions are maximum and close to *2* g/km; as speed increases, emissions decrease until values lower than *0.5* g/km for speeds above *80* km/h.

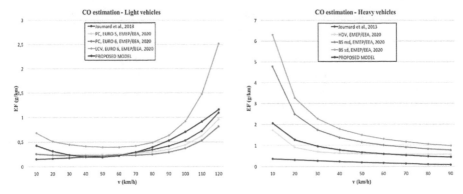

Fig. 1. Models for estimating the CO emission factor for light and heavy vehicles.

2.2   Standard Emission Models for NOx

For the estimation of the NOx pollutant, three models have been considered as the values of the parameters provided by the *EMEP/EEA* [9] are equal for EURO 5 and EURO 6 vehicles. For both light and heavy vehicles, the proposed model has an exponential expression:

$$EF = a\,v^b$$

where *EF* is the emission factor in *g/km*; *v* is the average speed in *km/h*; *a* and *b* are two parameters, appropriately calibrated, which respectively assume the values 0.250 gh/sq.km and −0.599 for light and *39.6* gh/sq.km and −0.815 for heavy vehicles.

Figure 2 shows the comparison between the basic models and the models estimated by interpolation (red lines) for light and heavy vehicles respectively. The proposed models follow an equilateral hyperbola trend with an evident difference in the order of magnitude; in the case of light vehicles, the curve starts from a value around *0.06* g/km and decreases versus the speed and falls below *0.02* g/km for speeds above *90* km/h; the curve relating to heavy vehicles (right) starts from a value close to *6* g/km and decreases to approximately *1* g/km at *90* km/h.

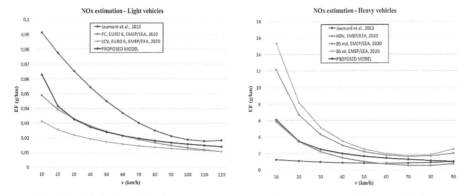

Fig. 2. Models for estimating the NOx emission factor for light and heavy vehicles.

2.3   Standard Emission Models for PM

Concerning Particle emission, in other terms Particulate Matter (PM), *Joumard et al.* [8] provided a model for estimating emissions for diesel vehicles only. This model assumes values not comparable with those of the *EMEP/EEA* [9]. This is due to the abatement of fine dust that the anti-pollution regulations of recent years produced, setting very strict maximum concentration levels for PM. For this reason, only the models for *EMEP/EEA* [9] light diesel vehicles have been considered here.

For the estimation of PM emissions, both for light and heavy vehicles, the following model is proposed for the *EF* emission factor (in *g/km*):

$$EF = a\,v^b$$

where v is the average speed in *km/h*; a and b are the model parameters. In relation to the calibration carried out, the latter assume, respectively, the values *0.0139* gh/sq.km and *−0.566* for light and *0.158* gh/sq.km and *−0.438* for heavy vehicles.

In Fig. 3 it is possible to observe the trend of the models considered for light and heavy vehicles respectively; the synthesis derivate models are identified in red. The trends of the proposed models are typically of equilateral hyperbola.

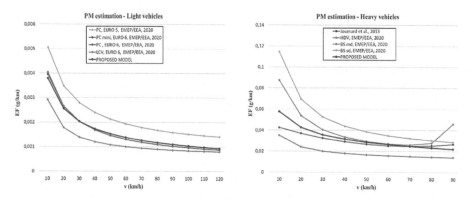

Fig. 3. Models for estimating the PM emission factor for light and heavy vehicles.

2.4   Summary of Standard Emission Models

Table 6 shows the calibrated parameters values for the proposed models relating to the different pollutants and for the two vehicle categories.

The proposed models give the emission factor EF_i, which provides the grams of pollutant by a category i vehicle (light, heavy) over a road unit length (*1* km). The total emission is, consequently, given by the sum on the network considering the flows (f_1) on all links of a road network and the relative lengths of the links (L_i):

$$E = \sum_l \left(\sum_i EF_i * f_{li} \right) * L_l$$

Table 6. Parameters values in the proposed models.

Vehicle type	Pollutant	a	b	c
Light	CO	1.79E−04	−0.0167	0.571
	NOx	0.250	−0.599	
	PM	0.0139	−0.566	
Heavy	CO	10.3	−0.701	
	NOx	39.6	−0.815	
	PM	0.158	−0.438	

3   A Case Study

The new models have been applied to a real network located in the urban area of Reggio Calabria, Italy (Fig. 4). The main network consists of four parallel urban streets (via Aschenez: Urban 1; via Tripepi: Urban 2; via Matteotti: Urban 3; via Falcomatà: Urban 4), a transversal urban road (via Argine destro Calopinace: Urban 5) and two ring roads (in the south direction: Tang. Sud; in the north direction: Tang. Nord). For each section, the speeds and flows relating to three specific time windows are known: morning rush hour, soft hour and night hour (Table 7). The estimation of emissions for CO, NOx and PM pollutants was carried out on each section by applying the models previously presented. The results are in Table 8.

Fig. 4. Reggio Calabria City road network ad traffic flows by simulation.

Table 7. Characteristics of road sections and traffic volumes of the study area. Current scenario

Road section	L (km)	Rush hour				Soft hour			Night hour		
		v (km/h)	f_{light} (veh/h)	f_{heavy} (veh/h)		v (km/h)	f_{light} (veh/h)	f_{heavy} (veh/h)	v (km/h)	f_{light} (veh/h)	f_{heavy} (veh/h)
Urb. 1	1.9	15	776	24		35	310	10	50	78	2
Urb .2	1.8	15	582	18		35	233	7	50	58	2

(*continued*)

Table 7. (*continued*)

Road section	L (km)	v (km/h)	f_{light} (veh/h)	f_{heavy} (veh/h)	v (km/h)	f_{light} (veh/h)	f_{heavy} (veh/h)	v (km/h)	f_{light} (veh/h)	f_{heavy} (veh/h)
		Rush hour			Soft hour			Night hour		
Urb. 3	1.6	20	1410	90	35	564	36	50	141	9
Urb. 4	1.6	20	1504	96	35	601	39	50	150	10
Tang. Sud	2.6	50	2295	405	70	935	165	70	234	41
Tang. Nord	2.6	50	1785	315	70	850	150	70	212	38
Urb. 5	1.8	40	1656	144	60	662	58	80	166	14

Table 8. Estimated emissions (in grams) in the study area

Road section	E_{CO}	E_{NOx}	E_{PM}	E_{CO}	E_{NOx}	E_{PM}	E_{CO}	E_{NOx}	E_{PM}
	Rush hour			Soft hour			Night hour		
Urb. 1	603.58	271.22	6.64	138.75	58.95	1.73	56.12	9.75	0.55
Urb. 2	428.86	192.71	4.72	98.00	39.94	1.20	40.15	8.38	0.42
Urb. 3	880.46	589.72	11.90	236.83	152.55	3.60	79.45	28.92	0.99
Urb. 4	939.16	629.04	12.70	253.24	164.79	3.87	84.97	31.88	1.07
Tang. Sud	1,814.79	1,862.92	39.12	916.92	580.31	13.62	169.68	144.28	3.55
Tang. Nord	1,411.50	1,448.94	30.43	833.57	527.55	12.38	154.89	133.48	3.27
Urb. 5	773.68	589.34	13.30	320.67	172.55	4.39	69.43	33.46	1.10
Total	6,852.03	5,583.90	118.81	2.797.98	1.696.65	40.80	654.70	390.15	10.95

The emissions estimation has been developed both in the current state (scenario 0) of the area and in a project scenario (scenario 1) in which a reorganization of the transport supply is designed in order to guarantee a sustainable mobility aims reducing the number of motorized vehicles on the inner city.

Specifically, it was decided to block the access to the two streets of the city's waterfront, via Matteotti (Urb.3) and via Falcomatà (Urb.4), and provide exchange parking lots at the ends that allow users to stop their car and move using another more sustainable mode of transport. This change in the transport supply has consequences on the demand of the whole city; many users, in fact, are forced to change their trips. In the project scenario the flows are redistributed as in Table 9.

Table 9. Characteristics of road sections and traffic volumes of the study area. Project scenario.

Road section	L (km)	Rush hour v (km/h)	f_{light} (veh/h)	f_{heavy} (veh/h)	Soft hour v (km/h)	f_{light} (veh/h)	f_{heavy} (veh/h)	Night hour v (km/h)	f_{light} (veh/h)	f_{heavy} (veh/h)
Urb. 1	1.9	15	823	25	35	497	15	50	124	4
Urb. 2	1.8	15	815	25	35	419	13	50	105	3
Urb. 3	1.6	0	0	0	0	0	0	0	0	0
Urb. 4	1.6	0	0	0	0	0	0	0	0	0
Tang. Sud	2.6	30	2,866	506	70	1,044	184	70	288	51
Tang. Nord	2.6	30	2,550	450	70	1,258	222	70	340	60
Urb. 5	1.8	20	1,656	144	60	662	58	80	166	14

The carried out emissions in the project scenario are shown in Table 10; in Table 11 a comparison is made between scenario 0 and 1. The closure of the two main streets allows a green space with benefits for people in centre city but, at the same time, causes an increase in global emissions due to a traffic redistribution. In scenario 1, a part of users chose to use more sustainable means; but other chose a longer alternative route and are forced in congestion and giving more emissions.

Table 10. Estimated emissions (in grams) in the study area in the project scenario

Road section	Rush hour E_{CO}	E_{NOx}	E_{PM}	Soft hour E_{CO}	E_{NOx}	E_{PM}	Night hour E_{CO}	E_{NOx}	E_{PM}
Urb. 1	638.81	283.90	7.00	220.77	90.22	2.71	90.25	18.05	0.93
Urb. 2	599.98	268.25	6.59	176.87	73.45	2.19	71.94	13.34	0.72
Urb. 3	0.00	0.00	0.00	0.00	0.00	0.00	0.00	0.00	0.00
Urb. 4	0.00	0.00	0.00	0.00	0.00	0.00	0.00	0.00	0.00
Tang. Sud	2,984.43	3,500.62	62.08	1,023.49	647.20	15.19	243.97	179.32	4.79
Tang. Nord	2,654.85	3,113.30	55.21	1,233.68	780.78	18.33	287.73	211.02	5.64
Urb. 5	1,250.20	1,016.77	18.66	320.67	172.55	4.39	104.60	33.46	1.35
Total	8,128.26	8,182.84	149.54	2,975.49	1,764.21	42.80	798.50	455.19	13.42

Table 11. Scenarios comparison

	Scenario 0 (g)			Scenario 1 (g)			Scen 1 − Scen 0 (Δ%)		
Total emission	Rush hour	Soft hour	Night hour	Rush hour	Soft hour	Night hour	Rush hour	Soft hour	Night hour
CO	6,852.0	2,798.0	654.7	8,128.3	2,975.5	798.5	18.6	6.3	22.0
NOx	5,583.9	1,696.7	390.1	8,182.8	1,764.2	455.2	46.5	4.0	16.7
PM	118.8	40.8	11.0	149.5	42.8	13.4	25.9	4.9	22.5

4 Conclusions

The paper has dealt with the issue of emissions and, specifically, a framework of literature models has been presented that allow the estimation of gas emissions from vehicular traffic for the pollutants CO, NOx and PM. The proposed models have a general character, they have been obtained from the envelope of literature models and are applicable in different traffic conditions.

The models have been developed for both light and heavy vehicles. An application has been presented on a real road network, located in the Reggio Calabria city, in three different traffic conditions (rush, soft, night hours). The application has been referred to two network conditions: the current scenario (scenario 0) and a project scenario (scenario 1). In the last one, a reorganization of the transport supply has been envisaged, blocking access to two central main road sections.

The case study analysis suggests that the effect of a change in the transport supply has negative impacts on emissions as the cutting of flows on two important roads involves an overload on other peripheral streets, which are forced to conditions of high congestion at peak times.

The impacts on emissions can be compared with the effects on consumption addressed in the study proposed by *Gattuso et al.* [10]. The urban regeneration policy translates into an increase in emissions but at the same time produces savings in terms of global consumption; in fact, there was an overall reduction in consumption of about 4% in the rush hours and of about 3% in the soft hours.

In-depth studies are underway for the modelling of emissions in the case of vehicular traffic with the presence of green vehicles (electric, hybrid, etc.).

References

1. Gattuso, D., Cassone, G.C., Pellicanò, D.S.: Modelli per la stima delle emissioni di inquinanti da traffico. Strade&Autostrade, n° 151 (2022)
2. Smit, R.: Development and performance of a new vehicle emissions and fuel consumption software (PΔP) with a high resolution in time and space. Atmos. Pollut. Res. **4**, 336–345 (2013)
3. Smit, R., Ntziachristos, L., Boulter, P.: Validation of road vehicle and traffic emission models – a review and meta–analysis. Atmos. Environ. **44**, 2943–2953 (2010)

4. Wang, X., Meng Li, M., Peng, B.: A study on vehicle emission factor correction based on fuel consumption measurement. In: IOP Conference Series: Earth and Environmental Science, vol. 108, p. 042049 (2018)
5. Sharma, S., Mathew, T.V.: Developing speed dependent emission factors using on-board emission measuring equipment in India. Int. J. Traffic Transp. Eng. **6**(3), 265–279 (2016)
6. Song, Y.Y., Yao, E.J., Zuo, T., Lang, Z.F.: Emissions and fuel consumption modeling for evaluating environmental effectiveness of ITS strategies. Discret. Dyn. Nat. Soc. **2013**, 581945 (2013)
7. Corvalán, R.M., Osses, M., Urrutia, C.M.: Hot emission model for mobile sources: application to the metropolitan region of the city of Santiago Chile. J. Air Waste Manag. Assoc. **52**(2), 167–174 (2011)
8. Joumard, R., Andre, J.M., Rapone, M., Zallinger, M., Kljun, N., et al.: Emission factor modelling and database for light vehicles - Artemis deliverable 3. https://hal.archives-ouvertes.fr/hal-00916945 (2007)
9. EMEP/EEA: Appendix 4 to chapter '1.A.3.b.i-iv Road transport', of Air pollutant emission inventory guidebook 2019. https://www.eea.europa.eu/emep-eea-guidebook (2020)
10. Gattuso, D., Cassone, G.C., Malara, M., Pellicanò, D.S.: Energy costs related to traffic on urban road networks - future perspectives of sustainable mobility. In: XI International Conference on Transport Sciences (2021). ISBN 978-615-5837-86-9

Structural Equation Modelling for Detecting Latent "Green" Attributes in Real Estate Pricing Processes

Elena Fregonara⬥ and Alice Barreca(✉)⬥

Architecture and Design Department, Politecnico di Torino, Viale Mattioli 39, 10125 Turin, Italy
alice.barreca@polito.it

Abstract. The presence of latent factors in the residential real estate market emerges by measuring and evaluating the effects of energy performance on price changes. The effects of green attributes on the price formation process are particularly interesting to explore, being closely linked to the energy performance of buildings. This last is regulated by increasingly stringent energy and environmental policies both internationally and nationally, translated into a thorough regulatory framework which ranges from the Directive 2002/91/EC - Energy Performance of Buildings Directive (EPBD) up to the recent Communication EU - A Renovation Wave for Europe 2020. Thus, this work aims to propose Structural Equation Models (SEMs) as a suitable approach to the detection and estimation of the effects of "green" attributes on residential property prices. This work, therefore, presents a survey of SEMs methodology, highlighting its theoretical and operational principles and suggesting future insights and applications.

Keywords: Structural equation modelling · Multiple regression analysis · Real estate market · Pricing processes · Green qualities · Latent variables · Renovation wave for Europe

1 Introduction

As emerges from recent studies (see Sect. 2), special attention is addressed towards physical-technical and qualitative characteristics of dwellings offered on the market, and more specifically towards latent variables able to influence consumers' behaviour in choosing and orienting their real estate investments. Precisely, the presence of latent factors in the residential real estate market emerges while measuring and assessing the effects of energy performance on price. The effects of green attributes on process pricing are particularly interesting to be explored, being strictly related to energy and environmental policies and related norms and regulations, these last considered both at an international and national level [1]. Coherently with these premises, this piece of work aims to propose Structural Equations Modelling (SEM) as a suitable approach for the detection and estimation of the effects of "green" attributes on residential real estate prices. Overall, the paper aims to enrich the body of literature that investigates the relationships between "green" attributes and residential units' prices suggesting that this

© The Author(s), under exclusive license to Springer Nature Switzerland AG 2022
F. Calabrò et al. (Eds.): NMP 2022, LNNS 482, pp. 1610–1620, 2022.
https://doi.org/10.1007/978-3-031-06825-6_155

issue could also be investigated by the means of SEM, i.e., a multivariate statistical approach that assumes causality links between variables, capitalizes on the concept of latent
variables and incorporates techniques such as factor analysis, path analysis and simultaneous regressions. More precisely, SEMs are statistical approaches that are usually
employed as an alternative to the Hedonic approach traditionally resolved through the
Multiple Regression Analysis (MRA) with Ordinary Least Squares (OLS) estimators.
In fact, SEMs are considered to be more effective than standard Multiple Regression
Models (MRMs) for their ability to detect and measure explicative variables, define the
functional form and control variables' multicollinearity. Contrarily to MRMs, they can
treat qualitative and latent variables, and, assuming the possible dependence between
explicative variables, they are able to define and measure their relationships. Latent
variables can be measured through multiple indicators or more observed variables for
each latent variable. Additionally, the analysis of the measurement errors represents a
particularly interesting aspect of the approach overcoming the stochastic term of the MR
equation: in fact, the weight of the error is quantified concerning the relations among the
model variables. Developed in Econometrics, SEM approaches have also been applied
in social sciences and economics, but their potentialities were underlined also for what
concerns the real estate estimative context [2–6]. Regarding this point it must be mentioned that the paper also aims to enrich this particular field of literature: in fact, SEMs
have usually not included "green" attributes and future experimentation that follows this
direction may thus represent an advancement. Overall, the work presented here can be
considered as the second part of ongoing research. In a first contribution [7], the literature recognition is illustrated, considering the methods and approaches recently used for
exploring how the buildings' energy efficiency may affect real estate: results highlight
the increasing importance of concepts such as latent variables and green attributes in
the real estate pricing process, and, methodologically, SEM is proposed as a suitable
and particularly promising research approach. In this second contribution, a literature
background insight and a methodological recognition are presented, highlighting the
theoretical and operative principles of the approach. The paper is articulated in the following parts: Sect. 2 presents the literature background. Section 3 illustrates the proposed
methodology (i.e. SEM); finally, Sect. 4 concludes the work and outlines future steps of
research.

2   A Literature Background

The permanent existence of a certain slowness in the execution of energy efficiency
interventions on buildings has contributed to increasing the urgency of the environmental question concerning the built heritage. The recent socio-economic crisis due to the
Covid-19 pandemic has led to the evaluation of energy redevelopment and green buildings not only as necessary actions for decarbonisation, but also as an economic lever
for the restart of the real estate market. In fact, the European Commission has recently
strengthened its commitment to achieving a real "green" transition with the Communication of the European Commission of 10 October 2020 entitled "A wave of renewal
for Europe: making our buildings greener, create jobs, improve lives". In this context,
the analysis of the real estate market price formation processes, user preferences for

purchasing and the variables underlying the changes in the real estate market are very useful for guiding redistributive policies. Implementing measures that consider multiple variables simultaneously can probably be an effective tool to understand which are latent variables that are underpinning the market changes. At the EU level, the European New Green Deal which aims to make Europe the first continent with zero climate impact represents the reference framework for green policies and will be financed by one-third of the 1.8 trillion euros of the Next Generation EU Recovery Plan and the EU seven-year budget. As regards national regulations, Italy has established, as part of the initiatives and long-term action plans of the PNRR, a series of emergency tax incentives aimed at favouring the retrofit and adaptation (including seismic) interventions of buildings, with a "Super Eco bonus" (Law n. 77, July 2020), dedicated to the energy efficiency of buildings. Contextually, with the evolution of the regulatory framework and the methods for their application, a growing literature has been produced concerning the analysis of the green attributes of buildings. From the perspective of the real estate economy, many studies aim to detect the impact of EPC labels and green features on pricing processes and market dynamics. Operationally, along the same lines as the widespread literature produced, a large number of these studies use the spatial regression models [8–12]; in other studies, spatial analyses are proposed to detect the presence of spatial dependence between different kinds of indicators and to manage the spatial latent variables in the property price determination process [13, 14]. In some recent studies, the spatial specification is tackled in relation with hedonic models [9, 15] or only by hedonic price model [16–21] in other studies the Geographically Weighted Regression [22, 23] or Quantile Regression [24, 25] were applied to understand the influence of the EPC labels on the price formation process and, to our knowledge, in only one paper employing SEMs [26]. As regards the analysis of the real estate market with particular reference to the analysis of latent variables other methodologies are applied in literature, but the one we want to focus on in this paper are the SEMs, for its potentialities, although it still does not very studied in relation to the market [2–6, 27] and the green attributes. In particular, Mendelsohn [27] demonstrates that the data of a single market with non-linear prices are consistent with a large set of underlying structural equations. By limiting the allowed functional form of structural equations, marginal price non-linearity can be used to identify the price and displacement parameters of a single member from the set. However, since the true form of supply and demand functions is often unknown, the identification approach must be used with great caution, as the necessary restrictions may be unwarranted. The work presented by Bravi and Fregonara [2] aims to identify and test out the structural equation model as an alternative to the application of multiple regression and in some respects include and extend the applicability of traditional (OLS) methods and bias estimators in the real estate market. Results of the application of SEM on a sample including different intrinsic and extrinsic building characteristics found some correlations and covariances between couples of variables and confirms that SEM has a better fit and prediction power than OLS models. The paper from Manganelli [3], considering some elements of the classical literature published in those years in Italy, he proposes as a step forward for the study of the real estate market: the use of structural equation models for mass appraisal. In his study, the author states that this methodology can overcome the problems of collinearity and the identification of the "latent" variables explaining the price. The results define that thanks to SEMs it is possible to obtain better and more reliable estimation due to their ability to represent and interpret the market price formation process. The paper of Manganelli and Morano [5] discusses a comparison between the performances of two different techniques: a system of structural equations and a neural network, applied on a concrete case of real estate valuation. The sample for the experimentation of the models consisted of n. 91 residences located in a homogeneous area of the city of Naples, Italy. The homogeneous area defines an urban boundary in which the mechanisms that regulate the price formation process are uniform. The observations of the sample include the distinctive characteristics of properties in the real estate market. The applied SEM removed errors due to multicollinearity between variables. The model was found to be suitable for the interpretation of market mechanisms because it provides consistent indications regarding the influence exerted by each variable on the price formation process. Hence, the unit price appears as a synthesis of contributions provided by selected features. In the work by Liu and Wu [6] they focused on the need for many developers to see which features are truly valuable to consumers when choosing a home. Hence, they focused on how to assess the value of the residential product regardless of its price. The paper then classified the characteristics of the residential product into four parts: location, neighbourhood, house structure and neighbourhood environment, and measured the value of the home with two variables: price and buyer satisfaction. The author performs comprehensive qualitative and quantitative analyses, with the use of the structural equation model and AMOS, 100 samples are selected to perform a case study. From the point of view of consumers and based on the utility and preferences of the consumer, the latter then measured the value of the house by judging the benefits that consumers obtain, together with the degree of consumer satisfaction. Also, Freeman and Zhao [2] agree that although hedonic regression remains the most popular technique for estimating property values, the SEM is increasingly seen as a realistic analytical alternative. This article presents a SEM analysis of a historical data set for a large Canadian real estate agent. An iterative approach is adopted for modelling. The first phase focused on the internal relationships between the intrinsic characteristics of the houses and the second on the living values and their determinants. In the final phase, the authors focused on advertised list prices and location details. A comprehensive evaluation of the resulting holistic model revealed a wealth of significant structural relationships, particularly between the home's style, structure, and attributes. Before SEM modelling, exploratory factorial analysis (EFA) was conducted to help determine the underlying structure of the data and gain insight into possible latent constructs. In the work of Nasrabadi and Hataminejad [30], the SEM was applied together with the Delphi Method to propose a sustainable building model (SH) and mitigation measures for housing in the city of the Mashhad region in Iran. The authors applied a descriptive-analytical method with a statistical sample of 384 people who answered a questionnaire. The responses were then analysed using Structural Equation Modelling (SEM) to illustrate the research SH model in the AMOS software. The results indicate that the adequacy of the housing model to meet needs together with the quality of health in housing has the greatest influence and, conversely, the number of housing units with the greatest amount of domestic green space has the least influence on sustainability of accommodation in the region. The original aspect of the study is to analyse the SH model of SEM, which has not been studied in nearly all other recent research. Bollen K.A. [28], defines the models of structural equations in relation to the models of multiple regression or ANOVA (Analysis of Variance) explaining their different nature based on the covariance differences. Instead of minimizing the difference between observed and predicted values, the covariance of the observed sample and the one predicted by the model is minimized. He puts as a basic hypothesis of this model that the covariance matrix of the observed variables is a function of a set of parameters. Analyses with structural equations allow you to make fewer assumptions in the premise and to "let the data give the answer" without reducing them to the premise. Goldberger A.S. [29], defines SEMs as stochastic models in which each equation represents a causal link, rather than an empirical association between data. The models arise in non-experimental situations and are characterized by simultaneity and/or errors in the variables. There may be errors in the variables because the measurable quantities do not correspond to the relevant theoretical quantities. Structural parameters do not coincide with regression coefficients between observable variables, but the model imposes constraints on these regression coefficients. As a result, he tackles subtle identification issues and draws on elaborate methods of statistical inference. From the point of view on application tools Hayduc L.A. [30], presented in his work SEM as a theory-driven analytical approach for the a priori evaluation of a research hypothesis relating to causal relationships between observed and latent variables. The hypothesis to be analysed can be expressed in many different forms, but the most common is to measure it through path analysis models, confirmatory factor models and latent variable path analysis models. The most common applications of SEM do not have descriptive or exploratory purposes but allow one or more theories on causality to be combined with more classes and correlation analysis. In the work of Blunch [31], the aims of the SEM are defined as two main ones. The first is due to the need to reduce the dimensionality of the data; when, in fact, the information contained in the interrelationships between many variables can be traced back to a smaller set, it becomes easier to identify a structure underlying the data. This is the idea that also characterizes the analysis. The second reason lies in the fact that the concept of the latent variable is found in many application fields, especially in the social sciences. Sarstedt et al. and Wong [32–34] instead, focus on defining the path model that they consider to be composed of two elements: the structural model represents the structural paths between the constructs, whereas the measurement models represent the relationships between each construct and its associated indicators. They present a quite different methodology: the Partial least squares structural equation modeling (PLS-SEM) where structural and measurement models are also referred to as inner and outer models.

3 Structural Equation Modelling: Methodological Aspects

As anticipated in Sect. 1, SEM builds on factor analysis, regression analysis and path analysis [35–38]. Specifically, the factor analysis is implemented in presence of direct and indirect effects, and with the inclusion of latent variables and multiple indicators. Operatively, SEMs consist of multi-equational stochastic models (or regression equations systems). In these systems, each regression equation is named "structural equation" and, in causal terms, it represents a causal link instead of an association [29]. The relations between dependent and independent variables are expressed in terms of deviation from the respective means; thus, the constant term disappears. It is important to underline that multi-equations systems are able to measure the error terms related to dependent and independent variables. More precisely, the method is based on a series of preliminary concepts/operative steps listed below:

1. The causation concept, which is different from the covariance one. In fact, we are in presence of covariance/correlation/association when two variables show concomitant variations, whilst we are in presence of causation when the variation in a variable can produce a variation in a second variable. The presence of causation is frequent in socio-economic analyses, where the presence of a stochastic term is due to the non-experimental nature of the analyses themselves. Thus, the verification of the causation is a preliminary fundamental step for the approach, developed through factor analysis.
2. Graphic modelling, which founds on path analysis. In Fig. 1a a graphic schematization of a multiple regression equation is illustrated, whilst in Fig. 1b the schematization of a structural equation according to the path analysis is reported. In Fig. 1b the regression coefficients—named structural parameters- are indicated on the arrows of the causal links among the variables.

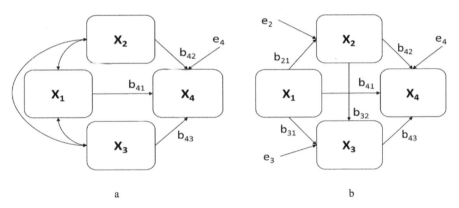

a   b

Fig. 1. Regression model graphic modelling (1a), and structural equations model graphic modelling with path diagram (1b). (Source: authors' elaboration from A. Lubisco, Analisi statistica multivariata, Università di Bologna, 2005–2006)

3. Formalization, which founds on the correspondence between the graphic representation and the following Equations system (1):

$$X_2 = \beta_{21}X_1 + \varepsilon_2$$
$$X_3 = \beta_{31}X_1 + \beta_{32}X_2 + \varepsilon_3 \tag{1}$$
$$X_4 = \beta_{41}X_1 + \beta_{42}X_2 + \beta_{43}X_3 + \varepsilon_4$$

where X_2 is caused by X_1; X_3 is caused by X_1 and X_2; X_4 is caused by all. According to the Econometric terminology, the variables X_2, X_3 and X_4 are "endogenous" variables, caused by other variables considered in the system; X_1 is an exogenous variable instead, and it can be only independent, being external to the model. This "independence" indicates that the variable is not caused by other variables considered in the model. Furthermore, SEM distinguishes among direct/indirect effects, pure associations and total effects (the sum of the direct and indirect effects). Notice that MRA considers only the direct effects through the regression coefficients. Then, notice that the effect of a variable over another one is defined as "direct" when this is not mediated by any other one; on the contrary, it is "indirect" when in presence of a mediating variable. In the equations above, X_2 and X_3 are mediating variables, i.e., they are both the result of a predictor and a predictor themselves; following the path diagram graphic representation, mediating variables present at least one arrow pointing to them and one arrow exit from them. Finally, notice that the error is divided into diverse components, furtherly analysed.

4. Latent variables detection. This step is fundamental: latent variables are theoretical constructs that cannot be directly observed and measured, and they are defined by their relations with observable variables (indicators) and/or other theoretical constructs. Observable variables present error terms, whereas not observable variables do not present measuring error terms.

On the basis of these preliminary steps, the method develops on four phases: model definition, measurement model definition, model parameters estimation, model testing and residuals analysis. These phases are summarized as follows:

1. Definition of the model. It consists in defining the model, by means of the observed/latent variables detection, the causal links between endogenous/exogenous variables definition, and a number of parameters of unknown/zero value (zero when in presence of non-attendance of a causal link). The model is formalized according to the following equation, known as the first basics Eq. (2):

$$\eta = B\eta + \Gamma\xi + \zeta \tag{2}$$

where η, ξ and ζ are in compact form the three vectors which stand for, respectively, the endogenous variables (η), the exogenous variable (ξ), the error terms (ζ); B stands for the square matrix of structural coefficients of the link between exogenous variables with diagonal equal to 0, Γ stands for the matrix of the structural coefficients of the link between endogenous variables. These matrices are furtherly specified by their covariance matrices: Φ between exogenous and Ψ between the endogenous ones.

2. Definition of the measurement model. It consists of the link between latent variables and measured variables (these are assumed perfectly correlated). Then, a scale of measurement is assigned to the latent variables to read the estimated coefficient (parametrization). Formally, the definition of the measurement model can be represented by the second and third basics Equations of the model, as in (3):

$$X = \lambda x\xi + \delta$$

$$Y = \lambda y\eta + \varepsilon \tag{3}$$

where X represents the causal variables indicators of the exogenous latent variables ξ, Y represents the causal variables indicators of the endogenous latent variables η, δ and ε represent the error terms, and λ coefficients represent the systematic differences in the measurement scales. Finally, the matrices θ_δ and θ_ε represent the covariance matrices of the measurement errors. In these matrices, the main diagonal contains the variances errors associated with the indicators: thus, it is possible to measure the relations between latent variables and to fix the error limits. To clarify the steps of the methodology, the graphic synthesis of the model is presented in Fig. 2.

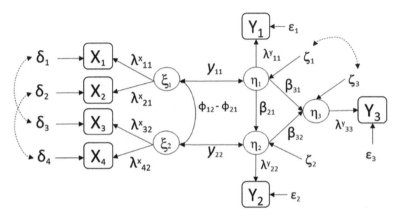

Fig. 2. A graphic synthesis of the structural equations model (Source: authors' elaboration from A. Lubisco, Analisi statistica multivariata, Università di Bologna, 2005–2006)

Notice that the model definition must observe specific limitations: 1) the variables are measured in terms of deviation from the mean; 2) independent variables and errors are not correlated; 3) stochastic terms of the different equations are not correlated; 4) the matrices related to the structural equations are not singular and invertible; 5) other relations between the variables, not included in the model, are not possible.

3. Model parameters estimation. This is produced through the Maximum Likelihood Method (MLE) approach (in place of the OLS approach), which, as known, founds on minimizing the distances between theoretical and observed covariance matrices (based on the theoretical and observed data respectively), and on adopting a variance-covariance matrix instead of the single observations. Thus, it is necessary to shift from the equation parameters estimation to the system parameters estimation. Formally, these last consist of the links between the theoretical model and the covariance matrix of the observed variables (sample covariance matrix). The model estimation can be formalized through the following Eq. (4):

$$\Sigma = \Sigma(\theta) \qquad (4)$$

where Σ stands for the covariance matrix for Y and X, and $\Sigma(\theta)$ represents the covariance matrix as a function of the free parameters of the model in θ. In other words, this last includes the covariances of Y, of X and of X and Y [28]. Notice that according to this formula each element of the covariance matrix has to be a function of at least one of the parameters of the model.

4. Model testing and residuals analysis. As a final step, the reliability of the model and of single parameters is tested and the residuals analysis is produced, by means of statistical methods (Chi-square based tests). According to the results of this step of the analysis, the model can be improved.

4   Conclusions and Future Steps of the Research

The present article has outlined that the evolution of research questions has led scholars to apply different methods, ranging from the Hedonic Price Model to different regression approaches. Given that effects on real estate prices might be exerted not only by energy performance but also by other characteristics, this study assumes the importance to explore the application of SEM. Particularly, in this step of the research, after an in-depth study of the literature, a recognition of the method was proposed. Concerning future research, the perspective is the application of SEMs on a case study at the building scale. Assuming an "Energetic Quality" latent variable (encompassing comfort, as well as savings and environmental benefits) employing SEM this variable will be investigated through measurable indicators, to detect the effects produced by identified variables on the real estate pricing process.

References

1. Del Giudice, D.E., Massimo, F., Salvo, P., De Paola, M., Ruggiero, M.D., Musolino, M.: Market price premium for green buildings: a review of empirical evidence. Case study. In: Smart Innovation, Systems and Technologies (2021)
2. Freeman, J., Zhao, X.: An SEM approach to modeling housing values. In: Data Analysis and Applications 1: Clustering and Regression, Modeling-estimating, Forecasting and Data Mining (2019)
3. Bravi, M., Fregonara, E.: Structural equations models in real estate appraisal. In: International Real Estate Conference - American Real Estate and Urban Economics Association AREUEA, 23–25 May 1996 (1996)
4. Manganelli, B.: Un sistema di equazioni strutturali per la stima di masse di immobili. Genio Rurale, vol. 2, Bologna (2001)
5. Manganelli, B., Morano, N.: Comparative performance of structural equations system and neural networks for real estate appraisal. In: New Logics for the New Economy, Atti del VII Sigef Congress 2001 (2001)
6. Liu, Y., Wu, Y.X.: Analysis of residential product's value based on structural equation model and hedonic price theory. In: 2009 International Conference on Management Science and Engineering - 16th Annual Conference Proceedings, ICMSE 2009 (2009)
7. Fregonara, E., Rubino, I.: Buildings' energy performance, green attributes and real estate prices: methodological perspectives from the European literature. Aestimum 79, 1–18 (2021)
8. Bottero, M., Bravi, M., Mondini, G., Talarico, A.: Buildings energy performance and real estate market value: an application of the spatial auto regressive (SAR) model. Green Energy and Technology (2017)
9. Dell'Anna, F., Bravi, M., Marmolejo-Duarte, C., Bottero, M.C., Chen, A.: EPC green premium in two different European climate zones: a comparative study between Barcelona and Turin. Sustainability (Switzerland) 11, 5605 (2019)

10. Morano, P., Tajani, F., Locurcio, M.: Multicriteria analysis and genetic algorithms for mass appraisals in the Italian property market. Int. J. Hous. Mark. Anal. **11**, 229–262 (2018)
11. Barreca, A., Fregonara, E., Rolando, D.: Epc labels and building features: spatial implications over housing prices. Sustainability (Switzerland) **13**, 2838 (2021)
12. De Paola, P., Del Giudice, V., Massimo, D.E., Del Giudice, F.P., Musolino, M., Malerba, A.: Green building market premium: detection through spatial analysis of real estate values. A case study. In: Smart Innovation, Systems and Technologies (2021)
13. Barreca, A., Curto, R., Rolando, D.: Urban vibrancy: an emerging factor that spatially influences the real estate market. Sustainability (Switzerland) **12**, 346 (2020)
14. Barreca, A., Curto, R., Rolando, D.: Housing vulnerability and property prices: spatial analyses in the Turin real estate market. Sustainability (Switzerland) **10**(9), 3068 (2018)
15. Bisello, A., Antoniucci, V., Marella, G.: Measuring the price premium of energy efficiency: a two-step analysis in the Italian housing market. Energy Build. **208**, 109670 (2020)
16. Cespedes-Lopez, M.F., Mora-Garcia, R.T., Perez-Sanchez, V.R., Perez-Sanchez, J.C.: Meta-analysis of price premiums in housing with energy performance certificates (EPC). Sustainability (Switzerland) **11**, 6303 (2019)
17. Fuerst, F., McAllister, P., Nanda, A., Wyatt, P.: Energy performance ratings and house prices in Wales: an empirical study. Energy Policy **92**, 20–33 (2016)
18. Jensen, O.M., Hansen, A.R., Kragh, J.: Market response to the public display of energy performance rating at property sales. Energy Policy **93**, 229–235 (2016)
19. Olaussen, J.O., Oust, A., Solstad, J.T.: Energy performance certificates – informing the informed or the indifferent? Energy Policy **111**, 246–254 (2017)
20. Stanley, S., Lyons, R.C., Lyons, S.: The price effect of building energy ratings in the Dublin residential market. Energ. Effi. **9**(4), 875–885 (2015). https://doi.org/10.1007/s12053-015-9396-5
21. Fregonara, E., Rolando, D., Semeraro, P.: Energy performance certificates in the Turin real estate market. J. Eur. Real Estate Res. **10**, 149–169 (2017)
22. Marmolejo-Duarte, C., Chen, A., Bravi, M.: Spatial implications of EPC rankings over residential prices. In: Green Energy and Technology (2020)
23. McCord, M., Lo, D., Davis, P.T., Hemphill, L., McCord, J., Haran, M.: A spatial analysis of EPCs in The Belfast Metropolitan Area housing market. J. Prop. Res. **37**, 25–61 (2020)
24. Wilhelmsson, M.: Energy performance certificates and its capitalization in housing values in Sweden. Sustainability (Switzerland) **11**, 6101 (2019)
25. McCord, M., Haran, M., Davis, P., McCord, J.: Energy performance certificates and house prices: a quantile regression approach. J. Eur. Real Estate Res. **13**, 409–434 (2020)
26. Nasrabadi, M.T., Hataminejad, H.: Assessing sustainable housing indicators: a structural equation modeling analysis. Smart Sustain. Built Environ. **8**, 457–472 (2019)
27. Mendelsohn, R.: Identifying structural equations with single market data. Rev. Econ. Stat. **67**, 525–529 (1985)
28. Bollen, K.A.: Structural equation modeling with latent variables. Structural Equation Modeling (1989)
29. Goldberger, A.S.: Structural equation methods in the social sciences. Econometrica **40**, 979–1001 (1972)
30. Hayduc, L.A.: Structural Equation Modelling with LISREL. J. Hopkins University, Baltimore (1987)
31. Blunch, N.: Introduction to Structural Equation Modelling Using SPSS and AMOS (2012)
32. Sarstedt, M., Ringle, C.M., Hair, J.F.: Partial least squares structural equation modeling. Handbook of Market Research (2017)
33. Ringle, C.M., Sarstedt, M., Mitchell, R., Gudergan, S.P.: Partial least squares structural equation modeling in HRM research. Int. J. Hum. Resour. Manag. **31**, 1617–1643 (2020)

34. Wong, K.K.-K.: Partial least squares structural equation modeling (PLS-SEM) Techniques using SmartPLS. Mark. Bull. **24**, 1–32 (2013)
35. Wang, J., Wang, X.: Structural equation modeling: applications using mplus (2019)
36. Ullman, J.B., Bentler, P.M.: Structural equation modeling. In: Handbook of Psychology, 2nd edn., pp. 661–690. Wiley (2013)
37. Clogg, C.C., Bollen, K.A.: Structural equations with latent variables. Contemp. Sociol. **45**, 289–308 (1991)
38. Corbetta, P.: Metodi di analisi multivariate per le scienze sociali. Il Mulino, Bologna (1994)

Designing Forms of Regeneration. Spatial Implication of Strategies to Face Climate Change at Neighborhood Scale

Kevin Santus[1]([⊠]) [iD], Emilia Corradi[1] [iD], Monica Lavagna[2] [iD], and Ilaria Valente[1] [iD]

[1] DASTU Department, Politecnico di Milano, 20133 Milan, Italy
kevin.santus@polimi.it
[2] ABC Department, Politecnico di Milano, 20133 Milan, Italy

Abstract. Climate change and the built environment have a double relation, in which the one influence the other. However, if on the one hand the built environment's impact on climate-altering emissions, is already evident, on the other it still lacks a proper understanding of the design morphological modification undergoing due to the resilience and decarbonization objectives we are facing. Indeed, the climate crisis requires the rethinking of the design action to tackle the increasing environmental fragility. This new condition should address the adaptation of the urban environment and, at the same time, the mitigation of carbon emission, to curb climate-change effects.

Thus, the contribution aims to present Nature-Based Solution and Circular Economy as a unitary design approach to display adaptation and mitigation actions to structure a long-term resilience and decarbonization. In this perspective, these two approaches are considered critical drivers for an ecological transition, already fostered by several international design competitions and actions.

Starting from a theoretical approach, the argumentation will present the impacts of Nature-Based Solutions and Circular Economy within urban regeneration projects. It will highlight the centrality of the neighborhood scale in implementing adaptation and mitigation toward a method that could enhance a renewed preparedness to climate change events and limit the carbon emission of the project. Finally, the article reflects the morpho typological sphere of the project, where the application of Nature-Based Solution and Circular Economy implies a design discourse bond to the spatial new character of the project.

Keywords: Urban regeneration · Resilient design · Zero-carbon project

1 Introduction

Several contributions highlighted the uncertainty as the main feature of our contemporaneity, also in changing the morpho-typological character of the architectural practice [1]. In this regard, the climate and the environmental crisis seems to have a crucial role [2, 3], sharpening spatial fragilities but also implying a series of modification related to temporal and formal connotations [4]. The changing condition of the environment is

© The Author(s), under exclusive license to Springer Nature Switzerland AG 2022
F. Calabrò et al. (Eds.): NMP 2022, LNNS 482, pp. 1621–1630, 2022.
https://doi.org/10.1007/978-3-031-06825-6_156

increasing the number of hazards, and climate change is "acting as a catalyst for this set of events" [5], thus making urgent a radical reformulation of the design vision. The visible consequences of climate change are pointing out new risks for the built environment [6], that is increasingly exposed to climate-related disasters and hazards (Fig. 1), and at the same time has a prominent role regarding the impacts of climate-altering emissions derived from the whole building life cycle [7]. Therefore, an ecological transition should address strategies to foster a resilient and zero-carbon project, considering the depletion of resources, so implementing logics of circularity to tackle on a broader level the human impacts on the environment. Because of these reasons, architecture must implement adaptive and mitigating strategies, considering climate threads, enabling the project to face issues such as urban floods, rainstorms, heat island, etc. At the same time, methods to reduce the carbon emissions, operating a keen use of the resources, should define a priority in rethinking the design process.

Starting from that, the contribution probes this design condition, considering Nature-Based Solution (NBS) and Circular Economic (CE) as approaches that could help define a complimentary design method to adapt the urban space and mitigate carbon emissions toward a post-carbon city. Hence, the aim of the contribution is a critical reflection about this complementarity, resonating on the scale of intervention and the impact that could have on the morpho-typological configuration of the project.

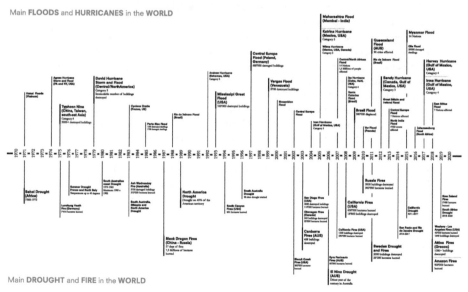

Fig. 1. Main climate-disasters in the last 50 years, data collected through the International Disaster Database. Drawing by K. Santus.

2   Research Methodology

This contribution focuses on regeneration processes that operate within the condition of obsolescence and decay, present in many peripheral urban areas in Europe. The spatial fragility is considered an opportunity for a development that considers the necessity of decarbonization of the built environment reconsidering the role of nature and usage of resources in the current practice. In this term, urban regeneration is framed as a design application to preserve virgin grounds from new urban expansion and enhance the already existing built environment through its transformation. To survey this statement, the neighborhood scale is selected as a spatial dimension the research will focus on. Specifically, the research identifies a synergy of design actions that interact with the ongoing environmental crisis exacerbated by climate change.

The article starts with an essential cultural background (partially stated in the introduction) that critically refers to a literature review regarding the ongoing climate crisis, setting the context in which the design practice is. The literature review has been conducted considering nearly 100 articles, searched both through online peer-reviewed articles and through web of science like SCOPUS and Google Scholar. At the same time, virtual databases as JSTOR and books deducted from the scientific articles have been examined. The research considered the following keywords: nature-based solutions, circular economy, climate change, resilient design, urban regeneration, climate adaptation, and climate mitigation. Furthermore, some international experiences and competition, such as Reinventing Cities, will be portrayed to examine scales and strategies for a resilient and zero-carbon regeneration.

Thus, the article portrays the NBS and CE as tools that could restore neglected urban areas, helping to reduce carbon emissions and define interventions toward an ecological horizon. The two strategies were selected starting from their international relevance, detected through a critical analysis of some international documents that established the European agenda, and the attention raised due to their impacts on adaptation, mitigation, and acting toward a broad idea of resilience. The two strategies will be critically described in their multiscalar perspective, highlighting the main applications within the design practice. These data have been detected analyzing documents present in DocsRoom of the European Commission website. Once these tools are defined for the project, the article will identify the research gaps related to the scale and the morpho-typological implications.

Regarding the scale, the neighborhood will be selected as crucial to implement the design action to achieve the connection between building and city, also in connection to some contemporary experimentations that are trying to re-set the scale of intervention to structure a resilient and zero-carbon project. The final remark of the contribution will focus on the spatial implication. This survey is conducted through some reasoning regarding which main elements of the project are entangled with the application of NBS and CE, understanding their influence in modifying some spatial/ontological aspects of the project. In these terms, the contribution tries to fill the current lack regarding a proper reflection upon the morphological. Indeed, the final aim is to stimulate a debate on how the post-carbon and resilient city could affect the design practice and the spatial dimension of the project, characterizing the morphological scenario of future cities.

3   Cultural Background

The depletion of resources, new climate risks and other factors are nowadays weighing on our society, and the project of the space should take into consideration these new challenges both from a cultural and technical perspective. Indeed "Current trends – including the implementation of evaluation standards such as LEED, BREEAM and C2C certification [...] – suggest that it is an opportune moment to reconsider and reevaluate what sustainability means to the discipline of architecture" [4].

Considering the way we design in relation to these factors means understanding the perils to the conservation of the environment, but also the risks that climate change brings to the space we inhabit. Besides, cities are particularly exposed to climate risks [8]. In this perspective, urban regeneration appears to have a leading role in contrasting the continuous and limitless usage of resources for new urban expansions, in the mitigation of the climate-altering emissions, but also to implement mitigation and adaptation for the built environment.

The issue of a necessary rethinking of the ecological responsibility and subsequently of design action is not a recent theme. Already in the second half of the XX century, B. Fuller exposed how the presence of systemic crises was directly connected to a lack of design [9], which implies the responsibility for the practice to transform and modify the space in relation to the changing condition of the environment.

It is evident how the concerns about the planet's survival have acquired growing importance, and Climate change has undoubtedly emerged as a crucial issue. Since 1988 the International Panel for Climate Change has been working to define the relation between human action and climate change. The forecasts presented [10] shows how the phenomena linked to climate variability will intensify in the coming decades, increasingly constituting a social and ecological risk.

The outlined urgency clarifies that the climate crisis could not be considered just as an occurrence, so defined as a temporary emergency; rather, it should be framed as a discontinuity from the previous global climate-environmental system. We could argue that climate change could be framed as a macro-catastrophe [11]. This concept allows understanding the event's temporal feature, which implies the necessity to act both with a short and long-term approach.

In this context, architecture plays an important role. Indeed, the built environment counts around the 38% of the annual emissions [12], and in a perspective of future decarbonization, we should consider that approximately 2/3 of the nowadays global building stock will still exist in 2040 [13]. It is evident the importance that regeneration could play in this panorama and the necessity to identify strategies to drive it.

We could foresee two main actions to tackle the environmental and climate crisis: on the one hand, acting to curb the climate-altering emissions and reduce the resources used for the building cycle; on the other hand, adapting the built environment to make the space able to resist to the climate threads. In this framework, the research studies NBS and CE as the main climate strategies affecting the design practice. These are well-known approaches aimed at making urban systems more resilient but still lack a proper reflection regarding the application at an intermediate scale and about a morphological impact on the project.

4   Envisioning New Design Opportunities

4.1   Nature-Based Solution and Circular Economy Synergy

The climate issue is being addressed in many of the design competitions and initiatives of the last few years. Hence, regeneration could be a specific area of interest to rethink fragile built fabrics. Here, an interconnection between adaptation and mitigation actions is required to restore urban spaces, making them adapted to the new climate condition and, at the same time, mitigating the increasing climate change acceleration [14]. Therefore, the design action to decarbonize and adapt the urban environment requires a complementarity of tools to structure a resilient post-carbon built environment. In this direction, looking at the European agenda, we could gaze explicit attention to strategies based on nature and an effort to transform the whole building and urban cycle into a circular one. Dealing with the built environment, two important European documents could be seen in COM/2020/98 [15] and COM/2020/662 [16]. The first one relates the CE as one of the main paradigms to implement in design processes to reduce greenhouse gas emissions and waste. The attention to these key themes is also increased in documents such as "Circular Economy - Principles for Building Design" [17] and other international guidelines such as the "Roadmap to 2050. A Manual for Nations to Decarbonize by Mid-Century" [18]. COM/2020/662, instead, refers to the Renovation Wave for Europe, promoting the greening of buildings, with more attention to the whole life cycle of the built environment. With these, going through national and international documents, such as "The New Green Deal", NBS, and CE seem to cover an essential role for the ecological transition, working as approaches to implement in the design process.

The NBS concept can be defined as solutions that are inspired and supported by nature, providing environmental benefits towards resilience [19]. The scope in applying NBS is not only related to a better living environment, rather a necessity to restore a damaged urban environment and making it adaptive to the new climate hazards.

Circularity is a design approach that characterized several projects along the architectural history, reducing the amount of new material usage and reusing existing components or buildings. It could embodies an important role, being both a mitigation tool to curb the building process emissions and help in tackling the depletion of resources, characterizing the current state of the environment [20]. This concept assumes a perspective concerning the life cycles, in a "rediscovery of the cyclical dimension of historical time related to the need to curb the dissipation of resources and to promote sustainable development" [21]. Accordingly, reflecting on this circular approach requires to examine the extent of the reuse/recycling strategies, understanding how it could be defined as a resilient approach, to regenerate the built environment, decarbonize the building process, and limit resource's usage.

Considering the current panorama that foresees the application of the two strategies, we can frame the city and the building scale as the principal ones in which the two tools are implemented. NBS are often displayed as an act of urban forestation or implementation of green corridors, while at the building scale, they are synthesized with generic abacus of interventions. CE, instead, present a series of approaches from the urban mining scale to the idea of building as material bank (Fig. 2).

As stated, to structure a project that could work both toward an urban resilience and a zero-carbon project, the research suggests an approach where the complementary application of CE and NBS can define a long term-project. Hence, capable of being beneficial for current climate pressure but also reduce the carbon impacts on a longer environmental perspective.

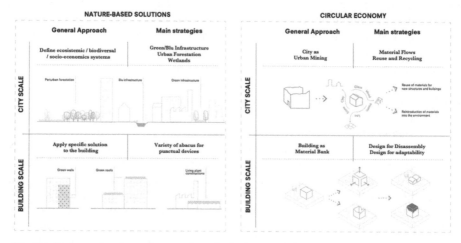

Fig. 2. Scales and actions of nature-based solutions and circularity. Drawing by K. Santus.

4.2   An Intermediate Scale Between the City and the Building

Working toward the post-carbon built environment means understanding a possible scale of intervention that could address a synergy of the tools presented. As expressed in the previous chapter, it is possible to identify how the two main scales in which we can see the application of NBS and CE are either the city scale or the building one. However, this points out an intermediate gap, where the project needs a reconsideration to understand better how to interlock the urban and the plot scales.

This consideration refers to a necessity to refocus on the neighborhood scale as a project working field, looking at it through the architectural lens and not by the planning one, thus operating with the morphology of the project and its features.

Exemplary is the Reinventing Cities initiative, by the C40 network, that have the intermediate scale as a primary action site to develop resilient strategies for the city of the future. With the last edition of competitions, Reinventing Cities proposed 31 areas to regenerate with a resilient and decarbonized intervention. Looking at the "Guidance to Design a Low-Carbon, Sustainable and Resilient Project" [22], it is clear the willingness of the network to work with what is already in the city, renovating it, making the new project environmentally and energetically sustainable, curbing the carbon footprint of the intervention. This idea is also visible in some experimentations by Lacaton&Vassal, where the intermediate scale regeneration deals with the reuse and renovation of the existing building stock, and work with nature for a renewed ground project. From the unbuilt projects "Neighborhood Development" and "Ecological neighborhood La Vecquerie" we can see the willingness to reduce the resources employed, with a high attention to nature as design material.

All these initiatives highlight strategies that simultaneously work for mitigation and adaptation of the built environment, displaying circular logic and NBS. In these terms, the intermediate scale is about the effectiveness of the presented project tools, especially when dealing with adaptation strategies, which requires an intervention beyond the single plot. Consequentially, it becomes central to focus on the intermediate scale since it assumes that semantic unity that identifies specific relational fields, once revealing sensitivity towards the territory's resources, identifying the neighborhood as a point of inflexion between plan and project [23]. Working at the intermediate scale allows setting the projects considering neither the whole urban/land scale nor the single building, rather a relevant intermediate portion of space, capable of working within a minimum urban cycle [24].

5 Morpho-Typological Inquiry

Understanding how the ecological transition is transforming the morpho-typological dimension of the project means having a glimpse of the qualitative solutions that we could gaze in the future resilient post-carbon city. As stated by K. Frampton, "There is no manifest reason why environmentally responsive and sustainable design should not be culturally stimulating and aesthetically expressive" [25]. Indeed, if adaptation and mitigation take the core of the design answer to the climate crisis, it is still vague which could be the morpho-typological effects of the ongoing transition.

Some recent projects that worked with regeneration strategies applying NBS and CE allow to identify some recurring themes that could be assumed as starting points for further analysis and critical reflections. A first issue to survey could be the impact of reuse for regeneration processes. Considering the intermediate scale, reuse could be defined as a form of architectural regeneration able to entangle the building and the city, revealing the relation between the site and the surroundings in its formal features. In this sense, the bearing structure of the building represents the core in reusing, producing a typological permanence. An example could be found in the regeneration of the Milanese Torre Bonnet by PLP architects, where the existing structure of the former tower remained as an unmodified element. This, on the one hand, halved the carbon footprint of the building process since it did not require demolition nor a reconstruction and, at the same time, kept a typological identity from the previous building (Fig. 3a). Hence, the architectural work is closer to an act of repairing than of reshaping [26], where the theme of circularity is intimately related to the concept of duration, revealing an attitude to consolidate the morphological building dimension.

A second layer of morpho-typological analysis, implying NBS and CE, could be seen focusing on the thresholds of the building. Especially when dealing with an intermediate scale, thresholds are the elements of the relation between the building and the city. A renewed importance in this perspective could be addressed by the base of the building and its crowning element. The base acquires high importance being a permanent trace of the project. Specifically, the reuse of low-rise buildings, similarly to the previous concept, can define a permanence of the previous urban settlement, as visible in the redevelopment of "Klaprozenbuurt neighborhood" by BETA office, where the regeneration of a former industrial site kept the bases and so the fabric structure (Fig. 3b). At the same time, thinking at the base as the floor in contact with possible climate risks requires shaping forms able to resist extreme events. Because of this, a spatial response could consist in the elevation of the ground floor, also working with the external ground section, implementing drainage solutions. An example is visible in the "Iseldoks" project, by the studio De Urbanisten, where the typological section of the buildings and the neighborhood's ground was rethought to define a spatial resiliency to floods (Fig. 3c).

The last element that we could display as crucial is the ground. This could be framed as other typological elements to connect the urban scale and the plot. Regenerate the ground means to restore the ecological potential of urban areas. Here NBS can play a central role: as visible in projects such as De Ceuvel by Space & Matter, or Bottière Chenaie by Atelier de Paysages Bruel Delmar, where nature could enhance the soil reclamation and produce a new form of the urban landscape (Fig. 3d).

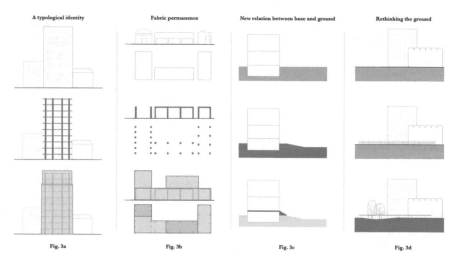

Fig. 3. Representation of morpho-typological modification. Drawing by K. Santus.

6   Conclusion

This initial morpho-typological consideration reveals the necessity for a critical description of the interventions approaching the ecological transition and decarbonization. Indeed, if on one hand we must consider the quantitative impacts of our design action, on the other we should understand the qualitative aspects of them to give shape to the transition and understand the spatial issue of it.

Given the above, one of the main contributions of this work is to open the discussion upon design strategies that could be used together to design and regenerate the urban built environment, increasing the resilience and decarbonization of metropolitan areas. Specifically, the usage of NBS envisions the possibility of tackling the adaptation of the urban space, with nature as a regeneration vehicle to restore urban grounds, and making the urban space capable of resisting extreme climate events such as floods. Moreover, NBS can produce a systemic urban adaptation, reducing the climate pressure in the urban environment, contrasting effects such as the heat island. Instead, the circular perspective focuses on the project's resources, stressing the necessity to deal with what has been built in the previous decades and imposing new reflection regarding what we can do with what we already have, enhancing the resilience for the availability of raw materials. Moreover, buildings' reuse and renovation could drive the practice to reduce carbon emissions derived from the demolition and reconstruction of new fabrics, decreasing the overall life cycle's emissions and mitigating the impacts of the built environment. Hence, the complementarity of nature and circularity exemplifies the possibility for spatial regeneration toward a new paradigm of urban ecology [27] portrayed in the resilient and post-carbon city.

The necessity of having this double approach envisions the neighborhood scale, an intermediate level between city and building, as the main scale to focus for the ecological transition. This makes the project work both on the building form and on its relationship with the open space and the ground.

Finally, the reflection regarding the morpho typological aspects highlighted the qualitative aspects that should be considered when resilient tools are applied, understanding the new challenges and possibilities that the projects could enhance. Indeed, if it is a necessary transition toward a built environment able to curb its carbon footprint, to reduce the usage of resources, and to adapt to the climate threads, the way we are going to do it should be addressed not only through a quantitative lens but considering the cultural and morphological implications. Because of this, architecture should investigate the role of the form in this transition, as a result and driver toward a project entangled with the environment, in its limits, hazards, and possibilities.

References

1. Gregotti, V.: Incertezze e simulazioni. Architettura tra moderno e contemporaneo. 1st edn. Skira editore, Milan (2012)
2. Wallerstein, I., Rojas, C.A., Lemert, C.C.: Uncertain Worlds World-Systems Analysis in Changing Times, 1st edn. Routledge, New York (2013)
3. Kelman, I., Mercer, J., Gaillard, J.C.: The Routledge Handbook of Disaster Risk Reduction Including Climate Change Adaptation, 1st edn. Routledge, New York (2020)
4. Lee, S. (ed.): Aesthetics of Sustainable Architecture. 1st edn. NAI/010 Publsher, Rotterdam (2011)
5. Fabian, L., Bertin, M.: Italy is fragile: soil consumption and climate change combined effects on territorial heritage maintenance. Sustainability 13(11), 6389 (2021). https://doi.org/10.3390/su13116389
6. Wilby, R.: A review of climate change impacts on the built environment. Built Environ. 33(1), 31–45 (2021). https://doi.org/10.2148/benv.33.1.31
7. Röck, M., et al.: Embodied GHG emissions of buildings – the hidden challenge for effective climate change mitigation. Appl. Energy 258, 114107 (2020). https://doi.org/10.1016/j.apenergy.2019.114107

8. Climate Change Impacts in Europe. https://www.eea.europa.eu/highlights/why-does-europe-need-to/climatechangeimpactineurope.pdf/view. Accessed 28 Dec 2021
9. Fuller, B.: The Year 2000. In: Architectural Design (February 1967)
10. IPCC: Climate Change: The Physical Science Basis. Cambridge University Press (2021)
11. Bertin, M., Marango, D., Musco, F.: Pianificare l'adattamento al cambiamento climatico come gestione di una macro-emergenza locale. Territorio **89**, 138–144 (2019)
12. UN: 2020 Global status report for buildings and construction. 1st edn. Nairobi (2020)
13. IEA: Energy Technology Perspectives 2020. 1st edn. IEA Publications, France (2020)
14. Steffen, W., Broadgate, W., Deutsh, L., Gaffney, O., Ludwig, C.: The trajectory of the Anthropocene: the great acceleration. Anthropocene Rev. **2**(1), 81–98 (2015)
15. European Commission: COM 98, A new Circular Economy Action Plan. For a cleaner and more competitive Europe. Brussels (2020)
16. European Commission: COM 662, A Renovation Wave for Europe - greening our buildings, creating jobs, improving lives. Brussels (2020)
17. European Commission - Circular Economy: Principles for Building Design. https://ec.europa.eu/docsroom/documents/39984. Accessed 12 Oct 2021
18. Roadmap to 2050 A Manual for Nations to Decarbonize by Mid-Century. https://roadmap2050.report. Accessed 12 Oct 2021
19. European Research Executive Agency: Nature-based solutions: Horizon 2020 research projects tackle the climate and biodiversity crisis. Publications Office (2021)
20. Ellen Macarthur Foundation - Circular economy in cities. https://ellenmacarthurfoundation.org/circular-economy-in-cities. Accessed 20 Dec 2021
21. Valente, I.: 'Durata/Duration'. In: Marini, S., Corbellini, G. (eds.) Recycled Theory. Quodlibet, Macerata (2016)
22. Reinventing Cities Guidance to Design a Low-Carbon, Sustainable and Resilient Project. https://www.c40reinventingcities.org/en/guidelines/. Accessed 27 Dec 2021
23. Rossi, A.: L'architettura della città. 1st edn. Marsilio Editore, Padova (1966)
24. van Timmeren, A.: Climate Integrated Design and Closing Cycles. Solutions for a Sustainable Urban Metabolism. In: van Bueren, E.M., van Bohemen, H., Itard, L., Visscher, H. (eds.) Sustainable Urban Environments, An Ecosystem Approach. Springer, Dordrecht (2012)
25. Frampton, K.: Urbanization and discontents: megaform and sustainability. In: Lee, S. (ed.) Aesthetics of Sustainable Architecture. 1st edn. NAI/010 Publsher, Rotterdam (2011)
26. García Germán, J.: Estrategias operativas en arquitectura. Nobuko, Buenos Aires (2012)
27. Morin, E.: L'entrée dans l'ère écologique. 1st edn. Edition de l'Aube, La Tour-d'Aigues (2020)

Investigating the Effect of Form and Material of Spatial Structures on Energy Consumption in Hot and Dry Climates
Case Study: Kerman City

Zinat Javanmard$^{(\boxtimes)}$ (iD) and Consuelo Nava

Mediterranean University of Reggio Calabria, Reggio Calabria, Italy
zinat.javanmard@unirc.it

Abstract. Climate and environmental conditions have a special and inevitable physical and psychological effect on humans that should be considered in the design of buildings depending on the heating or cooling situation (Yoglaucp 1972: 251). Climate design is a way to achieve bioclimatic comfort. Actually, the basic solutions to bioclimatic design are commonly found in vernacular building. Some modifications and improvements can be also observed while particular methods differ depending on the regional traditions, available materials, developed techniques etc. (Widera 2015). The building outer skin is the interface between the interior and exterior. The complex design of the skin can bring the issue of energy consumption optimization closer to the ambitious goals of energy saving through environmental considerations (Zemella and Faraguna 2014). An important function of the outer skin of the building (decking of space frames) is to control environmental and physical factors such as light, heat, sound, etc. to achieve comfort conditions with minimal energy consumption (Oral et al. 2004). Various simulation programs have been developed for this purpose that can lead the design solutions to the optimization of the building by simulating the building and measuring the exact energy consumption (Myers and Pohl 1992). In this research, after researching and studying articles and researches in the field of comparing energy simulation software, it has been tried to use the latest and most-up-to-date simulation software.

Keywords: Hot and dry climate · Climate design · Spatial structure's cladding · Energy consumption · Kerman

1 Introduction

Energy consumption is an important and complex issue in today's designs and in the construction industry. Traditional construction equipment and consequently not paying attention to the specific climatic conditions of each project will create a great challenge in terms of energy consumption.

Climate and environmental conditions have a special and inevitable physical and psychological impact on humans that should be considered in the design of buildings depending on the heating or cooling situation. (Yoglaucp 1972: 251) The basic design of buildings according to the climatic conditions of the area and the correct use of solar energy can play an important role in reducing energy consumption (Taban, Naghsh Jahan Magazine, No. 3). Choosing the right building materials is closely related to the climatic design of the building. In today's world, paying attention to the use of local materials is one of the concerns of the people of architecture and construction and the formation of proportional facades. To be an effective builder with local materials, you need to think about the principles of local design and use it in the design of new buildings (Zolfaghari 2015).

In this study, by introducing the four main categories of climatic division in Iran and finally by examining hot and dry climate as a case study, we seek to offer available solution as a cladding for spatial structures in a way that is suitable for hot and dry climates.

2   Research Questions

1. Considering the climatic conditions of the site of each project, can the use of suitable claddings for spatial structures in large projects cause cost savings and optimization in the consumption of materials and energy or not?
2. Can the attitude towards spatial structures as structures with sustainable architectural potential be introduced as environmentally friendly structures from the perspective of energy consumption?

3   Research Method

Method The present article is applied research using field, library and analytical studies to collect data to analyze environmental impacts. The software used for modeling and analysis is design builder. In this regard, first the variables studied in the research are introduced and then the dome roof as a suitable form for hot and dry climate with 2 types of metal and glass cover in Kerman as a case study in hot and dry climate by software design builder is simulated. Therefore, the relationship between energy consumption and different materials of glass and aluminum as the studied samples among the existing claddings in hot and dry climates is the main basis of the simulations.

4   Literature Review and Research Background

The chosen topic is an interdisciplinary topic and considering that no study has been done directly on Claddings and energy in spatial structures so far, It is necessary to indirectly study and review the research conducted in several areas related to the subject that form the main structure of research response.

For this purpose, to find the research background, a review of research records in the following research fields has been done:

1. **Factors affecting energy consumption:** (Javanmard and Nasser Alavi 2017), (Golabchi and Soroushnia 2012)
2. **Knowledge of spatial structures:** (Jiang et al. 2017; Zhang et al. 2020; Yuan and Yang 2019),
3. **Energy efficiency management:** (Chen and Yang 2017; Shoaib and Torkashvand in 2014; Khayyam Bashi and Taki 2016)
4. **Use of renewable energy:** (Sabzi Parvar 2001; Mohammadifar 2009).
5. **Climatic design and characteristics of hot and dry climate:** (Saeedi et al. 2017; Purdihimi 2008; Ghobadian 2005; Kasmaei 2010; Hashemi Rafsanjani and Heidari 2018).
6. **Bioclimatic Desing:** (Battisti 2020; Widera 2015; Nava 2021)
7. **Use of modern materials:** (Belinda et al. 2017; Ghazi Vakili et al. 2018; Alyyami and Regui 2012)
8. **Sustainable architecture:** (Reed et al. 2012; Raisi et al. 2016; Mahmoudi and Jedi 2009; Boer et al. 2015).
9. **Energy simulation software:** (Mortazavi and Roshan 2015; Dury et al. 2008; Ghiaei et al. 2013; Zumrodian and Tehsildoost 2015).

5   Theoretical Foundation

1. spatial structures: Spatial structure Structures that are basically three-dimensional in behavior so that their overall behavior can in no way be approximated using one or more independent two-dimensional sets are called spatial structures.
2. cladding: It is a verbal claddings that is used to surround the exterior of buildings to prevent and take care of the weather. In light buildings, the claddings itself bears its weight, but in general, wind and snow load and the weight of the cover itself to the frame of the structure. It can be transferred. In some cases, the thickness of the cover may also be used as a cover for lateral loads, but in general, it is used as insulation for indoor and outdoor spatial.

6   Discussion and Results

2 types of studies were performed on the amount of energy consumption in each of the coatings (metal and glass):

– Investigation between energy consumption in flat roof and dome roof with metal materials.
– Investigation between energy consumption in flat roof and dome roof with glass materials.

Variables

The energy consumption variable depends on the continuous variables of roof shape and material used in the claddings: a. Roof shape, b. Cover used.

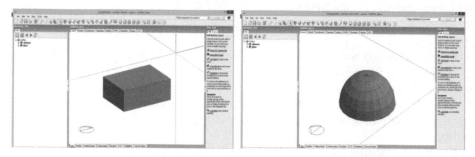

Fig. 1. The flat roof and dome modeled in the software

6.1   Roof Shape

Two types of flat and dome roof shapes were investigated as the basic shape of the simulations.

Figure 1 A figure caption is always placed below the illustration. Short captions are centered, while long ones are justified. The macro button chooses the correct format automatically.

The area of the spatials in both types of roof shapes was assumed to be fixed in order to better investigate the relationship between energy consumption and the shape of the roof. The floor area was defined as 25 * 25.

6.2   Cladding Used

Two types of metal and glass mixers were selected as the materials. The following tables specify the infrastructure, type, number and thickness of layers used.

6.2.1   Metal Materials (Aluminum)

The material used as the metal cladding is aluminum sheet in six layers, including its infrastructure, which is described in the table below. Layers in order from outside to inside: 1) Calzip with a thickness of 00001 m. 2) Polystyrene with a thickness of 0.001 m. 3) Airtight (glass-fiber) with a thickness of 0.0005 m. 4) Aluminum with a thickness of 0.0005 m. 5) Vacuum layer (air) with a thickness of 0.0065 m. Aluminum with a thickness of 0.0005 m (due to material limitations in this software, these three layers are used instead of corrugated sheets).

Table 1. Specifications of glass material infrastructure and Infrastructure of glass materials

Layers	
Number of layers	4
Outermost layer	
Material	Glass-cellular sheet
Thickness (m)	0.0010
Layer 2	
Material	Air gap 10 mm
Thickness (not used in thermal calcs) (m)	0.0020
Layer 3	
Material	Glass-foam
Thickness (m)	0.0050
Innermost layer	
Material	Glass-foam
Thickness (m)	0.0050
Inner surface	
Convective heat transfer coefficient (w/m^2−k)	4.460
Radiative heat transfer coefficient (w/m^2−k)	5.540
Surface resistance (m^2−k/w)	0.100
Outer surface	
Convective heat transfer coefficient (w/m^2−k)	19.870
Radiative heat transfer coefficient (w/m^2−k)	5.130
Surface resistance (m^2−k/w)	0.040
No bridging	
U-Value surface to surface (w/m^2−k)	2.754
R-Value (m/m^2−k)	0.503
U-Value (w/m^2−k)	**1.988**
With bridging (BS EN ISO 6946)	
Thickness (m)	0.0130
Km-internal heat capacity (kJ/m^2−k)	0.0000
Upper resistance limit (m^2−k/w)	0503
lower resistance limit (m^2−k/w)	0503
U-Value surface to surface (W/m^2−k)	2.754
R-Value (m^2-k/w)	0.503
U-Value (W/m^2-k)	**1.988**

Table 2. Specifications of aluminum material infrastructure and Infrastructure of aluminum

Layers	
Number of layers	6
Outermost layer	
Material	Aluminium
Thickness (m)	0.00010
Layer 2	
Material	XPS Extruded Polystyrene-HDC BIo
Thickness (not used in thermal calcs) (m)	0.0010
Layer 3	
Material	R-10 Glass-fiber Batt Insulation
Thickness (m)	0.00050
Layer 4	
Material	Metal aluminium cladding
Thickness (m)	0.00050
Layer 5	
Material	Air gap 10mm
Thickness (not used in thermal calcs)(m)	0.0065
Innermost layer	
Material	Metals-aluminium cladding
Thickness (m)	0.00050
Inner surface	
Convective heat transfer coefficient (w/m^2−k)	2.152
Radiative heat transfer coefficient (w/m^2−k)	5.540
Surface resistance (m^2−k/w)	0.130
Outer surface	
Convective heat transfer coefficient (w/m^2−k)	23.290
Radiative heat transfer coefficient (w/m^2−k)	1.710
Surface resistance (m^2−k/w)	0.040
No bridging	
U-Value surface to surface (w/m^2−k)	5.129
R-Value (m^2−k/w)	0.365
U-Value (W/m^2-k)	**2.740**
With bridging (BS EN ISO 6946)	
Thickness (m)	0.0091
Km-internal heat capacity (kJ/m^2−k)	1.6128
Upper resistance limit (m^2−k/w)	0.365

(*continued*)

Table 2. (*continued*)

Layers	
Lower resistance limit (m^2–k/w)	0.365
U-Value surface to surface (W/m^2–k)	5.129
R-Value (m^2–k/w)	0.365
U-Value (W/m^2-k)	**2.740**

6.2.2   Glass Materials

The glass claddings is considered in 4 layers, which are as follows: Layer 1) Security with a thickness of 0.0001 m. 2) Vacuum layer (air) with a thickness of 0.0020 m. 3) Glass with a thickness of 0.005 m. 4) Glass with a thickness of 0.005 m.

6.3   Case Study

The hot and dry climate of Iran has been studied as a base climate to study the effect of the type of materials and climatic characteristics with the amount of energy consumption. The shape of the roof is also considered as a variable in the form of dome and flat. Also, Kerman as a selected city in hot and dry climate is the basis of studies. In addition to the above, sports use (due to the widespread use in the use of spatial structures) was selected as the study to measure energy consumption. The standard of energy consumption in sports use is according to Table 3.

Table 3. Egyptian energy standard for stadium usage source: www.energystar.gov/PMGlossary.

Broad category	Primary function	Further breakdown (where needed)	Source EUI (kBtu/ft^2)	Site EUI (kBtu/ft^2)	Reference data source peer group comparison
Entertainment/Public Assembly	Stadium	Indoor Arena	85.1	45.3	CBECS-Public Assembly
		Race Track			
		Stadium (Closed)			
		Stadium (Open)			
		Other - Stadium			
	Other	Aquarium			
		Bar/Nightclub			
		Casino			

(*continued*)

Table 3. (*continued*)

Broad category	Primary function	Further breakdown (where needed)	Source EUI (kBtu/ft^2)	Site EUI (kBtu/ft^2)	Reference data source peer group comparison
		Zoo			
		Other - Entertainment/Public Assembly			

6.4   Data Analysis

Analyzes performed by the software after modeling: Flat roof in Kerman (Figs. 2 and 3).

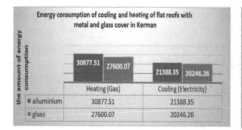
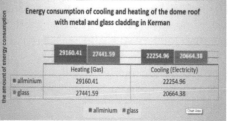

Fig. 2. The diagram shows that energy consumption will be lower in both cooling and heating when using a "glass cover" on a flat roof and that the amount of energy consumption in both cooling and heating will be lower when using a "glass cover" on the dome roof.

Descriptive Analysis

According to the selected area and the type of roof and materials in different climates, the following results were obtained from the analysis:

Hot and dry climate Comparing two metal and glass coverings with flat roofs in hot and dry climate (Kerman city), it was found that:

- the energy consumption of metal cladding is higher in both cooling and heating than glass cladding:
   Therefore, a flat roof with a glass cover is a better condition than a metal cover in this climate. And by comparing two metal and glass coverings with a domed roof in hot and dry climate (Kerman city), it was found that:
- the energy consumption of metal coatings in cooling and heating is higher than glass coatings

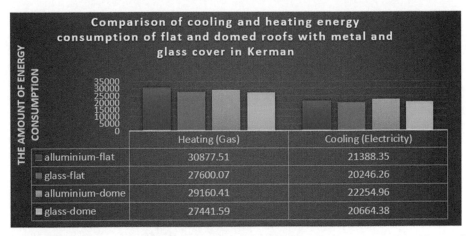

Fig. 3. Comparison of annual cooling and heating energy consumption of domed and flat roofs in Kerman

Therefore, Glass Cover Is Better Than Metal Cover in This Type of Roof and This Climate.

In general, the comparison of two metal and glass coverings with flat and domed roofs in hot and dry climate (Kerman city) showed that:

Dome roof with glass claddings is the best type of roof in this climate according to the amount of cooling and heating energy consumption, followed by dome roof with metal claddings, flat roof with glass cladding and finally flat roof with metal cladding with priority. The next design is in this climate. The general result of a hot and dry climate:

1. Comparing defined materials; The glass cladding has a better performance than the metal cover due to the energy consumption in the cooling and heating sector.
2. Comparing the performance of two types of flat and dome roofs, dome roofs have better efficiency in terms of energy consumption than flat roofs in this climate.

Finally, the global standard for energy consumption in stadiums was determined and proposed as a criterion for comparing the analyzes performed. In order to drive the amount of energy consumption to the relevant standard, some hypotheses were raised.

Hypotheses for improving performance and reducing energy consumption in relevant buildings are typically tested in hot and dry climates:

Theories

1. Changing materials and using more thermal insulation leads to reduced energy consumption.
2. Creating openings in the walls of structures reduces energy consumption.
3. The use of a combination of glass and metal materials in the roof leads to a reduction in energy consumption.

In the following, each of these hypotheses will be examined.

1- Changing the materials and increasing the air layer leads to reducing energy consumption. (For example, in a dome roof with a glass cover in hot and dry climates)
    The analysis showed that with increasing the thickness of the air layer in the dome roof, the amount of energy consumption decreases and a correct hypothesis is predicted (Fig. 4).

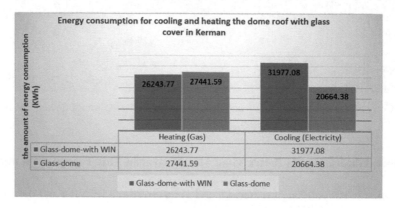

Fig. 4. Investigations of air layer thickness change

2- Creating openings in the walls of structures reduces energy consumption (Fig. 5)

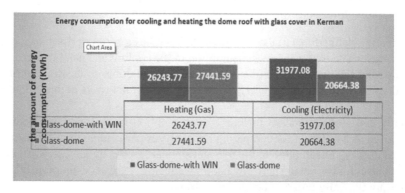

Fig. 5. Comparison of energy consumption with the hypothesis of creating openings

Creating openings in the dome roof structure reduces energy consumption in the heating sector and increases energy consumption in the cooling sector. Therefore, creating valves with a specific schedule (summers open and winters closed) helps reduce energy consumption.

3- The use of a combination of glass and metal materials in the roof leads to a reduction in energy consumption (Fig. 6).

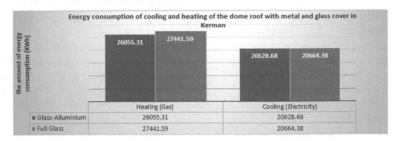

Fig. 6. Use a combination of glass and metal materials

The results showed that the simultaneous use of two types of coatings is a way to reduce energy consumption and the amount of energy required for heating can be reduced to some extent in the cold season.

7 Conclusions

In general, the following factors can be considered in choosing a folder suitable for each climate:

1- cladding material, 2- cladding form, 3- Proportion of each material used in the substructure of the cladding, 4-The presence or absence of air vents in the ceiling or wall, 5- Paying attention to the hot and cold days of the study area and using composite materials suitable for the cold and hot seasons, 6- Use insulation layers on the roof.

References

Battisti, A.: Bioclimatic architecture and urban morphology. Studies on intermediate urban open spaces (2020)

Ghiaei, M.M., Mahdavi Nia, M., Tahabaz, M., Mofidi Shemirani, S.M.: Methodology for selection of energy simulation applications in the field of architecture. City Identity **13**, 55–45 (2013). Year 7, Spring 2013

Ghobadian, V.: Principles and Concepts in Western Architecture, 29th edn. Farhang va Memari Publications, Tehran (2015)

Golabchi, M., Soroushnia, E.: Architectural Enhancement Details, 1st edn. University of Tehran Press, Tehran (2012)

Hashemi Rafsanjani, L.S., Heidari, S.: Evaluation of adaptive thermal comfort in hot and dry climate residential case study: Kerman. Q. J. Hot Dry Climate Archit. **6**(7), 65–43 (2018). Spring and Summer 1397

Javanmard, Z., Nasser Alavi, S.S.: Cladding design of spatial structures in the direction of sustainable architecture (2017)

Jiang, C., Tang, C., Seidel, H.P., Wonka, P.: Design and volume optimization of spatial structures. ACM Trans. Graph. (TOG) **36**(4), 1–14 (2017)

Kasmaei, M.: Climate and Architecture, 6th edn. Khak Publishing, Tehran (2010)

Khayyam Bashi, F., Taki, D.: Energy saving strategies in hot and dry climates in optimal building design. In: National Conference on New Research and Educational Findings in Iranian Civil Engineering, Architecture, Urban Planning and Environment, Tehran (2016)

Mahmoudi Kohneh Rudposht, A., Jedi, S.: Principles of design of sustainable buildings. In: The First Conference on Sustainable Architecture, Hamedan (2009)

Mohammadifar, L.: The use of solar energy with photovoltaic systems along with solar architecture to generate electricity and reduce consumption, the first national conference to improve the pattern of electricity consumption, Ahvaz, Khuzestan regional electricity network (2009)

Mortazavi, S.H., Roshan, M.: A study and comparison of building energy simulation software. In: Second National Conference on Architecture, Civil Engineering and Modern Urban Development, Urmia (2015)

Nava, C.: Advanced Sustainable Design (ASD) for resilient scenarios. In: Chiesa, G. (ed.) Bioclimatic approaches in urban and building design. PSS, pp. 255–274. Springer, Cham (2021). https://doi.org/10.1007/978-3-030-59328-5_13

Oral, G., Yener, A., Bayazit, N.: Building envelope design with the objective to ensure thermal, visual and acoustic conditions. Build. Environ. **39**, 281–287 (2004)

Green Building Strategy Supported by PostgreSQL and ArcGis System

Carlo Bernardo[✉]

Mediterranea University of Reggio Calabria, 89124 Reggio Calabria, CB, Italy
gevaul2@gmail.com

Abstract. The majority of existing buildings are old, in a state of decay and in need of rehabilitation interventions. When selecting the type of possible intervention to be applied, the choice falls between two alternatives: simple unsustainable ordinary maintenance versus ecological retrofitting i.e., an increase in the quality of the indoor environment and building energy saving using local bio-natural materials and products. The present research seeks to respond to the requests of recent comprehensive reviews which ask for the retrofitting of the world's huge existing building stocks and portfolios by proposing an approach and pre testing it in a prototype reference building repeated in the future on a larger urban scale. The proposed test achieved the important outcome and goal of a Green Building strategy and post-carbon city framework i.e. the significant enhancement of the thermal performance of the buildings as a result of a few targeted key external works and the consequent saving of energy. All the above show that all buildings can be ecologically retrofitted at an affordable cost, although initially slightly more expensive than the cost of ordinary unsustainable construction. However, this difference is offset by the favorable pay-back period, which is fast, acceptable and of short duration.

Keywords: Appraisal · Valuation · Valuation of green building · Valuation of post-carbon city strategy · Valuation of building energy · Energy performance simulation programs (EPSPs) · Data base management system · DBMS

1 Introduction

Climate change and the consequential global warming and the destructive side effects: higher atmospheric temperatures notably at the Poles and in Greenland; permafrost, glacier and snowfield melting; the consequential rising ocean levels and salt water surface warming. It also includes more frequent storms - hurricanes in the Atlantic, typhoons in the Pacific and cyclones in the Indian Ocean. These often cause huge floods, droughts due to the lack of regular rainfall and scarcity of fresh water.

It has to be especially noted that one important cause of climate change is the huge over consumption of energy in the world (the latest data expressed in million tonnes of oil equivalent, Mtoe [1 Mtoe = 11,63 TWh]) - the world total is 13.970, China uses 3.063, USA 2.155, Europe 1.828, India 882, Africa 812, Middle East 750, Russia 732

F. Calabrò et al. (Eds.): NMP 2022, LNNS 482, pp. 1643–1657, 2022.
https://doi.org/10.1007/978-3-031-06825-6_158

and Italy 125. This energy is largely coming from fossil fuels- coal, gas, oil and biomass, the burning of those fuels, the subsequent greenhouse gas emissions. The atmosphere is being polluted at all levels and this causes climate change as well as heavy health impacts.

It is crucial to recognize that the whole construction sector consumes over 40% of total world energy from fossil sources for the thermal management of residential as well as non-residential units. This percentage is and even higher in densely populated urban areas especially the world's megalopolises e.g. (2015 UN estimates) North East USA 52 million (m), Mexico City 25 m, Rio de Janeiro and SE Brazil 51m, Bogota 29 m, Tokyo 38 m, Seoul 25 m, Kolkata 65 m, Delhi 46 m, Mumbai 80.

Thus, the civil sector is the biggest consumer of fossil energy and consequently the biggest polluter and the biggest cause of planet climate change [1]. Reducing fossil energy consumption by retrofit works in this sector is one of the most effective steps in reducing greenhouse gas emissions. It is also a smart investment for families and homes given the immediate multiple impacts just at the end of retrofit works. Fossil fuel energy consumption must be dramatically reduced to ensure the planet's survival [2–8] and the solution is bio-ecological enhancement of a building's energy efficiency. This is a feasible strategy for overcoming these issues- a strategy for Green Buildings, Green Urban Districts/ Quartiers/ Neighborhoods, Post Carbon Cities and Green Regions. Regrettably this strategy is often seen as an additional bad cost for retrofitting of older buildings and maintenance in new constructions.

A scientific valuation must be rapidly performed and completed to correctly understand, demonstrate and implement the reduction of energy consumption in buildings to show a resulting gain in income and efficiency and not just as a passive bad cost. In fact, as soon as a building or an elementary unit is retrofitted and made energy efficient the owner and the users start to save forever on relevant expenses for indoor micro climate management and operating costs.

Therefore, the owner and the users of a bio-ecologically retrofitted unit or building gain immediately on higher technical value in terms of lower energy consumption, economically in terms of higher productivity [9, 10], in wellbeing with a healthier indoor environment and ecologically in terms of lower outdoor CO_2 emissions as well as a much-improved air quality in the atmosphere.

The above results show that retrofitting is not just an increased cost but ensures money is saved with higher productivity and lower running costs.

Appraisal and economic research addresses some of the benefits stemming from energy efficiency derived from Ecological Retrofitting [18, 19, 20, 21, 22, 23, 24, 27, 31, 32, 33, 34, 37] and calls for more case studies to consolidate the data and better implement the strategies of Post Carbon City - experimenting first at the bottom levels of residential units and, if the results are positive, implementing it further in a building and in an urban districts/neighborhoods and finally in an entire city as well as in an entire region.

2 Green Building Strategy Supported by PostgreSQL and ArcGis System

2.1 Research Purpose

The present research performed the thermal and energy assessment [11–16] of the Urban residential Block #102 (see Fig. 1) and the purpose is to experiment reduction of building energy consumption by implementing Ecological Retrofitting.

Fig. 1. Urban Block #102. Source: free Google Maps 2D. 2021.

2.2 GeoDataBase

The data to be managed will be of considerable quantity when retrofitting is extended to many real estate <u>units</u> or <u>sub-parcel</u> (often: <u>apartments</u>) (so called: "*Subalterni Catastali*"), buildings and blocks. Information Technology tools are needed for simultaneous management of large amount of geometric, numeric and alphanumeric data.

The state of the arts recommends using the DBMS enterprise in association with GIS for coordinated management of the above cited large amount of spatial and economic information. This part introduces and represents the adopted procedure, based on state of the arts DBMS PostgreSQL and ArcGis Spatial Engine as first step of a decision system to supporting the energy diagnosis and retrofitting of buildings, wards and quartiers. This computer system can be used to build the knowledge system of one's own building heritage and carry out interventions relating to energy efficiency, it is a contribution to systematize a set of services, data, software applications, procedures and methodologies.

2.3 Case Study

The prototype work is intended to compose a georeferenced database of certified information relating to the urban block and its building under study, where the geodatabase becomes an aggregator and a container of diversified information sources.

The pilot project, with the appropriate additions, can then be extended to the entire building stock of a city also through a web geoportal, therefore the bottom up model was adopted in this prototype study. A 3D model of the individual blocks has been built-up and extended, by replicating the operation, to the buildings of the whole entire 'Latin Quartier' and the whole City of Reggio Calabria (see Fig. 2a,b).

(a)   **(b)**

Fig. 2. 'Latin Quartier' (North, East beside city port) in the City of Reggio Calabria context. (**a**) Aerial bird's-eye view. Source: free Wikipedia 2020; (**b**) 3D in ArcMap - AscScene. Source: Author's elaboration.

The data of cartographic interest, thermal energy evaluation, pollution avoided, relating to prototype Block # 102 and its subordinates, have been transferred to the Geodatabase, so we have carried out all the necessary elaborations allowed by the software applications and the results of which will be presented below.

So, in the near future it will be possible to replicate the study on every single block and its subordinates to the whole city of Reggio, a modeling that has already begun. In order to make the processing even more realistic, some sample buildings were modeled in greater depth and detail with accurate photo texturing using the SketchUp software. This software was used as there is the possibility of exporting the three-dimensional file directly in the GIS environment (see Fig. 3).

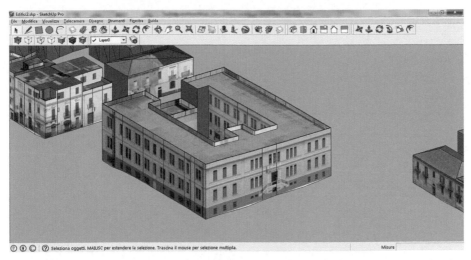

Fig. 3. Detail of the three-dimensional modelling and texturing of Block #82 in the *SketchUp* environment. Source: Author's Survey and elaboration.

2.4   The Thermal and Energy Assessment of Block #102

As described above, the data of cartographic interest, thermal and energy evaluation, relating to Block #102 and its subordinates, have been transferred to the Geodatabase so below we will describe the data transfer procedure and then the methodology used to query the geodatabase and the information obtained.

In this work about ecological retrofitting, DBMS PostgreSQL is adopted and the data of interest are transferred in the geodatabase "reggio".

The basic cartographic data were represented in Cad dwg format and converted the data into shapefile, check and add any missing data and then transfer this data to the PostgreSQL geodatabase, for all this we used the ArcGIS/ArcMap/Catalog applications (Esri). It is possible to transfer directly Cad files to geodatabase but in this case we had to choose a different path to enrich our base data (see Fig. 4).

The necessary thermal and energy assessment data to enrich our database were in Excel, data which are shown in below in the associated Table. So we can view this data in the next tables images from PostgreSQL where there are one table for one Building.

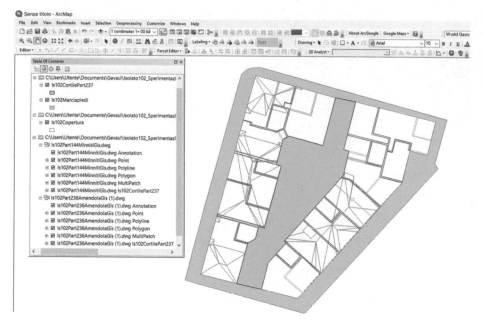

Fig. 4. CAD data represented in ArcMap. Source: Author's Survey and elaboration.

2.5   General Description About the Data

All the tables, within data base, used to collect the Sub-units' ordinates Geodatabase are composed of GIS-oriented dataset attributes, that is connected to geometric information. For the tracking of sub-parcel (often: apartments) (so called: "*Subalterni Catastali*"), a polygonal theme was used which allows to obtain the size of the real estate unit and, consequently, the surface in square meters. The attributes table of the "reggio" Geodatabase can be viewed by right-clicking on the layer name in the Table Of Contents (TOC), with the "Open Attribute Table" command.

The structuring of the spatialized Geo Data Base of the cadastral real estate units (Sub-units = "*subalterni*") of Block #102 of the post-earthquake 1908 liberty reconstruction will have the purpose of providing and maintaining an accessible data repertoire by ensuring access to information without limitations (compatibly with data licenses) and ensuring a continuous updating process. The Geodatabase, for each sub-parcel (so called: "*Subalterni Catastali*"), inserted, contains the following fields:

Objectid. This numeric field, the unique identifier of the row, is a progressive ID containing the progressive numbering of the real estate units in the sequence corresponding to their spatial tracing. It is a field created by default (like the shape field) in the associated table and contains a sequential order starting from number 1.

Layer. This column, together with Entity and Elevation, is automatically derived following the import of the DraWinG file from the "AutoCAD" assisted drawing software and contains the name of each subordinate reference, consisting of the floor level on which it is located and its cadastral numbering.

Elevation. Automatically derived after importing the DraWinG file, it contains the elevation already set in the "AutoCAD" software.

Piano (Floor). Indicates the floor level of the building.

Superficie (Surface). Field containing the dimensions (areas) of the real estate units expressed in square meters (Square Meters). The Area is obtained automatically using the Calculate geometry function applied to the specific field of the GDB.

Volume. Field containing the volumes of the real estate units expressed in cubic meters (Cubic Meters).

kWh_Y_SCom (kWh per Year. Common Scenario). Field containing the quantities of Energy consumed (measured in kWh), for each real estate unit, measured over the calendar year, in the intervention scenario hypothesized for its passivation, therefore aimed at its energy containment, through the use of commonly used materials and techniques and, for this reason, referred to as the Common Scenario.

kWh_Y_Sost (kWh per Year. Sustainable Scenario). Field containing the quantities of Energy consumed (measured in kWh), for each real estate unit, measured over the calendar year, in the intervention scenario hypothesized for its passivation, therefore aimed at its energy containment, through the use of innovative materials and techniques that allow greater containment of consumption together with a minimum environmental impact and, for this reason, indicated as a Sustainable Scenario.

Var_kWh_Y (Variation (Δ) in kWh per Year). Field containing the difference between the values indicating the consumption of Energy (measured in kWh) between the 2 reference scenarios for the restoration project hypothesized and aimed at the passivation of the building.

Perc_kWh_Y (Percentage (%) in kWh per Year). Field containing the percentages of Energy savings (measured in kWh) between the 2 reference scenarios for the restoration project hypothesized and aimed at the passivation of the building.

kWh_Y_Mq_C (kWh per Year and per Mq. Common Scenario). Field containing the quantities of Energy consumed (measured in kWh), for each square meter of the real estate unit, measured over the calendar year, in the intervention scenario hypothesized for its passivation, therefore aimed at its energy containment, by means of '' use of commonly used materials and techniques, and for this reason, referred to as the Common Scenario.

kWh_Y_Mq_S (kWh per Year and per Mq. Sustainable Scenario). Field containing the quantities of Energy consumed (measured in kWh), for each square meter of the real estate unit, measured over the calendar year, in the intervention scenario hypothesized for its passivation, therefore aimed at its energy containment, by means of use of innovative materials and techniques that allow greater containment of consumption together with a minimum environmental impact and, for this reason, indicated as a Sustainable Scenario.

VarkWhYmq (Variation (Δ) in kWh per Year and per mq). Field containing the difference between the values indicating the consumption of Energy per square meter (measured in kWh) between the 2 reference scenarios for the restoration project hypothesized and aimed at the passivation of the building.

Per_kWhYmq (Percentage (%) di kWh per Year and per mq). Field containing the percentages of Energy savings per square meter (measured in kWh) between the 2 reference scenarios for the restoration project hypothesized and aimed at the passivation of the building.

CO_2_Y_Scom (CO_2 Emission per Year. Common Scenario). Field containing the quantities of carbon dioxide (CO_2) emitted (measured in kg), for each real estate unit, measured over the calendar year, in the intervention scenario hypothesized for its passivation, therefore aimed at its energy containment, by means of the use of materials and techniques used, and for this reason, referred to as the Common Scenario.

CO_2_Y_Sost (CO_2 Emission per Year. Sustainable Scenario). Field containing the quantities of carbon dioxide (CO_2) emitted (measured in kg), for each real estate unit, measured over the calendar year, in the intervention scenario hypothesized for its passivation, therefore aimed at its energy containment, by means of the "use of innovative materials and techniques that allow greater containment of consumption together with a minimum environmental impact and, for this reason, indicated as a Sustainable Scenario.

Var_CO_2_Y (Variation (Δ) of CO_2 emissions per Year). Field containing the difference between the values indicating the amount of carbon dioxide (CO_2) emitted (measured in kg) between the 2 reference scenarios for the restoration project hypothesized and aimed at the passivation of the building.

Perc_CO_2_Y (Percentage (%) of CO_2 emissions per Year). Field containing the percentages of reduction in the amount of carbon dioxide (CO_2) emitted (measured in kg) between the 2 reference scenarios for the restoration project hypothesized and aimed at the passivation of the building.

CO_2_Y_Mq_C (CO_2 Emission per Year and per mq. Common Scenario). Field containing the quantities of carbon dioxide (CO_2) emitted (measured in kg), for each square meter of the real estate unit, measured over the calendar year, in the intervention scenario hypothesized for its passivation, therefore aimed at its energy containment, through the use of commonly used materials and techniques, and for this reason, referred to as the Common Scenario.

CO_2_Y_Mq_S (Emissioni di CO_2 per Year e per Mq. Scenario Sostenibile). Field containing the quantities of carbon dioxide (CO_2) emitted (measured in kg), for each square meter of the real estate unit, measured over the calendar year, in the intervention scenario hypothesized for its passivation, therefore aimed at its energy containment, through the use of innovative materials and techniques that allow greater containment of consumption together with a minimum environmental impact and, for this reason, indicated as a Sustainable Scenario.

VarCO$_2$Ymq (Variation (Δ) of CO$_2$ emissions per Year e per mq). Field containing the difference between the values indicating the amount of carbon dioxide (CO$_2$) emitted (measured in kg) between the 2 reference scenarios for the restoration project hypothesized and aimed at the passivation of the building.

Per_CO$_2$Ymq (Percentage (%) of CO$_2$ emissions per Year and per mq). Field containing the percentages of reduction in the amount of carbon dioxide (CO$_2$) emitted (measured in kg) between the 2 reference scenarios for the restoration project hypothesized and aimed at the passivation of the building.

2.6   Trasfering Shapefile to DBMS PostgreSQL

Now it was necessary to transfer the shapefiles into the PostgreSQL geodatabase. Data transfer in the PostgreSQL geodatabase was possible by using the Tool "Export Data" data (see Fig. 5), one shapefile at time. Joining the CAD data and the Excel data, ArcMap produced the two fundamental shapefiles identified in the geodatabase with the name:

- reggio.sde.Is102Part144Minniti3mPolygon;
- reggio.sde.Is102Part236Amendola3mPolygon.

The following files have also been transferred to the geodatabase:

- reggio.sde.Is102Part236AmendolaGisPolyline;
- reggio.sde.Is102Copertura;
- reggio.sde.Is102Part144MinnitiGisPolyline;
- reggio.sde.Is102Marciapiedi;
- reggio.sde.Is102CortilePart237;

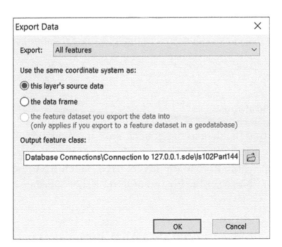

Fig. 5. Dialog box used to transfer shapefiles to Geodatabase PostgreSQL. Source: ArcMap.

After joining and transferring the data from shapefiles to DBMS PostgreSQL it is now possible to view both the Excel data and the cartographic data in one Table (see Fig. 6 and Fig. 7).

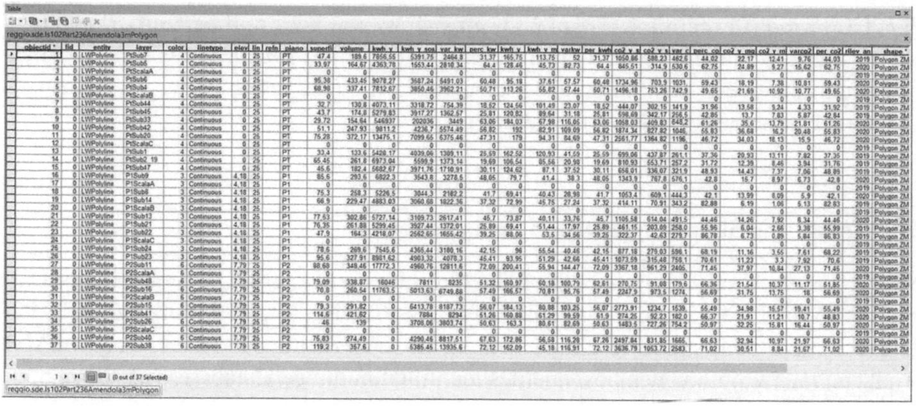

Fig. 6. Geodatabase table in PostgreSQL – Building facing Amendola Street. Source: ArcMap.

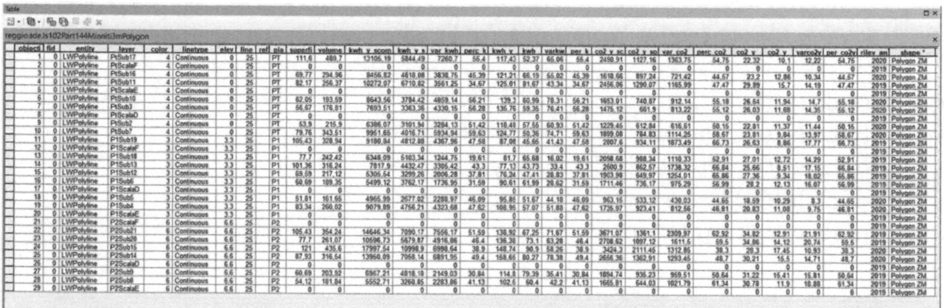

Fig. 7. Geodatabase table in PostgreSQL – Building facing Minniti Street. Source: ArcMap.

2.7 Query the Geodatabase

After having transferred the data to the geodatabase it is now possible to query it, as in the plan image of the Block #102 building on the ArcMap screen (see Fig. 8).

In the TOC it's possible to read the layers used, taken from the PostgreSQL Geodatabase. It's also possible to see the PostgreSQL data source because a part name layer's in the TOC is 'sde'.

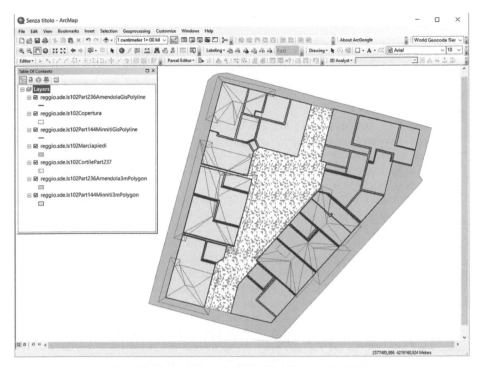

Fig. 8. Plan of Block# 102 in Gis. Source: ArcMap.

3   Conclusion

The research attempts to contributing TO:

- an existing building retrofit implementation approach by adopting natural, bio-ecological, historical, recyclable/renewable and local/regional materials in the framework of the Circular Economy;
- a simulation of bio-ecological retrofitting of the prototype building such as the second Scenario;
- an assessment (ecological as well as financial) of energy saving and CO_2 emission mitigation;

The experimentation in the provided prototype or reference small building achieved the important goal of the Green Building and Post Carbon City strategies [17, 25, 26, 28–30, 35, 36, 38–44] i.e. the significant enhancement of building thermal performance resulting from a few targeted keys works (="Lavorazioni") and the consequent permanent structural saving of energy in those existing constructions.

Indeed, the data obtained is very encouraging and the following observations can be made:

- energy saving amounts to 47%;

- avoided CO_2 pollution is 57%.

All of the above show that the buildings can be bio-ecologically built or retrofitted.

References

1. Massimo, D.E., Del Giudice, V., Malerba, A., Bernardo, C., Musolino, M., De Paola, P.: Valuation of ecological retrofitting technology in existing buildings: a real-world case study. Sustainability **13**, 7001 (2021). https://doi.org/10.3390/su13137001
2. Statistical Review of World Energy Archived 6 January 2009 at the Wayback Machine. Workbook (xlsx), London (2016)
3. Renewable Energy Employs 8.1 Million People Worldwide: United Nations Framework Group on Climate Change. 26 May 2016. Archived from the original on 18 April 2019. Accessed 18 Apr 2019
4. Energy Transition Investment Hit $500 Billion in 2020 – For First Time: Bloomberg NEF. (Bloomberg New Energy Finance). 19 January 2021. Accessed 19 Jan 2021
5. The European Power Sector in 2020/Up-to-Date Analysis on the Electricity Transition (PDF). ember-climate.org. Ember and Agora Energiewende (25 January 2021). Accessed 25 Jan 2021
6. Chrobak, U., Chodosh, S.: Solar power got cheap. So why aren't we using it more?. Popular Science. Archived from the original on 29 January 2021. {{cite news}}: lfirst1=has generic name (help) Chodosh's graphic is derived from data in "Lazard's Levelized Cost of Energy Version 14.0" (PDF). Lazard.com. Lazard (19 October 2020). Accessed 28 Jan 2021
7. Majority of New Renewables Undercut Cheapest Fossil Fuel on Cost. IRENA.org. International Renewable Energy Agency. 22 June 2021. Accessed 22 June 2021. Infographic (with numerical data) and archive thereof
8. Data: BP Statistical Review of World Energy, and Ember Climate (3 November 2021). "Electricity consumption from fossil fuels, nuclear and renewables, 2020". OurWorldInData.org. Our World in Data consolidated data from BP and Ember. Accessed 3 Nov 2021
9. Massimo, D.E.: Valuation of urban sustainability and building energy efficiency. A case study. Int. J. Sustain. Dev. **12**, 223–247 (2010). https://doi.org/10.1504/IJSD.2009.032779
10. Massimo, D.E.: Green building: characteristics, energy implications and environmental impacts. case study in Reggio Calabria, Italy. In: Coleman-Sanders, M. (ed.) Green Building and Phase Change Materials: Characteristics, Energy Implications and Environmental Impacts, vol. 1, pp. 71–101. Nova Science Publishers: New York (2015)
11. Crawley, D.B., Lawrie, L., Winkelmann, F.C., Pedersen, C.O.: Energy Plus: New capabilities in a whole-building energy simulation program. Build. Simul. **33**, 51–58 (2001)
12. Yilmaz, A.Z.: Evaluation of energy efficient design strategies for different climatic zones: comparison of thermal performance of buildings in temperate-humid and hot-dry climate. Energy Build. **39**, 306–316 (2007). https://doi.org/10.1016/j.enbuild.2006.08.004
13. Crawley, D.B., Hand, J.W., Kummert, M., Griffith, B.T.: Contrasting the capabilities of building energy performance simulation programs. Build. Environ. **43**, 661–673 (2008). https://doi.org/10.1016/j.buildenv.2006.10.027
14. Rallapalli, H.S.: A comparison of EnergyPlus and eQuest whole building energy simulation results for a medium sized office building. Master's Thesis, Arizona State University, Tempe, AZ, USA (2010)
15. Sousa, J.: Energy simulation software for buildings: review and comparison (2012). https://www.semanticscholar.org/paper/Energy-Simulation-Software-for-Buildings-%3A-Revie-Sousa/b4b6593df77024a585b68d066bf2bd668838f852. Accessed 25 May 2021

16. Malerba, A., Massimo, D.E., Musolino, M., Nicoletti, F., De Paola, P.: Post carbon city: building valuation and energy performance simulation programs. In: Calabrò, F., Della Spina, L., Bevilacqua, C. (eds.) ISHT 2018. SIST, vol. 101, pp. 513–521. Springer, Cham (2019). https://doi.org/10.1007/978-3-319-92102-0_54

17. Del Giudice, V., De Paola, P., Manganelli, B., Forte, F.: The monetary valuation of environmental externalities through the analysis of real estate prices. Sustain. Build. Environ. **9**, 229 (2017). https://doi.org/10.3390/su9020229

18. Del Giudice, V., Massimo, D.E., De Paola, P., Forte, F., Musolino, M., Malerba, A.: Post carbon city and real estate market: testing the dataset of Reggio Calabria market using spline smoothing semiparametric method. In: Calabrò, F., Della Spina, L., Bevilacqua, C. (eds.) ISHT 2018. SIST, vol. 100, pp. 206–214. Springer, Cham (2019). https://doi.org/10.1007/978-3-319-92099-3_25

19. Massimo, D.E., Del Giudice, V., De Paola, P., Forte, F., Musolino, M., Malerba, A.: Geographically weighted regression for the post carbon city and real estate market analysis: a case study. In: Calabrò, F., Della Spina, L., Bevilacqua, C. (eds.) Smart Innovation, Systems and Technologies; New Metropolitan Perspectives. ISHT 2018, vol. 100, pp. 142–149. Springer, Cham (2019). https://doi.org/10.1007/978-3-319-92102-3_17

20. De Paola, P., Del Giudice, V., Massimo, D.E., Forte, F., Musolino, M., Malerba, A.: Isovalore maps for the spatial analysis of real estate market: a case study for a central urban area of Reggio Calabria, Italy. In: Calabrò, F., Della Spina, L., Bevilacqua, C. (eds.) ISHT 2018. SIST, vol. 100, pp. 402–410. Springer, Cham (2019). https://doi.org/10.1007/978-3-319-92099-3_46

21. Del Giudice, V., Massimo, D.E., De Paola, P., Del Giudice, F.P., Musolino, M.: Green buildings for post carbon city: determining market premium using spline smoothing semiparametric method. In: Bevilacqua, C., Calabrò, F., Della Spina, L. (eds.) NMP 2020. SIST, vol. 178, pp. 1227–1236. Springer, Cham (2021). https://doi.org/10.1007/978-3-030-48279-4_114

22. Del Giudice, V., Massimo, D.E., Salvo, F., De Paola, P., De Ruggiero, M., Musolino, M.: Market price premium for green buildings: a review of empirical evidence. case study. In: Bevilacqua, C., Calabrò, F., Della Spina, L. (eds.) NMP 2020. SIST, vol. 178, pp. 1237–1247. Springer, Cham (2021). https://doi.org/10.1007/978-3-030-48279-4_115

23. De Paola, P., Del Giudice, V., Massimo, D.E., Del Giudice, F.P., Musolino, M., Malerba, A.: Green building market premium: detection through spatial analysis of real estate values. a case study. In: Bevilacqua, C., Calabrò, F., Della Spina, L. (eds.) NMP 2020. SIST, vol. 178, pp. 1413–1422. Springer, Cham (2021). https://doi.org/10.1007/978-3-030-48279-4_132

24. Spampinato, G., Massimo, D.E., Musarella, C.M., De Paola, P., Malerba, A., Musolino, M.: Carbon sequestration by cork oak forests and raw material to built up post carbon city. In: Calabrò, F., Della Spina, L., Bevilacqua, C. (eds.) Smart Innovation, Systems and Technologies; New Metropolitan Perspectives, ISHT 2018, vol. 101, pp. 663–671. Springer, Cham (2019)

25. Manganelli, B., Morano, P., Tajani, F.: The risk assessment in Ellwood's financial analysis for the indirect estimate of urban properties. AESTIMUM **55**, 19–41 (2009)

26. Manganelli, B., Tajani, F.: Optimised management for the development of extraordinary public properties. J. Prop. Invest. Financ. **32**(2), 187–201 (2014)

27. Morano, P., Tajani, F.: Least median of squares regression and minimum volume ellipsoid estimator for outliers detection in housing appraisal. Int. J. Bus. Intell. Data Min. **9**(2), 91–111 (2014)

28. Morano, P., Locurcio, M., Tajani, F., Guarini, M.R.: Fuzzy logic and coherence control in multi-criteria evaluation of urban redevelopment projects. Int. J. Bus. Intell. Data Min. **10**(1), 73–93 (2015). ISSN: 1743-8187

29. Tajani, F., Morano, P., Locurcio, M., D'Addabbo, N.: Property valuations in times of crisis: artificial neural networks and evolutionary algorithms in comparison. In: Gervasi, O., et al.

(eds.) ICCSA 2015. LNCS, vol. 9157, pp. 194–209. Springer, Cham (2015). https://doi.org/10.1007/978-3-319-21470-2_14

30. Tajani, F., Liddo, F.D., Guarini, M.R., Ranieri, R., Anelli, D.: An assessment methodology for the evaluation of the impacts of the COVID-19 pandemic on the Italian housing market demand. Buildings **11**(12), 592 (2021)

31. Massimo, D.E., Musolino, M., Fragomeni, C., Malerba, A.: A green district to save the planet. In: Mondini, G., Fattinnanzi, E., Oppio, A., Bottero, M., Stanghellini, S. (eds.) SIEV 2016. GET, pp. 255–269. Springer, Cham (2018). https://doi.org/10.1007/978-3-319-78271-3_21

32. Bentivegna, V.: The evaluation of structural-physical projects in urban distressed areas. In: Mondini, G., Fattinnanzi, E., Oppio, A., Bottero, M., Stanghellini, S. (eds.) SIEV 2016. GET, pp. 17–36. Springer, Cham (2018). https://doi.org/10.1007/978-3-319-78271-3_2

33. Massimo, D.E., Fragomeni, C., Malerba, A., Musolino, M.: Valuation supports green university: case action at Mediterranea campus in Reggio Calabria. Procedia Soc. Behav. Sci. **223**, 17–24 (2016). https://doi.org/10.1016/j.sbspro.2016.05.278

34. Musolino, M., Massimo, D.E.: Mediterranean urban landscape. integrated strategies for sustainable retrofitting of consolidated city. In: Sabiedriba, Integracija, Izglitiba [Society, Integration, Education], Proceedings of the Ispalem/Ipsapa International Scientific Conference, Udine, Italy, 27–28 June 2013, vol. 3, pp. 49–60. Rezekne Higher Education Institution, Rezekne (2013)

35. Calabrò, F., Cassalia, G., Lorè, I.: The economic feasibility for valorization of cultural heritage. the restoration project of the reformed fathers' convent in Francavilla Angitola: The Zibìb Territorial Wine Cellar. In: Bevilacqua, C., Calabrò, F., Della Spina, L. (eds.) NMP 2020. SIST, vol. 178, pp. 1105–1115. Springer, Cham (2021). https://doi.org/10.1007/978-3-030-48279-4_103

36. Calabrò, F., Mafrici, F., Meduri, T.: The valuation of unused public buildings in support of policies for the inner areas. the application of SostEc model in a case study in Condofuri (Reggio Calabria, Italy). In: Bevilacqua, C., Calabrò, F., Della Spina, L. (eds.) NMP 2020. SIST, vol. 178, pp. 566–579. Springer, Cham (2021). https://doi.org/10.1007/978-3-030-48279-4_54

37. Mallamace, S., Calabrò, F., Meduri, T., Tramontana, C.: Unused real estate and enhancement of historic centers: legislative instruments and procedural ideas. In: Calabrò, F., Della Spina, L., Bevilacqua, C. (eds.) ISHT 2018. SIST, vol. 101, pp. 464–474. Springer, Cham (2019). https://doi.org/10.1007/978-3-319-92102-0_49

38. Barrile, V., Malerba, A., Fotia, A., Calabrò, F., Bernardo, C., Musarella, C.: Quarries renaturation by planting cork oaks and survey with UAV. In: Bevilacqua, C., Calabrò, F., Della Spina, L. (eds.) NMP 2020. SIST, vol. 178, pp. 1310–1320. Springer, Cham (2021). https://doi.org/10.1007/978-3-030-48279-4_122

39. Spampinato, G., Malerba, A., Calabrò, F., Bernardo, C., Musarella, C.: Cork oak forest spatial valuation toward post carbon city by CO_2 sequestration. In: Bevilacqua, C., Calabrò, F., Della Spina, L. (eds.) NMP 2020. SIST, vol. 178, pp. 1321–1331. Springer, Cham (2021). https://doi.org/10.1007/978-3-030-48279-4_123

40. Fregonara, E., Curto, R., Grosso, M., Mellano, P., Rolando, D., Tulliani, J.-M.: Environmental technology, materials science, architectural design, and real estate market evaluation: a multidisciplinary approach for energy-efficient buildings. J. Urban Technol. **20**, 57–80 (2013). https://doi.org/10.1080/10630732.2013.855512

41. Fregonara, E., Giordano, R., Rolando, D., Tulliani, J.-M.: Integrating environmental and economic sustainability in new building construction and retrofits. J. Urban Technol **23**, 26 (2016). https://doi.org/10.1080/10630732.2016.1157941. ISSN: 1063-0732

42. Fregonara, E., Ferrando, D.G.: How to model uncertain service life and durability of components in life cycle cost analysis applications? The stochastic approach to the factor method. Sustainability **10**, 3642 (2018). https://doi.org/10.3390/su13052838

43. Fregonara, E., Ferrando, D.G., Pattono, S.: Economic – environmental sustainability in building projects: introducing risk and uncertainty in LCCE and LCCA. Sustainability **10**, 1901 (2018). https://doi.org/10.3390/su10061901
44. Fregonara, E., Ferrando, D.G., Chiesa, G.: Economic valuation of buildings sustainability with uncertainty in costs and in different climate conditions. In: Bevilacqua, C., Calabrò, F., Della Spina, L. (eds.) NMP 2020. SIST, vol. 178, pp. 1217–1226. Springer, Cham (2021). https://doi.org/10.1007/978-3-030-48279-4_113

Climate Change Adaptation of Buildings Using Nature-Based Solutions: Application in Alentejo Central - LIFE myBUILDINGisGREEN

Teresa Batista[1,2(✉)] ⓘ, Ricardo Barros[2] ⓘ, José Fermoso Domínguez[3] ⓘ,
Raquel Marijuan Cuevas[3] ⓘ, Jordi Serramia Ruiz[4], and Salustiano Torre Casado[5] ⓘ

[1] Instituto Mediterrâneo para a Agricultura, Ambiente e Desenvolvimento, (MED) Universidade de Évora, Pólo da Mitra, Ap. 94, 7006-554 Évora, Portugal
mtfb@uevora.pt
[2] Comunidade Intermunicipal do Alentejo Central (CIMAC), R. 24 de Julho, n°1, 7000-673 Évora, Portugal
{tbatista,ricardo.barros}@cimac.pt
[3] CARTIF Technology Center, Parque tecnológico de Boecillo, p. 205, 47151 Boecillo, Valladolid, Spain
[4] SingularGreen, C. Francisco Carratalá Cernuda, 34, 03010 Alicante, Spain
[5] Real Jardín Botánico, CSIC, Pl. Murillo 2, 28014 Madrid, Spain

Abstract. Climate change has been recognized as one of the most serious environmental, social and economic challenges facing the world today. The Intermunicipal Plan for Climate Change Adaptation in Alentejo Central (PIAAC-AC), has already identified the tendencies and future scenarios of climate change in Alentejo Central until the end of the XXI century, namely the increase in the number of days with very high temperatures, the number of tropical nights (above 20 °C) and heat waves and the general decrease in annual rainfall. In this scenario, the concerns with school communities and users of social services increase. The project "LIFE-myBUILDINGisGREEN" - "Application of Nature-Based Solutions for local adaptation of educational and social buildings to Climate Change", developed in partnership by CIMAC (Portugal), CARTIF Technology Center (Spain), Diputación de Badajoz and CSIC - Consejo Superior de Investigaciones Cientificas (Real Jardin Botanico – Spain – Project Leader) and the Porto City Council (Portugal), focuses on the construction sector, in particular on education and social services buildings in cities in Europe. It aims to implement prototypes (building adaptation) of nature-based solutions (NBS) on walls, roofs, exterior surfaces and parking on three pilot buildings. The overall objective is to contribute to improve resilience in these buildings using autochthone natural vegetation. This paper presents the project, its objectives and expected results.

Keywords: Climate Change Adaptation · Nature Based Solutions · Green roofs · Green walls · Educational buildings

F. Calabrò et al. (Eds.): NMP 2022, LNNS 482, pp. 1658–1663, 2022.
https://doi.org/10.1007/978-3-031-06825-6_159

1 Introduction

Climate change has been recognized as one of the most serious environmental, social and economic challenges facing the world today [1, 2]. in Alentejo Central (south of Portugal), the Intermunicipal Community of Municipalities developed in 2018 the Intermunicipal Plan for Climate Change Adaptation (PIAAC-AC) [3]. This study identified the actual tendencies and future scenarios for climate change until the end of the XXI century, which are mainly the increase in the number of days with very high temperatures, the increase of the number of tropical nights (above 20 °C) and heat waves, and the general decrease in annual rainfall. In this scenario, the concerns with school communities and users of social services increase, and should be addressed.

Actually, europe's educational and social service buildings will face multiple challenges in the coming decades, and climate change will add pressure to it. The project LIFE myBUILDINGisGREEN focuses on the building sector, specifically on the public buildings dedicated to education and social services existing in all cities and towns in Europe. The impacts of climate change (heat waves, changes in annual and seasonal precipitation patterns) are affecting the health and well-being of children and elderly people who are the main users of these types of centers. In this sense, classroom temperatures should not exceed 27 °C according to law regulations in many countries [4]. However, indoor temperatures could reach higher values than 38 °C in the Mediterranean area during last Spring and Summer time.

Initially, LIFE myBUILDINGisGREEN focused their efforts on predicting the impact of nature-based solutions (NBS) implementation. The main concern of this initial stage was to analyze the benefits on both interior comfort and energy efficiency [5–8].

Chronic low prioritization of funds and resources to support environmental health in schools and lack of clear regulatory oversight undergird the new risks from Climate Change (CC). NBS offer an exciting prospect for resilience building and advancing urban planning to address complex urban challenges simultaneously [9–12]. NBS can replace or compliment air conditioning for heat risk reduction by reducing outdoor temperature and isolating buildings envelope (green roofs and façades or shading structures) [13–17].

The project LIFE myBuildingisGreen aims to strengthen and support cities, and EU within the shift of the old urban paradigm based on grey surfaces, and works to contribute with specific and innovative solutions with nature-based (more efficient, more sustainable, greener), to the adaptation of buildings improving the bioclimatic comfort and quality of life of users, but also act (in each school - Évora, Oporto and Solana de Los Barros), as an example, inspiration and a starting point to shift and adapt to climate change in cities.

This article explains the most innovative NBS developed for the Évora pilot building and the expected results. The project is in implementation until December 2023.

2 Nature Based Solutions (NBS): Prototypes and Pilot Solutions

2.1 Prototype Roofs mBiGWTray for Application in Évora Pilot

This system was created by the CARTIF Team and SingularGreen [18] and consists of a multilayer tray to maintain cover vegetation that is encapsulated with a white waterproof sheet to collect rainwater and reduce water loss. The design of the system, including the selection of the appropriate plant species, is done so as not to require the installation of auxiliary irrigation. In the upper part of the encapsulation, there are some holes for planting plant species.

Schematically, the system would be a system of extensive vegetative cover pocketed to make it more resistant to rainfall shortages (Fig. 1).

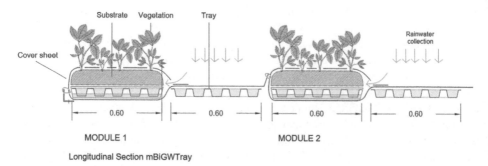

Fig. 1. Prototype mBiGWTray: schema [5, 18].

This system allows the installation of the trays on the roof directly occupying 50% of the surface with the vegetal part initially and the rest with a white surface that avoids the excessive capture of thermal energy. The installation is carried out using a checked system in which the planting surfaces alternate with the water collection surfaces.

In the joining area of each module, some holes allow the entrance of water from the zone of collected water towards the zone with the vegetal system.

The bags have a drainage hole at the surface of the tray to allow the greatest amount of water to be stored avoiding the pooling of the substrate in which the roots of the plants are found.

Both the weight of the tray itself and its flat design mean that, in principle, an auxiliary roof anchor system is not necessary. However, if the prevailing winds in the area were very strong, the system could be weighed down using draining aggregate as partial filling of the holes in the tray.

On the other hand, the system is compatible with drip irrigation that can be integrated into the base structure. More homogeneous vegetation maintenance would be achieved throughout the year. However, the initial design has been carried out so that the implementation of an irrigation system is not necessary and the previous tests that are going to be carried out are aimed in this direction.

2.2   Prototype Facades mBiGToldo for Creating Vertical Green Surfaces

This system has been designed to create vertical surfaces with vegetation of very low thickness to create shading with a contribution of humidity to the environment [6]. The system consists of an impermeable sheet on which a non-woven felt is adhered and on which a semi-woven substrate is projected. Due to the low substrate thickness, a hydroponic irrigation is integrated that is distributed by gravity across the surface of the substrate (Fig. 2). In the lower zone, a channel for collecting excess irrigation is integrated and returned to the irrigation station.

Irrigation is done through hydroponic irrigation station with adequate programming to cover the needs of the vertical garden at all times of the year and the environmental conditions that are to be generated. It is necessary to connect the irrigation station, to be installed under the level of the mBiGToldo, with all the gardens, both to provide the irrigation and to collect.

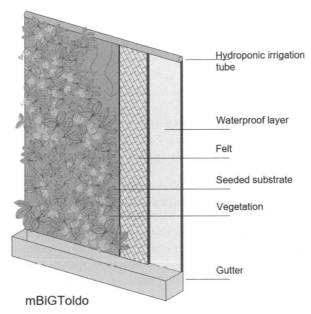

Fig. 2. mBiGToldo schema [6, 19].

The components of the system are:

- Support frame. Structure with the appropriate dimensions made with the material with which the vertical wall structure that will support the prototypes of the facade is built.
- Waterproof support. It must be a sheet-shaped material resistant to punching and tearing to facilitate fixing. The currently recommended material is a PVC awning sheet but more sustainable materials are being sought.
- Root fixing sheet based on non-woven felt or rock wool.
- Mix of substrate and compatible seeds that is applied by projection.
- The characteristics of the selected species will be easily propagated by seeds.
- Drip irrigation tube in the upper part and water collection gutter and injected to leftover water collection tube.

The system is only compatible with hydroponic drip irrigation that can be integrated into the support structure of the awning. The design includes a system for collecting excess irrigation water and returning it to the tank for optimization of water consumption.

Among the species that can be used are the *Festuca rubra*, *Agrostis stolonifera* and *Sagina subulata*, but others can be evaluated depending on the location and suggestions made by the experts from Royal Botanic Garden.

3   Conclusions

With the execution of this project, the following results are expected:

- Climate adaptation of the pilot buildings (three educational centers) through the implementation of NBS prototypes.
- Well-being improvement and thermal comfort for around 1.000 citizens in the 3 pilot buildings and communities.
- The elaboration of Reference Reports and Good Practice Manuals on the application of NBS in public buildings of education services for the 3 climatic risk areas of the EU.
- For the entire project it is expected to reduce the production of 27 Ton of CO_2/year and 144 kg of NO_x (reduction 20% and 7% respectively).
- It will be collected about 2.700 m^3/project of rainfall that represents 30% of each building.
- Increase the area of green areas in each building by approximately 0.5 ha.
- Reduction of 50% of the energy costs for cooling and 10% for heating. This amounts to 1000€/building.
- Integration of NBS into regulations, action plans and environmental programs.
- Execution of expert meetings, demonstration workshops, expert workshops, online seminars and transnational conferences, stakeholders training, website, publication of articles, connection with media associations and videos.

The overall conclusion is that NBS can be a technical and scientific solution for these types of buildings allowing a more affordable and energyefficient solution for the adaptation to climate change in Mediterranean regions.

References

1. del Río, S., et al.: Modelling the impacts of climate change on habitat suitability and vulnerability in deciduous forests in Spain. Ecol. Indic. **131**, 108202 (2021). https://doi.org/10.1016/j.ecolind.2021.108202
2. del Río, S., Álvarez-Esteban, R., Cano, E., Pinto-Gomes, C., Penas, Á.: Potential impacts of climate change on habitat suitability of *Fagus sylvatica* L. Forests in Spain. . Plant Biosyst. **152**, 1205–1213 (2018)
3. CIMAC: Plano Intermunicipal de Adaptação às Alterações Climáticas do Alentejo Central (PIAAC-AC) (2017)
4. RD 485/1997. https://www.boe.es/buscar/act.php?id=BOE-A-1997-8668
5. Gómez, G., et al.: Selection of nature-based solutions to improve comfort in schools during heat waves. Int. J. Energy Prod. Mgmt. **6**(2), 157–16 (2021)
6. Gómez, G., et al.: Prediction of thermal comfort and energy behavior through NatureBased Solutions implementation: a case study in Badajoz (Spain). WIT Trans. Built Environ. **195** (2020). https://doi.org/10.2495/ARC200021

 7. Fermoso, J., et al.: Improvement of classroom conditions and CO_2 concentrations through natural ventilation measures reinforced with NBS implementation. In: Ksibi, M., et al. (eds.) EMCEI 2019. ESE, pp. 2305–2309. Springer, Cham (2021). https://doi.org/10.1007/978-3-030-51210-1_361

 8. Batista, T.: Soluções baseadas na natureza (NBS Nature-Based Solutions) para a adaptação local de edifícios públicos às alterações climáticas no Alentejo Central - LIFE MyBuildingis-Green. Presentación en el Seminario "Soluciones naturales para la adaptación de edificios al cambio climático". Badajoz, 13 de Noviembre 2019 (2019)

 9. Frantzeskaki, N., et al.: Nature-based solutions for urban climate change adaptation: linking science, policy, and practice communities for evidence-based decision-making. BioScience **69**, 455–466 (2019) https://doi.org/10.1093/biosci/biz042

10. Spampinato, G., Malerba, A., Calabrò, F., Bernardo, C., Musarella, C.: Cork oak forest spatial valuation toward post carbon city by CO_2 sequestration. In: Bevilacqua, C., Calabrò, F., Della Spina, L. (eds.) NMP 2020. SIST, vol. 178, pp. 1321–1331. Springer, Cham (2021). https://doi.org/10.1007/978-3-030-48279-4_123

11. Barrile, V., Malerba, A., Fotia, A., Calabrò, F., Bernardo, C., Musarella, C.: Quarries renaturation by planting cork oaks and survey with UAV. In: Bevilacqua, C., Calabrò, F., Della Spina, L. (eds.) NMP 2020. SIST, vol. 178, pp. 1310–1320. Springer, Cham (2021). https://doi.org/10.1007/978-3-030-48279-4_122

12. Massimo, D.E., Del Giudice, V., Malerba, A., Bernardo, C., Musolino, M., De Paola, P.: Valuation of ecological retrofitting technology in existing buildings: a real-world case study. Sustainability **13**, 7001 (2021). https://doi.org/10.3390/su13137001

13. Connop, S., et al.: Renaturing cities using a regionally focused biodiversity-led multifunctional benefits approach to urban green infrastructure. Environ. Sci. Policy **62**, 99–111 (2016)

14. Panno, A., et al.: Nature-based solutions to promote human resilience and wellbeing in cities during increasingly hot summers. Environ. Res. **159**, 249–256 (2017)

15. Musolino, M., Malerba, A., De Paola, P., Musarella, C.M.: Building efficiency adopting ecological materials and bio architecture techniques. ArcRHistoR S **6**, 706–717 (2019)

16. Del Giudice, V., Massimo, D.E., De Paola, P., Del Giudice, F.P., Musolino, M.: Green buildings for post carbon city: determining market premium using spline smoothing semiparametric method. In: Bevilacqua, C., Calabrò, F., Della Spina, L. (eds.) NMP 2020. SIST, vol. 178, pp. 1227–1236. Springer, Cham (2021). https://doi.org/10.1007/978-3-030-48279-4_114

17. Massimo, D.E., Musolino, M., Fragomeni, C., Malerba, A.: A green district to save the planet. In: Mondini, G., Fattinnanzi, E., Oppio, A., Bottero, M., Stanghellini, S. (eds.) SIEV 2016. GET, pp. 255–269. Springer, Cham (2018). https://doi.org/10.1007/978-3-319-78271-3_21

18. Fermoso, J.: Prototipo fachadas mBiGToldo para creación de superficies verdes verticales. CARTIF. Life - MyBuildingisGreen Technical report (2019)

19. Fermoso, J.: Prototipo cubiertas mBiGWTray mediante bandejas encapsuladas con alta capacidad de retención de agua. CARTIF. Life - MyBuildingisGreen Technical report (2019)

The Use of Plants for Building Purposes in the Popular Tradition

Miriam Patti[1], Carmelo Maria Musarella[1]([envelope]) [iD], Valentina Lucia Astrid Laface[1] [iD],
Ana Cano-Ortiz[2,3] [iD], Ricardo Quinto-Canas[4,5] [iD], and Giovanni Spampinato[1] [iD]

[1] Department of Agriculture, Mediterranean University of Reggio Calabria, Reggio Calabria,
Italy
carmelo.musarella@unirc.it
[2] Department of Animal and Plant Biology and Ecology, Section of Botany, University of Jaén,
Jaén, Spain
[3] Institute for Secondary Education and Baccalaureate, Ministry of Education Madrid, Madrid,
Spain
[4] Faculty of Sciences and Technology, University of Algarve, Faro, Portugal
[5] Centre of Marine Sciences (CCMAR), University of Algarve, Faro, Portugal

Abstract. Designing, building and operating a "green" building means that it considerably reduces the negative impacts and increase the positive ones on the climate and the natural environment. Thanks to green buildings it is possible to preserve natural resources and improve the quality of people's life.

Ethnobotany is a science that describes the relationships between humans and plants, in particular by describing and analyzing the traditional uses that are made of them. In addition to the most common food, medicinal and religious uses, there are artisanal ones and in particular for building purposes.

The main goal of this work is to present a synthesis of the traditional uses of plants as a building material, useful for constructing ecologically sustainable buildings.

Among the most common species for this last purpose are two species belonging to the Poaceae family: *Ampelodesmos mauritanicus* (Mauritania grass) and *Arundo donax* (giant reed). These species have very strong fibres and a high potential in carbon sequestration too: therefore, in addition to representing a valid natural resource that can be re-evaluated in modern and more technological terms, they could guarantee a valid aid for the abatement of greenhouse gases.

Even ethnobotany, therefore, can actively contribute to determining environmental policy choices that guarantee an effective fight against climate change.

Keywords: Ethnobotany · Green buildings · Green districts · Post-carbon city · Natural products

1 Introduction

1.1 For a Green Planet: From the Past to the Future

"Green" is a word that is used a lot nowadays to mean anything or action that is environmentally friendly [1]. The issue of "Green buildings" is very topical and has been

F. Calabrò et al. (Eds.): NMP 2022, LNNS 482, pp. 1664–1670, 2022.
https://doi.org/10.1007/978-3-031-06825-6_160

dealt with in various scientific contributions about both their ecological retrofitting and valuation [2–5] and price market premium [6–12]. In order to be able to deal with these issues in a broader and more integrated way, the concept of "Green district" is increasingly affirming itself [13–17], as a generalization of the "Green Buildings" strategy, up to the creation of the "post-carbon city", thanks to the consolidated naturalistic characterization of energy efficiency obtained with natural materials which come from plant species very useful for carbon sequestration [18–23]. Furthermore, a city can be considered sustainable if it is built with materials whose extraction and processing do not have a strong impact on the environment and whose ecological footprint can be easily mitigated [24].

For this purpose, there are numerous examples of building interventions carried out with panels of granulated cork: they are very useful to prevent mould, provide insulation from the cold and the warm (reducing at the same time energetic costs and CO_2 emissions) and act as igro-regulators [3, 4, 14, 17, 25]. Furthermore, it has been demonstrated that "healthier", "greener" and with better energy performance buildings can be sold at higher prices [9–11, 26, 27].

This model today followed is the same of the past, ac-cording to which peoples have always resorted to natural resources to meet their primary needs, including housing: all this knowledge is studied by Ethnobotany.

1.2 What is Ethnobotany?

Ethnobotany is the scientific study of the complex and dynamic relationships between peoples and the plant heritage that characterizes their natural environment [28]. People established a relationship with plants by looking first and foremost those with which to feed himself, cure himself of diseases, manufacture tools, tools for daily use in the fields and at home, create artifacts, toys and hobbies [29–34].

This work aims to review the traditional uses of plants as a building material, with a low environmental impact and the ability to reduce the concentration of CO_2 in the atmosphere, in order to demonstrate how it is possible to plan modern buildings using "green" materials.

2 Traditional Uses of Plants in Buildings

There are many traditional uses of plants that have been found for different building purposes. Among them, we consider here two species belonging to the Poaceae family: *Ampelodesmos mauritanicus* (Poir.) T.Durand & Schinz (Mauritania grass) and *Arundo donax* L. (giant reed). *Ampelodesmos mauritanicus* is very common throughout the Mediterranean basin, from the Iberian Peninsula to Greece, including Northern Africa, growing from the coastal to the hilly areas (up to 1,200 m.a.s.l.) [35]. *Arundo donax* is a perennial grass growing spontaneously in temperate and tropical zones almost all over the world and can act as an invasive species [36].

There are very few building uses found for *A. mauritanicus*. Once dried, the stems of *A. mauritanicus* were mowed and cleaned of the leaves, cut to size and used to build domestic shutters [37]. Its stems were used until the 1950s to make roller shutters [38].

In general, its fibers were used in the construction of the roofs of the huts due to their durability, strong resistance to water and heat insulation [39].

The roofs of the rural houses were also made with the stems of *A. donax*: after cutting the stems to the appropriate extent and cleaned from the leaves, they were tied together with thin iron wire in order to obtain large mats. The stems were placed laterally on the skeleton of the roof which was made up of load-bearing wooden beams. A mixture of plaster was spread on the cane mats and finally the tiles were laid in an imbrice [39, 40]. The barrel roofs of town houses were also made with the reeds [40]. In southern Iraq, "Mudhif", an imposing building used as a meeting place for ceremonies and for the reception of foreign guests, were built with large and thick columns realized with *A. donax* stems folded up to arches. This building system creates a pre-stressing of the arches that are initially inserted into the soil at opposing angles [41, 42]. In Calabria, after the earthquake of 1783, a reconstruction plan of the collapsed buildings was initiated by providing a load-bearing structure framed in wood and a sheath which, in the "poorer" solution, adopted above all in rural areas, consisted of two layers of *A. donax* covered by a plaster layer: this building system is still evident in some abandoned house [43]. In the XIX century, *A. donax* was introduced from the Mediterranean area into North America for roof thatching [44]. In Sicily, some ethnobotanical studies reported, through interviews carried out to the rural population, that the stems of *A. donax* were used for the construction of raftered ceilings: from the ridge of the roof, some beams were putted on the perimeter walls. Above them were applied bundles of reeds which previously had been well cleaned of the residues of the leaves and cut to size and, then, were arranged tying them together with string or wire. A layer of lime was spread over this roof and the tiles rested on it [45, 46].

Arundo donax and *A. mauritanicus* are very good for carbon sequestration [47, 48]. It has been shown that *A. donax* has carbon accumulation rates (3.7 to 4.9 t has year^{-1}) far greater than those in tree biomass of other species, e.g. *Eucalyptus cladocalyx* F.Muell. of the south of Australia [49]. However, *A. donax* is native to western or southern Asia, cultivated for millennia in the Mediterranean regions and in others with similar climate, where it is now completely naturalized and sometimes invasive: this allowed it to out-compete native plant species, dramatically altering riparian habitats [50]. It is considered as one of the 100 world's worst invasive alien species [51]. Its widespread use is due to its robustness which has also been demonstrated experimentally by various authors: in fact, the *A. donax* fiber has a very high tensile strength [52]. The production of panels made with particles obtained from the grinding of stems and rhizomes of *A. donax* with low thermal conductivity and excellent mechanical properties is demonstrated [53].

3   Conclusions

Numerous scientific evidences confirm that natural resources increasingly represent a precious source of useful resources for various purposes, including construction. Thanks to these works, with our review we can confirm: a) the technical efficiency of different materials of plant origin for the energy efficiency of buildings; b) the considerably reduced environmental impact they have; c) Ethnobotany can be considered as an useful tool for investigating the traditional uses of plants for building purposes. Therefore, we can again affirm that man must move more and more convinced towards a sustainable use of natural resources. Only in this way will we be able to obtain both significant energy savings in economic and ecological terms, and respect for nature that will allow us to slow down the ongoing process of climate change, which is already causing irreversible damage to our planet, including habitat and biodiversity erosion [54–58].

References

1. Panuccio, M.R., Mallamaci, C., Attinà, E., Muscolo, A.: Using digestate as fertilizer for a sustainable tomato cultivation. Sustainability **13**, 1574 (2021). https://doi.org/10.3390/su13031574

2. Massimo, D.E., Del Giudice, V., Malerba, A., Bernardo, C., Musolino, M., De Paola, P.: Valuation of ecological retrofitting technology in existing buildings: a real-world case study. Sustainability **13**, 7001 (2021). https://doi.org/10.3390/su13137000

3. Malerba, A., Massimo, D.E., Musolino, M., Nicoletti, F., De Paola, P.: Post carbon city: building valuation and energy performance simulation programs. In: Calabrò, F., Della Spina, L., Bevilacqua, C. (eds.) ISHT 2018. SIST, vol. 101, pp. 513–521. Springer, Cham (2019). https://doi.org/10.1007/978-3-319-92102-0_54

4. Massimo, D.E.: Green building: characteristics, energy implications and environmental impacts. Case study in Reggio Calabria, Italy. In: Coleman-Sanders, M. (ed.) Green Building and Phase Change Materials: Characteristics, Energy Implications and Environmental Impacts, vol. 1, pp. 71–101. Nova Science Publishers, New York (2015)

5. Massimo, D.E.: Valuation of urban sustainability and building energy efficiency. A case study. Int. J. Sustain. Dev. **12**, 223–247 (2010). https://doi.org/10.1504/IJSD.2009.032779

6. Del Giudice, V., Massimo, D.E., De Paola, P., Del Giudice, F.P., Musolino, M.: Green buildings for post carbon city: determining market premium using spline smoothing semiparametric method. In: Bevilacqua, C., Calabrò, F., Della Spina, L. (eds.) NMP 2020. SIST, vol. 178, pp. 1227–1236. Springer, Cham (2021). https://doi.org/10.1007/978-3-030-48279-4_114

7. Del Giudice, V., Massimo, D.E., Salvo, F., De Paola, P., De Ruggiero, M., Musolino, M.: Market price premium for green buildings: a review of empirical evidence. Case study. In: Bevilacqua, C., Calabrò, F., Della Spina, L. (eds.) NMP 2020. SIST, vol. 178, pp. 1237–1247. Springer, Cham (2021). https://doi.org/10.1007/978-3-030-48279-4_115

8. De Paola, P., Del Giudice, V., Massimo, D.E., Del Giudice, F.P., Musolino, M., Malerba, A.: Green building market premium: detection through spatial analysis of real estate values. A case study. In: Bevilacqua, C., Calabrò, F., Della Spina, L. (eds.) NMP 2020. SIST, vol. 178, pp. 1413–1422. Springer, Cham (2021). https://doi.org/10.1007/978-3-030-48279-4_132

9. Massimo, D.E., Del Giudice, V., De Paola, P., Forte, F., Musolino, M., Malerba, A.: Geographically weighted regression for the post carbon city and real estate market analysis: a case study. In: Calabrò, F., Della Spina, L., Bevilacqua, C. (eds.) ISHT 2018. SIST, vol. 100, pp. 142–149. Springer, Cham (2019). https://doi.org/10.1007/978-3-319-92099-3_17

10. Del Giudice, V., Massimo, D.E., De Paola, P., Forte, F., Musolino, M., Malerba, A.: Post carbon city and real estate market: testing the dataset of Reggio Calabria market using spline smoothing semiparametric method. In: Calabrò, F., Della Spina, L., Bevilacqua, C. (eds.) ISHT 2018. SIST, vol. 100, pp. 206–214. Springer, Cham (2019). https://doi.org/10.1007/978-3-319-92099-3_25

11. De Paola, P., Del Giudice, V., Massimo, D.E., Forte, F., Musolino, M., Malerba, A.: Isovalore maps for the spatial analysis of real estate market: a case study for a Central Urban Area of Reggio Calabria, Italy. In: Calabrò, F., Della Spina, L., Bevilacqua, C. (eds.) ISHT 2018. SIST, vol. 100, pp. 402–410. Springer, Cham (2019). https://doi.org/10.1007/978-3-319-92099-3_46

12. Del Giudice, V., De Paola, P., Manganelli, B., Forte, F.: The monetary valuation of environmental externalities through the analysis of real estate prices. Sustain. Build. Environ. **9**, 229 (2017). https://doi.org/10.3390/su9020229

13. Malerba, A., Massimo, D.E., Musolino, M.: Valuating historic centers to save planet soil. In: Mondini, G., Fattinnanzi, E., Oppio, A., Bottero, M., Stanghellini, S. (eds.) SIEV 2016. GET, pp. 297–311. Springer, Cham (2018). https://doi.org/10.1007/978-3-319-78271-3_24

14. Massimo, D.E., Musolino, M., Fragomeni, C., Malerba, A.: A green district to save the planet. In: Mondini, G., Fattinnanzi, E., Oppio, A., Bottero, M., Stanghellini, S. (eds.) SIEV 2016. GET, pp. 255–269. Springer, Cham (2018). https://doi.org/10.1007/978-3-319-78271-3_21

15. Musolino, M., Massimo, D.E.: Mediterranean urban landscape. Integrated strategies for sustainable retrofitting of consolidated city. In: Sabiedriba, Integracija, Izglitiba [Society, Integration, Education], Proceedings of the Ispalem/Ipsapa International Scientific Conference, Udine, Italy, 27–28 June 2013, vol. 3, pp. 49–60. Rezekne Higher Education Institution, Rezekne, Latvija (2013)

16. Massimo, D.E., Battaglia, L., Fragomeni, C., Guidara, M., Rudi, G., Scala, C.: Sustainability valuation for urban regeneration. The Geomatic Valuation University Lab research. Adv. Eng. Forum 594–599 (2014). https://doi.org/10.4028/www.scientific.net/AEF.11.594

17. Massimo, D.E., Fragomeni, C., Malerba, A., Musolino, M.: Valuation supports green university: case action at Mediterranea campus in Reggio Calabria. Procedia Soc. Behav. Sci. **223**, 17–24 (2016). https://doi.Org/10.1016/j.sbspro.2016.05.278

18. Spampinato, G., Massimo, D.E., Musarella, C.M., De Paola, P., Malerba, A., Musolino, M.: Carbon sequestration by cork oak forests and raw material to built up post carbon city. In: Calabrò, F., Della Spina, L., Bevilacqua, C. (eds.) ISHT 2018. SIST, vol. 101, pp. 663–671. Springer, Cham (2019). https://doi.org/10.1007/978-3-319-92102-0_72

19. Spampinato, G., Malerba, A., Calabrò, F., Bernardo, C., Musarella, C.: Cork oak forest spatial valuation toward post carbon city by CO2 sequestration. In: Bevilacqua, C., Calabrò, F., Della Spina, L. (eds.) NMP 2020. SIST, vol. 178, pp. 1321–1331. Springer, Cham (2021). https://doi.org/10.1007/978-3-030-48279-4_123

20. Musolino, M., Malerba, A., De Paola, P., Musarella, C.M.: Building Efficiency Adopting Ecological Materials and Bio Architecture Techniques, pp. 707–717. ArcRHistoR (2019)

21. Barrile, V., Malerba, A., Fotia, A., Calabrò, F., Bernardo, C., Musarella, C.: Quarries renaturation by planting Cork oaks and survey with UAV. In: Bevilacqua, C., Calabrò, F., Della Spina, L. (eds.) NMP 2020. SIST, vol. 178, pp. 1310–1320. Springer, Cham (2021). https://doi.org/10.1007/978-3-030-48279-4_122

22. Nunes, L.J.R., Raposo, M.A.M., Meireles, C.I.R., Pinto-Gomes, C.J., Almeida Ribeiro, N.M.C.: Carbon sequestration potential of forest invasive species: a case study with Acacia dealbata Link. Resources **10**, 51 (2021). https://doi.org/10.3390/resources10050051

23. Quinto-Canas, R., et al.: Cork oak vegetation series of Southwestern Iberian Peninsula: diversity and ecosystem services. In: Bevilacqua, C., Calabrò, F., Della Spina, L. (eds.) NMP 2020. SIST, vol. 178, pp. 1279–1290. Springer, Cham (2021). https://doi.org/10.1007/978-3-030-48279-4_119

24. Pérez-García, F.J., Salmerón-Sánchez, E., Martínez-Hernández, F., Mendoza-Fernandez, A., Merlo, E., Mota, J.F.: Towards an eco-compatible origin of construction materials. Case study: Gypsum. In: Bevilacqua, C., Calabrò, F., Della Spina, L. (eds.) NMP 2020. SIST, vol. 178, pp. 1259–1267. Springer, Cham (2021). https://doi.org/10.1007/978-3-030-48279-4_117

25. Massimo, D.E.: Valuation of urban sustainability and building energy efficiency. A case study. Int. J. Sustain. Dev. **2–4**(12), 223–247 (2009)

26. Massimo, D.E.: Emerging Issues in Real Estate Appraisal: Market Premium for Building Sustainability, Aestimum, pp. 653–673 (2013)

27. Massimo, D.E.: Stima del green premium per la sostenibilità architettonica mediante Market Comparison Approach. Valori e Valutazioni (2011)

28. Voeks, R.: Ethnobotany. In: Richardson, D., Castree, N., Goodchild, M.F., Kobayashi, A., Liu, W., Marston, R.A. (eds.) International Encyclopedia of Geography: People, the Earth, Environment and Technology (2017). https://doi.org/10.1002/9781118786352.wbieg0300

29. Novais, M.H., Santos, I., Mendes, S., Pinto-Gomes, C.J.: Studies on pharmaceutical ethnobotany in Arrábida Natural Park (Portugal). J. Ethnopharmacol. **93**, 183–195 (2013)

30. Maruca, G., Spampinato, G., Turiano, D., Laghetti, G., Musarella, C.M.: Ethnobotanical notes about medicinal and useful plants of the Reventino Massif tradition (Calabria region, Southern Italy). Genet. Resour. Crop Evol. **66**(5), 1027–1040 (2019). https://doi.org/10.1007/s10722-019-00768-8

31. Musarella, C.M., Paglianiti, I., Cano-Ortiz, A., Spampinato, G.: Ethnobotanical study in the Poro and Preserre Calabresi territory (Vibo Valentia, S-Italy). Atti della Società Toscana di Scienze Naturali Memorie Serie B **126**, 13–28 (2019). https://doi.org/10.2424/ASTSN.M.2018.17

32. Abdul Aziz, M., Ullah, Z., Pieroni, A.: Wild Food Plant Gathering among Kalasha, Yidgha, Nuristani and Khowar Speakers in Chitral, NW Pakistan. Sustainability **12**, 9176 (2020). https://doi.org/10.3390/su12219176

33. Singh, B., et al.: Exploring plant-based ethnomedicine and quantitative ethnopharmacology: medicinal plants utilized by the population of Jasrota Hill in Western Himalaya. Sustainability **12**, 7526 (2020). https://doi.org/10.3390/su12187526

34. Bhat, M.N., Singh, B., Surmal, O., Singh, B., Shivgotra, V., Musarella, C.M.: Ethnobotany of the Himalayas: safeguarding medical practices and traditional uses of Kashmir Regions. Biology **10**, 851 (2021). https://doi.org/10.3390/biology10090851

35. Minissale, P.: Studio fitosociologico delle praterie ad *Ampelodesmos mauritanicus* della Sicilia. Colloques Phytosociologique **XXI**, 615–652 (1995)

36. Coffman, G.C., Ambrose, R.F., Rundel, P.W.: Wildfire promotes dominance of invasive giant reed (*Arundo donax*) in riparian ecosystems. Biol. Invasions **12**, 2723–2734 (2010)

37. Arcidiacono, S., Costa, R., Marletta, G., Pavone, P., Napoli, M.: Usi popolari delle piante selvatiche nel territorio di Villarosa (Enna, Sicilia Centrale). Quaderni di Botanica Ambientale e Applicata **21**, 95–118 (2010)

38. Savo, V.: Analisi etnobotanica della costiera amalfitana e valutazione dei risultati da un punto di vista scientifico ed economico (2009). http://hdl.handle.net/2307/604

39. Zergane, H., et al.: Habibi *Ampelodesmos mauritanicus* a new sustainable source for nanocellulose substrates. Ind. Crop. Prod. **144** (2020). https://doi.org/10.1016/j.indcrop.2019.112044

40. Martínez Francés, V., et al.: *Arundo donax* L. In: Pardo de Santayana, M., Morales, R., Aceituno, L., Molina, M. (eds.) Inventario Español de los Conocimientos Tradicionales relativos a la Biodiversidad. Primera Fase: Introducción, Metodología y Fichas. Ministerio de Agricultura, Alimentación y Medio Ambiente (2014)

41. Barreca, F., Fichera, C.R.: Wall panels of *Arundo donax* L. for environmentally sustainable agriculture buildings: thermal performance evaluation. J. Food Agric. Environ. **11**, 1353–1357 (2013)

42. Barreca, F., Martinez, A., Flores, J.A., Pastor, J.J.: Innovative use of giant reed and cork residues for panels of buildings in Mediterranean area. Resour. Conserv. Recycl. **140**, 259–266 (2019). https://doi.org/10.1016/j.resconrec.2018.10.005

43. Barreca, F.: Use of giant reed *Arundo donax* L. in rural constructions. Agric. Eng. Int. CIGR J. **14**(3), 46–52 (2012)

44. Pilu, R., Bucci, A., Badone, F.C., Landoni, M.: Giant reed (*Arundo donax* L.): a weed plant or a promising energy crop? Afr. J. Biotechnol. **11**, 9163–9174 (2012)

45. Aleo, M., Cambria, S., Bazan, G.: Tradizioni etnofarmacobotaniche in alcune comunità rurali dei Monti di Trapani (Siciliaoccidentale). Quad. Bot. Amb. Appl. **24**, 27–48 (2013)

46. Arcidiacono, S., Napoli, M., Oddo, G., Pavone, P.: Piante selvatiche d'uso popolare nei territori di Alcara Li Fusi e Militello Rosmarino (Messina, NE Sicilia). Quad. Bot. Ambient. Appl. **18**, 103–144 (2007)

47. Williams, C.M.J., Biswas, T.K., Marton, L., Czako, M.: *Arundo donax*. In: Singh, B.P. (ed.) Biofuel Crops: Production, Physiology and Genetics, p. 249. CABI, Wallingford (2013)

48. Corona, P., Badalamenti, E., Pasta, S., La Mantia, T.: Carbon storage of Mediterranean grasslands. Anales Jard. Bot. Madrid **73**(1), e029 (2016)

49. Paul, K.L., Jacobsen, K., Koul, V., Leppert, P., Smith, J.: Predicting growth and sequestration of carbon by plantations growing in regions of low-rainfall in Southern Australia. For. Ecol. Manag. **254**, 205–216 (2008)

50. CABI: *Arundo donax*. In: Invasive Species Compendium. CAB International, Wallingford, UK. https://www.cabi.org/isc/datasheet/1940. Accessed 05 Jan 2022

51. ISSG: Global Invasive Species Database (GISD). Invasive Species Specialist Group of the IUCN Species Survival Commission (2007)

52. Manniello, C., Cillis, G., Statuto, D., Di Pasquale, A., Picuno, P.: Experimental analysis on concrete blocks reinforced with *Arundo donax* fibers. J. Agric. Eng. (2021). https://doi.org/10.4081/jae.2021.1288

53. Andreu-Rodriguez, J., et al.: Agricultural and industrial valorization of *Arundo donax* L. Commun. Soil Sci. Plant Anal. **44**, 598–609 (2013). https://doi.org/10.1080/00103624.2013.745363

54. Del Río, S., Álvarez-Esteban, R., Cano, E., Pinto-Gomes, C.J., Penas, Á.: Potential impacts of climate change on habitat suitability of *Fagus sylvatica* L. Forests in Spain. Plant Biosyst. **152**, 1205–1213 (2018). https://doi.org/10.1080/11263504.2018.1435572

55. del Río, S., et al.: Modelling the impacts of climate change on habitat suitability and vulnerability in deciduous forests in Spain. Ecol. Ind. **131**, 108202 (2021). https://doi.org/10.1016/j.ecolind.2021.108202

56. Cano-Ortiz, A., Piñar Fuentes, J.C., Quinto Canas, R.J., Pinto Gomes, C.J., Cano, E.: Analysis of the relationship between bioclimatology and sustainable development. In: Bevilacqua, C., Calabrò, F., Della Spina, L. (eds.) NMP 2020. SIST, vol. 178, pp. 1291–1301. Springer, Cham (2021). https://doi.org/10.1007/978-3-030-48279-4_120

57. Raposo, M.A.M., et al.: *Prunus lusitanica* L.: an endangered plant species relict in the Central Region of Mainland Portugal. Diversity **13**, 359 (2021)

58. Raposo, M.A.M., del Río, S., Pinto-Gomes, C.J., Lazare, J.J.: Phytosociological ANALYSIS of *Prunus lusitanica* communities in the Iberian Peninsula and South of France. Plant Biosyst. (2021). https://doi.org/10.1080/11263504.2021.1998242

Urban Transformation of the Coastline from a Landscape Perspective. Analysis of Cases on the Costa del Sol (Spain)

Alessandro Malerba[1]($\boxtimes$), Hugo Castro Noblejas[2],
Juan Francisco Sortino Barrionuevo[2], and Matías Mérida Rodríguez[2]

[1] Mediterranea University, 89124 Reggio Calabria, Italy
gevaul02@gmail.com
[2] Departamento de Geografía, Universidad de Málaga, 29071 Malaga, Spain

Abstract. The Spanish Mediterranean coastline has been one of the regions with the fastest urbanization process since the mid-20th century. This study analyzes the urban colonization of the first coastline in two municipalities of the Costa del Sol in southern Spain, Manilva and Marbella, from the mid-twentieth century to the present. For this purpose, pre-existing data were processed by GIS using combined photointerpretation and remote sensing techniques from the Urban Atlas and alphanumeric information from the Spanish Cadastre. The unavailable information was elaborated ad-hoc by means of photo-interpretation techniques and field work. The results show an intense process of urbanization, in which the N-340 road, which has ended up clogging the coastline of Marbella, while in Manilva there are still spaces free of buildings. If the study is carried out for the different periods, we can see how the urban development model has evolved, from an initial phase with a predominantly extensive residential model, passing through construction profiles of greater height and building density in the following decades to, from the beginning of the 21st century, return to an extensive model in a saturated territory.

The integration of different photointerpretation and remote sensing techniques with geo-referenced data shows a very high efficiency to analyze the territory from a landscape perspective, to which historical information can be added in the future in graphic or written formats, especially at a scale of detail that does not reach the historical aerial photography.

Keywords: Photointerpretation · Land cover and use · Urbanization · GIS tools

1 Introduction

The European Environment Agency (EEA) has described sprawl as the physical pattern of low-density expansion of large urban areas, under market conditions, mainly towards the surrounding agricultural areas [1]. Uncontrolled sprawl is the spearhead of urban growth and implies a lack of control over land subdivision planning. Development is fragmentary and dispersed, with a tendency to discontinuity: cities are full of empty

F. Calabrò et al. (Eds.): NMP 2022, LNNS 482, pp. 1671–1682, 2022.
https://doi.org/10.1007/978-3-031-06825-6_161

spaces that indicate the inefficiencies of development and reveal the sequences of uncontrolled growth. Although sprawl has been considered an urban manifestation rooted in the Anglo-Saxon context [2–4], urban sprawl has progressively spread to Europe [5–7], starting from the northwest [8], expanding since the end of the 20th century along the Mediterranean arc [9, 10] or, more recently, in Eastern Europe [11, 12].

Urban expansion in Europe was accompanied by a reinterpretation of this urban model, based on a "change of scale" within metropolitan regions [13, 14]. In this way, cities expanded following a general "urban life cycle" based on different phases of growth, from urbanization to suburbanization and from counter-urbanization to re-urbanization. Within this scope, an "economistic" view of the relationship between cities and the surrounding territory was developed, which has become one of the most common perspectives from which to analyze sprawl in Europe [15, 16]. In this context of territorial planning, the case of the Spanish Mediterranean coast is of interest, where the N-340 highway has been a historical axis, going beyond its infrastructural function or its function as a support for lucrative activities. It has been common for the implementation of this type of infrastructure to generate an immediate effect on land prices, but the effects on the real development of the surrounding area, on its demography or economic activities, occur gradually or even fail to bear fruit. The direct repercussion of the implementation of this type of infrastructure is the reordering of the territorial hierarchical structure of the settlements, favoring greater development of the settlements that benefit directly from its service. However, as Fariña et al. [17] point out, for a road infrastructure to become a territorial axis, it must be accompanied by complementary political, economic and social measures. On the Costa del Sol, the passage of this road and the construction of the airport in Malaga contributed to a very rapid urban expansion that, as will be analyzed later, has gone through several urban planning models. To analyze this and other processes of urban expansion at different territorial scales, the use of photointerpretation has been common.

This technique is applied either by direct interpretation of orthophotography [18] or by relying on previous remote sensing work, such as the information provided on land use and occupation by the CORINE Land Cover project of the European Union (scale 1:100,000) [19], the Urban Atlas project of urban areas of more than 100,000 inhabitants (scale 1:10,000) [20] or the use of other satellites [21]. In the case of Spain, the geographic databases of the SIOSE (Land Occupation Information System of Spain) project (scale 1:25,000) and other regional databases of greater detail are available [22]. These studies of land use evolution, even those that carry out multiscale studies, usually adopt the municipality or other administrative units as the unit of study, which sometimes limit the understanding of territorial processes. However, it is considered that the coastal strip has entity and characteristics to be analyzed specifically, through the specific land uses: beach, dunes, etc.; its scarce population and with an agrarian economic activity, such as fishing).

The objectives of this study are: a) to analyze the landscape transformations experienced in the first line of the coast due to the urbanization process, showing the evolution of the urbanization process, b) to characterize the main dominant construction typologies in each urbanization stage and c) to analyze the land uses that have lost presence and, in particular, the environmental impacts that have occurred.

2 Methodology

2.1 Study Area

The study focuses on the coastal strip of the western Costa del Sol, in the province of Malaga, southern Spain. Specifically, these are the cases of Marbella and Manilva (see Fig. 1).

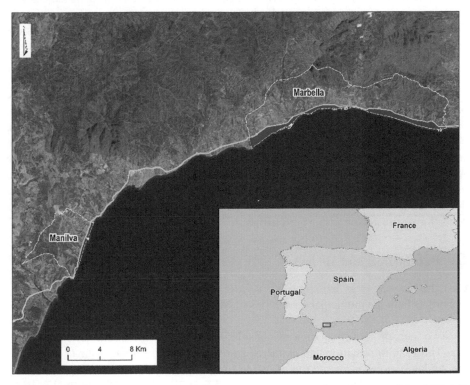

Fig. 1. Situation of the study cases. In red, the analyzed area. In yellow, the route of the old N-340 road.

Until the 1950s, this coastal zone was a rural area based on a traditional agricultural and fishing production model. In the mid-twentieth century, with the pioneering appearance of high-end hotels, a productive model focused on tourism and construction began to be implemented. The area had a great tourist potential based, in addition to its coastal character, on the existence of an attractive landscape, with an important presence of vegetation, a very comfortable climate, the proximity to the international airport of Malaga and the road connection through the N-340 [23]. This change in the productive model broke with the traditional configuration of the area, which was linked to an urban development process marked by urban growth that led to a real estate boom. The ex-tourist expansion brought with it a hypertrophic and uncontrolled urban development of the pre-existing nuclei [24], which has barely taken into account the conservation or creation of green areas in the interior of the cities. Currently, the municipalities of this coastal area form a conurbation, the result of the intense urban development experienced during the last half century. Marbella has its own brand that keeps it in the market as a luxury tourist destination. This is due, on the one hand, to the legacy that still remains of the glamour of the first luxury hotels, continued with the construction, in the 1970s, of Puerto Banús, a high standing marina that would become one of the main local tourist landmarks of the national wealthy classes; on the other hand, the adoption of its own construction model is key to its tourist image, with the introduction of resources of a landscape nature for the real estate business, such as the numerous golf courses [25]. For its part, Manilva, on the western edge of the region, has experienced a much more recent urban growth, between the 90s of the last century and the present, preserving an important agricultural sector (vineyards), which is a singularity in its spatial context.

Both municipalities have been connected by transport infrastructure since the construction of the N-340 road. The construction of the road between the border of the province of Cadiz and Malaga was extended between 1863 and 1929, leaving, as it passed through the municipalities of Manilva and Marbella, a strip of land ranging between 70 and 1000 m. Currently, both municipalities have an unfolded connection with subsequent infrastructures, which surround the urban centers in the interior, the AP-7 freeway. The N-340, which is the boundary of the study area, forms a clear and useful limit, as it runs largely parallel to the coastline. It constitutes both an edge (border) of the urbanization process and, at the same time, an articulating element due to its condition as the main transport infrastructure.

2.2 Data Sources and Methods

First, the study area was delimited with a Geographical Information System (GIS), taking as borders both the municipal boundaries of Manilva and Marbella and the route of the N-340.

Two sources of information were considered for the evolution of land cover and land use. The 1956 information was extracted from the vector information layer offered by the Andalusian government (Spain), while for the information concerning 2018, only the information for Marbella could be obtained from the *Urban Atlas* of the European Union. For the case of Manilva, a layer of information equivalent to that available in the case of Marbella was generated through a photo-interpretation work, using as a basis the most recent aerial orthophotography of the Spanish *National Aerial Orthophotography Plan* (PNOA). Subsequently, the work consisted of processing the information in GIS, aggregating the land uses according to territorial logic and performing statistical calculations to extract the most notable patterns. With respect to the information concerning building typologies, it was based on information provided by the Spanish Cadastre, using the plot layout in vector format and alphanumeric information, available at a building scale. A final step consisted of grouping the plots according to date of age, using time intervals with a historical criterion.

3 Results

The results show different patterns of urban expansion at different scales, with different construction models and asymmetric disappearance of other land uses.

As can be seen in Tables 1 and 2, the only area that increases substantially is built-up area. The case of Marbella is particularly striking, where the growth rate of the urbanized area in the period 1956–2018 increased by 2637.8%, which can be explained by a process of urbanization that began prior to that of Manilva, when there were still no mechanisms to regulate the urbanization process or the preservationist consideration of the coastline. There are two types of vegetation that, exceptionally, have increased, apart from urban landscaped areas. These are the riparian vegetation of the riverbeds in Manilva, although a miniscule increase that can be explained by small deviations in the measurements. This is not the case for the area of conifers in Marbella, where it has increased by 25%. This can be explained by the reforestation policy and the objective of fixing the littoral dunes in part of the municipality.

Table 1. Evolution of the main land covers on the Manilva coastline.

Type of ground cover	Surface 1956 (ha)	Surface 1956 (%)	Surface 2018 (ha)	Surface 2018 (ha)	Rate difference 2018–1956 (%)
Urban fabric	14.4	9	34.5	38,9	139.6
Crops	44.9	28,1	0	0,0	−100
Predominant scrubland	43.9	27,4	17.0	19,1	−61.2
Beaches, dunes, sands	54.2	33,9	34.6	39,0	−36.2
Rivers and natural streams	2.6	1,6	2.7	3,0	1.5
Total	160	100	88.8	100	

The dynamics shown in Tables 1 and 2 are spatially translated in Figs. 1 and 2.

On the left side of Fig. 2, the expansion of urbanized land can be seen, which takes place almost continuously in space. Only in the southwestern part of the study area are extensions of unbuilt land observed due to the difficulty of the terrain as it

Table 2. Evolution of the main land covers on the Marbella coastline.

Type of ground cover	Surface 1956 (ha)	Superficie 2018 (ha)	Rate difference 2018–1956 (%)
Urban fabric	37.7	1031.0	2637.8

(continued)

Table 2. (*continued*)

Type of ground cover	Surface 1956 (ha)	Superficie 2018 (ha)	Rate difference 2018–1956 (%)
Crops	560.4	11.6	−91.4
Predominant scrubland	222.3	78.6	−64.6
Beaches, dunes, sands	428.7	131.9	−69.2
Rivers and natural streams	33.6	10.8	−67.7
Coniferous forest	20.4	25.5	25.3
Eucalyptus forest	30.7	5.5	−82.0
Mixed forest	32.6	26.5	−18.9

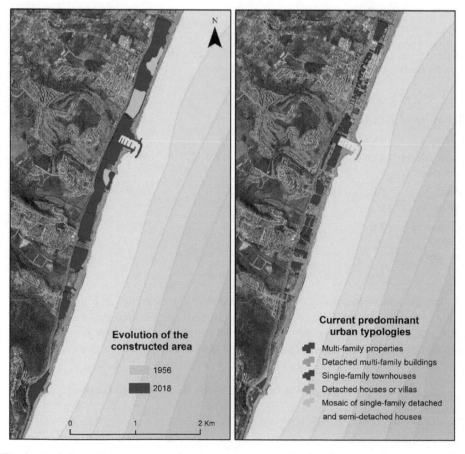

Fig. 2. Evolution of the constructed area and current predominant constructive types on the Manilva coastline.

is a cliff area. On the right side of Fig. 1, it can be seen, as the main construction pattern, how in the north central area, where some buildings already existed in 1956, an area of mainly multi-family buildings has spread, including the surroundings of a sports marina that corresponds to the incoming form in the sea and that now make up a compact nucleus. Progressively towards the southwest, there is a predominance of more extensive construction models, of detached houses and single-family townhouses.

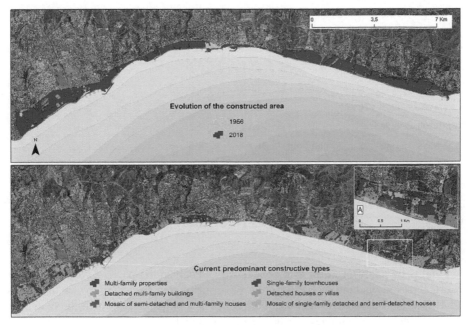

Fig. 3. Evolution of the constructed area and current predominant constructive types on the Marbella coastline.

In the case of Marbella (see Fig. 3), it can be seen in the image above how the evolution of the construction stain is much more intense, going from a small expansion to the south of the historic center in the central area of the map and some scattered construction in the western zone (orange stain), to a predominance in all the segments of the coast of the municipality. Regarding the lower image, a similar pattern can be observed in the case of Manilva. In the continuation of the main nuclei of Marbella (center) and San Pedro Alcántara-Puerto Banús (West), the predominance of higher-rise multi-family construction units is observed, which generate a compact coastal continuation of the historic nuclei. As the observer moves towards the outskirts, the preponderance of extensive construction types is identified, alternating the urbanizations of townhouses and villas. The peripheral zones (see zoom A), however, come to present varied constructive compositions, with a logic of patches that is explained by their lack of contemporaneity, being built interspersed, as is usual in the sprawling process. An alternation of architectural styles, volumes and dimensions and even functionality can be observed, since many buildings are hotels. As can be seen in Table 3, 83.9% of the study area has been transformed in terms of land cover and land use. The type of surface area that has remained relatively stable is that corresponding to beaches and dune formations, accounting for more than half of these areas, despite having lost more than 60% of its extension (see Tables 1 and 2), followed by the built-up land, due to the first urban pieces already existing at the initial date (see Tables 1 and 2).

Table 3. Ground cover that has remained unchanged in the study area.

Type of ground cover	Surface (ha)	% remained surface	% of total surface
Eucalyptus forest	2.04	0.8	0.13
Crops	4.97	2.0	0.32
Rivers and natural streams	8.54	3.4	0.54
Coniferous forest	10.50	4.2	0.67
Highways, freeways & road junctions	21.49	8.5	1.37
Predominant scrubland	22.70	9.0	1.45
Urban fabric	49.76	19.7	3.17
Beaches, dunes, sands	132.51	52.5	8.45
Total	252.51	100	16.1

Finally, by analyzing Fig. 4, it is possible to analyze the pace of construction by historical periods and the successive preferred construction models. Between the 1950s and 1970s, the first part of the construction boom took place, in which the extensive model of single-family homes was the dominant model.

Subsequently, in the 1980s, a much more compact construction model became more important, focused on multi-family housing, as the land considered developable was running out. This trend reached its peak in the following decade, under the influence of local politicians who sought to obtain maximum economic profitability from the land, above other values, such as landscape. The construction of buildings of more than five storeys high on the coastline, in front of pre-existing single-family homes, was common, especially in the expansion of the center of Marbella. Subsequently, with the filling of the available land, especially in the case of Marbella, and the tightening of urban planning legislation, which has sought to preserve the environment of the coastal strip, the rate of construction has slowed down, with a new focus on low-rise housing for the medium-high housing segment.

Similar studies can be found in Italy [26–32] bringing convergent and useful results in the general field, and also leading the consequent pioneer and key research concerning the relationship among environmental features, settlements, buildings, values and real estate market prices [33–38].

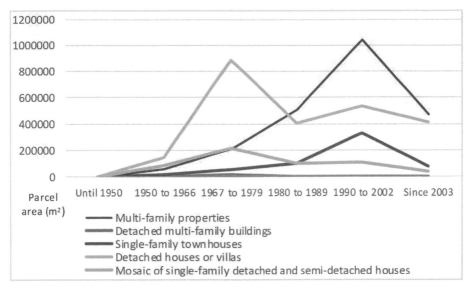

Fig. 4. Evolution of constructive development types by period.

4 Discussion and Conclusions

This study has been able to analyze the process of urban expansion in the coastal strip of two municipalities of the Spanish Mediterranean coast, from the mid-twentieth century to the present, characterizing the predominant construction models in each historical stage. Among the most striking results, the absolute loss of traditional land uses stands out, since there are practically no more agricultural plots left, as well as the great environmental degradation due to the invasion of beaches and dune formations by construction land. On the other hand, the study allows us to identify how the constructive models that have been following one after the other have been clogging the coastline. Comparing the two municipalities studied, it can be seen that in the case of Marbella the construction process has been more intense, due to that urban expansion began earlier, and that the layout of the highway generally leaves a larger strip of land towards the sea. Overall, this radiography helps to explain the complete landscape disconnection of the coastal strip with its past.

The main limitation of the study, with respect to the presentation of the results, is that the analog support prevents the display of more complex cartography, which would combine several variables, such as the introduction of the period of antiquity to which the plots belong, differentiated by their urban typology.

The integration of photointerpretation databases with cadastral information makes it possible to evaluate the evolution of a territorial unit at different scales: from a region such as the Costa del Sol to an urban piece, such as a man-block. At a detailed scale, this methodology can integrate other techniques, such as historical photographs and other types of documents to complete the information observed from the air. The consideration of a road axis (N-340) as a delimiting element of the study area is evaluated as an opportunity to deepen in the territorial processes that concern this now urban space. Similar studies can be found in Italy [26–32], and also leading the pioneer research concerning the relationship among environmental features, buildings, values and real estate market prices [33–38].

References

1. EEA: Urban Sprawl in Europe-The Ignored Challenge. European Environment Agency, Copenhagen, Denmark (2006)
2. Peiser, R.: Decomposing urban sprawl. Town Plann. Rev. 275–298 (2001). https://doi.org/10.3828/tpr.2001.72.3.275
3. Squires, G.D. (ed.): Urban Sprawl: Causes, Consequences, & Policy Responses. The Urban Institute (2002)
4. Hamidi, S., Ewing, R.: A longitudinal study of changes in urban sprawl between 2000 and 2010 in the United States. Landsc. Urban Plann. 128, 72–82 (2014). https://doi.org/10.1016/j.landurbplan.2014.04.021
5. Hennig, E.I., Schwick, C., Soukup, T., Orlitová, E., Kienast, F., Jaeger, J.A.: Multi-scale analysis of urban sprawl in Europe: towards a European de-sprawling strategy. Land Use Policy 49, 483–498 (2015). https://doi.org/10.1016/j.landusepol.2015.08.001
6. Triantakonstantis, D., Stathakis, D.: Examining urban sprawl in Europe using spatial metrics. Geocarto Int. 30(10), 1092–1112 (2015). https://doi.org/10.1080/10106049.2015.1027289
7. Smiraglia, D., Salvati, L., Egidi, G., Salvia, R., Giménez-Morera, A., Halbac-Cotoara-Zamfir, R.: Toward a new urban cycle? A closer look to sprawl, demographic transitions and the environment in Europe. Land 10(2), 127 (2021). https://doi.org/10.3390/land.10020127
8. Couch, C., Petschel-Held, G., Leontidou, L. (eds.): Urban Sprawl in Europe: Landscape, Land-Use Change and Policy. Wiley, New York (2008)
9. Díaz-Pacheco, J., García-Palomares, J.C.: Urban sprawl in the Mediterranean urban regions in Europe and the crisis effect on the urban land development: Madrid as study case. Urban Stud. Res. (2014). https://doi.org/10.1155/2014/807381
10. Zambon, I., Serra, P., Sauri, D., Carlucci, M., Salvati, L.: Beyond the 'Mediterranean city': socioeconomic disparities and urban sprawl in three Southern European cities. Geografiska Annaler: Ser. B Hum. Geogr. 99(3), 319–337 (2017). https://doi.org/10.1080/04353684.2017.1294857
11. Garcia-Ayllon, S.: Urban transformations as indicators of economic change in post-communist Eastern Europe: territorial diagnosis through five case studies. Habitat Int. 71, 29–37 (2018). https://doi.org/10.1016/j.habitatint.2017.11.004
12. Kovács, Z., et al.: Urban sprawl and land conversion in post-socialist cities: the case of metropolitan Budapest. Cities 92, 71–81 (2019). https://doi.org/10.1016/j.cities.2019.03.018
13. Salvati, L., Gargiulo Morelli, V.: Unveiling urban sprawl in the Mediterranean region: towards a latent urban transformation?. Int. J. Urban Reg. Res. 38(6), 1935–1953 (2014). https://doi.org/10.1111/1468-2427.12135
14. Morote, Á.F., Hernández, M.: Urban sprawl and its effects on water demand: a case study of Alicante. Spain. Land Use Policy 50, 352–362 (2016). https://doi.org/10.1016/j.landusepol.2015.06.032
15. Scott, A.J., Storper, M.: The nature of cities: the scope and limits of urban theory. Int. J. Urban Reg. Res. 39(1), 1–15 (2015). https://doi.org/10.1111/1468-2427.12134
16. Lux, G.: Minor cities in a metropolitan world: challenges for development and governance in three Hungarian urban agglomerations. Int. Plan. Stud. 20(1–2), 21–38 (2015). https://doi.org/10.1080/13563475.2014.942491

17. Fariña Tojo, J., Pozueta Echavarri, J., Lamíquiz Daudén, F.J.: Efectos territoriales de las infraestructuras de transporte de acceso controlado. Departamento de Urbanística y Ordenación del Territorio, Madrid, Spain (2000)
18. Díaz-Pacheco, J., García-Palomares, J.C.: A highly detailed land-use vector map for Madrid region based on photo-interpretation. J. Maps 10(3), 424–433 (2014). https://doi.org/10.1080/17445647.2014.882798
19. Cieślak, I., Biłozor, A., Szuniewicz, K.: The use of the CORINE land cover (CLC) database for analyzing urban sprawl. Remote Sens. 12(2), 282 (2020). https://doi.org/10.3390/rs.12020282
20. Pazúr, R., Feranec, J., Štych, P., Kopecká, M., Holman, L.: Changes of urbanised landscape identified and assessed by the urban atlas data: case study of Prague and Bratislava. Land Use Policy 61, 135–146 (2017). https://doi.org/10.1016/j.landusepol.2016.11.022
21. Checa, J.: Urban intensities. The urbanization of the Iberian Mediterranean coast in the light of nighttime satellite images of the earth. Urban Sci. 2(4), 115 (2018). https://doi.org/10.3390/urbansci2040115
22. Membrado, J.C., Huete, R., Mantecón, A.: Urban sprawl and Northern European residential tourism in the Spanish Mediterranean coast. Via. Tour. Rev. (10) (2016). https://doi.org/10.4000/viatourism.1426
23. García Manrique, E.: La Costa occidental malagueña. In: Málaga, vol. 1, pp. 229–260. Anel (1984)
24. Ferre, E., Ruíz-Sinoga, J.D.: Algunos aspectos del impacto del turismo en la Costa del Sol Occidental: el caso de Marbella. Baetica. Estudios de Arte, Geografía e Historia 9, 57–73 (1986)
25. Villar Lama, A.: La mercantilización del paisaje litoral del mediterráneo andaluz: El caso paradigmático de la Costa del Sol y los campos de golf. Revista de estudios regionales 96, 215–242 (2013)
26. Massimo, D.E., Musolino, M.: Mediterranean urban landscape. Integrated strategies for sustainable retrofitting of consolidated city. In: Society, Integration, Education Utopias and Dystopias in Landscape and Cultural Mosaic. Visions Values Vulnerability. Proceedings of the International Scientific Conference, 27–28 June 2013. Sabiedriba, Integracija, Izglitiba. III, 49–60 (2013). ISSN 1691-5887
27. Massimo, D.E., Musolino, M., Barbalace, A., Fragomeni, C.: Landscape quality valuation for its preservation and treasuring (2014). https://doi.org/10.4028/www.scientific.net/AEF.11.625. (In: 1st International Symposium "New Metropolitan Perspectives. The Integrated Approach of Urban Sustainable Development through the Implementation of Horizon/Europe 2020" (ISTH 2020). ISBN 978-3-03826-486-6. Adv. Eng. Forum. ISSN 2234-991X, vol. 11, pp. 625–633 (2020)
28. Massimo, D.E., Musolino, M., Fragomeni, C., Malerba, A.: A green district to save the planet. In: Mondini, G., Fattinnanzi, E., Oppio, A., Bottero, M., Stanghellini, S. (eds.) SIEV 2016. GET, pp. 255–269. Springer, Cham (2018). https://doi.org/10.1007/978-3-319-78271-3_21
29. Malerba, A., Massimo, D.E., Musolino, M.: Valuating historic centers to save planet soil. In: Mondini, G., Fattinnanzi, E., Oppio, A., Bottero, M., Stanghellini, S. (eds.) SIEV 2016. GET, pp. 297–311. Springer, Cham (2018). https://doi.org/10.1007/978-3-319-78271-3_24
30. Massimo, D.E., Musolino, M., Malerba, A.: Valuation to foster-up landscape preservation. Treasuring new elements through landscape planning. ArcHistoR6, 674–687 (2019). ISSN 2384-8898, https://doi.org/10.14633/ahr190
31. Spampinato, G., Massimo, D.E., Musarella, C.M., De Paola, P., Malerba, A., Musolino, M.: Carbon sequestration by Cork Oak forests and raw material to built up post carbon city. In: Calabrò, F., Della Spina, L., Bevilacqua, C. (eds.) ISHT 2018. SIST, vol. 101, pp. 663–671. Springer, Cham (2019). https://doi.org/10.1007/978-3-319-92102-0_72

32. Massimo, D.E., Del Giudice, V., Malerba, A., Bernardo, C., Musolino, M., De Paola, P.: Valuation of ecological retrofitting technology in existing buildings: a real-world case study. Sustainability **13**, 7001 (2021). https://doi.org/10.3390/su13137001.c0

33. Del Giudice, V., Massimo, D.E., De Paola, P., Forte, F., Musolino, M., Malerba, A.: Post carbon city and real estate market: testing the dataset of Reggio Calabria market using spline smoothing semiparametric method. In: Calabrò, F., Della Spina, L., Bevilacqua, C. (eds.) ISHT 2018. SIST, vol. 100, pp. 206–214. Springer, Cham (2019). https://doi.org/10.1007/978-3-319-92099-3_25

34. De Paola, P., Del Giudice, V., Massimo, D.E., Forte, F., Musolino, M., Malerba, A.: Isovalore maps for the spatial analysis of real estate market: a case study for a central urban area of Reggio Calabria, Italy. In: Calabrò, F., Della Spina, L., Bevilacqua, C. (eds.) ISHT 2018. SIST, vol. 100, pp. 402–410. Springer, Cham (2019). https://doi.org/10.1007/978-3-319-92099-3_46

35. Massimo, D.E., Del Giudice, V., De Paola, P., Forte, F., Musolino, M., Malerba, A.: Geographically weighted regression for the post carbon city and real estate market analysis: a case study. In: Calabrò, F., Della Spina, L., Bevilacqua, C. (eds.) ISHT 2018. SIST, vol. 100, pp. 142–149. Springer, Cham (2018). https://doi.org/10.1007/978-3-319-92102-3_17

36. Del Giudice, V., Massimo, D.E., De Paola, P., Del Giudice, F.P., Musolino, M.: Green buildings for post carbon city: determining market premium using spline smoothing semiparametric method. In: Bevilacqua, C., Calabrò, F., Della Spina, L. (eds.) NMP 2020. SIST, vol. 178, pp. 1227–1236. Springer, Cham (2021). https://doi.org/10.1007/978-3-030-48279-4_114

37. Del Giudice, V., Massimo, D.E., Salvo, F., De Paola, P., De Ruggiero, M., Musolino, M.: Market price premium for green buildings: a review of empirical evidence. Case study. In: Bevilacqua, C., Calabrò, F., Della Spina, L. (eds.) NMP 2020. SIST, vol. 178, pp. 1237–1247. Springer, Cham (2021). https://doi.org/10.1007/978-3-030-48279-4_115

38. De Paola, P., Del Giudice, V., Massimo, D.E., Del Giudice, F.P., Musolino, M., Malerba, A.: Green building market premium: detection through spatial analysis of real estate values. a case study. In: Bevilacqua, C., Calabrò, F., Della Spina, L. (eds.) NMP 2020. SIST, vol. 178, pp. 1413–1422. Springer, Cham (2021). https://doi.org/10.1007/978-3-030-48279-4_132

The Teaching of Environmental Sciences in Secondary Education, High School and University to Fight Against Climate Change

Ana Cano-Ortiz[1,2] (ID), Carmelo Maria Musarella[3] (ID), José Carlos Piñar Fuentes[1] (ID),
Ricardo Quinto-Canas[4,5] (ID), Jehad Igbareyeh[6], Valentina Lucia Astrid Laface[3] (ID),
and Eusebio Cano[1(✉)] (ID)

[1] Department of Animal and Plant Biology and Ecology, Section of Botany, University of Jaén,
Jaén, Spain
ecano@ujaen.es
[2] Institute for Secondary Education and Baccalaureate, Ministry of Education Madrid, Madrid,
Spain
[3] Department of Agriculture, Mediterranean University of Reggio Calabria, Reggio Calabria,
Italy
[4] Faculty of Sciences and Technology, University of Algarve, Faro, Portugal
[5] Centre of Marine Sciences (CCMAR), University of Algarve, Faro, Portugal
[6] Department of Plant Production and Protection, Hebron, Palestine

Abstract. The environmental contents in Compulsory Secondary Education (ESO) and Baccalaureate are reviewed, being scarce and not adapted to the environmental reality. Faced with the profound climatic changes that have occurred and that increase day by day, causing natural disasters, in terms of loss of life, and in terms of great economic losses. For this reason, it is essential to incorporate new content in the teaching of Natural Sciences (Biology, Geology, Physics, Chemistry) in the different teachings, both university and non-university.

Keywords: Pollution · Green city · Environment · Mitigation · CO_2

1 Introduction

The contents taught in this subject are aimed at students acquiring the foundations of scientific culture, with special emphasis on the unity of the phenomena that structure the natural environment, on the laws that govern them and on the mathematical expression of these laws, thus obtaining a rational and global vision of our environment with which they can face current problems related to life, health, the environment and technological applications, with special emphasis on climate change.

In the Secondary Education, Baccalaureate and University stages, among others, the physical-chemical, biological-geological aspects of nature are addressed.

In today's society, science is an indispensable instrument for understanding the world around us and its transformations, as well as for developing responsible attitudes about aspects related to natural resources and the environment.

© The Author(s), under exclusive license to Springer Nature Switzerland AG 2022
F. Calabrò et al. (Eds.): NMP 2022, LNNS 482, pp. 1683–1691, 2022.
https://doi.org/10.1007/978-3-031-06825-6_162

Knowledge about natural sciences must be consolidated and expanded during secondary and high school, by incorporating practical activities, and adapting the contents to the social and environmental reality of the historical moment, so it is essential to include in the contents various environmental aspects, such as land [1–4], water [5–7], fire [8–10], climate change [11–13], green cities [14–17]. This will allow to adopt critical attitudes based on knowledge to analyze, individually or in groups, scientific and technological environmental issues, which allow us to face the risks of today's society in aspects related to the environment. Understand the importance of using the knowledge of the natural sciences to satisfy human needs and participate in the necessary decision-making around local and global problems that we face. Know and value the interactions of science and technology with society and the environment, with particular attention to the problems that humanity faces today and the need to search and apply solutions, subject to the precautionary principle, to moving towards a sustainable future. Recognize the tentative and creative nature of the natural sciences, as well as their contributions to human thought throughout history, appreciating the great debates overcoming dogmatisms and the scientific revolutions that have marked the cultural evolution of humanity and its conditions of life [18, 19].

Consequently, it is necessary to establish as fundamental objectives in the teaching of Natural Sciences [20–22]: 1) understand and use the strategies and basic concepts of the natural sciences to interpret natural phenomena, as well as to analyze and assess the repercussions of techno-scientific developments and their applications; 2) apply, in problem solving, strategies consistent with the procedures of science, such as the discussion of the interest of the problems posed, the formulation of hypotheses, the elaboration of resolution strategies and experimental designs, the analysis of results the consideration of applications and repercussions of the study carried out and the search for global coherence; 3) understand and express messages with scientific content using oral and written language properly, interpret diagrams, graphs, tables and elementary mathematical expressions, as well as communicate arguments and explanations in the field of science to others; 4) obtain information on scientific topics, using different sources, including information and communication technologies, and use them, evaluating their content, to support and guide work on scientific topics. All this in order for the student to acquire certain capacities, which allow them to acquire specific competences to combat climate change [23–26]. As would be the 1) knowledge and interaction with the physical world; 2) mathematical competence, since it is necessary to use mathematical language to quantify natural phenomena, analyze causes and consequences, and express data and ideas about nature; 3) competence in the treatment of information and communication, as well as the acquisition of digital competence; 4) social and civic competence, the contribution of Natural Sciences to this competence is linked to the role of science in the preparation of democratic, participatory and active citizens in decision-making, in addition, it contributes to a better understanding of issues that are important to understand the evolution of society in past times and analyze current society; 5) competence on linguistic communication, in this case this competence is achieved through the construction of scientific discourse, aimed at arguing or making explicit their relationships taking care of the precision of the terms used, properly chaining ideas or in verbal expression and acquisition of the specific terminology on living beings, objects and natural phenomena; 6) it is necessary for the student to learn the contents associated with the way of constructing and transmitting scientific knowledge constitute an opportunity for the development of this competence. Lifelong learning, in the case of knowledge of nature, is produced by the incorporation of information that comes sometimes from one's own experience and on other occasions from written or audiovisual media. Finally, personal initiative is required, placing special emphasis on the formation of a critical spirit, capable of questioning dogmas and challenging prejudices, allowing to contribute to the development of autonomy and personal initiative. In this sense, it is important to point out the role of science as an enhancer of the critical and participatory spirit in the search for solutions.

The objective of this study is to highlight the importance of adapting the environmental reality in teaching, for which an adaptation of the contents of Natural Sciences is proposed, and an increase in practical learning over theoretical.

2   Methodology

Learning is conceived as a change in conceptual schemes on the part of the learner. It is based, therefore, from the acceptance that the students have previous schemes of interpretation of reality.

The organization of the contents takes into account the very nature of science as a constructive activity and in permanent revision. Construction of knowledge is favored through active learning, so that the student cannot be a mere passive receiver [26].

In this way, what is learned depends fundamentally on what has already been learned (previous knowledge), and on the other hand, the learner constructs the meaning of what has been learned from his own experience; that is, from their activity with the learning content and its application to family situations.

A work process is put into practice, which allows the use of the didactic elements that make up the different learning situations. Therefore, it is about applying different methods: a) inductive: starting from the particular and close to the student, to finish in the general, through increasingly complex conceptualizations, such as the knowledge of some plant species; b) deductive: starting from the general, to conclude on the particular, in the environment close to the student, for this, satellite images are used on the landscape; c) inquiry: through the application of the scientific method; d) active: based on the performance of activities by the student, laboratory activities; e) explanatory: based on explanation strategies; f) participatory: inviting debate; g) mixed: tending to unite in the same didactic unit the practice of more than one of the previous methods. When using these methods, it should be taken into account that adolescence is a stage in which important and great changes occur, not only in the individual himself and in his way of interacting with his peers and other people, but also in the acquisition of new ways of thinking.

As a general strategy, the motivation, curiosity and interest of the student must be aroused, through the relevance and presentation of the information.

To achieve the objectives stated above, the teacher must pay attention to students with specific educational needs, being multiculturalism in the classroom an adaptive teaching model [27].

The teaching of Natural Sciences requires a development of the ability to observe and interpret reality, which subsequently enables the student to make proposals for environmental improvement [28–32].

Teaching that is necessary as a result of the increase in CO2 due to climate change, which is due to an excessive use of fossil fuels, making it necessary for governments to promote renewable energies.

Global warming is now an unequivocal fact, as reflected in the fourth report of the Intergovernmental Panel on climate change. The third evaluation report revealed that there had been an increase in temperature of approximately 0.6 °C [between 0.4°C and 0.8°C], affecting physical and biological systems in different parts of the globe; The linear trend at 100 years is estimated at 0.74 °C [between 0.56 °C and 0.92 °C], as opposed to 0.6°C in the third report.

The greenhouse effect caused by global warming can and should be mitigated by applying sustainable development, which is achieved through bioclimatic agricultural and forest management, and using herbaceous plant covers (Stellarietea mediae) in agriculture.

3 Results and Discussion

The conceptual contents of Environmental Sciences are established in the common teaching curriculum, where a connection is established between Physics and Chemistry and Biology and Geology, with respect to the work and scientific method, as well as an assessment of the contributions of the natural sciences to development of humanity. These contents were established in Spain in 2007, which should force a revision of contents, which better adapt to the current environmental reality [34]. The contamination of land and water by the indiscriminate use of chemical products, has caused a profound change in cultivation techniques, this being the cause of the loss of plant species, with a decrease in the soil seed bank. The same phenomenon occurs in most of the Mediterranean areas treated with herbicides; the increase in CO_2 as a result of the excessive use of fossil fuels, forest fires, has led to climate change on the planet. Change that is being accelerated, since CO_2 continues to increase at the same time that the vegetation cover decreases, which could mitigate said change, by acting as a CO_2 sink [35–39].

In the case of the majority crops, it is evident that until a few years ago the cultivation has been something traditional, with life models acquired throughout history, so that the crops have always been somewhat plural, insofar as it has been given a heterogeneity and not a monoculture that has been reached in certain cases, which has put the territories in the hands of speculators.

Management models were acquired that were not standard models, but in each territory there was a specific model, therefore phenomena such as pruning, tillage, type of plantation, obtaining agricultural product, was typical of each town / region, all of this it is lost in an inordinate desire to increase production, and it moves to a sustained agriculture in which there is a loss of biodiversity and stability of the agroecosystem.

When we talk about sustainability, we are talking about the ability to endure over time without detriment to natural capital (soil, diversity, fauna, etc.) and culture, such as types of management, which enable production and renewal over time. From the different definitions given for "Sustainable Agriculture", we can choose that of the American Agronomy Society: "sustainable agriculture is one that in the long term improves the quality of the environment and the basic resources on which it depends, provides the necessary food and fiber for humanity, it is economically viable and improves the quality of life of the farmer and of society as a whole". In reality what the farmer must pursue is to abandon the criteria of maximum production per hectare, improving the productivity of the farm, through better management of productive factors.

Although for some authors traditional and sustainable agriculture is the same thing, it should not be treated in this sense, since traditional agriculture involves a type of knowledge acquired over millennia, and very empirically contrasted, so it cannot be substituted by current organic farming practices, since this represents a very simplistic model. It can be admitted that the traditional forms of management constitute coherent eco-compatible systems, while the current sustainable cultivation is not or does not follow 100% the traditional model.

The consequence of practicing sustained agriculture causes profound transformations, which in many cases causes loss of biodiversity, which is taking place in a gigantic way. The loss of floristic biodiversity brings with it fauna losses, biological control mechanisms are affected and an insect-plague explosion arises. This loss of floristic diversity is due, among other causes, to inappropriate tillage techniques, loss of plant cover, lack of organic fertilizers, use of rollers (soil compaction), indiscriminate use of herbicides and as a consequence, increased erosion and loss of soil [5, 40]. For this reason, it is convenient to use non-aggressive agricultural practices with the environment, among which would be traditional cultivation and sustainable cultivation, at least for marginal areas. For non-marginal high production areas, it is convenient to mix both techniques, that is, planning the cultivation on an ecological / botanical basis [41].

At this time a non-aggressive cultivation is possible, since there are advanced botanical investigations, both bioclimatic and edaphic [42], through which we can know the nutritional status of the soil, which allows us to use appropriate doses of fertilizer, avoiding the excess, which only affects the contamination of aquifers and economic losses for the farmer [43–45].

In the last 50 years a new agriculture has been introduced, which with full economic and social acceptance has survived to this day. A consequence of this has been the excessive use of pesticides, herbicides and phytosanitary products, causing a great environmental impact, with the consequent damage to the population. A crop is unproductive when the cost of production is equal to or greater than the income received by the farmer. In said expense is included the entire set of activities necessary for production and the environmental cost, which in most cases is not quantifiable. Considering as unproductive that whose cost is higher than the income it generates, even those apparently productive crops with modern technology should be considered unproductive, if the environmental cost is excessively high, and all those situations in which the cultivation causes irreversible losses of a certain resource. However, those cases of "contamination" in which the system itself is capable of self-regeneration can be allowed. Therefore, the expression "the polluter pays" is not correct, since resources can be lost that cannot be priced, as they are high ecological-natural values. Furthermore, cultivation is not valid at the expense of irreversible losses of soil, biodiversity, contamination of water and land, the latter being the ones that must be considered essential to carry out the implantation of a crop. All the situations contemplated within any of the previous cases, must pass to agrarian reform of a technical nature. Carrying out agricultural planning in which the crop is maintained respecting the above principles, is productive and can even increase said production, it is also possible that said production is of quality, since quantity and quality are compatible, this implies having taking into account the bioclimatic aspects of the territory.

It can be affirmed that living plant covers are therefore beneficial, not only to prevent erosion and maintain productivity, but to mitigate climate change, by acting as a sink for CO_2. On the other hand, the use of plant covers is not such a big deal novel, since they were already used by the Romans in the cultivation of vineyards, having been studied and used since the beginning of the 20th century.

The study on the teaching of natural sciences in the classroom, in relation to climate change, increase in CO2, pollution phenomena, forest fires, anthropic action in general, reveals the need for more practical than theoretical learning, which will imply that the curricula of the various countries must adapt to the new world reality.

4 Conclusions

As a consequence of the analysis of the current environmental problems and the proposal of new contents in the teaching of ESO and Baccalaureate, basic competences are acquired, which will allow decision-making, regarding what to do to provoke sustainable development and avoid that being given the climate change. To achieve this, it is necessary that teaching be given in accordance with the environmental needs of the moment. Therefore, the teacher must not only teach, but also motivate learning in this field, making the student aware of the importance of this learning, so that it in turn radiates to society. Teaching that must be structured with content adapted to the environmental reality, since at this time they are relatively scarce.

References

1. Singh, S., Singh, B., Surmal, O., Bhat, M.N., Singh, B., Musarella, C.M.: Fragmented forest patches in the indian himalayas preserve unique components of biodiversity: investigation of the floristic composition and phytoclimate of the unexplored bani valley. Sustainability **13**, 6063 (2021). https://doi.org/10.3390/su13116063
2. Quinto-Canas, R., et al.: Quercus rotundifolia Lam. Woodlands of the Southwestern Iberian Peninsula. Land. 10, 268 (2021). https://doi.org/10.3390/land10030268
3. Cano-Ortiz, A., et al.: Forest and arborescent scrub habitats of special interest for SCIs in central Spain. Land, **10**, 183 (2021). https://doi.org/10.3390/land10020183
4. Musarella, C.M., Brullo, S., del Galdo, G.G.: Contribution to the orophilous cushion-like vegetation of central-southern and insular Greece. Plants **9**, 1678 (2020). https://doi.org/10.3390/plants9121678
5. Cano-Ortiz, A., et al.: Indicative value of the dominant plant species for a rapid evaluation of the nutritional value of soils. Agronomy **11**, 1 (2021). https://doi.org/10.3390/agronomy11010001

6. Spampinato, G., et al.: Contribution to the knowledge of Mediterranean wetland biodiversity: Plant communities of the Aquila Lake (Calabria, Southern Italy). Plant Sociol. **56**, 53–68 (2019)

7. Cano-Ortiz, A., Musarella, C., Fuentes, C., Pinto-Gomes, C., del Río, S., Cano, E.: Diversity and conservation status of mangrove communities in two areas of Mesocaribea biogeographic region. Curr. Sci. **115**, 534–540 (2018). https://doi.org/10.18520/cs/v115/i3/529-534

8. Nunes, L.J.R., Raposo, M.A.M., Meireles, C.I.R., Gomes, C.J.P., Ribeiro, N.M.C.A.: Energy recovery of shrub species as a path to reduce the risk of occurrence of rural fires: a case study in Serra da Estrela Natural Park (Portugal). Fire **4**, 33 (2021). https://doi.org/10.3390/fire4030033

9. Nunes, L.J.R., Raposo, M.A.M., Meireles, C.I.R., Gomes, C.J.P., Ribeiro, N.M.C.A.: The impact of rural fires on the development of invasive species: analysis of a case study with *Acacia dealbata* Link. In Casal do Rei (Seia, Portugal). Environments **8**, 44 (2021). https://doi.org/10.3390/environments8050044

10. Nunes, L.J.R., Raposo, M.A.M., Pinto-Gomes, C.J.: A Historical perspective of landscape and human population dynamics in Guimarães (Northern Portugal): possible implications of rural fire risk in a changing environment. Fire **4**, 49 (2021). https://doi.org/10.3390/fire4030049

11. Nunes, L.J.R., Meireles, C.I.R., Pinto-Gomes, C.J., Ribeiro, N.M.C.A.: The impact of climate change on forest development: a sustainable approach to management models applied to mediterranean-type climate regions. Plants **11**, 69 (2022). https://doi.org/10.3390/plants11010069

12. Nunes, L.J.R., Meireles, C.I.R., Pinto-Gomes, C.J., Almeida Ribeiro, N.M.C.: Forest contribution to climate change mitigation: management oriented to carbon capture and storage. Climate **8**, 21 (2020). https://doi.org/10.3390/cli8020021

13. del Río, S., et al.: Modelling the impacts of climate change on habitat suitability and vulnerability in deciduous forests in Spain. Ecol. Indicators **131** (2021). https://doi.org/10.1016/j.ecolind.2021.108202

14. Massimo, D.E., Del Giudice, V., Malerba, A., Bernardo, C., Musolino, M., De Paola, P.: Valuation of ecological retrofitting technology in existing buildings: a real-world case study. Sustainability **13**, 7001 (2021). https://doi.org/10.3390/su13137001

15. Malerba, A., Massimo, D.E., Musolino, M., Nicoletti, F., De Paola, P.: Post carbon city: building valuation and energy performance simulation programs. In: Calabrò, F., Della Spina, L., Bevilacqua, C. (eds.) ISHT 2018. SIST, vol. 101, pp. 513–521. Springer, Cham (2019). https://doi.org/10.1007/978-3-319-92102-0_54

16. Massimo, D.E.: Green building: characteristics, energy implications and environmental impacts. case study in Reggio Calabria, Italy. In: Green Building and Phase Change Materials: Characteristics, Energy Implications and Environmental Impacts; Coleman-Sanders, Mildred, Ed.; Nova Science Publishers: New York, NY, USA, vol. 1, pp. 71–101 (2015)

17. Massimo, D.E.: Valuation of urban sustainability and building energy efficiency. A case study. Int. J. Sustain. Dev. **12**, 223–247 (2010). https://doi.org/10.1504/IJSD.2009.032779

18. Cañas, A., Martín-Díaz, M., Nieda, J.: Competencia en el conocimiento y la interacción en el mundo físico. La competencia científica. Madrid. Alianza Editorial. (2007)

19. Miño, M.H., Toia, S.N., Pérez, G.M., Gutierrez, T.N., González Galli, L.M., Meinardi, E.M.: Comparación del conocimiento metacognitivo sobre la lectura de textos de Biología entre estudiantes de la Ciudad de Buenos Aires. Revista Electrónica de Enseñanza de las Ciencias **20**(1), 114–134 (2021)

20. Cano-Ortiz, A., Piñar Fuentes, JC., Ighbareyeh, J.M.H., Quinto-Canas, R., Cano, E.: Aspectos Didácticos en la Enseñanza de Conceptos Geobotánicos. IJHSSE **8**(4), 1–6 (2021). https://doi.org/10.20431/2349-0381.0804008

21. Cano-Ortiz, A., Piñar Funtes, J.C., Cano, E.: Didactics of natural sciences in higher scondary education. IJHSSE **8**(9), 1–5 (2021). https://doi.org/10.20431/2349.0809001
22. Cano Ortiz, A., Piñar Fuentes, JC., Ighbareyeh, J.M.H., Quinto-Canas, R., Cano, E.: Didactic aspects in the teaching of vegetation in secondary and high school education. IJHSSE **8**(6), 1–7 (2021). https://doi.org/10.20431/2349-0381.0806002
23. Bello Benavides, L.O., Cruz Sánchez, G.E., Meira Cartera, P.A., González Gaudiano, E.: El cambio climático en el bachillerato. Aportes pedagógicos para su abordaje. Enseñanza de las Ciencias, 39–1, 137–156 (2021). https://doi.org/10.5565/rev/ensciencias.3030
24. Cañal, P.: Cómo evaluar la competencia científica. Investigación en la Escuela **78**, 5–17 (2012)
25. García Barros, S., Martínez Losada, C., Rivadulla López, J.: Actividades de textos excolares. Su contribución al desarrollo de la competencia científica. Enseñanza de las Ciencias, 39–1, 219–238 (2021). https://doi.org/10.5565/rev/ensciencias.3099
26. Fernández, I.M., Pires, D.M., Villamañán, R.M.: Educación científica con enfoque ciencia-tecnología-sociedad-ambiente: construcción de un instrumento de análisis de las directrices curriculares. Formación Universitaria **7**(5), 23–32 (2014). https://doi.org/10.4067/S0718-500 62014000500004
27. Tovar-Gálvez, J.C.: Design of intercultural teaching practices for science education based on evidence. Enseñanza de las Ciencias **39–1**, 99–115 (2021). https://doi.org/10.5565/rev/enscie ncias.2891
28. Cano-Ortiz, A.: El uso de la tecnología para la enseñanza de estructuras foliares. In: 5th International Conference on Educational Research and Innovation- CIVINEDU, 29–30 September, Madrid (2021)
29. Cano-Ortiz, A.: Propuestas para el aprendizaje de la diversidad vegetal. In: 5th International conference on educational research and innovation- CIVINEDU, 29–30 September, Madrid (2021, in press)
30. Cano-Ortiz, A., Cano, E: On the virtual teaching of plant species and communities in High School and University Education. In: VI Virtual International Conference on Education, Innovation and ICT- EDUNOVATIC-2021, 1–2 December (2021, in press)
31. Cano-Ortiz, A., et al.: Botanical gardens and natural and national parks and biosphere reserves: tools for teaching botanical-ecological values. In: XIV International Seminar Biodiversity Management and Conservation: Biodiversity and Sustainability: Two Importants Keywords for the Future. Reggio Calabria, 5–9 Junio (2022)
32. Cano-Ortiz, A., et al.: Measuring and teaching biological diversity. In: XIV International Seminar Biodiversity Management and Conservation: Biodiversity and Sustainability: two importants keywords for the future. Reggio Calabria, 5–9 Junio (2022)
33. Andrade, J.: El proceso de diseño del plán de estudios. Educ. Med. Saldud **5**, 20–39 (1971)
34. Spampinato, G., Massimo, D.E., Musarella, C.M., De Paola, P., Malerba, A., Musolino, M.: Carbon sequestration by cork oak forests and raw material to built up post Carbon City. In: Calabrò, F., Della Spina, L., Bevilacqua, C. (eds.) ISHT 2018. SIST, vol. 101, pp. 663–671. Springer, Cham (2019). https://doi.org/10.1007/978-3-319-92102-0_72
35. Cano, E., et al.: Mitigating climate change through bioclimatic applications and cultivation techniques in agriculture (Andalusia, Spain). In: Jhariya, M.K., Banerjee, A., Meena, R.S., Yadav, D.K. (eds.) Sustainable Agriculture, Forest and Environmental Management, pp. 31–69. Springer, Singapore (2019). https://doi.org/10.1007/978-981-13-6830-1_2
36. Cano, E., et al.: Climatología, Bioclimatología y Cubiertas Vegetales: Herramienta para Mitigar el Cambio Climático. Simposium Homenaje al Prof. Dr. Hc. Salvador Rivas Martínez, León (2021) (2021, in press)
37. Cano, E., et al.: Bioclimatology and Botanical Resources for Sustainability in Natural Resources Conservation and Advances for Sustainability, First Edition, ed. Manoj K Jhariya, pp. 377–387. Elsevier (2021). ISBN: 978–0–1282296–7. https://www.elsevier.com/books-and-journals

38. Barrile, V., Malerba, A., Fotia, A., Calabrò, F., Bernardo, C., Musarella, C.: Quarries renaturation by planting Cork oaks and survey with UAV. In: Bevilacqua, C., Calabrò, F., Della Spina, L. (eds.) NMP 2020. SIST, vol. 178, pp. 1310–1320. Springer, Cham (2021). https://doi.org/10.1007/978-3-030-48279-4_122

39. Piñar Fuentes, J.C., et al.: Impact of grass cover management with herbicides on biodiversity, soil cover and humidity in olive groves in the Southern Iberian. Agronomy **11**, 412 (2021). https://doi.org/10.3390/agronomy11030412

40. Cano Ortiz, A., Piñar, J.C., Pinto-Gomes, C.J., Musarella, C.M., Cano, E.: Expansion of the Juniperus genus due to anthropic activity. In: Weber, R.P. (ed.) Old-Growth Forest and Coniferous Forests, pp. 55–65. Nova Science Publishers, New York (2015)

41. Cano-Ortiz, A.: Bioindicadores ecológicos y manejo de cubiertas vegetales como herramienta para la implantación de una agricultura sostenible. Universidad de Jaén, Tesis Doctoral (2007)

42. Cano, E., Cano-Ortiz, A.: Bioclimatología aplicada a la agronomía in Nuevas Tendencias en Olivicultura. Serv. Publ. Univ. Jaén, 7–64. Material para las asignaturas: "Diseño de Plantaciones y Técnicas de Cultivo" y "Bioindicadores y Cubiertas Vegetales en el Olivar" (2016). ISBN:978–84–8439–000–0

43. Cano-Ortiz, A., Piñar Fuentes, J.C., Quinto-Canas, R., Pinto-Gomes, C.J., Cano, E.: Analysis of the relationship between bioclimatology and sustainable developement. In: Bevilacqua, C., et al. (eds.) New Metropolitan Persperctives. NMP 2020, SIST, vol. 178, pp. 1291–1301. Springe (2020)

44. Cano-Ortiz, A.: Bioindicadores y cubiertas vegetales en el olivar in Nuevas Tendencias en Olivicultura. Serv. Publ. Univ. Jaén, 69–115. Material para las asignaturas: "Diseño de Plantaciones y Técnicas de Cultivo" y "Bioindicadores y Cubiertas Vegetales en el Olivar" (2016). ISBN:978–84–8439–000–0

45. Cano, E.: Una agricultura respetuosa con el medio ambiente y productiva es posible: la bioclimatología como herramienta viabilizadora. Foro **5**(2), 17–23 (2021). www.revistaforo.com/2021/0502-03

Urban Spaces as a Phytogenetic Reserve

Mauro Raposo[1,2(✉)], Maria da Conceição Castro[1,2], and Carlos Pinto-Gomes[1,2,3]

[1] Department of Landscape, Environment and Planning, School of Science and Technology, University of Évora. Rua Romão Ramalho, nº 59, 7000-671 Évora, Portugal
mraposo@uevora.pt
[2] MED - Mediterranean Institute for Agriculture, Environment and Development, Universidade de Évora, Edifício dos Regentes Agrícolas, 7006-554 Évora, Portugal
[3] Institute of Earth Sciences, University of Évora. Rua Romão Ramalho, nº 59, 7000-671 Évora, Portugal

Abstract. Plants, plant communities and habitats are exposed to serious conservation threats in their natural environment. The negligent man's action contributes to endangered the biodiversity in Mediterranean regions. It has been considered that urban spaces (green spaces) can help to mitigate these threats as they have good potential for implementation of phytogenetic reserves, due to the lowest incidence of threats in plants and communities, such as rural fires, invasive plants, intensive agriculture and the wild herbivory. As green spaces consume huge quantities of water, it is important to highlight the use of vegetation well adapted to Mediterranean climate and those areas could also be elected for raising awareness among citizens. So, in this document we attempt to explore the importance of green spaces as a way of mitigate the decrease of occurrence of some species. The analysis used for this study was based on bibliographic sources, as well as data collected by the authors of this article. It was also developed the concept of using native plants, with resources at local ecotypes, as a tool for adaptation and resilience to climate change, increasing the identity of the landscape and reducing green spaces maintenance costs. In this way, green spaces can be as sustainable as possible with a large seasonal change, depending on the diversity of species introduced in landscape design. They also allow to get positive profits on economy, environment and society at local level.

Keywords: Endangered plants · Green spaces · Native plants · Re-forestation · Climate change mitigation

1 Introduction

The anthropic action on landscape has contributed to deepen the changes in patterns of biological diversity on the world [1–3]. The Mediterranean basin, as one of the great biodiversity hot spots worldwide, has suffered a significant reduction on number of species. Some endangered species, are on the red list, according to the International Union for the Conservation of Nature [4, 5], with several continuous updates on a local and global scale [6]. Since land use has been changing, mainly due to the increase of a monoculture, there is a reduction of flora, vegetation and quality of natural and semi-natural habitats [7–9]. In addition to these, a set of threats such as the abandonment of some rural areas, the expansion of invasive species and the increase of fire risk in short periods of time speed up the loss of biodiversity [10]. Long-term fires are in some cases suitable for biodiversity, such as the coast redwood (*Sequoia sempervirens* Endl.) in California or eucalyptus (*Eucaplyptus* spp.) forests in Australia, promoting their natural regeneration [11, 12]. However, fire in short cycles prevents the maturation of forest species, which are more demanding from the point of view of ecological succession, which avoids the formation of new seeds and, consequently, natural regeneration.

One of the plant groups that has been suffering most from this deepened and continued anthropic action is the natural forest environment which, in the Mediterranean region, is mainly dominated by oaks with persistent and marcescent leaves, such as the holm oak (*Quercus rotundifolia* Lam.), the cork oak (*Q. suber* L.) and the Portuguese oak (*Q. broteroi* (Cout.) Rivas Mart. & Sáez de Rivas) [13, 14]. The characteristic plants in these forests have suffered a strong reduction and are currently been transformed into the montado/dehesa agrosystem, in the western Mediterranean region [15]. Examples of these plants are lianas such as honeysuckle (*Lonicera* spp.), wild madder (*Rubia peregrina* L.), rough bindweed (*Smilax aspera* L.), asparagus (*Asparagus* spp.), as well as various geophytes and hemicryptophytes that are currently rare and with strong difficulties resulting from changes in habitat conditions [16]. However, Man forgets that native plants can contribute to the mitigation of climate change [17–21]. So, it is urgent to adopt management and conservation strategies for vegetation cover, in order to reduce the threats which go-ahead in specific protected areas.

Thus, this article aims to identify ways of protection and mitigation the decrease and extinction of plants in nature, contribute to increase the resilience of green spaces to climate change and reduce the maintenance costs in these areas.

2   Results and Discussion

Green spaces are a good solution to safeguard species with extinction problems in nature [22] as they are less susceptible to a set of threats, such as fire, invasive plants, intensive agriculture and the wild herbivory. The implementation of a proper management, the improvement of firefighting access, and the rational organization of different volumes of vegetation (given by the contrast between open and closed areas), make the urban environment suitable for the conservation of species of several ecological types. On the other hand, the use of RELAPE plants (it means in Portuguese "rare, endemic, localized, threatened or in danger of extinction") in an urban context, allows their dissemination and awareness to a wider population [16, 23]. The inhabitants of large cities sometimes live far from the rural reality, despite depending on it, with regard to the supply of food and the practice of a wide range of passive or active activities where the citizens also have the possibility to contact with several cycles of the nature, among others [24]. However, some of the RELAPE plants can also be found in areas with less accessibility to humans. So, their introduction in urban areas can become easier their knowledge by general public which reduces the impact of human activities in areas of natural occurrence. Since it is only possible to conserve what is known, it is essential to raise awareness of population about the floristic values in each territory, otherwise they will disappear through negligence actions. Unfortunately, society's involvement with local botanical values is sometimes reduced or non-existent, even with species with high ornamental value, such as the Portuguese-laurel (*Prunus lusitanica* L.) which has green foliage throughout the year and produces an exuberant and fragrant flowering. This is an example of a plant threatened with the category of Endangered at a global level which is clearly decreasing in its natural occurring areas [25, 26]. However, it has a high potential in urban space, especially in Central and Northern of the Iberian Peninsula (Fig. 1).

Fig. 1. Portuguese-laurel (*Prunus lusitanica* L.) in a hedge (a) or with its natural shape (b) in an urban space.

The main advantages of using native plants in green spaces are to contribute for the decrease of establishing and maintaining costs of these areas, not only due to pruning and fertilization, but also due to the reduction of irrigation [27]. Water is an essential resource for our lives, despite being very abundant on the planet, in some regions, is, however, increasingly scarce in Mediterranean regions and sometimes with poor quality. There is a global trend towards increased fecal bacteria, lead, nitrates, chloride, arsenic and cadmium in groundwater [28]. This issue becomes more relevant in the climate change scenarios as there is a reduction in precipitation and a significant increase in temperature at a global level [29]. According to that, the investment about using native plants is a reality, as irrigation costs can be reduced, being only necessary during the first or second year after planting (Fig. 2). On the other hand, the use of native plants could give identity to a landscape where it is worth highlighting the importance of tourism. Certainly, an Australian citizen will not feel like visiting Portugal as a tourist seeing extensive eucalyptus plantations.

Thus, indigenous plants should be valued in green spaces according to their ecology [30, 31]. For this, the organization of the territory in homogeneous areas, from a bioclimatic, geological and pedology, among others, are defined in a hierarchical way through biogeographic boundaries, as they relate these factors to the distribution of species, communities, habitats, biocenoses and natural ecosystems of the Earth [32]. The selection of plants according to the characteristics of the soil, either in its texture or in its chemical composition, is essential for a good adaptation of plants to the environment [33]. In addition to the aesthetic qualities of plants such as color, texture, shape, but it should not be forgotten the limitations of plants adaptation to certain soils such as acidic or alkaline (it is important to distinguish acidophilous plants from calciferous plants), among other factors. Therefore, planting design should not be seen as a simple exercise as there are many factors to consider in order to be successful.

Fig. 2. Use of native plants in the Calouste Gulbenkian Fundation's Garden (Lisbon - Portugal) in different systems of ways. Portuguese-laurel (*Prunus lusitanica* L.) and myrtle (*Myrtus communis* L.), on the right and Laurustinus (*Viburnum tinus* L.), on the left (c) Box (*Buxus sempervirens* L.) (d).

However, one of the biggest constraints is still the lack of availability of plants in local nurseries, sometimes hindering the use of native plants. In this sense, the political power must encourage the production of native plants in order to represent each Bio-geographic Sector, as it is at this level that there is the largest representation of the series of climatophilous vegetation, where endemic species are included [32]. In this way, genetic contamination between ecotypes of different geographical origins or hybridization resulting from the introduction of other taxa within a given genus is avoided, as happened with the introduction of red oak (*Quercus rubra* L.) in territories with potential for pedunculate oak (*Quercus robur* L.), resulting in forest formations dominated by this hybrid. Despite the high ornamental value of certain plants, it is necessary to ensure that their introduction does not result in negative impacts [33]. The use of exotic plants in urban spaces has evolved in line with the thinking of societies [34, 35]. As a result of the Age of Discovery, especially from 16th century, a large number of exotic plants were introduced into European gardens. However, some plants did not adapt soil and climatic conditions and others became invasive. Several examples can be cited of the negative consequences about the introduction of exotic species, such as mimosa wattle (*Acacia dealbata* Link.), which has become one of the most invasive species in mainland Portugal and it is very difficult to control it [36] or pampas grass (*Cortaderia selloana* Ach. & Graebn.) throughout some areas of the Mediterranean region [37].

3   Conclusions

There is a growing concern about the reduction of establishment and management costs in green spaces mainly the irrigation as water is a limited resource in Mediterranean regions. However, the selection of plants must be made taking into account the local edaphocli-matic conditions and/or their limitations in adapting to certain situations. Native plants, from local ecotypes, can be a good strategy to overcome problems related with climate change as they are resilient to some situations. Green spaces are suitable for conservation of endangered species, due to a proper management. They also have the possibility to show the beauty of the native flora to population, which can be used by schools or environmental awareness groups to highlight its importance. Thus, even if a certain taxon, in nature, becomes extinct there is a genetic heritage from an alive rescue which allows the production of new seeds and, consequently, makes it possible the production of new individuals and the recovery of potential areas of occurrence.

References

1. Magurran, A.E., Dornelas, M.: Biological diversity in a changing world. Philosoph. Trans. Roy. Soc. B: Biol. Sci. **365**, 3593–3597 (2010). https://doi.org/10.1098/rstb.2010.0296
2. Muller, N., Werner, P., Kelcey, J.G.: Urban Biodiversity and Design. Wiley (2010)
3. Musolino, M.; Massimo, D.E. Mediterranean Urban Landscape. Integrated Strategies for Sustainable Retrofitting of Consolidated City. In Sabiedriba, Integracija, Izglitiba [Society, Integration, Education], Proceedings of the Ispalem/Ipsapa International Scientific Conference, Udine, Italy, 27–28 June 2013; Rezekne Higher Education Institution: Rezekne, Latvija, 2013, vol. 3, pp. 49–60 (2013)
4. Myers, N., Mittermeier, R.A., Mittermeier, C.G., da Fonseca, G.A.B., Kent, J.: Biodiversity hotspots for conservation priorities. Nature **403**, 853–858 (2000). https://doi.org/10.1038/350 02501
5. Rodrigues, A.S.L., Pilgrim, J.D., Lamoreux, J.F., Hoffmann, M., Brooks, T.M.: The value of the IUCN Red List for conservation. Trends Ecol. Evol. **21**, 71–76 (2006). https://doi.org/10.1016/j.tree.2005.10.010
6. Mendoza-Fernández, A.J., et al.: Red List Index application for vascular flora along an altitudinal gradient. Biodivers. Conserv. **28**(5), 1029–1048 (2019). https://doi.org/10.1007/s10531-019-01705-y
7. del Río, S., et al.: Modelling the impacts of climate change on habitat suitability and vulnerability in deciduous forests in Spain. Ecol. Ind. **131**, 108202 (2021). https://doi.org/10.1016/j.ecolind.2021.108202
8. García, F.M., Pardo, F.M.V., Pozo, L.F.F., González, M.Á.R., Fernández, J.C., Batista, M.T.F.: Ensayo para la determinación del estado de conservación de la vegetación. Folia Botanica Extremadurensis. 5 (2010)
9. Musarella, C.M., Cano-Ortiz, A., Quinto Canas, R. Introductory chapter: habitats of the world. In: Musarella, C.M., Cano-Ortiz, A., Quinto Canas, R. (eds.) Habitats of the World - Biodiversity and Threats, IntechOpen, pp. 3–8 (2020). https://doi.org/10.5772/intechopen. 86454
10. Gonçalves, C., et al.: On the development of a regional climate change adaptation plan: Integrating model-assisted projections and stakeholders' perceptions. Sci. Total Environ. **805**, 150320 (2021). https://doi.org/10.1016/j.scitotenv.2021.150320
11. Burrows, N., Gardiner, G., Ward, B., Robinson, A.: Regeneration of *Eucalyptus wandoo* following fire. Aust. For. **53**, 248–258 (1990). https://doi.org/10.1080/00049158.1990.106 76084
12. Kilgore, B.M., Taylor, D.: Fire history of a sequoia-mixed conifer forest. Ecology **60**, 129–142 (1979). https://doi.org/10.2307/1936475
13. Costa, J.C., et al.: Vascular plant communities in Portugal (continental, the Azores and Madeira). Global Geobotany **2**, 1–180 (2012)

14. Quinto Canas, R., et al.: *Quercus rotundifolia* Lam. Woodlands of the Southwestern Iberian Peninsula. Land **10**, 268 (2021). https://doi.org/10.3390/land10030268

15. Pinto-Correia, T., Ribeiro, N., Potes, J.M.: Livro Verde dos Montados. Universidade de Évora, Évora (2013)

16. Carapeto, A., Francisco, A., Pereira, P., Porto, M.: Lista Vermelha da Flora Vascular de Portugal Continental. Sociedade Portuguesa de Botânica, Associação Portuguesa de Ciência da Vegetação – PHYTOS e Instituto da Conservação da Natureza e das Florestas (coord.). Imprensa Nacional-Casa da Moeda., Lisboa (2020)

17. Musolino, M., Malerba, A., De Paola, P., Musarella, C.M.: Building Efficiency Adopting Ecological Materials and Bio Architecture Techniques, pp. 707–717. ArcRHistoR (2019)

18. Spampinato, G., Malerba, A., Calabrò, F., Bernardo, C., Musarella, C.: Cork oak forest spatial valuation toward post carbon city by CO_2 sequestration. In: Bevilacqua, C., Calabrò, F., Della Spina, L., (eds.) New Metropolitan Perspectives, NMP 2020; Smart Innovation, Systems and Technologie, vol. 178. Springer, Cham (2021). https://doi.org/10.1007/978-3-030-48279-4_123

19. Spampinato, G., Massimo, D.E., Musarella, C.M., De Paola, P., Malerba, A., Musolino, M.: Carbon sequestration by cork oak forests and raw material to built up post carbon city. In: Calabrò, F., Della Spina, L., Bevilacqua, C., (eds.) Smart Innovation, Systems and Technologies; New Metropolitan Perspectives, ISHT 2018, vol. 101, pp. 663–671. Springer, Cham (2019)

20. Barrile, V., Malerba, A., Fotia, A., Calabrò, F., Bernardo, C., Musarella, C.: Quarries renaturation by planting Cork oaks and survey with UAV. In: Bevilacqua, C., Calabrò, F., Della Spina, L., (eds.) New Metropolitan Perspectives, NMP 2020; Smart Innovation, Systems and Technologies, vol. 178. Springer, Cham (2021). https://doi.org/10.1007/978-3-030-48279-4_122

21. Massimo, D.E., Del Giudice, V., Malerba, A., Bernardo, C., Musolino, M., De Paola, P.: Valuation of ecological retrofitting technology in existing buildings: a real-world case study. Sustainability **13**, 7001 (2021). https://doi.org/10.3390/su13137001

22. Kendal, D., et al.: The importance of small urban reserves for plant conservation. Biol. Cons. **213**, 146–153 (2017). https://doi.org/10.1016/j.biocon.2017.07.007

23. Primack, R.B., Ellwood, E.R., Gallinat, A.S., Miller-Rushing, A.J.: The growing and vital role of botanical gardens in climate change research. New Phytol. **231**, 917–932 (2021)

24. Ribeiro Telles, G.: Um novo conceito de cidade: a paisagem global. Contemporânea Editora, Matosinhos (1996)

25. Raposo, M., Nunes, L.J.R., Quinto-Canas, R., del Río, S., Pardo, F.M.V., Galveias, A., Pinto-Gomes, C.J.: *Prunus lusitanica* L.: An endangered plant species relict in the central region of mainland Portugal. Diversity **13**, 359 (2021). https://doi.org/10.3390/d13080359

26. Vivero, J.L.: IUCN Red List of Threatened Species: *Prunus lusitanica* subsp. *Lusitanica*. https://www.iucnredlist.org/en

27. Helfand, G.E., Sik Park, J., Nassauer, J.I., Kosek, S.: The economics of native plants in residential landscape designs. Landsc. Urban Plan. **78**, 229–240 (2006). https://doi.org/10.1016/j.landurbplan.2005.08.001

28. Smith, R.A., Alexander, R.B., Wolman, M.G.: Water-quality trends in the nation's rivers. Science (1987). https://doi.org/10.1126/science.235.4796.1607

29. IPCC: Climate Change 2007: The Physical Science Basis. Contribution of Working Group I to the Fourth Assessment Report of the Intergovernmental Panel on Climate Change. Cambridge University Press, Cambridge, United Kingdom and New York (2007)

30. Cabral, F.C., Telles, G.R.: A árvore em Portugal. Assírio & Alvim (1999)

31. Martínez, P.A.: Diseño de áreas verdes con criterios ecológicos. Estudio de dos casos en la comunidad de Castilla-La Mancha, España. Cuadernos de Investigación Urbanística **101**, 80 (2015). https://doi.org/10.20868/ciur.2015.101.3188

32. Rivas-Martínez, S., et al.: Biogeographic units of the iberian peninsula and baelaric islands to district level. A concise synopsis. In: Loidi, J. (ed.) The Vegetation of the Iberian Peninsula: Volume 1, pp. 131–188. Springer, Cham (2017). https://doi.org/10.1007/978-3-319-547 84-8_5

33. Laface, V.L.A., Musarella, C.M., Cano Ortiz, A., Quinto Canas, R., Cannavò, S., Spampinato, G.: Three new alien taxa for europe and a chorological update on the alien vascular flora of calabria (Southern Italy). Plants. **9**, 1181 (2020). https://doi.org/10.3390/plants9091181

34. Musarella, C.M.: *Solanum torvum* Sw. (Solanaceae): a new alien species for Europe. Genet. Resour. Crop Evol. **67**, 515–522 (2020)

35. Spampinato, G., Laface, V.L.A., Posillipo, G., Cano Ortiz, A., Quinto Canas, R., Musarella, C.M.: Alien flora in Calabria (Southern Italy): an updated checklist. Biological Invasions (IN PRESS)

36. Raposo, M.A.M., Pinto Gomes, C.J., Nunes, L.J.R.: Evaluation of species invasiveness: a case study with *Acacia dealbata* Link. on the Slopes of Cabeça (Seia-Portugal). Sustainability **13**, 11233 (2021). https://doi.org/10.3390/su132011233

37. Domènech, R., Vilà, M., Pino, J., Gesti, J.: Historical land-use legacy and *Cortaderia selloana* invasion in the Mediterranean region. Glob. Change Biol. **11**, 1054–1064 (2005). https://doi.org/10.1111/j.1365-2486.2005.00965.x

Improving the Efficiency of District Heating and Cooling Using a Geothermal Technology: Underground Thermal Energy Storage (UTES)

Jessica Maria Chicco[1] (ID), Dragi Antonijevic[2] (ID), Martin Bloemendal[3,7] (ID),
Francesco Cecinato[4] (ID), Gregor Goetzl[5] (ID), Marek Hajto[6] (ID), Niels Hartog[7] (ID),
Giuseppe Mandrone[1] (ID), Damiano Vacha[1(✉)] (ID), and Philip J. Vardon[8] (ID)

[1] Interuniversity Department of Regional and Urban Studies and Planning, University of Turin,
Viale Pier Andrea Mattioli 39, 10125 Turin, Italy
`{jessica.chicco,giuseppe.mandrone,damiano.vacha}@unito.it`
[2] Innovation Center of Faculty of Mechanical Engineering, University of Belgrade, Kraljice
Marije16, Belgrade, Serbia
`dantonijevic@mas.bg.ac.rs`
[3] Department of Water Management, Delft University of Technology, PO Box 5048, 2600 GA
Delft, The Netherlands
`j.m.bloemendal@tudelft.nl`
[4] Dipartimento di Scienze della Terra "A. Desio", Università degli Studi di Milano, Via
Mangiagalli, 34, 20133 Milan, Italy
`francesco.cecinato@unimi.it`
[5] Department of Hydrogeology and Geothermal Energy, Geological Survey of Austria,
Neulinggasse 38, 1030 Wien, Austria
`gregor.goetzl@geologie.ac.at`
[6] Faculty of Geology, Geophysics and Environmental Protection, AGH University of Science
and Technology, 30-059 Krakow, Poland
`mhajto@agh.edu.pl`
[7] KWR Water Research Institute, P.O. Box 1072, 3430 BB Nieuwegein, The Netherlands
`niels.hartog@kwrwater.nl`
[8] Department of Geoscience and Engineering, Delft University of Technology, P.O. Box 5048,
2600 GA Delft, The Netherlands
`p.j.vardon@tudelft.nl`

Abstract. For efficient operation of heating and cooling grids, underground thermal energy storage (UTES) can be a key element. This is due to its ability to seasonally store heat or cold addressing the large mismatch between supply and demand. This technology is already available and there are many operational examples, both within and outside a district heating network. Given the range of available UTES technologies, they are feasible to install almost everywhere. Compared to other storage systems, UTES have the advantage of being able to manage large quantities and fluxes of heat without occupying much surface area, although the storage characteristics are always site specific and depend on the geological and geothermal characteristics of the subsoil. UTES can manage fluctuating production from renewable energy sources, both in the short and long term, and fluctuating demand. It can be used as an instrument to exploit heat available

F. Calabrò et al. (Eds.): NMP 2022, LNNS 482, pp. 1699–1710, 2022.
https://doi.org/10.1007/978-3-031-06825-6_164

from various sources, e.g., solar, waste heat from industry, geothermal, within the same district heating system. The optimization of energy production, the reduction in consumption of primary energy and the reduction in emission of greenhouse gases are guaranteed with UTES, especially when coupled with district heating and cooling networks.

Keywords: Geothermal · UTES · District heating/cooling · Energy transition · Buildings energy retrofitting

1 Introduction

Europe is entering a decade of decarbonization. Referring to the latest policies of the EU commission, such as the "Clean Energy for all European" and the "European Green Deal" [1], Europe is aiming towards the decarbonization of the energy sector. Energy production by renewable energy sources (RES) and improved energy efficiency of buildings are considered key drivers in the pathway from a fossil fuel-based to a carbon neutral society and economy.

A substantial effort and focus are currently placed on electricity, industry and traffic, but half of the energy consumption in the EU is spent on heating and cooling in total [2]. Direct delivery of heating/cooling to consumers via district heating/cooling is seen as an important option to allow the decarbonization of our heating/cooling systems. Optimizing the performance of a sustainable and renewable district heating/cooling grid is becoming an increasingly important topic to both reduce greenhouse gas (GHG) emissions and ensure affordable heating/cooling.

Societal dependence upon energy has increased significantly in the last few decades. Air conditioning systems have increased worldwide from about 4 TW in 1990 to 11 TW (and more) in 2016, and the energy consumption for space heating and cooling is expected to more than triple by 2050 [3]. Since global energy demand for heating and cooling is growing rapidly, good economic and environmental performance are extremely important. Global energy demand is set to grow by more than a quarter to 2040 and the share of generation from renewables is projected to rise from 25% today to around 40%. This is expected to be achieved by promoting the accelerated development of clean and low carbon renewable energy sources and improving energy efficiency. At present, buildings in Europe account for 41% of the final energy consumption, more than transport (32%), and industry (25%), hence the integration of renewable energy technologies is extremely necessary [4]. In addition, cooling needs for warm climates such as Mediterranean areas are increasing. As incomes rise and populations grow, the use of air conditioners is becoming increasingly common, especially in commercial buildings and high-density residences of the hottest world regions. They currently account for about a fifth of the total electricity in buildings around the world [3].

The pressure is high for finding solutions to reduce energy imports, enable low carbon sources and fight against critical heat waves due to climate change. An already feasible and sustainable solution is geothermal energy. It has been used for decades in Europe, and its range of technological solutions can provide electricity, heating and cooling.

As proposed by the Heat Roadmap Europe project [5], heating and cooling grids are crucial for reaching the decarbonization of the heating and cooling sector covering at least 25% of the future end user supply. "Geothermal-DHC", a research network for including geothermal technologies into decarbonized heating and cooling grids, is addressing these topics [6].

Renewable heat can come from many sources, each of which exhibit different characteristics. However, none can be typically generated on-demand and they are often available either continuously or at different times than the demand. To overcome the temporal mismatch in supply and demand of thermal energy, storage facilities are needed.

Therefore, in this paper, the potential of a particular technology concerning geothermal energy and district heating and cooling is depicted: Underground Thermal Energy Storage (UTES). It can serve as seasonal storage or help overcoming short-term peaks of energy requests.

2   UTES in the Energy Transition of District Heating/Cooling Grids

Approximately 1.4 million GWh could be saved and 400 million tons of GHG could be reduced annually by the application of thermal energy storage (TES) in Europe [7].

One of the major sources of renewable heat is solar thermal energy, which harnesses energy from the sun using solar thermal collectors. Most of the solar thermal energy is produced in small-scale systems for domestic space and water heating, but large-scale solar thermal systems for district heating are also common. The supply of solar thermal energy varies seasonally resulting in more energy being available in the summer and less in winter. However, the peak space heating demand is during the winter months, thus creating a seasonal mismatch between the supply and demand. A solution can be to collect energy in the summer, store it seasonally, and use it to cover demand in the winter. One of the methods for seasonal storage of energy is UTES. It includes several different technologies, each one with its specific characteristics. For instance, UTES systems based on sensible heat storage, typically offer a storage capacity of 10–50 kWh/t and storage efficiencies between 50–90%, depending on the specific heat of the storage medium, storage size and thermal insulation technologies [7, 8].

Various UTES systems can be easily integrated into a district heating system in a centralized (large capacity units for neighbourhoods) or decentralized (usually smaller scale units in public and office buildings or households) manner. The utilization of UTES for coupling and the integration of decentralized renewable heat sources contribute to the overall efficiency, flexibility, and response time of a district heating system. The coupling of local renewable energy subsystems and installations to an onsite thermal energy storage system reduces the overall heat consumption of the district heating system. In the case of combined district heating and cooling systems, cold storage is also used. Cold storage may have common storage units with heat storage (operating in seasonal modes) or be designed separately as, for example, cold-water storage or ice storage.

This paper addresses the integration of geothermal energy into multivalent decarbonized district heating and cooling (DHC) networks operating at temperature levels between less than 30 °C (5th generation DHC) and approximately 100 °C (3rd generation DHC). Currently, Geothermal energy (Fig. 1) can be used for baseload supply due to its low operational costs (OPEX). Typically, peak load systems are designed to have a low capital expenditure (CAPEX) with higher OPEX as they are designed to run for only a limited time. UTES can be used to reduce the amount of time peak load systems operate by using low OPEX energy (e.g., from geothermal or other local available renewable or recyclable heat sources). UTES can therefore be used to both fill peak demand loads as well as provide backup supply.

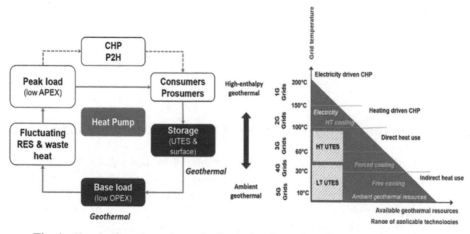

Fig. 1. Sketch of geothermal uses in district heating and cooling networks, from [6].

UTES can provide large-scale seasonal storage of cold and heat in the underground in different ranges of temperature. Commonly, literature refers to Low Temperature (LT) and High Temperature (HT) UTES. The main technical differences between them are:

- LT-UTES uses fluids at temperatures lower than ~25–30 °C; it is usually coupled with geothermal heat pump systems and improves the overall efficiency. It is the most widespread UTES application.
- HT-UTES uses higher temperatures, up to >90 °C, and typically deeper reservoirs; it can be easily coupled with traditional (higher temperature) and innovative district heating, with or without heat pumps to increase temperature.

Moreover, heat pumps at various temperature levels and positions inside the network can also help to modulate and stabilize the DHC network, and waste heat from providing cooling could be recycled through storage. These concepts could be applied to new networks but permit also upgrading existing heating only networks, increasing their overall efficiency.

3   UTES State of the Art in a DHC Context

3.1   Different Types of UTES

UTES falls in the family of technologies based on sensible heat, which is dependent on the mass, specific heat and temperature. The available UTES technologies (Fig. 2, a-f)

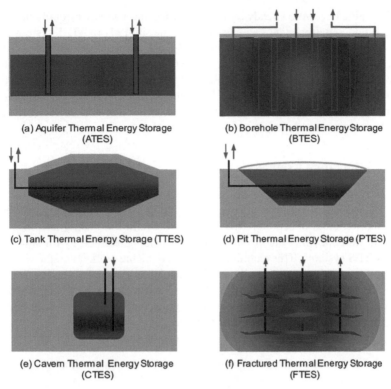

(a) Aquifer Thermal Energy Storage (ATES)

(b) Borehole Thermal Energy Storage (BTES)

(c) Tank Thermal Energy Storage (TTES)

(d) Pit Thermal Energy Storage (PTES)

(e) Cavern Thermal Energy Storage (CTES)

(f) Fractured Thermal Energy Storage (FTES)

Fig. 2. Sketch of most common UTES technologies, from [12].

include: Aquifer (ATES), Borehole (BTES), Tank (TTES), Pit (PTES), Cavern (CTES), and Fractured (FTES) thermal energy storage [9–11].

ATES uses naturally occurring groundwater bodies at depth between <30 m and up to 2–3 kms. In general, all ATES concepts are based on at least two wells (well doublet) for the extraction and injection of groundwater. These wells are connected to a heat exchanger to transfer the heat between the geofluid circuit and the DHC circuit. For technical, physical and legislative/environmental reasons, the maximum storage temperatures of ATES systems are usually limited to around 90 °C.

BTES can be considered as an improvement on conventional closed-loop Ground Source Heat Pump (GSHP) systems. They consist of an array of vertical borehole heat exchangers (BHE) installed in wells drilled at certain distances and depths depending on geological, hydrogeological and thermo-physical conditions of the underground. BHEs are designed in a way such that heat or cold energy are stored or extracted seasonally from a cylindrical volume of soil or rock (typical distance of heat exchangers: 2–3 m) while GSHPs typically provide dissipation of thermal energy into the subsurface (typical distance of heat exchangers: 6–8 m). Due to their closed-loop technology, they can be installed in almost all locations, but can store less heat for the same CAPEX than ATES systems.

Tank and Pit TES were already used in the 1950s and 1960s, especially in Denmark. They use water containing tanks and pits and operate similarly to ATES systems. Cavern and Fractured systems, again operate similarly, but use natural caverns or fractured formations to store heat, although so far have not been commonly used and their use is restricted to specific sites where the geological and mining conditions allow it.

Currently, ATES and BTES seem to be solutions that guarantee large-scale use and modularity, ranging from storage for small complexes to integration with district heating and cooling networks of large cities.

3.2   UTES Operating Examples

In DHC networks, geothermal energy technology will be a key energy source both in smart cities and smart rural communities, in addition to supplying energy for industry, services and agricultural sectors. This is due to its ability to supply not only heating, cooling and water heating, but also to be a key balancing technology for smart thermal grids via UTES systems. This technology is already on the market and there are several examples of its use, both within and outside a district heating network. The Technology Readiness Level (TRL) provides a useful metric to be used which ranges from technology validated in relevant environment (4–5) to actual system proven in operational environment (9) and ready for the market (Table 1).

Table 1. UTES systems at a glance and their Technology Readiness Level.

Item	Tank	Pit	Borehole	Aquifer
TRL	8–9	Up to 2 GWh: 7–9 above: 3–4	8–9	5–6 (HT) 7–8 (LT)
Storage depth	Surface	Surface to 30 m	30–1,000 m	10–1,000 m
Temperature range	Atmospheric: <100 °C pressurized: >100 °C	<100 °C	Up to 30 °C for shallow and 100 °C for deep systems	Up to 20 °C for shallow and 100 °C for deep systems
Thermal storage capacity	30–80 kWh/m^3	30–50 kWh/m^3	15–30 kWh/m^3	30–40 kWh/m^3
Strengths	Applicable in any place; low development risk	Applicable in any place; low development risk	Low development risk; small surface footprint	High efficiency rate; small surface footprint
Weaknesses	High investment cost; visible landmark	High surface footprint; low efficiency rate	High investment costs; lower efficiency of thermal output	Only applicable in aquifers; development risks

ATES systems provide sustainable heating and cooling energy for different building typologies and can be integrated at a district/urban level. They require a suitable sub-surface which allows water to flow easily and can store water (i.e. an aquifer). In [13] is reported that there were around 3,000 ATES applications worldwide by 2017, mostly concentrated in Europe. They are mostly applied for single buildings and small building complexes in the Netherlands with over 2,500 sites, and Nordic Countries such as Sweden and Denmark with 220 and 55 examples, respectively. A more limited number of examples are in Great Britain, China, Japan, Germany, North America and Turkey. The total amount of heat and cold produced by ATES is currently estimated to be 2.5 TWh per year.

The growing number of systems are mainly focused on LT-ATES systems, probably due to market incentive programs and the authorities supporting these kinds of systems. Most of the LT-ATES are in the Netherlands, are shallow and operate with well depths ranging between 25 and 250 m, with temperatures lower than 25 °C, while in Germany the current temperature threshold for these depths is at a maximum of 20 °C for heating and 5 °C for cooling. The TRL of LT-ATES can range between 7 and 8. The applicability of LT-ATES in other European countries is high based on the characteristics of the subsurface. Despite this, the current level of implementation outside the Netherlands lags at the European and worldwide levels [13]. Some large-scale ATES systems are integrated into district heating and cooling networks [14]. The low-temperature range has caused some problems with the integration of ATES systems, as the majority of DHC networks operate at higher temperatures.

A good example of the HT-ATES concept has been developed in Delft, NL (Fig. 3) [15].

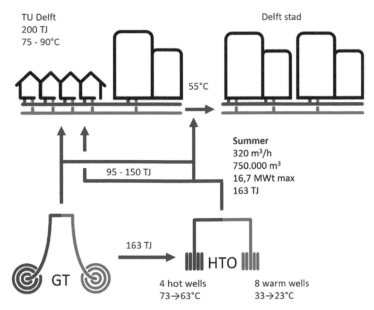

Fig. 3. Delft, the Netherlands: a summary of the HT-ATES at University campus.

The HT-ATES system is proposed to be connected to the existing district heating network of the TU Delft university campus. Developments are ongoing, with plans to realize a geothermal system to supply the heat, and the network will be extended to part of the city of Delft.

Currently, there is a significantly lower implementation of HT-ATES systems, partly due to lack of regulation and partially due to poor technology recognition. In fact, according to [13], only 5 HT-ATES systems are in operation worldwide. The TRL is considered to be between 5 and 7. Generally, several market barriers in the energy market are often preventing the development of such a system in specific countries, where this technology is not yet developed.

BTES applications can be considered to range between a TRL of 7 and 9. They are becoming popular because of their suitability for seasonal storage thanks to slow thermal response and large storage capacities. In recent years, experimental facilities have been used to better understand the underground thermal behavior using these systems, where for example various charging and discharging strategies have been tested [16].

BTES require only a small amount of space to tap into a large volume of subsurface rocks at a relatively low cost. There is no exchange of groundwater like in ATES systems, which increases the geographical applicability. Moreover, a literature review [17] reveals that the energy efficiency of BTES is best when diurnal and seasonal storage are used in conjunction.

An example of this technology can be seen at Crailsheim Hirtenwiesen, Germany (Fig. 4) [18], where 7,300 m^2 of solar thermal collectors provide 50% of the heat for a housing area with 260 units. Heat is stored in two water tanks (100 and 480 m^3) and a seasonal 37,500 m^3 borehole storage. A collector area of 9,700 m^2 (6.8 MWth) and a 75,800 m^3 borehole storage were foreseen. A 489 kWth high-temperature heat pump transfers heat from the larger buffer storage to the smaller one, when necessary, to ensure there is always hot water at 70 °C available.

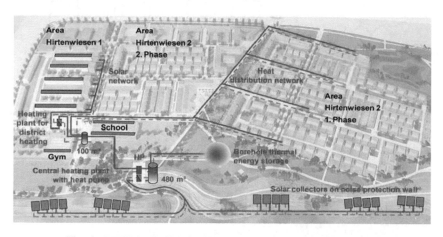

Fig. 4. BTES in Crailsheim Hirtenwiesen, Germany (after [18]).

CTES plants have reached a TRL between 5 and 7. CTES systems require a suitable natural or man-made cavern. Abandoned mines with stable caverns can provide such a facility, especially when located near urban areas (there are over one million abandoned mines worldwide). Depending on the geology and local groundwater conditions, polluted mine water may have to be perpetually treated, becoming a long-term economic burden on current and future generations [19]. Hence, environmental remediation costs could be potentially compensated by exploiting mine water for CTES.

One of the few pilot examples is at Fraunhofer IEG Bochum, Germany [20] (Fig. 5). It is a fully functional high-temperature mine thermal energy storage (HT-MTES) pilot plant, for the energetic reuse of an abandoned coal mine. The seasonal surplus heat available during the summer from solar thermal collectors is stored within the mine workings and is used during the winter season for heating the institute buildings of the "Fraunhofer Institut für Energieinfrastruktur und Geothermie" (IEG).

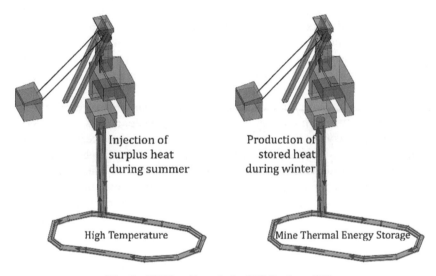

Fig. 5. CTES at Fraunhofer IEG Bochum [20].

4 Discussion

Existing housing infrastructure represents a considerable share of the heating and cooling demand, that can be efficiently and sustainably supplied by geothermal heat pumps and geothermal district heating systems. New materials and designs of the system have also produced promising results to reduce costs and to increase efficiency, but here further work is needed.

Furthermore, geothermal district heating will be increasingly targeted at existing buildings and old inner cities in dense urban areas. At the same time, the concept of UTES is attracting increasing interest from industries, research institutions, and public authorities. It is expected to gain acceptance and market uptake as it will provide a solution to partially replace the use of fossil fuels and to reduce the costs of heating and cooling. It will deliver the combination of geothermal energy with underground storage which will constitute a powerful tool in the context of sector coupling. Geothermal energy combined with small thermal grid systems offers one of the most effective options for this market, both in terms of carbon footprint and economics.

The European Parliament approved an Own Initiative Report on energy storage. The report highlights the role of geothermal as a provider of energy storage and flexibility services, including seasonal thermal energy storage in the underground, batteries and generation of electricity from flexible renewable sources such as geothermal [21].

Stakeholders in district heating and cooling projects such as developers, local authorities, utilities, consumers, and housing associations can be disheartened in investing in UTES technologies, due to non-technical aspects, despite that their technical feasibility has been widely proven. Investment mechanisms, clear guidelines and regulations for planning, building standards and environmental protection, can contribute to accelerate the deployment of UTES projects in the district heating and cooling context [22]. Policy makers should work on removing unnecessary barriers to project progress by ensuring robust planning procedures and assisting potential stakeholders. This stresses the need to implement the current EU Strategy [23] that aims to have flexible procedures in the district heating and cooling market, but still requires special planning for UTES technologies.

The important work of policy makers in this field can be also linked to funding R&D and demonstrations in order to prove the system benefits as well as to promote media campaigns encouraging consumer uptakes. Price support mechanisms are one of the important pillars between the main outlook until 2050; they can help drive the competitiveness of decarbonizing district heating and cooling on the whole, helping to increase demand for thermal storage. Increases in reliability given by thermal storage, and the ability to improve flexibility of using multiple sources, could also be included in such mechanisms.

5 Conclusions

The environmental objectives of the European Green Deal represent an innovation-driven development strategy for Europe, making sustainable development a priority. The new strategy raises several multilevel governance challenges involving not only prosperous cities or capitals but also suburban and rural areas. Energy production by RES and improved energy efficiency of buildings are considered key drivers in the pathway from a fossil fuel-based to a carbon neutral society and economy. Investment decisions taken by agents involved in energy retrofitting of buildings and the development of energy systems, alongside the value of related investments, can determine the success or failure of a fast energy transition.

One of the major challenges for this future energy systems is to overcome the mismatch between supply and demand through the development of energy management tools achieved thanks to new information and communication technologies and a new smart energy system approach.

Underground Thermal Energy Storage enables the utilisation of various sources of heating and cooling and the integration of such renewable energy sources in urban areas.

It is a ready-to-market technology, extensively tested in labs, in numerical models and real scale test sites, with many already operating systems, although some versions of the technology require further development. Compared to other storage systems it has the advantage of being able to store large quantities of heat without occupying a significant surface area, although the storage characteristics are always site-specific and depend on the geological and geothermal characteristics of the subsoil.

Between the most important outlooks until 2050, the International Renewable Energy Agency shows the key attributes of UTES technologies [22], identifying priorities for ongoing research and development such as:

- investment to drive technological development and measures are needed to enhance a market pull, together with well-defined and favourable energy policies aimed at scaling up the use of geothermal energy in the district heating and cooling sector also combined with other renewables;
- UTES systems can contribute to the energy transition investment package available to Countries for post-COVID recovery; this can strengthen health and economic infrastructure and align the energy development with global climate and sustainability goals.

Furthermore, the RHC and EGEC Agenda [23] sets forth certain technologies showing how the well-established, low temperature heat pump supported applications, energy produced from low temperature air source, water source and solar thermal energy could be stored underground and used for heating and cooling purposes. Based on the findings and conclusions reported in this agenda, these systems could become an important provider for heating and cooling for individual houses, industry and utility buildings, as well as for district heating and cooling.

UTES applications reveal an effective technology to bridge the gap between energy demand and supply, seasonally for both district heating and cooling networks. This improves efficiency, techno-economic feasibility and reduces the impact on the environment. Storage is currently the only way to use volatile renewables, such as solar-thermal, and make use of waste heat to enhance the overall energy efficiency of the heating and cooling market, which is an EU goal.

Acknowledgements. This work is part of the COST project "Geothermal DHC. Research Network for Including Geothermal Technologies into Decarbonized Heating and Cooling Grids (https://www.geothermal-dhc.eu/)". It is the result of a joint cooperation of the WP1 – technologies, Ad Hoc WG2 - UTES systems.

References

1. European Commission: A European Green Deal. https://ec.europa.eu/info/strategy/priorities-2019-2024/european-green-deal_en. accessed 16 Dec 2021
2. European Commission: Heating and cooling (2020). https://ec.europa.eu/energy/topics/energy-efficiency/heating-and-cooling_en. Update 25 Oct 2021, Accessed 16 Dec 2021
3. IEA: Global Energy Review 2019, Report Extract Electricity (2019). https://www.iea.org/reports/global-energy-review-2019/electricity. Accessed 16 Dec 2021

4. Todorov, O., Alanne, K., Virtanen, M., Kosonen, R.: A method and analysis of aquifer thermal energy storage (ATES) system for district heating and cooling: a case study in Finland. Sustain. Cities Soc. **53**, 101977 (2020)
5. Heat Roadmap Europe. https://heatroadmap.eu. Accessed 16 Dec 2021
6. Goetzl, G., Milenic, D., Schifflechner C.: Geothermal-DHC, European research network on geothermal energy in heating and cooling networks. In: Proceedings World Geothermal Congress 2020+1. IGA, Reykjavik, Iceland (2021)
7. IEA ETSAP & IRENA: Thermal Energy Storage, Technology Brief **E17**, 1–20 (2013)
8. Wild, M., Lüönd, L., Steinfeld, A.: Experimental investigation of a thermochemical reactor for high-temperature heat storage via carbonation-calcination based cycles. Front. Energy Res. **9**, 748665 (2021)
9. Hellström, G., Larson, S.: Seasonal thermal energy storage – the HYDROCK concept. Bull. Eng. Geol. Environ. **60**, 145–156 (2001)
10. Novo, A.V., Bayon, J.R., Castro-Fresno, D., Rodriguez-Hernandez, J.: Review of seasonal heat storage in large basins: water tanks and gravel-water pits. Appl. Energy **87**(2), 390–397 (2010)
11. Pavlov, G.K., Olesen, B.W.: Thermal energy storage - a review of concepts and systems for heating and cooling applications in buildings: Part 1. Seasonal Storage Ground **18**(3), 515–538 (2012)
12. Janiszewsky, M.: Techno-economic aspects OD seasonal underground storage of solar thermal energy in hard crystalline rocks. Doctoral dissertation, Aalto University (2019)
13. Fleuchaus, P., Godschalk, B., Stober, I., Blum, P.: Worldwide application of aquifer thermal energy storage. A review. Renew. Sustain. Energy Rev. **94**, 861–876 (2018)
14. Schmidt, T., Pauschinger, T., Sørensen, P.A., Snijders, A., Thornton, J.: Design aspects for large-scale pit and aquifer thermal energy storage for district heating and cooling. Energy Procedia **149**, 585–594 (2018)
15. Bloemendal, M., et al.: Feasibility study: HT-ATES at the TU Delft campus. TU Delft/ENGIE (2020). https://www.warmingup.info/documenten/window-fase-1---a1---ver kenning-hto-tud---feasibilityht_ates_tudelft.pdf
16. Nilsson, E., Rohdin, P.: Empirical validation and numerical predictions of an industrial borehole thermal energy storage system. Energies **12**(12), 2263 (2019)
17. Lanahan, M., Tabares-Velasco, P.C.: Seasonal thermal-energy storage: a critical review on BTES systems, modeling, and system design for higher system efficiency. Energies **10**(6), 743 (2017)
18. Miedaner, O., Mangold, D., Sørensen, P.A.: Borehole thermal energy storage systems in Germany and Denmark – construction and operation experiences. In: Greenstock Beijing 2015, Sensible TES, vol. C-1, pp. 1–8 (2015)
19. Menéndez, J., Ordóñez, A., Álvarez, R., Loredo, J.: Energy from closed mines: underground energy storage and geothermal applications. Renew. Sustain. Energy Rev. **108**, 498–512 (2019)
20. Hahn, F., et al.: The reuse of the former Markgraf II Colliery as a mine thermal energy storage. In: European Geothermal Congress 2019. EGEC, Den Haag, The Netherlands (2019)
21. European Parliament: On a comprehensive European approach to energy storage (2019/2189(INI)) (2019). https://www.europarl.europa.eu/doceo/document/A-9-2020-0130_EN.html
22. IRENA: Innovation Outlook: Thermal Energy Storage. International Renewable Energy Agency, Abu Dhabi (2020). ISBN 978-9260-279-6, www.irena.org/publications
23. European Commission: Towards a smart, efficient and sustainable heating and cooling sector. EU to fight energy waste with the first Heating and Cooling Strategy (2016). https://ec.eur opa.eu/commission/presscorner/detail/en/MEMO_16_311

A Multi-criteria Assessment of HVAC Configurations for Contemporary Heating and Cooling Needs

Ilaria Abbà$^{(\boxtimes)}$ (ID) and Giulia Crespi (ID)

TEBE-IEEM Research Group, Energy Department, Politecnico di Torino, Corso Duca degli Abruzzi 24, 10129 Turin, Italy
ilaria.abba@polito.it

Abstract. Due to the significant impact that HVAC systems have on the overall building consumption, there is the need to encourage consumers to invest in increasingly more efficient and sustainable technologies to accelerate the transition of the building sector. Moreover, new occupants' habits, higher environmental awareness, and the increment of external air temperatures due to climate change are varying buildings energy demands, which are highly experiencing also simultaneous heating and cooling needs. Consumers' investment decisions in the building sector are usually driven by financial convenience. However, the energy investment decision-making process is a multi-dimensional problem, characterized by different and often conflicting aspects, belonging to energy, technological, financial, environmental, and social domains. In the light of the above, the work aims to compare different electric HVAC configurations, capable of meeting the same contemporary heating and cooling loads, from a multi-perspective standpoint, ranking their performances according to a set of criteria that can potentially influence the choice of the most appropriate HVAC solution in line with consumers' needs. To this purpose, a multi-criteria analysis is developed, in the form of the PROMETHEE II method, with the final scope of ranking the selected alternatives according to different and conflicting criteria.

Keywords: HVAC configurations · Contemporary heating and cooling loads · Multi-criteria decision analysis

1 Introduction

It is well known that buildings are among the most environmentally impacting economies at global level, with the HVAC sector playing a crucial role in the attempt of reducing its consumptions and emissions. Indeed, due to the significant impact that HVAC systems have on overall building consumption [1], there is the need to encourage consumers to invest in increasingly more efficient and sustainable technologies to accelerate the transition of the building sector, which will be largely shaped by electrification, identified as a key pillar of this changeover, in line with European trajectories [2]. The investigation on buildings retrofit solutions and efficient HVAC technologies needs to face

© The Author(s), under exclusive license to Springer Nature Switzerland AG 2022
F. Calabrò et al. (Eds.): NMP 2022, LNNS 482, pp. 1711–1720, 2022.
https://doi.org/10.1007/978-3-031-06825-6_165

the changes in buildings energy demand due to new occupants' habits and behaviors (i.e., the diffusion of smart working activities because of the COVID-19 pandemic) and due to climate change and global warming consequences, which are mostly reflected in a severe increment of external air temperatures [3]. All these aspects influence building energy demands, varying typical heating and cooling profiles and increasing air conditioning needs and consumptions. Moreover, all these considerations profoundly affect energy systems reflections, asking for the adoption of more efficient technological solutions, able to satisfy in a cost-effective way also simultaneous heating and cooling requests [4, 5]. Despite the interesting solutions already present in the market to meet new buildings energy needs, in the process of selecting the most appropriate HVAC configurations for a specific building, the comparison is usually done according to their technical or financial performances, assessed separately, without deepening any possible trade-off between the different perspectives [5]. From a consumer standpoint, indeed, decisions are usually made according to a purely financial perspective, selecting the solution to be installed based on financial convenience, rather than on energy or environmental considerations. However, decision-making in the energy field is by definition a multi-dimensional problem, which asks for proper methods to study and rank the alternative solutions at disposal [6]. In this framework, evaluation tools are in the spotlight for supporting the decision-making process, being instruments capable of studying energy issues by integrating different elements, belonging to diverse and often contrasting domains [6]. Among these tools, multi-criteria decision analysis (MCDA) methods are commonly used for supporting investment and design decisions in the building sector, considering the judgements or preferences of the different stakeholders potentially involved in the decision-making process. In line with this, the work aims to compare different HVAC systems, capable of meeting the same contemporary space heating and cooling loads, not focusing on a mere energy comparison between the considered alternative configurations but enlarging the discussion to a multi-perspective standpoint, to consider a richer set of criteria that can potentially influence the choice of the most appropriate HVAC configuration. To this purpose, a multi-criteria analysis is developed, in the form of the PROMETHEE II (Preference Ranking Organization METHod for Enrichment of Evaluations) method, with the final scope of ranking the selected alternatives according to different and conflicting multi-dimensional criteria.

2   Materials and Methods

In recent years, dealing with energy issues or energy transition concept has been no longer a mere energy matter, but has increasingly involved environmental, financial, and social aspects [6]. When dealing with complex issues, a MCDA approach allows to consider different and often contrasting standpoints. For this reason, MCDA tools are particularly useful to help decision-makers in expressing rational and consistent preferences, needed to take confident decisions [7] and are acknowledged as beneficial in providing to interested stakeholders an instrument to select the best strategical option or to rank the studied alternatives, in accordance with their needs and goals [8, 9]. Generally speaking, a typical MCDA follows specific methodological steps: (i) to frame the research context, defining scope, target and key involved stakeholders; (ii) to identify the alternatives to be compared and evaluated; (iii) to define the evaluation criteria; (iv) to score the performances of each alternative against the criteria; (v) to weight each criterion to reflect its relevance with respect to the others in the decision process; (vi) to combine weights and scores for each alternative to get an overall value; and (vii) to examine results and to perform proper sensitivity analyses to evaluate if and how changes in scores or weights can affect the overall results [7].

Despite the general characteristics of the multi-criteria approach, there exists a variety of MCDA techniques, characterized by different definitions of the decision context, nature of information (e.g., qualitative, quantitative, or mixed), disaggregation of complex problems, level of compensation and weighting technique for estimating the final scores of the considered alternatives. Specifically, for the scope of this research, the outranking PROMETHEE II method was selected [10], allowing to manage, select, and rank a finite set of alternatives according to diverse criteria. In detail, once selected the proper alternatives and criteria for the study, according to the general MCDA framework, the following methodological steps need to be carried out [10]:

1. Establishment of a double-entry impact matrix, connecting alternatives and criteria.
2. Application of a proper preference function to each criterion to identify how much an alternative is preferred to another; each is a normalized function between 0 and 1, where 1 corresponds to a huge preference of an alternative over the other, while 0 means that the decision-maker is indifferent between the two alternatives.
3. Weighting of each criterion, according to experts' preferences, to define its importance with respect to the others in the decision process; the weights assignment allows to calculate the overall preference index for each alternative, which represents the intensity of one alternative over the others.
4. Computation of the outranking flows (i.e., leaving and entering flows) for the different alternatives; the higher the leaving flow and the lower the entering flow, the better the alternative is.
5. Comparison of outranking flows by calculating the net flows between the alternatives to define the complete ranking of the alternatives and to identify the best solution(s).

The weighting process is here performed involving experts with different expertise and background, potentially affected by or influencing the decision-making process; based on each expert's preferences, different scenarios of analysis can be defined. In this work, the Simos-Roy-Figueira method (SRF) was exploited for the experts' involvement [11]. All other steps were performed using the user-friendly Visual PROMETHEE software.

3   Case Study

The multi-criteria assessment was performed to compare and rank different HVAC systems, able to meet the same thermal loads, for supporting energy investments decision-making. In recent years, attention is mainly devoted to electric technologies, which deployment in new and retrofitted buildings can guarantee energy efficiency improvements, reduced environmental impacts and energy consumptions [12]. For this reason, in this work, five all-electric HVAC configurations were selected as alternatives, modeling their operation modes using the characteristics of real commercial units. All systems were compared for equal load profiles, defined as Gaussian-shaped curves (see Fig. 1) [4, 5], having fixed an average percentage of contemporaneity of space heating and cooling needs, equal to 52%. Maximum heating and cooling loads were set equal to 640 kW and 630 kW, respectively [5]. The external air temperature distribution of Strasbourg was considered for the analysis.

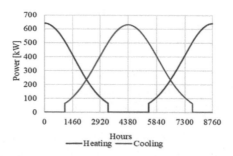

Fig. 1. Gaussian-shaped load profiles for a 52% percentage of contemporaneity [4, 5].

Due to the presence of contemporary heating and cooling requests during the year, different units were coupled in multi-unit systems able to match the required loads; specifically, 4 configurations consider the integration of reversible heat pumps (HPs), electric boilers (EBs) or chillers (CHs) to guarantee the satisfaction of contemporary needs. Conversely, the last configuration is characterized by the presence of a single unit, the polyvalent heat pump (PHP), which is recognized as a promising, but still not widespread solution for buildings [4, 5], able to provide space heating and cooling simultaneously and independently, and not only seasonally, as traditional HPs. From a technical perspective, the PHP can be defined as a heat pump equipped with a heat recovery unit, allowing the machine to operate in three different modes: heating only (as a traditional HP), cooling only (as a traditional chiller) and combined heating and cooling (which represents the main strength of this technology). The considered configurations

Table 1. Selection of alternatives.

	Primary unit	Secondary unit
HP + EB	Reversible HP (660 kW) with priority on space cooling	Electric boiler (500 kW) as backup for space heating contemporary needs
HP + CH	Reversible HP (660 kW) with priority on space heating	Chiller (520 kW) as backup for space cooling contemporary needs
HP + HP	Reversible HP (660 kW) with hourly priority on the highest need	Reversible HP (370 kW) as backup for the non-served contemporary needs
EB + CH	Electric boiler (640 kW) for space heating needs	Chiller (660 kW) for space cooling needs
PHP	Polyvalent heat pump (660 kW) for space heating and cooling needs	

are summarized in Table 1, dividing them between primary and secondary units and according to the loads they primarily match.

Once alternatives were defined, the following step concerned the identification of the evaluation criteria and the definition of their corresponding preference functions. For the sake of simplicity, criteria were divided into 4 main dimensions (i.e., technical, energy, financial and environmental), as shown in Fig. 2; a total of 10 criteria was considered, being all quantitative, with the sole exception of TECH.2.

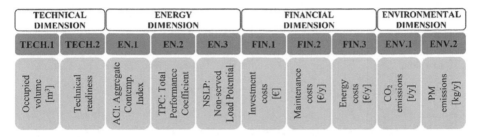

Fig. 2. Set of evaluation criteria.

Going into detail, within the technical dimension, occupied volume (TECH.1) and technical readiness (TECH.2) criteria were identified. TECH.1 is a quantitative criterion evaluating the overall encumbrance (in m^3) of each HVAC configuration, extrapolated from technical datasheets; on the other hand, TECH.2 is a qualitative criterion expressing the level of maturity and deployment of the considered technologies. Concerning the energy dimension, three quantitative criteria were identified and computed using a numerical model, developed in order to couple the considered load profiles with the units operation modes [4, 5]: Aggregate Contemporary Indicator (ACI), Total Performance Coefficient (TPC) and Non-served Load Potential (NSLP). ACI (EN.1) aims to evaluate the units performances only in contemporaneity hours, calculating the ratio between the requested contemporary heating and cooling loads and the corresponding electricity consumption [5]. Conversely, TPC (EN.2) is an annual indicator used to evaluate the total energy performance of the units, computed as the ratio between the sum of all requested loads and the total yearly electricity consumption [4]. Finally, NSLP (EN.3) indicates in percentage terms the quota of contemporary load that the primary unit would not be able to satisfy alone, without integration (NSLP is null for the PHP, meaning that all contemporary loads can be satisfied by the PHP unit) [5]. Moving to the financial dimension, investment (FIN.1), maintenance (FIN.2) and energy (FIN.3) costs were included. For FIN.1 estimation, real investment costs of commercial units were considered, with the sole exception of the cost of the electric boiler, derived from [3]. Annual maintenance costs were computed as percentages of the units investment costs, in line with [13], while energy costs were computed considering non-domestic electricity prices from [14] for the year 2019 (only variable quota was considered). Finally, CO_2 (ENV.1) and PM (ENV.2) emissions were identified as quantitative criteria for the environmental dimension, computed using appropriate emissions factors for electricity [15, 16]. Based on alternatives and criteria definition, the impact matrix can be built by computing, per each alternative, the value of each criterion. Figure 3 shows the input parameters of the double-entry impact matrix (with alternatives on the rows and criteria on the columns) and the direction of preference, indicating if criteria should be maximized or minimized, according to their definition. At first glance, it is possible to note that there is not a priori dominant alternative.

	Criteria	TECHNICAL DIMENSION		ENERGY DIMENSION			FINANCIAL DIMENSION			ENVIRONMENTAL DIMENSION	
		TECH.1	TECH.2	EN.1	EN.2	EN.3	FIN.1	FIN.2	FIN.3	ENV.1	ENV.2
	Direction of preference	min	max	max	max	min	min	min	min	min	min
Alternatives	HP + EB	79.6	Good	1.64	2.29	50.4%	186'240	5'282	371'624	874.6	8.6
	HP + CH	73.1	Very good	3.91	3.76	49.6%	298'000	8'940	226'991	534.2	5.2
	HP + HP	66.9	Very bad	3.95	3.78	27.1%	272'000	8'160	225'882	531.6	5.2
	EB + CH	79.6	Average	1.66	1.65	50.0%	142'240	3'962	515'831	1214.0	11.9
	PHP	39.8	Average	5.41	4.24	0.0%	210'000	6'300	201'128	473.4	4.6

Fig. 3. Impact matrix input data.

Based on the performed calculations, a proper preference function needs to be assigned to each criterion, choosing among six types of preference functions, differing in terms of shape and threshold values (i.e., preference and indifference thresholds, if present). In the current application, the linear function was applied to EN.1 and EN.2 and to all financial criteria; the U-shape (i.e., quasi criterion) preference function was selected for both environmental criteria and for EN.3, while the V-shape (i.e., criterion with liner preference) was chosen for TECH.1. Finally, the qualitative criterion concerning the technical readiness (TECH.2) was modeled using the level preference function. Afterwards, the weighting procedure was carried out involving and interviewing three experts with different backgrounds and expertise, relevant for the scope of the analysis, allowing to build three different scenarios: (i) an Energy scenario, according to a building physics expert; (ii) an Environmental scenario, based on the opinions of an expert in the sustainability field for industry; and finally (iii) a Financial scenario, accounting for an economic expert's standpoint. Due to the COVID-19 pandemic, experts' interviews were performed using the online DecSpace tool [17].

4   Results and Discussion

Before going into the details of the scenarios results, attention should be paid to the outcomes of the experts' interviews, summarized in Fig. 4. The weights, which are inserted as input to the impact matrix, are coherent with the experts' fields; moreover, it is interesting to note that, despite the different experts' areas of belonging, TECH.2 and EN.2 are considered highly relevant for all three experts. After completing the impact matrix, the outranking flows of the different alternatives can be computed using Visual PROMETHEE software, allowing the definition of the final ranking of the considered alternatives per each developed scenario. In the current analysis, the first two positions of the ranking result to be the same according to all three scenarios, as well as the last one. In particular, the best alternative is always the PHP, followed by the HP+CH configuration, while the worst alternative is represented by EB+CH. On the other hand, configurations HP+HP and HP+EB cover the third and fourth positions for the environmental and energy scenarios, while they experience an inversion in the financial scenario.

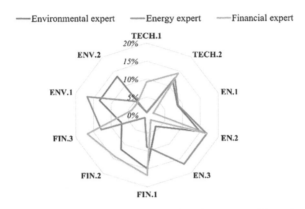

Fig. 4. Normalized weights assigned based on experts' interviews.

To critically analyse results and investigate the impacts of the weighting procedure on the final ranking of the alternatives, a sensitivity analysis was performed by assigning equal weights to all criteria (considering a 10% weight for all 10 criteria). In the Equal Weights scenario, the final rank of alternatives reflects the outcomes of the energy and environmental scenarios; however, the reciprocal distance in terms of outranking flows between the alternatives changes.

Furthermore, starting from this scenario with equal distribution of weights, the robustness of the assessment was addressed by checking the stability range of each criterion. Through Visual PROMETHEE, it is possible to graphically visualize how the alternatives ranking would change because of the variation of the weights assigned to each criterion. In particular, the stability analysis allows to identify the largest range in which a variation of the weight of a criterion does not affect the outcome of the scenario. Table 2 summarises the stability ranges for the Equal Weights scenario, showing the minimum and maximum weights that could be assigned to the different criteria in this scenario to maintain the final ranking of the alternatives unaltered.

Table 2. Stability intervals for the Equal Weights scenario.

Criterion	Min weight [%]	Equal weight [%]	Max weight [%]
TECH.1	0.00	10.00	40.15
TECH.2	3.63	10.00	24.90
EN.1	0.00	10.00	86.09

(*continued*)

Table 2. (*continued*)

Criterion	Min weight [%]	Equal weight [%]	Max weight [%]
EN.2	0.00	10.00	98.10
EN.3	0.00	10.00	100.00
FIN.1	0.00	10.00	31.28
FIN.2	0.00	10.00	29.20
FIN.3	0.00	10.00	97.43
ENV.1	0.00	10.00	100.00
ENV.2	0.00	10.00	100.00

EN.3, ENV.1 and ENV.2 criteria experience the greatest stability intervals, while TECH.2, FIN.1 and FIN.2 result to be the most sensitive criteria. For the sake of exemplification, FIN.1 criterion is explored in Fig. 5, in which x- and y- axes represent the normalized weights and the net flow rankings, respectively, while the light blue lines represent the net flow trend of each alternative according to the weight variation. The green/red vertical line is positioned in correspondence of FIN.1 weight in the Equal Weights scenario (10%), while the blue dotted vertical lines represent the minimum and maximum weights for which the final rank of all alternatives remains unaltered. It is interesting to note that if a weight higher than 75% is assigned to FIN.1, the ranking experiences a complete overturning of results, affecting not only the intermediate positions, but also the first and last ones. In particular, if considering a 100% weight for FIN.1 criterion, EB+CH configuration (i.e., the worst in all scenarios) would become the best alternative, having the lowest investment cost. In addition, in this extreme scenario, there is also an inversion of net flow signs for some alternatives.

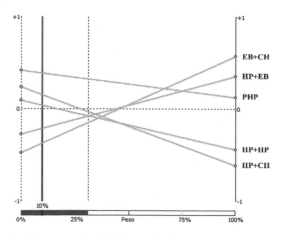

Fig. 5. Stability range for FIN.1 criterion in the Equal Weights scenario.

Based on these considerations on the instability of the financial criteria and bearing in mind that energy investments are usually made by consumers, whose decisions are still mainly driven by the financial convenience of the compared solutions to be installed in their buildings, an additional sensitivity scenario was developed, named Financial Extreme. Specifically, it was built assigning a 25% weight to each criterion belonging to the financial dimension, while the remaining percentage is equally distributed among the other criteria (to guarantee that the sum of all weights is equal to 100%).

Figure 6 shows a snapshot of the net flow ranking of all five considered scenarios (i.e., Environmental, Energy, Financial, Equal Weights and Financial Extreme). Also in the Financial Extreme scenario, PHP remains the best alternative, even though this scenario induces some relevant variations in the other positions; specifically, it is interesting to note how HP+CH configuration (second-best solution for all other scenarios) reaches the last position in the ranking, while other solutions (and specifically those using the electric boiler) rise, thanks to their lower investment and maintenance costs (e.g., EB+CH and HP+EB), despite their lower environmental and energy performances.

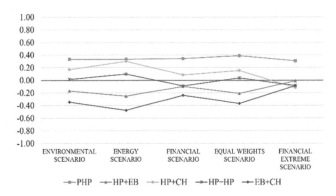

Fig. 6. Final ranking of the HVAC configurations according to the five developed scenarios.

5 Conclusions

The paper presented a multi-criteria assessment, using PROMETHEE II outranking method, to compare and rank different all-electric HVAC configurations, capable of meeting the same contemporary heating and cooling loads, according to diverse experts' preferences. Based on the developed scenarios, for all experts, PHP appeared to be the most promising solution, thanks to its capability of providing heating and cooling services simultaneously and independently. The scenarios results allow to highlight the potentialities of the units from different standpoints and to support consumers' choices in investing on more efficient and sustainable HVAC technologies. Indeed, on the one side, the outcomes can be useful for decision-makers to sustain the evolution of proper financial mechanisms to push consumers' investments; on the other side, technical sales professionals may use the results to support the proposal of still not widespread technologies, such as PHPs, to match consumers' energy needs. Future work will be devoted to enlarging the set of alternatives, including also non-electric technologies, more commonly used in buildings, and to testing other multi-criteria techniques.

Acknowledgements. The work was conducted thanks to the cooperation with industrial partner Rhoss S.p.A.

The work of Ilaria Abbà has been financed by the Research Fund for the Italian Electrical System under the Contract Agreement between RSE S.p.A. and the Ministry of Economic Development - General Directorate for the Electricity Market, Renewable Energy and Energy Efficiency, Nuclear Energy in compliance with the Decree of April 16th, 2018.

References

1. Li, H., Hong, T., Lee, S.H., Sofos, M.: System-level key performance indicators for building performance evaluation. Energy Build. **209**, 109703 (2020)
2. European Commission: The European Green Deal. COM (2019) 640 final (2019)
3. Witkowski, K., Hearing, P., Seidelt, S., Pini, N.: Role of thermal technologies for enhancing flexibility in multi-energy systems through sector coupling: technical suitability and expected developments. IET Energy Syst. Integr. **2**, 69–79 (2020)
4. Abbà, I., Crespi, G., Corgnati, S.P., Morassutti, S., Prendin, L.: Sperimentazione numerica delle dinamiche di funzionamento di sistemi polivalenti. In: Proceedings of AICARR Conference 2020: 37° Convegno Nazionale AiCARR (2020)
5. Crespi, G., Abbà, I., Corgnati, S.P., Morassutti, S., Prendin, L.: HVAC polyvalent technologies to balance contemporary loads in buildings. In: Proceedings of 16th Conference on Sustainable Development of Energy, Water and Environment Systems, SDEWES 2021 (2021)
6. Miller, C.A., Ritcher, J., O'Leary, J.: Socio-energy systems design: a policy framework for energy transitions. Energy Res. Soc. Sci. **6**, 29–40 (2015)
7. Department for Communities and Local Government: Multi-criteria analysis: a manual (2009)
8. Blanco, G., Amarilla, R., Martinez, A., Llamosas, C., Oxilia, V.: Energy transitions and emerging economies: a multi-criteria analysis of policy options for hydropower surplus utilization in Paraguay. Energy Policy **108**, 312–321 (2017)
9. De Felice, F., Petrillo, A.: Multi-criteria decision making: a mechanism design technique for sustainability. In: Luo, Z. (ed.) Mechanism Design for Sustainability: Techniques and Cases, pp. 15–35 (2013)
10. Brans, J.P., Vincke, P., Mareschal, B.: How to select and how to rank projects: the PROMETHEE method. Eur. J. Oper. Res. **24**, 228–238 (1986)
11. Figueira, J., Roy, B.: Determining the weights of criteria in the ELECTRE type methods with a revised Simos' procedure. Eur. J. Oper. Res. **139**, 317–326 (2002)
12. Singh Gaur, A., Fitiwi, D.W., Curtis, J.: Heat pumps and our low-carbon future: a comprehensive review. Energy Res. Soc. Sci. **71**, 101764 (2021)
13. European Committee for Standardization: EN 15459:2017: Energy performance of buildings - Economic evaluation procedure for energy systems in buildings - Part 1: Calculation procedures, Module M1-14 (2017)
14. ARERA Database. https://www.arera.it/it/prezzi.html. Accessed 21 July 2020
15. International Organization for Standardization: ISO 52000-1. Energy Performance of Buildings – Overarching EPB assessment, Part 1: General framework and procedures (2017)
16. Istituto Superiore per la Protezione e la Ricerca Ambientale (ISPRA): Fattori di emissione atmosferica di gas a effetto serra e altri gas nel settore elettrico. Rapporto 280 (2018)
17. DecSpace (Pre-Alpha). http://app.decspacedev.sysresearch.org/#/. Accessed 05 July 2021

Vanadium Redox Flow Batteries: Characteristics and Economic Value

Cinzia Bonaldo[1]([✉]) and Nicola Poli[2,3]

[1] Department of Management and Engineering, University of Padova, Padova, Italy
cinzia.bonaldo@phd.unipd.it
[2] Department of Industrial Engineering, University of Padova, Padova, Italy
[3] Interdepartmental Centre Giorgio Levi Cases for Energy Economics and Technology, University of Padua, Padova, Italy

Abstract. One of the main goal of energy transition is the decarbonization of global electricity networks. Toward this aim, the integration of Variable Renewable Energy Sources with the electricity grid has increased dramatically over the last ten years. However, the desire to obtain large fractions of electricity from VER has encountered many challenges mainly due to their random nature. The Vanadium Redox Flow Battery represents one of the most promising technologies for large stationary applications of electricity storage. It has an independent power and energy scalability, together with long life cycle and low long-term self-discharge process, which make it useful in applications where batteries need to remain charged for long periods of time while maintaining a state of readiness for use. The usage of such battery, coupled with variable renewable energy sources, allows to overcome the constraints linked to the limited dispatchability of the latter, thus favoring a large penetration of variable renewable energy source in the power grid. This article proposes to study the energy storage through Vanadium Redox Flow Batteries as a storage system that can supply firm capacity and be remunerated by means of a Capacity Remuneration Mechanism. We discuss a real option model to evaluate the value of investment in such technology.

Keywords: Energy storage systems · Renewable energy · Electrical grid · Vanadium redox flow batteries · Real options · Capacity markets

1 Introduction

The international scientific community agrees that climate change is a consequence of human activities and a real threat to future generations.[1] This growing awareness is the result of the commitment of several stakeholders, including non-governmental organizations, governments, institutions and companies, which in recent years have proposed policies and concrete actions to protect climate and environment at a global level. According to the 2018 special report by the Intergovernmental Panel on Climate Change, achieving a warming scenario of 1.5 °C will require that global CO_2 emissions

---

[1] See for instance IPCC 2013 (https://www.ipcc.ch/report/ar5/wg1/).

© The Author(s), under exclusive license to Springer Nature Switzerland AG 2022
F. Calabrò et al. (Eds.): NMP 2022, LNNS 482, pp. 1721–1731, 2022.
https://doi.org/10.1007/978-3-031-06825-6_166

decline by about 45% from their 2010 levels by 2030, reaching net zero around 2050, while achieving a warming scenario of 2 °C will require that CO_2 emissions decline by about 25% by 2030 and reach net zero around 2070.[2] Several countries have committed to curb greenhouse gas emissions (GHG). Since the first United Nations Conference, held in 1992 as part of the "Earth Summit" in Rio de Janeiro, followed by the Tokyo Protocol in 1997, the European 20–20-20 package of 2007, and the Paris Agreement of 2015 on the reduction of harmful emissions, we arrived at the European Green Deal. Nowadays, with the European Green Deal, European Union has decided to make its target for reduction of net harmful emissions even more ambitious, moving from a 40% CO_2 cut (with the Clean Energy Package) to a 55% cut from 1990 levels by 2030. In addition, the European Commission aims to reduce net GHG emissions to zero by 2050. To achieve such targets, progressive decarbonization of the energy system is foreseen, as well as a development of energy from renewable sources with the simultaneous elimination of coal.[3] Toward this aim, the integration of Renewable Energy Sources (RES) into the electricity system has increased over the last ten years, and it is expected to further rise.[4] The growth and integration of utility-scale renewable energy technologies have changed the economic and operational dynamics that traditionally underpinned power sector management and planning [1]. Electric grid operators around the world are rethinking the way they plan and operate their systems and markets, in order to accommodate various forms of policy that are promoting investment in Variable Energy Resources (VER). The definition of VER refers to any generation resource whose output is not completely controllable by a transmission system operator, since it depends on the supply of a primary energy source whose availability is random and that cannot be directly stored or stockpiled. Wind and solar power generation are the most common VER. The random nature of VER poses increasing challenges to the management of electric grids. Indeed, system operators need to maintain the grid balanced in real time, matching power supply and demand on a continuous basis. If variability between VERs and demand were completely synchronous, so that VERs would increase (or decrease) in output right at the moments when demand increases or decreases in output, then there would be no unbalances. But this is not the case. The problem is that although wind and solar energy can be forecasted [2, 3], depending on local features and climate, they are intermittent, opening spatial and temporal gaps between the availability of energy and its consumption by end users [4, 5]. Thus the random nature of VER and the cyclical structure of load poses challenges to the grid management [6], which will increase as VER will rise their penetration into the power mix.

Energy Storage Systems (ESS), or batteries, can address this issue by providing a solution to improve the stability and reliability of the grid. There is a growing research studies on the identification and development of new technological solutions to cut

---

[2] https://www.ipcc.ch/sr15/download/

[3] The production and use of electricity in the various economic sectors represent over 75% of greenhouse gas emissions in the EU (https://ec.europa.eu/commission/presscorner/detail/en/IP_20_1599).

[4] The RES in the Europe energy mix in 2019 were 19.7%, while the new 2030 target requires to double this share and reach a 40% of RES in the energy mix. See https://ourworldindata.org/renewable-energy.

storage costs, increase storage capacity and ensure the environmental and economic sustainability of batteries. One of the most promising ESS technology in this sense is represented by the Redox Flow Batteries (RFB). These batteries allow to convert electrical energy into chemical energy by means of electrochemical cells and store it in fluid electrolytes in external tanks. Several chemicals, based on different active species, such as iron, vanadium, zinc-bromine, iron-chromium and vanadium-chromium, have been investigated so far [7–9]. In particular, among RFB, the all-vanadium ones (VRFB) have a series of unique advantages when compared to other RFB such as: no cross-contamination problems, very low capacity-loss, long life-cycle and a negligible self-discharge during extended standby. Such feature makes VRFB attractive to support VER penetration, allowing them to provide firm capacity, i.e. the possibility to supply energy when needed even with short advanced notice and without the risk of being idle or not available. This opens to VER, coupled with VRFB, the opportunity to participate to capacity markets [10]. The aim of this paper is to show how to evaluate electricity storage accruing from VER and stored in VRFB. As mentioned above, VRFB can provide firm capacity, i.e. the capacity to deliver energy when needed, which has the characteristics of an option on the energy market, and thus can be analyzed as a Real Option (RO). In this paper, we propose a RO model that allows to evaluate the participation of VRFB to the capacity market. In order to do so, we first provide a literature review about VRFB, including their main characteristics and advantages, then we examine how to assess the value of capacity supplied by VRFB (fed by VER) in a capacity market. This paper is organized as follows: in Sect. 2 a general overview on technical and economic aspects of VRFB is reported and in Sect. 3 an economic model based on RO analysis is introduced.

2 The Vanadium Redox Flow Battery

At present, most ESS used for portable devices, electric vehicles and large-scale storage are based on electrochemical storage systems, in particular lithium-ion and lead-acid batteries. However, implementing long-term storage requires very low self-discharge rates and sufficiently large numbers of charging cycles and long life time, which are not the characteristic of lithium-ion or lead-acid batteries. For these reasons, despite some attempts to provide grid services, to date such ESS were never actively exploited to provide energy for long-term storage [11]. Instead, the VRFBs can represent an effective solution. These batteries store the chemical energy by using a mixture of vanadium compounds dissolved in sulfuric acid. The main component of this battery is the stack, made by several electrochemical cells where the electrical energy is converted in chemical energy and vice-versa.

2.1 Technical Specificities and Advantages of VRFB

The main technical features of ESS include:

- Roundtrip cycle efficiency. This is an indicator of the power loss during charge and discharge processes. In most studies, the VRFB roundtrip cycle efficiency is about 80% [11, 12].

- Cycle life. This corresponds to each round of full discharge and then full recharge. Usually, a battery cycle life ranges from 500 to 1200, meaning a life cycle of 1.5–3 years for conventional batteries. VRFB have longer life cycles as they can operate for decades without deterioration or need for replacement [14].

Table 1. Advantages and disadvantages of VRFB

Advantages	References
1. Flexibility: in VRFB energy capacity and power are independent, allowing higher flexibility in the energy-power relationships	[12, 21, 23]
2. Low environmental impact: unlike other energy storage systems, such as lead-acid and lithium-ion batteries, VRFB have a very low environmental impact. Both the anolyte and catholyte can be easily recycled, due to high reversibility of the redox reaction of VRFB	[24, 30]
3. High safety: no risk of fire and explosion. In addition, it operates at low pressure and room temperature	[26, 27]
4. Response time: VRFB have a fast response time, less than 1 min and typically in the order of seconds;	[19, 36]
5. Stability. The VRFB system is considered to be very stable, offering a long lifetime and unlimited electrolyte cycle time. In addition, the same reactive element is used in both half-cells, limiting cross-contamination. This results in virtually no operational costs over the life time of the VRFB	[26, 28, 30]
Disadvantages	
1. A relatively poor energy density in comparison with standard batteries, making them useful just for stationary applications. Generally, the energy density of VRFB is 10–50 Wh/kg, while for lithium-ion battery it is between 100–200 Wh/kg	[4]
2. A limited operating temperature between −5 °C and 40 °C	[4]
3. Cross-over and side reactions, which decrease the capacity of the battery during charge-discharge cycles	[34]
4. Higher capital cost in comparison to other ESS, in the range of about 500 $/kWh for VRFB	[4, 23]
5. Shunt currents, which take place in the hydraulic circuits of the stack, due to the conductivity of the vanadium electrolyte	[35]

- Depth of discharge (DoD). This indicates the percentage of the battery that has been discharged relative to the overall capacity of the battery. Usually, it is not recommended to discharge a battery entirely, as that dramatically shortens the working life of the battery. Lithium-ion batteries are damaged when discharged below 20% of their state of charge, and they typically need to work between 20% and 100% DoD [12]. VRFB can reach 0% DoD with no detrimental effects [11, 15]. In addition, VRFB can be left completely discharged for long periods of time without damage [16].

- Self-discharge. This often occurs due to chemical reactions, consumption of the electronic system or high temperatures. In general, all batteries are affected by self-discharge. One of the most interesting features of VRFB is that, even though they can suffer self-discharge due to the crossover phenomenon [17–19], in standby condition, where the tanks are hydraulically isolated from the stacks, this self-discharge is totally negligible. This makes them useful in applications where batteries are left charged for long periods of time, with little or no maintenance, while still keeping a ready-to-use state [22, 26]. This is why VRFB, unlike lithium and lead-acid batteries, are often indicated as ideal batteries for long-term storage [22].

Finally, a number of advantages and disadvantages of VRFB emerged from the literature reviewed, as reported in Table 1.

2.2   Economic Models Used for VRFB Investment Assessment

The most widely used economic methodologies for studying VRFB in literature are:

1. Levelized Cost Of Energy (LCOE). This corresponds to the ratio between the sum of the discounted costs and the discounted value of the energy stored over the expected lifetime of the project [23, 36, 37].
2. Net Present Value (NPV), a static methodology to assess the profitability of the investment [20, 37, 38]. The methodology consists of discounting cash flows for each year of operation of the project and subtracting the initial cost of investment in the technology. NPVs are often associated with optimization methodologies. The goal of the optimization method is to achieve the highest profit ratio in the shortest payback period.

VRFB have a number of advantages and flexibilities that contribute to value creation, which should not be ignored. Interestingly enough, in a recent conference proceeding [39], the authors provided an attempt to study the profitability of VRFB on the basis on the NPV. They stated that it is not worthwhile to invest in this technology due to a negative NPV level, concluding that ".. if the additional flexibility given from this technology to the system could be valued, vanadium batteries could be attractive for the performance of the grid. A good technique that may be.. Real Options analysis" [39].

Moreover, their characteristics imply they can provide firm capacity to an electricity system, namely the possibility to generate electricity when needed. This allow avoiding problems of security of supply, providing to the system the possibility to be balanced whenever there is a spike in the load.

3   A Real Option Model for VRFB

Real options analysis can capture the flexibility in future operations where the value of the operation is uncertain and allows for decision making under this uncertainty. We propose to study the value of capacity supplied by a VRFB that is fed by VER and that participates to the capacity market (with a Capacity Remuneration Mechanism – CRM) by means of real option analysis. As said before, VRFB have the technical characteristics to participate in such a market. Basically, VRFB have the main advantages that they do not discharge when they are left idle even for long periods of time and that they have a long life time. This implies that they have no risk of not being available when they are needed to provide energy, which makes the capacity provision of VRFB similar to the capacity offered by a thermal power plant. The former stores energy into external tanks and converts it into electricity by means of stacks. The latter stores energy into the primary source (that can be guarded as inventory), and then converted into electricity by means of power generators. However, there are two key differences between the capacity supplied by a thermal power plant and by a VRFB: (1) the marginal cost of the energy provided by a VRFB fed by VER is null, while the marginal cost of the hydrocarbons is positive; (2) the initial investment cost per unit of VRFB is generally higher than the investment costs of thermal power plants per unity of capacity provided.

Generally speaking, the day-ahead prices in electricity markets coincide with the marginal cost of the marginal technology[5] that often is a thermal-fired power plant[6]. When a CRM is introduced in the market it has the effect of avoiding the electricity price to rise above the marginal cost of the marginal. Therefore, it acts as a price cap on the electricity price: the electricity price is free to fluctuate reflecting the marginal cost of the marginal technology, avoiding undesired cases in which the day-ahead electricity price would rise above the marginal cost of the marginal technology[7]. Following this scheme, we can assume that there are two stochastic variables in the energy market: the day-ahead electricity price and the price cap (which would coincide with the marginal cost of fuel that feeds the plant, being marginal at time t). In addition, there is the remuneration provided by the CRM, which for the sake of simplicity we assume being known and constant. The profit function at time $t \geq 0$ for a VRFB supplying capacity in the CRM can be written as:

$$\pi_t^c = k + P_t - \max(P_t - C_t, 0) \tag{1}$$

which can be simplified as:

$$\pi_t^c = \begin{cases} \pi_t^1 = k + C_t \ \text{if} P_t \geq C_t \\ \pi_t^2 = k + P_t \ \text{if} P_t < C_t \end{cases} \tag{2}$$

In Eq. (1), π_t^c is the instantaneous profit, k is the fix remuneration for participating to the CRM, P_t is the day-ahead electricity price, and C_t is the marginal cost of the marginal technology, which is acting as a price cap. Note that Eq. (2) makes clear that there are two regimes for the instantaneous profit, depending on whether the price of electricity

---

[5] Except in the presence of a load shedding.

[6] When this is the case the day-ahead electricity price coincides with the marginal cost of the fuel of the thermal plant.

[7] In other words, by introducing a CRM, a market regulator would be able to set a cap to the maximum energy price, avoiding price spikes that are generated by undesired stress conditions of the system.

is either below or above the marginal cost of the marginal plant. In Eq. (2), π_t^1 is the instantaneous profit in the first region, where $P_t \geq C_t$, while π_t^2 is the instantaneous profit in the second region, where $P_t < C_t$. Finally, in both equations there are no marginal cost of energy provision from VRFB, as it is assumed that the ESS is fed by VRE, whose marginal cost is obviously null. In Eq. (2), the transition point of the profit function between the first region and the second region is given by $P_t = C_t$. The solution to the investment decision problem is not a straightforward integration because the profit function, π_t^c, freely switches between two different functional forms in the two regions, since both the day-ahead electricity price, P_t, and the price cap, C_t, are stochastic.[8] This makes clear that the option value for this problem is similar to a spread option.[9] The value of capacity supply can be determined by stochastic dynamic programming approaches and the final value of plant that participate to the CRM can be summarized as follows:

$$V(P,C) = \begin{cases} V^1(P,C) = \dfrac{\pi^1(C)}{r} + OV(V^1) & if P \geq C \quad (3) \\[2mm] V^2(P,C) = \dfrac{\pi^2(P)}{r} + OV(V^2) & if P < C \quad (4) \end{cases}$$

Where $V^1(P, C)$ is the value in the first region and $V^2(P, C)$ is the value in the second region. r is the risk-adjust discount rate. The valuation Eqs. (3) and (4) are composed of two parts. The first terms represent the present value of expected profits if the process is staying in the boundary forever. The second terms are the option value of hitting the boundary and entering in the other region. This second term could be positive or negative. Indeed, even if ex ante it is normal to conceive a positive option value, there can be cases in which the possibility to switch from one regime to the other one reduces the value of the investment, which yields negative option values (see for example [40–42]). Determine the magnitude and the sign of these option values go beyond the scope of this paper, they should be studied using HJB equation[10] as they depend on two stochastic underlying processes, i.e. P and C.[11]

---

[8] A further assumption is that P_t and C_t are not correlated.

[9] In financial option theory, spread options are options to exchange one risky asset for another [43]. In this case, the two risky assets would be P_t and C_t.

[10] The HJB equation is an arbitrage condition that divides the entire evaluation problem in two parts at each time point: the immediate state (represented by immediate cash flow) and the future possible state (capital gain). The stochastic dynamic programming is solved by using Ito's lemma and partial differential equation. The detailed explanation of all theorem and mathematical steps are omitted, as they are beyond the aim of this study.

[11] Calculating the exact OV for VRFB goes beyond the scope of the present study, and it depends on assumptions about the specific values of the parameters as well as on the assumed underlying stochastic process. Explicit or semi-explicit functional forms can be obtained just for specific stochastic processes, and simulations can be implemented to derive approximation of the option value.

4   Conclusions

For the energy transition goal to be efficient and effective, specific means of energy storage will be required, given the uncontrollability of VER. In this study, we first provide a literature review about VRFB, a new technology for long-term electricity storage, in order to investigate the technical features of VRFB and the economic model used to study VRFB value. We show that the economic evaluation is performed following the LCOE or the NPV approaches, and that currently in the literature there is just one conference paper that links real option analysis and VRFB. In addition, so far there is no study in literature that considers the possibility to exploit this technology in the long term electricity markets, like the capacity ones. We model the possibility of coupling VRFB with a VER, which can participate to a capacity market providing electricity at null or negligible variable costs. We show that the value of capacity provision by VRFB can be calculated by means of the real option theory. The evaluation is not trivial, since there are two regimes that represent the flow of operating profits accruing from VRFB that participate to a CRM. The value of a VRFB participating to a CRM is similar to a spread option, since the final value depends on two stochastic prices, namely the electricity price and the price cap (or the marginal cost of fuel that feeds the marginal plant). The final value is composed by two regions. The first regime is when the electricity price is not affected by the supply of energy remunerated by the CRM, and the second corresponds to a regime in which the electricity price is capped by the effect of the extra energy supplied by the technology that receives the CRM. These two regimes generate two option values embedded into the possibility to fall into the other regime when the VRFB is being operated into a given one. These option values can have either a positive or a negative value, and therefore add or subtract value to the simple net present value of the investment in VRFB. This is an aspect neglected so far in the literature that we highlight with this research. More details of this analytical model will be develop by future works. Another step is then to assess the analytical framework by adopting real market data and providing some numerical results.

References

1. Say, K., John, M.: Molehills into mountains: Transitional pressures from household PV-battery adoption under flat retail and feed-in tariffs. Energy Policy **152**, 112213 (2021). https://doi.org/10.1016/j.enpol.2021.112213
2. Foley, A.M., Leahy, P.G., Marvuglia, A., McKeogh, E.J.: Current methods and advances in forecasting of wind power generation. Renew. Energy **37**, 1–8 (2012). https://doi.org/10.1016/j.renene.2011.05.033
3. Yang, D., Quan, H., Disfani, V.R., Rodríguez-Gallegos, C.D.: Reconciling solar forecasts: temporal hierarchy. Sol. Energy **158**, 332–346 (2017). https://doi.org/10.1016/j.solener.2017.09.055
4. Alotto, P., Guarnieri, M., Moro, F.: Redox flow batteries for the storage of renewable energy: a review. Renew. Sustain. Energy Rev. **29**, 325–335 (2014). https://doi.org/10.1016/j.rser.2013.08.001

5. Sun, C., Negro, E., Vezzù, K., et al.: Hybrid inorganic-organic proton-conducting membranes based on SPEEK doped with WO3 nanoparticles for application in vanadium redox flow batteries. Electrochim. Acta. **309**, 311–325 (2019). https://doi.org/10.1016/j.electacta.2019.03.056

6. Agostini, M., Bertolini, M., Coppo, M., Fontini, F.: The participation of small-scale variable distributed renewable energy sources to the balancing services market. Energy Econ. **97**, 105208 (2021). https://doi.org/10.1016/j.eneco.2021.105208

7. Lourenssen, K., Williams, J., Ahmadpour, F., et al.: Vanadium redox flow batteries: a comprehensive review. J. Energy Storage 25 (2019). https://doi.org/10.1016/j.est.2019.100844

8. Khor, A., Leung, P., Mohamed, M.R., et al.: Review of zinc-based hybrid flow batteries: from fundamentals to applications. Mater Today Energy **8**, 80–108 (2018). https://doi.org/10.1016/j.mtener.2017.12.012

9. Zhang, H., Sun, C.: Cost-effective iron-based aqueous redox flow batteries for large-scale energy storage application: a review. J. Power Sources **493**, 229445 (2021). https://doi.org/10.1016/j.jpowsour.2020.229445

10. Cretì, A., Fontini, F.: Economics of Electricity Markets, Competition and Rules (2019)

11. Lucas, A., Chondrogiannis, S.: Smart grid energy storage controller for frequency regulation and peak shaving, using a vanadium redox flow battery. Int. J. Electr. Power Energy Syst. **80**, 26–36 (2016). https://doi.org/10.1016/j.ijepes.2016.01.025

12. Baldinelli, A., Barelli, L., Bidini, G., Discepoli, G.: Economics of innovative high capacity-to-power energy storage technologies pointing at 100% renewable micro-grids. J Energy Storage 28 (2020). https://doi.org/10.1016/j.est.2020.101198

13. Lucas, A., Chondrogiannis, S.: Electrical Power and Energy Systems Smart grid energy storage controller for frequency regulation and peak shaving, using a vanadium redox flow battery. Int. J. Electr. Power Energy Syst. **80**, 26–36 (2016). https://doi.org/10.1016/j.ijepes.2016.01.025

14. Zsiborács, H., Baranyai, N.H., Zentkó, L., et al.: Electricity market challenges of photovoltaic and energy storage technologies in the European union: regulatory challenges and responses. Appl. Sci. 10 (2020). https://doi.org/10.3390/app10041472

15. Li, M.-J., Zhao, W., Chen, X., Tao, W.-Q.: Economic analysis of a new class of vanadium redox-flow battery for medium- and large-scale energy storage in commercial applications with renewable energy. Appl. Therm. Eng. **114**, 802–814 (2017). https://doi.org/10.1016/j.applthermaleng.2016.11.156

16. Coronel T, Buzarquis E, Blanco GA (2017) Analyzing energy storage system for energy arbitrage. In: Analyzing energy storage system for energy arbitrage. Institute of Electrical and Electronics Engineers Inc., Grupo de Investigación en Sistemas Energéticos, Facultad Politécnica - U.N.A., San-Lorenzo, Paraguay, pp 1–4

17. Sun, C., Zlotorowicz, A., Nawn, G., et al.: [Nafion/(WO3)x] hybrid membranes for vanadium redox flow batteries. Solid State Ionics **319**, 110–116 (2018). https://doi.org/10.1016/j.ssi.2018.01.038

18. Zeng, Y.K., Zhao, T.S., An, L., et al.: A comparative study of all-vanadium and iron-chromium redox flow batteries for large-scale energy storage. J. Power Sources **300**, 438–443 (2015). https://doi.org/10.1016/j.jpowsour.2015.09.100

19. Tang, A., Bao, J., Skyllas-Kazacos, M.: Thermal modelling of battery configuration and self-discharge reactions in vanadium redox flow battery. J. Power Sources **216**, 489–501 (2012). https://doi.org/10.1016/j.jpowsour.2012.06.052

20. Resch, M., Bühler, J., Schachler, B., et al.: Technical and economic comparison of grid supportive vanadium redox flow batteries for primary control reserve and community electricity storage in Germany. Int. J. Energy Res. **43**, 337–357 (2019). https://doi.org/10.1002/er.4269

21. Bryans, D., Amstutz, V., Girault, H.H., Berlouis, L.E.A.: Characterisation of a 200 kw/400 kwh vanadium redox flow battery. Batteries 4 (2018). https://doi.org/10.3390/batteries4040054

22. Lai, C.S., McCulloch, M.D.: Levelized cost of electricity for solar photovoltaic and electrical energy storage. Appl Energy **190**, 191–203 (2017). https://doi.org/10.1016/j.apenergy.2016.12.153

23. Rodby, K.E., Carney, T.J., Ashraf Gandomi, Y., et al.: Assessing the levelized cost of vanadium redox flow batteries with capacity fade and rebalancing. J. Power Sources 460 (2020). https://doi.org/10.1016/j.jpowsour.2020.227958

24. He, H., Tian, S., Tarroja, B., et al.: Flow battery production: materials selection and environmental impact. J. Clean Prod. **269**, 121740 (2020). https://doi.org/10.1016/j.jclepro.2020.121740

25. Kear, G., Shah, A.A., Walsh, F.C.: Development of the all-vanadium redox flow battery for energy storage: a review of technological, Financial and policy aspects. Int. J. Energy Res. **36**, 1105–1120 (2012). https://doi.org/10.1002/er.1863

26. Yuan, X.Z., Song, C., Platt, A., et al.: A review of all-vanadium redox flow battery durability: degradation mechanisms and mitigation strategies. Int. J. Energy Res. **43**, 6599–6638 (2019). https://doi.org/10.1002/er.4607

27. Geurin, S.O., Barnes, A.K., Balda, J.C.: Smart grid applications of selected energy storage technologies. In: 2012 IEEE PES Innovation Smart Grid Technology. ISGT 2012, pp. 1–8 (2012). https://doi.org/10.1109/ISGT.2012.6175626

28. Sun, C., Vezzù, K., Pagot, G., et al.: Elucidation of the interplay between vanadium species and charge-discharge processes in VRFBs by Raman spectroscopy. Electrochim Acta **318**, 913–921 (2019). https://doi.org/10.1016/j.electacta.2019.06.130

29. García-Quismondo, E., Almonacid, I., Martínez, M.Á.C., et al.: Operational experience of 5 kW/5 kWh all-vanadium flow batteries in photovoltaic grid applications. Batteries 5 (2019). https://doi.org/10.3390/batteries5030052

30. Poullikkas, A.: A comparative overview of large-scale battery systems for electricity storage. Renew. Sustain. Energy Rev. **27**, 778–788 (2013). https://doi.org/10.1016/j.rser.2013.07.017

31. Leung, P.K., Xu, Q., Zhao, T.S., et al.: Preparation of silica nanocomposite anion-exchange membranes with low vanadium-ion crossover for vanadium redox flow batteries. Electrochim Acta **105**, 584–592 (2013). https://doi.org/10.1016/j.electacta.2013.04.155

32. Mohammadi, T., Chieng, S.C., Kazacos, M.S.: Water transport study across commercial ion exchange membranes in the vanadium redox flow battery. J. Memb. Sci. **133**, 151–159 (1997). https://doi.org/10.1016/S0376-7388(97)00092-6

33. Zhang, H., Tan, Y., Luo, X.D., et al.: Polarization effects of a rayon and polyacrylonitrile based graphite felt for iron-chromium redox flow batteries. Chem. Electr. Chem. **6**, 3175–3188 (2019). https://doi.org/10.1002/celc.201900518

34. Chen, N., Zhang, H., Luo, X.D., Sun, C.Y.: SiO2-decorated graphite felt electrode by silicic acid etching for iron-chromium redox flow battery. Electrochim Acta **336**, 135646 (2020). https://doi.org/10.1016/j.electacta.2020.135646

35. Trovò, A., Marini, G., Sutto, A., et al.: Standby thermal model of a vanadium redox flow battery stack with crossover and shunt-current effects. Appl. Energy **240**, 893–906 (2019). https://doi.org/10.1016/j.apenergy.2019.02.067

36. Zakeri, B., Syri, S.: Electrical energy storage systems: a comparative life cycle cost analysis. Renew. Sustain. Energy Rev. **42**, 569–596 (2015). https://doi.org/10.1016/j.rser.2014.10.011

37. Poli, N., Bonaldo, C., Trovò, A., Guarnieri, M.: Optimal energy storage systems for long charge/discharge duration. In: ICAE 2021 International Conference on Applied Energy 2021 (2022)

38. Lujano-Rojas, J.M., Zubi, G., Dufo-López, R., et al.: Novel probabilistic optimization model for lead-acid and vanadium redox flow batteries under real-time pricing programs. Int. J. Electr. Power Energy Syst. **97**, 72–84 (2018). https://doi.org/10.1016/j.ijepes.2017.10.037

39. Coronel, T., Buzarquis, E., Blanco, G.A.: Analyzing feasibility of energy storage system for energy arbitrage. Institute of Electrical and Electronics Engineers Inc., Grupo de Investigación en Sistemas Energéticos, Facultad Politécnica – U.N.A., San Lorenzo, Paraguay, pp 1–6 (2017)

40. Hamed, G.: Belending under uncertainty. Energy Econ. **61**, 110–120 (2017)

41. Fontini, F., Vargiolu, T., Zormpas, D.: Investing in electricity production under a reliability options scheme. SSRN Electron. J. 1–21 (2019). https://doi.org/10.2139/ssrn.3465196

42. Bonaldo, C.: Benefits of blending mandate in sustainable economies. In: Bevilacqua, C., Calabrò, F., Della Spina, L. (eds.) NMP 2020. SIST, vol. 178, pp. 526–535. Springer, Cham (2021). https://doi.org/10.1007/978-3-030-48279-4_50

43. Margrabe, W.: The value of an option to exchange one asset for another. J. Financ. **33**, 177–186 (1978). https://doi.org/10.2307/2326358

The Role of Quality Management Services (QMSs) in Aligning the Construction Sector to the European Taxonomy: The Experience of the QUEST Project

Marta Bottero⬤ and Federico Dell'Anna⁽⊠⁾⬤

Interuniversity Department of Regional and Urban Studies and Planning, Politecnico di Torino,
Viale Mattioli 39, 10125 Turin, Italy
federico.dellanna@polito.it

Abstract. The European Green Deal defines a series of strategies to address climate and environmental problems through a sustainable economy. To achieve the goals set by this ambitious plan significant investments are required, not only in the public sector but also in the private sector. These investments, often characterized by the application of innovative technologies developed on a large scale, are subject to technical risks that may increase the possibility of errors in the predictions of the energy performance of buildings. In this context, the Horizon 2020 project QUEST (Quality Management Investments for Energy Efficiency) aims to identify the potential of investments in sustainable and energy-efficient buildings, proposing to reduce the risk attributed by investors to this type of intervention. The QUEST project aims to assess the impacts of innovative Quality Management Services (QMSs) in reducing the technical and financial risks affecting the return on investment for energy-efficient buildings and in aligning the construction industry with the requirements of the European Taxonomy. The research activities of the QUEST project are leading to the definition of a predictive tool to assess the risks associated with building and energy investments, integrating effective QMSs, such as technical monitoring, sustainability protocols, and commissioning certificates. This work focuses on the role of QMS within the construction industry to align it with the EU Taxonomy by proposing a review. In addition, the experience of the QUEST project is recounted, presenting the basis and purpose of the predictive tool that is being developed.

Keywords: Risk analysis · Green building certificates · Technical monitoring · Building commissioning

1 Introduction

Climate change and environmental degradation are a huge threat to Europe and the world. It is crucial to reduce emissions over the next decade to make Europe the first climate-neutral continent by 2050 and to make the European Green Deal a concrete reality [1]. To overcome these challenges, the European Green Deal will transform the

F. Calabrò et al. (Eds.): NMP 2022, LNNS 482, pp. 1732–1741, 2022.
https://doi.org/10.1007/978-3-031-06825-6_167

EU into a modern, resource-efficient, and competitive economy, ensuring that by 2050 no more net greenhouse gas emissions are generated [2, 3]. Furthermore, the European Green Deal is a useful strategy to overcome the problems arising from the COVID-19 pandemic. One-third of the €1,800 billion investment in the Next Generation EU recovery plan and the EU's seven-year budget will fund the European Green Deal. The transition to a climate-neutral economy by 2050 requires clear tools and guidelines, reflecting scientific evidence and market experience, to give companies and investors' confidence to act [4, 5]. Of course, public sector investment alone cannot guarantee the achievement of the targets set. To address this, the EU's executive body launched the Action Plan for Sustainable Finance [6] to create a body of rules around sustainable finance in March 2018. This led to the formulation of a tool to include the private sector in this green transition process, the European Taxonomy [7]. The Taxonomy entered into force in 2021 and can be understood as a kind of vocabulary describing all activities that can significantly reduce environmental impacts and can contribute to the achievement of decarbonization goals by directing capital flows already in the market towards sustainable investments. For the implementation of the Taxonomy, the EU has set up a Technical Expert Group (TEG) that aims to define strategies and actions to be taken in the different sectors. Under the recently approved EU Taxonomy regulation, investors and companies will disclose the environmental performance of the assets they invest in, in line with the green and sustainable economy.

The building sector now accounts for 38% of total energy-related emissions. Building energy needs to be halved by 2030 to achieve a net zero-emission building stock by 2050. Various directives and plans have been implemented to decarbonize the building sector, such as the Renovation Plan [8] and the Energy Performance Building Directive and recasts [9–11]. The expected wave of renovation will involve the implementation of innovative, high-cost technologies applied on a large scale, especially in the construction sector, representing a huge challenge for Europe [12]. For the building sector, the main objective is to reduce emissions through retrofitting of the building stock, and the implementation of renewable energy. However, literature has shown that a lack of quality in the management of systems can result in a loss of energy and environmental performance of 10–20% every 10 years [13]. This phenomenon can of course be considered not only as a technical risk but above all as a financial risk. Referring to the funding established by the Green Deal due to the lack of quality in the systems, it is possible to estimate a loss of around 25 billion euros.

The EU Taxonomy has identified as a prerequisite for significant real estate assets, the need to be able to include appropriate planning in the management of buildings to limit these losses in environmental and financial performance. In this context, the Quality Management Investments for Energy Efficiency (QUEST) project, funded by the European Union's Horizon 2020 Programme, is examining the role of quality management services in achieving better and more efficient buildings [14]. In particular, the QUEST project is investigating the role of Quality Management Services (QMSs), such as technical monitoring, commissioning management and certification of Green Buildings, in achieving the objectives and aligning the building sector with the requirements of the EU Taxonomy [15, 16]. These QMSs are currently partially implemented and have not yet become standard in the building sector. Thus, understanding their usefulness could facilitate their commissioning, inclusion in contracts and obtaining reliable empirical data on the impact of quality management services.

This contribution aims to focus on the role of QMS within the construction industry to align it with the EU Taxonomy by proposing a review. In addition, the experience of the QUEST project is recounted, presenting the basis and purpose of the predictive tool that is being developed.

2 Research Background

2.1 European Taxonomy

EU Regulation 2020/852 introduced into the European regulatory system the Taxonomy of eco-friendly economic activities; a classification of activities that can be considered sustainable based on their alignment with EU environmental objectives and compliance with certain social clauses. Six are the strategic objectives of this European legislation, ranging from mitigating the effects of climate change to preventing pollution, the sustainable use of natural resources and protecting ecosystems and biodiversity, as well as promoting the transition to a circular economy. The EU taxonomy is a tool that supports multiple public and private actors. The EU Taxonomy is a guide for companies to assess their activities, to define corporate policies towards greater environmental sustainability and to report to stakeholders more comprehensively and comparably. The EU taxonomy is an important tool for banks, which have to take environmental, social and corporate governance (ESG) criteria into account when assessing loans. The EU Taxonomy helps investors to integrate sustainability issues into investment policies and to understand the environmental impact of the economic activities in which they invest or might invest. Public institutions can use the EU Taxonomy to define and improve their green transition policies.

To be environmentally friendly, an activity must contribute positively to at least one of the six environmental objectives and must not produce negative impacts on any other objective. To achieve these objectives, the TEG has selected the sectors with the highest levels of CO_2 emissions and can contribute to reducing greenhouse gas emissions. Within these sectors, 67 activities were selected, which today account for 93.5% of EU emissions. For each of these, the TEG identified technical criteria to determine the quantitative thresholds within which there is a contribution to mitigation objectives and no negative impacts on other environmental objectives. By thus providing criteria for activities in these areas, it is expected that the EU Taxonomy will greatly expand the market's understanding of the sustainable financing opportunities available today.

2.2 The Financial Sustainability of the Construction Sector

Within the EU Taxonomy, four economic activities related to real estate have been identified and regulated according to CO_2 emission reduction targets: construction of new buildings, renovation of existing buildings, spot renovation measures, purchase and ownership of real estate [17, 18]. The TEG has identified technical screening criteria for all building-related activities. In particular, for purchase and ownership activity, buildings constructed before December 2020 must be at least EPC class A, or within the top 15% of the national building stock expressed in primary energy demand. Buildings constructed after December 2020 must meet the EU Taxonomy criteria for the construction of new buildings, i.e. the primary energy demand of new construction should be at least 10% lower than the near-zero building energy requirements in the national measures. In addition, large non-residential buildings with HVAC power >290 kW should operate efficiently through energy performance monitoring and assessment. In this perspective, QMSs can improve energy performance and reduce the gap between the calculated and estimated energy performance of buildings and their actual performance. It is now recognized that many problems that arise with the energy performance of buildings are not caused by inadequate technology, but by the absence of technical monitoring and quality management throughout the building's life cycle, which inevitably affects the actual performance of the building [19]. QMSs include the installation of instruments and devices to measure, regulate and control the energy performance of a building. Others consist of technical consultancy (energy advice, energy simulations, project management, production of energy performance contracts, dedicated training) related to improving the energy performance of buildings, accredited energy audits and building performance assessments, energy management services, energy performance contracts and energy services provided by energy service companies (ESCOs). Within the QUEST project, therefore, attention has been paid to identifying the effects of implementing the correct QMS to minimize the technical and financial risk that may occur.

2.3 Quality Management Services (QMSs)

QMSs are considered an essential part of any investment in Europe's built heritage. Being a rare exemption at this stage, QMSs need to scale up massively to be applied in the millions of projects that need to be facilitated to achieve the goals of the EU Green Deal: a climate-neutral building stock. This section gives an overview of common QMSs applied in the building sector.

Technical Monitoring
Technical monitoring (TMon) is a QMS that focuses on the functionality of the building services. A definition of the service is provided by the Mechanical and Electrical Engineering Working Party of National, Regional and Local Authorities (AMEV) [20]. The service includes a precise definition of functional requirements for the whole building and for individual systems and a specified way of testing before delivery and in the normal operation of a building. AMEV recommends that the TMon is always provided either by the owner (or by the public building authority itself) or by an independent third-party service provider. The TMon process can be set up largely as a digital service, from specification to data management to reporting.

Commissioning
According to REHVA guidance [13], commissioning (Cx) is a quality-focused process to improve the delivery of a new building or major refurbishment of an existing building. The process focuses on verifying and documenting that all commissioned systems and assemblies are planned, designed, installed, tested, operated and maintained to meet the

owners' project requirements. Cx includes a variety of services ranging from verification of design documents, operability (e.g. accessibility of air handling units for maintenance services to allow functional testing of systems), life cycle costing, operation and maintenance (O&M) documentation and supervision of the training of building maintenance staff. The commissioning process allows the owner to have a detailed check on whether the delivered building meets the project requirements. The methodological steps of Cx involve successive phases, including the definition of the owner's design requirements for the building, planning of the quality management process, the design phase, the construction and testing phase, O&M and systems manuals, training, and information. Commissioning has proven its effectiveness in achieving targets. Energy savings alone typically generate payback times of less than 3 years. The availability of building HVAC performance monitoring software and the ability to integrate measurements with Building Energy Monitoring Systems (BEMS) allows the automated creation of large data sets that can be used for ongoing evaluation and reporting.

Green Building Certification
The performance of buildings plays a key role in the global energy scenario. In recent years, many countries have developed certification procedures to assess the environmental sustainability of buildings, aiming to reduce energy consumption and environmental impacts during the construction, management, and operation of a building [21–23]. The main international schemes are Leadership in Energy and Environmental Design (LEED) and Building Research Establishment (BREEAM) [24, 25]. A number of national schemes such as Deutsche Gesellschaft für Nachhaltiges Bauen (DGNB) in Germany, Haute Qualité Environnementale (HQE) in France, Green Mark (Singapore), offer certification in their domestic markets and abroad. Green Building (GB) certifications vary significantly based on their goal, administrator, and importance in the industry. The sustainability of buildings can be assessed through different labelling tools, which are characterized by different calculation methods, credits, weights, and issues considered in each of these protocols. The differences between various GB protocols can have a major impact on final scores, which turn out to be very different. Several benefits of GB certification schemes have been recognized, including increased rental or resale value, increased occupant and tenant satisfaction, increased worker productivity and pride of place, lower operating costs, and opportunities for local and national recognition [26, 27].

Self-assessment
Self-assessment (SA) tools aim at measuring the environmental, social and governance performance of individual assets or portfolios grounding on self-reported data. Different tools and metrics are available across Europe, including for example Global ESG Benchmark for Real Assets (GRESB) [28] or European Public Real Estate (EPRA) [29]. The purpose of this assessment is to collect, validate, evaluate, and compare building environmental, social, and governance data to make recommendations on optimizing performance (e.g., energy consumption) and to report on their performance over the years. The assessment is usually developed using questions and checklists. The data is self-reported and could be validated by a third party and evaluated before being used to generate the final benchmarks. The credibility of sustainability data increases when

third-party assurance is performed by an objective, independent assurance provider. Most SA tools are available for free online; however, some systems require an application fee and others require an assessment fee.

Individual Expert Services

Individual expert services include procedures that, when implemented, allow buyers, sellers, and lenders of assets to quantify performance and energy savings. Examples of these systems include design consultations or specific services such as Efficiency Valuation Organization (EVO) International Performance Measurement & Verification Protocol (IPMVP) [30]. The purpose of these services is to provide an assessment of actual building performance in terms of, for example, energy and water consumption and an estimate of potential demand savings. Since energy, water, or demand savings cannot be measured directly, because savings represent the absence of energy/water consumption or demand, these are determined by comparing measured consumption or demand before and after program implementation. Assessment procedures and methodologies used for calculation may vary depending on the type of services required and the intended use of the building.

Performance Contracting

According to Directive 2012/27/EU [31], the energy performance contract (EPC) is defined as a contractual agreement between the beneficiary and the provider of an energy efficiency improvement measure, verified and monitored throughout the life of the contract, in which investments (work, supply or service) are paid in relation to a contractually agreed level of energy efficiency improvement or other agreed energy performance criteria, such as financial savings [32]. In an EPC, an ESCO engages in a project to provide energy efficiency improvements on the client's premises and uses the revenue stream from the savings to repay the project costs. The approach is based on transferring technical risks from the customer to the ESCO based on performance guarantees provided by the ESCO. The resulting savings are used to partially or fully pay for investments made. After the contract ends, the cost benefits brought by the energy savings remain with the client. Once the installation of the energy efficiency measures is complete, the project moves to the new performance evaluation phase. The specific nature of the service provided will depend on the contract. Energy savings is a key benefit that should be achieved as the EPC service is paid for by the energy cost savings realized. The contract between the ESCO and the client contains guarantees for the cost savings and assumes the financial and technical risks of implementation and operation for the life of the project, typically 5 to 15 years. Energy savings are typically between 10 and 30% of the baseline.

3   QUEST Project

The QUEST project aims to identify the potential of investments in sustainable and energy-efficient buildings by reducing the risk investors attribute to this type of intervention. Through an international consortium of universities and key players in this field such as engineering companies, insurance companies and trade associations, the QUEST project aims to use innovative QMSs to reduce the gap between the calculated and actual energy performance of buildings, which is a source of risk affecting the return on investment for energy-efficient buildings. To solve this problem and reduce investment risks, the QUEST project is developing a predictive tool to assess the risks associated with building and energy investments by integrating effective QMSs.

The QUEST project is based on two main reasons. The first relates to the fact that new buildings and buildings where major modifications have been made have experienced poor energy performance and additional operating costs in the first years due to poor quality. Poor quality is primarily related has an inadequate quality of work in all phases of the project, from requirements definition to design, construction, and operations. Other market studies and previous research, on the other hand, provide numerous examples of properly certified buildings where better technical building performance correlates with better financial performance, including high market value. However, to date, these two performance categories have not been sufficiently well-linked to reliably predict or visualize the financial benefits of quality management in buildings based on their technical performance. The QUEST project is trying to close this gap by examining the causal relationships between very specific aspects of technical performance of buildings, such as the implementation of QMSs, and very specific aspects of financial performance of buildings, such as cash flow, rent level, market value, or risk level [33]. The main objective is to establish a clear report that is understandable to both engineers and lenders, especially for investors.

To help financial stakeholders to manage the risks associated with low-quality building management, QUEST has identified three QMS that can help reduce risk exposure on construction projects related to new construction and major renovations; technical monitoring, building commissioning and Green Buildings certification. To ensure that the services are well-defined and reliably applied, the QMS provided reference to standardized building sustainability certification systems. In addition, the three selected QMSs are all provided by third parties.

The predictive tool QUEST will provide a first predictive assessment for defining the sustainable viability of investments in large real estate portfolios. The basic idea of the tool involves a tool set up on two registers. A first register refers to the input data entered by the user that relates to the characteristics of the building in question. The second register instead refers to the output data, mainly aimed at defining the added value provided by the implementation of QMS not only in terms of the market but also in terms of operating costs (energy, maintenance) and revenues (rental income, occupancy rate) [34–36]. The first step in building the tool is to consult statistical applications that relate building sustainability to operational savings. The second step considers the perception of risk related to technicians' trust with the system under consideration. The third step is to capitalize on energy savings rather than rent increases over the time under consideration so that the value-added relative to investment costs can be calculated. As such, QUEST plans to develop a tool to predict the value-added generated by these QMS for different project profiles and is structuring a process to continually collect data to create and subsequently improve a reliable database on their impact and improve the value-added prediction.

4   Conclusions and Future Perspectives

The EU Taxonomy will guide future green investments in the financial marketplace towards ESG criteria. In this new regulatory environment, Quality Management Services can improve the energy performance of buildings. The QUEST project will propose a tool for de-risking investments by closing the gap between expected and actual energy performance through the implementation of QMSs. The tool has two key objectives; first, to answer questions related to investment decisions by industry stakeholders in terms of the added value generated by QMS systems. Second, to help investors understand how green the investment under consideration can be. Answering these questions seems of utmost importance in the context of the new issues that are emerging in the context of EU Taxonomy. The predictive tool in this context would allow the investor to optimize credit and ensure access to financing.

The QUEST tool is currently undergoing calibration, integration, and simplification, both in terms of input data and visualization of results. As a future perspective, the revision of the algorithms and default data with reference to the data feedback from the pilot cases of the QUEST project. The evaluation model will include the reduction of input variables to be entered by the decision-maker to simplify the calculation phase. Furthermore, given the variation of input and output variables in the different member country cases, it will be ensured that sensitivity analyses can be developed to validate the robustness of the results.

References

1. European Commission: The European Green Deal. European Commission 53, 24 (2019)
2. Napoli, G., Barbaro, S., Giuffrida, S., Trovato, M.R.: The European Green Deal: new challenges for the economic feasibility of energy retrofit at district scale. In: Bevilacqua, C., Calabrò, F., Della Spina, L. (eds.) NMP 2020. SIST, vol. 178, pp. 1248–1258. Springer, Cham (2021). https://doi.org/10.1007/978-3-030-48279-4_116
3. Assumma, V., Datola, G., Mondini, G.: New cohesion policy 2021–2027: the role of indicators in the assessment of the SDGs targets performance. In: Gervasi, O., et al. (eds.) ICCSA 2021. LNCS, vol. 12955, pp. 614–625. Springer, Cham (2021). https://doi.org/10.1007/978-3-030-87007-2_44
4. Ruggeri, A.G., Calzolari, M., Scarpa, M., Gabrielli, L., Davoli, P.: Planning energy retrofit on historic building stocks: a score-driven decision support system. Energy Build. **224** (2020). https://doi.org/10.1016/J.ENBUILD.2020.110066
5. Dell'Anna, F., Marmolejo-Duarte, C., Bravi, M., Bottero, M.: A choice experiment for testing the energy-efficiency mortgage as a tool for promoting sustainable finance. Energy Effic. **15**, 27 (2022). https://doi.org/10.1007/s12053-022-10035-y
6. European Commission: The European Commission's Action Plan on Financing Sustainable Growth. https://ec.europa.eu/info/publications/sustainable-finance-renewed-strategy_en
7. European Commission: Sustainable finance taxonomy - Regulation (EU) 2020/852. https://ec.europa.eu/info/law/sustainable-finance-taxonomy-regulation-eu-2020-852_en
8. European Commission: A Renovation Wave for Europe - greening our buildings, creating jobs, improving lives, Brussels (2020)
9. European Commission: Directive 2018/844/UE, Energy Performance of Building Directive 2nd Recast (new EPBD). https://ec.europa.eu/info/news/new-energy-performance-buildings-directive-comes-force-9-july-2018-2018-jun-19_en

10. European Commission: Directive 2002/91/CE, Energy Performance of Building Directive (EPBD). https://eur-lex.europa.eu/legal-content/EN/TXT/HTML/?uri=CELEX:32002L0091&from=EN

11. European Commission: Directive 2010/31/UE, Energy Performance of Building Directive Recast (EPBD recast). https://eur-lex.europa.eu/legal-content/EN/TXT/HTML/?uri=CELEX:32010L0031&from=it

12. Becchio, C., Corgnati, S.P., Crespi, G., Pinto, M.C., Viazzo, S.: Exploitation of dynamic simulation to investigate the effectiveness of the Smart Readiness Indicator: application to the Energy Center building of Turin. Sci. Technol. Built Environ. **27**, 1127–1143 (2021). https://doi.org/10.1080/23744731.2021.1947657

13. Plesser, S., Teisen, O., Ryam, C.: Quality Management for Buildings. Improving Building Performance through Technical Monitoring and Commissioning. REHVA (2019)

14. Quest project: Project Quest. https://project-quest.eu/

15. Mehnert, J., Reiß, D., Plesser, S., Hannen, M.: An algorithmic module toolkit to support quality management for building performance. In: E3S Web of Conferences. EDP Sciences (2019)

16. Loureiro, T., Gil, M., Desmaris, R., Andaloro, A., Karakosta, C., Plesser, S.: De-risking energy efficiency investments through innovation. In: Proceedings 2020, vol. 65, page 3. 65, 3 (2020). https://doi.org/10.3390/PROCEEDINGS2020065003

17. Caprioli, C., Bottero, M., De Angelis, E.: Supporting policy design for the diffusion of cleaner technologies: a spatial empirical agent-based model. ISPRS Int. J. Geo Inf. **9**, 581 (2020). https://doi.org/10.3390/ijgi9100581

18. Ruggeri, A.G., Gabrielli, L., Scarpa, M.: Energy retrofit in European building portfolios: a review of five key aspects. Sustainability **12**, 7465 (2020). https://doi.org/10.3390/su12187465

19. Alencastro, J., Fuertes, A., de Wilde, P.: The relationship between quality defects and the thermal performance of buildings. Renew. Sustain. Energy Rev. **81**, 883–894 (2018). https://doi.org/10.1016/J.RSER.2017.08.029

20. Mechanical and Electrical Engineering Working Party of National, Regional and Local Authorities (AMEV): Technical Monitoring as an Instrument for Quality Assurance. https://www.quantum-project.eu/fileadmin/Publications/AMEV_Tmon_2017_eng.pdf

21. Nguyen, B.K., Altan, H.: Comparative review of five sustainable rating systems. Procedia Eng. **21**, 376–386 (2011). https://doi.org/10.1016/j.proeng.2011.11.2029

22. Mattoni, B., Guattari, C., Evangelisti, L., Bisegna, F., Gori, P., Asdrubali, F.: Critical review and methodological approach to evaluate the differences among international green building rating tools. Renew. Sustain. Energy Rev. **82**, 950–960 (2018). https://doi.org/10.1016/J.RSER.2017.09.105

23. Brambilla, A., Lindahl, G., Dell'Ovo, M., Capolongo, S.: Validation of a multiple criteria tool for healthcare facilities quality evaluation. Facilities **39**, 434–447 (2020). https://doi.org/10.1108/F-06-2020-0070

24. USGBC: U.S. Green Building Council - Green building rating system. https://www.usgbc.org/leed

25. BRE Global: GreenBook Live: Certified BREEAM Assessments. https://www.greenbooklive.com/search/scheme.jsp?id=202

26. Dell'Anna, F., Bottero, M.: Green premium in buildings: evidence from the real estate market of Singapore. J. Clean. Prod. **286**, 125327 (2021). https://doi.org/10.1016/j.jclepro.2020.125327

27. Mangialardo, A., Micelli, E., Saccani, F.: Does Sustainability affect real estate market values? Empirical evidence from the office buildings market in Milan (Italy). Sustainability **11**, 12 (2018). https://doi.org/10.3390/su11010012

28. GRESB: Real Estate Refernce Guide. https://documents.gresb.com/generated_files/real_e state/2021/real_estate/reference_guide/complete.html
29. European Public Real Estate Association (EPRA): EPRA Sustainability Best Practices Recommendations Guidelines (2017)
30. Efficiency Valuation Organization: IPMVP. Generally accepted M&V principles. International Performance Measurement and Verification Protocol (2018)
31. European Commission: Directive 2012/27/EU of the European Parliament and of the Council of 25 October 2012 on energy efficiency, amending Directives 2009/125/EC and 2010/30/EU and repealing Directives 2004/8/EC and 2006/32/EC. https://eur-lex.europa.eu/legal-content/ EN/TXT/?uri=CELEX:32012L0027
32. Boza-Kiss, B., Zangheri, P., Bertoldi, P., Economidou, M.: Practices and opportunities for Energy Performance Contracting in the public sector in EU Member States (2017). https://op.europa.eu/en/publication-detail/-/publication/5627bb1c-809b-11e7-b5c6-01aa75ed71a1/language-en
33. Leskinen, N., Vimpari, J., Junnila, S.: A review of the impact of green building certification on the cash flows and values of commercial properties. In: Sustainability 2020, vol. 12, page 2729. 12, 2729 (2020). https://doi.org/10.3390/SU12072729
34. Bisello, A., Antoniucci, V., Marella, G.: Measuring the price premium of energy efficiency: a two-step analysis in the Italian housing market. Energy Build. **208**, 109670 (2020). https:// doi.org/10.1016/j.enbuild.2019.109670
35. Bragolusi, P., D'Alpaos, C.: The valuation of buildings energy retrofitting: a multiple-criteria approach to reconcile cost-benefit trade-offs and energy savings. Appl. Energy **310**, 118431 (2022). https://doi.org/10.1016/j.apenergy.2021.118431
36. Dell'Anna, F., Bottero, M., Becchio, C., Corgnati, S.P., Mondini, G.: Designing a decision support system to evaluate the environmental and extra-economic performances of a nearly zero-energy building. Smart Sustain. Built Environ. **9**, 413–442 (2020). https://doi.org/10. 1108/SASBE-09-2019-0121

A Multi-dimensional Decision Support System for Choosing Solar Shading Devices in Office Buildings

Maria Cristina Pinto[1,2(✉)] ⓘ, Giulia Crespi[1] ⓘ, Federico Dell'Anna[2] ⓘ, and Cristina Becchio[1] ⓘ

[1] TEBE-IEEM Research Group, Energy Department, Politecnico di Torino, Corso Duca degli Abruzzi 24, 10129 Turin, Italy
mariacristina.pinto@polito.it
[2] Interuniversity Department of Regional and Urban Studies and Planning, Politecnico di Torino, Viale Mattioli 39, 10125 Turin, Italy

Abstract. The smart building revolution requires constructions to be equipped with innovative and advanced management systems able to improve energy performance and to guarantee adequate comfort conditions to occupants. In this framework, properly designed solar shading devices coupled with automation and control systems are of interest, being able to improve energy performance and indoor lighting conditions, especially during summer. The selection of the most adequate devices and related control mechanisms depends on various and potentially conflicting factors, concerning financial, environmental, socio-economic, energy, and architectural aspects. The present study proposes a multi-step methodological approach, which integrates EnergyPlus-based dynamic energy simulations with PROMETHEE II-based multi-criteria analysis to obtain a ranking of different alternative solutions. The work identifies a set of multi-domain indicators to give a snapshot of the different domains that can potentially influence the selection of shading solutions to be installed in office buildings and defines the actors to be involved in the decision-making process.

Keywords: Solar shading devices · Smart building management · Multi-criteria decision analysis

1 Introduction

Due to the well-recognized energy and environmental impacts of buildings, actions able to reduce their energy demand and improve their efficiency are crucial in the energy transition process [1]. Moreover, the rise of external air temperatures because of climate change and global warming issues are affecting the way energy is used within buildings, increasing air conditioning demands; these new considerations strongly ask buildings to be equipped with efficient HVAC systems for heating and cooling provision [2].

In this context, solar shading devices are recognized as an effective building design strategy, leading to a reduction of space cooling demands and an improvement of natural lighting conditions of indoor spaces [3]. Generally identified as systems able to dynamically respond to external solar stresses, if adequately installed and managed, solar shadings can help maximizing thermal gains during winter and reducing loads during summer; indeed, properly designed shadings, coupled with adequate automatic mechanisms, can improve the energy performance of buildings, allowing to achieve significant energy savings [4]. Moreover, besides energy-related considerations, the exploitation of automation and connectivity in smart buildings can enhance the role of solar shadings also for guaranteeing occupants' thermal and visual comfort [5].

Despite the attention on the benefits guaranteed by such systems, the investment decision-making process is still mainly influenced by financial aspects. For this reason, diverse incentive mechanisms are in place in favor of shading devices installation in new and existing buildings, aiming to increase the awareness about the impacts that these installations can have on buildings and occupants. Among them, in Italy, it is worth mentioning that, thanks to "2020 Relaunch Decree", energy efficiency interventions in buildings, including the installation of shading systems, can access a 110% tax deduction [6].

However, besides purely financial considerations, the decision-making process for the selection of appropriate solar shading systems and relative control strategies should take into consideration not only single aspects, but the wide set of criteria that can influence the choice of a specific system to be installed. Indeed, the selection of the proper shading device and related control mechanism (if present) to be installed depends on various factors, not limited to design properties (e.g., building orientation, window size, etc.), and can be characterized by several and potentially conflicting economic, environmental, social, energy, architectural and daylighting aspects.

In line with this, the paper highlights the need for developing proper methodologies able to support decision-makers for identifying the solutions representing the best compromise between the different conflicting criteria and proposes a multi-step methodological approach to support the comparison of different alternative solutions, in accordance with decision-makers' preferences and judgements. Specifically, the work focuses on office buildings, recognizing that the role of solar shadings is particularly interesting for this building category. Indeed, given the impact that solar shading devices have on occupants' comfort, their efficient design can be essential to increase employees' productivity, in turn connected to favorable indoor environmental conditions.

2 Decision-Making Frameworks for Optimum Shading Devices

This section presents a literature review on the use of decision-making models in the energy sector, focusing mainly on the context of shading systems investment decisions. Specifically, the review has the scope of highlighting the most recent trends and key issues relevant to this topic and obtaining an overview of decision-making tools to support the evaluation of alternative design solutions. The literature showed that the most frequently used approaches in the domain of energy investment decision problems are Life Cycle Assessment (LCA) and Life Cycle Cost (LCC), Discounted Cash Flow Analysis (DCFA) and Cost-Benefit Analysis (CBA), Environmental Impact Assessment (EIA) and Multi-Criteria Decision Analysis (MCDA) [7, 8]. Although the spread of assessment tools makes it possible to analyze all aspects of sustainability in the energy sector, the different nature of the approaches leads to assessment results that are not always consistent [9]. In this framework, it would be appropriate to define what are the advantages and disadvantages of these approaches in the context of shading devices investments. Deepening this field of application, the literature review showed that the most used approaches are mainly LCA, LCC, and MCDA.

Specifically, MCDA is an evaluation tool for simultaneously considering a multiplicity of qualitative and quantitative criteria to bring out the different points of view of the actors involved in the decision-making process [10, 11]. According to MCDA theory, the criteria weighting makes it possible to consider stakeholders' opinion in the decision process, which otherwise would be omitted using a traditional evaluation procedure, such as LCA and LCC [12–14].

Several applications of MCDA techniques used to compare diverse shading systems are present in literature. Stamatakis et al. [15] proposed an MCDA based on the PROMETHEE (Preference Ranking Organization METHod for Enrichment Evaluation) method to evaluate 13 alternative fixed shading systems combined with PV, according to four quantitative (i.e., PV power generation, Heating, Cooling, and Lighting) and three qualitative criteria (i.e., Aesthetic, View, Glare). Furthermore, the analysis included a moonlighting of social and economic stakeholders that are generally involved in either the use (users of office buildings), the construction (installers of PV systems), the study (architects, academic researchers) or the control (building and energy inspectors) of the building construction, to collect different points of view. Fontenelle and Bastos [16] applied a model based on ELECTRE III to select one of six proposed window solutions (differing in size, glass type and solar protection devices), according to three criteria: landscape view, daylight level on work plan and energy efficiency. Jalilzadehazhari et al. [17] proposed an integrated model to optimize the selection of windows and blinds in buildings, considering the existence of potential conflicts between visual comfort, thermal comfort, energy consumption and life cycle cost. The authors developed a decision support model by integrating dynamic energy simulations performed with EnergyPlus with the Analytic Hierarchy Process (AHP) algorithm [18]. In particular, a set of combinations of external shading systems and window glazing types were compared considering criteria of visual comfort, thermal comfort, energy consumption, and LCC [17].

One of the main elements that emerges from the literature review is the combination of financial and economic evaluation techniques with energy simulation models. Some authors emphasized the need to create integrated frameworks that could combine the simulated energy performance of the investment solutions with their environmental, economic, and social performances. On the one hand, the use of energy dynamic simulation models, such as EnergyPlus or TRaNsient SYstem Simulation (TRNSYS) software, allows to evaluate also visual comfort (e.g., amount of light, glare, uniformity, light intensity distribution), thermal comfort (e.g., Fanger's thermal comfort model, long-term percentage dissatisfaction, temperature) and LCC (investment, consumption, maintenance) performances [17]. On the other hand, MCDA results as a flexible evaluation model, allowing criteria of different nature (both qualitative and quantitative) to be integrated, with the aim of supporting decision-makers with the preliminary information available. In addition, the development of a MCDA model, and specifically the criteria weighting phase, allows for an evaluation of the criteria considered by the different stakeholders (e.g., building owner, building manager, occupants, etc.), to obtain a final evaluation that is as shared as possible.

To conclude, in the light of the findings of the developed literature review, the combination of multi-criteria approaches with energy dynamic modelling and simulation seems to be adequate for supporting investment decisions related to shading devices installations in buildings.

3 Methodological Proposal

The integration of MCDA methods with energy modelling and simulation tools can be beneficial for supporting the identification of the best alternative among the options at disposal for the decision-makers. Focusing on the investment decisions related to the selection of shading devices and relative control strategies, it is undoubted that energy simulation tools are powerful instruments to evaluate their performances in energy, environmental and comfort terms, when installed in a specific building. However, their usage for supporting the decision-making is usually restricted to the analysis of a single criterion at once, without identifying the solution representing the best trade-off between different multi-domain criteria [17]. To this purpose, energy simulation tools can be beneficially coupled with MCDA methods. In line with this, the work aims to develop a multi-step methodological proposal to support decision-makers in selecting an adequate solar shading device and relative control strategy to be installed in an office building, considering and integrating a wide set of energy, environmental, economic and social criteria [19, 20]. Specifically, the methodological proposal combines EnergyPlus-based energy dynamic simulations, to estimate the performances of the considered devices, and the PROMETHEE II MCDA method, to evaluate them according to diverse multi-domain indicators.

The workflow of the developed assessment model is reported in Fig. 1, outlining the progressive methodological steps needed to assess and rank the different alternatives at disposal, including the wide set of criteria that can potentially influence the decision-making process. After the identification of the building under study, the first step of the methodological approach consists in the assessment of the building energy performance in its current state (i.e., without any shading device installed), by means of energy dynamic simulations through EnergyPlus software (see Sect. 3.1). Then, once selected the different combinations of solar shading device and automation and control strategy to be potentially installed in the building, a set of multi-domain criteria is defined and computed, to compare the performances of the alternative solutions in energy, environmental, economic, and social terms. As previously mentioned, the PROMETHEE II multi-criteria method is used for the analysis, to weight the criteria, involving different experts, and to score and rank the selected alternatives from a multi-dimensional standpoint (see Sect. 3.2). Finally, proper sensitivity analyses are developed to estimate the stability of the final ranking and to determine its changes in accordance with the variation of the relative importance of the considered criteria; specifically, weighting coefficients are changed by exaggerating the criteria belonging to a domain at once, while criteria are validated by examining the visual stability intervals. To conclude, the methodology is developed with the final scope of providing guidelines and recommendations to the involved decision-makers.

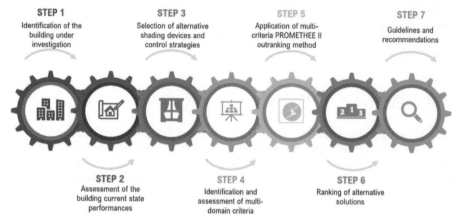

Fig. 1. Proposal of multi-step methodological workflow.

3.1 EnergyPlus Software

EnergyPlus software is used for the estimation of the energy performances of the building under investigation, in both its current state and in case different shading devices and control strategies are installed. EnergyPlus, developed and funded by the U.S. Department of Energy (DOE), is an open-source whole-building energy modelling engine, deeply used by engineers, architects and academics to model the energy and water consumption of a building [21]. The tool allows to perform the simulation of the building under analysis, providing energy- and comfort-related (both thermal and visual) outcomes, with an hourly or sub-hourly resolution, using location-specific information of solar radiation and external air characteristics. EnergyPlus is selected for the developed assessment model since, as underlined by Jalilzadehazhari et al., it can be defined as "the most frequently used simulation tool for evaluating thermal comfort, energy consumption and LCC" [17].

3.2 PROMETHEE II Method

The multi-criteria method chosen to address the problem investigated in this study is PROMETHEE II [22, 23]. It is an outranking method that performs a pairwise comparison of comparable alternatives to rank them against a set of often conflicting criteria. The method is chosen mainly because it does not require prior normalization of the quantitative data of the evaluation items, which can be used in their original units. Being an outranking method, PROMETHEE II does not manifest the danger of compensation between criteria. Moreover, the judgements made by each stakeholder can be easily integrated. The use of Visual PROMETHEE software facilitates the implementation procedure and the visualization of the obtained results [24, 25].

The first essential requirement for the implementation of the model concerns the importance of each specific criterion assumed by the decision-maker, through the definition of importance weights. The numerical weights are assigned to define, for each criterion, the relative ratings according to the chosen scale and reflect the priorities of the decision-makers. This information can be deduced using the SRF (Simos-Roy-Figueira) method [14], which requires the decision-maker to order the evaluation criteria (represented on cards) from the most to the least important. The card-playing technique gives the opportunity to indirectly obtain valuable information through numerical values, transforming the ranking into importance weights. Moreover, the second piece of information needed for the use of the PROMETHEE II is the definition of a preference function for each considered criterion, used to determine how much an alternative is preferred to another in terms of each criterion. Each preference function takes values between 0 and 1, where 0 is used in the case of indifference, while 1 means that a narrow preference is revealed. Six choices for preference functions are suggested by Brans and Vincke [26]: usual criterion, quasi-criterion, criterion with linear preference, level criterion, criterion with linear preference and indifference area, and Gaussian criterion. Finally, to define the complete ranking of alternatives, PROMETHEE II provides the calculation of the net outranking flow for each alternative at disposal.

4   Results

4.1   Identification and Assessment of Criteria

To take care of the potential conflicting aspects involved in the selection of the most adequate solar shading option and relative control strategy for an office building, four different domains are involved in the assessment, to reflect the main aspects that can potentially drive and influence the selection of these systems, in line with current sustainability and energy transition pillars. Based on a review of the possible indicators to be included in the assessment, a total of 9 criteria of interest was selected, as summarized in Table 1, through experts' panels and interviews. All selected criteria are quantitative, with the sole exception of the visual impact (S3), which is assessed through a qualitative scale.

Table 1. Considered multi-domain criteria for the assessment.

Domain	Criterion	Unit of measure
Energy	Cooling needs savings (EN1)	%
	Primary energy for heating and cooling (EN2)	kWh/m^2
Environment	CO_2 emissions for heating and cooling (ENV1)	kg/m^2
	PM emissions for heating and cooling (ENV2)	g/m^2

(*continued*)

Table 1. (*continued*)

Domain	Criterion	Unit of measure
Society	Percentage of occupied hours in thermal comfort (S1)	%
	Percentage of occupied hours in visual comfort (S2)	%
	Visual impact (S3)	Qualitative scale 1–5
Economy	Investment cost (EC1)	€/m^2
	Maintenance cost (EC2)	€/m^2

Starting from the energy domain, both indicators are calculated through the energy dynamic simulation of the building under investigation, through which it is possible to simulate the presence of the different shading systems and relative control strategies. In particular, the cooling needs savings (EN1) criterion is assessed in percentage terms with respect to the cooling needs of the building simulated in its current state, without any shading system applied. The focus is conducted specifically on summer loads, because of the stronger influence that solar shadings have on them, as well as considering the higher impact that climate change is having on building cooling demands. EN2 criterion, instead, is related to both heating and cooling performances, in terms of primary energy consumption of the building, expressed in kWh/m^2.

To tackle the environmental impacts of the different alternatives to be compared, CO_2 (ENV1) and PM (ENV2) emissions are elaborated using appropriate emissions factors, applied to the energy consumptions extracted from the building model.

The social domain is investigated through the identification of three different criteria, aiming to assess the impacts of shading devices on the thermal and visual comfort of occupants, as well as the visual aesthetical impact of the installed systems for the building under analysis. Thermal comfort is evaluated considering the Predicted Percentage of Dissatisfied (PPD), defined by Fanger [27] as the percentage of occupants not satisfied by the thermal conditions of the indoor spaces [28]; specifically, S1 considers the percentage of annual occupied hours in which PPD is lower than 10% (representing an indication of thermal comfort condition [27]). Moreover, to consider the effects of solar shading on the internal lighting conditions of the offices, S2 criterion is defined as the percentage of occupied hours in visual comfort, the latter defined according to the Discomfort Glare Index (DGI) value; in particular, visual comfort is assumed to be guaranteed in case DGI is lower than 22 [21]. Both S1 and S2 are computed based on the hourly results from the EnergyPlus energy dynamic simulations of the building model. Then, the third social criterion (S3) is qualitatively evaluated by experts' assessment, estimating the visual impact of shading systems through a qualitative scale ranging from 1 to 5; a lower score means a better aesthetics of the alternative, with respect to the urban context in which the building is inserted.

Finally, attention is devoted also to the economic perspective; in particular, EC1 considers the investment cost required for the installation of the shading devices and the relative accessors for control strategies (if required), while EC2 refers to the maintenance cost, calculated over the entire lifetime of the systems under analysis. Both economic criteria are evaluated in €/m^2 and are determined based on standardized national and regional price lists or market prices.

4.2 Importance of Weighting Criteria

One of the key steps of multi-criteria decision-making models concerns the assignment of weights to criteria. Generally, the subjective determination of weights is based on experts' opinions, which can help improving the quality of decisions, rendering the decision-making process more explicit, rational, and efficient [29, 30]. In this research context, to design appropriate shadowing systems in line with the views of different economic and social actors, it is necessary to take into account their interests and suggestions. Different actors can be involved, using their judgements to create separate scenarios for the multi-criteria analysis. In this work, the following actors may be accounted [15]. Firstly, it is important to consider the average user of the office building, since he/she spends the entire working life within the hypothetical office and knows the needs to be satisfied; a general user can be represented by a group of civil servants selected as a representative sample for this scenario. Their attention can be focused on visual and thermal comfort criteria. Then, planners are responsible for the architectural design of office buildings and at the same time participate to the construction phase by organizing, supervising, and financing these projects. They control most of the criteria included in the model. Finally, the building manager supervises the performance of the building throughout its life cycle. In this sense, fulfilling different objectives related to the rational management of economic and energy resources is crucial.

5 Conclusions and Future Perspectives

To achieve the decarbonization targets, the building sector is experiencing a transition phase towards smartness, giving relevance to the adoption of more efficient and smarter technologies and on the improvement of the envelope characteristics. In this framework, the building envelope is becoming more and more dynamic, and solar shading devices and their control strategies play a crucial role, making use of automation technologies and advancement in connectivity.

Due to the diverse solutions present in the market and to the multi-domain criteria that can influence the investment decision-making process, the work aims to propose a multi-dimensional decision support system for selecting the best solar shading solution coupled with adequate control strategies for an office building, aiming to maximize the internal comfort of occupants and, at the same time, to reduce the operational energy and environmental costs of the building under analysis. The methodological proposal integrates a Multi-Criteria Decision Analysis technique, in the form of the PROMETHEE II method, with energy dynamic modelling and simulations, using EnergyPlus tool. The paper wants to stress how the synergy and integration of different instruments and methodologies gives the possibility to extract valuable results and informed solutions. Specifically, the work identifies and evaluates a set of multi-domain indicators, selected to give a snapshot of the several (and often conflicting) domains influencing the selection of the solutions to be installed. Moreover, thanks to the involvement of diverse experts for the weighting procedure, the approach allows to consider and combine the suggestions of different actors, potentially influenced by or influencing the decision-making process.

Future work will allow to deepen the presented study, applying the developed integrated assessment model on several case studies of interest, to test its capability of supporting investment decision-making processes in the building sector.

Acknowledgments. The work of Maria Cristina Pinto has been financed by the Research Fund for the Italian Electrical System under the Contract Agreement between RSE S.p.A. and the Ministry of Economic Development - General Directorate for the Electricity Market, Renewable Energy and Energy Efficiency, Nuclear Energy in compliance with the Decree of April 16th, 2018.

References

1. European Commission: 'Fit for 55': delivering the EU's 2030 Climate Target on the way to climate neutrality, 14 July 2021. https://eur-lex.europa.eu/legal-content/EN/TXT/?uri=CELEX%3A52021DC0550
2. Crespi, G., Abbà, I., Corgnati, S.P., Morassutti, S., Prendin, L.: HVAC polyvalent technologies to balance contemporary loads in buildings. In: Proceedings of 16th Conference on Sustainable Development of Energy, Water and Environment Systems SDEWES2021 (2021)
3. Prowler, D: (FAIA): Sun control and shading devices, 08 September 2016. https://www.wbdg.org/resources/sun-control-and-shading-devices
4. Becchio, C., Corgnati, S.P., Crespi, G., Pinto, M.C., Viazzo, S.: Exploitation of dynamic simulation to investigate the effectiveness of the Smart Readiness Indicator: application to the Energy Center building of Turin. Sci. Technol. Built Environ. **27**, 1127–1143 (2021)
5. De Luca, F., Voll, H., Thalfeldt, M.: Comparison of static and dynamic shading systems for office building energy consumption and cooling load assessment. Manage. Environ. Qual. Int. J. **29**, 978–998 (2018)
6. Ministry of Economic Development: Superbonus e Sismabonus 110% - Decreti attuativi. https://www.mise.gov.it/index.php/it/incentivi/energia/superbonus-110
7. Bottero, M., Dell'Anna, F., Morgese, V.: Evaluating the transition towards post-carbon cities: a literature review. Sustainability **13** (2021)
8. Strantzali, E., Aravossis, K.: Decision making in renewable energy investments: a review. Renew. Sustain. Energy Rev. **55**, 885–898 (2016)
9. Hoogmartens, R., Van Passel, S., Van Acker, K., Dubois, M.: Bridging the gap between LCA, LCC and CBA as sustainability assessment tools. Environ. Impact Assess. Rev. **48**, 27–33 (2014)
10. Figueira, J., Greco, S., Ehrogott, M.: Multiple criteria decision analysis: state of the art surveys. Springer, New York (2005)
11. Figueira, J.R., Greco, S., Roy, B., Słowiński, R.: ELECTRE methods: main features and recent developments. In: Zopounidis, C., Pardalos, P. (eds.) Handbook of Multicriteria Analysis. Applied Optimization, pp. 51–89. Springer, Berlin, Heidelberg (2010). https://doi.org/10.1007/978-3-540-92828-7_3
12. Mustajoki, J., Hamalainen, R.P., Salo, A.: Decision support by interval SMART/SWING-Incorporating imprecision in the SMART and SWING methods. Decis. Sci. **36**, 317–339 (2005). https://doi.org/10.1111/j.1540-5414.2005.00075.x
13. Tsoukias, A., Keeney, R.L., Raiffa, H.: Decisions with multiple objectives: preferences and value tradeoffs. J. Oper. Res. Soc. **45**, 1093 (1994). https://doi.org/10.2307/2584151

14. Figueira, J., Roy, B.: Determining the weights of criteria in the ELECTRE type methods with a revised Simos' procedure. Eur. J. Oper. Res. **139**, 317–326 (2002)
15. Stamatakis, A., Mandalaki, M., Tsoutsos, T.: Multi-criteria analysis for PV integrated in shading devices for Mediterranean region. Energy Build. **117**, 128–137 (2016)
16. Fontenelle, M.R., Bastos, L.E.G.: The multicriteria approach in the architecture conception: defining windows for an office building in Rio de Janeiro. Build. Environ. **74**, 96–105 (2014)
17. Jalilzadehazhari, E., Johansson, P., Johansson, J., Mahapatra, K.: Developing a decision-making framework for resolving conflicts when selecting windows and blinds. Archit. Eng. Des. Manage. **15**, 357–381 (2019)
18. Saaty, T.L.: The Analytic Hierarchy Process (1980)
19. Dell'Anna, F., Bottero, M., Becchio, C., Corgnati, S., Mondini, G.: Designing a decision support system to evaluate the environmental and extra-economic performances of a nearly zero-energy building. Smart Sustain. Built. Environ. **9**, 413–442 (2020)
20. Sward, J., et al.: Integrating social considerations in multicriteria decision analysis for utility-scale solar photovoltaic siting. Appl. Energy **288**, 116543 (2021)
21. Department of Energy (DOE): EnergyPlus software. https://energyplus.net/
22. Brans, J.P., Mareschal, B., Vincke, P.: PROMETHEE: a new family of outranking methods in multicriteria analysis. In: Operational Research (1984)
23. Brans, J.P., Vincke, P., Mareschal, B.: How to select and how to rank projects: the Promethee method. Eur. J. Oper. Res. **24**, 228–238 (1986)
24. Andreopoulou, Z., Koliouska, C., Galariotis, E., Zopounidis, C.: Renewable energy sources: using PROMETHEE II for ranking websites to support market opportunities. Technol. Forecast. Soc. Chang. **131**, 31–37 (2018)
25. Visual PROMETHEE software. https://www.promethee-gaia.net
26. Brans, J., Vincke, P.: A preference ranking organization method: the PROMETHEE method for MCDM. Manage. Sci. **31**, 647–656 (1985)
27. International Standard Organization, ISO 7730 - Ergonomics of the thermal environment — Analytical determination and interpretation of thermal comfort using calculation of the PMV and PPD indices and local thermal comfort criteria (2005)
28. Liu, M., Wittchenm, K., Heiselberg, P.: Control strategies for intelligent glazed façade and their influence on energy and comfort performance of office buildings in Denmark. Appl Energy **145**, 43–51 (2015)
29. Bottero, M., Assumma, V., Caprioli, C., Dell'Ovo, M.: Decision making in urban development: the application of a hybrid evaluation method for a critical area in the city of Turin (Italy). Sustain. Cities Soc. **72**, 103028 (2021). https://doi.org/10.1016/j.scs.2021.103028
30. Bottero, M., Datola, G.: Addressing social sustainability in urban regeneration processes. An application of the social multi-criteria evaluation. Sustainability **12**, 7579 (2020). https://doi.org/10.3390/su12187579

A Multi-criteria and Multi-domain Model to Support the Comprehensive Assessment of Building Performance

Giulia Vergerio[1]([✉]) [iD], Giulio Cavana[2], Federico Dell'Anna[2] [iD], Cristina Becchio[1] [iD], Sara Viazzo[1] [iD], and Marta Bottero[2] [iD]

[1] TEBE-IEEM Research Group, Energy Department, Politecnico di Torino, 10129 Turin, Italy
giulia.vergerio@polito.it

[2] Interuniversity Department of Regional and Urban Studies and Planning, Politecnico di Torino, 10125 Turin, Italy

Abstract. Buildings can play a significant role in reducing energy consumptions and related emissions. However, the renovation process towards better-performing buildings is slowed down also by non-technical barriers, like the lack of accounting for indirect impacts of building renovation. Benefits of building retrofit that are not yet reflected by their market value, e.g., increase in people comfort and well-being, reduction of environmental impacts, must be quantified to make them tangible and, thus, to become leavers towards new habits and investment decisions. Thanks to the spreading of smart sensing in the building sector and the increasing availability of monitored data coming from indoor spaces, the control of such aspects in building design and management practices will become more and more feasible.

The paper aims at proposing a multi-criteria and multi-domain (energy, environmental, financial, and socio-economic) methodology for building performance assessment, exploiting monitored data. The set of the evaluation criteria per each domain, namely the Key Performance Indicators (KPIs), and the parameters needed for their quantification represent the main results of this research. Moreover, the research contributes to identifying an evaluation framework based on the Multi Attribute Value Theory, built on the standardization of the various indicators through so-called value functions. This integrated approach enables the translation of monitored data into useful and easy-to-understand metrics to support building design and management. Future works will include the translation of such approach into informative services and the deepening of the relative weights of KPIs and domain of impacts in the evaluation schema.

Keywords: Indoor Environmental Quality (IEQ) · Monitoring system · Key Performance Indicator (KPI) · Multi-criteria analysis

F. Calabrò et al. (Eds.): NMP 2022, LNNS 482, pp. 1752–1761, 2022.
https://doi.org/10.1007/978-3-031-06825-6_169

1   Introduction

Buildings have a great potential in terms of energy savings and subsequent emissions reduction; thus, they are considered key contributors in reaching European targets on energy efficiency and carbon neutrality by 2050. Furthermore, being the place where people spend 90% of their time, buildings can also play a significant role to reach many Sustainable Development Goals (SDGs), and in particular SDG3 about good health and well-being of people.

However, the renovation process towards better-performing buildings is slowed down by both technical and non-technical barriers. Among the latter, it is recognized that buildings market value does not fully reflect an increase in their energy performance, and it accounts even less for non-monetary benefits related to building renovation, such as an increased Indoor Environmental Quality (IEQ). Therefore, positive consequences on people health and well-being are neglected. Thus, there is the need for an integrated approach addressed to decision-makers, buildings managers, and/or owners and buildings occupants aiming at raising awareness on the multiple benefits coming from buildings renovation [1]. To do that, benefits must be quantified, to make them tangible (thus, easy to communicate) and to be able to account for them in the framework of building performance assessment procedures adopting a comprehensive and multi-criteria approach.

The attention to a multi-criteria approach to building performance assessment is well reflected in the evolution of the regulation on building energy performance in the last ten years. Indeed, the Energy Performance of Building Directive (EPBD) 2010/31/EU represented a first shifting in the attention from only energy efficiency concerns to financial ones in the definition of building retrofit, by introducing the concept of Cost-optimality. The Cost-optimal level corresponds to a level of performance of a building that minimizes primary energy consumptions and global cost at once. The latter is computed as the actualized sum of all the cost items occurring over the lifespan of the building, including not only investments, but also operation and maintenance, energy expenses, substitutions, and residual values. More recently, the new EPBD 2018/844/UE highlights a further shift of the paradigm of building performance, by stressing on the inclusion of the benefits of buildings renovation associated with the improvement of the IEQ and the subsequent levels of occupants' comfort. In the last two decades, huge attention for the quantification and monetization of the impacts of IEQ on occupants' productivity was given, with particular interest on office and school buildings, as summarized by [2]. Nowadays, the control of IEQ is studied under an even broader paradigm, paying attention to its impact on health and general well-being of the occupants [3], as it is stressed by the new EPBD. Not by chance, the most recent protocols for the certification of buildings quality (e.g., WELL) focus, among the aspects that building design and management should control, also on criteria related to health and well-being of occupants. The topic received a new boost due to the Covid-19 pandemic. Among the scientific evidence of the impacts of IEQ on people health and well-being, the prevalence of Sick Building Syndrome (SBS) symptoms is widely studied [4–7].

In face of this trend and of the scientific evidence about the impacts of IEQ on people health and well-being, a methodology for the evaluation of buildings performance and their alternative intervention strategies cannot fail in measuring co-benefits related to these aspects, as done in [8].

To complete the picture on the problem of evaluating building performances to support building design and management towards a more sustainable future, it is worthy to point to the role of data, as stressed by the European Green Deal. According to the latter, targeting sustainable goals requires exploiting the potential of data and digital platforms. In the domain of building management, it is expected to have an increase in data availability, which are going to be more and more monitored data, rather than simulated or self-reported. Monitored data are also related to IEQ parameters, whose data-driven control is still poorly concerning building design and management practices.

In this context, the paper aims at proposing a multi-criteria and multi-domain (energy, environmental, financial, and socio-economic) methodology for building performance assessment, exploiting monitored data collected from standing monitoring systems. Particular attention is given to the problem of the identification of an evaluation framework that enables the translation of monitored data into useful and easy-to-understand metrics to support building design and management. The set of the evaluation criteria per each domain, namely the Key Performance Indicators (KPIs), and the parameters needed for their quantification represent the main results of this research.

The paper is structured as follows. Section 2 clarifies the research gaps in the context depicted in the introduction. In Sect. 3, the methodology proposed, and the methods adopted to bridge the gap are reported. Finally, Sect. 4 summarizes the results in terms of identified metrics for a comprehensive evaluation of buildings performance. Conclusion and expected future developments are provided in Sect. 5.

2   Research Gap

The most common method to monitor and control a process or a project in many fields refers to the use of KPIs. Research on the evaluation of buildings performance and its retrofit alternatives have been widely pursued, but studies focusing on the identification of appropriate KPIs for holistic evaluation remain limited [9]. A recent literature review on such KPIs has identified four categories of aspects that are addressed, namely economic, environmental, users' perspective, and health and safety [9]. The authors identified fifty-two indicators, which were shortlisted based on the opinion of experts collected through dedicated focus groups to define a set of nineteen relevant KPIs. From this work [9], three interesting points are identified and addressed in the research presented in this paper.

Firstly, the authors concluded that the reasons for the exclusion of some of the KPIs were mostly about impracticality for their computation, in terms of excess of resources needed, like time, data or money. Secondly, the authors point out the fact that indicators must be simple to understand, which is an important aspect when seeking funding support or approval from parties joining retrofit projects. Finally, in the shortlisted KPIs, the ones related to health and safety are poorly represented (two out of nineteen). The authors suggest that this does not corroborate the relative importance of the different domains and that further work is needed to determine the weightings of the indicators that constitute the intended performance evaluation scheme.

The spreading of smart sensing technologies, including reliable low-cost air quality sensors (LCAQS), might partially overcome the barrier mentioned as the first important aspect (i.e., the need for resources to compute KPIs). For this reason, in this research, the use of monitored data is maximized. Furthermore, the issue of comprehensibility is addressed by proposing a multi-criteria model to aggregate KPIs into easy-to-understand domain and super-domain indices, which are expressed in a number from 0 to 1, reflecting their preferability. Finally, regarding the relative importance of different domains, particular attention is given to the KPIs related to people's well-being in indoor spaces, since this is an aspect to leverage on to trigger new habits and investment decisions [10]. Furthermore, to address the need for further work on this topic, the use of a multi-criteria model is proposed. It enables to account for the preferences of the actors involved in the building design or management process, whose opinion must be collected to assign different priorities (in terms of weighting systems) to the domains of impacts of a building performance (e.g., energy, financial, environmental, etc.).

Finally, it is worthy to mention that the continuous monitoring of the running performances of a building, the inclusion of personalized weights, and the engagement of occupants' and building managers in the evaluation process are the aspects in which the methodology presented in this paper differs from the existing protocols for building performance certification schemes.

All these aspects are deepened and clarified in the methodological proposal (Sect. 3) and in the results (Sect. 4), as follow.

3 Methodological Proposal

A multi-criteria and multi-domain (energy, environmental, financial, and socio-economic) methodology for building performance assessment is presented in this section. The methodology is based on the following steps:

1. Statement of the evaluation problem and identification of performance domains and related KPIs.
2. Monitoring of parameters per each domain KPI.
3. Calculation of KPIs from monitored data.
4. Aggregation of domain KPIs into domain index.
5. Aggregation of the domain indices to assess the overall building performance (Building Performance index) (Fig. 1).

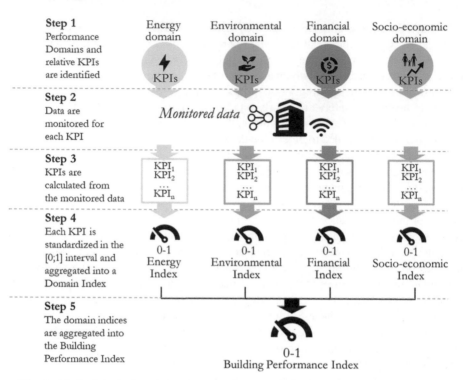

Fig. 1. Methodological workflow for comprehensive building performance assessment.

Steps 1–3 are answering the question "what to assess/measure?". Step 4–5 proposes to adopt specific methods from literature belonging to the domains of multi-criteria analysis to answer the question "how to assess?", meaning how data are treated as part of an evaluation model accounting for a full range of criteria. The final objective of these stages is to provide for a single metric, coming from the progressive aggregation of the domain specific KPIs, which represents the overall performance of the building and/or of alternative strategies that have been adopted. Indeed, describing and assessing complex problems as buildings sustainability requires not only to identify a system of indicators describing all the different aspects (e.g., energy performance, occupants' satisfaction), but also an aggregation strategy. The latter allows a more straightforward communication, benchmarking and ranking process, but opens the challenge of finding the balance between synthesis and loss of information in the aggregation. There are various methods that allow to score the performance of a project/process/system/building. Among monetary methods, the Cost-Benefit Analysis allows to translate all impacts into monetized costs and benefits to compute a final monetary index. However, multi-criteria analyses are considered preferred, since they do not require to monetize all the indicators (which is not always feasible) and they embed stakeholder preferences in the model. They are an established tool for the evaluation multidimensional problems and allow to face the trade-off between synthesis and loss on information mentioned before. So, in this research a multi-criteria model is proposed, specifically the Multi Attribute Value

Theory (MAVT) [11]. Among the advantages of this method, it is worthy to mention that:

- It is based on a standardization process of each of the evaluation criteria (i.e., KPI), allowing to provide for easy-to-understand metric, ranging from 0 to 1, where 0 corresponds to the worst performance while 1 to the most desirable one. The standardization enables the aggregation of data.
- It is able to account for preferences by means of value functions, which are enabling the aggregation of data, and weighting systems, accommodating for the relative importance of KPIs and domains.
- It can accommodate for metrics that differ in type (i.e., continuous, dichotomic, qualitative, etc.) and nature (monetary and non-monetary), which are all standardized before being aggregated.
- It can deal with many elements to be assessed: once the general evaluation framework (value functions) is identified, the increase in the number of configurations that have to be evaluated is not followed by an increase in the computational burden on the evaluator (some distinctions have to be done for certain construction methods of the value functions).

MAVT is a specific type of Multi-Criteria Analysis (MCA) introduced by Keeney and Raiffa [11]. MAVT is a quali-quantitative discrete MCA that can be used to evaluate heterogeneous dimensions of a decision problem, assuming the existence of a real value function that represents the preferences of the decision-maker and is used to formulate the evaluation of different possible alternatives into a single value [12].

In particular, the evaluation matter is decomposed into attributes that are assessed separately in their performance against a stated goal and then aggregated to score a unique value that allows the decision-maker to rank the different alternatives in their capability to reach the objective of the decision problem.

The aggregation of the attributes score could be done following different models, the simplest one is the additive one [13], defined by the Eq. (1):

$$V(a) = \sum w_i * v_i(a_i) \tag{1}$$

where $V(a)$ is the overall score of alternative a, $v_i(a_i)$ is the specific attribute performance, and w_i is a weighting factor that represents the importance of the attribute in reaching the stated goal.

As it could be noted from Eq. (1), MAVT works within the weak sustainability concept, allowing the compensation of a weak performance of one criterion by a good performance of another one.

In the development of a MAVT model some crucial steps have to be done: the first one is devoted establishing a representation of the human judgment on the worthiness of each impact of the alternatives under investigation (that defines $v_i(a_i)$), and the second one is the definition of the relative importance of the attributes in the calculation of the final score of each alternative (which decides the weight w_i for each attribute) [12].

The first step is modelled via a value function [14]. The value function standardizes the performance of the alternative on the specific attribute in a dimensionless score in the [0; 1] interval that represents the degree to which an alternative satisfies a decision objective: a score of 1 means a high achievement of a specific part of the overall goal, while, on the contrary, 0 represents a poor achievement of the goal [15]. In this sense, value functions are mathematical representations of the human judgment [12] and from the different shapes in which these functions are drawn, one could infer the value system of the individual involved in the evaluation process. The standardization of the attributes performances into a dimensionless value is a fundamental passage that allows to aggregate the heterogeneous dimensions of the decision problem.

The second fundamental step, as was mentioned before, is the definition of the weighting vector of the different attributes to prioritize different objectives. Many methods have been proposed to assess the different weights. To name a few, swing-weights, rating, pairwise comparison, trade-off, qualitative translation are among the most used [15]. In particular, the final value that each alternative will score is deeply influenced by the different weights that will be assigned to each attribute considered in the evaluation process, therefore the crucial role of the weighting vector elicitation process has been often stressed [12].

4   Results

The set of the evaluation criteria, the KPIs, per the identified domains of impact represents the main result of this work. Moreover, the research contributes to identifying an evaluation framework based on the MAVT. Results are reported in terms of the definition of the evaluation problem and the domain of impacts, and of the identification of the relevant KPIs per domain. The latter are reported in Table 1 with the parameters involved in their quantification.

The object of the assessment that the KPIs support is the building performance in terms of impacts on energy, environmental, financial, and socio-economic dimensions related to the use of the building itself. Thus, KPIs should be able to measure the impacts that the building performance has in terms of consumptions (energy domain), environmental burdens due to the energy consumptions (environmental domain), monetary burdens and benefits connected to building operation (financial domain), and outcomes of the guaranteed IEQ on people comfort, health, and well-being (socio-economic domain). So, the answer to the question "what to assess/measure?" posed in Sect. 3 is that building performances are measured by proposing several KPIs that in turn allow to answer the following questions per each domain:

- Energy: 'what are the energy consumptions of the building?'
- Environmental: 'what are the environmental impacts due to the energy consumptions of the building?'
- Financial: 'what are monetary value and maintenance costs of the building?'
- Socio-economic: 'which is the IEQ of the building in terms of its impacts on people comfort, health, and well-being?'.

Table 1. Key Performance Indicators and needed parameters for their quantification.

Domain	KPI	Parameters [unit] (Collection method)
Energy	– Primary energy consumption for space heating per square meter[a] – Final energy consumption for space cooling per square meter[a] – Final energy consumption for appliances per square meter[a] – Final energy consumption for lighting per square meter[a]	– Thermal energy [kWh] (Monitored) – Electric energy [kWh] (Monitored) – Systems efficiencies [-] (Reported) – Conditioned square meter [m^2] (Reported)
Environmental	– CO_{2eq} emissions per square meter[a] – NO_x emissions per square meter[a] – SO_x emissions per square meter[a] – $PM_{2.5}$ emissions per square meter[a] – PM_{10} emissions per square meter[a]	– Emission per each pollutant [g] (Computed, as consumptions per energy carrier times emission factors) – Conditioned square meter [m^2] (Reported)
Financial	– Management costs per square meter – Add-value due to QMS[b]	– Management cost [€] (Reported) – Presence of QMS [yes/no] (Reported) – Total square meter [m^2] (Reported)
Socio-economic	– Performance at work [16] – Sick leaves [17] – Hygro-thermal quality – Indoor air quality – Lighting quality – Acoustic quality	– Indoor air temperature [°C] (Monitored) – Ventilation rate [h^{-1}] (Measured) – Indoor relative humidity [%] (Monitored) – CO_2 concentration indoor [ppm] (Monitored) – $PM_{2.5}$ concentration indoor [$\mu g/m^3$] (Monitored) – TVOC concentration indoor [$\mu g/m^3$] (Monitored) – Illuminance on desks [lux] (Monitored) – Noise indoor [dBA] (Monitored) – Occupancy [1/0] (Monitored)

[a]Referred to conditioned area. [b]QMS: quality management strategies (e.g., adoption of technical monitoring, building commissioning, etc.).

For this purpose, a set of KPIs is proposed based on their relevance in office buildings.

Computing all these KPIs on office buildings allow to monitor and control their performances against multiple criteria, whose aggregation is enabled by MAVT framework.

5 Conclusions and Future Perspectives

In this paper, a set of KPIs to measure the overall performance of office buildings is proposed, specifying the parameters needed for their quantification. Indeed, the KPIs are thought as part of a single system, represented by an evaluation framework based on the MAVT. According to the latter method, each KPI must be standardized according to so-called value functions (currently under study), representing the favourability of each KPI status measured on a scale from 0 to 1, and, later, aggregated according to specific weighting factors (i.e., weights of each domain specific KPI against the others, and per each domain against the remaining domains).

The advantage of the proposed methodological framework is that it is possible to get a single key figure of the multidimensional performance of a building based on real data, also accommodating for stakeholders' preferences by means of the systems of weights. Moreover, the final aggregated index, being a value ranging from a minimum (0) to a maximum (1), is easy-to-understand and communicate through user-friendly informative services (e.g., Dashboard). Readiness for communication and suitability for ranking alternative strategies are the main reasons for choosing an approach based on aggregation. However, the method also offers the opportunity to monitor specific indices.

Among the future developments of this research, two actions are expected. Firstly, a survey to collect stakeholder preferences on the relative importance of single KPIs and domains will be conducted. The objective is to build the weighting systems needed to apply the MAVT. Secondly, the translation of the model into a Web App is also expected. It would play as an informative system to support the design and management of buildings in achieving better performance against multiple criteria, as allowed by the MAVT.

The spreading of such integrated approaches in building performance evaluation might have implications on policy and investment choices by making non-monetary outcomes of building renovation more tangible, and also by better controlling the risks of investments through the accounting of their non-financial implications. Regarding private investments, the deployment of evaluation tools which are based on multiple domains (e.g., environmental, financial, social) is in line with the adoption of ESG (i.e., Environment, Social, Governance) criteria. ESG represents an approach to the evaluation of the sustainability of investments and of the performance of enterprises in terms of their impacts on ESG dimensions. The adopting of such multi-domain approach to the evaluation of buildings and their interventions might co-create the conditions for the alignment of the real estate portfolio to ESG principles (especially E and S).

References

1. ENEA: Report Annuale Efficienza Energetica (RAEE) (2020). https://www.efficienzaen ergetica.enea.it/component/jdownloads/?task=download.send&id=453&catid=40%20&Ite mid=101. Accessed 22 Dec 2021
2. Wargocki, P., Ten, W.D.P.: Questions concerning thermal and indoor air quality effects on the performance of office work and schoolwork. Build. Environ. **112**, 359–366 (2017)
3. BPIE: How to integrate indoor environmental quality within national long-term renovation strategies. https://www.bpie.eu/publication/policy-paper-how-to-integrate-indoor-environme ntal-quality-within-national-long-term-renovation-strategies/. Accessed 22 Dec 2021

4. Erdmann, C.A., Steiner, K.C., Apte, M.G.: Indoor Carbon Dioxide Concentrations and Sick Building Syndrome Symptoms in the Base Study Revisited: Analyses of the 100-Building Dataset. Lawrence Berkeley National Laboratory, Berkeley (2002)
5. Mendell, M.J., Mirer, A.G.: Indoor thermal factors and symptoms in office workers: findings from the US EPA BASE study. Indoor Air **2009**(19), 291–302 (2009)
6. Fisk, W.J., Mirer, A.G., Mendell, M.J.: Quantitative relationship of sick building syndrome symptoms with ventilation rates. Indoor Air **19**, 159–165 (2009)
7. Wang, B., Takigawa, T., Yamasaki, Y., Sakano, N., Wang, D., Ogino, K.: Symptom definitions for SBS (sick building syndrome) in residential dwellings. Int. J. Hyg. Environ. Health **211**, 114–120 (2008)
8. Fisk, W.J., Black, D., Brunner, G.: Benefits and costs of improved IEQ in U.S. offices. Indoor Air **21**, 357–367 (2011)
9. Ho, A.M.Y., Lai, J.H.K., Chiu, B.W.Y.: Key performance indicators for holistic evaluation of building retrofits: systematic literature review and focus group study. J. Build. Eng. **43**, 102926 (2021)
10. H2020 MOBISTYLE project. https://www.mobistyle-project.eu/en/mobistyle/project. Accessed 22 Dec 2021
11. Keeney, R., Raiffa, H.: Decisions with Multiple Objectives: Preferences and Value Trade-offs. Wiley, New York (1976)
12. Ferretti, V., Bottero, M., Mondini, G.: Decision making and cultural heritage: an application of the Multi-Attribute Value Theory for the reuse of historical buildings. J. Cult. Herit. **15**(6), 644–655 (2014)
13. Belton, V., Stewart, T.J.: Multiple Criteria Decision Analysis: An Integrated Approach. Kluwer Academic Press, Boston (2002)
14. Montibeller, G., Franco, A.: Decision and risk analysis for the evaluation of strategic options. In: O'brien, F.A., Dyson, R.G. (eds.) Supporting Strategy: Frameworks, Methods and Models. Wiley, Chichester (2007)
15. Beinat, E.: Value Functions for Environmental Management. Kluwer Academic Publishers, Dordrecht (1997)
16. Seppänen, O., Fisk, W.J., Lei, Q.: Effect of Temperature on Task Performance in Office Environment. Ernest Orlando Lawrence Berkley National Laboratory, Berkley (2006)
17. Fisk, W.J., Seppänen, O., Faulkner, D., Huang, J.: Economic Benefits of an Economizer System: Energy Savings and Reduced Sick Leave. Ernest Orlando Lawrence Berkley National Laboratory, Berkley (2004)

Evaluating Positive Energy Districts: A Literature Review

Tiziana Binda[1,2(✉)], Marta Bottero[1] ⓘ, and Adriano Bisello[2]

[1] Interuniversity Department of Regional and Urban Studies and Planning,
Politecnico di Torino, Viale Mattioli 39, 10125 Turin, Italy
{tiziana.binda,marta.bottero}@polito.it
[2] EURAC Research, Viale Druso, 1, 39100 Bolzano, Italy
adriano.bisello@eurac.edu

Abstract. To achieve the climate goals in the Paris Agreement and clean energy transition, positive energy districts must be promoted. A positive energy district is focused on increasing the efficiency of the buildings within it, using the renewable energy it produces, favouring electric and hybrid cars, and storing all the energy produced, in order to make clean energy for the whole city. Positive energy is a concept that takes into account not only the energy aspect, but also the environmental, social, and economic sphere. In order to be effective, this transformation requires the intervention of the community and the local decision-makers. The aim of the paper is to investigate the scientific literature, through the scientific dataset SCOPUS, in order to develop an evaluation framework for energy transition to support the decision-makers. Since the positive energy district is a recent paradigm, the investigation is extended to consider energy fields and takes into account different levels of urban scale. Specific keywords are used in order to find different economic methods in the literature, which can be used to support positive energy transition.

Keywords: Energy districts · Positive energy district · Economic evaluation · Economic valuation · Economic assessment · Economic analysis

1 Introduction

The concept of a positive energy district (PED) is becoming a possible solution for clean energy and meeting the climate goals set by Agenda 2030 and the Paris Agreement [1–3]. Given that 70% of people are expected to be living in cities by 2050, it is fundamental to find a sustainable solution at the urban level in order to create a healthy habitat for citizens. This concept is underlined by the United Nations in the definition of the 17 sustainable development goals (SDGs) within the 2030 Agenda, in particular by SDG 11, which aims to make cities inclusive, safe, resilient and sustainable [2].

In the last few years PEDs have received much attention as a possible solution for the global situation [3–5]. A PED could be defined as an energy-efficient and energy-flexible urban area aimed at creating a surplus of clean energy for the city by using renewable energy, producing an annual net-zero energy import and net-zero CO_2 emissions in a certain time frame [3].

F. Calabrò et al. (Eds.): NMP 2022, LNNS 482, pp. 1762–1770, 2022.
https://doi.org/10.1007/978-3-031-06825-6_170

To increase PEDs it is important to involve the local decision-makers and the community. In order to support this complex decision, evaluation methodology can help to quantify the benefit derived from the application of positive energy districts. Until now, the main evaluation methods used to monetize the environmental impact in the field of energy decision-making problems have been life cycle cost (LCC) [6], and cost-benefit analysis (CBA) [7]. In recent years, the analysis of multi-criteria decisions (MCDA) [8] has also been used to support decisions-makers. MCDA has the aim to involve decisions-makers in focusing not only on the energy aspect but also taking a wide range of perspectives into account. There are already various evaluation methods in use for PEDs, but to support the decision-maker more of them need to be identified in the literature. The aim of this research is to conduct a literature review to find more evaluation methods, in this way the concept of the PED is not confined to academic areas but can be put into practice. The literature review was carried out by using the scientific database SCOPUS.

The concept of the PED is new in literature. In fact, the period of publication regarding this concept only started in 2018 and by the end of 2021 there were fewer than 100 documents which mentioned it. In detail, the economic evaluation methods applied for PEDs are life cycle costs (LCC) [9, 10], cost-benefit analysis (CBA) [11], multi-criteria decision analysis (MCDA) [12, 13], and life cycle assessment (LCA) [14, 15]. More recently, also other approaches have been considered, such as the estimation of environmental and social impacts or the sensitivity analysis.

The rest of the paper is structured as follows: the Sect. 2 regards the research methodology, aiming to show the steps used for the literature review analysis. The third part is dedicated to the results, the outcome of the analysis. The last part is the conclusion, summarizing the overall research and pointing out a future perspective on this topic.

2   Research Methodology

The literature bibliography analysis was conducted by using the SCOPUS database.

All the periods present in the literature, from 1975 to 2021, were taken into consideration. The analysis was conducted by using a multistep research approach (see Fig. 1). In detail, it was developed in three steps. The first step was the general analysis, using the keywords *("economic evaluation" OR "economic valuation" OR "economic assessment") AND ("energy")*. *"All fields"* gave a result equal to 85'043, instead of with the selection *"title, abstract, keyword"* 7'633 results.

Starting from the general analysis, the second step was to reduce the research area by adding specific words and creating two groups. For the first analysis, the words related the scale of application were added, creating the group B; in detail:

- *(("economic evaluation" OR "economic valuation" OR "economic assessment") AND ("energy") AND ("Urban"))* = 215 results
- *(("economic evaluation" OR "economic valuation" OR "economic assessment") AND ("energy") AND ("District"))* = 211 results
- *(("economic evaluation" OR "economic valuation" OR "economic assessment") AND ("energy") AND ("Neighborhood"))* = 14 results

Research method

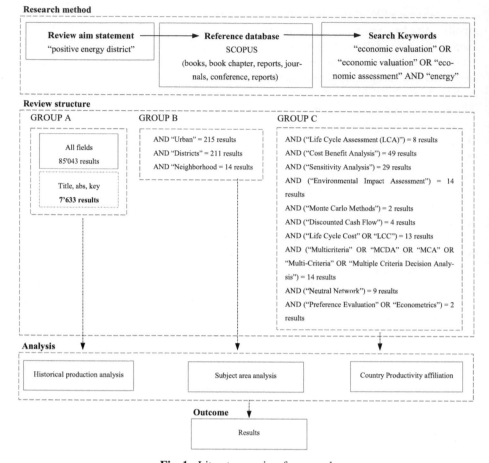

Fig. 1. Literature review framework

For the second analysis, specific keywords regarding the evaluation methods were added, creating group C; in detail:

- *AND ("Life Cycle Assessment (LCA)")* = 8 results
- *AND ("Cost Benefit Analysis")* = 49 results
- *AND ("Sensitivity Analysis")* = 29 results
- *AND ("Environmental Impact Assessment")* = 14 results
- *AND ("Monte Carlo Methods")* = 2 results
- *AND ("Discounted Cash Flow")* = 4 results
- *AND ("Life Cycle Cost" OR "LCC")* = 13 results
- *AND ("Multicriteria" OR "MCDA" OR "MCA" OR "Multi-Criteria" OR "Multiple Criteria Decision Analysis")* = 12 results
- *AND ("SROI" OR "Social Return on Investment")* = 0 results
- *AND ("Preference Evaluation" OR "Econometrics")* = 2 results
- *AND ("Quantitative Analysis")* = 4 results

The last step was the analysis of the groups created. Specifically, the analysis was conducted in: *Historical production, Subject area analysis* and *Country productivity affiliation*. The *historical production* allowed the production activity regarding the research analysis created during the time to be understood. The second analysis, *country productivity affiliation,* showed the geographical area where the documents were published. The last analysis, *subject area analysis,* according to the SCOPUS database, show the area of publication, for example, energy, environmental sciences, and social area.

3   Results and Discussions

3.1   Group A: Analysis About Search Fields

The first analysis was the general research in the energy area in the framework of economic methods. The outcome shows a total of 85′043 documents for *"all field"*, and 7′633 documents for the limited research to *"title, abstract, keywords"*. The documents start to be published in 1951 (see Fig. 2). The most intense period of production for the area *"all fields"* is between the 2008 and 2021, with 77′171 documents, equal to 91%. In particular, the last four years produced 34,196 documents, equal to 40% of the total production.

For the area referred to *"title, abstract and keywords"*, the most intense periods of production were the last 15 years, with a production of 6′040 documents, equal to 79% of total. Especially, the last four years have produced 2′349 documents, which represent the 31% of total. Figure 2 reports the *historical production analysis* comparing the two areas, *"all fields"* and the limited area *"title, abstract, and keywords"*.

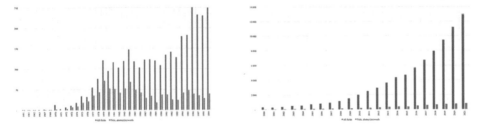

Fig. 2. *Historical production* for *"all fields"* and research limited *"title, abs and key"*.

The second analysis regards the examination of the *subject areas* (see Fig. 3). The database SCOPUS identify 27 subject areas for the field *"all fields"* and the limited research *"title, abstract and keywords"*. In particular, for both the fields the main subject areas are energy, environmental sciences, engineering and chemical engineering. In detail for *"all fields,"* 60% of all documents are produced in the four subject areas identified, instead of the *"title, abstract, and keywords"* research limitation 71% of documents are produced from the four-subject area.

The last analysis regards the country of production. For the research *"all fields"* it emerges that there are 160 countries which produced documents, instead of for the limit research *"title, abstract and keywords"* there are 132. From the dataset SCOPUS, the countries which have mostly produced documents are China, United States, United Kingdom and Italy.

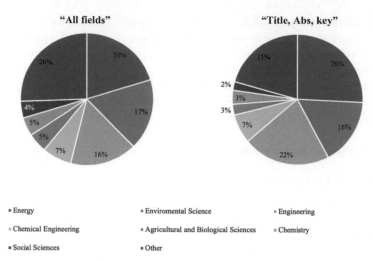

Fig. 3. *Subject area analysis* for "*all fields*" research on the left and "*title, abstract and keywords*" on the right.

3.2   Group B: Economic Method and Energy View at Territorial Scale

The second analysis focused on the territorial scale. In detail, the research was conducted in "*title, abstract, keywords*" research limitation, adding new keywords related to the territorial scale. Specifically, the words used were *urban*, *district* and *neighbourhood*. The results shows that the keyword *urban* has produced 215 documents, instead *neighborhood* has produced 14 documents. From the *historical production analysis*, the word *urban* was present since 1972, instead, the word *neighborhood* has appeared just since 2013 (see Fig. 4).

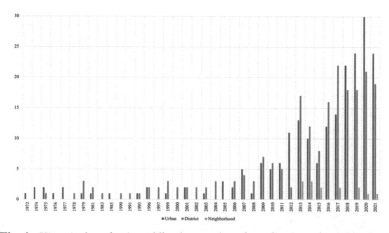

Fig. 4. *Historical production* adding keywords, *urban, district* and *neighborhood*.

For the analysis of the *subject area* specific keywords have been added related to the territorial scale. The areas with the highest number of documents were the same as the previous group in detail: *energy* with 201 documents, *engineering* with 189 and *environmental sciences* with 180.

The third analysis regards the country which produced documents. The SCOUP dataset shows that 64 countries have published documents. From the results it emerges that Italy is the country that has been producing the most part of documents. In detail, the keyword *"urban"* appears in 35 documents and for *"district"* 34 documents have been produced. Instead, Germany is the country that has produced the most part of documents with the keyword *"neighbourhood"*, 14 documents (see Fig. 5). Also, the analysis shows that countries of Northern Europe use more the keyword *"district"* such as the United Kingdom with 21 documents, Sweden with 18 documents and Germany with 12 documents. Figure 5 reports the countries which have the keywords mentioned with a minimum of 2 documents.

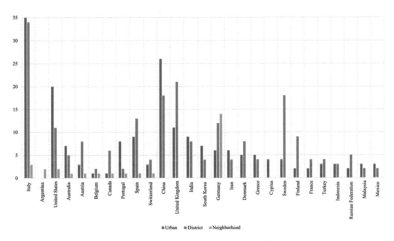

Fig. 5. Country of production focused on the keywords, *urban, district* and *neighborhood.*

3.3 Group C: Evaluation Approach

The last analysis, the most interesting, focused on the evaluation approach. The keywords added to Group B are related to different evaluation economic methods. Some words were suggested by the SCOPUS dataset, as Cost-benefit analysis (CBA) (49 documents), but others were added in order to find different evaluation economic methods, from the most common methods to monetize the energy as the Life Cycle Cost (LCC) to the methods used to estimate the social or environmental impact such as the Multicriteria analysis (MCA). Figure 6 reports all the economic methods added to the research.

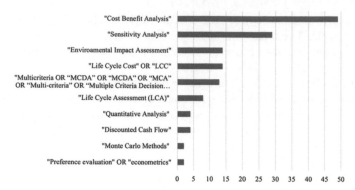

Fig. 6. Number of documents related to evaluations approaches.

The *historical analysis* shows that in 1975 the economic methods used are the preference evaluation and the costs benefits analysis. The most intense period of documents productions was in 2018. The sensitivity analysis started to be present since 2005 and it increased in the period between 2010 and 2014 (see Fig. 7). Results of the *subject area analysis* shows an important production in the fields of environmental sciences, energy and engineering, as reported in Fig. 8.

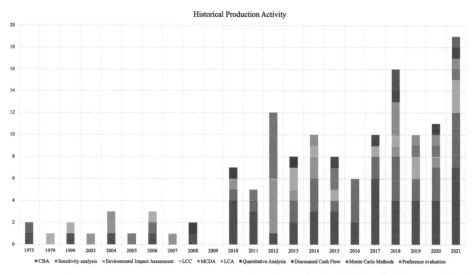

Fig. 7. Historical production activity for different evaluation approaches.

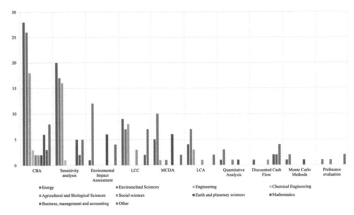

Fig. 8. *Subject areas* related to the different evaluation approaches and methods.

4 Conclusion

The present paper allowed the application of different economic methods in the field of energy at different territorial scales to be understood better, in order to support the benefits resulting from positive energy districts (PEDs).

The transformation of a city into PEDs is complex. In fact, this type of urban transformation consists in increasing the efficiency of buildings, using renewable energy, favoring the use of electric and hybrid cars and storing all the energy produced. To achieve this goal, a literature review was conducted in order to find different economic methods to support the decision making in order to develop PEDs. The PED paradigm is new, so the literature review was extended to take into consideration energy fields in the framework of economic evaluations at different urban scales.

From the analysis it emerged that the concept of urban scale is present throughout the literature dating back to 1975, whereas the district scale started to appear from 2009. Also, it is clear from the analysis that Italy and China are the countries most interested in working on and developing research on the urban and district scale. The evaluation approach analysis shows that the economic methods mostly used are Cost of Benefits Analysis (CBA), sensitivity analysis and environmental impact.

From a future perspective, it would be interesting to investigate integrated approaches, such as the application of the evaluation methods CBA and MCA together in order to increase the creation of PEDs, since they are two approaches that complement each other. In fact, CBA is based on a detailed cost study, whereas MCA takes into account a wide range of perspectives, such as the human behaviors area.

The outcome of this study could support local decision-making in order to transform urban areas into PEDs. In this way, the concept of PED will not just be related to the academic area, but it will have a practical application.

Acknowledgements. This work was developed within the context of the International Energy Agency (IEA) Energy in buildings and construction (EBC) Annex 83 workgroup "Positive Energy Districts". This research has received funding from the European Union's Horizon 2020 research and innovation programme under grant agreement No. 101036723.

References

1. S&P Global Sustainable Homepage. https://www.spglobal.com/esg/about/index. Accessed 22 Dec 2021
2. United Nations Department of Economic and Social Affairs: The sustainable development goals repost 2016. United Nations Department of Economic and Social Affairs, New York, NY, USA (2016)
3. Bottero, M., Dell'Anna, F., Morgese, V.: Evaluation the transition towards post-carbon cities: a literature review. Sustainability **13**, 567 (2021)
4. Brozovsky, J., Gustavsen, A., Gaitani, N.: Zero emission neighborhoods and positive energy districts-A state-of-the-art review. Sustain. Cities Soc. **72**, 103013 (2021)
5. Guarino, F., et al.: State of the art on sustainability assessments of positive energy districts: methodologies, indicators and future perspectives. Smart Innov. Syst. Technol. **263**, 479–492 (2022)
6. Cellura, M., Fichera, A., Guarino, F., Volpe, R.: Sustainable development goals and performance measurement of positive energy district: a methodological approach. Smart Innov. Syst. Technol. **263**, 519–527 (2022)
7. International Organization for Standardization: Building and Constructure Assets, Service-life Planning, Part 5: Life Cycle Costing. ISO 15686:2008. International Organization for Standardization, Geneva, Switzerland (2008)
8. European Commission: Guide to Cost-benefit Analysis of Investment Projects: Economic appraisal tool for cohesion policy 2014–2020. European Commission, Brussels, Belgium (2014)
9. Strantzali, E., Aravossis, K.: Decision making in renewable energy investments: a review. Renew. Sustain. Energy Rev. **55**, 885–898 (2016)
10. Salom, J., et al.: An evaluation framework for sustainable plus energy neighborhoods: moving beyond the traditional building energy assessment. Energies **14**(14), 4314 (2021)
11. Laitinen, A., Lindholm, O., Hasan, A., Reda, F., Hedman, A.: A techno-economic analysis of an optimal self-sufficient district. Energy Convers. Manage. **236**, 114041 (2021)
12. Kiwan, S., Venezi, L., Montagnino, F.M., Parede, F., Damseh, R., Damseh, R.: Techno-economic analysis of a concentrated solar polygeneration plant in Jordan. Jordan Journal of Mechanic and Industrial Engineering **12**(1), 1–6 (2018)
13. Alpagut, B., Romo, A.L., Hernandez, P., Tabanoglu, O., Martinez, N.H.: A GIS-based multicriteria assessment for identification of positive energy district boundary in cities. Energy **14**(22), 7517 (2021)
14. Bisello, A.: Assessing multiple benefit of housing regeneration and smart city development: the European project Sinfonia. Sustainability (Switzerland) **12**(19), 1–28 (2020)
15. Marotta, I., Guarino, F., Longo, S., Cellura, M.: Environmental sustainability approaches and positive energy district: a literature review. Sustainability (Switzerland) **13**(23), 13063 (2021)

Neighbourhood Energy Community: Norms, Actors and Policies. The Case of Pilastro-Roveri

Federica Rotondo[(✉)], Giancarlo Cotella, and Isabella M. Lami

Interuniversity Department of Regional and Urban Studies and Planning (DIST), Politecnico di Torino, Viale Mattioli, 39, 10125 Turin, Italy
federica.rotondo@polito.it

Abstract. The decarbonization of cities implies a complex transition where energy solutions are conceived as social configurations which have emerged contingently in particular institutional and spatial contexts. Within this framework, neighbourhood energy communities have recently gained attention due to their multi-agent character and the deriving governance implications. Their study calls for a renovated attention to public policies and spatial governance arrangements, to better analyse the peculiar characteristics of energy communities' processes on specific territories. Acknowledging the above, the chapter explores the main features and governance practices that characterise the experience of Pilastro-Roveri energy community in the city of Bologna. More specifically, it looks into: (i) the spatial conditions that led to focus the attention on a specific area and selected project proposal; (ii) the actors involved in decision-making and their relations at different stages of the process (iii) the implications for urban governance and policies. The analysis draws on a mixed methodology that combines documental review, semi-structured interviews with privileged actors and on-site observations. Overall, the collected evidence contributes to a better understanding of the main levers and catalysts that trigger process innovation in shaping norms and practices of neighbourhood energy communities in the Italian context.

Keywords: Energy transition · Energy community · Urban governance · Sustainable neighbourhoods · Italy

1 Introduction

Energy communities represent an alternative model of clean energy production, consumption and distribution that is based on local energetic, environmental and social needs (De Vidovich et al. 2021). Differently from the concept of *sustainable neighbourhood* (Bottero et al. 2019) this model implies the presence of innovative forms of organization that include citizens, public entities, businesses, and other actors that operate within a defined geographic area with a view to self-consumption and energy self-sufficiency (Lopez et al. Lopez et al. 2019). In the context of the energy transition and the global Sustainable Development Goals[1] (SDG 7 and SDG 11), the role of energy communities is recognized for the environmental, economic and social benefits addressed to the

---

[1] Defined in the framework of 2030 Agenda for Sustainable Development (United Nations General Assembly 2017).

F. Calabrò et al. (Eds.): NMP 2022, LNNS 482, pp. 1771–1779, 2022.
https://doi.org/10.1007/978-3-031-06825-6_171

urban and territorial contexts within which they are located (De Vidovich et al. 2021). In particular, the SDG Target 11a underlines the role that spatial governance and planning can play in the process (Berisha et al. 2022).

Aiming at shedding light on the above, the chapter investigates the role of energy communities from the urban governance and spatial policies perspective (Sonetti et al. 2020; Valkenburg and Cotella 2016). To this aim the authors consider the energy community of Pilastro and Roveri neighborhoods (Bologna, Italy) to retrospectively and diachronically analyse its development against the background of the more recent European and national legislative framework (Heldeweg and Saintier 2020). More in detail, the research focuses on three main dimensions: (i) the spatial conditions that lead to focus the attention on a specific area and selected project proposal; (ii) the actors involved in the decision-making process and their relations at different stages of the latter (iii) the implications for urban governance and policies. The analysis draws on a mixed methodology that combines documental review, semi-structured interviews with privileged actors, and on-site observations.

After this brief introduction, the following section sketches out the main coordinates concerning the normative framework concerning energy transition and energy communities in the European and Italian context; secondly, the selected case study is presented in the light of the main research question animating the research. Finally, a conclusive section rounds off the contribution, highlighting a number of relevant aspects related to neighbourhood energy community practices and regulations in the national and European context. Overall, the collected evidence contributes to a better understanding of the main levers and catalysts that trigger process innovation in shaping norms and practices of neighbourhood energy communities in the Italian context.

2   Energy Communities Between Norms and Practices

The liberalization of the energy market pushes towards the incremental decentralization of energy production and distribution networks (Agazzi in Ruggieri and Acanfora 2021). In this context, the European Union (EU) promotes new and renovated organisation arrangements that imply the active role of citizens in the production, consumption, and exchange of clean energy towards forms of collaboration and self-consumption (Cotella et al. 2016a)[2]. This issue is particularly relevant in spatial governance and urban policies because it widens the public arena of public decision-making and opens up possibilities for new alliances among various interests and actors (Rotondo et al. 2022; Lami and Mecca 2020; Valkenburg and Cotella 2016).

The European Clean Energy Package (CEP), adopted in 2018, includes the directives 2018/2001/EU (Renewable Energy Directive - RED II) and 2019/944/EU (Internal Electricity Market Directive - IEM) that together represent the regulatory reference framework at the European level to enhance energy transition and promote energy communities. More specifically, the European directives promote different forms of energy communities (e.g. the Renewable Energy Community-REC and the Citizen Energy Community-CEC), meant as legal entities based on voluntary and open participation

---

[2] For a more detailed overview of the implications of the EU for the evolution of domestic territorial governace and planning see: Cotella (2020), Cotella and Dabrowski (2021).

and composed by shareholders that are individuals, public entities, and small enterprises operating on a non-profit basis. In addition to this, energy facilities are owned by the community and located in a specific area (Cunha et al., 2021). In Italy, the Milleproroghe Decree-Law[3] extended the timeframe for the transposition of the European directives t to allow for a longer experimentation phase (Zulianello et al., 2020). While the transposition was eventually concluded at the end of 2021[4], in the last two years the legislation concerning energy community legislation has further evolved at the national level as a consequence of the recent efforts put in place to face the ongoing pandemic and its impact on urban areas[5] (e.g., the Italian National Recovery and Resilience Plan[6]). Accordingly, the most recent legislation allows for an increase in the power limit of plants eligible for incentive mechanisms, in addition to an enlargement of the spectrum of potential subjects eligible to become community members.

Drawing on the introduced legislative framework, a number of energy communities have recently developed and consolidated in Italy, altogether contributing to framing the transition toward net-zero carbon society as a more inclusive and spatially-sensitive process (Agazzi in Ruggieri and Acanfora 2021). These experiments of Renewable Energy Community (REC) have emerged throughout the country at different scales (from the neighbourhood to the metropolitan city), and called into play different governance levels (municipal, regional, state) (De Vidovich et al. 2021; Zulianello et al. 2020). Among them, the chapter focuses on REC concerning the neighbourhood level and, more in detail, on their spatial governance and planning implications. It does so by illustrating the experience of the Pilastro-Roveri energy community; this case was chosen for several reasons: (i) it is one of the first experiments of neighbourhood energy community in Italy; (ii) it involves a variety of public, private and third sector actors acting on multiple urban governance levels; (iii) its replicability has been discussed on a national and international level.

3 Pilastro-Roveri Energy Community

Since the beginning of the 2000s, the city of Bologna elaborated urban plans and policy programs in the direction of promoting energy transition and contrasting climate change[7]. In more recent times, the new General Urban Plan (Italian acronym: PUG) pays particular attention to the promotion of energy transition and circular economy processes and identifies eight neighbourhood in the city as priority areas of intervention. In synergy

---

[3] The decree-law n.187 of 30.12.2020 is available at: https://www.gazzettaufficiale.it/eli/id/2020/12/31/20G00206/sg [accessed on December 2021].

[4] The decree-law n.199 of 08.11.2021 is available at https://www.gazzettaufficiale.it/atto/serie_generale/caricaDettaglioAtto/originario?atto.dataPubblicazioneGazzetta=2021-11-30&atto.codiceRedazionale=21G00214&elenco30giorni=true [accessed on February 2022].

[5] For a more detailed examination of the impaction that the COVID-19 pandemic may have on cities see: Cotella and Vitale Brovarone (2020, 2021).

[6] An English version is available here: https://www.mise.gov.it/images/stories/documenti/it_final_necp_main_en.pdf [accessed on December 2021].

[7] With reference to: Municipal Energetic Program PEC (2007), the Sustainable Energy Action Plan (2012) and the adapting climate change plan BLUEAP (2015).

with the PUG, the last Sustainable Energy and Climate Action Plan (SECAP) introduced the theme of renewable energy communities at the neighbourhood scale also through the Green Energy Community (GECO) pilot project.

The GECO project[8], funded by the European partnerships EIT Climate-KIC, has been launched in 2019 to implement the first neighbourhood energy community in the city of Bologna (Italy) and is still ongoing. The European project supports the role or prosumers in the community production, distribution and self-consumption of renewable energy by explicitly focusing on the involvement of residents, businesses, and enterprises and on the integration of social, environmental and economic dimensions.

The pilot project proposes to implement a Renewable Energy Community (REC) within the existing neighbourhoods of Pilastro and Roveri, located in the north-eastern part of Bologna (Fig. 1). The project is promoted by the Energy and Sustainable Development Agency (Italian acronym: AESS)[9], the Italian National Agency for New Technologies, Energy and Sustainable Economic Development (Italian acronym: ENEA), and the University of Bologna (UNIBO) and it is co-funded by the project promoters (50%) and the European partnership EIT Climate-KIC (50%). The three-year project started in September 2019 and it has been delayed due to pandemic conditions and the evolution of the legislative national framework on energy communities.

Fig. 1. Bologna metropolitan context and the neighbourhoods of Pilastro and Roveri (authors' elaboration)

---

[8] https://www.gecocommunity.it/ [accessed on December 2021].

[9] AESS is a non-profit organization involved in the promotion of renewable energy sources, energy efficiency and reduction of energy consumption among Local Authorities, SMEs, schools and consumers.

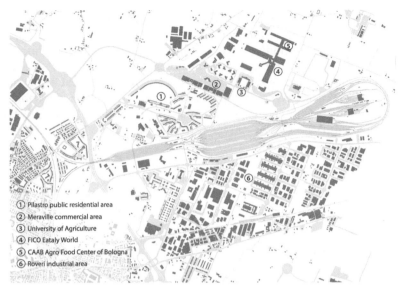

① Pilastro public residential area
② Meraville commercial area
③ University of Agriculture
④ FICO Eataly World
⑤ CAAB Agro Food Center of Bologna
⑥ Roveri industrial area

Fig. 2. Zoom on the neighbourhoods of Pilastro and Roveri and its urban fabric (authors' elaboration)

3.1 Spatial Conditions and Project Proposal

The districts of Pilastro and Roveri are located in the north-eastern part of the city of Bologna, marked by the presence of the rail yard of San Donato, the ring road to the west, and an agricultural area to the east. North to the rail yard of San Donato there is a mixed urban fabric composed of the public residential area of Pilastro, the commercial area of Meraville, the University of Agriculture, FICO Eataly Word and the Agro Food Center of Bologna (CAAB). South to the rail yard of San Donato there is the Roveri industrial area (Fig. 2).

Taking into account the spatial conditions that lead to focus the attention on the neighbourhoods of Pilastro and Roveri, it is important to consider: (i) the presence of a mixed urban fabric composed by residential, commercial, and production areas; (ii) the stratification over time of energy and urban regeneration projects targeting the area of interest; (iii) the presence of an already existing energy infrastructure, that was considered as a potential asset upon which the GECO project could build. The energy and urban regeneration projects previously proposed in the districts of Pilastro and Roveri include the energy renovation and promotion of sustainable mobility services targeting Pilastro public residential area, the energy renovation and the photovoltaic plant installation in the Meraville commercial area, the promotion of sustainable mobility and electric charging for the university area and the energy renovation targeting Roveri industrial area (Cunha et al. 2021; Sustainable Energy and Climate Action Plan 2012).

In 2018, when the GECO project is launched, the districts of Pilastro and Roveri stand out for the presence of an already existing photovoltaic system located in the proximity on the rooftop of the CAAB building and of a district heating network located in the Pilastro residential area. In this direction, the GECO project intends to integrate the existing energy infrastructure with new photovoltaic systems to be located on the rooftops of both public and private buildings. More specifically the project proposal mainly includes: (i) the integration of the existing photovoltaic system on the rooftop of CAAB building (ii) the implementation of a new biogas plant for organic waste disposal and (iii) a new photovoltaic system on the rooftops of residential public buildings. In addition to this, the energy community project aims to use part of the renewable energy generated by the new and integrated photovoltaic systems to the benefit of both the companies located in the CAAB surroundings and the people living in the Pilastro public residential area.

The above-mentioned spatial conditions are considered favourable to the development and consolidation of a neighbourhood energy community due to the possibility to partially benefit from an existing clean energy production system, as well as to produce clean energy on-site also to the benefit of surrounding residential areas.

3.2 Actors and Relations

When it comes to the actors involved in the decision-making process and their relations the main elements of discussions are: (i) the leading role of AESS and ENEA throughout the entire process as actors that were previously involved in the past energy projects for the neighbourhoods of both Pilastro and Roveri and (ii) the involvement of the municipality of Bologna and some local actors already active in the territory (e.g. the CAAB centre located in the northern part of the neighbourhood or the former Agency for Local Development of Pilastro located in the public residential area).

Not only because they are promoters of the GECO project; AESS and ANEA assume a key role in support and coordination among different actors and interests. They both could benefit from certain familiarity with the local resources and urban features of Pilastro and Roveri neighbourhoods thanks to their previous experience at the occasion of projects and initiatives dealing with smart and energy efficiency issues in 2018[10]. Furthermore, they both organize focus groups to discuss the integration of energy community European directives in the Italian context together with national actors.

From the very beginning of the process implementation, the project promoters favour the engagement of local actors in combination with the promotion of education activities by involving the Municipality of Bologna, the residents, and the local agencies and citizens' associations based in the Pilastro public residential district. More specifically, the Municipality of Bologna acts as a facilitator of an open discussion table involving the representatives of the different associations operating in the Pilastro public residential district. In addition to this, the Agency for Local Development of Pilastro has a primary role in connecting the different initiatives in the area and promoting opportunities for dialogue and exchange with the residents. Among the local citizens' associations, the so-called "neighbourhood house" organizes education and awareness-raising activities targeting children and young people (such as urban trekking or readings on the theme of sustainability and energy).

---

[10] In 2018 AESS is involved in the Neighborhood Economics project, funded by EIT Climate-KIC that concerns the Pilastro public residential area. In the same year, ENEA is involved in the Roveri Smart Village project that concerns the Roveri industrial area.

At the moment, the specific business model of the GECO project and the composition of the Pilastro-Roveri energy community are still under definition, also as a consequence of the delays that occurred in the preliminary phase of feasibility checks and the recent final transposition of the EU directive into the national legislation. In this context, the network of involved actors and their relations are considered particularly important for the functioning of the neighbourhood energy community, in particular in relation to the creation of new and renovated alliances between actors belonging to different spheres as well as to favour the development of shared strategies through which to spatially channel the available resources on the area.

3.3   Urban Policies and Governance Implications

The selected case of neighbourhood energy community emphasizes urban policy and governance implications related to (i) the replicability of the pilot project on a wider scale and (ii) the reshaping of the national regulatory framework for promoting energy communities.

The pilot project of the Pilastro-Roveri energy community opens to prospects of replicability in other peripheral contexts of the city of Bologna thanks to the numerous information and facilitation activities organized by the Municipality itself (Bologna 2021). More specifically, the specific pilot project could potentially be replicated in other neighbourhoods characterized by similar features such as a mixed urban fabric in terms of residential, commercial and industrial areas, the presence of existing photovoltaic plants and other energy infrastructures and the presence of a consolidated network of local associations.

In addition to this, the pilot project enriches the Italian panorama of the first Renewable Energy Community (REC) practices offering some insights for the Italian regulatory framework still (De Vidovich et al. 2021). In fact, according to the previous regulatory framework, existing photovoltaic plants could not be included in the creation of the energy community and, at the same time, associations could not be recognized as members of the energy community. The reshaping of the current national regulatory framework took into consideration ongoing pilot projects of neighbourhood energy communities in existing contexts and push to the active involvement of different local actors already present on the territory.

4   Conclusion and Future Perspectives

The chapter presents and discusses the case of an Italian energy community (Pilastro-Roveri), to highlights its contextual characteristics and reflect on what conditions might favour or inhibit the replicability of energy community creation at the neighbourhood level. It does so also in the light of the evolving European legislative framework on the matter, and how the latter has been transposed in the Italian context (Vernay and Sebi 2020). More in detail, the investigation provided a number of evidence concerning the spatial conditions that favoured the development of the discussed project proposal, the actors involved in the decision-making process and the implications for urban governance and policies.

First of all, recent practices place particular emphasis on neighbourhood energy communities in metropolitan contexts, thus initiating some first experimentation at the local communitarian level with a view to subsequent replicability on a wider scale in the metropolitan area (Cotella et al. 2016b). Secondly, the selected case highlights the importance of taking into account the existing contexts by initiating a recon phase of past and ongoing projects and local resources already existing in the selected area to integrate strategies towards the production of clean energy together with the reduction of energy consumption (Rotondo et al. 2020). Finally, the ongoing processes in the city of Bologna emphasizes the importance of fostering sharing processes with local actors and with the inhabitants also through initiatives promoted by the municipality in the direction of stimulating co-design paths for community management of a common resource.

Overall, the collected evidence contributes to a better understanding of the main levers and catalysts that trigger process innovation in shaping norms and practices of neighbourhood energy communities in the Italian context. At the same time, it outlines a number of avenues for future research in the field of energy communities' practices and regulations in Italy and in Europe, that may help and understand in more detail the context- and path-dependence of these initiatives.

References

Agazzi, D.: La transizione è un processo politico. In: Ruggieri, G., Acanfora, M. (eds.) Che cos'è la transizione ecologica clima, ambiente, disuguaglianze sociali: Per un cambiamento autentico e radicale. Altreconomia (2021)

Cunha, F.B.F, Carani, C., Nucci, C.A., Castro, C., Santana Silva, M., Andrade Torres, E.: Transitioning to a low carbon society through energy communities: lessons learned from Brazil and Italy. Energy Res. Soc. Sci. **75**, 101994 (2021). https://doi.org/10.1016/j.erss.2021.101994

Berisha, E., Caprioli, C., Cotella, G.: Unpacking SDG target 11. a: what is it about and how to measure its progress? City Environ. Interact. 100080 (2022). https://doi.org/10.1016/j.cacint.2022.100080

Bottero, M., Caprioli, C., Cotella, G., Santangelo, M.: Sustainable cities: a reflection on potentialities and limits based on existing eco-districts in Europe. Sustainability **11**(20), 5794 (2019). https://doi.org/10.3390/su11205794

Cotella, G.: How Europe hits home? The impact of European Union policies on territorial governance and spatial planning. Géocarrefour 94/3 (2020). https://doi.org/10.4000/geocarrefour.15648

Cotella, G., Crivello, S., Karatayev, M.: European Union energy policy evolutionary patterns. In: Lombardi, P., Gruenig, M. (eds.) Low-Carbon Energy Security from a European Perspective, pp. 13–42. Elsevier, Amsterdam (2016a). https://doi.org/10.1016/B978-0-12-802970-1.00002-4

Cotella, G., Janin Rivolin, U., Santangelo, M.: Transferring 'good' territorial governance across Europe: opportunities and barriers. In: Schmitt, P., Van Well, L. (eds.) Territorial Governance Across Europe: Pathways, Practices and Prospects. Routledge, New York (2016b)

Cotella, G., Dabrowski M.K.: EU cohesion policy as a driver of europeanisation: a comparative analysis. In: Rauhut, D., Sielker, F., Humer, A. (eds.) EU Cohesion Policy and Spatial Governance. Edward Elgar Publishing (2021)

De Vidovich, L., Tricarico, L., Zulaniello, M.: Community energy map. Una ricognizionedelle prime esperienze di comunità energetiche rinnovabili. Franco Angeli Editore (2021). https://francoangeli.it/Ricerca/scheda_libro.aspx?id=27647

General Urban Plan: Comune di Bologna (2021). http://dru.iperbole.bologna.it/pianificazione?fil ter=Piano%20Urbanistico%20Generale%20(PUG). Accessed Dec 2021

Heldeweg, M., Saintier, S.: Renewable energy communities as 'socio-legal institutions': a normative frame for energy decentralization? Renew. Sustain. Energy Rev. **119**(April 2019) (2020). https://doi.org/10.1016/j.rser.2019.109518

Lami, I.M., Mecca, B.: Assessing social sustainability for achieving sustainable architecture. Sustainability **13**, 142 (2020). https://doi.org/10.3390/su13010142

Lopez, F., Pellegrino, M., Coutard, O.: Local Energy Autonomy. ISTE Ltd./Wiley, New York (2019). ISBN 978-1-786-30144-4

Rotondo, F., Abastante, F., Cotella, G., Lami, I.M.: Questioning low-carbon transition governance: a comparative analysis of European case studies. Sustainability **12**(24), 10460 (2020). https://doi.org/10.3390/su122410460

Rotondo, F., Abastante, F., Cotella, G., Lami, I.M.: Investigating "sustainable neighborhoods" in the Italian context: a diachronic approach. In: Abastante, F., et al. (eds.) Urban Regeneration Through Valuation Systems for Innovation. Green Energy and Technology. Springer, Heidelberg (2022, forthcoming)

Sonetti, G., Arrobbio, O., Lombardi, P., Lami, I.M., Monaci, S.: «"Only Social Scientists Laughed"»: reflections on social sciences and humanities integration in European energy projects. Energy Res. Soc. Sci. **61**, 101342 (2020)

Sustainable Energy and Action Plan: Comune di Bologna (2012). http://www.comune.bologna.it/media/files/paes_12maggio2012_approvato_1.pdf. Accessed Dec 2021

Sustainable Energy and Climate Action Plan: Comune di Bologna (2021). http://www.comune.bologna.it/media/files/piano_azione_per_energia_sostenibile_e_clima_paesc_2.pdf. Accessed Dec 2021

United Nations General Assembly: Global Indicator Framework for the Sustainable Development Goals and Targets of the 2030 Agenda for Sustainable Development. United Nation, New York (2017)

Vernay, A.L., Sebi, C.: Energy communities and their ecosystems: a comparison of France and the Netherlands. Technol. Forecast. Soc. Chang. **158**(May), 120123 (2020). https://doi.org/10.1016/j.techfore.2020.120123

Valkenburg, G., Cotella, G.: Governance of energy transitions: about inclusion and closure in complex sociotechnical problems. Energy, Sustain. Soc. **6**(1), 1–11 (2016). https://doi.org/10.1186/s13705-016-0086-8

Zulianello, M., Angelucci, V., Moneta, D.: Energy community and collective self consumption in Italy. In UPEC 2020 - 2020 55th International Universities Power Engineering Conference, Proceedings (2020). https://doi.org/10.1109/UPEC49904.2020.9209893

Dimensions of Social Acceptance in Energy Transition

Paolo Bragolusi$^{(\boxtimes)}$ and Maria Stella Righettini

Interdepartmental Centre Giorgio Levi Cases, University of Padova, Francesco Marzolo Street, 9, 35131 Padova, Italy
paolo.bragolusi@unipd.it

Abstract. The European Union fixed medium and long-term targets to reduce GHG emissions up to 80–95% compared to 1990 levels by 2050. The governments' ambitious goal to increase the share of renewable energy fosters technological innovation. Nevertheless, the innovation results could be jeopardized by the shortage of and the lack of local communities' social acceptance towards renewable energy plants (REPs). The lack of social acceptance of REPs may be a significant barrier to success in energy transition policies. Therefore, it is paramount to analyze social acceptance dimensions to avoid energy policy implementation gaps. The present paper investigates in deep such dimensions concerning REPs to fill this gap. We first investigate the distinction between acceptance and acceptability definitions; then, we analyze the dimensions of social acceptance, i.e., socio-political acceptance, community acceptance, and market acceptance. We finally propose a new analytical framework to map social acceptance in concrete case studies empirically.

Keywords: Social acceptance · Acceptability · Energy transition · Renewable energy plants · Trust

1 Introduction

In 2019, the European Union adopted the Clean Energy package to promote in member states energy transition policies, substitute fossil fuels, and reduce Greenhouse Gas (GHG) emissions. Contextually, European Union also has fixed medium and long-term targets, i.e., 2030 Climate and Energy Framework and 2050 Long Term Strategy. Member states have to reduce GHG emissions by up to 80–95% concerning 1990 levels by 2050. To reach these targets, governments and energy operators need to adopt and implement renewable energies policies by installing REPs. In the local context, the social acceptance of REPs plays a central role in achieving climate protection goals, and, fundamentally, local communities and residents support them. Thus, the non-acceptance of REPs becomes the most significant barrier toward achieving renewable energy targets [1] as many developments of REPs have been hampered by non-acceptance phenomena. Social acceptance has a multidimensional character [2], and various factors may contribute to shaping it [1]. In this respect, there are few studies in literature focused on

© The Author(s), under exclusive license to Springer Nature Switzerland AG 2022
F. Calabrò et al. (Eds.): NMP 2022, LNNS 482, pp. 1780–1789, 2022.
https://doi.org/10.1007/978-3-031-06825-6_172

social acceptance dimensions related to REPs, and it is paramount to fill this gap. In this contribution, we conducted a literature review to deepen the main dimensions of social acceptance defined as follows: socio-political acceptance, community acceptance, and market acceptance [2]. The social acceptance framework we derive from literature can be helpful to provide policymakers with a map of critical points to be monitored and may support REP policy design in anticipating and preventing oppositions and conflicts.

2 Acceptance and Acceptability Definitions

In literature, authors provided a different definition of acceptance related to a technology or a policy. A general and simple definition is that acceptance may be defined as an active or passive approval of a given technology/product or policy [1, 3, 4]. In technology use field, ref. [4] argues that acceptance is a subjective measure related to people's willingness to accept a facility independently of rational judgments. Ref. [5] argues that acceptance is mainly a descriptive notion and refers to a specific condition in which users, other levels of governments involved, the general public, or formal and informal stakeholders express their endorsement to a given technology. Ref. [6] defines social acceptance as the appropriation of a given technology by a social body that is defined in a proper space (the community) and politically established. In this context, the social body is mobilized to address the challenge of energy issues (e.g., air pollution, scarcity of energy resources, etc.). Some authors [3, 6, 7] have defined four levels of acceptance and non-acceptance depending on people's attitudes and actions: passive acceptance i.e., "approval"; active acceptance, i.e., "support"; non-passive acceptance, i.e., "rejection"; active non-acceptance, i.e., "resistance". Regarding acceptability, ref. [4] argues that it concerns the judgments of experts regarding the construction of a particular facility by assessing whether the construction of that facility is reasonable considering quantifiable criteria (e.g., the impact on health, noise generated, etc..). Another definition provided by authors is that acceptability is a normative notion, i.e., the expression of the assessment that something is acceptable according to a normative standard (e.g., ethical codes, moral norms, and values, some public values, or moral standards established by-laws) [5]. Refs. [8, 9] defined acceptability as a decision-making framework, i.e., a contested and dynamic process informed by collective norms (with a geographical and historical connotation) on which different evaluation processes are involved as well as struggle policies; the process is more generally recognized as legitimate. The distinction between "acceptance" and "acceptability" is essential because sometimes, the two terms are misused as synonyms [4].

3 Dimensions of Social Acceptance

In literature, authors agree with the dimensional classification of social acceptance suggested by [2]. According to this classification, social acceptance is divided into three dimensions, as follows:

- Socio-political acceptance (of technologies and policies, by the public, by stakeholder, by policymakers);

- Community acceptance (procedural justice, distributional justice, trust);
- Market acceptance (consumers, investors, intra-firm).

In this section, all these dimensions are deepened and analysed.

3.1 Socio-political Acceptance

Socio-political acceptance is defined as social acceptance at the broadest level [2]. Some authors [10, 11] argue that people support a policy depending on the perceived fairness and effectiveness of the policy itself. Ref. [12] state that there is less effort to form an attitude than acting on it once it is assumed by people, and political acceptance is necessary for its support. Ref. [13] assert that exist two mechanisms by which policy acceptance may be increased, these mechanisms concern information activities for changing people's attitudes and effective participatory engagement that may increase people's perceptions of policy legitimacy [10, 13, 14]. Socio-political acceptance is also related to other sub-dimensions of community acceptance theorized by ref. [2] i.e., procedural justice and distributive justice. In particular, socio-political acceptance may be increased by adopting decision-making processes that are perceived as fair, respectful, transparent and inclusive [15]. By focusing on social acceptance for renewable energy policies and related technologies, some surveys revealed that people are usually in favour of renewable energies but at local level, the so-called NIMBY ("Not In My Yard") phenomenon may be the main barrier for the development of REPs [2, 16]. As for socio-political acceptance of technologies, ref. [17] argue that energy infrastructures are proposed or implemented by authorities or companies and individuals do not contest or actively oppose them. However, this top-down perspective is criticized in literature as researchers argue that sustainable development should not be imposed on the community, and it cannot take place until all people are actively engaged in it [2, 17, 18]. Renewable energy technologies implementation may be properly planned by promoting and developing participatory processes on which social actors are actively engaged [13]. During years, different technology acceptance models (TAM) have been developed in order to analyse technologies that are not large-scale ones as renewable energy technologies (see ref. [19] for details); a gap of literature is thus found, and future research developments may be focused on this field.

3.2 Community Acceptance

Community acceptance refers to the specific acceptance of siting decisions and renewable energy projects by local stakeholders, in particular, residents and local authorities [2]. In this dimension of acceptance, the NIMBY phenomenon is one of the most critical issues that may involve non-acceptance of REPs. In particular, the NIMBY phenomenon means that people have a positive attitude towards something, but then they assume oppositional behaviours for personal reasons [16, 20]. According to ref. [21] this phenomenon is strongly correlated to the place where there is the intention to locate the plant (or where the plant is located) but it is characterized by having nature, spatial scale and importance which vary according to the local context as well as use and non-use economic values of the territory; the author also argues that people who have a sense of identity from particular rural landscapes are more likely to oppose against REPs [21, 22]. The non-acceptance of community may have more complex explanations than NIMBY phenomenon, in particular, it may be caused by the sentimental value that people give to places (i.e., "place attachment"), by concerns about potential loss of landscape economic value due to REPs (i.e., "landscape concerns") and lack of procedural and distributive justice [23]. Other important factors are related to perceived risks and benefits [24], personal emotions and values [25], perceived trust in the owners and operators of technologies, fairness of decision-making and the distribution of associated costs and benefits [26, 27]. An interesting study conducted by ref. [20] indicates that community acceptance may have a temporal dimension. In particular, it follows a U-shaped trend starting from the phases before the implementation of the project up to the final phase on which the plant is in service [20]. The acceptance level is therefore higher in the phase prior to the design stage and then falls in the phase in which the location of the facility is chosen, finally, once the system is in service, the acceptance level increases again [20]. To avoid protests in the planning phase of REPs and in the phase of siting decisions, it is necessary to involve all interested actors in the decision-making process [2, 3, 28, 29]. This aspect defines the so-called "procedural justice" which is one of the sub-dimensions of community acceptance defined by [2]. The most important elements of procedural justice concern participation rights, transparency, access to information and equity [1, 3, 26]. In order to respect procedural justice principles, ref. [28] has postulated six essential criteria: equal treatment of actors (coherence), completeness and correctness of information (accuracy), the possibility of retracting decisions (fairness), the involvement of all actors in the decision-making process (representation) and compliance with elementary moral and ethical values (ethics). The inclusion of all the actors in decision-making processes allows them to be part of the project creation, to assume certain responsibilities and being able to influence important decisions [3, 26]. Another sub-dimension of community acceptance defined by ref. [2] is the so-called "distributive justice". Distributive justice concerns the equal distribution of costs and benefits generated by REPs [3]. As regards benefits, they can be classified into direct or indirect monetary ones (e.g., shares of stocks, purchase of bonds, compensation payments, reduction of taxes and energy costs, creation of jobs for the community, etc…) or non-monetary ones (e.g., environmental compensations or landscaping) [1, 3]. As for the benefits, costs may have or not monetary nature, e.g., the decrease in market prices of real estate assets near the plant, the increase in the pollution level, the increase in vehicular traffic, noise, the modification of spatial planning, etc. [3]. An important aspect of distributive justice is that it cannot be measured in an objective manner as it has a strong character of subjectivity, in particular, each individual may consider equitable or not different levels of costs and benefits [3]; this fact suggests the difficulty of achieving distributive justice at the aggregate level. As regards the interaction between procedural justice and distributive justice, ref. [30] argue that the perception of procedural justice has a greater influence than distributive justice. The authors claim that if procedural justice is respected and outcome distribution is unfair or disadvantageous, people still accept the outcome because they feel that are treated equally and fairly [30]. The third sub-dimension of community acceptance is trust. The concept of trust is very broad and focusing on REPs context, many authors argue that the lack of trust of people towards local government, project developers, operators and stakeholders may involve to the non-acceptance of REPs [26, 31, 32]. Ref. [33] theorize that when people have little knowledge about the technology, acceptance mainly depends on the trust in actors that are responsible for that technology as this trust serves as a heuristic or alternative basis on which people form their own opinion. Trust is also closely related to procedural justice and distributive justice, in particular, the participation in the decision-making processes and the fair distribution of benefits may involve people's perception of trust [34]. A sustainable energy transition requires governance structures that need to be both participatory and inclusive, as well as enable people to become stakeholders of the process while also sharing benefits [34]. In addition to trust, this approach also helps to satisfy broader principles that they concern the legitimacy and the democracy of the entire processes [34]. The majority of existing literature on social acceptance in the context of REPs is based on studies which adopt survey methodologies to elicit people preferences related to specific characteristics of renewable energy or policies, as well as specific schemes used to promote REPs [21]. These studies analyze the acceptance from a socio-economic point of view. It was found that acceptance may also depend on the level of education attainment, age and income of individuals. In particular, higher levels of environmental awareness are positively correlate with higher levels of education attainment, income and lower age levels [4]. However, in some cases, people's acceptance of REPs emerged from surveys was followed by phenomena of local opposition, these studies did not investigate on these aspects and a literature gap was found [21]. In order to provide good practices for social acceptance of REPs, researchers should not adopt only qualitative and quantitative approaches based on surveys. In particular, policy approaches should be also considered as they integrate the passivity linked to economic models. Policy approaches have to analyze how social acceptance or acceptability is built within the processes (e.g., modification of people's behavior, rational and non-rational analysis of people decision-making processes).

3.3 Market Acceptance

The third dimension of social acceptance defined by ref. [2] is market acceptance. In this dimension, the actors are consumers, investors and companies. Market acceptance refers to the diffusion of technologies for producing renewable energies within the market and to the extent to which its participants, such as consumers and companies, accept them [7]. Market acceptance operates at an intermediate level between national politics and local communities, it involves consumers (who adopt the technology) and investors (who want to support technology production and use) [35]. It is important to distinguish between large-scale technologies (i.e., REPs) and small-scale ones (such as microgeneration systems, solar thermal or photovoltaic domestic power plants, etc.) [2]. As regards the latter, social acceptance may be interpreted as the acceptance of the technology by the market or its adoption process (i.e., market acceptance is achieved with the purchase of a technology by users) [2]. As for large-scale technologies, they are linked to infrastructures that make them intrinsically more complex than other products, both in terms of innovation and implementation aspects [2]. Market acceptability of renewable energies also depend on social, technical, political and economic factors such as national regulations, barriers to green energy implementation and others that affect local acceptance/opposition [35]. Ref. [36] analyzed broad factors for which 18 countries adopted different typologies of renewable policy mechanisms, and they found that those who are "pro-renewable" may be explained by political and cultural structure, energy security and the absence of a strong conventional energy industry [35, 36]. In particular, countries characterized by a market-oriented culture and liberal governments have often chosen proper mechanisms for market penetration as well as regulatory structures; they also provided optimal incentives schemes for investors [35, 36]. In addition, countries that are more concerned about energy security and/or climate change issues were more likely to adopt renewable energies, in particular, those who have scarce coal and oil reserves [35, 36]. As for barriers, they may be related to the lack of proper information campaigns on green technologies, distorted price signals that did not take into consideration the advantages of renewable energy adoption, costs of conventional energy sources, regulations at political level that favored historical companies and energy suppliers as well as disinformation on how renewable energy technologies work [35]. Other barriers are related to subsidies schemes for conventional energy production (i.e., by using fossil fuels), high initial investment costs of REPs, imperfect markets, financial risks and uncertainties of investments in REPs as well as a variety of institutional factors [37]. Ref. [35] have identified three factors that may involve acceptance of renewable energies in the market:

- competitive installation and/or production costs: renewable energy technologies may produce electricity with competitive market prices compared to others if they are supported by government incentive schemes, a wide allocation of resources and/or a strong local manufacturing availability;
- information and feedback mechanisms: investors and users/producers must have access to reliable information on policies, market prices and opportunities offered by renewable energies;
- access to financing: producers and users must have access to low-cost national sources of financing and/or must be able to benefit from specific government financing schemes.

By considering the supply and demand of renewable energy market, through the green energy marketing users have the opportunity to switch to renewable energy supply without actually being involved in energy generation, in this context, barriers to diffusion of renewable energy are reduced as market adoption may be isolated from the broader framework of social acceptance [2, 38]. Another issue of renewable energy market is related to international green energy trade, countries which have large resources for producing renewable energy may not be willing to export them to others as they want to limit the use of their landscape from the installation of REPs [2]. Ref. [2] also highlight issues related to the way in which social acceptance is developed within the companies that produce renewable energies, in this context, there may be cognitive barriers related to environmental and sustainability issues [2, 39]; there is a lack of literature in this field and future research developments are needed [2, 39].

Therefore, we go forward providing a general framework to map the dimensions of social acceptance (Fig. 1) and, in the conclusions, we summarize the most important findings.

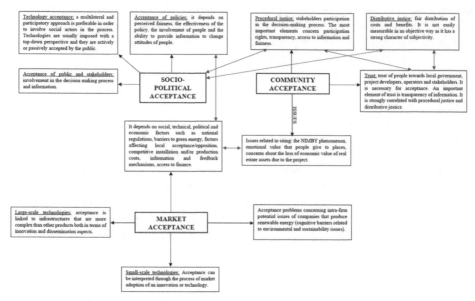

Fig. 1. Social acceptance framework (arrows in blue show the links between acceptance dimensions

4 Conclusions

In this contribution, we conducted a literature review to analyze the dimensions of social acceptance, i.e., socio-political acceptance, community acceptance, and market acceptance. In the socio-political acceptance dimension, the policy's endorsement may be reached through the principle of fairness and by providing adequate information to individuals. Policies may be designed to change people's attitudes to increase acceptance of renewable energies. To increase renewable energy technologies acceptance, a multilateral and participatory approach in technology design and implementation is needed. Regarding community acceptance, the NIMBY syndrome turns out to be one of the most critical issues. Other factors may hinder acceptance, such as the emotional value that people give to places or concerns about potential real estate economic value losses due to REP's installation. Procedural justice involves stakeholders' participation in the decision-making process by respecting principles of transparency, access to information, fairness, and participation rights. Distributive justice is appreciated through the fair distribution of costs and benefits that REPs generate. Trust of people towards local government, project developers, operators, and stakeholders are essential for REP's acceptance; information transparency is crucial for trust. Market acceptance depends on many technical, financial and economic factors, and it changes by considering large-scale technologies and small-scale ones. As for the latter, acceptance may be interpreted through the market adoption of technology, while it is linked to more complex infrastructures for large-scale technologies. The most important findings of our literature review were summarized in a conceptual framework which may be useful for developers in anticipating and overcoming non-acceptance barriers related to the implementation of REPs. An important literature gap was emerged from our study, in particular researchers usually adopt qualitative and quantitative approaches based on surveys to investigate on how social acceptance or acceptability is built within the processes of REPs implementation. Thus, future developments of this study may analyze alternative policy approaches to favor social acceptance of REPs, such as behavioral approach.

References

1. Segreto, M., et al.: Trends in social acceptance of renewable energy across Europe-a literature review. Int. J. Environ. Res. Public Health **17**(24), 9161 (2020). 1–19
2. Wüstenhagen, R., Wolsink, M., Bürer, M.J.: Social acceptance of renewable energy innovation: an introduction to the concept. Energy Policy **35**(5), 2683–2691 (2007)
3. Schweizer-Ries, P.: Energy sustainable communities: environmental psychological investigations. Energy Policy **36**(11), 4126–4135 (2008)
4. Bertsch, V., Hall, M., Weinhardt, C., Fichtner, W.: Public acceptance and preferences related to renewable energy and grid expansion policy: empirical insights for Germany. Energy **114**, 465–477 (2016)
5. Van de Poel, I.: A coherentist view on the relation between social acceptance and moral acceptability of technology. In: Franssen, M., Vermaas, P., Kroes, P., Meijers, A. (eds.) Philosophy of Technology after the Empirical Turn. Philosophy of Engineering and Technology, vol. 23. Springer, Cham (2016). https://doi.org/10.1007/978-3-319-33717-3_11
6. Rau, P., Schweizer-Ries, J., Hidelbrand, J.: The silver bullet for the acceptance of renewable energies? In: Kabisch,, S., Kunath,, A., Schweizer-Ries,, P., Steinfuhrer, A. (eds.) Vulnerability, Risks and Complexity: Impacts of Global Change on Human Habitats, pp. 177–191. Hogrefe Publishers, Gottingen, Germany (2012)
7. Schumacher, K., Krones, F., McKenna, R., Schultmann, F.: Public acceptance of renewable energies and energy autonomy: a comparative study in the French, German and Swiss Upper Rhine region. Energy Policy **126**, 315–332 (2019)
8. Szarka, J.: Wind Power in Europe. Politics, Business and Society. Palgrave Macmillan, London (2007)
9. Fournis, Y., Fortin, M.-J.: From social 'acceptance' to social 'acceptability' of wind energy projects: towards a territorial perspective. J. Environ. Plann. Manag. **60**(1), 1–21 (2017)
10. Drews, S., van den Bergh, J.C.J.M.: What explains public support for climate policies? A review of empirical and experimental studies. Clim. Policy **16**(7), 855–876 (2016)
11. Dreyer, S.J., Walker, I.: Acceptance and support of the Australian carbon policy. Soc. Justice Res. **26**(3), 343–362 (2013)
12. Dreyer, S.J., Polis, H.J., Jenkins, L.D.: Changing tides: acceptability, support, and perceptions of tidal energy in the United States. Energy Res. Soc. Sci. **29**, 72–83 (2017)
13. PytlikZillig, L.M., Hutchens, M.J., Muhlberger, P., Gonzalez, F.J., Tomkins, A.J.: Policy acceptance. In: Deliberative Public Engagement with Science. Springer Briefs in Psychology. Springer, Cham (2018). https://doi.org/10.1007/978-3-319-78160-0_5
14. Wallner, J.: Legitimacy and public policy: seeing beyond effectiveness, efficiency, and performance. Policy Stud. J. **36**(3), 421–443 (2008)

15. Brockner, J.: Understanding the interaction between procedural and distributive justice: the role of trust. In: Kramer, R.M., Tyler, T.R. (eds.) Trust in organizations: Frontiers of theory and research, pp. 390–413. Sage, Thousand Oaks, CA (1996)
16. Bell, D., Gray, T., Haggett, C.: The 'Social Gap' in wind farm citing decisions: explanations and policy responses. Environ. Polit. **14**, 460–477 (2005)
17. Batel, S., Devine-Wright, P., Tangeland, T.: Social acceptance of low carbon energy and associated infrastructures: a critical discussion. Energy Policy **58**, 1–5 (2013)
18. Barr, S.: Strategies for sustainability: citizens and responsible environmental behaviour. Area **35**, 227–240 (2003)
19. Alomary, A., Woollard, J.: How is technology accepted by users? A review of technology acceptance models and theories. In: Conference: The IRES 17th International Conference, London, UK (2015)
20. Wolsink, M.: Invalid theory impedes our understanding: a critique on the persistence of the language of NIMBY. Trans. Inst. Br. Geogr. **31**, 85–91 (2006)
21. Van der Horst, D.: NIMBY or not? Exploring the relevance of location and the politics of voiced opinions in renewable energy siting controversies. Energy Policy **35**(5), 2705–2714 (2007)
22. Woods, M.: Deconstructing rural protest: the emergence of a new social movement. J. Rural. Stud. **19**, 309–325 (2003)
23. Delicado, A.: Local responses to renewable energy development. In: Oxford Handbook of Energy and Society, pp. 343–360 (2018)
24. Visschers, V.H.M., Siegrist, M.: Find the differences and the similarities: relating perceived benefits, perceived costs and protected values to acceptance of five energy technologies. J. Environ. Psychol. **40**, 117–130 (2014)
25. Truelove, H.B.: Energy source perceptions and policy support: image associations, emotional evaluations, and cognitive beliefs. Energy Policy **45**, 478–489 (2012)
26. Gross, C.: Community perspectives of wind energy in Australia: the application of a justice and community fairness framework to increase social acceptance. Energy Policy **35**, 2727–2736 (2007)
27. Wolsink, M.: Wind power implementation: the nature of public attitudes: equity and fairness instead of 'backyard motives. Renew. Sustain. Energy Rev. **11**, 1188–1207 (2007)
28. Leventhal, G.S.: What should be done with equity theory? New approaches to the study of fairness in social relationships. In: Gergen, K.J., Greenberg, M.S., Willis, R.H. (eds.) Social Exchange: Advances in Theory and Research, pp. 27–55. Plenum Press, New York (1980)
29. Bauwens, T., Devine-Wright, P.: Positive energies? An empirical study of community energy participation and attitudes to renewable energy. Energy Policy **118**, 612–625 (2018)
30. Cropanzano, R., Folger, R.: Procedural justice and worker motivation. In: Steers, R.M., Porter, L.W., Bigley, G.A. (eds.) Motivation and Leadership at Work, pp. 72–83. McGraw-Hill, New York (1996)
31. Titov, A., Kövér, G., Tóth, K., Gelencsér, G., Horváthné Kovács, B.: Acceptance and potential of renewable energy sources based on biomass in rural areas of Hungary. Sustainability (Switzerland) **13**(4), 2294 (2021). 1–19
32. Mazzanti, M., Modica, M., Rampa, A.: The biogas dilemma: an analysis on the social approval of large new plants. Waste Manag. **133**, 10–18 (2021)
33. Huijts, N.M.A., Molin, E.J.E., Steg, L.: Psychological factors influencing sustainable energy technology acceptance: a review-based comprehensive framework. Renew. Sustain. Energy Rev. **16**, 525–531 (2012)
34. Lennon, B., Dunphy, N.P., Sanvicente, E.: Community acceptability and the energy transition: a citizens' perspective. Energy Sustain. Soc. **9**(1), 1–18 (2019). https://doi.org/10.1186/s13705-019-0218-z

35. Sovacool, B.K., Lakshmi Ratan, P.: Conceptualizing the acceptance of wind and solar electricity. Renew. Sustain. Energy Rev. **16**(7), 5268–5279 (2012)
36. Haas, R., et al.: Promoting electricity from renewable energy sources–lessons learned from the EU, United States, and Japan. In: Competitive Electricity Markets, pp. 419–468 (2008)
37. Beck, F., Martinot, E.: Renewable energy policies and barriers in encyclopedia of energy cutler Cleveland (ed.). Academic Press/Elsevier Science (2004)
38. Bird, L., Wüstenhagen, R., Aabakken, J.: A review of international green power markets: recent experience, trends, and market drivers. Renew. Sustain. Energy Rev. **6**(6), 513–536 (2002)
39. Bansal, P., Roth, K.: Why companies go green: a model of ecological responsiveness. Acad. Manag. J. **43**(4), 717–736 (2000)

Strategies for the Valorisation of Small Towns in Inland Areas: Critical Analysis

Emanuela D'Andria⬤, Pierfrancesco Fiore⬤, and Antonio Nesticò⁽⊠⁾⬤

Department of Civil Engineering, University of Salerno, Fisciano, SA, Italy
{emdandria,pfiore,anestico}@unisa.it

Abstract. Many EU Countries are facing the depopulation of inland areas and small towns. In the light of such a pressing and intense exodus from rural areas, the cities are experiencing increasing overcrowding. This leads to a rapid growth in urban congestion, as well as the consumption of land and resources. It is only in recent years, due to the crisis of the metropolitan model, also accentuated by the current pandemic emergency, that a radical change in trend has taken place, which sees the valorisation of inland areas as an indispensable action for the sustainable development of Countries. The presence of large green spaces, clean air, quality of the built environment and landscape reveal the great potential of these areas, which have proved to be extremely resilient to sudden changes. For these reasons, both at European and national level, strategies are being implemented to recover small towns and inland areas.

Starting from European data on rural demographic contraction, the paper presents a critical examination of the main valorisation actions adopted both in the EU and in some Countries, with the aim of highlighting their criticalities and potentials. This is an essential premise for designing economic policy models that can have a significant impact on local realities.

Keywords: Inland areas · Small towns · Valorisation strategies · Investment projects

1 Introduction

Since the Industrial Revolution, Europe witnesses the gradual and continuous depopulation of its inland areas in favour of medium-sized and large cities. Places that once held the economies of Countries are downgraded for the more industrialised areas. Work in the fields is rapidly replaced by new machines, forcing most of the population to migrate in search of a different occupation. Moreover, between the two World Wars, many marshes and peat bogs were drained, encouraging the establishment of settlements on the plains, which were easier to access and closer to the main trade hubs. This has led to a general impoverishment of small municipalities in inland areas which, from being pivotal elements in territorial planning, have become 'marginal places'. However, the phenomenon of abandonment of these realities, if on the one hand has encouraged their degradation,

---

E. D'Andria, P. Fiore and A. Nesticò—Contributed equally to this work.

F. Calabrò et al. (Eds.): NMP 2022, LNNS 482, pp. 1790–1803, 2022.
https://doi.org/10.1007/978-3-031-06825-6_173

on the other hand has allowed the preservation of multiple identity and community values. In fact, the small towns are testimonies of a vast intangible and tangible heritage that translates into ancient traditions, architectural and artistic heritage, local building techniques, and production activities handed down from generation to generation. Therefore, the preservation of these places is an essential and indispensable tool to safeguard the peculiarities and history of Nations. In this respect, it is only recently that an awareness of the importance of inland areas and small municipalities has been raised, with the development of strategies for their recovery.

In the light of the above, the aim of this work is to illustrate the main valorisation actions carried out at European and national level, highlighting critical points and potentials.

The paper is structured in 5 sections. Following the introduction, Sect. 2 provides an overview of the demographic decline trend in European inland areas. Section 3 presents the most significant initiatives implemented by the EU and some European Countries. Results and discussions are in Sect. 4. Section 5 outlines conclusions and research developments.

2 The Demographic Contraction of Inland Areas in Europe

The demographic contraction of inland areas, especially rural ones, is a wide-ranging and complex issue that affects many countries both in Europe and abroad. Although this is not a recent phenomenon, it is only in recent years that concrete questions have been asked about the future of these realities and the role they could play in the sustainable development of territories [1]. The climate and metropolitan crisis, together with the current health emergency, are leading to a rethink of spatial planning, highlighting the pressing need to rebalance the available resources and prepare new settlement scenarios [2]. A new and possible 'form of living' is gradually emerging, aimed at repopulating the countryside and reconstructing the ancient city-rural environment dialogue. In this sense, there are numerous researches and studies that not only investigate the reasons behind depopulation, but also provide strategies and guidelines for the enhancement of inland areas and small towns [3].

Among the most significant research projects there is the *European Shrinking Rural Areas: Challenges, Actions and Perspectives for Territorial Governance* (ESCAPE), developed under the European Commission's ESPON programme, which presents relevant data on the status quo of rural areas. At present, these areas occupy around 40% of the EU's land, housing only a third of its population. The recent report published by ESCAPE [1] lists the main countries suffering the most from rural depopulation: Finland, Sweden, Estonia, Lithuania, Latvia, Poland, Hungary, Romania, Greece, Spain, Italy, Ireland and Portugal. In addition, forecasts predict an exponential growth of the phenomenon involving other Countries by 2033. These studies are corroborated by analyses carried out by the Spanish Data Agency, published by Europa Press, which ranks European Countries in the light of the demographic contraction of rural municipalities, with Estonia, Finland, Latvia and Spain at the top [4].

Further information is provided by the *European Network for Rural Development* (ENRD), which supports the implementation of Rural Development Programmes, promoting not only the exchange of information and best practices, but also cooperation between Countries. ENRD studies show that only 22% of rural residents have a high level of education, while 59% of homes have broadband connections [5].

The reasons for this demographic contraction are largely the same in all European Countries: lack of adequate infrastructure, insufficient essential services and job opportunities, distance from the main cities, vulnerability to natural risks [6]. To confirm this, the European Union has recently concluded a public consultation on the future of rural areas: more than 50% of the participants indicated infrastructure as a primary sector on which to intervene; 43% stressed the importance of increasing access to basic services, such as electricity, water, banks, post offices; there was a need to improve digital connectivity, as well as environmental performance in agriculture [5].

The ESPON programme also investigates the main causes of rural depopulation, grouping them into three 'vicious circles', subdivided into sub-themes. The first circle is about accessibility to Services of General Interest (SGI); the second is about employment; the third is about the vitality and presence of young people [7]. The effects of migration are also the same across Europe. The decay of the built heritage, the cultural, economic and social isolation, the increase in the number of elderly people, the merging/contraction of some basic services, are only some of the direct consequences of the gradual abandonment of inland areas and small towns (Fig. 1).

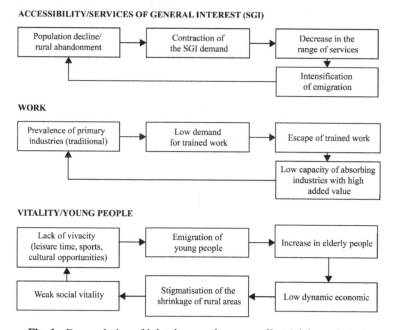

Fig. 1. Depopulation of inland areas: the cause-effect 'vicious circles'.

3 Strategies for the Valorisation of Inland Areas and Small Towns

In response to the marginalisation and abandonment phenomenon, many strategies have been put in place in the last decade to valorise inland rural areas and small municipalities. The recovery of these realities promotes the decrease of land consumption, the mitigation of urban decongestion and overcrowding, and the implementation of healthy and sustainable lifestyles. In addition to all this, there is the high quality of the air, water and soil, the value of the historical, architectural and artistic heritage, the presence of identity traditions, as well as the availability of large green spaces and valuable natural-landscape contexts [8, 9].

In the light of these considerations, the following sub-sections describe plans, actions and valorisation strategies undertaken both at European level and at national and local scale by associations, bodies and administrations.

3.1 EU Plans and Actions

Among the adopted policies for the valorisation of inland rural areas and small towns, of crucial importance is the LEADER (*Liaison Entre Actions de Développement de l'Économique Rurale*) approach, presented in 1990 by a group of European Commission officials. The aim is to bring together authorities, citizens and administrations to facilitate rural development through the creation of public-private partnerships. The main tool is the LAGs (Local Action Groups: currently 2,800 across Europe), which are responsible for drawing up and implementing participatory local strategies, optimising the allocation of available EU resources. The driving idea is that there are multiple possible development models, whose peculiarities necessarily derive from the specificities of each place, considering not only the economic components, but also and above all the intangible ones, closely linked to the culture of the territory.

The LEADER method is, in fact, «the first Community tool that promoted the creation of development agencies, providing them with specific financial resources for the administration and execution of animation actions, whose purpose, beyond the management of the planned actions with the EU contribution, is to strengthen cohesion, the sense of belonging and local identity» [10, p.12].

In 2013, the approach is officially included among the tools of European rural development policies, involving 2,402 inland areas. Over time, the LEADER system required a stronger and more extensive implementation network. To this end, National Rural Networks (NRNs) are established for each Member State to support networking and partner search activities.

More recent is *A Long-term Vision for the EU's rural areas*, presented by the European Commission in the summer of 2021 to offer a concrete response to the depopulation of inland rural areas [5]. Starting from the analysis of the potential and criticalities of these places, the Vision proposes the definition of a Rural Pact between several EU actors and an Action Plan based on pilot initiatives. The intervention fields, which correspond to the four objectives of the Plan, are subdivided into thematic sub-areas as shown in Table 1.

Table 1. The areas and sub-areas of intervention, with their objectives, of *A Long-Term Vision for the EU's rural areas.*

	Intervention areas	Intervention sub-areas
Goal	Making inland areas *STRONGER*	Empowering communities
		Accessing services
		Social innovation
	Making inland areas *CONNECT*	Digital connection
		Transport links and new mobility patterns
	Making inland areas *PROSPEROUS*	Diversifying economic activities
		Sustainable food production
	Making inland areas *RESILIENT*	Climate change resilience
		Environmental resilience
		Social resilience

Guide projects are proposed for each goal. In particular:

- *Stronger inland areas.* A *Rural Revitalisation Platform* is developed containing data and information on ongoing interventions and possibilities for funding and collaboration between different stakeholders [5];
- *Connected inland areas.* The concept of connection is extended both to the technological-digital sphere with the spread of the broadband and to the infrastructure-mobility sector, considering new forms of sustainable transport and the strengthening of the existing connections. A pilot initiative is *Rural Digital Futures* which aims to guarantee digital accessibility to all services by 2030, achieving 100% fast internet coverage in the inner regions of Europe [5];
- *Prosperous inland areas.* The intention is to implement economic diversification not only in the agri-food sector, but also in the commercial and industrial fields through the use of digital technology. In this regard, the European Commission will carry out targeted actions to support the establishment of medium and small enterprises in the framework of the *Entrepreneurship and Social Economy in Rural Areas* project. The hope is to encourage the setting up of an effective network between the different business realities [5];
- *Resilient inland areas.* The commitment is to improve not only environmental and climate resilience, helping small municipalities in the energy transition, but also social resilience, promoting gender and race equality. There are three 'flagship' initiatives in this field: *Climate action in peatlands through the "carbon farming"* which aims to reduce the level of carbon in the soil by restoring and conserving peat bogs and marshes; The *EU mission on soil health and food* which focuses on restoring soil for agricultural purposes; *Social resilience and women in rural areas* which is committed to supporting women in entrepreneurship [5].

Also relevant is the Horizon2020 project ROBUST (*Rural-Urban Outlooks: Unlocking Synergies*), which aims to strengthen relations between rural, peri-urban and urban areas through the development of innovative governance processes. ROBUST is currently working in 11 rural areas belonging to the urban districts of Ede, Frankfurt, Gloucestershire, Helsinki, Lisbon, Ljubljana, Lucca, Styria, Tukums and Valencia. In these territories, discussion and collaboration between policy-makers, researchers, companies, service providers and citizens are key elements in developing effective solutions in response to the specific needs of the places [11].

ENRD also proposes initiatives to support rural areas and small towns. These include *Smart and Competitive Rural Areas*, which works in line with the EU's three rural development goals, i. e. promoting the transmission and dissemination of knowledge and innovative practices in agriculture, forestry and rural areas; improving the profitability and competitiveness of the agricultural sector in all regions with the use of new technologies; optimising and encouraging the organisation of food supply chains. To date, the project is structured according to three different strategies: *Food and Drink Supply Chains*; *Rural Businesses*; *Smart Villages*. The latter is based on the transformation of small towns into 'smart villages' through the use of digital technology and the active participation of citizens: residents are the real protagonists of the transformation and, therefore, they must be ready for change, proposing valid solutions to the most pressing problems.

In the field of transport, there is the *SMARTA - Rethinking Rural Mobility* initiative, sponsored by the European Parliament, which analyses rural mobility needs in order to draw up action plans that take into account the policies and resources of each Country. Thus, *ad hoc* strategies are proposed, supported by the use of digital technologies. At the same time, the percentage of EU residents, divided into 'share of inhabitants living in cities', 'share of inhabitants living in rural areas' and 'share of inhabitants living in suburbs', is monitored and constantly updated [12].

In summary, the main actions of the *SMARTA* project are: Analysis of population and mobility in rural areas, as well as the political context of the countries examined; Identification of the main best practices and their level of innovation; Proposal of pilot projects involving the use of digital; Monitoring of projects and collection of results; Online sharing of acquired data.

Social issues are addressed within the *SIMRA-Social Innovation in Marginalised Rural Areas* project, which works to: Develop a theoretical and operational framework useful for cataloguing, understanding and implementing social innovation strategies across multiple scales of intervention; Prepare a 'Diversity Catalogue' of social innovation experiences in European rural areas; Assess the degree of social innovation and the impact of these experiences; Analyse the success factors of some of these projects; Create a network of innovative strategies that have been or are being implemented in order to facilitate collaboration and the exchange of information between different rural realities.

To date, *SIMRA*'s open access database collects numerous social innovation strategies in Europe, organising them according to the following purposes: environmental preservation; forest management and fire prevention; sustainable agriculture; employment; local development; social agriculture; service provision; tourism; integration of migrants; access to land; local foods; fishing/aquaculture; energy; crowdfunding [13].

All the analysed actions are in line with the goals of the *Next Generation EU (NGEU)* programme which, approved by the Council of Europe in July 2020, has allocated EUR 750 billion to promote the Member States' economic recovery. In this programme, a decisive role is played by the *Recovery and Resilience Plan (RRP)*, drawn up and approved by each country on the basis of the *NGEU*. This tool is designed to mitigate the social and economic impacts that the Covid19 pandemic has generated, making territories more resilient and prepared for the challenges of the green and digital transition. Considering the six pillars on which the *Plan* is articulated (Green transition; Digital transformation; Smart, sustainable and inclusive growth; Social & territorial cohesion; Health, and economic, social and institutional resilience; Policies for next generation), it is clear that the revaluation of rural areas is a crucial issue for the revitalisation of territories. In fact, Italy, France and Spain, as well as many other EU countries, have included measures aimed at enhancing historic villages and rural areas in their plans. The goal is to improve accessibility, not only in terms of sustainable mobility, but also in terms of digital infrastructure and essential services, in order to ensure equal opportunities for economic and social growth.

3.2 National Strategies

In addition to the EU's actions, the multiple strategies adopted by individual countries, associations and bodies are also relevant. A significant example is Portugal which, in 2015, approved the first *Programa Nacional para a Coesão Territorial* (PNCT), later renamed *Programa de Valorização do Interior*. The aim of this policy is to contrast the demographic decline of inland areas by identifying effective actions for the valorisation of small municipalities. This is done by attracting business investment to create jobs and protect endogenous resources. At present, the measures adopted have not succeeded in slowing down the marginalisation of rural areas. In fact, only 22% of the population lives in these places, which cover about 70% of the national territory.

On the other hand, the last three years have seen a change in the Government's focus on the problems of small rural towns. Tangible evidence of this shift is the adoption of specific programmes to promote the creation of local businesses. These include the *Programa de Captação de Investimento para o Interior*, which simplifies the administrative procedures required to carry out an investment. This ensures not only a reduction in costs, but also in the time needed for the financing process. With regard to the tourism sector, of interest is the *Linha de Apoio à Valorização Turística do Interior*, which supports the creation of hospitality and hotel accommodation projects [14].

In this context, Italy also plays a key role with the *Strategia Nazionale per le Aree Interne* (SNAI), a national policy for territorial development and cohesion born in 2013 to stem the demographic decline of rural areas. The aim of the *Strategia* is to develop new ways of multilevel local governance, focused on the protection and valorisation of the territories' cultural and natural heritage. Starting from this goal, SNAI has mapped 72 pilot areas, defined as 'inland' in view of the travel time needed to reach the major poles. The allocation of funding (ERDF, ESF and EAFRD European funds) is divided into three phases: *selection of areas* through a public inquiry procedure; *approval of the Area Strategy* by the Department for Cohesion Policies; *signing of the Framework Programme Agreement* between the central administrations of the Inner Areas Technical Committee, the Regions and the Municipalities.

To date, the *Area Strategy* has been approved for 59 pilot areas, while the *Framework Programme Agreement* has been signed for only a few. This is due to the very long lead time of the allocation procedure, with several years passing between each phase.

More recent is the project *RIPROVARE - Abitare i Paesi. Strategie Operative per la Valorizzazione e Resilienza delle Aree Interne*, funded in 2020 by the former Ministry of the Environment and coordinated by the University of Campania "Luigi Vanvitelli", with the collaboration of the University of Salerno and the University of Basilicata. This project is structured in three research phases with three corresponding goals [15].

At present, the work of *RIPROVARE* has completed the first two steps, identifying three pilot areas: the Matese and Ufita in Campania Region, the Val d'Agri in Basilicata. For these territories, phase 3 foresees the definition of targeted strategies that take into account the resilience degree and the peculiarities of each municipality.

Finally, there are also actions carried out by associations and bodies involved in the valorisation of small municipalities. Among these, it is necessary to mention the *Global Ecovillage Network* (GEN), a worldwide organisation that promotes the recovery of small towns and inland areas through projects based on eco-sustainability and sharing. The adopted model is that of the *Ecovillage*, defined by Robert Gilman as a place on a human scale in which anthropogenic activities are harmlessly integrated into the natural world [16].

GEN is flanked by the GEN Europe which supports EU Ecovillages, encouraging their cooperation and promoting the dissemination of experiences. There are also national initiatives, such as the *Associazione Rete Italiana Villaggi Ecologici* (RIVE), which catalogues and maps the various settlements and supports their growth and management. Similar work is carried out by the Spanish *Red Ibérica de Ecoaldeas* (RIE), which is currently involved in the *Rehabitar* project, aimed at defining a pilot programme to encourage the repopulation of abandoned villages in rural areas.

Around the world, there are currently about 10,000 Ecovillages, of which 250 are on the European continent alone [GEN data, 2020].

Another widely adopted model is that of the *Albergo Diffuso* (AD - Diffused Hotel). Born in Italy in 2004 from an idea of Giancarlo Dall'Ara, the strategy is based on an innovative form of tourist hospitality, which welcomes visitors in some of the historic buildings of the village, appropriately recovered and restored. This concept derives from a reinterpretation of traditional Japanese inns, the *Ryokan*, which particularly fascinated Dall'Ara because of the functional and spatial flexibility of the interiors, used as bedrooms at night and as dining rooms during the day. Moreover, «each *Ryokan* was different, but all were rooted in Japanese culture and history of hospitality» [17, p.3]. In ten years, the *Albergo Diffuso* model has rapidly spread throughout Europe, with 80 accommodation facilities in Italy alone.

Alongside the *Ecovillage* and the *Albergo Diffuso* there are also other strategies with different purposes: productive, cultural, social. Among these, it is important to mention the *Borgo del Benessere* model, which focuses on welcoming and caring for third-age users, and the *Borgo dell'Arte*, in which workshops are set up and artists are hosted, recovering the existing historical, architectural and social heritage.

4   Discussions

Strategies for the valorisation and recovery of inland areas and small towns are numerous. However, the many actions carried out at both European and national level seem to have little effect in tackling the complex issue of the marginalisation of inland villages and hamlets. What emerges is the general lack of an organic vision able to consider different and complex problems – social, environmental, economic, cultural – to be examined on a broad territorial scale. The programmes and models implemented to date, although initially based on a multi-sectoral/multi-disciplinary approach, generally take the form of mono-directional valorisation interventions, which address only some of the aspects and problems affecting small towns. One example is the Smart Village model which, even if based on the active participation of residents in the definition of concrete projects and on the use of digital technologies to enhance services, is often reduced to the design of applications for mobile phones or tablets. While the model of the Albergo Diffuso itself has the merit of recovering part of the built heritage, it focuses exclusively on tourist-economic issues, which do not encourage the effective repopulation of places. In addition, there is a lack of local services, which limits the stay of tourists, who are not very attracted by the territory's offering.

Moreover, municipalities are often unprepared and not very proactive in targeting European funding effectively. This is probably due to cultural factors, as well as to weak planning of interventions. The result is a superficial reading of the places, which leads to punctual interventions of secondary importance. Indeed, the funds designed to enhance these realities are not lacking: just think of the Fund for Development and Cohesion, which allocates 50 billion euro for the 2021–2027 cycle, useful for financing projects of national, interregional and regional importance; or the MIC Fund (Ministry of Culture), which directs 30 million euro to the recovery of municipalities in southern Italy. On the contrary, there is no an overall vision that combines the need to create networks with the demands of the single municipalities.

Although the plans carried out by the EU are clear and valid, in practice they are 'dematerialised' at national and even more at local level. There is a need not only for integrated and organic strategies, firmly supported by long-term policies, but also for actions aimed at raising public consciousness on the valorisation of small rural towns. Starting with people is the first, indispensable step towards revolutionising the way in which inland areas are seen and experienced, laying the concrete bases for an effective return [18].

Table 2 summarises the strengths and weaknesses of each analysed strategy.

Table 2. Criticalities and potentials of the analysed strategies.

	Strategy	Criticality	Potentiality
European context	LEADER	Generic approach, mainly focused on social issues, managed through 'bottom-up' processes	Importance given to human resource. The aim is to reconnect people to places, rebuilding a sense of belonging and safeguarding different local identities
	A Long-term Vision for the EU's rural areas	Strategy too recent to highlight critical issues	Strategy too recent to highlight potential issues
	ROBUST	'Sector-based' strategy, which does not look at a comprehensive system of interventions, but at individual 'thematic' actions	Emphasis on relations between rural, peri-urban and urban areas to be strengthened through innovative governance processes and through collaboration between different stakeholders
	SMARTA	Strategy that looks only at infrastructure issues	A careful analysis of infrastructure in inland areas is carried out. This offers the possibility of proposing targeted strategies that 'listen' to the territory, exploiting innovative digital technologies
	SIMRA	Strategy that looks only at social issues	An open access database of social innovation initiatives in European rural areas is provided. Best practices and impact assessments are reported for each initiative

(*continued*)

Table 2. (*continued*)

	Strategy	Criticality	Potentiality
National context	Programa de Valorização do Interior	It contains actions broken down by intervention areas. It lacks an overall view	Encourages the creation of local businesses by simplifying the required administrative procedures
	Strategia Nazionale per le Aree Interne	Excessively long timeframe for approval of Strategy documents. Many pilot areas are so heterogeneous that the implementation of the strategies is not feasible	First national mapping of inland areas. In addition, the Strategy directed part of the political attention to the issues of the hinterland depopulation and the urgency of intervening with valorisation actions
	RIPROVARE	Strategy too recent to highlight critical issues	Strategy too recent to highlight potential issues
	Ecovillage	The risk is to further marginalise small towns, encouraging the emergence of enclaves	It promotes the recovery of existing buildings. The model is based on eco-sustainable communities, on sharing and collaboration
	Albero Diffuso	A model that does not encourage a real return to the inland areas because it does not act on essential services and facilities for tourist reception	It promotes the recovery of existing buildings and knowledge of local traditions

(*continued*)

Table 2. (*continued*)

Strategy	Criticality	Potentiality
Smart Village	Sometimes the model is reduced to mobile phone applications or punctual interventions	At the heart of the strategy there is the citizen. The use of digital is fundamental. There is a careful knowledge of the territories and communities where the intervention takes place
Wellness Village	The model is addressed to only one type of user	It promotes the recovery of existing buildings, as well as the natural and landscape heritage. There is also a recovery of ancient traditions
Art Village	The model is addressed to only one type of user	It promotes the recovery of existing buildings, of the natural and landscape heritage as well as of local productive traditions

5 Conclusions

From the analysis of the many initiatives implemented in Europe, it is clear that the valorisation of inland areas and small municipalities is a shared and urgent issue. Providing a concrete answer to urban overcrowding and diffuse pollution has become an imperative goal, the pursuit of which is crucial to the sustainable development of Countries. UN forecasts for 2018 predict exponential population growth in metropolises over the next few years, to the point that by 2050 large cities will be hosting around 70% of the world's population [19].

In this scenario, rethinking the role of inland areas as a driving force for the reorganisation of territories is essential and possible. In fact, with reference to the 2030 Agenda, the recovery of small towns and rural territories is a decisive action within multiple goals: Good health and Well-being; Quality education; Affordable and clean energy; Decent work and economic growth; Industry, Innovation and Infrastructure; Reduced inequalities; Sustainable cities and communities; Responsible consumption and production; Climate action; Life below water; Life on land. In this respect, it is useful to emphasise that rural areas with their small towns are a cultural and historical palimpsest of great value, characterised by a tangible and intangible heritage that can be seen in the quality of the landscape and architecture, in the presence of green spaces, in the 'restance' of community and identity values, as well as local traditions and sustainable production models. In the light of the coexistence of these multiple aspects, there is a clear need to implement strategies that do not look at a single field of intervention, but consider issues of different nature (social-anthropological, cultural, environmental, economic) [20, 21].

With these considerations, in the light of a critical analysis, the work identifies the main limits and potentials of Community and national actions for inland areas and small towns. A fundamental precondition for research perspectives aimed at characterising economic policy models able to achieve the valorisation goals. These goals must necessarily take into account the wide range of social, cultural, environmental and economic problems that often affect inland areas. Therefore, work developments will concern the setting up of a valorisation model that combines the different fields of intervention, considering issues both at the scale of the single village and at that of the territorial context in which it is located. This is done in the light of the potential and criticalities that have emerged from the analysis of the actions implemented to date. The characterisation of such a model would make it possible to act effectively and organically on the recovery of small municipalities in inland areas, assessing their real needs and opportunities.

References

1. ESPON: ESCAPE European Shrinking Rural Areas: Challenges, Actions and Perspectives for Territorial Governance, Final Report (2020). https://www.espon.eu/sites/default/files/attachments/ESPON%20ESCAPE%20Main%20Final%20Report.pdf. Accessed 26 Nov 2021
2. Troisi, R., Alfano, G.: Is regional emergency management key to containing COVID-19? A comparison between the regional Italian models of Emilia-Romagna and Veneto. Int. J. Public Sect. Manag. **35**(2), 195–210 (2021). https://doi.org/10.1108/IJPSM-06-2021-0138
3. Bencardino, M., Nesticò, A.: Demographic changes and real estate values. A quantitative model for analyzing the urban-rural linkages. Sustainability **9**(4), 536 (2017). https://doi.org/10.3390/su9040536
4. Agencia de Datos, Europa Press. www.epdata.es. Accessed 26 Nov 2021
5. ENRD: Long term vision for rural areas. 1st edn. Publications Office of the European Union: Luxembourg (2021)
6. Troisi, R., Castaldo, P.: Technical and organizational challenges in the risk management of road infrastructures. J. Risk Res. 1–16 (2022). https://doi.org/10.1080/13669877.2022.2028884
7. ESPON: ESPON Policy Brief on the Future of Rural Areas (2021)
8. D'Andria, E., Fiore, P., Nesticò, A.: Small towns recovery and valorisation. An innovative protocol to evaluate the efficacy of project initiatives. Sustainability **13**, 10311 (2021). https://doi.org/10.3390/su131810311
9. D'Andria, E., Fiore, P., Nesticò, A.: Historical-architectural components in the projects multi-criteria analysis for the valorization of small towns. In: Bevilacqua, C., Calabrò, F., Della Spina, L. (eds.) NMP 2020. SIST, vol. 178, pp. 652–662. Springer, Cham (2021). https://doi.org/10.1007/978-3-030-48279-4_61
10. Rete Rurale Nazionale 2014–2020, L.E.A.D.E.R. nei Programmi di Sviluppo Rurale 2014-2020, Settembre 2016
11. ROBUST, ROBUST Project. https://rural-urban.eu/. Accessed 28 Nov 2021
12. SMARTA: The SMARTA project in a nutshell. https://ruralsharedmobility.eu/about/. Accessed 28 Nov 2021
13. SIMRA: Project. http://www.simra-h2020.eu/index.php/project/. Accessed 28 Nov 2021

14. Costa Gonçalves, P., Tralhão, M.: Lo spopolamento dell'entroterra portoghese e l'utilizzo degli appalti pubblici come strategia di contrasto delle asimmetrie tra l'entroterra e la costa. Istituzioni del Federalismo, no. 2, pp. 517–536 (2020)

15. Galderisi, A., Fiore, P., Pontrandolfi, P.: Strategie operative per la valorizzazione e la resilienza delle aree interne: il Progetto RI.P.R.O.VA.RE. BDC-Università degli Studi di Napoli Federico II **20**(2), 297–316 (2020)

16. Gilman, R.: The eco-village challenge. In: Context, no. 10 (1991)

17. Dall'Ara, G.: Albergo Diffuso. Un modello di ospitalità italiano nel mondo. 1st edn. ADI, Campobasso (2019)

18. Della Spina, L., Giorno, C., Galati Casmiro, R.: Bottom-Up processes for culture-led urban regeneration scenarios. In: Misra, S., et al. (eds.) ICCSA 2019. LNCS, vol. 11622, pp. 93–107. Springer, Cham (2019). https://doi.org/10.1007/978-3-030-24305-0_8

19. United Nations: Department of Economic and Social Affairs, Population Division: World Population Prospects 2019: Highlights (ST/ESA/SER.A/423). UN, New York (2019)

20. Calabrò, F., Cassalia, G., Lorè, I.: The economic feasibility for valorization of cultural heritage. the restoration project of the reformed fathers' convent in francavilla angitola: the zibìb territorial wine cellar. In: Bevilacqua, C., Calabrò, F., Della Spina, L. (eds.) NMP 2020. SIST, vol. 178, pp. 1105–1115. Springer, Cham (2021). https://doi.org/10.1007/978-3-030-48279-4_103

21. Dolores, L., Macchiaroli, M., De Mare, G.: Sponsorship's financial sustainability for cultural conservation and enhancement strategies: an innovative model for sponsees and sponsors. Sustainability **13**(16), 9070 (2021). https://doi.org/10.3390/su13169070

From Condominium to Energy Community: Energy and Economic Advantages with Application to a Case Study

Concettina Marino, Antonino Francesco Nucara, Maria Francesca Panzera(✉),
Matilde Pietrafesa, and Federica Suraci

Diceam, Mediterranea University, Reggio Calabria, Italy
matilde.pietrafesa@unirc.it

Abstract. The energy transition is now extremely necessary and urgent: for it to take place changes based on energy saving, consumption efficiency and use of renewable energy sources must be triggered. In such a scenario, the activation of new forms of collective action and collaborative economies, combined with the opportunities offered by new digital technologies, constitute the cornerstones, as well as represent an opportunity for new green economy models. In this new development scenario the topic of the work arises, which aims to analyze the satisfaction of the electricity needs of a residential condominium located in the Municipality of Reggio Calabria, comparing, in terms of energy and economic benefits, its electric supply obtained separately for each user through a photovoltaic system, with that obtainable as members of an energy community. The analysis shows that even if there is no marked reduction in fluxes exchanged with the grid, thanks to the incentives a significant reduction of expenses is obtained (–50%), with NPV benefiting from a saving of about € 14,000 at the end of plant life.

Keywords: Energy transition · Energy community · Energy district

1 The Energy Transition and Economic Programs

The transition to more sustainable modes of production and consumption has become one of the great challenges of the contemporary world. The energy transition, intended as the transition from the use of fossil energy sources to renewable ones, is part of the more extensive transition to sustainable economies [1, 2]. It leads to eliminate dependence on fossil fuels and improving energy efficiency, both of energy production and of user consumption, as well as in its distribution and conservation [3]. The emergence of a high-carbon society has generated instability, inequality and social inequity. Global warming, climate change, loss of biodiversity, environmental and social injustices require a profound rethinking of the way in which governments, businesses, financial systems and individuals interact with our planet. In this frame, the 2030 Agenda [4] by *United Nations Organization* is a very important document. In it 17 *Sustainable Development Goals* define a vision, detailed in concrete actions, through which communities can

© The Author(s), under exclusive license to Springer Nature Switzerland AG 2022
F. Calabrò et al. (Eds.): NMP 2022, LNNS 482, pp. 1804–1817, 2022.
https://doi.org/10.1007/978-3-031-06825-6_174

modify their organization and relationships, becoming synergistic and sustainable. At European level, one of the most important actions is the *Green New Deal* [5], which requires the EU to reduce overall climate-changing emissions of 40% by 2030 in order to achieve carbon neutrality by 2050.

The regulation of the *NGEU (Next Generation EU)* [6], the 750 billion € European fund proposed by the European Commission in 2020 in response to the economic crisis triggered by the pandemic, provides that at least 37% of the budget of the national plans will support climate objectives set by the *Green New Deal*. Furthermore, Europe aims to tackle the broad global challenges concerning climate change also with the new research and innovation plan *Horizon Europe 2021–2027*.

At national scale, in Italy the *National Recovery and Resilience Plan* [7] is the programmatic document aimed to achieve the climate neutrality objectives in the country. It is divided into 6 missions, one of the most important is *Green Revolution and ecological transition*, where almost 70 billion € of the approximately 235 total are allocated, making Italy the first beneficiary of the NGEU fund among the European countries. The largest portion of the mission will be used to guide the energy transition (*Renewable energy, hydrogen, grid and energy transition and sustainable mobility*, 25 billion €); significant is also the portion dedicated to energy efficiency and the redevelopment of public and private buildings (22 billion €).

The current Italian target for 2030, set in the *Integrated National Energy and Climate Plan* [8], is to reach 30% of energy produced from RES in final consumption and 35% of emission reduction; the plan supports *energy communities*, based on collective self-production and self-consumption of renewable energy. Moreover, the Legislative Decree n. 48/2020 [9] aligns the Italian legislation on the energy performance of buildings with the new European rules envisaged by the EU directive 2018/844 [10]. With this act, fully consistent with the New Green Deal strategy, the era of *smart buildings* and *nZEBs* (*nearly Zero Energy Building*) begins in Italy as well.

Cities consume two thirds of the energy supply and produce 70% of CO_2 emissions; among them the construction sector is the most responsible, with a share in Europe of 40% of energy consumption and 36% of CO_2 emissions. The role of construction is fundamental for existing buildings: the European Commission has estimated that the envelope redevelopment can lead to a reduction in consumption of 5–6%, while a plant with an intelligent control system of 15–20%.

2 Energy Communities

Today, self-consumption can be implemented not only individually [11, 12], but also collectively within condominiums or *energy communities* [13, 14]. The increase in distributed generation, especially through the spread of photovoltaic systems, has made the integration of energy production and consumption within neighborhoods and districts relevant (*Energy Districts*). *Energy communities* implement innovative forms of *presumption*, maximizing the consumption of energy produced within the community itself. Each user owns his energy production plant, of which he consumes a part, while the remaining portion can be exchanged with consumers physically close, accumulated and returned to the consumption units at the most appropriate time, and only at the end of the process is fed into the network. This makes it possible to reduce grid exchanges, energy lost for transport, transport costs and system charges, to enjoy economic benefits and to contribute to the energy transition and sustainable development. Primary purpose of the community is not the generation of financial profits, but the achievement of environmental, economic and social security for its members. In an energy community all interested users belonging to the same medium/low voltage electrical substation have the right to join it. Participants retain their rights as end customers, including that to choose their supplier and to leave the community. It is possible to create distinct categories of members: users, who do not participate in the investment and user/investors, who financially support it.

Regulations are contained within the EU package *Clean Energy for all Europeans*:

1) the Renewable Energy Directive (EU Directive 2018/2001) [16], containing the definitions of collective self-consumption and **Renewable Energy Community (CER)**
2) the Directive on the internal electricity market (EU Directive 2019/944) [17] which defines the **Energy Community of Citizens (CEC)**.

The main differences between CERs and CECs are:

1) the CER is based on the principle of autonomy among the members and on the need for proximity to the generation plants; it can manage energy in different forms (electricity, heat, gas) generated from a renewable source
2) the CEC does not provide for the principles of autonomy and proximity and can only manage electricity, produced from both renewable and fossil sources.

In Italy the national law for the transposition of the two Directives has not yet promulgated, nevertheless a phase of experimentation has initiated on the first. To date, the Italian regulation on collective self-consumption and renewable energy communities consists of Art. 42-bis of the law no. 8/2020 and Decree 16/9/2020 [18].

Many are the economic advantages related to energy communities:

- **tax concessions** (deductions or super-depreciation): for private individuals, the construction of a PV system falls within the scope of building renovations, provided for by the *Inland Revenue*, for access to tax concessions. It is possible to deduct 50% of the construction costs from the income tax; for businesses, the super-amortization of 130% of the investment value is envisaged
- **savings in the bill**: the more energy is self-consumed, the more the costs of its variable components are reduced (energy quota, network charges and related taxes such as excise duties and VAT)
- **source of income**: incentive mechanisms, namely the *Dedicated Withdrawal;* moreover, for systems installed after 1/3/2020, two more types of incentive rate, cumulative with tax deductions, are provided:

a) shared energy as part of collective self-consumption (same building or condominium): € 100/MWh;

b) shared energy within the renewable energy communities (same medium/low voltage electrical substation): 110 €/MWh;

The law also provides for the return of some items in the bill due to the avoided transmission of energy to the grid, the remuneration of the energy fed into the grid at an *Hourly Zonal Price*, equal to approximately 50 €/MWh, with a consequent relief quantified at 10 €/MWh for the Collective Self-consumption and in 8 €/MWh for CERs on shared energy. The sum of all the benefits amounts to around € 150–160/MWh.

3 Analysed Case Study

Within the frame of energy sharing, the paper analyses the consumption of four users belonging to a same condominium located in the *Gallico* district of the municipality of *Reggio Calabria* (Italy) (Fig. 1), served by PV systems, considering them first as individual users and then as members of an energy community, in order to identify its energetic and economic benefits. The condominium analysed, dating back to the 1980s, is in the marine area, mainly residential, with a high urbanization index and a high population density. The approach used for the analysis assessed, for each condominium user, the daily and monthly energy needs relating to electricity in the various seasons of the year, estimating the annual one. Subsequently, for each user, a grid connected photovoltaic system has been installed on the available surfaces of the building, each with different power based on the loads, making an estimate of produced and consumed energy. An economic evaluation was made, taking into consideration the incentives in force, such as the tax deduction and the on site-exchange mechanism. Subsequently, the case of individual users was compared with that in which they are part of an energy community, sharing and exchanging energy produced by a condominium PV system.

Fig. 1. Top view of the intervention area (Google Earth)

3.1   Single Users

3.1.1   Users Load

The condominium considered consists of 4 houses, with different daily energy consumption profiles, dictated by the different habits of the occupants (Table 1). For each user the yearly consumption of the individual devices was determined through a data sheet, starting from the knowledge of their absorption and their daily functioning hours in the 4 annual periods (winter, summer and autumn/spring). The daily, monthly and annual total energy requirement was determined from the sum of the energy consumption of the individual appliances, calculating the percentages of day and night load. Table 2 shows the annual needs obtained for the four users with their respective day/night shares: it can be seen that users B and C have an almost entirely daytime consumption, unlike users A and D which show equal loads in daytime and evening.

Table 1. Users' description

User	A	B	C	D
Description	Family (4 persons)	Office (2 persons)	Family (1 person)	Family (4 persons)

Table 2. Yearly energy load, nighttime and daytime rates for the four users

Consumption (%)	A	B	C	D
Daytime	47	97	79	47
Nighttime	53	3	21	53
Yearly load (kWh)	**4,164**	**1,822**	**2,869**	**7,141**

3.1.2   Photovoltaic System: Sizing and Energy Production

To satisfy the load of the various users photovoltaic systems have been by installed, determining for each of them the peak powers and the necessary areas for installation. Monocrystalline silicon photovoltaic modules were chosen, with a surface area of 1.66 m^2, power of 360 W$_p$ and technical characteristics shown in Table 3. Table 4 shows the sizing of the PV systems: it can be seen that user D, showing the greatest need, is consequently equipped with a PV system of greater power.

To evaluate the productivity of each photovoltaic system, E, it is necessary to know the installed surface S, the panel efficiency η, net of losses, and the incident solar radiation I_s, function of the panel exposure and orientation:

$$E = S \times \eta \times I_s \tag{1}$$

The examined building has a pitched roof: to calculate the productivity of the generator the correction coefficient for solar radiation was taken into account for a 30° pitch

Table 3. Technical characteristics of the selected photovoltaic panel

Type of panel	η_r (%)	NOCT (°C)	β (%/°C)	t_r (°C)
Monocrystalline	20.5	38.0	0.43	25

Table 4. PV powers and covered surfaces for the four users

User	N. of panels	Generator surface (m²)	Power (kWp)
A	8	13.28	2.88
B	4	6.64	1.44
C	6	9.96	2.16
D	14	23.24	5.04

slope, with a south orientation. The energy produced by each plant was then calculated, obtaining the available energy in both direct current (DC) and alternating one (AC), considering the system losses and the inverter efficiency. Subsequently, the values of energy fed into the grid and taken from it were determined (Table 5); such rates depend on PV production, load and self-consumption. By comparing the users in energy terms, we observe that user D, due to the greater peak power of its PV system, shows the highest energy production, and the greatest exchanges with the grid.

Table 5. Energy rates for the four users

User	Energy production in CA (kWh)	Load (kWh)	Self-consumption (kWh)	Energy fed (kWh)	Energy withdrawn (kWh)
A	4,460	4,164	1,862	2,598	2,302
B	2,230	1,822	1,604	626	218
C	3,345	2,869	2,111	1,234	758
D	7,806	7,141	3,104	4,702	4,037

3.1.3 Energy Comparison

Figure 2 shows users loads whereas Fig. 3 reports energy productions, shared as:

$$Load = Self-consumed\ Engery + Energy\ withdrawn\ from\ the\ grid$$
$$Engery\ production = Self-consumed\ Engery + Energy\ fed\ into\ the\ grid$$

For users with higher daytime consumption, B and C, self-consumption covers a much higher rate, minimizing the contribution of the grid. Utilities A and D show in

winter months a higher amount of energy drawn from the network than those of utilities
B and C, as in that period of the year their load is high and distributed more at night.

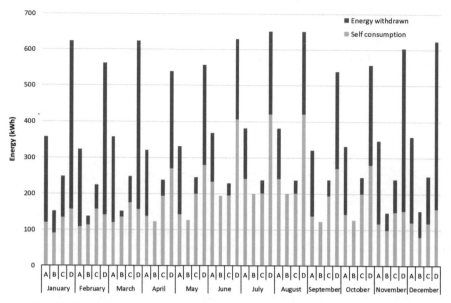

Fig. 2. Load shares: self-consumption and withdrawal from the utility network

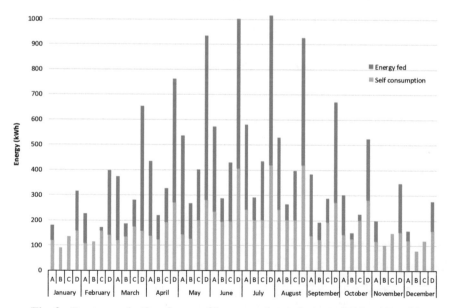

Fig. 3. Energy production shares: self-consumption and energy fed into the grid

3.1.4 Economic Comparison

An economic analysis was carried out, determining the costs and benefits deriving from the installation of the PV systems serving the users. The *Net Present Value* (*NPV*) was used, assessing the discounted value of the net costs over the useful plant life (20 years), comparing it with that referring to pre-existing case of absence of PV:

$$NPV = \sum_{j=1}^{n} \frac{C_j - B_j}{(1 + r)^j} + I_0 \tag{2}$$

where:

- j year
- n years of plant life
- cost of investment and replacement of components [€]
- Cj operating costs [€/year]
- Bj benefits [€/year]
- r interest rate [%]

The adopted value of the interest rate (0.51) is provided by the *Italian Ministry of Economic Development* [19]. The investment cost was estimated at € 1,800/kW$_p$, also considering the maintenance costs that the system needs to operate efficiently.

The annual operating costs were obtained adopting an all-inclusive tariff equal to 0.214 €/kWh (average cost of electricity in the last quarter) [20]. The benefits are a function of the incentives currently in force: tax deductions, valid for the first 10 years of life of the plant and equal to 50% of the investment cost, and the incentivizing mechanism of the *on site-exchange* for energy production, normed by an *Integrated Text* issued by the *Electricity and Gas Authority* [21]. This service allows user to offset the electricity produced annually and fed into the grid with that taken and consumed at a different time, allowing fictitious *refund* for the energy he has fed. The mechanism can be summarized in expression (3), where the first term is the *energy* share and the second the *service* one:

$$C_S = \min[O_E; \; C_{EI}] + C_{US} \times E_S \tag{3}$$

in which:

- Cs: amount in exchange account paid to the user
- OE: energy charges, obtained as the product of energy purchased from the network and the Single National Price, PUN (0.0214 €/kWh), the wholesale reference price of electricity purchased on the Italian Power Exchange Market, i.e. the average of the zonal prices of the Day-Ahead Market weighted with total purchases, net of pumping
- CEI: economic value of injected energy, given by the product between the amount of energy injected into the grid and the Zonal Energy Price on the Day Ahead Market, PZ (0.0210 €/kWh).
- CUS: lump sum exchange fee, equal to 0.065 €/kWh
- ES: energy exchanged, equal to the minimum between kWh injected and withdrawn.

Surpluses are also reimbursed, if at the end of the calendar year energy input value C_{EI} is greater than the corresponding withdrawn one O_E.

Considering energy costs, contributions and surpluses, an overall exchange contribution on site was obtained for the four utilities, shown in Table 6 together with the investment benefit and the cost items. Table 7 shows NPV for the various users: it is evident how the presence of a PV system, thanks also to its incentivizing mechanisms, allows for significant savings, depending on self-consumption and ranging from a minimum of 55% to a maximum of 78%.

Table 6. Economic rates for calculating the NPV

User	Investment cost (€)	Cost (€/year)	Investment benefit (€)[a]	On site exchange benefit (€)	Total benefit (€)
A	5,184	493	259	200	464
B	2,592	47	130	19	157
C	3,888	162	194	66	270
D	9,072	864	454	350	816

[a]for the first ten years

Table 7. Comparison of NPV with and without PV system for the four users

NPV		With FV (€)	Without FV (€)	Difference (€)	Difference (%)
User	A	7,919	17,823	−9,904.0	56
	B	1,713	7,798	−6,085.6	78
	C	4,260	12,282	−8,021.9	65
	D	13,823	30,562	−16,739.3	55

3.2   Energy Community

In this paragraph the installation of a condominium system, capable of managing all the loads of the various users, connected as an energy community, will be considered. The functioning of the energy community is obviously more complex, as all energy flows will be primarily exchanged within the building, withdrawing or feeding energy into the grid only when all users have already satisfied their load.

3.2.1   Energetic Analysis

First of all, the overall load (15,996 kWh, Table 2) of the four users was determined and the photovoltaic system was sized, estimating a peak power of 10.8 kW_p, a number of 30 panels and an occupied area of approximately 50 m^2. The energy rates are reported in Table 8 and shown in Figs. 4, 5.

Table 8. Energy community: energy rates

Load (kWh/year)	Self consumption (kWh/year)	Energy fed (kWh/year)	Energy withdrawn (kWh/year)
15,996	8,939	7,787	7,057

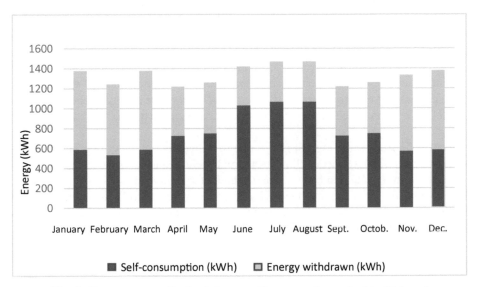

Fig. 4. Energy community–load shares: self-consumption and grid withdrawal

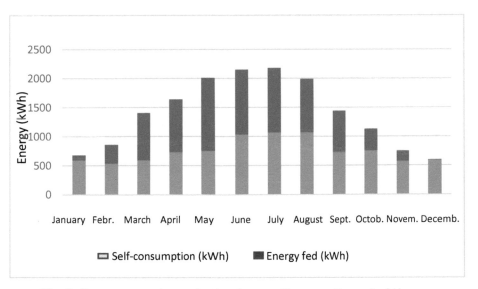

Fig. 5. Energy community–production shares: self-consumption and grid income

3.2.2   Economic Analysis

The same economic analysis carried out for individual users is applied, adopting the same functioning and investment (including tax benefit) cost, but different incentive rates, due to energy communities. Three types of incentives are envisaged:

- a premium rate on all shared energy with the grid (equal to the minimum, on an hourly basis, between the electricity fed and the taken one): it is paid € 0.10/kWh for groups of self-consumers or € 0.11/kWh for renewable energy communities
- a unit fee equal to the sum of the transmission tariff (€ 7.61/MWh) for low voltage users and of the highest value of the distribution variable component for other uses ones (0.61 €/MWh). For groups of renewable energy self-consumers an additional contribution is envisaged due to avoided grid losses, equal to about 1.3 €/MWh
- the incentive of the *Dedicated Withdrawal* (*RID*) on all the energy injected, equal to € 0.05/kWh, in replacement of the on-site exchange mechanism.

Among the incentives the highest rate is due to shared energy with the grid: it follows that a storage system, while entailing an energy benefit, reducing grid exchanges, would generate a minor economic benefit, due to the battery and the system oversizing cost, also reducing the incentive rate due to shared energy with the grid. The investment and functioning costs, together with total benefits, are reported in Table 9; Table 10 shows NPV, compared with the value referred to the case without PV.

Table 9. NPV cost and benefit items for the energy community

User	Investment cost (€)	Functioning Cost (€/year)	Investment Benefit (€/year)[a]	EC Benefit (€/year)	Total Benefit (€/year)
EC	19,440	1,510	972	1,315	2,287

[a]for the first ten years

Table 10. Comparison of NPV with and without PV system for the energy community

NPV with FV	NPV without FV	Savings
€ 13,750	€ 68,465	€ 54,715

3.3   Comparison Between Single Users and Energy Community

Table 11 shows a comparison between the energy community and the individual users, both powered by PV. As it can be seen, in energy terms, compared to the case of individual users, the energy community benefits from a reduction in incoming and output energy, in particular of that injected (−1,373 kWh, −15%) against a reduced withdrawal of 250 kWh. From an economic point of view, the case of single users presents a double expense compared to the energy community, which thanks to the incentives benefits from a saving of € 13,965 on NPV over 20 years. Both energy community and individual users benefit of very conspicuous reductions with respect to the case of absence of PV system, saving respectively € 54,715 and € 68,680.

Table 11. Energy rates and NPV for single users and energy community

User	Self-consumpt. (kWh/year)	Energy fed (kWh/year)	Energy withdrawn (kWh/year)	NPV (€)
Energy community	8,939	7,787	7,057	€ 13,750
A + B + C + D	8,681	9,160	7,315	€ 27,715
Difference	*+258*	*−1,373*	*−258*	*−€ 13,965*

4 Conclusions

The purpose of the work was to apply to a residential case study the concept of energy community, highlighting the energy and economic benefits entailing individual users. Collective self-consumption allows to pool electricity and to maximize the consumption of energy produced within the community itself, reducing withdrawals from the grid and energy lost in the power lines, but also transport costs and system charges; in addition conspicuous incentives are envisaged.

In the present case, 4 single users were analyzed, comparing in terms of energy and economic benefits their electric supply obtained separately through a PV system, with that obtainable as members of an energy community. In the latter case, if there is no marked reduction in the energy annually fed into the grid (−15%) and that withdrawn is practically unchanged, on the contrary, from an economic point of view, thanks to the incentives, a significant reduction of expenses (−50%) is obtained, benefiting from a saving of € 13,965 on NPV over 20 years. Compared to the case of absence of PV system, savings on NPV of both energy community and individual users are much more conspicuous, respectively equal to € 54,715 and € 68,680.

References

1. Barbaro, G., Foti, G., Labarbera, F., Pietrafesa, M.: Energetic and economic comparison between systems for the production of electricity from renewable energy sources (hydro-electric, wind generator, photovoltaic), Atti Accademia Peloritana dei Pericolanti, Classe di Scienze Fis., Mat. e Nat. , vol. 97, N. S2, A31 (2019). Edizioni Scientifiche Italiane, ISSN 1825-1242

2. Gattuso, D., Greco, A., Marino, C., Nucara, A., Pietrafesa, M., Scopelliti, F.: Sustainable mobility: environmental and economic analysis of a cable railway, powered by photovoltaic system. Int. J. Heat Technol. **34**(1), 7–141 (2016)

3. Marino, C., Nucara, A., Pietrafesa, M.: Electrolytic hydrogen production from renewable source, storage and reconversion in fuel cells: the system of the Mediterranea University of Reggio Calabria. Energy Procedia **78**, 818–823 (2015). 6th International Building Physics Conference, Torino 14–17 june 2015. Elsevier Science. ISSN: 1876-6102. https://doi.org/10.1016/j.egypro.2015.11.001

4. UN General Assembly: Transforming our world: the 2030 Agenda for Sustainable Development, October 2015. A/RES/70/1, https://www.refworld.org/docid/57b6e3e44.html

5. H.Res.109–Recognizing the duty of the Federal Government to create a Green New Deal. 116th Congress (2019–2020)

6. European Commission 2020: Communication to the European Parliament. Europe's moment: Repair and Prepare for the Next Generation. https://eur-lex.europa.eu/legal-content/EN/TXT/PDF/?

7. Presidenza del Consiglio dei Ministri, Piano Nazionale di ripresa e resilienza, 5 maggio 2021. https://www.governo.it/it/articolo/piano-nazionale-di-ripresa-e-resilienza/16782

8. MISE: MATTM, MIT, Piano Nazionale Integrato per l'Energia e il Clima (2030). https://www.mise.gov.it/index.php/it/energia/energia-e-clima-2030

9. DL 10/6/2020, n. 48, Attuazione della direttiva UE 2018/844 del Parlamento europeo e del Consiglio, del 30/5/2018, che modifica le direttive 2010/31/UE sulla prestazione energetica nell'edilizia e 2012/27/UE sull'efficienza energetica. GU Serie Generale n.146 del 10/6/2020

10. Directive (EU) 2018/844 of the European Parliament and of the Council of 30 May 2018 amending Directive 2010/31/EU on the energy performance of buildings and Directive 2012/27/EU on energy efficiency, http://data.europa.eu/eli/dir/2018/844/oj

11. Marino, C., Nucara, A., Panzera, M.F., Pietrafesa, M.: Towards the nearly zero and the plus energy building: Primary energy balances and economic evaluations. Therm. Sci. Eng. Prog. **13** (2019). art. n. 100400

12. Malara, A., Marino, C., Nucara, A., Pietrafesa, M., Scopelliti, F., Streva, G.: Energetic and economic analysis of shading effects on PV panels energy production. Int. J. Heat Technol. **34**(3), 465–472 (2016)

13. Gjorgievski, V.Z., Cundeva, S., Georghiou, G.E.: Social arrangements, technical designs and impacts of energy communities: a review. Renew. Energy **169**, 1138–1156 (2021). ISSN 0960-1481, https://doi.org/10.1016/j.renene.2021.01.078

14. Lowitzsch, J, Hoicka, C.E., van Tulder, F.J.: Renewable energy communities under the 2019 European Clean Energy Package–Governance model for the energy clusters of the future? Renew. Sustain. Energy Rev. **122**, 109489 (2020). ISSN 1364-0321,https://doi.org/10.1016/j.rser.2019.109489

15. Reis, I.F.G., Gonçalves, I., Lopes, M.A.R., Antunes, C.H.: Business models for energy communities: a review of key issues and trends. Renew. Sustain. Energy Rev. **144**, 111013 (2021). ISSN 1364-0321,https://doi.org/10.1016/j.rser.2021.111013

16. EU Directive 2018/2001 of the European Parliament and of the Council of 11 December 2018 on the promotion of the use of energy from renewable sources. https://eur-lex.europa.eu/legal-content/EN/TXT/?

17. EU Directive 2019/944 of the European Parliament and of the Council of 5 June 2019 on common rules for the internal market for electricity and amending Directive 2012/27/EU. http://data.europa.eu/eli/dir/2019/944/oj

18. MISE, Decreto 16/9/2020, Individuazione della tariffa incentivante per la remunerazione degli impianti a fonti rinnovabili inseriti nelle configurazioni sperimentali di autoconsumo collettivo e comunità energetiche rinnovabili, in attuazione dell'articolo 42-bis, comma 9, del

decreto-legge n. 162/2019, convertito dalla legge n. 8/2020, GU Serie Generale n. 285 del 16 November 2020. https://www.gazzettaufficiale.it/eli/id/2020/11/16/20A06224/sg

19. MISE: DM 27/12/2021, Tasso da applicare per le operazioni di attualizzazione e rivalutazione ai fini della concessione ed erogazione delle agevolazioni in favore delle imprese, GU n. 309, 30 December 2021. https://www.mise.gov.it/index.php/it/incentivi/impresa/strumenti-e-pro grammi/tasso-di-attualizzazione-e-rivalutazione

20. ARERA–Autorità di Regolazione per Energia Reti e Ambiente: Andamento del prezzo dell'energia elettrica per il consumatore domestico tipo in maggior tutela, II semestre (2021)

21. Deliberazione dell'Autorità per l'energia elettrica e il gas ARG/elt 74/08, Testo Integrato per lo Scambio sul Posto, 26 July 2012

Energy Requalification of a Neighbourhood of Reggio Calabria with a View to an Energy District

C. Marino, A. Nucara, M. F. Panzera[✉], M. Pietrafesa, and A. Votano

DICEAM, Università Mediterranea Reggio Calabria, Reggio Calabria, Italy
francesca.panzera@unirc.it

Abstract. The energy transition, intended as the construction of a new energy and social organization model based on the production and consumption of energy from renewable sources, with a view to distributed generation, cannot be further postponed. This transformation no longer represents a choice, but a necessity, which also becomes an opportunity to generate new production models and embrace new eco-sustainable behaviours. In this scenario, the activation of new forms of collective action and collaborative economies, such as energy districts and communities, are at the heart of the European Union's development programs and constitute the cornerstones in which to trigger the green revolution in response to the climate crisis, but also to economic inequalities and socio-environmental injustices. In this context, the objective of the work is the redevelopment of an area in the district of S. Caterina in the Municipality of Reggio Calabria from an energy district perspective. The analysis showed that by drawing from the district's virtual energy reservoir, the balance between energy productivity and needs is almost equal.

Keywords: Energy district · Energy community · Energy management

1 Introduction

The transition to more sustainable ways of producing and consuming energy represents one of the great contemporary challenges. A low-carbon energy system is based on the transition from fossil to renewable sources and the transformation of the electricity generation paradigm from centralized to distributed. To a large extent today, electricity generation takes place mainly in large thermoelectric, nuclear and hydroelectric plants, with the transport of energy to end users through extensive transmission and distribution lines, with significant losses. This system, which has established itself by virtue of its high efficiency, has triggered a dependence on fossil fuels, which today, thanks to the extraordinary development of renewables and the overwhelming advancement of digital technologies, can be overcome. The new model will bring with it wide repercussions on the electricity system: energy independence, greater efficiency, flexibility and reliability of the grid. Furthermore, the entry into the market of many operators, increasing the level

© The Author(s), under exclusive license to Springer Nature Switzerland AG 2022
F. Calabrò et al. (Eds.): NMP 2022, LNNS 482, pp. 1818–1829, 2022.
https://doi.org/10.1007/978-3-031-06825-6_175

of competition, will bring economic and social benefits, and small local units, becoming producers/consumers, will acquire greater awareness of management. and energy consumption, favouring its more efficient use. However, this transformation requires a regulatory adaptation aimed at simplifying procedures: bureaucratic barriers are in fact an obstacle that slows down the process, due to the difficulties encountered in the authorization phases. In this scenario the subject of the work arises, consisting in the energy redevelopment of an area of the S. Caterina neighbourhood in the city of Reggio Calabria with a view to an energy district.

2   The Civil Sector: The Most Energy-Demanding One

Energy is an asset and a strategic service for the growth and development of a country: implementing energy policies to reduce consumption, increase energy efficiency and integrate self-production renewable plants is extremely urgent to tackle climate change and reduce energy dependence. However, so far such contents have not found much space, due to an oppressive bureaucracy and economic-political interests present in fossil fuel supplier countries. Today, the largest contribution to consumption is provided by the civil sector (tertiary and residential) (36%), followed by that of transport (32%) and industry (23%). Italy is among the first countries in Europe for primary energy consumption and the first for CO_2 emissions for energy uses in the civil sector; in particular, cities are responsible for two thirds of energy supplies and produce 70% of emissions. The Italian building stock has 13.7 million buildings, of which 12.1 for residential use, corresponding to approximately 27 million homes, of which 22 are permanently inhabited and heated. The average energy requirement of a conventional building is 100 kWh/m^{2}year, while for an old one (before Legislative Decree 373/76) it is 150 kWh/m^{2}year (ENEA) [1]: this derives from the presence of many buildings (70%) made before the entry into force of the directives on energy efficiency (1976), moreover 25% of them never underwent extraordinary maintenance. Compared to a low-consumption building (in Europe the average requirement is 25–60 kWh/m^{2}year) Italian buildings have much higher consumption, with excessive use of fuels and high emissions. The legislation on the subject in the EU through the *Renovation Wave* (2020) [2] aims to double the energy requalification rate of buildings compared to current levels; in Italy, improving the energy performance of buildings is one of the main objectives of the *Integrated National Plan for Energy and Climate* (PNIEC) [3].

3   Energy Districts and Smart Cities: European Programs

The European Directive 2018/844 [4] on energy efficiency includes, among other things, distributed generation, i.e. the decentralized production of energy through many small-sized units that exploit local resources [5, 6], as a political and economic strategy priority for the coming years. Unlike large power plants, with large production units located in individual areas, distributed generation uniformly spreads production over the territory, exploiting the forms of energy present at the source [7, 8]. This model is particularly relevant on a neighbourhood scale, in medium and low voltage networks, where the energy districts are located. They shift the focus from the single block construction of buildings, creating the interaction between condominiums, infrastructures, mobility and ICT through the integration of renewable energy systems and the optimization of flow management with smart strategies, obtaining reductions in consumption higher than 50%. The District is *Energy Positive* if the neighbourhood is self-sufficient, it does not generate net CO_2 emissions, and produces energy in excess of its own needs, putting it back into the grid. The districts can affect contexts with different types of users: residential, tertiary (hospitals, hotels, schools, shopping centres), transport, industries and are an integral part of the *smart city*, a new model of city and urbanization that is efficient and sustainable from the social, economic and environmental point of view. The area of *Smart City & Communities* was a priority and strategic in the previous European Horizon 2020 program, but it is also currently present in the 17 sustainable development goals of the UN and the *2030 Agenda* [9]. Broader global challenges affecting cities and societies are addressed by the new *Horizon Europe* 2021–2027 plan, which focuses on health and safety, digitalisation, energy and climate change. Finally, sustainable urbanization strategies are contained in the *Urban Europe* program, particularly aimed at energy districts; in addition, 8 Directives on energy issues, approved in 2019, *Clean energy for all Europeans* [10], put in place legal frameworks aimed at promoting the energy transition, regulating the energy performance of buildings, energy efficiency, renewable energy and electricity market and assigning a leading role for citizens in the energy sector.

3.1   Structure and Technology of the Energy District

Smart Micro Grid. An energy district is characterized by a variable number of Smart Buildings interconnected through a Smart Micro Grid that allows the exchange of electricity between multiple users at local level; it is intelligent as it optimizes its distribution, minimizing overloads and variations in electrical voltage. Smart grids are used for the management of medium and low voltage distribution, in which production gaps and peaks are repeated, which require intelligent management to protect the extremes. Figure 1 shows the structure of the district.

Smart Buildings produce energy through RES systems, a portion of which is self-consumed for the energy needs of the housing units, another can be stored through domestic storage systems and finally the remaining part is exchanged with other users through the *Smart District MicroGrid*: this creates a capillary diffusion of energy that travels along multiple nodes and reaches large basins of users. The production plants are

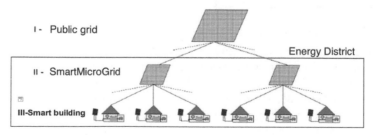

Fig. 1. Hierarchical structure of an energy district

of lesser power, require less space and less investment and are therefore very widespread in the area. The district energy surplus is stored in district energy accumulators or fed into the public grid.

Energy Community. Self-consumption allows users to consume the electricity produced by a local generation plant on site to meet their energy needs. Prosumer is the user who is not limited to the passive role of consumer, but actively participates in the production process (producer) and in the management of energy flows, enjoying autonomy and economic benefits. Today self-consumption can be implemented not only individually but also collectively within energy communities, by implementing innovative forms of prosumption. Energy communities are coalitions of users who, through the voluntary adhesion to a contract, share the development of a project to produce renewable energy and the economic and social benefits that derive from it. With due distinctions, they all share the same goal: to provide affordable renewable energy to their members, rather than prioritizing economic profit like a traditional energy company. Figure 2 illustrates the conformation of individual, collective condominium self-consumption and through energy communities. The two European directives that regulate them are contained in the aforementioned Clean Energy Package:

- EU Directive 2018/2001 on renewable energy, which contains the definitions of collective self-consumption and of the *Renewable Energy Community* (CER) [11]
- EU Directive 2019/944 on the internal electricity market, which defines the *Energy Community of Citizens* (CEC) [12].

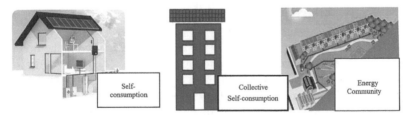

Fig. 2. Types of self-consumption: individual, collective and energy community

The *Nearly Zero Energy Building.* In the transition to the districts, a particularly important role is played by highly energy efficient buildings: the nearly Zero Energy Buildings (nZEB). They combine two fundamental aspects: improvement of the energy performance of the envelope and the system and consequent reduction of needs (a), satisfaction of needs with renewable energy sources technologies (b). In addition to the nZEB, the role of the Plus Energy Building is fundamental, a residential or commercial building connected to the public grid, which in a calendar year produces more energy from renewable sources than it consumes, becoming a small power plant capable of selling the energy that it does not consume and to profit from its sustainability. Smart buildings cover an added value, whose shell components and systems are managed in an intelligent and automated way, allowing energy consumption to be minimized.

The Storage Systems. Promoting the use of renewable sources and encouraging the spread of local distributed generation plants requires overcoming the two problems they suffer from: the difficult integration into the network and the inability to make the most of the available sources. The storage systems play a fundamental role for this purpose, being able to be used for multiple applications, including acting as a buffer for exchanges with the grid and allowing the storage of excess energy produced, to return it when production fails to satisfy demand: very often, in fact, the maximum production occurs in periods of low demand from domestic users, as in the case of photovoltaic systems. The technologies that are integrated in renewable generation plants are substantially two: electrochemical storage through batteries, for small plants, and hydroelectric storage through pumping stations, for larger sizes, the only systems capable of interfacing with small local production facilities are batteries. They allow greater exploitation of the energy produced, as well as a reduction in peaks, often the cause of imbalances, due to the random nature of renewables, levelling the power profiles fed into the grid. A different technology in full development is that of hydrogen storage systems, considered the energy vector of the future for its environmental sustainability and versatility in different fields of use [13, 14]. They are based on fuel cells, electrochemical devices that transform the chemical energy of hydrogen into electricity through a process in which it combines with oxygen, forming water.

Hydrogen must be produced using green systems, such as electrolysis combined with RES. In the coming years a significant rise in these technologies is estimated.

4 Case Study. Analysis Methodology

The study was focused on a part of the *S. Caterina* district, located in the north of the Municipality of *Reggio Calabria*, adjacent to the city centre. The neighbourhood is one of the most populous, mainly residential, with a high urbanization index, and a high population density. Most buildings date back to the fascist period, during which they were rebuilt following the devastating earthquake of 1908. Its urban aspect is essentially a checkerboard type, the main streets being orthogonal to each other. The area considered extends into the quadrilateral with the perimeter outlined in Fig. 3. The applied methodology started with a comparison between the district energy productivity, that is energy that could be produced installing photovoltaic systems on the available surfaces of the buildings, and their total needs. Subsequently, with a view to an energy district, the analysis has been carried out at the building, block and district scale, applying the concept of shared and exchanged energy among users.

Fig. 3. Top view of the intervention area (Source: Google Earth)

4.1 Classification of the Building Stock Under Study

The selected area is characterized by a high heterogeneity of the buildings. Figure 4 shows the map of the district with the relative numbering of the buildings. Preliminarily, in the study, a classification of the building stock was carried out according to the supporting structure of the building, the number of floors and interiors, the type of roofs and the presence of commercial activities (Fig. 5). The area consists of 12 blocks for a total of 66 buildings, formed by single and semi-detached or multi-family houses and condominiums. The formers are characterized by 1 or 2 interiors on 1 or 2 floors above ground; of them, most have a load bearing structure in masonry and only a few are in reinforced concrete. Multi-family buildings are characterized by several interiors ranging from 2 to 6, on a maximum of 3 floors above ground, with both load-bearing masonry and reinforced concrete structures. Finally, the condominiums have a number of interiors greater than 6 and are exclusively reinforced concrete structures, with a minimum number of 4 floors above ground. All load-bearing masonry structures (53%

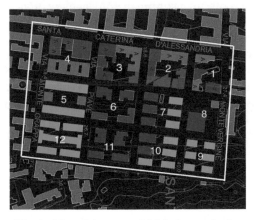

Fig. 4. Map of the area with blocks numbering

of the total, as shown in Fig. 5 in pink and light green) are characterized by pitched roofs, while those in reinforced concrete (47%, in blue, green and orange) have flat roofs.

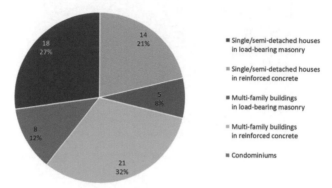

Fig. 5. Classification of the buildings in the neighbourhood. In brackets the n. of businesses

4.2 Photovoltaic Plant

The productivity of a photovoltaic system depends on the installed surface S, the efficiency of the system net of losses η and the incident solar radiation I_s, in turn a function of the exposure and orientation of the panels:

$$E = S \times \eta \times I_s \tag{1}$$

The photovoltaic panel selected is a SUNPOWER model in monocrystalline silicon, with a power of 315 W_p, which uses 96 solar cells with back-contact technology, providing a conversion efficiency of 19.3%. The module reduced voltage-temperature coefficient, anti-reflective glass and low-light performance ensure high energy production. In the case of flat roofs, to maximize the productivity of the systems, installations of panels facing South with a tilt angle of 28° were considered; in the case of modules installed on pitched roofs, a coplanar arrangement with the inclination, equal to about 18° and the orientation of the roof, was considered. For each calculated area, 10% was reduced as a safety margin, obtaining the results shown in Table 1. The number of modules that can be installed and the total productivity of the district was then determined, net of shading. In the case of a flat roof, the shading between the modules was considered, determining the optimal distance between the rows.

Considering the angle of inclination (28°) and that of azimuth (0°), a solar height of 28.5°, an optimal distance of 2.72 m has been estimated. Given the size of the panel (1.37 m²), their number was 1,518. In the case of pitched roofs, the modules follow the slope of the roof (about 18°) and the exposure is constrained by the orientation of the building (E/W or S/N): corrective factors have been considered to take this into account. For these buildings, the number of panels that can be installed was 3,813.

Table 1. Total surfaces available for each type of building

	Total surface (m^2)
Single/semi-detached houses in load-bearing masonry	3,355
Single/semi-detached houses in reinforced concrete	765
Multi-family buildings in load-bearing masonry	6,290
Multi-family buildings in reinforced concrete	2,150
Condominiums	5,540

4.3   Estimation of the Electrical Load

Known the number of houses and the presence of activities per building and per block, the energy needs have been estimated starting from the estimate of the load of the individual houses. According to ARERA (Regulatory Authority for Energy Networks Environments) the average expenditure for electricity per year of a family in Southern Italy is equal to € 639: considering the average cost of energy in 2020 (€ 0.2083/kWh), the average consumption resulted equal to 3,068 kWh/year. For commercial activities, on the other hand, a higher consumption was estimated, equal to 5,000 kWh/year, due to the greater number of electrical appliances and their higher absorption. The total estimated demand for the entire district was 1,475 MWh/year.

4.4   Energy Comparison

Energy productivity and average needs were determined for the entire district, by comparing them as the scale varies. The building scale was first considered, in which two units, one residential and one commercial, exchange energy between them before taking/sending it into the grid. Subsequently, we moved on to the block scale, evaluating for each building the difference between the load and the energy that can be produced; finally, it was considered on a district scale, considering all the blocks.

A) Building Scale: In the first case, two users were analysed, the first of a residential type, served by a photovoltaic system with a power of 2 kW$_p$, the second a commercial activity, with a complementary load to the previous one, having a predominantly daytime consumption, served by a 3 kW$_p$ plant. The utilities were initially considered distinct and then connected to each other as if they constituted an elementary energy community, sending to the second the flows not used by the first and vice versa: this interaction makes it possible to avoid the greater withdrawals/injections into the grid since, before taking advantage of it, the energy will be supplied or withdrawn by nearby users. Figures 6–7 show the rates of energy injected into the grid and withdrawn from it, both determined through a calculation datasheet, in the two cases: by exploiting the energy before sending or withdrawing it, a difference of about 10% is observed between the incoming and outgoing fluxes.

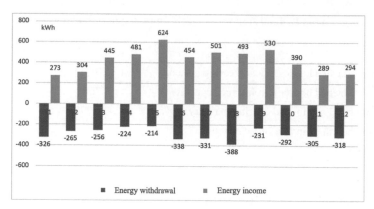

Fig. 6. Incoming and outgoing energy flows in the different months for separate users

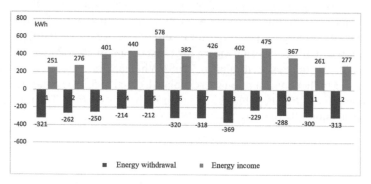

Fig. 7. Incoming and outgoing energy flows in the different months for connected users

B) Isolate Scale. Subsequently, the production and energy requirements for each block were compared. If the need has been assumed to be constant for each dwelling and strictly depends on the number in the buildings, the productivity depends on their type: buildings with greater development in height, such as multi-family ones and condominiums, occupy a small surface area, which translates into a smaller available surface and therefore in a smaller number of installable panels; conversely, the other types, mainly older buildings, develop in length or width over a small number of floors and this increases the area available for installation. Figure 8 shows the comparison between the productivity and the needs of the buildings of three blocks, 10, 11 and 12, respectively characterized by an almost zero energy balance (annual surplus 0.93 MWh), by the maximum consumption and by the maximum productivity.

Block 11, in addition to being the one with the highest electricity consumption, is the penultimate in terms of productivity, since all the buildings from which it is composed are condominiums. Opposite results are obtained for block 12, which has the greatest difference between production and consumption (+163 MWh/year), covering negative differentials such as those of block 11 adjacent to it and allowing an energetic surplus to be reintroduced on the neighbourhood micro grid.

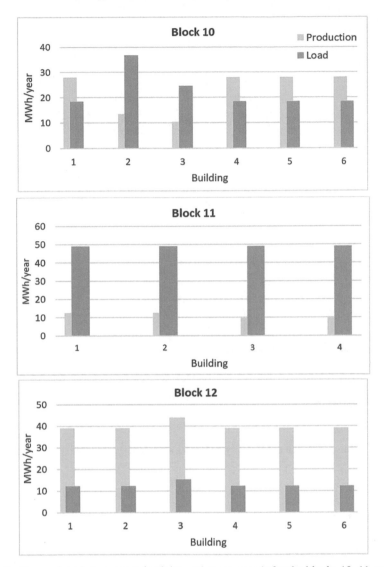

Fig. 8. Comparison between productivity and energy needs for the blocks 10, 11 and 12.

C) District Scale. On a district scale, taking into consideration all twelve blocks, we have what is reported in Fig. 9; Fig. 10, on the other hand, shows the difference between the productivity and the requirement of each isolate. It is possible to note that the blocks with the highest consumption peaks are 3 and 11 (in red) while those with the highest productivity are 5 and 12 (in green): these conditions are closely linked to their confor-mation, in particular the 3rd and the 11th block have the highest population density and therefore the greatest needs. The productivity and the needs of the entire district show balance, with an energy surplus of only 9.57 MWh/year, which in turn could be stored, reusing it within the same district.

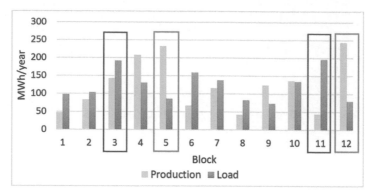

Fig. 9. Comparison between productivity and energy requirement for each block (Color figure online)

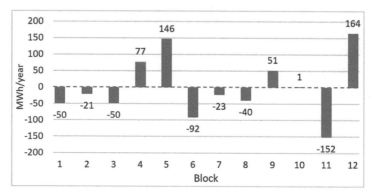

Fig. 10. Differential between productivity and energy needs for each block (Color figure online)

5 Conclusions

The work tackles the urgent issue of the energy transition, intended as the construction of a new model of energy and social organization based on the production and consumption of energy from renewable sources. This transformation is no longer a choice, but a necessity, which also becomes an opportunity to create new production models and embrace new eco-sustainable behaviours. In this scenario, the activation of new forms of collective action and collaborative economies, such as *energy districts*, now at the centre of the European Union's programs, are the cornerstones. This concept has been applied to a case study, an area of the *S. Caterina* neighbourhood in the Municipality of Reggio Calabria (Italy), in which a more rational management of energy is implemented with a view to an energy district. By applying the concept of sharing and exchanging energy between users, the benefits in terms of energy have been identified at various residential scales (building, block, district), given by the reduction of surplus/deficit of energy for the units or buildings, drawing on a virtual energy reservoir given by the community itself. Analysing 12 blocks belonging to the district in question, it was observed that in terms of district it is possible to obtain an almost net energy balance (+9.57 MWh/year) between productivity and loads.

References

1. Agenzia nazionale per le nuove tecnologie, l'energia e lo sviluppo economico sostenibile (ENEA) e Comitato Termotecnico Italiano (CTI) – Rapporto annuale 2020. Certificazione Energetica degli Edifici (2020). ISBN 978-88-8286-399-9
2. European Commission: A Renovation Wave for Europe–Greening our buildings, creating jobs, improving lives Communication COM/2020/662 final
3. MISE: MATTM, MIT, Piano Nazionale Integrato per l'Energia e il Clima (2030). https://www.mise.gov.it/index.php/it/energia/energia-e-clima-2030
4. Directive (EU) 2018/844 of the European Parliament and of the Council of 30 May 2018 amending Directive 2010/31/EU on the energy performance of buildings and Directive 2012/27/EU on energy efficiency. http://data.europa.eu/eli/dir/2018/844/oj
5. Marino, C., Nucara, A., Panzera, M.F., Pietrafesa, M.: Towards the nearly zero and the plus energy building: Primary energy balances and economic evaluations. Therm. Sci. Eng. Prog. **13** (2019). art. no. 100400
6. Bevilacqua, P., Morabito, A., Bruno, R., Ferraro, V., Arcuri, N.: Seasonal performances of photovoltaic cooling systems in different weather conditions. J. Clean. Prod. **2721**. Art. N. 122459, ISSN 0959–6526, https://doi.org/10.1016/j.jclepro.2020.122459
7. Narayanan, A., Mets, K., Strobbe, M., Develder, C.: Feasibility of 100% renewable energy-based electricity production for cities with storage and flexibility. Renew. Energy **134**, 698–709. https://doi.org/10.1016/j.renene.2018.11.049
8. Malara, A., Marino, C., Nucara, A., Pietrafesa, M., Scopelliti, F., Streva, G.: Energetic and economic analysis of shading effects on PV panels energy production. Int. J. Heat Technol. **34**(3), 465–472 (2016). https://doi.org/10.18280/ijht.340316
9. UN General Assembly: Transforming our world: the 2030 Agenda for Sustainable Development, October 2015. A/RES/70/1, https://www.refworld.org/docid/57b6e3e44.html
10. European Commission: Clean Energy for all Europeans Package (2019). https://ec.europa.eu/energy/topics/energy-strategy/clean-energy-all-europeans_en
11. EU Directive 2018/2001 of the European Parliament on the promotion of the use of energy from renewable sources. https://eur-lex.europa.eu/legal-content/EN/TXT/?
12. EU Directive 2019/944 of the European Parliament and of the Council of 5 June 2019 on common rules for the internal market for electricity and amending Directive 2012/27/EU. http://data.europa.eu/eli/dir/2019/944/oj
13. Carbone, R., Marino, C., Nucara, A., Panzera, M.F., Pietrafesa, M.: A case-study plant for a sustainable redevelopment of buildings based on storage and reconversion of hydrogen generated by using solar energy. ArcHistoR **6**, 596–615 (2019). 14633/AHR184
14. Marino, C., Nucara, A., Panzera, M.F., Pietrafesa, M., Varano, V.: Energetic and economic analysis of a stand-alone PV system with hydrogen storage. Renew. Energies **142**, 316–329. Elsevier (2019). https://doi.org/10.1016/j.renene.2019.04.079

Environmental Assessment of a Hybrid Energy System Supporting a Smart Polygeneration Micro-grid

Giovanni Tumminia[1], Davide Aloisio[1], Marco Ferraro[1], Vincenzo Antonucci[1], Maurizio Cellura[1,2(✉)], Maria Anna Cusenza[2], Francesco Guarino[2], Sonia Longo[2], Federico Delfino[3], Giulio Ferro[4], Michela Robba[4], and Mansueto Rossi[3]

[1] CNR-Istituto di Tecnologie Avanzate per L'Energia "Nicola Giordano" (ITAE), Salita S. Lucia Sopra Contesse, 5, 98126 Messina, Italy
maurizio.cellura@unipa.it

[2] Engineering Department, University of Palermo, Viale Delle Scienze, Building 9, 90128 Palermo, Italy

[3] Department of Naval, Electrical, Electronic and Telecommunication Engineering, University of Genoa, 16145 Genoa, Italy

[4] Department of Informatics, Bioengineering, Robotics and Systems Engineering (DIBRIS), University of Genoa, Via Opera Pia 13, 16145 Genoa, Italy

Abstract. To support the global transition towards a climate-neutral economy by 2050, countries all over the world are implementing low carbon and energy efficiency policies. This is leading to a rapid increase in the installations of distributed generation technologies. A hybrid system consisting of two or more energy sources could provide a more reliable supply of energy and mitigate storage requirements. In this context, this paper develops an environmental early analysis by the employment of a simplified Life Cycle Assessment approach. The above-mentioned approach is used to investigate the environmental performances of the hybrid energy system of the smart polygeneration micro-grid of the University Campus of Savona, which integrates photovoltaics and combined heat and power generators with an electricity storage system. Moreover, two further comparison scenarios are analyzed: electricity demand met by importing energy from the electricity grid (Scenario 1) and management of cogeneration plants to primarily satisfy thermal demand (Scenario 2). An early environmental assessment analysis shows that the cogeneration systems have a predominant weight on the majority of the environmental impact categories analyzed. On the other hand, the PV systems are the most responsible for the impacts on human toxicity cancer effects (35%), freshwater ecotoxicity (60%), and resource depletion (72%); while the energy imported from the grid has a predominant weight on freshwater eutrophication (55%). Finally, the results show that the alternative scenarios investigated are responsible for higher environmental impacts than the case study. The only exceptions are represented by the resource depletion for Scenario 1 and the global warming potential for Scenario 2. In fact, for these indicators, the base case shows higher environmental impacts than those of the alternative scenarios.

Keywords: Environmental impacts · Energy storage system · Renewable energy systems · Smart grid

© The Author(s), under exclusive license to Springer Nature Switzerland AG 2022
F. Calabrò et al. (Eds.): NMP 2022, LNNS 482, pp. 1830–1841, 2022.
https://doi.org/10.1007/978-3-031-06825-6_176

1 Introduction

The Paris Agreement has set out the ambition to keep global temperature rise this century below 2 °C above pre-industrial levels and to pursue efforts to limit the temperature increase even further to 1.5 °C [1]. This target requires zero global human-caused CO_2 emissions by 2050. Moreover, according to a recent Intergovernmental Panel on Climate Change report, 45% of global human-caused CO_2 emissions need to be reduced by 2030 compared to 2010 levels to achieve this objective [2]. To support this global transition towards a climate-neutral economy, renewable energy systems are becoming more common, both for large and small-scale applications. Moreover, given that the worldwide energy demand is expected to grow, renewable energy will continue to bear significance in the global energy mix [3, 4].

However, the presence of non-programmable renewable sources into the existing power grids can result in the difficulty of management and balancing them [5–7]. Moreover, the usual design of renewable energy systems, based on rules of thumbs, the new needs will require further considerations including the temporal alignment of energy uses and generation [8–11] and the environmental point of view [12–14]. Micro-grids can be particularly effective in this domain, by balancing services for the safe operation of the power system [15, 16]. In detail, micro-grids, are usually defined as an aggregation of generation systems, storage systems and loads that are managed by a central controller. They can alleviate the management of the grid by clustering several distributed energy generation systems in a single entity, but they require flexible and reliable energy management systems, which automatically schedule the plants [17, 18].

In recent literature, the optimization of micro-grids has been approached from different points of view [19, 20]. Different approaches have been developed both for planning and operational purposes, in islanded or grid-connected micro-grids [21, 22]. However, further studies are needed for the management of real case studies in order to reduce costs, improve energy performance, and limit system uncertainties [22]. In this context, [23] proposes an optimization model for low voltage micro-grids that can optimize both active and reactive power flows defining the optimal schedule of generation and storage systems that minimizes daily operational costs. The optimization procedure has been integrated within the supervisory software that monitors the Savona Campus Smart Polygeneration Microgrid (SPM) and tested employing an experimental campaign on the field.

Although the use of smart micro-grids could have a significant impact on electricity grids in terms of stability and resilience, the transition towards a more efficient and sustainable energy system implies the evaluation of the environmental impacts connected to these systems. Previous studies already analyzed the environmental impacts of smart grids, their components, and energy systems [24–27]. However, when dealing with the energy sectors, where supply chains are long and complex, an analysis of the environmental consequences should be carried out with a life cycle perspective. Therefore, to consider the entire environmental footprint of these systems, a reliable methodology is represented by the Life Cycle Assessment (LCA) [28, 29].

In this context, this paper aims at contributing to the body of knowledge on environmental performances of hybrid energy systems by investigating the environmental impacts of the Savona Campus Smart Polygeneration Micro-grid, through an environmental early analysis by the employment of a simplified LCA approach. The study could represent an early analysis of the environmental performances of the systems investigated, which although being able to give some preliminary results, will need to be strengthened and expanded in future studies.

In detail, the present study is developed from previously published researches [30, 31] in which a novel multidisciplinary design approach, based on LCA methodological principles, to investigate the environmental performances of renewable energy systems, considering a wide range of environmental impacts categories, is proposed. Moreover, to verify whether the control of technologies based on daily management cost minimization is also the one that allows reducing environmental impacts, two comparison scenarios are investigated.

2 Materials and Methods

This section describes the methodology and data used to carry out the analysis. In detail, the following paragraphs show how the methodology previously developed in [30, 31] is used to analyze the environmental impacts of the SPM of the University Campus of Savona. Section 2.1 briefly shows the case study. Section 2.2 reports the methodology used to perform the environmental analysis, while Sect. 2.3 shows the comparison scenarios investigated.

2.1 Case Study

The Savona Campus Smart Polygeneration Micro-grid is equipped with renewable energy systems (3 different photovoltaic systems) and combined heat and power systems (2 gas microturbines) integrated with electrical storage systems and connected to the electricity grid. The overall system is monitored and controlled by a central controller (based on an optimization model presented and validated in [23]) to find the optimal technologies schedule that minimizes daily costs.

Figure 1 shows the system configuration analyzed. In particular, the SPM is equipped with the following systems for the generation or the storage of electricity:

- 2 combined heat and power (CHP) systems (CHP 1 and CHP 2) (electrical nominal power of 65 kW_p, thermal nominal power of 112 kW_{th});
- 3 photovoltaic systems (PV 1, PV 2 and PV 3) (PV 1 nominal power = 80 kW_p, PV 2 nominal power = 15 kW_p and PV 3 nominal power = 21.25 kW_p);
- electric energy storage system (ESS) based on sodium-nickel chloride batteries (nominal capacity = 141 kWh).

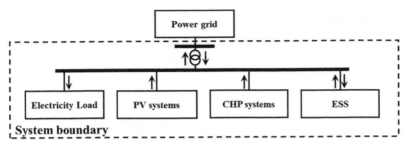

Fig. 1. The system configuration of the case study.

The electricity demand of the SPM consists of several buildings with different intended uses (laboratories, classrooms, offices, library, student residences, etc.). Table 1 shows the monthly electricity generation and use of the case study. Moreover, it also reports the exported energy as well as the imported energy. In detail, the annual electricity requirement of the University Campus of Savona is approximately 1,140 MWh$_e$, while on a monthly scale it varies between approximately 74 MWh$_e$ (August) and 114 MWh$_e$ (July). The yearly electricity produced by all on-site systems is about 950 MWh$_e$, while monthly it is between 73 MWh$_e$ (January) and 90 MWh$_e$ (July). The only months with exported energy are August (5.2 MWh$_e$) and June (1.5 MWh$_e$). In particular, the month of July does not show exported energy due to the high electrical load for the buildings' air conditioning (during August most of the buildings of the campus are not occupied). Finally, the yearly energy imported from the power grid is equal to about 216 MWh$_e$ and monthly it varies between about 2 MWh$_e$ (June) and 35 MWh$_e$ (December).

Table 1. Monthly electricity generation and use of the case study.

	Elec. Dem.	CHP 1	CHP 2	PV 1	PV 2	PV 3	Export.	Import.
	MWh$_e$	MWh$_e$	MWh$_e$	MWh$_e$	MWh$_e$	MWh$_e$	MWh$_e$	MWh$_e$
Jan	96.7	33.1	33.1	4.9	0.9	1.3	0	24.5
Feb	97.4	29.8	29.8	6.5	1.2	1.7	0	29.5
Mar	110.6	33.0	33.0	10.6	2.0	2.8	0	30.3
Apr	92.7	32.0	32.0	13.5	2.6	3.6	0	10.1
May	94.9	33.1	33.1	13.7	2.6	3.6	0	9.9
Jun	82.7	29.7	29.7	16.3	3.1	4.3	1.5	2.2
Jul	113.6	32.6	32.6	16.9	3.2	4.4	0	25.0
Aug	73.5	28.5	28.4	15.7	3.0	4.1	5.2	0.0
Sep	92.0	31.9	31.8	11.6	2.2	3.1	0	12.5
Oct	89.1	33.1	33.1	8.9	1.7	2.4	0	11.1
Nov	97.8	32.0	32.0	6.7	1.3	1.8	0	25.2
Dec	102.2	32.6	32.5	2.0	0.4	0.5	0	35.3
Year	1,143.1	381.3	381.0	127.3	24.1	33.6	6.7	215.6

2.2 Environmental Analysis

Table 2 shows the environmental impact indicators used in the analysis. They are evaluated using the ILCD 2011 Midpoint + method [32].

Table 2. Environmental impact categories investigated.

Impact category	Acronym	Unit
Global warming potential	GWP	kg CO_{2eq}
Ozone depletion potential	ODP	kgCFC-11$_{eq}$
Human toxicity – cancer effects	HT-c	CTUh
Human toxicity – non-cancer effects	HT-nc	CTUh
Particulate matter	PM	kg PM2.5$_{eq}$
Photochemical ozone formation	POF	kg NMVOC$_{eq}$
Acidification	Ac	molc H^+_{eq}
Terrestrial eutrophication	TE	molc N_{eq}
Freshwater eutrophication	FE	kg P_{eq}
Marine eutrophication	ME	kg N_{eq}
Freshwater ecotoxicity	FET	CTUe
Resource depletion, mineral, fossil, renewable	RD	kg Sb_{eq}

The environmental impacts (I_{Tot}) of the investigated system is evaluated using Eq. 1.

$$I_{Tot} = I_{CHP} + I_{PV} + I_{GRID} + I_{ESS} = El_{CHP} \cdot i_{CHP} + El_{PV} \cdot i_{PV} + El_{GRID} \cdot i_{GRID} + \frac{\eta \cdot C_{ESS}}{\nu} \cdot i_{ESS}$$

$$(1)$$

where:

- El_{CHP} = yearly electric energy produced by the CHP systems (kWh$_e$/year);
- i_{CHP} = specific impact due to the production of a kWh of electricity from the CHP systems (environmental impact/kWh$_e$);
- El_{PV}= yearly energy produced by the PV systems (kWh$_e$/year);
- i_{PV} = specific impact due to the production of a kWh of electricity from the PV systems (environmental impact/kWh$_e$);
- El_{grid} = yearly energy imported from the electricity grid (kWh$_e$/year);
- i_{Grid} = specific impact due to the import of a kWh from the grid (environmental impact/kWh$_e$);
- n = number of the installed electrical energy storage system replacement during the system useful life;
- ν = system useful life (year);

- C_{ESS} = nominal capacity of the installed electrical energy storage system (kWh$_e$);
- i_{ESS} = specific impact due to the production of the electrical energy storage system, referred to 1 kWh of capacity (environmental impact/kWh$_e$).

As reported by Eq. 1, the term I_{ESS} takes into account the environmental impacts associated with the production of the electrical storage system. Moreover, the impact of the electricity storage is annualized taking into account the useful life of the PV system (25 years) [5]. Finally, it is assumed that the replacement of each battery will occur after 3,000 charge-discharge cycles. The electrical grid impact (I_{GRID}) includes the environmental impacts associated with the energy withdrawn from the national electricity grid. It is evaluated by taking into account the Italian national energy consumption mix [2]. Since the energy produced on-site and fed into the power grid is beyond the system boundaries, the impact potentially avoided due to the surplus energy fed into the electricity grid is not taken into account.

Table 3 shows the values of the specific impacts used to calculate the environmental impacts. These background data are taken from the Ecoinvent database [33].

Table 3. Specific environmental impacts [33].

Impact category	Unit	i_{CHP}	i_{PV}	i_{GRID}	i_{ESS}
GWP	kg CO_{2eq}/kWh	6.15E−1	6.87E−2	4.62E−1	5.27E+1
ODP	kgCFC-11$_{eq}$/kWh	5.32E−8	7.44E−9	5.66E−8	4.91E−6
HT-c	CTUh/kWh	1.93E−8	8.68E−8	6.07E−8	1.64E−4
HT-nc	CTUh/kWh	4.77E−9	1.08E−8	1.34E−8	1.82E−5
PM	kg PM2.5$_{eq}$/kWh	4.24E−5	7.37E−5	1.35E−4	3.15E−1
POF	kg NMVOC$_{eq}$/kWh	6.25E−4	2.94E−4	9.71E−4	6.50E−1
Ac	molc H$^+_{eq}$/kWh	7.07E−4	5.83E−4	2.29E−3	5.93E+0
TE	molc N$_{eq}$/kWh	1.38E−3	9.00E−4	4.37E−3	1.36E+0
FE	kg P$_{eq}$/kWh	1.00E−5	5.78E−5	1.12E−4	8.43E−2
ME	kg N$_{eq}$/kWh	1.27E−4	8.85E−5	3.26E−4	1.08E−1
FET	CTUe/kWh	1.03E+0	1.42E+1	3.46E+0	1.78E+4
RD	kg Sb$_{eq}$/kWh	1.32E−6	2.24E−5	1.80E−6	1.60E−2

2.3 Comparison Scenarios

To compare the environmental impacts due to the use of the case study two further comparison scenarios are analyzed:

- Scenario 1: absence of electricity generation systems and electricity storage systems within the SPM. In particular, it is assumed that the entire electricity demand is satisfied with energy imported from the national electricity grid. Moreover, it is considered that the thermal energy produced by the two cogeneration systems and used to meet the thermal energy needs of the university campus, is produced, also evaluating the environmental impacts (Table 4 shows the specific impacts used), through a natural gas boiler.
- Scenario 2: management of the two CHP systems according to the heat requirement. In detail, unlike the current state in which the control of CHP systems takes place by favoring the electricity production, in this scenario, the CHP systems are considered as thermal generators that produce electricity as a function of the thermal load. Moreover, it is considered that the electricity demand no longer satisfied by these systems is satisfied through energy imported from the grid

Table 4. Specific environmental impacts due to 1 kWh of thermal energy produced by a natural gas boiler [33].

	Unit	Specific impact		Unit	Specific impact
GWP	kg CO_{2eq}/kWh	2.57E−1	Ac	molc H^+_{eq}/kWh	3.30E−4
ODP	kgCFC-11$_{eq}$/kWh	2.28E−8	TE	molc N_{eq}/kWh	6.89E−4
HT-c	CTUh/kWh	1.06E−8	FE	kg P_{eq}/kWh	1.24E−5
HT-nc	CTUh/kWh	2.90E−9	ME	kg N_{eq}/kWh	6.52E−5
PM	kg PM2.5$_{eq}$/kWh	1.92E−5	FET	CTUe/kWh	8.28E−1
POF	kg $NMVOC_{eq}$/kWh	2.83E−4	RD	kg Sb_{eq}/kWh	2.57E−1

3 Results

Table 5 and Fig. 2 show the environmental impacts (I_{Tot}) associated to the case study. The contribution of each of the two CHP systems on the environmental impacts examined varies between 8.7% (RD) and 40.3% (GWP). The contribution of the PV 1 (nominal power equals to 85 kW$_p$) is between 1.5% (GWP) and 49.6% (RD), while the incidence of the PV 2 (nominal power equals to15 kW$_p$) and of the PV 3 (nominal power equals to 21.25 kW$_p$) vary, respectively, from 0.3% (GWP) to 9.4% (RD) and from 0.4% (GWP) to 13.1% (RD). Finally, the contribution to the environmental impacts of the ESS is between 0.1% (GWP and ODP) and 6.7% (Ac), while the incidence of energy imported from the grid is between 6.7% (RD) and 55.2% (FE).

Table 5. Base case environmental analysis results.

	Unit	$I_{CHP\,1}$	$I_{CHP\,2}$	$I_{PV\,1}$	$I_{PV\,2}$	$I_{PV\,3}$	I_{ESS}	I_{GRID}	I_{Tot}
GWP	kg CO_{2eq}/year	2.34E+5	2.34E+5	8.75E+3	1.66E+3	2.31E+3	7.21E+2	9.96E+4	5.82E+5
ODP	kgCFC-11$_{eq}$/year	2.03E−2	2.03E−2	9.47E−4	1.80E−4	2.50E−4	6.73E−5	1.22E−2	5.42E−2
HT-c	CTUh/year	7.36E−3	7.35E−3	1.10E−2	2.10E−3	2.91E−3	2.25E−3	1.31E−2	4.61E−2
HT-nc	CTUh/year	1.82E−3	1.82E−3	1.37E−3	2.61E−4	3.62E−4	2.49E−4	2.88E−3	8.77E−3
PM	kg PM2.5$_{eq}$/year	1.62E+1	1.62E+1	9.38E+0	1.78E+0	2.47E+0	4.31E+0	2.91E+1	7.94E+1
POF	kg NMVOC$_{eq}$/year	2.38E+2	2.38E+2	3.74E+1	7.10E+0	9.86E+0	8.90E+0	2.09E+2	7.49E+2
Ac	molc H$^{+}_{eq}$/year	2.69E+2	2.69E+2	7.42E+1	1.41E+1	1.95E+1	8.12E+1	4.94E+2	1.22E+3
TE	molc N$_{eq}$/year	5.25E+2	5.25E+2	1.15E+2	2.17E+1	3.02E+1	1.87E+1	9.41E+2	2.18E+3
FE	kg P$_{eq}$/year	3.83E+0	3.82E+0	7.36E+0	1.40E+0	1.94E+0	1.15E+0	2.41E+1	4.36E+1
ME	kg N$_{eq}$/year	4.85E+1	4.84E+1	1.13E+1	2.14E+0	2.97E+0	1.47E+0	7.03E+1	1.85E+2
FET	CTUe/year	3.91E+5	3.91E+5	1.80E+6	3.42E+5	4.75E+5	2.44E+5	7.46E+5	4.39E+6
RD	kg Sb$_{eq}$/year	5.02E−1	5.02E−1	2.86E+0	5.42E−1	7.53E−1	2.19E−1	3.88E−1	5.76E+0

As reported in Fig. 2, the two CHP systems together have a predominant weight on eight (GWP, ODP, HT-nc, PM, POF, Ac, TE, and ME) of the twelve environmental impact categories, reaching values higher than above 60% in the case of GWP (81%), ODP (75%) and ME (52%). On the other hand, the energy generated by the PV systems is the most responsible for the impacts on HT-c (35%), FET (60%), and RD (72%) while the energy imported from the grid has a predominant weight on FE (55%).

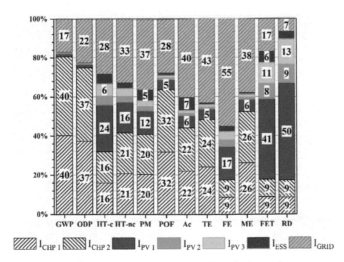

Fig. 2. Base case environmental analysis results.

Table 6 shows the comparison between the environmental impacts of the case study and those of the alternative scenarios investigated.

Except for the RD environmental impact category, Scenario 1 is responsible for greater environmental impacts than the base case. In detail, Scenario 1 shows an increase of the environmental impacts between +5% (FET) and +214% (FE) compared to the base case. Moreover, for 7 (HT-c, PM, POF, Ac, TE, FE, and ME) of the 12 environmental impact categories investigated, the increase compared to the base case is higher than 70%. However, Scenario 1 shows a reduction compared to the base case of the RD environmental impact category equal to 52%.

On the other hand, except for the GWP environmental impact category, Scenario 2 shows an increase of the environmental impacts between +2% (ODP) and +81% (FE) compared to the base case (except for the FE impact category, the increase is lower than 50%). However, Scenario 2 shows a reduction compared to the base case of the GWP environmental impact category equal to 9%.

Table 6. Environmental analysis results: scenarios comparison.

	Unit	SPM	Sc. 1	Sc. 1 Var.	Sc. 2	Sc. 2 Var.
GWP	kg CO_{2eq}/year	5.82E+5	7.27E+5	25%	5.29E+5	−9%
ODP	kgCFC-11$_{eq}$/year	5.42E−2	8.23E−2	52%	5.53E−2	2%
HT-c	CTUh/year	4.61E−2	7.76E−2	68%	6.04E−2	31%
HT-nc	CTUh/year	8.77E−3	1.75E−2	100%	1.17E−2	33%
PM	kg PM2.5$_{eq}$/year	7.94E+1	1.69E+2	113%	1.11E+2	40%
POF	kg NMVOC$_{eq}$/year	7.49E+2	1.33E+3	78%	8.68E+2	16%
Ac	molc H$^+_{eq}$/year	1.22E+3	2.88E+3	136%	1.77E+3	45%
TE	molc N$_{eq}$/year	2.18E+3	5.53E+3	154%	3.21E+3	47%
FE	kg P$_{eq}$/year	4.36E+1	1.37E+2	214%	7.87E+1	81%
ME	kg N$_{eq}$/year	1.85E+2	4.24E+2	129%	2.54E+2	37%
FET	CTUe/year	4.39E+6	4.60E+6	5%	5.23E+6	19%
RD	kg Sb$_{eq}$/year	5.76E+0	2.74E+0	−52%	5.93E+0	3%

4 Discussion and Conclusions

The study investigates the environmental impacts due to the use of the energy systems of the Savona Campus Smart Polygeneration Micro-grid, through an environmental early analysis by the employment of a simplified LCA approach. Moreover, two further comparison scenarios are analyzed.

The early environmental analysis shows that the cogeneration systems have a predominant weight on the majority of the environmental impact categories analyzed (GWP, ODP, HTc, PMP, POCP, AP, TEP, and MEP). In this context, future works will focus on the environmental assessment of these technologies fully through the application of the LCA methodology, to investigate their environmental life cycle impacts and also to analyze the alternative solution to reduce them, such as the use of biogas as fuel.

On the other hand, the PV systems are the most responsible for the impacts on human toxicity cancer effects (35%), freshwater ecotoxicity (60%), and resource depletion (72%) while the energy imported from the grid has a predominant weight on freshwater eutrophication (55%).

The scenarios comparison results show that both alternative scenarios investigated are responsible for higher environmental impacts than those of the base case. The only exceptions are represented by the RD for Scenario 1 and the GWP for Scenario 2. In fact, for these indicators, the base case shows higher environmental impacts than those of the alternative scenarios.

Finally, the study demonstrates that a renewable-energy-based system controlled using strategies that minimized management costs is affordable for a smart polygeneration micro-grid. However, to reduce the environmental impacts necessary for limiting a rise in global average surface temperatures to less than 2 °C, future energy systems will have to be controlled by strategies that also minimize environmental as well as economic performance. Moreover, to avoid the shift of burdens between environmental impacts categories, environmental performances should be considered as holistically and integrated as possible. Therefore, future works will focus on improving the system control strategies to find the optimal technologies schedule that at the same time minimizes costs and environmental performances.

Moreover, the next work developments foresee a more in-depth analysis of the individual technologies investigated through the application of the LCA methodology, in order to fully take into account both the embodied energy and the embodied environmental impacts as well as the end-of-life energy and environmental impacts, which were not investigated in this study.

References

1. Dimitrov, R.S.: The Paris agreement on climate change: behind closed doors. Glob. Environ. Polit. **16**, 1–11 (2016)
2. Masson-Delmotte, V., et al.: Global warming of 1.5 C. An IPCC Spec. Rep. Impacts Glob. Warm 1 (2018)
3. International Energy Agency: World energy outlook 2020 (2020)
4. Cusenza, M.A., Guarino, F., Longo, S., Mistretta, M., Cellura, M.: Environmental assessment of 2030 electricity generation scenarios in Sicily: an integrated approach. Renew. Energy **160**, 1148–1159 (2020)
5. Obi, M., Bass, R.: Trends and challenges of grid-connected photovoltaic systems–a review. Renew. Sustain. Energy Rev. **58**, 1082–1094 (2016)
6. Cusenza, M.A., Guarino, F., Longo, S., Ferraro, M., Cellura, M.: Energy and environmental benefits of circular economy strategies: the case study of reusing used batteries from electric vehicles. J. Energy Storage **25**, 100845 (2019)
7. Fichera, A., Pluchino, A., Volpe, R.: From self-consumption to decentralized distribution among prosumers: a model including technological, operational and spatial issues. Energy Convers. Manag. **217**, 112932 (2020)
8. Airò Farulla, G., et al.: A review of key performance indicators for building flexibility quantification to support the clean energy transition. Energies 14 (2021)

9. Cellura, M., et al.: Analysis of load match in nearly zero energy buildings: a parametric analysis of an italian case study. In: IEEE 4th International Forum Research and Technology for Society and Industry. RTSI 2018 – Proceedings (2018)

10. Guarino, F., Longo, S., Tumminia, G., Cellura, M., Ferraro, M.: Ventilative cooling application in Mediterranean buildings: impacts on grid interaction and load match. Int. J. Vent. 16 (2017)

11. Cellura, M., et al.: A net zero energy building in Italy: design studies to reach the net zero energy target. Proc. Build. Simul. 649–655 (2011)

12. Beccali, M., Cellura, M., Finocchiaro, P., Guarino, F., Longo, S., Nocke, B.: Life cycle assessment performance comparison of small solar thermal cooling systems with conventional plants assisted with photovoltaics. Energy Procedia **30**, 893–903 (2012)

13. Longo, S., Cellura, M., Guarino, F., Brunaccini, G., Ferraro, M.: Life cycle energy and environmental impacts of a solid oxide fuel cell micro-CHP system for residential application. Sci. Total Environ. **685**, 59–73 (2019)

14. Cellura, M., La Rocca, V., Longo, S., Mistretta, M.: Energy and environmental impacts of energy related products (ErP): a case study of biomass-fuelled systems. J. Clean. Prod. **85**, 359–370 (2014)

15. Zia, M.F., Elbouchikhi, E., Benbouzid, M.: Microgrids energy management systems: a critical review on methods, solutions, and prospects. Appl. Energy **222**, 1033–1055 (2018)

16. Fichera, A., Fortuna, L., Frasca, M., Volpe, R.: Integration of complex networks for urban energy mapping. Int. J. Heat. Technol. **33**, 181–184 (2015)

17. Soshinskaya, M., Crijns-Graus, W.H.J., Guerrero, J.M., Vasquez, J.C.: Microgrids: experiences, barriers and success factors. Renew. Sustain. Energy Rev. **40**, 659–672 (2014)

18. Lidula, N.W.A., Rajapakse, A.D.: Microgrids research: a review of experimental microgrids and test systems. Renew. Sustain. Energy Rev. **15**, 186–202 (2011)

19. Gamarra, C., Guerrero, J.M.: Computational optimization techniques applied to microgrids planning: a review. Renew. Sustain. Energy Rev. **48**, 413–424 (2015)

20. Fathima, A.H., Palanisamy, K.: Optimization in microgrids with hybrid energy systems – review. Renew. Sustain. Energy Rev. **45**, 431–446 (2015)

21. Zhang, J., Li, K.-J., Wang, M., Lee, W.-J., Gao, H., Zhang, C., et al.: A bi-level program for the planning of an islanded microgrid including CAES. IEEE Trans. Ind. Appl. **52**, 2768–2777 (2016)

22. Parisio, A., Rikos, E., Glielmo, L.: A model predictive control approach to microgrid operation optimization. IEEE Trans. Control Syst. Technol. **22**, 1813–1827 (2014)

23. Delfino, F., Ferro, G., Robba, M., Rossi, M.: An energy management platform for the optimal control of active and reactive powers in sustainable microgrids. IEEE Trans. Ind. Appl. **55**, 7146–7156 (2019)

24. Adefarati, T., Bansal, R.C.: Reliability, economic and environmental analysis of a microgrid system in the presence of renewable energy resources. Appl. Energy **236**, 1089–1114 (2019)

25. Gildenhuys, T., Zhang, L., Ye, X., Xia, X.: Optimization of the operational cost and environmental impact of a multi-microgrid system. Energy Procedia **158**, 3827–3832 (2019)

26. Peppas, A., Kollias, K., Politis, A., Karalis, L., Taxiarchou, M., Paspaliaris, I.: Performance evaluation and life cycle analysis of RES-hydrogen hybrid energy system for office building. Int. J. Hydrogen Energy **46**, 6286–6298 (2021)

27. Bortolini, M., Gamberi, M., Graziani, A., Pilati, F.: Economic and environmental bi-objective design of an off-grid photovoltaic–battery–diesel generator hybrid energy system. Energy Convers. Manag. **106**, 1024–1038 (2015)

28. ISO 14040:2006 Environmental management – Life cycle assessment – Principles and framework, p. 20 (n.d.)

29. ISO 14044: 2006 Environmental management – Life cycle assessment. RequirGuidel, p. 20 (n.d.)

30. Tumminia, G., et al.: Towards an integrated design of renewable electricity generation and storage systems for NZEB use: a parametric analysis. J. Build. Eng. 103288 (2021)
31. Tumminia, G., Guarino, F., Longo, S., Aloisio, D., Cellura, S., Sergi, F., et al.: Grid interaction and environmental impact of a net zero energy building. Energy Convers. Manag. **203**, 112228 (2020)
32. European Commission: Joint Research Centre: Characterisation Factors of the ILCD Recommended Life Cycle Impact Assessment Methods: Database and Supporting Information (2012)
33. Wernet, G., Bauer, C., Steubing, B., Reinhard, J., Moreno-Ruiz, E., Weidema, B.: The ecoinvent database version 3 (part I): overview and methodology. Int. J. Life Cycle Assess. **21**(9), 1218–1230 (2016). https://doi.org/10.1007/s11367-016-1087-8

Drainage Layer in Green Roofs: Proposal for the Use of Agricultural Plastic Waste

Stefano Cascone(✉) 🄳

Mediterranea University of Reggio Calabria, 89124 Reggio Calabria, Italy
stefano.cascone@unirc.it

Abstract. Green roof promises to become an increasingly important option for building owners and community planners because it can address many of the challenges facing urban residents. The success of the roof depends on the specific build-up of the green roof, therefore, special attention was focused on the identification and application of more environmentally friendly materials in green roofs. However, except the substrate, very few studies have denoted the potential to use reused materials for drainage layers, despite its important role in determining thermal and hydraulic performance of green roofs. This study analyzes the scientific literature on the use of innovative materials from the recovery/recycling of products for the drainage layer and propose the use of low-density polyethylene (LDPE) granules coming from the recycling of waste films used in agriculture for greenhouse roofing and mulching as innovative and sustainable drainage material in green roofs. In addition, the proposal concerns the possibility to enclose the plastic granular material into micro-perforated bags made of recycled polyethylene and to use the soil coming from washing of agricultural films as substrate in green roofs. This proposal will be further investigated by laboratory assessment on the thermo-physical properties of several polyethylene granules and by experimental set-up having the goal of assessing the thermal performance of different green roof technologies, in order to represent a suitable alternative to materials commercially used from an environmental, economic, and social point of view.

Keywords: Greenhouse covering · Low-density polyethylene · Recycled materials

1 Introduction

On a hot, sunny summer day, the temperature of black roof surfaces is much higher than the ambient air temperature. This is because non-reflective roofs absorb and retain solar energy as heat, which contributes not only to a hotter roof, but also to uneven thermal expansion/contraction and aging of the roof, and sometimes to heat gain within the rest of the building [1]. The top floors of the building underneath can be heated up by the hot roof and cause either discomfort for the building inhabitants or increased local cooling loads, particularly in older buildings, which tend to have less insulation [2, 3].

© The Author(s), under exclusive license to Springer Nature Switzerland AG 2022
F. Calabrò et al. (Eds.): NMP 2022, LNNS 482, pp. 1842–1849, 2022.
https://doi.org/10.1007/978-3-031-06825-6_177

Green roofs can help to mitigate some of the negative effects of the urban hardscape by reintroducing a natural landscape into urban environments without making major changes to city's infrastructure [4]. There are two main types of green roofs: extensive roofs, which are relatively inexpensive to install and are used mainly for environmental benefit, and intensive roofs which allow a greater variety and size of plants such as shrubs and small trees, but which are usually more expensive to install and maintain, partly due to the need for irrigation. Green roofs consist of waterproofing and anti-root membrane, drainage layer, filter layer, substrate (soil or growing medium) and vegetation (plants), overlying a traditional roof [5] (see Fig. 1).

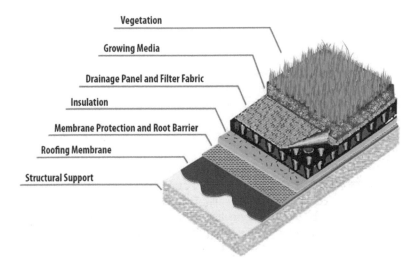

Fig. 1. Example of green roof assembly.

Drainage layer must allow excess water to be shed quickly from the roof surface, whilst holding sufficient to support the vegetation. Insufficient water storage means that either additional irrigation will be required, or the vegetation will struggle to flourish. There are three main types of drainage layers used on green roofs: simple geocomposite drain layers, reservoir sheets, and granular mineral medium. Granular medium is the best at slowing and delaying runoff. Reservoir sheets are designed to capture rainwater in indentations in their surface. Geocomposites, or multi-layered materials made from a combination of synthetic polymers, are adapted for green roofs, and designed to allow water to flow easily through the material, draining excess water away from a roof. However, drainage layer is still based on conventional materials, in some cases non-environmentally friendly materials, such as polypropylene or polyester geotextile membranes, polyethylene or polystyrene panels, bitumen or PVC membranes.

In the literature, special attention was focused on the identification and application of more environmentally friendly materials in different layers of green roofs. A growing trend regarding the construction of green roof systems has been the use of recycled and reused materials in their drainage and substrate layers, which can generate environmental, technical, economic, and aesthetic benefits while also providing the possibility of incorporating waste into the construction production chain [6]. Therefore, there is a need to replace current green roof materials by more environmentally friendly and sustainable products.

Except the substrate, very few studies have denoted the potential to use reused materials for green roofs, such as drainage layer. This study analyzes the scientific literature on the use of innovative materials from the recovery/recycling of products for the drainage layer in green roofs. In addition, the use of low-density polyethylene (LDPE) granules from waste films used in agriculture for greenhouse roofing and mulching as a drainage material in green roofs is proposed.

2 Literature Review

This section includes a review of research papers to provide insight into recycled materials used in green roofs. In particular, the literature review explores the possibility of using recycled materials in the drainage layer.

About a decade ago, the first important studies on the possibility of using alternative materials for the drainage layer in green roofs were carried out at the GREA Innovació Concurrent research center of the University of Lleida, Spain. Researchers studied the possibility of using rubber crumbs from tires as drainage layer in green roofs, substituting the porous stone materials currently used (such as expanded clay, expanded shale, pumice, and natural pozzolana). Vila et al. [7] and Pérez et al. [8] studied the ability for draining of recycled rubber granules and compared them with that one offered by pozzolana in a laboratory set-up and tracked the new solution mounted in trays to see if recycled rubber affected the plants development. In addition, the new solution using rubber crumbs was also studied to test if it would keep the same insulating properties that the green roof with stone materials presented in previous studies in Continental Mediterranean climate, by comparing three experimental cubicles made of the same construction system and only varying the composition of the roof. Then, Rincón et al. [9] evaluated the environmental performance of green roofs in which the drainage layer was made of rubber crumbs, by applying the Life Cycle Assessment (LCA) methodology. The operational phase of the LCA was obtained from an experimental set-up consisting of four house-like cubicles with each type of roof, located in Lleida, where the energy consumptions of the heating and cooling period were measured. Finally, Coma et al. [10] compared the thermal performance and electrical energy consumption by the heat pumps of the three cubicles in an experimental facility during the cooling period. The authors concluded that extensive green roofs can be a good tool to save energy during summer in Continental Mediterranean climate, and the use of rubber crumbs instead of pozzolana as drainage layer material in extensive green roofs is possible and should not arise any problem for its good operation. In addition, extensive green roofs with recycled rubber as drainage layer is an environmentally friendly constructive system and it should be recommended for the use in buildings.

Recently, other studies were carried out at the GeMMe Building Materials, Urban and Environmental Engineering (UEE) research center of the University of Liege, in Belgium, where green roofs with the substrate and drainage layer incorporating coarse recycled materials were tested and assessed following ISO 9869-1. In a first study,

Kazemi and Courard [11] simulated specific rubber crumbs and volcanic gravel on the base of the temperature and humidity transfers within green roof systems and the thickness of substrate and drainage layer was optimized and adapted for dry summer in Mediterranean climate. Kazemi et al. [12] assessed the heat resistance of green roof systems with a drainage layer of incinerated municipal solid waste aggregate (IMSWA) and a substrate layer with recycled tiles and bricks in wet and dry states and investigated the effect of media mixtures' moisture content on thermal performance of green roof. Moreover, the drainage and substrate layers were separately exposed to the temperature to measure their heat resistance. The authors concluded that the use of IMSWAs for the drainage layer in green roof systems represents an interesting solution for helping water evacuation and, additionally, dry the substrate layer. This implies a high thermal resistance and better insulating properties of the green roof system. Then, Kazemi and Courard [13] and Kazemi et al. [14] experimentally assessed the thermal resistance of green roof layers with coarse recycled materials following ISO 9869-1 standard. Thereafter, the green roof layers' thermal properties were utilized for the simulation of green roof systems and the modeling data were validated with experimental data. Finally, the temperature variation within green roof layers was evaluated, where the substrate and drainage layer' thickness was changed to achieve an optimum design of green roof systems with an adequate thermal performance. The authors concluded that both control and proposed green roofs provided nearly the same thermal resistance for rooftops and applied the same dead load on the top of buildings. However, to reduce the burden on the environment and save natural resources, the use of green roof with recycled materials and optimized thicknesses of the different layers is recommended.

Other researchers investigated the usage of recycled materials as alternative to commercial materials. Nagase [15] choose materials (cocopeat, PET bottle, and bamboo) based on the following factors: easily obtainable, reminiscent of environmental issues, long durability, and ability to maintain water for plant growth, focusing on plant growth and moisture uptake implications. The results showed that reused materials were observed to function well as commercial green roof materials. Naranjo et al. [16] evaluated scale prototypes of three semi-intensive green roof systems with different types of drainage systems made of recycled and reused materials (rubber, trays, and bottles) and compared them to a traditional green roof system using natural aggregates (gravel) as drainage layer. In terms of hydraulic performance, the results showed that the granular drainage systems (gravel and rubber) were very efficient because they retained all the precipitation, while the systems composed of module containers (bottles and trays) retained approximately half of the water supplied at that level of precipitation. With respect to the thermal behavior, the use of green roof systems with drainage layers made of recycled and reused materials had, like gravel roofs, the potential to reduce the consumption of electrical energy in buildings derived from artificial cooling. Finally, Pushkar [17] evaluated alternative materials for each layer of green roofs to identify their potential environmental impact, concluding that Perlite substitution with byproducts in the substrate and drainage layers was a sensitive issue for the different byproduct evaluation approaches because no difference was noted between perlite-based and byproduct-based roofs.

The following Table 1 gives a summary of the papers analyzing recycled materials in green roof drainage.

Table 1. Summary of papers related to the use of recycled materials in green roof drainage.

Author	Material	Scope
Vila et al. [7]	Rubber crumbs	Hydraulic behavior Thermal behavior
Pérez et al. [8]	Rubber crumbs	Hydraulic behavior Thermal behavior
Rincón et al. [9]	Rubber crumbs	Environmental performance
Coma et al. [10]	Rubber crumbs	Energy consumption
Kazemi and Courard [11]	Rubber crumbs	Heat and moisture transfers
Kazemi et al. [12]	Incinerated municipal solid waste aggregate	Heat transfer
Kazemi and Courard [13]	Coarse recycled materials	Heat resistance Hygrothermal conditions Optimizing thickness
Kazemi et al. [14]	Coarse recycled materials	Heat resistance Optimizing thickness
Nagase [15]	Cocopeat, PET bottle, and bamboo	Plant growth Moisture uptake
Naranjo et al. [16]	Rubber, trays, and bottles	Hydraulic behavior Thermal behavior
Pushkar [17]	Byproducts	Environmental performance

3 Plastic Granule from Recycled Greenhouse Covering Films

Plastic films are used in greenhouse cultivation system as covering materials, in the form of transparent sheets for under-tarp moisture collection or as black sheets for crop mulching. At the end of their useful life, these films are taken off and treated as waste by disposing of them in landfills or by recycling them into secondary raw materials for a wide range of applications, including rubbish bags and boxes. Unfortunately, by now, around 50% of plastic wastes generated by agricultural activities is treated in landfills, so emphasizing upon the urgent need to find and follow alternative and more sustainable routes.

In Italy, each year, more than 350,000 t of plastic materials are used in agricultural activities with a consequent post-use material flow of about 200,000 t. In addition, more than 2 million hectares are used for greenhouse cultivation system, i.e. approximately 21% of the whole cultivated surface in Italy.

A previous study conducted by Cascone et al. [18] reported upon a combined evaluation of environmental issues associated with the manufacture of plastic granules from agricultural waste as a zero-burden material input.

Proposing low-density polyethylene (LDPE) granules from waste films used in agriculture for greenhouse roofing and mulching as a drainage material in green roofs can increase the sustainability and can represent an innovative way of using this material.

The production of polyethylene granules complies with the UNI 10667-1 standard on the recycling and recovery of plastic waste. In order to be further processed, this standard provides that the particle size of the recycled granule is between 2 and 3.5 mm. In the previous section, studies on rubber crumbs have shown that the suitable diameter is between 3.5 and 7 mm to optimize energy and hydraulic performance. Furthermore, to make the polyethylene lighter than the one currently produced, it is necessary to reverse the degassing step during the manufacturing process. The low weight of the drainage layer in green roofs is essential as the material can be used in existing buildings constructed before the entry into force of the regulation on the reduction in energy consumption, representing more than 80% of existing buildings.

Currently, in order to reduce installation times and costs, the granular material commercially used for the drainage layer (perlite, expanded clay, etc.) is closed in non-woven fabric bags, thus avoiding its dispersion into the environment during the installation phase. The costs of both materials and non-woven fabric bags are high and comparable to that of plastic panels representing the commercial alternative to granular material for the drainage layer in green roofs, as mentioned in Sect. 1.

Making both bags and granules in recycled polyethylene would reduce costs compared to commercial solutions (perlite with non-woven fabric bags). Finally, the micro-perforated bags will be designed to allow water to flow through the drainage layer and, at the same time, to prevent the fine particles of the substrate occluding the drainage layer. Finally, the possible use of soil coming from the washing of agricultural films as a substrate in green roofs will be verified, replacing commercial substrates.

4 Future Research Development

The next step of the research will concern the laboratory assessment on the thermo-physical properties (i.e. particle size, specific weight, minimum and maximum density, permeability, thermal conductivity, etc.) of several polyethylene granules (different in the production process, in the particle size, etc.), and they will be compared with those of perlite, the granular material commercially used for the drainage layer in green roofs. The test will be performed at different water content conditions, i.e. moist, saturated and dried, at which the tested materials can be found most commonly.

In addition, the thermal performance of the proposed system will be compared with those of commercial green roofs and traditional roof. To this end, a set-up will be installed having the goal of assessing the thermal performance of different green roof technologies. The analysis of thermal behavior will involve both thermo-physical parameters (surface temperatures and heat flows) and dynamic parameters (decrement factor and time lag).

5 Conclusions

Based on the extensive review of scientific literature on the use of innovative materials from the recovery/recycling of products for the drainage layer in green roofs these studies, it seems well worth developing as simple green roof systems as possible, without several artificial layer materials, suggesting finding alternative materials for the components with the greatest environmental impacts, or leaving them away, when possible, in order to have a minimum environmental impact while performing the desired green roof functions.

The proposed solution, starting from already tested industrial productions (recycling of polyethylene from agricultural films, production of both small diameter granules and plastic bags) aims to increase innovation through the creation of a finished product such as polyethylene with a diameter greater than 3.5 mm and lighter through the inversion of the current degassing phase, the creation of micro-perforated plastic bags and the use of the soil coming from the washing of agricultural films. The proposed solution may represent an alternative to the commercial solutions used for green roofs from an environmental, economic and social point of view.

Acknowledgement. This research is carried out in collaboration with I.LA.P. s.p.a., a Sicilian company producing granules through the regeneration of low-density polyethylene from the recovery of waste films used in agricultural greenhouses.

References

1. Detommaso, M., Cascone, S., Gagliano, A., Nocera, F., Sciuto, G.: Cool roofs with variable thermal insulation: UHI mitigation and energy savings for several Italian cities. In: Littlewood, J., Howlett, R.J., Capozzoli, A., Jain, L.C. (eds.) Sustainability in Energy and Buildings. SIST, vol. 163, pp. 481–492. Springer, Singapore (2020). https://doi.org/10.1007/978-981-32-9868-2_41
2. Oke, T.R., Mills, G., Christen, A., Voogt, J.A.: Urban Climates. Cambridge University Press, Cambridge (2017). https://doi.org/10.1017/9781139016476
3. Evola, G., Cascone, S., Sciuto, G., Parisi, C.B.: Performance comparison between building insulating materials made of straw bales and EPS for timber walls. In: IOP Conference Series: Materials Science and Engineering (2019). https://doi.org/10.1088/1757-899X/609/7/072020
4. Jim, C.Y., Chen, W.Y.: Bioreceptivity of buildings for spontaneous arboreal flora in compact city environment. Urban For. Urban Green. **10**, 19–28 (2011). https://doi.org/10.1016/j.ufug.2010.11.001
5. Vijayaraghavan, K.: Green roofs: a critical review on the role of components, benefits, limitations and trends. Renew. Sustain. Energy Rev. **57**, 740–752 (2016). https://doi.org/10.1016/j.rser.2015.12.119
6. Koroxenidis, E., Theodosiou, T.: Comparative environmental and economic evaluation of green roofs under Mediterranean climate conditions – extensive green roofs a potentially preferable solution. J. Clean. Prod. **311** (2021). https://doi.org/10.1016/j.jclepro.2021.127563
7. Vila, A., Pérez, G., Solé, C., Fernández, A.I., Cabeza, L.F.: Use of rubber crumbs as drainage layer in experimental green roofs. Build. Environ. **48**, 101–106 (2012). https://doi.org/10.1016/j.buildenv.2011.08.010
8. Pérez, G., Vila, A., Rincón, L., Solé, C., Cabeza, L.F.: Use of rubber crumbs as drainage layer in green roofs as potential energy improvement material. Appl. Energy **97**, 347–354 (2012). https://doi.org/10.1016/j.apenergy.2011.11.051

9. Rincón, L., Coma, J., Pérez, G., Castell, A., Boer, D., Cabeza, L.F.: Environmental performance of recycled rubber as drainage layer in extensive green roofs. A comparative Life Cycle Assessment. Build. Environ. **74**, 22–30 (2014). https://doi.org/10.1016/j.buildenv.2014.01.001

10. Coma, J., Pérez, G., Castell, A., Solé, C., Cabeza, L.F.: Green roofs as passive system for energy savings in buildings during the cooling period: use of rubber crumbs as drainage layer. Energ. Effi. **7**(5), 841–849 (2014). https://doi.org/10.1007/s12053-014-9262-x

11. Kazemi, M., Courard, L.: Modelling thermal and humidity transfers within green roof systems: effect of rubber crumbs and volcanic gravel. Adv. Build. Energy Res. 1–26 (2020). https://doi.org/10.1080/17512549.2020.1858961

12. Kazemi, M., Courard, L., Hubert, J.: Heat transfer measurement within green roof with incinerated municipal solid waste aggregates. Sustainability **13** (2021). https://doi.org/10.3390/su13137115

13. Kazemi, M., Courard, L.: Modelling hygrothermal conditions of unsaturated substrate and drainage layers for the thermal resistance assessment of green roof: effect of coarse recycled materials. Energy Build. **250**, 111315 (2021). https://doi.org/10.1016/j.enbuild.2021.111315

14. Kazemi, M., Courard, L., Hubert, J.: Coarse recycled materials for the drainage and substrate layers of green roof system in dry condition: parametric study and thermal heat transfer. J. Build. Eng. **45**, 103487 (2022). https://doi.org/10.1016/j.jobe.2021.103487

15. Nagase, A.: Novel application and reused materials for extensive green roof substrates and drainage layers in Japan – plant growth and moisture uptake implementation. Ecol. Eng. **153**, 105898 (2020). https://doi.org/10.1016/j.ecoleng.2020.105898

16. Naranjo, A., Colonia, A., Mesa, J., Maury-Ramírez, A.: Evaluation of semi-intensive green roofs with drainage layers made out of recycled and reused materials. Coatings **10**, 525 (2020). https://doi.org/10.3390/COATINGS10060525

17. Pushkar, S.: Modeling the substitution of natural materials with industrial byproducts in green roofs using life cycle assessments. J. Clean. Prod. **227**, 652–661 (2019). https://doi.org/10.1016/j.jclepro.2019.04.237

18. Cascone, S., Ingrao, C., Valenti, F., Porto, S.M.C.: Energy and environmental assessment of plastic granule production from recycled greenhouse covering films in a circular economy perspective. J. Environ. Manag. **254**, 109796 (2020). https://doi.org/10.1016/j.jenvman.2019.109796

Towards the Environmental Sustainability of the Construction Sector: Life Cycle Environmental Impacts of Buildings Retrofit

Simona Rosaria La Mantia[1], Roberta Rincione[1], Francesco Guarino[1], Sonia Longo[1], Marina Mistretta[2]([⊠]), and Maurizio Cellura[1]

[1] Department of Engineering, University of Palermo, 90128 Palermo, Italy
[2] Department of Heritage, Architecture, Urbanism, University Mediterranea of Reggio Calabria, Salita Melissari - Feo di Vito, Reggio Calabria, Italy
marina.mistretta@unirc.it

Abstract. In the context of the need for carbon emissions reduction, the building sector, as one of the most energy intensive one, needs tools and approaches towards carbon neutrality and the increase of the buildings overall energy performances. The paper proposes an integrated approach towards the environmental performances analysis of a small neighborhood having as goal the achievement of the Positive Energy District target and the assessment of its environmental impacts. The methodology proposed includes building modeling and dynamic energy simulation using the Energy Plus engine and a simplified Life Cycle Assessment approach. Two scenarios are investigated: i) the existing neighborhood, ii) the redesigned one achieving Positive Energy District levels. Electricity use is computed in both scenarios, embodied impacts for the additional materials and systems used are also computed in the second one. Results highlight significant improvements (up to 50%) in most environmental indicators, strengthening the need for holistic approaches when discussing the feasibility of carbon neutrality objectives, specifically when the general inclination is to choose just one indicator as metric.

Keywords: Positive Energy Districts · Neighborhoods · Building simulation · Life Cycle Assessment · Environmental performances

1 Introduction

With COP 26 concluded in the past months, the way towards a more sustainable energy sector is a shared need among policies and technical approaches, which – although being accepted quantitatively at a slower pace than what is probably needed – is becoming increasingly present and achieving an always more convincing awareness among stakeholders.

However, achieving the trajectory for 1.5 °C maximum increase in the next decades is a major accomplishment that would require a significant increase in the efforts throughout the world and cannot overlook a substantial rethinking of the concept of urban agglomerations themselves.

F. Calabrò et al. (Eds.): NMP 2022, LNNS 482, pp. 1850–1859, 2022.
https://doi.org/10.1007/978-3-031-06825-6_178

The buildings sector, as one of the most energy intensive ones, needs to aim to an effective and true decarbonization [1], including a reliable paradigm shift to the concept of zero energy and carbon districts and cities, from mere users of energy and resources to a more complex entity including smart management (and generation) of electricity and heat, generation and re-use of resources.

In the context of the European New Bauhaus initiative [2], and in a more general framework in the EU centered on the improvement of energy performance of buildings [3, 4] and on circular economy [5], the focus is shifting from the single building perspective towards the districts/neighborhood level. This is due to several reasons including the possibility to achieve the management of energy generation through renewable sources, balance generation and load through smart grids, optimize energy planning, and, ultimately, leading to poles able to act as aggregators and examples leading the change throughout the city also through social innovation and living labs.

The Positive Energy Districts (PED) [6, 7] cultural change developed throughout Europe under these premises: under the umbrella of the European Strategic Energy Technology Plan (SET PLAN Action 3.2), a PED is defined as a "district with annual net zero energy import and net zero CO_2 emission working towards an annual local surplus of renewable energy".

The programme "Positive Energy Districts and Neighborhoods for Sustainable Urban development" [8] supports the planning, deployment and replication of 100 positive energy neighbourhoods by 2025 and is conducted by JPI Urban Europe. Also from a more research-oriented perspective several initiatives at different levels are available either at the international (IEA EBC Annex 83 "Positive Energy Districts) [9] and at the EU level (i.e. European Energy Research Alliance on PED, EU PED NET COST Action etc.) [10].

The paper proposes an energy/environmental analysis of a small group of buildings located in the Mediterranean area (Sicily) to investigate the potential for the achievement of the PED level in the area through retrofit [11] of the existing buildings and deployment of renewable energy generation techniques. As per the PED definition, the environmental cost of the fulfillment of the positive energy target will be investigated in the use stage and with some life cycle considerations [12].

2 Methodology

The study was developed on the analysis of two different buildings located in Sicily (Southern Italy). The main methodological stages followed in the analysis are briefly recapped in the following:

- Site analysis: the necessary documentation for the buildings energy analysis was acquired. This includes geometrical and energy information on both the envelope and the indoor spaces, occupation data for all indoor areas, HVAC, lighting and other technical information and more than one year of energy bills for both electricity and methane;

- Energy modeling was performed through Google Sketchup [13] for the Energy Plus engine [14]. The thermal zoning included the development of more than twenty zones per building in order to achieve the necessary accuracy. All available energy and physical data acquired were input to the model including thermal-physical properties of the materials;
- Building energy simulation in non-steady state was performed through Energy Plus and energy uses for the simulation for both buildings were compared to the baseline from the energy bills. Model calibration was performed on the more uncertain parameters;
- A set of redesign solutions, including HVAC and envelope modification, is developed and implemented for each of the buildings as well as for deployment of PV systems to offset the energy uses towards a net positive annual primary energy balance;
- A simplified Life Cycle Assessment approach is employed to quantify the environmental impacts of the materials and systems used for the energy retrofit, developed according to the ISO 14040 standards and while using EPD impact indicators. The approach includes in its system boundaries the stages of production and use and compares two different scenarios: i) the existing district, ii) the retrofitted district, computing the environmental impacts of the materials used in the redesign and the reduced impacts in the use stage of the buildings. Data from the Ecoinvent database are used for the materials and systems computed in the analysis [15].

The energy efficiency solutions investigated are briefly summarized in the list below and are implemented within both buildings:

1. Insulation of the opaque structure through the use of polystyrene. In both buildings the new U values will be coherent with the current legislation in the domain of energy performance of buildings in Italy i.e. the U value for the opaque vertical walls is 0.34 $W/(m^2 K)$ while for the roof this value is equal to 0.33 $W/(m^2 K)$ in both buildings;
2. Glazings and frames are substituted with higher performance ones (Double glazings, with $U < 2.20$ $W/(m^2 K)$);
3. The old energy systems are substituted with higher efficiency water heat pumps;
4. The lighting system is expected to be substituted through the use of LED solutions;
5. Photovoltaics systems are expected to be planned either on the roof of the buildings or on adjacent structure;
6. A 200 kWh electricity storage system based on the lithium-ion technology is expected to be used to improve the load match of the system.

3   Description of the Buildings

The two investigated office buildings are located in Monreale in the province of Palermo, Italy (Latitude: 38° 4′ 48,6" N Longitude: 13° 17′ 20" E Altitude (m): 314) and for the sake of simplicity will be defined in the following as building A and B.

Building A houses the local municipal police command, is composed of three levels, one of which is partially underground. Each floor has an area of 500 m². From the main entrance it is possible to reach the municipal police command by passing through a concierge. The right side houses the staircase, toilets, archives and a large room used as a dressing room. The first floor plan has a "C" shape, characterized by a long central corridor and two lateral arms. The remaining area is occupied by a large terrace (Fig. 1).

Fig. 1. Office A, (North façade (left)

Office B is the main venue of the municipal offices, it develops on three levels, all above ground. Each floor measures 371 m². The main façade overlooks the street, while the west one borders the municipal police command from which it is separated from by an avenue intended for pedestrians only (Fig. 2).

Fig. 2. Office B, (North façade)

Building A was built around the 1960, the envelope is based on bricks (50 cm thick with a U value of 1,086 W/(m² K), with single glazed windows (average glazing U value: 5,894 W/(m² K). Building B instead has 30 cm thick brick walls and a U value: 1,574 W/m² K). Figure 3 and 4 show the plans for the first level of both buildings.

Occupancy is variable according to the zone considered but a maximum number of 60 occupants in building A and 45 in building B is considered, distributed among the different areas of the building., with variable occupancy rates between 9 a.m. until 1 p.m. slowly decreasing until 7 p.m.

Artificial lighting is based mostly on fluorescent tubes and its power installed varies slightly within the different building areas between 10 W/m² to 12 W/m².

The energy systems used to ensure the covering of heating and cooling are based on low efficiency single split air conditioning solutions.

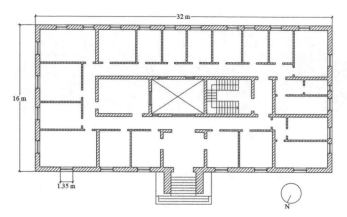

Fig. 3. Office A, first level plan

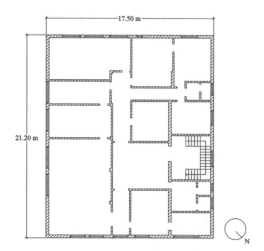

Fig. 4. Office B, first level plan

4 Results

The preliminary evaluation of the quantitative outputs of the models is performed by comparing the available models results with the available energy bills data and by performing model calibration to reduce the deviations among the two data sets. The results of this procedure are reported in Fig. 5 and 6 showing electricity aggregated energy data for heating, cooling, lighting, appliances and any other auxiliary electricity use. In all cases the deviation between monitored electricity use and simulated ones are lower than 10% on a monthly base, with the total at the annual scale equal to 1% deviation for building A and 1.94% for building B.

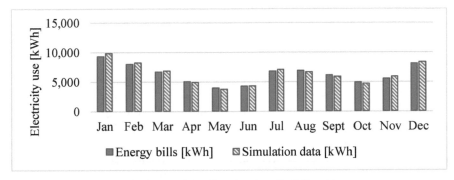

Fig. 5. Model calibration data (Building A)

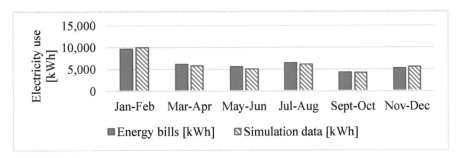

Fig. 6. Model calibration data (Building B)

The results for the calibrated building simulation are briefly recapped in the following:

- Building A is characterized by a use of electricity equal to around 76,000 kWh per year (50.66 kWh/m^2),
- Building B is characterized by an electricity use of 36,303 kWh/year, or 32.62 kWh/m^2).

In this context the potential for energy efficiency is investigated by implementing the aforementioned energy efficiency solutions separately, to investigate their potential for improvement in building performances.

The main results of this scenario analysis are reported in Table 1.

Table 1. Redesign solutions for the analysed

	Building A [kWh]	Building B [kWh]	Variation Building A [%]	Variation Building B [%]
1. Insulation	62063.03	29900.23	−18.34%	−17.64%
2. Glazings and frames	73835.92	36170.44	−2.85%	−0.37%
3. Heat pumps	68170.89	33280.94	−10.30%	−8.32%
4. LED Lighting	65957.35	32559.13	−13.21%	−10.31%

Table 1 shows a predominant weight in the impact of the insulation, by the substitution of the heat generators as well as the implementation of the LED lighting solutions. Glazing and frames substitution would have only a limited impact due to the limited window to wall ratio of all facades on both buildings (around 8–10%).

Figure 7 reports the monthly trends of energy use between the existing small neighborhood and the redesigned one. The highest difference is traced for December equal to 50.74%, the lowest for May equal to 22.38%.

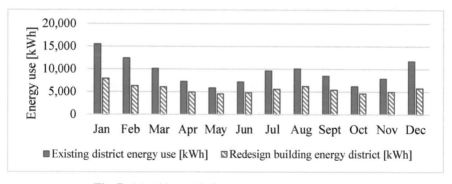

Fig. 7. Monthly trend of energy use, redesign scenario

The installation of photovoltaics is investigated throughout several south oriented surfaces on both buildings as shown in Fig. 8. The PV is considered also to be installed on a structure on top of a vehicle parking area adjacent to building A for a total of 52 kW peak power and an overall area equal to 407.18 m^2.

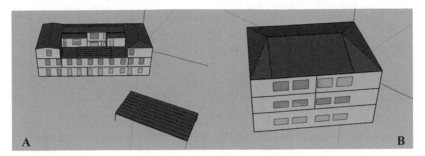

Fig. 8. Installation of PV modules (purple areas) within the two buildings

If all the buildings are considered and the overall generation per year is concerned, the overall neighborhood energy uses of the redesign district are equal to 67,052 kWh, while the overall generation is equal to 74,123 kWh (see Fig. 9 for the monthly distribution of energy generation and consumption), thus per the definition of Positive Energy District as a neighborhood able to generate more energy than the one that is used within its geographical boundaries, the PED target can be considered as achieved for the small neighborhood investigated.

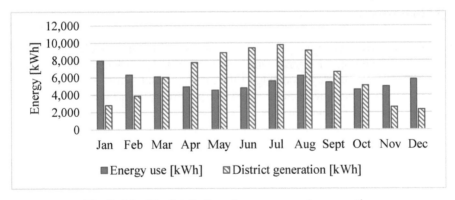

Fig. 9. Monthly distribution of energy use and consumption

The last step of the analysis is based on the quantification of environmental impacts caused by the scenarios described to achieve the positive energy target.

Table 2 shows the main results of this comparative analysis which computes electricity use for scenario 1 and embodied contributions, as well as the residual electricity use for the redesign scenario 2.

Results clarify a significant reduction in most indicators with global warming being reduced by a half in the redesign scenario as well as energy use being nearly three times lower. Only the abiotic depletion shows an increase, which is to be explained with the use of photovoltaic panels in scenario 2.

The percentage variations are in the range of 28,14%(Eutrophication) to 56.03% (Abiotic depletion- fossil fuels).

Table 2. Environmental impacts of the neighborhood (annualized data)

Impact category	Unit	Scenario 1	Scenario 2–Redesign *	Environmental impacts/Embodied energy (Scenario 2)
1. Acidification	kg SO2 eq	176.69	113.33	78.12
2. Eutrophication	kg PO4--- eq	49.22	35.37	25.56
3. Global warming	kg CO2 eq	46,501.50	20,584.54	11,318.95
4. Photochemical oxidation	kg NMVOC	105.48	66.78	45.76
5. Abiotic depletion, elements	kg Sb eq	0.44	1.83	1.75
6. Abiotic depletion, fossil fuels	MJ	607,946.12	267,325.16	146,189.77
7. Water scarcity	m3 eq	23,930.18	14,409.36	9,641.19
8. Ozone layer depletion	kg CFC-11 eq	0.0060	0.0040	0.0030

5 Conclusions

The paper has investigated the performances of a small group of buildings under the assumptions of the Positive Energy Districts framework. In particular, the analysis has included the detailed energy modeling of the building, calibration efforts and a set of redesign scenarios that have demonstrated the possibility to achieve the PED target within the case study in southern Italy.

While the technical feasibility was evaluated as possible, the main results from the simplified life cycle oriented environmental study show that the deployment of all technologies and materials required to achieve the PED level is justified in terms of comparison of embodied energy and impacts increase, if compared to the substantial energy use/environmental impacts associated to the use stage only.

This leads to some final considerations:

- While it is understandable to try and develop simple formulations of the PED definition to increase its potential for diffusion, it should always be kept in mind that embodied contributions have a weight in the energy balance and in its emissions that is not negligible and is often comparable (as is the case for this paper) with the use stage of conventional districts. Thus in itself the definition proposed by the Setplan (which includes carbon neutrality), is much more difficult to achieve than a simple use stage balance would anticipate;
- Neglecting the life cycle perspective can actually lead to the choice of wrong solutions sets and actually steer further away from carbon neutrality if materials with large embodied impacts are selected;

- Not all indicators chosen for the study identify clear and shared reductions in overall environmental impacts. This reinforces the need for holistic approaches when discussing the feasibility of such objectives, specifically when the general inclination is to choose just one indicator as metric.

References

1. Cusenza, M.A., Guarino, F., Longo, S., Cellura, M.: An integrated energy simulation and life cycle assessment to measure the operational and embodied energy of a mediterranean net zero energy building. Energy Build. **254** (2022). https://doi.org/10.1016/j.enbuild.2021.111558
2. https://europa.eu/new-european-bauhaus/index_it. Accessed 27 Dec 2021
3. https://eur-lex.europa.eu/legal-con-https://eur-lex.europa.eu/legal-con-tent/EN/ALL/;ELX_SESSIONID=FZMjThLLzfxmmMCQGp2Y1s2d3TjwtD8QS3pqdkhXZbwqGwlgY9KN!2064651424?uri=CELEX:32010L0031. Accessed 27 Dec 2021
4. https://eur-lex.europa.eu/legal-con-tent/EN/TXT/?qid=1399375464230&uri=CELEX:32012L0027. Accessed 27 Dec 2021
5. https://ec.europa.eu/environment/strategy/circular-economy-action-plan_it. Accessed 27 Dec 2021
6. Marotta, I., Guarino, F., Longo, S., Cellura, M.: Environmental sustainability approaches and positive energy districts: a literature review. Sustainability (Switzerland) **13**(23), 13063 (2021)
7. Brozovsky, J., Gustavsen, A., Gaitani, N.: Zero emission neighbourhoods and positive energy districts – a state-of-the-art review. Sustain. Cities Soc. **72**, 103013 (2021)
8. https://jpi-urbaneurope.eu/ped/. Accessed 27 Dec 2021
9. Hedman, Å., et al.: IEA EBC Annex83 positive energy districts. Buildings **11**, 130 (2021). https://doi.org/10.3390/buildings11030130
10. https://pedeu.net/. Accessed 27 Dec 2021
11. Cellura, M., Ciulla, G., Guarino, F., Longo, S.: Redesign of a rural building in a heritage site in Italy: towards the net zero energy target. Buildings **7**(3), 68 (2017)
12. Cusenza, M.A., Guarino, F., Longo, S., Ferraro, M., Cellura, M.: Energy and environmental benefits of circular economy strategies: The case study of reusing used batteries from electric vehicles. Journal of Energy Storage **25**, 100845 (2019)
13. https://www.sketchup.com/it. Accessed 27 Dec 2021
14. https://energyplus.net/. Accessed 27 Dec 2021
15. https://ecoinvent.org/the-ecoinvent-database/. Accessed 27 Dec 2021

Reversible, Sustainable and Circular Constructive Systems: Buildability Conditions

Tecla Caroli[✉] ⓘ, Andrea Campioli ⓘ, and Monica Lavagna ⓘ

ABC Department, Politecnico di Milano, 20133 Milano, Italy
tecla.caroli@polimi.it

Abstract. The paper reports the results of an investigation of the Reversible Technologies oriented to solve the problem of the huge quantity of building components that once disassembled have still residual performances. Thus, using Reversible Technologies, it is possible to reach the extension of the use of constructive systems, activate the potential circularity of the disassembled systems and obtain the environmental impacts optimization of the buildings, demonstrated by the analysis of LCA studies taken from literature.

In practice, the potentialities of the reversibility are not always exploited, due to the lack of effective reversible products on the market and competent operators. To face it, the research aimed to define the Buildability Conditions on technical and operative levels, for the actualization of reversible technologies and the activation of circular solutions.

Thanks to the design and manufacturing practices (Pre-conditions) that generate the reversible transition in the construction sector, the Buildability Conditions are applicable. Specifically, the conditions defined are the operative actions (Soft Conditions), regarding the actions to develop by the operators involved and the definition of tools and methods, that support the development and verification of the technical characteristics (Hard Conditions) of the constructive systems.

The next step of the research will be the application of the Buildability Conditions on practical experiences, to verify their feasibility and actual contribution.

Keywords: Reversible Technologies · Sustainability · Circular economy · LCA

1 The Potentiality of Reversible Technologies

In recent years, life cycle design has been promoted by implementing the design for disassembly and deconstruction (Design for Disassembly - DfD) of prefabricated constructive systems to facilitate the reuse, remanufacturing and recycling of building elements at end of their service life [1]. Thus, the brief use of reversible and long-lasting construction systems makes it possible to recover them with different circular strategies by reintroducing components that are functionally obsolete (end of service life) but not yet considered waste (end of life) for their residual performances.

To ensure the efficient use of resources and the verification of the environmental effects of design and construction choices, it is necessary to use assessment methods

© The Author(s), under exclusive license to Springer Nature Switzerland AG 2022
F. Calabrò et al. (Eds.): NMP 2022, LNNS 482, pp. 1860–1869, 2022.
https://doi.org/10.1007/978-3-031-06825-6_179

that evaluate the benefits during the whole life cycle of the object analyzed. Despite the degree of uncertainty due to the different assumptions on the retrieval of the data used, that do not respond to the wide range of technological solutions present, the LCA (Life Cycle Assessment) is useful and fundamental for the analysis of environmental sustainability [2].

1.1 LCA Studies: Verification of Environmental Benefits

Kibert [3] suggested the basic rules for implementing a closing-loop construction process: the buildings are disassembled; the building components are composed by disassembly layers; and, finally, materials are recoverable. Taking into account these rules, reversibility has been identified as the necessary means to promote circular building processes [4]. Reversible technologies are therefore essential for the disassembly and recovery of end-of-life components to minimize emissions.

Evidence of this reduction is the comparative LCA study carried out by Eckelman et al. [5] in which 8 concrete and steel buildings were compared, 7 designed to be disassembled (DfD) and recovered at the end of service life and 1 consisting of conventional construction technologies that allow only demolition and landfill at the end of life. It has been possible to verify that DfD buildings made of durable materials and disassembly components allow to activate reusing solutions, obtaining a reduction in environmental impacts ranging from 60% to 70% compared to a conventional building. As mentioned above, this variation depends on the composition of the materials used (e.g. recycled content), the method of manufacture (partially or totally prefabricated), the transport of materials and components, the method of assembly of components that allows the activation of different circular strategies.

Design for disassembly has proved to be a necessary but not sufficient condition for reducing end-of-life impacts: the design phase should not be limited to the appropriate selection of materials and components but must also provide for a second use of the same at the end of the first useful life. Predicting a new life for the building would reduce the values of different environmental impact categories by up to 40% [6]. On the contrary, if the second use is not foreseen, the economic cost of the disassembly operation could become very high as well as the recovery of components.

By comparing the impacts on constructive systems composed by disassembly and reused components with new components, it is possible to achieve a reduction in impacts during the entire life cycle of up to 40% [7], considering the structure, space plan, skin and stuff building layers [8].

Furthermore, by comparing the application of the most diffused circular strategies (reusing, recycling and remanufacturing) for the second use of building components (skin, stuff, space plan) at the end of useful life, recycling generates the lower environmental optimization:

- the reusing allows a reduction of the CO_2 emissions about 53%, compared to the recycling [9];
- the remanufacturing uses the 18% less energy than recycling [10].

The study carried out by Assefa et al. [11] on the recovery scenarios of reuse and regeneration of a disassembly building, shows a potential reduction that varies between 20 and 40%, in six of the seven environmental categories evaluated. The highest reduction is obtained for eutrophication of 37%, followed by acidification (29%). Carbon dioxide emissions and fossil fuel consumption show an avoided impact of 33% and 34% respectively as a result of the decision to resort to selective deconstruction rather than complete demolition and new construction. Ozone depletion in the stratosphere was the only impact category that was not strongly influenced by the reuse of buildings (2% reduction).

1.2 Hard and Soft Technologies Towards Circularity

Extending the life of a building by adapting it to the new needs for different uses can potentially save materials, energy and carbon impacts compared to the new construction. Such reductions are only possible if the building is designed and constructed to be reversible and if its whole life cycle is planned.

To this end, the constructive systems must design and build considering the operative actions that generate the reversible constructive process (Soft Technologies) and the reversible constructive technics that characterized the materials, elements, systems levels of the building (Hard Technologies) [12].

Currently, the importance of developing operational aspects, including methodologies and tools, to support the effective reversibility of construction technologies has not been perceived. This lack has led to not fully exploiting the potential of reversible technologies, which allow both to reduce environmental impacts but also to activate circular solutions on disassembled construction systems.

To overcome the barriers and limits that hinder the actualization of the reversible technologies, it is fundamental to define the conditions to facilitate the construction as well as the adoption of construction techniques (Hard) and operative features (Soft) that allow an effective and efficient constructive process [13]: Buildability Conditions.

2 Pre-conditions for the Transition Towards Reversibility

The State of Art analysis, about the use of Reversible Technologies in the construction sector, highlighted that the design and manufacturing practices are the drivers for the implementation of Reversible Technologies on Hard and Soft levels. For the actualization of Buildability Condition, it is necessary the realization of the Pre-conditions with which the transition towards reversibility can be activated. The transition consists both in the design of the extension of the life of constructive systems and in the manufacturing of the value retention of resources, exploiting their potential circularity. The aim is to reuse the resource available, with residual performances, without the production of waste and use of new materials.

2.1 The Extension of Life of the Project

Conventionally, an architectural project starts when the client wants to invest in a new building. After, the definition of the client's request and resources, time and costs available, it is chosen the professional figure who will take care of the project.

Design reversible system is not simple. The interaction with stakeholders testified that to realize reversible and circular buildings it is fundamental to plan all Hard and Soft aspects that a project must follow to design the extended life of constructive systems. It means that the potential circularity of all its components can be activated at 100% without losing the value of materials, systems and products. To generate reversible constructive systems, it is necessary:

- the definition of project requirements towards reversibility, sustainability and circularity, to setting up the project design;
- the coordination and creation of a consortium among all operators involved in the constructive process, to extend the responsibility among operators;
- the planning the project life cycle, to manage and monitor the development of the project;
- the integration of BIM and LCA/LCC as a design support tool, to assess and verify the project during its development.

2.2 The Value Retention of Products

The task of the manufacturing operations is to offer to the other fields the material resources (products) for the realization of reversible projects and processes. How the products are conceived and manufactured influences the effective reversibility of a constructive system. As said before, in order to obtain the reversibility for the activation of potential circularity, it is necessary to develop simultaneously Hard and Soft conditions referred to technical and operational requirements. It is possible if there is the direct involvement of the manufacturer in the construction process, making the manufacturing supply chain (manufacturing consortium) aware of product responsibility.

Exploiting the extended life of reversible products and conceiving the used products as resources and not as waste, the products can be reused and counted in different life cycles and resources remain in the closing-loop processes. In this way, a circular economy model is created in which nothing is lost that occurs thanks to the design and construction of reversible technologies.

The manufacturing practices to offer the product performances for the success of the reversible and circular constructive systems are:

- the identification of the manufacturing consortium as responsible for the product, to extend the producer responsibility;
- the conceiving of the product from good to service, to offer the right and effective performances;
- the design and production of the reversible performances of the product, to develop the Hard characteristics of the product;
- the elaboration of Material Passport of the product, to track and maintain the product value;
- the creation of second use program in the company, to facilitate circularity.

The practices described are all voluntary. The radical change can be developed thanks to their mandatory application. The transition will be reached if all operators involved in the process believe and activate the change. For this reason, the paper defines the Buildability Conditions on Hard and Soft levels to have guidelines to design and build reversible, sustainable and circular constructive systems.

3 Towards Reversibility, Sustainability and Circularity: Buildability Conditions

The way in which a constructive system is designed is strictly connected to the way it will be constructed. The Buildability Conditions defined are the technical characteristics (Hard) and operative actions (Soft) that generate reversible, sustainable and circular constructive processes, projects and products. The Reversibility starts with the design and so, to achieve technical requirements, it is essential to have an operative support. To this end, firstly, the conditions regard the role of the operators involved in the different phases of the constructive process and in the support tools. Secondly, the conditions refer to the technical and additional characteristics of the different building layers [14] of the different constructive systems (material, element, system, building).

3.1 Phases and Actors for Reversible Process Management: Soft Conditions

The previous chapter has defined the practices to realize the context, the Transition towards Reversibility, Sustainability and Circularity, in which the Buildability Conditions can be actualized.

The consortium is the driver for the development of a reversible construction process. To this end, on the operative level, it is essential to plan the life cycle of the constructive process taking into account the operators involved and the tool and methods that will support, guide and verify the choices.

Pre-design Phase: Definition of Requirements and Project Goals. The client thanks the support of the development manager to create the consortium with all operators involved in the process. Once defined the project requirements, the consortium elaborates the Strategic Plan: for the description of the role of each member of the consortium that deals with their specific expertise; for the collection of information about the choices taken during the process and the requirements of the project for upgrading, adaptability and flexibility operations during the time (different use phases); for the realization of the constructive project; and for the planning of the second use of the constructive systems.

Design Phase: Selection of Products and Project Development. In this phase, the relation between the designer and the manufacturer is strictly connected. Indeed, the designer seeks advice from manufacturers on whether, and how, product value can best be maintained through reuse and how products can be certified for reuse. Where it is possible, an idea should be identifying reusable elements from other buildings (or other fields), incorporating these in the new project and following the reversible strategy. Furthermore, through comparative assessments carried out with the aid of Life Cycle Assessment and Life Cycle Costing methodologies, it is possible to choose the constructive systems with low environmental and economic impacts.

After the selection of all products, their information is collected in an inventory tool, the Material Passport and inserted in the BIM model of the product or project.

Operative Design and Construction Phase: Assembly and Reversibility Plan. Following the instructions contained in the BIM model, the constructors start the site. The detailed drawings are implemented to not compromise the element/system integrity and to ensure maintenance and disassembly purposes. To this end, the reversibility plan and the second-use plan are developed by the designer, manufacturer, constructor, deconstructor and facility manager. The plans contain the inventory of the products selected, their assembly/disassembly method, the maintenance operations according to the manual, the specifications for the application of circular solutions. The information described is updated in the Material Passport of the BIM model of the project.

Use Phase: Product Extended Use. The creation of a maintenance plan integrated in the BIM model, is crucial for the future costs of managing the property, providing the information useful for the proper management operations, optimizing costs, services and transformations of the building over time.

In addition, the maintenance plan allows to maintain systems in the best conditions and to extend their useful life. This is the starting point for the transformation in the time of the components to adapt to the changing needs, in terms of use, safety and integration of functions.

Deconstruction and Second-Use Phase: Disassembly and Circular Strategies Activation. The reversibility and flexibility of the constructive systems offer a positive risk investment: for example, the potential second life cycle of a building can guarantee cheaper financing and loan insurance.

The information about the possible second use of a constructive system is collected in the Material Passport, which allows to track the product data even if the owner of the product change over time. In the best vision in which it is activated the extended responsibility strategy, the ownership is maintained by the manufacturer. The latter will facilitate the management of the second use and will avoid the loss of resources and information.

3.2 Design Guidelines on Different Building Levels: Hard Conditions

After the planning of a reversible process, it is fundamental to define how the constructive system must be designed and built to be reversible. For this reason, in the following paragraphs are described the Buildability Conditions on hard level respectively of materials, elements, systems and buildings levels, to achieve reversible, sustainable and circular goals.

Material Level: Selection Criteria and Characteristics. The achievement of Hard Reversibility of constructive systems starts with the selection of materials during the design phase. The characteristics follow both sustainable and circular goals, to reduce emissions and to produce zero waste. The materials represent the first level of resource of a reversible constructive system, they are considered as the input for the realization of reversible elements, systems and buildings (output). Their characteristics regarding the composition: renewable materials, such as timber and derivatives, with a controlled extraction; recycled, such as the steel with recycled content processed thanks to the recovering of residual steel after the production processes; mono-material, since the composite materials hinder the activation of circular solutions; origins, preferring local products for the optimization of transport costs; no toxic materials, reducing the potential worker and occupant exposure to environmental and health impacts; low environmental and economic impacts materials, assessing the with the LCA and LCC methodology. Finally, the materials selected must be codified collecting all the information in the Material Passport.

Element Level: Manufacturing Features ad Technical Characteristics. The elements are made by the materials that have partially or totally the characteristics cited, and they could be new or reused. The latter derive from disassembly operations of other buildings or from other fields in which the elements have still residual performances. Specifically, the elements must be prefabricated and the manufacturing process should follow sustainable features, producing no waste, consuming low energy and water, using few and local materials and generating low carbon emissions. They must be standard, modular, lightweight and well-dimensioned for easy transportation and reuse, as they generate flexible and multifunctional constructive systems suitable for different locations and buildings and spaces with different uses. These characteristics are enhanced by the reversibility between the parts that compose the element; for example, a dry system should be preferred for the union of the parts of a laminated panel (e.g. sandwich panel) or to use adhesives in small quantities and compatible with the other material elements, in order to apply circular solutions on the entire panel or parts of it (if layered).

In addition, the products that leave the production company and are then transported to the site to be assembled, must be finished elements that do not need wet or secondary finishes which may cover connections, facilitating the identification of the position of the joint points. In this way, the elements result suitable and accessible for maintenance activities (e.g. cleaning and/or replacement).

Focusing on the characteristics of the connections, they must be composed of ductile materials, providing adequate mechanical tolerances for assembly and disassembly operations and minimizing the damages on other elements.

Moreover, it is necessary to minimize the number of fasteners and connectors, speeding the disassembly operations, and use small joints that are material-compatible for the potential recycling at the end of their technical life.

The trackability and the impact transparency (EPD certifications) should be achieved by digitalizing and codifying the elements information in the Material Passport added in the BIM model, provided by the manufacturer.

System Level: Assembly Method ad Second-Use Planning. The system is generated by the assembly of elements. For this reason, the most Hard Conditions referred to the elements are transferable to the system level.

In addition, in order to facilitate maintenance operations during the use phase, systems of elements related to a specific layer of the building (the same function) must be designed and built as independent systems. This feature also allows partial disassembly of systems for different reasons such as spatial expansion (structure and space plan), aesthetic renewal (skin and stuff), malfunction or implementation of plants (service).

Regarding the structure system, it is necessary that it is over-dimensioned to accommodate different functions during the time, in case of disassembly of the other systems and reuse of the structure for other building functions. Hence, the modularity of elements that compose the different systems facilitates the operation of assembly/disassembly and transportation of the different parts.

Finally, as for the previous levels, all the features related to the systems must be collected within the Material Passport that will be completed with the information related to the entire building.

Building Level: Investments and Second-Use Potentiality. The last level is the building. A building is defined as reversible on a Hard level if the elements and the systems that compose it are dry assembled and considering the characteristics fixed before.

Hence, the building reversibility increased the open building systems, composed of flexible elements and adaptive spaces, with the minimal cost for the client. Meeting the user needs means accommodating future changes, from adaptation to additions.

For this reason, the initial investment in reversible buildings is balanced by the maintenance of the value of the building over time, providing the resale to future users or the potential second or infinite use.

Once again, also the building is digitalized since the design phase. Following the Second-use Plan collected in the Material Passport, it is possible to provide all information for the activation of potential circularity.

4   Findings and Future Developments

If in some fields, such as the food industry or the automation industry [14, 15], circular economy proposals have been successfully adopted and generated significant results [16], in the construction sector the development is slow, due to the lack of reversible products and competent operators.

The built environment needs an important transformation. The approach to design and build must change considering the whole life cycle of the building: to conceive the products at the end of service life as resources; to develop reversible technologies, activating the potential circularity of disassembled constructive systems; to generate new recovery markets to create sustainable networks among stakeholders.

To this end, the Pre-Conditions towards the reversible transition and the Buildability Conditions are strictly connected: without the realization of the firsts cannot be realized the seconds, although the Hard conditions cannot be activated without the realization of the Soft ones. Thus, the buildability of reversible constructive systems is achieved thanks to the support of management and assessment methods and tools that facilitate their actualization.

Extending the use of constructive systems with the extension of their technical life and activating the potential circularity of the disassembled systems, it is possible to generate investment risk reduction and environmental optimization. Thus, the analyzed LCA studies testify the reduction of environmental impacts with the use of the disassembly constructive systems compared to the conventional ones; the possibility to activate circular economy solutions, on the end of use components, compared to landfill; and finally, the application of reusing and remanufacturing solutions, through the total reversibility of constructive systems, compared to the recycling, due to the partial reversibility. In addition, the potential of reversible technologies can be really exploited through their dissemination and implementation at a theoretical and practical level.

The future development may be achieved by both technical (Hard) and operative (Soft) innovation, thanks to the experimentation of the Buildability Conditions. The practical application will highlight the feasibility and actualization of the conditions and the diffusion of the reversible practices in the construction sector. Only by combining the theoretical research with practice, it will be possible to realize the transformation and generate reversible, sustainable, and circular processes, projects, and products.

References

1. Jaillon, L., Poon, C.S.: Life cycle design and prefabrication in buildings: a review and case studies in Hong Kong. Autom. Constr. **39**, 195–202 (2014)
2. Jrc, E.C.: European Commission, Joint Research Center: Level(s) – A Common EU Framework of Core Sustainability Indicators for Office and Residential Buildings. Publications Office of the European Union, Luxembourg (2017)
3. Kibert, C.J.: Sustainable Construction, Green building Design and Delivery. Wiley, Hoboken (2008)
4. Durmisevic, E.: Circular Economy in Construction. Design Strategies for Reversible Buildings, BAMB, European Union's Horizon 2020, Netherlands (2019)
5. Eckelman, M.J., Brown, C., Troup, L.N., Wang, L., Webster, M.D., Hajjar, J.F.: Life cycle energy and environmental benefits of novel design-for-deconstruction structural systems in steel buildings. Buil. Environ. **143**, 421–430 (2019)
6. Arrigoni, A., Zucchinelli, M., Collatina, D., Dotelli, G.: Life cycle environmental benefits of a forwards-thinking design phase for buildings; the case study of a temporary pavilion built for an international exhibition. J. Clean. Prod. **87**, 974–983 (2018)
7. Krystofik, M., Luccitti, A., Parnell, K., Thurston, M.: Adaptive Remanufacturing for multiple lifecycles: a case study in office furniture. Resour. Conserv. Recycl. **135**, 14–23 (2018)
8. Brand, S.: How Buildings Learn: What Happens After They're Built. Penguin, USA (1994)
9. Rasmussen, F.N., Birkved, M., Birgisdòttir, H.: Upcycling and Design for Disassembly – LCA of building employing circular design strategies. SBE19 Brussels BAMB-CIRCPATH. IOP Conf. Series: Earth and Environmental Science, vol. 225 (2019)
10. Eberhardt, L.C., Charlotte, L.: Potential of circular economy in sustainable buildings. In: Proceeding Paper, IOP Conference Series: Materials Science and Engineering, vol. 471 (2019)
11. Assefa, G., Ambler, C.: To demolish or not demolish: Life Cycle consideration of repurposing buildings. Sustain. Cities Soc. **28**, 146–153 (2017)
12. Cowan, H.J.: Handbook of Architectural Technology. Van Nostrand Reinhold, New York (1991)
13. Glass, J.: Encyclopaedia of Architectural Technology. Wiley Academy, Chichester (2002)
14. Ardente, F., Talend Peirò, L., Mathieux, F., Polverini, D.: Accounting for the environmental benefits of remanufactured products: method and application. J. Clean. Prod. **198**, 1545–1598 (2018)

15. Fegade, V., Shrivatsava, R.L., Kale, A.V.: Design for remanufacturing: methods and their approaches. In: 4th International Conference on Materials Processing and Characterization, Proceedings, vol. 2, pp. 1849–1858 (2015)
16. European Commission: Towards a circular economy: A zero waste programme for Europe. COM 398 (2014a)

Advanced Circular Design, a Life Cycle Approach

Methods and Tools for an Eco-Innovative Life Cycle Approach for Buildings Energy and Resource Optimization

Domenico Lucanto[(✉)] [iD]

Università degli Studi Mediterrnea di Reggio Calabria, 89122 Reggio Calabria, Italy
domenico.lucanto@unirc.it

Abstract. The built environment with its operational phase, its production chain, and its maintenance, is responsible for over a third of global greenhouse gas emissions, the validation of these data positions the construction sector as the largest responsibility for producing long-term greenhouse gas emissions.

The evolutionary scenarios developed by the IPCC see five possible conditions, based on different levels of future emissions, which will result in a warming of 1.5° or more by 2040, thus exceeding the limit threshold for maintaining biodiversity.

The study conducted retracing the advancement proposed in Regenerative Design, proposes a comparison of perspective in the definition of the fundamental principles of regenerative design, and underlines the importance of considering the aspects concerning the sustainability of buildings already in the early design phase, intercepting in the responses offered by alternatives the need to find solutions that in the medium-term make the transition towards a carbon-free system for the construction industry, increasing the well-being of those who use the goods produced and minimize the need for natural resources such as use soil, biodiversity, water, air, and energy.

Keywords: Advanced circular design · Parametric LCA · GWP optimization

1 Introduction

The built environment with its operational phase, its production chain, and its maintenance, is responsible for over a third of GHG emissions, the validation of these data positions the construction sector as the largest responsibility for producing long-term greenhouse gas emissions.

The studies conducted during the doctoral research period on the Life Cycle Analysis applied to buildings reveal that the energy incorporated in the energetic and environmental balance can represent more than 30% [1] of the primary energy requirement during the life cycle of a common residential building.

Based on this we can estimate that the energy incorporated by the extraction, to the production, use phase, and disposal of building materials represents the 60% of the energy consumed for heating [2].

© The Author(s), under exclusive license to Springer Nature Switzerland AG 2022
F. Calabrò et al. (Eds.): NMP 2022, LNNS 482, pp. 1870–1878, 2022.
https://doi.org/10.1007/978-3-031-06825-6_180

Greenhouse gas emissions, soil consumption, loss of biodiversity, and the exponential increase in materials to be disposed of in landfills are considered the key factors on which to act to counter the effects of climate change.

The scenario outlined in the latest report by the IPCC [3], consolidated by its recent government recognition within COP 26, has no doubt regarding the history and possible evolutions of this issue. The increase in CO2 emissions and other Green House gases (GHG) is certainly the key factor that most determines the increasing of temperature in our climate changing.

The study carried out on **Regenerative Design** [4], assume his comparison of different methodologies and approach based on Life Cycle Assessment to propose a new methodological approach based on the fundamentals of parametric design that underlines the importance of considering the sustainability of buildings already in the Early Design Phase.

The discussed methodology wants to intercept the Characters of Responsivity, defining the alternatives that buildings need to find solutions to operate the transition towards a carbon-free system for the construction industry, by also increasing the well-being of those who use the goods produced and minimizing the need for natural resources such as land use, biodiversity, water, air, and energy.

2 Life Cycle Design: Approach, Methods, and Tools

2.1 Advanced Circular Design and Life Cycle Approach

Advanced Circular Design thinking finds its form as a theoretical framework of sustainable concepts, traducing that into the realization of a digital exportable algorithmic model.

The proposed experimental field can work as a sustainable design approach capable of defining the Life Cycle design evolution, as an innovation of the technological system which interfaces their model for optimizing architectural components design.

The key steps that need to be followed to achieve the Advanced Circular Design Method are refered to the PLCA model, this one are listed under the design and development section (Fig. 1).

The methodological framework according to the correlation between the inputs and the design and development phase, the CAD-CAM incorporates the development of a list of design solutions for the ecological improvement based on the Life Cycle Impact.

The topic proposed in the discussion aims to illustrate the method, the project, and the theory [5] of advanced circular design in an experimental context, underlining the possible aspects of innovations that link the optimization strategy of the life cycle through a methodology of simplified-parametric type [6] with a performance-based approach operated in the same parametric environment.

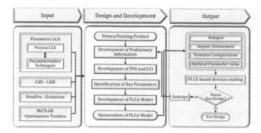

Fig. 1. Framework of methodology Source: S.Kamalakkannan; A.K.Kulatunga

2.2 Life Cycle Thinking: The Methodological Approach for Controlling Environmental Impacts and Energy Balances

Life cycle assessment using the LCA method is certainly the most effective approach to assessing the quantification of the environmental impacts of production processes, it is also regulated by the UNI EN ISO 14040 standards. Studies conducted on simplified approaches to assessing environmental impacts [7, 8], made it possible to re-examine the construction of LCA tools with a parametric approach.

Despite some gaps in the availability of sources and case studies due to the innovative character and recent constitution of the specific sector, an attempt was made to introduce the topics to be taken into consideration for a possible experimental presentation that is understandable and comparable with the results.

The methodology proposed in this document aims to overcome the existing difficulties of architects and engineers to manage the complexity of LCA.

The lack of studies on simplified approaches highlights the experimental character on the study conducted in applying this type of methodology.

During the design of the optimization phase, the study was performed in a parametric environment by comparing the operation of the Grasshopper *Bombyx* (Fig. 2) [9] and *Cardinal LCA* (Fig. 3) [10] plug-ins.

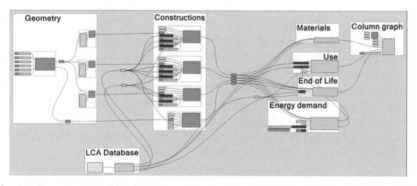

Fig. 2. Configuration algorithm based on artifacts using the Bombyx 2.0 plugin for the simplified assessment of the impacts of the life cycle of buildings on Grasshopper.

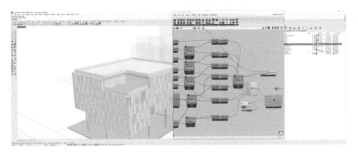

Fig. 3. Data-based Parametric Visualization is processed through the Cardinal LCA plugin for the simplified assessment of the life cycle impacts of buildings on Grasshopper. Elaborations on HARVARD Studies on Cardinal

LCA results are normally presented in the form of aggregate environmental loads or impacts relating to the functional unit, in both cases values can be reproducibly measured using the unit CO2eq/Kg.

Considering that it should be noted that the characteristics of this parametric approach to LCA can include different variations based on the methodological choices.

This approach can simplify in the future the development for a detailed analysis for the Life Cycle Assessment, in this case, oriented to the certification of environmental performance.

The studies are mostly comparative according to UNI EN ISO 14040, the general categories of environmental impacts to be considered in an LCA resume the normal phases of use according to regulations, integrating all those aspects related to the digitization of impacts, to innovation in building materials and the possible implications on trials for digital fabrication.

2.3   Goal and Scope Definition

The phases of the life cycle analysis identify the operational processes that make it possible to define the boundaries of the study for the assessment of environmental impacts.

Whose succession of phases takes up the conventional definitions from Life Cycle Assessment: Production stage; Construction phase; Use; End of Life, in our case we only consider, in its execution, the production stage (from cradle-to-gate) [11].

The goal is to open the building to different design alternatives for the test object, parameterizing its data on the Life Cycle Assessment considering the performance capabilities, and trying to maximize the potential of the innovative components.

Improving the sustainability performance of the building must start already in the design phase, as the evaluation of the optimization potential in the early stages of the project when the impacts of building modifications and construction costs are lower.

3 Life Cycle Based Indicators for Early Design Phase

3.1 Life Cycle Assessment and Computational Design

The context in which the design experimentation operates defines the reasons for the choice of methods and proposes the integration with high computational tools operated as per the prospectus, a holistic approach to advanced design [15].

The parametric algorithmic design approach (Grasshopper) was presented in this study to optimize the geometric capabilities of the building, in particular increasing the ability to manage and interface in a formal way material, components, and systems to evaluate their ability to interface both with biological systems to promote the increase of biodiversity by operating a positive impact on the biosphere, both with components with high-performance capabilities, thus optimizing the impacts relating to the technosphere in the life cycle.

The aspects related to the circularity of the project, which concerns the interaction with the processes that interfere with the biosphere, the hybrid approach to architecture proposes in the articulation of the projects studied, a new configuration between nature and the built environment, biosphere and technosphere.

3.2 Life Cycle Methodologies Declined on Computational Design

Different methodologies, standards, and research projects were analyzed to determine the final set of key indicators that should be considered in this methodology, supporting the evaluation and management of the project process during the early design stages [16].

Taking this into account, the two groups of indicators described in the previous paragraphs have been defined: (Par. 3) Performance-Based Design Indicators (Par. 4) Life Cycle Based Design Indicators representing in all their complexity the set of indicators concerning the environmental footprint.

The basic indicators that identify the characters for the functional co-optation of the parts can be used in the conceptual phase of the project, the performance indicators can be used in the pre-design phases for the optimization of passive operation, the indicators on the life cycle as well they are used to optimize the use of the technologies chosen by intervening on the aspects that influence the energy incorporated in the system-product (Fig. 4).

Fig. 4. Parametric display of the results of the simplified evaluation of the life cycle of buildings based on the data processed by the Bombyx 2.0 plugins (left) Cardinal LCA (right) on Grasshopper.

The data obtained released in the form of an algorithm will be the subject of subsequent studies for design optimization in all phases of the Advanced Circular Design with an inter-scalar approach.

This exit is made possible by the balance on which the entire algorithmic sequence, applied in the experimentation, is based, that is the balance of the impacts relating to the energy incorporated and the operational phase of the possible architectural system under study.

4 Conclusions

4.1 Algorithm Based Performance Optimization

With this approach designers and architects will be able to focus on the layers that most influence the environmental footprint of the building to identify design solutions that will improve the eco-performance.

As is known, the Life Cycle Assessment (LCA) is a methodology used to analyze and evaluate the environmental impact of a material, product or service during the entire life cycle, usually from the extraction of raw materials to disposal for recovery.

The proposed design-operational methodology allows conducting direct experimentation on the factors that contribute both to the increase of the Global Warming Potential [17] throughout the life cycle, referring to the studies conducted on the parameters of 'LCA' that insist on the environmental benefits that the reuse of materials brings in buildings (non-renewable energy and GHG emissions).

It is also possible to identify all those possible implications that allow intervening in the operational phase not only by optimizing the performance aspects but by identifying the ability to integrate techniques and technologies for the storage of CO_2.

As demonstrated above, the methods for obtaining the main indicators aim to predict the sustainability performance of the building in the Early Design Phase, where the available data is scarce, and where the project lends itself more to setting an open-source configuration [18].

4.2 Advanced Circular Design: New Methodological Framework

The aspects related to the decision-making phase for construction parameters of buildings, therefore, become a key element in the characterization of the advanced circular design processes, to represent the capacity offered by the building technologies evolutions by studying frontier research in the sector in the field of circular economy in architecture.

The first iteration is run considering the algorithm-aided design [16], the second one is based on the performance-driven design approach [14]. The synthesis finds its natural evolution in applications to an advanced circular design, designed as a regenerative algorithm to be used for the up-cycling of materials, components, and construction systems innovated for architecture.

The performance-based design approaches naturally flow into a regenerative process of responsive optimization of building materials, which achieves advanced circular design [19].

The proposed approach aims to demonstrate that a sufficient comparison can be provided for the optimization of the parts in an advanced circular perspective already in the early design phase, adding to the performative tools, the parametric evaluation tools of the impacts [20].

By accessing international opensource databases, they carry out an L.C.A. of a simplified type and allow you to quantify, view, and optimize the impacts on the entire life cycle of buildings in an algorithmic sequence, having an objective comparison of the values in terms of CO2eq/Kg, can be the key to determine the optimization of the Global Warming Potential.

The case study conducted on a generic test project demonstrates the different solutions proposed on the performative models by the ADC model for the parametric optimization of the life cycle (Fig. 5).

The environmental optimization of both the parametric software and the two variable objective functions shows that the impact of a unitary surface considering the traditional construction techniques is about 3.0 kg CO2 eq. CO2 (eq) can reach through the combined Galapagos optimization the result of 1.2039 kg of CO2 (eq). The optimization of the product and process parameters assumes optimal values and allows to obtain a reduction of about 70% by carrying out all the phases of the reiterative model.

Both software used opt for the mitigation of design and process parameters that have the total environmental impact, intervening on the characteristics of recyclability, project circularity, sustainable production and low energy consumption. Such performance of both scenarios is better than the existing product scenario and does not affect the design requirements.

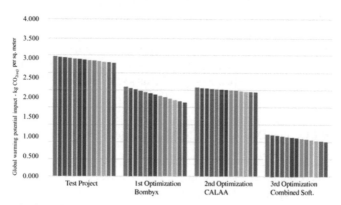

Fig. 5. Comparisation of the results of the simplified evaluation of the life cycle of a generic building based on the data processed by the Bombyx 2.0 plugins Cardinal LCA optimized in Grasshopper.

Acknowledgements. This text is part of the scientific activities carried out by ABITAlab, as a contribution to the theoretical and experimental discussion of TREnD (Transition with Resilience for Evolutionary Development) research interest, which has received funding from the European Union's Horizon 2020 research and innovation program under the Marie Skłodowska-Curie grant agreement No. 823952.

References

1. UNEP SBCI: Building Design and Construction: Forging Resource Efficiency and Sustainable Development (2012)
2. Bribián, I.Z., Usón, A.A., Scarpellini, S.: Life cycle assessment in buildings: state-of-the-art and simplified LCA methodology as a complement for building certification. Build. Environ. **44**(12), 2510–2520 (2009). ISSN 0360-1323
3. IPCC 2021: Climate Change (2021): The Physical Science Basis. Contribution of Working Group I to the [IPCC Sixth Assessment Report|Sixth Assessment Report] of the Intergovernmental Panel on Climate Change (2021)
4. Naboni, E., Havinga, L.C.: Regenerative Design in Digital Practice: A Handbook for the Built Environment. Eurac, Bolzano (2019)
5. Nava, C.: Ipersostenibilità e tecnologie abilitanti. Teoria, metodo e Progetto, vol. 1. Aracne Editrice (2019)
6. Suppipat, S., Teachavorasinskun, K., Hu, A.H.: Challenges of applying simplified LCA tools in sustainable design pedagogy. Sustainability **2021**(13), 2406 (2018). https://doi.org/10.3390/su13042406
7. Kiss, B., Szalay, Z.S.: The applicability of different energy performance calculation methods for building life cycle environmental optimization. Int. Rev. Appl. Sci. Eng. IRASE **9**(2), 115–121 (2018). https://akjournals.com/view/journals/1848/9/2/article-p115.xml. Accessed 3 Jan 2022
8. Basic, S., et al.: IOP Conf. Ser. Earth Environ. Sci. **323**, 012112 (2019)
9. Montana, F., et al.: Multi-objective optimization of building life cycle performance. A housing renovation case study in Northern Europe. Sustainability **12**, 7807 (2020). https://doi.org/10.3390/su12187807
10. Westfall, L.A., Davourie, J., Ali, M., McGough, D.: Cradle-to-gate life cycle assessment of global manganese alloy production. Int. J. Life Cycle Assess. **21**(11), 1573–1579 (2016). https://doi.org/10.1007/s11367-015-0995-3
11. Bogenstätter, U.: Prediction and optimization of life-cycle costs in early design. Build. Res. Inf. **28**(5/6), 376–386 (2000)
12. Melis, A.: Urban Exaptation: Towards a new taxonomy of the ecological city. Research output: Chapter in Book/Report/Conference proceeding › Other chapter contribution (2018)
13. Kocak, C.: An algorithm aided design approach for using daylight in early phases of architectural design Mahmut Can Koçak Department of Informatics Architectural Design Computing Programme (2019). https://doi.org/10.13140/RG.2.2.12253.77289
14. Oskam, P., Mota, J.: Design in the Anthropocene: Intentions for the Unintentional (2020). https://doi.org/10.1007/978-3-030-61671-7_26
15. D'Urso, S., Cicero, B.: From the efficiency of nature to parametric design. A holistic approach for sustainable building renovation in seismic regions. Sustainability **11**(5), 1227 (2019). https://doi.org/10.3390/su11051227
16. Hollberg, A.: A parametric method for building design optimization based on Life Cycle Assessment. Bauhaus University Weimar (2016)
17. Hollberg, A., Habert, G., Schwan, P., Hildebrand, L.: Potential and limitations of environmental design with LCA tools. Creat. GAME Theory Pract. Spat. Plan. **5**, 34–455 (2015)
18. Meex, E., Hollberg, A., Knapen, E., Hildebrand, L., Verbeeck, G.: Requirements for applying LCA-based environmental impact assessment tools in the early stages of building design. Build. Environ. **133**, 228–236 (2018). ISSN 0360-1323

19. Pranjale-Bokankar, P.: ECO DESIGN: DESIGN WITH NATURE Biophilic Design -A Sustainable Approach (2019)
20. Hollberg, A., Kaushal, D., Basic, S., Galimshina, A., Habert, G.: A data-driven parametric tool for under-specified LCA in the design phase online. In: World Sustainable Built Environment Conference, p. 052018. IOP Publishing, Bristol (2020)

Design Strategies Toward Low-Carbon Buildings and Neighborhoods. The Use of LCA to Support a Project Proposal for Reinventing Cities

Monica Lavagna$^{(\boxtimes)}$ ⓘ, Andrea Campioli ⓘ, Anna Dalla Valle ⓘ, and Serena Giorgi ⓘ

ABC Department, Politecnico di Milano, 20133 Milano, Italy
monica.lavagna@polimi.it

Abstract. Within building design competitions, the use of LCA method is increasingly required to demonstrate the lower impact of project choices. However, in response to the European Green Deal and decarbonisation objectives, the attention is generally focused only on carbon footprint. In the context of C40 Reinventing Cities call, the article highlights the chosen design and energy strategies to achieve low-carbon buildings and neighbourhoods, with specific reference to a project proposal. In particular, the LCA methodological approach aimed at quantifying the reduction of carbon footprint and the results obtained are discussed, as well as some critical issues faced by the authors in responding to the call requests related to carbon footprint reduction. The paradoxes that derive from stressing the use of low-carbon without a wider environmental and LCA attention and the unbalanced trends towards low-carbon design solutions which fail to target material resource efficiency are argued.

Keywords: Life cycle assessment · Carbon footprint · Building design

1 The Goal of Carbon Neutrality and the Role of Life Cycle Environmental Impact Assessment

The European Union commitment for the reduction of climate change has led to the definition of multiple action strategies, operational measures and Framework Programs setting out common long-term objectives regarding a climate-neutral society by 2050. This objective has recently been relaunched by European Green Deal (2020), stressing a particular attention to the decarbonization and reduction of CO_2 emissions. The first steps towards a more sustainable building sector have been taken by the directives on the energy efficiency of buildings (2002/91/EC, 2010/31/EU, 2012/27/EU) during the use phase. However, to achieve the goal of decarbonization, the focus on the sole use phase is not enough. It is crucial to consider the carbon footprint of the entire life cycle of buildings and to control all design choices.

To establish a common methodological approach to allow Member States to assess and compare the environmental performance of products and services over their life

© The Author(s), under exclusive license to Springer Nature Switzerland AG 2022
F. Calabrò et al. (Eds.): NMP 2022, LNNS 482, pp. 1879–1888, 2022.
https://doi.org/10.1007/978-3-031-06825-6_181

cycle, the European Commission has identified Life Cycle Assessment (LCA) as the most suitable tool for measuring environmental impacts (European Commission 2015). Although Life Cycle methodologies are internationally recognized and recalled by the European Commission, their integration in the process of policy definition by public authorities appears to be very limited (Allen et al. 1995). To enhance LCA practices, an important initiative for the building supply chain is represented by national and international building design competitions that require the use of LCA as a design supporting tool. Nevertheless, design competitions usually call for only low-carbon strategies (to comply with decarbonization objectives), limiting the use of LCA to the evaluation of the single environmental indicator of Global Warming Potential (GWP), as carbon footprint. If, on one side, the biogenic carbon assessment approach is still discussed in the scientific debate (Levasseur et al. 2013; Breton et al. 2018), on the other side, it risks to create burden shifting, by orienting design choices without considering the environmental impacts in an holistic way (Fouquet et al. 2015; Sandin et al. 2014).

In this context, the paper aims to highlight the critical issues related to the evaluation of only carbon footprint and the paradoxical design trends that emerge from the maximization of GWP reduction. To this end, the paper shows an example of low-carbon buildings and neighborhoods designed to meet the requirements of an international building competition – C40 Reinventing Cities – in which Authors have taken part as environmental expert consultants for the assessment of the project carbon footprint (including buildings and urban spaces). The pursued design and energy strategies are thus presented in connection with the methodological assumption adopted in the LCA study.

2   Case Study of Low-Carbon Buildings and Neighborhoods Design: LOreto VErde Project

The case study presented in this paper has been submitted to the second edition (2020) of Reinventing Cities call, an international competition organized by C40 Cities Climate Leadership Group with the support of Climate KIC. Key aim of the competition is to drive carbon neutral and resilient urban regeneration, through the involvement of a multidisciplinary team, composed as a minimum by developers, architects, environmental experts, in charge of collaborating with other design and construction operators to design zero-carbon buildings and neighborhoods.

Among C40 competition sustainability requirements, there are mandatory accomplishments of: i) Energy efficiency and low-carbon energy; ii) Life cycle assessment and sustainable materials management. To comply the first challenge (energy issues) the design team have to show: the energy consumption of project in $kWh/m^2/year$ broken down into energy source and by usage; and the carbon footprint of the energy consumption in $kgCO_2e/m^2/year$. To meet the second challenge (LCA issues) the team have to display: the carbon footprint of the project across the whole life cycle in tCO_2e or tCO_2e/m^2; and the quantity of low-carbon construction material used for building in m^3/m^2. In both cases, the competition requires to compare project results with "Business As Usual" (BAU), obtained by modelling a standard building (same shape, function and location, but conventional technological solutions and performances).

The presented case study refers to LOreto Verde project[1] and it concerns the requalification of Piazzale Loreto area in Milan (Italy). The project involves a student housing building with a gross surface of 2.839 m^2 and the public square of about 17.000 m^2, offering commercial spaces both underground for about 6.000 m^2 and aboveground providing retail pavilions for about 2.000 m^2. In particular, LOreto VErde project reorganizes the square plan for encouraging direct movement of pedestrians, cyclists and public transport, integrating green areas and reactivating the neighborhood with various commercial activities. Moreover, the existing building is transformed and expanded to host a student housing, improving architectural quality and environmental performance.

3   Life Cycle Design Strategies to Reduce Carbon Footprint

In view of low-carbon C-40 competition goals, all design and energy choices of LOreto VErde project have been targeted at the reduction of carbon footprint, involving from the preliminary phases all parties and expertise. Significant efforts have been made to evaluate different alternative design options in terms of material and energy flows along the whole life cycle for finding the best solutions towards low-carbon emissions, adopting a multiscale approach (from materials to constructive subsystems up to the entire settlement). Such strategies concern the whole building life cycle and range from material properties, including bio-based materials, recycled content, recyclability, durability, and constructive principles, such as prefabrication, reversibility, modularity, to potential technological processes oriented to life extension. Moreover, energy strategies focuses on the reduction of operational energy use, with the aim of achieving Net Zero Energy Buildings (NZEB). Here, special emphasis is placed on envelope efficiency, the production in situ of renewable energy and the green energy supply.

During design, the carbon footprint of LOreto VErde project has been constantly calculated according to LCA method (EN15978) and progressively monitored along the design process for verifying the effective reduction in terms of CO$_2$eq compared to BAU model. The assessment has been gradually performed from materials to building level "from cradle to cradle", including all life cycle phases and assuming 60 years of reference service life. Specifically, the carbon footprint of the project considers: production phase, namely the procurement of raw materials (A1), their transport to the manufacturing plant (A2) and production process (A3); construction phase, namely the transport to the construction site (A4); use phase, namely the replacement operations (B4), energy consumption (B6) and water consumption (B7); and the end-of-life phase, namely the transport to landfill (C2) and waste treatment (C3).

The following paragraphs provide an overview of the design choices pursued for achieving low-carbon buildings and neighborhood. In particular, the carbon footprint is explained separately for "cradle to gate" and "cradle to cradle" phases, focusing on the most significant parts in terms of carbon impacts.

---

[1] Name of team/project: LOreto VErde. Team representative: VICOM S.R.L. Architect/Urban designer: CITTERIO - VIEL & PARTNERS S.R.L. Environmental expert: POLITECNICO DI MILANO – DABC.

3.1 Carbon Footprint Cradle to Gate: The Role of Material Choices

The constructive and material choices of LOreto VErde project are based on LCA studies of alternative design options oriented towards the reduction of the carbon footprint. Comparisons have been conducted, on one hand, taking into consideration several materials and, on the other, exploring the environmental profile of different products by looking into various Environmental Product Declaration (EPDs) and encouraging the use of local materials. Project impacts are calculated from EPD primary data, selecting for each material supply chain the third best product from a carbon emission point of view (not the absolute environmentally favorable products) to allow broader selection during the procurement stage. Instead, in order to better reflect current construction market, BAU impacts are assumed as average values collected from database secondary data (including Ecoinvent, Ökobaudat and ICE). Note that building materials solutions in some case differ between project and BAU models, while in others they are the same but considering different emissions.

As shown in Fig. 1, thanks to low carbon design choices, the carbon footprint of LOreto VErde project presents during the production phase an overall saving in term of $kgCO_2eq$ of −64% compared to BAU model. In particular, the proposed student housing contributes to the achievement of 116% carbon footprint reduction, while carbon footprint is reduced to only 18% in the case of square.

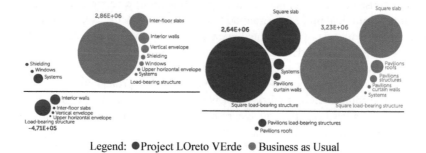

Legend: ● Project LOreto VErde   ● Business as Usual

Fig. 1. Carbon footprint "cradle to gate" [$kgCO_2eq$] per constructive subsystems of the student housing (left) and of square (right) respectively.

Due to the high impacts generally associated to load-bearing structure, special attention has been paid on it. Compared to the traditional BAU solution, consisting of reinforced concrete frame, a series of load-bearing walls and brick-concrete slabs, a set of structural alternative solution[2] have been evaluated. At last the solution proposed for the project combines steel for frame and wood for slabs, composed of not full section with external Xlam panels (60 mm thick) and internal stiffening beams in laminated wood (CLT), in order to optimize the use of material and lighten the structure. In this way, it allows a carbon footprint reduction of −127% mainly thanks to the wood-based slabs inclusion (Fig. 2).

---

[2] Structural and envelope consultant: FACES ENGINEERING S.R.L.

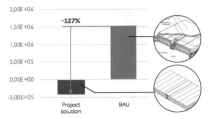

Fig. 2. Comparison of emissions [kgCO$_2$eq] of load-bearing structure solutions (1165 m^2).

Concerning student housing envelope, preliminary comparative LCA studies have been developed in order to select for each layer the materials distinguished by low impacts in terms of CO$_2$eq. For this purpose, a wide range of thermal insulation materials are analyzed in depth (Fig. 3), disclosing how wood fibre insulation and cork insulation have a smaller footprint (design choice rests on wood fibre insulation, since more cost-effective than cork).

At material level, another core study concerns the exploration of façade cladding (Fig. 4), selecting fibrecement since compliant with the requirements of reversibility, disassembly, durability and recyclability. Here, despite the low emission, wood-based cladding materials have been excluded because of subject to material degradation, due to atmospheric exposure, and not aligned with the architectural language of Milan.

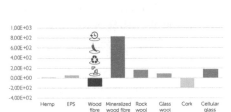

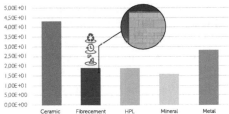

Fig. 3. Comparison of emissions [kgCO$_2$eq/m^3] of insulation materials.

Fig. 4. Comparison of emissions [kgCO$_2$eq/m^2] of cladding materials.

These studies in conjunction with many others allowed to define the best technological-constructive solution for both the vertical envelope and upper horizontal envelope. Project envelope composed of prefabricated modules with wood substructure, wood fibre insulation (20 cm), OSB panels and internal plasterboard achieves a reduction of -132% carbon footprint compared to BAU hollow brick block walls with EPS insulations (Fig. 5). Note that the vertical envelope enables a reduction of the project carbon footprint, obtained along with an optimization of the thermal transmittance, equal to 0.17 W/m^{2}K in the project vertical envelope, while at about 0.26 W/m^{2}K in BAU.

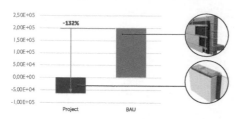
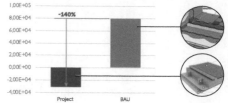

Fig. 5. Comparison of emissions [kgCO$_2$eq] of vertical envelope solutions (1884 m^2).

Fig. 6. Comparison of emissions [kgCO$_2$eq] of roof solutions (776 m^2).

The carbon footprint reduction of student housing has been attained also through a careful selection of roof solutions, corresponding to -140% compared to BAU (Fig. 6). These benefits result primarily from the design choice of dry constructive systems and wood-based materials, since the structural slab is completed by wood fibre insulation (21 cm), wooden slats, OSB panels, waterproofing and then mostly finished as a green roof and the remaining with wood flooring. BAU solution differs not only for the use of traditional wet constructive systems, consisting of lightweight concrete screed, EPS insulation (13 cm) and waterproofing, but also for the use of sintered stone flooring, assuming instead an equal green roof solution to the project.

As previously evidenced, another building parts that significantly affects the carbon footprint reduction of the proposed student housing are partitions. Interior wall, made of dry assembled lightweight partitions (wood substructure, wood fibre insulation, external OSB panels and internal plasterboard) enables a reduction of -197% of CO$_2$eq compared to BAU hollow brick blocks with finishes in plaster. In addition, wood finishes of inter-floor slabs allow to reduce project CO$_2$eq emissions by -115% compared to BAU solution consisted of lightweight concrete screed and ceramic flooring.

Besides optimized design solutions for construction materials, it is important to highlight that project systems reveal an increase in the carbon footprint. This is caused by the inclusion of the production impacts of photovoltaic panels, expected in greater number in project student housing (548 m^2) than in BAU scenario (88 m^2), since the same extent of sanitary and distribution systems are estimated for both models.

As regards the re-layout of Piazzale Loreto, particular attention has been given to the design of the structural elements of the new deck. Two alternative options have been compared: a composite steel-concrete structure and a prestressed reinforced concrete structure. On the basis of multi-criteria analysis (including environmental, mechanical, functional and aesthetic issues), the project team decided to select the composite structure, because of structural performance and compatibility with functional and architectural needs (the composite steel-concrete structure allows a free plan underground space, particularly suited to retail activities and to the subway line).

Accounting square project, the most consistent reductions in terms of carbon footprint are attributable to pavilions load-bearing structure and pavilion roofs, reaching negative values of kgCO$_2$eq since made of wood-based materials, as opposed to BAU model composed of concrete frame and brick-concrete slabs. Wood is also established for the window frame of both Loreto pavilion and student housing, embedding the lowest carbon footprint (Fig. 7) and guaranteeing durability requirements since designed protected from atmospheric agents. Among the other low-carbon design strategies, it is worth mentioning the comparative assessment performed for external paving, selecting stone and calcestre as alternative materials to the traditional concrete flooring (Fig. 8).

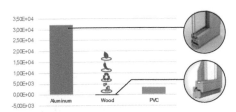

Fig. 7. Comparison of emissions [kgCO$_2$eq/m^3] of windows frame materials.

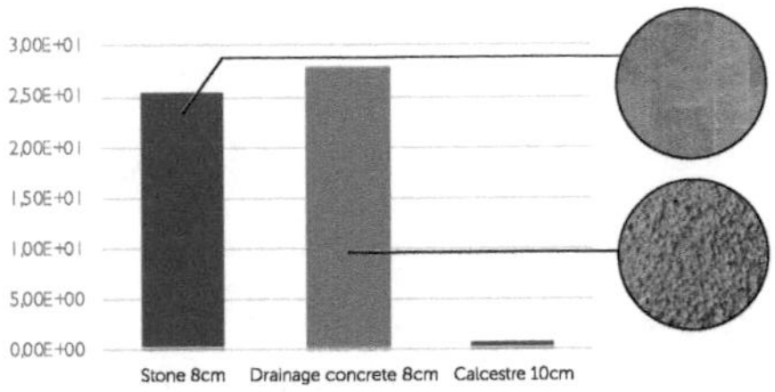

Fig. 8. Comparison of emissions [kgCO$_2$eq/m^2] of external paving materials.

Finally, as disclosed for student housing, results show an increase in the square carbon footprint solely for plant systems, as the production impacts of photovoltaic panels are included to a greater extent in the proposed project (720 m^2) compared to BAU scenario (126 m^2). Note that, if for the production phase this design choice appears environmentally disadvantageous, it entails significant benefits during the use phase.

3.2 Carbon Footprint Cradle to Cradle

Extending the carbon footprint assessment from the production phase to the entire life cycle, it is possible to figure out how material selection enables limited reduction in relation to the overall carbon footprint savings of LOreto VErde project (Fig. 9). Indeed, the proposed technologies and materials corresponds to about −12% reduction of the total, including all low-carbon strategies implemented at the different life cycle stages. Besides production phase, previously discussed, they cover: the construction phase, by prioritizing the use of local materials and prefabricated modules; the use phase, by selecting long-lasting products; and the end-of-life phase, by opting for dry solutions.

As evidenced in Fig. 9, the most remarkable achievements in the reduction of carbon footprint concerns the use phase exclusively in terms of energy consumption. In particular, it has been simulated per 60 years of service life using IES-ve 2019 software, by modelling separately the student housing and the square with retail pavilions, both for project and BAU scenario[3]. BAU model has been set up following the most updated regional energy legislation (Decree 18456/2019 published on 4 January 2020), considering minimum transmittance values defined by laws and reference building systems

---

[3] Energy consultant: ARIATTA INGEGNERIA DEI SISTEMI S.R.L.

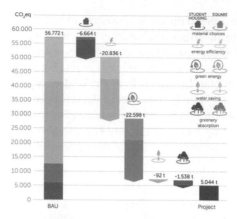

Fig. 9. Emission balance in 60 years [tCO$_2$eq].

with average efficiency values, as well as the integration of the mandatory photovoltaic panels (D.lgs 28/2011 Annex 3). LOreto VErde project is designed with highly efficient systems, linked to intervention resilience to climate change at local and global scale and comprising air conditioning system that combines the heat exchange efficiency with groundwater and the accuracy of HVRF systems in terms of metering consumption. In addition, with the aim to achieve Net Zero Energy Building (NZEB), the project includes the installation of photovoltaic panels both on roof and South façade of student housing and on roofs of retail pavilions in Piazzale Loreto. Furthermore, since photovoltaics are not enough to achieve NZEB goals, the whole proposed settlement is supplied with green energy entirely procured from renewable sources in agreement with Edison (partner of the project). In this way, the operational carbon footprint reductions can be summarized into three main energy strategies. First, the optimization of project energy consumption (heating, cooling, DHW, ventilation, lighting and people transport). Second, the on-site production of energy through the installation of photovoltaic panels. Third, the supply of green electricity from renewable sources, totally derived from hydroelectric (about 70%) and wind-powered stations (about 30%), both present to a very limited extent in the national energy mix assumed for BAU model. Accordingly, the carbon reduction paired with energy efficiency solutions stands at about −37%, further increased to an additional −40% due to the green energy supply, widely decreasing the entire carbon footprint of BAU scenario.

Other design strategies implemented for the project regards the operational water management and the increase of vegetation. With regard to both drinking water consumption and waste water treatment, the reduction are confined to −0,2%, primarily thanks to the recovery of rainwater but also to rainwater absorption, such as the installation of perlite under the vegetative layers, leading to sewage reduction. Moreover, LOreto VErde project provides an increase of vegetation[4] in the intervention area, ensuring qualitative benefits as well as the absorption of CO$_2$. The proposal in fact includes 78 trees in the student housing on the rooftop and on the common terraces at the different floors and

---

[4] Landscape consultant: P'ARCNOUVEAU.

the planting of 184 new trees in Piazzale Loreto. Along with contributing to the reduction of temperature, heat island effect and energy consumption at the project scale (summer cooling), the greenery allows to decrease the carbon footprint by -3% compared to BAU scenario.

4   The Criticality of the Decarbonization Target

In the previous paragraphs, the low carbon strategic choices aimed at achieving the reduction of carbon footprint, as required by the design competition, are evidenced. Nevertheless, it is important to highlight some critical issues relating to the partial application of LCA method, based on the observation of the single indicator GWP.

The presented case study, as happened in many other competing projects, chose a particular strategy to reduce CO_2, namely the use of wood as main building material. In fact, to obtain a zero-carbon intervention, wood and bio-based products usage are encouraged, since represent the only materials able to reach negative impact in A1–A3 phases, providing significant environmental benefits. Nonetheless, this benefit is only apparent, deriving from the methodological approach adopted for LCA calculation.

Currently, there is an open debate on the methodological assumption to assess the CO_2 profile of construction products. The more widely used approach (Hoxha et al. 2020), called "$-1/+1$ approach", considers both the absorption of biogenic CO_2 (-1), as a negative impact (or avoided) in Module A, and the release at the end of building life ($+1$), as a positive impact in Module C. If the life cycle balance of CO_2 stored in wood is respected, including the release at the end-of-life, there would be no advantage in terms of carbon footprint in the use of wood.

However, the bias in the assessment occurs because EPDs developed in compliance with EN15804:2012+A1 standard not always show end-of-life impacts (Module C). In EPDs, Module D is also misleading, because of GWP indicator shows negative emissions linked to the energy recovery of wood products at the end-of-life (subtracting the biogenic carbon declared in C3) and the benefit of energy recovery which replace electrical and thermal energy (produced by European average scenario).

On account of this methodological approach, there is the risk of a design paradox: to use more material, rather than optimize the amount of material to achieve more "sustainable" buildings. For example, with the same performance, it is advantageous to use a solid wood slab in Xlam rather than a lighter joist with panel floor, since the heavier solution stores more CO_2 and contributes further to carbon footprint reduction (Fig. 10).

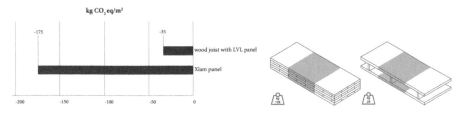

Fig. 10. GWP comparison of Xlam panel and wood joist with LVL panel for structural floor.

Another enhanced strategy implemented for achieving a low-carbon project is the green energy supply. The promotion of renewable sources (including biomass) without considering other impacts (e.g. NO_x emissions) beyond CO_2 balance and neglecting the impacts for the production of system plants necessary to transform renewable energy (e.g. panels photovoltaic, wind turbines, batteries for energy storage, energy distribution network), risks to underestimate the overall impacts along the entire life cycle.

5   Conclusion

The paper shows the application of design strategies to meet the low-carbon target set for a building competition, that requires LCA as a support and verification tool. The analysis of methodological approach demonstrates that the promotion of carbon zero design choices, focused on a single environmental indicator of GWP, can trigger incorrect design choices that move away from material and resource efficiency. The evaluation of single environmental indicator leads to a limited use of LCA as support tool, without exploiting its key peculiarity that is the evaluation of a wide set of environmental indicators, intended for avoiding burden shifting among impacts.

In the light of this, it is crucial to avoid an improper and limited use of LCA, in order to not promote uncontrolled design trends (such as the use of wood regardless of the context and material optimization).

References

Allen, D.T., Consoli, F.J., Davis, F.J., Davis, G.A., Fava, J.A., Warren, J.L.: Public policy applications of life-cycle assessments. In: Proceedings from the Workshop on Application of Life-Cycle Assessment to Public Policy, SETAC Technical Publications Series (1995)

Breton, C., Blanchet, P., Amor, B., Beauregard, R., Chang, W.: Assessing the climate change impacts of biogenic carbon in buildings: a critical review of two main dynamic approaches. Sustainability 10(6), 1–30 (2018)

EC (European Commission): COM 98 A new Circular Economy Action Plan For a cleaner and more competitive Europe – final, Brussels (2020)

EU Commission: Science for environment Policy, improving resource efficiency: new method identifies key areas of product improvement (2015)

Fouquet, M., et al.: Methodological challenges and developments in LCA of low energy buildings: application to biogenic carbon and global warming assessment. Build. Environ. 90, 51–59 (2015)

Hoxha, E., et al.: Biogenic carbon in buildings: a critical overview of LCA methods. Build. Cities 1(1), 504–524 (2020)

Levasseur, A., Lesage, P., Margni, M., Samson, R.: Biogenic carbon and temporary storage addressed with dynamic life cycle assessment. J. Ind. Ecol. 17(1), 117–128 (2013)

Sandin, G., Peters, G.M., Svanström, M.: Life cycle assessment of construction materials: the influence of assumptions in end-of-life modelling. Int. J. Life Cycle Assess. 19, 723–731 (2014)

Regenerative Design and Hybrid Buildings to Address Climate Change

Consuelo Nava[✉] [ID]

ABITAlab, dArTe, Mediterranean University of Reggio Calabria, Via dell'Università.25, 89124 Reggio Calabria, Italy
cnava@unirc.it

Abstract. Is it possible to redefine the space of action, of project and experimental type, of the regenerative design, following the scenarios of climate change and for the questions related to the incidence of the construction sector and the production of architecture and urban spaces? Does the hybrid building model respond to the need to achieve climate neutrality performance levels by 2050? This text, as a contribution to the ongoing research on advanced sustainable design, conducted within ABITAlab, proposes some possible trajectories in the field of frontier research, opening up the possibility of re-founding some theoretical paradigms, after having gone through applied experiences on "prototype cases", activities with a strong innovative character with the generation of new operational models, interpretation and management of data and resources, simulation and prototyping of solutions and evaluation of results with advanced tools of ecological and digital design. In this contribution are provided some important references of literature and critical reconstruction of the new principles on "zero impact" and "zero net buildings", in the need to refound a metric and assessment tools in a circular key, in order to experiment with the approaches of advanced design and its prototyping on case studies, the real possibility to respond to the great challenge posed by environmental and climate issues in the transition scenarios '30, '50, '85 (IPPC). The discussion frames the research trajectory, investigating between radial and continuous innovation, as an operational space on which to conduct experiments, to produce design methodologies that even in the context of transition scenarios, are able to produce predictive models whose role becomes that of founding multiple ecological and digital assets towards climate neutrality.

Keywords: Regenerative design · Hybrid buildings · Climate neutrality

1 Introduction

Under climate change regime, even advanced hybrid building design processes must be conducted according to a new approach to innovation and sustainability. Therefore, in the field of frontier research, there is a need for applied experimentation in innovation-driven design in transition scenarios, through the expression of more advanced spatial and functional configurations, con-tributing to the discussion of sustainability in its most radical version, with a building design that projects its goals beyond green and efficient

F. Calabrò et al. (Eds.): NMP 2022, LNNS 482, pp. 1889–1901, 2022.
https://doi.org/10.1007/978-3-031-06825-6_182

performance, towards climate neutrality, overcoming the concept of "zero carbon" and through regenerative type systems of positive and hybrid response. It is appropriate to argue that, to address the design and management of hybrid buildings under climate change, regenerative-type design is the most complex yet highest performing translation of advanced sustainable design. What is experimented to achieve "the resilience of buildings in the built environment", in fact, changes the very concept of hybrid building, in the architecture of its possible forms and materials and spaces of context, in its functions of urban device, of ecological and technological device, or even as a regenerative device to "live" or "perform activities".

All regenerative systems are themselves "hybrids", since in order to produce performances with a "positive/productive balance", they must be able to function both on high levels of efficiency and effectiveness. Thus, for example, the passive-active behavior of any technological system designed for a level of thermo-hygrometric well-being, is the result of the integration of physical-environmental devices of technical elements and digital ones, of regulation and control of natural/passive behaviors [1].

The environmental issues related to the effects of climate change and the incidence of the field of production of architecture and urban strategic planning, remain the operational space in which, the design of transformations on transition scenarios, becomes "a new physical space" on which to act. The architecture of transition, in its highest ambition of circular process and resilient project, must achieve zero-impact configurations, without waste and able to overcome the application areas of "low carbon" and "net-zero building", as efficient responses to a restorative level of the transformation project and more reach the states of "climate neutrality", through a regenerative model of effective response with "zero carbon" and "hybrid building".

In this direction, it is necessary to ask if there is a real "climate dimension of resilience" [2] that, measuring itself on the local contexts, for the impacts that must be triggered by mitigation or adaptation actions, can affect in a global way the causes and the effects that they cause, through an approach as much strategic as design.

In this sense all the recent literature, in theoretical terms, advances with the international contributions on the issues of interdisciplinary dialogue in the relationship between architecture and resilience, new regulations and assessment tools and related policies on the issues of governance of buildings in relation to climate change, zero impact as a predictive capacity of the project; in experimental and applicative terms on ecological design, as a condition for founding new responses to the design of transformations in architecture, landscape and urbanism, the bioclimatic approach to urban and building design, strategies and tools and practices of product design for sustainability.

In this socio-productive context, the enabling technologies, assume a further role of acceleration of transformation processes but participate, with wide spaces of performance and innovation of systems, in the hypersustainable design of living systems and production of new ecological and digital qualities of transition [3]. The predictive and forecasting characters of the project and its design levels, find themselves new fields of experimentation, in which the relationship between available material and im-material resources and phases of consumption and production, respond in metrics and evaluation to codifiable "life cycles", according to an open and circular model.

Sustainable design becomes a precondition available to the processes of repair and reclaim activated by enabling technologies, on the profiles of zero impact and zero energy, able to trigger a new approach in which the role of hybrid buildings becomes fundamental to make "transit" from a present scenario to the future, the new energy communities (see Fig. 1).

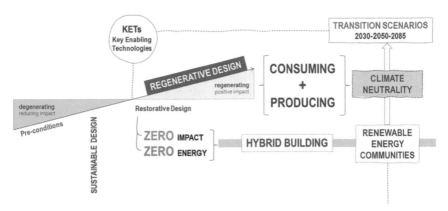

Fig. 1. Relationship between design, presuming, technologies and REC towards climate neutrality in transition scenario. (source: elaboration by G.Mangano on C.Nava's note - Design Culture Technology Course – dArTe, 09-11-2021)

In this model of regenerative design, emerges that character of innovativeness that F.Tucci [4] entrusts to the possible interaction between "Self-reliant approach (whose reference principles are: reflexivity, self-organization and inclusivity)", "Error-friendliness approach (whose reference principles are: robustness, flexibility and adaptability)" and "Dynamic-responsive approach (whose reference principles are: integration, connectivity and responsiveness)".

2   Background and Literature Review

2.1   First Issue: Technologies, Advanced Design, Regulation

The contribution to the first theoretical question discusses "how to ensure that the contribution that technologies and advanced design can make can be made on a large scale and in a timely manner" [5], actually recovering from the same sector that is responsible for the greatest impact on carbon emissions and therefore on global warming, new levels of quality of environmental and energy requirements. The impacts of the last decade have been produced by the construction sector, which accounts for 40% of global energy and 35% of global carbon emissions in the construction, maintenance and use of buildings, 80% of which relates to consumption in the operational phase for heating, cooling and ventilation of buildings and for the operation of household appliances. The mitigation potential of the buildings themselves can be enhanced with renewable energy production systems, but even more so the same potential can reach high levels of adaptation if, in addition to active technologies, passive building design and performance from free gains in heat and cold and special devices contribute to a zero-impact, ecological hybrid building design.

There is no doubt that in order to promote the spread of practices useful to the design of low-carbon and zero-impact buildings, new governance tools are needed at all levels, so as to involve the public sector but also the tried and tested. The most advanced level of experimentation in recent years, has been that produced by the systems of energy and environmental certification and qualification, or even by classification tools (BREEAM, LEED, etc.), in benchmarking or labeling mode, in which they are essentially mapped, questioned and evaluated performance, on frameworks of requirements related to the quality of the physical parts of buildings and measured on frameworks of national and/or international standards. The shift from the model of sustainable design to the regenerative one, marks a different participation of users to contribute to the production of new levels of efficiency and energy and environmental production. The systems of energy autonomy, the smart grid of storage and redistribution, the models of distributed production affect both buildings, as new nodes in a smart grid, and the behavior of prosumers (consumers/producers).

It is clear that the systems that have regulated a model of efficiency related only to the performance of buildings, in different local, national and international regulations, have concerned the achievement of standards through mandatory regulations on energy efficiency and often voluntary on the sustainability of buildings. With the advancement of the concept of "resilience" and the need to involve users in the processes of reducing

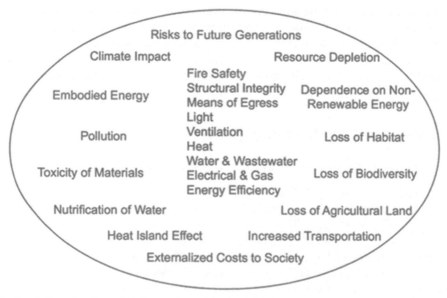

Fig. 2. Addressing these risks is not about either/or, it is about a balanced approach. With the full set of impacts and hazards in view, it becomes possible to consider them all together, across all regulatory boundaries and address them in a process that balances risks and seeks the best overall risk performance (source: D.A.Eisenberg, p.43, 2018)

impacts and responding to climate change, it is necessary to codify, map and address all levels of risk for future generations and reduce environmental, physical and economic costs for new models of society, thinking in a single system to be governed and regulated and long-term. Hawken, Lovins, and Lovins (1999) state that "optimizing components (of a system) in isolation tends to pessimize the whole systems." [6] It is therefore necessary to have an approach that is not exclusive to the relationship between impact and risk, but one that considers a risk-balance approach capable of assessing direct and indirect effects on the systems involved, with the best overall performance, both local and global [7] (see Fig. 2).

2.2 Second Question: Shared Definition of "Building Resilience In The Built Environment"

The discussion contribution to the second theoretical question, seeks a definition of the shared term "resilience", according to a radical and transformative meaning [8], which directly concerns the field of transformations for the built environment. It is of interest to investigate whether it is possible, an effective definition to share and support on "building resilience in the building environment". Too often, in fact, in transferring the meanings of the term from other disciplines, we operate field reductions or assimilation of context that forget the reference condition of the scenario of our interest: the aggregate vectors of "risk", "exposure", "vulnerability" to which we refer any action of mitigation or adaptation in relation to climate change. It is not easy to operate in this condition, even answering the question of "who should do it" when it comes to act for "designing resilient building". Conventional building processes based on the life cycle and durability of buildings must be put before the need to prioritize the design of energy-environmental performance capable of making buildings as responsive as possible, open in their functions, and regenerative with respect to the contexts in which they are settled. Susan Roaf (2018) in this regard, brings out the unpredictability of phenomena con-connected precisely to the complexity of risks to be considered, as exposed in the discussion of the first issue, stating "we live in a non-linear and unpredictable world of boom and bust, of war and infrastructural failures, where rapid change an uncon-trolled growth and driving the collapse of stablished system" [9] (see Fig. 3).

In this sense are configured two characteristics of the hybrid building, in addition to the physical-environmental performance, which concern its "temporal dimensions" as a building that changes over time and its "functional extensions" as a building-device, which is able to integrate its responses in terms of performance. It connects the inside and the outside of its spaces, through the system of "frontier", characterizes its envelop as an evolved technological system and mediates and interacts with the systems of urban functioning, with other devices and spaces able to improve further the overall performance, at each change of scenario (climatic, physical, functional and social).

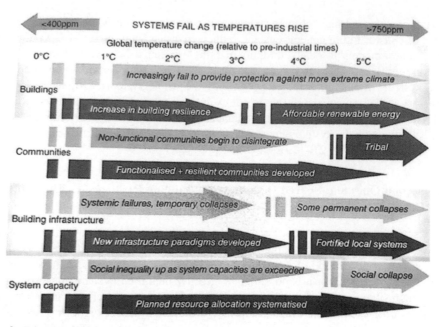

Fig. 3. Diagram charting warming climate and rate of collapse of social and physical systems in the built environment (source: S. Roaf, p.144, 2018)

On the first character we can say that the phase of "use" becomes as important as the phase of "design and construction", its character becomes "neutral" with respect to the typology and spatial organization, the possibility of modification of parts, recycling materials and systems in other arrangements. This is an approach to the life cycle of the building, which goes beyond the concept of reuse and recycling, through the strategies and practices of upcycling, from the scale of the building to that of the component. The second character, refers to the concept of "frontier" as an integrated system of several technological systems, of devices for activation, service and control of the performance of the building and the environment, but also of level of operation "open-to", able to express all the innovative and advanced characters of the experimentation of ecodesign. The two characters participate in equal measure to recognize the systemic approach of regenerative design scenario, through the most resilient expressions towards climate neutrality (2050), as "design for innovative and transitional systems". An approach, the latter, that allows to design/plan the flow of materials and energy from one component of the system to another, reducing the amount of waste and scrap and transforming them into inputs and resources for other systems. [10] In this area of transformation of the built environment, the concept of "building resilience" produces "the architecture of transition, non-extractive 'on designing without Depletion.'"[11] (Fig. 4).

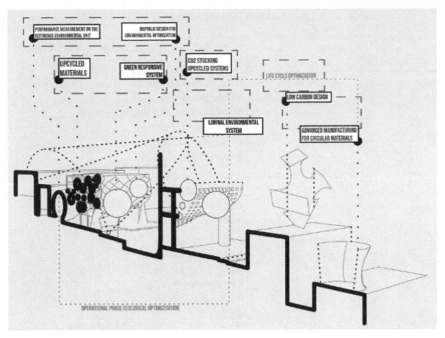

Fig. 4. Diagram section of LES - liminal environmental system (source: D. Lucanto, Upcycling + GRS Sillabus, NYIT, 2021)

3   Approach Method – Metrics and Assessment

The goal of climate neutrality has to be answered in the production of transition scenarios and advanced sustainability strategies, founding through experimentation "the principles of zero". Therefore it is important, in the methods used to apply tools of design, digital simulation and prototyping, to search for the meanings and evidence of a more complex and comprehensive definition of zero impact, even when referring to "zero energy building". This means pursuing the goals of "net zero site", "net zero source", "net zero cost", "net zero emissions". A new metric and innovative assessment tools must be investigated, following the evolution of the "net zero" definition [12, a] (see Table 1) but also aiming at a new framework based on life cycle assessment, including in theory and practice the need to measure all "impacts of ecological systems and user behaviors". [12, b] (see Table 2).

Table 1. Net zero building definitions (source: Ming Hu, 2019)

	Name	Definition	Year	Metric of balance
USA	Net zero site	A net zero produces as much energy as it uses, as measured at the site. Onsite generation within the building footprint is preferable	2006	Site energy
	Net zero source	A net zero source produces as much energy as it uses, as measured at the source	2006	Source energy
	Net zero cost	A net zero cost building receives as much financial credit for exported energy as it is charged on the utility bills	2006	Cost
	Net zero emissions	A net zero emissions building produces at least as much emissions-free renewable energy as it uses from emissions-producing energy sources	2006	Carbon emissions
France	Positive Energy building	A positive energy building (BEPOS) in a building whose overall energy balance in positive	2009	
EU	Nearly Zero energy building	Nearly zero energy buildings have very high energy performances. The low amount of energy that these buildings require comes mostly from renewable resources	2010	Source energy

(*continued*)

Table 1. (*continued*)

	Name	Definition	Year	Metric of balance
Switzerland	Minergie - A	The plug energy building: expenses for space heating, water heating, and air renewal; all electrical appliances and lighting are covered by specially produced renewable energies	2011	Source energy (includes embodied energy)
Norway	Zero emissions building	A zero emissions building produces enough renewable energy to compensate for the building's greenhouse gas emissions over its life span	2013	Carbon emissions
Sweden (Skanska)	Net primary zero energy building	A net primary zero energy building needs to fulfill a net zero primary energy balance between the total energy demand and total energy generation	2013	Source energy
Sweden (SCNH)	Zero energy building	A zero energy building has a net balance of energy imported onsite and energy exported from the site	2015	Source energy
Sweden (Boverket)	Nearly zero energy building	A nearly zero energy building as a near net balance of energy imported onsite and energy exported from the site	2015	Source energy
Japan	Zero energy building and zero energy house	A net zero energy house has a zero annual net consumption of primary energy (METI)	2015	Source energy

(*continued*)

Table 1. (*continued*)

	Name	Definition	Year	Metric of balance
Germany	Efficienchauz Plus	Effcienhauz Plus house have both a negative annual primary energy requirement and a negative annual final energy requirement	2016	Site energy
Korea	Zero energy house/zero energy building	n/a	n/a	n/a
New Zealand (Zero carbon act)	Net zero energy building	A net zero building is low-energy and offsets any energy that is generated from greenhouse gas-emitting fuels with renewable energy generation, such ad hydro, solar and wind	2018	Site energy

Table 2. Impact categories and measurement units (source: Ming Hu, 2019)

Impact	Sub-categories	Negative impact	Mitigation	Measurement
Energy	Embodied Energy	Energy consumption and carbon emissions	Renewable energy generation onsite	Average GHG emissions per unit floor area (metric tons/mq)
	Operating energy			
	Transport energy			
	Introduced energy			
Water	Quantity		Capture rainwater, recycle and reuse grey and black water	Liters per floor area (liters/mq)

(*continued*)

Table 2. (*continued*)

Impact	Sub-categories	Negative impact	Mitigation	Measurement
	Quality *Acidification*	Energy/emissions related to generating clean water, treating grey/black water	Use low impact materials Capture rainwater onsite	Measurement of increase in hydrogen ion (H) concentration in the water (Kg SO2 eq/mq)
	Quality *Eutrophication*	Emissions lead to acidifying effects on the environment		Kg N ew/mq
Environment (MACRO)	Global Warming Potential	Carbon emissions	Reduce carbon emissions	GHG emissions per unit floor area (metric tons/mq)
	Smog potential	NOx e VOCs generated in the presence of sunlight lead to lung damage and other impacts on the ecosystems	Use less harmful materials and optimize construction methods Reduce energy consumption	Kg O3 eq/mq
	Ozone depletion potential	Primary sources of ozone precursors are electricity generation and motor vehicle	Use less harmful materials and optimize construction methods Reduce energy consumption	Kg CFC eq/mq
Environment (MICRO)	Heat Island effect	Temperature difference between the building site and the city average	Vegetated roof, use less reflective exterior materials	°C
	Wind flow	Speed difference between the building site and until site	Use building massing, design to optimize window	m.p.s or m/s
	Biodiversity	Quantify the biodiversity of an urban habitat (building site)	Use landscape design to create habitats for diverse species	Simpson index of diversity (0–1)

(*continued*)

Table 2. (*continued*)

Impact	Sub-categories	Negative impact	Mitigation	Measurement
Health (physical and psychological)	Ability	Enable occupants to perform certain tasks	Light level Nose level Thermal comfort Cleanness Air quality	Lux (lx) Decibel (dB) °C Parts of millions (ppm) Indoor environmental quality (scale 0–10)
	Motivation	Assess whether a person in willing to perform a certain task	Create a good environment for high productivity and enhance social interaction	Indoor spatial quality (scale 0–10) Accessibility to views Personal control
	Opportunity	Provide conditions that reduce health and safety risks	Create a healthy, friendly indoor environment	Indoor habitats Quality (scale = −10)

4 Implications and Research in Progress

From the proposed framework of method, it emerges that in the last ten years the definition of "zero impact building", on the issues of performance qualification, has marked the transition from a conservative model of "zero energy" to a regenerative model of "positive energy". The definitions (Table 1) provided by USA, EU, Switzerland, Holland, have actually guided the transition towards a more productive concept, already referred to the energy not only saved and consumed, but to the energy incorporated and produced, already present in the French experience, in addition to references on the use of renewable energy. The reference to the dimension of impacts on the whole set of categories and subcategories (Table 2) has a matrix of reference in the assessment of the life cycle of each production system. The hybrid building model and regenerative design build their framework on these approaches. Within the experiences of ABITALab the case studies on experimentation activities conducted with competitive research, measure their level of technological maturity (TRL) with reference to the experimental and pre-industrial transfer, conducted between the design and the laboratory environment, between the prototyping system and the demonstrator system [13].

References

1. Nava, C.: Advanced sustainable design (ASD) for resilient scenarios. In: Chiesa, G. (ed.) Bioclimatic Approaches in Urban and Building Design. PSS, pp. 255–274. Springer, Cham (2021). https://doi.org/10.1007/978-3-030-59328-5_13

2. Losasso, M.: Progetto, Ambiente, Resilienza. In: Technè nr. 15, pp. 16–20. FUP ed. (2018)
3. Nava, C.: Ipersostenibilità e Tecnologie abilitanti. Teoria, Metodo, Progetto, 336. Aracne editrice, Roma (2019)
4. Tucci, F.: Resilienza ed economie green per il future dell'architettura e dell'ambiente costruito. In: Technè nr. 15, pp. 153–164. FUP ed. (2018)
5. Van der Heijden J.: The new governance for low carbon buildings: mapping, exploring, interrogating. In: Lorch, R., Laubscher, J., Chan, E., Visscher, H. (eds.) Building Governance and Climate Change, pp. 191–200. Routledge, New York (2018)
6. Hawken, P., Lovins, A., Lovins, H.: Natural Capitalism Creating the Next Industrial Revolution. Little Brown, Boston (1999)
7. Eisenberg, D.E.: Transforming building regulatory systems to address climate change. In: Lorch, R., Laubscher, J., Chan, E., Visscher, H. (eds.) Building Governance and Climate Change, pp. 41–46. Routledge Publ., New York (2018)
8. Nava, C., Melis, A.: Un paradigma radicale bioecologico per le tecnologie progettanti con approccio transdisciplinare. In: Technè nr. 21, pp. 103–111. FUP ed. (2021)
9. Roaf, S.: Building resilience in the built environment. In: Trogal, K., Bauman, I., Lawrence, R., Petrescu, D. (eds.) Architecture & Resilience. Interdisciplinary Dialogues, pp. 143–157. Routledge Publ., New York (2019)
10. Ceschin, F., Gaziulusoy, I.: Evolution of design for sustainability: from product design to design for system innovations and transitions. Des. Stud. **47**, 118–163 (2016)
11. Grima, J.: Space Caviar. https://www.artribune.com/progettazione/architettura/2021/05/intervista-joseph-grima-vac-foundation-venezia/
12. Hu, M.: Net Zero Building. Predicted and Unintended Consequences, pp. 19–20 (10, a), pp. 123–124 (10, b). Routledge Publ., New York (2019)
13. Researches. www.abitalab.unirc.it

Printed by Printforce, the Netherlands